DATE DUE

DEMCO 38-296

HANDBOOK OF FLUID DYNAMICS AND FLUID MACHINERY

HANDBOOK OF FLUID DYNAMICS AND FLUID MACHINERY

VOLUME III: APPLICATIONS OF FLUID DYNAMICS

Edited by

Joseph A. Schetz and Allen E. Fuhs

A WILEY-INTERSCIENCE PUBLICATION

JOHN WILEY & SONS, INC.

New York • Chichester • Brisbane • Toronto • Singapore

This text is printed on acid-free paper.

This publication is designed to provide accurate and
authoritative information in regard to the subject
matter covered. It is sold with the understanding that
the publisher is not engaged in rendering professional
services. If legal accounting, medical, psychological, or any other
expert assistance is required, the services of a competent
professional person should be sought.

Library of Congress Cataloging in Publication Data:
Handbook of fluid dynamics and fluid machinery/editors, Joseph A.
  Schetz and Allen E. Fuhs.
      p.   cm.
    Includes index.
    ISBN: 0-471-14090-2 (Volume III)
    ISBN: 0-471-87352-7 (set)
      1. Fluid mechanics. 2. Hydraulic machinery. I. Schetz, Joseph A.
  II. Fuhs, Allen E.
    TA357.H286   1996
    620.1′06—dc20   95-5671
Printed in the United States of America

10 9 8 7 6 5 4 3 2 1

EDITORIAL REVIEW BOARD

LIST OF CONTRIBUTORS

DANIEL F. ANCONA, U.S. Department of Energy, Washington, D.C.

JOHN D. ANDERSON, JR., University of Maryland, College Park, MD

W. KYLE ANDERSON, NASA Langley Research Center, Hampton, VA

TAKAKAGE ARAI, Muroran Institute of Technology, Muroran, Japan

GEORGE D. ASHTON, U.S. Army Cold Regions Research and Engineering Laboratory, Hanover, NH

FRITZ H. BARK, The Royal Institute of Technology, Stockholm, Sweden

ALAN BEERBOWER (Deceased), University of California at San Diego, La Jolla, CA

MAREK BEHR, University of Minnesota, Minneapolis, MN

R. BYRON BIRD, University of Wisconsin–Madison, Madison, WI

RODNEY D. W. BOWERSOX, Air Force Institute of Technology, Dayton, OH

BOBBIE CARR, Naval Postgraduate School, Monterey, CA

CAMPBELL D. CARTER, Systems Research Laboratories, Dayton, OH

MICHAEL V. CASEY, Sulzer Innotec/Sulzer Hydro, Winterthur/Zurich, Switzerland

JACK E. CERMAK, Colorado State University, Fort Collins, CO

CHAU-LYAN CHANG, High Technology Corporation, Hampton, VA

HERBERT S. CHENG, Northwestern University, Evanston, IL

PASQUALE CINNELLA, Mississippi State University, Starkville, MS

FAYETTE S. COLLIER, JR., NASA Langley Research Center, Hampton, VA

J. ROBERT COOKE, Cornell University, Ithaca, NY

MALCOLM J. CROCKER, Auburn University, Auburn, AL

CLAYTON T. CROWE, Washington State University, Pullman, WA

NICHOLAS A. CUMPSTY, University of Cambridge, Cambridge, UK

WILLIAM G. DAY, JR., David Taylor Model Basin, Carderock, MD

REINER DECHER, University of Washington, Seattle, WA

JAMES D. DE LAURIER, University of Toronto, Toronto, Canada

Note: All Contributors to Volumes I, II, and III are mentioned in this Section of the handbook.

SERGE T. DEMETRIADES,   STD Research Incorporated, Arcadia, CA

AYODEJI O. DEMUREN,   Old Dominion University, Norfolk, VA

THOMAS E. DILLER,   Virginia Polytechnic Institute and State University, Blacksburg, VA

PANAYIOTIS DIPLAS,   Virginia Polytechnic Institute and State University, Blacksburg, VA

LOUIS V. DIVONE,   U.S. Department of Energy, Washington, D.C.

FRANKLIN T. DODGE,   Southwest Research Institute, San Antonio, TX

EARL H. DOWELL,   Duke University, Durham, NC

DONALD J. DUSA (Retired),   General Electric Aircraft Engines, Evandale, OH

HARRY A. DWYER,   University of California, Davis, CA

PETER R. EISEMAN,   Program Development Corporation, White Plains, NY

GEORGE EMANUEL,   University of Oklahoma, Norman, OK

JOHN F. FOSS,   Michigan State University, East Lansing, MI

ALLEN E. FUHS (Emeritus),   U.S. Naval Postgraduate School, Monterey, CA

SUSAN E. FUHS,   Rand Corporation, Santa Monica, CA

WALTER S. GEARHART (Retired),   Pennsylvania State University, State College, PA

WILLIAM K. GEORGE,   State University of New York at Buffalo, Buffalo, NY

ALFRED GESSOW,   University of Maryland, College Park, MD

K. N. GHIA,   University of Cincinnati, Cincinnati, OH

U. GHIA,   University of Cincinnati, Cincinnati, OH

CARL H. GIBSON,   University of California—San Diego, La Jolla, CA

RICHARD J. GOLDSTEIN,   University of Minnesota, Minneapolis, MI

SANFORD GORDON,   Sanford Gordon & Associates, Cleveland, OH

ROBERT A. GREENKORN,   Purdue University, West Lafayette, IN

EDWARD M. GREITZER,   Massachusetts Institute of Technology, Cambridge, MA

BERNARD GROSSMAN,   Virginia Polytechnic Institute and State University, Blacksburg, VA

FREDRICK G. HAMMITT (Deceased),   University of Michigan, Ann Arbor, MI

EVERETT J. HARDGRAVE, JR. (Retired),   Applied Physics Laboratory, The Johns Hopkins University, Laurel, MD

A. GEORGE HAVENER,   U.S. Air Force Academy, Colorado Springs, CO

STEPHEN HEISTER,   Purdue University, West Lafayette, IN

ROBERT E. HENDERSON,   Pennsylvania State University, State College, PA

JACKSON R. HERRING, National Center for Atmospheric Research, Boulder, CO

JOHN L. HESS (Retired), Douglas Aircraft Company, Long Beach, CA

RAYMOND M. HICKS, NASA Ames Research Center, Moffett Field, CA

GUSTAVE J. HOKENSON (Deceased), Air Force Institute of Technology, Dayton, OH

TERRY L. HOLST, NASA Ames Research Center, Moffett Field, CA

DAVID P. HOULT, Massachusetts Institute of Technology, Cambridge, MA

THOMAS J. R. HUGHES, Stanford University, Stanford, CA

EVANGELOS HYTOPOULOS, Automated Analysis Corp., Ann Arbor, MI

TAKAO INAMURA, Tohoku University, Sendai, Japan

DAVID JAPIKSE, Concepts ETI, Incorporated, Wilder, VT

SAM P. JONES, TCOM, L.P., Columbia, MD

HELMUT KECK, Sulzer Innotec/Sulzer Hydro, Winterthur/Zurich, Switzerland

JAMES L. KEIRSEY (Retired), Applied Physics Laboratory, The Johns Hopkins University, Laurel, MD

LAWRENCE A. KENNEDY, University of Illinois at Chicago, Chicago, IL

JOHN KIM, University of California Los Angeles, Los Angeles, CA

ANJANEYULU KROTHAPALLI, Florida State University, Tallahassee, FL

PAUL KUTLER, NASA Ames Research Center, Moffett Field, CA

K. KUWAHARA, Institute of Space and Astronautical Science, Sagamihara, Japan

E. EUGENE LARRABEE (Emeritus), Massachusetts Institute of Technology, Cambridge, MA

J. GORDON LEISHMAN, University of Maryland, College Park, MD

PETER E. LILEY (Retired), Purdue University, West Lafayette, IN

RAINALD LÖHNER, The George Mason University, Fairfax, VA

LUIZ M. LOURENCO, Florida State University, Tallahassee, FL

MUJEEB R. MALIK, High Technology Corporation, Hampton, VA

JAMES F. MARCHMAN, III, Virginia Polytechnic Institute and State University, Blacksburg, VA

CRAIG SAMUEL MARTIN, Georgia Institute of Technology, Atlanta, GA

HUGH R. MARTIN, University of Waterloo, Waterloo, Ontario, Canada

C. D. MAXWELL, STD Research Incorporated, Arcadia, CA

UNMEEL B. MEHTA, NASA Ames Research Center, Moffett Field, CA

JOHN E. MINARDI, University of Dayton, Dayton, OH

ALBERT M. MOMENTHY,   Boeing Commercial Airplane Company, Seattle, WA

THOMAS B. MORROW,   Southwest Research Institute, San Antonio, TX

HANY MOUSTAPHA,   Pratt and Whitney Canada, Montreal, Canada

S. NAKAMURA,   Ohio State University, Columbus, OH

RICHARD NEERKEN (Retired),   The Ralph M. Parsons Company, Pasadena, CA

WAYNE L. NEU,   Virginia Polytechnic Institute and State University, Blacksburg, VA

NEIL OLIEN,   National Institute of Standards and Technology, Boulder, CO

BHARATAN R. PATEL,   Fluent Incorporated, Lebanon, NH

VICTOR L. PETERSON (Retired),   NASA Ames Research Center, Moffett Field, CA

J. LEITH POTTER (Retired),   Vanderbilt University, Nashville, TN

THOMAS H. PULLIAM,   NASA Ames Research Center, Moffett Field, CA

SAAD RAGAB,   Virginia Polytechnic Institute and State University, Blacksburg, VA

RICHARD H. RAND,   Cornell University, Ithaca, NY

EVERETT V. RICHARDSON (Emeritus),   Colorado State University, Fort Collins, CO

DONALD O. ROCKWELL,   Lehigh University, Bethlehem, PA

COLIN RODGERS,   Los Angeles, CA

JOHN P. ROLLINS,   Clarkson University, Potsdam, NY

PHILIP G. SAFFMAN,   California Institute of Technology, Pasadena, CA

MANUEL D. SALAS,   NASA Langley Research Center, Hampton, VA

P. SAMPATH,   Pratt and Whitney Canada, Montreal, Canada

TURGUT SARPKAYA,   Naval Postgraduate School, Monterey, CA

JOSEPH A. SCHETZ,   Virginia Polytechnic Institute and State University, Blacksburg, VA

LEON H. SCHINDEL,   Naval Surface Warfare Center, Silver Spring, MD

JOHN E. SCHMIDT (Retired),   Boeing Commercial Airplane Company, Seattle, WA

WILLIAM B. SHIPPEN (Retired),   Applied Physics Laboratory, The Johns Hopkins University, Laurel, MD

TERRY W. SIMON,   University of Minnesota, Minneapolis, MI

HELMUT SOCKEL,   Technical University of Vienna, Vienna, Austria

GEOFFREY R. SPEDDING,   University of Southern California, Los Angeles, CA

PHILIP C. STEIN, JR.,   Stein Seal Company, Kulpsville, PA

WILLIAM G. STELTZ (Retired),   Westinghouse Electric Company, Orlando, FL

KENNETH G. STEVENS,   NASA Ames Research Center, Moffett Field, CA

PAUL N. SWARZTRAUBER,   National Center for Atmospheric Research, Boulder, CO

ROLAND A. SWEET,   University of Colorado at Denver, Denver, CO

JULIAN SZEKELY,   Massachusetts Institute of Technology, Cambridge, MA

JIMMY TAN-ATICHAT,   California State University, Chico, Chico, CA

RICHARD S. TANKIN,   Northwestern University, Evanston, IL

TAYFUN E. TEZDUYAR,   University of Minnesota, Minneapolis, MN

JAMES L. THOMAS,   NASA Langley Research Center, Hampton, VA

CHANG LIN TIEN,   University of California, Berkeley, CA

EUGENE D. TRAGANZA,   Naval Postgraduate School, Monterey, CA

STEVENS P. TUCKER,   Naval Postgraduate School, Monterey, CA

ERNEST W. UPTON,   Bloomfield Hills, MI

MICHAEL W. VOLK,   Oakland, CA

JAMES WALLACE,   University of Maryland, College Park, MD

CANDACE WARK,   Illinois Institute of Technology, Chicago, IL

FRANK M. WHITE,   University of Rhode Island, Kingston, RI

JOHN M. WIEST,   Purdue University, West Lafayette, IN

JAMES C. WILLIAMS, III,   Auburn University, Auburn, AL

SCOTT WOODWARD,   State University of New York at Buffalo, Buffalo, NY

TERRY WRIGHT,   University of Alabama at Birmingham, Birmingham, AL

GEORGE T. YATES,   Consultant, Boardman, OH

H. C. YEE,   NASA Ames Research Center, Moffett Field, CA

HIDEO YOSHIHARA (Retired),   Boeing Company, Seattle, WA

TSUKASA YOSHINAKA,   Concepts ETI, Incorporated, Wilder, VT

VIRGINIA E. YOUNG,   Virginia Polytechnic Institute and State University, Blacksburg, VA

JAMES L. YOUNGHANS,   General Electric Aircraft Engines, Evandale, OH

HENRY C. YUEN,   TRW Space and Technology Group, Redondo Beach, CA

EDWARD E. ZUKOSKI,   California Institute of Technology, Pasadena, CA

CONTENTS

23 Fluid Dynamic Related Technologies

HERBERT S. CHENG
Northwestern University
Evanston, IL

MALCOLM J. CROCKER
Auburn University
Auburn, AL

EDWARD E. ZUKOSKI
California Institute of Technology
Pasadena, CA

JULIAN SZEKELY
Massachusetts Institute of Technology
Cambridge, MA

RAYMOND M. HICKS
NASA Ames Research Center
Moffett Field, CA

EARL H. DOWELL
Duke University
Durham, NC

E. EUGENE LARRABEE (Emeritus)
Massachusetts Institute of Technology
Cambridge, MA

HELMUT SOCKEL
Technical University of Vienna
Vienna, Austria

SAM P. JONES
TCOM, L.P.
Columbia, MD

JAMES D. DE LAURIER
University of Toronto
Toronto, Canada

Handbook of Fluid Dynamics and Fluid Machinery, Edited by Joseph A. Schetz and Allen E. Fuhs
ISBN 0-471-12598-9 Copyright © 1996 John Wiley & Sons, Inc.

FAYETTE S. COLLIER, JR.
NASA Langley Research Center
Hampton, VA

FREDRICK G. HAMMITT (Deceased)
University of Michigan
Ann Arbor, MI

WILLIAM G. DAY, JR.
David Taylor Model Basin
Carderock, MD

FRANKLIN T. DODGE
Southwest Research Institute
San Antonio, TX

ALLEN E. FUHS
U.S. Naval Postgraduate School
Monterey, CA

A. GEORGE HAVENER
U.S. Air Force Academy
Colorado Springs, CO

CONTENTS

23.1 LUBRICATION
Herbert S. Cheng

Lubrication with oil or other fluids is used mainly to reduce friction and wear between sliding surfaces. The effectiveness of lubrication depends upon how far apart the *surface asperities* are separated by fluid pressure generated by wedging a viscous lubricant through a convergent gap, by squeezing the lubricant out of a thin gap, or by a high pressure lubricant supplied externally. Fluid pressure generated internally is known as *hydrodynamic lubrication* and supplied externally is known as *hydrostatic lubrication*.

When the surfaces are totally separated by hydrodynamic or hydrostatic action, the process is known as *full-film lubrication*. In this Section, emphasis is given to the basic principles of full-film lubrication assuming perfectly smooth surfaces, its application to sliding bearings and rolling element bearings, and the methods used in determining the lubrication performance of these bearings.

23.1.1 The Reynolds Equation

In hydrodynamic lubrication, the main concern is to determine the fluid pressure generated between two moving surfaces separated by a thin gap, as shown in Fig. 23.1. For most lubrication conditions, the following assumptions are valid: 1) the fluid inertia is negligible, 2) the flow is laminar, 3) the pressure is constant across

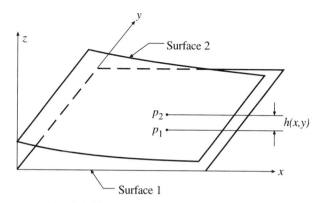

FIGURE 23.1 Geometry of a thin fluid film.

the thin gap, 4) lubricant viscosity variation normal to the surface is negligible, and 5) velocity gradient parallel to the surface is negligible.

Based on these assumptions, it is shown in typical lubrication texts [Szeri (1980) and Cameron (1966)] that the mass flows, q_x and q_y, through a unit segment in the x and y direction are

$$q_x = -\frac{\rho h^3}{12\mu} \frac{\partial p}{\partial x} + \left(\frac{U_1 + U_2}{2}\right) h \tag{23.1}$$

$$q_y = -\frac{\rho h^3}{12\mu} \frac{\partial p}{\partial y} + \left(\frac{V_1 + V_2}{2}\right) h \tag{23.2}$$

In the above, μ, p, and ρ are the lubricant viscosity, pressure, and density respectively, h is the local film thickness, U_1 and V_1 are the velocity components of a point p_1 on surface 1 in the x, y direction, U_2 and V_2 are the velocity components of the point p_2 directly across the film from p_1 on surface 2 in the x, y direction. The first term is the mass flow due to the pressure gradient and is known as the *Poiseuille flow*. The second term is the mass transport by the surface velocities in the x, y coordinates and is known as the *Couette flow*.

Considering a typical element bounded by Δx, Δy and equating the mass flow into the element to the increase in mass within Δx, Δy, one obtains

$$-\left(\frac{\partial q_x}{\partial x} + \frac{\partial q_y}{\partial y}\right) = \frac{\partial \rho h}{\partial t} \tag{23.3}$$

Substituting Eqs. (23.1) and (23.2) into Eq. (23.3), one obtains

$$\frac{\partial}{\partial x}\left(\frac{\rho h^3}{12\mu} \frac{\partial p}{\partial x}\right) + \frac{\partial}{\partial y}\left(\frac{\rho h^3}{12\mu} \frac{\partial p}{\partial y}\right) = \left(\frac{U_1 + U_2}{2}\right)\frac{\partial(\rho h)}{\partial x} + \left(\frac{V_1 + V_2}{2}\right)\frac{\partial(\rho h)}{\partial y} + \frac{\partial(\rho h)}{\partial t} \tag{23.4}$$

For liquids with a constant density, the above equation becomes the well-known, incompressible *Reynolds equation* for determining the pressure in a thin gap.

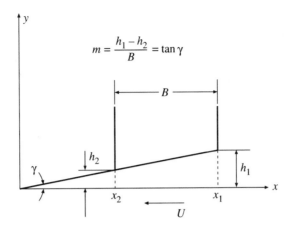

FIGURE 23.2   Geometry of a plane slider.

23.1.2 Hydrodynamic Sliders

The *plane slider* is the most basic form of the hydrodynamic bearings. Figure 23.2 shows the cross section of a fixed plane slider which forms a straight convergent film with a moving flat surface. The nondimensional Reynolds equation for the plane slider under a static load is obtained by setting $V_1 = V_2 = U_2 = 0$, $V_1 = -U$, and $\partial h / \partial t = 0$.

$$\frac{\partial}{\partial \bar{x}} \left(\bar{x}^3 \frac{\partial \bar{p}}{\partial \bar{x}} \right) + 4 \left(\frac{B}{L} \right)^2 \frac{\partial}{\partial \bar{y}} \left(\bar{x}^3 \frac{\partial \bar{p}}{\partial \bar{y}} \right) = -1 \tag{23.5}$$

where $m = (h_1 - h_2)/B$, $h = mx$, $\bar{x} = x/B$, $\bar{y} = 2y/L$, and $\bar{p} = Bpm^2/6\mu U$. For the infinitely long slider, $\partial \bar{p} / \partial \bar{y} = 0$. The second term on the left side vanishes, and the pressure can be directly integrated for $\bar{p} = 0$ at $\bar{x} = \bar{x}_2$ and $\bar{x} = \bar{x}_1$. The resulting pressure and total load [Szeri (1980)] become

$$\bar{p}_\infty = \frac{(\bar{x} - \bar{x}_2)(\bar{x}_2 + 1 - \bar{x})}{(2\bar{x}_2 + 1)\bar{x}^2} \tag{23.6}$$

$$\overline{W}_\infty = 6\bar{x}_2 \left[\ln \left(1 + \frac{1}{\bar{x}_2} \right) - \frac{2}{2\bar{x}_2 + 1} \right] \tag{23.7}$$

where $\bar{p}_\infty$ and $\overline{W}_\infty$ are the dimensionless pressure and load for the infinitely long plane slider, $\overline{W}_\infty = W_\infty h_2^2/\mu ULB^2$, and W_∞/L is the load per unit length.

For plane sliders of a finite length, L, analytical solution for the pressure distribution becomes more involved but is still attainable [Hays (1958) and Szeri and Powers (1970)]. However, with the numerical techniques for solving the two-dimensional Reynolds equation so highly developed in recent times, calculation of the pressure distribution for the sliders becomes a fairly routine task even for sliders with a curved or pocketed surface. Numerical solution for a finite plane slider [Rai-

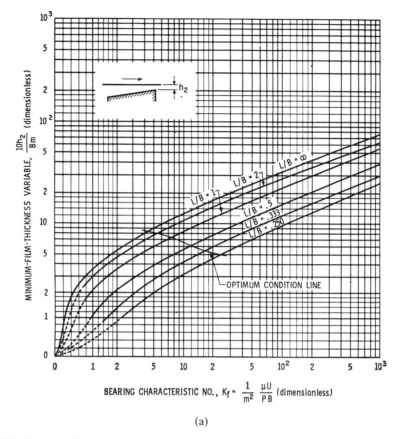

(a)

FIGURE 23.3 Performance of a plane slider bearing. (a) Load curves for a finite plane slider and (b) coefficient of friction of a finite plane slider. (From Raimondi, 1968.)

mondi and Boyd (1955) and Raimondi (1968)], in Fig. 23.3(a) and (b), show how the minimum film thickness and coefficient of friction increases with the bearing characteristic parameter $\mu U/m^2 pB$ for the length to width ratio, L/B, from 1/4 to infinity. Other design charts can be found in Raimondi (1968) to determine oil flow, and oil temperature rise.

The straight taper of a plane slider is geometrically simple, but it is not the optimum profile for the maximum load based on a given design minimum film. The optimum profile was shown to be the *step pad* [Fig. 23.4(a)]. However, the step pad has a major drawback, because its nondimensional load capacity for film thicknesses smaller than the design film thickness is greatly inferior to that of the other geometrical profiles.

The load capacity of a fixed plane slider at small film thicknesses can be improved considerably by allowing the slider to *pivot* at a point corresponding to the optimum load [Fig. 23.4(b)]. This pivot is always closer to the exit edge of the slider. The slider geometries shown in Figs. 23.2 and 23.4 work only for unidirectional sliding. If the sliding direction is reversed in these cases, hydrodynamic lift would not result and contact would occur.

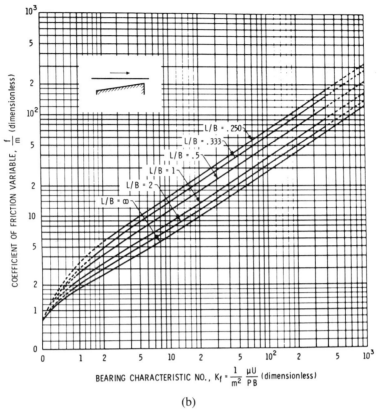

(b)

FIGURE 23.3 (*Continued*)

For bi-directional applications, the profile geometry must be symmetric with respect to the center. Typical profiles may include those shown in Fig. 23.5(a), (b), and (c). In the first two fixed pads, the load capacity is a result of the *film cavitation* in the divergent section. For a discussion of cavitation see Sec. 23.12. The tilt of the *pivoted crowned pad* is automatically adjusted to give a higher load capacity and is the best compromise for sliders to operate in both directions.

23.1.3 Thrust Bearings

A thrust bearing generally consists of a series of loaded pads distributed over an annular area as shown in Fig. 23.6. In most cases, the inside and outside radii are

(a)   (b)

FIGURE 23.4 Geometry of step or pivoted pads. (a) Step pad and (b) pivoted pad.

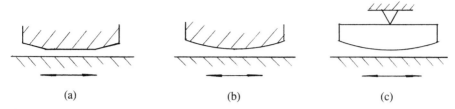

FIGURE 23.5 Bi-directional sliders. (a) Bidirectional tapered pad, (b) crowned pad, and (c) centrally pivoted and crowned pad.

sufficiently close to each other so that the solutions for a finite slider with a rectangular bearing area can be used to determine the lubrication performance of the entire bearing for completely parallel surfaces or for nonparallel surfaces resulting from a slight angular misalignment. For cases where the difference of the two radii is large enough so that the radius effect must be taken into account, the solution of a sector shaped bearing must be used [Pinkus (1958)].

23.1.4 Journal Bearings

The geometry shown in Fig. 23.7 for a full journal bearing consists of a shaft of a radius R rotating inside a bearing of a radius $R + C$, where C is the radial clearance. For $C/R \ll 1$, the gap distribution along the circumference can be approximated by

$$
\begin{aligned}
h &= C + e \cos \theta \\
&= C(1 + \epsilon \cos \theta)
\end{aligned}
\tag{23.8}
$$

where e is the eccentricity, and ϵ is the eccentricity ratio, e/C.

For analyzing the pressure generation, the influence of curvature of the surfaces is extremely small and can be neglected. Thus, the cylindrical surfaces can be unwrapped to form a rectangular region bounded by $-\pi R < x < \pi R$ and $-L/2 < y < L/2$, where $x = R\theta$.

The Reynolds equation can now be expressed in the reference frame fixed to the line joining the journal and bearing centers. If the angular speed of the shaft and the

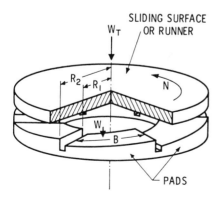

FIGURE 23.6 Geometry of a fixed-pad thrust bearing.

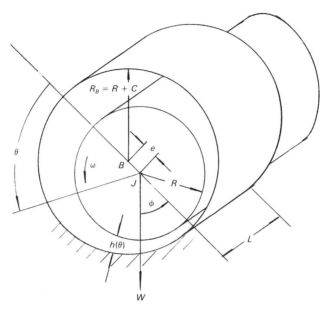

FIGURE 23.7   Geometry of a journal bearing.

angular speed of the line of centers with respect to the bearing are ω and $\dot{\phi}$, respectively, then the tangential velocities of the two surfaces in the frame containing the line of centers become

$$U_1 = -\dot{\phi}R, \qquad U_2 = (\omega - \dot{\phi})R \qquad (23.9)$$

and

$$V_1 = V_2 = 0, \qquad \frac{\partial h}{\partial t} = \dot{e}\cos\theta \qquad (23.10)$$

Using the above velocities in Eq. (23.4) for a constant ρ, one obtains

$$\frac{1}{R^2}\frac{\partial}{\partial\theta}\left(\frac{h^3}{12\mu}\frac{\partial\rho}{\partial\theta}\right) + \frac{\partial}{\partial y}\left(\frac{h^3}{12\mu}\frac{\partial p}{\partial y}\right) = -\frac{(\omega - 2\dot{\phi})}{2}e\sin\theta + \dot{e}\cos\theta \quad (23.11)$$

The above Reynolds equation is applicable for journal bearings lubricated by an incompressible lubricant under a dynamic load. For statically loaded bearings, $\dot{e} = \dot{\phi} = 0$, the right-hand side becomes $-(\omega/2)e\sin\theta$. The solutions without the time dependent terms are known as the *static journal theory.*

The following are results of several well known static journal theories.

Infinitely Long Journal Bearing. For very large L/D, the axial pressure distribution is nearly constant, and $\partial p/\partial y$ can be neglected in Eq. (23.11) to give

$$\frac{\partial}{\partial\theta}\left(H^3\frac{\partial\bar{p}}{\partial\theta}\right) = -\epsilon\sin\theta \qquad (23.12)$$

where

$$\bar{p} = pC^2/6\mu\omega R^2, \quad H = h/C \tag{23.13}$$

The solution of Eq. (23.12) depends on the boundary conditions. If the oil of an inlet pressure, p_i, is fed from an inlet axial groove at the largest film thickness $\theta = 0$, then the boundary conditions are $\bar{p} = \bar{p}_i$ at $\theta = 0$ and $\theta = 2\pi$. Equation (23.12) can be readily integrated through the use of the *Sommerfeld transformation*. The resulting pressure distribution is

$$\bar{p}(\theta) = \frac{\epsilon \sin \theta \, (2 + \epsilon \cos \theta)}{(2 + \epsilon^2) \, (1 + \epsilon \cos \theta)^2} + \bar{p}_i \tag{23.14}$$

This is the Sommerfeld solution for 360° long journal bearings and is shown in Fig. 23.8. It is only valid for a very high inlet oil pressure. The pressure distribution is antisymmetric with respect to $\theta = \pi$. Because of this antisymmetry, the force components due to the fluid pressure parallel to the line of centers cancel completely. Therefore, the resultant fluid force vector is always normal to the line of centers, as shown in Fig. 23.8. The angle between the resultant and the line of centers, which is known as the *attitude angle*, ϕ, is always $\pi/2$ for the Sommerfeld long bearing.

The negative pressure shown in Fig. 23.8 for a 360° Sommerfeld bearing may not exist completely if the lubricant inlet pressure is low since liquid cannot sustain much tension. Thus, the 360° solution is only valid for a very high lubricant supply

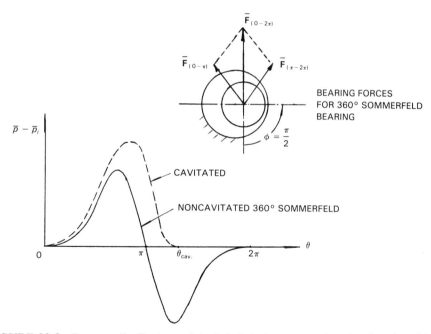

FIGURE 23.8 Pressure distributions of the infinitely long noncavitated and cavitated journal bearing.

pressure at $\theta = 0$. For a low inlet pressure, the oil film in the divergent section ($\theta > \pi$) breaks up into *streamers* shortly after the minimum film at $\theta = \pi$ and re-emerges into a continuous film in the convergent section. In the region of the oil streamers, the lubricant pressure is slightly below the ambient pressure and is never as large as the negative pressure predicted by the 360° Sommerfeld solution. The point where the streamers appear is often referred to as the *caviation point* and can be determined by using a floating boundary condition, $p = p_{cav.}$ and $dp/d\theta = 0$ at $\theta = \theta_{cav.}$, where $p_{cav.}$ is a prescribed subambient pressure and $\theta_{cav.}$ is the place where the film breaks.

The pressure distribution for $p_{cav.}$, the same as the inlet pressure, is also shown in Fig. 23.8. Since this pressure profile is not antisymmetric, the resultant force is not normal to the line of centers like the 360° bearing. The attitude angle between this force vector and the line of centers is less than 90° depending on the eccentricity ratio. Figure 23.9 shows the locus of the journal center and the relation between ϕ and ϵ for a cavitated bearing using $p = dp/d\theta = 0$ as the cavitation boundary condition.

Finite Length Journal Bearing. For bearings with a finite L/D ratio under a static load, the pressure can be determined by solving Eq. (23.11) numerically excluding the terms containing $\dot{e}$ and $\dot{\phi}$. A complete set of curves are available [Raimondi and Boyd (1958)] for determining the load capacity and frictional load force covering a wide range of L/D ratio for full journal as well as for partial arc bearings. Figure 23.10(a), (b) plot the *Sommerfeld number*, S (a nondimensional parameter representing reciprocal of the load parameter) and the nondimensional frictional force, F, against ϵ, the eccentricity ratio for 360° finite bearings with L/D from 1/4 to ∞.

23.1.5 Thermal Effects

In most sliding bearings, the viscosity in the lubricant film is not uniform due to increases in temperature. However, the isoviscous solutions described above can

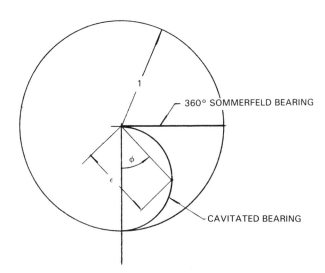

FIGURE 23.9   Locus of journal center for 360° Sommerfeld bearing and cavitated bearing.

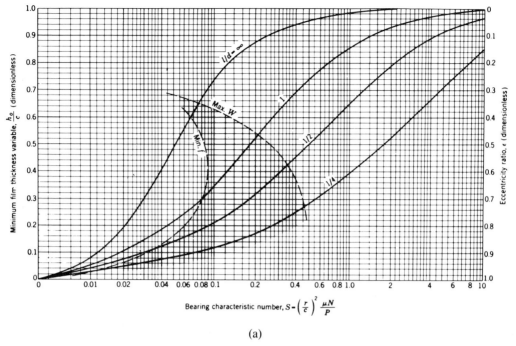

FIGURE 23.10 Charts for finite journal bearings. (a) Chart for minimum film thickness for finite journal bearings. The left boundary of the shaded area defines the optimum h_0 for minimum friction; The right boundary is the optimum h_0 for maximum load. r = Radius of the journal, C = Radial clearance, μ = Average lubricant viscosity in the film, N = Shaft rotational speed in rev. per sec., and P = The load per unit projected bearing area. (b) Chart for coefficient of friction. (From Shigley, 1977.)

still be applied for predicting the lubrication performance if a uniform effective lubricant viscosity corresponding to an average lubricant temperature is assumed in the entire film. The *average film temperature* is determined from the following gross heat balance, assuming that all heat generated by sliding is convected away by the lubricant

$$T_{ave.} = T_{in.} + \frac{KUf}{\rho C_p Q} \tag{23.15}$$

where U = the sliding speed, ρ = lubricant density, C_p = specific heat of the lubricant, f = frictional force, Q = lubricant flow, and K = a factor between 1/2 and 1. Since f and Q are dependent on the average temperature through the average viscosity in the Reynolds equation, Eq. (23.15) is solved together with the Reynolds equation iteratively [Shigley (1977)]. For high speed sliding bearings, the use of a uniform effective viscosity is inadequate, because the temperature variation in the film is excessive. For these cases, accurate pressure predictions must be obtained by solving the generalized Reynolds equation and the energy equation [Dowson (1962) and Ezzat and Rohde (1973)].

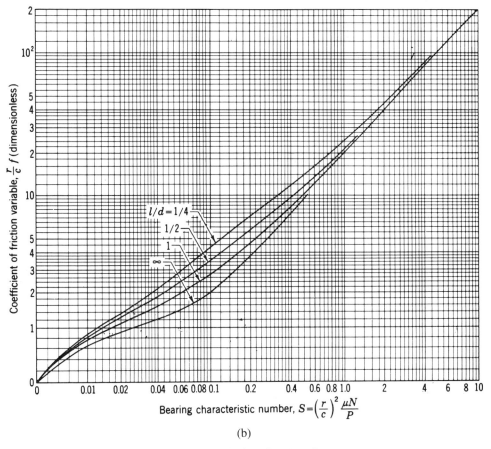

Bearing characteristic number, $S = \left(\dfrac{r}{c}\right)^2 \dfrac{\mu N}{P}$

(b)

FIGURE 23.10   (*Continued*)

23.1.6   Effect of Turbulence

When the Reynolds number, Re, defined as $CR\omega/\nu$ for journal bearings and Uh/ν for sliders, becomes very large, the flow ceases to be laminar and becomes turbulent. For conventional journal bearings with a rotating shaft and a fixed bearing, the transition may not be sudden. It may be preceded by the formation of layers of toroidal vortices in the axial direction as shown in Fig. 23.11. These are the well known *Taylor vortices* between rotating cylinders [Taylor (1923)]. The transition from laminar flow to the formation of toroidal vortices between concentric cylinders takes place when the term $\mathrm{Re}^2\,(C/R)$, known as the *Taylor number*, T, reaches a critical number $T_{cr} = 1707.8$ or $\sqrt{T_{cr}} = 41.3$ [Rosenhead (1963)]. As the speed increases further, the boundaries between vortex cells become wavy, and finally when the Reynolds number reaches a critical value around 2000, the flow suddenly becomes turbulent. However, for bearings of a small C/R ratio, the critical Reynolds number may be reached before the critical Taylor number. For these cases, the transition from laminar to turbulence is direct without the formation of Taylor vortices.

For cases of a rotating bearing wall and a stationary shaft, the transition from laminar to turbulence is also sudden with absence of any vortices. The effect of a

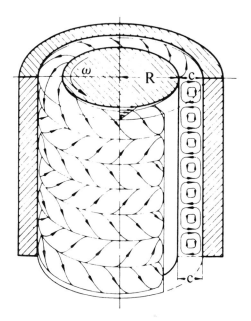

FIGURE 23.11 Taylor vortices between concentric rotating cylinders with inner cylinder rotating. (Reprinted with permission, Szeri, A. Z., *Tribology: Friction, Lubrication, and Wear*, Taylor & Francis, Washington, D.C., Copyright © 1980.)

fully turbulent film can be incorporated by using effective turbulent viscosities μ_x and μ_y, defined by [Constantinescu (1959), Ng and Pan (1965), Elrod (1967), and Hirs (1974)]. $\mu_x = \mu/12G_x$ and $\mu_y = \mu/12G_y$. With these effective viscosities, the Reynolds equation for a turbulent film becomes

$$\frac{\partial}{\partial x}\left(\frac{G_x h^3}{\mu}\frac{\partial p}{\partial x}\right) + \frac{\partial}{\partial y}\left(\frac{G_y h^3}{\mu}\frac{\partial p}{\partial y}\right) = \frac{U}{2}\frac{\partial h}{\partial x} \tag{23.16}$$

The effect of turbulence is not isotropic; the value of G_x in the direct of Couette flow is different than G_y normal to the Couette flow. G_x and G_y vary with the local Reynolds number, Uh/ν, and are plotted in Fig. 23.12. Empirical expressions [Szeri (1980)] for G_x and G_y as functions of Re $= Uh/\nu$ can be used to obtain the numerical solutions of Eq. (23.16) for the pressure distribution in a turbulent film.

23.1.7 Effects of Inertia

In deriving the Reynolds equation, the effect of fluid inertia is neglected. This assumption is shown to be valid as long as a modified Reynolds number defined as follows does not exceed unity [Szeri (1980)]: (Re · C/R) for journal bearings and (Re · h_2/B) for sliders, where Re $= \rho\omega RC/\mu$ for journal bearings, and Re $= \rho Uh_2/\mu$ for sliders.

If the modified Reynolds number is larger than unity, the method of an average inertia across the film [Osterle and Saibel (1955) and Osterle *et al.* (1957)] may be used to account for the inertia effect shown on the left-hand side of the momentum equation in the x-direction for infinitely long bearings

$$\rho\left(u\frac{\partial u}{\partial x} + w\frac{\partial u}{\partial z}\right) = -\frac{\partial p}{\partial x} + \mu\frac{\partial^2 u}{\partial z^2} \tag{23.17}$$

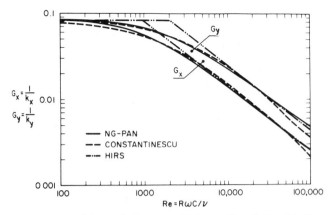

FIGURE 23.12 Variation of the turbulence coefficients G_x and G_y with Reynolds number. Comparison of various theories. (Reprinted with permission, Szeri, A. Z., *Tribology: Friction, Lubrication, and Wear*, Taylor & Francis, Washington, D.C., Copyright © 1980.)

By averaging the left-hand side across the film, it is no longer a function of z. The solution for pressure can be handled in a similar manner as the inertialess Reynolds equation but with an additional term to account for the average inertia force.

23.1.8 Effect of Surface Roughness

When the film thickness in sliding bearings becomes sufficiently small, the surface roughness will have a significant influence on the load capacity and friction. This effect is due mainly to the ratio of the film thickness divided by the *composite roughness height*, σ, for both sliding surfaces (the composite σ is defined by $\sqrt{\sigma_1^2 + \sigma_2^2}$, where σ_1 and σ_2 are the r.m.s. roughness height of the two sliding surfaces (see Fig. 23.13). This ratio is commonly known as the *specific film parameter*, λ. For rough surfaces with a Gaussian height distribution, contact between asperities begins when λ reaches three. The roughness effects become increasingly important as λ approaches three. Machined or ground surface roughness has protruding asperities like ridges running parallel in one direction. Height profiles taken in the direction of the ridges show wavelength structure much longer than those taken normal to that di-

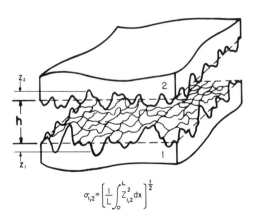

FIGURE 23.13 Geometry of film lubrication between two rough surfaces. (From Dowson and Higginson, 1977.)

$$\sigma_{1,2} = \left[\frac{1}{L} \int_0^L Z_{1,2}^2 \, dx \right]^{\frac{1}{2}}$$

rection. The orientation of these lays with respect to the flow clearly has an influence on hydrodynamics when λ is near to or less than three.

Surface roughness of long ridges can be treated as one-dimensional roughness whose height only runs in the direction normal to the ridge. Purely longitudinal roughness with ridges running parallel to the flow and purely transverse roughness with ridges normal to the flow have been analyzed extensively in the past for sliders [Christensen (1969), Chow and Cheng (1976), and Rhow and Elrod (1974)], journal bearings [Christenson and Tonder (1973)], and concentrated contacts [Chow and Cheng (1976)]. Analyses of one-dimensional roughness have been reviewed by El-rod (1978). Two-dimensional, ridge-like roughness has asperities which are long and thin. A parameter γ, known as the *surface pattern parameter*, can be used to indicate the ratio of the characteristic asperity dimension along the flow to that normal to the flow. For ridges running in the direction of flow $\gamma > 1$; if the ridges are turned around 90° normal to the flow, then $\gamma < 1$. For $\gamma = 1$, the profiles are independent of directions, and this is known as *isotropic roughness* [Constantinescu (1959)] (see Fig. 23.14). Mathematically, γ is the ratio of the correlation length of the profile auto correlations in two mutually perpendicular directions.

There have been two major approaches to analysis of the two-dimensional roughness lubrication problems. Elrod (1979) adopted the method of small-parameter expansion for two-dimensional roughness and developed a general theory for representing the mass flow between two sliding rough surfaces. The flows calculated by this approach agree well with an alternative approach known as the *average flow method* [Patir and Cheng (1978)].

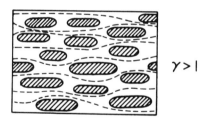

$\gamma > 1$

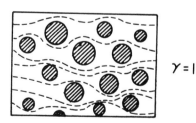

$\gamma = 1$

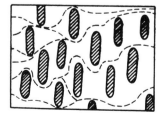

$\gamma < 1$

FIGURE 23.14 Typical contact areas for longitudinally oriented ($\gamma > 1$), isotropic ($\gamma = 1$) and transversely oriented ($\gamma < 1$) surfaces. (From Patir and Cheng, 1978.)

23.1.9 Hydrostatic Bearings

When the wedging or squeezing mechanism is insufficient to generate a lubricant film to separate the sliding asperities, a high pressure lubricant from an external source can be introduced to separate the asperities. These bearings are known as *hydrostatic bearings*. A hydrostatic bearing usually consists of a flat or curved pad extending over a circular, rectangular, cylindrical, or sector-shaped area. The high pressure is introduced into a recessed central pocket. The pump can either supply a constant volume of lubricant or a constant pressure by means of a regulator. For a constant pressure bearing, a restrictor must be used between the lubricant supply and the recess in order to allow the recess pressure to vary in response to load change.

There are four basic requirements in hydrostatic bearings: 1) a film thickness sufficiently thick to separate the asperities, 2) a low combined frictional and pump loss, 3) a sufficiently high overload capacity to prevent asperity overload, and 4) a high stiffness to prevent a large excursion of film thickness under fluctuating loading.

The analyses for determining the above quantities may be illustrated with the three basic hydrostatic systems for a very long rectangular pad shown in Fig. 23.15. These systems include: 1) constant volume system, 2) constant pressure system with a laminar restrictor, and 3) constant pressure system with an orifice restrictor. The total flow, Q, through the film thickness for a bearing length, L, is

$$Q = 2 \frac{h^3 L}{12\mu} \frac{p_r}{l} \tag{23.18}$$

For a constant volume system, the variation of the recess pressure or the load is inversely proportional to h^3. If h is nondimensionalized with a design film thickness, h_0, and p_r with p_s, which is the maximum allowable recess pressure, then

$$\bar{p} = \bar{Q}/\bar{h}^3 \tag{23.19}$$

where $\bar{p} = p_r/p_s$, $\bar{h} = h/h_0$, and $\bar{Q} = 6\mu l Q/h_0^3 p_s L$ = constant. For a constant supply pressure with a laminar restrictor such as a capillary tube, the flow between the pressure supply and the restrictor is

$$Q = \frac{\pi d^4}{128\mu} \left(\frac{p_s - p_r}{l_t} \right) \tag{23.20}$$

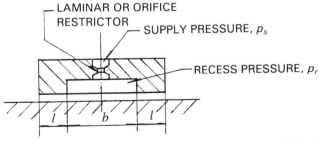

FIGURE 23.15 Schematic of an infinitely long hydrostatic bearing.

where d and l_t are the diameter and length of the capillary tube. Equating Eqs. (23.18) and (23.20), one obtains

$$\bar{p} = \frac{p_r}{p_s} = \frac{1}{1 + \bar{h}^3/K_c}$$ (23.21)

$K_c = 3\pi d^4 l/64 l_t L h_0^3$, and the flow for the capillary restrictor is

$$\bar{Q} = \bar{p}\bar{h}^3 = \frac{\bar{h}^3}{1 + \bar{h}^3/K}$$ (23.22)

For a constant supply pressure with an orifice restrictor with a sharp-edged orifice, the flow between the supply pressure and the recess for a liquid is

$$Q = \frac{\pi}{4d} C d^2 \left(\frac{2(p_s - p_r)}{\rho}\right)^{1/2}$$ (23.23)

where C_d is the discharge coefficient and d the orifice diameter. If Eqs. (23.18) and (23.23) are combined, the nondimensional recess pressure, $\bar{p}$, is governed by a quadratic equation

$$\bar{h}^6 \bar{p}^2 - K_0(1 - \bar{p}) = 0$$ (23.24)

where $K_0 = 9\pi^2 (C_d d^2 \mu l/h_0^3 L)^2/2\rho p_s$. The solution of the quadratics in Eq. (23.24) gives

$$\bar{p} = \frac{K_0}{2\bar{h}^6}\left[\left(1 + 4\left(\frac{\bar{h}^6}{K_0}\right)\right)^{1/2} - 1\right]$$ (23.25a)

and

$$\bar{Q} = \bar{p}\bar{h}^3$$ (23.25b)

Using these equations, variations of flow and the recess pressure with respect to $\bar{h}$ can be readily determined. Typical curves are shown in Fig. 23.16 for all three types of hydrostatic bearings [Davies and Howarth (1978)].

For full hydrostatic journal bearings, more recesses are required in order to provide a radial stiffness. Details of the hydrostatic journal bearing can be found in Raimondi (1968) and Elrod and Ng (1967).

23.1.10 Gas-Lubricated Slider

For high speed, lightly loaded, bearings, gas and air can be used as lubricants because gases are readily available, friction is extremely low, and viscosity increases with temperature. However, *gas* and *air bearings* also have many disadvantages: 1) the load capacity for self-acting bearings is low and does not increase linearly with speed or viscosity, 2) the bearings require high precision in surface geometry, and 3) the performance is highly susceptible to mechanical and thermal distortion.

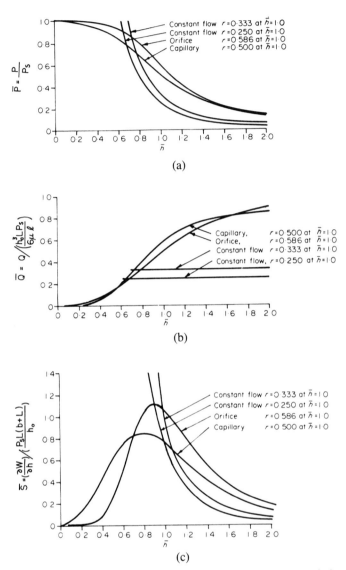

FIGURE 23.16 Comparison of characteristics of three types of hydrostatic bearings. (From Davies and Howarth, 1978.) (a) Load capacity, (b) flow rate, and (c) stiffness.

Nevertheless, these bearings have been used in gyro-rotors, high speed spindles, computer pick-up heads, and other precision instruments.

Considering an infinitely long plane slider, the equation governing the pressure is

$$\frac{\partial}{\partial x}\left(h^3\rho\,\frac{\partial p}{\partial x}\right) = 6\mu u\,\frac{\partial(\rho h)}{\partial x} \tag{23.26}$$

For most operating conditions, it was shown [Elrod and Burgborfer (1958)] that the gas film is isothermal. Using the isothermal relation, $p/\rho = $ constant, and introducing the following nondimensional variables, $p = p_a$ where p_a is the ambient pressure,

$H = h/h_2$, where h_2 is the minimum film thickness at the exit end, $\bar{x} = x/B$, and $\Lambda = 6\mu UB/p_a h_2^2$, Eq. (23.26) becomes

$$\frac{\partial}{\partial x}\left(H^3 P \frac{\partial p}{\partial \bar{x}}\right) = \Lambda \frac{\partial(PH)}{\partial \bar{x}} \tag{23.27}$$

Exact and approximate analyses of this nonlinear equation have been obtained [Harrison (1913) and Gross (1962)]. The principal difference between the gas and liquid slider is the pressure as affected by the compressibility parameter, Λ, which is a measure of the hydrodynamic pressure relative to the ambient pressure. When $\Lambda \rightarrow 0$, the film pressure is small compared to p_a, and the problem is essentially incompressible. For this case, the maximum pressure is somewhere in the middle along the slider, as shown in Fig. 23.17. For very large Λ, the maximum pressure moves to the edge of the slider. Results of load capacity for infinitely long sliders [Gross (1962)] are plotted in Fig. 23.18.

23.1.11 Gas-Lubricated Journal Bearings

The 360° full journal configuration has been analyzed extensively. The equation governing the pressure in a finite gas journal bearing is

$$\frac{\partial}{\partial \theta}\left(PH^3 \frac{\partial P}{\partial \theta}\right) + \frac{1}{\left(\dfrac{L}{D}\right)}\frac{\partial}{\partial y}\left(PH^3 \frac{\partial P}{\partial \bar{y}}\right) = \Lambda \frac{\partial(PH)}{\partial \theta} \tag{23.28}$$

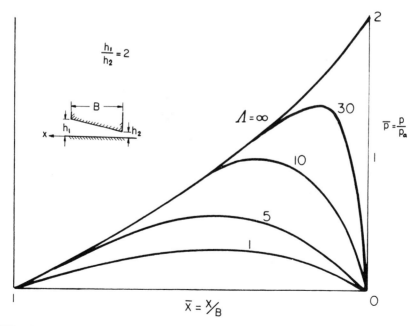

FIGURE 23.17 Pressure distribution in a gas lubricated slider. (Reprinted with permission, Gross, W. A., *Gas Film Lubrication*, John Wiley & Sons, New York, 1962.)

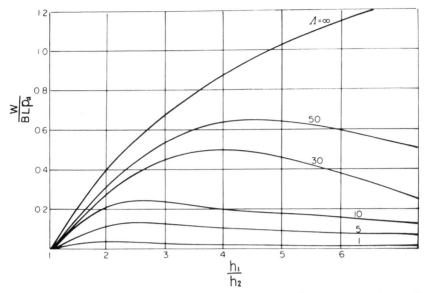

FIGURE 23.18   Load capacity of a gas lubricated slider. (Reprinted with permission, Gross, W. A., *Gas Film Lubrication*, John Wiley & Sons, New York, 1962.)

where $\bar{y} = y/(L/2)$ and Λ for journal bearing is $6\mu\omega(R/C)^2/p_a$. Approximate analytical solutions to this equation have been obtained by perturbing either P or PH about the eccentricity ratio [Ausman (1961) and Ng (1965)], ϵ, or by the method of Galerkin [Cheng and Pan (1965)]. However, numerical calculations using the finite difference method [Raimondi (1961)] still give the most accurate solution for the steady state problem. Figure 23.19 gives the load capacity curves for a finite bearing

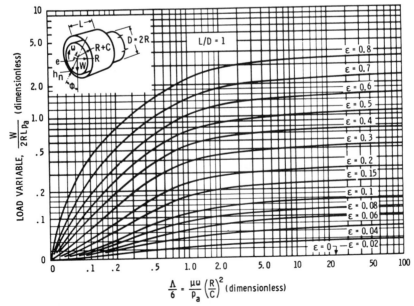

FIGURE 23.19   Load capacity for full bearing, $L/D = 1$. (From Raimondi, 1961.)

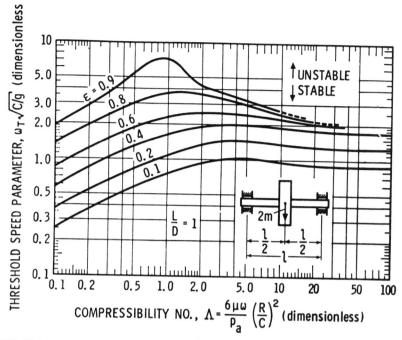

FIGURE 23.20 Stability (half frequency whirl) chart for plain cylindrical bearing, $L/D = 1$. (From Cheng and Pan, 1965.)

of $L/D = 1$ as a function Λ and ϵ. Figures for other L/D and additional performance characteristics can be found in Raimondi (1961).

One of the major problems with the *gas journal bearing* is its susceptibility to a self-excited instability known as the *fractional frequency whirl*. When whirling occurs, the amplitude becomes unbounded, and this eventually leads to metallic contact and failure. For a centrally located rigid rotor, a series of charts is available to predict the threshold speed at which whirling begins [Cheng and Pan (1965)]. A typical chart for $L/D = 1$ is shown in Fig. 23.20 which plots the variation of a nondimensional threshold speed against Λ for ϵ from 0.1 to 0.9.

23.1.12 Rotor-Bearing Interaction

For a liquid journal bearing, the time dependent pressure is governed by Eq. (23.11). The time dependency is caused by the journal motions, $\dot{e}$ and $\dot{\phi}$, which are controlled by the dynamics of the rotor-shaft system under a time-varying load. If one considers a centrally located rigid rotor supported on two liquid bearings as shown in Fig. 23.21, the equations of motion become

$$m(\ddot{e} - e\dot{\phi}^2) = 2 \iint p \cos\theta R \, d\theta \, dy + W \cos\phi$$

$$m(e\ddot{\phi} + 2\dot{e}\dot{\phi}) = 2 \iint p \sin\theta R \, d\theta \, dy - W \sin\phi \tag{23.29}$$

where $(\cdot)$ stands for d/dt.

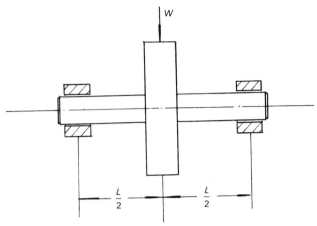

FIGURE 23.21 Schematic of a centrally located rigid rotor supported by two fluid film bearings.

Since the Reynolds equation, Eq. (23.11), is linear and contains $(\omega - 2\dot\phi)$ and $\dot e$ on the right-hand side, the pressure contributed by $\dot e$ and $(\omega - 2\dot\phi)$ can be solved separately. By superimposing these components together, the film resultant forces appear as

$$\int\int p \cos \theta R \, d\theta \, dy = F_{rr}\dot e + F_{rt}(\omega - 2\dot\phi)$$

$$\int\int p \sin \theta R \, d\theta \, dy = F_{tr}\dot e + F_{tt}(\omega - 2\dot\phi) \qquad (23.30)$$

where $\epsilon = e/c$, and F_{rr}, F_{rt}, F_{tr}, and F_{tt} are functions of ϵ. These functions can be solved directly from Eq. (23.11) without any coupling with the equations of motion. Equations (23.29) with the film forces represented by Eqs. (23.30) are two coupled nonlinear equations which can be solved readily by the Runge–Kutta method for the trajectory of the journal center with a given dynamic loading function.

The above method is valid for small or large journal motion. For small oscillations, the rotor-bearing interaction can be analyzed by perturbing the dynamic pressure about the equilibrium solution. The perturbed pressure gives rise to the dynamic film forces which can be represented by

$$\begin{Bmatrix} \Delta F_r \\ \Delta F_t \end{Bmatrix} = -\begin{bmatrix} k_{\epsilon\epsilon} & k_{\epsilon\phi} \\ k_{\phi\epsilon} & k_{\phi\phi} \end{bmatrix} \begin{Bmatrix} \Delta G \\ \epsilon_0 \Delta\phi \end{Bmatrix} - \begin{bmatrix} C_{\epsilon\epsilon} & C_{\epsilon\phi} \\ C_{\phi\epsilon} & C_{\phi\phi} \end{bmatrix} \begin{Bmatrix} \Delta\dot e \\ \epsilon_0 \Delta\dot\phi \end{Bmatrix} \qquad (23.31)$$

where ϵ_0 is the equilibrium eccentricity, and $k_{\epsilon\epsilon} = -\partial F_r/\partial\epsilon$, $k_{\epsilon\phi} = -\partial F_r/\epsilon\, \partial\phi$, etc., are the *stiffness coefficients*, and $C_{\epsilon\epsilon} = -\partial F_r/\partial\dot e$, $C_{\epsilon\phi} = -\partial F_r/\epsilon_0\, \partial\dot\phi$, etc., are the *damping coefficients* in the rotating coordinates. These coefficients can be readily transformed into a set of stiffness and damping coefficients in the fixed coordinates if the equations of motion can be more conveniently solved in these coordinates.

Procedures for calculating these coefficients for journal bearings can be found in Lund and Sternlicht (1962).

The use of the stiffness and damping coefficients facilitates the application of standard techniques in vibration and control of linear systems to determine the dynamic response and stability thresholds of complicated rotor-bearing systems.

23.1.13 Contact Lubrication

When the bearing surfaces are both convex, such as in contact between spheres or rollers, the lubrication is confined to a very small area under a very high pressure. For these cases, the surface deformation must be taken into account in determining the film thickness. Research in this area is known as *elastohydrodynamic lubrication* (EHL).

Developments in EHL have greatly enhanced the understanding of lubrication and failure mechanisms in rolling bearings and gears. It is now widely recognized that the most significant EHL parameter in controlling fatigue, scuffing, and wear is the lubricant film thickness to surface roughness ratio, $h_{min}/\sqrt{\sigma_1^2 + \sigma_2^2}$, known as λ, where h_{min} is the minimum film thickness, and σ_1 and σ_2 are the r.m.s. roughness of the two surfaces.

The minimum film thickness between two deformed rollers can be determined by [Dowson and Higginson (1977)]

$$\frac{h_{min}}{R} = 2.65 \frac{G^{0.54} U^{0.7}}{W^{-0.13}} \tag{23.32}$$

where $G = \alpha E$, α is the pressure viscosity exponent in $\mu = \mu_0 e^{\alpha p}$, $U = \mu_0 u/ER$, μ_0 is the ambient viscosity, $u = (u_1 + u_2)/2$, $W = w/ER$, and w is the load per unit length of the rollers.

For well-lubricated point contacts having a circular or elliptical contacting area as shown in Figs. 23.22 and 23.23, the minimum film thickness can be determined by [Hamrock and Dowson (1977)]

$$H_{c,F} = 2.69 U^{0.67} G^{0.53} W^{-0.067} (1 - 0.61 e^{-0.73k})$$

$$H_{min,F} = 3.63 U^{0.68} G^{0.49} W^{-0.073} (1 - e^{-0.68k}) \tag{23.33}$$

Where

$$H_{c,F} = h_{c,F}/R_x$$
$$h_{c,F} = \text{central film thickness for flooded contacts}$$
$$H_{min,F} = h_{min,F}/R_x$$
$$h_{min,F} = \text{minimum film thickness for flooded contacts}$$
$$k = \text{elliptical parameter } (k = 1.03(R_y/R_x)^{0.64})$$
$$R_x = R_{x1}R_{x2}/(R_{x1} + R_{x2})$$
$$R_y = R_{y1}R_{y2}/(R_{y1} + R_{y2})$$
$$R_{x1}, R_{x2} = \text{principal radii for body 1 and 2 in the direction of rolling}$$
$$R_{y1}, R_{y2} = \text{principal radii for body 1 and 2 normal to the direction of rolling}$$
$$W = w/ER_x^2$$

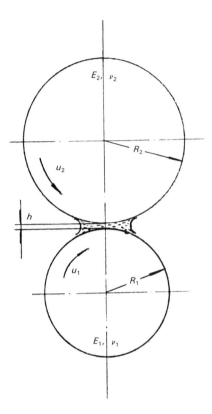

FIGURE 23.22 Geometry of a lubricated roller contact.

$$U = \mu_0(u_1 + u_2)/2ER_x$$
$$G = \alpha E$$

23.2 ACOUSTICS
Malcolm J. Crocker

23.2.1 Introduction

Acoustics may be defined as the science of sound. We are all familiar with atmospheric sound which we can hear. It is less known that sound is a motion comprised of compressional waves. Of course, compressional waves can travel through any fluid and even solids too. Besides the atmospheric sound waves, which we can hear, there are other types of waves in fluids and solids (some of which are discussed in Chap. 6). However, here we shall mainly be concerned with compressional waves in air, usually described as sound waves. Much sound such as speech and music is wanted. Noise is usually defined as any unwanted sound.

23.2.2 Wave Motion

Although wave motion is discussed separately in Chap. 6, it will be necessary for some discussion to be presented here to aid in the understanding of acoustics. Wave

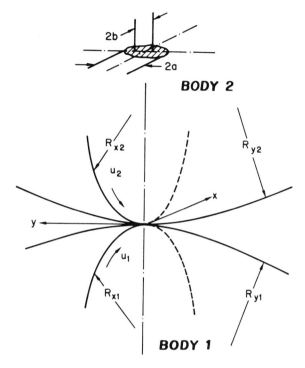

FIGURE 23.23 Geometry of an elliptical EHL contact.

motion is easily observed in the waves on stretched strings and as the ripples on the surface of water. Waves on strings and surface water waves are very similar to sound waves in air (which we cannot see), but there are some differences which are useful to discuss.

If we throw a stone into a calm lake, we observe that the water waves (*ripples*) travel out from the point where the stone enters the water. The ripples spread out circularly from the source at a wave speed which is virtually constant and independent of the wave height. Like the water ripples, sound waves in air travel at a constant speed, almost independent of their strength. Like the water ripples, sound waves in air pass by transferring momentum and energy between air particles. There is no net flow of air away from a source of sound, just as there is no net flow of water away from the source of water waves. Of course, the waves on the surface of a lake are circular or two-dimensional, while in air, sound waves in general are spherical or three-dimensional.

As water waves move away from a source, their curvature decreases, and the *wavefronts* may be regarded almost as straight lines. Such waves are observed in practice as *breakers* on the seashore. A similar situation occurs with sound waves in the atmosphere. At large distances from a source of sound, the spherical wave front curvature decreases, and the wave fronts may be regarded almost as plane surfaces.

Plane sound waves may be defined as waves which have the same acoustic properties at any position on a plane surface drawn perpendicular to the direction of propagation of the wave. Such plane sound waves can exist and propagate along a long straight tube or duct (such as an air conditioning duct). In such a case, the waves propagate in a direction along the duct and the plane waves are perpendicular to this direction (and are represented by duct cross sections). Such waves in a duct are one-dimensional like the waves on a long string or rope under tension (or like the ocean breakers described above).

Although there are so many similarities between one-dimensional sound waves in air, waves on strings and surface water waves, there are some minor differences. In a fluid such as air, the fluid particles vibrate back and forth in the same direction as the direction of wave propagation; such waves are known as *longitudinal*, *compressional*, or *sound waves*. On a stretched string, the particles vibrate at right angles to the direction of wave propagation; such waves are usually known as *transverse waves*. The surface waves described are also mainly transverse waves, with the complication that the water particles move up and down and to a lesser extent back and forth horizontally (describing elliptical paths). The wave direction is, of course, horizontal.

23.2.3 Acoustic Wave Equation

Acoustic Plane Waves. If a disturbance in a thin element of fluid in a duct is considered, a mathematical description of the motion may be obtained by assuming that: 1) the amount of fluid in the element is conserved, 2) the net longitudinal force is balanced by the inertia of the fluid in the element, and 3) the process in the element is adiabatic (i.e., there is no flow of heat in or out of the element). Then the following equation of motion may be derived [Crocker and Price (1975)]

$$\frac{\partial^2 p}{\partial x^2} - \frac{1}{c^2}\frac{\partial^2 p}{\partial t^2} = 0 \tag{23.34}$$

This equation is known as the one-dimensional equation of motion or *wave equation*, and it relates the second rate of change of the sound pressure with the coordinate x with the second rate of change of the sound pressure with time t through the square of the speed of sound c. Identical wave equations may be written if the sound pressure, p, in Eq. (23.34) is replaced with the particle displacement, ξ, the particle velocity, u, fluctuating density, ρ', or the fluctuating temperature, T'. However, the wave equation in terms of the sound pressure in Eq. (23.34) is perhaps most useful, since the sound pressure is the easiest acoustic quantity to measure (using a microphone) and is the acoustic perturbation we sense with our ears. The sound pressure, p, is the acoustic pressure perturbation or fluctuation about the time-averaged or undisturbed pressure, p_0.

The *speed of sound waves*, c, is given for a perfect gas by

$$c = (\gamma R T)^{1/2} \tag{23.35}$$

The speed of sound is proportional to the square root of the absolute temperature, T. The ratio of specific heats, γ, and the gas constant, R, are constants for any particular gas. Thus Eq. (23.35) may be written as

$$c = c_0 + 0.6t \tag{23.36}$$

where, for air, $c_0 = 331.6$ m/s, the speed of sound at 0 C, and t is the temperature, C. Note that Eq. (23.36) is an approximate formula valid for t near room temperature. The speed of sound does not depend on the atmospheric pressure, although it does vary slightly with the humidity [Kurze and Beranek (1971)].

A solution to (23.34) is

$$p = f_1(ct - x) + f_2(ct + x) \tag{23.37}$$

where f_1 and f_2 are arbitrary functions such as sine, cosine, exponential, log, etc. It is easy to show that Eq. (23.37) is a solution to the wave equation, Eq. (23.34), by differentiation and substitution into Eq. (23.34). Varying x and t in Eq. (23.37) demonstrates that $f_1(ct - x)$ represents a wave traveling in the positive x-direction with wave speed c, while $f_2(ct + x)$ represents a wave traveling in the negative x-direction with wave speed c (see Fig. 23.24).

The solution given in Eq. (23.37) is usually known as the *general solution*, since, in principle, any type of sound wave form is possible. In practice, sound waves are usually classified as impulsive or steady in time. One particular case of steady wave is of considerable importance. Waves created by sources vibrating sinusoidally in time (e.g., a loudspeaker, a piston, or a more complicated structure vibrating with a discrete angular frequency ω) vary both in time t and space x in a sinusoidal manner

$$p = p_1 \sin(\omega t - kx + \phi_1) + p_2 \sin(\omega t + kx + \phi_2) \tag{23.38}$$

At any point in space, x, the sound pressure, p, is simple harmonic in time. The first expression on the right of Eq. (23.38) represents a wave of amplitude p_1 traveling in the positive x-direction with speed, c, while the second expression represents a wave of amplitude p_2 traveling in the negative x-direction. The symbols ϕ_1 and ϕ_2 are *phase angles*, and k is the *acoustic wavenumber*. It is observed that the wavenumber $k = \omega/c$ by studying the ratio of x and t in Eqs. (23.37) and (23.38). At some instant, t, the sound pressure pattern is sinusoidal in space, and it repeats itself each time kx is increased by 2π. Such a repetition is called a *wavelength*, λ. Hence, $k\lambda = 2\pi$ or $k = 2\pi/\lambda$. This gives $\omega/c = 2\pi f/c = 2\pi/\lambda$, or

$$\lambda = c/f \tag{23.39}$$

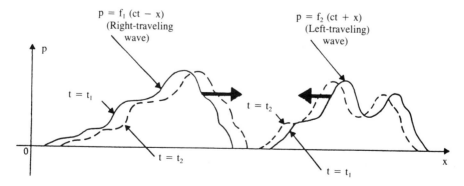

FIGURE 23.24 Plane waves of arbitrary wave form.

The wavelength of sound becomes smaller as the frequency is increased. At 100 Hz, $\lambda \simeq 3.5$ m ≈ 10 ft. At 1000 Hz, $\lambda \approx 0.35$ m ≈ 1 ft. At 10,000 Hz, $\lambda \approx 0.035$ m $\simeq 0.1$ ft $\simeq 1$ inch.

At some point x in space, the sound pressure is sinusoidal in time and goes through one complete cycle when ωt increases by 2π. The time for a cycle is called the *period*, T. Thus, $\omega T = 2\pi$, $T = 2\pi/\omega$, and

$$T = 1/f \tag{23.40}$$

Wave Impedance and Sound Intensity. We see that the sound wave disturbances travel with a constant wavespeed, c, although there is no net, time-averaged movement of the air particles. The air particles oscillate back and forth in the direction of wave propagation (x-axis) with velocity u. We may show that for any right-traveling plane wave, at any instant

$$p/u = \rho c \tag{23.41}$$

and for any left-traveling plane wave

$$p/u = -\rho c \tag{23.42}$$

The quantity, ρc, is called the *characteristic impedance* of the fluid, and for air, $\rho c = 428$ kg/m^2 s at 0°C and 415 kg/m^2 s at 20°C.

The *sound intensity* is the rate at which the sound wave does work on an imaginary surface of unit area in a direction perpendicular to the surface. Thus, it can be shown that the instantaneous sound intensity is obtained by multiplying the sound pressure, p, by the particle velocity, u. Therefore

$$I = pu \tag{23.43}$$

and for a right-traveling plane wave this becomes

$$I = p^2/\rho c \tag{23.44}$$

The time-averaged sound intensity for a right-traveling plane wave $\langle I \rangle_t$ is

$$\langle I \rangle_t = \langle p^2 \rangle_t/\rho c \tag{23.45}$$

and for the special case of a sinusoidal (pure tone) wave

$$\langle I \rangle_t = \langle p^2 \rangle_t/\rho c = p_1^2/2\rho c \tag{23.45a}$$

where p_1 is the pressure amplitude and the mean square pressure is thus $\langle p^2 \rangle_t = p_{rms}^2 = p_1^2/2$.

Three-Dimensional Wave Equation. In most acoustic fields, sound propagation occurs in two or three dimensions. The three-dimensional version of Eq. (23.34) in Cartesian coordinates is

$$\frac{\partial^2 p}{\partial x^2} + \frac{\partial^2 p}{\partial y^2} + \frac{\partial^2 p}{\partial z^2} - \frac{1}{c^2}\frac{\partial^2 p}{\partial t^2} = 0 \tag{23.46}$$

This equation is useful if sound wave propagation in rectangular spaces such as rooms is being considered. However, it is helpful to recast Eq. (23.46) in spherical coordinates if sound propagation from sources of sound in free space is being considered. It is a simple mathematical procedure to transform Eq. (23.46) into spherical coordinates, although the resulting equation is quite complicated. However, for propagation of sound waves from a spherically symmetric source (such as the idealized case of a pulsating balloon) then the equation becomes quite simple (since there is no angular dependence)

$$\frac{\partial^2(rp)}{\partial r^2} - \frac{1}{c^2}\frac{\partial^2(rp)}{\partial t^2} = 0 \tag{23.47}$$

Here, r is the distance from the origin, and p is the sound pressure at that distance [Crocker and Price (1975)].

Equation (23.47) is identical in form to Eq. (23.34) with p replaced by rp and x by r. The general and simple harmonic solutions to Eq. (23.47) are thus the same as Eqs. (23.37) and (23.38) with p replaced by rp and x with r. The general solution is

$$rp = f_1(ct - r) + f_2(ct + r) \tag{23.48}$$

or

$$p = \frac{1}{r}f_1(ct - r) + \frac{1}{r}(ct + r) \tag{23.49}$$

The first term on the right of Eq. (23.49) represents a wave traveling outward from the origin; the sound pressure p is seen to be proportional to the distance r. If the distance, r, is doubled, the sound pressure level [Eq. (23.59)] decreases by $20 \cdot \log_{10}(2) = 20(0.301) \cong 6$ dB. This is known as the *inverse square law*. The second term in Eq. (23.49) represents sound waves traveling inward toward the origin, and in most practical cases these can be ignored (if reflecting surfaces are absent).

The simple harmonic solution of Eq. (23.47) is

$$p = \frac{p_1}{r}\sin(\omega t - kr + \phi_1) + \frac{p_2}{r}\sin(\omega t + kr + \phi_2) \tag{23.50}$$

The second term on the right of Eq. (23.50), as before, represents sound waves traveling inward to the origin and is of little practical interest. However, the first term represents simple harmonic waves of angular frequency ω traveling outward from the origin and this may be rewritten as [Crocker and Price (1975)]

$$p = \frac{\rho ckQ}{4\pi r}\sin(\omega t - kr + \phi_1) \tag{23.51}$$

where Q is termed the *strength* of an *omnidirectional* (*monopole*) *source* situated at the origin, and $Q = 4\pi p_1/\rho ck$. The mean square pressure p_{rms}^2 may be found by averaging the square of Eq. (23.51) over a period T

$$p_{rms}^2 = (\rho ck)^2 Q^2/32\pi^2 r^2 \qquad (23.52)$$

From Eq. (23.52), the mean square pressure is seen to decrease with the inverse square of the radius, for such an idealized omnidirectional point source, everywhere in the sound field. Again, this is known as the *inverse square law*. We may show that Q has units of volume flow rate (m³/s).

The particle velocity in a spherically spreading sound field is

$$u = -\frac{1}{\rho}\int \frac{\partial p}{\partial r}\, dt \qquad (23.53)$$

and substituting Eqs. (23.51) and (23.53) into (23.43) gives the intensity in such a field as

$$\langle I \rangle_t = p_{rms}^2/\rho c \qquad (23.54)$$

the same result as for a plane wave. The sound intensity decreases with the inverse square of the radius. The same result as Eq. (23.54) is found to be true for any source of sound as long as the measurements are made sufficiently *far* from the source. The intensity is *not* given by the simple result of Eq. (23.54) *close* to sources such as dipoles, quadrupoles, or more complicated sources of sound. Close to such sources, Eq. (23.43) must be used for the instantaneous intensity, or

$$\langle I \rangle_t = \langle pu \rangle_t \qquad (23.55)$$

for the time-averaged intensity.

Sound Power. The *sound power*, W, of a source is given by integrating the intensity over any closed surface S around the source

$$W = \int_S \langle I_r \rangle_t \, dS \qquad (23.56)$$

The intensity, I_r, must be measured in a direction perpendicular to the elemental area, dS. If a spherical surface is chosen, then the sound power of an omnidirectional (monopole) source is [Crocker and Price (1975)]

$$W = \langle I_r \rangle_t \, 4\pi r^2 \qquad (23.57a)$$

$$W = (p_{rms}^2/\rho c) \, 4\pi r^2 \qquad (23.57b)$$

and from Eq. (23.52)

$$W = \rho ck^2 Q^2/8\pi \qquad (23.57c)$$

It is apparent from Eq. (23.57c) that the sound power of an idealized (monopole) source is independent of radius, a result required by conservation of energy and also to be expected for all sound sources.

Equation (23.57b) shows that for an omnidirectional source (in the absence of reflections) the sound power can be determined from measurements of the mean square pressure made with a single microphone. Of course, for such a source, measurements should really be made with a *reflection-free (anechoic) environment* or very close to the source where reflections are presumably less important.

In practical situations with real directional sound sources and where reflections are important, use of Eq. (23.57b) becomes difficult and less accurate, and then sound power is more conveniently determined from Eq. (23.57a) with a sound intensity measurement system.

Decibels and Levels. The range of sound pressure magnitude and sound power experienced in practice is very large (see Fig. 23.25). Thus, logarithmic rather than linear measures are often used for sound pressure and power. The most common is the *decibel*. The decibel represents a relative measurement or ratio. Each quantity in decibels is expressed as a ratio relative to a *reference sound pressure, power,* or *intensity.* Whenever a quantity is expressed in decibels the result is known as a *level.*

The decibel (dB) is the ratio r given by

$$\log_{10} r = 0.1$$

$$10 \log_{10} r = 1, \text{ dB} \tag{23.58}$$

Thus, $r = 10^{0.1} = 1.26$. The decibel is seen to represent the ratio 1.26. A larger ratio, the *bel* is sometimes used. The *bel* is the ratio R given by $\log_{10} R = 1$. Thus, $R = 10^1 = 10$. The bel represents the ratio 10.

The *sound pressure level, L_p,* is given by

$$L_p = 10 \log_{10} \left(\frac{\langle p^2 \rangle_t}{p_{\text{ref}}^2} \right) = 10 \log_{10} \left(\frac{p_{\text{rms}}^2}{p_{\text{ref}}^2} \right) = 20 \log \left(\frac{p_{\text{rms}}}{p_{\text{ref}}} \right), \text{ dB} \tag{23.59}$$

where p_{ref} is the reference pressure $p_{\text{ref}} = 20 \ \mu\text{Pa} = 0.00002 \ \text{N/m}^2 \ (\simeq 0.0002 \ \mu\text{bar})$. This reference pressure was originally chosen to correspond to the quietest sound (at 1000 Hz) that the average young person can hear.

The sound power level, L_W, is given by

$$L_W = 10 \log_{10} \left(\frac{W}{W_{\text{ref}}} \right), \text{ dB} \tag{23.60}$$

where W is the sound power of a source source, and W_{ref} is the reference sound power $= 10^{-12} \ \text{W}$.

The sound intensity level, L_i, is given by

$$L_i = 10 \log_{10} \left(\frac{I}{I_{\text{ref}}} \right), \text{ dB} \tag{23.61}$$

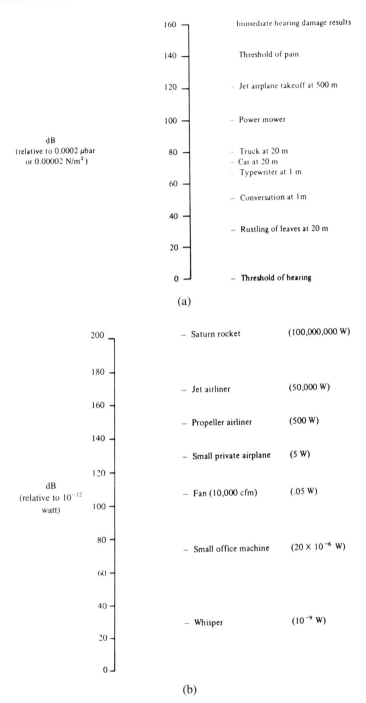

FIGURE 23.25 Some typical sound levels: (a) sound pressure and (b) sound power.

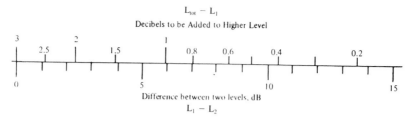

$L_{tot} - L_1$

Decibels to be Added to Higher Level

Difference between two levels, dB

$L_1 - L_2$

FIGURE 23.26 Chart for combination of decibels (uncorrelated sources).

where I is the sound intensity in a given direction, and I_{ref} is the reference sound intensity $= 10^{-12}$ W/m^2.

Some typical sound pressure and sound power levels are given in Fig. 23.25(a) and (b).

If two sound sources are independent (*incoherent*), then the total sound pressure level at some point in space, or the total sound power level may be determined using Fig. 23.26. *First Example:* If two independent sound sources create sound pressure levels each of 80 dB, what is the total level? *Answer.* The difference in levels is 0 dB, thus the total sound pressure level is $80 + 3 = 83$ dB. *Second Example:* If two independent sound sources are 70 and 73 dB, what is the total level? *Answer.* The difference in levels is 3 dB, thus the total sound power level is $73 + 1.8 = 74.8$ dB.

Reflection, Refraction, and Diffraction. If a sound wave is incident on a fluid medium of different characteristic impedance, ρc, then both reflected and transmitted waves are formed (see Fig. 23.27).

From energy considerations (provided no losses occur at the boundary) the sum of the reflected intensity, I_r, and transmitted intensity, I_t, equals the incident intensity, I_i

$$I_i = I_r + I_t \tag{23.62}$$

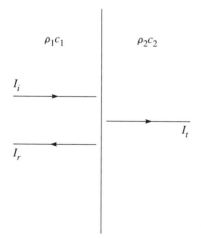

FIGURE 23.27 Plane wave incident on boundary between two different fluids.

and dividing throughout by I_i

$$I_r/I_i + I_t/I_i = R + T = 1 \tag{23.63}$$

where R is the *reflection coefficient* and T is *the transmission coefficient*. For plane waves at normal incidence on a plane boundary between two fluids (see Fig. 23.27)

$$R = (\rho_1 c_1 - \rho_2 c_2)^2/(\rho_1 c_1 + \rho_2 c_2)^2 \tag{23.64}$$

and

$$T = 4\rho_1 c_1 \rho_2 c_2/(\rho_1 c_1 + \rho_2 c_2)^2 \tag{23.65}$$

Some interesting facts can be deduced quickly from Eqs. (23.64) and (23.65). Both the reflection and transmission coefficients are independent of the direction of the wave, since interchanging $\rho_1 c_1$ and $\rho_2 c_2$ does not effect the values of R and T. For example, for sound waves traveling from air to water or water to air, almost complete reflection occurs, independent of direction, and the reflection coefficients are the same and the transmission coefficients are the same for the two different directions.

As discussed before, when the characteristic medium ρc of a fluid medium changes, incident sound is both reflected and transmitted. It can be shown that the wave transmitted into the changed medium changes direction. This effect is called *refraction*. In practice, temperature changes and wind speed changes in the atmosphere are the main causes of refraction.

Wind speed normally increases with altitude and Fig. 23.28 shows the refraction effects to be expected for an idealized wind speed profile. Atmospheric temperature changes alter the speed of sound, c, and temperature gradients can also produce sound shadow and focussing effects as seen in Figs. 23.29 and 23.30.

23.2.4 Subjective Response to Sound

The human ear is a marvelous mechanism for detecting sound. It is very sensitive. If it were only slightly more so, people would be able to hear the Brownian (random)

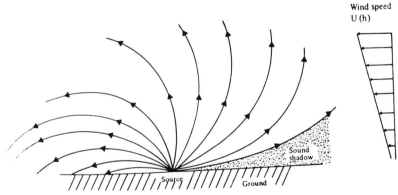

FIGURE 23.28 Refraction of sound in air with wind speed increasing with altitude.

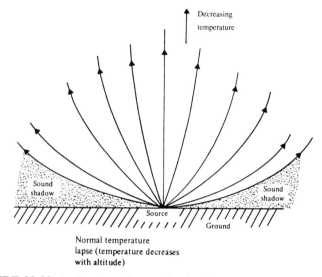

FIGURE 23.29 Refraction of sound in air with normal temperature lapse.

motion of the air molecules and have a perpetual buzz in their ears! The ear has a wide frequency response from about 15 to about 16,000 Hz. Also, the ear has a large dynamic range; the ratio of the loudest sound pressure which can be tolerated to the faintest which can be heard is about ten million (10^7).

There are three essential reasons to consider human hearing. Sound levels are now so high in industrialized societies that many individuals are exposed to intense noise and permanent damage results. Large numbers of other people are exposed to noise from aircraft, surface traffic, construction equipment or machines, and appliances, and disturbance and annoyance results. Lastly, there are subjective reasons. An understanding of people's subjective response to noise allows environmentalists and engineers to reduce noise in more effective ways. For example, noise should be

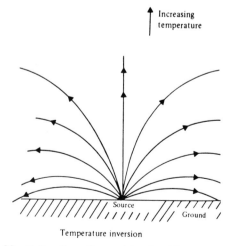

FIGURE 23.30 Refraction of sound in air with temperature inversion.

reduced in the frequency range in which the ear is most sensitive. Noise reduction should be by a magnitude which is subjectively significant.

Construction of the Human Ear.   The ear can be divided into three main parts (Fig. 23.31)—the *outer*, *middle*, and *inner* ear. The outer ear consisting of the fleshy *pinna* and ear canal conducts the sound waves onto the eardrum. The middle ear converts the sound waves into mechanical motion of the auditory *ossicles*, and the inner ear converts the mechanical motion into neural impulses which travel along the auditory nerves to the brain. The anatomy and functioning of the ear are described more completely in various other references and textbooks [Crocker and Price (1975), Kinsler *et al.* (1982), and Parker (1972)].

The function of the fleshy appendage on the side of the head (the *pinna*) is to focus sound into the ear canal. It helps to localize the source of sound, particularly in the vertical direction, and is more effective at higher frequencies. The ear canal is about 25 mm long and ends at the tympanic membrane (*eardrum*).

The eardrum is connected to the *malleus*, the first of the three small bones known as the *auditory ossicles*. The middle ear air cavity is connected to the back of the mouth by the *Eustachian tube*. The smallest of the ossicles, the *stapes*, is connected to a small oval window in the *cochlea*. The cochlea consists of spiral liquid-filled cavities inside the bone of the skull. The cochlea is comprised of a passageway which makes two and one half turns rather like a snail shell. Connected to the cochlea are the semicircular canals which are the balance mechanism and are unrelated to hearing.

The passageway of the cochlea is separated into a lower and upper gallery by a membranous duct (the *organ of Corti*). The upper and lower galleries are connected together only at the apex of the cochlea.

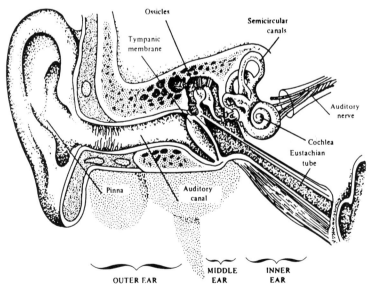

FIGURE 23.31   Simplified cross section through the human ear. (From Palmer, 1972.)

How the Ear Works. When a sound wave reaches the ear, it travels down the auditory canal until it reaches the eardrum. It sets the eardrum in motion, and this vibration is transmitted across the 2 mm gap to the oval window by the lever system comprised of the three auditory ossicles. It is thought that this mechanical system of eardrum, ossicles, and window is an impedance-matching device.

Movement of the footplate of the stapes which is connected to the oval window causes pressure waves in the fluid of the upper gallery of the cochlea. The fluid in the lower gallery is separated from that in the upper gallery by the cochlear duct containing the organ of Corti. The organ has about 35,000 sensitive hair cells distributed along its length which are connected in a complicated way to about 18,000 nerve fibers which combine into the auditory nerve which runs into the brain. The pressure waves cause the *basilar membrane* to deflect, and a shearing motion occurs between the basilar and *tectorial* membranes. The hair cells situated between the two membranes sense the shearing motion, and, if the stimulus is great enough, the neuron to which each hair cell is attached sends an impulse along the nerve fiber to the brain cortex. The brain must interpret the neural impulses to produce the sensation of hearing.

Loudness. The way in which the brain interprets signals from the ear mechanism is still a matter for research. Various experiments have been conducted on groups of people to determine the average sensation of loudness. Since every person's hearing is somewhat different, statistical results must be sought for the loudness of sounds. Figure 23.32 shows equal loudness contours for pure tone sounds. The lowest curve in Fig. 23.32 is labeled MAF, *mean audible field.* This is the hearing threshold which is the quietest sound at each frequency that the average young person can hear.

The equal loudness contours show that people are most sensitive to sound at about 4000 Hz. The equal loudness contours are not flat with frequency. For example, notice that a pure tone of 40 dB at 1000 Hz appears equally as loud as one of about 80 dB at 30 Hz. The intensity of this tone at 30 Hz is 40 dB higher than at 1000 Hz in order for it to appear equally as loud. This represents a 10,000-fold increase in sound intensity or equivalently a 100-fold increase in sound pressure. As the sound pressure level is increased, the equal loudness contours become flatter.

Other subjective experiments show that the ear is a nonlinear device. It has been found that for most people with pure tones at 1000 Hz, when the sound pressure level is increased by about 10 dB the sound appears twice as loud, and when the sound pressure level is increased by 20 dB the sound appears four times as loud and so on. Since equal loudness contours join all pure tones with the same apparent loudness level, then this result may be generalized to say that a doubling in *loudness* occurs when the *loudness level* is increased by 10 *phon*, or

$$S = 2^{(P-40)/10} \tag{23.66}$$

where S = loudness, *sone;* and P = loudness level, *phon.*

So far, the loudness of pure tones has been described. However, many noises experienced are predominantly *broad-band* in nature. Similar schemes to rate the loudness of broad-band noise have been devised by Zwicker and Stevens and have been standardized by the International Organization for Standardization [ISO (1965),

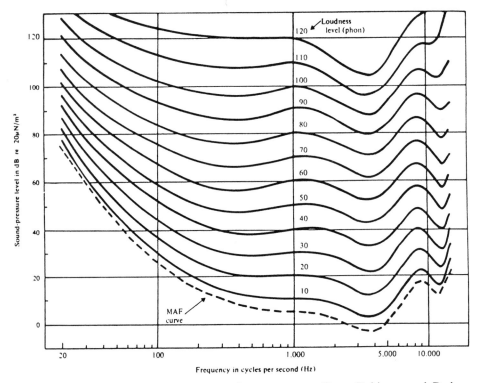

FIGURE 23.32 Equal loudness contours for pure tones. (From Robinson and Dadson, 1956.)

(1967)]. The schemes are quite complicated and the reader should consult ISO (1965), (1967), and Stevens (1961) for more details. Again the linear loudness, S, and loudness level, P, are related by Eq. (23.66).

Masking. Masking is a fairly familiar phenomenon to most people. In normal situations there is always some ambient noise from machinery, vehicles, or aircraft which interferes with people's ability to understand speed. This *masking* phenomenon has been studied experimentally by a number of workers and has been shown to be quite complicated. Results from different experiments, although showing some differences seem to agree on the following:

i) The masking effect is different when the masking noise is a pure tone from when it is a band of noise (of the same level). The band of noise causes more masking around its center frequency than does the pure tone.

ii) At low levels of masking noise, the masking effect is limited to a narrow frequency range around the center masking frequency. As the level of the masking noise is increased, the frequency range of the masking effect increases.

iii) The masking effect is asymmetrical with frequency. Sounds at frequencies above the masking noise or tone are more easily masked than those below.

Further discussion on masking can be found in Crocker and Price (1975).

Sound Levels. As already discussed, methods to rank the loudness of noises based on the work of Zwicker and Stevens have been standardized [ISO (1975), ANSI (1965), and Stevens (1961)]. However, since these methods are complicated, some have sought simpler methods to rate the loudness of noise. This has led to simple frequency weighting filters related to the equal loudness contours (see Fig. 23.32). The weighting filters which are most commonly used date back to the 1930's and to the early Fletcher–Munson equal loudness contours. The *A*-, *B*-, and *C*-frequency weightings shown in Fig. 23.33 approximately correspond to the inverse of the Fletcher–Munson, 40-, 70-, and 100-Phon contours after some frequency smoothing is applied.

The sound pressure level measurements made with the frequency-weighting filters are described as sound levels. The readings then are known as A-weighted sound levels, B-weighted sound levels, and C-weighted sound levels. Since the A-weighting filter is the one most commonly used, if the description sound level is given without qualification, it is assumed to be A-weighted. The most complete description would be to say that the A-weighted sound level was 50 dB, for example. A common abbreviation is to say the sound level is 50 dB(A).

Hearing Loss. Some people are born with the severe handicap of deafness. Some others suffer sudden hearing loss later, or, more commonly, gradually lose their hearing. There are several common causes of deafness, [Crocker and Price (1975)] and these may be grouped in two main types—*conduction deafness* and *sensory-neural deafness*. Conduction deafness is related to disorders in the outer and middle ear and normally manifests itself as a fairly uniform loss over most frequencies. Some common causes include blockage of the ear canal with wax, middle ear infections, and calcification of the smallest ossicle, the stapes. Sensory-neural deafness is associated with disorders in the inner ear, the auditory nerves, the auditory cortex in the brain, or combinations of all three. Unlike conduction deafness, sen-

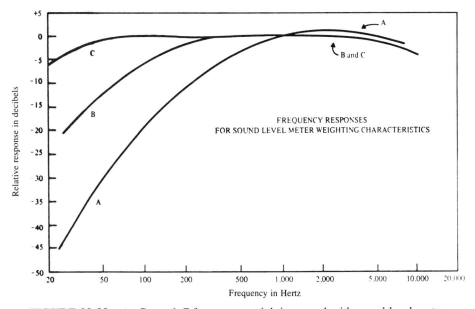

FIGURE 23.33 *A*-, *B*-, and *C*-frequency weightings used with sound level meters.

sory neural deafness is often most severe at the higher frequencies. Sensory-neural deafness can be inherited or caused by various viral infections. Another cause which seems to occur in all western societies, is *presbycusis*, which is believed to be caused by deterioration of the central nervous system with aging and by changes in the hearing mechanism in the inner ear. Presbycusis mainly effects the higher frequencies (above about 2000 or 3000 Hz), and men more than women.

Another cause of sensory-neural deafness is noise-induced deafness. It is well known that if very intense noise above about 135 or 140 dB is experienced that, immediate hearing damage often results. However, it is now known that much lower sound pressure levels about 90 or 100 dB(A), if experienced for long periods, (weeks, months, or years) can cause permanent deafness. Noise-induced deafness often appears as a *notch* (loss) at about 4000 Hz. It is, however, difficult to differentiate conclusively between noise-induced deafness and presbycusis. Noise-induced deafness has been shown to occur mainly by the gradual destruction of the hair cells in the middle ear and is thus irreversible.

23.2.5 Measurement of Sound and Vibration

There is a great variety of sound and vibration measuring equipment available. With rapid advances in electronics, the sophistication of this equipment seems to increase each year. A brief attempt will be made here to classify the equipment available and to describe the most important instruments available in acoustics. In most acoustical measurements, it is usually necessary to combine several types of instruments into one measurement system. A typical system is shown in Fig. 23.34. In order to make useful measurements, it is necessary to have an understanding of the acoustical phenomena being investigated and of the functioning of the instrumentation.

The first item in any measuring system is the transducer (1) in which the signal is converted from one physical form to another, e.g., sound pressure into electrical voltage. The electrical signal is not normally suitable at this stage for direct read out, and a signal conditioner (2) must be used to amplify, attenuate or transform the signal. The signal is normally passed through a signal processor (4) of some form before it is displayed (5). The processing may be done with analog or digital devices. In addition it is optional to store data at several stages in the system, (3) and (5).

The first item in a measurement system, the transducer, suffers from two main

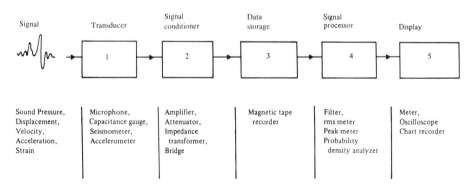

FIGURE 23.34 Idealized measurement system.

shortcomings: 1) it will respond slightly to other variables in addition to the primary variable of interest (e.g., a microphone will be most sensitive to sound pressure but also slightly to temperature, humidity, magnetic field, and vibration), and 2) the transducer will interfere in some way with the measurement medium. The most common transducers in acoustics are microphones and accelerometers.

Condenser Microphones. Condenser microphones are the ones most commonly used in acoustical measurements. They consist of a thin metal diaphragm stretched under tension and spaced a short distance from an insulated back plate (see Fig. 23.35). This type of microphone has uniform sensitivity over a wide frequency range and is very stable. However, it suffers from several disadvantages, principally, its rather low sensitivity, high internal impedance, and the need for a polarizing voltage.

Piezoelectric Microphones. Much thicker diaphragms are used with piezoelectric than condenser microphones (usually about 50 times greater) (see Fig. 23.36). This leads to a lower resonance frequency for the diaphragm and usually a poorer high frequency response. However, the thicker diaphragm makes the piezoelectric more robust than the condenser microphone. Thus, it is usually preferred in field measurements, while the condenser microphone is preferred in the laboratory.

Electret Microphones. The *electret microphone* is basically a form of condenser microphone (Fig. 23.37). However, a polarization voltage is not needed, since the electret foil diaphragm is permanently polarized during manufacture. In addition, the electret microphone has the advantage that it is more rugged, has a higher capacitance and a lower cost than the condenser microphone.

Displacement Transducers. Transducers can be built to measure displacement, velocity, or acceleration in fluids or vibrating structures. Ultrasonic transducer arrays have been developed to measure particle velocity [Nordby and Bjor (1987)]. Several noncontacting magnetic and capacitance transducers are available to measure vibra-

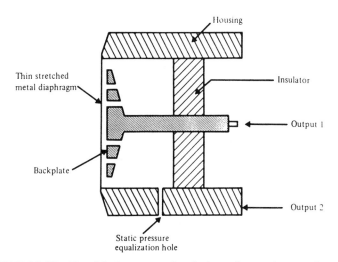

FIGURE 23.35 Simplified cross-sectional view of a condensor microphone.

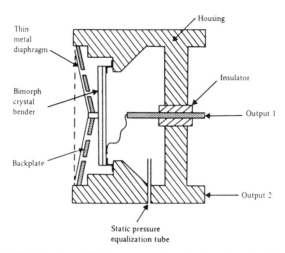

FIGURE 23.36   Simplified cross section of piezoelectric microphone.

tion of structures. The most common transducer used to measure structural vibration, however, is the piezoelectric accelerometer. Although accelerometers are very convenient and can be made very lightweight, they then become less sensitive to vibration. More massive accelerometers are more sensitive but may load a lightweight structure such as a thin pane [Starr (1971)].

Transducers for Intensity Measurement. Transducers have recently been developed for directly measuring sound intensity [see Eq. (23.55)]. Measurements of sound intensity are very useful in determining the sound power of sources such as machinery [see Eq. (23.56)], the sound transmission loss of panel and wall structures and the absorption coefficient of acoustic absorbing materials. In principle, it is also possible to measure the vibration intensity of flexural energy flowing in structures.

Sound Intensity Transducers. Two-microphone probes have become widely used to measure sound intensity [Waser and Crocker (1984)]. Several microphone arrangements have been used (see Fig. 23.38). The particle velocity needed in Eq.

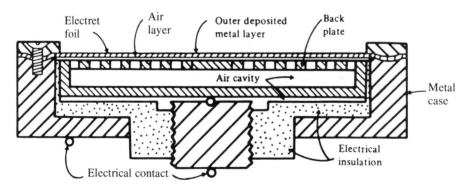

FIGURE 23.37   Cross section of electret microphone. (From Sessler and West, 1966.)

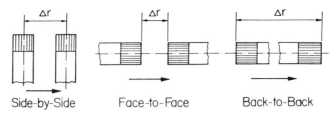

Side-by-Side   Face-to-Face   Back-to-Back

FIGURE 23.38 Different two-microphone probe arrangements. The geometrical separation distance is denoted by Δr; the arrow indicates the direction in which the pressure gradient component, and hence the intensity, is measured.

(23.55) is estimated by the sound pressure gradient which is obtained from the sound pressure difference and the microphone separation. The sound pressure needed is obtained from the average sound pressure measured by the two microphones. Instruments are available to process the signals from the two microphones, both by analog and digital methods, to obtain the sound intensity. Information on noise sources of machinery and sound power can be obtained by moving the two-microphone probe over a control surface enclosing the source or sources of noise [see Eq. (23.56)]. The microphone probe may be placed at a discrete number of points on the control surface or traversed in a continuous manner sometimes called *sweeping* [Waser and Crocker (1984)].

23.2.6 Machinery and Fluid Sources of Sound

Noise is usually defined as *unwanted sound*. Most machines produce noise. Here are some of the most common machinery noise sources [Bridgewater and Mumford (1979)]: 1) pipe and duct systems, 2) control valves, 3) steam turbines, 4) gearboxes, 5) steam or air leaks, 6) centrifugal and screw compressors, 7) electric motors, 8) airfin coolers, 9) furnaces, 10) fans, 11) reciprocating compressors, and 12) gasoline and diesel engines.

These machinery sources generate noise through a variety of mechanisms. The noise of most of these sources is broad band, although many contain pure tone sources. The pure tones are normally high (up to 4000 Hz) for the first sources in the list and low (down to 30 Hz) for the last sources in the list.

Flow noise is usually generated in pipes and ducts where pressure reduction occurs. Turbulence is created, and its interaction with the pipe or duct wall gives rise to internally transmitted fluid-borne sound or structural vibration resulting in external airborne sound. Control valves create sound in a similar manner. Shock waves can also be created at the valves. Many control valve suppliers manufacture low noise valves or else give information on methods to predict noise from their valves. Steam turbines produce broadband noise caused by turbulence but also tonal noise which is related to the product of the number of blades and the rotational speed and harmonics (integer multiples). Gearboxes produce *tonal noise* at both meshing frequencies and harmonics. Electric motors produce tonal noise caused by magnetic forces. The magnetic force noise occurs at multiples of twice the line frequency. Electric motors usually contain cooling fans.

Fan noise is broadband but also contains tonal components at integer multiples of the blade passing frequency (number of fan blades multiplied by rotational speed in revolutions/second). Reciprocating compressors and gasoline and diesel engines produce some broadband noise but also have tonal components at pumping or firing frequencies and harmonics. These tonal components are related to integer multiples of the pumping or firing strokes each cycle multiplied by the rotational speed (in revolutions per second).

23.3 FLUID DYNAMICS OF COMBUSTION
Edward E. Zukoski

23.3.1 Introduction

The aim of this Section is to acquaint the uninformed reader with a range of combustion phenomena which will be discussed in a qualitative manner. Glassman (1987), Strehlow (1968), Stull (1977), Kuo (1986), and Williams (1965) are texts which discuss combustion in great detail. Older texts such as Jost (1946), Lewis and von Elbe (1951), and Lewis *et al.* (1956) contain extensive data as well as theoretical material and remain useful works. In addition, the *Combustion Symposium* volumes and a number of journals, e.g., *Combustion and Flame, Combustion Science and Technology,* and *Combustion, Explosion, and Shock Waves,* should be consulted for current research results and reviews. Although not specifically oriented to combustion, the *AIAA Journal,* the *Journal of Spacecraft and Rockets,* the *Journal of Propulsion and Power,* and the *Physics of Fluids* contain combustion articles of interest. Several specific articles from the combustion literature are discussed in this section [Field (1982), Fay (1962), Launder and Spalding (1972), and Gordon (1982)].

23.3.2 Heat Release Calculations

An important part of any calculations dealing with chemical reactions is the calculation of the heat released in the reaction. This subsection deals with some of the definitions and ideas used to calculate heat release. Refer to Chap. 22 and Sec. 2.7 for complementary information.

Stoichiometry. For systems involving air, the mixture ratio is called the *fuel–air ratio* and is often specified by fixing a quantity called the *equivalence ratio* which is defined as the ratio of the actual fuel–air ratio and the *stoichiometric fuel–air ratio.* The stoichiometric fuel–air ratio is the value of the fuel–air ratio for which sufficient oxygen is present in the mixture to completely oxidize the fuel atoms. The same nomenclature is used for systems with oxidizers other than air with the word *air* replaced by the word *oxidizer.*

As an illustration of these definitions, consider the stoichiometric methane–oxygen system

$$2(O_2) + (CH_4) = (CO_2) + 2(H_2O) \tag{23.67}$$

To achieve the stoichiometric fuel–oxidizer ratio, sufficient oxygen must be present to reduce the hydrogen to water and the carbon to carbon dioxide. Thus, for the methane–oxygen system, 2 moles of oxygen are required for each mole of methane and the fuel–oxygen ratio, given in terms of the mass, will be about 0.25. If the methane–air system is considered, the stoichiometric fuel–air ratio is reduced to about 0.058, because the mass fraction of oxygen in air is only about 23%. For most gaseous hydrocarbons, the value of the stoichiometric fuel–air ratio is between 0.065 and 0.070. Thus, about 16 kg of air are required to completely oxidize 1.0 kg of hydrocarbon fuel.

The equivalence ratio, ϕ, is a convenient expression for the mixture ratio, because most of the heat release and reaction rate parameters of combustion systems depend in a similar manner on the fuel–air ratio when it is expressed in terms of the equivalence ratio. For example, the temperature of the combustion products and the flame speeds of most hydrocarbons and other fuels have maximum values near an equivalence ratio of 1.1.

Heat of Combustion. The enthalpy of a gas is defined in terms of a differential relationship, hence it is only defined to within an additive constant. When gases are treated which can undergo a chemical reaction, these constants can no longer be arbitrary but must be selected in conformity with general thermodynamic laws. This can be done by using the following set of definitions.

The *enthalpy* is defined as

$$h_i(T) = \int_{T_s}^{T_{c_{pi}}} dT + H_{fi}(T_s) \tag{23.68}$$

where T_s is the reference or standard state and H_{fi} is the heat of formation of species (i) from the elements comprising (i) evaluated at the reference temperature and pressure, T_s and p_s. In addition, the heat of formation of the elements in their standard states are taken to be zero.

To illustrate this definition, consider the following simple problem: A thermodynamic system consists of a gas contained within a box whose volume can be changed to maintain the pressure within the box constant. At the start of the experiment, state (1), the gas is a stoichiometric mixture of oxygen and hydrogen at reference conditions of temperature and pressure, T_s and p_s. The gas is then allowed to react completely to form water vapor in a constant pressure and *adiabatic* process. The water vapor will be at pressure p_s but at a higher temperature, T_2, defined as the *adiabatic flame temperature*. Finally, heat is added to the box to reduce the temperature to the initial value T_s. Thus in state (3), the gas is water vapor at the reference values for the temperature and pressure. The enthalpies of the system in states (1) and (2) are identical because heat was not transferred to the system and the pressure was held constant. The heat added to the system, between states (2) and (3), to return the water vapor to the reference temperature is called the *heat of formation*, H_f, of the water. In this example, it is clearly a negative number.

Thus, the use of the definition for the enthalpy, which includes the heat of formation, keeps the enthalpy of a system undergoing an isobaric and adiabatic chemical reaction unchanged, as it must. It is convenient to think of the heat of formation defined in this manner as the *chemical potential energy* of the molecule. Note that for water, carbon dioxide, and other oxides, the heat of formation will be a large negative number, consequently the enthalpies of these species defined in this manner will also be negative for temperatures near room temperature. Heats of formation

TABLE 23.1   Heats of Formation at 25 C

Compound	Formula	Phase	Heat of Formation, H_{fi}, kJ/gram-mole
Carbon	C	solid	0
Carbon	C	gas	+529
Various gases	N_2, O_2, H_2	gas	0
Methane	CH_4	gas	− 75
Acetylene	C_2H_2	gas	+227
n-Octane	C_8H_{18}	liquid	−250
Methyl alcohol	CH_3OH	liquid	−239
Water	H_2O	liquid	−286
Carbon monoxide	CO	gas	−111
Carbon dioxide	CO_2	gas	−394
Hydroxyl radical	OH	gas	+ 42
Hydrogen atom	H	gas	+218
Oxygen atom	O	gas	+248

and specific heats are the basic chemical information needed to make thermochemical calculations when the species involved are known. Table 23.1 lists a few representative values. Note the large positive heats of formation for atomic hydrogen and oxygen, and the hydroxyl radical. The fuel acetylene has a positive heat of formation and is unstable when shocked or heated.

Given this definition of the enthalpy, the heat of reaction is built into the definition of the enthalpy and need not be treated as heat added from an outside source. Thus, if a steady, inviscid, one-dimensional flow of a compressible, combustible, ideal gas is considered in which reactions are allowed to occur as the gas flows from states (1) to (2), the energy equation is reduced to the statement that the total enthalpy is conserved across the reaction zone. This result can be written as

$$\sum_r Y_r \left(\int_{T_s}^{T_1} c_{pr} \, dT + H_{fr} \right) + V_1^2/2 = \sum_p Y_p \left(\int_{T_s}^{T_2} C_{pp} \, dT + H_{fp} \right) + V_2^2/2 \quad (23.69)$$

Here, Y_i is the mass fraction of reactants or products, V is the velocity of the mixture, and subscripts r and p refer to reactants and products. Thus, in the simple case that the specific heats are about equal and are not functions of the temperature, Eq. (23.69) reduces to

$$[c_p(T_2 - T_1)] = \left[\sum_r (Y_r H_{fr}) - \sum_p (Y_p H_{fp}) \right] + (V_1^2 - V_2^2)/2 \quad (23.70)$$

The first term with square brackets on the right-hand side is the chemical energy which has been released in the reaction and which is available to increase the temperature of gas; the second term is the change in the kinetic energy. The first term is sometimes called the *heat of reaction* or *heat of combustion* in our case. Note that the temperature increase is maximized when the reactants have large positive heats of formation, and the products have large negative heats of formation.

For hydrocarbon fuel reacting with air at an initial temperature of 300 K, the heat of reaction is in the range of 40 to 50 MJ/kg when the water vapor in the products of combustion is assumed to remain in the vapor state. This value, called the *lower heat of formation*, is slightly less than the *upper heat of formation* which is based on the assumption that the water vapor condenses to a liquid state. For pure hydrogen, the heating value is about 120 MJ/kg and for partially oxidized compounds, such as carbon monoxide or methyl alcohol, the heating value is greatly reduced. Values for the lower heat of combustion are listed in Table 23.2 for a range of fuels burning in air at a pressure of one atmosphere. The initial temperature of the mixture is 25 C. Note, that as the carbon-to-hydrogen ratio increases, the heating value decreases. In general, the heating value per unit mass of a liquid hydrocarbon decreases as the density of the fuel increases, since the mole fraction of hydrogen in the molecule also decreases.

The fuel–air mass ratio (mass of fuel to mass of air) for the stoichiometric mixture for most of the hydrocarbon fuels listed in Table 23.2 is in the range 0.05 to 0.10; the value for hydrogen is only 0.0292. The adiabatic flame temperatures for hydrocarbon fuels burning in air have maximum values for fuel–air ratios in the range 110 to 130% of the stoichiometric value. For most of the hydrocarbon fuels listed in Table 23.2, the adiabatic flame temperature is close to 2200 K; the corresponding values for hydrogen, carbon monoxide, and acetylene are slightly higher.

When the initial mixture temperature is between 300 and 1300 K, an increase in the initial temperature of one degree produces an increase in the adiabatic temper-

TABLE 23.2   Lower Heats of Combustion, MJ/kg of Fuel, and Maximum Adiabatic Flame Temperatures (Fuel at 25 C and 1 atm, with Gaseous Water in Products)

Compound	Formula	Heat of Combustion	Phase of Fuel	Maximum Adiabatic Flame Temperatures
Hydrogen	H_2	120.0	gas	2450 K
Carbon	C	32.8	solid	
Carbon monoxide	CO	10.1	gas	2350
Methane	CH_4	50.0	gas	2210
Ethane	C_2H_6	47.5	gas	2170
Propane	C_3H_8	44.4	gas	2200
n-Butane	C_4H_{10}	45.7	gas	2170
n-Pentane	C_5H_{12}	45.5	gas	
n-Hexane	C_6H_{14}	45.1	gas	
n-Heptane	C_7H_{16}	44.9	gas	2290
n-Octane	C_8H_{18}	44.8	gas	
Benzene	C_6H_6	40.6	gas	
Ethylbenzene	C_8H_{10}	41.3	gas	
n-Butylbenzene	$C_{10}H_{14}$	42.8	gas	
Ethylene	C_2H_4	47.2	gas	2250
Acetylene	C_2H_2	40.6	gas	2600
Naphthalene	$C_{10}H_8$	40.3	solid	
Methyl alcohol	CH_4O	26.8	liquid	
Ethyl alcohol	C_2H_6O	38.9	liquid	
Crude oil	—	43 to 45	liquid	
Bituminous coal	—	25 to 34	solid	~2000
Wood	—	16 to 19	solid	

ature for the stoichiometric mixture of about 1/2 degree. This result occurs because the specific heats of the products of combustion increase rapidly with temperature for temperatures close to the adiabatic flame temperature. The heat of combustion also depends on the pressure, because the degree of dissociation decreases as the pressure increases. However, this dependence is weak.

Equilibrium Calculations. The chemical equation given above in Eq. (23.67) is of course not an accurate description of what actually occurs in combustion reactions even when the stoichiometric fuel–air ratio is used. The species present at thermodynamic equilibrium in a mixture which is formed by complete combustion of fuel in air in an adiabatic system also can include oxygen, hydrogen, carbon monoxide, the oxides of nitrogen, and other species such as the hydroxyl radial and monatomic oxygen and hydrogen which are the result of dissociation reactions. The mass fraction of these species increases with increasing temperature and decreasing pressure. Since these product species have zero or positive heats of formation, their presence in the mixture reduces the temperature of the gas. The data listed in Table 23.3 illustrates the magnitude of this effect. Equilibrium values for the mole fraction ratios are given as a function of the temperature T and pressure p (in atmospheres).

For the data of Table 23.3, the fuel has the composition $C_n H_{2n}$, and the stoichiometric fuel–air ratio was used. Note the rapid increase in all the terms as the pressure decreases and temperature increases. At atmospheric pressure and temperatures below about 1500 K (2700 R), the products of dissociation can be ignored.

The equilibrium composition of a mixture can be found from a thermodynamic calculation which requires a knowledge of the *Gibbs function* for all of the species present in the products and reactants (see Chap. 7 and Sec. 2.7). Thus, the species present and the pressure and temperature of the mixture must be known. The calculation is based on the fact that the magnitude of the Gibbs function for the mixture is a minimum value when the composition of the gas reaches its equilibrium value.

A number of computer programs are available which calculate compositions and temperatures for simple gas mixtures using a minicomputer. When solid or liquid products exist, more complex programs are required. The code by Gordon and McBride from NASA Lewis Research Center is highly recommended.

Reaction Rates. In a reacting gas mixture, reactions can occur when the molecules collide under certain circumstances. The collisions which produce reactions are typically of the form

TABLE 23.3   Ratios of Products for Equilibrium Compositions

T, K	p, atm	CO/CO_2	H_2/H_2O	H/H_2O	O_2/H_2O
1500	1.0	0.0064	0.0024	0.0000	0.0003
2050	0.1	0.086	0.0183	0.0024	
	1.0	0.040	0.0085	0.0005	0.0143
	10.0	0.0015	0.0040	0.0001	
2500	1.0	0.316	0.0516	0.0146	0.0878
3000	1.0	1.889	0.259	0.254	0.3068

Adapted from Gordon (1982).

$$OH + H_2 \rightarrow H_2O + H$$

$$CO + OH \rightarrow CO_2 + H$$

$$H_2O + M \rightarrow OH + H + M$$

$$OH + H + M \rightarrow H_2O + M \tag{23.71}$$

Reactions of the type used to describe the stoichiometry of the reaction, see Eq. (23.67) for example, do not occur in a single collision but are the result of many individual reaction steps of the type shown above. Note that the reverse reactions can also occur (see the last two equations). Reverse reactions are usually considered to be separate reactions, and they proceed at different rates.

The rate at which reactions occur in a gas mixture depends very strongly on the temperature and composition of the mixture. For a second-order reaction (which involves two molecules),

$$N_1 + N_2 \rightarrow 2N_3 + N_4 \tag{23.72}$$

the rate at which molecules of type N_1 are consumed per unit volume due to the reaction given in Eq. (23.72) is proportional to the rate at which collisions occur between the two species N_1 and N_2 and thus is proportional to the product of the molar concentrations of species 1 and 2, $[N_1]$ and $[N_2]$. (Here, $[N_2]$ is the number of moles of species 2 per unit volume.) In addition, because most molecules have a force field about them which repels other molecules when their separation distance is small, the collisions which can produce reactions are restricted to those which have a kinetic energy (in a center of mass system) greater than the limiting value required to overcome the effects of this potential field. This limiting energy is called the *activation energy*, A, and results from the kinetic theory of gases show that the fraction of collisions which have energy greater than A is proportional to the factor, $\exp(-A/RT)$ where R is the gas constant and T the temperature. This exponential term is called the *Arrhenius factor*.

Given this background, it is reasonable to expect that the rate at which molecules of species (1) are created (moles per unit volume and per unit time) could be written as

$$d[N_1]/dt = -K(T)\,[N_1]\,[N_2] = -f(T)\,[\exp(-A/RT)]\,[N_1]\,[N_2] \tag{23.73}$$

Here, K is the *rate coefficient*, and f is a dimensional function called the *frequency factor* which varies weakly with the temperature and has a value which lies between one and zero. Thus, the rate equation can be viewed as indicating that the reaction may only occur between two molecules when they collide with an energy greater than A and that the fraction of times that they actually react is f.

When the stoichiometric coefficients of the reaction equation are different from one, as is the case for the species N_3 in the above equation, the collision rate still depends on the product of the molar concentrations of the molecules involved in the reaction on the molecular scale, thus the rate of production of N_3 in the reaction

$$2N_3 + N_4 \rightarrow N_1 + N_2 \tag{23.74}$$

has the form

$$d[N_3]/dt = 2(-K_r(T) [N_3] \cdot [N_3] \cdot [N_4]) \tag{23.75}$$

or

$$d[N_3]/dt = -2(f_r(T) [\exp(-A_r/RT)] [N_3]^2 [N_4]) \tag{23.75a}$$

The subscript r is used on A, f, and K to show that the reverse reaction is involved.

The exponential nature of the reaction rate coefficient is responsible for many of the features of flames discussed below. In a typical case, the term A/R has a value between 5,000 and 25,000 K, consequently the reaction rate is a very sensitive function of the temperature. This sensitivity leads to problems in numerical calculations involving reactions. Rates are known for many reactions from experimental work but cannot yet be predicted from basic principles for arbitrary reactions [see Strehlow (1968)].

To illustrate the process of calculating the production rates of a species in a more general reacting mixture consider the reactions

$$aA + bB \rightarrow cC + dD, \qquad \text{forward rate constant } K_f(T) \tag{23.76}$$

and the reverse reaction

$$cC + dD \rightarrow aA + bB, \qquad \text{reverse rate constant } K_r(T) \tag{23.77}$$

where a, b, c, and d are the stoichiometric coefficients for the reactions for species A, B, C, and D. The rate of production of species C (*mass* per unit volume per unit time) is called w_c and is given by

$$w_c/M_c = d[N_c]/dt = c(K_f([N_a]^a \cdot [N_b]^b) - K_r([N_c]^c [N_d]^d)) \tag{23.78}$$

and in general,

$$w_a/aM_a = w_b/bM_b = w_c/cM_c = w_d/dM_d \tag{23.79}$$

M_i is the molecular weight of the ith species and is used to convert production rates in units of mass per unit volume to moles per unit volume. When more than a single, reversible, reaction step is involved, the rate of production of a given species is obtained by summing the production rates for all the reactions which produce or consume the species.

23.3.3 Ignition

Consider a small mass of hot gas created suddenly within a combustible mixture of fuel and oxidizer. Heat and species will diffuse from this hot core into adjacent elements of the unburnt mixture. As the adjacent combustible material is heated, the oxidization reactions will become important, and heat will be released by the chemical reactions. If the heat addition rate, due to conduction and chemical reactions is sufficiently larger than the heat loss rate, the reaction can continue to completion in that element and can spread on to adjacent elements to form a propagating flame.

It is clear from this simple model and confirmed from experimental results that the minimum energy required for ignition is a sensitive function of the mechanism used to form the original *ignition source*. Typical sources include: open flames, arc or spark heated gas, hot surfaces, or gas compression heating by shock waves, etc. The energy required for ignition depends on the apparatus and heat source used, on the fuel–air ratio of the mixture, and on the pressure and temperature. Minimum ignition energy is usually required when the mixture is close to the stoichiometric mixture. When differences in diffusion rates between fuel and oxygen are large, the composition of the gas in the combustion region can be quite different from that in the mixture itself, therefore the initial mixture ratio for minimum ignition energy can be far from the stoichiometric value. Thus, for example, the *minimum ignition energy* for heptane occurs near twice a stoichiometric mixture, whereas that for methane is near 0.8 of stoichiometric [Lewis and von Elbe (1951)]. The diffusion coefficient for oxygen is much greater than that for heptane, but is smaller than that for methane. The ignition process also exhibits lean and rich limits beyond which ignition is not possible. These limits are quantitatively similar to the flammability limits discussed below for laminar flame propagation.

Heating a large mass of combustible material can also cause ignition; this ignition process is called *spontaneous ignition* and occurs at a surprisingly low temperature for many common fuels. For hydrocarbon fuels burning in air, the *minimum spontaneous ignition temperature* is about 700 K for fuels with low specific gravity (in the range 0.6 to 0.7) and falls to values between 500 and 600 K for fuels with high specific gravity (in the range about 1.0). Ignition occurs at the lowest temperature for fuel–air ratios near stoichiometric, and the ignition limits spread to a wider range as the temperature is increased.

Ignition phenomena are associated with another effect called *quenching*. If a burning gas is cooled convectively, for example, the flame can be extinguished. This quenching process is often studied by forcing a flame front, which is propagating through a premixed mixture, to pass through a set of plates set parallel to the gas stream. The spacing between the plates which first causes the flame to be extinguished is called the *quenching distance*. The dependence of minimum ignition energies and quenching distance on the properties of a combustion system are qualitatively similar.

23.3.4 Modes of Combustion

Combustion is the process of oxidation of a fuel by an oxidizer, and this process can occur with a large variety of fuel and oxidizer types and with a large number of combustion mechanisms. A few mechanisms will be considered here and will be restricted to those processes in which an appreciable heat release results from the chemical reactions. Combustion systems can be characterized by a number of properties. These include: 1) the phases of the fuel and oxidizer, 2) the degree to which the fuel and oxidizer are mixed prior to the initiation of combustion, 3) the intensity of turbulence in the flow, and 4) the influence of buoyancy on the combustion process.

To illustrate these categories consider a few examples. The typical home gas furnace involves the combustion of laminar jets of premixed fuel gas and air. The *Otto cycle* for the internal combustion engine is approximately a premixed, prevaporized, turbulent fuel–air system whereas combustion in the *Diesel cycle* occurs more nearly in a turbulent diffusion flame between a vaporizing fuel jet and air. Both premixed and droplet combustion can be important in gas turbine engines. Finally, the combustion of coal in an industrial furnace occurs in both gaseous diffusion flames, involving volatile components of the coal and air, and diffusion flames between a gas and a solid fuel particle.

High turbulence levels occur in most of the combustors described here, and the influence of buoyancy forces may be important in several. In the following paragraphs a few of these processes will be described in more detail to introduce the subject and to develop the nomenclature used in this field.

Premixed Gaseous Flames. In many interesting applications, gaseous fuel and oxidizer are mixed together before combustion is allowed to occur. Combustion can occur in these premixed systems in a wave-like process called a *deflagration wave* or more commonly a *flame*. In this process, chemical reactions and heat release occur in a well defined region called a *flame front*. The front is usually very thin (a fraction of a millimeter or less for hydrocarbon–air flames burning at a pressure of one atmosphere), and it separates a cool, unburnt fuel–air mixture on one side from hot products of combustion on the other. The products may be at temperatures of several thousand degrees Kelvin, and the pressure jump across the flame for deflagration waves is often negligible.

Most of the chemical reaction is restricted to this thin region, however, in some flames, slow chemical reactions may continue for several flame thicknesses downstream of the principal reaction region. For example, in flames involving the combustion of carbon, the oxidation of carbon monoxide to carbon dioxide usually extends an appreciable distance downstream of the front.

Radiation generated as a result of the chemical reactions often makes the flame front visible. In hydrocarbon systems, visible radiation is also produced in very fuel-rich mixtures due to the production of radiation from *soot* generated in the reaction. For nonsooting systems, the fraction of the heat released which is lost due to radiation is usually negligible, except perhaps at some limiting condition such as a *flammability limit*.

In a very crude sense, the processes occurring in the flame involve the transport of heat and species from the hot burnt gas forward into unburnt material. When the reaction rate in the unburnt gas heated by this process is fast enough, sufficient heat will be released to maintain a high temperature and thus sustain the chemical reaction, and the flame will move forward. If the reaction rate is too slow, the transport processes will dominate, the temperature in the unburnt gas will not rise fast enough, and the reaction process will die out. This balance between reaction and transport rates is important in many processes involving premixed fuel–air systems.

Laminar Flame Speed: Although diffusive processes fix the rate of transport of chemical species and heat which occur within the laminar flame front, premixed laminar flames propagate through the mixture with a constant velocity (relative to the unburnt mixture). This unexpected behavior results in large part from the exponential dependence of the reaction rates on the temperature.

The propagation velocity or *flame speed* in a quiescent mixture is fixed by transport properties of the mixture, such as the diffusivity and thermal conductivity, and also by the chemical properties such as the heat of combustion and the chemical reaction rates for critical reactions. The chemical properties depend strongly on the mixture ratio and more weakly on the temperature and pressure.

The flame speed has a maximum near the stoichiometric fuel–air ratio where the adiabatic flame temperature is a maximum, and it approaches zero as the ratio approaches values on either side of stoichiometric ratio, called the *flammability limits*. Flame speeds increase with increasing temperature. The dependence on pressure is weak, and flame speeds can increase or decrease with increasing pressure.

Mass Consumption Rate: The flame speed is one of the important characteristics of a fuel–oxidizer system, however the *mass consumption rate*, the mass flow rate per unit area through the flame front or the product of flame speed and mixture density, is perhaps a more useful parameter. For the methane–air system at room temperature and pressure, the maximum value for the flame speed is about 0.4 m/s and consumption rates are less than 0.5 kg/m^{2}s. For hydrogen–air flames, a maximum flame speed of about 3.0 m/s and consumption rates of 4 kg/m^{2}s are observed. Flame speeds in mixtures of pure oxygen and hydrocarbon fuels are much higher. For example, for methane–oxygen mixtures the maximum speed and consumption rates are about 3 m/s and 4 kg/m^{2}s, and the corresponding values for hydrogen–oxygen mixtures are 9 m/s and 11 kg/m^{2}s.

Flammability Limits: The *flammability limits* mentioned above are the fuel–air ratios, both on the *lean* (or excess-oxygen) side of stoichiometric and on the *rich* (or excess-fuel) side, outside of which a propagating flame cannot be sustained in the premixed mixture. The values reported often depend slightly on the apparatus used in the experiment, and considerable scatter occurs in reported values. A general rule for hydrocarbons is that the flammability limit corresponds to the fuel–air ratio at which the temperature of the flame drops to a value in the range of 1100 to 1300 K. Typical lean and rich flammability limits for hydrocarbons burning in air are about 0.5 and 3.0 times the stoichiometric fuel–air ratio. The addition of an inert diluent such as nitrogen, helium, or even products of combustion to a fuel–air mixture will reduce the flame temperature and flame speed and also will move both flammability limits toward the stoichiometric value.

In general, given sufficient effort, the characteristics of a laminar flame can be deduced accurately from calculations provided the thermal properties and the chemical reaction rate constants of the gases involved are known. Thus, even the production of trace species such as nitric oxides can be calculated satisfactorily for a laminar, premixed methane–air flame, however about 100 reaction rate equations and accompanying constants are required to achieve accurate predictions.

Table 23.4 lists some characteristics of several fuel–air mixtures which are typical of laminar flames. The flame speeds reported in Table 23.4 are the maximum velocities (which occur with fuel–air ratios near stoichiometric) with air at room conditions. The flammability and ignition limits are given in volume percent of the fuel vapor in the fuel–air mixture and are for room conditions of pressure and temperature.

Detonation Waves. A *detonation wave* propagating into a premixed fuel–air mixture can be usefully thought of as the combination of a conventional shock wave which

TABLE 23.4   Laminar Flame Properties

Fuel	Flame Speed cm/s	Flammability Limits		Ignition Limits	
		Lean	Rich	Lean	Rich
H_2	260	12	75	4	74
CH_4	40	5	14	5	15
C_3H_8	46	2	10	—	—
Heptane	32	1	6	1	6
NH_4	26	15	28	—	—

Note: The oxidizer is air.

is followed by an *incubation zone* for the chemical reaction and then by a combustion zone. The high temperature and pressure produced by the shock wave (see discussion given in Chap. 8) keep the incubation zone very short and allow the mixture to burn at the very high speeds characteristic of the detonation process. Typically the whole region is several millimeters thick [Fay (1962)].

This system of waves propagates through premixed and gaseous mixtures with velocities of one to two thousand meters per second and with Mach numbers as high as 5 for fuel–air mixtures. The wave speed observed at a specific time and in a particular system depends in a complex manner on the geometry of the system and the ignition point, and it may be strongly time dependent.

To illustrate this process, calculated properties of a detonation wave in a mixture of hydrogen (20% by volume) and air are presented in Table 23.5. In Table 23.5, station (1) is located upstream of the shock wave, (2) is in the incubation zone, and (3) is downstream of the wave. The Mach number and velocity are values seen by an observer moving with the detonation wave. Note the large pressures and the pressure drop associated with the heat addition process [station (2) to (3)].

Most detonations in the laboratory or in accidents occur in premixed fuel–air mixtures as the result of a transition from a deflagration process. When the combustion process starts in a partially confined space, it can cause an appreciable pressure rise which is then available to accelerate the unburnt mixture. The resulting high speed flow produces a shock wave in the combustible mixture which then develops into the detonation wave. Early in the development of the detonation, the original deflagration front may be separated from the detonation front by unburnt material which burns later.

Under certain circumstances a single, well-defined speed is identified, and it is called *the detonation velocity*. This velocity is the *Chapman–Jouguet speed* and has

TABLE 23.5   Properties for a Typical Detonation

Station	1	2	3
Velocity, m/s	1500	300	540
Mach number	4.5	0.42	1.0
Pressure, atm	1	23	13
Temperature, K	294	1350	2425

Adapted from Glassman, 1977.

the property that the velocity downstream of the wave is sonic with respect to the wave. Thus, weak pressure waves will not overtake the detonation front, and the wave speed in a one-dimensional flow can be constant. The example given above in Table 23.5 has this property.

The picture given above of the detonation wave as a plane shock followed by a combustion zone is over-simplified; the shock front in a real detonation may not be a plane wave, but it is often a combination of curved fronts which intersect to form *lambda shocks*. This complex wave structure appears to play an essential role in fixing the development of the wave. However, for computational purposes, the simple picture leads to reasonable estimates of the wave speed.

Detonations have many of the properties of laminar flames. Thus, detonation velocities have a maximum value at fuel–oxidizer ratios close to the stoichiometric value and exhibit lean and rich limits beyond which no detonations are possible. For example, detonations can occur in hydrogen–air mixtures when the percent of hydrogen in the mixture is in the range, 18% to 60%; the corresponding range for deflagrations is about 4% to 74%. The stoichiometric ratio for this system is 2.92%.

Note that although the fuel vapor and air system is the best understood detonation process, the fuel need not be premixed on the molecular scale; detonations have been observed to propagate through mixtures of air and either fuel droplets or dust particles [Field (1982)]. Thus, for example, detonations have been observed in mixtures of air and either starch or coal dust, and detonations are a dangerous hazard in systems involving the transport of pulverized coal, wheat, and corn.

In strong contrast with the premixed laminar flames described above, very large pressure increases can occur across the detonation wave as a result of the normal shock, and many accidental explosions are associated with detonations. However, deflagrations produce large temperature increases in the combustion products, consequently a deflagration in a closed volume can also cause a large pressure rise in the system which will produce many of the characteristics of an explosion. Combustion in an internal combustion engine occurs in a premixed mixture of fuel and air as a deflagration wave when the system is operating properly. However, under some operating conditions and when *knocking* is heard, a detonation occurs during some part of the combustion process.

Detonations can also occur in premixed fuel and oxidizer systems in which the components are in the liquid or solid phases. In explosive compounds such as TNT, the fuel and oxidizer can be contained within the same molecule.

Diffusion Flames. Diffusion flames are the second major category of combustion processes and involve combustion in systems in which the fuel and oxidizer are initially completely separated.

Gaseous Diffusion Flames: Consider first diffusion flames which occur at a front between unmixed masses of gaseous fuel and oxidizer. Combustion occurs in a relatively thin region of chemical reaction and heat release which is again called a *flame front*. The transport of fuel and oxygen to the front and that of the products away from it are accomplished by a diffusion process. Because the reaction rates are highest in a stoichiometric mixture, the flame front sits at the position for which the ratio of mass fluxes of fuel and oxygen is the stoichiometric ratio. Hence, the dependence on fuel–air ratio which was so strong for the premixed flames is absent here. As a first approximation, the temperature in the flame front is the adiabatic flame temperature for the stoichiometric ratio.

The flame does not propagate through the mixture in the same sense as the premixed flame, rather the fuel and oxidizer diffuse into the flame and products of combustion and heat diffuse away from it. Depending on the stoichiometry of the system, the flame moves slowly into either the fuel or oxidizer gas mass, however the products of combustion remain around the flame and form a layer on either side of it. As the thickness of this region increases, the diffusive gradients are reduced, consequently the flux of fuel and oxidizer to the front decrease. The mass consumption rate of the flame also decreases and falls approximately as the reciprocal of the square root of the time. Thus, the diffusion flame is inherently a nonsteady process, and for times of a millisecond or less, the consumption rate is comparable to that of the premixed flame.

As the fuel molecules diffuse toward the flame, they are exposed to the products of combustion which contain little if any free oxygen and to temperatures approaching the adiabatic flame temperature. Under these conditions, hydrocarbon fuels undergo a complex set of reactions called *pyrolysis* reactions which result in the production of very long chains of solid particles. These particles, called *soot*, consist of high carbon to hydrogen ratio compounds, and the particles grow to sizes in the 1/10 micron range. In a diffusion flame, the particles are burned in a surface reaction between solid fuel and gaseous oxygen or carbon monoxide. In a large diffusion flame, the combustion of these soot particles keeps the particle surface temperature in the 1200 to 1400 K range, and the resulting thermal radiation produces the yellow light characteristic of hydrocarbon diffusion flames. This thermal radiation from the burning soot results in a radiation flux from the flame which can amount to 20 to 40% of the total heat released by the chemical reactions.

A steady diffusion flame can be created in a region of strong shear in which the flow and the flame are strained at a steady rate. Flows such as that which occur at a stagnation point can be used to study this phenomena, and a steady state flame with a constant thickness can be produced.

In many industrial processes, diffusion flames occur when fuel is injected into air in discrete jets. When the momentum of the jet is small enough, the *buoyancy* of the resulting hot gas formed in the combustion process will dominate the motion of the gas and the combustion process itself. Buoyancy appears to dominate when the initial momentum flux of the jet is smaller than the buoyant force acting on the whole flame region. Properties of buoyancy and momentum dominated flames such as flame lengths and entrainment rates of air are quite different.

Heterogeneous Diffusion Flames: Diffusion flames can also exist when one of the components is gaseous and the other is either solid or liquid. For example, the combustion of a single liquid fuel droplet injected into air can occur in a diffusion flame which surrounds the droplet. Heat transferred from the flame vaporizes the droplet or particle, and the vapor feeds a spherical diffusion flame located a number of droplet radii from the surface. When a cloud of droplets is present, the diffusion flame may at times surround the whole cloud rather than individual droplets. In this example, the combustion still occurs as a diffusion flame between gaseous fuel and air.

A similar process is involved in the early stages of the combustion of a solid coal particle; heat transferred to the solid surface produces a flow of volatile material from the coal which feeds a diffusion flame that surrounds the particle. However, in a later stage of the combustion of a coal particle, the primary reaction occurs at the surface of the particle between a gaseous oxidizer and a solid or liquid carbon residue which contains little hydrogen. This process is still called a diffusion flame, because the oxidizer must diffuse to the surface. Surface reactions of this type are also important in the combustion of soot particles.

In the limit that the solid or liquid fuel is dispersed in very small diameter particles through air, the process has some of the characteristics of a premixed system even though the combustion itself may occur as a diffusion flame lying about each particle. Consider for example the combustion of aluminum dust in the size range of 1 to 30 microns dispersed in air at concentrations of 0.1 to 0.3 kg/m^3. This mixture has a well defined laminar flame speed with propagation velocities of 10 to 30 cm/s, a lean flammability limit and minimum ignition energy [see discussions in Stull (1977)]. Similar situations arise for corn starch or pulverized coal. These systems differ from gaseous flames because the transfer of energy by radiation becomes very important in the solid fuel case and because separation or addition of the solid or liquid fuel particles from the air can occur as a result of motion of the air. The latter effect is particularly important in *dust explosions* [Fleid (1982)]. For example, dust can be removed because the dense particles sink to the floor due to gravitational forces, however motion of the gas can pick up particles from floors or walls and return them to the gas phase where they can burn. Thus, the gas motion resulting from the combustion process causes additional particles to be picked up from floors and walls. Combustion of this additional fuel causes increased heat addition and an increase in gas velocities which can cause more dust to be picked up. Thus, the *dust pick-up process* furnishes a strong feed back path which can lead to rapid heating of the gas and even to detonations.

Turbulent Flames. In most combustion systems the influence of turbulence [Launder and Spalding (1972)] on the process is marked. Turbulent motion in the gas surrounding a flame has several effects on either a premixed or diffusion flame. First, if the scale of the motions in the gas is large compared with the thickness of the front, we can expect that the major effect of the turbulence will be to distort the front and hence increase the rate of consumption by increasing the area of the flame surface. Second, when the turbulent velocity fluctuations become comparable with the flame speed, the stretching or straining of the flame surface resulting from the fluctuations will act to increase the rate of consumption by decreasing the thickness of the flame front and hence increasing the gradients of temperature and species. However, if the straining rate becomes too large, it can cause the temperature gradients in the reaction zone to increase. Then, the heat transfer rate from the reaction zone will overwhelm the heat added by the chemical reaction, and the temperature in the zone will decrease. Consequently, the chemical reaction will stop, and the flame will be quenched. Finally, if the scale of the turbulence becomes smaller than that of the flame front, the transport properties of the gas will be increased above their molecular values, hence the flame speed and consumption rates will increase.

Calculations of turbulent flame properties is complex because of the requirement for the prediction of a local and instantaneous chemical reaction rate. The rate depends on the *instantaneous* values of the temperature, pressure, and the local concentrations of the species involved in the reaction which are mixed at the molecular level. Thus, the mean and fluctuating values of these parameters are needed as usual for a turbulent flow calculation, but, in addition, a prediction is needed for the correlations between the fluctuations of the parameters. An accurate description of the fluctuations is necessary because the concentration fluctuations can be very large and because, in most reactions, temperature enters in a rapidly varying exponential term. Hence, the common k-ϵ treatment of a turbulent flow (see Sec. 4.5) is not sufficient for cases in which reactions are important. In these cases, the probability distribution functions for the various variables must be known before a calculation can be made.

23.3.5 Equations of Motion

The discussion presented above gives a qualitative description of some combustion phenomena. Before closing this section, it is appropriate to present the equations of motion for a system of reacting gases and to discuss briefly some of the new terms which appear in the equations when combustion is considered. The complications which are encountered in these problems arise because the gas is a mixture of gases whose composition may change due to either chemical reactions or diffusion of the species relative to one another.

The continuity and momentum equations for the mixture as a whole are identical to those for a nonreacting, single species gas with the exception that the transport properties of the system such as viscosity and thermal conductivity depend on the local composition of the gas in addition to the temperature and pressure. Thus, derivatives of these coefficients must also contain terms involving the gradients of the species.

When chemical reactions and diffusion must be considered, the concentrations of the individual species must be known at every point, and equations for the conservation of mass for each species must also be used. These equations contain the mass-averaged velocity vector for each species which may differ from the mass-averaged velocity for the mixture as a whole. In a multicomponent gas mixture, the molecules of one species move in a mass-averaged sense relative to the other species in response to gradients of concentration, pressure, and temperature. The three diffusion coefficients depend on the properties of the molecule in question and also on the properties of the other species present in the mixture. The mass-averaged velocity of species (i) is usually written as the sum of the mass-averaged velocity for the whole mixture, $\bar{\mathbf{v}}$, and a diffusion velocity for species (i), $\bar{\mathbf{V}}_i$. Both are vectors, and $\bar{\mathbf{V}}_i$ can be either positive or negative. Thus, $(\bar{\mathbf{v}} + \bar{\mathbf{V}}_i)$ is, by definition of $\bar{\mathbf{V}}_i$, the mass averaged velocity for the ith species.

When, ρ is the density and C_i is the mass fraction of the ith species in the mixture, ρY_i is the density of the ith species and the continuity equation for this species can be written as

$$\partial(\rho C_i)/\partial t + \nabla \cdot (\rho C_i(\bar{\mathbf{v}} + \bar{\mathbf{V}}_i)) = w_i \qquad (23.80)$$

Here, w_i is the rate at which mass of species (i) is generated per unit volume per unit time by the chemical reaction. Thus, w_i depends on the kinetic constants described above and is, thus, usually a strong function of the temperature and species concentrations.

The diffusion velocity in a laminar flow depends on the concentration, temperature and pressure gradients, and on the concentrations of the other species present at the position in question. However, in many applications only the dependence on concentration gradients need be considered; for this case and for a simple binary gas mixture it can be expressed approximately as

$$\overline{V}_i = -D_{ij}(\nabla C_i) \tag{23.81}$$

which is usually identified as *Ficks' Law*. The *diffusion coefficient*, D_{ij}, depends in a known manner on the properties of the two species in question and on the pressure and temperature of the mixture. See Sec. 4.13 for further discussions and Sec. 2.16 for some data for diffusion coefficients.

In the energy equation, account must be taken of the mass averaged velocity for each species in the enthalpy flux terms, but there is no need to add an explicit term for the heat released by chemical reactions when the enthalpy and internal energy are defined in terms of the heat of formation as discussed above. Thus, for gases, the enthalpy, h, and internal energy, e, per unit mass for the whole system can be defined as

$$h = \sum_i (C_i h_i) \tag{23.82}$$

$$h_i = H_{fi} + \int_{T_s}^T c_{pi}\, dT \tag{23.83}$$

$$e = h - p/\rho \tag{23.84}$$

where H_{fi} is the heat of formation evaluated at T_s and c_{pi} is the specific heat at constant pressure of the ith species. The energy equation for flows in which the body forces, dissipation, and kinetic energy can be neglected, can be put in the simple form

$$\partial \rho e/\partial t + \nabla \cdot \sum_i (\rho C_i [\overline{v} + \overline{V}_i] h_i) = \nabla \cdot (k\nabla T) \tag{23.85}$$

The steady state form of the continuity and energy equations can be expressed as

$$\nabla \cdot (\rho \overline{v}) = 0 \tag{23.86}$$

$$\nabla \cdot \left(\sum_i \rho C_i [\overline{v} + \overline{V}_i] \right) = w_i \tag{23.87}$$

$$\nabla \cdot \left(\sum_i \rho C_i [\overline{v} + \overline{V}_i] h_{ti} \right) = \nabla \cdot (k\nabla T) - \sum_i (w_i H_{fi}) \tag{23.88}$$

where

$$h_{ti} = \int_{T_0}^T c_{pi}\, dT \tag{23.89}$$

is the *thermal enthalpy* of the species (i). The momentum equation is in its standard form.

Three new terms appear here when reactions in a multi-component gas mixture must be considered. First, the diffusion velocity term, $\overline{V}_i$, which was discussed above appears in the last two equations. Second, the chemical production rate term, w_i, appears in the continuity equations, and it appears in a term which corresponds to the energy addition due to chemical reaction in the last term on the right-hand side of the energy equation. The production rate term, w_i, may be calculated when the local temperature, pressure, species concentrations, and rate constants are known for each chemical reaction which occurs in the mixture.

The calculation of the properties of a laminar flow which involves combustion can be carried out satisfactorily by use of numerical techniques, although considerable computation is required. Care must be taken to avoid numerical problems which can arise because of the exponential dependence of the rate equations on the temperature (see Chap. 22).

The corresponding calculation for a turbulent flow cannot be treated yet from first principals, and calculation schemes in use now must rely on strong and often unverified *ad hoc* assumptions. The primary new problem which arises here is concerned with the calculation of the instantaneous values of the variables which enter the reaction rate equations as described earlier.

23.3.6 Dimensionless Parameters

Although a number of dimensionless groups have been defined in the course of the study of various combustion problems, only a few are used as regularly as, for example, the Reynolds Number is used in fluid dynamics. The complexity and wide range of combustion problems and the fact that many parameters are often required to characterize a single system prevent one or even a few dimensionless parameters from being generally useful. Several examples are given below.

Damkohler [Glassman (1977)] introduced a number of parameters which are now called the *Damkohler Numbers* or *Similarity Groups*. As an example, the first group is the ratio of two time scales—a residence time in a combustion region and a time required for a chemical reaction to reach completion at some temperature, pressure, etc. The *residence time* could be defined as the residence time in the laminar flame and hence would be the ratio of the laminar flame thickness, δ, and the flame speed, δ/s. The *chemical time* can be taken to be the ratio of a typical mass per unit volume, i.e., the density, ρ, and a generation rate per unit volume, w, evaluated at some characteristic condition of the flow in question. Thus, the dimensionless group is

$$D_1 = (\delta/s)/(\rho/w) \tag{23.90}$$

when D_1 is greater than one, the reaction in question is expected to reach an equilibrium state.

A second parameter arises in simple models for calculation of the laminar flame speeds and flammability limits. The dimensionless group is the ratio of the heat transported by the gas flow, given by $\rho c_p s \Delta T$, to that conducted forward into the unburnt gas, approximated by $k \Delta T / \delta$. In these expressions, s is the laminar flame speed and ΔT is a characteristic temperature difference. The ratio of these quantities is $\alpha/\delta s$, where α is the thermal diffusivity, $k/\rho c_p$. When this ratio is relatively constant for a given chemical reaction, it can give guidance concerning the effects of the thermal diffusivity on the flame speed or thickness.

As a final example, a third dimensionless parameter arises in the combustion of liquid or solid fuels, and it is called the *transfer number* or *B number*. It is approximately the ratio of the heat released in a reaction involving fuel vapor, ΔH, to the latent heat, L, required to vaporize the fuel. Thus, when $B = \Delta H/L$ is large, only a small fraction of the heat generated in the reaction must be returned to the fuel surface to supply the energy required to vaporize the fuel and produce the fuel flow rate. Consequently, a high fuel consumption rate is expected when B is large, since usually a considerable fraction of the heat released in the combustion of a solid does get returned to the surface of the fuel. This parameter plays an important role in fixing the burning rates of liquid pools or droplets and solid fuels for which combustion occurs in the gas phase near the fuel surface [Glassman (1977)].

In addition to parameters of this type, several groups are used which deal with the transport properties of the flow. These include,

$$\textit{Prandtl number: } \mathrm{Pr} = c_p\mu/k = \nu/\alpha \tag{23.91}$$

$$\textit{Lewis number: } \mathrm{Le} = k/\rho D_{ij}c_p = \alpha/D_{ij} \tag{23.92}$$

$$\textit{Schmidt number: } \mathrm{Sc} = \mu/\rho D_{ij} = \nu/D_{ij} \tag{23.93}$$

Here, ν is the kinematic viscosity, μ/ρ. Note, that in the simple kinetic theory of ideal gases, these ratios are all one; for air, the Prandtl and Schmidt numbers are about 0.7.

23.4 FLUID DYNAMICS IN MATERIALS PROCESSING
Julian Szekely

23.4.1 Introduction

Until recent years, materials science and engineering has concentrated on the structure and properties of materials and on the thermodynamics that governed equilibrium between phases. During the past two decades, there has been a growing interest in understanding the factors that govern the rates at which materials are processed and the associated heat flow, mass transfer, and fluid flow phenomena. Thus, the materials community should develop knowledge of fluid mechanics and at the very least an awareness of the types of problems that can be effectively treated. At the same time, fluid mechanicians may regard the materials field as an interesting source of problems that are of both practical and fundamental interest. Our aim in presenting the material in this section seeks to serve this dual purpose.

Materials processing covers a very broad field of activity that ranges from the manufacture of tonnage materials, such as steel, aluminum, and cement, on the one hand, and the production of specialty materials, such as diamond films, metal matrix composites, electronic chip substrates, on the other. These systems have both distinguishing and common features. The important distinguishing features are the scale of operation.

The production scale of tonnage materials is measured in the millions or at least hundreds of thousands of tons per annum, and the linear dimensions of the equipment are measured in meters, sometimes tens of meters. As shown in Fig. 23.39, an iron blast furnace is about thirty meters high and ten meters in diameter. A copper or nickel converter, shown in Fig. 23.40, is a few meters in diameter and may be 20 meters long. Both these are excellent examples of quite complex fluid flow problems.

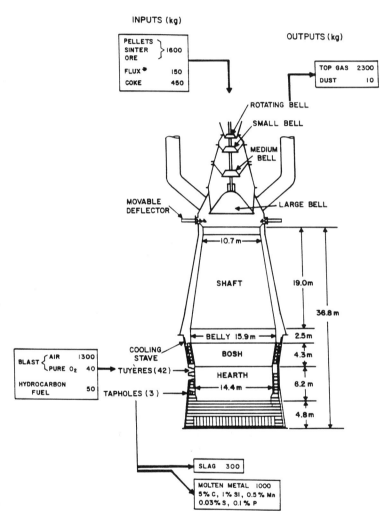

FIGURE 23.39   Vertical profiles of Fukuyama number 5 blast furnace. [Adapted from Peacey and Davenport (1979) and Sugarawa *et al.* (1976).] Two features are notable—the quadruple bell system which indicates that the furnace is, like most modern units, being operated under pressure, and the movable deflectors around the throat which are used to obtain uniform burdening across the furnace. The working volume, tuyeres to stockline, and the internal volume, hearth bricks to stockline, of this furnace are 3930 and 4620 m³, respectively. Not shown in the figure are the descending burden of sinter (ore) coke and limestone and the ascending gas stream, injected through the tuyeres.

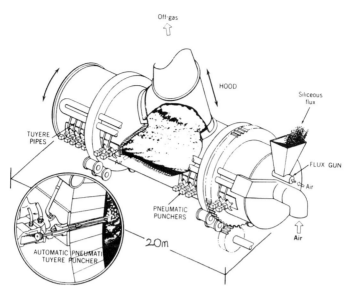

FIGURE 23.40 Cutaway view of horizontal side-blown copper converter. [Adapted from Boldt (1967) and Rosenqvist (1983).] In this system, which conceptually has not changed over many decades, a mixture of molten copper and iron sulphides (matte) is reacted with an air stream, blown into the melt through tuyeres. The iron sulphide is converted to iron oxide (slag) while the copper sulphide is converted to *blister copper*, i.e., copper containing oxygen, which has to be further refined.

In the iron blast furnace, solid iron ore, coke, and fluxing (slag forming) materials are fed at the top, while preheated air is blown in through side mounted *tuyeres* at the bottom. The air reacts with the coke and generates thermal energy and a hot reducing gas. This ascending gas heats and reacts with the descending solid charge, which is eventually melted and partially consumed to produce molten pig iron and a molten metal oxide (slag) stream. Both of these are tapped from the furnace periodically. Thus, using fluid flow terminology, we are dealing with spatially non-uniform flow through porous media in the upper part of the furnace, with the counter current flow of slag and metal and gas streams through a porous matrix in the lower part, and the flow of slag and metal streams at the bottom. In view of the obvious economic importance of blast furnaces, some 500 million tons of pig iron of a total value of about $75 billion is being produced annually in the world, these problems have been studied to a considerable extent [Peacey and Davenport (1979), Sugarawa *et al.* (1976), and Iron and Steel Inst. of Japan (1987)].

In copper converting, a molten copper sulfide is reacted with air that is blown into the copper converter through tuyeres to produce *blister copper*. This is an interesting submerged-jet problem, coupled with two-phase, bubbly flow phenomena, complicated by the existence of multiple, possibly interacting jets and by the fact that the liquid is of two or three phase (matte, metal, and slag) [Rosenqvist (1983)].

Many other large scale metals and materials processing operations present similar problems, i.e., the handling of physically large systems, although the details of the geometry and the contacting arrangements may differ. Table 23.6 provides a summary of some key primary metals and materials processing operations.

TABLE 23.6 Fluid Flow Problems in Some Key Primary Metals and Materials Processing Operations

Description	Fluid Flow Problems	Current Status
1. Ironmaking in the blast furnace A large (say 30 m high and 5–10 m in diameter) shaft furnace, with a counterflow of a solid charge and air injected through tuyeres at the bottom.	• Spatially nonuniform gas flow through a porous medium accompanied by chemical reactions, heat and mass transfer. • counterflow of molten metal and slag and gas through a porous medium • flow of molten steel and slag on tapping.	Extensive modeling efforts. References: Peacey and Davenport (1979), Sugarawa *et al.* (1976), Rosengrist (1983), and Iron and Steel Inst. Japan (1987), (1990)
2. Steel Processing Liquid steel is made in converters, where a supersonic oxygen jet is blown at a molten iron bath, say 60–300 tons in size. The oxygen reacts with the carbon and the other alloying elements contained in the iron charge.	• Supersonic oxygen flow through nozzles • cooling water flow through nozzles • complex multiphase flow situation in the bath, involving a slag–metal and gas mixtures. There are important fluid flow problems in the subsequent treatment of molten steel, including the flow of steel through "tundishes" i.e., launders, and during its casting.	Extensive modeling of the lance phenomena in the 1960s and 1970s. Details of the flow in the steelmaking vessel only partially understood. But not thought to be critical for good process control. References: Harabuchi and Pehlke (1988), McPherson and McLean (1988), and Iron and Steel Inst. of Japan (1990)
3. Aluminum Metallurgy Aluminum is produced by the electrolysis of molten alumina in the so-called Hall Cell, which consists of a shallow flat bath into which large graphite electrodes are immersed. By passing a DC current through the system, the metal oxide (which is dissolved in molten kryolite) is decomposed and molten aluminum is collected at the bottom.	Aluminum electrolysis is a rich source of fluid flow problems and has been extensively studied. Bath circulation, which has a profound effect on the cell efficiency is driven by a combination of electromagnetic forces, buoyancy and also by the gas bubble streams that are generated. Furthermore, the free surface waves that are generated can have a major deleterious effect on the overall performance of the system.	Hall Cell electrolysis has been very extensively studied, not only through numerical techniques, but some very elegant analytical solutions have also been generated. References: Voller *et al.* (1991), Szekely *et al.* (1987), (1988), Sohn and Geskin (1988), and Sahai and St. Pierre (1992)

4. Glass Manufacture

   Glass is produced by melting together soda ash, silica (sand), and return glass, called cullet, a large, shallow reverbatory furnace, some 20 m long, 5 m wide and 1 m deep.

   A critical issue in the melting process is the "fining" or refining, in course of which the small gas bubbles occluded in the glass are allowed to float out. Since molten glass is highly viscous, of the order of 1,000 poise or more, the flow in glass tanks is linear and the flotation of the gas bubbles takes a very long time. The actual movement of the molten glass in these tanks is quite complex, driven by a combination of the bulk motion due to discharge and thermal natural convection.

   A great deal of work has been done on both the mathematical and the physical modeling of flow phenomena in glass tanks, but much of this effort has been kept confidential by the glass companies. The melting of fiberglass entails similar problems, except these systems are perhaps even more sensitive to disturbances.

   References: CMP (1990), EPRI (1988)

5. Cement manufacture

   Cement is produced in large rotary kilns, typically 4–6 m in diameter and 100–150 m long. Lime, silica, alumina, and other additives are charged into the kiln, which is fired, using natural gas, coal, or oil as a fuel.

   The important fluid flow problem in cement manufacture is the control of the flame and its interaction with the charge.

   A great deal of work has been done on the mathematical modeling of cement kilns, but this has been mainly focussed on the heat transfer aspects of the system.

   References: Gorog et al. (1981), (1983), Barr et al. (1989), Brinacombe (1989)

6. Manufacture of copper, nickel, and lead and other nonferrous metals.

   The production of these metals takes place on a much smaller scale than steel.

   A whole range of fluid flow problems. These are in general much less well explored than steel or aluminum, although interesting results have been published on copper.

   Some work has been done in different aspects of these processes, but a great deal of problems still remain to be tackled.

   References: Rosenquist (1983), Reddy and Weizenbach (1993)

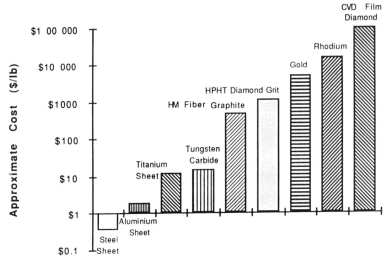

FIGURE 23.41   Price comparison between a range of materials. (From Bush, 1994.)

The arrangements for the production of specialty materials contrasts quite sharply with this picture. The production rate of specialty materials is much smaller, but at the same time, the intrinsic values per unit weight are much greater. Figure 23.41 shows a price comparison between a range of materials; it is seen that from a low of a few cents per pound for cement to many thousands of dollars per pound for CVD diamond, a very broad range is being covered. The actual scale of the processing operations is also much smaller.

In contrast to these, silicon wafers used as substrates are usually 6–8 inches (150–200 mm) in diameter and 1 mm thick, ceramic components may have a characteristic dimension of a few tens of millimeters, and solder beads connecting circuitboard components are measured in tens of microns. The equipment in which they are usually processed is also characteristically smaller, with typical dimensions being about 1 meter or less. Figure 23.42 provides a sketch of the Czochralski (CZ) system used for the growing of single crystals of silicon; here, melt is contained in a heated, rotating crucible that may be 20–40 inches in diameter, a solid single crystal of silicon is being slowly withdrawn from the melt (at a typical rate of 35 mm/hr) and this single crystal is used as the starting point for the wafers in semi-conductor technology. Also shown in the figure are some computed stream line patterns, velocity vectors and concentration isopleths.

The growing of silicon single crystals is also a very rich source of fluid flow problems and provides an excellent illustration of the need for a better fundamental understanding of the phenomena associated with crystal growth. The crucible rotation and the counter-rotation of the crystal that is being withdrawn impose a rotational flow field on the system, which necessarily interacts with the buoyancy driven vertical circulation loops produced by the temperature gradients between the hot walls and the colder crystal surface. In addition, the horizontal temperature gradients at the free surface may give rise to surface tension (*Marangoni*) convection. Within the context of producing high quality wafers, (an absolute necessity as the perfor-

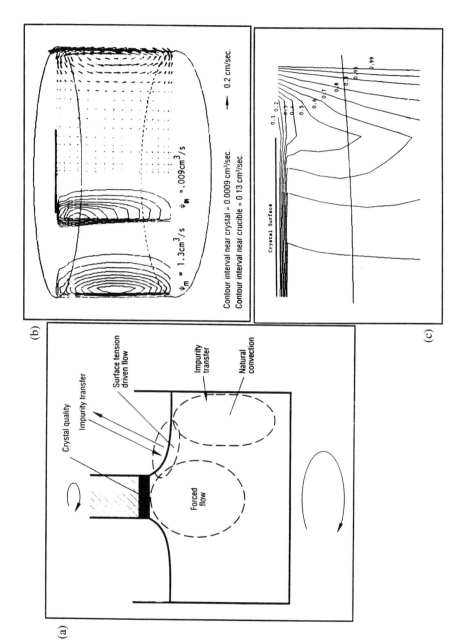

(b)

$\psi_m = 1.3 \, cm^3/s$ $\psi_m = .009 \, cm^3/s$

→ 0.2 cm/sec.

Contour interval near crystal = 0.0009 cm³/sec.
Contour interval near crucible = 0.13 cm³/sec.

Crystal Surface

0.1 0.2   0.3   0.5   0.6   0.7   0.8   0.9   0.99

(c)

Crystal quality

Impurity transfer

Surface tension driven flow

Impurity transfer

Natural convection

Forced flow

(a)

FIGURE 23.42 Magnetic Czochralski (CZ) system. (a) Schematic sketch showing the different flow regimes. (b) Computed flow pattern in a CZ system. (c) Computed dopant distribution near the crystal in a CZ system with a magnetic field of 2 kilogauss. We note here that the imposition of magnetic fields is a recent, but popular development in the crystal growing field. (Reprinted with permission, Ilebugsi, O. J. and Szekely, J., *Computational Mechanics in Metals Processing*, I&SM, p. 29, 1991.)

mance of computer chips is being continually improved) the convective flow in the melt has to be precisely controlled for the following reasons:

i) Convection determines the temperature gradients at the melt-crystal surface, which in turn govern the all important solid structure.

ii) Convection will also determine the shape of the melt-solid interface, which will affect crystal quality.

iii) Most important, convection will affect the oxygen content of the crystal (due to the convective transport of oxygen from the walls to the crystal surface), hence, the all important doping level concentration of trace impurities such as oxygen which has to be controlled within strict limits.

It follows that the precise knowledge and control of a highly complex fluid flow situation would be highly desirable.

Other specialty processing applications include the coating of wafers by Chemical Vapor Deposition (CVD), as depicted in Fig. 23.43, where a mixture of hydrogen and an organo–metallic compound, such as silane, is passed over a wafer with which it reacts, forming a deposit. Here, the issue is to control the gas flow such that the deposit is of the correct properties and uniformity. Most CVD equipment has been developed from laboratory scale experience. From a materials standpoint, the critical issue is to produce wafers or coatings of the desired structure and property, which are primarily controlled by the surface temperature and the composition of the reactant gas. However, in an engineering sense, especially if we are dealing with larger surfaces or multiple wafers, it becomes very important to arrange for spatially uniform temperatures, gas flows, and deposition rates. Here, fluid flow issues may take on a special importance. Similar considerations apply to the production of many coatings, including that of synthetic diamond. In all these cases, the equipment tends to be small, at least the section where the wafers are contained.

Physical Vapor Deposition (PVD) is another example of specialty materials processing, where, as sketched in Fig. 23.44, a molten metal pool is struck with an electron beam, such that vaporization occurs, then this vapor is deposited onto the substrate. There are also numerous fluid flow problems associated with this operation, including the circulation of the melt, which is driven by a combination of surface tension and buoyancy forces and the behavior of the vapor cloud which is being transferred. This latter problem has received little attention up to the present time.

It is readily appreciated that the flow will tend to be turbulent, at times very highly turbulent in many tonnage materials operations, while the flow is likely to remain laminar, but with a strong influence of surface tension effects and *end effects* in specialty materials processing.

A common characteristic feature of all materials processing operations is that the end product is a solid of a desired shape (steel sheet, silicon wafer, extruded polymer component) chemical composition, structure and properties. Shape, structure and properties (such as strength, toughness, corrosion resistance, permeation resistance, electrical conductivity, etc.) are the key features of solid materials that distinguish them from fluids. In the majority of cases, the materials person's main interest is not in fluid mechanics *per se*, but on the actual effect of fluid flow phenomena on the process, such as it may influence convective heat transfer, mixing, mass transfer,

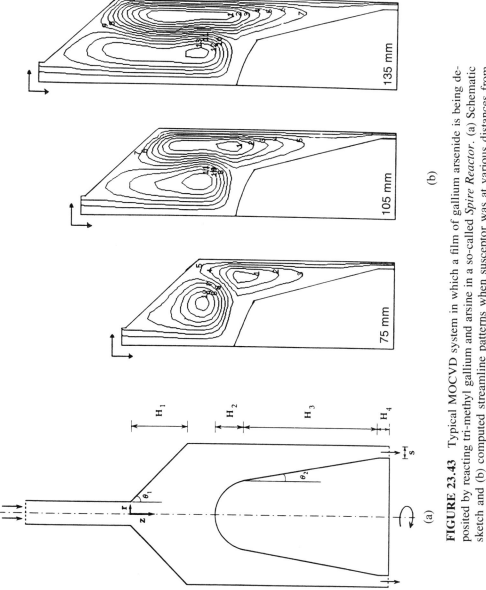

FIGURE 23.43 Typical MOCVD system in which a film of gallium arsenide is being deposited by reacting tri-methyl gallium and arsine in a so-called *Spire Reactor*. (a) Schematic sketch and (b) computed streamline patterns when susceptor was at various distances from the inlet. (Reprinted with permission, Dilawari, A. H. and Szekely, J., "Computed Results for the Deposition Rates and Transport Phenomena for an MOCVD System With a Conical Rotating Substrate," *J. Crystal Growth*, Vol. 97, pp. 777–791, 1989.)

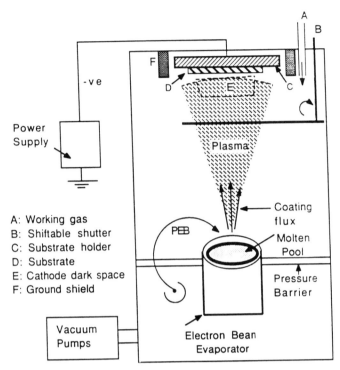

A: Working gas
B: Shiftable shutter
C: Substrate holder
D: Substrate
E: Cathode dark space
F: Ground shield

FIGURE 23.44   Schematic diagram of an ion plating apparatus.

and the like. However, even within this restriction, virtually every chapter in this handbook could be of potential interest to scientists or engineers engaged in the materials field, even though the actual materials applications may not be clearly spelled out.

In view of the great breadth of fluid flow applications in materials processing, it is impractical within the confines of this section to provide a comprehensive description, so the approach adopted will be to present a classification of fluid flow situations most frequently encountered in materials processing, with illustrations of typical examples, frequent cross references to the other chapters in this text, and references to the materials literature for further reading.

We may classify fluid flow problems encountered by materials engineers into two broad categories:

i) Conventional fluid flow problems that could be normally treated by the mechanical, chemical, or civil engineering community, such as flow through pipes, cooling water circuits, fuel lines, sedimentation of sludges, filtration and electrostatic precipitation, and the like. These tend to be quite *clean* fluid flow problems, where the materials processing context does not really play a major role in the problem formulation, although the ultimate success of the project will depend on having found the correct fluid flow solution. A selection of such problems is given in Table 23.7.

ii) Fluid flow problems closely tied to the materials system itself, such as turbulent mixing in steelmaking vessels, flow instabilities in crystal growing sys-

TABLE 23.7   Some Pure Fluid Flow Problems in Materials Processing

- Cooling circuits (water, air, mist, oil, etc.,) to lance, furnace, casting, solidification, deformations, electronic devices, and heat treatment components.
- Fluid delivery equipment (pumps, compressors, heat exchangers, etc.)
- Supply of reactant gases in metallurgical operations (combustion, stirring, etc.)
- Supply and delivery of gases and liquids in semiconductor processing.
- The operation of waste gas and waste water treatment circuits.
- The operation of supersonic jets (lances in steel or copper processing, burners in heating operations, torches in thermal spraying processing, etc.)

tems, melt circulation in induction furnaces, plasma processing, etc. In this latter group, even the identification of the fluid flow issues may be of some difficulty to the practicing materials engineer, while the fluid flow expert may find it quite troublesome to establish the connection between the relatively clean fluid flow problems with which he is familiar and the rather ill-defined, complex problems presented by most materials processing operations. It is the intent of this brief section to help in establishing these connections between the fluids and the materials communities.

23.4.2   Classification of Fluid Flow Problems in Materials Processing

The Tonnage Materials Processing Field. In the tonnage field, we can consider single-phase systems, which are perhaps the most easily tractable. In the liquid phase, good examples are the flow of molten metals or slags through troughs, launders, or tundishes, as metal or slag streams are being transferred between furnaces or being fed to casting machines, where solidification occurs. Other single phase examples include the pouring of melts into molds (this is a topic of major interest, because of the CAD-CAM design of casting operations) [Doyama *et al.* (1990), (1993), Kim and Kim (1991), Sou and Mehrabian (1986), Dantzig and Berry (1984), Rappaz *et al.* (1991), and Piwonka and Katgerman (1993)] and the operation of induction furnaces, combustion flames, and plasma jets.

Most of these problems tend to be three dimensional (occasionally, two-dimensional or axisymmetric), and they are now being handled numerically, using a combination of general purpose fluid flow packages, such as PHOENICS, FIDAP, FLUENT, FLOW 3-D, or specific, custom developed codes. A typical example of such calculations is shown in Fig. 23.45, which shows the computed velocity profiles and the trajectories of inclusion particles in the operation of a tundish in steel processing [Szekely and Trapaga (1994) and Szekely and Ilgebusi (1989)].

Consideration of single-phase problems is a gross idealization (the inclusion particles were assumed not to affect the fluid flow); indeed, the vast majority of fluid flow problems in tonnage processing involve highly complex multi-phase situations. The behavior of gas bubble plumes is of considerable interest in steel processing, because the injection of gas streams provides an excellent means for agitation.*
These problems have been tackled by the metals community by a combination of

---

*For obvious reasons, it is impractical to agitate molten steel mechanically, no stirrer would survive, so induction stirring or agitation with gas bubble streams represent the only practical alternatives.

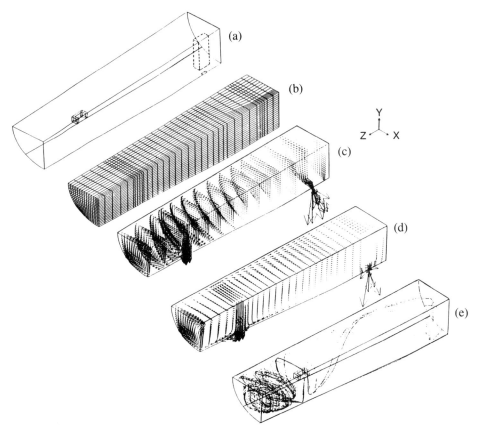

FIGURE 23.45 Computed results representing the modeling of a tundish. Such tundishes are being used in all cases when steel is being continuously cast. The molten steel is being poured into the tundish from which it is fed into the molds of the continuous casting machine. The study of these problems is of interest, because melt behavior in tundishes will affect its cleanliness. (a) Computational domain, (b) grid (15 × 18 × 35 cells along the x, y, and z directions) employed in the axisymmetric calculations (i.e., only half the actual tundish), (c) velocity vectors on selected vertical-transverse planes along the tundish length, and (d) velocity vectors along the faces of the figure.

physical (water) modeling work and also by numerical computation [Ilegbusi *et al.* (1993), Woo *et al.* (1990), Iguchi *et al.* (1988), and Schwarz and Turner (1988)]. Figure 23.46 shows a comparison between the theoretical predictions, using a numerical code and experimental measurements obtained with an air–water model, and the agreement seems to be quite good.

  The majority of real tonnage processing systems involve the contacting of several phases, e.g., a slag–metal–gas and suspended solids, as is the case of steelmaking. Many of these problems have defied a fully quantitative description, although a great deal of progress has been made in recent years.

  Another important group of problems pertains to solidification processing. Here, Fig. 23.47 shows a schematic sketch of the arrangements used for the continuous casting of steel, i.e., the transformation of molten steel to a solid slab. It is seen that molten steel is being poured into a (usually copper) mold, where a solid skin is

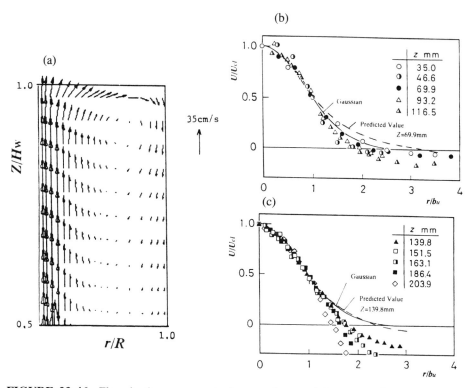

FIGURE 23.46 Flow in the upper part of a vessel, containing water, into which an air stream is being blown through the bottom; such systems model argon—stirred ladles of molten steel—for the experimental conditions of data by Ilegbusi *et al.* (1993). (a) Predicted liquid velocity vectors, (b) radial variation of normalized axial liquid velocity at $z = 69.9$ mm, and (c) radial variation of normalized axial liquid velocity at $z = 139.8$ mm. (Reprinted with permission, Ilebugsi, O. J., Szekely, J., Iguchi, M., Takeuchi, H., and Morita, Z., "A Comparison of Experimentally Measured and Theoretically Calculated Velocity Fields in a Water Model of a Argon Stirred Ladle," *ISIJ International*, Vol. 33, pp. 474–478, 1993.)

formed and then this solid skin containing a molten core is being continuously withdrawn. The outer skin or shell is being further cooled, until the whole slab is solidified; this solid material is then rolled to obtain the final desired form. Various shapes of steel, copper, and aluminum are being cast this way, including slabs (1–2 m wide and 200 mm thick), thin slabs, (1–2 m wide and 50 mm thick), billets (100–200 mm square), and *blooms* (300–400 mm square). Important fluid flow issues in casting are the distribution of the molten metal through appropriately shaped and sized nozzles, as it is being fed into the sump and the interaction of the circulating molten metal pool with the solidifying surface. A great deal of useful work has been and is being done in this area also, particularly in connection with steel processing.

Specialty Materials Processing. As discussed earlier, the scale of operations in specialty materials processing is much smaller, so that the flow will tend to be laminar rather than turbulent. The smaller scale of the operations also means that interfacial phenomena will tend to be much more important, so that meniscus problems and

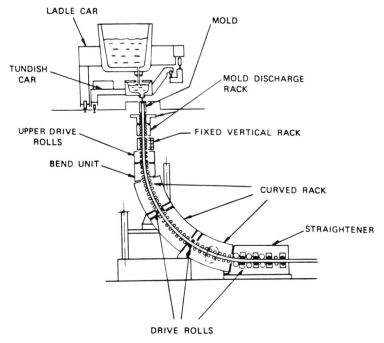

FIGURE 23.47 Sketch of a system used for the continuous casting of steel. Typically, the sections produced by this means range from say 100 mm × 100 mm or larger billets or slabs that may be 1–2 m wide and 200–250 mm thick. [Adapted from Lankford *et al.* (1985).]

surface tension driven flows (*Marangoni Convection*) will tend to play a much more important role here.

Specialty processing operations tend to be even more diverse than the tonnage systems, and, since they have evolved from laboratory-scale experiments, the geometries tend to be rather complex. Another important distinction is the fact that in specialty materials procssing systems, the quality of the finished product is extremely important. Indeed, major premiums may be obtained for extra high performance wafers or coatings. Since this product quality may be markedly influenced by the flow conditions (e.g., flow uniformity, absence of instabilities, avoidance of transitional flows, etc.), fluid flow phenomena may play a critical role here. At the same time, most of these technologies have been evolved very recently, with little input from fluid mechanicians, so this may be a very ripe field for research.

In the following, we shall show some illustrations of recent work in specialty materials processing. Figure 23.48 shows the computed stream lines in a Chemical Vapor Deposition (CVD) system, which is seen to result in a quite uniform flow, because of the modified *stagnation flow* arrangement. The velocity field involved here will give spatially uniform deposition rates and is the result of a design that has evolved over the years.

Figure 23.49 shows a schematic representation of the incomplex phenomena involved during welding operations, while Fig. 23.50 shows the computed velocity fields in a weldpool. Such weldpool problems have been extensively studied over the years, and it has been shown that metal circulation is driven by a combination

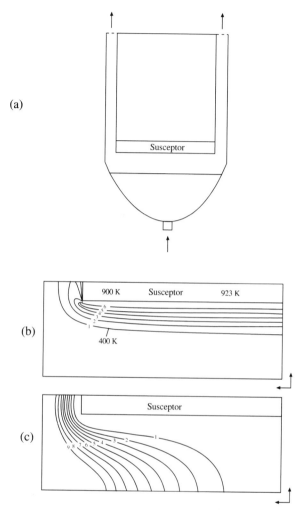

FIGURE 23.48 Stagnation flow CVD system (reactor diameter = 200 mm; susceptor diameter = 180 mm; distributor diameter = 160 mm), for the production of gallium arsenide or silicon wafers. (a) Schematic sketch, (b) computed temperature profiles, and (c) computed streamline patterns (#1 corresponds to 6.63×10^{-5} kg/s and #9 to 5.967×10^{-4} kg/s). (Reprinted with permission, Dilawari, A. H. and Szekely, J., "Computed Results for the Deposition Rates and Transport Phenomena for an MOCVD System With a Conical Rotating Substrate," *J. Crystal Growth*, Vol. 97, pp. 777–791, 1989.)

of factors, including surface tension gradients (so-called *Marangoni flow*) which are the dominant and also electromagnetic and buoyancy forces. Such weldpool circulation is important in practice, because the metal flow will affect the shape of the molten regions which in turn will influence the structure and properties of the solidified welded joint.

Figure 23.51 shows a comparison of the experimentally measured and the theoretically calculated shapes of a solder bead. In these situations, surface tension is the dominant force. The knowledge of solder bead shapes is important in the design

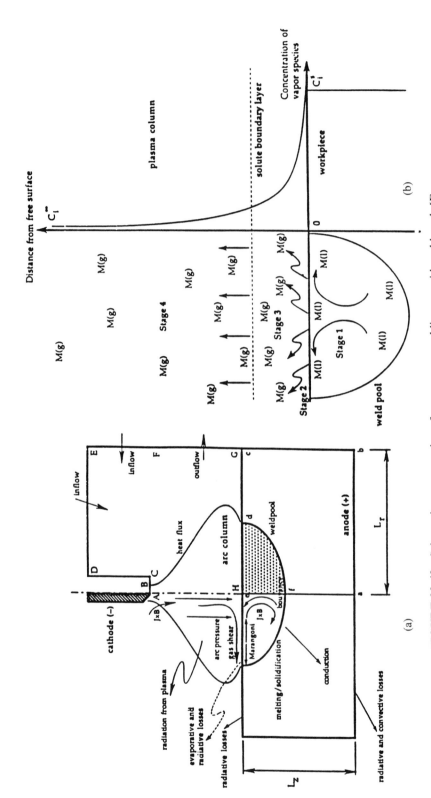

FIGURE 23.49 Schematic representation of gas tungsten welding arc with weld pool. [From Choo *et al.* (1992).] (a) The left portion indicates the various physical phenomena occurring with the workpiece while the right portion shows the computational domain. The origin for the weld pool is at point *e*, while the origin for the welding arc is at point *A*. (b) Vaporization stages of volatile species from the weld pool surface. The diagram on the left shows the mechanism by which high vapor pressure species escape from the weld pool, while the diagram on the right shows the concentration profile of that vapor species as a function of distance from the free surface.

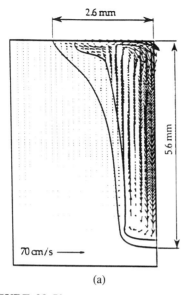

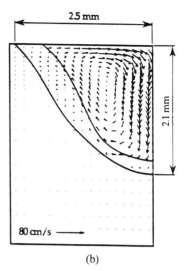

(a)   (b)

FIGURE 23.50 Calculations for a weld pool (the current is 100 A). [From Choo *et al.* (1992).] (a) Weld pool shape and melt velocity profile for a spot welding (i.e., stationary heat source) application, assuming laminar physical properties and (b) weld pool shape and flow profile calculated based on the effective viscosity and thermal conductivity that are 30 times the molecular values.

of circuit boards (surface mount technology), because we must have joints with adequate strength, but at the same time short circuiting between the joint has to be avoided [Szekely and Trapaga (1994), Racz *et al.* (1993), and Racz and Szekely (1993)].

In contrast to many other applications of fluid mechanics, the scale of the system and the nature of the materials involved (nontransparent melts at high temperatures) make the direct measurement of velocities and other transport properties rather difficult, although some quite ingenious systems have been devised for this purpose. One way of addressing this problem is to work with model systems [Szekely *et al.* (1988)]. The ammonium chloride–water system has served as an excellent model for the solidification of alloys, because the aqueous phase is transparent, and the solidifying system does form a eutectic mixture analogous to a large number of metallic alloy systems. Figure 23.52 shows interesting results reported by Incropera and Bennon, indicating the velocity fields, the composition field, and the macrosegregation patterns in such a system. It should be noted that macrosegregation would be a topic of primary interest to the materials community, but one must tackle the fluid flow problem to reach this desired answer.

Figure 23.53 illustrates one successful application of modeling and experiment, showing the implication, flattening, and solidification of a metal droplet on a surface. The numerical solution of this problem was generated by solving the laminar, Navier–Stokes equations for an axisymmetric situation, but for a free boundary, using the FLOW 3-D software package. This is, of course, an idealization of a spray forming operation, where a very large number of droplets are made to fall on a surface, nonetheless the example is of some interest.

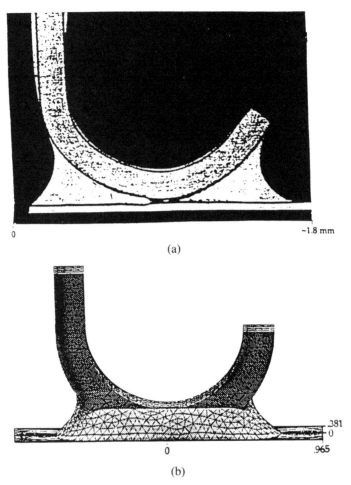

FIGURE 23.51   The mathematical model of a solder joint, that binds a lead wire to a chip on a circuit board. The calculation represents the equilibrium shape of a *j*-bend lead with surrounding solder joint. (a) photomicrograph showing the actual cross section. (Reproduced with permission of K. Patel, Digital Equipment Corp.) and (b) computed results (contact angle 45°, volume: 6.29 × 10^{-5} cm^3).

23.4.3   Current State of Fluid Flow Studies in Materials Processing

The materials processing field is rich in applications of fluid mechanics, but this area has been only recently *discovered*, and major advances may be expected in the forth-coming years. A characteristic feature of fluid flow problems in materials processing is that we are dealing with coupled problems, so the understanding of the fluid flow behavior will only provide a partial answer. Of course, fluid flow phenomena are readily represented in terms of the conventional Navier–Stokes equations. As might be expected the Navier–Stokes equations are familiar to the reader; the main differences in solving these equations in a materials processing context are the more complex geometries, the coupled nature of the problems, and the more complex boundary conditions. For multi-phase systems some success has been achieved by writing

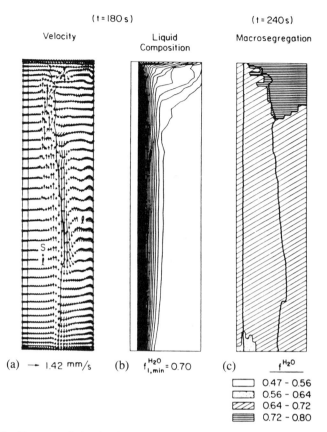

FIGURE 23.52 Transport behavior in a solidifying 30 pct NH_4Cl–H_2O ingot (i.e., a physical model of a real metals solidification problem) [From Bennon and Incropera, (1987a) and (1987b).] (a) Velocity field at $t = 180$ s, (b) liquid isocomposition lines at $t = 180$ s, and (c) macrosegregation at $t = 240$ s.

down a set of equations for each phase and coupling them through so-called *friction coefficients*.

The following comments should be made:

i) In many specialty materials processing applications, the flow is laminar.

ii) In the majority of tonnage materials applications, the flow is turbulent, so that one has to invoke an appropriate turbulence model. Most of the work done to-date has used the k-ϵ model or some variation of it. It is appreciated that the k-ϵ model has to be regarded as a first approximation and that we must await the development of more sophisticated models and more computational horsepower.

iii) Most of the work has involved the representation of steady-state two- or three-dimensional systems; transients, transitional flows, wave motion, and free boundary problems have received much less attention.

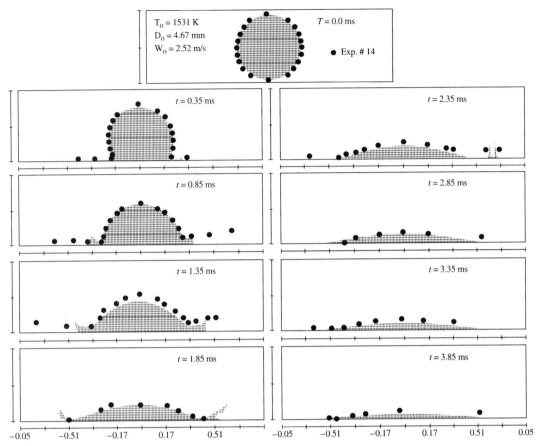

FIGURE 23.53 Spreading and solidification of a Cu droplet on a Cu substrate. Comparison between experimental data and predicted droplet shape (cross section) as function of time. $h = 2 \times 10^5$ W/m^2 K. [From Trapaga *et al.* (1992).]

iv) Melt-solid slurries such as encountered in metals processing and aqueous slurries used in ceramics processing pose very interesting non-Newtonian fluid flow problems, but these have been explored only to a limited extent.

v) Since many melts (e.g., molten metals, molten metal oxides, and halides) of interest in materials processing are electric conductors, electromagnetically driven flows are of considerable interest to the materials community.

vi) Vortex flows, wave motion, flow instabilities, and flow bifurcation are of major potential interest to the materials community, especially in their application to specialty materials processing systems, but these areas have not been well explored.

One may conclude by stating that fluid flow phenomena have become an important part of the materials processing field. A useful start has been made in the application of fluid flow knowledge to the improved design and operation of many materials processing systems, but a great deal more could and should be done. It is hoped that this brief section has stimulated interest in this fascinating field.

23.5 LIFTING SURFACES
Raymond M. Hicks

23.5.1 Introduction

An aircraft in flight is supported by surfaces which develop forces created by the motion of the vehicle through the air. These surfaces may be called *wings*, *canards*, or *tails* depending on their purpose and location on the aircraft. Canard and tail surfaces are usually smaller than the wing and are located near the front or rear, respectively, of the aircraft (see Fig. 23.54). Canard and tail surfaces are used primarily to balance and control the aircraft; the main difference is that the balancing force is directed upward for the canard and usually downward for the tail if the aircraft is statically stable. Hence, the canard helps support the aircraft, while the tail may cause the wing to carry an additional load. If the center of gravity is located well aft, the balancing force on the tail will be directed upward. Such an aircraft will have *reduced static longitudinal* stability and may require a stability augmentation system to be safely flown. Some modern aircraft are designed to operate this way because of improved aerodynamic efficiency. Other aircraft have multiple wings, such as *biplanes*, or wings which flap or rotate, such as ornithopters (see Sec. 27.10) or helicopters (see Sec. 27.3).

The forces which are most important to flight are *lift* and *drag* which act perpendicular and parallel, respectively, to the flight direction. Efficient flight is possible if the lift is much larger than the drag. Lift and drag may be considered to act through a fixed point on the aircraft and may generate a moment about this point. The moment tends to change the angle between the airplane's reference axis and the relative wind and is called the *pitching moment*. This angle is usually referred to as the *angle of attack*. The lift, drag, and pitching moments are reduced to coefficient form by

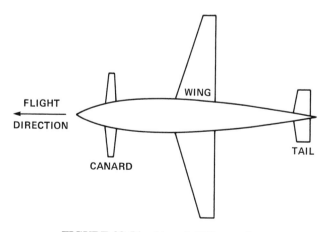

FIGURE 23.54 Aircraft lifting surfaces.

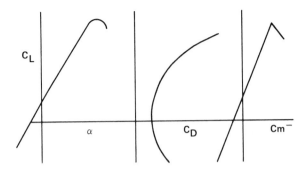

FIGURE 23.55 Typical nondimensional aerodynamic coefficients.

dividing by the dynamic pressure and appropriate reference areas and lengths. The resulting nondimensional coefficients are usually presented as shown in Fig. 23.55. The wing aerodynamic characteristics are determined, in part, by the characteristics of the airfoil sections (cross-section profiles) of the wing.

23.5.2   Aircraft

Subsonic Flight. The aerodynamic characteristics of lifting surfaces are determined by both the planform and the airfoil sections. Lifting surfaces for aircraft operating at low subsonic speeds are generally developed by first selecting the airfoil sections from handbooks or reports giving the coordinates and aerodynamic characteristics of the airfoils. A typical cambered airfoil section with pertinent nomenclature is shown in Fig. 23.56. The most widely used low speed airfoil sections in the United States are the 4-digit, 5-digit, 6-series, and 16-series sections developed by the National Advisory Committee for Aeronautics (NACA) during the 1930's and 1940's [Abott and von Doenhoff (1945)].

The numerical designations for airfoils are usually related to the aerodynamic and geometric properties of the airfoils. For example, the digits of the 6-series airfoils have the following meaning:

first digit—family designation

second digit—chordwise position of minimum pressure in tenths of chord, when the airfoil is operated near the design lift coefficient (also possible extent of laminar flow)

third digit (usually a subscript)—range of lift coefficients in tenths above and below the design lift coefficient for which there exist a favorable pressure gradient on both surfaces (often referred to as the *drag bucket*)

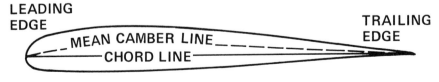

FIGURE 23.56 Cambered airfoil.

fourth digit—design lift coefficient in tenths

fifth and sixth digits—airfoil thickness in percent chord

NACA 63_2-412 is a 6-series section with minimum pressure at 30% chord, a *drag bucket* over a lift coefficient range from 0.2 to 0.6, a design lift coefficient of 0.4 and a maximum thickness which is 12% of the chord.

The development of the 4-digit family of airfoils began in 1929 with a systematic study of a large class of airfoils with thickness distributions given by a 4th degree polynomial with a square root term. These airfoils had higher ratios of maximum lift to minimum drag than had been achieved previously. The development was later extended to consideration of airfoils with the same thickness distribution but with the position of maximum camber located nearer to the leading edge. Airfoils in this class are identified by 5-digits. Some of the 5-digit airfoils give better maximum lift coefficients than the 4-digit airfoils due to the more forward location of maximum camber. However, these airfoils also exhibit an abrupt loss of lift at *stall*. Abrupt loss of lift is a characteristic which can degrade the safety of an aircraft using this type of airfoil. The section stalls when the flow separates over a substantial portion of the upper surface and is felt by the pilot as a loss in lift which may be abrupt. Separated flow is discussed in Sec. 4.9.

The NACA 6-series sections were designed to have extensive regions of *laminar flow* over both surfaces for a limited range of lift coefficients. Hence, these sections can achieve lower drag than either the 4-digit or 5-digit airfoils. The low drag of the 6-series airfoils results from a favorable pressure gradient which exists over a larger percentage of the airfoil chord and over a wider range of lift coefficients than was possible with the 4- and 5-digit airfoils. The 6-series airfoils maintain laminar flow only if the airfoil surface is accurate and smooth. These are conditions which are difficult to achieve with riveted aluminum construction and for aircraft in regular service.

The high lift characteristics of the 6-series airfoils are not as good as those of the 4- or 5-digit airfoils and may be affected by roughness near the leading edge. This has led to the development of several modifications designed to improve the maximum lift coefficients of these sections [Hicks *et al.* (1975) and Kelly (1950)]. One modification consists of drooping the leading edge and increasing its radius. Another modification increases the thickness of the forward 30–40% of the upper surface. Both modifications delay the stall to a higher angle of attack but may change the pressure gradient over the forward region of one or both surfaces from favorable to adverse at low lift coefficients. An adverse pressure gradient will decrease the extent of laminar flow over the airfoil surfaces (see Sec. 4.3), thereby degrading the aerodynamic efficiency of such airfoils.

The thickness and pressure distributions of the NACA 6-series airfoils were derived by conformally transforming a circle to an airfoil. Such transformations relate the known flow about a circle to the flow about an airfoil. Mean-line distributions which give a uniform chordwise loading from the leading edge to a preselected point along the chord and a linearly decreasing load from that point to the trailing edge are combined with the thickness distributions to produce the airfoil sections. Experience has shown that serious degradation of airfoil performance at high speed can result from scaling the coordinates of a 6-series airfoil to obtain an airfoil with a different thickness.

In recent years, many modern airfoil sections have been designed with the aid of high speed computers. Some of the most widely used are those developed by McGhee, Beasley, and Whitcomb (1973), (1975), (1976), (1977), (1978), (1979), Somers (1981a), (1981b), Eppler (1969), (1979), Wortmann (1960), (1972a), (1972b), (1974), Liebeck (1973), (1978), and Liebeck and Ormsbee (1973).

Presently, the most widely used modern airfoils in this country are the low speed (LS-XXXX), medium speed (MS-XXXX), and the *natural laminar flow* (NLF-XXXX) sections. As the names imply, the low-speed airfoils were designed to meet the needs of advanced *General Aviation* aircraft which operate at low speeds. The medium-speed airfoils were intended for use on aircraft which cruise near Mach 0.7 and also need good low speed performance and handling qualities. The first series of natural laminar flow airfoils were designed for low speed flight and achieved laminar flow over approximately 40% of the upper surface and nearly 60% of the lower surface when flown near the design lift coefficient. The Eppler sections have been used on sailplanes and ultralight airplanes. Wortmann has designed airfoils for both low- and high-speed flight including sections for helicopter rotor blades. Liebeck has developed a class of high lift, low speed airfoils which are characterized by low pitching moments. These airfoils have been used on modern sailplanes, ultralight aircraft, racing cars, and axial flow fans.

The aerodynamic characteristics of lifting surfaces are determined by the planform geometry as well as the airfoil sections used. The planform variables of most interest are *area, aspect ratio, taper ratio, twist, sweep,* and *span*. These variables are depicted in Fig. 23.57. The *twist* is the difference in angle of incidence between the root and tip.

It is not uncommon for a lifting surface to consist of more than one airfoil section. Low speed wings sometimes use two sections, one at the root and the other at the tip with linear lofting between sections. The use of different sections at the root and tip is useful in controlling stall progression or span load distribution. It is usually desirable for the stall of a lifting surface to begin at or near the root and propagate

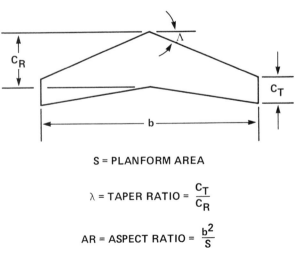

S = PLANFORM AREA

$$\lambda = \text{TAPER RATIO} = \frac{c_T}{c_R}$$

$$AR = \text{ASPECT RATIO} = \frac{b^2}{S}$$

FIGURE 23.57 Typical wing planform.

toward the tip so that the flow over the control surfaces (*ailerons*) will remain attached, and roll control will be maintained. Wing twist also affects stall progression. The tip section may be set at a smaller angle-of-attack than the root section to promote early root stall. Stall progression control is particularly critical for tapered wings, since decreasing chord length from root to tip means the tip region will operate at lower Reynolds numbers than the root and hence will tend to stall at a lower angle of attack. As mentioned above, this can be counteracted by twist.

The aspect ratio is an important parameter in determining the magnitude of the drag-due-to-lift coefficient (this coefficient is usually denoted by C_{D_i} and is given by $C_{D_i} = C_L^2/\pi e AR$ where e is the *span efficiency factor* which has a theoretical value of 1 for a wing with an elliptical span load distribution). The span load distribution can be controlled by twist, taper, or a spanwise variation of airfoil section or a combination of all three parameters. If the planform is rectangular, then twist and spanwise variation of airfoil section can be used to approximate an elliptical span load distribution. Since C_{D_i} is inversely proportional to aspect ratio, it is particularly important for an airplane to have a large aspect ratio if it flies at large lift coefficients.

Most low speed wings have little or no leading edge *sweep*, because lift is greatest for a given angle-of-attack when the angle of sweep is zero. If the leading edge is swept back, the aerodynamic influence of the root on the tip may cause the tip to stall before the root, pitching the nose of the airplane up and further aggravating a stall which may be difficult to recover from. Tip stall can usually be alleviated for sweptback wings in a manner similar to that for tapered wings, i.e., by twisting the wing so that the tip airfoil section operates at a lower angle of attack than the root section. A wing with forward sweep will usually have natural stall progression from root to tip and hence will not need twist to promote early root stall.

Nearly all wings are fitted with *high lift devices*, such as trailing-edge flaps, to increase the maximum lift coefficient. A higher maximum lift coefficient gives a lower stall speed, hence a shorter runway is required for landing and takeoff. High lift devices installed along the leading edge of the wing are called *slats*, slots, or *leading edge flaps*. Leading edge devices increase the maximum lift coefficient of the wing by increasing the angle-of-attack for stall without appreciably affecting the camber of the wing or the angle for zero lift. When trailing edge devices are deflected, the wing camber increases causing a decrease in the angle for zero lift without appreciably changing the angle-of-attack at stall.

Transonic Flight. A flow field is called *transonic* if subsonic and supersonic zones are present in the flow simultaneously (Chap. 9 discusses Transonic Flow in more detail). Shock waves are a feature of transonic flow which are responsible for energy loss and rapid drag increase with increasing Mach number. Early encounters with transonic flow were baffling to engineers and pilots who were only familiar with subsonic flow. It was during this time that the mysterious *sound barrier* became popular.

As the free stream Mach number increases from low subsonic values, a speed is finally reached where the local flow over a lifting surface accelerates to supersonic speed. Further increases in Mach number cause the supersonic zone to grow in both horizontal and vertical extent, moving the shock wave downstream and increasing its strength. When the free stream Mach number reaches unity, another shock wave appears far upstream and approaches the leading edge with further increases in Mach number. The flow over the wing is called *supersonic* when this second shock wave becomes attached to the leading edge and the lowest local Mach number in the flow field exceeds the speed of sound. The progression from subsonic to transonic to supersonic flow is continuous with no insurmountable barriers as once were thought to exist. The drag increases rapidly as the transonic range is encountered and then decreases again in the supersonic range. Lift and pitching moments also increase through the transonic range and then stabilize at lower values at supersonic speeds. The speed for *maximum aerodynamic efficiency* is encountered at the beginning of the rapid drag rise, hence this is a speed where many transonic aircraft are designed to operate.

Early airfoil sections operating near the transonic drag-rise were designed to delay the onset of supersonic flow over the airfoil surfaces to as high a Mach number as possible (i.e., to increase the *critical Mach number*). The methods which have been used to delay the drag increase associated with the critical Mach number include: 1) the use of thin airfoils and airfoils with thickness and camber distributions which delay the development of supersonic flow, 2) sweeping the leading and trailing edges of the lifting surface, 3) injection of air into the boundary layer to control boundary layer separation, 4) vortex generators, and 5) multielement airfoils. Each method has disadvantages which may partially or completely offset the gains in aerodynamic efficiency; for example, thin airfoils may have poor stall characteristics and low strength to weight ratios, sweep decreases the maximum lift coefficient available for landing and may cause pitch-up due to tip stall, and mechanical boundary layer control requires complex plumbing which decreases useful load.

The first airfoils developed in the United States with the goal of increasing the critical Mach number were the NACA 1 series airfoils. These airfoils had critical Mach numbers greater than that of the NACA 4- and 5-digit airfoils. However, the 1-series sections exhibited poor low speed, high lift characteristics due to small leading edge radii. The NACA 6-series sections also had higher critical Mach numbers than the NACA 4- and 5-digit sections but gave somewhat lower maximum lift coefficients at low speed.

Modern transonic airfoil sections are designed with different criteria than the older sections. Less emphasis is placed on increasing the critical Mach number and more attention is devoted to delaying the drag-rise which occurs above the critical Mach number. Drag-rise can be delayed by tailoring the shape of the supersonic zone through careful contouring of the airfoil, which minimizes the strength of the shock wave terminating the supersonic zone. This design philosophy recognizes the fact that it is not the supersonic flow which degrades airfoil efficiency but rather the losses associated with the shock waves and the boundary layer separation that is often caused by shock waves.

The first transonic airfoils designed to operate at the cruise Mach number with large regions of supersonic flow without shock waves were called *supercritical*. Such *shock-free airfoils* sometimes experienced a gradual drag increase at Mach numbers below the critical value. This premature drag increase is often called *drag-creep* and is associated with adverse pressure gradients caused by transient shock waves which appear at Mach numbers below the design Mach number. Such shock waves can disappear at the design Mach number allowing for a reduction in drag before the transonic drag-rise. Further research indicated that the *aerodynamic efficiency* of the airfoil, as measured by the *lift-drag ratio*, is often greater for airfoils designed to operate with weak shock waves rather than shock-free, hence most modern transonic airfoils exhibit large supersonic zones terminated by weak shock waves at the design Mach number.

Three of the most important parameters for determining the flow conditions over a transonic airfoil are: 1) Mach number, 2) maximum value of thickness-chord ratio, and 3) the lift coefficient. It has been observed that an airfoil section can operate at transonic speeds without shock waves if the following inequality holds

$$100M + 10C_l + 100(t/c)_{\text{max}} < 100 \tag{23.94}$$

Garabedian and his associates developed a family of shock-free airfoils with numerical designations based on the above inequality [Bauer, Garbadian, Korn, and Jameson 1975)]. For example, airfoil 79-03-10 is designed to operate at a Mach number of 0.79 and a lift coefficient of 0.3 with a maximum thickness/chord ratio of 0.1.

The first airfoils designed to tailor the development of the supersonic flow over the upper surface of the airfoil and hence delay the drag rise are the *peaky airfoils* designed by Pearcey (1962). These airfoils increased the Mach number for drag-rise as much as 0.03 when compared with an NACA 6-series airfoil of the same thickness and design lift coefficient but had slightly poorer low speed performance. Another family of transonic airfoils was developed by Whitcomb and Harris and are characterized by flat curvature of the middle region of the upper and lower surfaces with large positive camber near the trailing edge [Whitcomb (1974) and Harris (1971), (1974), (1975a,b,c)].

Transonic drag-rise may also be controlled by *wing sweep;* the appearance of supersonic flow over the wing is delayed by the use of sweep. The flow characteristics over a swept wing are determined by the velocity normal to the leading edge (i.e., $V \cos \Lambda$) rather than the flight speed, V. Hence, for a wing with $45°$ of sweep, the velocity approaching the leading edge is only 71% of the flight speed.

Supersonic Flight. A lifting surface traveling faster than the speed of sound causes an abrupt compression of the air ahead of its leading edge. The compression propagates in all directions as spherical waves at a velocity near the speed of sound. Since the vehicle is traveling at a speed greater than the speed of sound, the compression waves accumulate along the leading edge forming a shock wave whose angle approaches the limiting value of $\mu = \sin^{-1}(1/M)$ far from the source. The angle μ is called the *Mach angle*, and the associated compression front is called a *Mach line* (see Chap. 8). These quantities are shown in Fig. 23.58. A force called *wave drag* is associated with the shock wave.

Since all pressure disturbances generated by the approaching wing propagate at velocities close to the speed of sound, the space ahead of the shock front is a *zone of silence* with respect to the wing. This phenomenon is demonstrated by bullets traveling at speeds greater than that of sound. This lack of warning associated with supersonic flight is very different from the conditions of subsonic flight where all space is aware of a vehicle traveling at low speed.

An interesting characteristic of supersonic flow is its ability to turn through large angles without separating. Such expanding flows are treated analytically by the *Prandtl–Meyer expansion* theory [Pope (1958)]. The physical mechanism which re-

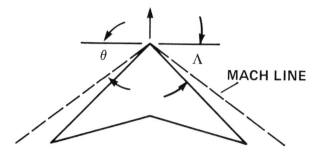

FIGURE 23.58 Supersonic wing.

duces the possibility of flow separation is the pressure reduction associated with supersonic expansion. This expansion takes place without loss of energy or momentum, but the compression which follows to restore the pressure to the free stream value gives rise to wave drag.

The wave drag associated with supersonic compressions can be reduced, as stated previously, by the use of sweep which produces a smooth flow over the surfaces of the wing similar to that found at subsonic speeds. If V is greater than the speed of sound, a, but $V \cos \Lambda$ is less than a, the sweep angle is greater than the Mach angle and the wing will lie behind the Mach lines along which pressure disturbances propagate. Under these conditions, a point ahead of the Mach line is in the zone of silence, but points behind the Mach line and ahead of the leading edge receive signals from the wing causing the flow to behave as though it were subsonic. The wing is then said to have a *subsonic leading edge*. The streamlines will follow paths similar to those found in subsonic flow.

The most widely used airfoils for supersonic flight are NACA 6-series, double wedge, hexagonal, and circular arc sections. These sections are shown in Fig. 23.59. The sharp edged airfoils have poor low-speed characteristics, hence high lift devices are usually required to obtain acceptable landing speeds. The type of airfoil used on a supersonic aircraft is partially determined by the type of flow normal to the leading

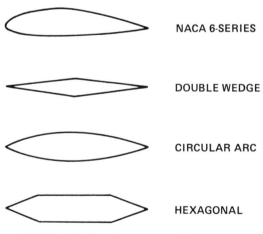

FIGURE 23.59 Supersonic airfoil sections.

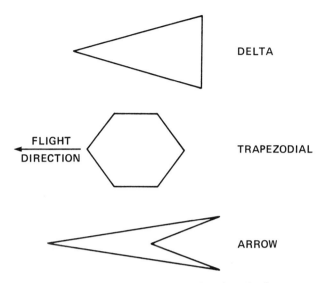

FIGURE 23.60 Basic supersonic wing planforms.

edge. If the sweep is large enough to produce a subsonic leading edge, it may be appropriate to use an airfoil with a round leading edge, such as an NACA 6-series section.

Past research has shown that in some cases sweepback can increase the form drag more than it reduces wave drag at high supersonic speeds making an unswept leading edge more desirable for some supersonic applications. The three basic planforms which have been most widely used on supersonic aircraft are shown in Fig. 23.60. Other planforms which have been used on supersonic aircraft and can be considered as variations of the three basic planforms are illustrated in Fig. 23.61. The planforms shown in Figs. 23.60 and 23.61 have bilateral symmetry, which, as shown by Jones

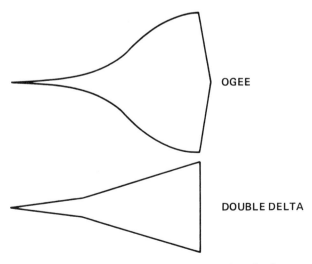

FIGURE 23.61 Advanced supersonic wing planforms.

(1972) is required only for subsonic flight. The small disturbance theory for supersonic flight indicates that the mutual interference of two bodies (or lifting surfaces) flying side by side, i.e., with bilateral symmetry, creates four times the drag of the same two bodies flying in staggered formation at Mach numbers near 1. An antisymmetric lifting surface with its lateral axis swept behind the Mach lines emanating from its most forward point will have substantially less wave drag than a symmetric lifting surface at the same supersonic speed. Examples of each type are shown in Fig. 23.62.

23.5.3 Helicopter Rotor Airfoils

Until about 1965, the airfoil sections used on helicopter rotor blades were nearly always symmetrical NACA 4-digit airfoils with maximum thickness/chord ratios ranging from 0.10 to 0.15. Cambered airfoils became acceptable during the late 1960's. The first cambered sections used were often modifications of the NACA 5-digit airfoil sections, since these airfoils have very small pitching moments about their aerodynamic centers. This is a characteristic required to reduce rotor torsional loads during forward flight. The 5-digit sections also have larger maximum lift coefficients than most other sections with the same amount of camber. However, the 5-digit sections often exhibit an abrupt stall and relatively low critical Mach numbers making them less desirable for the rotor tip region than the NACA 16-series sections. The characteristics of good rotor sections can be summarized as follows: 1) high maximum lift coefficients near $M = 0.4$, 2) high drag-rise-Mach-number and gradual drag increase beyond drag rise, 3) low drag at a Mach number and lift coefficient both equal to 0.6, and 4) pitching moments near zero measured about the aerodynamic center at all speeds up to drag rise. The first characteristic is related to the stalling performance of the retreating blade. The second characteristic is asso-

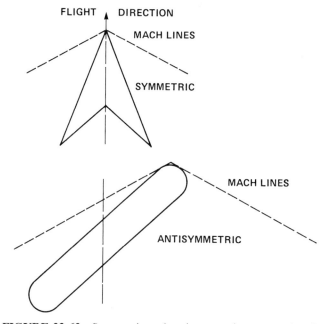

FIGURE 23.62 Symmetric and antisymmetric supersonic wings.

ciated with the drag rise of the advancing blade. In contrast with the retreating blade, the advancing blade must operate at high subsonic Mach numbers with a large amount of supersonic flow over the tip section of the blade. The rate of drag increase beyond the drag rise Mach number is important, because many rotor tips operate at speeds above drag rise. The third and fourth characteristics are important for good hover performance and low hub weight, respectively.

Another characteristic which should be considered during rotor blade design is the stall type. Three types of stall are possible on lifting surfaces: 1) leading edge stall, which is abrupt and is usually associated with thin airfoils or airfoils with small leading edge radii, 2) trailing edge stall, which is usually characterized by a gradual progression of separation forward from the trailing edge as the angle of attack is increased, and 3) a combined leading-trailing edge stall, which usually consists of the separation beginning almost simultaneously at the leading and trailing edges and moving rapidly toward each other. The most desirable type of stall from the point of view of structural considerations and handling qualities is gradual, trailing edge stall, since the loss of lift is not abrupt and the load excursions imposed on the rotor blade are minimized. The greatest difficulty associated with designing airfoils to exhibit a particular stall type is lack of theoretical methods available to calculate separated flows. Section 4.9 discusses separated flows. Hence, the design for stall type must be carried out primarily by experimental methods such as wind tunnel testing with theory providing only qualitative guidance.

The first four rotor characteristics discussed above can be readily applied to rotor section design because of the rapid advances in high speed scientific computers and computational fluid dynamic (CFD) software. It is now possible to eliminate many unacceptable blade shapes analytically before wind tunnel or flight testing. The rotor design problem is more complicated than that for a fixed wing aircraft due to the wide range of speed and angles-of-attack which exist simultaneously on a rotor blade in forward flight. The results of initial CFD studies led to the development of two distinct classes of rotor blade sections. The first class of sections had good high lift characteristics giving good retreating blade performance at Mach numbers near 0.4, but less than desirable performance at the high Mach numbers associated with the advancing blade tip. The second class of sections had high critical Mach numbers but poor high lift performance at moderate Mach numbers. CFD design methods have been developed which permit the designer to consider both speed regimes simultaneously during the design process, producing airfoils which have good advancing and retreating blade performance [Hicks and McCroskey (1980)].

Many advanced rotor sections have been designed by use of the new CFD methods [Drees and Wortmann (1969), Blackwell and Hinson (1977), Sloof *et al.* (1975), Dadone and Fukushima (1975), and Bingham and Noonan (1979)]. These airfoil sections show substantial progress has been made toward developing the desirable characteristics discussed above. Helicopters using the new sections have shown increased efficiency because of their ability to use higher rotor tip speeds without encountering compressibility limitations or to attain higher forward speed with a given rotor tip speed without the need for additional power.

23.5.4 Watercraft

The selection of hydrofoil sections for hydrofoil ships is often governed by the requirement that the hydrofoil be nearly cavitation-free at the design condition. *Cav-*

itation occurs when the pressure at some point on the hydrofoil surface falls to a value near the vapor pressure of water. The cavitation process is very complex and, among other things, is a function of the contaminants present in the water; see Sec. 23.12 for more information on cavitation. Cavitation may cause structural damage due to large pressure fluctuations and may be responsible for losses in the performance of the hydrofoil lifting surface.

Hydrofoil sections that delay the onset of cavitation are those sections which minimize the acceleration of the flow over the hydrofoil surfaces. Modern theoretical methods have been applied to the design of advanced hydrofoil sections with good results. Such sections maintain pressures below the vapor pressure of water (avoid cavitation) over a wider range of lift coefficients than was possible with the older NACA sections [Shen and Eppler (1981)].

Hydrofoil ships using noncavitating hydrofoil sections are, however, limited to speeds of approximately 55 knots. To permit higher speeds, it is necessary to use *supercavitating hydrofoils* which permit speeds approaching 80 knots [Rowe (1979)]. A hydrofoil ship using this type of section may operate economically at cruising speeds near 45 knots with the hydrofoil in a subcavitating condition and at much higher speeds with the hydrofoil operating at a supercavitating condition. The higher speeds require a substantial increase in propulsive power and hence give lower operating efficiencies and less range.

Hydrofoils do not use trailing edge flaps to achieve large changes in lift coefficients, because the corresponding change in induced velocities over the suction surface of the hydrofoil may cause cavitation. Hydrofoils which operate over a wide lift coefficient range (speed range) may require a change in the amount of surface wetting (i.e., the extent of cavitation).

Hydrofoil sections for high speed ships operating with large attached vapor cavities are often wedge shaped with sharp leading edges. Such sections are known to exhibit a hydroelastic instability called *leading-edge flutter* [Waid and Lindberg (1957)] which is similar to the flutter generated by the separated flow over an airfoil in stalled flight (see Sec. 23.6). This flutter manifests itself as a chordwise bending which results in a wave pattern on the cavity surface originating at the leading edge of the hydrofoil. Such vibrations have resulted in structural failures in supercavitating propeller blades. This chordwise bending differs from the wing bending of aircraft; only a chordwise mode has been observed on hydrofoils, whereas aircraft wings may exhibit a torsional mode along with a bending mode. Hydrofoils with vapor cavities that close near the trailing edge exhibit flow instabilities at nearly all speeds. The lift changes rapidly as the cavity closure oscillates from the suction side of the section to a point downstream of the trailing edge.

23.6   AEROELASTICITY AND HYDROELASTICITY
Earl H. Dowell

23.6.1   Introduction

The subject has been defined as encompassing those physical phenomena for which aerodynamic, elastic (structural), and inertial (dynamic) forces interact in a significant way [Bisplinghoff and Ashley (1962), Bolotin (1975), Dowell (1975), Dowell *et al.* (1979), Forsching (1974), Fung (1955), Petre (1966), Scanlan and Rosenbaum

(1951), and Smirnou (1980)]. The famous Collar aeroelastic triangle of forces is a graphical definition of the field [Bisplinghoff *et al.* (1955)].

Such phenomena are now so widespread in engineering practice, that in the space available for this section some critical choices must be made with respect to subject coverage. Fortunately, there are well written textbooks (cited above) which give much of the background for the now classical aspects of the field, and, moreover, elsewhere in this volume several aspects of aeroelasticity and hydroelasticity are discussed *inter alia*.

In this Section, the emphasis will be on: 1) the basic physical phenomena of aeroelasticity and hydroelasticity with a special focus on fluid mechanics, and 2) providing an entree into the advanced literature beyond the standard texts. Among important topics of aeroelasticity *not* treated in this article are: 1) flutter of plates and shells, 2) feedback control of flutter, and 3) structural optimization subject to aeroelastic constraints.

The latter two are beyond the scope of an article in a handbook devoted primarily to fluid mechanics. For the first, there is a volume available [Dowell (1975)] to which the reader is referred.

23.6.2 Static Aeroelasticity of the Typical Section Airfoil

Consider the geometry of Fig. 23.63. This is the *typical section* airfoil, so-called because it is a representation of some chordwise wing section at a typical spanwise location. The translational and rotational springs simulate the bending and torsional stiffnesses of the wing, respectively. In the following two sections more accurate, but more complex, representations of the structure will be considered.

The equations of motion for the typical section model are [Bisplinghoff *et al.* (1955)]

$$M[\ddot{h} + \omega_h^2 h] + S_\alpha \ddot{\alpha} = -L \tag{23.95}$$

$$I_\alpha[\ddot{\alpha} + \omega_\alpha^2 \alpha] + S_\alpha \ddot{h} = M_y \tag{23.96}$$

where

$$\omega_h^2 \equiv K_h / M \tag{23.97}$$
$$\omega_\alpha^2 \equiv K_\alpha / I_\alpha$$

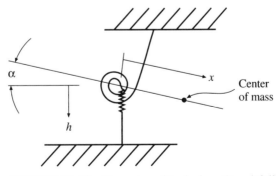

FIGURE 23.63   Geometry of typical section airfoil.

L is the lift, K_h and K_α are the bending and torsion spring stiffness, and M, I_α, and S_α are the mass, moment of inertia, and mass static unbalance with respect to the point of spring location. Since static aeroelasticity is a special case of dynamic aero-elasticity, one could solve the above dynamic model directly and assess both its static and dynamic aeroelastic behavior. However before doing so, it is instructive first to set the time derivatives to zero and consider the static equilibria conditions which are possible. Hence Eqs. (23.95) and (23.96) reduce to

$$M\omega_h^2 h = -L \qquad\qquad (23.95a)$$

$$I_\alpha \omega_\alpha^2 \alpha = M_y \qquad\qquad (23.96a)$$

Now, consider the aerodynamic force and moment. For small angles of attack, α, it is known that for a streamlined aerodynamic shape one has

$$L = qc \frac{\partial C_L}{\partial \alpha} \alpha \qquad\qquad (23.98)$$

$$M_y = M_{AC} + Le \qquad\qquad (23.99)$$

where $q = 1/2\rho_\infty U_\infty^2$, is the airfoil chord (the wing is taken to have unit span), and $\partial C_L/\partial \alpha$ is the lift curve slope (a nondimensional aerodynamic parameter which depends primarily on Mach number and Reynolds number). For small α, $\partial C_L/\partial \alpha > 0$. M_{AC} is the moment about the aerodynamic center axis (which by definition does not depend upon α) and e is the distance from the aerodynamic center to the point where the springs are attached (effectively the elastic axis; see Sec. 23.6.3).

Since L and M_y are independent of h, the second of the two equations of static equilibrium Eq. (23.96) is of primary interest. Once α is determined from Eq. (23.96), h readily follows from Eq. (23.95).

Using Eqs. (23.98) and (23.99) in (23.96), one obtains

$$K_\alpha \, \alpha = M_{AC} + eqc \frac{\partial C_L}{\partial \alpha} \alpha \qquad\qquad (23.100)$$

Solving for α

$$\alpha = \frac{M_{AC}}{K_\alpha - eqc \dfrac{\partial C_L}{\partial \alpha}} \qquad\qquad (23.101)$$

The most interesting aspect of this result is that $\alpha \to \infty$ as

$$q \to q_D \equiv \frac{K_\alpha}{ec \dfrac{\partial C_L}{\partial \alpha}} \qquad\qquad (23.102)$$

This condition of infinite twist is called *divergence*. q_D is the dynamic pressure at which divergence occurs and is an important result of any aeroelastic investigation.

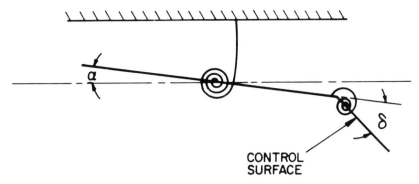

FIGURE 23.64 Typical section airfoil with control surface.

From Eq. (23.102) it is seen that as $q \to q_D$, since $\alpha \to \infty$, then also $L \to \infty$. Clearly for $q < q_D$, aeroelastic effects may increase the airfoil lift. However, of more concern is that aeroelastic effects may also lead to a decrease in airfoil lift under some circumstances. For example, consider the typical section airfoil with a control surface attached to the main airfoil as in Fig. 23.64. Now it is possible that when the control surface itself is rotated through an angle, δ, to increase lift, the lift on the control surface will tend to twist the main airfoil nose down a sufficient amount that the decrease in lift due to decreasing α just offsets the increase in lift due to δ. This condition is called *reversal*, and, in this simple form or its more elaborate counterparts for more realistic structural models, it is also a matter of great importance to the aeroelastician.

Divergence and reversal are the central physical phenomena which may be treated by *static* aeroelastic models. These are thoroughly discussed in standard texts listed earlier. Rather than discussing these further here, we now turn to dynamic models and solution methods which include these static phenomena as well as others.

For further study of static aeroelasticity, consult Bisplinghoff *et al.* (1955), Bisplinghoff and Ashley (1962), Dowell *et al.* (1979), and Fung (1955).

23.6.3 Bending/Torsion Motion of a Wing

Consider the geometry of Fig. 23.65. For simplicity, the airfoil structure is represented as a beam-rod with a straight *elastic axis*, e.a. An elastic axis is an ideal-

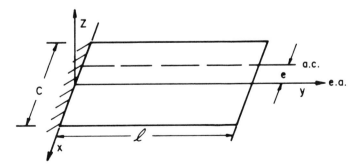

FIGURE 23.65 Beam-rod representation of a wing.

ization of a wing structure whereby: 1) static forces applied on such an axis give rise to bending of [but no twisting (*torsion*) about] the axis, and 2) static moments about the axis give rise to torsion (but no bending). This idealization is often reasonably accurate for high aspect ratio wings, turbine blades or rotor blades, and provides a useful prototype for more complex aeroelastic models.

It can be shown [Scanlan and Rosenbaum (1951)] that the equations of motion for such a structural model can be written as (if one accounts for bending by a single bending natural mode and torsion by a single torsion natural mode)

$$M_{mm}[\ddot{h}_m + \omega_{h_m}^2 h_m] + S_{\alpha mn}\ddot{\alpha}_m = -L_m \tag{23.103}$$

$$I_{\alpha nn}[\ddot{\alpha}_n + \omega_{\alpha n}^2 \alpha_n] + S_{\alpha mn}\ddot{h}_n = M_{yn} \tag{23.104}$$

where $h_m(t)$ and $\alpha_n(t)$ are generalized coordinates in the sense of Lagrange, and the physical motion has been taken to be of the form

$$h(y, t) = h_m(t)f_m(y)$$
$$\alpha(y, t) = \alpha_n(t)g_n(y) \tag{23.105}$$

and f_m, g_n are the mth and nth natural bending alone and torsion alone modes of a beam-rod. ω_{h_m}, $\omega_{\alpha n}$ are the corresponding natural frequencies. If several modes are included to represent h and α then suitable summations over m and n will appear in Eqs. (23.103) to (23.105). By restricting ourselves to a single mode in bending and torsion the results are more transparent and retain the essential features of the problem.

In Eqs. (23.103) and (23.104) the several coefficients have the following meaning

$$M_{mm} \equiv \int_0^l Mf_m^2 \, dy$$

$$I_{\alpha nn} \equiv \int_0^l I_\alpha g_n^2 \, dy$$

$$S_{\alpha mn} \equiv \int_0^l s_\alpha f_m g_n \, dy \tag{23.106}$$

$$L_m \equiv \int_0^l Lf_m \, dy$$

$$M_{yn} \equiv \int_0^l M_y g_n \, dy$$

where M, I_α, S_α, L, and M_y are all per unit distance variables along the y axis and are respectively, mass, moment of inertia, static unbalance (mass times distance from center of mass axis to elastic axis), aerodynamic lift and aerodynamic moment. (Note that if one chooses $f_m = 1$ and $g_n = 1$, then the equations reduce to those of the *typical section* model of the previous discussion, a common pedagogical device used in many texts.)

M_{mm}, $I_{\alpha nn}$, and $S_{\alpha mn}$ are generalized mass coefficients in the sense of Lagrange and L_m and M_{y_n} are generalized forces. If one uses a simple, quasi-steady, strip-theory representation for the aerodynamics, see the subsequent discussion in Sec. 23.6.5, e.g.,

$$L = qc \frac{\partial C_L}{\partial \alpha} \left[\frac{\dot{h}}{U} + \alpha \right] \tag{23.107}$$

and

$$M_y = Le + M_{AC} \tag{23.108}$$

then L_m and M_{y_n} become

$$L_m = q \frac{\dot{h}_m}{U} C_{L_{\alpha ff}} + q\alpha_n C_{L_{\alpha gf}} \tag{23.109}$$

$$M_y = \int_0^l M_{AC} g_n \, dy + q \frac{\dot{h}_m}{U} C_{M_{\alpha efk}} + q\alpha_n C_{M_{\alpha egg}}$$

where

$$C_{L_{\alpha ff}} \equiv \int_0^l \frac{\partial C_L}{\partial \alpha} c f_m^2 \, dy$$

$$C_{L_{\alpha gf}} \equiv \int_0^l \frac{\partial C_L}{\partial a} c g_n f_m \, dy$$

$$C_{M_{\alpha efk}} \equiv \int_0^l ec \frac{\partial C_L}{\partial \alpha} f_m g_n \, dy$$

$$C_{M_{\alpha egg}} \equiv \int_0^l ec \frac{\partial C_L}{\partial \alpha} g_n^2 \, dy \tag{23.110}$$

Here, M_{AC} is the aerodynamic moment about the aerodynamic center axis and by definition is independent of h and α. e is the distance from the aerodynamic center axis to the elastic axis and c the total wing chord at some spanwise location.

Using Eqs. (23.108) and (23.109) with Eqs. (23.103) and (23.104) gives two linear, ordinary differential equations for h_m and α_n. Of central importance to the aeroelastician is the question of *dynamic stability*. That is, for a given set of initial conditions, will the solutions of these equations grow or decay (exponentially) with time? In particular, the flow velocity, U_F, at which a growing solution first appears is the *flutter velocity*. Without elaborating on the details, the stability analysis is carried out as follows: Let

$$h_m = \bar{h}_m e^{pt}$$

$$\alpha_n = \bar{\alpha}_n e^{pt} \tag{23.111}$$

(We may set M_{AC} to zero, since it will only influence the static equilibrium of the wing and not its stability.) Substitution of Eq. (23.111) into Eqs. (23.103) and (23.104) using Eqs. (23.108) and (23.109) and cancelling out common factors e^{pt} gives two homogeneous algebraic equations in $\bar{h}_m$ and $\bar{\alpha}_n$. p may be thought of as a Laplace transform variable and $\bar{h}_m$ as the Laplace transform of h_m, etc. For nontrivial (be they stable or unstable) solutions the determinant of the coefficient matrix must be zero, viz.

$$
\begin{vmatrix}
M_{mm}(p^2 + \omega_{b_m}^2) & S_{\alpha_{mn}} p^2 \\
+ \dfrac{qp}{U} C_{L_{\alpha_f}} & + q C_{L_{\alpha_f}} \\
\\
S_{\alpha_{mn}} p^2 & I_{\alpha_{nn}}(p^2 + \omega_{\alpha_n}^2) \\
+ C_{M_{\alpha_{efk}}} \dfrac{qp}{U} & + q C_{M_{\alpha_{egg}}}
\end{vmatrix} = 0 \qquad (23.112)
$$

This is clearly a fourth order polynomial in p. The four roots of p determine the stability of the wing. In general, the p are complex, i.e.,

$$
p = p_R + i p_I \qquad (23.113)
$$

Each of the p_I give a possible natural (aeroelastic) frequency of wing motion and the p_R a rate of decay (if $p_R < 0$) or growth ($p_R > 0$) of the motion. If all $p_R < 0$, the wing is *stable*. If any one of the four p_R is positive the wing is *unstable*, i.e., it flutters. The principal result of a flutter analysis is the determination of the roots of p as a function of flow velocity, U. Again, the smallest U for which one of the p_R is positive is the *flutter velocity*.

Sometimes for computational reasons or because of its interest from a design point of view, it is assumed that one of the roots is *neutrally stable*, $p_R = 0$, and one solves for the flow velocity, U_F, and one of the other parameters, say air density, ρ, for which this is the case. This procedure is particularly attractive computationally when more elaborate aerodynamic representations are used which are only strictly valid or readily available for neutrally stable simple harmonic motion, i.e.,

$$
p_R = 0, \ p = i p_I \qquad (23.114)
$$

and

$$
\begin{aligned}
h_m &= \bar{h}_m e^{i p_I t} \\
\alpha_n &= \bar{\alpha}_n e^{i p_I t}
\end{aligned} \qquad (23.115)
$$

There is considerable current work, however, on methods for generalizing harmonic motion aerodynamic representations to arbitrary motions ($p_R \neq 0$). This is discussed further in Sec. 23.6.

As a notational matter, it should be pointed out that as an alternative to p a complex frequency, ω, can be introduced which is simply related to p by

$$p \equiv i\omega \tag{23.116}$$

Hence

$$p_R = -\omega_I \tag{23.117}$$
$$p_I = +\omega_R$$

The results of a flutter analysis are sometimes presented in terms of p_R, p_I or ω_R, ω_I or even ω_R, p_R. They are all related as above. Note that ω_R is the frequency associated with an aeroelastic root and $\omega_I < 0$ or > 0 denotes unstable or stable motion, respectively. See, e.g., Dowell *et al.* (1979) for further details.

23.6.1 Planar Motion of a Lifting Surface

For lower aspect ratio wings, the wing (lifting surface) may bend as a plate in two directions (x and y) rather than as a (one-dimensional) beam-rod. Nevertheless, the basic equations of motion and solution procedures are similar to those discussed in the last section for a beam-rod. If one expresses the wing (plate) deflection normal to its undeformed surface in series form using its natural modes, $\phi_m(x, y)$, and associated generalized coordinates, q_m, viz.,

$$z_a(x, y, t) = \sum_m q_m(t)\phi_m(x, y) \tag{23.118}$$

then from Lagrange's equations one may obtain the modal equations of motion as

$$M_m[\ddot{q}_m + \omega_m^2 q_m] = Q_m \tag{23.119}$$

where the aerodynamic generalized forces are given by

$$Q_m(t) \equiv \int_{\text{over wing area}} \int \Delta p(x, y, t)\phi_m(x, y) \, dx \, dy \tag{23.120}$$

and the structural generalized masses by

$$M_m = \int_{\text{over wing area}} \int m(x, y)\phi_m^2(x, y) \, dx \, dy \tag{23.121}$$

Δp is the net aerodynamic pressure (force/area) normal to the lifting surface and is determined by methods discussed in Sec. 23.6.5.

If one selects

$$z_a = -h(y, t) - x\alpha(y, t) \tag{23.122}$$

or

$$z_a = -h_m(t) f_m(y) - x\alpha_n(t)g_n(y) \tag{23.123}$$

and identifies

$$q_1 = h_m$$
$$\phi_1 = -f_m(y)$$
$$q_2 = \alpha_n$$
$$\phi_2 = -xg_n(y) \tag{23.124}$$

then from Eq. (23.119) one may retrieve essentially Eqs. (23.103) and (23.104) recognizing that

$$L \equiv \int_0^c \Delta p \, dx$$

$$M_y \equiv \int_0^c \Delta p x \, dx \tag{23.125}$$

The correspondence is exact if for the beam-rod, $S_\alpha = 0$, i.e., the elastic and center of mass axes coincide and thus f_m and g_n are natural modes of the beam-rod when both bending and torsion are permitted. For $S_\alpha \neq 0$, f_m, and g_n are so-called *uncoupled natural modes*, i.e., those for which only bending or torsion separately are permitted. Starting from f_m and g_n, one may determine the coupled natural modes of the beam-rod. However, a discussion of this topic is more suitable for a handbook on structural mechanics and the reader is therefore referred to a reference such as Bisplinghoff and Ashley (1962).

As is shown in Sec. 23.6.5, Δp may be expressed formally as

$$\Delta p(x, y, t) = \int_0^t \int_{\text{over wing area}} \int A(x - \xi, y - \eta, t - \tau)w(\xi, \eta, \tau) \, d\xi \, d\eta \, d\tau$$

$$\tag{23.126}$$

where the *downwash*, w, is

$$w \equiv \dot{z}_a + U \frac{\partial z_a}{\partial x} \tag{23.127}$$

and t is time and A is the *aerodynamic influence function*.

Using Eqs. (23.118) and (23.127) in Eq. (23.126) and the results in Eq. (23.120), one obtains

$$Q_m(t) = \sum_r \int_0^t [\dot{q}_r(\tau)I_{rm\dot{q}}(t - \tau) + q_r(\tau)I_{rmq}(t - \tau)] \, d\tau \tag{23.128}$$

where the *aerodynamic impulse functions* are

$$I_{rm\dot{q}}(t - \tau) \equiv \iiint\int A(x - \xi, y - \eta, t - \tau)\phi_r(\xi, \eta)\phi_m(x, y) \, dx \, dy \, d\xi \, d\eta$$

$$I_{rmq}(t - \tau) \equiv \iiint\int A(x - \xi, y - \eta, t - \tau)U \frac{\partial \phi_r}{\partial \xi} (\xi, \eta)\phi_m(x, y) \, dx \, dy \, d\xi \, d\eta$$

$$(23.129)$$

Equation (23.119) with Eq. (23.128) is a set of linear, differential-integral, coupled (because of the aerodynamic generalized forces) equations. By taking a Fourier or Laplace transform, however, they can be reduced to a set of algebraic equations and the stability of the lifting surface determined by methods similar to those discussed for the bending/torsion of a wing. In general, one sets the determinant of the matrix of coefficients to zero, and (because of the aerodynamic transfer functions) this leads to a transcendental equation whose roots determine the stability of the system.

A Laplace transform would lead to search for p roots and a Fourier transform to a search for ω roots.

From an alternative, but equivalent, point of view, it is worth noting that the Fourier transforms of $I_{rm\dot{q}}$ are *aerodynamic transfer functions*. When an external force is added, the Fourier transform of Eq. (23.119) using Eq. (23.128) leads to the identification of *aeroelastic transfer functions* (which include both aerodynamic and structural parameters) whose poles in ω determine the stability of the lifting surface.

For further study, see Dowell *et al.* (1979).

23.6.5 Unsteady Aerodynamics and Hydrodynamics of Lifting Surfaces

A compact, self-contained discussion of this subject pertinent to aeroelasticity is available in Chap. 4 of Dowell *et al.* (1979). The basic, underlying assumptions which are usually invoked are that the fluid flow is: 1) inviscid, 2) isentropic, and 3) irrotational. This leads to a *potential flow* fluid model expressed in terms of a differential equation for the (perturbation from freestream) velocity potential, ϕ. For small disturbances, i.e., fluid velocity perturbation which are small compared to the oncoming mean stream velocity, U_∞, this equation is

$$\phi_{xx}[1 - M_L^2] + \phi_{yy} + \phi_{zz} - \frac{1}{a_\infty^2} [2U_\infty\phi_{xt} + \phi_{tt}] = 0 \qquad (23.130)$$

where

$$M_L^2 = M_\infty^2 \left[1 + \frac{(\gamma + 1)}{U_\infty} \phi_x \right] \qquad (23.131)$$

M_∞ is the freestream Mach number at a position far from the lifting surface, M_L is the local Mach number at some point in the flow and γ is the ratio of specific heats.

Equation (23.130) is valid for subsonic, supersonic, and transonic flow. For the former two flow regimes,

$$\left\| \frac{\phi_x}{U_\infty} \right\| \ll |M^2 - 1| \tag{23.132}$$

and Eq. (23.130) simplifies (it is linearized) by noting that to this approximation, $M_L = M_\infty$. Equation (23.130) with $M_L = M_\infty$ is recognized as the (convected) *wave equation*, and there is a vast literature devoted to its solution.

Equation (23.130) is to be solved for ϕ subject to boundary conditions. These are: 1) on the lifting surface (actually at its mean plane position, $z = 0$)

$$\phi_z|_{z=0} = \dot{z}_a + U_\infty \frac{\partial z_a}{\partial x} \equiv w(x, y, t) \tag{23.133}$$

and 2) far from the airfoil ($z \rightarrow \infty$) ϕ must be finite and represent outwardly propagating waves (if waves exist).

Subsonic or Supersonic Flow, $M_L = M_\infty$. A variety of methods has been developed for solving Eq. (23.130) subject to Eq. (23.133) when M_L is approximated by M_∞. All involve converting Eqs. (23.130) and (23.133) to an integral equation.

Because of the linearity of the model, one can divide the solution into lifting (antisymmetric in ϕ with respect to z) and nonlifting (or thickness) (symmetric with ϕ with respect to z) solutions. Noting that the (perturbation from ambient) pressure, p, is given in terms of ϕ through Bernoulli's equations,

$$p = -\rho_\infty \left[\dot{\phi} + U_\infty \frac{\partial \phi}{\partial x} \right] \tag{23.134}$$

one may show that:

for the lifting case

$$w(x, y, t) \equiv \int_0^t \int_{\text{over wing area only}} \int K(x - \xi, y - \eta, t - \tau)p(\xi, \eta, \tau) \, d\xi \, d\eta \, d\tau \tag{23.135}$$

Note that $\Delta p = 2p$ at $z = 0^-$ for the lifting case.

for the thickness case (nonlifting)

$$p = \int_0^t \int_{\text{over wing area only}} \int A(x - \xi, y - \eta, t - \tau)w(\xi, \eta, \tau) \, d\xi \, d\eta \, d\tau \tag{23.136}$$

where K and A are known (influence or Green's) functions. Clearly, the nonlifting case is much easier to treat, as the desired quantity p is given as a quadrature of the known downwash, w. By contrast for the lifting problem, the desired quantity, p is expressed in terms of the known, w, through an integral equation, whose solution turns out to be quite subtle.

K, sometimes called the *Kernel function*, is highly singular in $y - \eta$, and this poses great difficulties for the various numerical solution methods which have been proposed. Nevertheless, successful solution procedures have been developed (in some quarters, they would be called boundary integral methods). These are fully described in the literature [see Dowell *et al.* (1979) for an introduction]. Most solution techniques are based upon the assumption of simple harmonic motion. While, in principle, arbitrary time dependent motion solutions can be synthesized from those for simple harmonic motion, only relative recently have practical procedures have been developed for doing so and these are still under development. See Dowell (1980) for a representative method and a discussion of the literature.

Formally, the solution procedures for Eq. (23.135) may be thought of as allowing its inversion to the form for the nonlifting problem [Eq. (23.136)]. Hence, $p(x, y, t)$ is then known as a *convolution integral* of $w(\xi, \eta, \tau)$ over the wing (and previous times) for the lifting problem as well. This inversion must usually be done by numerical means and involves using a suitable computer code, several of which are available. See NASTRAN for representative nonproprietary codes.

Transonic Flow, $M_L^2 = M_\infty^2 [1 + (\gamma + 1) \varphi_x/U_\infty]$. For the transonic problem, entirely different solution procedures have been developed for the most part, although some effort has been expended in extension of the integral equation procedures [Nixon (1978), Tseng and Morino (1982)] to transonic flow. Returning to Eq. (23.130) with Eq. (23.131), one sees this equation is nonlinear. Successful solutions have been obtained by a variety of finite difference procedures. See, for example, [Ballhaus and Goorjian (1977), Rizzella and Yoshihara (1980), and Ehlers and Weatherill (1982)]. Such procedures are substantially more expensive in computer costs than integral equation methods.

One of the fundamental difficulties for the aeroelastician is that even when such solutions can be obtained, the use of Eq. (23.130) in combination with a suitable structural model leads to a *nonlinear aeroelastic model*, thereby precluding conventional methods of *linear aeroelastic analysis*.

There are three approaches to dealing with this difficulty:

i) use of a finite difference solution procedure to treat the combined aerodynamic and structural (aeroelastic) equations. This is costly in terms of computation [Borland and Rizzetta (1981) and Yang *et al.* (1981)].

ii) Use nonlinear transfer (describing) function ideas to represent the $p \to w$ relationship, and thereby reduce the aeroelastic analysis to an equivalent linear one [Ueda and Dowell (1982)]. This retains in an approximate way the nonlinear $p \to w$ relationship and, in particular, permits the calculation of limit cycle oscillations associated with flutter in addition to the velocity at which flutter begins. It greatly reduces the cost of aeroelastic analysis over a finite difference procedure.

iii) Separate the velocity potential into a steady, ϕ_s, and dynamic, ϕ_d, part, *viz.* $\phi = \phi_s + \phi_d$, where ϕ_s represents the steady (nonuniform) flow about a stationary airfoil and ϕ_d the velocity potential associated with the airfoil dynamic oscillation. If the oscillation is small, then one may substitute this definition into Eq. (23.130) and Eq. (23.131) and linearize in ϕ_d. The resulting equation for ϕ_d is linear, but with variable coefficients (because of the dependence of ϕ_s on x, y, z) [Dowell et al. (1978)]. It may be solved by finite difference procedures [Ehlers and Weatherill (1982)] or, more approximately, by integral equation methods [Cockey (1983) and Pi and Liu (1982)]. For the aeroelastician, the advantage of this procedure is that the model is inherently linear, and conventional, linear aeroelastic methods of analysis may be used. Williams (1980) (following Eckhaus) has developed a particularly efficient, but more approximate, aerodynamic method in this spirit. Of course, one may also use the nonlinear aerodynamic methodology for sufficiently small oscillations to extract a linear aerodynamic model. This approach has been followed by several authors [Yang and Batina (1982) and Edwards et al. (1982)]. In Dowell et al. (1981), discussion is given of the circumstances under which a linear aerodynamic model may be used in transonic flow.

One aspect of transonic flow and the use of a linear aerodynamic model deserves special emphasis, as it is a source of considerable confusion in the literature. Some respected experts, noting that in general shocks will be present for transonic flow, have sometimes stated erroneously that a linear aerodynamic theory can never be used when shocks are present. Sometimes, more explicitly, it is stated that shock movement cannot be accounted for by a linear aerodynamic theory. In fact, as is perhaps shown most clearly by Williams (1979), a linear aerodynamic theory can allow for shock movement *if* the shock movement is sufficiently small. Ballhaus and Goorjian (1977) and Dowell et al. (1983) suggest the shock movement should be less than 5% of the airfoil chord for a *transonic*, *linear* theory to be valid. To be sure, it should be emphasized, this transonic, linear theory which allows for (sufficiently small) shock movement is substantially more complex than the classical subsonic or supersonic linear theories. Nevertheless, there is growing evidence that *transonic*, *linear* aerodynamic theory will prove a very valuable model for the aeroelastician.

Separated Flows (Stall). When flow separation occurs, the complications for aerodynamic modeling are considerable. McCroskey has written several informative articles on this subject; McCroskey (1982) also contains an overview of unsteady aerodynamics, in general. As McCroskey notes one of the most hopeful approaches is that of Tran and Petot (1980). These authors, using static large angle of attack experimental data combined with data taken from small angle of attack oscillations about these large static angles, have devised a nonlinear differential equation modeling procedure (a form of *system identification*) for relating say lift, L, to angle of attack, α. Such a relationship is of the form

$$A_L(\alpha)L + A_{\dot{L}}(\alpha)\dot{L} + A_{\ddot{L}}(\alpha)\ddot{L} + \cdots = F_{L\alpha}(\alpha) + F_{L\dot{\alpha}}(\alpha)\dot{\alpha} + \cdots \quad (23.137)$$

Note that this equation is only nonlinear in α *per se* and not in L or $\dot{\alpha}$, $\ddot{\alpha}$, etc. This has important practical advantages in determining the functions A_L, $A_{\dot{L}}$, $A_{\ddot{L}}$, $F_{L\alpha}$, and $F_{L\dot{\alpha}}$ from the experimental data, since these are assumed to depend on α alone and not its time derivatives. Dat and Tran (1981) have carried out a flutter analysis using this type of aerodynamic modeling.

Chi (1980) following Dowell, Williams, and Chi (1980) has developed a conceptually simple, theoretical model for airfoils undergoing small oscillations about a mean, separated flow which has shown some success in turbomachinery applications. It may prove useful to combine the method of Tran with that of Chi.

Hydrodynamics, Cavitation, and the Kutta Condition. Hydroelasticity involves basically the same phenomena as aeroelasticity except there are certain possible fluid mechanical complications. Two will be discussed here, cavitation and the Kutta condition.

Cavitation: When cavitation bubbles are formed over some portion of a hydrofoil surface, the fluid forces on an oscillating hydrofoil can be significantly modified from their noncavitating counterparts. Only for two-dimensional flows, have theoretical models for determining such forces been developed. These are viewed critically and one of them used in a flutter analysis in Brennan *et al.* (1980). From a theoretical modeling viewpoint, the cavitation problem is broadly analogous to that of separated flows for aeroelastic applications. See also Sec. 23.12 of this handbook for a discussion of cavitation.

Kutta Condition: Several authors [McCroskey (1982), Tran and Petot (1980), Dat and Tran (1980), Chi (1980), Brennan *et al.* (1980), and Abramson (1969)] have expressed doubt as to the adequacy of the Kutta condition in describing the flow at the trailing edge of an oscillating hydrofoil. This question is still not answered satisfactorily, though it has been suggested that a modified form of the Kutta condition be used with an empirically chosen parameter. Also, some authors have questioned the use of the Kutta condition for aerodynamic flows over airfoils oscillating at high frequencies. However, no operationally superior alternative has been proposed.

Local Aerodynamic Approximations in Space and Time. In general, the pressure at a given point and a given time on an oscillating lifting surface depends upon the motion of the surface at all other points and at all previous times. However, under some circumstances, it is an acceptable approximation to describe the aerodynamic pressure solely in terms of the motion at the same point and/or same time. We briefly review when such a local approximation may be used.

Piston Theory: For small disturbance potential flow theory, when either the Mach number, $M \to \infty$, or the frequency of oscillation, $\omega \to \infty$* it may be shown that the perturbation pressure (from the freestream value) on the lifting surface is given by $p(x, y, t) = \rho_\infty a_\infty w(x, y, t)$, where a_∞ is the speed of sound in the free stream. This is clearly an approximation which is local in space and time. Moreover, the relationship between pressure, p, and downwash, w, is precisely that for a piston in an infinitely long tube, hence the name, *piston theory*.

The asymptote, $\omega \to \infty$ (for any M_∞) is perhaps the more meaningful, since small disturbance theory itself will become invalid when $M_\infty \to \infty$. However, historically this approximation was developed for high Mach number flows.

---

*Normally one considers the frequency on a nondimensional basis, e.g., $k \to \infty$ where $k \equiv \omega c / U_\infty$.

Quasi-Steady Theory: For small oscillation frequencies, a local approximation in time (but not generally in space) can normally be found. As $k \to 0$, the fluid reacts instantaneously to the slow surface motions, and the aerodynamic pressure is that determined from standard aerodynamic steady flow theory.

Note that piston theory ($k \to \infty$) and quasi-steady theory ($k \to 0$) may be thought of as the fluid mechanical analogs of adiabatic and isothermal thermodynamic processes. Also note that in the time domain (in contrast to the frequency domain), piston theory is asymptotically correct as $s \to 0^*$ ($k \to \infty$) and quasi-steady theory is the result as $s \to \infty$ ($k \to 0$).

Strip Theory: If the aerodynamic surface is of high aspect ratio, then one may treat each span location as aerodynamically independent of all others. Thus, each such chordwise *strip* at each span location may be considered an independent two-dimensional airfoil.

For further discussion of these matters, see Dowell *et al.* (1979) and McCroskey (1982).

23.6.6   Flutter

Flutter is a subject rich in physical detail whose variety is sometimes only dimly perceived by examination through the linear dynamical models discussed previously. Nevertheless, it is useful to examine first those types of flutter for which linear models provide a reasonably accurate description. The following discussion is exemplary rather than exhaustive. The experienced flutter practioner will have identified many other varieties of flutter behavior.

As discussed in Sec. 26.6.3 the movement of the roots, $p = p_R + ip_I$ or $\omega = \omega_R + i\omega_I$, with increasing flow velocity, U, allows a determination of dynamic stability. One format for displaying these roots is a locus of roots with U treated as a gain as is often done by control system engineers. Here a similar, but somewhat different, format is used. The following discussion closely follows that of Dowell *et al.* (1979). For definiteness, consider bending-torsion motion of a beam-rod wing model. We shall show the roots in terms of ω (rather than p).

Types of Flutter

Coalescence or *Merging Frequency Flutter:* In this type of flutter (also called *coupled mode* or *bending-torsion flutter*) the distinguishing feature is the coming together of two (or more) frequencies, ω_R, near the flutter condition, $\omega_I \to 0$ and $U \to U_F$ (see Fig. 23.66). For $U > U_F$ one of the ω_I becomes large and positive (stable pole) and the other which gives rise to flutter becomes large and negative (unstable pole), while the corresponding ω_R remain nearly the same. Although one usually speaks of the torsion mode as being unstable and the bending mode stable, the airfoil normally is undergoing a flutter oscillation composed of important contributions of both h and α. For this type of flutter, the out-of-phase or damping forces are not qualitatively important. Often one may neglect structural damping entirely

---

*The appropriate nondimensional time is $s \equiv tU_\infty/c$, consistent with the previous nondimensionalization of frequency.

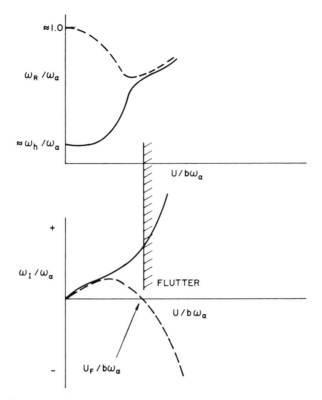

FIGURE 23.66 Real and imaginary components of frequency vs. air speed: "Coalescence" or "Merging Frequency" Flutter.

in the model and use a quasi-steady or even a quasi-static aerodynamic assumption. This simplifies the analysis and, perhaps more importantly, leads to generally accurate and reliable results based on theoretical calculations.

Single-Degree-of-Freedom Flutter: In this type of flutter, the distinguishing feature is the virtual independence of the frequencies, ω_R, with respect to variations in airspeed, $U/b\omega_{\alpha}$, where b is the half chord (see Fig. 23.67). Moreover, the change in the true damping, ω_I, with airspeed is also moderate. However, above some airspeed, one of the modes (usually torsion) which has been slightly positively damped becomes slightly negatively damped leading to flutter. This type of flutter is very sensitive to structural and aerodynamic out-of-phase or damping forces. Since these forces are less well described by theory than the in-plane forces, the corresponding flutter analysis generally gives less reliable results. One simplification for this type of flutter is the fact that the flutter mode is virtually the same as one of the system natural modes at zero airspeed and thus the flutter mode and frequency (though not flutter speed) are predicted rather accurately by theory. Airfoil blades in turbomachinery and bridges in a wind usually encounter this type of flutter.

Divergence or Zero Frequency Flutter: This is also a one-degree-of-freedom type of flutter, but of a very special type (see Fig. 23.68). The flutter frequency is zero,

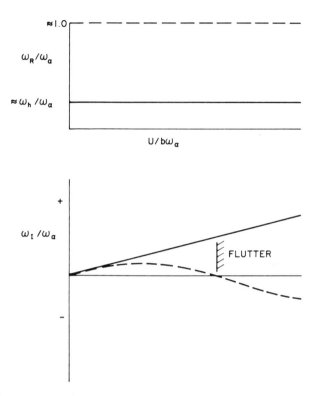

FIGURE 23.67 Real and imaginary components of frequency vs. air speed: "Single-Degree-of-Freedom" Flutter.

hence this represents the static instability which we have previously analyzed in our discussion of static aeroelasticity under the name of *divergence*. Because it is a static type of instability, out-of-phase forces are again unimportant, and the theory is generally reliable.

We note that in all of the above we have considered only positive ω_R even though there are negative ω_R as well, and these are physically meaningful. There are at least two reasons why this practice is usually followed. For those models where the aerodynamic transfer function can be (approximately) expressed as a polynomial in $p \equiv i\omega$, the negative ω_R plane is (nearly) the mirror image of the positive ω_R plane and the ω_I are identical, i.e., all poles are complex conjugates in p. Secondly, some of the structural damping models employed in flutter analysis are only valid for $\omega_R > 0$, hence the $\omega_R < 0$ in such cases cannot be interpreted in a physically valid way. However, there are some types of traveling wave flutter in plates and shells for which a consideration of negative ω_R is essential. In such cases, a change in sign of ω_R represents a change in direction of the traveling wave.

Flutter Calculations in Practice. At this point it should be emphasized that, in practice, one or another of several indirect methods is often used to compute the flutter velocity, e.g., the so-called U *(or V)-g method*. In this approach, structural damping is introduced by multiplying the structural natural frequencies squared, ω_h^2, ω_α^2 by 1 $+ ig$ where g is a structural damping coefficient and pure sinusoidal motion is as-

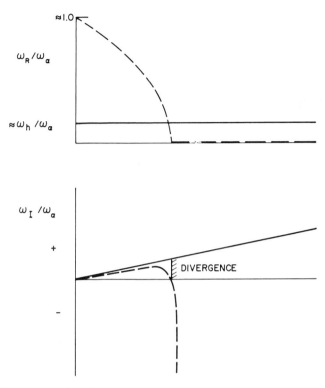

FIGURE 23.68 Real and imaginary components of frequency vs. air speed: ''Divergence'' or ''Zero Frequency'' Flutter.

sumed, i.e., $\omega = \omega_R$ with $\omega_I \equiv 0$. For a given U, the g required to sustain pure sinusoidal motion for each aeroelastic mode is determined. The computational advantage of this approach is that the aerodynamic forces only need to determined for real frequencies. The disadvantage is the loss of physical insight. For example, if a system (with no structural damping) is stable at a given airspeed, U, all the values of g so determined will be negative, but these values of g cannot be interpreted directly in terms of ω_I. Moreover, for a given system with some prescribed damping, only at one airspeed $U = U_F$ (where $\omega = \omega_R$ and $\omega_I \equiv 0$) will the mathematical solution be physically meaningful. The limitations of the *V-g method* are fully appreciated by experienced practitioners, and it is a measure of the perceived difficulty of determining the aerodynamic forces for other than pure sinusoidal motion, that this method remains very popular. For further discussion of this matter, including possible pitfalls, see Dowell *et al.* (1979).

Aerodynamic Roots. One discovery, which has come about through improved flutter analysis procedures that allow one to solve directly for $p = p_R + ip_I$ or $\omega = \omega_R + i\omega_I$, is that aerodynamic roots may sometimes contribute directly and significantly to flutter. The transcendental nature of the aerodynamic transfer function gives rise to the possibility of aerodynamic poles in the *aeroelastic transfer function* which are in addition to those whose origins are in the structural dynamics model. Historically, particularly with the *V-g method*, only those roots or poles which can be identified with structural natural modes at small U (or V) have been studied. Thus, a flutter analysis by the *V-g method* is an investigation of how only the structural mode poles change with flow velocity.

However, Dashcund (1980) and Rodden and Bellinger (1982) have shown that aerodynamic roots are sometimes present and determinate for stability. Three aspects of Dashcund's work deserve special mention: 1) use of the exact Theodorsen aerodynamic representation for two-dimensional incompressible flow (as well as more approximate ones), 2) verification of the appearance (and disappearance) of additional aerodynamic roots at certain flow velocities by utilizing Cauchy's argument principle, and 3) performance of experiments which verified the portions of the theoretical stability boundary attributed to the aerodynamic roots.

Nonlinear Flutter Behavior. There are two other types of flutter which are of importance, *transonic buzz* and *stall flutter*. Both involve significant aerodynamic nonlinearities and are therefore, not describable by our previous linear models. Indeed, both are imperfectly understood theoretically and recourse to experiment and/or empirical rules-of-thumb is usually made.

Transonic Buzz: Typically, an oscillating control surface gives rise to an oscillating shock which produces an oscillating pressure field which gives rise to an oscillating control surface which gives rise to an oscillating shock and so on and so forth. The airfoil profile shape is known to be an important parameter, and this fact plus the demonstrated importance of the shock means that any aerodynamic theory which hopes to successfully predict this type of flutter must accurately account for the nonuniform mean steady flow over the airfoil and its effect on the small dynamic motions which are superimposed due to control surface and shock oscillation. Aerodynamic methods which include a model of viscous boundary layer–shock wave interaction provide hope for better understanding of this phenomenon [see Steger and Bailey (1980) and Chyu and Kuwahara (1982)].

Stall Flutter: An airfoil oscillating through large angles of attack will create a time lag in the aerodynamic moment which may give rise to negative aerodynamic damping in pitch and, hence, flutter, even though for small angles of attack the aerodynamic damping would be positive. Compressor and helicopter blades are particularly prone to this type of flutter, since they routinely operate through large ranges of angle of attack. An earlier section discussed the aerodynamic modeling associated with this type of flutter [see Sec. 4.9 on Separated Flows *(stall)*].

Parameter Trends. Coalescence flutter is perhaps the most common for airfoils under conventional flow conditions (no shock oscillation and no stall). It is certainly the best understood. Hence, for this type of flutter, a good understanding of the effects of various structural and aerodynamic parameters on flutter is available. This is less true for other types of flutter [Dowell *et al.* (1979)].

23.6.7   Forced Vibration in Response to Gusts

In addition to the critical question of dynamic stability or flutter, the response of an aeroelastic system to external time dependent forces is of considerable interest.

Foremost among these are *gusts*, be they atmospheric disturbances encountered by aircraft, missiles or rotorcraft, or inlet distortions in jet engines. Here, the essence of the issue and associated analysis are sketched. Often the modeling of such disturbances treats them as random, hence random dynamic analysis and the use of power spectral density concepts have become widespread.

For definiteness, first consider an aircraft wing whose deformation is dominated by a single bending mode. Indeed for simplicity consider the beam-rod model of Sec. 23.6.3 with $h \neq 0$, $\alpha = 0$.

The lift force as given by a quasi-steady aerodynamic theory and including the effect of a vertical gust velocity component, w_G, is

$$L = qc \frac{\partial C_L}{\partial \alpha} \left[\frac{\dot{h}}{U} + \frac{w_G}{U} \right] \tag{23.138}$$

The corresponding generalized force is given by

$$L_m = q \frac{\dot{h}}{U} C_{L_{\alpha_{ff}}} + q \frac{w_G}{U} C_{L_{\alpha_{wf}}} \tag{23.139}$$

with

$$C_{L_{\alpha_{ff}}} = \int_0^l \frac{\partial C_L}{\partial \alpha} cf_m^2 \, dy$$

$$C_{L_{\alpha_{wf}}} = \int_0^l \frac{\partial C_L}{\partial \alpha} cf_m \, dy \tag{23.140}$$

where, for simplicity, we have assumed w_G depends only upon time, t, and is independent of y. The equation of motion is then

$$M_{mm}[\ddot{h}_m + \omega_{h_m}^2 h_m] + q \frac{\dot{h}_m}{U} C_{L_{\alpha_{ff}}} = q \frac{w_G}{U} C_{L_{\alpha_{wf}}} \tag{23.141}$$

It is straightforward to determine the solution of Eq. (23.141) when the gust time history is a sinusoidal function of time. Let

$$w_G = \overline{w}_G e^{i\omega t} \tag{23.142}$$

hence

$$h_m = \overline{h}_m e^{i\omega t} \tag{23.143}$$

Substituting Eqs. (23.142) and (23.143) into Eq. (23.141) and solving for $\overline{h}_m / \overline{w}_G$ gives the *aeroelastic transfer function*,

$$H_{hw} \equiv \frac{\overline{h}}{\overline{w}_G} = \frac{\dfrac{q}{U} C_{L_{\alpha_{wf}}}}{M_{mm}[-\omega^2 + \omega_{h_m}^2] + q \dfrac{i\omega}{U} C_{L_{\alpha_{ff}}}} \tag{23.144}$$

With Eq. (23.144) as a basic building block, one may now determine the mean square of h when w_G is a random function of time. In standard texts [Bisplinghoff and Ashley (1962)] the following key relationships are defined and deduced.

The correlation function of w_G is defined as

$$\phi_{w_G w_G}(\tau) \equiv \lim_{T \to \infty} \frac{1}{2T} \int_{-T}^{T} w_G(t) w_G(t + \tau) \, dt \qquad (23.145)$$

The power spectral density of w_G is defined as the Fourier transform of the correlation function, i.e.,

$$\Phi_{w_G w_G}(\omega) \equiv \frac{1}{\pi} \int_{-\infty}^{\infty} \phi_{w_G w_G}(\tau) e^{-i\omega\tau} \, d\tau \qquad (23.146)$$

From Eq. (23.146) one has immediately the companion Fourier relationship,

$$\phi_{w_G w_G}(\tau) = \tfrac{1}{2} \int_{0}^{\infty} \Phi_{w_G w_G}(\omega) e^{i\omega\tau} \, d\tau \qquad (23.147)$$

or

$$\phi_{w_G w_G}(\tau) = \int_{0}^{\infty} \Phi_{w_G w_G} \cos \tau \, d\tau \qquad (23.148)$$

Note that from Eq. (23.145), $\phi_{w_G w_G}(\tau)$ is an even function of τ, hence from Eq. (23.146), $\Phi_{w_G w_G}$ is an even real function of ω and Eq. (23.148) follows. Finally, note that from Eq. (23.145), the mean square of w_G is given by

$$\overline{w_G^2} = \phi_{w_G w_G}(\tau = 0) \qquad (23.149)$$

Thus from Eq. (23.148)

$$\overline{w_G^2} = \int_{0}^{\infty} \Phi_{w_G w_G}(\omega) \, d\omega \qquad (23.150)$$

Analogous to Eq. (23.150) is a similar relationship for $\overline{h^2}$, i.e.,

$$\overline{h^2} = \int_{0}^{\infty} \Phi_{hh}(\omega) \, d\omega \qquad (23.151)$$

where Φ_{hh} is the power spectral density of h. Finally, there is a fundamental relationship between the two power spectral densities which can be derived [Dowell et al. (1979)].

$$\Phi_{hh}(\omega) = |H_{hw_G}(\omega)|^2 \Phi_{w_G w_G} \qquad (23.152)$$

Using Eq. (23.152) in Eq. (23.151), one has the following expression for the mean square of h,

$$\overline{h^2} = \int_0^\infty |H_{hw_G}(\omega)|^2 \Phi_{w_G w_G}(\omega) \, d\omega \qquad (23.153)$$

Actually Eq. (23.152) is valid for any two input–output variables which are related by a set of linear equations for an aeroelastic system with time invariant properties. In particular, it is valid for a system with many degrees-of-freedom, provided the transfer function is appropriately generalized and determined.

In addition to determining $\overline{h^2}$, other responses such as acceleration or strain may be calculated. Also, one can, by assuming Gaussian statistics, determine other quantities such as the probability of exceeding certain response levels, etc., which are useful in fatigue studies. An excellent reference on this subject is the report by Houbolt, Steiner, and Pratt (1964).

23.6.8 Gaining an Entry to the Literature Including Available Computer Codes

Bibliography

Books: In addition to the books cited earlier, these can be recommended.

Simiu, E. and Scanlan, R. H., *Wind Effects on Structures—An Introduction to Wind Engineering*, John Wiley & Sons, New York, 1978.

Survey Articles: Among many possibilities, the following are generally well written and authoritative.

Landahl, M. T. and Stark, V. J. E., "Numerical Lifting Surface Theory-Problems and Progress," *AIAA J.*, No. 6, No. 11, pp. 2049–2060, 1968.

Aeroelastic Effects From a Flight Mechanics Standpoint, AGARD, Conference Proceedings No. 46, 1969.

Garrick, I. E., "Aeroelasticity—Frontiers and Beyond," *J. of Aircraft*, Vol. 13, No. 9, pp. 641–657, 1976.

Unsteady Aerodynamics Contribution of the Structures and Materials Panel to the Fluid Dynamics Panel Round Table Discussion on Unsteady Aerodynamics, AGARD Report R-645, 1976.

Rodden, W. P., *A Comparison of Methods Used in Interfering Lifting Surface Theory*, AGARD Report R-643, 1976.

McCroskey, W. J., "Some Current Research in Unsteady Fluid Dynamics—The 1976 Freeman Scholar Lecture," *J. Fluids Eng.*, pp. 8–39, 1977.

Aeroelastic Problems in Aircraft Design, Lecture Series, Von Karman Institute of Fluid Dynamics, 1979.

Ashley, H., Lehman, L. L., and Nathman, J. K., "The Constructive Uses of Aeroelasticity," AIAA 80-0877, 1980.

Tijdeman, H. and Seebass, R., "Transonic Flow Past Oscillating Airfoils," *Ann. Rev. Fluid Mech.*, pp. 181–222, 1980.

Noll, T. E., Huttsell, L. J., and Cooley, D. E., "Wing/Store Flutter Suppression Investigation," *J. Aircraft*, Vol. 18, No. 11, pp. 963–968, 1981.

Garrick, I. E. and Reed, W. H., III, "Historical Development of Aircraft Flutter," *J. Aircraft*, Vol. 18, No. 11, pp. 897–912, 1981.

Reed, W. H., III, "Aeroelasticity Matters: Some Reflections on Two Decades of Testing in the NASA Langley Transonic Dynamics Tunnel," presented at the International Symposium on Aeroelasticity, Nuremberg, Germany, 1981.

Abel, I. and Newsom, J., "Overview of Langley Activities in Active Controls Research," NASA TM 83149, 1981.

McCroskey, W. J., "Unsteady Airfoils," *Ann. Rev. Fluid Mech.*, pp. 285–311, 1982.

AIAA Journal, ASCE Transactions, Engineering Mechanics Division, Journal of Applied Mechanics, Int. Journal of Solids and Structures, Journal of Aircraft, Journal of Ship Research, and *Journal of Sound and Vibration.*

Design Literature: The impact of aeroelasticity on design is not discussed in any detail in this article. For insight into this important area the reader may consult the volumes prepared by the National Aeronautics and Space Administration in its series on Space Vehicle Design Criteria. Although these documents focus on space vehicle applications, much of the material is relevant to aircraft as well. The depth and breadth of coverage varies considerably from one volume to the next, but each contains at least a brief state-of-the-art review of its topic as well as a discussion of Recommended Design Practices. Further, some important topics are included which have not been treated at all in the present article. These include: 1) aeroelasticity of plates and shells (panel flutter) (NASA SP-8004), 2) aeroelastic effects on control system dynamics (NASA SP-8016, NASA SP-8036, NASA SP-8079), 3) structural response to time-dependent separated fluid flows (*buffeting*) (NASA SP-8001), 4) fluid motions inside elastic containers (*fuel sloshing*) (NASA SP-8009, NASA SP-8031), and 5) coupled structural-propulsion instability (*POGO*) (NASA SP-8055)

Availability of Computer Codes.

Many computer codes exist for calculating aerodynamic forces on oscillating lifting surfaces, flutter solutions, and gust response solutions. Among those which are nonproprietary and accessible the following, which are available from the National Aeronautics and Space Administration, are representative. Such codes are continually improved or superceded, of course.

NASTRAN. This program set was created originally to permit finite element analysis of structures. It now includes a capability for carrying out flutter and random gust response analysis. Various classical aerodynamic theories are available for subsonic and supersonic (but not transonic) flow.

DYLOFLEX [Perry *et al.* (1980)]. This code is designed for determining dynamic loads of flexible aircraft in response to gusts and control system inputs. Again the aerodynamic theories used are for subsonic and supersonic flow.

LTRAN2 [Reed (1981)]. This is one of the more useful codes for two dimensional transonic flow around an oscillating airfoil. Its counterpart for three-dimensional flow past oscillating lifting surfaces, LTRAN3, has also been developed [Borland and Rizzetta (1981)].

CAP-TSD. A popular three-dimensional, unsteady aerodynamic code is CAP-TSD developed by Batina *et al.* (1989). It is based on transonic potential, small disturbance theory.

A very readable survey of the computational state-of-the-art is given in Edwards and Malone (1991).

Experimental Methods. Although such methods are widely used and many specific papers present the results obtained by such methods, the literature on experimental methodology per se is relatively sparse. For the novice or experienced practitioner, the discussion of Halfman in Bisplinghoff *et al.* (1955) is still unsurpassed as an introduction or review. For a review of the accomplishments of the most active aeroelastic test facility in the United States, the NASA Langley Research Center Transonic Dynamics Wind Tunnel, and an incisive discussion of modern experimental methods, the paper by Reed (1981) is highly recommended.

23.7 ROAD VEHICLE AERODYNAMICS
E. Eugene Larrabee

23.7.1 Overview

This section describes typical aerodynamic problems of surface vehicles, primarily cars, trucks, and buses. Some of these problems are encountered also with trains. Those of ships will not be discussed. The reader can refer to Sec. 23.8 for train aerodynamics. The principal aerodynamic problems of road vehicles are the effects on performance and fuel economy of aerodynamic resistance to motion and the necessity for providing adequate ventilation for passengers, power plants, and equipment under noise-free conditions. In high speed vehicles, especially race and competition cars, the tendency to develop lift must be carefully controlled. Also, lateral response to cross wind disturbances and the velocity fields of adjacent vehicles should be minimized.

23.7.2 Effects on Vehicle Straight Line Performance

Practical design constraints make most road vehicles (and trains) bluff bodies, whose flow field is dominated by extensive regions of separation, which are triggered by sharp edged windshield pillars, rear quarter panels, uncovered wheel openings, and completely unfaired undersides. The drag coefficient of such a body is not Reynolds number dependent.

Speed squared variation of drag of this kind needs to be considered in relation to the other main source of vehicle resistance, namely the rolling friction of pneumatic tires. Tire friction is caused by hysteresis of the tread material which makes the pressure distribution in the foot print region unsymmetrical in the fore-and-aft direction. For normal inflation pressure and loading, the coefficient of tire rolling friction is typically between 0.014 (steel belted radials) and 0.02 (bias ply fabric), and it may be speed dependent, particularly for bias ply tires. By contrast, for steel wheels running on steel rails, the hysteresis lies in the ground supporting the rails. The vertical deflection of the rails is not quite symmetrical fore-and-aft, and the coefficient of friction is much lower, e.g., 0.002. Journal bearing friction has a similar level.

Assuming for simplicity that aerodynamic drag is proportional to speed squared, that rolling friction is independent of speed, and that the vehicle powerplant operates at constant power (except at very low speeds, where it is torque or traction limited), it is possible to derive simple analytic expressions which may be used to predict the effects of aerodynamic drag on vehicle straight line performance or to determine vehicle drag by application of parameter identification methods to road test data.

Coasting and Braking Performance. A vehicle is brought to high speed on a level road under calm air conditions, and its transmission is shifted into neutral. The decay of speed with either time or distance may then be observed. The equation of motion is

$$m(dv/dt) = mv(dv/ds) = -(\rho/2)v^2 A C_D - \mu mg \qquad (23.154)$$

where

m = vehicle mass (kg)
g = acceleration of gravity, 9.806 m/s^2
v = vehicle speed (m/s)
t = time (s)
s = distance (m)
ρ = air density, 1.225 kg/m^3 at 760 mm Hg and 15°C
A = vehicle frontal area (m^2)
C_D = vehicle drag coefficient (dimensionless)
μ = coefficient of rolling friction (dimensionless)

This equation is integrated to yield

$$(t_2 - t_1)/t_r = \arctan(r_1) - \arctan(r_2) \qquad (23.155a)$$

or

$$(s_2 - s_1)/s_D = \left(\frac{1}{2}\right) \ln\left[\frac{r_1^2 + 1}{r_2^2 + 1}\right] \qquad (23.155b)$$

for time and distance solutions, respectively. The solutions have been normalized with $s_D = 2m/\rho A C_D$ (m), the *aerodynamic penetration*, $v_r = \sqrt{\mu g s_D}$ (m/s), the *reference speed*, where aerodynamic drag equals rolling friction, and $t_r = s_D/v_r$ (s), the *reference time*. The quantities $r_1 = v_1/v_r$ and $r_2 = v_2/v_r$ are dimensionless speed ratios at t_1 (or s_1) and t_2 (or s_2), respectively. The use of these equations can be illustrated by an example.

Example 23.1 Coasting Performance of a Subcompact Sedan

$$A = 1.9 \text{ m}^2; \ C_D = 0.45$$

$$\rho = 1.225 \text{ kg/m}^3 \ (760 \text{ mm Hg}, \ 15°C)$$

$$m = 1000 \text{ kg}; \ \mu = 0.015; \ g = 9.806 \text{ m/s}^2$$

$$s_D = 2m/\rho A C_D = 1909.5 \text{ m}$$

$$v_r = \sqrt{\mu g s_D} = 16.759 \text{ m/s} \ (37.490 \text{ mph})$$

$$t_r = s_D/v_r = 113.937 \text{ s}$$

TABLE 23.8 Solution for Example 23.1 (the results are plotted in Fig. 23.69)

v (mph)	r	t (s)	s (m)	s (ft)
70	1.8672	0	0	0
60	1.6004	7.609	220.4	732.2
50	1.3337	17.281	457.4	1500.5
40	1.0670	29.774	707.5	2321.0
30	0.8002	46.056	960.6	3151.6
20	0.5335	67.111	1194.0	3917.3
0	0	122.950	1433.1	4701.8

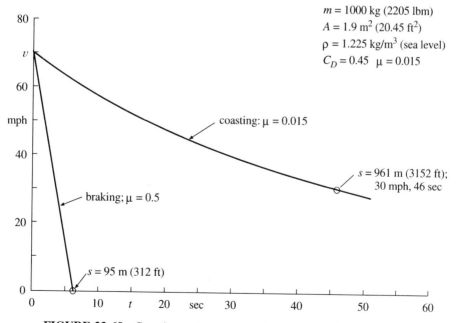

FIGURE 23.69 Coasting and braking performance for Example 23.1.

These equations may be elaborated to account for the additional apparent mass associated with angular momentum of the wheels and drive line, variation of rolling friction with speed, and slope of the roadway, but the example gives a clear idea of the time and distance scales for the process.

The same equations can be used to predict braking performance if the coefficient of friction is appropriately increased to about 1.0 for rubber tires on a dry road at low speed or 0.5 at high speed or about 0.2 for steel wheels on the point of sliding on steel rails. The effect of airbrakes also can be studied if the appropriate increase in drag coefficient is known. Figure 23.69 presents comparative speed-time histories for coasting and braking deceleration of a subcompact car. The braking distance from 70 mph is decreased from 99.9 m (327.8 ft) to 95.0 m (312 ft) when aerodynamic drag is accounted for.

Figure 23.70 illustrates a method of determining C_D and μ from an experimental speed-time history, in this case modeled by the results of Example 23.1. The reference speed, v_r, is estimated, and the arctangents of the speed ratios corresponding to the experimental data are plotted as a function of time. The correct reference speed will produce a linear relation; too small a value will produce an upwardly convex line and too large a value the opposite. The curvature is easily obscured by noise in actual data, however, but the slope of the line, (Δ arctangent γ)/Δt, is little affected by the choice of v_r. In any event

$$\mu = (v_r/g)[(-\Delta \text{ arctan } r)/\Delta t] \tag{23.156}$$

and

$$C_D = [(2m/\rho A v_r)][(-\Delta \text{ arctan } r)/\Delta t] \tag{23.157}$$

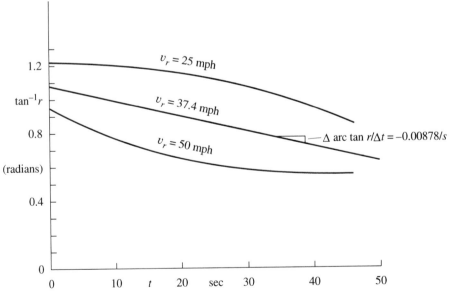

FIGURE 23.70 Determination of C_D and μ from coasting data: $v_r = 16.7$ m/s (37.4 mph), $C_D = 0.45$, $\mu = 0.015$.

showing that too small a value of v_r will tend to reduce the extracted value of μ and increase the value of C_D.

Acceleration Performance. A standard performance test is full throttle acceleration through gears to a speed near top speed; the time and distance to specified speeds are the quantities of interest. The engine-driver-driveline combination is modeled as a constant power device above the *intermediate speed*, v_i; below v_i, the driving force is limited to a value $\lambda\gamma mg$. Here, λ is the fraction of vehicle weight on the driven wheels and γ is the limiting coefficient of friction; torque limiting amounts to specifying a maximum value of the $\lambda\gamma$ product. The driveline efficiency, η, accounts for the mechanical losses in the driveline, the inability of the (driver) transmission combination to hold the engine at peak power as the vehicle speed changes, and the surprisingly large moments of the normal forces on the driven tires about the axles due to circumferential deflection of the driven tires.

The equations of motion are for the traction or torque limited range

$$m(dv/dt) = mv(dv/ds) = (\lambda\gamma - \mu)mg - (\rho/2)v^2AC_D \tag{23.158}$$

and for the power limited range

$$m(dv/dt) = mv(dv/ds) = (\eta P/v) - \mu mg - (\rho/2)v^2AC_D \tag{23.159}$$

Here, P is the engine power. Averre [see Larrabee (1974)] has integrated these equations with the following results for the traction or torque limited range

$$\frac{(t_2 - t_1)}{s_D/v_f} = \left(\frac{1}{2}\right)\ln\left[\frac{1 + \xi_2}{1 - \xi_2} \cdot \frac{1 - \xi_1}{1 + \xi_1}\right] \tag{23.160a}$$

and

$$\frac{(s_2 - s_1)}{s_D} = \left(\frac{1}{2}\right)\ln\left[\frac{1 - \xi_1^2}{1 - \xi_2^2}\right] \tag{23.160b}$$

Here

$$v_i = \eta P/\lambda\gamma mg \quad \text{(m/s)} \tag{23.161a}$$

$$v_f = \sqrt{(\lambda\gamma - \mu)gs_D} \quad \text{(m/s)} \tag{23.161b}$$

v_f would be the top speed if there were only *friction* limiting, and $\xi_1 = v_1/v_f$ and $\xi_2 = v_2/v_f$ are dimensionless speed ratios at t_1 (or s_1) and t_2 (or s_1), respectively, for $0 < v < v_i$.

For the power limited range

$$\frac{t_4 - t_3}{s_D/v_t} = \left[\frac{1}{3 + \phi}\right]\left[\Delta A + \frac{\Delta B}{2} - \left(\frac{3 + 2\phi}{\sqrt{3 + 4\phi}}\right)\Delta C\right] \tag{23.162a}$$

$$\frac{s_4 - s_3}{s_D} = \left[\frac{1}{3 + \phi}\right]\left[\Delta A - \left(\frac{2 + \phi}{2}\right)\Delta B - \left(\frac{\phi}{\sqrt{3 + 4\phi}}\right)\Delta C\right] \tag{23.162b}$$

where

$$\Delta A = \ln \left[\frac{1 - r_3}{1 - r_4} \right] \tag{23.163}$$

$$\Delta B = \ln \left[\frac{1 + \phi + r_4 + r_4^2}{1 + \phi + r_3 + r_3^2} \right] \tag{23.164}$$

$$\Delta C = \arctan \left(\frac{1 + 2r_4}{\sqrt{3 + 4\phi}} \right) - \arctan \left(\frac{1 + 2r_3}{\sqrt{3 + 4\phi}} \right) \tag{23.165}$$

Here, v_t is the top speed, and ϕ is the ratio of rolling friction to air drag at top speed. They may be calculated as

$$v_t^3 + \mu g s_D v = (s_D/m)\eta P \tag{23.166}$$

$$\phi = (\mu m g)/((\rho/2)v_t^2 A C_D) \tag{23.167}$$

The quantities $r_4 = v_4/v_t$ and $r_3 = v_3/v_t$ are dimensionless speed ratios at t_4 (or s_4) and t_3 (or s_3), respectively. Again, the use of these equations is best illustrated by an example.

Example 23.2   Acceleration Performance of a Subcompact Sedan

$$A = 1.9 \text{ m}^2; \ C_D = 0.45; \ \rho = 1.225 \text{ kg/m}^3$$

$$m = 1000 \text{ kg}; \ \mu = 0.015; \ \lambda\gamma = (0.62)(0.5)$$

$$P = 52 \text{ kW (70 hp)}; \ \eta = 0.85; \ g = 9.806 \text{ m/s}^2$$

$$s_D = 2m/\rho A C_D = 1909.5 \text{ m (as before)}$$

$$v_i = \eta P/\lambda\gamma m g = 14.540 \text{ m/s (32.525 mph)}$$

$$v_f = \sqrt{(\lambda\gamma - \mu)g s_D} = 74.323 \text{ m/s (166.255 mph)}$$

$$t_i = (s_D/v_f) \left(\frac{1}{2} \right) \ln \left[\frac{1 + (v_i/v_f)}{1 - (v_i/v_f)} \right] = 5.0918 \text{ s}$$

$$s_i = s_D(1/2) \ln [1/1 - (v_i/v_p)^2] = 37.258 \text{ m}$$

$$v_t^3 + \mu g s_D v_t = (\eta P/m)s_D v_t = 41.732 \text{ m/s (93.352 mph)}$$

$$\phi = (\mu m g)/(\rho/2)v_t^2 A D_D = 0.1613$$

TABLE 23.9   Solution for Example 23.2 (the results are plotted in Fig. 23.71)

v (mph)	r	ΔA	ΔB	ΔC	Δt (s)	t (s)	Δs (m)	s (m)
32.525	0.34841	0	0	0	0	5.092	0	37.26
50.0	0.53561	0.33868	0.19575	0.09551	3.8122	8.904	71.722	104.98
60.0	0.64273	0.60092	0.30696	0.14829	7.1840	12.276	155.041	192.3
70.0	0.74985	0.95735	0.41635	0.19197	12.034	17.126	296.706	334.0
80.0	0.85697	1.51637	0.52332	0.23117	19.912	25.004	562.545	600.0

These equations may be used also to predict train acceleration.

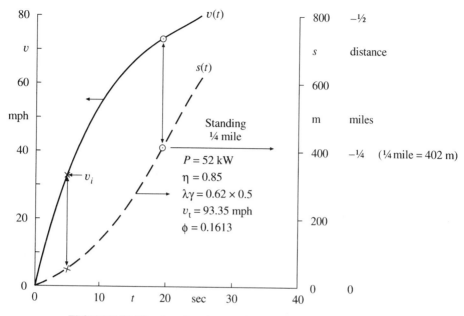

FIGURE 23.71 Acceleration performance for Example 23.2.

Fuel Economy. Above speeds of 40 mph (65 km/hr), the power required to maintain speed on a level road is dominated by aerodynamic drag, which therefore has important consequences on fuel consumption. The level road power requirement is given by

$$\eta P = (\rho/2)v^3 A C_D + \mu mgv \qquad (23.168)$$

and the fuel rate is found by multiplying P by the *specific fuel consumption*. The fuel rate per unit distance is found by dividing the fuel rate by the vehicle speed, as shown in the following example.

Example 23.3 Fuel Economy of a Subcompact Sedan specific fuel consumption = 0.423 ltr. gasoline/kW-hr

$$A = 1.9 \text{ m}^2; \; C_D = 0.45; \; \rho = 1.225 \text{ kg/m}^3$$

$$m = 1000 \text{ kg}; \; \mu = 0.015; \; g = 9.806 \text{ m/s}^2; \; \eta = 0.85$$

TABLE 23.10 Solution for Example 23.3 (1 U.S. gallon = 3.7854 lit)

v (mph)	v (km/hr)	v (m/s)	P (kW)	Fuel Rate lit/hr	lit/100 km	m/gal
40	69.374	17.882	6.617	2.799	3.348	54.1
50	80.467	22.352	10.748	4.546	5.650	41.6
60	96.561	26.822	16.531	6.992	7.242	32.5
70	112.654	31.293	24.294	10.277	9.122	25.8
80	128.746	35.763	34.370	14.539	11.292	20.8

Examples 23.2 and 23.3 illustrate the ordinary concerns of the vehicle designer. The analytic methods given here permit quantitative evaluation if aerodynamic drag can be estimated or determined from experiment. A typical method of estimating automobile drag may be found in the *Handbook of Tables for Applied Science* (1969).

23.7.3   Sources of Vehicle Drag and Lift

Figure 23.72 shows the flow field about the symmetry plane of a typical subcompact car. The only extensive regions of attached flow are above the roof, on the windshield, and perhaps on the sides (not shown). On these, a small skin friction drag is exerted by viscosity acting at the bottom of a thin boundary layer. Most of the vehicle drag is due to unbalanced surface pressure loads in the direction of motion caused by extensive regions of separated flow. In particular, separation of the flow field in high velocity regions, such as that produced by rain gutters on the side edges of windshield pillars, leads to very low pressures in the separation bubble (or cavity), and these tend to be communicated to aft facing panels, such as the rear window. On the other hand, positive pressure regions are produced on the forward face of the radiator core and at the base of the windshield by local deceleration of the flow. A thick sheet of eddying, turbulent flow exists between the vehicle underside and the moving road surface which serves to maintain the average velocity of the fluid in the eddies at stream velocity near the road. The rough underside of the vehicle, its wheels, and its suspension elements all tend to be buffeted by the rearward moving eddies, and they experience an average downstream force.

The flow pattern shown tends to produce an aerodynamic lift approximately equal in magnitude to the drag; it is a consequence of the relatively high velocities over the roof and the relatively low velocities beneath the floor. Ordinarily, the lift is of no consequence, amounting to about 5% of the vehicle weight at 70 mph (113 km/

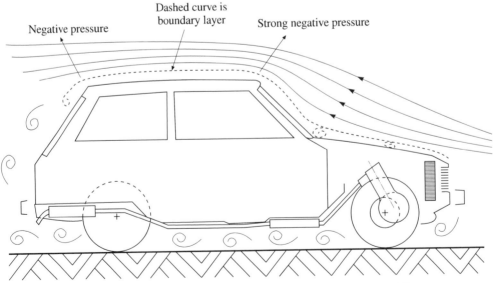

FIGURE 23.72   Flow about the symmetry plane of an automobile.

hr). If the vehicle is to run at higher speeds, on the Autobahnen in Germany, for example, where there is no posted speed limit, or in racing, then the lift, which tends to increase as the square of the speed, becomes much more important. It can even be reversed by the addition of external wings, or the use of faired belly pans which form one half of a venturi passage above the road surface. The tire rolling friction is of course increased by aerodynamic downloading devices. The use of downloading devices on competition cars is closely regulated by competition rules.

Figure 23.73 shows several lift reducing fixes and downloading devices which have been developed in racing. The lift producing capacity of exposed wheels fitted with wide, flat tread profile tires has been investigated by Fackrell and Harvey (1974). Tire contact with the road maintains near stagnation pressure from the most forward tire contour to the leading edge of the footprint; the region of negative pressure above the tire is influenced by the size of the separation bubble, which in turn is governed by the forward movement of the tread in that region.

The success of *chin spoilers* (sometimes called *air dams*) in reducing front end lift seems to be due to the formation of a separation bubble characterized by negative pressures behind the dam. The separation bubble covers almost the entire underbody. Cooling flow is increased through the radiator core, but it is decreased around the catalytic convertor, usually mounted under the floor. Adding a chin spoiler to a car with an *unfaired underbody* usually has little effect on its drag.

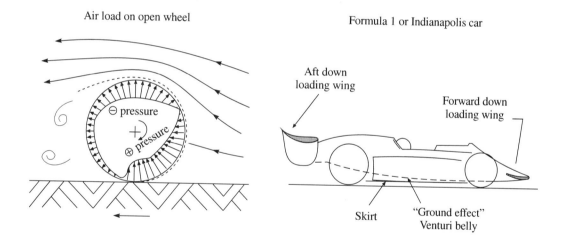

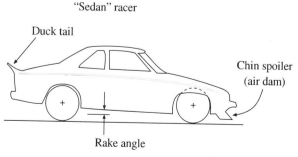

FIGURE 23.73 Lift producing and down loading devices.

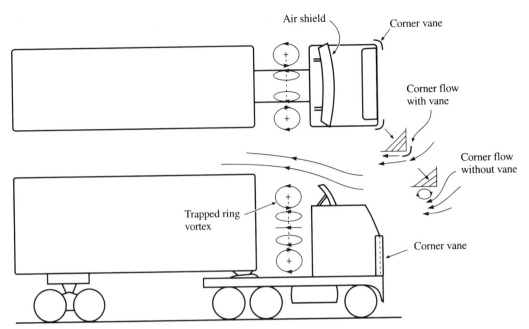

FIGURE 23.74   Aerodynamic devices used on trucks.

Concern for fuel economy also has led to the addition of drag reducing devices to trucks, as shown in Fig. 23.74. The simple square forward corners of a box-like cab often are fitted with semicircular vanes which reduce the high velocity separation that would otherwise occur. The separation bubble on the cab side is greatly reduced in intensity, as is the edge separation vortex, which has the potential for dirtying side windows with muddy water in wet weather conditions.

Another common device is the *air shield* mounted on the cab roof ahead of the taller, flat fronted trailer box. An air shield of proper geometry traps a ring vortex in the space between the cab and the trailer box so as to lead the air smoothly onto the trailer roof without creating large stagnation pressures on the forward wall of the trailer. The device was invented by Seldon W. Saunders and originally distributed by the Rudkin–Wiley Corporation. The corporation has filed many suits for patent infringement. Air shields have been demonstrated to reduce vehicle air resistance by as much as 15%.

Research conducted by Salzman and Steers (1977) at NASA demonstrated that an add-on smooth contour nose fairing for the tractor in combination with a smooth articulated joint cover between tractor and trailer had the capacity of reducing the air resistance to 63% of the usual value. The modified vehicle was drivable but could not be completely jack-knifed for parking in restricted spaces.

23.7   Wind Tunnel Testing of Land Vehicles

Wind tunnels were developed to determine the aerodynamic characteristics of complete airplane models and airplane components (such as wings); techniques for wind tunnel testing were well established by 1920 (Chap. 16 discusses wind tunnels).

These same tunnels were used from the beginning to determine the aerodynamic characteristics of automobiles and trains and land vehicle components (for example rail passenger car roof ventilators). Whenever automobile and train models were tested, serious doubts arose as to the validity of the flow field beneath the model underside. Sometimes the model was mounted in the middle of the jet of air (as an airplane model would be) and the flatness of the flow field at the vehicle wheel bottoms would be enforced by a smooth flat plate, either in contact with the wheels, or slightly separated from them. In either case, the flat plate would have its own boundary layer which would tend to mingle with and distort the turbulent sheath of eddying flow which naturally forms beneath the vehicle. These difficulties were magnified when the tunnel jet was confined by surrounding walls and the model rested on the test section floor, or on a balance platform flush with the floor. A solution to the problem is a *moving ground board* in the form of a roller supported carpet carried past the model at tunnel airspeed. Such a device has been installed in some large wind tunnels for tests of VTOL and STOL aircraft in ground proximity, and tests of a $\frac{3}{8}$ scale automobile [Turner (1967)] have shown that lack of ground board movement usually has only small effects on vehicle aerodynamics. Most automobile testing is conducted with fixed ground planes and corrections based on experience are applied to the data.

Another problem arises from the temptation to test an actual automobile in a test section with cross sectional area no more than three times the frontal area of the vehicle. If the jet is unconfined, the vehicle deflects the flow much more than it would an *infinite* jet, but if it is confined by walls, the straightening effect of the walls greatly increases the tendency for flow separation on the test vehicle. In either case serious questions arise as to what *infinite* stream velocity is being represented (if any) when the pressure difference along the tunnel jet nozzle is measured to determine the tunnel airspeed.

Wind tunnel balances are designed to resolve the airload acting on the model into six components—three each for the orthogonal components of a general force and moment vector referred to an origin determined by the balance design. Figure 23.75 shows two common schemes for air load resolution on automobile models. The

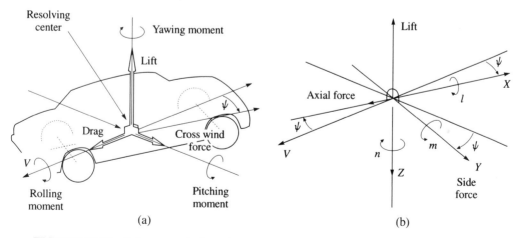

FIGURE 23.75 Balance resolution of air load. ψ = yaw angle, $\beta = -\psi$ = sideslip angle. (a) Wind axes and (b) stability axes.

forces are commonly made dimensionless with the dynamic pressure, $(\rho v^2/2)$, and a reference area which corresponds to the front view of the vehicle, including tires, but omitting external rear view mirrors, radio antennas, etc. The moments are made dimensionless with the dynamic pressure, the frontal area, and a characteristic length, commonly the wheelbase. Figure 23.76 presents some typical data obtained in wind tunnel tests of an actual subcompact automobile [Turner (1967)]. In these tests, the yaw angle, ψ, between the wind direction and the vehicle plane of symmetry was varied continuously to simulate the effect of sideslip of the vehicle, or of a cross wind uniform with height above the road surface.

23.7.5   Aerodynamic Lateral Disturbances

Prediction of the aerodynamic lateral disturbance of a vehicle by cross winds or the pressure and velocity field of adjacent vehicles requires the mathematical modeling of the following effects: 1) the side force of each tire caused by rolling along a line on the road surface misaligned with a plane determined by the wheel rim centerline (the slip angle), 2) differential changes in road wheel steer and camber angles in response to vehicle roll angle on its suspension caused by aerodynamic side load or centrifugal force, 3) driver activity at the steering wheel in response to perceived departure from the desired track, 4) aerodynamic effects of the cross wind gradient with height above the road surface and along the vehicle (cross wind), 5) aerodynamic effects of vehicle sideslip (the only aerodynamic effect represented by wind tunnel tests of the vehicle in yaw [in studies of aircraft and road vehicle lateral dynamics *yaw*, ψ, refers to motion in inertial space and *sideslip*, β, to misalignment between the vehicle symmetry plane and the relative wind]), and 6) aerodynamic effects of the lateral velocity and pressure field of an adjacent vehicle which may be moving at a speed different from the disturbed vehicle.

Studies [Larrabee (1974)] of the lateral response of a typical driver-vehicle combination to strong cross winds accounting for these effects show that the disturbances are commonly at the annoyance rather than the danger level, a fact borne out by driver experience. Some vehicles, notably the Type I Volkswagen (Beetle) may exhibit disturbance levels ten times greater than those of large American automobiles of the 1960's, but even these disturbances are readily handled by alert drivers. The extreme sensitivity of the Type I Volkswagen is due to its aft center of gravity location, the roll steer characteristics of its trailing arm front and swing axle rear suspensions, and its relatively high roof line. More recent front wheel drive subcompact cars are insensitive to cross wind disturbances.

23.7.6   Yaw Stability of Competition Cars

Vertical irregularities of record courses give rise to road wheel and sprung mass accelerations proportional to speed squared; at speeds above 300 mph (483 km/hr) it is not unusual for land speed record cars to lose ground contact momentarily. Under these conditions, the vehicle ought to have near zero *weathervane stability* so that it will remain aligned with its velocity vector while in flight. The natural tendency of a slender, streamlined body, pointed at both ends, is to have negative weathervane stability. According to the airship theory of Max Munk, the unstable yawing moment due to sideslip is given by

$$dN/d\beta = -(\rho/2)v^2(2 \times \text{body volume}) \ (\text{Nm/rad}) \qquad (23.169)$$

A small vertical fin of area, S_v (m^2), will yield a stabilizing moment,

$$dN/d\beta = +(\rho/2)v^2 S_v l_v (dC_L/d\alpha) \ (\text{Nm/rad}) \qquad (23.170)$$

where l_v (m) is the distance from the vehicle center of gravity to the aerodynamic center of the fin, and $dC_L/d\alpha$ is its lift curve slope, about 2 or 3 per radian for the low aspect ratios commonly employed.

Ordinary sedan bodies also have negative weathervane stability as shown in Fig. 23.76 (positive $dC_n/d\psi$ corresponds to negative $dC_n/d\beta$). According to low aspect ratio wing theory a body in sideslip develops a sideforce

$$dY/d\beta = -(\rho/2)v^2 \pi h_{max}^2 \qquad (23.171)$$

where h_{max} (m) is the maximum height of the vehicle. This side force acts at the centroid of side area ahead of the location corresponding to h_{max}; the centroid is commonly ahead of the vehicle center of mass. Lines corresponding to the estimate of Eq. (23.171) have been added to Fig. 23.76; the aerodynamic instability would

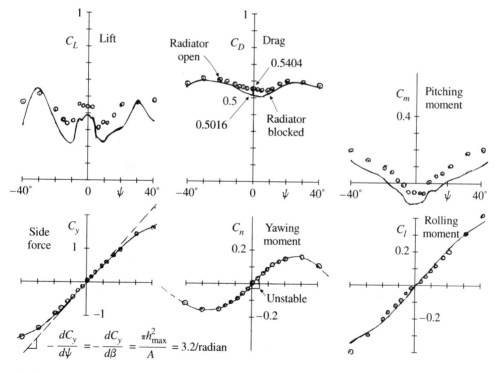

FIGURE 23.76 Wind tunnel data for a 1979 Chevette sedan. $A = 1.765$ m^2, $l = 2.477$ m, and $h_{max} = 1.341$ m. (Reproduced with permission, Kurtz, D. W., "Aerodynamic Characteristics of Sixteen Electric, Hybrid Subcompact Vehicles," JPL Publication 75-59, June, 1979.)

not be apparent to a driver except possibly in ice racing, where the vehicle might seek a stable aerodynamic sideslip angle at $\beta = \pm 60°$ with all four wheels sliding at high speed.

23.7.7 Cooling and Ventilation

Figure 23.76 shows that cooling flow through the radiator and engine compartment of a subcompact car increases the drag coefficient from 0.5016 to 0.5404, an increase of 7.74 %. The drag coefficients are based on a vehicle frontal area of 1.765 m^2; if we assume that the frontal area of the radiator is 0.387 m^2, the increase of drag coefficient, based on the radiator area, is 0.1770, which shows that the total head loss in the cooling flow is considerably less than stream dynamic pressure. Low cooling drag is insured by providing radiators of large frontal area in combination with inlet and exit passages which decelerate the external flow to a low velocity at the radiator face and accelerate it behind with little loss of total pressure. The cooling capacity of a radiator is proportional to the core area and the velocity of the cooling flow through the core; the cooling drag is proportional to the core area and the square of the the the velocity through the core, assuming no total head loss in the inlet and exit flow. At low speeds and high power, as in hill climbing, a powered fan will be required to provide adequate cooling. Substantial savings in fan power are possible if the fan is not driven at level road cruising speeds.

Passenger ventilation flow should be adequate and noiseless. An adequate flow at cruising speeds is provided by an inlet in a region of high positive pressure, typically at the base of the windshield, and exits in a region of negative pressure, for example the sides of the rear quarter panels between the rear side window and the rear window [Sovram et al. (1978)]. Exit flow is made noiseless by diffusing it through a porous cloth roof lining and damping material in the exit passages. An electrically driven fan is required for adequate ventilation at low speeds.

Ventilation by opening a side window or a sun roof may lead to an excitation of disagreeable cavity resonance pressure pulsations in the passenger compartment. These can be alleviated by fitting a rake-like spoiler to the upstream side of the opening. They thicken the boundary layer and tend to decouple the external flow from the cavity resonance flow within, but the spoiler also increases the vehicle drag.

23.7.8 Conclusion

Compared with aircraft, road vehicles are insensitive to aerodynamic characteristics associated with design practice since 1950. The excessive drag of sharp edged, box-like bodywork of the 20's and 30's; the excessive lift (and induced drag) of long, sloping, sharp edged front fenders seen on sports cars of the 30's; and the intense aerodynamic buffeting of passengers in open roadsters is a thing of the past except in period replicas created for nostalgia buffs. Recently, the increased cost of fuel has led to *aerodynamic* body work with well rounded corners, contoured, and flush mounted windows, belly pans blended into nose and side panels, smooth wheel covers merging with tire profiles, and cooling and ventilating ducts producing low total head loss [Zygmont (1985)]. Drag coefficients of practical sedans may approach 0.25 with zero lift coefficient at the expense of uncompromising design detail. The fuel economy benefits of such design are easily estimated by the methods of Sec. 23.7.2.

23.8 AERODYNAMICS OF TRAINS
Helmut Sockel

23.8.1 Introduction

Faster operating speeds and lightweight construction have increased the need for research in the field of train aerodynamics in order to maintain standards of comfort and safety and to reduce the aerodynamic drag. There are a number of other reasons for interest in the aerodynamics of trains. Wind effects on trains become important, and there is a risk of overturning at high wind speeds. Passing trains induce static pressure changes and airflows at the track site that may be a risk for people and objects. The distance between parallel tracks has to be increased to keep the pressure amplitudes caused by passing of trains at an acceptable level. Aerodynamic forces on the pantograph may cause problems at high operating speeds; wind induced vibrations of overhead lines have also been recorded. A major problem of a tunnel passage is the pressure fluctuations which may cause discomfort to the ears of passengers. Moreover, the aerodynamic drag of the train and the aerodynamic loads of the cars are increased at high speeds.

23.8.2 Methods of Investigation

Numerical Methods. For the symmetric flow over an aerodynamic well-shaped nose, the agreement between experimental results and the results calculated by using a distribution of surface singularities to solve the equations of steady inviscid flow is quite good [Stoffers (1979), Mackrodt *et al.* (1980), and Gawthorpe (1972)]. But, if the train nose is more blunt or if side-wind effects are considered, we find separations and fully developed wake flow with large vortices, and there is always a wake flow in the region of the underbelly. Empirical inputs, such as the measured separation lines, must be used with a panel method that includes these effects [Copley (1987)]. Additionally, the vortex strength on the underside due to the bogies is much less than that on the roof, which was assumed to be smooth. Recently, Reynold's equations were solved by modeling the turbulence using a standard k-ϵ model. Air flows over a two-dimensional cross section of a central coach with a cross-wind perpendicular to the train axis were investigated. The practical application of numerical methods to open air railway aerodynamics shows considerable potential as long as its limitations are appreciated and new applications are validated [Gaylard and Johnson (1993)]. So one can say that experiments in railway aerodynamics are very important today and will remain so in the near future, at least.

Much better agreement between theoretical and experimental results has been achieved in the field of tunnel aerodynamics. Pressure histories and the aerodynamic resistance for a passage of one or more trains through a tunnel can be calculated accurately using a computer program.

Model Experiments. The main advantages of model experiments in comparison over full-scale experiments are repeatability and that measures for improving the flow conditions can be investigated easily. For wind tunnel testing, the Reynolds number and the Mach number for full scale and model scale should be equal. If wave propagation problems are not considered and the Mach number, $M < 0.2$ for full scale, it is only important to have $M < 0.2$ in model scale. The flow speed in the wind tunnel will be approximately equal to the full scale speed of the train, while Re is much less than in full scale. This can influence the positions of separation lines and may have a severe influence on the flow details (as with the pantograph for instance) due to a change of the flow from a transcritical region of Re for full scale to a subcritical one for the model. The model scale depends on the cross-sectional area of the wind tunnel; the ratio of the area of the model to that of the tunnel, the so-called *blockage ratio*, should be less than 0.03 for a plate normal to the flow [Sockel (1985)], but can be higher for less blunt bodies. For train models at a yaw angle in the tunnel, the situation is similar to that of a plate, and a ratio less than 0.05 is recommended.

In a stationary wind tunnel model, the shear flow between the vehicle underbelly and the ground is not represented, but is replaced by a ground board boundary layer. Removing this boundary layer by suction or blowing has been tried, but that involves control problems in producing the correct shear flow [Gawthorpe (1978)]. Comparing measurements of the aerodynamic resistance of a model locomotive made with and without suction have shown only a minor difference if a height correction was made for displacement thickness of the boundary layer beneath the model [Mackrodt (1978) and Gawthorpe (1972)]. The *image method* using a double model is normally not adequate, since the different wake flow can cause a pressure distribution very much changed compared to that of the same body on the ground [Ackeret (1965)]. This method will be adequate if there are only minor wakes in the flow field. Moving ground belts have been used, but the lift of the belt and edge effects both influence the flow field. This measure seems to be important only if detailed flow problems in the region of the underbelly are investigated. Most experiments are done in a conventional wind tunnel, and simple methods to reduce the influence of the ground boundary layer are applied.

For a vehicle in a wind, we have a superposition of the relative motion of the vehicle and the atmospheric boundary layer. A correct modeling of these conditions is possible only with moving models. For fixed models in a wind tunnel, one can introduce some turbulence in the flow to see if this changes the measured values. Work on the aerodynamics of structures has shown that forces and moments of many body shapes are rather sensitive to the intensity and structure of the turbulence [Sockel (1985)].

Many model experiments dealing with aerodynamic problems of trains in the open air and during the tunnel entrance phase have been performed in the water tunnel of the Institute of Shipbuilding at Hamburg University [Neppert and Sanderson (1978)]. The models at a scale of $\frac{1}{25}$ were mounted on a carriage upside down and moved at the speed 0.7m/s in a 35 m long water tank. The Reynolds number was of the same order as in wind tunnel experiments.

British Rail has built a facility for moving models consisting of two 132 m long parallel tracks; trains at a scale of $\frac{1}{25}$ coast over a 46 m long test section [Pope (1991) and Schultz (1990)]. The remainder of the track is set aside for acceleration and braking. Rubber launchers positioned at the end of each track propel the train models as shown in Fig. 23.77. When a model reaches the end of the test section, a hook on the chassis engages a Kevlar cable which draws a piston through a tube which is deformed mechanically causing the model to brake. This facility is used to investigate open air problems including the passing of trains on adjacent tracks and train passage through short tunnels. If wave propagation problems in tunnels are inves-

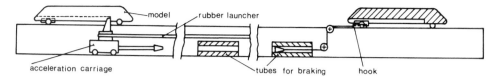

FIGURE 23.77 Moving model test facility of British Rail Research, Derby, UK.

tigated, the model speed has to be equal to the full scale speed, which means the Mach numbers must be equal.

A test rig was built at the University of Nottingham to investigate side-wind effects by propelling vehicle models across an environmental wind tunnel in which an atmospheric boundary layer had been simulated [Cooper (1984), Baker (1986), and Humphreys and Baker (1992)]. A strain gage balance is installed in the model, but since aerodynamic forces and mass forces are not measured separately, there is always mechanical noise in the results, and mean values have to be taken over many runs. Moreover, only wind normal to the train direction is simulated.

British Rail built a 900 m long test track to investigate side-wind effects, and a $\frac{1}{5}$ scale model of the British APT-P leading vehicle was connected to a half length vehicle which acted as an aerodynamic fairing and equipment housing [Gawthorpe (1978)]. The load cell information and the onset flow parameters from the model nose probe were recorded aboard the propelling vehicle. Further measurements of wind conditions were made at the tracksite with gust anemometers.

Full-Scale Tests. If a new train is tested, no fundamental design changes will be possible. Therefore, these full-scale tests are used primarily to check the validity of numerical calculations and the results of model experiments. The main purposes of such investigations are: 1) pressure distribution on the surface, 2) environmental effects on the tracksite, 3) aerodynamic resistance, 4) effects of passing of trains on adjacent tracks, 5) stability of the train and wagons in high winds, 6) pressure waves, and 7) aerodynamic resistance for a tunnel passage. Tests in the open air are severely influenced by the wind conditions.

23.8.3 Open-Air Aerodynamics

Flow Field near the Train Head. For a high speed train, an aerodynamically well-shaped head reduces the aerodynamic resistance and the intensity of the so-called *head wave* at the track side. A typical pressure distribution for such a nose shape is shown in Fig. 23.78, where experimental values from a wind tunnel and results from a numerical panel method are compared [Mackrodt *et al.* (1980)]. The agreement for this symmetric flow field is rather good, but it becomes worse with increasing yaw angle, especially for the pressure distribution in the cross-section. For a more blunt shaped nose, as with the German DB 103, the pressure differences taken from wind tunnel experiments are much higher, especially for a yaw angle of 30 degrees [Mackrodt (1978b)]. The large effects of side winds are shown in Fig. 23.79.

Flow along the Train. The turbulent boundary layers and the skin friction coefficients vary considerably over the perimeter and length of the train due to the effects

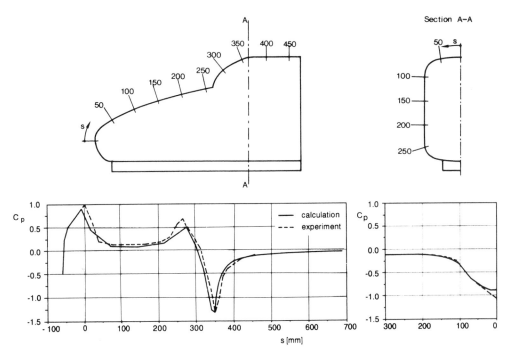

FIGURE 23.78   Pressure distribution on a train head model. (Reprinted with permission, Mackrodt, P. A., Steinheuer, J., and Stoffers, G., "Aerodynamisch optimale Kopfformen für Triebzüge," DFVLR Göttingen, IB 152 79 A27, 1980.)

of bogies, inter-car gaps, etc. For a typical conventional coach, full-scale measurements on the sides by British Rail [Gawthorpe (1972)] have shown, that the friction coefficient is 0.0031 (Re = 4.5 to 8.0 × 10^7) for a conventional train, and 0.0021 (Re = 2 × 10^8) for the British APT-E, corresponding to fully developed rough boundary layers. Values from German measurements of a train with an E 103 locomotive show a decrease of the friction coefficient from 0.006 at 100 m from the head to 0.003 at 300 m from the head [Sonntag *et al.* (1976)]. British Rail performed full-scale tests with an MK3 laboratory coach inserted into a service British HST; values in the range of $C_f = 0.0011$–0.0013 were obtained, which are well below the values for an aerodynamical smooth plate. As demonstrated at two different model scales, the train flow regime is substantially three-dimensional, so the assumption of a two-dimensional boundary layer is an unrealistic simplification [Brockie and Baker (1990)]. Based on results from model experiments and theoretical considerations, these authors expect a full scale friction coefficient between 0.002 and 0.004.

The growth of the boundary layer along the train side from different full-scale experiments is shown in Fig. 23.80. The problem is that the boundary layer thickness, δ, has been obtained from the extrapolation of measured velocity values. Comparisons of measured values are often made applying the formula for a turbulent boundary layer on a flat plate

$$\frac{\delta}{x} = C \left(\frac{V_T x}{\nu} \right)^{-1/5}$$ (23.172)

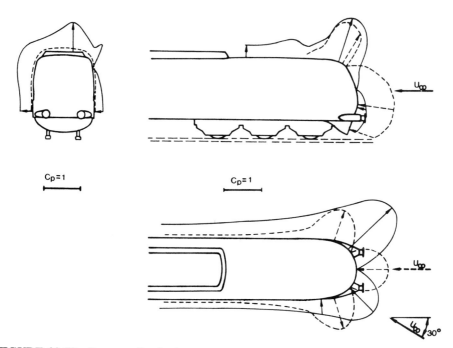

$C_p = 1$

$C_p = 1$

FIGURE 23.79 Pressure distribution on locomotive DB 103 for yaw angles 0° and 30°. (Reprinted with permission, Mackrodt, P. A., "Windkanalunterversuchungen am Modell der Schrellfahrlokomotive 103 der Deutschen Bundesbahn," DFVLR-AVA Bericht 251 78 A 08, 1978b.)

where the coefficient C is fitted to the experimental values and depends on the height above the ground for a train side [Sonntag *et al.* (1976)].

For the train passing problem, the displacement thickness, δ^*, is relevant, and it can be obtained more accurately from experiments than δ. Firchau *et al.* (1980) show that δ^* is in the range of $\delta/12$ to $\delta/8$. Sidewind effects have a strong influence on

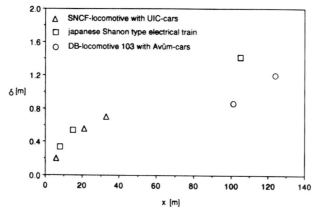

FIGURE 23.80 Boundary layer thickness on the walls of different trains at the height of the window. Experimental data from Firchau *et al.* (1980).

the growth of the boundary layer, causing the thickness of the layer on the leeside to be much higher than that on the windward side, thus the flow is unsymmetric. For the aerodynamic resistance, the boundary layer on the underbelly and the drag of the bogies and the wheels is at least as important as the friction forces on the train sides and the roof.

Aerodynamic Resistance. There are two terms of the total resistance force, F_R, caused by aerodynamic effects.

$$F_R = \dot{m}V_A + \tfrac{1}{2}\varrho C_D A_T V_A^2 \qquad (23.173)$$

The first one is the air momentum drag associated with the acceleration of air ingested for combustion, cooling, or air conditioning. The second term is the aerodynamic resistance, where the drag coefficient C_D depends on Re, and the yaw angle β. For train speeds higher than today, Ma will also become important. V_A is the train air-speed and is equal to the train speed, V_T, only for still air conditions (see Fig. 23.81). A_T is the train frontal area, but sometimes C_D is referred to a standard area of 10 m^2 [Voss *et al.* (1972)]. C_D is extremely sensitive to the yaw angle β; for a train of the German ICE type the following formula is appropriate [Peters (1990)].

$$C_D(\beta) = C_D(0)\,[1 + 0.02|\beta°|] \qquad |\beta°| < 30° \qquad (23.174)$$

For a motor-coach train consisting of 4 coaches, C_D was found to be constant up to $\beta = 4°$ [Glück (1985)]. For several freight wagons, there was an increase of C_D with increasing yaw angle [Watkins *et al.* (1992)]. An estimation of the relevant angle subtended by the annual average wind and the wagon speed is typically about 6° for freight wagons [Watkins *et al.* (1992)]. Therefore, C_D values for this yaw angle should be taken for the calculation of the aerodynamic resistance.

Figure 23.82 shows a breakdown of the aerodynamic drag coefficient $C_D(0)$ into its components (pressure drag and skin friction terms) for two typical train types [Gawthorpe (1978a)]. The body side and roof skin friction drag are the only components which decrease slightly due to the increasing thickness of the boundary layer. Since these terms represent only about 25% of the whole drag, they can be assumed to be constant along the train for simplicity. Therefore, $C_D(0)$ can be broken down into several terms

$$C_D(0) = C_{DL} + C_B + \lambda_T \frac{l_T - l_L}{\sqrt{A_T}} \qquad (23.175)$$

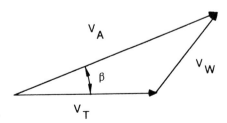

FIGURE 23.81   Train air-speed, V_A, diagram.

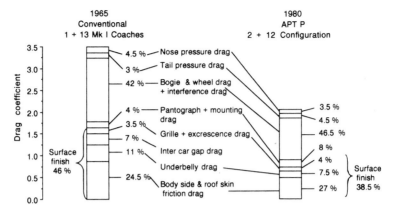

FIGURE 23.82 Breakdown of aerodynamic drag. (Reprinted with permission, Gawthorpe, R. G., "Aerodynamics in Railway Eng., Part I: Aerodynamics of Trains in the Open Air," *I. Mech. E.*, Railway Eng. Int., pp. 7–12, 1978a.)

where C_{DL} is the drag coefficient of the leading car or locomotive, C_B takes into account the back pressure at the end, and λ_T is the friction along the train, including bogies, wheels, interference, underbelly effects, etc., l_T is the total length, and l_L is the length of the leading car. C_B is in the range of 0.09–0.11 [Hoerner (1965)]. Wind tunnel tests show differences in the drag between a middle car and the last car in the range of 0.05 to 0.12 [Mackrodt (1980)]. The values in Table 23.11 have been taken or calculated from other values of different references given in the table. The C_B values for the German ICE and the Japanese Shinkansen have been calculated from other data. For all the other trains in the table, C_B was set 0.11. All values with an asterisk are estimated values needed for the calculation of λ_T. For mixed freight trains, λ_T values can be higher than for a container train. Bendel (1990) shows that a reduction of the aerodynamic resistance by 25–45% is possible by smoothing the sides and roofs of freight cars or containers and by other similar measures. Watkin *et al.* (1992) made experiments in a wind tunnel with hopper and gondola rail cars. They found that a practical wagon design, that was based on an existing wagon and allowed current methods of loading and unloading, would reduce average drag coefficients by 27% for an empty wagon and that savings up to 50% could be made with more substantial wagon redesign.

The so-called *Hannover-formula* [Voss *et al.* (1972)] takes into account, that for locomotive-hauled trains, the drag coefficient C_{D1} for the first car is higher than for

TABLE 23.11 Typical Drag and Friction Coefficients for Trains

Train Type	$C_D(0)$	$A_T\,[m^2]$	l_T	l_L	C_B	C_L	λ_T	Ref
APT-P	2.05	8.05	300	13.0	0.11*	0.2*	0.0172	Gawthorpe (1978a)
HST	2.11	9.12	300	17.4	0.11*	0.2*	0.0192	Gawthorpe (1978a)
Conv.Pass.Tr. MKII	2.75	8.8	300	20*	0.11*	0.3*	0.0248	Gawthorpe (1978a)
Container 80% loaded	6.5	8.8	300	20*	0.11*	0.5*	0.0624	Gawthorpe (1978a)
Shinkansen 200	1.52	13.3	300	24.5	0.11	0.2	0.0160	Maeda *et al.* (1989)
ICE	0.69	10.2	115	20.9	0.12	0.2	0.0125	Peters (1990)

a middle car C_{Dm}. The back pressure is accounted for by a higher value C_{Dn} for the last car. The values are referred to a normalized area $A = 10 \text{ m}^2$, therefore the coefficient C_K for the locomotive is different from the value C_{DL} in Eq. (23.175)

$$F_R = [C_K + C_{DL} + C_{Dn} + (n - 2)C_{Dm}] \frac{\rho}{2} V_t^2 A \qquad (23.176)$$

This equation does not take into account a momentum of ingested air.

Train Stability in High Winds. Increased speeds and lighter vehicles have necessitated stability checks for trains and cars. It is current practice to reduce the speed or to stop the train if high winds are blowing. For high-speed trains the highest overturning moments are experienced at the front of the train [Gawthorpe (1978a)]. But, as full-scale and model experiments show, there is an accident risk for freight cars too, especially in exposed regions as on bridges and embankments. If for a special car with known mass, the aerodynamic rolling moment coefficients for different wind angles have been evaluated, the train speed limit for a given wind speed can be calculated. Figure 23.83 shows these limits for a freight car and a container on a flat wagon on a bridge based on measurements in a turbulent flow in a wind tunnel. The mean values of the coefficients are referred to the mean of the flow velocity, and the extreme values are referred to the extreme of the flow velocity. For the extremes, the 99% fractile of a sample has been used. The extremes seem to be more adequate for practical application, since at the track side there are usually measured gust speeds with anemometers. For the container, there is a significant reduction in the allowable train speed if the extremes are taken. For the freight car, there is only a reduction for high train speeds. The comparison shows that rolling

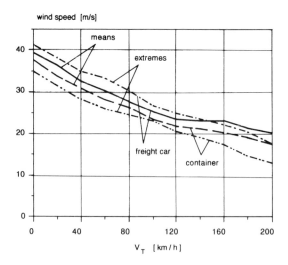

FIGURE 23.83   Train speed limits as a function of the wind speed for a container and a freight car. (Reprinted with permission of Elsevier Science Publishers, Krönke, I. and Sockel, H., "Model Tests About Cross Wind Effects on Containers and Wagons in Atmospheric Boundary Layers," *J. Wind Eng. and Ind. Aerodyn.*, Vol. 52, pp. 109–119, 1994.)

moment coefficients from smooth flow experiments [Peters (1989), (1990)], which are comparable to the means of turbulent flow, should be applied with care. In experiments with moving models across an environmental wind tunnel, accident wind speeds based on extreme values have been derived and compared to full-scale results. As mentioned earlier, it is difficult to separate the aerodynamic forces from the total values measured with the balance on the moving vehicle including mass forces.

Environmental Effects at the Track Site. A passing train produces rapid static pressure changes and strong induced air flows in the train boundary layer and wake at the track side. For a given train speed, the effects depend on the clearance from the track, on the profile of the train, and the ambient wind conditions. The static pressure changes are most pronounced as the head of the train passes and less for the passing of the tail. One method for comparing qualities of nose shapes is to measure the pressure distribution on a vertical wall at a certain distance from the track in a wind tunnel experiment as in Fig. 23.84. The amplitude, ΔC_p, referred to the dynamic pressure of the train speed depends mainly on the slenderness of the nose. The length, Δx, corresponds for the moving vehicle to a time scale $\Delta t = (\Delta x)/V_T$. For slender nosed trains, Δx compares closely to the nose shape length [Gawthorpe (1978a)]. As a result of full-scale measurements at mid train body height, amplitudes as a function of distance from the track centerline for two categories of slenderness of train nose shapes have been proposed and are plotted in Fig. 23.85. Results for container trains show an increase in amplitude up to 20% over the bluff nosed value and are associated with the pulse produced by the front of a passing group of isolated containers. For slender nose, the pressure pulses can be predicted well applying potential flow methods [Mackrodt *et al.* (1980) and Sonntag *et al.* (1976)].

Figure 23.86 shows the results of air velocity measurements 1 m above a station platform and 2 m from the edge for different types of trains measured under similar

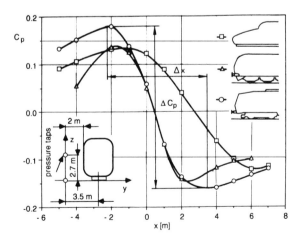

FIGURE 23.84 Head waves caused by different train noses. (Reprinted with permission, Mackrodt, P. A., Steinheuer, J., and Stoffers, G., "Aerodynamish optimale Kopfformen für Triebzüge," DFVLR Göttingen, IB 152 79 A27, 1980.)

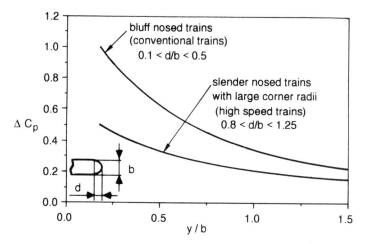

FIGURE 23.85 Lateral decay of train nose pressure perturbation as a function of distance from track center line. (Reprinted with permission, Gawthorpe, R. G., "Aerodynamics in Railway Eng., Part I: Aerodynamics of Trains in the Open Air," *I. Mech. E.*, Railway Eng. Int., pp. 7–12, 1978a.)

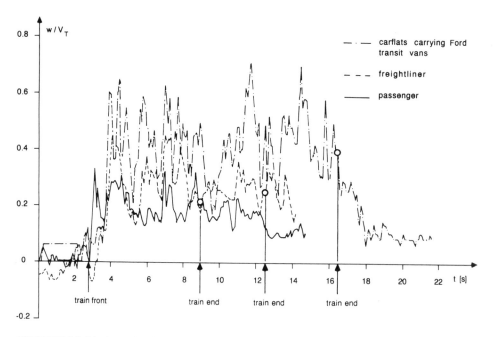

FIGURE 23.86 Longitudinal air velocities at the trackside for different passing trains, 1 m above station platform, 2 m from the edge. (Reprinted with permission, Glöckle, H. and Gawthorpe, R. G., "Aerodynamics–Trackside Safety Environment," *I. Mech. E.*, The Railway Division, 1991.)

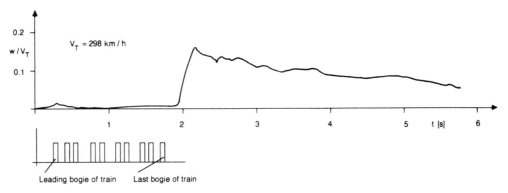

FIGURE 23.87 Longitudinal air velocities at the trackside for a passing ICE/V, 1.3 m above rail, 3.5 m from track center line. (Reprinted with permission, Glöckle, H. and Gawthorpe, R. G., "Aerodynamics–Trackside Safety Environment," *I. Mech. E.*, The Railway Division, 1991.)

conditions, and Fig. 23.87 gives results for the German ICE 3.5 m from track centerline and 1.3 m above the rail. In Fig. 23.86, disturbances are induced when the train is passing. In Fig. 23.87, the disturbances start when the train has already passed. The effect of ambient wind conditions on this flow field can be considerable. The train's boundary layer and wake are swept across the track by a sidewind, therefore its effects are felt more strongly on the leeside of a passing train.

Due to these aerodynamic effects, significant aerodynamic forces can be exerted on people and objects at the trackside. The peak values of the transient forces cause most of the problems, but the time scale is important too, since inertial effects of the objects are relevant for stability. One technique adopted to study these aspects has been to measure the static pressure and the air flow at the trackside, as mentioned above, and to calculate the effects of this on the people and objects concerned. The other method is to set the items or dummies into position and measure the forces. Results of forces on circular cylinders with a diameter of about 0.5 m and different lengths from full- and model-scale experiments and from theoretical consideration are reported in Neppert and Sanderson (1981). In Fig. 23.88, the results for the force coefficients, C_f, based on the dynamic pressure of the train speed for such a cylinder for the passing of the French TGV are plotted. The scatter of the points may have been caused by wind effects, but this is typical for full-scale measurements. Up to a distance of about 1.8 m from the rail, the higher forces are caused by the passage of the tail. For conventional trains, the higher values up to a distance of 2 m are caused by the head passage. Penwarden (1974) considers wind speeds around 20 m/s dangerous for people, particularly if they are old or infirm. His investigations in a wind tunnel concerning wind forces on 331 people [Penwarden *et al.* (1978)] show that the mean drag coefficient for people is 1.16 when facing the wind, and 1.01 when sideways to the wind. The mean projected area of the persons when facing the wind was about 0.6 m^2 and for sidewind was about 0.4 m^2. Facing the wind we get, with the speed limit of 20 m/s, a force limit of 167*N*, much higher than the 100*N* recommended by French Rail [Montagné (1973)]. Recent surveys by German Rail found 16 m/s (corresponding to 100*N* for facing the wind) to be ranked 3 on a scale of 1 (not unpleasant) to 7 (extremely unpleasant).

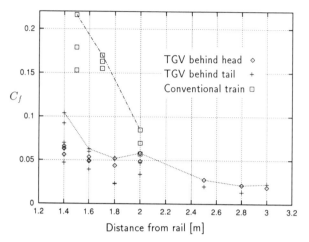

FIGURE 23.88   Force coefficients for a cylinder at the trackside for different passing trains. Experimental data from Montagné (1973).

The Train Passing Problem. If a train passes another one at rest, impulsive pressure changes occur on the passing sides similar to those on a vertical wall. Again, the main parameters are the speed, the distance of the train walls, and the shape of the nose. The pressure amplitude occuring on the resting train due to the nose passage of the passing train is nearly constant along the train (Fig. 23.89). If both trains are moving at the same speed, we find an increase of the amplitude near the end of the train. The increasing boundary layer thickness along the train, where the measure-

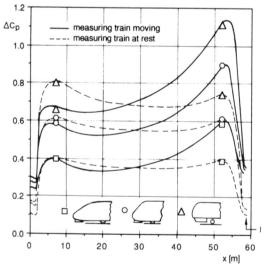

FIGURE 23.89   Pressure amplitudes ΔC_p along a train wall caused by a passing train; wall distance 0.65 m. [Reprinted with permission, Neppert, H. and Sandersen, R., "Untersuchungen zur Zugbegegnung, Bauwerks- und Personenpassage im Wasserkanal," Steinheuer, J. (Ed.), DFVLR IB 129-81/11, 1981.]

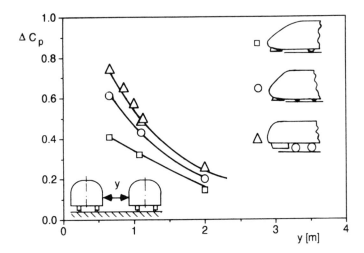

FIGURE 23.90   Pressure amplitudes ΔC_p on the train wall caused by passing as a function of the wall distance y. [Reprinted with permission, Neppert, H. and Sandersen, R., "Untersuchungen zur Zugbegegnung, Bauwerks- und Personenpassage im Wasserkanal," Steinheuer, J. (Ed.), DFVLR IB 129-81/11, 1981.]

ments are taken, causes an increase of the amplitudes near the train tail. Neppert and Sanderson (1981) have shown that the relevant distance is $y_{\text{eff}} = y - \delta^*$, where δ^* is the displacement thickness of the boundary layer. In Fig. 23.90, the passing maximum amplitudes as a function of the distance y between the train walls, if the measuring train is at rest, are plotted for model- and full-scale experiments.

Aerodynamic Problems with Overhead Line Equipment. Since the aerodynamic characteristics of the pantograph should be neutral, the mechanically produced contact force should be influenced by aerodynamic forces as little as possible. When testing a pantograph, the influence of Re should be regarded. Cross winds may also have an influence on the uplift [Gawthorpe (1978a)]. Aerodynamically induced vibrations, such as *galloping*, may occur in overhead lines. To check the stability of a profile to galloping, it is sufficient to measure the cross force coefficient as a function of the incident angle, but this has to be done in a turbulent flow [Sockel (1985)]. The most common type of failure which occurs in gails is the *blow-off devironment*, where the cross wind blows the wire out laterally so far that it comes back underneath the head of the pantograph.

23.8.4   Tunnel Aerodynamics

The Differential Equations. The unsteady turbulent flow caused by the passage of a train is strictly three-dimensional in space. But, theoretical and experimental results show the pressure disturbances are approximately constant over the tunnel cross-section, even for a short tunnel [Schultz and Sockel (1991), Ottitsch et al. (1994)]. Pressure transients may cause ear discomfort for passengers, an additional load on the cars, and an increase of the aerodynamic drag. Therefore, we are mainly interested in calculating pressure disturbances, which can be done by applying a theory for one-dimensional flow in space.

The governing equations for the conservation of mass, momentum, and energy for adiabatic flow are [Steinrück (1984)]

$$\frac{\partial(\varrho A)}{\partial t} + \frac{\partial(\varrho A w)}{\partial x} = 0 \tag{23.177}$$

$$\frac{\partial w}{\partial t} + w \frac{\partial w}{\partial x} = -\frac{1}{\varrho} \frac{\partial p}{\partial x} + R_f \tag{23.178}$$

$$T\left(\frac{\partial s}{\partial t} + w \frac{\partial s}{\partial x}\right) = P_T + w R_f \tag{23.179}$$

Two different flow regions are distinguished: the empty tunnel area A and the annular space between train and tunnel wall with the area $A - A_T$. R_f accounts for the friction force on train and tunnel wall respectively, and a quasisteady model is usually applied

$$R_f = \frac{1}{8(A - A_T)} \underbrace{U\lambda w|w|}_{\text{tunnel}} + \underbrace{U_T\lambda_T(w - V_T)|w - V_T|}_{\text{train}} \tag{23.180}$$

Here, w is the velocity, R_f is the frictional force on the train and tunnel walls, P_T is the work done by the train walls, and A is the tunnel cross-sectional area. The influence of unsteady friction has been investigated by Vardy (1980) and Schultz and Sockel (1988). For short tunnels (but for other tunnels too) the equations can be simplified [Schultz and Sockel (1991)] by replacing the energy equation with the constant entropy condition, $s = $ constant, and by considering friction in the momentum equation only [Woods and Pope (1979), Harwarth and Sockel (1979)].

The system in Eqs. (23.177)–(23.179) is a hyperbolic system of nonlinear differential equations, and three families of characteristic lines exist.

$$\left.\begin{array}{ll} \left(\dfrac{dx}{dt}\right)_\xi = w + a & \xi = \text{const.} \\[1.5em] \left(\dfrac{dx}{dt}\right)_\eta = w - a & \eta = \text{const.} \end{array}\right\} \text{Mach lines} \tag{23.181}$$

$$\left(\frac{dx}{dt}\right)_\epsilon = w \qquad \epsilon = \text{const.} \qquad \text{path lines} \tag{23.182}$$

There are two families of Mach lines, the propagation curves of weak pressure disturbances, and one family of path lines, which are the propagation curves of entropy disturbances. If Eq. (23.179) is replaced by $s = $ constant, we have the Mach lines in Eq. (23.181) only. Transformation of Eqs. (23.177)–(23.179) to the new variables ξ, η, ϵ gives the so-called *compatibility conditions* along the characteristic lines [Steinrück (1984)]

$$\left(\frac{dw}{dt}\right)_\xi + \frac{1}{\varrho a}\left(\frac{dp}{dt}\right)_\xi + R_f - \frac{a}{c_p}\frac{wR_f + P_T}{T} + \frac{aw}{A}\frac{\partial A}{\partial x} = 0 \qquad \xi = \text{const.}$$

(23.183)

$$-\left(\frac{dw}{dt}\right)_\eta + \frac{1}{\varrho a}\left(\frac{dp}{dt}\right)_\eta - R_f - \frac{a}{c_p}\frac{wR_f + P_T}{T} + \frac{aw}{A}\frac{\partial A}{\partial x} = 0 \qquad \eta = \text{const.}$$

(23.184)

$$\left(\frac{ds}{dt}\right)_\epsilon - \frac{wR_f + P_T}{T} = 0 \qquad \epsilon = \text{const.}$$

(23.185)

Very effective numerical characteristic methods have been developed for the integration of these equations [Fox and Vardy (1973), Harwarth and Sockel (1979), Steinrück (1984), and Waclawiczek and Sockel (1982)].

The Boundary and Initial Conditions. There are always several flow regimes illustrated in Fig. 23.91. The boundaries (train ends, tunnel portals) are normally assumed to be discontinuities of the flow area, where quasi-steady, one-dimensional, compressible flow is assumed and viscous flow effects are simulated using pressure loss coefficients. According to this theory, the entrance of the train head into the tunnel induces a propagating pressure discontinuity [Schultz and Sockel (1991)], but results of measurements [Woods (1973)] show that there is a steep (but profiled) rise over a finite time period. Because the grid size of about 50 m is used for the computation of long tunnels ($L > 1000$ m) an assumed pressure discontinuity is *smeared out* in the numerical result. For shorter tunnels, the effect of the actual pressure rise can be taken into account by replacing the actual cross-section of the train by an empirically derived area distribution. If a wave is reflected at the portal of a tunnel, the mean value of the pressure at the portal adjusts to the ambient pressure over a finite time period. For short tunnels, these two points must be taken into account for numerical results to agree with the experimental values [Schultz and Sockel (1991)]. Inside the tunnel, initial conditions can be either the air at rest, or a fully developed flow [Steinrück (1984)].

Pressure Histories. In Fig. 23.92(b) the pressure history for a point on the train 20 m behind the head is plotted. Figure 23.92(b) shows the $x - t$ plane, where the paths of the train ends and of the just mentioned point are shown. Furthermore, the

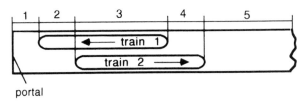

FIGURE 23.91 Flow regimes for the case of two trains passing in a tunnel.

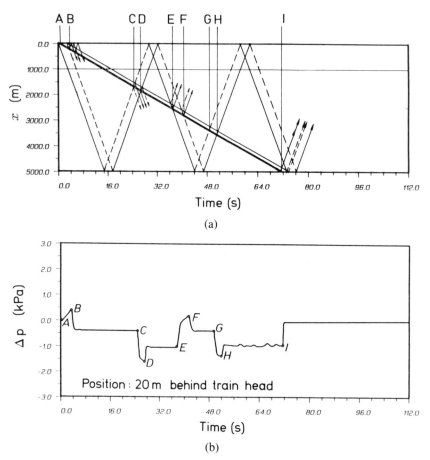

FIGURE 23.92   Pressure transients on a train during the passage in a tunnel $L = 5000$ m; $L_T = 185$ m; $V_T = 70$ m/s; $R = 0.12$. (a) paths of the train ends and propagation lines of waves; (b) pressure history for a point 20 m behind train head. [Reprinted with permission, Schultz, M. and Sockel, H., "Pressure Transients in Railway Tunnels," *Trends in Applications of Mathematics to Mechanics*, Schneider, W., Troger, H., and Ziegler, F. (Eds.), Longman Scientific & Technical, pp. 33–39, 1989.]

propagation lines of the waves due to the train ends entering the tunnel are included. The compression wave caused by the nose entrance is reflected on the opposite portal as a rarefaction wave. The entrance of the train tail causes an expansion wave reflected at the opposite portal as a compression wave. The pressure increase between points *A* and *B* is due to the wall friction in the gap between train and tunnel wall. The drop at *B* is caused by the entering of the train end. Then, the pressure is nearly constant until the reflected entrance wave arrives at *C*. All further main changes (*D, E, F, G, H*) are the effects of the two main waves. Minor changes are caused by reflections at the train ends. The increase at *I* is due to the passage of the train head at the portal.

Figure 23.93 is a comparison between calculated data and results of measurements by British Rail [Gawthorpe and Pope (1976)]. At the beginning, the agreement is very good, therefore the highest pressure changes are predicted rather well.

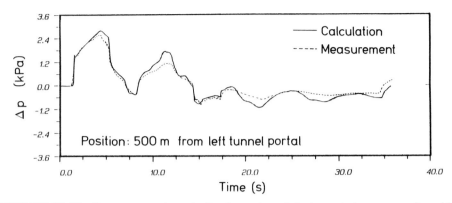

FIGURE 23.93 Pressure transients in Patchway tunnel during a train passage; $L = 1140$ m, $L_T = 100.3$ m, $V_T = 34.7$ m/s, and $R = 0.363$. [Reprinted with permission, Schultz, M. and Sockel, H., "Pressure Transients in Railway Tunnels," *Trends in Applications of Mathematics to Mechanics*, Schneider, W., Troger, H., and Zielger, F. (Eds.), Longman Scientific & Technical, pp. 33–39, 1989.]

But with increasing time, the attenuation of the pressure waves is underestimated by the computation. Possible reasons for this include an unfavorable superposition of several minor effects, such as unsteady friction [Vardy (1980) and Schultz and Sockel (1988)], assumption of quasi-steady boundary conditions [Schultz and Sockel (1991)] neglecting heat transfer, porosity of the ballasted track and of the train [Pope and Woods (1988)]. The agreement between measured and calculated results can be improved by using a numerical damping procedure [Sockel and Schultz (1990)]. The numerical calculation gives only a mean value for the flow velocity; the local measured values can be very different.

A measure for the intensity of the pressure fluctuations caused by a passing train is the intensity of the train head entrance wave, given by [Pope (1975)]

$$\frac{p - p_a}{p_a} = \frac{\dfrac{\kappa}{2} M^2 \left[1 - \dfrac{(1 - R)^2}{1 + \zeta_H} \right]}{\dfrac{(1 - R)^2}{1 + \zeta_H} + M \left[1 - \dfrac{(1 - R)^2}{1 + \zeta_H} \right] - \kappa M^2 \left[1 - \dfrac{(1 - R)^2}{2(1 + \zeta_H)} \right]} \qquad (23.186)$$

where $R = A_T/A$, κ is the isentropic exponent and $\zeta_H = C_1 \zeta_K R^2$; ζ_K is a loss coefficient and ζ_H is the loss coefficient for the train head [Steinrück (1984)].

$C_1 = 1$ to 2 for entrance phase

$C_1 = 1$ for passage

$\xi_K = 0$ to 4 depending on the train nose shape

ζ_H is negligible for aerodynamically well shaped train noses and small values of R. For this case, the pressure increase is a function of the train Mach number, M, and the blockage ratio $R = A_T/A$ only. Investigations concerning a reduction of pressure pulses by flared or perforated extensions of the tunnel show that these extensions have to be rather long for high speed trains [Dayman and Vardy (1979), Vardy and Dayman (1979), and Steinrück and Sockel (1985)]. These measures are effective if the head entrance wave, reflected from the exit portal, passes the train head in the region of the entrance extension thus reducing intensity. Assuming symmetrical tunnel extensions, the minimum length, l_{min}, for the extension required to reduce the wave intensity is given by

$$\frac{l_{min}}{l} = \frac{2M}{1 - M} \tag{23.187}$$

Improvements due to tunnel extensions are insignificant in double track tunnels where passing of trains causes rather high pressure changes. Airshafts can reduce the strength of pressure amplitudes. The optimum position depends on the speed and length of the train, but not on the length of the tunnel [Waclawiczek and Sockel (1982) and Vardy (1976)]. The cross-sectional area of the shaft and its lengths have an influence also. A train passing a shaft causes a pressure wave, and unfavorable superposition of waves caused by the air shafts in a double track tunnel can make the situation worse.

In a single-bore, two-way traffic tunnel in which one train may pass another, much larger pressure fluctuations are possible than those experienced with one-way operation. The magnitudes of pressure changes are very sensitive to the relative times of entry of the vehicles. As shown for a special case in Vardy and Anendarajah (1982), a shift of the time difference of 4 s can more than double the worst-pressure change. If two trains enter a tunnel at the same instant, the pressure change in the middle of the tunnel is doubled compared to the single train passage; for a point on the train the increase is less [Steinrück (1984)]. It is impossible to consider all possible cases, because the problem has so many parameters. The probability for such worst cases is low in practical applications.

Steinrück (1984) and Steinrück and Sockel (1985) describe a method for the approximate calculation of pressure fluctuations at the moment two train heads cross in a tunnel.

Effects of Pressure Transients. The pressure transients can cause considerable discomfort to train occupants and tunnel workers creating excessive pressure differences across the eardrum. Experiments with volunteers subjected to a variety of idealized pressure wave forms in a pressure chamber were made by British Rail [Gawthorpe (1985)]. The choice of a maximum pressure difference limit is not straightforward, because the perceived discomfort varies from one person to another and also depends on the journey characteristic [Gawthorpe (1991)]. According to tests and operating experience, the maximum pressure change for intercity route operation for British Rails is 4 kPa within any 4 s period. For the route London to the Channel Tunnel with approximately 30% in tunnels, there are proposed limits of 2.5 kPa and 4.0 kPa within 4 s for single and double track tunnels, respectively. As a result of comfort investigations for tunnel runs on the new line Würzburg–Fulda in Germany, a preliminary comfort criteria was published −0.5 kPa in 1 s; 0.8 kPa in 3 s and 1.0 kPa in 10 s [Glöckle (1991)]. In high-speed trains and for usual blockage ratios, these values can be achieved only with sealed coaches. Since pressure waves pass through the ventilating system of a sealed car, a new ventilating system which controls the charged and discharged flow rate by detecting the pressures outside and inside of a Japanese Shinkansen train was tested successfully [Kobayashi *et al.* (1991)]. A similar solution of the problem was proposed by German Rail [Klingel (1988)]. With mixed traffic in a double track tunnel, passengers in unsealed coaches will experience pressure transients almost the same as those outside the coach. As experiments with freight cars by German Rail have shown, pressure differences on a side wall can be so high that the risk of buckling occurs. An analogous risk exists for trucks on carflats.

If the train entrance wave is reflected at the exit portal, a pressure wave is radiated to the environment. The intensity of this radiated wave depends on the gradient of the entrance wave and may cause problems to houses near the track. The prevailing measure for Japanese Shinkansen was entrance hoods (longest 49 m) to decrease the pressure gradient of the head entrance wave [Ozawa *et al.* (1991)].

Aerodynamic Resistance in a Tunnel. The aerodynamic resistance for a tunnel passage can be calculated by applying the momentum equation if all data of the flow field have been calculated [Vardy (1980) and Steinrück (1984)]. During the entrance and exit phases, the part of the train in the open air has to be considered as discussed earlier. Due to the unsteadiness of the flow field, the resistance fluctuates during the passage, which is naturally more pronounced for short tunnels. For long tunnels, the drag and the aerodynamic power are approximately constant for long time intervals during a passage with constant speed (Fig. 23.94). For this long tunnel case, it is possible to estimate a so-called *tunnel factor*, T_f, the ratio of a typical drag coefficient, C_{DT}, for the tunnel passage to the drag coefficient, $C_D(0)$, in the open air as previously mentioned. For this estimation, the train is assumed to be accelerated from rest to the speed V_T in the middle of the tunnel, producing in this way a compression wave propagating to the exit portal and a rarefaction wave propagating to the entry portal. The flow over the train is assumed to be incompressible and

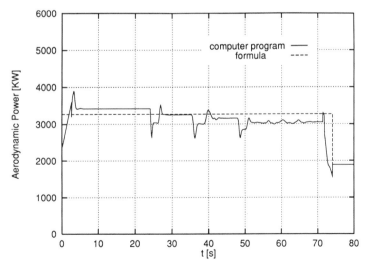

FIGURE 23.94 Aerodynamic power of a train during a tunnel passage; $L = 5000$ m, $L_T = 185$ m, $V_T = 70$ m/s, $R = 0.12$, and $C_D(0) = 0.92$.

quasi-steady. This yields the following equations for the calculation of T_f

$$T_f = \frac{C_{DT}}{C_D(0)} = \frac{1}{C_D(0)(1-R)^2}$$
$$\cdot \left\{ C_1' \left(1 - \frac{w}{V_T}\right)^2 - C_2'\sqrt{R} \left(\frac{w}{V_T} - R\right) \left|\frac{w}{V_T} - R\right| \right\} \qquad (23.188)$$

where w is a solution of

$$4\frac{a}{V_T}\frac{w}{V_T} + \frac{w^2}{V_T^2}\lambda\frac{L-L_T}{d_T}\sqrt{R}$$
$$= \frac{1}{(1-R)^2} \left\{ C_1'R \left(1 - \frac{w}{V_T}\right)^2 - C_2'\sqrt{R} \left(\frac{w}{V_T} - R\right) \left|\frac{w}{V_T} - R\right| \right\}$$

$$C_1' = R(1 + \zeta_K) + C_D(0) + \frac{L_T\lambda_T}{d_T}\frac{R}{1-R}$$

$$C_2' = \frac{L_T\lambda_T}{(1-R)d_T}; \qquad d_T = \frac{4A_T}{U_T} \qquad (23.189)$$

where a is the speed of sound and U_T is the circumference of the train. In Fig. 23.94 the aerodynamic power for the passage calculated with C_{DT} from Eq. (23.188) is compared to the varying power calculated with a computer program.

Figure 23.95 shows T_f values calculated with Eq. (23.188) for the passage of an aerodynamically well-shaped high speed train for different parameters as a function of the blockage ratio R, which is the most important parameter. T_f is practically

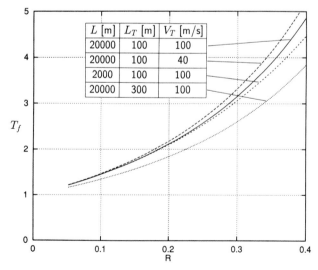

L [m]	L_T [m]	V_T [m/s]
20000	100	100
20000	100	40
2000	100	100
20000	300	100

FIGURE 23.95 Tunnel factor, T_f, as a function of the blockage ratio, R, for long tunnels for an aerodynamically well-shaped high speed train.

independent of the tunnel length, L, and the train speed, V_T, especially for the usual R values in the range of 0.1 to 0.2. This has been verified by full scale experiments [Peters (1990)]. T_f decreases with increasing train length, which is in accordance with full-scale results. In comparing full-scale values of T_f from coasting tests with a value calculated with Eq. (23.188), it should be considered that coasting may have an influence on the flow field due to the deceleration of the train. Therefore, German Rail made tests in a tunnel with a grade in the downhill direction so that the train speed V_T remained nearly constant.

The basic configuration of the Channel Tunnel consists of two single-track running tunnels connected at 250 m intervals with cross-connecting tubes called pressure relief ducts. The concept of these ducts results in considerable traction energy savings, lower pressure levels, and reduced cooling load. But, exhaust flows through the ducts, which can be substantial (50 m/s), impinge on the sides of passing trains. The induced forces can be a considerable risk for less, robust containers and freight wagons and for cars on the open upper deck of car carriers. These effects can be reduced by fitting deflectors on the connecting tubes [Gawthorpe (1992)].

23.9 LIGHTER-THAN-AIR VEHICLES
Sam P. Jones and James D. DeLaurier

23.9.1 Aerostatics for Aerostats and Airships

Introduction. The long and romantic history of airships and their contribution to the development of air transportation is well known. Less familiar is the development in recent years of the tethered aerostat as an important vehicle in communications, intelligence gathering, and surveillance. The development of the power tether which transmits electrical power from the ground has been a factor in the success of the latter. The largest of the tethered aerostats exceeds 500,000 cubic feet in volume and flies routinely at altitudes of 15,000 ft [Krausman (1991)]. The design and operation of these vehicles, both powered airships and tethered aerostats, depend upon environmental and other factors which fall under the general heading of *Aerostatics*. The few references on this subject include Brown and Speed (1962), Durand (1934), Myers (1969), Wright (1976), and Jones (1986). Much of this discussion is taken from the last of these. It is the objective of this section to present an overview of this field sufficient for the engineer to understand the important factors affecting these vehicles and to make elementary calculations of their static performance.

Strictly speaking, aerostat is a generic term applying to both airships and tethered lighter-than-air craft. In the context of this discussion, the airship is considered to be a powered, controlled, ligher-than-air vehicle, while the tethered aerostat, although of aerodynamic shape, is unpowered and tethered to the ground. The same principles of aerostatics apply to both, however the latter is more sensitive to environmental factors, since it flies much higher than the airship and, having no active controls, must be both statically and dynamically stable. The airship seldom flies above 5,000 ft and can overcome inherent instability with active controls.

The Atmosphere. All lighter-than-air vehicles must allow for the expansion of the lifting gas (usually helium) which is determined by the temperature and pressure or density of the external air. An accurate model of the atmosphere is, therefore, essential to prediction of their performance. The effect of ambient conditions is so strong that it is usually not sufficient to use Standard Atmosphere Tables [NOAA (1976)]. A fairly accurate model of the atmosphere is derived from the hydrostatic equation and the usually linear temperature gradient, with the surface conditions as a boundary.

$$dp = -\gamma \, dH \qquad (23.190)$$

$$T = T_s - aH \qquad (23.191)$$

Here, H is the altitude, p is the pressure, T is the absolute temperature, γ is the specific weight of air, a is the atmospheric lapse rate, and the subscript s denotes surface conditions.

Integration yields

$$p = p_s \left[\frac{T}{T_s} \right]^n \qquad (23.192)$$

$$\gamma = \gamma_s \left[\frac{T}{T_s} \right]^{n-1} \qquad (23.193)$$

where

$$n = \frac{\gamma_0 T_0}{p_0 a} \qquad (23.194)$$

Standard conditions at sea level are the following.

$p_0 = 1013.25 \; mbar = 29.921$ in. Mercury $= 2116.22$ lb/sq ft

$T_0 = 15 \; C = 59 \; F = 288.15 \; K = 518.67 \; R$

$a_0 = .0065 \; C/m = 0.0035662 \; F/ft$

Under these conditions,

$\gamma_0 = 1.2250$ kg/cu m $= .076474$ lb/cu ft

$\lambda_0 = 1.0559$ kg/cu m $= 0.06592$ lb/cu ft

$n_0 = 5.2558$

where λ_0 is the specific lift of pure helium.

If the surface is at sea level, then H is the *geopotential altitude* and Eqs. (23.191), (23.192), and (23.193), with n_0 will generate the Standard Atmosphere tables.

The presumption is dry air. The effect of humidity is to reduce the air density which is accounted for by use of *virtual temperature*, T_v. This is the temperature at which the dry air would have the same density as the moist air at the same pressure. It can be calculated using the following formula

$$T_v = \frac{T}{1 - 0.3779p_w/p} \tag{23.195}$$

where p_w is the partial pressure of water vapor.

Ballonets. An essential feature of both types of lighter-than-air craft is the presence of one or more *ballonets*, which are internal air chambers into which air is pumped or from which it is expelled to maintain pressurization as the temperature and pressure change. Tethered aerostats usually have one large ballonet, while airships more often have two or more. In the latter, they are frequently located fore and aft and used for trim purposes by shifting air from one to the other.

The size of the ballonet or ballonets determines the maximum altitude of the aerostat, since the helium expansion from the ground to altitude must be accommodated. The total ballonet size can be calculated using

$$\frac{V_b}{V_h} = 1 - \frac{(T_s)_{min}(p_H)_{min}}{(T_H)_{max}(p_s)_{max}} \tag{23.196}$$

Here, V_b is the ballonet volume, V_h is the volume available for helium when the ballonets are empty, $(T_s)_{min}$ is the minimum temperature on the ground, $(T_H)_{max}$ is the maximum temperature at maximum altitude and $(p_H)_{min}$ and $(p_s)_{max}$ are the helium pressures.

The large TCOM* 71 meter aerostat designed to carry a 3500 lb payload to 15,000 ft has a ballonet volume of 56% of the hull volume, while the Westinghouse** Sentinel 1000 airship's ballonets are 24% of the total volume.

The large ballonets of tethered aerostats present a special problem in the calculation of static trim and center of gravity. The ballonet air, being much heavier than the helium, tends to form a level surface, as shown in Fig. 23.96. This causes a variable center of gravity depending upon the volume of air and pitch angle. The problem is very minor in the much smaller ballonets of airships and is usually ig-

---

*TCOM, L.P., Columbia, MD.
**Westinghouse Airships, Inc., Baltimore, MD.

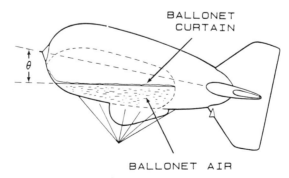

BALLONET
CURTAIN

BALLONET AIR

FIGURE 23.96 Leveling of air in the ballonet of a tethered aerostat.

nored. For large tethered aerostats, however, a family of curves must be prepared for each design showing the centroid of the ballonet air as a function of fullness and pitch angle. These are usually expressed as a family of polynomials used in both static and dynamic analyses.

Buoyancy Model. For dynamic analysis, the *Archimedean buoyancy* of the entire body and the total mass, including the enclosed gases, must be considered. This presents the difficulties due to air in the ballonets of large aerostats discussed in the preceding paragraph. A simplified model, commonly used for steady-state analysis, envisions the lifting gas as free to expand by expelling neutrally-buoyant air from the ballonet. In this model, the gross buoyant lift is that of an independent parcel of helium. Neglecting the small pressure differential (*superpressure*) and with no temperature differential relative to the ambient (*superheat*), the lift of such a parcel is independent of temperature and pressure and therefore of altitude. This is due to the fact that the specific lift of the gas decreases in exactly the same proportion as the volume increases. This model has the added advantage that the gross weight is that of the structure alone, which has a fixed weight, and center of gravity. The variable is the center of buoyancy, the centroid of the helium volume, calculated from the ballonet polynomials.

This simple model requires corrections for deviations from the simplifying assumptions that the internal gases are at the same temperature and pressure as the surrounding air and that both are dry. The gross lift of the helium thus corrected is referred to as the *Standard Gross Lift*.

Correction for Superpressure: The superpressure is the differential of the pressure inside the hull to the ambient and is responsible for maintaining the rigidity of the hull. Common practice for both airships and tethered aerostats is to maintain about 2 in. of water gauge above dynamic pressure. This compression of the enclosed gases results in a loss of lift, ΔL_p, given by

$$\Delta L_p = -\gamma V_t \frac{\Delta p T}{p T_g} \qquad (23.197)$$

which reduces to

$$\Delta L_p = -\frac{\gamma_0 T_0 V_t}{p_0} \frac{\Delta p}{T_g} \qquad (23.198)$$

where V_t and T_g are the total volume and temperature of the enclosed gases.

Since the absolute temperature does not vary greatly, the lift loss is almost independent of altitude. For a 500,000 cu ft vehicle with a superpressure of 2 in. water gauge, the lift loss would be about 195 lb.

Correction for Superheat: The term *superheat* in common usage refers to the temperature differential of the internal gases to the ambient air. It results from the radiation environment and is usually positive in the daytime due to solar radiation and negative at night when the aerostat radiates to a clear sky. The effect is maximum during clear, calm conditions when the heat exchange with the environment is by natural convection. Relative wind diminishes superheat due to the forced convection heat transfer. For that reason, the effect is much less important for airships except when moored, since they create their own relative wind.

Superheat has two effects. It increases the volume of helium, expelling air from the ballonets, and increases the lift by decreasing the density of the enclosed gases. Assuming the enclosed gases are at a uniform temperature, the net effect on buoyant lift is analogous to that for superpressure.

$$\Delta L_t = \gamma V_t \frac{\Delta T}{T_g} \tag{23.199}$$

For a 500,000 cu ft vehicle at 5000 ft altitude, the effect amounts to a gain of about 65 lb per degree F of superheat.

The magnitude of superheat depends upon many factors including the hull material, time of day, cloud cover, season, and relative wind. Figure 23.97 presents data taken on a TCOM 250 tethered aerostat at Grand Bahama Island during an unusual period of clear, calm conditions for two days in January 1974. The wind velocity was less than 5 knots, and the aerostat's altitude was maintained between 8500 and 10,800 ft. The shielded temperature probe was hanging about 15 ft from the top of the hull at the major diameter [Jones (1986)]. Like all subsequent TCOM aerostats, the laminated hull material included an outer layer of white Tedlar® film. During the day, subjected to solar radiation, the helium temperature reached about 25°F higher than the ambient air, while at night, due to radiation to space, the differential was negative by more than 10°F. These results are fairly typical of observations with other aerostats under similar conditions, irrespective of location and season.

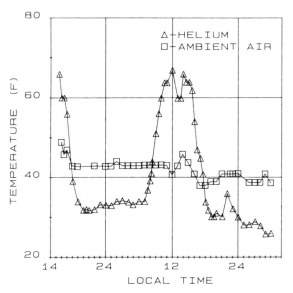

FIGURE 23.97 Helium and air temperature on a TCOM 250 aerostat during calm conditions, Grand Bahama Island, Jan. 18–20, 1974.

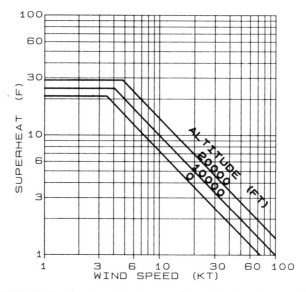

FIGURE 23.98 Semiempirical curves for estimating day superheat.

Based upon empirical data and heat transfer theory, the curves of Figs. 23.98 and 23.99 have been derived to estimate the superheat for day and night conditions, respectively. At low winds, free convection is assumed, and at higher winds cooling is primarily by forced convection.

The Effect of Humidity: Water vapor is lighter than air, so moist air provides less buoyant force on immersed bodies than dry air. This may be offset, however, by

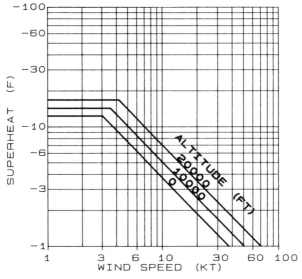

FIGURE 23.99 Semiempirical curves for estimating negative superheat at night.

water vapor inside the aerostat. A little known fact is that, although relatively impermeable to helium, most modern tethered aerostat and airship materials transmit water vapor rather well. Thus, tests on TCOM laminated hull material using ASTM E-398 gave a water vapor transmissivity of 1.3 grams per 100 sq in. per 24 hr at 100 F [Ashford *et al.* (1983)]. This is over 10 times the helium permeability measured at the same temperature when compared on the same basis.

While moisture in the ambient air decreases lift, water vapor inside the hull, being lighter than air, increases lift. The water vapor content inside the hull will approach that of the outside air, in which case, there will be no net effect on lift. For that reason and the difficulty of measuring the humidity in helium, this effect is usually ignored. However, ambient humidity fluctuates widely and water vapor change inside the hull lags by many hours so that the concentrations are seldom the same. Humidity should be taken into account when making precise measurements such as lift loss rate.

Figure 23.100 shows the effect of water vapor in both the air and the helium on the gross lift factor or ratio to the lift of dry helium [Jones (1986)]. The curves are given in terms of the dew point, which is a measure of water vapor content with a single number.

In addition to the effect on buoyant lift, water vapor in the helium affects the maximum altitude of the aerostat. Since it occupies volume, the altitude at which the ballonets are empty will be lowered.

Aerostatic Flight Envelopes. Both airships and tethered aerostats are limited in their maximum altitude and the temperature and pressure when moored by the helium fill (gross lift) and size of the hull and ballonets. In airship parlance, the maximum altitude is known as the *pressure height*, H_p. It is the altitude at which the helium has expanded to expel all the air in the ballonets and any further increase in altitude would cause a dangerous increase in hull pressure or venting of the helium. Thus, it is sometimes called the *vent ceiling*.

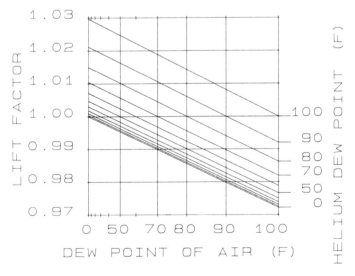

FIGURE 23.100 Nomograph showing the effect of water vapor on helium gross lift.

Making use of Eq. (23.193) and the gas laws, and neglecting the small superheat and superpressure, the pressure height of an airship is given by

$$H_p = \frac{T_s}{a} \left[1 - \left(\frac{Lp_0 T_s}{V_h \lambda_0 f_p p_s T_0} \right)^{1/(n-1)} \right] \tag{23.200}$$

where L is the standard gross lift of helium, and f_p is the helium purity expressed as volume fraction.

At low altitudes, as when moored, and low temperatures, the helium may contract to such a small volume that the ballonets are completely filled with air. This is the lower limit of helium volume.

$$V_f = V_h - V_b \tag{23.201}$$

This altitude can be found by substituting V_f for V_h in Eq. (23.200).

While the same equations apply to tethered aerostats, their ceiling altitude must take into account the weight of the tether and the much higher superheat and relative superpressure at the higher altitudes. In addition, tethered aerostats require significant free lift or excess buoyancy over that required to lift the system including the tether. It is essential to know how high the aerostat can safely fly under calm conditions and what helium fill must be provided. The operating altitude is determined by two factors, illustrated in Fig. 23.101, which is a plot of altitude as a function of helium fill in terms of gross lift. The curves of positive slope represent the lift limit for the gross weight, including tether and free lift. The greater the altitude, the greater the tether weight and thus the more lift required. The slope is the inverse of the unit weight of the tether. At night, negative superheat reduces lift, as shown by the displaced curve.

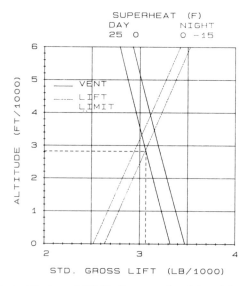

FIGURE 23.101 Vent altitude and lift limit curves for determining optimum altitude and lift (dashed lines).

The second factor determining operating altitude is the pressure height or vent ceiling, represented by the curves of negative slope. This curve is very sensitive to superheat, which reduces the pressure height.

The intersection of the lift limit and vent ceiling curves gives the altitude that can be maintained both day and night with a minimum specified free lift and allowances for superheat. This is known as the *optimum altitude*, and the corresponding gross lift is the *optimum helium fill*. Of course, wind reduces superheat and provides aero-dynamic lift, so it will be possible to fly higher than the optimum altitude under some conditions, but that will be at the expense of free lift based on helium buoy-ancy. For flights of many days, it may be necessary to compromise altitude by overfilling to account for helium leakage.

Figure 23.101 is specific for a given set of ambient conditions. In particular, the ambient temperature and pressure will affect the vent ceiling. Figure 23.102 shows how the optimum altitude varies with sea level temperature for the TCOM 31-meter tethered aerostat, with payload weight as a parameter. Standard sea level barometric pressure and lapse rate are assumed. The free lift is 20% of the gross lift, and the daytime superheat is taken to be 25°F. The negative superheat at night is not taken into account. It is a matter of strategy to ignore negative superheat, which only exists under calm conditions.

23.9.2 Aerodynamics for Aerostats and Airships

For a traditional configuration consisting of an axisymmetric body with aft fins, as shown in Fig. 23.103, a lighter-than-air vehicle's steady-state normal force, axial force, and pitching moment may be calculated from a modified cross-flow analysis. In this, it is assumed that the vehicle is divided into two distinct aerodynamic re-gions—the hull, which extends from the nose to the hull-fin intersection point, l_h,

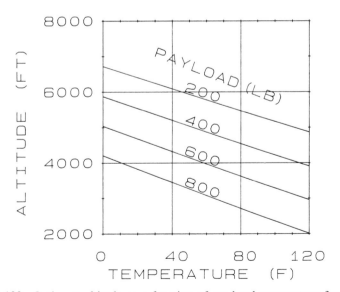

FIGURE 23.102 Optimum altitude as a function of sea-level temperature for the TCOM 31 meter aerostat with various payloads. Free lift is 20% of the gross lift.

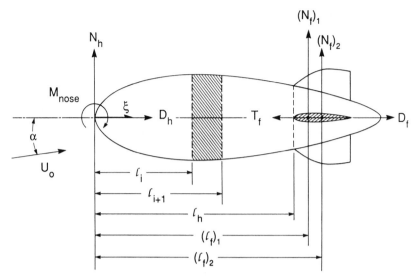

FIGURE 23.103 Schematic of analytical model.

and the fins, which continue from that point to their trailing edge. For nonseparated flow, the hull forces and moment are obtained from Allen and Perkins (1951), and the fins' normal force is obtained from Wardlaw (1979), with efficiency factors, η_k and η_f, introduced to account for mutual interference between hull and fins. From Jones and DeLaurier (1983), the resulting equations are for the normal force

$$N = Cnq_0S$$

$$Cn = [(k_3 - k_1)\eta_k \hat{I}_1 + 0.5(Cn_\alpha^*)_f\eta_f\hat{S}_f] \sin(2\alpha) \qquad (23.202)$$
$$+ [(Cd_c)_h \hat{J}_1 + (Cd_c)_f\hat{S}_f] \sin\alpha \sin|\alpha|$$

and for the axial force

$$D = Cdq_0S$$

$$Cd = [(Cd_h)_0\hat{S}_h + (Cd_f)_0\hat{S}_f] \cos^2\alpha$$
$$- (k_3 - k_1)\eta_k\hat{I}_1 \sin(2\alpha) \sin(\alpha/2) - (Ct)_f\hat{S}_f \qquad (23.203)$$

The pitching moment (about the nose) is

$$M_{nose} = Cm_{nose}q_0 S\bar{c}$$

$$Cm_{nose} = -[(k_3 - k_1)\eta_k\hat{I}_3 + 0.5(\hat{I}_f)_1(Cn_\alpha^*)_f\eta_f\hat{S}_f] \sin(2\alpha)$$
$$- [(Cd_c)_h\hat{J}_2 + (Cd_c)_f(\hat{I}_f)_2\hat{S}_f] \sin\alpha \sin|\alpha| \qquad (23.204)$$

The terms in these equations are defined as follows: q_0 = free-stream dynamic pressure = $\rho(U_0)^2/2$, S = (total hull volume)$^{2/3}$, the reference area; $\bar{c}$ = total hull length, the reference length; α = the angle of attack of the vehicle's center-line to the free stream flow direction; k_1, k_3 = axial and lateral apparent-mass coefficients, respectively, given by

$$k_1 = \gamma/(2 - \gamma); \qquad k_3 = \delta/(2 - \delta)$$

$$\gamma = 2 \left(\frac{1 - e^2}{e^3} \right) \left[\frac{1}{2} \ln \left(\frac{1 + e}{1 - e} \right) - e \right]$$

$$\delta = \frac{1}{e^2} - \frac{1 - e^2}{2e^3} \ln \left(\frac{1 + e}{1 - e} \right)$$

$$e = \left(\frac{f^2 - 1}{f^2} \right)^{1/2} \tag{23.205}$$

where f, the hull's fineness ratio, is f = total hull length/hull maximum diameter The nondimensional hull integrals, $\hat{I}_1$, $\hat{I}_3$, $\hat{J}_1$, and $\hat{J}_2$, are given by

$$\hat{I}_1 = \frac{1}{S} \int_0^{lh} \frac{dA}{d\xi} \, d\xi = \frac{A_h}{S}; \qquad \hat{I}_3 = \frac{1}{S\bar{c}} \int_0^{lh} \xi \frac{dA}{d\xi} \, d\xi$$

$$\hat{J}_1 = \frac{1}{S} \int_0^{lh} 2r \, d\xi; \qquad \hat{J}_2 = \frac{1}{S\bar{c}} \int_0^{lh} 2r\xi \, d\xi \tag{23.206}$$

where A = hull cross-sectional area, ξ = axial distance along the hull, measured from the nose, and r = hull radius.

Also,

$\hat{S}_f = S_f/S$, nondimensional fins' reference area (see Fig. 23.104 for S_f).

$\hat{S}_h = S_h/S$, the nondimensional hull reference area [usually $S_h = S$, depending how $(Cd_h)_0$ is defined].

$(Cd_h)_0$ = hull zero-angle axial drag coefficient, referenced to S_h. [This includes skin friction, base drag, and excrescence drag. A good reference for estimating these values is Hoerner (1965).]

$((Cd_f)_0)$ = fin's zero-angle axial drag coefficient, referenced to S_f. (This involves the profile drag of the exposed surfaces, as well as the interference drag where the roots of the fins intersect the hull.)

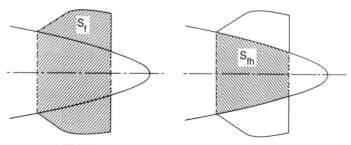

FIGURE 23.104   Fin planform definition.

$(Cd_c)_h$ = hull cross-flow drag coefficient, referenced to $\hat{J}_1 S$. [From Wardlaw (1979), this value is typically ≈ 0.3.]

$(Cd_c)_f$ = fin's cross-flow drag coefficient, references to S_f.

Plots for $(Cd_c)_f$ are presented in Wardlaw (1979), which are summarized in the following curve-fitted equations

$$(Cd_c)_f = A_0 + A_1 \times AR + A_2 \times AR^2 + A_3 \times AR^3 \tag{23.207}$$

for $0 < AR \leq 3$ and where

$$A_0 = 2.008 + 3.99\lambda \qquad\qquad A_1 = -0.832 - 3.289\lambda + 0.7885\lambda^2$$

$$A_2 = 0.185 + 2.028\lambda - 1.15\lambda^2 \quad A_3 = -0.0178 - 0.44\lambda + 3.2\lambda^2$$

$$\tag{23.208}$$

In these, AR = the fins' aspect ratio, defined by $AR = b^2/S_f$ (it is important to note that, for fins with a dihedral angle, Γ, that b is twice the semi-span length from the hull's center-line to the fin's tip. In other words, it is *not* the projected length on the horizontal plane) and λ = the fins' taper ratio. (Since the effective planform, shown by S_f in Fig. 23.104, does not generally have a simple straight taper, it has been found adequate to chose an equivalent straight-tapered planform as shown in Fig. 23.105.) For the purposes of initial estimations, a representative $(Cd_c)_f \approx 2.0$ for the AR and λ values typical for airship and aerostat fins.

$(Cn_\alpha^*)_f$ is the derivative of the isolated fins' normal-force coefficient with respect to α, at $\alpha = 0$ and referenced to S_f. A useful equation for estimating this is obtained from Lowry and Polhamus (1957), with a correction factor to account for the horizontal fins' dihedral angle

$$(Cn_\alpha^*)_f = \frac{2\pi AR \cos^2 \Gamma}{2 + \left[\dfrac{(AR)^2}{\kappa^2} (1 + \tan^2 \Lambda_{c/2}) + 4 \right]^{1/2}} \tag{23.209}$$

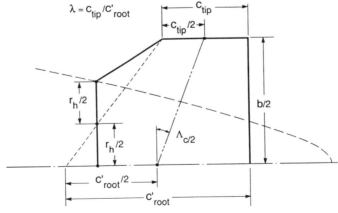

FIGURE 23.105   Equivalent straight-tapered planform.

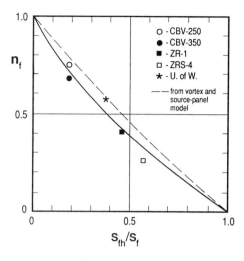

FIGURE 23.106 Fin-efficiency factor.

where

$\Lambda_{c/2}$ = sweep of the half-chord line of the equivalent straight-tapered planform, shown in Fig. 23.105.

κ = ratio of the airfoil section's lift-curve slope to 2π (a reasonable estimate is $\kappa \approx 0.95$).

Γ = fins' dihedral angle.

η_f = fin-efficiency factor accounting for the effect of the hull on the fins, given as a function of S_{fh}/S_f in Fig. 23.106 (S_{fh} = portion of the fins' area covered by the hull, as shown in Fig. 23.104). η_k = hull-efficiency factor accounting for the effect of the fins on the hull, given as a function of $S_f \cos^2 (\Gamma)/(J_1)_{\text{total}}$ in Fig. 23.107 [$(J_1)_{\text{total}}$ = the value of $\hat{J}_1 S$ when the integral for $\hat{J}_1$ extends to the end of the hull. Note that this is the hull's complete horizontal projected area (*shadow area*)].

Curves for η_k and η_f, shown in Figs. 23.106 and 23.107, were obtained from a semi-empirical method based on wind tunnel data for several airship and aerostat

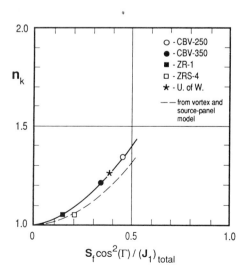

FIGURE 23.107 Hull-efficiency factor.

models as described in Jones and DeLaurier (1983). Subsequently, these curves were rederived by Wong, Zhiyung, and DeLaurier (1985) by using a vortex and source-panel computational model. The results are co-plotted on Figs. 23.106 and 23.107, where it is seen that they are a reasonably close match. Good results have been obtained from using the semi-empirical curves.

$(C_t)_f$ = the fins' leading-edge suction coefficient, given by

$$(Ct)_f = (Cn_\alpha^*)_f \eta_f \eta_t \frac{\sin (2\alpha)}{2} \tan \alpha - \frac{[(Cn_\alpha^*)_f \eta_f \sin (2\alpha)]^2 \eta_t}{4\pi AR} \qquad (23.210)$$

where

η_t = the leading-edge suction efficiency (For fins incorporating airfoils with rounded leading edges, $\eta_t \approx 1.0$. However, fins with sharp leading edges may give an $\eta_t \approx 0$)

and finally

$(\hat{l}_f)_1 = (l_f)_1/\bar{c}$, nondimensionalized distance from the hull's nose to the fins' aerodynamic center (based on the S_f planform)

$(\hat{l}_f)_2 = (l_f)_2/\bar{c}$, nondimensionalized distance from the hull's nose to the fins' center of cross-flow force (the area center of the exposed fins, $S_f - S_{fh}$)

This method was applied by Badesha and Jones (1993) to the CBV 71M aerostat shown in Fig. 23.108. This incorporates three identical fins arranged in an inverted-Y configuration, where $\Gamma = -45°$. The values of the parameters for this aerostat are: $\hat{S}_h = 1.0$, $\hat{S}_f = 0.8369$, $\hat{I}_1 = 0.2090$, $\hat{J}_1 = 1.5542$, $\hat{I}_3 = -0.1691$, $\hat{J}_2 = 0.5498$, $(Cn_\alpha^*)_f = 2.7101$, $(\hat{l}_f)_1 = 0.8108$, $(\hat{l}_f)_2 = 0.8798$, $(k_3 - k_1) = 0.7169$, $\eta_f = 0.6684$, $\eta_k = 1.0034$, $(Cd_c)_f = 1.1145$, $(Cd_c)_h = 0.1200$, $S_{fh}/S_f = 0.2851$, $S_f \cos^2 \Gamma/(J_1)_{total} = 0.2329$ and the resulting normal force and pitching coefficients are co-plotted with results from wind tunnel tests in Figs. 23.109 and 23.110.

The aerostat in Fig. 23.108 has a very small portion of hull extending aft of the fins' trailing edge. However, blimps and airships are characterized by larger aft hull portions (commonly called the *boat tail*), as drawn in Fig. 23.103. It has been found that, for this analysis, only the boat tail's skin friction and base drag contribution

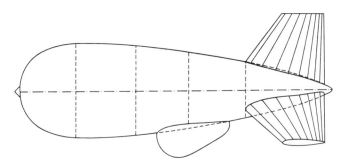

FIGURE 23.108 Aerostat CBV 71M.

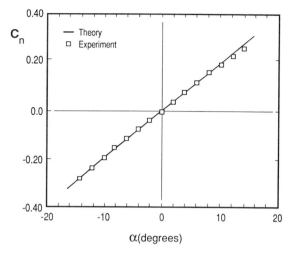

FIGURE 23.109 Results for Cn vs. α.

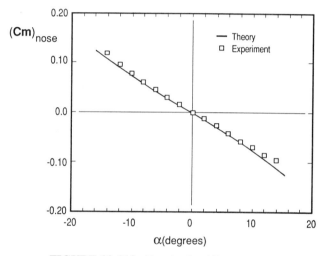

FIGURE 23.110 Results for $(Cm)_{nose}$ vs. α.

[through the $(Cd_h)_0$ term] need be accounted for. The flow straightening effect of the fins essentially eliminates any cross flow in the boat-tail region.

23.10 LAMINAR FLOW CONTROL ON AIRCRAFT
Fayette S. Collier, Jr.

23.10.1 Introduction

Drag reduction in the form of laminar flow control applied to future, advanced commercial, and military transports across the speed regime offers potential breakthrough opportunities in terms of reductions in takeoff gross weight (TOGW), operating empty weight (OEW), block fuel for a given mission, and significant

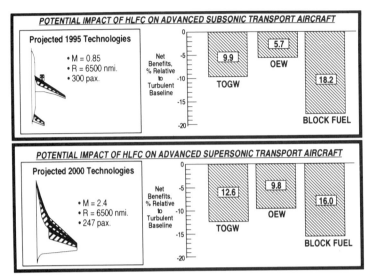

FIGURE 23.111   Potential impact of laminar flow control applied to advanced aircraft configurations.

improvements in cruise lift-to-drag ratio (*L/D*). NASA studies [Arcara *et al.* (1991)] of an advanced subsonic, twin-engine commercial transport with anticipated 1995 engine, structure, and aerodynamic technology improvements incorporated into the design indicate that the application of laminar flow to the wing upper surface, empennage, and engine nacelles may result in TOGW and OEW reductions of about 10% and 6%, respectively, when compared to the turbulent baseline shown in Fig. 23.111. The analysis included conservative estimates of the hybrid laminar flow control (HLFC) system weight and engine bleed air (to drive the suction device) requirements. Satisfaction of all operational and Federal Aviation Regulations (FAR) regulatory requirements, such as fuel reserves and balanced field length, was achieved. Also shown in Fig. 23.111, a laminar version of an advanced high-speed civil transport (HSCT) designed to carry 247 passengers at Mach 2.4 a distance of 6500 nautical miles with laminar flow over 40% of the wetted wing area results in impressive improvements in TOGW, OEW, and fuel burn of 12.6%, 9.8%, and 16.0% respectively, when compared to the turbulent version of the concept [Anon. (1990)]. Based upon a TOGW of 750,000 lb for the turbulent baseline HSCT aircraft, the projected reduction in TOGW for the laminar airplane is roughly equivalent to the payload fraction.

   The optimistic picture painted in the cursory system studies described in the last paragraph is tempered somewhat by a frank discussion of the practical issues associated with the development of a transonic, commercially viable, laminar aircraft. Lynch *et al.* (1991) and Robert (1992) have categorized the minimum prerequisites for inservice status of an HLFC aircraft as:

1)   resolution of potential performance penalties (assessed against the HLFC benefits) due to use of full-span leading-edge Krueger system as opposed to the usual leading-edge slat system

2) development of HLFC compatible ice-protection system for the leading-edge high-lift/insect shield Krueger system, as well as the main wing element

3) demonstration of the ability to achieve laminar flow in the wing root area, and downstream of the nacelle/pylon area, especially with the advent of advanced very high BPR propulsion system installations

4) development of viable high Reynolds number, wind-tunnel test techniques for HLFC configuration development

5) demonstration of acceptable reliability, maintainability, and operational characteristics for an HLFC configuration, and the ability to predict realistic total airplane related operating cost (TAROC) improvements for the customer and for the airlines.

In addition to the practical, implementation issues listed above, in order to compute potential improvements in direct operating costs or improvements in aerodynamic efficiency as a function of total laminarized surface area for laminar aircraft, the designer must be able to accurately predict the location of boundary layer transition on complex, three-dimensional geometries. Pressure gradient, surface curvature, wall temperature, wall mass transfer, and unit Reynolds number are known to influence the stability of the boundary layer. For practical HLFC designs, it is imperative to be able to accurately predict the required amount, location, and distribution of wall suction to attain a given (*designed for*) transition location. To further complicate matters, there exist many environmental disturbances, any one of which can cause immediate transition to turbulent flow if *supercritical* in nature. For example, although rare, ice crystals of sufficient size and density in cirrus clouds encountered during flight are known to *trip* a laminar boundary layer causing premature transition. Under such conditions, conventional linear stability theory is not applicable. Linear stability theory represents the current state-of-the-art for transition onset prediction for flows past complex, three-dimensional geometries at transonic, supersonic, and hypersonic speeds, but a great deal of effort is being devoted to the ability to account for nonlinear and receptivity issues. Reviews of recent advances in transition prediction methods are available and can be found in Arnal (1992), Saric (1992), and Reed *et al.* (1989) and Sec. 4.3 of this handbook.

The balance of this section will present an overview of wind tunnel investigations and flight research activities devoted to advancing the state-of-the-art and reducing the risk associated with the application of laminar flow control (LFC) technology to subsonic/transonic commercial and military transports. We will focus on LFC research conducted within the last 5 years and highlight activities devoted to the attack of the challenges listed earlier.

23.10.2 B757 Hybrid Laminar Flow Control Flight Experiment

Early laminar flow flight research demonstrated that manufacturing techniques needed to obtain the stringent surface smoothness and waviness criteria required for laminar flow aircraft presented a major challenge. Today, conventional production aircraft wing surfaces can be built to meet these design constraints. The most significant advance made in the development of laminar flow technology is the concept of hybrid laminar flow control (HLFC), an idea which integrates the concepts of natural laminar flow (NLF) and full-chord laminar flow control (LFC), and which

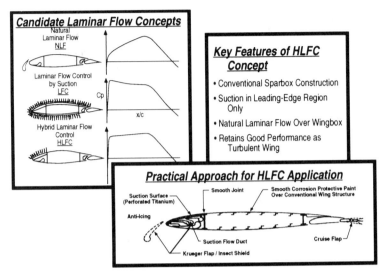

FIGURE 23.112   Candidate laminar flow concepts and schematic of the HLFC concept.

avoids the undesirable characteristics of both (see Fig. 23.112). NLF is sweep limited, and full-chord LFC is very complex. The key features of HLFC are: 1) conventional spar box construction techniques are utilized, 2) boundary layer suction is required only in the leading edge, 3) natural laminar flow is obtained over the wing box through proper tailoring of the geometry, and 4) the HLFC wing design has good performance in the turbulent mode. The Leading-Edge Flight Test on the NASA Jetstar [Powell *et al.* (1987)] initially addressed HLFC leading-edge system integration questions and set the stage for a large, commercial transport demonstration of HLFC.

Objectives. The B757 High Reynolds Number HLFC Flight Experiment was designed to meet three objectives: 1) perform high Reynolds number flight research on the HLFC concept, 2) develop a data base on the effectiveness of the HLFC concept applied to a large, subsonic commercial transport, and 3) develop and flight validate an integrated, practical high-lift, anti-ice, and HLFC system.

Technical Approach. A 20-foot span segment of the leading-edge box of the B757's port wing outboard of the engine nacelle pylon was replaced with an all metal construction, HLFC leading-edge box as shown schematically in Fig. 23.113. This new leading edge consisted of a microperforated titanium outer skin, subsurface suction flutes and collection ducts to allow for boundary layer suction to control cross flow disturbance growth aft to the front spar of the B757. The leading edge consisted of fully integrated high-lift Krueger/insect shield and hot air de-icing systems. The wing-box portion of the test area consisted of the original, production B757 surface and contour, and only required minor rework to meet surface waviness and smoothness requirements for the achievement of laminar flow. The design point for the flight tests was chosen as $M = 0.80$ at a lift coefficient of 0.50. Parametric variations around the design point were performed to investigate extent of laminar flow as a function of Mach number, unit Reynolds number and lift coefficient.

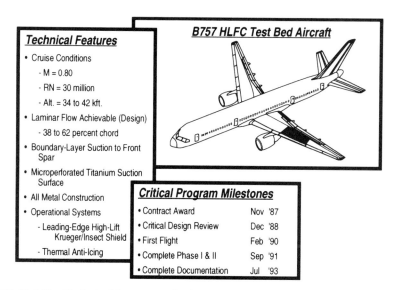

Technical Features

- Cruise Conditions
  - M = 0.80
  - RN = 30 million
  - Alt. = 34 to 42 kft.
- Laminar Flow Achievable (Design)
  - 38 to 62 percent chord
- Boundary-Layer Suction to Front Spar
- Microperforated Titanium Suction Surface
- All Metal Construction
- Operational Systems
  - Leading-Edge High-Lift Krueger/Insect Shield
  - Thermal Anti-Icing

B757 HLFC Test Bed Aircraft

Critical Program Milestones

• Contract Award	Nov '87
• Critical Design Review	Dec '88
• First Flight	Feb '90
• Complete Phase I & II	Sep '91
• Complete Documentation	Jul '93

FIGURE 23.113 Technical features and schematic layout of B757 HLFC test bed aircraft.

Instrumentation. The instrumentation package on board the aircraft included: 1) flush-mounted (in the perforated leading edge) and strip-of-tube static pressure measurement capability for external C_p distribution, 2) hot-film gages for transition detection of the wing box and attachment line boundary layers, 3) infrared camera for boundary layer transition detection (phase II only), and 4) wake survey probe for inferred local drag-reduction determination. A schematic of the instrumentation setup is shown in Fig. 23.114.

Results. During the conducting of the flight tests, it was shown that the HLFC concept was very effective in delaying boundary-layer transition to the rear spar around the design point. The state of the boundary layer is shown in Fig. 23.115 at a sample test condition with most of the hot-film gages indicating laminar flow beyond 65% chord. In fact, the suction rates required to routinely achieve laminar flow to 65% chord were about one third of those predicted during the design phase. As shown in Fig. 23.115, the wake-rake measurements indicated a local drag reduction on the order of 29% with the HLFC system operational, which results in an integrated 6% drag reduction for the aircraft.

23.10.3 Hybrid Laminar Flow Nacelle Demonstration

Another project was directed toward the aerodynamic flight demonstration of the hybrid laminar flow control concept applied to the external surface of large, turbofan engine nacelles. The friction drag associated with modern turbofan nacelles may be as large as 4 to 5% of the total aircraft drag for a typical commercial transport, and studies indicate potential specific fuel consumption reductions on the order of 1–1.5% for advanced nacelles designed to achieve laminar boundary layer flow.

Objectives. The main objective of the project was "to demonstrate the feasibility of laminar flow nacelles for wide-body aircraft powered by modern high-bypass

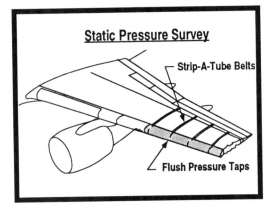
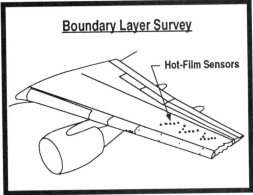
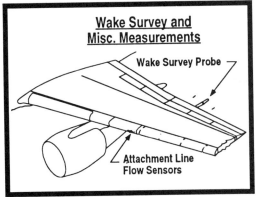

FIGURE 23.114 Schematic of the instrumentation layout for the B757 HLFC flight tests.

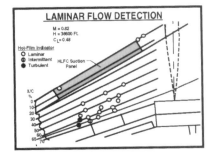

FIGURE 23.115 B757 high Reynolds number hybrid laminar flow control program results.

engines'' . . . and was ''geared to investigate the influence of aerodynamic characteristics and surface effects on the extent of laminar flow'' according to Bhutiani *et al.* (1993).

Technical Approach. A production GEAE CF6-50C2 engine nacelle installed in the number two (on the starboard wing) position of a Airbus A300/B2 commercial transport test-bed aircraft was modified to incorporate two hybrid laminar flow control panels, one inboard and one outboard as shown in Fig. 23.116. The panels were built to very stringent surface waviness specifications and were fabricated of a microperforated composite material. The design was capable of providing suction from the *highlight* aft to the outer barrel/fan cowl juncture. Suction was applied to the surface utilizing subsurface circumferential flutes, a design proven in previous LFC programs, and was collected and ducted to an industrial turbocompressor (TC) unit (the suction source) driven by engine bleed. For convenience, the TC unit was located in the storage bay of the aircraft. The flow through each flute was individually metered. The laminar flow contour extended aft over the fan cowl door and was accomplished through the use of a nonperforated *scab on* composite structure which was blended back into the original nacelle contour ahead of the thrust reverser. No provisions were made for anti-icing or insect contamination avoidance systems.

Instrumentation. The instrumentation package on the aircraft included: 1) static pressure taps on the external surface and in the flutes, 2) a boundary layer rake to quantify boundary layer buildup, 3) hot-film gages for boundary-layer transition detection, 4) surface embedded microphones to assess noise field influence on the state

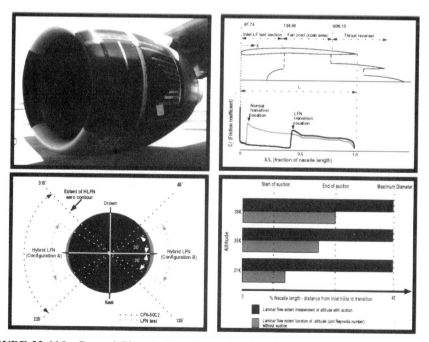

FIGURE 23.116 General Electric CF6-50C2 hybrid laminar flow nacelle program. (From Bhutiani *et al.*, 1993. Reprinted with permission of GE Aircraft Eng.)

of the boundary layer, 5) a charge patch for measurement of atmospheric particle concentration, and 6) infrared imaging for global laminar boundary layer transition detection.

Results. As shown in Fig. 23.16, the HLFC concept was extremely effective over the range of cruise altitudes and Mach numbers tested, resulting in laminar flow to 43% of the nacelle length (the design objective) independent of altitude. At this location, the static pressure sensors indicated the beginning of the pressure recovery region, which effectively transitions the laminar boundary layer to a turbulent boundary layer. Without suction, significant laminar flow was achieved, however the extent of laminar flow decreased as altitude decreased (unit Reynolds number increased) for the case of no suction.

23.10.4 Dassault Falcon 50 Hybrid Laminar Flow Flight Demonstrator

Objectives. The main objective of the flight tests on the Falcon 50 aircraft was the development of a new, laminar flow wing design in the highly three-dimensional region near the fuselage which could provide for leading-edge boundary layer suction aft to 10% of the chord on the upper surface, anti-icing and insect contamination avoidance, and fuselage turbulence contamination avoidance along the attachment line [Bulgubure *et al.* (1992) and Arnal (1991)]. Achievement of 30% chord laminar flow in this region was the design objective.

Technical Approach. A perforated stainless steel suction article was *gloved* over the existing inboard wing structure in close proximity to the fuselage. The glove was faired into the existing wing. Boundary layer suction was generated with the use of an ejector/plenum arrangement and was distributed chord-wise through six spanwise flutes (see Fig. 23.117). In addition, an anti-icing system was integrated into the design, and it performed the additional task of insect contamination avoidance. Monopropylene Glycol (MPG) was the fluid chosen.

The flight test phase was conducted with and without a *Gaster bump* styled turbulence diverter installed in the inboard region on the leading edge of the suction panel. Initially, without the Gaster bump, the primary objective of the flight investigation was the assessment of the anti-icing/insect avoidance system. In addition, the location of the attachment line was measured for proper placement of the Gaster bump. In the second phase, the effectiveness of the Gaster bump for turbulence contamination avoidance along the attachment line was assessed, as well as the effect of boundary layer suction and sweep angle on the chordwise extent of laminar flow. The flight tests were conducted such that the chord Reynolds number variation in the region of the test article was between 12 and 20 million. The leading-edge sweep angle of the test article was nominally 35°, however additional testing was conducted at sideslip of 5°, yielding a resultant leading-edge sweep angle of 30 degrees.

Instrumentation. The installed instrumentation package included: 1) three rows of pressure taps embedded in the suction article between the flutes for external pressure distribution, 2) three rows of 12 hot films each for transition detection flush mounted downstream of the suction article, 3) two multi-element hot-film sensor arrays oriented spanwise on either side of the attachment line for attachment line boundary

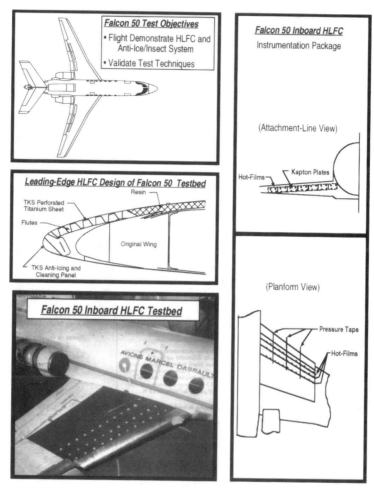

FIGURE 23.117 Dassault Falcon 50 hybrid laminar flow flight demonstrator. (From Bulgubure *et al.*, 1992 and Arnal *et al.*, 1991.)

layer state detection (used only during the leading-edge transition measurements), 4) infrared and video capability, 5) sensors for free-stream turbulence measurements, and 6) velocimeters coupled with static pressure taps for flute mass flow inference.

Results. It was demonstrated that the system was very effective for insect avoidance. During low-altitude flight tests over insect infested areas, the port (untreated) side of the aircraft had 600 insects per square meter impact the leading edge in the region of interest, whereas on the starboard side treated with the MPG fluid, no insect contamination was noted.

The effectiveness of the Gaster bump for turbulence contamination avoidance along the attachment line is illustrated in Fig. 23.118. With boundary layer suction and without the bump, the whole test article was observed to be turbulent. For various Reynolds number and sweep angle combinations, the best case revealed only a very small area of intermittent boundary layer outboard on the test article (Fig.

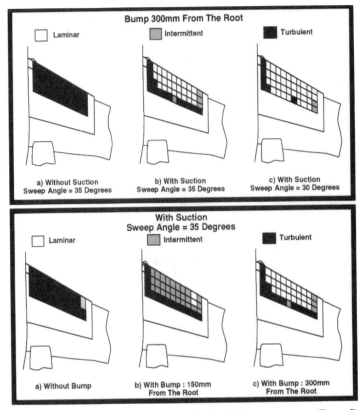

FIGURE 23.118   Flight test results on Falcon 50 HLFC demonstrator. (From Bulgubure *et al.*, 1992.)

23.118). With the Gaster bump installed on the leading edge at 150 mm from the side of fuselage and with the same suction rates as in the case of no bump, it was observed that the boundary layer was mostly intermittent, a significant improvement over the first case (Fig. 23.118). Finally, when the bump was moved to a position 300 mm from the root, most of the test article became fully laminar, as shown in Fig. 23.118. In this configuration, when the sweep angle was reduced to 30° by sideslipping the aircraft, a further slight improvement was observed in the extent of laminar flow (Fig. 23.118). As expected, when the boundary layer suction was eliminated, the flow over the test article became completely turbulent.

23.10.5   European Natural and Hybrid Laminar Flow Nacelle Demonstrator

A program was conducted with the goal of investigating the prospects of achieving extensive laminar flow on aircraft engine nacelles in flight [Barry (1994)]. The test vehicle chosen was the VFW-614/ATTAS aircraft which has twin Rolls/Snecma M45H Turbofans, as shown in Fig. 23.119. The program had the goals of: 1) demonstration of drag reduction with NLF and HLFC applied to a nacelle in flight, 2) verification of CFD design methodology, 3) verification of manufacturing techniques for laminar flow surfaces, and 4) validation of insect contamination avoidance by transpiration. For the natural laminar flow portion of the test program, two new

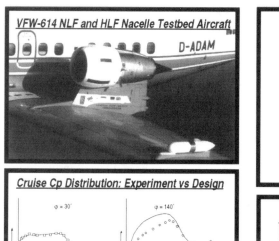

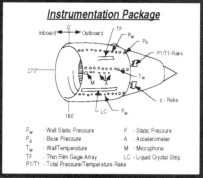

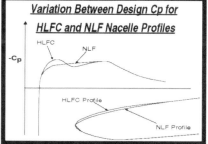

FIGURE 23.119 NLF and HLFC nacelle flight demonstration on the VFW-614 testbed. (From Barry, 1994.)

composite nacelles were constructed—one which consisted of the baseline nacelle lines, and another with a new set of aerodynamic lines. The flight test portion of the program consists of two phases. The first tested the composite NLF design, and the second tested the new composite baseline nacelle.

A third nacelle designed for validation of the hybrid laminar flow concept was built which included a liquid transpiration insect contamination avoidance system [see Humphreys (1992)]. The comparison between the NLF and HLF nacelle pressure distribution is also shown in Fig. 23.119 along with the required nacelle aero lines for each concept. Both profiles were carefully designed with advanced CFD methods, and they were required to meet acceptable low-speed performance characteristics for takeoff and landing. Stringent laminar flow design criteria in terms of surface waviness and steps were applied in each case.

The flight test portion of the program demonstrated that laminar boundary-layer flow was achievable over 60% of the nacelle length in the installed environment over a large range of flight conditions for both laminar flow concepts tested. It was noted that noise and vibration had little or no effect on the ability to achieve laminar flow for this design. The liquid transpiration-styled insect contamination avoidance system was operated successfully during the course of the flight testing.

23.10.6 European Laminar Flow Investigations

The European Laminar Flow Investigation (ELFIN) project, consisted of four primary elements, each of which concentrated on the development of laminar-low technology for application to commercial transport aircraft: 1) a transonic wind tunnel evaluation of the hybrid laminar flow concept on a large scale model, 2) the development of a boundary layer suction device, as well as the development of new wind tunnel and flight test techniques, 3) development of improved computational methods for laminar to turbulent flow predictive capability, and 4) a partial-span flight demonstration of natural laminar flow utilizing a foam and fiberglass glove over an existing wing structure [Birch (1992)].

Tasks 1 and 2 of the program called for a transonic wind tunnel evaluation of the hybrid laminar flow concept, evaluation of wind tunnel test techniques and development of viable boundary layer suction devices. A 1:2 scale model of one of the VFW-614 wings capable of leading-edge suction was built and tested in the French S1MA transonic tunnel [Schmidt *et al.* (1993)]. The model had a span of 4.7 m and a mean chord of 1.58 m. A schematic of the test setup is shown in Fig. 23.120, including a detailed view of the ejector system used to provide the boundary layer suction.

The perforated leading edge was built into the midspan region of the wing and was about 0.95 m in span and provided boundary-layer suction aft chord wise to about 15% chord on both the upper and lower surface. The titanium outer skin was 0.9 mm thick and had holes which were 40 microns in diameter spaced 0.5 mm apart. The suction surface had an average porosity of 44%. As shown in Fig. 23.120, the leading edge consisted of 38 suction flutes ganged to 17 collection ducts. The suction flow rate through each collection duct was individually controlled and measured.

In Fig. 23.120, the chord-wise transition location measured with infrared ther-

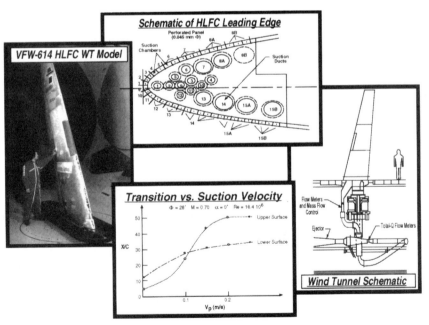

FIGURE 23.120 ELFIN large-scale HLFC wind-tunnel investigation in ONERA S1MA. (From Ledy *et al.*, 1993.)

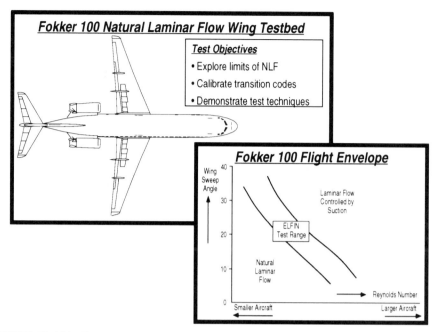

FIGURE 23.121 ELFIN Fokker 100 natural laminar flow test objectives and flight envelope. (From Birch, 1992, Mecham, 1992, Moxon, 1991, and Hemker, 1991. Portions reprinted with permission of Reed Business Publ.)

mography as a function of suction flow velocity for a given transonic test condition is illustrated. As the suction flow velocity is increased from zero, the transition front moves aft, an indication of successful control of the boundary layer disturbance growth rates.

Building upon the VFW-614/ATTAS NLF wing flights tests, the ELFIN project team embarked upon Task 4, choosing a Fokker 100 (Fig. 23.121) transport aircraft for the partial-span NLF demonstration [Birch (1992), Mecham (1992), Moxon (1991), and Hemker (1992)]. The main objectives of this task were to measure the drag reduction associated with a natural laminar flow wing design, validate laminar flow CFD methodology, and establish the upper limits (transition Reynolds number for a given leading-edge sweep angle) of NLF. As seen in Fig. 23.121, the geometry and flight envelope of the Fokker 100 allows for the exploration of the NLF boundary.

The starboard wing was modified with a full-chord, partial-span natural laminar flow glove which was bonded to the original wing surface and located between the flap track fairings as shown in Fig. 23.122. Glove construction was of foam and fiberglass; it was prefabricated in a very accurate mold and designed to provide over 50% chord laminar flow. A wake rake was mounted aft of the NLF test section so that local drag reduction could be inferred from wake pressure profile measurements. The aircraft was instrumented with two infrared cameras for boundary layer transition detection, one above and one below the wing (see Fig. 23.122). To facilitate transition detection with the IR cameras, a carbon-fiber heater mat was embedded in the glove to enhance the heat transfer process on the surface. The glove was further instrumented with chordwise, flush static pressure taps and embedded hot

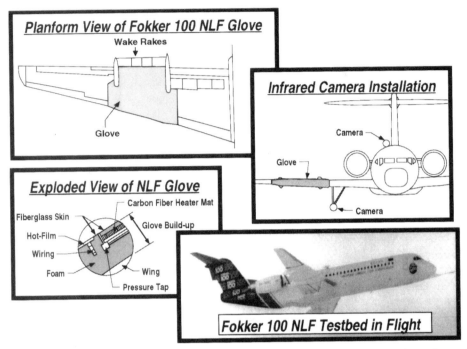

FIGURE 23.122 ELFIN Fokker 100 NLF instrumentation layout, glove schematic, and test bed. (From Birch, 1992, Mecham, 1992, Moxon, 1991, and Hemker, 1992. Portions reprinted with permission of Reed Business Publ.)

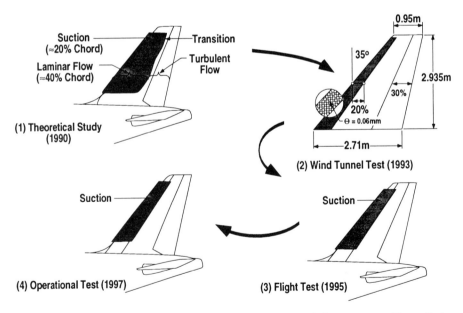

FIGURE 23.123 A320 hybrid laminar flow control vertical fin program. (From Robert, 1992.)

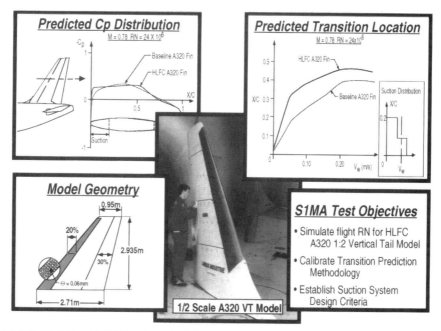

FIGURE 23.124   A320 HLFC vertical fin analysis and wind-tunnel investigation in ONERA S1MA (From Robert, 1992, Redeker *et al.*, 1992, and Thibert *et al.*, 1990).

films for boundary layer flow physics measurements—an additional means of detecting transition. The Fokker 100 flight-test phase produced results which confirmed predictions of 15% drag reduction, thereby validating preflight high-speed, wind tunnel investigations.

Finally, the vertical fin of an A320 aircraft was chosen as a candidate for evaluation of the feasibility of HLFC because of the availability of an aircraft for flight testing, simple installation, no de-icing system, attainment of flight Reynolds number in an existing wind-tunnel facility (S1MA at Modane), and minimized cost [Robert (1992), Redeker *et al.* (1992), and Thibert *et al.* (1990)]. The arrangement is given in Fig. 23.123. Analysis of the pressure distribution of the existing A320 vertical tail and a proposed HLFC A320 vertical tail are shown in Fig. 23.124. Results indicate that laminar flow is achievable to about 40% chord of the baseline A320 fin, and about 50% chord of the HLFC A320 fin for reasonable amounts of boundary layer suction. Further analysis of these results revealed that 1.0–1.5% aircraft drag reduction is possible by laminarizing the vertical fin.

23.11   CAVITATION
Frederick G. Hammitt

23.11.1   Nature of Cavitation

General Background.   An important problem for liquid handling devices with low pressure regions, is *cavitation*. Coined by R. E. Froude, but suggested by Euler (1754), it is similar to local *boiling*, providing bubbles and vapor regions in a liquid, due to local pressure reduction below the vapor pressure rather than increased temperature. There are two important consequences: 1) damage to surrounding structure from bubble collapse, and 2) deterioration of machine performance (decrease of output or efficiency) due to modification of flow streamlines around vapor regions. Two basic questions are: will cavitation occur, and if unavoidable, can a device still perform its function acceptably?

Bubble growth and collapse may be at a nominal rate, if caused by gas diffusion or gradual pressure change. However, cavitation bubbles may be explosive if caused by vaporization effects. Collapse occurs implosively if pressure is increased rapidly (position change or vibration effects) with small gas content, and less so for larger gas content. Such collapse can damage nearby structure.

If pressure is maintained for sufficient time below a *critical pressure*, depending on liquid flow conditions, cavitation will occur. If not, cavitation will not occur. Cavitation involves appearance and disappearance of cavitites (or *bubbles*) in a liquid. Since the bubbles are *empty*, the bubbles can play no part in the phenomenon, which then depends entirely on liquid behavior. This model is not entirely valid for initial *nucleation* and final *collapse*.

Cavitation may occur for liquids in motion or at rest, since pressure oscillations may also be due to surface vibration or acoustic radiation. It may occur in the body of liquid or along walls.

Cavitation erosion occurs at the point of bubble collapse rather than inception. These points sometimes coincide.

Many reviews of the very extensive literature and books exist [Arndt (1981), Acosta (1974), Thiruvengadam (1974), Roberston and Wislicenus (1969), Knapp *et al.* (1980), Hammitt (1980a), Acosta and Parkin (1975), and Pearsall (1972)].

Caviation Effects. Major effects are discussed here according to their importance.

Performance Effects: A substantial quantity of cavitation (*degree* or extent) in flow passages will alter streamlines and thus machine performance. The effects of cavitation are minimal for conditions near inception (nucleation), since only a few bubbles will not affect the overall flow. For extensive (*well-developed*) cavitation, a large portion of the low-pressure region becomes vapor filled. The velocity in the remainder of the passage is then increased, and pressure is further decreased giving added evaporation. Distortion of flow pattern and energy losses caused by the two-phase regime can then cause a sudden decrease in overall machine performance (head, efficiency, etc). Figure 23.125 shows the performance of a typical centrifugal pump for 3 flow coefficients as functions of inlet suppression pressure, and Fig. 23.126 shows impeller inlet passage flow patterns.

The most common nondimensional parameter to correlate cavitation performance is the *cavitation number, K,* or *cavitation sigma,* σ. Then

$$K = \sigma = p_{sv}/(\rho V^2/2) = \text{NPSH}/(V^2/2) \qquad (23.211)$$

where V is reference velocity; ρ, liquid density; p_{sv}, suppression pressure; NPSH net positive suction head. Note that $p_{sv} = \rho \cdot \text{NPSH} = p - p_v$, where p_v is vapor pressure.

Another important parameter applied especially to pumps is *suction specific speed,*

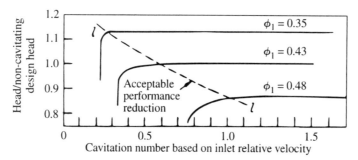

FIGURE 23.125 Effect of cavitation on performance of centrifugal pump. (Reprinted with permission of Simon & Schuster, Inc., from the Merrill/Prentice Hall text, Sabersky, Rolf H., Acosta, Allan J., and Hauptman, Edward G., *Fluid Flow, A First Course in Fluid Mechanics*, 3rd ed., Macmillan College Publishing Co., 1989; Data computed from Blom, C., "Development of Hydraulic Design for the Grand Coules Pumps," *Trans. ASME*, 72, 1950.)

S. The larger S, the more probable is cavitation. In English units, $S > 8000$ for centrifugal pumps implies possibility of cavitation. S is analogous to *specific speed*, except NPSH is substituted for head rise. S reflects pump inlet conditions rather than entire pump performance, assuring similarity for the inlet. $S = NQ^{1/2}\text{NPSH}^{-3/4}$, where N is the rotating speed (RPM), and Q is the volumetric flow rate (GPM). S is not nondimensional unless converted to consistent units.

Because of so-called *scale effects* and [see Hammitt (1980a), (1972) and Bonnin *et al.* (1981)], conventional similarity parameters cannot be applied precisely, for reasons not well understood. Divergence from classical scaling laws pertinent to single-phase flow is due to great complexity and the lack of understanding of any multiphase flow, and cavitating flows in particular. Scale effects occur for velocity, size, p_{sv}, fluid property, and temperature changes. Fluid property and temperature effects, termed *thermodynamic effects*, are discussed later.

Figure 23.127 from Hammitt (1980a) shows effects on inception and also well-developed cavitation *sigma*, σ, for water orifice flows, of orifice shape, Reynolds

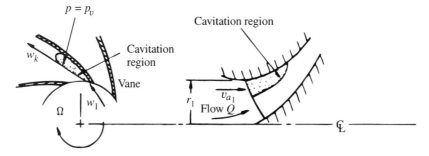

FIGURE 23.126 Sketch showing probable location of cavitation in a centrifugal pump. (Reprinted with permission of Simon & Schuster, Inc., from the Merrill/Prentice Hall text, Sabersky, Rolf H., Acosta, Allan J., and Hauptman, Edward G., *Fluid Flow, A First Course in Fluid Mechanics*, 3rd ed., Macmillan College Publishing Co., 1989; Data computed from Blom, C., "Development of Hydraulic Design for the Grand Coules Pumps," *Trans. ASME*, 72, 1950.)

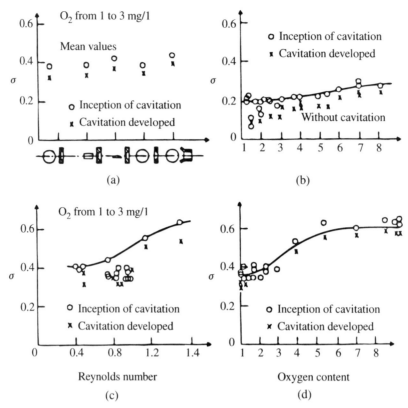

FIGURE 23.127   Influence on inception sigma of Reynolds number and oxygen content for various orifice shapes. Curve (a) shows various orifice shapes, and curves b, c, and d are for circular orifices. Tests reported J. Duport, SOGREAH, Grenoble, France. (Reprinted with permission, Hammitt, F. G., *Cavitation and Multiphase Flow Phenomena*, Adv. Book Series, McGraw-Hill, New York, 1980a.)

number, Re, and air content. The Re effect is from changes in velocity and diameter. Inception σ increases by ~50% for an increase of Re by $4X$, but σ for well-developed cavitation is little affected.

Figure 23.128 from Hammitt (1980a) shows velocity effects on *inception* σ for flow of water or sodium in a Venturi for various gas contents. For well deaerated water, sigma increases substantially for increased velocity, consistent with Fig. 23.127 for orifices. For high gas content, σ first decreases strongly, then increases for continued velocity increase. The curves converge at high velocity. Gas content effects are expected, but not those for velocity. Tests with sodium gave similar results [Hammitt (1980a)]. Turbopump results are also often similar. Optimum performance (maximum S) may occur at intermediate speeds, and S often decreases for higher speeds. Such results occur for any test liquid. Figure 23.129 shows data from a low specific speed pump for mercury and water. Strong S increase ($2X$) occurred for $4X$ RPM increase, with little difference between mercury and water, when plotted against RPM rather than Re. Thus, Re was not a good correlating parameter. No data is available separating the effects of velocity and diameter.

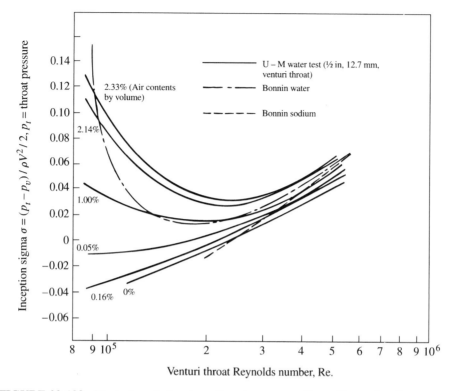

FIGURE 23.128 Venturi cavitation inception sigma, σ vs. Reynolds number for water and molten sodium. Water tests at Univ. Mech. Sodium tests, Elec. de France, J. Bonnin. Air content effect shown for water. (Reprinted with permission, Hammitt, F. G., *Cavitation and Multiphase Flow Phenomena*, Adv. Book Series, McGraw-Hill, New York, 1980a.)

Substantial *thermodynamic scale effects* for sigma and damage exist [Bonnin *et al.* (1981), Stahl and Stepanoff (1956), and Stepanoff (1961)] for changes in fluid properties or temperature. The mechanism is the large increase in vapor density with temperature for any fluid. Taking water, e.g., for *cold* water, vapor density, and bubble thermal content are negligible. Thus, bubble growth and collapse are inertially dominated, as in the classical Rayleigh (1917) analysis, and heat transfer restraints are negligible. For *hot* water (>100 C, e.g.), vapor density is increased ~40X, and the thermal content of a bubble is increased by a similar ratio. Rapid growth or collapse is then restrained by heat transfer as well as inertial effects. Both are reduced compared to the Rayleigh model. *Hot-water cavitation* is thus very similar to sub-cooled boiling. Boiling is, therefore, not damaging due to *thermodynamic effects* and the lack of high pressures to drive collapse.

Similar effects occur comparing *cold* water with fluids such as cryogenics, petroleum products, chemical process fluids, freons, liquid metals, etc. Comparative temperatures depend on physical properties, e.g., low for cryogenics, high for liquid metals, and intermediate for petroleum products, freons, etc. Cavitation effects on performance and erosion are strongly reduced. Thus, 200 C water may not be damaging at all. Experimental data confirms this hypothesis [Hammitt (1980a)].

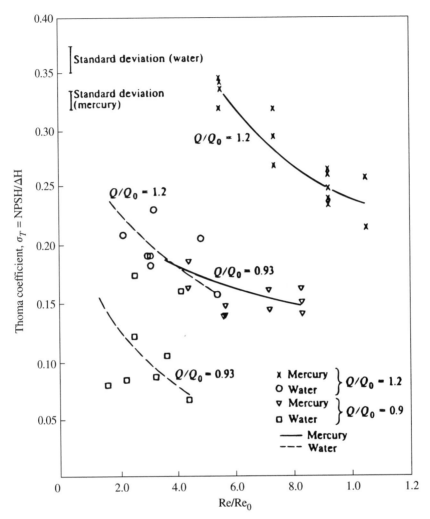

FIGURE 23.129 Thomas cavitation parameter vs. normalized Reynolds number for low specific speed centrifugal pump, Univ. Mich. tests, for mercury and water. Re_0 for pump design condition. (Reprinted with permission, Hammitt, F. G., *Cavitation and Multiphase Flow Phenomena*, Adv. Book Series, McGraw-Hill, New York, 1980a.)

Erosion Damage Effects: Cavitation damage has been important since the early 1900s, when first seen on British torpedo-boat propellors [Hammitt (1980a)]. The cause was high propeller speeds when turbines replaced reciprocating engine drives. The work of Rayleigh (1917) showed that spherical collapse of vapor cavities in *ideal* liquids caused *infinite* pressures and velocities at the site of collapse. Later analyses [see Hammitt (1980a)] showed that the Rayleigh model is not entirely valid, especially concerning spherical symmetry near a wall, where the Rayleigh model still shows correctly the damage potential. The point of damage is thus that of collapse instead of inception.

In the past 50 years, studies have been made to study spherical bubble collapse in *real* liquids, and nonspherical collapse near a wall, in pressure gradients, and with viscosity [Ivany and Hammitt (1965), Hickling and Plesset (1964), Plesset and Chapman (1971), and Mitchell and Hammitt (1973)]. The effects in real fluids are shown in Fig. 23.130. Bubble shapes (from high-speed movies) for collapse near a wall are shown in Fig. 23.131.

It is generally agreed that damage is due to both liquid *microjet* impact for non-symmetrical collapse (see Fig. 23.131) and shock waves from *rebounding* bubbles [Hammitt (1980a) and Hickling and Plesset (1964)] which grow again after collapse due to compressed internal gas and vapor. Such rebounds have been often observed

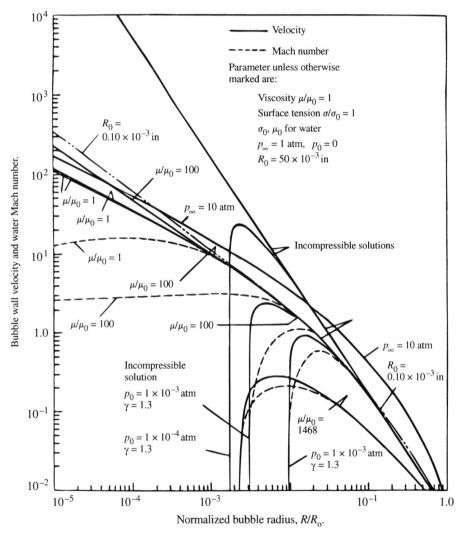

FIGURE 23.130 Spherical bubble collapse velocity and Mach number vs. normalized bubble radius. Compressibility, viscosity, and surface tension effects. (Reprinted with permission, Ivany, R. D. and Hammitt, F. G., "Cavitation Bubble Collapse in Viscous Compressible Liquids," *J. Basic Eng.* Vol. 87, D, pp. 977–985, 1965.)

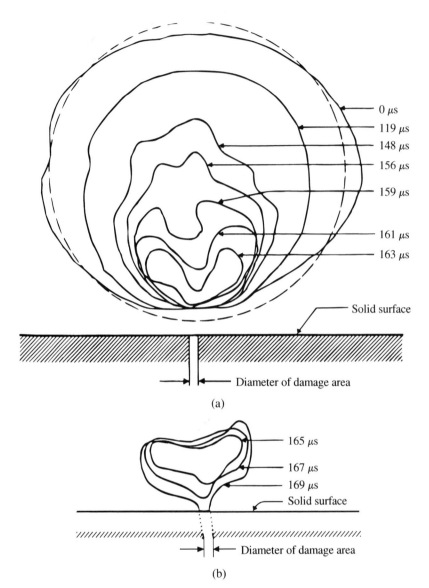

FIGURE 23.131   Venturi bubble collapse near wall of single (spark-generated) originally spherical bubble. High-speed cinematography. (Reprinted with permission, Kling, C. L. and Hammitt, F. G., "A Photographic Study of Spark Induced Cavitation Bubble Collapse," *J. Basic Engr.*, Vol. 94, No. 4, pp. 825–833, 1972.) (*a*) Initial portion bubble collapse. (*b*) Final portion bubble collapse showing microjet.

in water tunnel tests [Kling and Hammitt (1972)]. This disagrees with the Rayleigh (1917) model, which predicts damage from shock waves from spherically-symmetric collapse. High-speed movies show that, near walls, spherical collapses do not occur, and this is confirmed by computer studies [Kling and Hammitt (1972), Plesset and Chapman (1971), and Mitchell and Hammitt (1973)]. Potential intensity of such mechanisms can explain damage for most materials.

All materials (even the hardest and strongest) can be damaged by cavitation. Resilient materials such as elastomers may have a good resistance, preventing microjet formation [Hammitt (1980a)]. Still in most cases, harder materials are more resistant. This effect may be reduced if corrosion occurs. The combination of corrosion with *mechanical* cavitation can produce much more damage than each alone.

The best, most available, and easiest to apply material property is hardness (Brinell, *BHN*, or other). It has been traditionally so used. It appears, however, [Hammitt (1980a), (1980b) and Hammitt and Heymann (1975)] that better correlations are often found with more complex parameters as *ultimate resilience*, *UR*, for which theoretical backing and dimensional consistency exists, since *UR* is failure energy per volume for brittle fracture. $UR \cong (TS)^2/2E$, where *TS* is tensile strength and *E* is the elastic modulus. Then $MDPR^{-1} \propto UR^n$. MDPR is *mean depth of penetration rate* or volume loss rate/exposed area. *n* must be determined experimentally, but *n* $\cong$ 1 is expected. Data shows that $n \cong 1$ is often the best-fit exponent [Hammitt (1980a), (1980b), Hammitt and Heymann (1975), and Heymann (1968), (1969)]. Since $BHN \propto TS$ for metals, $n = 1$ corresponds to hardness exponent $= 2$, whereas 1.89 is reported [Hammitt (1980a), (1980b), Hammitt and Heymann (1975), and Heymann (1968), (1969)]. Discrepancy is not significant. Figure 23.132 shows data for this large group of materials, combining cavitation and liquid impact data. Standard deviation is $\sim 3X$, typical of fluid erosion data. Since liquid impact and cavitation are very similar, data fits for metals may be applied to both.

The discrepancy from best-fit curves for some materials is $\sim 10X$, due to micro-characteristics not reflected by conventional mechanical properties. Satellite 6-B, e.g., is much more resistant than expected from *BHN* or *UR* values. Microstructures are important, and mechanical properties, due partially to their relatively low loading rates, are not fully pertinent, since the damage mechanism occurs in a few microseconds. Little confidence can be placed in damage rate correlations for untested materials. The large data scatter is not surprising, since the range of resistances, $MDPR^{-1}$, is 10^3–10^4, between the most and least resistant.

Vibrations and Cavitation: Major machinery vibrations, sometimes damaging to structures, can occur. These involve macro- rather than the micro-loading for usual erosion. The vibrations are due to large and varying forces on a structure due to the collapse of large cavities, or smaller bubbles in phase. The large noise emission can also prevent machine operation. Little published data exists.

In some cases (diesel engine liners, e.g.), vibration-induced cavitation is the important damage mechanism [Zhou *et al.* (1982a), (1982b), and Speller and LaQue (1950)]. Vibratory cavitation also provides an essentially zero-flow test device to compare materials and fluid conditions [Hammitt *et al.* (1970) and ASTM (1977)]. It consists of a *vibratory horn*, to which damage specimens (buttons) are attached. The Michigan unit shown in Fig. 23.133 can be used over large temperature and pressure ranges, though most are for room temperature and pressure only.

The *vibratory* test and others such as *rotating disc, Venturi*, etc., provide highly accelerated erosion so that corrosion is suppressed compared to *mechanical* cavitation. Thus, the results cannot be well applied for prediction of field damage rates which are usually much smaller. This is a major difficulty in the present state-of-the-art.

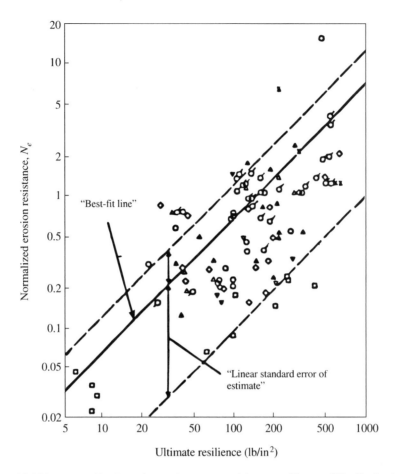

FIGURE 23.132   Normalized erosion resistance vs. ultimate resilience, UR. Cavitation and liquid impact data. UR = $TS^2/2E$. (Reprinted with permission, Hammitt, F. G., *Cavitation and Multiphase Flow Phenomena*, Adv. Book Series, McGraw-Hill, New York, 1980a.) (Also see Heymann, 1968, 1969.)

Noise: Noise inevitably accompanies cavitation and is important in naval applications such as submarines for detection avoidance. Cavitation also limits the transmitted power of sonar transducers. Cavitation produces essentially *white* noise to high frequencies (order of 10 MHz), since the damaging portion of bubble collapse often requires ~ 1 μs [Plesset and Chapman (1971), Mitchell and Hammitt (1973), and Kling and Hammitt (1972)]. Noise studies are limited by instrumentation so that maximum frequency is unknown. Most of the noise is from collapse rather inception. Thus, little noise occurs for boiling where violent collapse is absent.

Hydrophones can be used to detect inception and also to correlate noise with damage. A low frequency filter is useful to attenuate ordinary flow and machine noise, which do not usually have the very high frequency components of cavitation. A cut-off frequency of 40 kHz is suitable.

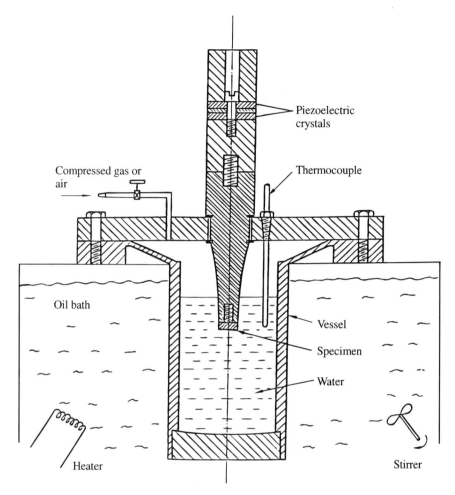

FIGURE 23.133 Univ. Mich. vibratory cavitation erosion test facility, 20 kHz, 14.3 mm (0.562 in.) specimen diameter. Amplitude to 3 mils (75 μm).

23.11.2 Causes and Prediction of Cavitation and Erosion

Cavitation Flow-Regime Prediction (Performance Effects). For stationary devices, inception and development (*degree* or *condition*) is best correlated by cavitation number, K [or *sigma* (σ)], i.e., the ratio of local suppression pressure, p_{sv} to kinetic pressure, $\rho V^2/2$, or the ratio of *net positive suction head, NPSH* to kinetic head, $V^2/2$, usually referred either to upstream values, or those values at the point of cavitation. Increased sigma means less cavitation.

For rotating machines other parameters are used, e.g., suction specific speed, S (already discussed) and *Thoma parameter*, σ_T, i.e., NPSH/ΔH. The Thoma parameter predates S. It is less useful, reflecting total performance rather than the low-pressure region only. It is, thus, affected by more variables, such as conventional specific speed.

Scale Effects and Modeling: Flow modeling, using similarity parameters discussed above, may be very imprecise due to substantial *scale effects*, involving *scale* (size), velocity, pressure, and temperature (*thermodynamic effects*). The mechanisms are only partially understood, and the details are too complex to present here. The mechanisms can best be evaluated from papers listed in Hammitt (1980a). Figures 23.127–23.129 are typical.

Without *scale effects*, inception σ and S would be constant for a given machine. In fact, inception σ and S usually vary substantially as shown in Figs. 23.127–23.129.

Impurities and Gas Content: The greatest effects are upon inception σ, through pressure thresholds for nucleation. For zero particulate and gas content, and perfect wetting, cavitation and boiling would not usually occur, since the tensile strength of pure liquids would prevent nucleation. That they do occur is proof of a *stress-relating* mechanism such that small under pressures do cause cavitation. Nucleation is assumed to be due to unwetted acute angle micro-crevices in which gas cavities exist. Nucleation occurs in the liquid from micro-particles with such nonwetted micro-crevices. This is the *Harvey mechanism* [Harvey *et al.* (1947)].

The Harvey mechanism is needed to provide a stable source of *nuclei*. Smaller bubbles would quickly dissolve, since the pressure in the bubbles is greater than for the liquid due to surface tension. Larger bubbles would be removed by buoyancy or centrigual effects. Other mechanisms such as a particle shell around a *nuclei*, or the combined effects of solution and vaporization [Chu (1981)] have been suggested. It is still assumed that the Harvey mechanism is of primary importance. Wetted particles have no effect, and nonwetted particles are similar to gas bubbles.

Dissolved gas has little effect, since solution proceeds slowly. Thus, only entrained gas is important, but measurement thereof is difficult, and no standard method exists. Presumably, entrained gas increases with total gas content. Laser-light scattering seems the best approach for its measurement [(Keller (1972)] but is complex. Entrained gas is only a small part of the total, so its measurement alone may be misleading. Total gas can be measured easily, e.g., Van Slyke apparatus [Hammitt (1980a)]. Suitable automated instruments are available.

Cavitation Damage Prediction: State-of-the-Art. Cavitation damage prediction until recently was seldom attempted. Rather, cavitation was to be avoided, or if that was impossible, performance of similar machines was applied to new designs to assure suitable operation. Damage prediction is then essentially empirical. Damage can often be repaired relatively easily (e.g., by welding in new material). However, in new applications, as nuclear power-plants, rocket pumps, etc., past procedures may not be suitable. Thus, more precise information is needed, and improved prediction capability is essential.

Zero damage is impossible to guarantee for the present the state-of-the-art, unless p_{sv} is set prohibitively high. Though NPSH vs. head curves can be measured accurately, the NPSH necessary to entirely eliminate bubble collapse is usually not known. It may be $10X$ (Fig. 23.134) that for usual inception point (3% head drop-off). In most cases, the needed NPSH for *zero cavitation* is 2–4X the *inception* value, and much greater in some cases. Prediction of the ratio is not now possible.

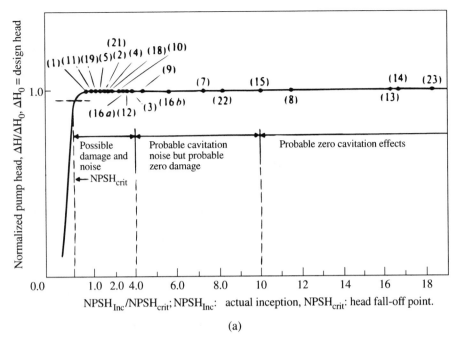

(a)

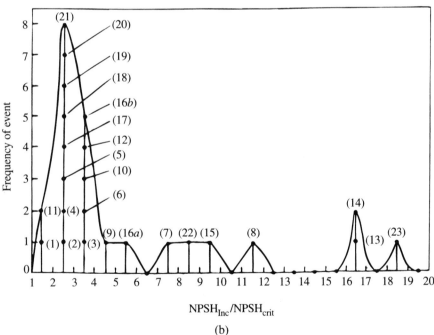

(b)

FIGURE 23.134 Normalized pump head vs. NPSH for actual inception (acoustic or visual) compared to head fall-off, and frequency of events. (a) Normalized pump head vs. NPSH. (b) Frequency of event vs. ratio of inception to critical NPSH. (Reprinted with permission, Hammitt, F. G., *Cavitation and Multiphase Flow Phenomena*, Adv. Book Series, McGraw-Hill, New York, 1980a.)

Damage Prediction Methods: Several quite imprecise methods exist. These are discussed in the order of probable utility.

Consider first the characteristic erosion curve and incubation period. Cavitation (or liquid impact) erosion often follows a *characteristic (S-shaped) curve*, as in Fig. 23.135. An *incubation period*, *IP* (damage rate very small) is followed by a zone of increasing rate, and then a maximum rate period ($MDPR_{max}$). This may persist for some time, depending on other parameters. Rate then usually decreases, sometimes reaching an eventual *steady-state* value. However, 2nd or 3rd maxima sometimes occur. Use of a characteristic curve for prediction is a rough approximation at best, unless the curve can be verified experimentally for each given case. An approximate relation, $MDPR_{max}^{-1} \propto IP^{n}$, exists [Hammitt (1980a), (1980b) and Hammitt and Heymann (1975)], where n ranges from ~ 0.7 and ~ 1.2. It depends on test parameters, type device, material, etc. Vibratory and Venturi tests [He and Hammitt (1981)] for 5 materials (carbon steels, aluminum alloys, and SS-316), showed $n = 0.94$ (± 0.1). If the unity exponent is approximately valid, the utility of the method would be increased.

Tests to determine *IP*, based on the appearance of significant pitting or other erosion, may be feasible in prototype machines using soft inserts (AL-1100-0, e.g.) in the expected damage region. Laboratory tests, such as vibratory, could compare *IP* and the remainder of the erosion curve, including $MDPR_{max}$, for a soft insert and prototype material. Assuming that the material resistance ratio between lab and pro-

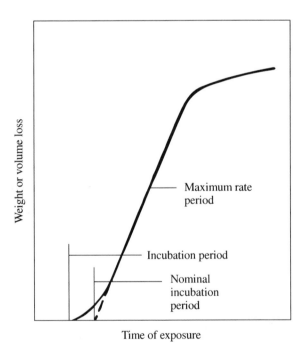

Time of exposure

FIGURE 23.135   Characteristic cavitation or liquid impact "S-shaped" erosion curve. (Reprinted with permission, Hammitt, F. G., *Cavitation and Multiphase Flow Phenomena*, Adv. Book Series, McGraw-Hill, New York, 1980a.) (Also see Hammitt, 1980b.)

totype is reasonably valid, *IP* and $MDPR_{max}^{-1}$ for prototype, and their time of occurrence, can be predicted. A *characteristic curve* can be measured in the laboratory and assumed valid for a prototype.

Consider second the predictions of scale and *thermodynamic effects*. Several important *scale effects* exist between model and prototype for erosion and performance. The most important involves velocity, *V*. Many tests show a strong damage increase with increased *V*. Assuming $MDPR_{max} \propto V^n$, the usual value is $n \sim 6$ [Hammitt (1980a) and (1980b)]. However, it obviously depends on the relation between p_{sv} and *V*, which depends on geometry, etc. The value of p_{sv} in the collapse region sometimes may not increase with *V*. Then, the damage increase with *V* will be a minimum. The range of *n* is usually $0 < n < 10$, but even negative values have been reported. Thus, the damage-exponent model is generally not valid. Still, increased velocity can produce catastrophic damage, where little existed before. Thus, damage tests at prototype velocity are vital.

Continuing the discussion consider suppression pressure, p_{sv}, effects. A maximum-damage p_{sv} exists (Fig. 23.136) for all machines, including vibratory, between low values (low collapse pressure, as in boiling) and high (cavitation suppressed). Conventional machines operate on the right side of the curve in Fig. 23.136, so increased p_{sv} usually reduces damage, but not always. Maximum damage often occurs near inception and then decreases for reduced NPSH. Vibratory test is on the left side of the curve, and damage increases about as p_{sv}^2.

Another factor is the diameter (size) effect. Little information is available, though it appears that diameter-damage exponents are in range 3–4 for total volume loss, giving an *MDPR* exponent of 1–2.

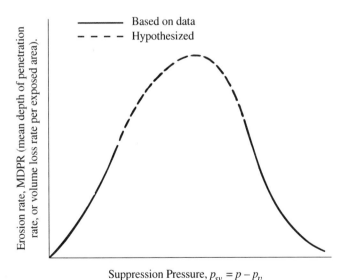

$$\text{Suppression Pressure, } p_{sv} = p - p_v$$

FIGURE 23.136 Cavitation erosion rate (volume or weight loss) vs. suppression pressure, p_{sv} ($= p - p_v$) or net positive suction head, NPSH. (Reprinted with permission, Hammitt, F. G., *Cavitation and Multiphase Flow Phenomena*, Adv. Book Series, McGraw-Hill, New York, 1980a.) (Also see Hammitt, 1980b.)

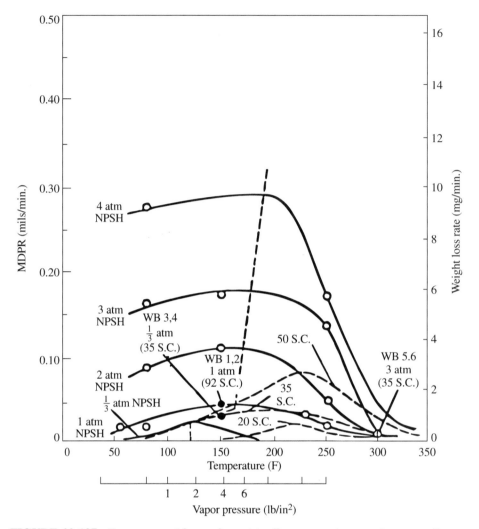

FIGURE 23.137 Temperature (*thermodynamic*) effect on erosion rate in water vibratory test. Maximum MDPR (mean depth of penetration rate) and weight loss rate vs. temperature and vapor pressure for bearing brass (SAE-660) at various static NPSH values (1, 2, 3, and 4 bar). Curves of constant subcooling (S.C.) also shown. (Reprinted with permission, Hammitt, F. G., *Cavitation and Multiphase Flow Phenomena*, Adv. Book Series, McGraw-Hill, New York, 1980a.) (Also see Hammitt, 1980b.)

Examine now thermodynamic (temperature) effects; a maximum damage temperature exists for any fluid (Fig. 23.137). Damage decreases for higher temperatures (thermodynamic effects), and less so for lower temperatures. The high temperature effects are best correlated by Stepanoff's B-factor. Figure 23.138 shows damage predictions with modified B for various fluids. Figure 23.137 shows that damage in water may not be important for $T \gtrsim 200$ C. Of course, temperature also affects both material properties and corrosion rates.

Information concerning gas content and impurity effects is now presented. En-

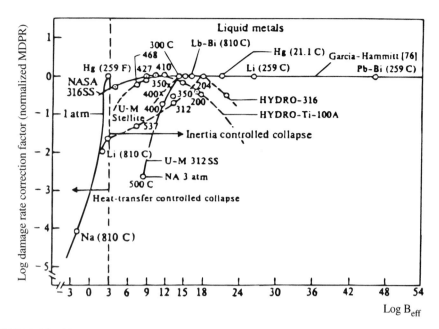

FIGURE 23.138 Effect of modified Stepanoff thermodynamic parameter, B_{eff} on cavitation erosion rate for various liquid metals and test materials, showing strong damage decrease at high temperature. (Reprinted with permission, Hammitt, F. G., *Cavitation and Multiphase Flow Phenomena*, Adv. Book Series, McGraw-Hill, New York, 1980a.) (Also see Hammitt, 1980b.)

trained gas facilitates nucleation, increasing bubble size and number. Added gas (entrained or total) increases erosion through this effect. However, increased gas in bubbles weakens collapse (Fig. 23.130) by a cushioning effect. Air injection into the collapse region has been used to reduce damage. While information is scanty, Fig. 23.139 shows the presumed overall effects [Hammitt (1972)]. Gas content near zero should reduce damage by suppressing inception by increased liquid tensile strength. Moderate gas reduces damage through the cushioning effect, and high gas more strongly. Tests exist verifying the high content effect especially [Rasmussen (1956)].

Additives could increase corrosion and other effects, and entrained particulates increase damage strongly. As an extreme case, cavitating slurries can be very damaging compared to a liquid alone, e.g., dredge pump sand-water slurries, and coal or ore-bearing pipelines. The pipeline itself and associated pumps, valves, or fittings present major erosion problems. It may then be required to design for *zero cavitation*, but little pertinent data exist.

A milder case is particulate erosion in sand-bearing rivers of hydroelectric plant components, e.g., the Yellow River in China. Vibratory tests of particulate and chemical additives on 1018 carbon steel showed a damage increase of 50% over tap water [Zhou and Hammitt (1983)].

The final approach to consider is predictions from noise measurements. Overall noise (or that in a set frequency band) has been used successfully [Lush and Hutton

(1976)] to predict and correlate change for both pumps and venturis. For pumps, maximum damage and noise often occur near conventional inception. This method seems limited to the geometry tested, since given overall noise (or noise in a fixed frequency band) could result from numerous collapses of nondamaging intensity or from a few which are much stronger and of damaging intensity.

An approach of more general applicability, tried in several labs with success [De and Hammitt (1983)] is to count and quantify individual collapses, using a suitable micro-transducer. A linear relation (Fig. 23.140) between collapse acoustic energy, so measured, and Venturi erosion rate was found. The damaging portion of bubble collapse lasts only few μs, and the loaded area is often few μm in size. A microprobe of adequate smallness and response rate is critical. State-of-the-art is 5 mm and 0.1 MHz. A 10 MHz probe and even smaller size is needed.

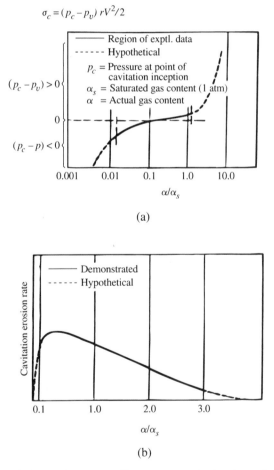

(a)

(b)

FIGURE 23.139 Air (or gas) content effects upon cavitation inception sigma and erosion rate—measured and hypothetical trends. (Reprinted with permission, Hammitt, F. G., *Cavitation and Multiphase Flow Phenomena*, Adv. Book Series, McGraw-Hill, New York, 1980a.) (Also see Hammitt, 1980b.)

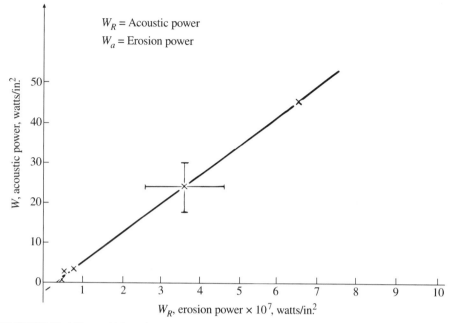

FIGURE 23.140 Bubble collapse acoustic power vs. measured erosion power for soft (1100-0) aluminum in a Venturi. *Erosion power* is the product of material failure energy rate per volume (*ultimate resilience*, UR) and eroded volume/exposed area (*mean depth of penetration rate*, MDPR). (Reprinted with permission, De, M. K. and Hammitt, F. G., "New Method for Monitoring and Correlating Cavitation Noise to Erosion Capability," *J. Fluids Eng.*, Dec. 1983.)

23.11.3 Occurrence of Cavitation

Cavitation can occur in any liquid-handling machine when p_{sv} is sufficiently low and velocity sufficiently high. There is, of course, no danger of cavitation in many cases. Important cases, rotating and stationary, follow.

Rotating Devices

Centrifugal and Axial Pumps: Pumps liable to cavitation problems are those in large powerplants, liquid-propellant rocket pumps, aircraft fuel pumps, and those driving high-speed ships. In some cases, liquids other than cold water are involved. Nuclear applications include molten sodium, and rocket pumps often use cryogenics.

Cavitating inducers (axial-flow) are often used to moderate effects for the main stage. These may involve *super cavitation*, where the collapse region is downstream of the blades. Such designs are also used for high-speed propellers (see Sec. 27.2).

Hydraulic Turbines: Cavitation is a major limitation for most hydraulic turbines, affecting the siting of the turbine and the rotating speed to obtain adequate σ. Thus, larger and more expensive units than otherwise required may be needed (refer to Sec. 27.9).

Marine Propellers: Design is limited by the need to avoid prohibitive cavitation from both performance and erosion viewpoints or effects on structure and bearings. Performance can be correlated by σ (see Sec. 27.2).

Corrosion, combined with *mechanical cavitation*, is often important. Special materials have been developed (e.g., propeller bronze) and used. Cavitation at increased depth (pressure), as for torpedoes and submarines, can cause increased damage.

Stationary Flow Devices. Erosion and performance effects occur. Damage can again be correlated by cavitation σ. Consider the following examples: 1) pipe fittings (elbows, tees, etc.) and relatively sharp pipe or tubing bends; 2) flow measuring devices such as orifices, nozzles, and Venturis, where cavitation is usually of the separated-flow type, so that while noise and performance effects exist, damage may be limited, since bubbles often collapse away from the surface. Figure 23.127 shows typical performance effects; 3) valves, which are similar to flow-measuring devices with regard to cavitation, but damage is often important, since the *collapse point* depends on geometries involved (refer to Sec. 25.3); and 4) hydrofoils, where performance and damage effects are important, and can be correlated with σ, velocity, etc.

23.12 SHIP HYDRODYNAMICS
William G. Day, Jr.

23.12.1 Introduction

The term *ship hydrodynamics* refers to the particular application of fluid mechanics to the solution of problems that are related to the motion of ships through water. As such, these problems can be treated as purely potential flows or purely viscous flows, but most often are a fully developed three-dimensional flows with both free surface and turbulent boundary layer characteristics. The solution to these problems is most often used in design and assessment of ship performance by naval architects, and most of the relevant literature can be found in the publications of that field [Comstock (1967) and Saunders (1953)].

This section will address briefly the principal areas of ship hydrodynamic resistance and flow, ship propulsion, and maneuvering and seakeeping. The latter subject areas are sometimes referred to as simply *ship dynamics*. In all cases, the complete solution to the fluid mechanics problems are not dealt with by analytical methods alone because of the complexity of the problems. However, the use of physical models and of empirical data from laboratory experiments and full-scale ship performance measurements has become an integral part of the disciplines of naval architecture.

The laboratory experiments in particular have formed the basis for understanding the physics of the fluid mechanics and have enabled the development of both empirical and purely theoretical analysis techniques. These analytical methods can then be applied to the solution of specific problems in the design and assessment of ship motion through the water.

This section will identify particular methods and experimental data that should enable the reader to understand the solution of these highly complex flows. Additional references are provided to assist those who are involved in more detailed study of ship flows.

23.12.2 Resistance and Flow

The resistance of a ship is comprised of forces due to viscous shear stress on the hull surface, a form drag component due to separation of the flow or due to interference flows between appendages and the hull, and a wavemaking component of force. In the prediction of resistance forces, all these components are calculated or estimated by empirical methods. These values for each component may be computed at various sizes or speeds, and then it is assumed that the total drag force is the sum of the components.

Viscous shear forces on the hull are referred to as *frictional resistance*. The values of frictional resistance are usually computed using empirical formulae derived from experiments in a tow tank or a wind tunnel using a long thin plank. The thin plank is assumed to produce little or no wave drag or separation drag, so the total drag is therefore due to frictional resistance. The values of frictional drag are nondimensionalized by the dynamic pressure ($\frac{1}{2}V^2$) and the wetted area. The coefficient derived is considered to be an inverse logarithmic function of the Reynolds number defined as

$$\text{Re} = \frac{VL}{\nu} \tag{23.212}$$

where V = speed, L = waterline length, and ν = kinematic viscosity. Calculations for most ship related problems use the 1957 ITTC ship-model correlation line to estimate a coefficient of frictional resistance

$$C_F = \frac{0.075}{(\log_{10} \text{Re} - 2)^2} \tag{23.213}$$

The form of the equation follows that of flat plate wall shear stress coefficients (see Chap. 4).

Form drag is usually estimated from empirical data either from wind tunnel or tow tanks. The most complete set of data are found in Hoerner (1965).

Wave resistance can be computed for some simple ship forms, but in most cases the wave resistance and form drag are combined by the Froude hypothesis into a single component known as *Residuary Resistance*. The assumption is made that the total resistance of the hull, R, is separated into frictional, C_F, and residuary, C_R, parts, i.e.,

$$C_T = C_F + C_R \tag{23.214}$$

and

$$C_T = \frac{R}{\frac{1}{2}\rho V^2 s} \tag{23.215}$$

where s = total wetted surface area. Traditionally, ship resistance has been estimated by using model towing tank data for C_R. The assumption is made that for geometrically similar forms at the same *Froude number* ($V/\sqrt{gL}$) that the residuary coefficient C_R is the same for ship and model scales. Therefore, a ship resistance is predicted by using C_R from either an experiment with a geometrically similar model or with historic data from series of systematically varied models such as the Taylor (1943) Standard Series.

It should be pointed out here that when the total resistance of a ship is estimated from model experimental data, a correlation allowance C_A is applied to account for scale effects, interactions between frictional and residuary resistance and for effects of hull roughness due to fouling or small appendages. The value of C_A is usually only five to ten percent of the total resistance coefficient (typically C_A is of the order of 0.0005 for a 400 foot ship).

Recent developments in experimental and computational methods have enabled the direct evaluation of wave-making drag and flows per past hulls and appendages. These Computational Fluid Dynamics (CFD) methods will eventually become routine tools in the design and evaluation process for ship resistance and flow problems. They have already been successfully applied in a number of ship hydrodynamic problems, most notably in the design of propellers [Morgan and Lin (1988)]. Other details of wave and wake flows have been computed for hull forms with bulbous bows in order to optimize performance at a particular Froude number. The increasing capabilities of both computer hardware and software will provide much more extensive and faster assessment of flows around ship hulls.

23.12.3 Ship Propulsion

For a ship operating at a steady speed in otherwise undisturbed water, the resistance that was discussed in the previous section must be overcome by a thrust force. The thrust force is generated by a propulsion device which converts the power from the main engines into a thrust power. Typically, the propulsion device is a marine screw propeller operating at the stern of the ship in order to take advantage of energy recovery of the boundary layer. Thrust is generated by accelerating the water around the hull to produce a net change in its momentum and thereby a force on the hull counter to the resistance force.

Many devices can be used to generate thrust for the ship, including ducted propellers, waterjets that use pumps internal to the hull and tandem, contra-rotating propellers that allow the aft propeller to gain efficiency in recovering rotational energy from the forward propeller. For simplicity, we will discuss only single rotation marine screw propulsion here. The emphasis will be on the propeller operating behind a ship; more details on marine propellers *per se* can be found in Sec. 27.2.

The efficiency of ship propulsion is defined by the ratio of effective power (resistance times speed) to the delivered power to the propeller (torque times revolutions)

$$\eta_D = \frac{R \cdot V}{2\pi Q \cdot n} \tag{23.216}$$

where R = resistance, Q = shaft torque, and n = revs. The propeller efficiency is generally calculated in terms of the thrust produced

$$\eta_D = \frac{T \cdot V_A}{2\pi Q \cdot n} \qquad (23.217)$$

where V_A = speed of flow into the propeller disk and V = speed of the ship. Because of the effect of propeller action on the hull surface, the thrust generated is usually greater than the hull resistance. The propeller changes the flow field around the hull and hence the *propelled resistance*. This interaction effect is called *thrust deduction* or *resistance augmentation* and is defined as

$$t = (T - R)/T \qquad \text{or} \qquad 1 - t = R/T \qquad (23.218)$$

Furthermore, the propeller operates in a flow field that is moving not necessarily at the same speed as the ship. The energy lost in the boundary layer due to frictional resistance slows the flow to the propellers. The effect is defined by a *Taylor wake fraction* as

$$w = (V - V_A)/V \qquad \text{or} \qquad V_A = (1 - w)V \qquad (23.219)$$

These two propeller-hull interaction quantities are combined into a *hull efficiency*

$$\eta_H = (1 - t)/(1 - w) \qquad (23.220)$$

and then the propulsive efficiency can be stated as

$$\eta_D = \eta_H \cdot \eta_B \qquad (23.221)$$

where η_B represents the propeller efficiency behind the hull, and η_H represents the propeller–hull (interaction) efficiency.

Estimates of propeller–hull interaction are obtained from towing tank experiments with a hull model of the design and a propeller model of similar characteristics. These data are then used to design a propeller specifically for the power and rpm of the engine to be installed in the ship. If model experiments are not available, historic data can be used to estimate $(1 - t)$ and $(1 - w)$.

Details of the flow into the propeller disk may be required for the specific propeller design. In that case, measurements are made using a towing tank model and either a Pitot tube or a laser doppler velocimeter for three components of flow in the exact location of the propeller plane. These velocity components enable the design of *wake-adapted propellers* that can achieve better performance with less vibration and cavitation than standard propellers.

If a standard propeller is deemed to be adequate for the particular ship, there are several sets of data from experiments with model propellers that have been performed using systematic variations of such parameters as blade number, blade area and pitch-diameter ratio. The most frequently used data were published by Troost (1951). These data have been analyzed, and a very useful regression analysis equation has been published by Oosterveld et al. (1969). Preliminary design and performance estimates can be easily made using a personal computer with Troost series regression equations.

Propeller design tools are one of the most successful applications of computational fluid dynamics. Potential flow lifting-line foil theory has been used to develop design algorithms that will produce specific pitch, camber, and loading that match the wake inflow and the thrust generation requirements of the ship. Further developments of the theory to include lifting surface effects are now available.

The use of both physical models and analytical predictions of propeller performance is commonplace in the determination of the powering performance of ships. Other constraints on the ship, such as draft limits that limit propeller diameter or engine operating characteristics that limit revolutions of the propeller may have a substantial impact on the efficiency of any given design. The cavitation performance must also be considered in any ship propulsion. Cavitating propellers usually cause vibration and noise as well as danger of blade erosion due to the bubble collapse. Propellers must therefore be designed not only to maximize efficiency but also to operate within acceptable bounds of cavitation performance. Typical guidance for these constraints can be found in Comstock (1967).

23.12.4   Maneuvering and Seakeeping

The motion of a ship due to its own control systems or due to the external forces of wind and waves is characterized chiefly as a potential flow problem. Few analytical tools exist for the prediction of maneuvering because of the complex interactions between hull, propeller, rudder, and the distorted flow field generated by a ship turning. The seakeeping performance of a ship, particularly the motions in the vertical plane-pitch and heave, are regularly predicted using *strip theory* approximations [Newman (1977)]. As part of a design and performance evaluation process, both physical models and simplified theoretical models are used.

The maneuvering characteristics of a ship are characterized by three particular maneuvers: tactical turn, zig-zag, and spiral. The tactical turn defines the forward most travel and greatest lateral travel (advance and transfer) for any given speed of approach and rudder angle. The zig-zag or overshoot maneuver characterizes how far a ship turns beyond a given heading when the rudder is periodically reversed. The spiral maneuver is used to determine if there are regions of operation of the rudder where the rudder forces are ineffective. All of these maneuvering characteristics are ascertained through model experiments in the design stage and are confirmed at sea during full-scale maneuvering trials once the ship has been built. With the necessity for ship control to avoid grounding and collision and the increase in ship size and speed, maneuvering characteristics have become an increasingly important design problem. Work on the analytical prediction of ship maneuvering performance is ongoing to take advantage of new capabilities in the area of computational fluid dynamics.

Seakeeping performance is a major consideration for both structural and human factors engineering. Slamming loads are frequently the most severe structural requirement for ships. In addition, more and more consideration is being paid to crew comfort and safety in new design.

The use of computational methods to predict ship motion in a seaway is regularly a part of the ship hydrodynamics assessment. As mentioned previously, pitch and heave motions are assessed for a variety of hull form shapes, and the trade-off between seakeeping performance and other hydrodynamic characteristics such as resistance performance are done using three analytical tools [Baitis *et al.* (1981)].

23.12.5 Summary

The complex hydrodynamics associated with ships has been addressed in a summary fashion here. The utility of these aspects of fluid mechanics is realized in ship designs that are efficient and seaworthy. The future of ship hydrodynamics will include more capable computational fluid dynamics methods coupled with validation exercises using laboratory experiments. Ultimately, the understanding of the detailed flow phenomena associated with ship hydrodynamics will be exploited in better ships.

23.13 DYNAMICS OF PARTIALLY FILLED TANKS
Franklin T. Dodge

23.13.1 Applications of Sloshing

Liquid in a partially full tank is easily caused to slosh about. The sloshing of fuel in airplanes and liquid-fuel rockets can affect flight stability, although adequate stability can be retained if the slosh natural frequencies can be made considerably higher than any control or disturbance frequency, say by partitioning a large tank into several smaller ones. If partitioning is not practical, various kinds of slosh-damping devices can be inserted in the tank. Liquid cargo trucks and railroad cars are also subject to slosh-induced stability problems, such as overturning on curves or rocking back and forth during starts or stops. Since it is not usually possible to change the disturbance frequencies, either the tanks must be partitioned to increase the slosh frequencies, or they must be kept nearly full to diminish the magnitude of the cargo shifting. Tanker ships transporting liquified natural gas also experience sloshing. In these large, generally unbaffled tanks, the impact of the slosh waves on the bulkheads and insulation can cause structural damage and failure. Again, the most practical solution is to keep the tanks as full as possible. Liquid storage tanks have collapsed as a result of sloshing induced by earthquakes. The sloshing of the water in the pressure suppression pool of a nuclear power plant of the boiling water type during an earthquake has also received attention. Two- and three-phase separation equipment for processing oil on flexible offshore platforms or barges are susceptible to a sloshing-induced loss of performance or capacity. It can be concluded, then, that sloshing may be a problem not only in moving vehicles but also in normally stationary tanks and pools.

Spacecraft may carry liquid propellants to supply the thrusters used for orbit correction and stationkeeping. In space, the gravity-like acceleration acting on liquid can be very small. Surface tension forces are then important, and the sloshing, now called "low-g" sloshing, is significantly different from what occurs when the gravitational body force is larger. When the spacecraft spins, nonsloshing forms of liquid motions, such as inertial wave resonances, may occur that have no counterpart in nonspinning tanks. Spinning projectiles and ballistic missiles are also subject to instabilities arising from these kind of inertial wave oscillations.

23.13.2 Basic Equations for Lateral Sloshing

Lateral sloshing in an *upright tank of rectangular cross-section* is discussed here; results for other tank shapes are given in Sec. 17.1.4 in the form of equivalent mechanical models. A number of simplifying but reasonable assumptions are made: the tank is rigid; the liquid is incompressible and nonviscous; the velocities and displacements are small; and the motion is irrotational (see Sec. 1.2). Viscous and other energy-dissipating effects are considered in Sec. 23.13.3. As shown in Fig. 23.141 (which also depicts a mechanical model that will be discussed in Sec. 23.13.4), the tank is subjected to small horizontal oscillations $x_0 \sin \omega t$ and small pitching oscillations $\theta_0 \sin \omega t$ in the same plane.

The velocity potential of the sloshing motions shows that there are certain natural sloshing or wave frequencies of the free surface that are given by

$$\omega_n = \left[\frac{(2n-1)\pi g}{w} \tanh (2n-1) \frac{\pi h}{w} \right]^{1/2}, \qquad n = 1, 2, 3 \ldots \quad (23.222)$$

Note that ω_n increases when the tank width, w, decreases, and becomes independent of the liquid depth, h, when $h/w > 1$. By integrating the unsteady pressure on the tank boundaries caused by the wave motions, the slosh forces and moments can be computed. The result for the horizontal force is

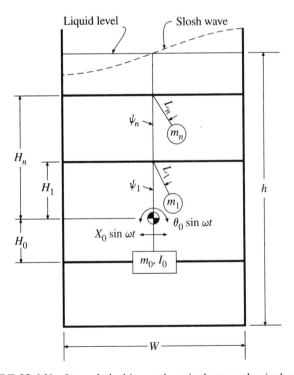

FIGURE 23.141 Lateral sloshing and equivalent mechanical model.

$$F_H = \rho x_0 \omega^2 \sin \omega t \left\{ bwh + \sum_{n=1}^{\infty} \frac{8bw^2}{(2n-1)^3 \pi^3} \left[\frac{\omega^2}{\omega_n^2 - \omega^2} \right] \tanh (2n-1) \frac{\pi h}{w} \right\}$$

$$+ \rho \theta_0 \omega^2 \sin \omega t \left\{ \sum_{n=1}^{\infty} \left[-\frac{8bw^2}{(2n-1)^3 \pi^3} \tanh (2n-1) \frac{\pi h}{w} \right] \right.$$

$$\left. \times \left[\left(\frac{\omega^2}{\omega_n^2 - \omega^2} \right) \left(\frac{h}{2} - \frac{2w}{(2n-1)\pi} \tanh (2n-1) \frac{\pi h}{w} \right) + \frac{g}{\omega_n^2 - \omega^2} \right] \right\}$$

(23.223)

Here, b is the breadth of the tank normal to the plane of the tank motion. Each term in the infinite series corresponds to one of the natural slosh modes. For $\omega \ll \omega_1$, the force due to the translation, x_0, is merely the inertial force of the rigidized mass of the liquid ρwbh, and that due to the pitching, θ_0, is the unbalanced force caused by pitching the tank when the free surface essentially remains flat. For $\omega \gg \omega_1$, the force due to x_0 is again that of an accelerating rigid body but in this case the liquid mass is only the smaller "apparent" mass

$$\text{Apparent Mass} = \rho bwh \left\{ 1 - \sum_{n=0}^{\infty} \frac{8w/h}{(2n-1)^3 \pi^3} \tanh (2n-1) \frac{\pi h}{w} \right\} \quad (23.224)$$

The net unsteady moment exerted on the tank walls and bottom can be derived similarly but is not written here explicitly. In an application, the moment can be derived from the equivalent mechanical models developed in Sec. 23.13.4. Both the force and moment take on indefinitely large magnitudes when ω approaches any ω_n. The exact magnitude of the forces and moments for that condition must be determined by consideration of viscous or damping effects.

23.13.3 Inclusion of Damping Effects

Damping can be included in the preceding results on an empirical basis by modifying the resonant denominators in Eq. (23.223). By analogy with vibration theory, the appropriate modification (in complex form with $i = \sqrt{-1}$) is

$$(\omega_n^2 - \omega^2) \rightarrow (\omega_n^2 + 2i\zeta_n \omega \omega_n - \omega^2) \quad (23.225)$$

where ζ_n is the critical damping ratio of the slosh mode in question. Empirical results for ζ_n are generally available only for the lowest frequency, fundamental mode, $n = 1$. In the absence of other data, ζ_1 can be used for all other important modes.

Dimensional analysis and empirical correlations [Abramson (1966)] both show that the damping arising from viscous boundary layers on the tank walls can be expressed as

$$\zeta_1 = C_1 (\nu/d^{3/2} g^{1/2})^{n_1} \quad (23.226)$$

where d, the significant width dimension of the tank, is equal to w for a rectangular tank, or to the radius R of a cylindrical or spherical tank, or to the radius of the free

TABLE 23.12   Parameters for Viscous Damping Correlation

Tank Shape		C_1	n_1
Circular Cylindrical: $h/R \geq 1.0$		0.79	0.5
	$h/R = 0.5$	1.11	0.5
	$h/R = 0.1$	3.36	0.5
Rectangular:	$h/w \geq 1$	≈ 1.0	0.5
Spherical:	3/4 Full	0.66	0.359
	1/2 Full	0.39	0.359
	1/4 Full	0.32	0.359
1.3/1.0 Oblate Spheroid: Half-full		1.11	0.5
Upright Conical		0.81	0.5
Inverted Conical		0.85	0.5

(From Abramson, 1966.)

surface of a general axisymmetric tank; ν is the liquid kinematic viscosity. Typical results for various tank shapes are listed in Table 23.12. Reasonable assumptions for other tank configurations are $C_1 \approx 1.0$ and $n_1 = 0.5$.

The damping provided by the viscous boundary layer is generally too small to diminish resonant forces and moments significantly. Various kinds of baffles generally attached along the vertical walls must then be added. For example, the damping provided by an annular ring in a cylindrical tank is [Abramson (1966)]

$$\zeta_1 = 2.83(A_r)^{3/2} \, (\bar{\eta}/R)^{1/2} \, \exp\left[-4.6d_s/R\right] \tag{23.227}$$

Here, A_r is the fraction of the tank area, πR^2, blocked by the ring, $\bar{\eta}$ is the maximum allowed wave amplitude at the wall, and d_s is the vertical distance to the ring below the free surface.

23.13.4 Development of Equivalent Mechanical Models for Lateral Sloshing

The form of Eq. (23.223) is similar to that of the forced response of a set of harmonic oscillators. In fact, the slosh force and moments can be duplicated identically by a system of pendulums, or spring-mass oscillators, if the parameters are chosen correctly. A pendulum analogy is chosen here, as shown in Fig. 23.141, but similar results can be obtained for a spring-mass system. One pendulum is chosen to represent each slosh mode, and a rigidly attached mass is chosen to represent the effect of any liquid mass that does not participate in the sloshing. The total lateral force due to the system of pendulums and the rigid mass for simple harmonic forced motions of the tank is

$$F = -\ddot{x}_0 \left\{ m_T + \sum_{n=1}^{\infty} m_n \left(\frac{\omega^2}{\omega_n^2 - \omega^2} \right) \right\} - \ddot{\theta}_0 \sum_{n-1}^{\infty} m_n \left[\frac{(H_n - L_n)\omega^2 + g}{\omega_n^2 - \omega^2} \right] \tag{23.228}$$

Here, $\omega_n = (g/L_n)^{1/2}$ is the natural frequency of the nth pendulum, and $m_T = m_0 + \sum_{n=1}^{\infty} m_n$ is the total mass of liquid. Term-by-term comparison with Eq. (23.223) shows that the analogy will be perfect if the pendulum parameters are chosen in accordance with

$$m_n = \frac{8\rho b w^2}{(2n-1)^3 \pi^3} \tanh (2n-1) \frac{\pi h}{w}$$

$$m_0 = \rho w b h - \sum_{n=1}^{\infty} m_n$$

$$L_n = \frac{w}{(2n-1)\pi} \coth (2n-1) \frac{\pi h}{w}$$

$$H_n = L_n + \frac{h}{2} - \frac{2w}{(2n-1)\pi} \tanh (2n-1) \frac{\pi h}{w} \qquad (23.229)$$

The rigidly-attached mass is assumed to have a centroidal moment-of-inertia of I_0, so the moment exerted on the tank by the mechanical system can be expressed as

$$M = -\ddot{x}_0 \sum_{n=1}^{\infty} m_n \left[\frac{(H_n - L_n)\omega^2 + g}{\omega_n^2 - \omega^2} \right] - \ddot{\theta}_0 \left\{ I_0 + m_0 H_0^2 + m_n (H_n - L_n)^2 \right.$$

$$\left. + \sum_{n=1}^{\infty} m_n \left[\frac{2(H_n - L_n)g + (g/\omega)^2 + (H_n - L_n)^2 \omega^2}{\omega_n^2 - \omega^2} \right] \right\} \qquad (23.230)$$

Term-by-term comparison with the sloshing results for the moment gives the same expressions for m_n, m_0, L_n, H_n, and ω_n, and the additional result that

$$I_0 = I_F - m_0 H_0^2 - \sum_{n=1}^{\infty} m_n (H_n - L_n)^2 \qquad (23.231)$$

where

$$I_F = \rho b h w \left\{ \frac{h^2}{12} + \frac{w^2}{16} - 2w^2 \sum_{n=1}^{\infty} \frac{16}{(2n-1)^4 \pi^4} \right.$$

$$\left. \times \left[1 - \frac{2w}{(2n-1)\pi h} \tanh (2n-1) \frac{\pi h}{2w} \right] \right\} \qquad (23.231a)$$

Damping can be added on an empirical basis as before. For example, with damping Eq. (23.228) becomes

$$F = -\ddot{x}_0 \left\{ m_T + \sum_{n=1}^{\infty} m_n \left[\frac{\omega^2}{\omega_n^2 - \omega^2 + 2i\zeta_n \omega \omega_n} \right] \right\}$$

$$- \ddot{\theta}_0 \sum_{n=1}^{\infty} m_n \left[\frac{(H_n - L_n)\omega^2 + g}{\omega_n^2 - \omega^2 + 2i\zeta_n \omega \omega_n} \right] \qquad (23.232)$$

It should be noted that the liquid moment-of-inertia, I_F, is increased by the damping, but unless the damping is very large, the effect is negligible [Dodge and Kana (1966)].

TABLE 23.13 Mechanical Model Parameters for a Cylindrical Tank

$$m_n = 2\pi\rho R^3 \tanh \frac{\xi_n h}{R} [\xi_n (\xi_n^2 - 1)]^{-1}$$

$$L_n = (R/\xi_n) \coth \frac{\xi_n h}{R}$$

$$H_n = h/w + (R/\xi_n) \coth \frac{\xi_n h}{R} - 2 \tanh \frac{\xi_n h}{R}$$

$$m_o = \pi\rho R^2 h - \Sigma m_n$$

$$H_0 = [\Sigma m_n(H_n - L_n)]/m_0$$

$$I_0 + m_0 H_0^2 + \Sigma m_n(H_n - L_n)^2 = \text{Fig. } 17.2$$

(From Abramson, 1966.)

The mechanical model parameters needed to represent sloshing in an *upright cylindrical tank* are shown in Table 23.13 and Fig. 23.142. The parameter ξ_n in Table 23.13 has the following values: $\xi_1 = 1.841$; $\xi_2 = 5.331$; $\xi_3 = 8.536$; $\xi_4 = 11.706$; $\xi_5 = 14.864$; and $\xi_n \approx \xi_5 + (n - 5)\pi$. The parameters needed to represent the fundamental mode ($n = 1$) for a *spherical tank* are shown in Fig. 23.143. Although results for some other tank shapes are available [Abramson (1966)], it is recommended that one of the available computer codes [Lomen (1965)] be used for tank shapes not covered by the above tabulations.

23.13.5 Analysis of Vertical Sloshing

When a partially full tank is vibrated vertically, the free surface moves relative to the tank (sloshes) only when the excitation frequency is a multiple (most commonly

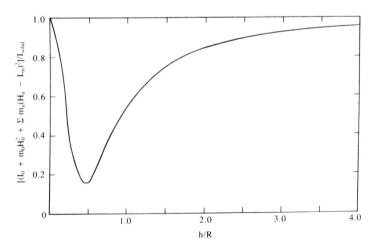

FIGURE 23.142 Ratio of effective liquid to rigid liquid moment of inertia for a cylindrical tank. (From Abramson, 1966.)

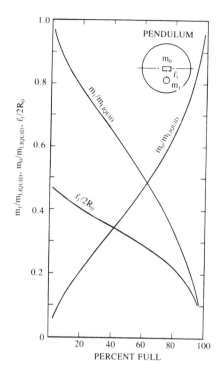

FIGURE 23.143 Mechanical model parameters for sloshing in a spherical tank. (From Abramson, 1966.)

twice) of one of the slosh natural frequencies. The frequency of the slosh waves is thus a fraction (most commonly $\frac{1}{2}$) of that of the forcing; such motions are called subharmonic. Further, the slosh mode can be antisymmetrical, as in lateral sloshing, in which one end of the wave moves upward while the other moves downward, or symmetrical, in which both ends move in phase. Over most of the forcing frequency range, the free surface remains flat and there are no slosh forces or moments. For that reason, vertical sloshing is usually not of any great concern.

The absence of liquid motion for most excitation frequencies is caused by a lack of direct forcing of the liquid; instead motion can only arise from an instability (of the Mathieu equation type) that is due to the oscillating body force produced in the liquid by the vertical excitation. Figure 23.144 shows a typical stability plot for an upright cylindrical tank having a diameter of 14.5 cm and $h/d = 1$, over a range of vertical excitation amplitude z_0 [Dodge *et al.* (1965)]. The fundamental antisymmetrical slosh mode for this tank is $\omega = 2.5$ Hz. As can be seen, there is only a very small range of frequencies for which a slosh response exists, even for large z_0. Further, the minimum z_0 needed to cause a motion is increased by damping, and the width of the unstable region is decreased.

In the design of storage tanks, it may be required to evaluate the loads due to the oscillating effective gravity for high frequency inputs. Since the sloshing is negligible, the resulting liquid pressure is simply $\rho g_{\text{eff}} H$, where H is the depth below the surface.

23.13.6 Analysis of Low Gravity Sloshing

When gravitational body forces are small, surface tension forces are important in determining the slosh characteristics. The static configuration of the liquid in the

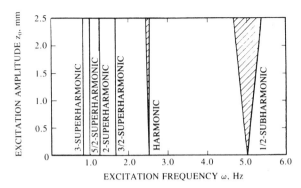

FIGURE 23.144 Stability chart of $n = 1$ mode for vertical sloshing in a cylindrical tank. (Reprinted with permission, Dodge, F. T., Kana, D. D., and Abramson, H. N., "Liquid Surface Oscillations in Longitudinally Excited Rigid Cylindrical Containers," *AIAA J.*, Vol. 3, No. 4, pp. 685–695, 1965.)

tank is also considerably different than it is in standard gravity conditions; for example, the free surface of liquid in a cylindrical tank in zero-*g* takes on the shape of a concave hemisphere. Some typical slosh results will be given here for liquids that have a zero-degree contact angle on the tank walls, which is the most common case in applications. (The contact angle is defined as the angle between the wall and the liquid surface, measured in the liquid; thus, a zero-degree contact angle means that the liquid surface is tangent to the tank wall at its contact with the wall.) Figure 23.145 shows some of the numerically computed results [Concus *et al.* (1967)] for the fundamental slosh frequency ($n = 1$) in a *cylindrical tank* as a function of liquid depth. The frequency ω_1 is presented in nondimensional terms, $\omega_1^2/[(1 + \text{Bo})\sigma/\rho R^3]$.

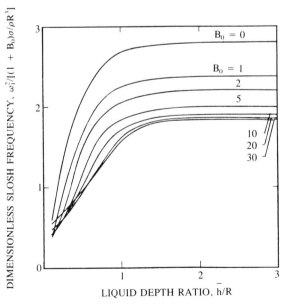

FIGURE 23.145 First mode frequency for low-g lateral sloshing in a cylindrical tank with hemispherical bottom. (From Concus *et al.*, 1967.)

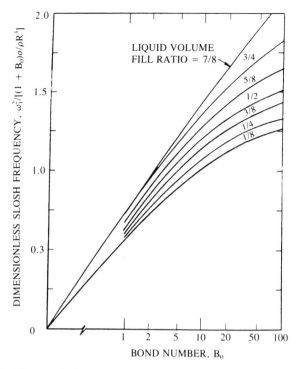

FIGURE 23.146   First mode frequency for low-g lateral sloshing in a spherical tank. (From Concus *et al.*, 1969.)

The parameter Bo is the Bond number, $\rho g R^2 / \sigma$, where σ is the surface tension and R the tank radius. The average liquid depth $\bar{h}$ is computed from the tank shape and the liquid volume. Figure 23.146 shows similar results for a *spherical tank* [Concus *et al.* (1969)]. Note that small values of g do not necessarily give small values of Bo or imply that the sloshing is low-g. For example, in a large rocket, R may be 300 cm; since $\sigma = 13$ dynes/cm and $\rho = 1.14$ gr/cm^3 for liquid oxygen, the value of Bo is about 80 even when the effective gravity (due to drag and other effects) is as small as 0.01 cm/sec^2. True low-g sloshing implies that Bo < 1.

The mechanical model parameters for low-g sloshing vary somewhat from those of normal sloshing [Dodge *et al.* (1991)]. The slosh mass can be approximated, in the absence of other data, from the data presented in Sec. 23.13.4. The pendulum must, however, incorporate a torsional spring K_θ to represent the effects of surface tension. The magnitude of K_θ can be found from matching the computed slosh natural frequency to that predicted from the mechanical model. The natural frequency of the spring-pendulum is

$$\omega_1 = \left[\frac{g}{L_1} + \frac{K_\theta}{m_1 L_1^2} \right]^{1/2} \tag{23.233}$$

while that given by the liquid slosh analysis, as shown in Figs. 23.145 and 23.146 is

$$\omega_1 = N[(1 + \text{Bo})\sigma/\rho R^3]^{1/2} \tag{23.234}$$

where N is the numerical value read off the Figs. 23.145 and 23.146 for the specified Bo and liquid filling ratio. By comparing the two expressions, it can be concluded that

$$L_1 = R/N \tag{23.235}$$

and

$$K_\theta = \sigma m_1/\rho RN \tag{23.236}$$

23.13.7   Liquid Motions in Spinning Tanks

Spacecraft and ballistic projectiles are sometimes made to spin to improve their stability, and the spinning liquid in them can undergo several kinds of motions. Because of the way the liquid equations of motion depend on the spin rate, Ω, sloshing (free-surface) motions will have a frequency greater than 2Ω and inertial (internal) waves will have a frequency less than 2Ω. The predominant form of excitation is the precession or wobbling of the spacecraft or projectile, which always has a frequency less than 2Ω. Thus, inertial waves are usually of the most concern in spinning tanks [Slabinski (1978)]. A few typical results for inertial wave frequencies will be given here.

For a *completely full spherical tank*, the inertial wave frequencies are the zeroes of certain Legendre polynomials. For the axisymmetric modes, the natural frequencies are [Aldridge and Toombe (1969)]

$$\omega_{nm} = 2\Omega x_{nm} \tag{23.237}$$

where $x_{nm} < 1$. Some of the predominant modes (i.e., those producing large disturbance moments on the tank) have the following values: $x_{11} = 0.6547$; $x_{21} = 0.4688$; $x_{31} = 0.3631$; and $x_{41} = 0.2958$. Note that $x_{n1} \to 0$ as n increases.

In a *partially-full cylindrical tank*, the free surface forms a cylindrical surface around the spin axis. Large disturbance torques are possible. Extensive tables are available to compute the inertial wave frequencies [Stewartson (1959)]. Table 23.14 has been extracted from those tables for two cases, a completely full tank and a 60% full tank. The parameters in the table are defined as c = axial length of tank; R = radius of tank; $\tau = \omega/\Omega$; and $j + 1/2$ = axial wave number. Any given resonance must correspond to an integer value of j. For example, for the full tank, consider a case with $c/R = 3$. For $j = 0$, $c/[(2j + 1)R] = 3$. There is no resonance for which $\tau < 0.2$. (It was assumed in preparing the tables that values of $\omega > 0.2\Omega$ were not of great interest.) For $j = 1$, $c/[(2j + 1)R] = 1$. There is a resonance at about $\tau = 0.00435$. For $j = 2$, $c/[(2j + 1)R] = 0.6$. There is a resonance at about $\tau = 0.171$.

Equivalent mechanical models for inertial waves are not available. In many cases, the stability of the motion can be determined solely from the magnitude of the energy dissipated by the liquid motions.

TABLE 23.14   Inertial Wave Frequencies for a Cylindrical Tank

τ	$c/[2j + 1)R]$		
a. Full Tank			
0.00	0.995	0.478	0.310
0.02	1.018	0.490	0.319
0.04	1.042	0.503	0.327
0.06	1.066	0.516	0.336
0.08	1.091	0.530	0.345
0.10	1.117	0.544	0.355
0.12	1.144	0.559	0.364
0.14	1.172	0.574	0.375
0.16	1.201	0.590	0.385
0.18	1.231	0.607	0.397
0.20	1.262	0.624	0.408
b. 60% Full Tank			
0.00	0.842	0.281	0.154
0.02	0.861	0.288	0.158
0.04	0.881	0.296	0.162
0.06	0.901	0.304	0.166
0.08	0.923	0.312	0.171
0.10	0.945	0.320	0.176
0.12	0.969	0.329	0.181
0.14	0.994	0.338	0.186
0.16	1.020	0.348	0.191
0.18	1.048	0.358	0.197
0.20	1.077	0.369	0.203

(From Stewartson, 1959.)

23.14   AERO- AND HYDRO-OPTICS

Allen E. Fuhs and A. George Havener

23.14.1   Introduction

Aero-optics is defined as the interaction between a refractive index field in gases and an electromagnetic wave. Hydro-optics has the same definition except the medium through which the electromagnetic wave propagates is a liquid. The words *refractive index field* mean a spatial variation of refractive index, $n(x, y, z)$. The spatial variation in refractive index results from variation in mass density, excitation of molecular internal degrees of freedom, and changes in concentrations of chemical species. For example, the process of combustion causes changes in mass density as well as the concentrations of various chemical species.

The familiar methods for flow visualization, i.e., schlieren, shadowgraph and Mach–Zehnder interferometry, are applications of aero-optics. These methods are discussed in Chap. 15. Aero- and hydro-optical phenomena are not restricted to the visible region of the spectrum; propagation of laser beams in the infrared is one example at longer wavelength.

In addition to flow visualization, which was mentioned earlier, the applications of aero- and hydro-optics include aerial photography, astronomical observations through the atmosphere, and celestial navigation for missiles and aircraft. Reconnaissance aircraft using photography may experience image degradation due to optical path distortion. Astronomy from airborne platforms may be affected by aero-optics. Missiles and high-speed aircraft using celestial navigation may have angular errors introduced by adverse effects of aero-optical phenomena.

Lasers have stimulated the investigation of aero-optical, and to a lesser degree, hydro-optical phenomena. Propagation of laser beams is degraded by turbulence, thermal blooming, a variety of boundary layers, shear layers, and wakes, and compressible flow over laser turrets mounted on high-speed aircraft. Blue-green lasers are being developed for communication to submerged submarines from aircraft or satellites. Propagation of the blue-green radiation within the ocean is an application of hydro-optics.

Figure 23.147 is a flow diagram for the analysis of hydro- and aero-optical phenomena. Each block in the flow diagram gives chapters which are relevant to the item within the block.

23.14.2 Wavefronts and Rays

The refractive index, n, is the ratio of velocity of a phase front in a particular medium, v, to the speed of light in a vacuum, c. As an equation, n is

$$n = \frac{c}{v} \tag{23.238}$$

As defined above, n is a real number. The refractive index can be defined as a complex number with the imaginary part leading to absorption [see Stratton (1941) or Jackson (1962)].

The optical path length, ζ, is defined as

$$\zeta = \int n(s) \, ds \tag{23.239}$$

where s is distance along a ray. As derived by Born and Wolf (1964), the optical path length is related to refractive index by

$$\left(\frac{\partial \zeta}{\partial x}\right)^2 + \left(\frac{\partial \zeta}{\partial y}\right)^2 + \left(\frac{\partial \zeta}{\partial x}\right)^2 = n^2 \, (x, \, y, \, z) \tag{23.240}$$

Equation (23.240) is known as the *eikonal equation* in geometrical optics. An alternate short-hand form for Eq. (23.240) is

$$(\text{grad } \zeta)^2 = n^2 \tag{23.241}$$

the surfaces

$$\zeta(\bar{\mathbf{r}}) = \text{constant} \tag{23.242}$$

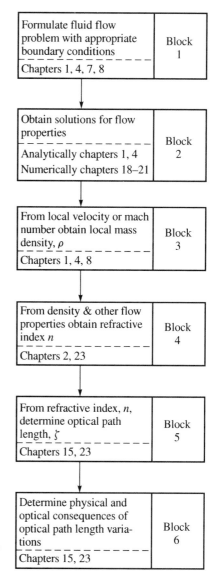

FIGURE 23.147 Flow diagram for analysis of hydro- and aero-optical problems.

are known as *geometrical wavefronts*. According to Eq. (23.239), ζ has the units of a length. On a wavefront, ζ has a constant value as noted by Eq. (23.242). If one moves along a ray in the direction of propagation by a distance ds, ζ increases by an amount

$$d\zeta = n \, ds \qquad (23.243)$$

which is the differential form of Eq. (23.239).

Rays are orthogonal to wavefronts. Define a unit vector $\bar{s}$ which is

$$\bar{s} = \frac{\text{grad } \zeta}{n} = \frac{\text{grad } \zeta}{|\text{grad } \zeta|} \qquad (23.244)$$

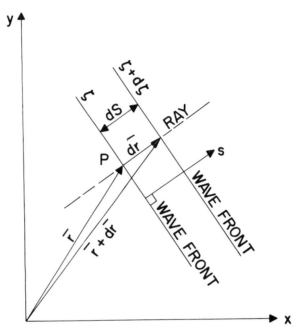

FIGURE 23.148   Geometry of wavefronts and rays.

The gradient of a function for a surface yields the normal to the surface. Hence, $\bar{s}$ is normal to wavefront and is parallel to a ray. The vector $\bar{r}$ defines a point, P, on a ray; $\bar{r}$ was used earlier in Eq. (23.242). Figure 23.148 illustrates the geometry of vector $\bar{r}$, the wavefront containing point P, and the ray passing through point P. Examination of Fig. 23.148 shows that

$$|d\bar{r}| = ds = dr \qquad (23.245)$$

Also one notices

$$d\bar{r} = \bar{s}\, dr = \bar{s}\, ds \qquad (23.246)$$

Consequently

$$\frac{d\bar{r}}{ds} = \bar{s} \qquad (23.247)$$

The increment of ζ can be obtained from the gradient of ζ and Eq. (23.243)

$$d\zeta = \bar{s} \cdot (\text{grad } \zeta)\, ds = n\, ds \qquad (23.248)$$

From Eq. (23.248) one finds

$$n\bar{s} = \text{grad } \zeta \qquad (23.249)$$

Combining Eqs. (23.247) and (23.249), the equation for rays is obtained

$$n \frac{d\bar{\mathbf{r}}}{ds} = \text{grad } \zeta = n\bar{\mathbf{s}} \qquad (23.250)$$

Equation (23.250) relates the rays to the corresponding wavefronts. A useful form for the ray equation uses the gradient of refractive index. The equation, which is derived in Born and Wolf (1964), is

$$\frac{d}{ds}\left[n \frac{d\bar{\mathbf{r}}}{ds} \right] = \text{grad } n \qquad (23.251)$$

A simple example can be studied as a method for gaining insight into Eq. (23.251). Refer to Fig. 23.149. Due to a positive gradient of refractive index in the y-direction, the wavefront is turned by an angle θ. For this simple case, the gradient of refractive index is

$$\text{grad } n = \bar{\mathbf{e}}_y \frac{\partial n}{\partial y} \qquad (23.252)$$

The distance Δs_1 is given by

$$\Delta s_1 = v\Delta t = \frac{c}{n} \Delta t \qquad (23.253)$$

The distance Δs_2 is given by

$$\Delta s_2 = \frac{c}{n + (\partial n/\partial y)\Delta y} \Delta t \simeq \frac{c}{n}\left(1 - \frac{1}{n}\frac{\partial n}{\partial y} \Delta y \right)\Delta t \qquad (23.254)$$

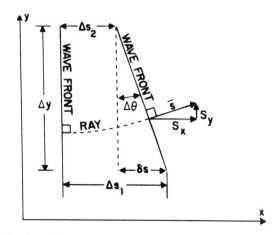

FIGURE 23.149   Turning of a wavefront by a positive gradient of refractive index, $\partial n/\partial y$.

The increment of distance which the lower end of the wavefront moves compared to the upper end is

$$\delta s = \Delta s_2 - \Delta s_1 = -\frac{c}{n^2}\frac{\partial n}{\partial y}\Delta y \Delta t \tag{23.255}$$

The angle $\Delta\theta$ is obtained from

$$\Delta\theta = \arctan\frac{\delta s}{\Delta y} \simeq \frac{\delta s}{\Delta y} = -\frac{c}{n^2}\frac{\partial n}{\partial y}\Delta t \tag{23.256}$$

Combining Eqs. (23.253) and (23.256), one finds

$$\frac{\Delta\theta}{\Delta s} = -\frac{1}{n}\frac{\partial n}{\partial y} \tag{23.257}$$

As developed here, $\Delta\theta$ is negative when the turn is counterclockwise.

23.14.3 Refractive Index of Gases and Liquids

The refractive index, n, can be expressed in terms of a *molar refraction*, $R(\lambda)$, using a formula developed independently by Lorentz and Lorenz. The Lorentz–Lorenz equation, which is discussed by Glasstone (1954) and by von Hippel (1959) is

$$R(\lambda) = \frac{M}{\rho}\frac{n^2 - 1}{n^2 + 2} \tag{23.258}$$

where M is molecular weight, gram/mole or kg/kmole; and ρ is mass density, kg/m^3. Molar refraction, which has the units of m^3/kmole, is applicable to substances in both the liquid and gaseous states. Molar refraction may be regarded as a property of a particular molecule. Only a few percent difference in $R(\lambda)$ for gaseous and for liquid states of a compound may be observed when density is varied by a factor of 1000 or so.

The molar refraction of a mixture, R_m, is obtained from

$$R_m = X_1 R_1 + X_2 R_2 + X_3 R_3 + \cdots = \Sigma X_i R_i \tag{23.259}$$

where X_1 is mole fraction of species 1 with molar refraction R_1, etc.

Molecules are categorized as being *nonpolar* or *polar* depending on whether or not the molecule has an electric dipole moment. Cole and Coles (1964) discuss the influence of electric dipole moment on refractive index. Nonpolar molecules have smaller values of $R(\lambda)$ than polar molecules. Water is a polar molecule. Values of refractive index for water and other substances are given in Chap. 2.

To understand the hydro-optical effects to be discussed shortly in this section, the change in refractive index of water as a function of density is needed. Equation (23.258) can be rearranged to yield

$$n^2 = \frac{1 + 2R\rho/M}{1 - R\rho/M} \qquad (23.258a)$$

Differentiation of Eq. (23.258a) gives

$$\frac{\partial n}{\partial(\rho/\rho_0)} = \frac{(3/2n)(R\rho_0/M)}{[1 - R\rho_0/M]^2} \qquad (23.260)$$

At 10°C and λ equal to the wavelength of the sodium D lines, n for water is 4/3. Using this value for n, Eq. (23.258) can be evaluated for $R\rho_0/M$. The result is $R\rho_0/M = 0.206$. For this calculation, the reference density, ρ_0, is that for water at 10°C and atmospheric pressure. Suppose the density of water changes by 0.1, i.e., $\rho/\rho_0 = 1.1$; what is the refractive index? Using Eq. (23.260) and the first terms of a Taylor series expansion

$$n = n_0 + \frac{\partial n}{\partial(\rho/\rho_0)} \Delta(\rho/\rho_0) + \cdots \qquad (23.261)$$

one finds $n = 1.333 + (0.367)(0.1) = 1.370$. For $R\rho_0/M = 0.206$ the value of $\partial n/\partial(\rho/\rho_0)$ is 0.367.

For gases, n is nearly unity. As a consequence, Eq. (23.258) can be simplified. Define

$$n = 1 + \delta \qquad (23.262)$$

where $\delta \ll 1.0$. Inserting Eq. (23.262) into Eq. (23.258) and eliminating terms of order δ^2 and smaller, one obtains

$$\delta = \frac{3}{2} \frac{R\rho/M}{1 - (R\rho/M)} \approx \frac{3}{2} \frac{R\rho}{M} \qquad (23.263)$$

For the refractive index, the appropriate equation becomes

$$n = 1 + \frac{3}{2} \frac{R\rho}{M} \qquad (23.264)$$

Equation (23.264) is of the form

$$n = 1 + \beta\rho \qquad (23.265)$$

where

$$\beta = \frac{3R(\lambda)}{2M} \qquad (23.266)$$

The quantity, β, is known as the *Gladstone–Dale constant*; the relationship between the Gladstone–Dale constant and the molar refraction is given by Eq. (23.266). An alternate form is

$$n = 1 + \kappa \rho / \rho_0 \tag{23.267}$$

In the form above, κ is nondimensional, and ρ_0 is a reference density. The relationships between the coefficients in Eqs. (23.264) and (23.265) are

$$\kappa = \beta \rho = \frac{3R\rho_0}{2M} \tag{23.268}$$

For problems involving aero-optics in the atmosphere, a useful form for refractive index is

$$n = 1 + \kappa_s \frac{\rho_\infty}{\rho_s} \frac{\rho}{\rho_\infty} \tag{23.269}$$

where κ_s is the value appropriate for sea level conditions (0.000292 at 589.3 nm), ρ_s is atmospheric density at sea level (1.226 kg/m^3), ρ_∞ is the ambient density at altitude, and ρ is the density at some point in a compressible flow. Equation (23.269) has the merit of explicitly showing the dependence of n on altitude through the term ρ_∞ / ρ_s.

Values of refractive index for liquids are tabulated in Chap. 2. Some values for κ_s appear in Table 23.15.

23.14.4 Hydro-Optics Applications and Examples

Two examples will be discussed in this section. One example involves the motion of a projectile in water. The other example is the hydro-optics associated with a metal jet from a shaped charge penetrating water.

Figure 23.150 illustrates the flow field for a 0.22 caliber projectile penetrating water. The photograph was obtained using a spark shadowgraph technique. The grid has lines spaced 2.5 mm by 12.7 mm (1.0 × 0.5 in.). Due to compression of the water as a result of fluid flow, the grid lines appeared to be curved. The displacement of the lines permits one to determine the spatial distribution of fluid density. The accuracy to which the fluid density is measured depends on the completeness of the

TABLE 23.15 Refractive Index Constant, κ_s

Gas	κ_s
Air	0.000292
Nitrogen	0.000297
Oxygen	0.000271
Carbon dioxide	0.000451
Helium	0.000036
Steam	0.000254
Argon	0.000281

Values are for λ = 589.3 nm.

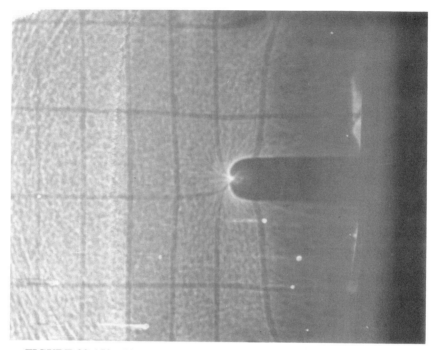

FIGURE 23.150 Photograph of a 0.22 caliber projectile penetrating water.

model used for ray tracing. A complete model starts with Eq. (23.251) and obtains a solution for n based on measured grid displacement. Here, for purposes of illustration, the model for obtaining grid displacement is simplified.

As derived in Born and Wolf (1964), the equation for a ray in a medium with spherical refractive index is given by

$$\theta = c \frac{dr}{r[n^2 r^2 - c^2]^{1/2}} \qquad (23.270)$$

The angle θ is measured from a reference direction to the radius vector for a point on the ray. Refer to Fig. 23.151. The refractive index is a function only of radius r; a surface of constant n has a spherical shape. The distance from the center of the sphere to a point, P, on the ray is r. The constant c is given by

$$c = nr \sin \phi \qquad (23.271)$$

As n, r, and ϕ change along a ray, the product given by Eq. (23.271) remains constant. The angle ϕ is measured from the radius vector, $\bar{r}$, to the unit vector tangent to the ray, $\bar{s}$. At point P_1 in Fig. 23.151, θ is 37°, and ϕ is 143°. At point P_2 in Fig. 23.151, θ_2 is 113°, and ϕ_2 is 72°. The deviation of a ray from its initial direction is δ. The value of δ is obtained from

$$\delta = \theta + \phi - 180° \qquad (23.272)$$

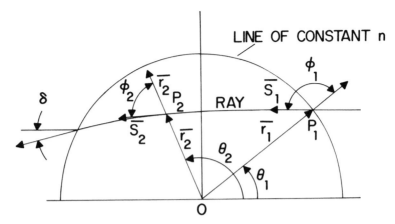

FIGURE 23.151   Diagram for geometry used to describe a ray.

or using degrees

$$\delta = \theta + \phi - 180° \tag{23.272a}$$

At point P_2, δ has a value of 5°. For the ray shown in Fig. 23.151, the refractive index increases toward the center which is point 0.

Equation (23.270) can be integrated for certain functions for $n(r)$. Table 23.16 summarizes several solutions. For the case at hand, the last example in Table 23.16 is appropriate. Consider a ray entering a region of spherical $n(r)$ as illustrated in Fig. 23.151. The initial values of θ and ϕ are as follows: $0 < \theta_i < 90°$ and $90° < \phi_1 < 180°$. The constant for the θ equation changes when argument of the arcsin becomes -1.0. This occurs near θ equal to 90°. Denote $\theta*$ as special value where the argument is -1.0. In the interval $0 \le \theta \le \theta*$, the equation for θ is

$$\theta = 180 - \phi_0 + \tfrac{1}{2}(\sin_0^{-1} - \sin^{-1}) \qquad 0 \le \theta \le \theta* \tag{23.273}$$

where

$$\sin^{-1} = \sin^{-1}\left[\frac{ax - 2(a + b) \sin^2 \phi_0}{x(a^2 + 4(a + b)b \sin^2 \phi_0)^{1/2}}\right] \tag{23.274}$$

and

$$\sin_0^{-1} = \sin^{-1}\left[\frac{a - 2(a + b) \sin^2 \phi_0}{(a^2 + 4(a + b)b \sin^2 \phi_0)^{1/2}}\right] \tag{23.275}$$

As indicated in Table 23.16, $x = (r/r_0)^2$. The equation for ϕ is

$$\phi = 180 - \sin^{-1}\left[\frac{(a + b) \sin^2 \phi_0)}{(a + bx)x}\right]^{1/2} \tag{23.276}$$

TABLE 23.16 Solutions for Ray Geometry for Spherically Symmetric Refractive Index

n	θ	ϕ	Remarks
$n = n_0 \left(\dfrac{r_0}{r}\right)^{1/2}$	$\theta_0 = \phi_0 - 2\tan^{-1}\left[\dfrac{r/r_0 - \sin^2\phi_0}{\sin^2\phi_0}\right]^{1/2}$	$\phi = \sin^{-1}\left[\left(\dfrac{r_0}{r}\right)^{1/2}\sin\phi_0\right]$	Refractive index increases as radius decreases.
$n = n_0 \left(\dfrac{r_0}{r}\right)$	$\theta = \phi_0 \pm \tan\phi_0 \ln\left(\dfrac{r}{r_0}\right)$	$\phi = \phi_0$	Angle ϕ does not change. Ray may spiral into center. Ray with circular shape possible.
$n = n_0 \left(\dfrac{r_0}{r}\right)^2$	$\theta = \phi_0 \pm \sin^{-1}\left(\dfrac{r}{r_0}\sin\phi_0\right)$	$\phi = \sin^{-1}\left[\dfrac{r}{r_0}\sin\phi_0\right]$	Ray may spiral into center.
$n = \left[a + b\left(\dfrac{r}{r_0}\right)^2\right]^{1/2}$	$\theta = \dfrac{1}{2}\sin^{-1}\left[\dfrac{ax - 2(a+b)\sin^2\phi_0}{x(a^2 + 4(a+b)b\sin^2\phi_0)^{1/2}}\right] +$ constant $x = (r/r_0)^2$	$\phi = \sin^{-1}\left[\dfrac{\dfrac{r}{r_0}\sin\phi_0\,(a+b)\sin^2\phi_0}{(a+bx)x}\right]^{1/2}$ $x = (r/r_0)^2$	By selecting magnitude of b, the refractive index gradient can be to any desired value.

In the interval $\theta* \leq \theta \leq \pi$ or 180, the equation for θ is

$$\theta = \phi_0 + \Delta\theta - \tfrac{1}{2}(\sin_0^{-1} - \sin^{-1}) \qquad \theta* \leq \theta \leq \pi \qquad (23.277)$$

where

$$\Delta\theta = 270 - 2\phi_0 + \sin_0^{-1} \qquad (23.278)$$

Also $\Delta\theta$ is related to $\theta*$ by (using θ in degrees)

$$\Delta\theta = 2(\theta* - 90) \qquad (23.279)$$

The equation for ϕ is

$$\phi = \sin^{-1}\left[\frac{(a + b)\sin^2\phi_0}{(a + bx)x}\right]^{1/2} \qquad (23.280)$$

If one wants only the maximum deflection of the ray, one needs to calculate δ when $x = (r/r_0)^2 = 1.0$ using Eqs. (23.272), (23.277), and (23.280).

Equation (23.258) can be used to calculate the refractive index for water as a function of density. Fluid dynamics can be used to find the pressure at points in the flow. To find density from pressure, one needs the equation of state of water. Richardson, Arons, and Halverson (1947) use

$$\frac{\rho}{\rho_1} = \left[1 + \frac{mp}{\rho_1 c_1^2}\right]^{1/m} \qquad (23.281)$$

where subscript 1 denotes a reference condition, which could be properties far from the body. The speed of sound in water is c_1 and m is a number. Richardson, Arons, and Halverson (1947) use $m = 7.15$.

To model the flow around the projectile, a Rankine 1/2-body is used. From Karamcheti (1966) along the stagnation streamline, the flow velocity, u, is

$$\frac{u}{U} = \left[1 - \frac{r_T^2}{2r^2}\right] = \left[1 - \left(\frac{r_S}{r}\right)^2\right] = \left[1 - \left(\frac{r_\infty}{2r}\right)^2\right] \qquad (23.282)$$

where U is the flow at infinity. The distance from the source to a point on the stagnation streamline is r. The three lengths are defined as follows: r_T is distance measured traversely from source to edge of body; r_S is distance from source to stagnation point; and r_∞ is radius of the Rankine 1/2-body at very large distance downstream from the source. Another description for r_T may be helpful; r_T is the distance from the source to the surface of the body measured along a line passing through the source at right angles to the main stream.

Knowing the velocity, the local pressure can be calculated from Bernoulli's law

$$p_0 = p + \rho u^2/2 \qquad (23.283)$$

which is valid only for $u \ll c_1$. Combining Eqs. (23.258), (23.282), and (23.283) one finds

$$\frac{\rho}{\rho_1} = \left\{ 1 + \frac{mp_1}{\rho_1 c_1^2} \left[\frac{p_0}{p_1} - \frac{\rho_1 U^2}{2p_1} \left(\frac{r_T^2}{2r^2} - 1 \right)^2 \right] \right\}^{1/m} \qquad (23.284)$$

and

$$n = \left[\frac{1 + 2\left(\dfrac{\rho}{\rho_1}\right)\left(\dfrac{\rho_1 R}{M}\right)}{1 - \left(\dfrac{\rho}{\rho_1}\right)\left(\dfrac{\rho_1 R}{M}\right)} \right]^{1/2} \qquad (23.285)$$

Equations (23.284) and (23.285) provide n as a function of distance along the stagnation streamline.

Having outlined the relevant equations, the appropriate values for the case shown in Fig. 23.150 will now be inserted into the equations. The density of water is $1000 \cdot \text{kg/m}^3$, and the velocity of the projectile was 340 m/s. The value of p at the projectile is nearly one atmosphere. Stagnation pressure is $p_0 = 1.013 \times 10^5 + 1/2(1000)(340)^2 = 5.7 \times 10^7 \text{ N/m}^2$. The speed of sound for the temperature of the water is $c_1 = 1445$ m/s. From Eq. (23.284)

$$\frac{mp_1}{\rho_1 c_1^2} = \frac{(7.15)(1.013 \times 10^5 \text{ Pa})}{(1000 \text{ kg/m}^3)(1445 \text{ m/s})^2} = 3.47 \times 10^{-4} \qquad (23.284a)$$

Equation (23.284) gives ρ/ρ_1 as a function of r. Using the values indicated above, the value of ρ/ρ_1 is 1.0117 at the stagnation point and 1.0 at upstream infinity. The corresponding values for n are obtained from Eq. (23.285). The value for R_ρ/M is 0.206 as noted earlier in the discussion following Eq. (23.260).

Equations (23.284) and (23.285) give the refractive index due to the compressible flow around a Rankine 1/2-body; denote this refractive index as n_R. The refractive index given by the fourth example in Table 23.16 is

$$n_S = [a + b(r/r_0)^2]^{1/2} \qquad (23.286)$$

The subscript "S" has been added to indicate that this is the refractive index due to the spherical distribution. The parameters a and b are open and can be selected so as to match n_R and n_S at two values of r/r_T. When n_R and n_S are matched at $r/r_T = 3.5$ and 7, the values of a and b are 1.781 and -4.37×10^{-5}, respectively. When n_R and n_S are matched at $r/r_T = 1.25$ and 5, $a = 1.793$ and $b = -5.522 \times 10^{-4}$. The equations in Table 23.16 have been derived so that $\phi = \phi_0$ when $r = r_0$. To determine the value of ϕ_0 to use in Eqs. (23.270) to (23.280), one wants to have $r/r_0 \leq 1.0$. Hence, select $r_0 \gg r_T$.

Figure 23.150 has been traced and appears as Fig. 23.152. The cavity diameter is 13 mm and $r_\infty = 6.5$ mm, also $r_T = 4.6$ mm and $r_S = 3.25$ mm. The amount of grid distortion is indicated below the figure caption. The measured ray deflection is calculated using

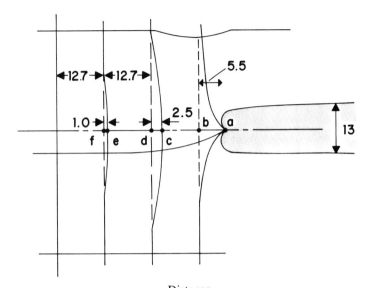

Point	Distance r, mm	r/r_T
a	0	0
b	5.5	1.19
c	15.7	3.42
d	18.2	3.96
e	29.9	6.51
f	30.9	6.72

FIGURE 23.152   Geometry of distorted grid lines.

$$\delta_m = \frac{r_b - r_a}{l} = \frac{5.5}{508}\frac{180}{\pi} = 0.62° \tag{23.287}$$

The example above is for points a and b in Fig. 23.152. The calculated ray deflection, δ_c, is obtained from Eqs. (23.272), (23.277), and (23.280). Results are shown in Table 23.17. The value for δ_c depends on the values of parameters a and b; the parameter b has a strong influence on δ_c. The simple model, which was selected to illustrate the principles, yields the correct magnitude for δ.

Figure 23.153 is a photograph of a metal jet from a shaped charge penetrating water. In contrast to Fig. 23.150, the flow is supersonic in water. A bow shock wave exists. Notice the concentration of radiation at the nose of the body in both Figs. 23.150 and 23.153. The concentrated radiation is due to hydro-optical effects.

TABLE 23.17   Comparison of Measured and Calculated Ray Deflection

r/r_T	r/r_0	a	b	ϕ_0	δ_m	δ_c
1.19	0.119	1.793	−0.05522	173.2	0.62	0.42
3.96	0.396	1.811	−0.01212	156.7	0.28	0.28
6.72	0.672	1.781	−0.00437	137.8	0.11	0.14

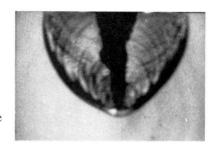

FIGURE 23.153 Metal jet from a shaped charge penetrating water.

23.14.5 Aero-Optics Applications and Examples

The summary of aero-optical applications given in Table 23.18 contains some examples for the causes of variable refractive index.

Flow of a compressible fluid causes density changes. According to Eq. (23.260), which is applicable to liquids or gases, a change in density causes a change in refractive index. All fluids are compressible to a greater or lesser extent. The important parameter is the Mach number. At low Mach number, compressibility effects are negligible relative to fluid flow properties and aerodynamic forces, however aero-optical phenomena may be important even at low Mach numbers. The discussion of the projectile penetrating water at low Mach number is an example.

An example of compressible flow causing variable n is flow over a laser turret. Figure 23.154(a) illustrates the flow geometry. Refer to Fig. 23.147. For Blocks 1 and 2, potential flow over a sphere was used. To obtain a compressibility correction, the Janzen–Rayleigh technique was employed [Fuhs and Fuhs (1980)]. For Block 3, the density was obtained using

$$\frac{\rho_0}{\rho} = \left[1 + \frac{\gamma - 1}{2} M^2 \right]^{1/(\gamma - 1)} \tag{23.288}$$

where ρ_0 is stagnation density, and ρ is static density. For Block 4 in Fig. 23.147, Eq. (23.267) was used. Equation (23.239) yielded the information required by Block 5. In both Fig. 23.154 and Eq. (23.239), distance along the beam is s. Figure 23.154(b) shows contours of constant phase across the beam after it has propagated

TABLE 23.18 Summary of Aero-Optical Applications

Cause for Variation of Refractive Index	Examples
Compressible flow	Flow over a laser turret
	Laser cavity flow
	Aerodynamic windows
Heat addition	Thermal blooming
	Kinetic cooling
	Laser cavity flow; laser quantum efficiency
	Laser cavity flow; electrical dissipation in electrical lasers
Excitation of molecules	Excimer lasers
	Excimer lasers
Change in chemical species	Combustion

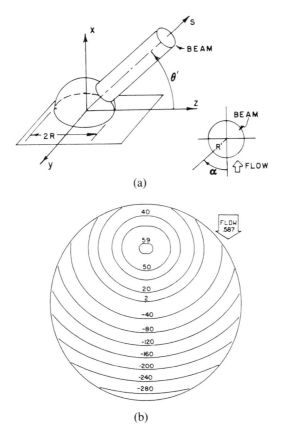

(a)

(b)

FIGURE 23.154 Laser beam degradation due to compressible flow over a laser turret. (a) geometry of hemispherical laser turret used for phase distortion calculations and (b) contours of constant phase shift for $\theta' = 18°$. The beam center is the reference.

a long distance from the turret. Figure 23.154(b) is an output of Block 5. Define

$$\delta = \zeta_2 - \zeta_1 \tag{23.289}$$

where ζ_1 is the reference optical path length. The beam center in Fig. 23.154 is the reference. The optical path length for a ray located at (α, R') in the beam is ζ_2. In Fig. 23.154(b), to avoid awkward decimal values, the phase distortion is the ratio δ/λ multiplied by 100. Hence, 250 from the graph is 2.5 wavelengths. The Mach number for the flow in Fig. 23.154 is $M = 0.587$ which is the critical Mach number for a sphere.

Consider now Block 6 in Fig. 23.147. From Eq. (23.289) one can determine the meaning of positive and negative contours. For a negative contour, $\zeta_1 > \zeta_2$. The beam is tilted forward into the freestream flow and is focused. A set of parallel lines would be pure tilt. A set of concentric circles would be a focus. If the numbers for contours increase toward the center of the concentric circles, a positive focus is obtained. Due to flow over the turret, the beam is partially focused and tilted into the wind.

Using the contours of constant phase, δ_{rms} can be calculated. When $\delta_{rms} \ll \lambda$, the *Strehl ratio* can be used to assess the significance of results such as that shown in Fig. 23.154. The *Strehl ratio* is

$$I/I_0 = \exp\left[-(2\pi\delta_{rms}/\lambda)^2\right] \tag{23.290}$$

where I_0 is beam intensity (watts/m^2) for $\delta_{rms} = 0$, and I is the beam intensity in the farfield for the value of δ_{rms}.

The flow in gas dynamic and chemical laser cavities is supersonic [Gross and Bott (1976)]. Due to waves in the cavity, the output beam may have a wrinkled wavefront, i.e., have a value for δ_{rms}. The beam quality is degraded as a result.

Laser cavities for gas dynamic lasers operate at a pressure of approximately 100 Torr. Combustion driven chemical lasers operate with a few tens of Torr pressure. To operate the laser in an ambient one-atmosphere environment, windows which can survive the intense radiation are not available. To avoid the difficulties of material windows, aerodynamic windows are used.

Aerodynamic windows use a turning of the flow to support a pressure gradient. The relevant equation is

$$\frac{\partial p}{\partial n} = \frac{\rho v^2}{R} \tag{23.291}$$

where n is distance normal to a streamline, ρ is density, p is pressure, v is velocity, and R is radius of curvature of the streamline. The aero-optical performance of aerodynamic windows is determined using the flow diagram of Fig. 23.147.

Heat addition in a flowing gas causes changes in density. Tsien and Beilock (1949) discuss line heat sources in both subsonic and supersonic flow. In subsonic flow, addition of heat causes the streamtube to expand. Considering the streamtube to be a solid body, which effectively it is, the adjacent streamtubes must move aside. The flow properties are altered throughout the flow as a result.

In supersonic flow, a line heat source causes compression and expansion waves which propagate outward at the Mach angle. Also, a wake extends downstream from the heat source. Both the waves and the wake cause variation in density and refractive index.

Thermal blooming is one application of the Tsien and Beilock paper. Thermal blooming occurs in laser beams due to absorption of laser radiation by the atmosphere. The energy removed from the beam heats the air which in turn changes density and refractive index. If the laser beam is slewing, the relative wind varies along the beam. Near the laser telescope, the wind velocity is low. Moving outward along the slewing laser beam, subsonic, transonic, and supersonic flow is encountered. Transonic flow is especially critical for laser propagation [Carey and Fuhs (1976)].

An unusual phenomenon occurs when CO_2 in the air absorbs radiation from a CO_2 laser. The phenomenon is termed *kinetic cooling*. Energy is added to the air and the (translational) temperature decreases, hence the name kinetic cooling. Details are given in Gebhardt and Smith (1972) and Sica (1973).

Laser quantum efficiency, η_Q, is defined as

$$\eta_Q = \frac{\epsilon_u - \epsilon_l}{\epsilon_u} \qquad (23.292)$$

where subscript u means upper level, and l, the lower level of the lasing transition. The energy of a level is ϵ. Due to laser quantum efficiency, which might be better termed laser quantum inefficiency, heat is added to the flow in a cavity, $\dot{Q}_c$. The equation is

$$\dot{Q}_c = P_L(1 - \eta_Q) \qquad (23.293)$$

P_L is laser power, watts, and $\dot{Q}_c$ is rate of heat addition, J/s or watts. To obtain the local value of heat addition $\dot{q}_c$ one uses

$$\dot{q}_c = \alpha I(1 - \eta_Q) \qquad (23.294)$$

where α is local gain coefficient (1/m) and I is local intensity (watts/m^2). Hence $\dot{q}_c$ has units watts/m^3. For Blocks 1 and 2 in Fig. 23.147, use the Tsien and Beilock equations along with Eq. (23.294) for the line heat sources within the laser cavity. Equation (23.288) is not applicable for this problem; hence for Block 3 one uses the formulas of Tsien and Beilock to find ρ. Blocks 4 to 6 are used according to previous discussion. For a detailed example of laser cavity flow with laser quantum efficiency, see Fuhs (1973).

For electrical lasers, heat is added due to the flow of current through an electrical potential. The electrical power per unit volume (watts/m^3) is

$$P_E = \bar{\mathbf{E}} \cdot \bar{\mathbf{J}} \qquad (23.295)$$

where $\bar{\mathbf{E}}$ is the electric field vector (volts/m) and $\bar{\mathbf{J}}$ is the current density (amperes/m^2). A fraction of P_E is used to pump the laser. Another fraction of P_E is dissipated as heat. Hence, for an electrical laser

$$\dot{q}_c = \alpha I(1 - \eta_Q) + f_d P_E \qquad (23.294a)$$

where f_d is the fraction of P_E dissipated as heat. An example of heating in an electrical laser cavity is given by Biblarz and Fuhs (1973).

The refractive index of a gas depends on the mole fraction of each specie and on the populations within each molecular specie. Using the formulation of Fano and Cooper (1968) one can show that the refractive index is given by

$$n - 1 = \sum_l X_l \left\{ \frac{2\pi e^2}{m} \sum_{i=0} \sum_{k=i+1} \frac{f_{ik} N_i}{\omega_{ik}^2 - \omega^2} \left[1 - \frac{N_k}{N_i} \frac{g_i}{g_k} \right] \right\}_l \qquad (23.296)$$

The symbols have the following meaning: X_l is mole fraction of species l; e is the charge of an electron; m is the mass of an electron; f_{ik} is the oscillator strength for transition between levels i and k; N_i is the number of molecules of species l in energy level i; ω_{ik} is the frequency for a transition between levels i and k; ω is the frequency at which the refractive index n is to be evaluated; g_i and g_k are the degeneracies of levels i and k, respectively. A variation in X_l changes n. A change in populations N_i or N_k changes n.

An example of transient refractive index in an XeF excimer laser is given by Fuhs, Cole, and Etchechury (1980). Quality for excimer lasers is discussed by Hogge and Crow (1978).

Combustion changes both populations and concentrations (mole fractions) of molecular species. Gross and Bott (1976) gives information about combustion driven chemical lasers.

23.14.6 Wave Front Variance

When wave flow interactions in turbulent flows are random, statistical analyses can be used to predict aero-optical distortion. Usual quantities sought are means, variances, and standard deviations, correlations, and length scales for the fluctuating quantities. Wave front variance, in particular, is an important statistic; it is fundamental to modeling other optical performance parameters like optical transfer functions, point spread functions, and Strehl ratios.

Wave front variance is the statistical expectation of the dispersion in the randomness about the mean, μ_Γ, of the optical path lengths, Γ, in a beam-aperture projected through the flow. Treating the fluctuations in optical path length (equivalently optical phase, Φ) as random variables belonging to the random function $\{\Gamma(t)\}$, wave front variance is,

$$\sigma_\Phi^2(t) = \text{Var}_\Phi(t) = E[(\Phi(t) - \mu_\Phi)^2]$$

$$= k^2 \, \text{Var}_\Gamma(t) = k^2 E[(\Gamma(t) - \mu_\Gamma)^2] \qquad (23.297)$$

where $\sigma^2(\ ) = \text{Var}(\ )$ is the variance, $E[\ \ ]$ denotes the statistical expectation, and k is the wave number, $k = 2\pi/\lambda$. A temporal variance pertains to the time dependent fluctuations at a specific point on the waves, whereas a spatial variance pertains to the instantaneous irregularities along a wave front. The temporal and spatial variances are equal when the randomness is *ergodic* (e.g., the turbulence is statistically isotropic and weakly stationary throughout the beam-aperture).

Models for specific aero-optical scenarios can be developed from Eq. (23.297) as, for example, the fundamental scenario illustrated by the ideal conditions of Fig. 23.155(a). Here, a turbulent fluid layer consisting of a thermally perfect gas flows uniformly across the flat surface of an optically perfect window. The optical waves enter the layer normal to the free surface ($z = L$) and thereafter propagate through the layer in the negative z-direction toward the interface between the fluid and the window ($z = 0$). The optical waves are not refracted by the flow density, ρ, and they experience no distortion outside L. The mean flow is a fully developed two-dimensional turbulent boundary layer with the mean velocity vector pointing in the positive x-direction. In the beam aperture, $\rho(z)_{x,y}$ is a random process, $\rho(z)_{x,y} = \{\rho_i(t)\}_{x,y}$, where i denotes N normally distributed sample functions as $z_{i=1} = 0, z_2,$ $z_3, \ldots, z_{i=N} = L$. Figure 23.156 illustrates typical temporal distributions for N-sample functions, and Fig. 23.157 shows a representative instantaneous spatial distribution compared to the mean, $\mu_\rho(z, t_1)$.

Owing to the flow turbulence, the change in optical phase across L is a random

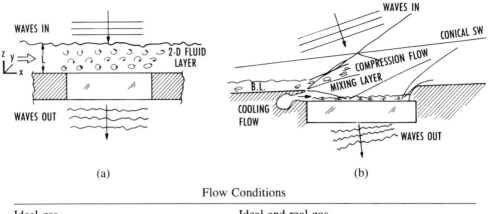

(a)   (b)

Flow Conditions

Ideal gas	Ideal and real gas
Developed 2nd flow	3rd asymmetric flow
No flow interactions	Strong flow interactions
No mixing	Strong viscous mixing
Homogeneous	Nonhomogeneous turbulence
Turbulence	
Small, constant l_ρ	Variable l_ρ
Subsonic flow validation	No validation
No refraction	Strong refraction
K_{GD} known	Species concentration gradients, K_{GD} unknown
No resonance effects	Resonance effects possible

FIGURE 23.155   Ideal and real flow aero-optic scenarios. (a) Ideal case and (b) real case.

variable with mean and fluctuating components,

$$\Phi(t)_{x,y} = \mu_\Phi(t) + \Phi'(t) \tag{23.298}$$

The prime denotes a fluctuation about the mean at (x, y). With $\Phi(t)$ assumed stationary, $\mu_\Phi(t)$ is constant. Combining Eqs. (23.297) and (23.298), the wave front

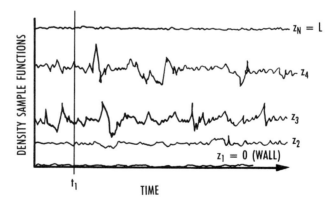

FIGURE 23.156   Illustration of normally distributed n-sample functions in l.

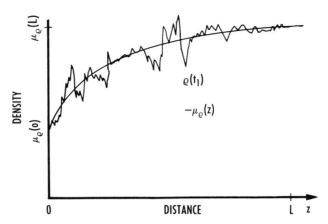

FIGURE 23.157 Illustration of representative instantaneous mean and spatial density distributions for one ray through l.

variance is,

$$\sigma_\Phi^2(\Delta L)_{x,y} = E[(\Phi')^2]_{x,y} = kE[(\Gamma')^2]_{x,y} \qquad (23.299)$$

where the dependence on t is hereafter understood, and ΔL signifies the variance over L.

At any instant, the optical phase (hence, $\sigma_\Phi^2(\Delta L)_{x,y}$) depends on $\Gamma(x, y)$ which, in turn, depends on the density fluctuations, $\rho'(z)_{x,y}$. Optical path length is the line integral of the index of refraction, $\Gamma(\Delta L)_{x,y} = \int_0^L n(z)_{x,y} \, dz$, which is also a random variable,

$$\Gamma(\Delta L)_{x,y} = \mu_\Gamma(\Delta L)_{x,y} + \Gamma'(\Delta L)_{x,y}$$

$$= \int_0^L \mu_n(z)_{x,y} \, dz + \int_0^L n'(z)_{x,y} \, dz \qquad (23.300)$$

Here, $n(z)_{x,y}$ is the local index of refraction, $n = \mu_n + n'$ for the (x, y)-ray. Considering wavelengths far from resonance, specific refractivity for an ideal gas is given by the Gladstone–Dale relationship, $K_{GD} = (n - 1)/\rho$, where $\mu_n(z) = 1 + \mu_\rho(z)K_{GD}$, and $n'_{x,y}(z) = \rho'_{x,y}(z)K_{GD}$. Combining Eqs. (23.299) and (23.300), and noting that $\Gamma'(\Delta L)_{x,y} = K_{GD} \int_0^L \rho'(z)_{x,y} \, dz$,

$$\sigma_\Phi^2(\Delta L)_{x,y} = \beta^2 E\left\{ \left[\int_0^L \rho'(z) \, dz \right]^2 \right\}_{xy} \qquad (23.301)$$

where $\beta = k \cdot K_{GD}$. So for an ideal gas, wave front variance is directly proportional to the statistical expectation of a density fluctuation line integral, squared. For high speed real gas flows, however, linking wave front variance to flowfield effects is not straightforward, because the wave flow interactions that account for specific refractivity and optical dispersion, or the density fluctuations along the path of wave propagation, are unknown.

Some statistical expectations are worthwhile here. First, $\rho'(z)$ is a zero-mean statistic since $E[\rho'(z)] = \mu_\rho'(z) = 0$ with respect to $\mu_\rho(z)$. Next,

$$E\left[\int_0^L \rho(z)\,dz\right]_{x,y} = \int_0^L E[\rho(z)]_{x,y}\,dz = \int_0^L \mu_\rho(z)_{x,y}\,dz \qquad (23.302)$$

and,

$$E\left[\left(\int_0^L \rho'(z)\,dz\right)^2\right]_{x,y} = E\left[\int_0^L \rho'(z_1)\,dz_1\int_0^L \rho'(z_2)\,dz_2\right]_{x,y}$$

$$= E\left[\int_0^L\int_0^L \rho'(z_1)\rho'(z_2)\,dz_1\,dz_2\right]_{x,y}$$

$$= \int_0^L\int_0^L E[\rho'(z_1)\rho'(z_2)]_{x,y}\,dz_1\,dz_2$$

$$= \int_0^L\int_0^L \mathrm{Cov}_\rho(z_1, z_2)_{x,y}\,dz_1\,dz_2 \qquad (23.303)$$

where $\mathrm{Cov}_\rho(z_1, z_2)_{x,y}$ is the *statistical covariance* between points z_1, z_2 along z. These expectations follow from the additive property of the *Normal Statistical Model:* The mean and variance of a summation of normally distributed random variables is the linear summation of the means and variances, respectively, of the individual random variables. Thus, Eq. (23.301) is,

$$\sigma_\Phi^2(\Delta L)_{x,y} = \beta^2\int_0^L\int_0^L \mathrm{Cov}_\rho(z_1, z_2)_{x,y}\,dz_1\,dz_2 \qquad (23.304)$$

which is the fundamental wave front variance model for Fig. 23.155(a) scenario. It is perfectly general for the assumptions stated thus far, and further modeling requires treatment for $\mathrm{Cov}_\rho(z_1, z_2)_{x,y}$. For fields with homogeneous turbulence, integral lengthscales, such as $l_\rho(z)$ for density, depend only on coordinate differences, $\Delta x = x_2 - x_1$, $\Delta y = y_2 - y_1$, $\Delta z = z_2 - z_1$. By definition $l_\rho(z)$ is the half plane area bounded by the correlation $\rho'(z_1) \otimes \rho'(z_2)$, $\otimes$ denotes correlation. Accordingly, two models proposed for the covariance are an exponential distribution after Steinmetz (1982),

$$\mathrm{Cov}_\rho(\Delta x, \Delta y, \Delta z) = \sigma_\rho^2(z)\exp - (\sqrt{(\Delta x/l_\rho)^2 + (\Delta y/l_\rho)^2} + (\Delta z/l_\rho)^2) \qquad (23.305)$$

and a Gaussian model after Wolters (1973),

$$\mathrm{Cov}_\rho(\Delta x, \Delta y, \Delta z) = \sigma_\rho^2(z)\exp - ((\Delta x/l_\rho)^2 + (\Delta y/l_\rho)^2 + (\Delta z/l_\rho)^2) \qquad (23.306)$$

Note that $\Delta x = \Delta y = 0$ in Eq. (23.304), and hereafter, the (x, y) subscript is understood.

It can be shown that substituting Eqs. (23.305) or (23.306), into Eq. (23.304) gives,

$$\sigma_\Phi^2(\Delta L) = 2\beta^2 \int_0^L \sigma_{\rho'}^2(z) l_\rho(z) W(z) \, dz \qquad (23.307)$$

where $W(z)$ is a weighting function,

$$W(z) = \begin{cases} 2(1 - \exp(-2z/l_\rho)), & 0 \le z \le L/2 \\ 2(1 - \exp(-2(L-z)/l_\rho)), & L/2 \le z \le L \end{cases} \qquad (23.308)$$

for Eq. (23.305), and,

$$W(z) = \begin{cases} \pi(\text{erf}\,(2z/l_\rho)), & 0 \le z \le L/2 \\ \pi(\text{erf}\,(2(L-z)/l_\rho)), & L/2 \le z \le L \end{cases} \qquad (23.309)$$

for Eq. (23.306), respectively; exp( ) and erf( ) are the *exponential* and *error functions*, respectively.

$W(z)$ is a *weighting* of $\rho'(z)$. Explicit definition of $W(z)$ depends directly on $\text{Cov}_{\rho'}(\Delta z)_{x,y}$, and for the simplified scenario here, it depends only on $l_\rho(z)$. As mentioned above, however, for real gas flows where specific refractivity and dispersion are variables, or for flows where $\rho'(z)$ results from the evolution of coherent flow structures, $W(z)$ is more complicated because treatment for $\rho'(z)$ is more complicated and most likely predictable only from Direct Numerical Simulations.

$W(z)$ influences are examined for three representative $l_\rho(z)$-distributions. Since Eqs. (23.308) and (23.309) produce similar results, only Eq. (23.308) is illustrated for: 1) $l_\rho(z)$ constant over L, 2) $l_\rho(z)$ parabolic over L, and 3) $l_\rho(z)$ exponentially piece wise continuous over L, this distribution being motivated by the experimental findings of Johnson and Rose (1976). First, $W(z)$ is normalized using $z^* = z/L$ and $\alpha = L/l_\rho$. Figure 23.158 shows $W(z^*)$ distributions for case 1, constant α for three different magnitudes. Note that $W(0) = W(1) = 0$, but elsewhere, $[0 < z^* < 1]$,

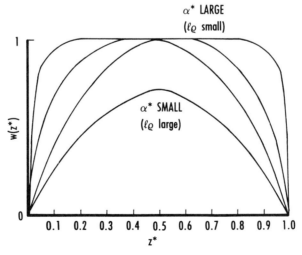

FIGURE 23.158 Illustration of $w(z^*)$ for constant α.

$W(z^*)$ approaches 1 at a rate dependent upon α. For $\alpha \gg 1$ (e.g., $20 \leq \alpha \leq 40$ in subsonic transonic flows), $W(z^*) \sim 1$ except at the boundaries. For other cases (e.g., $\alpha \leq 15$ in high speed flows), $W(z^*) \neq 1$, and to obtain accurate predictions for $\sigma_\phi^2(\Delta L)_{x,y}$, either Eqs. (23.308) or (23.309) must be included in evaluating Eq. (23.307). Analogous to optical wave transmission through materials with small-scale crystal structure that have numerous internal scattering surfaces, turbulent flows with small density length scales have more eddies that generate more random optical refraction than flows with relatively large length scales, so the predicted wave front distortion is larger, which means $W(z^*) = 1$. Otherwise, $W(z^*) < 1$ to account for fewer refraction sources. Figure 23.159 shows $W(z^*)$ distributions for all three cases.

To illustrate the influence of $W(z^*)$ on $\sigma_\phi^2(\Delta L)_{x,y}$, some additional normalization is helpful. First, the density variance and density length-scale ratios are normalized according to $\sigma_\rho^{2*}(z^*) = \sigma_\rho^2(z^*)/(\sigma_\rho^2)_m$, and $\alpha^*(z^*) = \alpha(z^*)/\alpha_m$, where the m subscript signifies the respective maximums over L. Then with $l_\rho(z^*) = L/\alpha^*(z^*)\alpha_m$, Eq. (23.307) becomes,

$$\sigma_\phi^2(\Delta L) = 2\beta^2 L^2 \frac{(\sigma_\rho^2)_m}{\alpha_m} \int_0^1 \frac{\sigma_\rho^{2*}(z^*)}{\alpha^*(z^*)} W(z^*) \, dz^* \qquad (23.310)$$

with,

$$W(z^*) = \begin{cases} 1 - \exp(-2\alpha^* z^*) & 0 \leq z^* \leq 1/2 \\ 1 - \exp(-2\alpha^*(1 - z^*)) & 1/2 \leq z^* \leq 1 \end{cases} \qquad (23.311)$$

Equation (23.310) is resolvable once the integrand factors are defined. Noting that the normalized density variance and density length scale functions are reciprocal factors bounded on [0, 1], a first-order simplification is that $\sigma_{\rho'}^{2*}(z^*)/\alpha^*(z^*) \approx 1$, arguing that $\sigma_{\rho'}^{2*}(z^*)$ is a maximum when $\alpha^*(z^*)$ is a maximum, this being the z^* where the $l_\rho(z^*)$ is a minimum, and that other relative extremes also coincide (note, $\sigma_{\rho'}^2(z)/\alpha(z) \neq 1$, necessarily). Thus,

$$\sigma_\phi^{2*}(\Delta L) = \int_0^1 W(z^*) \, dz^* \qquad (23.312)$$

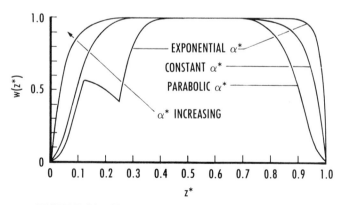

FIGURE 23.159   Illustration of $w(z^*)$ for variable α.

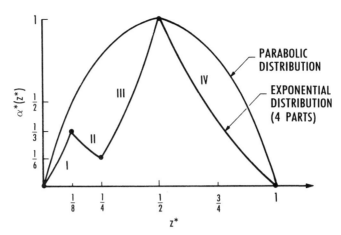

FIGURE 23.160 Normalized spatial distributions for the density length-scale parameter, l_ρ.

where $\sigma_\Phi^{2*}(\Delta L) = \sigma_\Phi^2(\Delta L)/2\beta^2 L^2(\sigma_{\rho'}^2)_m/\alpha_m$ is a normalized phase variance. So now for the same maximum conditions, the predicted wave front variance depends only on the integrated distribution of $W(z^*)$. Figure 23.160 shows $\alpha^*(z^*)$ distributions for the three cases. For case 1, $\alpha^*(z^*) = 1$, for case 2, $\alpha^*(z^*) = 4z^*(1 - z^*)$, and for case 3, $\alpha_I^*(z^*) = e^{2.30146^*} - 1$, $0 \le z^* \le 1/8$, $\alpha_{II}^*(z^*) = 2/3e^{-5.5452z^*}$, $1/8 \le z^* \le 1/4$, $\alpha_{III}^*(z^*) = 1/36e^{7.167z^*}$, $1/4 \le z^* \le 1/2$, and $\alpha_{IV}^*(z^*) = e^{1.38629(1-z^*)} - 1$, $1/2 \le z^* \le 1$. Using these models, Eq. (23.312) predicts $\sigma_\Phi^{2*}(\Delta L) = 0.3679$ for case 1, $\sigma_\Phi^{2*}(\Delta L) = 0.3075$ for case 2, and $\sigma_\Phi^{2*}(\Delta L) = 0.2077$ for case 3. The actual wave front variance is obtained from these values once the physical parameters (λ, K_{GD}, L, $\sigma_{\rho'}^2(\Delta L)_m$, and $l_\rho(z)_{min}$) are defined.

In summary, $W(z) \approx 1$ for $\sigma_\Phi^2(\Delta L)$ estimates when $L \gg l_\rho$ (large α), but omitting treatment for $W(z)$ when $l_\rho\, OL(\alpha \le 15)$ overestimates $\sigma_\Phi^2(\Delta L)$, sometimes by as much as 40% depending on the physical parameters of the scenario. So using $W(z) = 1$ is tantamount to *worst case* aero-optical estimates for the Fig. 23.155(a) scenario; elsewhere, it may have more significance.

Figure 23.155 presents a direct comparison between ideal (a) and real gas (b) aero-optics scenarios. Scenario b is for hypersonic flow over an aerodynamically cooled window where the interacting fluid streams create a viscous turbulent shear layer with concentration gradients, dissociation, chemical interactions, and thermal nonequilibrium states as possible real gas conditions. Effects owing to ablation, plasma, and resonance may also be present. When cast against Scenario a, the inappropriateness of Eq. (23.307) is recognized. All the assumptions needed to develop this model are now restrictions that render the applicability suspicious. To single out one—homogeneous turbulence, Chernov (1960) states that the density covariance models of Eqs. (23.305) and (23.306) are strictly valid only for flows with homogeneous, isotropic turbulence; homogeneous turbulence is unlikely in high speed complex flows like Scenario b. So how important is the homogeneous turbulence assumption? This question is unanswerable now, because there are too few data upon which to make conclusions.

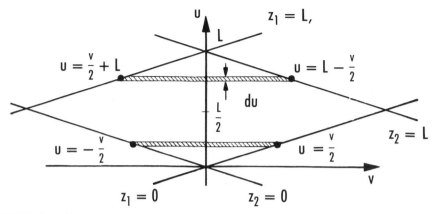

FIGURE 23.161   Coordinate transformation for evaluation of spatial density covariance functions.

Another concern is the proper choice for $l_\rho(z)$. Aside from characterizing the large-scale turbulent eddy effects, use of an integral length scale instead of a dissipation length scale appears to be a matter of convenience. However, in light of the $W(z)$ influences, the small scale eddies may well have the dominant effect on $\sigma_\Phi^2(\Delta L)_{x,y}$, in which case, dissipation length scales might be better characteristic dimensions for $\text{Cov}_{\rho'}(z_1, z_2)_{x,y}$ models. But, in practice, accurate dissipation length scales are difficult to determine, because they are inversely proportional to the second derivative of the $\rho'(z)$-correlation function thus small errors in the data are greatly magnified. Conversely, accurate integral length scales are relatively easy to determine, because small errors in the experimental data have little effect on the computation of the half-plane area bounded by the correlation function.

Other concerns for validity are discussed in Havener (1992), but in short, Eq. (23.307) should be used only heuristically when the governing assumptions are invalid.

REFERENCES

Abbott, I. H., von Doenhoff, A. E., and Stivers, L. S., Jr., "Summary of Airfoil Data," NACA Rept. 824, 1945.

Abramson, H. N. (Ed.), *The Dynamic Behavior of Liquids in Moving Containers*, NASA SP-106, 1966.

Abramson, H. N., "Hydroelasticity: A Review of Hydrofoil Flutter," *Applied Mechanics Review*, Vol. 22, No. 2, pp. 115–121, 1969.

Ackeret, J., "Anwendungen der Aerodynamik im Bauwesen," *Z. f. Flugwissenschaften*, Vol. 13, No. 4, pp. 109–122, 1965.

Acosta, A. J. and Parkin, B. R., "Cavitation Inception—A Selective Review," *J. Ship Res.*, Vol. 19, pp. 193–205, 1975.

Acosta, A. J., "Cavitation and Fluid Machinery," *Cavitation, Proc. Conf. Heriot-Watt Univ.*, Edinburgh, Scotland, Inst. Mech. Engrs., London, 1974.

AGARD Manual on Aeroelasticity, Vols. I–VII, beginning 1959 with continual updating.

Aldridge, K. D. and Toombe, A., "Axisymmetric Inertial Oscillations of a Fluid in a Rotating Spherical Container," *J. Fluid Mechanics*, Vol. 37, pp. 307–323, 1969.

Allen, H. J. and Perkins, E. W., "A Study of Viscosity on Flow Over Slender Inclined Bodies of Revolution," NACA Rep. 1048, 1951.

Anon., "Application of Laminar Flow Control to Supersonic Transport Configurations," NASA CR 181917, July, 1990.

Anon., "The NASTRAN Theoretical Manual," Section 17, Dec. 1977, with continual updates.

Arcara, P. C., Jr., Bartlett, D. W., and McCullers, L. A., "Analysis for the Application of Hybrid Laminar Flow Control to a Long-Range Subsonic Transport Aircraft," SAE Paper No. 91-2113, Aerospace Technology Conference and Exposition, Long Beach, CA, 1991.

Arnal, D. and Thibert, J. J., "Le Controle de L'Ecoulement Laminaire," *L'Aeronautique et l'Astronautique*, Nos. 148 and 149, Nos. 3 and 4, pp. 61–66, 1991.

Arnal, D., "Boundary Layer Transition: Prediction, Application to Drag Reduction," Special Course on Skin Friction Drag Reduction, AGARD Report 786, 1992.

Arndt, R. E. A., "Cavitation in Fluid Machinery and Hydraulic Structures," *Ann. Rev. Fluid Mech.*, Vol. 13, pp. 273–328, 1981.

Ashford, R. L., Bata, B. T., and Walsh, E. D., "Measurement of Helium Gas Through Aerostat Material," AIAA 83-1986, 1983.

ASTM, *ASTM Standard Method*, "Standard Method of Vibratory Cavitation Erosion Test," ANSI/ASTM G32-77, 1977.

Ausman, J. S., "An Improved Analytical Solution for Self-Acting, Gas Lubricated Journal Bearings of Finite Length," *J. Basic Eng.*, Vol. 83, pp. 188–194, 1961.

Badesha, S. and Jones, S. P., "The Aerodynamics of the TCOM 71M Aerostat," AIAA 93-4036-CP, 1993.

Baitis, A. E. *et al.*, "User's Manual for the Standard Ship Motion Program, SMP," DTNSRDC Report SPD-0936-01, Washington, D.C., 1981.

Baker, C. J. and Humphreys, N. B., "Aerodynamic Forces and Moments on Containers on Flat Wagons in Cross Winds From Moving Model Tests," Dept. of Civil Eng., Nottingham Univ., Rep. No. FR 91017, 1991.

Baker, C. J., "Train Aerodynamic Forces and Moments From Moving Model Experiments," *J. Wind Eng. Ind. Aerodyn.*, Vol. 24, pp. 227–251, 1986.

Ballhaus, W. F. and Goorjian, P. M., "Implicit Finite-Difference Computations of Unsteady Transonic Flow About Airfoils, Including the Effects of Irregular Shock Motions," *AIAA J.*, Vol. 15, No. 12, pp. 1728–1735, 1977.

Barr, P. V., Brimacombe, J. K., and Watkinson, A. P., "A Heat-Transfer Model for Pilot Kiln Trials," *Metallurgical Trans. B*, Vol. 20B, pp. 391–419, 1989.

Barry, B., Parke, S. J., Brown, N. W., Reidel, H., and Sitzmann, M., "The Flight Testing of Natural and Hybrid Laminar Flow Nacelles," ASME 94-GT-408, International Gas Turbine and Aeroengine Congress and Exposition, The Hague, The Netherlands, 1994.

Batina, J. T., Seidel, D. A., Bland, S. R., and Bennett, R. M., "Unsteady Transonic Flow Calculations for Realistic Aircraft Configurations," *J. Aircraft*, Vol. 26, No. 2, pp. 131–139, 1989.

Bauer, F., Garabedian, P., Korn, D., and Jameson, A., "Supercritical Wing Sections II," *Lecture Notes in Economics and Mathematical Systems*, Springer-Verlag, Berlin, 1975.

Bendel, H., "Untersuchungen zur Verringerung des aerodynamischen Widerstandes von Güterwagen," *ZEV-Glas. Ann.*, Vol. 114, No. 4, pp. 124–132, 1990.

Bennon, W. D. and Incropera, F. P., "A Continuum Model for Momentum, Heat, and

Species Transport in Binary Solid–Liquid Phase Change—I. Model Formulation,'' *Int. J. Heat Mass Trans.*, Vol. 30, No. 10, pp. 2161–2170, 1987a.

Bennon, W. D. and Incropera, F. P., ''A Continuum Model for Momentum, Heat, and Species Transport in Binary Solid–Liquid Phase Change—II. Application to Solidification in a Rectangular Cavity,'' *Int. J. Heat Mass Trans.*, Vol. 30, No. 10, pp. 2171–2187, 1987b.

Bhutiani, P. K., Keck, D. F., Lahti, D. J., and Stringas, M. J., ''Investigating the Merits of a Hybrid Laminar Flow Nacelle,'' *The Leading Edge*, General Electric Company, pp. 32–35, 1993.

Biblarz, O. and Fuhs, A. E., ''Laser Internal Aerodynamics and Beam Quality,'' *Developments in Laser Technology—II*, Vol. 41, Society of Photo-Optical Instrumentation Engineers, pp. 59–70, 1973.

Bingham, G. J. and Noonan, K. W., ''Experimental Investigation of Three Helicopter Rotor Airfoils Designed Analytically,'' NASA TP 1396, 1979.

Birch, S., ''Technology Update: Laminar Flow,'' *Aerospace Engineering, SAE*, Vol. 12, No. 3, pp. 45–47, March 1992.

Bisplinghoff, R. L. and Ashley, H., *Principles of Aeroelasticity*, John Wiley & Sons, New York, 1962. (Also available in Dover Edition.)

Bisplinghoff, R. L., Ashley, H., and Halfman, R. L., *Aeroelasticity*, Addison-Wesley, Boston, 1955.

Blackwell, J. A. and Hinson, B. L., ''The Aerodynamic Design of an Advanced Rotor Airfoil,'' NASA CR 2961, 1977.

Boldt, R., Jr., *The Winning of Nickel*, Van Nostrand, New York, 1967.

Bolotin, V. V., *Nonconservative Problems of the Elastic Theory of Stability*, Pergamon Press, Oxford, England, 1975.

Bonnin, J., Billet, M. L., Hammitt, F. G., and Chaix, B., ''Survey of Present Knowledge on Cavitation in Liquids Other than Cold Water (Thermodynamic Effect),'' *J. Hyd. Res. (IAHR)*, Vol. 19, No. 4, pp. 277–305, 1981.

Borland, C. J. and Rizzetta, D. P., ''Nonlinear Transonic Flutter Analysis,'' AIAA 81-0608-CP, 1981.

Born, M. and Wolf, E., *Principles of Optics*, Pergamon Press, Oxford, England, 1964.

Brennan, C., Oey, K. T., and Babcock, C. D., ''Leading-Edge Flutter of Supercavitating Hydrofoils,'' *J. Ship Research*, Vol. 24, No. 3, pp. 135–146, 1980.

Bridgewater, A. V. and Mumford, C. J., *Waste Recycling and Pollution Control Handbook*, Van Nostrand Reinhold, New York, 1979.

Brimacombe, J. K., ''The Extractive Metallurgist in an Emerging World of Materials,'' *Metallurgical Trans. B*, Vol. 20B, pp. 291–313, 1989.

Brockie, N. J. W. and Baker, J. W., ''The Aerodynamic Drag of High Speed Trains,'' *J. Wind Eng. Ind. Aerodyn.*, Vol. 34, pp. 273–290, 1990.

Brown, J. S. H. and Speed, L. A., ''Ballonet Kite Balloons Design, Construction and Operation,'' Royal Aircraft Establishment (Farnborough), Report No. Mech. Eng. 24, 1962.

Bulgubure, C. and Arnal, D., ''DASSAULT FALCON 50 Laminar Flow Flight Demonstrator,'' *Proceedings on the First European Forum on Laminar Flow Technology*, pp. 11–18, 1992.

Bush, John, IBIS, Inc., Wellesley, MA, private communication, 1994.

Cameron, A., *The Principles of Lubrication*, John Wiley & Sons, New York, 1966.

Carey, E. F. and Fuhs, A. E., ''Transonic Thermal Blooming due to an Intense Laser Beam,'' *J. Aircraft*, Vol. 13, pp. 974–980, 1976.

Cha, Y. S., "On Bubble Cavitation and Dissolution," *Intl. J. Heat and Mass Transfer*, 1983.

Cheng, H. S. and Pan, C. H. T., "Stability Analysis of Gas-Lubricated, Self-Acting, Plain, Cylindrical, Journal Bearings of Finite Length, Using Galerkin's Method," *J. Basic Engineering*, Vol. 87, No. 1, 1965.

Chernov, L. A., *Wave Propagation in a Random Medium*, McGraw-Hill, New York, 1960.

Chi, M. R., "Unsteady Aerodynamics in Stalled Cascade and Stall Flutter Prediction," ASME 80-C2-1, 1980.

Choo, R. T. C., Szekely, J., and David, S. A., "On the Calculation of the Free Surface Temperature of Gas-Tungsten-Arc Weld Pools from First Principles: Part II. Modeling the Weld Pool and Comparison with Experiments," *Met. Trans.*, Vol. 23B, pp. 371–384, 1992.

Chow, L. S. H. and Cheng, H. S., "The Effect of Surface Roughness on the Average Film Thickness between Lubricated Rollers," *J. Lubrication Technology*, Vol. 98, pp. 117–224, 1976.

Christensen, H. and Tonder, K., "The Hydrodynamic Lubrication of Rough Journal Bearings," *J. Lubrication Technology*, Vol. 95, No. 1, pp. 166–172, 1973.

Christensen, H., "Stochastic Models for Hydrodynamic Lubrication of Rough Surfaces," *Proc. Inst. Mech. Eng.*, Vol. 184, pp. 1013–1022, 1969.

Chyu, W. J. and Kuwahara, K., "Computations of Transonic Flow Over an Oscillating Airfoil with Shock-Induced Separation," AIAA 82-0350, 1982.

CMP, "Modeling of the Glass Melting Process," CMP Report No. 90-4, January 1990.

Cockey, W. D., *Panel Method for Perturbations of Transonic Flows With Finite Shocks*, Ph.D. thesis, Princeton University, Princeton, NJ, June 1983.

Cole, R. H. and Coles, J. S., *Physical Principles of Chemistry*, McGraw-Hill, New York, 1964.

Comstock, J. P. (Ed.), *Principles of Naval Architecture*, Society of Naval Architects and Marine Engineers, New York, 1967.

Concus, P., Crane, G. E., and Satterlee, H. M., "Small Amplitude Lateral Sloshing in a Cylindrical Tank with a Hemispherical Bottom Under Low Gravitational Conditions," NASA CR-54700, 1967.

Concus, P., Crane, G. E., and Satterlee, H. M., "Small Amplitude Lateral Sloshing in Spheroidal Containers Under Low Gravitational Conditions," NASA CR-72500, 1969.

Constantinescu, V. N., "On Turbulent Lubrication," *Proc. Inst. Mech. Eng.*, Vol. 173, No. 38, pp. 881–889, 1959.

Cooper, R. K., "Atmospheric Turbulence With Respect to Moving Ground Vehicles," *J. Wind Eng. Ind. Aerodyn.*, Vol. 17, pp. 215–238, 1984.

Copley, C., "The Three-Dimensional Flow Around Railway Trains," *J. Wind Eng. Ind. Aerodyn.*, Vol. 26, pp. 21–52, 1987.

Crocker, M. J. and Price, A. J., *Noise and Noise Control*, Vol. I, CRC Press, 1975.

Dadone, L. U. and Fukushima, T., "A Review of Design Objectives for Advanced Helicopter Rotor Airfoils," American Helicopter Society Symposium on Helicopter Aerodynamic Efficiency, 1975.

Dantzig, J. A. and Berry, J. T. (Eds.), *Engineering Foundation Conferences—Modeling of Casting and Welding Processes II*, The Metall. Soc., Warrendale, PA, 1984.

Dashcund, D. E., *The Development of a Theoretical and Experimental Model for the Study of Active Suppression of Wing Flutter*, Ph.D. thesis, MAE Technical Report 1496-T, Princeton University, Princeton, NJ, December 1980.

Dat, R. and Tran, C. T., "Investigation of the Stall Flutter of an Airfoil with a Semi-Empirical Model of 2D Flow,'' International Symposium on Aeroelasticity, Nuremberg, Germany, 1981.

Davies, P. B. and Howarth, R. B., ''Hydrostatic Lubrication,'' *Principles of Tribology,* Macmillan Press, 1978.

Dayman, B. and Vardy, A. E., ''Alleviation of Tunnel Entry Pressure Transients: 1. Experimental Program,'' *Proc. 3rd Int. Symp. Aero. Vent. Vehicle Tunnels,* Sheffield, BHRA Fluid Eng., pp. 343–362, 1979.

De, M. K. and Hammitt, F. G., ''New Method for Monitoring and Correlating Cavitation Noise to Erosion Capability,'' *J. Fluids Engr.,* Dec. 1983.

Dilawari, A. H. and Szekely, J., ''A Mathematical Representation of a Modified Stagnation Flow Reactor for MOCVD Applications,'' *J. Crystal Growth,* pp. 491–498, 1991.

Dilawari, A. H. and Szekely, J., ''Computed Results for the Deposition Rates and Transport Phenomena for an MOCVD System With a Conical Rotating Substrate,'' *J. Crystal Growth,* Vol. 97, pp. 777–791, 1989.

Dodge, F. T. and Kana, D. D., ''Moment of Inertia and Damping of Liquids in Baffled Cylindrical Tanks,'' *J. Spacecraft and Rockets,* Vol. 3, pp. 153–155, 1966.

Dodge, F. T., Green, S. T., and Cruse, M. W., ''Analysis of Small-Amplitude Low Gravity Sloshing in Axisymmetric Tanks,'' *Microgravity Science and Technology,* IV/4, pp. 228–234, 1991.

Dodge, F. T., Kana, D. D., and Abramson, H. N., ''Liquid Surface Oscillations in Longitudinally Excited Rigid Cylindrical Containers,'' *AIAA J.,* Vol. 3, pp. 685–695, 1965.

Dowell, E. H., ''A Simple Method for Converting Frequency Domain Aerodynamics to the Time Domain,'' NASA TM 81844, 1980.

Dowell, E. H., Bland, S. R., and Williams, M. H., ''Linear/Nonlinear Behavior in Unsteady Transonic Aerodynamics,'' *AIAA J.,* Vol. 21, No. 1, p. 38, 1983.

Dowell, E. H., Curtiss, H. C., Jr., Scanlan, R. H., and Sisto, F., *A Modern Course in Aeroelasticity,* Martinus Nijhoff, The Netherlands and Higham, MA, 1979.

Dowell, E. H., Williams, M. H., and Chi, M. R., ''An Assessment of Theoretical Methods for Viscous and Transonic Flow,'' presented at the AGARD Conference on Boundary Layer Effects on Unsteady Airloads, in Aix-en-Provence, France, Sept. 1980.

Dowell, E. H., *Aeroelasticity of Plates and Shells,* Noordhoff International Publishing, Leyden, 1975.

Dowson, D. and Higginson, G. R., *Elastohydrodynamic Lubrication,* Pergamon Press, Oxford, England, 1977.

Dowson, D., ''A Generalized Reynolds Equation for Fluid-Film Lubrication,'' *Int. J. Mechanical Sci.,* Vol. 4, pp. 159–170, 1962.

Doyama, M., Kihara, J., Tanaka, M., and Yamamoto, R. (Eds.), *Computer Aided Innovation of New Materials I and II—Proceedings of the Second International Conference and Exhibition on Computer Applications to Materials and Molecular Science and Engineering—CAMSE '92,* North-Holland, Amsterdam, 1993.

Doyama, M., Suzuki, T., Kihara, J., and Yamamoto, R. (Eds.), *Computer Aided Innovation of New Materials—Proceedings of the First International Conference and Exhibition on Computer Applications to Materials Science and Engineering—CAMSE '90,* North-Holland, Amsterdam, 1990.

Drees, J. M. and Wortmann, F. X., ''Design of Airfoils for Rotors,'' paper presented at CAL/AVLABS, 1969.

Durand, W. F., ''Aerostatics: Structure of the Atmosphere,'' *Aerodynamic Theory, Vol. 1,* Durand, W. F. (Ed.), Julius Springer, Berlin, pp. 216–223, 1934 (republished by Peter Smith, Gloucester, MA, 1976).

Edwards, J. H. and Malone, J. B., "Current Status of Computational Methods for Transonic Unsteady Aerodynamic and Aeroelastic Applications," presented at the AGARD Structures and Materials Panel Specialist's Meeting on Transonic Unsteady Aerodynamics and Aeroelasticity, San Diego, CA, 1991.

Edwards, J. W., Bennett, R. M., Whitlow, W., Jr., and Seidel, D. A., "Time Marching Transonic Flutter Solutions Including Angle-of-Attack Effects," AIAA 82-0685, 1982.

Ehlers, F. E. and Weatherill, W. H., "A Harmonic Analysis Method for Unsteady Transonic Flow and Its Application to the Flutter of Airfoils," NASA CR-3537, 1982.

Elrod, H. G. and Burgborfer, A., "Refinement of the Theory of Gas-Lubricated Journal Bearings of Infinite Length," *Proc. Int. Symp. on Gas-Lubricated Bearings*, ACR-49, pp. 93–118, 1958.

Elrod, H. G. and Ng, C. W., "A Theory for Turbulent Films and Its Application to Bearings," *Trans. ASME, Series F*, Vol. 89, pp. 346–362, 1967.

Elrod, H. G., "A General Theory for Laminar Lubrication with Reynolds Roughness," *Trans. ASME*, Vol. 101, pp. 8–15, 1979.

Elrod, H. G., "A Review of Theories for the Fluid Dynamic Effects of Roughness on Laminar Lubricating Films," *Surface Roughness Effects in Lubrication*, Dowson, D. *et al.* (Eds.), *Proc. 4th Leeds–Lyon Symp. on Tribology*, Mechanical Engineering Publications, pp. 11–28, 1978.

Eppler, R., "Laminar Airfoils for Reynolds Numbers Greater than 4×10^6," B-819-35, April 1969. (Available from NTIS as N69-28178.)

Eppler, R., "Some New Airfoils," *Science and Technology of Low Speed and Motorless Flight*, NASA CP-2085, Part I, 1979.

EPRI, Glass Industry Scooping Study, EPRI Report EM-5912, July 1988.

Euler, L., "More Complete Theory of Machines Driven by Hydraulic Reaction," (in French), *Historie de l'Academie Royale des Sciences et Belles Lettres, Classe de Philosophie Experimentale*, pp. 227–295, *Mem 10*, 1754, Berlin, 1765.

Ezzat, H. and Rohde, S., "A Study of the Thermohydrodynamic Performance of Finite Slider Bearings," *Trans. ASME, Series F*, Vol. 95, pp. 298–307, 1973.

Fackrell, D. E. and Harvey, J. K., "The Aerodynamics of the Isolated Road Wheel," *Proceedings of the 2nd AIAA Symposium on the Aerodynamics of Sports and Competition Cars*, Los Angeles, 1974.

Fay, J. A., "The Structure of Gaseous Detonations," *Eighth Symposium (International) on Combustion*, Williams and Wilkins Co., Baltimore, MD, 1962.

Field, P., *Dust Explosions*, Elsevier, Amsterdam, The Netherlands, 1982.

Fine, H. A. and Geiger, G. H., *Handbook on Material and Energy Balance Calculations in Metallurgical Processes (Revised Ed.)*, TMS, Warrendale, PA, 1993.

Firchau, W., Kiekebusch, B., and Neppert, H., "Grenzschichtdaten für moderne Züge," MBB Rep. No. TN-HE 211/212-22/, 1980.

Forsching, H. W., *Fundamentals of Aeroelasticity* (in German), Springer-Verlag, Berlin, 1974.

Fox, J. A. and Vardy, A. E., "The Generation and Alleviation of Air Pressure Transients Caused by the High Speed Passage of Vehicles Through Tunnels," *Proc. 1st Int. Symp. Aerod. Vent. of Vehicle Tunnels*, BHRA Fluid Eng., Paper G3, 1973.

Fuhs, A. E. and Fuhs, S. E., "Optical Phase Distortion Due to Compressible Flow over Laser Turrets," *Proceedings of the Aero-Optics Symposium on Electromagnetic Wave Propagation from Aircraft*, NASA Conf. Proc. 2121, pp. 287–325, 1980.

Fuhs, A. E., "Density Inhomogeneity in a Laser Cavity Due to Energy Release," *AIAA J.*, Vol. 11, pp. 347–375, 1973.

Fuhs, A. E., Cole, L., and Etchechury, J., "Transient Refractive Index in a XeF Laser," *J. Opt. Soc. Amer.*, Vol. 70, p. 1621, 1980.

Fung, Y. C., *An Introduction to the Theory of Aeroelasticity*, John Wiley & Sons, New York, 1955. (Also available in Dover Edition.)

Gaskell, D. R., *An Introduction to Transport Phenomena in Materials Engineering*, Macmillan, New York, 1991.

Gawthorpe, R. G., "Aerodynamic Aspects of Train Design for Operation Through the Channel Tunnel," Conf. "Train Technology for the Tunnel," Le Touquet, *I Mech. E*, C451/003, pp. 1–14, 1992.

Gawthorpe, R. G. and Pope, C. W., "The Measurement and Interpretation of Transient Pressures Generated by Trains in Tunnels," *Proc. 2nd Int. Symp. Aero. Vent. Vehicle Tunnels*, Cambridge, BHRA Fluid Eng., Paper C3, 1976.

Gawthorpe, R. G., "Aerodynamic Problems of High-Speed Trains Running on Conventional Tracks," *High Speed Ground Vehicles*, VKI Lecture Series 48, 1972.

Gawthorpe, R. G., "Aerodynamics in Railway Engineering Part I: Aerodynamics of Trains in the Open Air," *I. Mech. E.*, Railway Eng. Int., pp. 7–12, 1978a.

Gawthorpe, R. G., "Aerodynamics in Railway Engineering: Aerodynamic Problems with Overhead Line Equipment," *I. Mech. E.*, 1978, Railway Eng. Int., pp. 38–40, 1978b.

Gawthorpe, R. G., "Human Tolerance to Rail Tunnel Pressure Transients—A Laboratory Assessment," *Proc. 5th Int. Symp. Aero. Vent. Vehicle Tunnels*, Lille, BHRA The Fluid Eng. Centre, 1985.

Gawthorpe, R. G., "Pressure Comfort Criteria for Rail Tunnels Operation," *Aerodynamics and Ventilation of Vehicle Tunnels*, Haerter, A. (Ed.), Elsevier, London and New York, 1991.

Gaylard, A. P. and Johnson, T., "The Practical Application of CFD to Railway Aerodynamics," *I. Mech. E.*, C464/020, pp. 49–57, 1993.

Gebhardt, F. G. and Smith, D. C., "Kinetic Cooling of a Gas by Absorption of CO_2 Laser Radiation," *Appl. Phys. Letters*, Vol. 20, pp. 129–132, 1972.

Glassman, I., *Combustion*, Academic Press, New York, 1987.

Glasstone, S., *Textbook of Physical Chemistry*, 2nd ed., D. Van Nostrand, Princeton, NJ, 1954.

Glöckle, H. and Gawthorpe, R., "Aerodynamics—Trackside Safety Environment in Seminar Railways—Considerations of the Environment," *I. Mech. E.*, The Railway Division, 1991.

Glöckle, H., "Comfort Investigations for Tunnel Runs on the New Line Würzburg-Fulda," *Aerodynamics and Ventilation of Vehicle Tunnels*, Haerter, A. (Ed.), Elsevier, London and New York, 1991.

Glück, H., "Aerodynamik der Schienenfahrzeuge," TÜV Rheinland, 1985.

Gordon, S., *Thermodynamic and Transport Combustion Properties of Hydrocarbons With Air*, National Aeronautics and Space Administration, 1982.

Gorog, J. P., Adams, T. N., and Brimacombe, J. K., "Heat Transfer from Flames in a Rotary Kiln," *Metallurgical Trans. B*, Vol. 14B, pp. 411–424, 1983.

Gorog, J. P., Brimacombe, J. K., and Adams, T. N., "Radiative Heat Transfer in Rotary Kilns," *Metallurgical Trans. B*, Vol. 12B, pp. 55–70, 1981.

Gross, R. W. F. and Bott, J. F. (Eds.), *Handbook of Chemical Lasers*, John Wiley & Sons, New York, 1976.

Gross, W. A., *Gas Film Lubrication*, John Wiley & Sons, New York, 1962.

Guthrie, R. I. L., *Engineering in Process Metallurgy*, Clarendon Press, Oxford, 1992.

Hammitt, F. G., "Effects of Gas Content upon Cavitation Inception, Performance and Damage," *J. Hyd. Res. (IAHR)*, Vol. 10, No. 3, pp. 259–290, 1972.

Hammitt, F. G., *Cavitation and Multiphase Flow Phenomena*, Adv. Book Series, McGraw-Hill, New York, 1980a.

Hammitt, F. G., "Cavitation and Liquid Impact Erosion," *Wear Control Handbook*, Peterson, M. B. and Winer, W. O. (Eds.), ASME, New York, 1980b.

Hammitt, F. G. and Heymann, F. J., "Liquid-Erosion Failures," *Metals Handbook*, Vol. 10, 8th ed., Am. Soc. Metals, Metals Park, OH, 1975.

Hammitt, F. G., Chao, C., Kling, C. L., and Rogers, D. O., "ASTM Round-Robin Test with Vibratory Cavitation and Liquid Impact Facilities of 6061-T-6511 Aluminum Alloy, 316 Stainless Steel, and Commercially Pure Nickel," *ASTM Mat. Res. and Stds.*, Vol. 10, No. 10, pp. 16–36, 1970.

Hamrock, B. J. and Dowson, D., "Isothermal Elasto Hydrodynamic Lubrication of Point Contacts, Part III—Fully Flooded Results," *J. Lubrication Technology*, Vol. 99, No. 2, p. 264, 1977.

Handbook of Tables for Applied Engineering Science; Table 5-22, adapted from White, R. G. S., SAE Paper 690189, 1969.

Harabuchi, T. B. and Pehlke, R. D., *Continuous Casting Volume Four, Design and Operations*, The Iron and Steel Society, Warrendale, PA, 1988.

Harris, C. D., "Aerodynamic Characteristics of an Improved 10-Percent-Thick NASA Supercritical Airfoil," NASA TM X-2978, 1974.

Harris, C. D., "Aerodynamic Characteristics of the 10-Percent-Thick NASA Supercritical Airfoil 33 Designed for a Normal-Force Coefficient of 0.7," NASA TM X-72711, 1975a.

Harris, C. D., "Aerodynamic Characteristics of a 14-Percent-Thick NASA Supercritical Airfoil Designed for a Normal-Force Coefficient of 0.7," NASA TM X-72712, 1975b.

Harris, C. D., "Transonic Aerodynamic Characteristics of a 10-Percent-Thick NASA Supercritical Airfoil 31," NASA TM X-3203, 1975.

Harris, C. D., "Wind Tunnel Investigation of Effects of Trailing-Edge Geometry on a NASA Supercritical Airfoil Section," NASA TM X-2336, 1971.

Harrison, W. I., "The Hydrodynamic Theory of Lubrication with Special Reference to Air as a Lubricant," *Trans. Cambridge Philos. Soc.*, Vol. 22, pp. 37–54, 1913.

Harvey, E. N., McElroy, W. D., and Whitely, A. H., "On Cavitation Formation in Water," *J. Appl. Phys.*, Vol. 18, No. 2, pp. 162–172, 1947.

Harwarth, F. and Sockel, H., "Unsteady Flow Due to Trains Passing a Tunnel," *Proc. 3rd Int. Symp. Aero. Vent. Vehicle Tunnels*, Sheffield, BHRA Fluid Eng., pp. 151–160, 1979.

Havener, A. G., "Optical Wave Front Variance: A Study on Analytic Models in Use Today," AIAA 92-0654, 1992.

Hays, D. F., "Plane Sliders of Finite Width," *Trans. Amer. Soc. of Lubrication Eng.*, Vol. 1, pp. 233–240, 1958.

He, J. G. and Hammitt, F. G., "Velocity Exponent and Sigma for Venturi Cavitation Erosion," *Wear*, Vol. 80, No. 1, pp. 43–58, 1982.

Hemker, H., "Turbulenzen in Grenzen," *Flug Revue*, No. 2, pp. 59–61, 1992.

Hemker, H., "Laminare Triebwerksgondeln im Flugversuch, Verkleidungs-Kunster," *Flug Revue*, No. 2, pp. 74–77, 1993.

Heymann, F. J., "Erosion by Cavitation, Liquid Impingement, and Solid Impingement," Westinghouse Electric Engr. Rept. E-1460, Mar. 15, 1968.

Heymann, F. J., "Toward Quantitative Prediction of Liquid Impact Erosion," ASTM STP-474, pp. 212–248, 1969.

Hickling, R. and Plesset, M. S., "Collapse and Rebound of a Spherical Bubble in Water," *Phys. Fluids*, Vol. 7, pp. 7–14, 1964.

Hicks, R. M. and McCroskey, W. J., "An Experimental Evaluation of a Helicopter Rotor Section Designed By Numerical Optimization," NASA TM 78622, 1980.

Hicks, R. M., Mendoza, J. P., and Bandettini, A., "Effects of Forward Contour Modification on the Aerodynamic Characteristics of the NACA 641-212 Airfoil Section," NASA TM X-3293, 1975.

Hirs, G. G., "A Systematic Study of Turbulent Film Flow," *Trans. ASME, Series F*, Vol. 96, pp. 118–126, 1974.

Hoerner, S. F., *Fluid-Dynamic Drag*, published by the author, Midland Park, NJ, 1965.

Hogge, H. D. and Crow, S. C., "Flow and Acoustics in Pulsed Excimer Lasers," Paper II-4, AIAA Conference on Fluid Dynamics of High Power Lasers, Cambridge, MA, 1978.

Houbolt, J. C., Steiner, R., and Pratt, K. G., "Dynamic Response of Airplanes to Atmospheric Turbulence Including Flight Data on Input and Response," NASA TR R-199, June 1964.

Humphreys, B., "Contamination Avoidance for Laminar Flow Surfaces," *Proceedings of First European Forum on Laminar Flow Technology*, Hamburg, Germany, DGLR-Bericht 92-06, 1992.

Humphreys, N. D. and Baker, C. J., "Forces on Vehicles in Cross Winds From Moving Model Tests," *J. Wind Eng. Ind. Aerodyn.*, Vols. 41–44, pp. 2673–2684, 1992.

Iguchi, M., Tani, J., Uemura, T., Kawabata, H., and Takeuchi, H., "The Characteristics of Water and Bubbling Jets in a Cylindrical Vessel with Bottom Blowing," *Tetsu-to-Hagane*, vol. 74, p. 1785, 1988.

Ilebugsi, O. J. and Szekely, J., *Computational Mechanics in Metals Processing*, I&SM, p. 29, 1991.

Ilebugsi, O. J., Szekely, J., Iguchi, M., Takeuchi, H., and Morita, Z., "A Comparison of Experimentally Measured and Theoretically Calculated Velocity Fields in a Water Model of a Argon Stirred Ladle," *ISIJ International*, Vol. 33, pp. 474–478, 1993.

ISO, "Method for Calculating Loudness Level," ISO Rec. R532-1967, American National Standards Institute, New York, 1967.

ISO, "Method for Calculating Loudness," ASA 33-1965, American National Standards Institute, New York, 1965.

Ivany, R. D. and Hammitt, F. G., "Cavitation Bubble Collapse in Viscous Compressible Liquids," *J. Basic Engr.*, Vol. 87D, pp. 977–985, 1965.

Jackson, J. D., *Classical Electrodynamics*, John Wiley & Sons, New York, 1962.

Johnson, D. A. and Rose, W. C., "Turbulence Measurements in a Transonic Boundary Layer and Free Shear Layer Flow Using Velocimeter and Hot Wire Anemometer Techniques," AIAA 76-399, 1976.

Jones, R. T., "Reduction of Wave Drag by Antisymmetric Arrangement of Wings and Bodies," *AIAA J.*, Vol. 10, No. 2, pp. 171–176, 1972.

Jones, S. P. and DeLaurier, J. D., "Aerodynamic Estimation Techniques for Aerostats and Airships," *J. of Aircraft*, Vol. 20, No. 2, pp. 120–126, 1983.

Jones, S. P., "Tethered Aerostat Performance Modeling," AIAA 86-2567-CP, 1986.

Jones, S. P. and Thach, D., "The Transmission of Water Vapor through Aerostat Hull Material and the Effect on Buoyant Lift," AIAA 95-1619, 1995.

Jost, W., *Explosion and Combustion Processes in Gases*, McGraw-Hill, New York, 1946.

Karamcheti, K., *Principles of Ideal-Fluid Aerodynamics*, John Wiley & Sons, New York, 1966.

Keller, A., "The Influence of Cavitation Nucleus Spectrum on Inception, Investigated with a Scattered Light Counting Method," *J. Basic Engr.*, Vol. 94, No. 4, pp. 917–925, 1972.

Kelly, J. A., "Effects of Modifications to the Leading Edge Region on the Stalling Characteristics of the NACA 631-012 Airfoil Section," NACA TN 2228, 1950.

Kim, C. and Kim, C. (Eds.), *Numerical Simulation of Casting Solidification in Automotive Applications—Proceedings of the 18th Annual Automotive Materials Symposium*, The Metall. Soc., Warrendale, PA, 1991.

Kinsler, L. E., Frey, A. R., Coppens, A. B., and Sanders, J. V., *Fundamentals of Acoustics*, John Wiley & Sons, New York, 1982.

Kling, C. L. and Hammitt, F. G., "A Photographic Study of Spark Induced Cavitation Bubble Collapse," *J. Basic Engr.*, Vol. 94, No. 4, pp. 825–833, 1972.

Klingel, R., "Druckertüchtigte Reisezugwagen für Neubaustrecken; Grundlagen, Lösungen für Klimaanlagen," *ZEV-Glas. Ann.*, Vol. 112, No. 1, pp. 10–18, 1988.

Knapp, R. T., Daily, J. W., and Hammitt, F. G., *Cavitation*, 1970, McGraw-Hill, New York. (Also Iowa Inst. of Hyd. Res., Iowa City, 1980.)

Kobayashi, M., Suzuki, Y., Akutsu, K., Iida, M., and Maeda, T., "New Ventilating System of Shinkansen Cars for Alleviating Aural Discomfort of Passengers," *Aerodynamics and Ventilation of Vehicle Tunnels*, Haerter, A. (Ed.), Elsevier, London and New York, 1991.

Kou, S. and Mehrabian, R. (Eds.), *Modeling and Control of Casting and Welding Processes*, The Metall. Soc., Warrendale, PA, 1986.

Krausman, J. A., "An Overview of TCOM LTA Technology and Operations," AIAA 91-1271, 1991.

Krönke, I. and Sockel, H., "Model Tests About Cross Wind Effects on Containers and Wagons in Atmospheric Boundary Layers," *J. Wind Eng. and Ind. Aerodyn.*, Vol. 52, pp. 109–119, 1994.

Kuo, K., *Principles of Combustion*, Wiley–Interscience, New York, 1986.

Kurtz, D. W., "Aerodynamic Characteristics of Sixteen Electric, Hybrid, and Subcompact Vehicles," JPL Publication 79-59, 1979.

Kurze, U. and Beranek, L. L., "Sound Propagation Outdoors," *Noise and Vibration Control*, Beranek, L. L. (Ed.), McGraw-Hill, New York, 1971.

Lambourne, N. C., "Flutter in One Degree of Freedom," *AGARD Manual on Aeroelasticity*, Vol. V, AGARD, Paris, 1963.

Lankford, W. T., Jr., Samways, N. L., Craven, R. F., and McGammon, H. E. (Eds.), *The Making, Shaping, and Treating of Steel 10th ed./Latest Technology*, Association of Iron and Steel Engineers, Pittsburgh, PA, 1985.

Larrabee, E. E., "Aerodynamics of Road Vehicles, or Aerodynamics as an Annoyance," *Proceedings of the 2nd AIAA Symposium on the Aerodynamics of Sports and Competition Cars*, Los Angeles, 1974.

Launder, B. E. and Spalding, D. B., *Lectures in Mathematical Models of Turbulence*, Academic Press, New York, 1972.

Ledy, J. P., Charpin, F., and Garcon, F., "ELFIN (European Laminar Flow Investigation) Test in S1MA Wind Tunnel," *1992 Scientific and Technical Activities, ONERA*, pp. 60–61, 1993.

Lewis, B. and von Elbe, G., *Combustion, Flames, and Explosions of Gases*, Academic Press, New York, 1951.

Lewis, B., Pease, R. N., and Taylor, H. S. (Eds.), *Combustion Processes*, Vol. II, Princeton University Press, Princeton, NJ, 1956.

Liebeck, R. H. and Ormsbee, A. I., "Optimization of Airfoils for Maximum Lift," *J. Aircraft*, Vol. 7, No. 5, pp. 409–415, 1970.

Liebeck, R. H., "A Class of Airfoils Designed for High Lift in Incompressible Flow," *J. Aircraft*, Vol. 10, No. 10, pp. 610–617, 1973.

Liebeck, R. H., "Design of Subsonic Airfoils for High Lift," *J. Aircraft*, Vol. 15, No. 9, pp. 547–561, 1978.

Lomen, D. O., "Digital Analysis of Liquid Propellant Sloshing in Mobile Tanks with Rotational Symmetry," NASA CR-230, 1965.

Lord Rayleigh (John William Strutt), "On the Pressure Developed in a Liquid During the Collapse of a Spherical Cavity," *Phil. Mag.*, Vol. 34, pp. 94–98, 1917.

Lowery, J. G. and Polhamus, E., "A Method for Predicting Lift Increments Due to Flap Deflection at Low Angles of Attack in Incompressible Flow," NACA TN 3911, 1957.

Lund, J. W. and Sternlicht, B., "Rotor-Bearing Dynamics with Emphasis on Attenuation," *J. Basic Eng.*, Vol. 84, No. 4, pp. 491–502, 1962.

Lush, P. A. and Hutton, S. P., "The Relation Between Cavitation Intensity and Noise in a Venturi-Type Section," *Proc. Intl. Conf. on Pump and Turbine Design*, I. Mech. Engr., London, 1976.

Lynch, F. T. and Klinge, M. D., "Some Practical Aspects of Viscous Drag Reduction Concepts," SAE-912129, SAE Aerospace Technology Conference and Exposition, Long Beach, CA, 1991.

Mackrodt, P. A., "Messungen an der Lokomotive 103 im Windkanal und auf der freien Strecke," *Aerodynamisches Seminar spurgeführter Fernverkehr Rad/Schiene-Technik*, Göttingen, 1978a.

Mackrodt, P. A., "Windkanaluntersuchungen am Modell der Schnellfahrlokomotive 103 der Deutschen Bundesbahn," DFVLR-AVA Bericht 251 78A08, 1978b.

Mackrodt, P. A., "Zum Luftwiderstand von Schienenfahrzeugen," DFVLR-AVA IB 251 80C 08, 1980.

Mackrodt, P. A., Steinheuer, J., and Stoffers, G., "Aerodynamisch optimale Kopfformen für Triebzüge," DFVLR Göttingen, IB 152 79 A 27, 1980.

Maeda, T., Kinoshita, M., Kajiyama, H., and Tanemoto, K., "Aerodynamic Drag of Shinkansen Electric Cars (Series 0, Series 200, Series 100)," *QR of RTRI*, Vol. 30, No. 1, pp. 48–56, 1989.

McCroskey, W. J., "Unsteady Airfoils," *Ann. Review of Fluid Mechanics*, pp. 285–311, 1982.

McGhee, R. J. and Beasley, W. D., "Effects of Thickness on the Aerodynamic Characteristics of an Initial Low-Speed Family of Airfoils for General Aviation Applications," NASA TM X-72843, 1976.

McGhee, R. J. and Beasley, W. D., "Low-Speed Aerodynamic Characteristics of a 17-Percent-Thick Airfoil Section Designed for General Aviation Applications," NASA TN D-7428, 1973.

McGhee, R. J. and Beasley, W. D., "Low-Speed Wind-Tunnel Results for a Modified 13-Percent-Thick Airfoil," NASA TM X-74018, 1977.

McGhee, R. J. and Beasley, W. D., "Wind-Tunnel Results for an Improved 21-Percent-Thick Low-Speed Airfoil Section," NASA TM X-78650, 1978.

McGhee, R. J., Beasley, W. D., and Somers, D. M., "Low-Speed Aerodynamic Characteristics of a 13-Percent-Thick Airfoil Section Designed for General Aviation Applications," NASA TM X-72697, 1975.

McGhee, R. J., Beasley, W. D., and Whitcomb, R. T., "NASA Low and Medium Speed Airfoil Development," NASA TM-78709, 1979.

McPherson, N. A. and McLean, A., *Continuous Casting Volume Six, Tundish to Mold Transfer Operations*, The Iron and Steel Society, Warrendale, PA, 1988.

Mecham, M., "Europeans Test New Laminar Flow Design, Target Lower Transport Operating Costs," *Aviation Week*, Vol. 136, No. 5, p. 51, 1992.

Mitchell, T. M. and Hammitt, F. G., "Asymmetric Cavitation Bubble Collapse," *J. Fluids Engr.*, Vol. 95, No. 1, pp. 29–37, 1973.

Montagné, M. S., "Mesure du Souffle Provoqué par le Passage de la Rame TGV 001 a grande Vitesse," Inf. Techniques-SNCF-Equipment Paris, No. 12, 1973.

Moreau, R., *Magnetohydrodynamics*, Kluwer, Dordrecht, 1990.

Morgan, W. B. and Lin, W. C., "Ship Performance Prediction," *Computational Fluid Dynamics and Experiments*, Schiffstechnik, Hamburg, 1988.

Moxon, J., "Drag Down 15% with Laminar Flow Wing," *Flight International*, Vol. 140, No. 4299, p. 10, 1991.

Myers, P. F., *Tethered Balloon Handbook*, Goodyear Aerospace Corp. (AFCRL-69-0017), 1969.

Neppert, H. and Sanderson, R., "Untersuchungen zur Zugbegegnung, Bauwerks- und Personenpassage im Wasserkanal," in Steinheuer, J., *Aerodynamische Wirkungen von schnellfahrenden Schienenfahrzeugen auf die Umgebung*, DFVLR IB 129-81/11, 1981.

Neppert, H. and Sanderson, R., "Untersuchungen zur Zugbegegnung, Bauwerk- und Personenpassage im Wasserkanal," *Aerodynamisches Seminar Spurgeführter Fernverkehr Rad/Schiene-Technik*, Göttingen, 1978.

Newman, J. N., *Marine Hydrodynamics*, The MIT Press, Cambridge, MA, 1977.

Ng, C. W. and Pan, C. H. T., "A Linearized Turbulent Lubrication Theory," *Trans. ASME, Series D*, Vol. 87, pp. 675–688, 1965.

Ng, C. W., "Linearized PH Stability Theory for Finite Length, Self-Acting Gas-Lubricated, Plain Journal Bearings," *J. Basic Eng.*, Vol. 87, No. 3, pp. 559–603, 1965.

Nineteenth (and Earlier Volumes) Symposium (International) on Combustion, The Combustion Institute, Pittsburgh, PA, 1982.

Nixon, D., "Calculation of Unsteady Transonic Flows Using the Integral Equation Method," *AIAA J.*, Vol. 16, No. 9, pp. 976–983, 1978.

NOAA, *U.S. Standard Atmosphere, 1976*, NOAA-S/T 76-1562, Oct. 1976.

Nordby, S. A. and Bjor, O. H., "Measurement of Sound Intensity by Use of a Dual Channel Real-Time Analyser and a Special Sound Intensity Microphone," *Proceedings Inter-Noise 84*, 1984.

Oosterveld, M. *et al.*, "The Wageningen B-Screw Series," *Trans. Soc. Naval Architects and Marine Engineers*, New York, 1969.

Osterle, J. F. and Saibel, E., "On the Effect of Lubricant Inertia in Hydrodynamic Lubrication," *Z. Angew. Math. Phys.*, Vol. 6, p. 334, 1955.

Osterle, J. F., Chou, Y. T., and Saibel, E., "The Effect of Lubricant Inertia in Journal Bearing Lubrication," *Trans. ASME*, Vol. 79, pp. 494–496, 1957.

Ottitsch, F., Sockel, H., and Peiffer, A., "The Influence of Abrupt Changes in the Cross-Sectional Area of a Railway Tunnel on the Propagation of Pressure Waves Caused by Passing Trains," *Aerodynamics and Ventilation of Vehicle Tunnels*, Cockran, I. J. (Ed.), Mech. Eng. Publ. Ltd., London, 1994.

Ozawa, S., Maeda, T., Matsumura, T., Uchida, K., Kahiyama, H., and Tanemoto, K., "Countermeasures to Reduce Micro-Pressure Waves Radiating from Exits of Shinkansen Tunnels," *Aerodynamics and Ventilation of Vehicle Tunnels*, Haerter, A. (Ed.), Elsevier, London and New York, 1991.

Parker, J. W., *Anatomy for Speech and Hearing*, Harper & Row, New York, 1972.

Patir, N. and Cheng, H. S., "An Average Flow Model for Determining Effects of Three-Dimensional Roughness on Partial Hydrodynamic Lubrication," *Trans. ASME*, Vol. 100, pp. 12–18, 1978.

Peacey, J. P. and Davenport, W. G., *The Iron Blast Furnace*, Pergamon Press, New York, 1979.

Pearcey, H. H., "The Aerodynamic Design of Section Shapes for Swept Wings," *Adv. Aero. Sciences*, Vol. 3, pp. 277–322, 1962.

Pearsall, I. S., *Cavitation*, Mills & Boone, London, 1972.

Penwarden, A. D., "Acceptable Wind Speeds in Towns," BRE CP 1/74, 1974.

Penwarden, A. D., Grigg, P. F., and Rayment, R., "Measurements of Wind Drag on People Standing in a Wind Tunnel," BRE CP 74/78, 1978.

Perry, B. III, Kroll, R. I., Miller, R. D., and Goetz, R. C., "DYLOFLEX: A Computer Program for Flexible Aircraft Flight Dynamic Loads Analyses with Active Controls," *J. Aircraft*, Vol. 17, No. 4, pp. 275–282, 1980.

Peters, J. L., "Bestimmung des aerodynamischen Widerstandes des ICE/V im Tunnel und auf freier Strecke durch Auslaufversuche," *Eisenbahntechnische Rundschau*, Vol. 39, No. 9, pp. 559–564, 1990.

Peters, J. L., "Measurement of the Influence of Tunnel Length on the Tunnel Drag of the ICE/V Train," *Aerodynamics and Ventilation of Vehicle Tunnels*, Haerter, A. (Ed.), Elsevier, London and New York, 1991.

Peters, J. L., "Windkanaluntersuchung zum Verhalten von Schienenfahrzeugen unter Windeinfluß auf Dämmen und Brücken mit und ohne Schutzeinrichtungen," Krauss Maffei Verkehrstechnik GmbH, T.B. L333-TB-07/90, 1990.

Peters, J. L., "Bestimmung der Wind-und Geschwindigkeitsgrenzen für Containertransport anf Rungenwagen durch Messungen in DNW-Windkanal," Krauss Maffei Verkehrstechnik GmbH, B.AE-TB-0189, 1989.

Petre, A., *Theory of Aeroelasticity, Vol. I, Statics, Vol. II Dynamics* (in Romanian), Publishing House of the Academy of Socialist Republic of Romania, Bucharest, 1966.

Pi, W. S. and Liu, D. D., "A Unified Linear Pressure Panel Procedure for Oscillating Airfoils in Mixed Transonic Flow with Embedded Shocks," AIAA 82-0725, 1982.

Pinkus, O., "Solutions of the Tapered-Land Sector Thrust Bearing," *Trans. ASME*, Vol. 80, pp. 1510–1516, 1958.

Piwonka, T. S. and Katgerman, L. (Eds.), *Modeling of Casting, Welding, and Advanced Solidification Processing—VI*, The Metall. Soc., Warrendale, PA, 1993.

Plesset, M. S. and Chapman, R. B., "Collapse of an Initially Spherical Cavity in Neighborhood of a Solid Boundary," *J. Fluid Mech.*, Vol. 47, Pt. 2, p. 283, 1971.

Pope, A., *Aerodynamic of Supersonic Flight: An Introduction*, Pitman Publishing Corp., 1958.

Pope, C. W. and Woods, W. A., "Transient Pressures Inside Passenger Trains Travelling Through Tunnels," *Proc. 6th Int. Symp. Aero. Vent. Vehicle Tunnels*, Durham, BHRA The Fluid Eng. Centre, pp. 3–21, 1988.

Pope, C. W., "The Simulation of Flows in Railway Tunnels Using a 1/25th Scale Moving Model Facility," *Aerodynamics and Ventilation of Vehicle Tunnels*, Haerter, A. (Ed.), Elsevier, London and New York, 1991.

Pope, C. W., "Transient Pressures in Tunnels—A Formula for Predicting the Strength of the Entry Wave Produced by Trains with Streamlined and Unstreamlined Noses," British Railways & Development Division Tech Memo AERO 12, 1975.

Porier, D. R. and Geiger, G. H., *Transport Phenomena in Materials Processing*, The Metall. Soc., Warrendale, PA, 1994.

Powell, A. G. and Varner, L. D., "The Right Wing of the LEFT Airplane," NASA CP 2487, 1987.

Racz, L. and Szekely, J., "Determination of Equilibrium Shapes and Optimal Volume of Solder Droplets in the Assembly of Surface Mounted Integrated Circuits," *ISIJ International*, Vol. 33, No. 2, pp. 336–342, 1993.

Racz, L., Szekely, J., and Brakke, K. A., "A General Statement of the Problems and Description of a Proposed Method of Calculation of Some Meniscus Problems in Materials Processing," *ISIJ International*, Vol. 33, No. 2, pp. 328–335, 1993.

Raimondi, A. A. and Boyd, J., "A Solution for the Finite Journal Bearing and Its Application to Analysis and Design," *Trans. Amer. Soc. Lubrication Engineers*, Vol. 1, pp. 159–209, 1958.

Raimondi, A. A. and Boyd, J., "Applying Bearing Theory to the Analysis and Design of Pad-Type Bearings," *Trans. ASME*, Vol. 77, pp. 287–309, 1955.

Raimondi, A. A., "A Numerical Solution for the Gas-Lubricated Full Journal Bearing of Finite Length," *Trans. ASLE*, Vol. 4, pp. 131–155, 1961.

Raimondi, A. A., "Analysis and Design of Sliding Bearings," *Standard Handbook of Lubrication Engineering*, McGraw-Hill, New York, 1968.

Rappaz, M., Ozgu, M., and Mahin, K. (Eds.), *Modeling of Casting, Welding, and Advanced Solidification Process—V*, The Metall. Soc., Warrendale, PA, 1991.

Rasmussen, R. E. H., "Some Experiments on Cavitation Erosion in Water Mixed with Air," *Proc. 1955 NPL Symp. on Cavitation in Hydrodynamics*, Paper 20, H. M. Stationary Office, London, 1956.

Reddy, R. G. and Weizenbach, R. N. (Eds.), *Proceedings of The Paul E. Queneau International Symposium, Extractive Metallurgy of Copper, Nickel and Cobalt, Volume I: Fundamentals Aspects*, The Metall. Soc., Warrendale, PA, 1993.

Redeker, G., Quast, A., and Thibert, J. J., "Das A320 Laminar-SeitenleitWerks-Program," *Proceedings of JAHRBUSH 1992 Vol. III*, 29 Sept.–02 Oct., Deutsche Luft-und Raumfahrt-Kongeß, DGLR-Jahrestagung, Bremen, pp. 1259–1270, 1992.

Reed, H. L. and Saric, W. S., "Stability of Three Dimensional Boundary Layers," *Annual Review of Fluid Mech.*, Vol. 21, p. 235, 1989.

Reed, W. H. III, "Aeroelastic Matters: Some Reflections on Two Decades of Testing in the NASA Langley Transonic Dynamics Tunnel," International Symposium on Aeroelasticity, Nuremberg, Germany, 1981.

Rhow, S. K. and Elrod, H. G., "The Effects on Bearing Load-Carrying Capacity of Two-Sided Striated Roughness," *J. Lubrication Technology*, Vol. 96, No. 4, pp. 554–560, 640, 1974.

Richardson, J. M., Arons, A. B., and Halverson, R. R., "Hydrodynamic Properties of Sea Water at the Front of a Shock Wave," *J. Chemical Physics*, Vol. 15, pp. 758–794, 1947.

Rizzetta, D. P. and Yoshihara, H., "Computation of the Pitching Oscillation of a NACA 64A010 Airfoil in the Small Disturbance Limit," AIAA 80-128, 1980.

Robert, J. P., "Drag Reduction: An Industrial Challenge," Special Course on Skin Friction Drag Reduction, AGARD Report 786, 1992.

Robertson, J. M. and Wislicenus, G. F. (Eds.), *Cavitation State of Knowledge*, ASME, New York, 1969.

Robinson, D. W. and Dadson, R. S., "A Re-determination of the Equal Loudness Relations for Pure Tones," *Brit. J. Appl. Phys.*, Vol. 7, p. 166, 1956.

Rodden, W. P. and Bellinger, E. D., "Aerodynamic Lag Functions, Divergence and the British Flutter Method," *J. Aircraft*, Vol. 19, No. 7, pp. 596–598, 1982.

Rosenhead, L., *Laminar Boundary Layers*, Oxford University Press, London, 1963.

Rosenqvist, T., *Principles of Extractive Metallurgy*, McGraw-Hill, New York, 1983.

Rowe, A., "Evaluation Study of a Three-Speed Hydrofoil with Wetted Upper Side," *J. Ship Research*, Vol. 23, No. 1, pp. 55–56, 1979.

Sabersky, R. H., Acosta, A. J., and Hauptmann, E. G., *Fluid Flow*, Macmillan, New York, 1971.

Sahai, Y. and St. Pierre, G. R., "Advances in Transport Processes in Metallurgical Systems," *Transport Processes in Engineering*, Vol. 4, Elsevier, New York, 1992.

Salzman, E. and Steers, L., "Reduced Truck Fuel Consumption Through Aerodynamic Design," *J. of Energy*, Vol. 1, No. 5, 1977.

Saric, W. S., "Laminar-Turbulent Transition: Fundamentals," Special Course on Skin Friction Drag Reduction, AGARD Report 786, 1992.

Saunders, H. E., *Hydrodynamics in Ship Design*, Society of Naval Architects and Marine Engineers, New York, 1953.

Scanlan, R. H. and Rosenbaum, R., *Introduction to the Study of Aircraft Vibration and Flutter*, Macmillan, New York, 1951. (Also available in Dover Edition.)

Schmitt, V., Reneaux, J., and Priest, J., "Maintaining Laminarity by Boundary Layer Control," *1992 Scientific and Technical Activities*, ONERA, pp. 13–14, 1993.

Schultz, M. and Sockel, H., "Pressure Transients in Railway Tunnels," *Trends in Applications of Mathematics to Mechanics*, Schneider, W., Troger, H., and Ziegler, F. (Eds.), Longman Scientific & Technical, pp. 33–39, Harlow, 1989.

Schultz, M. and Sockel, H., "Pressure Transients in Short Tunnels," *Aerodynamics and Ventilation of Vehicle Tunnels*, Haerter, A. (Ed.), Elsevier, London, and New York, 1991.

Schultz, M. and Sockel, H., "The Influence of Unsteady Friction on the Propagation of Pressure Waves in Tunnels," *Proc. 6th Int. Symp. on the Aerod. Vent. Vehicle Tunnels*, Durham, BHRA The Fluid Eng. Centre, pp. 123–136, 1988.

Schultz, M., "Experimentelle und rechnerische Ermittlung von Druckwellenausbreitungsvorgängen bei der Fahrt eines Zuges durch einen Tunnel. Untersuchung des Einflusses der instationären Wandreibung auf die Wellendämpfung," Dissertation, Technical Univ. Vienna, 1990.

Schwarz, M. P. and Turner, W. J., "Application of the Standard k-ε Turbulence Model to Gas-Stirred Baths," *Applied Math. Modeling*, Vol. 12, p. 273, 1988.

Sessler, G. M. and West, J. F., "Foil Electret Microphones," *J. Acoust. Soc. Amer.*, Vol. 40, p. 1433, 1966.

Shen, Y. T. and Eppler, R., "Wing Sections for Hydrofoils—Part 2: Nonsymmetrical Profiles," *J. Ship Research*, Vol. 25, No. 3, pp. 191–200, 1981.

Shigley, J., *Mechanical Engineering Design*, McGraw-Hill, New York, 1977.

Sica, L., "Three Beam Interferometer for the Observation of Kinetic Cooling in Air," *Appl. Optics*, Vol. 12, pp. 2848–2854, 1973.

Simiu, E. and Scanlan, R. H., *Wind Effects on Structures—An Introduction to Wind Engineering*, John Wiley & Sons, New York, 1978.

Slabinski, V. J., "INTELSAT IV In-Orbit Liquid Slosh Tests and Problems in the Theoretical Analysis of the Data," *Comsat Technical Review*, Vol. 8, pp. 1–40, 1978.

Sloof, J. W., Wortmann, F. X., and Duhon, J. M., "The Development of Transonic Airfoils for Helicopters," 31st Annual National Forum of the American Helicopter Society, Washington, D.C., 1975.

Smirnov, A. I., *Aeroelastic Stability of Aircraft* (in Russian), Izdatel'stvo Mashinostroenie, Moscow, 1980.

Sockel, H. and Schultz, M., ''PC-Rechenprogramm für die aerodynamische Berechnung der Durchfahrt eines Zuges durch einen Tunnel,'' Bericht für DB, 1990.

Sockel, H., *Aerodynamik der Bauwerke*, Vieweg, Braunschweig, 1985.

Sohn, H. Y., and Geskin, E. S. (Eds.), *Metallurgical Process for the Year 2000 and Beyond*, TMS, Warrendale, PA, 1988.

Somers, D. M., ''Design and Experimental Results for a Natural-Laminar-Flow Airfoil for General Aviation Applications,'' NASA TP 1861, 1981a.

Somers, D. M., ''Design and Experimental Results for a Flapped Natural-Laminar-Flow Airfoil for General Aviation Applications,'' NASA TP 1865, 1981b.

Sonntag, H., Schmidt, M., and Voss, G., ''Aerodynamische Untersuchungen bei Geschwindigkeiten über 200 km/h,'' Bericht 2-76, Lehrstuhl und Institut für Schwerfahrzeuge und maschinelle Bahnanlagen der TU-Hannover, 1976.

Sovran, G., Morel, T., and Mason, W. T., Jr. (Eds.), *Aerodynamic Drag Mechanisms of Bluff Bodies and Road Vehicles*, Plenum Press, New York, 1978.

Speller, F. N. and LaQue, F. L., ''Water Side Deterioration of Diesel Engine Cylinder Liners,'' *Corrosion*, Vol. 6, pp. 209–215, 1950.

Stahl, H. A. and Stepanoff, A. J., ''Thermodynamic Aspects of Cavitating Centrifugal Pumps,'' *Trans. ASME*, Vol. 78, pp. 169–193, 1956.

Starr, E. A., ''Sound and Vibration Transducers,'' *Noise and Vibration Control*, Beranek, L. L. (Ed.), McGraw-Hill, New York, 1971.

Steger, J. L. and Bailey, H. E., ''Calculation of Transonic Aileron Buzz,'' *AIAA J.*, Vol. 18, No. 3, pp. 249–255, 1980.

Steinmetz, W. J., ''Optical Degradation due to a Thin Turbulent Layer,'' *AeroOptical Phenomena*, Progress in Astronautics and Aeronautics, Vol. 80, AIAA, 1982.

Steinrück, P. and Sockel, H., ''Further Calculations on Transient Pressure Alleviation and Simplified Formulae for Initial Tunnel Design,'' *Proc. 5th Int. Symp. Aero. Vent. Vehicle Tunnels*, Lille, BHRA The Fluid Eng. Centre, pp. 317–341, 1985.

Steinrück, P., ''Ein Verfahren zur Berechnung instationärer Strömungsvorgänge bei Fahrten von mehreren Zügen durch Tunnels mit und ohne Portalvorbauten,'' dissertation, Technical University of Vienna, 1984.

Stepanoff, A. J., ''Cavitation Properties of Liquids,'' *J. Engr. Power*, Vol. 83, pp. 195–200, 1961.

Stevens, S. S., ''Procedure for Calculating Loudness, Mark VI,'' *J. Acoust. Soc. Am.*, Vol. 33, p. 1577, 1961.

Stewartson, K., ''On the Stability of a Spinning Top Containing Liquid,'' *J. Fluid Mechanics*, Vol. 5, pp. 577–592, 1959.

Stoffers, G., ''Untersuchung der Umströmung der Schnellfahrlokomotive E103 bei Seitenwind,'' DFVLR IB 151-79A2, 1979.

Stratton, J. A., *Electromagnetic Theory*, McGraw-Hill, New York, 1941.

Strehlow, R. A., *Fundamentals of Combustion*, International Textbook Company, Pennsylvania, 1968.

Stull, D. R., *Fundamentals of Fire and Explosion*, AIChE, New York, 1977.

Sugarawa, T., Ikeda, M., Shimotsuma, T., Higuch, M., Izuka, M., and Kuroda, K., ''Construction and Operation of No. 5 Blast Furnace, Fukuyama Works, Nippon Kokan K. K.,'' *Ironmaking and Steelmaking*, Vol. 3, No. 5, p. 242, 1976.

Szekely, J. and Ilegbusi, O. J., *Physical and Mathematical Modelling of Tundish Operations*, Springer-Verlag, New York, 1989.

Szekely, J. and Themelis, N. J., *Rate Phenomena in Process Metallurgy*, Wiley–Interscience, Toronto, 1971.

Szekely, J. and Trapaga, G., "Some Perspectives on the Mathematical Modeling of Materials Processing Operations, Modeling and Simulation in Materials Science," in press, 1994.

Szekely, J. and Trapaga, G., *METEC Congress 94, Proceedings Volume 1, Continuous Casting, Near-Net-Shape Casting*, pp. 40–45, 1994.

Szekely, J., Evans, J. W., and Brimacombe, J. K., *The Mathematical and Physical Modeling of Primary Metals Processing Operations*, John Wiley, Toronto, 1988.

Szekely, J., Evans, J. W., Blazek, K., and El-Kaddah, N. (Eds.), *Magnetohydrodynamics in Process Metallurgy*, TMS, Warrendale, PA, 1991.

Szekely, J., Hales, L. B., Henein, H., Jarrett, N., and Rajamani, K. (Eds.), *Mathematical Modeling of Materials Processing Operations*, The Metallurgical Society, Inc., Warrendale, PA, 1987.

Szekely, J., *Fluid Flow Phenomena in Metals Processing*, Academic Press, New York, 1979.

Szeri, A. Z. and Powers, D., "Pivoted Plane Pad Bearings: A Variational Solution," *Trans. ASME, Series F*, Vol. 92, pp. 466–472, 1970.

Szeri, A. Z., *Tribology, Friction, Lubrication, and Wear*, Taylor & Francis, Washington, D.C., 1980.

Taylor, D. W., "Speed and Power of Ships," Government Printing Office, Washington, D.C., 1943.

Taylor, G. I., "Stability of a Viscous Liquid Contained Between Two Rotating Cylinders," *Philos. Trans. Roy. Soc. London, Ser. A.*, Vol. 223, pp. 289–343, 1923.

The Iron and Steel Institute of Japan, *Blast Furnace Phenomena and Modeling*, Elsevier, New York, 1987.

The Iron and Steel Institute of Japan, *The Sixth International Iron and Steel Congress*, Nagoya Congress Center, Nagoya, Japan, Vols. 1–5, 1990.

Thibert, J. J., Reneaux, V., and Schmitt, V., "ONERA Activities in Drag Reduction," *Proceedings of 17th Congress of the International Council of the Aeronautical Sciences*, pp. 1053–1064, 1990.

Thiruvengadam, A., "Handbook of Cavitation Erosion," Tech. Rep. 7301-1, Hydronautics, Inc., Laurel, MD, 1974.

Trans, C. T. and Petot, D., "Semi-Empirical Model for the Dynamic Stall of Airfoils in View of the Application of Responses of a Helicopter Blade in Forward Flight," Paper 48, Proc. 6th European Rotorcraft and Powered-Lift Aircraft Forum, Bristol, England, 1980.

Trapaga, G., Matthys, E., Valencia, J., and Szekely, J., "Fluid Flow, Heat Transfer, and Solidification of Molten Metal Droplets Impinging on a Substrate: Comparison of Numerical and Experimental Results," *Met. Trans. B*, Vol. 23B, No. 6, pp. 701–718, 1992.

Troost, L., "Open Water Test Series with Modern Propeller Forms," Parts I, II, and III, North East Coast Institute, United Kingdom, 1951.

Tseng, K. and Morino, L., "Nonlinear Green's Function Method for Unsteady Transonic Flows," *Transonic Aerodynamics*, Nixon, D. (Ed.), Vol. 81, AIAA, New York, pp. 565–603, 1982.

Tsien, H. S. and Beilock, M., "Heat Source in a Uniform Flow," *J. Aero. Sci.*, Vol. 16, p. 756, 1949.

Turner, T. R., "Wind Tunnel Investigation of a 3/8 Scale Automobile Model Over a Moving Ground Plane," NASA TN D-4229, 1967.

Ueda, T. and Dowell, E. H., "Flutter Analysis Using Nonlinear Aerodynamic Forces," AIAA 82-0728, 1982.

Vardy, A. E. and Anandarajah, A., "Initial Design Considerations for Rail Tunnel Aerodynamics and Thermodynamics," *Proc. 4th Int. Symp. Aero. Vent. Vehicle Tunnels*, York, BHRA Fluid Eng., pp. 353–366, 1982.

Vardy, A. E. and Dayman, B., "Alleviation of Tunnel Entry Pressure Transients: 2. Theoretical Modeling and Experimental Correlation," *Proc. 3rd Int. Symp. Aero. Vent. Vehicle Tunnels*, Sheffield, BHRA Fluid Eng., pp. 363–376, 1979.

Vardy, A. E., "The Use of Airshafts for the Alleviation of Pressure Transients in Railway Tunnels," *Proc. 2nd Int. Symp. Aero. and Vent. Vehicle Tunnels*, Cambridge, BHRA Fluid Eng., pp. 55–69, 1976.

Vardy, A. E., "Unsteady Airflows in Rapid Transit Systems," *Proc. Inst. Mech. Eng.*, Vol. 194, pp. 341–356, 1980.

Vardy, A. E., "Unsteady Flows: Fact and Friction," *Proc. 3rd Int. Conf. on Pressure Surges*, Canterbury, BHRA The Fluid Eng. Centre, pp. 15–26, 1980.

Voller, V. R., Stachowicz, M. S., and Thomas, B. G. (Eds.), *Materials Processing in the Computer Age*, TMS, Warrendale, PA, 1991.

von Hippel, A. R., *Dielectrics and Waves*, John Wiley & Sons, New York, 1959.

Voss, G., Gackenholz, L., and Wiebels, R., "Eine neue Formel (Hannover'sche Formel) zur Bestimmung des Luftwiderstandes spurgebundener Fahrzeuge," *ZEV-Glas. Ann.*, Vol. 96, No. 6, pp. 166–172, 1972.

Waclawiczek, M. and Sockel, H., "Pressure Transients and Aerodynamic Power in Railway Tunnels with Special Reference to Entropy and Airshafts," *Proc. 4th Int. Symp. Aero. Vent. of Vehicle Tunnels*, York, BHRA Fluid Eng., pp. 337–351, 1982.

Waid, R. L. and Lindberg, Z. M., "Experimental and Theoretical Investigations of a Supercavitating Hydrofoil," Cal. Inst. Tech. Eng. Rep. No. 47-8, 1957.

Wardlaw, A. B., "High Angle-of-Attack Missile Aerodynamics," *Missile Aerodynamics*, AGARD Lecture Series 98, 1979.

Waser, M. and Crocker, M. J., "Introduction to the Two-Microphone Cross-Spectral Method of Determining Sound Intensity," *Noise Control Eng. J.*, pp. 76–85, 1984.

Watkins, S., Saunders, J., and Kumar, H., "Aerodynamic Drag Reduction of Good Trains," *J. Wind Eng. Ind. Aerodyn.*, Vol. 40, pp. 147–178, 1992.

Whitcomb, R. T., "Review of NASA Supercritical Airfoils," ICAS Paper No. 74-10, 1974.

Williams, F. A., *Combustion Theory*, Addison-Wesley Publishing Company, MA, 1965.

Williams, M. H., "Linearization of Transonic Flows Containing Shocks," *AIAA J.*, Vol. 17, No. 4, pp. 394–397, 1979.

Williams, M. H., "Unsteady Thin Airfoil Theory for Transonic Flows with Embedded Shocks," *AIAA J.*, Vol. 18, No. 6, pp. 615–625, 1980.

Wolters, D. S. J., "Aerodynamic Effects on Airborne Optical Systems," MDC A2582, McDonnell Douglas Corporation, St. Louis, MO, 1973.

Wong, K. Y., Zhiyung, L., and DeLaurier, J., "An Application of Source Panel and Vortex Methods for Aerodynamic Solutions of Airship Configurations," AIAA 85-0875, 1985.

Woo, J. S., Szekely, J., Castillejos, A. H., and Brimacombe, J. K., "A Study on the Mathematical Modeling of Turbulent Recirculating Flows in Gas-Stirred Ladles," *Metallurgical Trans. B*, Vol. 21B, p. 269, 1990.

Woods, W. A. and Pope, C. W., "On the Range of Validity of Simplified One-Dimensional Theories for Calculating Unsteady Flows in Railway Tunnels," *Proc. 3rd Int. Symp. Aerod. Vent. Vehicle Tunnels*, Sheffield, BHRA Fluid Eng., pp. 115–150, 1979.

Woods, W. A., "Wave Action Associated with a Train Entering a Tunnel," *Proc. 1st Int. Symp. Aero. Vent. Vehicle Tunnels*, Canterbury, BHRA Fluid Eng., pp. 65–88, 1973.

Wortmann, F. X., "Experimental Investigations on New Laminar Profiles for Gliders and Helicopters," TIL/T 4906, British Ministry of Aviation, 1960.

Wortmann, F. X., "The Quest for High Lift," AIAA 74-1018, 1974.

Wortmann, F. X., *Design of Airfoils with High Lift at Low and Medium Subsonic Mach Numbers*, AGARD CP102, 1972a.

Wortmann, F. X., "Airfoils With High Lift Drag Ratios at a Reynolds Number of About One Million," NASA CR-2315, 1972b.

Wright, J. B., "Computer Programs for Tethered-Balloon System Design and Performance Evaluation," AFGL-TR-76-0195, 1976.

Yang, T. Y. and Batina, J. T., "Transonic Time-Response Analysis of Three D.O.F. Conventional and Supercritical Airfoils," AIAA 82-0688, 1982.

Yang, T. Y., Guruswamy, P., and Striz, A. G., "Application of Transonic Codes to Flutter Analysis of Conventional and Supercritical Airfoils," AIAA 81-0603, 1981.

Yang, W. J., Mochizuki, S., and Nishiwaki, N., *Transport Process in Engineering Series No. 6, Transport Phenomena in Manufacturing and Materials Processing*, Elsevier Science B.V., Amsterdam, The Netherlands, 1994.

Zhou, Y. K. and Hammitt, F. G., "Vibratory Cavitation Erosion of Aqueous Solutions," *Wear*, Vol. 87, No. 3, pp. 163–171, 1983.

Zhou, Y. K., He, J. G., and Hammitt, F. G., "Cavitation Erosion of Diesel Engine Wet Cylinder Liners," *Wear*, Vol. 76, No. 3, pp. 321–328, 1982a.

Zhou, Y. K., He, J. G., and Hammitt, F. G., "Cavitation Erosion of Cast Iron Diesel Engine Liners," *ibid.*, *Wear*, Vol. 76, No. 3, pp. 329–335, 1982b.

Zygmont, J., "Truck of the Future," *High Technology*, Vol. 5, pp. 28–33, 1985.

24 Fluid Dynamics in Nature

JACKSON R. HERRING
National Center for Atmospheric Research
Boulder, CO

JACK E. CERMAK
Colorado State University
Fort Collins, CO

WAYNE L. NEU
Virginia Polytechnic Institute and State University
Blacksburg, VA

EVERETT V. RICHARDSON
Colorado State University (Emeritus)
Fort Collins, CO

PANAYIOTIS DIPLAS
Virginia Polytechnic Institute and State University
Blacksburg, VA

RICHARD H. RAND
J. ROBERT COOKE
Cornell University
Ithaca, NY

GEORGE T. YATES
Consultant
Boardman, OH

GEOFFREY R. SPEDDING
University of Southern California
Los Angeles, CA

JAMES D. DE LAURIER
University of Toronto
Toronto, Canada

DAVID P. HOULT
Massachusetts Institute of Technology
Cambridge, MA

Handbook of Fluid Dynamics and Fluid Machinery, Edited by Joseph A. Schetz and Allen E. Fuhs
ISBN 0-471-12598-9 Copyright © 1996 John Wiley & Sons, Inc.

GEORGE D. ASHTON
U.S. Army Cold Regions Research and Engineering Laboratory
Hanover, NH

CONTENTS

24.1 ATMOSPHERIC FLOWS
Jackson R. Herring

24.1.1 Introduction

Atmospheric motions range in scale from millimeters to planetary scales (thousands of kilometers), yet all such scales play their roles in weather and climate. The largest are dominated by the effect of the earth's rotation, while the smallest are frequently buoyantly unstable and insensitive to rotation. The conventional meteorological terms used to distinguish between such scales of motion are: 1) *planetary scale*, those whose extent is a substantial fraction of the earth's diameter, 2) *synoptic scale*, ranging from 2000 (approximately continental scale) to 1000 km, 3) *mesoscale scale*, 1000 km to 100 km, and finally 4) *microscale*, less than 10 km. The driving force of all atmospheric motions is the combination of solar heating and the earth's rotation. Their effects lead to differing physics of the regimes noted above and to the distribution in latitude of various zones of the flow. How this comes about may be traced by considering an idealized atmosphere under the forces of gravity, $\bar{g}$, rotation, $\bar{\Omega}$, and solar heating. We write the equations of motion of the atmosphere as

$$\partial_t \rho + \nabla \cdot (\rho \bar{u}) = 0 \tag{24.1}$$

$$\rho(\partial_t + \bar{u} \cdot \nabla)\bar{u} = -\nabla p - 2\rho \bar{\Omega} \times \bar{u} + \bar{g}\rho + \nabla \cdot \bar{\bar{\tau}} \tag{24.2}$$

$$\rho(\partial_t + \bar{u} \cdot \nabla)(c_v T) = -p\nabla \cdot \bar{u} + \nabla \cdot k\nabla T + \phi \tag{24.3}$$

where the tensor τ_{ij} and scalar ϕ express the effects of viscosity,

$$\tau_{ij} = \mu(\partial_i u_j + \partial_j u_i) - (\tfrac{2}{3})\lambda \partial_n u_n \delta_{ij} \tag{24.4}$$

$$\phi = \tau_{ij}\partial_j u_i \tag{24.5}$$

and $\bar{\Omega}$ is the rotation rate of the earth, p the atmospheric pressure, and k the thermal diffusivity. In Eqs. (24.4) and (24.5), we have used an alternate notation to express $\bar{u}$ as (u_1, u_2, u_3), with the summation convention implied. We add to this set the ideal gas law,

$$p = \rho RT \tag{24.6}$$

and recall the relation between specific heat at constant volume, c_v, and that at constant pressure, c_p

$$R = c_p - c_v \tag{24.7}$$

Most of the incident solar radiation is in the visible range of the spectrum, and most of that penetrates to the ground, where it is partially adsorbed, with the remainder reradiated into the atmosphere and then eventually into space. The atmosphere's time-averaged temperature must then vary with height in such a way that the upward heat flux associated with its vertical gradient [through the conductive and radiative processes in Eq. (24.3)] balances the heat input from the incident solar radiation. Convective instability will ensue if the vertical temperature gradient exceeds the *adiabatic lapse rate*, which is*

$$(\partial T/\partial z)_{ad} \equiv -(T/p)\{(c_p - c_v)/c_p\}\partial p/\partial z \tag{24.8}$$

We know that solar heating is sufficient to cause a convective instability in the tropics, where it is most intense. Let us follow such a buoyantly induced motion, slightly north of the equator. After its initial vertical rise, continuity, Eq. (24.1), implies that it turn northward, as depicted in Fig. 24.1. But, as it begins to move northward, it must also move eastward faster than the earth's rotation rate, since its angular momentum

$$(|\bar{v}\Re| \cos(\vartheta)) \tag{24.9}$$

is approximately conserved, and since its moment-arm, $\Re \cos(\vartheta)$, is progressively shortened with its northward progress. Here, ϑ is the co-latitude of the parcel. (In these considerations, we assume an air parcel's distance from the center of the earth, $\Re$, changes negligibly during the course of its motion). Thus, near the tropics, we have an eastward wind (from the west) aloft, with a return westward wind (from the east) near the ground.** A necessary return flow from the east follows if we assume the heating cannot impart net angular momentum to the air. Such a single-cell system would imply a surface flow everywhere counter to the earth's rotation, which would act by surface drag to eventually slow the earth's rotation. In any case, this poleward motion persists only to about 30 degrees, at which point the flow descends and returns to the equator. The torus-like convective cell defined by this motion is called the *Hadley cell*. North of 30° (or southward in the southern hemisphere), the flow is a west wind, both at the surface and aloft. In the northern hemisphere, this band

---

*If we assume an atmosphere initially at rest and introduce a small perturbation (u_i', T') on this motionless state, then Eqs. (24.1)–(24.3) imply $c_v \partial_t T' = -c_v\{\partial T_0/\partial z - (\partial T/\partial z)_{ad}\}w$. If this is positive, a *nondissipative* $(\mu = k = 0)$ flow is unstable, since in that case vertical motion would be associated with an excess of temeprature. Our considerations here are for *dry* air, and if the air is moist, Eq. (24.8) must be modified.

**Custom designates motion that moves air from the west towards the east a west wind, and *vice versa*.

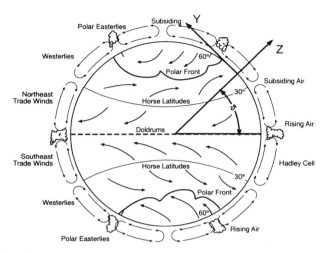

FIGURE 24.1 Schematic of large-scale atmospheric motion, depicting the three major flow zones: 1) the low-latitude Hadley cell (with the doldrums between the rising limbs of the cells; 2) the mid-latitude waves, extending from $\sim 30°$ to $\sim 60°$; and 3) the polar cell, $\sim 60°$ to 90°. (Reprinted with permission, Lutgen, F. K. and Tarbuck, E. J., *The Atmosphere, an Introduction to Meteorology*, Prentice-Hall, Englewood Cliffs, NJ, p. 478, 1982, p. 171.) Coordinate frame sketched here (x, y, z) is that used in Sec. 24.1.2. The direction x points into the page.

persists from roughly 30 degrees to about 60 degrees. Northward of that is the polar vortex, with westerlies aloft and eastward flow (mainly) as a surface flow.*

The net effect of the eastward motion in the temperate zones (~ 30 to $\sim 60°$) is to export heat from the tropics towards the poles. This is done by a sinuous motion, in which the flow picks up heat on approaching the lower-latitude of the Hadley cell, and transports it by advection to the lower-latitude of the polar vortex, ascending on its northward path, and descending on returning. Such motion is largely two-dimensional, nearly parallel to the earth's surface. On average, it is stable against vertical overturning, although from a certain perspective it may be viewed as a slantwise convection that exists as a result of an instability involving both rotation and an elastic response of the stable atmosphere (*baroclinic* instability). In the next section, we will develop some approximate but simple equations derived from Eqs. (24.1)–(24.3) that give insight into these large, temperate-zone eddies that are the essence of the weather patterns.

24.1.2 Idealized Atmospheric Flows and Some of Their Instabilities

The Quasi-Geostrophic Equations. Atmospheric flows are generally such that the full apparatus of Eqs. (24.1)–(24.3) are not necessary to understand their basic dynamics. The pressure scale height is much smaller than the earth's radius, and this constrains motions to be largely two-dimensional, especially at planetary scales. The Mach number of atmospheric motions is usually sufficiently small that the flow may

---

*For a more complete discussion of these zones and their physics, we refer the reader to the classic paper by C.-G. Rossby (1945).

be considered incompressible, or at least *anelastic*, so that the time derivative in Eq. (24.1) may be dropped. Here, we adopt the simplest possible approach to this issue, regarding the flow as incompressible. This means that the flows considered are confined to a depth, D, which is the order of the *pressure scale height*, RT/g. Thus, as our first simplifying assumptions

$$\nabla \cdot \overline{\mathbf{u}} = 0 \tag{24.10}$$

In nonequatorial regions, large-scale motions are controlled by rotation. This may be seen by estimating rotation and inertial effects as expressed in Eq. (24.2). The ratio of $|\overline{\mathbf{u}} \cdot \nabla \overline{\mathbf{u}}|$ to $|\overline{\mathbf{u}} \times \overline{\mathbf{\Omega}}|$ is the *Rossby number*

$$\mathrm{Ro} \equiv u/\Omega l \tag{24.11}$$

If l is a typical planetary scale of the motion field ($\sim 10^4$ km), we obtain for $u \sim$ 20 km/hr, Ro $\sim 1/20$. We have noted that in temperate zones, the flow is on average stable against convection, with convective events confined to storm systems of small extent. We may further estimate that for large scales, the gravitational term $\overline{\mathbf{g}}\rho$ dominates the inertial term, $\rho|\overline{\mathbf{u}} \cdot \nabla \overline{\mathbf{u}}|$, provided $\overline{\mathbf{u}}$ has a significant component parallel to $\overline{\mathbf{g}}$. The dominance of gravitational and rotational forcing suggests that a useful idealized problem to consider is the limit $\overline{\mathbf{g}}, \overline{\mathbf{\Omega}} \to \infty$, and to ask what characteristic motions remain finite in this limit. By examining equations simplified in this limit, we may hope to capture those motions which would obtain some time after arbitrary initial fields $\overline{\mathbf{u}}(\overline{\mathbf{x}}, 0)$, $T(\overline{\mathbf{x}}, 0)$ are introduced as initial data into Eqs. (24.1)–(24.3) and transients die out.

At this point, it is convenient to write the equations of motion, Eqs. (24.1)–(24.3), in a Cartesian frame, whose tangent plane is at latitude δ, in the neighborhood of which we wish to study the flow. We take the z coordinate as the radial direction, y as northward, and x as eastward, as shown in Fig. 24.1. Then $\overline{\mathbf{\Omega}} = \overline{\mathbf{j}}\Omega_y + \overline{\mathbf{k}}\Omega_z$, $\Omega_y = \Omega \cos(\vartheta)$, $\Omega_z = \Omega \sin(\delta)$. According to remarks in the previous paragraph, we may further neglect density variations, except where they are multiplied by large quantities, such as the magnitude of $\overline{\mathbf{g}}$. We thus arrive at equations of motion simpler than Eqs. (24.2)–(24.3) that have some validity for atmospheric flows. In addition to Eq. (24.10), these are

$$\rho_0(\partial_t + \overline{\mathbf{u}} \cdot \nabla)\overline{\mathbf{u}} = -\nabla p + \overline{\mathbf{g}}\rho - 2\rho_0\overline{\mathbf{\Omega}} \times \overline{\mathbf{u}} \tag{24.12}$$

and

$$(\partial_t + \overline{\mathbf{u}} \cdot \nabla)T = -\beta w \tag{24.13}$$

Here, $\beta = d\overline{T}/dz$, where $\overline{T}$ is the horizontal mean temperature, which we shall take to be constant. This is a plausible simplification, provided that $\beta > 0$ (convectively stable stratification). We assume a linear relation between ρ and T, appropriate for small fluctuations

$$\rho = \rho_0(1 - \alpha(T - T_0)) \tag{24.14}$$

and use Eq. (24.14) instead of Eq. (24.6). In the analysis that follows, we neglect the effects of Ω_y. Its contribution to the x-component of Eq. (24.12) occurs in the combination $(\Omega_y w - \Omega_z v)$ and we recall that $w \sim (D/\Re)\,(u, v)$. Its other appearance is in the z-component of Eq. (24.12), where its contribution is overshadowed by the hydrostatic balancing of the gravitational and pressure terms. This is equivalent to taking $\boldsymbol{\Omega}$ parallel to $\mathbf{k}$.

If $\bar{\mathbf{u}}(\bar{\mathbf{x}}, t)$ and $T(\bar{\mathbf{x}}, t)$ are to be finite as $\boldsymbol{\Omega}, \bar{\mathbf{g}} \to \infty$, it is clear that the flow must be dominantly perpendicular to $\bar{\mathbf{g}}$. We must further require both horizontal and vertical accelerations induced by $\bar{\mathbf{g}}$ and $\boldsymbol{\Omega}$ to be finite. The vertical dynamics are isolable by forming $\{\nabla \times \nabla \times \partial_t \bar{\mathbf{u}}\}_z$, which from Eq. (24.12) is

$$\{\nabla \times \nabla \times \partial_t \bar{\mathbf{u}}\}_z = f \partial_z \omega_z + \alpha g \nabla_\perp^2 \theta \tag{24.15}$$

with

$$\nabla_\perp^2 \equiv (\partial_x^2 + \partial_y^2)$$

$$\theta = T - \overline{T} \tag{24.16}$$

and

$$f = 2\Omega \sin (\vartheta) \tag{24.17}$$

In Eq. (24.15), $\omega_z = \{\nabla \times \bar{\mathbf{u}}\}_z$. To the extent that $\bar{\mathbf{u}}$ is two-dimensional, it may be specified by a stream function, ψ, defined as

$$(u, v) = (-\partial_y \psi, \partial_x \psi) \tag{24.18}$$

with

$$\omega_z = -\nabla_\perp^2 \psi \tag{24.19}$$

If the vertical velocity is now to remain finite as g and f become large, we must require the right-hand side of Eq. (24.15) to vanish

$$\{f \partial_z \omega_z + g \alpha \nabla_\perp^2 \theta\} = 0 \Rightarrow f \partial_z \psi - g \alpha \theta = 0 \tag{24.20}$$

Using the hydrostatic relation for the pressure perturbation field $(\partial_z \delta p = g \alpha \theta)$, and Eq. (24.20) results in

$$\delta p = \rho_0 f \psi \tag{24.21}$$

The horizontal dynamics $(\{\nabla \times \partial_t \bar{\mathbf{u}}\}_z)$ follow from Eq. (24.12) as

$$(\partial_t + \bar{\mathbf{u}} \cdot \nabla)\omega_z = (f - \omega)\partial_z w + (\partial_y f)v \tag{24.22}$$

The last term on the right-hand side of Eq. (24.22) is responsible for *Rossby waves*, whose existence is traced to the north-south dependence of the *Coriolis parameter*, *f*. This dependence is frequently linearized, $f = f_0 + \hat{\beta}y$, assuming the flow of interest is of a limited north-south extent. Large scale waves (on a sphere), for which such linearization is not valid are *Haurwitz waves* [Haurwitz (1940)]. We shall not include either of these in our present discussion, and refer to Pedlosky (1979), for a more complete discussion of their properties.

Returning to our discussion of Eq. (24.22) for $f \to \infty$, it is reasonable to neglect $\omega \partial w_z$ compared to $f \partial w_z$ on the right-hand side. Eliminating $\partial_z w$ between ∂_z of Eqs. (24.13) and (24.22), and using Eq. (24.20), then gives

$$\{\partial_t + \bar{\mathbf{u}} \cdot \nabla\} \{\omega_z - f^2/(g\alpha\beta)\partial_z^2\psi\} = 0 \tag{24.23}$$

Defining

$$q \equiv \omega_z - (f^2/(g\alpha\beta))\partial_z^2\psi \tag{24.24}$$

allows Eq. (24.23) to take the form

$$(\partial_t + \bar{\mathbf{u}} \cdot \nabla)q = 0 \tag{24.25}$$

We preserve the customary relation between vorticity, q, and stream function, ψ, provided we rescale the vertical coordinate to

$$\hat{z} \equiv f\sqrt{(g\alpha\beta)}z \tag{24.26}$$

$$-q = \hat{\nabla}^2\psi \tag{24.27}$$

where $\hat{\nabla}^2 \equiv (\partial^2/\partial x^2 + \partial^2/\partial x^2 + \partial^2/\partial \hat{z}^2)$ (we shall henceforth drop the caret on $\hat{\nabla}^2$). Here, q is synonymous with ω_z, as given by Eq. (24.19). Note that the advection operator $(\bar{\mathbf{u}} \cdot \nabla)$ in Eq. (24.25) is two-dimensional, whereas the Laplacian in $-q = \hat{\nabla}^2\psi$, Eq. (24.27), is three-dimensional. Finally, we need boundary conditions for Eqs. (24.24), and (24.25). For these, we suppose that the flow is confined by impenetrable plates at $z = (0, D)$, on which $w = 0$. Thus, from Eqs. (24.13), (24.14), (24.16), and (24.20),

$$(\partial_t + \bar{\mathbf{u}} \cdot \nabla)\partial_z\psi(x, y, z \in \mathcal{B}) = 0 \tag{24.28}$$

where $\mathcal{B}$ specifies the boundaries. Equation (24.25), with $\bar{\mathbf{u}}$ from Eq. (24.18), and q from Eq. (24.27), and boundary conditions in Eq. (24.28), is known as the *quasi-geostrophic approximation* and was proposed as a useful way of characterizing rapidly rotating, stratified flows by Charney (1947) and Eady (1949). The derivation presented here is but a sketch; a more rigorous derivation is to be found in Pedlosky (1979).

In our efforts to be brief in arriving at the quasi-geostrophic approximation, we have perhaps obscured the fact that it represents a systematic perturbation expansion, appropriate to small Rossby number, Ro. Since we will shortly need more details than have been as yet brought out, we record here some of the relations that result from introducing into Eqs. (24.12) and (24.13) the perturbation series in Ro for the velocity field $\bar{\mathbf{u}}$

$$\bar{u} = \bar{u}_g + \text{Ro } \bar{u}_a + \cdots \tag{24.29}$$

Here, $(\bar{u}_g, \text{Ro } \bar{u}_a)$ are the *geostrophic* and *ageostrophic* components of $\bar{u}$, and $\bar{u}_g = (-\partial_y, \partial_x)\psi$.

Introducing Eq. (24.29) into the Rossby number nondimensionalized form of Eq. (24.12) gives

$$\text{Ro } Du/Dt = v - \partial_x\hat{p}, \quad \text{Ro } Dv/Dt = -u - \partial_y\hat{p} \tag{24.30}$$

$$\text{Ro } Dv/Dt = -u - \partial_y\hat{p}, \quad \hat{p} \equiv p/f \tag{24.31}$$

we find

$$(u_g, v_g) = (-\partial_y, \partial_x)\hat{p} \tag{24.32}$$

and

$$D_g(u_g, v_g)/Dt = (-v_a, u_a) \tag{24.33}$$

Here, D_g/Dt is the *substantial derivative* following $\bar{u}_g$. Notice that the vertical velocity, w, is a component of $\bar{u}_a$ ($\bar{u}_g$ is entirely horizontal), the only appearance of w is in Eq. (24.13). Continuity for $\bar{u}_a$ implies equations for $\bar{u}_g$, which are summarized by Eqs. (24.20)–(24.25).

One peculiarity of the quasi-geostrophic equations is the inability to advect q in the vertical; its equation of motion, Eq. (24.25), contains only horizontal advection. This despite the fact that the vertical velocity is given by Eq. (24.13) as

$$-\beta w = (\partial_t + \bar{u} \cdot \nabla)\partial_z\psi \tag{24.34}$$

seems not to advect q.

The resolution of this seeming paradox lies at a higher order of the perturbation series in Eq. (24.29); a more precise equation of motion for q states that it is advected along a three-dimensional surface and that these surfaces are advected by (u, v, w). Finally, we note two quadratic inviscid constants of motion

$$E \equiv \tfrac{1}{2} \int d\bar{x} |\nabla\psi|^2 \tag{24.35a}$$

$$\mathcal{E} \equiv \tfrac{1}{2} \int d\bar{x} |\nabla^2\psi|^2 \tag{24.35b}$$

where the volume integral $d\bar{x}$ is bounded by surfaces on which $\bar{n} \cdot \nabla\psi(\bar{x})$ is zero. We will make particular use of the quadratic constants of motion as in Eq. (24.35), despite the fact that Eq. (24.25) implies an infinity of additional conservation laws. In spite of their structural simplicity, the quasi-geostrophic equations still present a formidable system in practice. We have noted, for example, that the dynamics leave the vertical distribution of q open. More precisely, this is determined by moleclar or dissipative processes, since, without vertical dissipation, the vertical distribution of q is fixed by its initial distribution. Modelers are reluctant to ascribe much physics to such diffusive terms, since, at practical numerical resolutions, these represent numerical artifacts rather than actual processes.

We will now describe the application of the quasi-geostrophic equations to large-scale flow fields. The first topic is an analysis of the sinuous motion that is responsible for the temperate-zone poleward heat transport (in the latitudes of ~ 30 to $\sim 60°$), and is known as the *Eady problem*, after the discoverer of the instability, the result of which are large-scale *weather waves* [Eady (1949)].

Baroclinic Instability. Consider a steady eastward flow $(\overline{\mathbf{u}} = (U(z), 0, 0))$ in hydrostatic and geostrophic balance. The zeroth order balance is then among the members of the right-hand side of Eq. (24.12), which remains as $\overline{\mathbf{g}}$ and $\overline{\mathbf{\Omega}} \to \infty$. This gives for U

$$\rho_0 f_0 U = -\partial_y \delta\rho \tag{24.36}$$

and

$$-g\delta\rho = \partial_z \delta p \tag{24.37}$$

so that, eliminating δp, we find

$$\partial_z U = (g/f_0)\partial_y \delta\rho \tag{24.38}$$

where, we recall [from Eq. (24.12) and the discussion after Eq. (24.14)]

$$f_0 \equiv 2\overline{\mathbf{\Omega}} \cdot \overline{\mathbf{k}} \tag{24.39}$$

Here, $\overline{\mathbf{k}}$ is the unit vector along z.

Equation (24.38) is the *thermal wind relation*, expressing the balance between vertical shear and density gradient in the northward direction. Equations (24.36)–(24.39) express the balance between Coriolis and gravitational forces needed for equilibrium. Assuming the domain of interest is sufficiently small to permit linearization, they imply a variation of total density

$$\rho(y, z) = \rho_0 - \rho_0 \alpha\beta[z - (f_0 U/D)/(g\alpha\beta)y] \tag{24.40}$$

Thus the isolevels of the total density field increase northward at a rate $(f_0 U/D)/(g\alpha\beta)$, where D is the vertical depth permitted to the flow (in our approximation, a pressure scale height), and $\overline{\mathbf{u}}$ is the mean over the depth.

We will now discuss briefly the stability of a constant sheart flow, indicated by Eqs. (24.36)–(24.40), to small perturbations. Our description follows Pedlosky (1979), except for minor notational changes. The simplest way to realize Eqs. (24.38)–(24.40) is through a stream function

$$\Psi = -\overline{U}zy/D \tag{24.41}$$

so that the total stream function ψ, including a small perturbation, φ, whose putative growth rate we wish to examine is

$$\psi = \Psi + \varphi \tag{24.42}$$

Introducing Eq. (24.42) into Eq. (24.23) yields

$$\{\partial_t + (z/D)\overline{U}\partial_x\} \{\nabla_\perp^2 + f^2/(g\alpha\beta)\partial_z^2\}\varphi = 0 \tag{24.43}$$

for which boundary conditions $w = 0$ [see Eq. (24.13)], with $\theta \sim \partial_z\psi$ at $z = 0$ and D imply

$$[\{\partial_t + (z/D)\overline{U}\partial_x\}\partial_z\varphi - (\overline{U}/D)\partial_x\varphi = 0]_{z=(0,D)} \tag{24.44}$$

If we assume

$$\varphi \sim \exp(i(kx - \sigma t) \sin(n\pi y/L)F(z) \tag{24.45}$$

$(i = \sqrt{-1})$, then we find from Eq. (24.43)

$$F(z) = A \sinh(Kz) + B \cosh(Kz) \tag{24.46}$$

where

$$K^2 = ((k^2 + (n\pi/L)^2)(N/f)^2 \tag{24.47}$$

The boundary condition Eq. (24.44), $z = 0$ implies

$$A/B = kU/(\sigma KD) \tag{24.48}$$

while that at $z = D$ yields,

$$\frac{A}{B} = \frac{\{K(kU - \sigma) \sinh(KD) - (kU/D) \cosh(KD)\}}{\{K(kU - \sigma) \cosh(KD) - (kU/D) \sinh(KD)\}} \tag{24.49}$$

Equating Eqs. (24.48) and (24.49) gives a quadratic equation for σ, whose solution is

$$\sigma = (kU/2) \{1 \pm \sqrt{(KD)^2 - 4KD \coth(KD) + 4/(KD)}\} \tag{24.50}$$

Thus, instability occurs for values of KD less than the root of

$$KD = 2 \coth(KD/2), \quad K = \sqrt{k^2 + (n\pi/L)^2} (N/f) \tag{24.51}$$

which is $KD = 2.399$. This may be turned into an absolute condition for instability: If

$$L_D \equiv D(N/f) < 0.764L \tag{24.52}$$

the flow is always unstable. We recall that L is the north–south extent of the flow [see Eq. (24.45)]. Here, L_D is called the *Rossby radius of deformation*. In general, notice that it is the large-scale meridional waves $\sim \exp{(ikx - \sigma t)}$ that are more unstable than the small waves.

Finally, note that according to Eqs. (24.13) and (24.23), the integrated time rate of change of kinetic energy, $\frac{1}{2}\int(u^2 + v^2)\,dx\,dy\,dz = \frac{1}{2}\int(|\partial_x\psi|^2 + |\partial_y\psi|^2)\,dx\,dy\,dz$, is given by

$$\frac{1}{2}\frac{d}{dt}\int (u^2 + v^2)\,dx\,dy\,dz = g\alpha \int dx\,dy\,dz(w\theta) \qquad (24.53)$$

Equation (24.53) implies that growth in kinetic energy is associated with a net positive correlation between vertical velocity and temperature fluctuations, the same association as in buoyantly unstable thermal convection. But, here the vertical stratification is stable [see Eq. (24.13)], although the *slantwise stratification* [see Eq. (24.40)] has been found to be unstable under condition of Eq. (24.52). This analogy of baroclinic instability with convective instability has been exploited by Tennekes (1977), who has proposed a more quantitative assessment of *slantwise convection* to explain the large-scale flows of the temperate zones.

Barotropic Instability. Even without the presence of vertical motion, two-dimensoinal flows suffer from a variety of instabilities of meteorological importance. We discuss here only the simplest of these, first explored by Rayleigh (1880, 1887). Our discussion here is brief, a more complete description is to be found in Drazin and Reid (1981).

To this end, consider an inviscid channel flow bounded by $-L \le y \le L$, with an arbitrary $U(y)$. The condition for stability is that $U(y)$ have no inflection point within $(-L \le y \le L)$. Conversely, if $d^2U(y)/dy^2 = 0$ for $(-L < y < L)$, the flow will be unstable. The following brief sketch [see Monin (1990) and Rayleigh (1887)], shows why this is so. As before, we assume

$$\psi(x, y) = \Psi(y) + \varphi(x, y) \qquad (24.54)$$

but now suppress any z-dependence. Introducing Eq. (24.54) into Eq. (24.23) [with $(f/(g\alpha\beta)) \to 0$], and retaining only first-order terms in φ gives

$$(\partial_t + U(y)\partial_x)\nabla_\perp^2\varphi + \phi_x(\partial_y^2 U(y)) = 0 \qquad (24.55)$$

where $U(y) = -\partial_y\Psi(y)$. Assuming $\varphi(x, y, t) = \exp{(ikx)}F(y, t)$, results in

$$\{(\partial_t + ikU)\,(D_y^2 - k^2) + ikD_y^2 U\}F(y) = 0 \qquad (24.56)$$

Boundary conditions are that $v(x, y = \pm L) = -\partial_x\varphi(x, y = \pm L, t) = 0 \Rightarrow F(y = \pm L) = 0$.

Rayleigh (1880, 1887) gave the simplest analysis of why such flows whose value of D_yU changes sign are unstable. He begins by noting that flows for which dU/dy consists of constant segments may be constructed from solutions of $(D_y^2 - k^2)F = 0$ in each of these segments. The constants of integation belonging to each segment are chosen by constraining the flow by continuity and by Eq. (24.56) interpreted as an integral constraint imposed across each segment. Growth rates of a disturbance of the form

$$F(y, t) = \exp{(i\sigma t)}G(y) \tag{24.57}$$

are then determined by the usual consistency condition. Finally, if an instability exists, the imaginary part of σ [denoted here as Im(σ)] is $\neq 0$. Then, from Eq. (24.55),

$$2\,\text{Im}\,(\sigma) \cdot k \int_{-L}^{L} dy (D_y^2 U)\,|\,G(y)\,|^2/|\,U + \sigma/k\,| = 0 \tag{24.58}$$

which implies that $D_y^2 U(y)$ must change signs on $-L \leq y \leq L$.

Before terminating our brief discussion of instability in rapidly rotating flows such as those considered here, we should note that such instabilities in quasi-two-dimensional flows are fundamentally different from those in nonrotating flows. The reason is a consequence of the existence of two inviscid constants of motion: energy and *enstrophy* [Eq. (24.35)]. From these constraints stems the fact that both energy and squared vorticity remain finite as dissipative processes approach zero (the limit of infinite Reynolds number). These constraints limit the type of unstable growth and their extent. As an example of the difference between instabilities in three dimensions and in two dimensions, we note that in three dimensions a constant shear suffices (depending on boundaries) to induce instability, whereas, as we have just seen, this is not so in two dimensions. Another difference is that gradients of velocity fields grow indefinitely with increasing Reynolds number in three dimensions, while a consequence of Eq (24.35) is that such gradients remain bounded by their initial distribution in quasi-geostrophic flows. A more subtle consequence of Eq. (24.35) is the fact that gradients of vorticity remain finite for finite times, if they are initially bounded [a theorem of Kato (1967)]. As we shall shortly see, in certain circumstances, these are unrealistic restrictions for many important large-scale features of the atmosphere, in particular, the formation of weather fronts, *frontogenesis*.

Beyond Instability: Structures and Turbulence. The study of flow instabilities in frequently thought to give some insight into the character of the nonlinear regime into which the unstable flow develops. Thus, it is argued that an examination of baroclinic wave instability leads to the anticipation that such waves (even those that are unstable) move eastward, with a phase velocity in accordance with Eq. (24.50). The early, exponential growth associated with the instability is eventually supposed to be stabilized by a sharpening of wave forms, transforming the initial sinusoidal perturbations into near shock-like structures. Another aspect of the fully nonlinear process is the breakdown of the organized linear instability profiles into turbulence, a disorganized motion which is thought to be best characterized statistically. Both sharp structures and the chaos in which they are imbedded are characteristics of the fully nonlinear regime. Neither can be neglected in atmospheric flows. We shall discuss here only briefly two of the aspects of atmospheric flow which display the structure-*chaos dichotomy*, the development of near discontinuities in baroclinic flows, and the development of intense, isolated vortices in barotropic flows. But, the discussion of frontogenesis (the first topic) will lead into more general issues than quasi-geostrophy as covered here.

The past two decades have witnessed a large number of initial value type studies of two-dimensional flows, largely devoted to an understanding of how the scale size distribution natural to two-dimensional flows is established. Such studies were largely numerical, and stressed the nonlinear dynamics implicit in equations such as Eq. (24.25). We will describe in the next section certain ideas that emerge from a scaling analysis of this equation, as well as how these may relate to the real atmosphere. Here, we simply stress that intense structural features such as compact vortices seem a generic feature of two-dimensional and quasi-two-dimensional flow. These features were largely unanticipated by the theorist and are not easily comprehended by concepts such as energy spectra or scale-size distributions. The first indication of the presence of such intense vortices was obtained by Fornberg (1977). Later Basdevant *et al.* (1981) and McWilliams (1984) demonstrated that such structures are an intrinsic feature of two-dimensional flow by means of a more refined numerical simulation, and they showed how such features modified the spectra of kinetic energy. Finally, Carnavale *et al.* (1991) showed how compact vortices could be viewed as comprising the essential particles of two-dimensional turbulence. A key feature of this dynamics is that, in regions where the magnitude of strain exceeds the magnitude of vorticity, vorticity is rapidly destroyed by stretching. Such a criterion for survival is known as the *Weiss criterion*, after the researcher who proposed it from an approximate analysis of the Lagrangian dynamics [Weiss (1981), (1991)], although a similar result may be deduced from a spectral form of rapid distortion theory [Herring (1975)]. The surviving vortices then move in their mutual self-induction field, until a pair merge, with a rather sudden decrease in their number.

Our discussion of nonlinear issues so far concerns features that are comprehended by the quasi-geostrophic dynamics. We should at least mention briefly a problem which leads to structures that may signal the breakdown of the assumptions made to derive the quasi-geostrophic dynamics, *frontogenesis*. Fronts are formed from growing baroclinic waves. Their dimension is the order of the Rossby radius, Eq. (24.52), in one direction, but much sharper in the transverse direction. They comprise the descent of a tongue of stratospheric air into the *tropopause*. Since the cross-stream length scale is small, the basic assumption of quasi-geostrophy that the Rossby number is small becomes suspect, so that some generalization may be necessary for this configuration. The proper generalization [see Hoskins (1982)] is to be found by including in the advection of the geostrophic flow $(\partial_t + \overline{\mathbf{u}} \cdot \nabla q)$ not only by the geostrophic flow $\overline{\mathbf{u}}_g$, but the ageostrophic component, $\overline{\mathbf{u}}_a$ as well.

24.1.3 Scale Size Distribution of Atmospheric Motions

Reynolds numbers of atmospheric flows (below the tropopause) are quite large, consequently a broad range of scale sizes must be thought of as comprising the flow. Since the flow is turbulent, different scales are related statistically and have a degree of independence from one another. To put this issue more quantitatively, we may resolve fields such as $(\overline{\mathbf{u}}, \theta)$ into an orthonormal set of function, $\phi_k(\overline{\mathbf{r}})$

$$\Psi(\overline{\mathbf{r}}) = \sum_k \phi_k(\overline{\mathbf{r}}) a(\overline{\mathbf{k}}) \tag{24.59}$$

where $\Psi(\overline{\mathbf{r}})$ represents either of the fields $(\overline{\mathbf{u}}, \theta)$. For $\phi_k(\overline{\mathbf{r}})$ we choose Fourier modes, which we designate generically as

$$\phi_k(\bar{\mathbf{r}}) = \exp{(i\bar{\mathbf{k}} \cdot \bar{\mathbf{r}})} \tag{24.60}$$

In terms of $\phi_k(\bar{\mathbf{r}})$, we define the scale-size distribution (or energy spectrum) through

$$U(\bar{\mathbf{k}}) = \langle |a(\bar{\mathbf{k}})| \rangle^2 \tag{24.61}$$

Here, the angular brackets represent a time or, more properly, an ensemble average. We ignore boundaries, assuming the flow to be reasonably homogeneous.*

Figure 24.2 shows the observed energy spectra, $2\pi k^2 U(k)$, of large-scale motion of the earth's atmosphere at a height near the tropopause. We recall the earth's circumference $\sim 1.8 \times 10^4$ km. Spectra for both zonal and meridional winds are shown, with the meridional spectrum shifted to the right for display purposes. (The horizontal motion field is not far from isotropy). Here, k is the horizontal wave number on a Cartesian tangent projection. At scales smaller than about 1000 km, the spectrum is not inconsistent with a $k^{-5/3}$ slope, but at larger scales, it is steeper, approximately $\sim k^{-3}$.

One explanation proposed for the observed scale-size distributions is in terms of quasi-two-dimensional turbulence, whose essential dynamics are described by Eqs.

---

*Specifying $U(\bar{\mathbf{k}})$ is equivalent to specifying the covariance, $\mathcal{C}(\mathbf{r}, \bar{\mathbf{r}}') \equiv \langle \Psi(\bar{\mathbf{r}})\Psi(\bar{\mathbf{r}}') \rangle$ only for homogeneous flows, for which $\mathcal{C}$ is a function of $\bar{\mathbf{r}} - \bar{\mathbf{r}}'$ only.

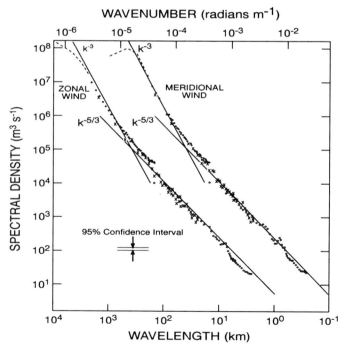

FIGURE 24.2   Observed kinetic energy spectra $2\pi k^2 U(k)$ (m³ sec⁻²) of large-scale motion of the earth's atmosphere at a height near the tropopause. Here, units of k are rad m⁻¹. (Reprinted with permission, Nastrom, G. D., Gage, K. S., and Jasperson, W. H., "Atmospheric Kinetic Energy Spectrum, 10^0–10^4 km," *Nature*, Vol. 310, p. 36, 1984.)

(24.23)–(24.27). We will use the observed spectrum (Fig. 24.2) to illustrate some of the traditional ideas concerning the statistical dynamics of large-scale atmospheric flows. First, for the largest scales, $E(k) \sim k^{-3}$, it is plausible that any vertical variation [as in Eq. (24.23)] may be neglected. Then, the quasi-geostrophic potential vorticity, q, is advected as a scalar in a very nearly conservative manner. If we assume that q may be regarded as a passive scalar advected by large scales, its spectrum would be determined wholly by the strain field associated with $\bar{\mathbf{u}}$, whose *rms* value we denote by ζ. The strain's dimensions are that of a frequency. Then, if we must make a spectrum whose integral over k has dimensions of q^2, it follows that

$$Q(k) \equiv 2\pi k \langle q(-\bar{\mathbf{k}}) q(\bar{\mathbf{k}}) \rangle = \zeta^2 k^{-1} \tag{24.62}$$

The factor $2\pi k$ converts the *modal enstrophy spectrum*, Q, to an *omnidirectional spectrum*, assuming isotropy. From Eq. (24.62) the omnidirectional energy spectrum (recall that $-q = \nabla^2 \psi$, $(u, v) = (-\partial_y, \partial_x)\psi$) is

$$E(k) \sim \zeta^2 k^{-3} \tag{24.63}$$

Notice that dissipation plays no explicit role in deducing Eq. (24.63). But a closer look at its derivation shows that the physics of Eqs. (24.62) and (24.63) is a straining out of the vorticity field into progressively thinner filaments, and there is a tacit assumption in the derivation of Eq. (24.63) that dissipative processes destroy filaments whose scale is smaller than a certain size set by the value of viscosity. We refer to the straining of vortex elements into progressively smaller scales as a *cascade* to smaller scales. Thus, Eq. (24.63) is characterized by a flux of enstrophy, $d\langle q^2 \rangle / dt$, whose value in the statistical mean is set by ζ.

We next inquire as to the dynamics of the spectrum in the range 1 km $\leq k \leq 16 \times 10^2$ km). Here, we mention two principal explanations: 1) a weakly interacting system of gravity waves [Van Zandt (1982)], and 2) inverse-cascading (*forced*) quasi-two-dimensional turbulence. The latter was proposed by Gage (1979) and Lilly (1983). The small-scale forcing (in the vicinity of 10^0 km) may then be attributable to anvil outflows from convective storm systems. We focus on alternative 2), to illustrate certain features of near two-dimensional dynamics in which energy is injected at small scales and migrates to larger scales. Gage (1979) proposed that the paradigm of this $k^{-5/3}$ range is that of inverse cascading two-dimensional turbulence, a theory described by Kraichnan (1967) and Lilly (1971). We mentioned above enstrophy cascade to small scales, and an eventual destruction of vorticity at extremely small scales by some form of viscous dissipation. But, here we are referring an unorthodox case in which the cascade of energy is to progressively larger scales, with its eventual destruction by frictional effects at the very largest of the planetary scales. To describe this dynamics, note that both energy and enstrophy are conserved for near two-dimensional inviscid flows [see Eq. (24.35)]. This suggests that a constant energy flux (of dimensions: $d\langle u^2 \rangle / dt \equiv -\epsilon$) determines $E(k)$. If $E(k)$ is comprised only of k and ϵ (the traditional scaling argument), we can only have

$$E(k) \sim \epsilon^{2/3} k^{-5/3} \tag{24.64}$$

But here, ϵ is a flux towards larger instead of smaller scales.

We now give a more quantitative setting to such ideas of a turbulent cascade by introducing a simple heuristic model that is only a slight modification of that proposed by Leith (1967). The starting point is the observation that the straining of vortex filaments constitutes the mechanism by which energy or enstrophy is transferred from large to small scales. If the flow is turbulent, the straining is quasi-random and of finite duration. Its lifetime [denoted here by $\tau(k)$] depends on the scale size. We picture the cascade as a series of random straining events, whose orientation and magnitude are somewhat random, but ultimately determined by the turbulence itself. It is a conservative process, with dissipation relegated to much smaller scales than those under consideration. This means that there is no loss of the cascaded quantity (such as energy) as it is passed from larger to smaller scales. Moreover, the strain transforms a given scale into one of only slightly smaller (or larger, for the case of inverse cascade) scale. We call such a process *local*, meaning that only neighboring scale-sizes are involved. Localness of transfer is rooted in the observation that very large scales amount to a uniform translation of the flow field structures, and hence is dynamically irrelevant. It is then plausible to suppose that the cascade is diffusive (in scale size, or wave number) and to write,

$$\partial_t \mathcal{V}(k, t) \equiv T(\{\mathcal{V}\}(k, t)) = \gamma \frac{\partial}{\partial k} k^{a+2} \left\{ \frac{\partial}{\partial k} k^{-a} \mathcal{V}(k) \, S(k)/\tau(k) \right\} - 2\nu k^2 \mathcal{V}(k)$$

(24.65)

where, $\mathcal{V}(k)$ is the spectral variance examined (enstrophy, energy, or scalar variance), and $S(k)$ is the square of the large-scale strain

$$S(k) = \int_0^k p^2 E(p) \, dp$$

(24.66)

Notice that the wave-number factors comprising $T(k)$ in Eq. (24.65) are homogeneous of degree zero and that the factor $S(k)/\tau(k)$ has dimensions of $1/t$. If the time scale, $\tau(k)$, is to be determined wholly by S, then there is no other choice than,

$$\tau(k) \sim 1/\sqrt{S(k)}$$

(24.67)

For two-dimensional turbulence, Eq. (24.65) must be consistent with energy and enstrophy conservation, Eq. (24.35), which means (for $\mathcal{V} = E$) that

$$\int_0^\infty dk T(k) = \int_0^\infty dk k^2 T(k) = 0$$

(24.68)

which implies $a = -3$, so that our model, Eq. (24.65), becomes,

$$\partial_t E(k) = \gamma \frac{\partial}{\partial k} \frac{1}{k} \frac{\partial}{\partial k} \{k^3 E(k) \sqrt{S(k)}\} - 2\nu k^2 E(k) + \mathcal{F}(k)$$

(24.69)

where γ is a constant of order unity. In Eq. (24.69) we have included a possible forcing term, $\mathcal{F}(k)$. If $T(k) = 0$, $\mathcal{F} = 0$, and $\nu = 0$, there are two solutions to Eq.

(24.69), namely $E(k) \sim k^{-5/3}$ and $E(k) \sim k^{-3}$. In practice, dissipation may be small, but its indirect effects are always manifest. Consider the solution $E(k) \sim k^{-3}$, in a range of k where $2\nu k^2 E(k)$ is small relative to the other terms of Eq. (24.69). Then, forming $\int_k^\infty dk k^2$ of Eq. (24.69), [with $\mathcal{F}(k) = 0$], there results

$$\frac{\partial}{\partial k} \{k^3 E(k) \sqrt{S(k)}\} = 2\nu \int_k^\infty k^4 E(k) \, dk - \int_k^\infty k^2 \partial_t E(k) \, dk \qquad (24.70)$$

In secular decay, the second member on the right-hand side is much smaller than the first, in the limit $\nu \to 0$. The first term,

$$\eta(k) = 2\nu \int_k^\infty k^4 \, dk E(k) \qquad (24.71)$$

is the viscous dissipation of enstrophy, and it may be taken to be approximately constant in k, for small k. Then, Eq. (24.70) may be solved as an equation for $S(k)$ by recognizing that $\gamma k^2 E(k) = dS(k)/dk$

$$E(k) \sim \eta^{2/3} k^{-3}/(\ln (k))^{1/3} \qquad (24.72)$$

The precise value of η can be known only by solving Eq. (24.68) not only in the inertial range where $E(k) \sim k^{-3}$, but also in the dissipation range, where $E(k)$ decreases sharply so that the integral in the definition of $\eta(k)$ converges. The wave number where this sharp decrease begins, k_η, is the *(enstrophy) dissipation scale*, and it may be estimated by equating the eddy turnover rate $\sqrt{S(k)}$ to the dissipation rate $2\nu k^2$.

Next, consider Eq. (24.68) for the case in which $\mathcal{F}(k)$ injects energy into some intermediate scale, k_0, and we are interested in the spectrum at scales larger than $2\pi/k_0$. If we form an equation for the total energy from Eq. (24.69) in that case, we see that there is no possibility for a steady state unless energy is extracted from large scales by frictional effects. If this is done, and we call ϵ the *rate of energy extraction by friction*, we may form $\int_0^k dk$, $(k < k_f)$ of Eq. (24.69) to get,

$$\epsilon \sim \gamma \frac{1}{k} \frac{\partial}{\partial k} \{k^3 E(k) \sqrt{S(k)}\} \qquad (24.73)$$

whose solution is,

$$E(k) \sim \epsilon^{2/3} k^{-5/3} \qquad (24.74)$$

For three-dimensions, we have only a single inviscid constant of motion, kinetic energy, and how to choose a [in Eq. (24.65)] is less clear. Here, we simply state that $a = 2$ gives consistency with the principle of equipartitioning of energy on a finite wave-number band, and hence is a plausible choice. In this case, a repeat of the analysis for the energy spectra, except now ϵ represents the forward transfer of energy to small scales (the large-scale friction is not needed) yields Eq. (24.74).

The formulas presented here for the energy spectrum and their transfers were derived from a more secure perspective by Kraichnan (1971), and a more complete account of such spectra (for both two and three dimensions) and their relationship to energy and enstrophy fluxes may be found elsewhere, for example, in the book by Lesieur (1990).

24.1.4 Predictability

Two flows that differ only slightly initially may be quite dissimilar at some later time. The inquiry into the length of time needed for significant differences to develop between two initially similar flows is known as the *predictability problem*, and it is of special significance for atmospheric sciences, since one of its tasks is to predict the weather from initial data specified at data situations. Consider, then a rectangular network of data stations, a distance l apart, and at each of which the stream function $\psi(x, y)$ is accurately measured. For simplicity, we restrict our discussion to two-dimensional turbulence, as described by Eq. (24.25), and the observation network to be a square lattice. The question is the extent to which the finite resolution, l, limits the ability to forecast $\psi(x, y)$.

Imagine a computer able to solve the equations of motion on a much smaller mesh than l. Since there is no data at the finer mesh points, we must assign values of $\psi(x, y)$ in a manner consistent with the known data specified at the coarser mesh points, and consistent with the spectrum of turbulent motions, such as that given by Eq. (24.63). One way to achieve this consistency is to make random assignments on the finer mesh that meet the consistency requirement noted above. In this way, we arrive at the following question whose solution would give an understanding of the problem posed. Given an ensemble of equivalent flows, and supposing that for any two members of the ensemble (a, b) we have, at $t = 0$

$$W(k) \equiv \langle \psi^a(\mathbf{k}, 0)\psi^b(\mathbf{k}, 0)\rangle = E(\mathbf{k}, 0), \ |k| \leq 2\pi/l \qquad (24.75a)$$

$$W(k) \equiv \langle \psi^a(\mathbf{k}, 0)\psi^b(\mathbf{k}, 0)\rangle = 0, \ |k| \geq 2\pi/l \qquad (24.75b)$$

Here, the angular brackets denote ensemble averages. Let L be the scale at which most of the energy resides, and suppose that $L \gg l$. The question is then: How long after the initial time will $W(2\pi/L)$ remain $\sim \mathcal{O}(1)$? Perhaps surprisingly, at large Reynolds numbers this time turns out to be independent of l. Its magnitude is found to be several (5 to 10) large-scale turnover times, L/u_{rms}, where u_{rms} is the rms velocity field [Lorenz (1969), Leith and Kraichnan (1972)]. To see how this happens, at least for a homogeneous domain, we write equations of motion for $W(k, t)$, defined by Eqs. (24.75a) and (24.75b). If we know the energy spectrum equation of motion such as Eq. (24.69),

$$\partial_t E(k, t) = T(\{E\}, t) \qquad (24.76)$$

where T is specified by, say, Eq. (24.65), then the statistical theory of turbulence [Leith and Kraichnan (1972), Herring (1984)] gives for $W(k, t)$

$$\partial_t W(k, t) = T(\{W\}, t) - \eta(\{E - W\})W(k) \qquad (24.77)$$

where η is an eddy viscosity, also computable from its functional argument *via* the statistical theory of turbulence. Equation (24.77) indicates that the fate of $W(k, t)$

depends on the competition between two effects: 1) the sweeping of correlation to ever smaller scales (the first term), and 2) the destruction of correlation by the same physics as eddy viscosity. But, here the eddy viscosity is constructed from the error energy $\Delta(k, t) \equiv E(k, t) - W(k, t)$, instead of from $E(k, t)$. It is not *a priori* clear which of these competing effects will win. However, given the instabilities to which Navier–Stokes is susceptible, it is plausible that eventually, $W(k, t)$ will be destroyed.

The *localness principle*, discussed in Sec. 24.1.3, carriers over to the correlation spectrum $W(k)$, and means that for the initial value problem considered here, Eqs. (24.75a) and (24.75b), $W(k, t)$ passes from a value near unity to zero as k increases by roughly an order of magnitude. We denote the *crossover* value of k at which $R(k, t) \equiv W(k)/E(k) = 1/2$ as $k_c(t)$. The equation for $R(k, T)$ is

$$\partial_t R(k, t) = R[T(W)/W - T(E)/E] - \eta(\Delta)R(k, t) \qquad (24.78)$$

A simple way of representing $R(k, t)$ is as a self similar function,

$$R(k, t) = R(k/k_c(t)) \qquad (24.79)$$

For orientation, we simplify the shape of R to a step function centered at k_c. Then, the left-hand side of Eq. (24.79) becomes

$$\dot{R}(k/k_c(t)) = -\delta(1 - k/k_c)\dot{k}_c/k_c \qquad (24.80)$$

The dimension of Eq. (24.80) is $1/t$, which, if localness of transfer is all that is needed, must be $\sim \delta(1 - k/k_c)/\tau(k)$, where $\tau(k)$ is given by Eq. (24.67). Then

$$-\dot{k}_c/k_c \sim \sqrt{k_c^3 E(k_c)} \qquad (24.81)$$

For $E(k) \sim \epsilon^{2/3}k^{-5/3}$,

$$k_c^{2/3}(t) \sim 1/(\Gamma\epsilon^{1/3}t + k_c(0)^{-2/3}) \qquad (24.82)$$

Here Γ is a constant of order unity. Note from Eq. (24.82) that even if $k_c(0) = \infty$, $k_c(t)$ is finite, so that the effects of subgrid uncertainties appear at a scale $2\pi/k_c$ at a time independent of the mesh size, no matter how small. Detailed computations both by way of direct idealized numerical simulations [Leith (1971) and Herring *et al.* (1973)] and those for the more realistic models of the general circulation of the atmosphere [Mullen and Baumhefner (1988)] confirm these estimates. This point is illustrated in Fig. 24.3, which shows the error energy spectrum ($E(k, t) - W(k, t)$) [see Eq. (24.78)], with the time (in days) after the initial spectrum labeling each curve. These calculations are done by a theory similar to that discussed here, but with added refinements. In general, computations using detailed equations of motion show quantitatively similar growth patterns [see Mullen and Baumhefner (1988)].

24.1.5 The Planetary Boundary Layer

As noted earlier, the smaller scales of atmospheric flows are insensitive to rotation and are fully three-dimensional and turbulent. They are in general buoyantly active (either stable or unstable). The velocity and temperature fields are determined by

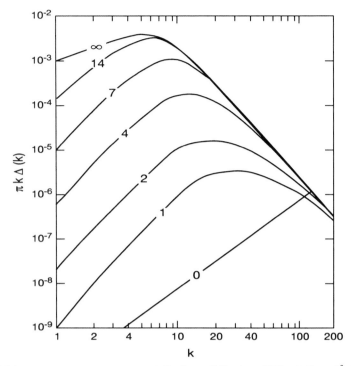

FIGURE 24.3 Error energy spectrum, $\Delta(k, t) \equiv E(k, t) - W(k, t)$ (in $\text{rad}^3 \, \text{day}^{-2}$) as a function of planetary wavenumber, k (rad^{-1}). Curves are labeled by days after high resolution observations with alias error (alias error spectrum is that labeled by 0). (Reprinted with permission, Leith, C. E., "Atmospheric Predictability and Two-Dimensional Turbulence," *J. Atmos. Sci.*, Vol. 28, No. 2, pp. 145–161, 1971.)

imposed fluxes of momentum and heat that are in turn set by solar heating and the large-scale flow context. The mean vertical distribution of velocity, $U(z, t)$, and of temperature, $\overline{T}(z, t)$, are of great interest. A universal similarity theory for these has been proposed by Monin and Obukhov (1958) and is widely used in the meteorological context. To develop these ideas, we form the horizontal average of the equations of motion for $\overline{\mathbf{u}}$, and T, Eqs. (24.12) and (24.13), and assume steady-state conditions. We further neglect $\mathbf{\Omega}$. There result two simple equations expressing the constancy of the flux of momentum and of heat

$$-\overline{uw} + \nu \partial_z U(x) = v^{*2} \tag{24.83}$$

$$\overline{w\theta} - k\partial_z \overline{T} = q \tag{24.84}$$

Here, v^{*2} and q are the flux of momentum and heat, which are constants of motion under steady-state conditions. It is plausible that far from the ground, where viscosity and conductivity are unimportant, these constants of motion determine the profiles of $U(z)$ and $\overline{T}(z)$. To make such a statement precise, we must note what distance is sufficient for viscous (or conductive) effects to be neglected. We denote this distance as z_0, the *roughness length scale*. It is the only distance comprised of ν and the constants of motion (v^* and q)

$$z_0 = \nu/v^* \tag{24.85}$$

The central idea is to utilize the constants of motions, $v*$ and q, to obtain scaling laws. To that end, we must comprise velocity, temperature, and length scales from these. The velocity scale is $v*$. A length scale, $\mathcal{L}$, is obtained by equating the inertial to buoyancy term in Eq. (24.12), with ∇ estimated by $\mathcal{L}^{-1}$, $u \sim v*$ and $g\rho \sim g\alpha q/v*$. Thus,

$$\mathcal{L} \equiv -v*^3/(g\alpha q) \tag{24.86}$$

and the temperature scale is

$$\mathfrak{I} = (q/v*) \tag{24.87}$$

Hence, for the universal form of mean gradients,

$$dU(z)/dz = (v*/\mathcal{L})\mathcal{G}(z/\mathcal{L}) \tag{24.88}$$

and

$$d\bar{T}(z)/dz = (\mathfrak{I}/\mathcal{L})\mathcal{G}_1(z/\mathcal{L}) \tag{24.89}$$

Here, $\mathcal{G}(\mathfrak{z})$ and $\mathcal{G}_1(\mathfrak{z})$ are universal functions of their arguments. Note that the heat flux, q, may be either positive, for convectively active regions, or negative, for buoyantly stable region. The sign convention in Eq (24.86) is chosen so that $\mathcal{L}$ is positive for stable stratification.

The experimental search for universal forms for functions such as $\mathcal{G}$ and $\mathcal{G}_1$ has been the topic of much meteorological research and is summarized in Monin (1990) and Azad (1993). Here, we draw out one simple example, the case in which $v* \rightarrow 0$, pure thermal convection. In that case, Eq. (24.89) for $\partial_z \bar{T}$ must be independent of $v*$, which implies

$$\mathcal{G}_1(\mathfrak{z}) \rightarrow \mathfrak{z}^{-4/3}, \quad \mathfrak{z} \rightarrow \infty \tag{24.90}$$

So that for pure convection,

$$\partial_z \bar{T} = q^{2/3}/(\alpha g)^{1/3} z^{-4/3} \tag{24.91}$$

This equation may be used to define a thermal length scale, z_t, by equating $\partial_z \bar{T}$ to q/k, the thermal gradient in the presence of pure conduction

$$z_t = (k^3/(\alpha g q))^{1/4} \tag{24.92}$$

Notice that a volume integral of Eq. (24.12) implies that $\epsilon \equiv \langle(\partial_i u_j)^2\rangle = \alpha g q$, so that Eq. (24.92) is just the *Kolmogorov length scale* times $\mathrm{Pr}^{-3/4}$, where Pr is the Prandtl number.

Finally, we can estimate the Nusselt number, Nu, defined as the efficiency of convection to conduction,

$$\mathrm{Nu} = q/(k\Delta T/L) \tag{24.93}$$

where ΔT is the total drop of temperature from the surface to the interior. Thus, we also have,

$$\text{Nu} = L/z_t \tag{24.94}$$

Equating Eq. (24.93) to (24.94) gives,

$$\text{Nu} = \{\alpha g \Delta T L^3/(k^2)\}^{1/3} \tag{24.95}$$

Equation (24.94) expresses the familiar *Priestly's Law* relating Nusselt number to Rayleigh number (Ra $\equiv \alpha g \Delta T L^3/(k\nu)$) [Priestley (1960)].

Although Eq. (24.95) seems very plausible, there is growing evidence that it is not correct, because of the effects of the large-scale context in which the flow is imbedded. Thus, recent laboratory experiments [Helsot *et al.* (1987)] imply Nu ∼ Ra$^{2/7}$, a result directly attributable to a dependence on large-scale circulation effects on the convection [Shraiman and Siggia (1991)]. Moreover, more careful analysis of the planetary boundary data also suggests certain deviations from the universal scaling discussed here [Yaglom (1993) and Zilitinkevich (1993)].

24.2 FLOW OVER BUILDINGS—WIND ENGINEERING
Jack E. Cermak

24.2.1 Introduction

Treatment of wind movement over and around buildings requires integration of knowledge on bluff-body fluid mechanics and boundary-layer meteorology and comprises one aspect of a new engineering discipline that was identified in 1970 as *Wind Engineering*. Other aspects of this fluid dynamic related technology, wind loads on structures (towers, bridges, solar energy collectors, etc.), atmospheric transport of mass (gaseous air pollutants, soil, sand, snow), and definition and control of wind characteristics, are discussed in Cermak (1975), (1979) and in proceedings of national and international conferences.

Flow over buildings results in various effects caused by transfer and transport of momentum, mass and heat. Since the transfer and transport mechanisms are sensitive to the approaching flow, information on wind characteristics, with emphasis on strong winds, is presented in Sec. 24.2.2.

The extreme complexity of wind characteristics and building geometry, coupled with great variability in site geometry, requires that physical modeling be used to obtain both qualitative and quantitative data for wind-engineering applications. However, computational methods show promise for future wind-engineering studies [Murakami (1992)]. Two basic flow domains must be accounted for in the modeling process, the *atmospheric boundary layer* (ABL) upwind and over the building site and flow over the exterior building surfaces. Similarity requirements and types of wind tunnels for physical modeling of atmospheric boundary layers are presented in Sec. 24.2.3. Following meteorological considerations, requirements and procedures for physical modeling of wind pressures, wind forces and moments, dynamic responses (peak wind loads, deflections and accelerations), local wind characteristics and local mass transport are discussed in Sec. 24.2.4.

Wind effects significantly influence the performance, safety, construction cost, and operation and maintenance costs of a building. Therefore, some basic information and guidance on major effects are presented in Sec. 24.2.5.

Section 24.2.6 provides a brief commentary on full-scale measurements of wind and wind effects.

24.2.2 Wind Characteristics

The nature of wind approaching a building depends, among many other factors, upon the type of meteorological event in progress. Severe storms such as tornadoes, thunderstorms, hurricanes, downslope winds, low-level jets, and extra tropical cyclones described in Cermak (1975) can be the cause of extreme winds. Although each meteorological event has distinguishing wind characteristics, in engineering practice, wind is commonly considered to have characteristics associated with the most frequent type of flow in the lower atmosphere, a planetary boundary layer [ASCE (1988) and Haugen (1979)].

Severe Storms. Wind speeds for the most distinctive storms decrease in the follwong order: tornadoes, hurricanes, downslope winds (also referred to as *chinooks, foehns* and *boras*). Table 24.1 gives some characteristics of these storms. Buildings other than nuclear reactor structures and some nuclear materials processing buildings are not designed to survive tornadoes without significant damage. The probability that a particular building will experience such a wind is low (10^{-5} to 10^{-7} per year), and the cost would be excessive. However, low-cost reinforcement of room without external walls can provide safety to occupants of housing units [Mehta *et al.* (1975)].

Variations of wind characteristics up to about 500 m for most strong winds are typical of turbulent boundary layers. Although flows near the eyewall of hurricanes and in downslope winds vary significantly from a zero pressure gradient boundary layer, they too are treated as a simple boundary layer for most wind-engineering applications. Fastest mile wind speeds with annual probabilities for being exceeded of 0.04, 0.02, and 0.01 at locations throughout the United States are tabulated in ASCE (1988). These reference wind speeds are given for an elevation of 10 m but

TABLE 24.1 Characteristics of Primary Severe Storms

Storm	Maximum Wind Speeds[a] (m/s)	Probability of Annual Occurrence	Diameter of Maximum Velocity Core (m)	Maximum Pressure Drop In Core (Pa)	References
Tornadoes[b]	300	10^{-7}	160	$\sim 1.4 \times 10^4$	Fujita *et al.* (1980)
	200	10^{-5}	110	$\sim 6 \times 10^3$	
Hurricanes	45^c	10^{-2}	10^4–10^5	1×10^4	Batts *et al.* (1980)
Downslope Winds	40	1^d	—	—	Miller *et al.* (1974)

[a]Approximate 15/s gust at height of 10 m.
[b]Magnitudes vary with location in the United States.
[c]Near coastlines.
[d]Local areas east of mountain ridges in the United States.

can, after selection of the appropriate *power-law* exponent defined in Sec. 24.2.2, be used to estimate extreme wind speeds at any height within the boundary layer.

Boundary Layer Winds. Characteristics of mean (15 to 30 minute average) wind velocity in the *atmospheric boundary layer* (ABL) depend upon temperature gradients, *aerodynamic roughness* of the surface, z_0, and the *Coriolis parameter, f*. In general, the mean wind velocity changes direction with height. For wind-engineering applications, a simplified form of the ABL is used. Wind approaching an isolated building is considered to be unidirectional, and the variation of mean wind speed with height is represented by a *power-law* relationship

$$U/U_g = (z/z_g)^{1/n} \tag{24.96}$$

In this formulation, z_g is the gradient wind height and corresponds to the boundary-layer thickness, δ, where surface frictional effects cease to influence the flow. The exponent $1/n$ ranges from about 0.1 to 0.4 for generally flat surfaces with profiles varying as shown in Fig. 24.4. For strong winds, the boundary-layer characteristics (power-law exponent, $1/n$, and gradient wind height, z_g) depend strongly upon the surface roughness, z_0. This dependence is shown in Fig. 24.5 along with the surface cross-isobar angle, α_0 (total change in wind direction through the ABL) and the surface drag coefficient, $v*/U_g$, for a wind speed of 10 m/s at $z = 10$ m.

In the lower 100 to 150 m (the atmospheric surface layer) where flow over most buildings occurs, the mean wind speed has a logarithmic distribution for neutral thermal stability

$$U/v* = k^{-1} \ln (z/z_0) \tag{24.97}$$

When wind speeds become weak, the atmospheric surface layer becomes thermally stratified, and the mean velocity profile and turbulence structure are modified. A

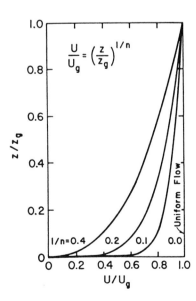

FIGURE 24.4 Mean velocity distribution for strong wind *power-law* profiles.

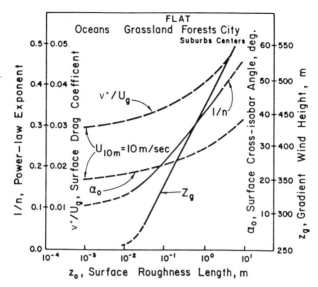

FIGURE 24.5   Dependence of gradient wind height, z_g, and *power-law* exponent, $1/n$, upon surface roughness length, z_0 (from Davenport, 1965) and dependence of surface cross-isobar angle, α_0, and surface drag coefficient, $v*/U_g$, upon surface roughness length z_0, for a 10 m/s mean wind at 10 m. (From Hanna, 1969.)

log-linear distribution

$$U/v* = k^{-1}[\ln (z/z_0) + \beta z/L] \qquad (24.98)$$

gives a good approximation for $U(z)$. For near neutral flow, $\beta \cong 5$. For stable stratification, the *Monin–Obukhov length*

$$L = -v*^3/[(g/T)H_0/(\rho C_p)] \qquad (24.99)$$

is positive, negative for unstable stratification, and it approaches an infinite magnitude as the flow becomes neutral.

24.2.3   Physical Modeling of Boundary Layer Wind

Requirements for physical modeling of the simplified ABL (without Coriolis effects) in wind tunnels are presented in *Similarity Requirements* below. A brief description of wind tunnels designed for this purpose is given in *Typical ABL Simulation Wind Tunnels* below. Refer, also, to Chap. 16.

Similarity Requirements. Detailed discussions of similarity requirements for physical modeling of the ABL are presented in Cermak (1975), (1979), and (1982). The essential requirements are established through dimensional analysis of appropriate equations for conservation of mass, momentum, and energy. In summary, requirements for *exact* similarity for vertical distributions of mean and turbulent fluctuations of velocity and temperature at a particular site may be stated as follows (neglecting Coriolis effects):

i) Undistorted scaling of upwind boundary geometry (buildings, trees, and topographic),
ii) equality of Reynolds number, $Re = U_g L_x / \nu$
iii) equality of gross Richardson number

$$Ri = [(\Delta T)_0 / T_0] \, (L_0 / U_0^2) g \tag{24.100}$$

iv) equality of Prandtl number, $Pr = \nu / (k / \rho c_p)$
v) zero pressure gradient in direction of mean flow, and
vi) similarity of local surface roughness, topographic relief, and surface temperature distribution.

For simulation of strong-wind ABL's, the model flow can be isothermal and requirements iii), iv), and vi) (surface temperature distribution) can be relaxed, since strong mechanical mixing inhibits thermal stratification.

Simulation of wind effects on buildings requires that the ratio of boundary-layer thickness, δ_{max}, to building height, H, be the same for model and prototype. Geometrical scaling ratios, to meet this requirement, ordinarily range from 1 : 5000 to 1 : 100. Therefore, simulation in unpressurized, subsonic wind tunnels results in unequal Re for model and prototype with their ratio approximately equal to the length scale. Fortunately, Re dependence of flow over rough boundaries is asymptotic in nature as shown in Fig. 24.6. Independence of Re, for a specified relative roughness, K_s / L_x, is achieved at a value much smaller than the prototype Re. Accordingly, requirement ii) can be relaxed to require that the model Re be sufficiently large to fall in the shaded region of Fig. 24.6.

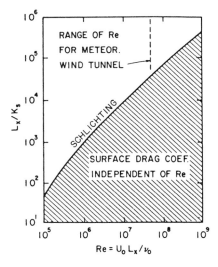

FIGURE 24.6 Reynolds number independence for surface drag coefficient in neutral flow as functions of relative roughness. (Reprinted with permission, Cermak, J. E., "Applications of Fluid Mechanics to Wind Engineering—A Freeman Scholar Lecture," *ASME J. Fluids Eng.*, Vol. 97, pp. 9–38, 1975.)

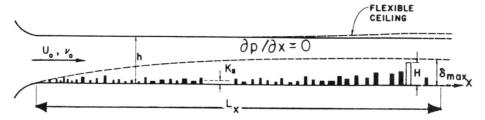

FIGURE 24.7   Long test-section wind tunnel ($L_x/h > 12$) for simulation of the atmospheric boundary layer.

Typical ABL Simulation Wind Tunnels. Wind tunnels of the most common type designed for ABL simulation and wind-engineering tests may be classified as subsonic boundary-layer wind tunnels. Two basic types in use are *recirculating* and *open circuit.* The former may incorporate temperature controls to develop thermal stratification. Figure 24.7 is a schematic representation of a typical *long test section* while Fig. 24.8 shows an interior view. Relative dimensions of a *long test section* are $L_x \cong 20\delta_{max}$, $h \cong 2\delta_{max}$ and width $w \cong 2\text{–}4\delta_{max}$. A desirable magnitude for δ_{max} is approximately 1 m.

Design details for a recirculating meteorological wind tunnel capable of simulating thermally stable and unstable boundary layers are given in Cermak (1981). The 30-m long test section can attain Ri values in the range -1 to 1 in the lower 10% of the boundary layer.

Short boundary-layer wind tunnels ($L_x \cong 10\delta_{max}$, $h \cong 2\delta_{max}$ and $w \cong 2\text{–}4\delta_{max}$) can be utilized for wind effect studies by placement of boundary-layer augmentation devices such as vortex generators, spires, and fences at the test-section entrance. Their effect on similarity of flow and precautions in their use are discussed in Cermak (1975), (1982).

FIGURE 24.8   Equitable Center West, New York (1:400 scale) and surrounding buildings in a boundary-layer wind tunnel viewed from downwind. (Courtesy Cermak/Peterka & Associates, Inc., Fort Collins, CO.)

24.2.4 Physical Modeling of Wind Effects on Buildings

Essentially all data on wind forces and diffusion around buildings are obtained by measurements on small-scale physical models. Therefore, in addition to simulation of the ABL, similarity criteria for flow on and near the building and the effects resulting from interaction with this flow must be considered. Criteria for similarity of wind pressures, dynamic responses and dispersion of effluents emitted from stacks and vents are given below.

Model Tests for Wind Pressures, Forces, and Moments. For buildings with sharp edges, Re invariance of local flow features (separation, reattachment, and vortex formation) is usually satisfied if Re $= U_H W/\nu$ is greater than 10^5. For buildings with curved surfaces, the separation location varies with Re within the usual test range of Re. In this case, separation locations corresponding to the full-scale Re can be achieved by roughening of the curved surface.

Mean and fluctuating pressure distributions are obtained by perforating the model with 500 to 1000 piezometer taps. The pressure measuring system must have good frequency response up to at least 150 Hz and an A/D conversion sampling rate of at least 250 per second to obtain an adequate resolution of pressure fluctuations.

Mean wind forces and moments on a building can be obtained by integration of the measured mean pressure distributions. However, fluctuating wind forces and moments on an entire building must be measured directly. This can be accomplished for buildings having small effects of aerodynamic damping by placement of the rigid model on a stiff high-frequency (200 to 400 Hz) multi-component balance [Boggs and Peterka (1989)]. A second method of measuring fluctuating moments that includes the effects of aerodynamic damping for slender buildings and, in addition, permits direct measurement of fluctuating deflections and accelerations is to use an aeroelastic model. See Sec. 23.6 for a discussion of aeroelasticity.

The simplest aeroelastic model reproduces primary bending modes about principal axes at the base of a tall building. Figure 24.9 is a schematic representation of the *stick model*. Criteria for dynamic response similarity are that the model and

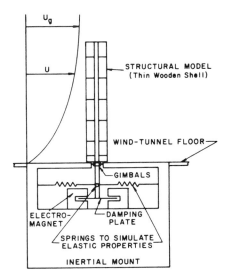

FIGURE 24.9 Schematic representation of primary two-sway-mode aeroelastic building model.

prototype have equal values of the natual frequency ratio, f_x/f_y, density ratio, $\rho_{structure}/\rho_{air}$, reduced velocity, $U_H/(f_{x \text{ or } y} W)$, and damping ratio, c. More complex aeroelastic models can be constructed to reproduce the torsional mode and higher modes for tall buildings of irregular geometry.

Atmospheric Transport Tests. Physical modeling of effluent concentration distributions on a building introduces an additional set of similarity criteria for the source characteristics. A commonly used set of criteria discussed in Cermak (1975) for a passive gaseous effluent is similarity of source geometry; equality of ratio of efflux speed to mean wind speed at height of stack or vent, V_E/U_S; equality of Froude number, $\text{Fr} = \rho_a U_s^2/[(\rho_a - \rho_E)/gD]$; and equality of internal Reynolds number, $\text{Re}_i \equiv V_E D/\nu$. The last criterion cannot be realized, however duct interiors can be roughened to produce turbulent flow of the exiting effluent. Equality of Fr is achieved by mixing helium with air to produce a large density difference while reducing the flow velocity U_S. Alternate sets of similarity criteria are discussed in Poreh (1981). Buoyant flows are discussed in Sec. 13.3.

Mean concentration fields are conveniently established by adding a known concentration of a *tracer gas* such as propane or ethane to the effluent. Samples withdrawn from an array of points on the building can be analyzed for tracer gas concentration with a gas chromatograph.

24.2.5 Wind Effects on Buildings

The primary manifestation of wind on a building is an unsteady, nonuniform distribution of pressures, $p(x, y, z, t)$, acting upon exterior surfaces (and interior surfaces when openings exist in the building envelope). Pressure fluctuations $p'(x, y, z, t)$ from the mean $\bar{p}(x, y, z, t)$ results in fluctuating forces on local areas $f'(x, y, z, t)$ as well as fluctuating forces $\mathbf{F}'$ and moments $\mathbf{M}'$ on the entire building. These, in turn, cause fluctuating deflections and accelerations. Another important effect of wind flowing over a building is recirculation and downwash of air pollutants emitted from vents or nearby stacks. A review of building aerodynamics is given in Cermak (1976).

Buildings encountered in practice vary greatly in size, shape, architectural details, and surroundings. The buildings range from low-rise buildings (including mobile homes) to high-rise structures (including skyscrapers in excess of 100 stories) in every conceivable setting from complete isolation to city centers as shown in Fig. 24.8. Buildings less than or equal to 18.3 m (60 ft) in height are classified as *low-rise* and those of greater height as *high-rise* [ASCE (1988)]. Most low-rise buildings and many taller buildings are designed for wind pressures that would exist on a reference building (a block-type building of equal height in an isolated setting) as specified in a city building code or a standard such as ASCE (1988). In order to understand why wind effects specified for a reference building may or may not be appropriate for a particular building, some general features of flow over buildings are identified in the next section.

Aerodynamic Features of Flow Over Buildings. Figure 24.10 illustrates a typical flow pattern for flow normal to one face of a low-rise building. Flow separation occurs on all faces excepting the upwind face. Particularly large negative mean pres-

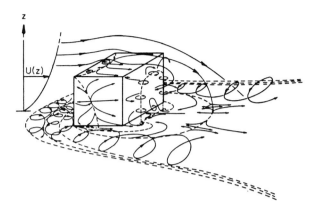

FIGURE 24.10 Mean flow pattern over a low block-type building. (From Woo *et al.*, 1977.)

sures, $\bar{p}$, and negative pressure fluctuations, p', occur on upstream portions of the sides and roof. Flow over an isolated high-rise block building with one face normal to the wind is illustrated in Fig. 24.11. In this flow, a vertical pressure gradient on the upwind face induces a strong downward movement of air that results in strong wind at street level. Flow around the same building with wind approaching at an angle that gives the maximum negative peak pressures (15° from the normal) is shown in Fig. 24.12.

A pair of attached vortices is shown in Fig. 24.13 over the roof of a high-rise building. This flow feature causes the largest negative peak pressures found on building envelopes as well as the most intense shear or scouring action and is responsible for initiation of most wind damage to roofs. Attached vortices also occur at geometrical discontinuities such as *setbacks* and intersections of low-rise buildings with a high-rise building. Shedding vortices that form alternately on opposite downstream edges of a tall building cause a quasi-periodic crosswind force that can result in severe crosswind oscillations when the shedding frequency and natural frequency of the building are nearly equal. Details of these vortices that form a von Kármán vortex street are given in Sec. 4.12.

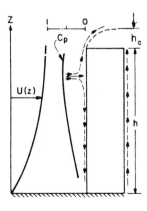

FIGURE 24.11 Mean vertical flow pattern over a high-rise block-type building. (Reprinted with permission, Cermak, J. E., "Aerodynamics of Buildings," *Annual Review of Fluid Mechanics*, Vol. 8, pp. 75–106, 1976.)

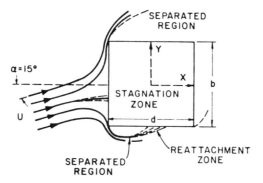

FIGURE 24.12   Mean horizontal flow pattern around a high-rise block-type building. (Reprinted with permission, Cermak, J. E., "Aerodynamics of Buildings," *Annual Review of Fluid Mechanics*, Vol. 8, pp. 75–106, 1976.)

Local Mean and Fluctuating Pressures on Exterior Surfaces. The instantaneous pressure at a tap for a given building geometry, wind direction, and wind characteristics can be expressed as the mean value plus the instantaneous fluctuation from the mean; i.e.,

$$p(x, y, z, t) = \bar{p}(x, y, z, t) + p'(x, y, z, t) \tag{24.101}$$

Four useful *pressure coefficients* may be formed as follows

$$\text{Mean:} \qquad\qquad C_{\bar{p}} = \bar{p}/(\rho U_g^2/2) \tag{24.102a}$$

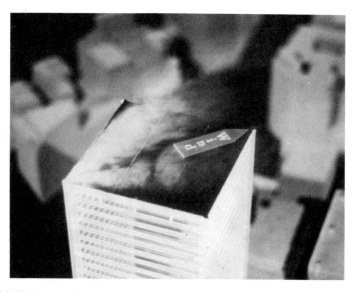

FIGURE 24.13   Vortex formation on flat roof of a block-type building for wind parallel to diagonal. (Courtesy of Fluid Dynamics and Diffusion Laboratory, Colorado State University, Fort Collins, CO.)

$$\text{Root-mean-square (rms):} \quad C_{p_{rms}} = p'_{rms}/(\rho U_g^2/2) \qquad (24.102b)$$

$$\text{Peak maximum:} \quad C_{\hat{p}} = p_{max}/(\rho U_g^2/2) \qquad (24.102c)$$

$$\text{Peak minimum:} \quad C_{\check{p}} = p_{min}/(\rho U_g^2/2) \qquad (24.102d)$$

A similar set of coefficients for an area averaged pressure can be obtained by either pneumatic averaging or simultaneous measurement of pressures at a set of taps distributed over the area. The reference pressure is usually taken to be $\rho U_g^2/2$, however $\rho U_H^2/2$ and $\rho U_{10m}^2/2$ are sometimes used. Therefore, pressure coefficients for which the *reference pressure* is not specified must be used with caution. Figure 24.14 illustrates a typical distribution of pressure coefficients around the periphery of an isolated high-rise building of square cross sections. Minimum peak pressure coefficients, $C_{\check{p}}$, near edges where separation occurs depend strongly upon the geometry of the building and nearby buildings. They range from -1.5 to -4.0 and may be substantially constant or highly variable along the edge. ASCE (1988) gives recommendations for both peak minimum and mean pressures on *reference-type* buildings but recommends wind tunnel tests for other types of buildings. Local values of $C_{\check{p}}$ measured near the vortex root have been found to be as large as -10 on a flat-top roof (without parapet as shown in Fig. 24.13), -5.5 below edges of a setback (such as on the Sears Tower, Chicago) and -4.0 near intersections of a tall building wall with the street or attached low-rise building. Peak positive pressure coefficients, $C_{\hat{p}}$, seldom exceed $+1$ unless flow is accelerated by the presence of nearby buildings or a local architectural feature.

Mean and Fluctuating Forces and Moments. Mean wind force and moment coefficients obained by integration of mean pressures over a family of isolated rectangular buildings for which W/L varies from 0.25 to 1.0 and H/W varies from 1 to 8, as defined by Fig. 24.15 have been reported [Akins *et al.* (1977)]. Table 24.2 gives

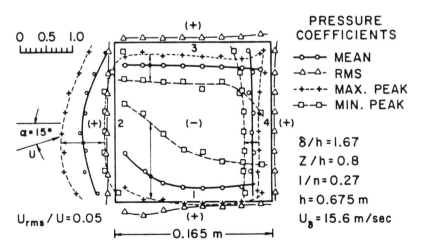

FIGURE 24.14 Distribution of pressure coefficients around periphery of an isolated high-rise block-type building. (Reprinted with permission, Cermak, J. E., "Aerodynamics of Buildings," *Annual Review of Fluid Mechanics*, Vol. 8, pp. 75–106, 1976.)

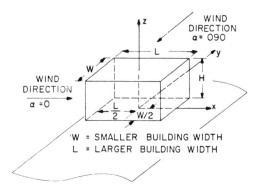

FIGURE 24.15 Coordinate system and nomenclature for family of rectangular block-type buildings.

details of buildings with $W/L = 1.0$ for which the coefficients are presented in Fig. 24.16 when submerged in four different boundary layers. Power-law exponents, $1/n$, for each of the four boundary layers were $0.12, 0.27, 0.34$, and 0.38 for boundary layers 1, 2, 3, and 4, respectively. The force and moment coefficients given in Fig. 24.16 are defined as

$$C_{FX} = F_x/(1/2\rho U_A^2 WH) \tag{24.103}$$

$$C_{FY} = F_y/(1/2\rho U_A^2 LH) \tag{24.104}$$

$$C_{FZ} = F_z(1/2\rho U_A^2 WL) \tag{24.105}$$

$$C_{MX} = M_x(1/2\rho U_A^2 LH^2) \tag{24.106}$$

$$C_{MY} = M_y/(1/2\rho U_A^2 WH^2) \tag{24.107}$$

$$C_{MZ} = M_z/(1/2\rho U_A^2 WLH) \tag{24.108}$$

By using a reference pressure $1/2\rho U_A^2$, where

$$U_A = \left(\int_0^H U(z) \, dz \right) \bigg/ H \tag{24.109}$$

TABLE 24.2 Building Geometry and Flow Parameters for Data in Fig. 24.16

Building	W (m)	L (m)	H (m)	γ	β	H/δ	Re $\times 10^{-4}$ Max	Re $\times 10^{-4}$ Min
A	0.127	0.127	0.254	1.0	2	0.2	9.5	6.5
B	0.127	0.127	0.508	1.0	4	0.4	9.0	9.0
C	0.254	0.254	0.254	1.0	1	0.2	19.0	13.0

$\gamma = W/L$, $\beta = H/W$.

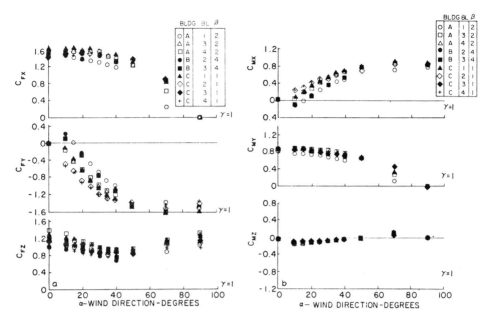

FIGURE 24.16 Mean wind force and moment coefficients as a function of wind direction for a building of square cross section. (Reprinted with permission, Akins, R. E., Peterka, J. A., and Cermak, J. E., "Mean Force and Moment Coefficients for Buildings in Turbulent Boundary Layers," *J. Industrial Aerodynamics*, Vol. 2, pp. 195–209, 1977.)

the force and moment coefficients for boundary layers 1 to 4 become nearly equal for a given wind direction and *aspect ratio*, γ. The coefficients are essentially independent of the height-to-width ratio, β.

The coefficients given in Fig. 24.16 are mean values averaged over a 30 to 60 minute period. A good estimate of 10 to 15 second peak values can be obtained by applying a *gust factor* of 1.3 to the wind speed, U_A. However, if the building is slender, $H/W > 5$, dynamic effects may result in even larger peak values.

Wind Induced Oscillations. Wind excites building oscillations primarily at the fundamental natural frequencies, f_x, f_y, and f_z, through three basic mechanisms: 1) buffeting by gusts, 2) quasi-periodic vortex shedding, and 3) interaction with building motion to produce a *galloping type* aerodynamic instability. Buffeting induced motion occurs to some degree on all buildings. Vortex-shedding excitation becomes significant primarily for isolated high-rise buildings with unbroken vertical lines. The galloping-type excitation would occur for an unusually slender building with very low damping and is not a common problem associated with flow over buildings.

Buffeting by Gusts: Estimates of the power spectral density for the drag (along-wind) coefficient $F(C_D)(n)$ can be obtained in terms of the mean drag coefficient C_D and a power spectral density for the along-wind component $F(u)(n)$ [Davenport (1961)]

$$F(C_D)(n) = 4|\chi(n)|^2(C_D/U_A)^2F(u)(n) \qquad (24.110)$$

Equation (24.110) includes the *aerodynamic admittance*, $\chi(n)$, for a building as a function of frequency, n. An empirical expression for the aerodynamic admittance [Vickery (1967)] is

$$\chi^2(n) = 1/[1 + (2n \sqrt{A}/U_A)^{4/3}] \tag{24.111}$$

where A is the area of building elevation projected on a plane perpendicular to U_A. The following approximation for $F(u)(n)$ may be used in Eq. (24.110) [Kaimal *et al.* (1972)]

$$nF(u)(z_A, n)/v^* = 200N(1 + 50\ N)^{-5/3} \tag{24.112}$$

where $N = nz_A/U_A$ is the nondimensional frequency and z_A is the height at which $U(z) = U_A$ [see Eq. (24.109)].

Discussions of *gust excited building motion* and methods for calculation of deflections and accelerations are presented in ASCE (1988) and Simiu and Scanlan (1978). No formulation has been proposed that adequately relates cross-wind oscillations to atmospheric turbulence.

Vortex-Excited Oscillations: Maximum dynamic response occurs when the dominant shedding frequency is near one of the fundamental natural frequencies, f_x, f_y, or f_z. The frequencies at which vortices are shed from a building depend upon the cross sectional shape, the ratio of building height to width, H/W, and the mean wind speed, U_A. A nondimensional frequency, the *Strouhal number*, $S = fW/U_A$, is essentially a constant for cross sections with sharp corners and a particular value of H/W. On the other hand, S for cross sections with rounded corners such as a circular cylinder varies with the Reynolds number, $U_A d/\nu$. Figure 24.17 gives values of S for a variety of rectangular cross-sections.

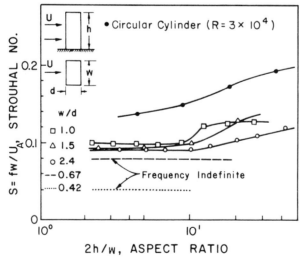

FIGURE 24.17 Strouhal number dependence upon aspect ratio 2h/w and building shape w/d. (Reprinted with permission, Cermak, J. E., "Aerodynamics of Buildings," *Annual Review of Fluid Mechanics*, Vol. 8, pp. 75–106, 1976.)

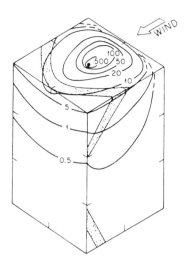

FIGURE 24.18 Distribution of concentration coefficients $K = \chi U_H A_p/Q$ for roof-centered source with velocity ratio $V_E/U_H = 0.3$ on a block building, $L/H = W/H = 0.67$. (From Hosker, 1983.)

Transport of Vent and Stack Emissions. Emissions from roof vents may enter either the upwind separation cavity or the leeward separation cavity following reattachment of the upwind separated flow shown in Fig. 24.10. For a roof centered vent, the cavity in which primary circulation will occur depends upon building geometry (W/L and H/W for block-type buildings), wind direction, atmospheric stability, and the geometry of surrounding buildings. The concentration, χ, of effluent from a source of given strength, Q, on various surfaces of a building is highly dependent upon which cavity receives the effluent. Figures 24.18 and 24.19 show that maximum concentration coefficients, $K = \chi U_H A_R/Q$, can be ten times larger for release into the upwind cavity than when released downwind of the reattachment region. Terms in the coefficient K are as follows: χ = concentration of pollutant (gm/m^3 or m^3/m^3), Q = source strength of pollutant in effluent (gm/s or m^3/s), and A_p = upwind projected area of the building. As wind direction varies, a given vent on a rectangular building discharges into either the upwind or the leeward cavity. For

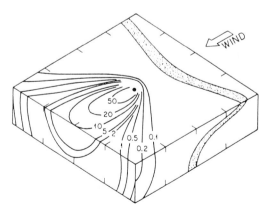

FIGURE 24.19 Distribution of concentration coefficients $K = \chi U_H A_p/Q$ for roof-centered source with velocity ratio $V_E/U_H = 0.39$ on a block building, $L/H = W/H = 4.00$. (From Hosker, 1983.)

cornering winds, effluent can be transported onto the roof surface by vortices shown in Fig. 24.13. Concentration coefficients may be decreased by increasing the velocity ratio, V_E/U_S, and/or the vent height. The ratio V_E/U_S should exceed 1.5 in order to prevent *stack downwash*. To avoid recirculation of effluents, the vent height must be greater than the height of the cavity, h_c, which, for a flat-topped rectangular building, is approximately $0.3A_p^{1/2}$.

Additional information on dispersion from building surface sources is available in ASHRAE (1993) and EPA (1979). Unfortunately, dispersion around buildings is so strongly dependent upon building geometry, source location and wind direction that predictive relationships for concentration coefficients have not been formulated, and physical modeling is necessary to predict concentrations.

Effluent from stacks located downwind of a building is subject to downwash if release is into flow with steamlines inclined downward into the building wake. The height above a building where streamlines no longer have a downward slope is approximately $1.5A_p^{1/2}$. A guideline for avoidance of downwash is that the ratio $H_s/H \geq 2.5$. The stack height, H_s, resulting from this guideline may be over conservative, therefore wind-tunnel tests are recommended for determination of a *good engineering practice* stack height that satisfies criteria established by the Environmental Protection Agency [EPA (1979)].

24.2.6 Full-Scale Measurements of Wind and Wind Effects

Full-scale testing can be used to advantage for several purposes:

   i) Diagnostic study of the relationship between wind and wind effects on an existing building
   ii) Exploratory study to establish effectiveness of local architectural, landscaping, or operational modifications as a solution for a particular problem, and
   iii) Acquisition of data to validate predictions made by small-scale physical modeling or numerical modeling

A major disadvantage of full-scale testing is that meteorological conditions cannot be controlled, therefore a long period of testing may be necessary to reach valid conclusions. Regardless of whether information on wind pressures, building motion, contaminant concentration or local wind speeds is being sought, a companion set of *reference meteorological data* measured at a representative location is necessary. In many cases, qualitative judgments can be made by observation of local flow patterns with visual tracers.

24.3 OCEAN FLOWS
    Wayne L. Neu

The only significant differences between geophysical fluid dynamics, including both ocean and atmospheric flows, and the engineering fluid mechanics considered in the other sections of the Handbook are boundary conditions and scale. The fact that the flows occur on a global scale means that we must consider the acceleration due to the rotation of the earth—the *Coriolis acceleration*. This can have a profound effect on the nature of the flow, e.g., velocity is now nearly perpendicular to horizontal pressure gradients. See Stommel (1987) for an excellent explanation of the Coriolis acceleration (in conjunction with the geostrophic flow discussed below) given from the view point of an observer fixed somewhere off the earth in space.

One must also appreciate that the oceans are extremely thin with depths on the order of one one-thousandth of typical width scales. The depth of the ocean has been likened to a coat of varnish on a globe or as having the dimensions of a very thin sheet of paper. This means that horizontal velocities are typically very much greater than vertical velocities, i.e., the flow is very nearly horizontal, or normal to the local gravity vector, $\bar{\mathbf{g}}$. Finally, as a consequence of the large scale, the flow is virtually always turbulent.

The point of departure for a discussion of ocean flows is the Reynolds averaged Navier–Stokes equation for an incompressible flow with the addition of the Coriolis acceleration which is treated as the apparent *Coriolis force*

$$\frac{D\bar{\mathbf{u}}}{Dt} = -\frac{\nabla p}{\rho} - 2\bar{\mathbf{\Omega}} \times \bar{\mathbf{u}} + \bar{\mathbf{g}} + \epsilon\nabla^2\bar{\mathbf{u}} + \text{tidal forces} \qquad (24.113)$$

Here, $\bar{\mathbf{u}} = u\bar{\mathbf{i}} + v\bar{\mathbf{j}} + w\bar{\mathbf{k}}$ is the mean* velocity vector, $\bar{\mathbf{\Omega}}$ is the earth's rotational velocity vector with magnitude 7.29×10^{-5} sec^{-1}, and ϵ is a kinematic eddy viscosity. The *tidal forces* term includes the gravitational forces which drive the tides. It is included for completeness but will not be further discussed in this section. We will be using a cartesian coordinate system with x directed to the east, y to the north and z vertically upward with origin at the water surface.

The Coriolis term can be expanded as,

$$-2\bar{\mathbf{\Omega}} \times \bar{\mathbf{u}} = (2\Omega v \sin\phi - 2\Omega w \cos\phi)\bar{\mathbf{i}} - 2\Omega u \sin\phi\bar{\mathbf{j}} + 2\Omega u \cos\phi\bar{\mathbf{k}} \qquad (24.114)$$

where ϕ is latitude. When the equation of motion is written as three component equations, the term $2\Omega w \cos\phi$ is usually neglected compared to other terms in the x direction equation since $w \ll u, v$. In the z equation, the term $2\Omega u \cos\phi$ is small compared with the pressure gradient and with g. Thus, this term is usually neglected. In the two remaining terms, the substitution $f \equiv 2\Omega \sin\phi$ is often used, with f being the *Coriolois parameter*.

The eddy viscosity is not a property of the fluid but rather depends on the local characteristics of the flow. In ocean flows, there are large differences in horizontal and vertical flow characteristics, thus we must make use of horizontal and vertical eddy viscosities, ϵ_H and ϵ_Z. The magnitudes differ widely with $\epsilon_Z = O(10^{-6} \epsilon_H)$.

The horizontal components of the equation of motion can now be written as,

$$\frac{Du}{Dt} = -\alpha\frac{\partial p}{\partial x} + fv + \epsilon_H\frac{\partial^2 u}{\partial x^2} + \epsilon_H\frac{\partial^2 u}{\partial y^2} + \epsilon_Z\frac{\partial^2 u}{\partial z^2}$$

$$\frac{Dv}{Dt} = -\alpha\frac{\partial p}{\partial y} - fu + \epsilon_H\frac{\partial^2 v}{\partial x^2} + \epsilon_H\frac{\partial^2 v}{\partial y^2} + \epsilon_Z\frac{\partial^2 v}{\partial z^2} \qquad (24.115)$$

---

*Mean, in the sense that the turbulent fluctuations are averaged out. Longer term time variation is retained as is the use of the total derivative, D/Dt.

where $\alpha = 1/\rho$. The vertical component can be reduced to simply the hydrostatic equation

$$\frac{\partial p}{\partial z} = -\rho g \tag{24.116}$$

as all other terms in this equation are negligible compared to these two. Equations (24.115) and (24.116), together with conservation of mass for an incompressible fluid, form the system of governing equations for most ocean flow problems. The conservation of energy equation will be important when considering the heat budget and distribution of temperature in the ocean. It should be noted that we have made use of the *Boussinesq approximation* which says that one can neglect the effect of density variations on the mass of the fluid but must retain their effect on its weight. This means that in Eq. (24.115) where the mass (or inertia) appears, density is treated as a constant, but in Eq. (24.116), where the weight appears, the density variations must be retained in order to calculate the pressure field.

In the interior of the ocean, away from boundaries, velocity gradients and accelerations are small. Thus, for the bulk of the ocean's mass, both the friction and the inertial terms can be neglected compared with the pressure gradient and Coriolis terms. Equation (24.115) becomes

$$\alpha \frac{\partial p}{\partial x} = fv$$

$$\alpha \frac{\partial p}{\partial y} = -fu \tag{24.117}$$

which, together with Eq. (24.116) form the *geostrophic equation*. Defining $V^2 \equiv u^2 + v^2$ and $(\partial p/\partial n)^2 \equiv (\partial p/\partial x)^2 + (\partial p/\partial y)^2$, Eq. (24.117) can be written as

$$\alpha \frac{\partial p}{\partial n} = fV \tag{24.118}$$

which is a balance between the horizontal pressure force and the Coriolis force. The Coriolis force ($fv\mathbf{i} - fu\mathbf{j}$) is normal to the velocity and directed to its right in the northern hemisphere (left in the southern hemisphere where $f < 0$). Thus, the pressure gradient in Eq. (24.118) is also normal to the velocity as in Fig. 24.20. In other words, the flow is along lines of constant pressure. In the northern hemisphere,

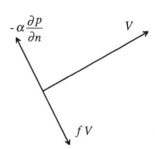

FIGURE 24.20   Balance of forces for geostrophic flow.

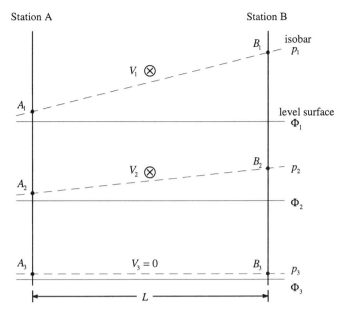

FIGURE 24.21 Isobaric and level surfaces for geostrophic flow.

looking in the direction of the flow, high pressure is to the right, or isobaric surfaces slope up to the right.

The slopes of these isobaric surfaces result from differences in the vertical density structure (due to temperature and salinity differences) at different horizontal stations. Referring to Fig. 24.21, if the density at station B between points B_1 and B_2 is less than that at station A between A_1 and A_2, then a larger column of water will be required at station B than at station A to create the pressure difference between isobars p_1 and p_2. These density differences give rise to variations in the slopes of the isobaric surfaces and, hence, variations in the horizontal pressure gradients with depth. The ocean surface, which is an isobar, need not be (and usually is not) level. It has many low mounds and valleys.

The slopes of these isobars are small, on the order of 1 in 10^5. There is presently no means of measuring these from a ship at sea. Fortunately, differences in these slopes, and thus differences in the geostrophic velocities at different levels, can be deduced from a knowledge of the density variation with depth. The density variation is obtained from the temperature and salinity variations which can be accurately measured. These are usually cast as functions of pressure rather than depth, since pressure is much easier to measure at sea.

Equation (24.118) can readily be solved together with Eq. (24.116) to yield

$$f(V_1 - V_2) = \alpha g \frac{\partial}{\partial n} \int_{z_1}^{z_2} \rho \, dz \qquad (24.119)$$

but Eq. (24.119) is not in a form which can be easily evaluated. To rework the mathematics in terms of the measurable quantities, the geopotential, Φ, is introduced. It is defined such that $d\Phi = g \, dz$. Lines of Φ = constant are level surfaces.

Making use of Eq. (24.116), we can write $d\Phi = -\alpha\,dp$. Integrating from level z_1 to z_2,

$$\Phi_2 - \Phi_1 = g(z_2 - z_1) = -\int_{p_1}^{p_2} \alpha\,dp \qquad (24.120)$$

It can further be shown that

$$\left.\frac{\partial p}{\partial n}\right|_{\Phi = \text{const.}} = \rho\left.\frac{\partial \Phi}{\partial n}\right|_{p = \text{const.}} \qquad (24.121)$$

Thus, from Eq. (24.118)

$$f(V_2 - V_1) = \frac{\partial}{\partial n}(\Phi_2 - \Phi_1) = -\frac{\partial}{\partial n}\int_{p_1}^{p_2} \alpha\,dp \qquad (24.122)$$

Notice that while the geopotential gradients cannot be calculated, differences in them between levels can be found in terms of the density structure. Further, α is now being integrated as a function of pressure, the form in which it is available from measurements, rather than as a function of depth. This leads to finding differences in the geostrophic velocities between levels rather than absolute velocities. Averaging Eq. (24.122) over the distance L between stations A and B in Fig. 24.21 we have

$$\overline{f(V_1 - V_2)} = \frac{1}{L}\left[\left(\int_{p_1}^{p_2} \alpha\,dp\right)_B - \left(\int_{p_1}^{p_2} \alpha\,dp\right)_A\right] \qquad (24.123)$$

If, between levels 1 and 2 in Fig. 24.21, α at station B is greater than α at station A (or ρ at B is less than ρ at A), the right-hand side of Eq. (24.123) is positive and $V_1 > V_2$. The water at level 1 is flowing, *relative to the water below it*, in a direction such that the less dense water is to its right. This *light on the right* rule is valid in the northern hemisphere. Relative directions will be reversed in the southern hemisphere.

To find absolute velocities, it is often assumed that a level of no motion (isobar p_3 in Fig. 24.21) exists at some deep depth, say 1000 m, or one may use a level of known current if it has been measured at some depth. Schott and Stommel (1978) proposed a method, called the *beta-spiral*, which will yield absolute velocities directly from the density data if the spacing of stations is dense enough.

Most of our current knowledge of ocean currents is based on the geostrophic equation, Eq. (24.118). Of course it will not be valid where friction is important (i.e., has a magnitude on the order of the Coriolis force or greater) such as near the bottom or at lateral boundaries where large velocity gradients are present. We are rarely completely free of friction. Figure 24.22 modifies the balance of forces in geostrophic flow to include a small friction force opposing the motion. Now the pressure force is balanced by the vector sum of the Coriolis and friction forces. To enable this balance, the velocity vector is slightly reduced and rotated slightly downslope. Thus, the flow slowly slides down the isobars as it moves along them. Were

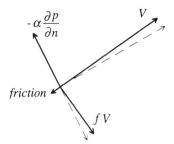

FIGURE 24.22 Balance of forces for geostrophic flow modified by friction. Dashed vectors are without friction.

this the only flow mechanism of the ocean, the isobars would eventually level out and the currents would stop. What keeps this from happening is the forcing provided by the wind on the surface.

We now return to Eq. (24.115) and consider the case where there is no horizontal pressure gradient, but there is friction in the form of a vertical velocity shear as would occur near the surface due to the imposition of a wind stress on the surface. The Coriolis force is now balanced by the friction force

$$0 = fv + \epsilon_z \frac{\partial^2 u}{\partial z^2}$$

$$0 = -fu + \epsilon_z \frac{\partial^2 v}{\partial z^2} \tag{24.124}$$

Considering a parcel of water at the surface, Fig. 24.23, there is wind stress acting on its upper surface and a water velocity shear stress acting on its lower surface. These friction forces cannot be collinear (as they would be in the absence of rotation), since there vector sum must balance the Coriolis force. Considering the parcel just below this surface parcel, the shear on its upper surface is equal and opposite to the bottom shear on the surface parcel. If the vector sum of this force and the shear on the bottom of the parcel is to balance the Coriolis force on this parcel, its velocity vector must be rotated off to the right of the velocity of the surface parcel. Descending farther into the water column, the velocity vector continues to rotate to the right.

Taking the wind to be in the y direction for simplicity, and considering it to exert a wind stress, τ_y, on the surface, surface boundary conditions for Eq. (24.124) can be written as,

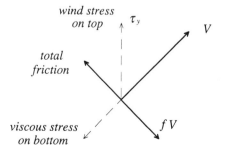

FIGURE 24.23 Balance of forces at the surface in the presence of wind stress.

$$\tau_x = \rho\epsilon_z \left.\frac{\partial u}{\partial z}\right|_{z=0} = 0$$

$$\tau_y = \rho\epsilon_z \left.\frac{\partial v}{\partial z}\right|_{z=0} \qquad (24.125)$$

The solution of Eq. (24.124), first presented by Ekman (1905), is then,

$$u = \pm V_0 e^{az/\sqrt{2}} \cos\left(\frac{az}{\sqrt{2}} + \frac{\pi}{4}\right)$$

$$v = V_0 e^{az/\sqrt{2}} \sin\left(\frac{az}{\sqrt{2}} + \frac{\pi}{4}\right) \qquad (24.126)$$

where $a^2 = |f|/\epsilon_z$ and $V_0 = \tau_y/\rho a\epsilon_z$ is the surface velocity. In the first equation, the $(+)$ applies in the northern hemisphere and the $(-)$ applies in the southern hemisphere. In the northern hemisphere at the surface, the velocity is 45° to the right of the wind. The magnitude of the velocity decreases as its direction turns to the right with increasing depth ($z < 0$). At the depth $D_E = \pi\sqrt{2}\epsilon_z/f$, the magnitude of the velocity has decayed to $e^{-\pi}$ or about 4% of the surface value, and the direction is opposite the direction of the surface flow. This solution, illustrated in Fig. 24.24, is called the *Ekman spiral*, and D_E is called the *Ekman depth* or *depth of frictional influence*. At this depth, the wind driven current has essentially decayed to zero. It is taken as the effective depth of this current and is sometimes called the *Ekman layer*. The Ekman depth is on the order of 10–100 m.

The Ekman spiral can be integrated vertically to find the net mass flow rates in the x and y directions. Integrating Eq. (24.126) over z and multiplying by the (assumed constant) density, gives,

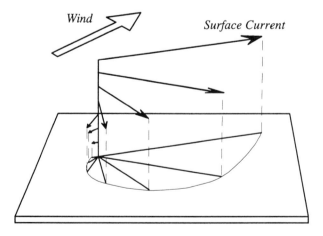

FIGURE 24.24   The *Ekman spiral*. The current direction rotates to the right (northern hemisphere) and its magnitude decreases with increasing depth.

$$M_x = \frac{\rho V_0}{a} = \frac{\tau_y}{f}$$

$$M_y = 0 \tag{24.127}$$

or, the mass transport is normal to the wind.

If the wind is blowing along a coastline which is to the left of the wind in the northern hemisphere, the Ekman layer will be drawn away from the coast. By continuity, this water must be replaced by water from below. This is called *upwelling*. It is common along the eastern sides of oceans during the summer when the wind tends to blow towards the equator. The region near the coast will see a slight depression of isobars and is said to be a region of *divergence* (see Fig. 24.25). This depression of the isobars induces a geostrophic current which flows along the slope in the direction of the wind. Note that once this system is set up, the small down slope component of the geostrophic current due to friction is balanced by the up-slope Ekman transport. We thus arrive at a steady state with the wind indirectly forcing a geostrophic current in the direction of the wind.

If the wind is blowing in the opposite direction along the coastline, the Ekman transport will cause water to pile up along the coast producing a region of *convergence*, a local rise in the isobars, and again a geostrophic current in the direction of the wind. Downwelling is not as easily accomplished as upwelling due to the large vertical density gradients common in coastal areas. The wind induced onshore-offshore currents are generally considerably slower than the resulting geostrophic currents and thus are difficult to measure.

Regions of divergence and convergence can also be produced in the open ocean away from boundaries. Consider a region where the wind speed varies with latitude as indicated by the open arrows in Fig. 24.26. Such a situation exists in the North Atlantic with the *westerlies* to the north and the *trade winds*, or *easterlies*, to the south. The varying Ekman transport creates regions of convergence and divergence, which in turn induce geostrophic flow. Note that the wind need not change direction, as illustrated here. A variation with latitude of the magnitude only will suffice. While the Ekman transport will be all in the same direction, convergences will occur where more flow is entering a region than is leaving, and divergences will occur when the opposite is true. In this case, there will be a region where the geostrophic current is opposite the wind.

A discussion of wind driven circulation is typically completed with a look at three more studies. The concepts discussed conceptually above were demonstrated analytically by Sverdrup (1947). He solved the equations of motion retaining the pres-

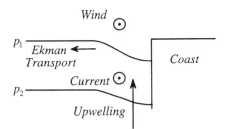

FIGURE 24.25 Ekman transport inducing a region of divergence and upwelling near a coast.

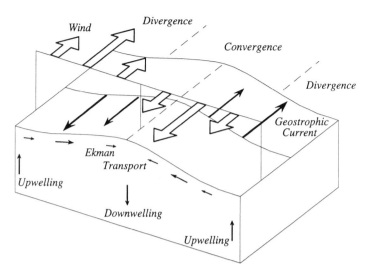

FIGURE 24.26   Variations in the wind profile, and thus in the Ekman transport, inducing regions of convergence and divergence, and resulting geostrophic current, and upwelling and downwelling in the open ocean.

sure, Coriolis and vertical friction terms in a vertically integrated form. Since the resulting equations are first-order in horizontal derivatives, he could only satisfy one lateral boundary condition. Stommel (1948) added a simple lateral friction term to Sverdrup's equations so that he could consider steady flow in a closed basin. He showed that the tendency for currents to be faster on the western sides of ocean basins (e.g., the *Gulf Stream*, the *Kuroshio*), called *westward intensification*, is due to the variation of the Coriolis parameter with latitude. Munk (1950) added the lateral friction terms to Sverdrup's equations and used the actual wind field to produce the first detailed account of the general ocean circulation.

These solutions are covered in many physical oceanography texts. Among them are Neumann and Pierson (1966), Pedlosky (1987), Pond and Pickard (1983), and von Schwind (1980), and, for a slightly less mathematical treatment, Knauss (1978). These texts will also contain more details on the material covered in this section.

24.4   SEDIMENTATION
Everett V. Richardson

24.4.1   Introduction

Sedimentation is the engineering science that is used to determine the yield of sediment from the land surface, the transport of sediment by streams, and reservoir sedimentation. It requires an understanding of fluvial geomorphology and river mechanics, sediment properties, hydraulics of open-channel flow, and bed forms in alluvial channels. This section describes sediment properties, the beginning of motion of sediment particles, bed forms and associated resistance to flow and concentration of sand size bed material, sediment yield from land surface, sediment transport, and reservoir sedimentation. The fluid dynamics of rivers and open-channel flow are described in Sec. 24.5.

24.4.2 Fluvial Geomorphology

Fluvial geomorphology is the science dealing with the profiles and planforms of streams and rivers. River mechanics is the quantification of fluvial geomorphology. The fluvial geomorphology of a stream determines its mode of sediment transport, response to change in climate, tectonic activity, man's activity, and changes in sediment and water discharge. Rivers are dynamic, always changing their shape, position, and other morphological characteristics with time.

Stream planform is broadly classified as meandering, straight, braided, or *anabranching* as shown in Fig. 24.27. Anabranching streams are similar to braided except the sand bars and islands are more permanent islands. These broad classifications have been subclassified by geomorphologists and engineers, but for sedimentation studies, these broad classifications will normally be sufficient. Geomorphic factors affecting stream planform and morphology are stream size, flow habit (ephemeral, perennial but flashy or perennial), bed and bank material, valley setting, floodplains, natural levees, apparent incision, channel boundaries, vegetation, stream planform, variability of width, and development of bars. These factors are illustrated in Fig. 24.28.

Braided streams are wide with many channels around sand bars or islands that interlace and change position with time. They are characterized by unstable banks, steep, and shallow watercourses. They have a large bed material discharge moving in contact with the bed and a relatively low suspended bed material discharge. There are two primary causes of a braided stream: 1) the stream may be supplied with more sediment than it can transport (overloaded) and is aggrading, or 2) the stream is steep, has easily eroded banks and a large range in discharge, which at its high flows forms a wide single channel, and at low flows forms many channels.

A *meandering* stream consists of an S-shaped channel with alternating bends. In

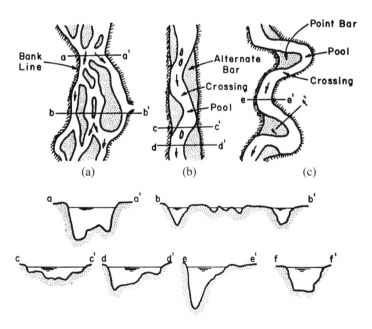

FIGURE 24.27 Braided, straight, and meandering planform.

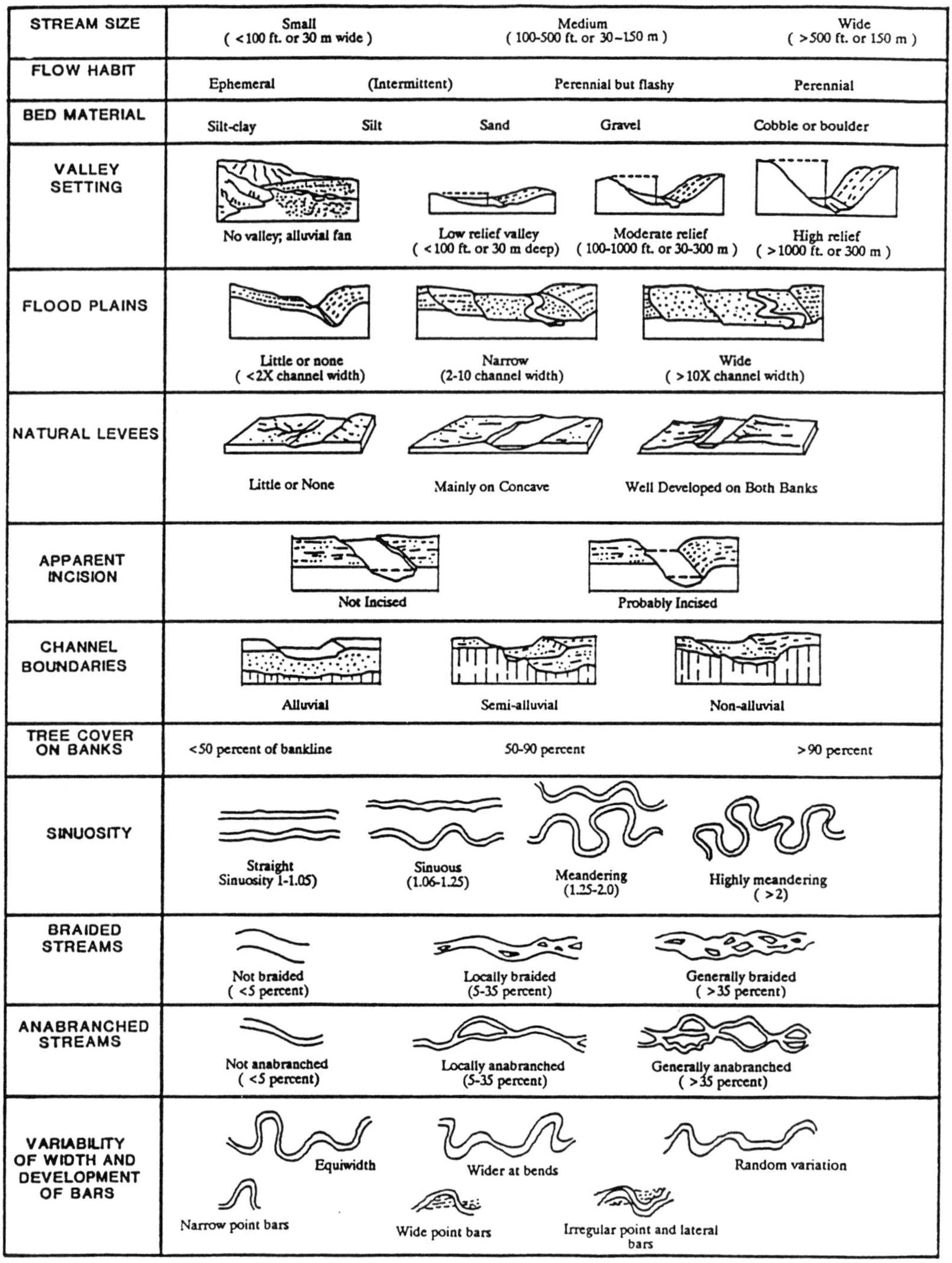

STREAM SIZE	Small (<100 ft. or 30 m wide)		Medium (100-500 ft. or 30–150 m)		Wide (>500 ft. or 150 m)
FLOW HABIT	Ephemeral	(Intermittent)	Perennial but flashy		Perennial
BED MATERIAL	Silt-clay	Silt	Sand	Gravel	Cobble or boulder

VALLEY SETTING

No valley; alluvial fan   Low relief valley (<100 ft. or 30 m deep)   Moderate relief (100-1000 ft. or 30-300 m)   High relief (>1000 ft. or 300 m)

FLOOD PLAINS

Little or none (<2X channel width)   Narrow (2-10 channel width)   Wide (>10X channel width)

NATURAL LEVEES

Little or None   Mainly on Concave   Well Developed on Both Banks

APPARENT INCISION

Not Incised   Probably Incised

CHANNEL BOUNDARIES

Alluvial   Semi-alluvial   Non-alluvial

TREE COVER ON BANKS

<50 percent of bankline   50-90 percent   >90 percent

SINUOSITY

Straight Sinuosity 1-1.05   Sinuous (1.06-1.25)   Meandering (1.25-2.0)   Highly meandering (>2)

BRAIDED STREAMS

Not braided (<5 percent)   Locally braided (5-35 percent)   Generally braided (>35 percent)

ANABRANCHED STREAMS

Not anabranched (<5 percent)   Locally anabranched (5-35 percent)   Generally anabranched (>35 percent)

VARIABILITY OF WIDTH AND DEVELOPMENT OF BARS

Equiwidth   Wider at bends   Random variation

Narrow point bars   Wide point bars   Irregular point and lateral bars

FIGURE 24.28 Geomorphic factors that affect stream morphology. (From Lagasse *et al.*, 1991.)

the bends, there is a deep section (pool) in the outer part and a shallow section in the inner part (often with a sand bar called a *point bar*, see Fig. 24.27). The main current (*thalweg*) flows from a pool through a crossing to the next pool. The channel slowly migrates downstream with time. In laboratory flumes with homogeneous sand, the sinuous channels are very regular, however, in nature because of differences in the erodibility of the bank material, the sinuous channel can be very irregular. At low flow, the thalweg is located close to the outside of the bend. At larger flows, the thalweg tends to straighten. In contrast to a braided stream, the meandering channel is more stable, deeper, and will have a larger portion of its bed material transport in suspension. *Sinuosity* is the ratio of stream length to valley length.

Lane [Richardson *et al.* (1990)] studying the relationship between slope S, discharge Q and channel planform observed that when $SQ^{1/4} < 0.0017$, a sand-bed stream had a meandering planform. Also, when $Q^{1/4} > 0.01$, a stream had a braided planform. A stream with a value between 0.0017 and 0.01, could have either a meandering or braided planform. A meandering stream that has its slope increased such that the value of $SQ^{1/4}$ was larger than 0.0017, might become a braided stream. Conversely, a braided stream that has its discharge or slope decreased, such that the value of $SQ^{1/4}$ was smaller than 0.01, might change to a meandering planform.

Lane [Schumm (1972)], in addition to developing the relation to determine if a stream would be braided or meandering, proposed the following qualitative qualitative relation to predict the response of a stream or river to changes in water discharge, Q, slope, S, sediment discharge, Q_s, and median diameter of the bed material D_{50}.

$$QS \sim Q_s D_{50} \qquad (24.128)$$

For example, this relation shows that with a decrease in Q and no changes in the quantities on the right, there would then be an increase in the slope. The increase in slope results from the deposition of Q_s. Changes in other quantities would induce a response to keep a balance between the left- and right-side of the relation. Richardson *et al.* (1990), Simons and Senturk (1992), and Lane [Schumm (1972)] give examples of the uses of this relation and also expand on the effect of additional variables on the response of streams to change. Leopold and Maddock, Mackin, and others [Schumm (1972), (1977)] give additional insight into the response of streams to change, qualitative relations between width, depth, and velocity with discharge, and the importance of fluvial geomorphology in sedimentation.

24.4.3 Sediment Properties

Sediment properties of importance in sedimentation are the physical size, D, fall velocity, ω, and density of the particles and the bulk properties of the bed and bank material, and sediment deposits. Sediments are composed of clay (0.0002 to 0.004 mm), silt (0.004 to 0.062 mm), sand (0.062 to 0.2 mm), gravel (2.0 to 64 mm), cobble (64 to 250 mm), or boulder (250 to 4,000 mm) material. Bulk properties are described by the size frequency distributions, specific weight, and porosity. Size distributions are determined by sieving, visual accumulation tube analysis, pebble count, and pipette methods. Size distributions are usually expressed as percent finer than a given size in the distribution.

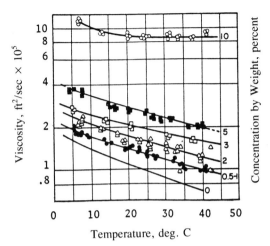

FIGURE 24.29   Apparent kinematic viscosity of water-bentonite dispersions. (From Richardson *et al.*, 1990.)

Of major importance is the viscosity of suspensions with large concentrations of silts and clays. Figure 24.29 shows the change in apparent kinematic viscosity of water-bentonite dispersions with temperature and concentration by weight of bentonite. For example, at a temperature of 24 C, the viscosity is 1 at 0 concentration of bentonite and is around 9 at 10 percent. This change in viscosity can have a significant effect on bed configuration, resistance to flow, and sediment transport. A concentration of silts and clay in a natural stream of 30 percent by weight can increase the transportation of sand-size bed material by a factor of 10.

Sediment properties and methods to determine these properties are discussed in detail by Brown (1950), Vanoni (1975), Richardson *et al.* (1990), and Simons and Senturk (1992).

24.4.4   Beginning of Motion

Knowledge of when fluid forces are large enough to move sediment particles is important in sediment transport, erosion and the design of *riprap*. Shields [Brown (1950), Vanoni (1975), and Simons and Senturk (1992)] experimentally determined a relation between the ratio of the critical shear stress, τ_c, to move a particle and its submerged weight expressed as $[(S_s - 1)\gamma D]$ to the shear velocity, size of particle Reynolds number, $(gRS)^{0.5}D/\nu$, at beginning of motion, where S_s = specific gravity of the particle, γ = unit weight of water, g = acceleration of gravity, R is the hydraulic radius, D = particle size, and ν = kinematic viscosity. The critical shear stress $\tau_c = K(S_s - 1)\gamma D$, where K ranges from 0.047 to 0.03 for sand sizes or larger. Lane, Fortier, and Scobey, Keown and others [Brown (1950), Vanoni (1975), Richardson *et al.* (1990), and Simons and Senturk (1992)] give values for the critical shear stress or velocity for the beginning of motion of silts, clays, sand, and coarser particles. Equations for determining the shear stress on a boundary are also given in the above references. Neil [Richardson *et al.* (1993)] gives the following equation for the critical velocity at beginning of motion of a particle

$$V_c = 1.58[(S_s - 1)\gamma D]^{1/2}(y/D)^{1/6} \qquad (24.129)$$

where V_c = critical velocity (ft/s) above which bed material of size D (ft) and smaller will be transported, y = depth of flow (ft). When S_s equals 2.65, a typical value for sand, Eq. (24.129) in English units reduces to

$$V_c = 11.52\,y^{1/6}D^{1/3} \qquad (24.129a)$$

24.4.5 Sediment Yield

Erosion results from overland flow processes and streamflow. Overland flow processes are sheetwash and rilling which constitutes the sediment yield from the land surface, whereas streamflow erosion is the removal of a stream's bed and bank material. A stream not only erodes its bank and bed, but also transports the product of overland flow erosion. This latter will be described in the Sediment Transport section below.

The *Universal Soil Loss Equation* (USLE), developed by the U.S. Department of Agriculture, is one of the most widely used regression equation for predicting sediment yield from overland flow [Vanoni (1975), Simons and Senturk (1992)].

$$A = R \cdot K \cdot LS \cdot C \cdot P \qquad (24.130)$$

where A = annual soil loss (U.S. tons per acre), R = a rainfall factor (inches per hour), K = a soil erodibility factor (tons per acre per unit of rainfall factor R), LS = topographic factor where S = land gradient and L = length of slope, C = a cropping and management factor, and P = an erosion control practices factor (LS, C, and P are dimensionless).

24.4.6 Bed Forms

The interaction between the flow in a stream and alluvial bed material consisting of sand or medium gravel sizes (D_{50} from 0.062 to 16 mm) creates different bed forms. These bed forms are divided into an upper and lower flow regime with a transition between them (see Fig. 24.30). The characteristics and magnitude of resistance to flow and sediment transport is related to the type of bed form as in Table 24.3. These bed forms depend on the discharge, slope, depth of flow, size of bed material, and viscosity of the fluid. A change in any of the above conditions can change the bed form and thus, resistance to flow and sediment transport which change flow depth, river stage, bed elevation, flow velocity, and sediment concentration.

Alluvial streams with steep slopes may flow in the upper flow regime all the time, whereas streams with lower slopes change from lower to upper flow regimes depending upon the discharge, bed material size, and fluid viscosity. This results in streams that, at low flow, are in the lower flow regime and, at high flow, are in the upper flow regime. Also, with the change in viscosity that occurs with a change in temperature, this can result in streams that have a dune bed form in the summer and washed out dunes, plane bed, or antidunes in the fall and winter. An example is the Missouri River as it flows along the border between Nebraska and Iowa. In the summer when the water temperature is high (70 to 80°F), the bed form is in the

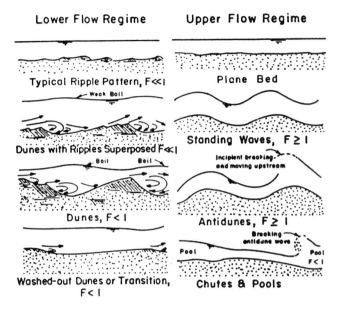

FIGURE 24.30 Regimes of flow and bed forms in alluvial sand and medium gravel stream. (From Simons and Richardson, 1966.)

lower flow regime and the Manning n is 0.020, and in the fall, water temperatures are much lower (35 to 65°F) and the bed form is in the transition and Manning n is 0.015. Consequently, this results in a lower depth and higher velocity for the same discharge [U.S. Corps of Engineers (1968)]. These changes in flow regime with discharge and temperature produce discontinuous or shifting rating curves for many sand channel streams (see Fig. 24.31) and affects the determination of discharge, depths of flow, velocity, and bed material discharge in these streams.

TABLE 24.3   Manning's n (see Sec. 25.5) and Sediment Concentration for Various Bed Forms

Alluvial Sandbed Channels	Manning's n	Concentration PPM
Lower Flow Regime		
plane bed	0.014	0
ripples $d \leq 0.7$ mm	0.018–0.028	10–200
dunes	0.020–0.040	200–3000
Transition		
washed out dunes	0.014–0.025	1000–4000
Upper Flow		
plane bed	0.010–0.013	2000–4,000
Antidunes		
standing waves Fr ≈ 1	0.010–0.015	2000–15,000
breaking Fr ≥ 1	0.012–0.020	5000–50,000
chute and pools	0.018–0.035	5,000–50,000

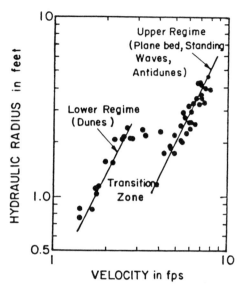

FIGURE 24.31 Relation of hydraulic radius to velocity, Rio Grande near Bernalillo, NM. (From Nordine, 1964.)

24.4.7 Sediment Transport

The quantity of sediment transported by a stream consists of the bed material discharge and fine material discharge (*wash load*) from the watershed and banks. Fine material discharge is that quantity of sediment that is not found in appreciable quantities in the bed of a stream. In sand channels, the fine sediment discharge will consist of silt- and clay-size sediments. In steep mountain streams with bed material of cobbles and boulders, the fine sediment discharge (wash load) consists of sand and/or small gravel-size material. The sediment is transported in suspension by the turbulence of the flow (suspended sediment discharge) and in contact with the bed by the shear force of the flow (contact sediment discharge of bed load). Also, because of the methods used to measure sediment discharge, the discharge is classified as measured or unmeasured sediment discharge. The separation of the sediment discharge by the source of the sediment, its mode of transport, and measurement is illustrated in Table 24.4.

Laursen [Richardson *et al.* (1993)] determined that when the ratio of the shear velocity, $(gRS)^{0.5}$, to the fall velocity, ω, of bed material is less than 0.5, the bed material discharge is mostly contact bed material discharge; between 0.5 and 2.0 has some suspended bed material discharge; and larger than 2.0 is mostly suspended bed material discharge. When the bed material discharge is mostly in contact with the bed, the Meyer–Peter and Muller equation is normally used to determine bed material discharge. When the sediment discharge is mixed mode, equations developed by Einstein, and others are used to calculate the total bed material discharge. A modified Einstein method developed by Colby and Hembree calculates the total bed material and fine sediment discharge of a stream using measured suspended sediment discharge measurements. These equations and methods are described by Vanoni (1975), Richardson *et al.* (1990), and Simons and Senturk (1992). The

TABLE 24.4   Classification of Sediment Discharge

	Classification of Sediment Discharge		
	Source of Sediment	Mode of Sediment Movement	Measurement of Sediment Discharge
Total Sediment Discharge	Fine Sediment (Wash Load) Discharge	Suspended Sediment Discharge	Measured Sediment Discharge
	Bed Material Discharge	Contact (Bed Load) Discharge	Unmeasured Discharge

Meyer–Peter and Muller equation and a graphical method of determining total bed material discharge developed by Colby (1964) is given in a later section.

Suspended sediment discharge is measured using suspended sediment samplers developed by the U.S. Inter-Agency Subcommittee on Sedimentation [Vanoni (1975)]. The samplers collect a velocity weighted concentration at their nozzle and either collect it at a point or integrate the velocity weighted sample through a vertical. The samplers come in various models, from hand-held or suspended by cable, and are point or depth integrating. The samplers only measure within a certain distance to the bed (0.3 to 0.4 ft) and, thus, do not measure the contact sediment discharge and suspended sediment discharge near the bed (unmeasured sediment discharge). However, in very turbulent streams or specially constructed turbulent flumes, where all the sediment transport is in suspension and uniformly distributed in the vertical or the sampler traverses the total depth, as at a weir, the suspended sediment samplers measure the total load. The suspended sediment discharge is calculated by the following equation

$$Q_s = KC_sQ \tag{24.131}$$

where Q_s = suspended discharge (U.S. tons per day); C_s = suspended sediment concentration (milligrams per liter); Q = water discharge (cubic feet per second); and K = coefficient to convert to the indicated units. K equals 0.0027 for sediment concentrations less than 36,000 milligrams per liter by weight.

Suspended sediment measurements are made by integrating the suspended sediment concentration in the vertical and across the stream using the suspended sediment sampler. To integrate across the stream, the sampler measures the concentration in equal-discharge increments (EDI) or at the centrodes of equal-length at an equal transit rate (ETR) across a stream. The EDI method requires the measurement of the water discharge prior to use, whereas ETR does not. The EDI measurement of suspended sediment concentration is the average of the concentration from several depth-integrated samples. The ETR concentration is the composite of several depth-integrated samples.

The distribution of suspended sediment in the vertical is given by

$$\frac{C}{C_a} = \left[\frac{y_0 - y}{y} \frac{a}{y_0 - a} \right]^Z \tag{24.132}$$

where C = sediment concentration at a point y, C_a = sediment concentration, at a point a above the bed, y_0 = the depth of flow, and Z = ratio of the fall velocity, ω, of the bed material to the shear velocity, $(gRS)^{0.5}$, times two coefficients. One coefficient is von Kármán's kappa in the logarithmic velocity distribution and the other is a coefficient for momentum exchange.

Sediment transport is predicted by the use of many equations. For streams with mostly coarse bed material and contact bed material discharge, the Meyer–Peter and Muller equation is often used [Vanoni (1970), Richardson *et al.* (1990), and Simons and Senturk (1992)]. A version is

$$q_s = [39.25q^{2/3}S_0 - 0.95D_{50}]^{3/2} \tag{24.133}$$

where q_s = bed material transport (pounds per second per foot of width), q = water discharge (cubic feet per second per foot of width), S_0 = slope, and D_{50} = median particle size (feet).

Einstein developed a procedure for calculating bed material transport by size fractions. In the method, he separates the shear (friction) on the boundary into a grain and form roughness, with the grain roughness being responsible for the transport of the contact load. His method first computes the contact bed material discharge by size fraction. It then uses this discharge to aid in the integration of the product of the suspended sediment concentration profile and velocity profile, to determine the suspended bed material discharge by size fraction. It then adds the two discharges by size fraction to obtain the total bed material discharge.

A modified Einstein procedure was developed by Colby and Hembree [Vanoni (1970), Richardson *et al.* (1990), and Simons and Senturk (1992)] which included the measurement of the suspended bed material discharge in the framework of the original Einstein method to calculate the bed material transport by size fractions. The modified Einstein method, because it is based on suspended sediment measurements which may constitute from 50 to 90 percent of the total sediment discharge, is the most accurate method of determining sediment discharge.

Colby (1964) developed a very useful graphical method of determining total bed material discharge. The method is given in Figs. 24.32 and 24.33 and Eqs. (24.134) and (24.135).

$$q_T = [1 + (K_1 K_2 - 1)K_3]q_n \tag{24.134}$$

$$Q_s = Wq_T \tag{24.135}$$

where q_T = total bed material discharge per unit width (U.S. tons/day/ft), $K's$ = correction coefficients determined from Fig. 24.33, q_n = discharge of bed material determined from Fig. 24.32, and Q_s = total bed material discharge for a channel of width W (ft)(U.S. tons/day). The uncorrected total bed material discharge, q_n, is determined from Fig. 24.32 for a given velocity, median diameter of the bed material, and two depths that bracket the desired depth and interpolating on a logarith-

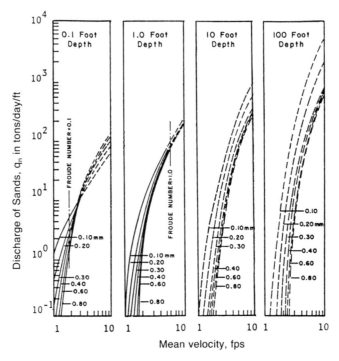

FIGURE 24.32 Colby's (1964) relation of discharge of sands to mean velocity for six median sizes of sand, four depths of flow, and at a water temperature of 60 F.

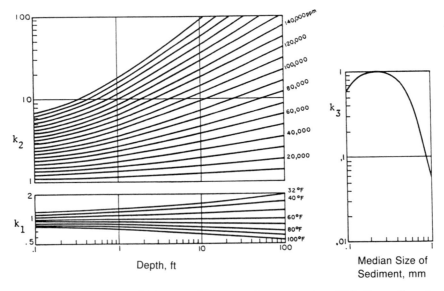

FIGURE 24.33 Colby's (1964) correction curves for temperature (K_1), fine sediment concentration (K_2), and sediment size (K_3).

mic graph of depth versus q_n to obtain the bed material discharge per unit width. The corrected bed material discharge per unit width, q_T, is determined using Eq. (24.134) and Fig. 24.33. The total bed material discharge is then obtained using Eq. (24.135).

24.4.8 Reservoir Sedimentation

The rate of depletion of reservoir storage from storage of sediment depends on: 1) the volume of the reservoir in relation to inflow, 2) sediment inflow, 3) reservoir trap efficiency, and 4) the specific weight (density) of the sediment deposits. Sometimes, the distribution of the sediment in the reservoir can be important. Sedimentation damages a reservoir over time if it decreases storage volume to such an extent that it no longer can serve its design function.

   Reservoirs with large storage volume in relation to average annual inflow of water have a lower rate of storage loss due to sedimentation than reservoirs that do not, even with same trap efficiency. Lake Mead on the Colorado River and Lake Nasser on the Nile River have storage volumes that are about double their averaged annual inflow. Their trap efficiencies are almost 100 percent, but their rate of sediment depletion of storage is so low that their useful life expectancy is measured in thousands of years. The reservoir behind Tarbella Dam on the Indus River has a storage volume that is less than 20 percent of the average annual flow. Its trap efficiency is also about 100 percent, and its rate of storage depletion is so large that its useful life as a storage reservoir is less than 100 years. All three reservoirs have approximately the same annual sediment inflow. Of interest is the fact that the reservoir behind old Aswan Dam on the Nile River downstream of Lake Nassar with a storage to inflow ratio of less than 3 percent has an infinite useful life. It stored water for over 50 years before Lake Nassar was created by the High Aswan Dam. Its trap efficiency is close to zero, because it has under sluices across the total dam width and most of the annual flood passes through the reservoir and only the tail of the annual flood is stored.

   The sediment inflow into a reservoir can be estimated by: 1) use of the recorded annual measured total sediment discharge of the stream, 2) computation by flow-duration-sediment-rating curve method, and 3) estimating the sediment yield from the watershed.

   The use of recorded annual measured total sediment discharge of a stream is limited to those sites that have a historical record based on frequency sampling to establish a reliable estimate of the sediment inflow into the reservoir. Normally, the record of sediment inflow is less than the streamflow record and is only suspended sediment discharge. However, the suspended sediment record can be adjusted to obtain the total sediment discharge using methods described in the previous section.

   The flow-duration-sediment-rating curve method extends the available total sediment discharge record to the historical streamflow record and is the most desirable. In this method, a sediment rating curve is established relating the daily sediment discharge (normally expressed in tons/day) to the daily discharge. There will be a larger scatter in the data, some of which may be seasonal, but a reliable set of curves can be developed. A flow-duration curve is also prepared for the entire streamflow record. From these two sets of relations, the average annual sediment discharge is determined [Vanoni (1975) and Simons and Senturk (1992)]. The sediment rating curve may be developed using measured suspended sediment discharge, measured total sediment discharge (suspended sediment plus unmeasured sediment discharge), or a measured suspended sediment discharge corrected for the unmeasured sediment discharge.

The sediment yield from a watershed is estimated using the USLE. This value is then used with the appropriate trap efficiency to determine sedimentation rates in a reservoir. This method is often used for small reservoirs.

Trap efficiency is the measure of the percentage of the sediment inflow retained (trapped) in the reservoir. It depends on the velocity of flow through the reservoir and sediment size. The velocity of flow depends on the size of the reservoir, type of dam, and operating procedures. Reservoirs that are formed from large embankments (soil, concrete, or rock), with large storage volumes, over year storage and outlets that normally discharge less flow than the incoming floods have large trap efficiencies (close to 100%). These reservoirs may have some sediment removed by density currents or large velocities when the reservoir is low, but in consideration of the other approximations and uncertainties in the determination of the sediment inflow, these amounts are ignored. Small reservoirs with large velocities or reservoirs with under sluices than can pass large inflows will have low trap efficiencies. Operation of the dam to have a low pool with high velocities part of each year will decrease trap efficiency.

To determine trap efficiency, Churchill, from a study of TVA reservoirs, developed a relation between the percent of incoming silt passing through the reservoir and the ratio of period of retention divided by mean velocity, and Brune developed a relation between percent of sediment trapped and a ratio of reservoir capacity divided by the mean annual inflow [Vanoni (1975) and Simons and Senturk (1992)]. In employing these relations, engineering judgment must be used to determine trap efficiency for a particular reservoir.

The specific weight of the sediment is needed to convert the estimate of sediment deposited in the reservoir, normally given in terms of weight, to volume. The specific weight of the sediment deposits in the reservoir increases with time as they consolidate. Coarse sediments (sand and gravels) will consolidate faster than the finer silts and clays and will reach their ultimate weight faster. Also, sediments consolidate faster if they are not always submerged. Vanoni (1975) and Simons and Senturk (1992) present methods for estimating the specific weight that take into account the type of sediment and degree of submergence.

The distribution of sediment in reservoirs depends on the size composition of the sediment flowing into a reservoir and management of the outflow. Coarse sediments are deposited in the upper reaches of the reservoir and finer sediments farther down. If the water storage in the reservoir is managed so as to have a low water level part of the year, then the coarser materials are moved farther down into the reservoir. Coarse sediment deposits at the upper end of the reservoir can increase the backwater effects upstream. Vanoni (1975) and Simons and Senturk (1992) discuss the location of sediment deposits and presents methods to evaluate the location.

Sediment deposits in reservoirs are measured using sonic sounders and surveying techniques to monitor the loss of storage with time. Vanoni (1975) describes methods for conducting these surveys.

24.5 RIVER FLOWS
Panayiotis Diplas

24.5.1 Introduction

Open channel flows are characterized by the presence of a free surface, and is the main feature that distinguishes them from pipe flows. The free surface is subject to atmospheric pressure, however its position is not known beforehand and is part of the solution of the problem. It is also likely to change its elevation with respect to time and space. The driving force in channel flows is typically, but not always, gravity. Although there are a lot of similarities between the two types of flow, the free surface has been found to affect the turbulence intensity and secondary currents of open channel flows [Nezu and Nakagawa (1993)]. Furthermore, the geometry and roughness characteristics of channel boundaries vary more widely, resulting in flow behavior that is more complex than is usually encountered in pipe flows. Since pipe flows have been studied more thoroughly, it is common to extend results from such studies to channel flows by using simple geometrical considerations, e.g., by replacing pipe diameter with hydraulic radius. In light of the partial list of differences outlined here, generalizations that are based on simplistic approaches should be viewed cautiously until they have been tested.

Rivers have always played an important role in man's civilization. Naturalists and philosophers were probably the first to attempt to explain general fluvial phenomena in the context of wide-ranging discussions on the origins of earth and its physical features. Aristotle made observations about the rapid silting up of navigable rivers around the shores of the Black Sea and on the Bosphorous, while the historian Herodotus characterized Egypt as the gift of the (Nile) river. From an engineering standpoint, river hydraulics might be one of the oldest research topics, as indicated by a network of canals found in Mesopotamia that is dated back to 4000 B.C. Rivers are not any less important today. The navigable waterways within the United States exceed 25,000 linear miles, serve 130 of the nation's 150 largest cities, and handle in excess of two billion tons of commerce each year. Similarly, the number of main bridge river crossings exceeds 500,000. Finally, the disastrous 1993 flood of the Mississippi speaks convincingly about the power of rivers.

Open channels can be man-made (artificial) or natural. Artificial channels have, typically, rigid boundaries and regular geometric shapes. Natural or *alluvial* streams possess an erodible boundary which interacts dynamically with the flow. The location and pattern exhibited by the channel boundary, in this case, is the result of solid-liquid interplay. Therefore, all boundaries of a river can be considered as free surfaces [Kennedy (1983)]. This behavior has prompted Leopold and Wolman (1957) to characterize alluvial rivers as "authors of their own geometry." Rigid boundary hydraulics are simpler, better understood, and well presented in several books [e.g., Chaudhry (1993), Mahmood and Yevjevich (1975), Henderson (1966), Chow (1959), and Stoker (1957)]. The present section will deal with some aspects of alluvial stream hydraulics. For a more complete account of this extensive subject, the relevant literature should be consulted [e.g., Leopold (1994), Simons and Senturk (1992), Raudkivi (1990), Ikeda and Parker (1989), Chang (1988), Garde and Ranga Raju (1985), Cunge *et al.* (1980), Jansen (1979), and Vanoni (1975)].

24.5.2 Physical Properties of Fluvial Sediments

The outcome of fluid-sediment interaction in river flows is greatly influenced by the properties of the sediment particles. Sediment is partly of mineral and partly of organic origin. The mineral part originates from the decomposition of rock. The mineral type that is most commonly encountered in the river and coastal environments is quartz (specific gravity $\gamma \simeq 2.65$). It is also the most durable. Other common rock types include limestone ($\gamma \simeq 2.6 \sim 2.8$), basalt ($\gamma = 2.7 \sim 2.9$), and granite ($\gamma = 2.7$). The properties used to characterized individual sediment particles are: 1) shape, 2) size, and 3) settling velocity.

The coarser grains are typically approximated as ellipsoids. The particle dimensions are measured along three mutually perpendicular axes, with a the length along the longest (major) axis, b along the intermediate, and c along the smallest (minor). The shape factor, $SF = c/\sqrt{ab}$, provides a measure of particle shape ($SF \approx 0.7$ is an average value for well-rounded natural sediments). Zingg used the ratios b/a and c/b to classify the particles as spheroids, disks, blades, and rollers [Vanoni (1975)].

Particle size (and shape to a lesser extent) have a direct effect on particle mobility. Common measures and definitions of particle sizes are: 1) *sieve diameter*, D, is the size of a square sieve opening through which the particle will just pass, 2) *sedimentation diameter* is the diameter of a sphere of the same specific weight and same terminal velocity as the given particle in the same fluid, and 3) *nominal diameter*, D_n, is the diameter of a sphere of the same volume as the given particle. The b-axis is the best single axis estimator of D_n, however a more accurate estimate is given by $D_n = (abc)^{1/3}$. For naturally worn particles, the sieve diameter is usually slightly smaller than the nominal diameter. Particle size is measured in mm or in *phi-scale*, $\phi = -\log_2 D$, with D in mm.

For very small grains, the *standard fall velocity*, which is the terminal fall velocity of a particle in quiescent distilled water at 24 C, is used for characterization.

Sediments are classified according to their size in the following way: boulders ($D > 256$ mm), cobbles (64 min $< D < 256$ mm), gravel (2 mm $< D < 64$ mm), sand (0.064 mm $< D < 2.0$ mm), silt (0.002 mm $< D < 0.064$ mm), and clay ($D < 0.002$ mm). Silt or coarser particles are typically produced by mechanical action, including fracture and abrasion. Clays are produced by chemical action, and they exhibit cohesive behavior which renders them more resistant to erosion. Noncohesive types of sediment will be considered in the present chapter. The present state of knowledge about the behavior of cohesive sediment is given by Mehta *et al.* (1989).

Methods for Measuring Particle Sizes. Because of the wide range of sediment sizes encountered in natural streams, there is not a single method suitable for measuring all of them. Instead, different approaches are employed depending upon the particle size. For very large particles ($D < 256$ mm), the intermediate axis b is measured. For 32 mm $< D < 256$ mm, templates with square openings are used. This technique provides a measure of size that is equivalent to conventional sieving. For 0.0625 mm $< D < 32$ mm, the regular sieve analysis is used. For $D < 0.0625$ mm, a hydraulic settling method is suggested (hydrometer analysis, visual accumulation tube, pipette method, bottom withdrawal) [Vanoni (1975)]. It should be mentioned that the size specifications obtained by measuring the b-axis, sieve diameter D, and settling velocity are not equivalent. Usually, b is close to D. However, the exact relation between b and D depends on the clast flatness ratio, c/b [Church *et al.* (1987)].

Sediment Size Distribution. Sediment deposits normally contain a range of grain sizes. It is, therefore, necessary to separate a sediment sample into a number of size classes, for example by means of a sieve analysis. The results are then presented in terms of a cumulative size-frequency distribution curve. It is customary to use a semi-log plot, because the size distribution provided by a sieve analysis is often close to log normal. The geometric mean diameter, D_g, the geometric standard deviation, σ_g, the mean diameter, D_m, the median size, D_{50}, the mode, D_{md}, D_{90}, and D_{10} are some measures of the size distribution commonly used to characterize various aspects of the sediment in two-phase, fluid-solid, flow phenomenon. In general, D_x indicates the grain size diameter that is coarser than $x\%$ of the material in the sediment sample in terms of weight. The general expressions are

$$D_g = \prod_i D_i^{f_i} \tag{24.136}$$

$$\sigma_g = \exp\left[\sum_i \left(\ln \frac{D_i}{D_g}\right)^2 f_i\right]^{1/2} \tag{24.137}$$

where f_i denotes the probability of occurrence of the value D_i. For log-normally distributed grain sizes the above expressions can be simplified to read

$$D_g = \sqrt{D_{16}D_{84}} \tag{24.136a}$$

and

$$\sigma_g = \sqrt{\frac{D_{84}}{D_{16}}} = \frac{D_{50}}{D_{16}} = \frac{D_{84}}{D_{50}} \tag{24.137a}$$

In this case, $D_g = D_{50} = D_{md}$. If the data deviate from log-normal, use

$$\sigma_g = \frac{1}{2}\left(\frac{D_{84}}{D_{50}} + \frac{D_{50}}{D_{16}}\right) \tag{24.137b}$$

Material with $\sigma_g < 1.3$ is designated as *well-sorted*, and for all practical purposes it can be treated as uniform. Material with $\sigma_g > 1.6$ is termed *poorly-sorted*.

24.5.3   Gravel- and Sand-Bed Streams

A river is classified as *gravel-bed* or *sand-bed* when its bed material, D_{50}, is coarser or finer than 2.0 mm, respectively. Both types often occur in a single system. The trend exhibited by the sediment sizes along the Saskatchewan River, shown in Fig. 24.34, is typical of natural streams. The bed material becomes progressively finer in the downstream direction. Usually, in the headwaters (mountains) where the river

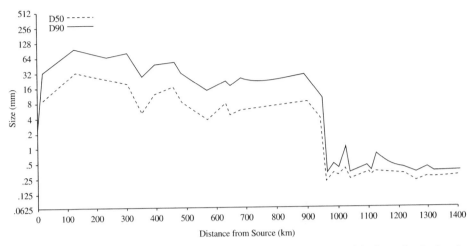

FIGURE 24.34 Downstream changes in the size of the bed material along the Saskatchewan River, Canada. (From Shaw and Kellerhals, 1982.)

slope is steeper, the bed material is in the gravel or coarser ranges, while in the valley it is within the sand or finer ranges. The transition from gravel to sand is quite abrupt (Fig. 24.34). Attempts have been made recently to explain and quantify these phenomena [Paola *et al.* (1992), Parker (1991), and Shaw and Kellerhals (1982)]. Sand- and gravel-bed streams exhibit certain distinctly different features that deserve close attention. First, the bed material in sandy streams is usually well sorted, therefore its D_{50} sufficiently describes the entire grain size distribution. In gravelly streams, however, it is poorly sorted, and its size distribution often exhibits bimodal behavior with one mode in the sand range and another one in the gravel range as in Fig. 24.35. Second, the bed shear stress in sandy streams is frequently considerably higher than the critical value required to mobilize the bed material. In gravel streams, the maximum bed shear typically does not exceed two to three times the critical value, and sediment transport is an infrequent event [Andrews (1994) and Parker (1978)]. Third, the bed material in sandy streams does not exhibit any stratification in terms of grain size. In a gravel river, however, a surface layer that is coarser than the subsurface material is typically present at the river bed (see Fig. 24.36). The surface layer is termed *pavement* or *armor layer*, depending on whether sediment motion takes place or not, while the layer below is called *subpavement* [Parker and Sutherland (1990) and Diplas (1987)]. Fourth, the dominant bed forms in sand-bed rivers are *ripples* and *dunes*, which are washed away during floods (see Fig. 24.30). For gravel rivers, the dominant bed forms are gravel bars, which are typically weakly formed [Diplas (1994) and Parker and Peterson (1980)]. Last, scour and fill of the channel bed are typically larger and faster in sandy streams.

24.5.4 Sediment Sampling

The characteristics of a stream are intimately linked to the material that comprises its channel boundary. It is, therefore, necessary to employ methods that would accurately sample and analyze the bed material to determine its composition. Simi-

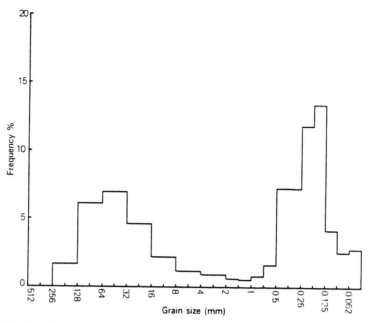

FIGURE 24.35 Size-frequency distribution of a bed material sample from the Athabasca River, a gravel stream in Canada. (From Shaw and Kellerhals, 1982.)

larly, it is important to accurately measure the type of sediment and the rate at which it is transported by the flow as bed load and in suspension. The point- and depth-integrating samplers, developed by the Interagency Committee, are commonly used for suspended sediment [Vanoni (1975)]. The Helley–Smith sampler seems to be the favored portable device for measuring bed load transport [Hubbell (1987)]. The vortex-tube and Birkbeck type samplers are permanent installations that have been used successfully in the field to provide continuous monitoring of bed load discharge at a station [Tacconi and Billi (1987) and Reid et al. (1980)]. While considerable information is available in the literature on the methods and equipment used for sampling moving sediments, limited attention has been paid to the problem of sampling bed material.

Volumetric, or *bulk*, sampling is considered to be the standard sampling procedure, because it provides unbiased results. It consists of the removal of a predetermined volume of material, sieve analysis, and interpretation of the results in terms of a grain-size frequency distribution by weight. The volume of the sample should be large enough to be independent of the dimensions of the individual particles [Kellerhals and Bray (1971)]. This is a suitable procedure for sampling sandy streams because of the absence of vertical grain size stratification within the channel bed. However, it would be inappropriate for sampling gravel-bed streams which, typically, possess three horizontal layers with each containing a different grain-size population. The top layer, or pavement, is coarser than the second layer, or subpavement, which is typically richer in fine sediment. The material in the third, or bottom, layer is similar in size to that of the subpavement but not as rich in *fines*. The thickness of each of the top two layers is close to the size of the largest grain, while the third layer does not have a predetermined thickness. The material within each

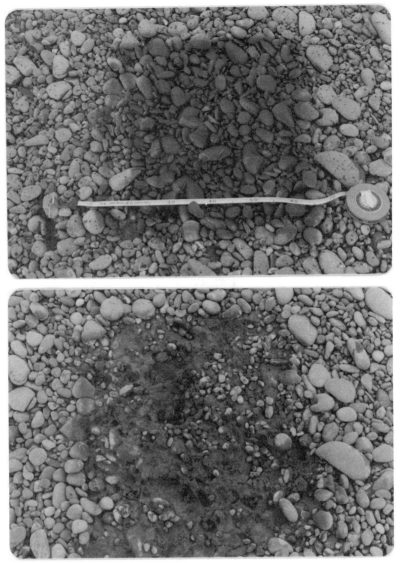

FIGURE 24.36 Coarser bed surface layer, pavement, at the Waimakariri River, New Zealand. In the bottom figure, part of the bed surface layer has been removed to highlight the contrast in size between the surface and subsurface materials.

layer influences the hydraulics of the river and its ecology in a distinct way and should be sampled and analyzed separately. The composition of the pavement material is important for determining the roughness and stability of the channel bed. The subpavement, especially its fines content, is crucial for evaluating the suitability of the stream for spawning by fish and other species.

Because of their limited thickness, the pavement and subpavement layers cannot be sampled volumetrically [Church *et al.* (1987)]. Instead, a surface-oriented sampling procedure should be employed. Such procedures, however, provide biased results. To convert a surface-oriented sample into its volumetric equivalent, Kellerhals and Bray (1971) suggested the formula

$$p(V)_i = C_1 p(S)_i D_i^x \qquad (24.138)$$

where $p(V)_i$ is the percentage of material retained on sieve size i based on a volumetric sample, $p(S)_i$ is the percentage of material retained on sieve size i based on a surface sampling method, D_i is the geometric mean of the sieve sizes i and $i + 1$, and C_1 is a proportionality constant. The exponent x depends on the type of surface sampling method being used.

The grid-by-number and area (clay)-by-weight are reliable procedures, suitable for sampling a single layer of bed material. The first, is the procedure most widely used in the field for sampling coarser sediments ($D > 15$ mm) [Fripp and Diplas (1993)]. A grid is placed over the surface of the channel bed with the grains lying below grid points constituting the sample. A simplified version is for an operator to pace off regular intervals and pick up the particles below the operator's toe. An acceptable sample size consists of 100 stones. The results of sieve analysis are presented in terms of a frequency by number distribution. This is the only approach that is equivalent to bulk sampling, e.g., $x = 0$ in Eq. (24.138). The only drawback is that in the presence of finer particles ($D < 15$ mm), it provides a sample that is truncated at its lower end.

The second procedure uses pottery clay pasted on a flat plate and subsequently pushed against the channel bed to remove all the grains that are on the surface of a specified area. The sample is sieved and the results are presented in terms of frequency-by-weight distribution. To convert this sample into an equivalent volumetric, Eq. (24.138) is used with $x = -1.0$ [Diplas and Fripp (1992)]. The adhesive strength of clay can consistently remove grains with sieve diameter of 40 mm or smaller. Then, in the presence of coarser sediments, the clay method provides a sample that is truncated at its upper end.

The advantages of these two surface-oriented procedures is that they can be used to sample both dry and under-water sediment deposits, they require minimal equipment, and they can be adequately carried out with one or two field workers. To overcome the truncation problems, a hybrid method has been proposed by Fripp and Diplas (1993) that combines the grid and clay procedures to provide the entire size distribution of the bed material. They also suggest a method for determining the required sample size for estimating various sediment characteristics (e.g., D_{50}, D_{90}) with a desirable accuracy. The results of sampling the bed material in the New River, Virginia, by using the hybrid technique are shown in Fig. 24.37.

The freeze-core method is another approach suitable for sampling smaller grains. In this case, liquid carbon dioxide or liquid nitrogen is injected into a single or multiple standpipe within which the substrate freezes. This technique provides undisturbed samples and is especially useful when the vertical variation of the bed material is of interest. This method may not sample the coarser particles of a sediment deposit in a representative way resulting in biased samples.

24.5.5 Velocity Profiles

The boundary of an open channel flow is classified according to the relation between the effective roughness height, k_s, and the thickness of the viscous sublayer, $5(v/v^*)$,

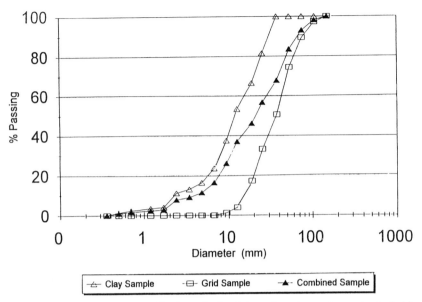

FIGURE 24.37 Particle size distribution obtained by using the under-water hybrid sampling technique at the New River, VA.

where $v* = \sqrt{\tau_0/\rho}$ is the *shear velocity*, τ_0 is the boundary shear stress, v is the kinematic viscosity, and ρ is the fluid density. For $k_s^+ = k_s v*/v < 5$ it is termed *hydraulically smooth*; for $k_s^+ > 70$ it is termed *completely rough*, and *transitional* for $5 \leq k_s^+ \leq 70$ the same as for pipe flows. Wall-bounded turbulent flows are traditionally split into inner, outer, and overlap (inertial) layers. Within the overlap layer, both inner and outer velocity profiles are valid with their expressions becoming logarithmic. The thickness of the inner layer is $y \simeq 0.2\,\delta$, where y is the distance from the channel boundary and δ is the thickness of the boundary layer. For sufficiently wide channels ($B > 5h$), $\delta = h$, where B is the channel width and h is the flow depth. For narrower channels, the location of the maximum velocity is below the free surface, resulting in the well known velocity dip, therefore $\delta < h$ [see, e.g., Ferro and Baiamonte (1994)]. For the case of the smooth boundary (a condition rarely encountered in river flows), the inner layer can be subdivided into three smaller layers. Within the viscous-dominated region adjacent to the wall, $y^+ = yv*/v \leq 5$, called *viscous sublayer*, the velocity profile is linear

$$u^+ = y^+ \tag{24.139}$$

where $u^+ = U/v*$, and U the time-averaged velocity at a distance y from the wall. For $y^+ \geq 50$ and $(y/h) < 0.2$ the velocity profile is described by the log-law

$$u^+ = \frac{1}{\kappa}\ln y^+ + A \tag{24.140}$$

where κ is the von Kármán constant with a value around 0.4 and $A \cong 5.5$. Both are near-universal constants for turbulent flow past a smooth boundary. The intermedi-

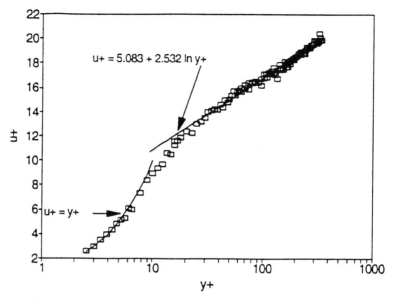

FIGURE 24.38 The time-averaged velocity profile within the inner layer of an open channel flow with smooth boundary. (From Balakrishnan and Dancey, 1994.)

ate region ($5 < y^+ < 50$) is called the *buffer layer* and provides a smooth transition between the other two layers. The velocity profile within the inner layer of an open channel flow measured with a 3D-LDA is shown in Fig. 24.38.

In the case of a rough boundary, the viscous sublayer is obliterated by the roughness elements, and the log-law describes the velocity profile within the inner region ($y \leq 0.2h$)

$$u^+ = \frac{1}{\kappa} \ln \frac{y}{k_s} + B \tag{24.141}$$

where $B \simeq 8.5$ [Diplas (1990)]. Estimates about the k_s value vary widely. Yen (1992) has attributed this variability to k_s dependence on flow depth. A reasonable value is $k_s = 2D_{90}$ [Diplas (1990)]. It should be mentioned, that for very shallow flows, $h < 5k_s$, the velocity profile deviates from the logarithmic expression in Eq. (24.141). The agreement between open channel and pipe flow velocity profiles in the wall region, for both smooth and rough boundaries, should not be surprising because this region is not affected by the presence of the free surface. The velocity profile in the outer region ($y > 0.2h$) is adequately represented by the velocity defect law for both smooth and rough boundaries

$$\frac{U_{max} - U}{v^*} = -2.5 \ln \left(\frac{y}{h} \right) + 5\Pi \cos^2 \left(\frac{\pi y}{2h} \right) \tag{24.142}$$

where U_{max} is the maximum velocity at the edge of the boundary layer and Π is Cole's wake strength parameter, which describes the deviation of the velocity profile from the logarithmic law in the outer region. For clear water, open channel flows,

Π mainly depends upon the boundary roughness. The Π values for free surface flows are lower than the corresponding values for zero pressure gradient boundary layers in closed conduits. Typical open channel Π values are 0.2 for smooth boundary [Nezu and Nakagawa (1993)] and 0.08 or smaller for rough boundaries [Ferro and Baiamonte (1994) and Diplas (1990)]. Therefore, Π is small in open channel flows, and the logarithmic expressions in Eqs. (24.140) and (24.141) can be considered to approximate the velocity profile over the entire flow depth.

It should be emphasized that the results presented here are based on experimental data collected from flumes with clear water flows, and either impermeable or permeable boundary but with small voids (e.g., the bed material consists of finer grains and large standard deviation). For flows carrying sediment in suspension, or with permeable boundaries and quite porous beds, it is customary to extend these results and attempt to fit the log law [Eqs. (24.140) and (24.141)] to the velocity profile for the entire flow depth by adjusting the constants κ, A, and B. However, this extension is contrary to the theory of boundary layers which specifies that Eqs. (24.140) and (24.141) are strictly valid in the wall region [White (1991), Zippe and Graf (1983), and Coleman (1981)]. Appropriate use of Eqs. (24.140) and (24.141) suggests that the von Karman κ value remains constant, equal to 0.4, for all flow conditions. The presence of suspended load, however, influences the velocity profile in the outer region by changing the value of the wake strength parameter. Based on experiments with a smooth boundary, Coleman (1981) found that Π was increasing with the amount of suspended load, ranging from $\Pi = 0.19$ for clear flow to $\Pi = 0.86$ for flows laden nearly to capacity with sediment. The net result is a reduction in the drag of the channel boundary with an increase in suspended load. On the other hand, Mendoza and Zhou (1992) and Zippe and Graf (1983) found that the drag of a permeable surface is higher than that of an impermeable surface having identical roughness. This is attributed to the continuous exchange of fluid between the flow and the upper layers of the permeable bed. The difference becomes more pronounced with increasing porosity. Furthermore, the friction factor for fully rough permeable boundaries is not constant but increases with the flow Reynolds number [Lovera and Kennedy (1969)].

It is worthwhile to mention that the velocity profiles discussed here pertain to 2-D or weakly curved open channel flows. For strongly 3-D flows, like flow through a well-developed *meander* bend, the profile of the streamwise velocity component can no longer be assumed to be logarithmic [Falcon (1984)].

24.5.6 Channel Flow Resistance

From an engineering standpoint, the accurate determination of the channel resistance characteristics is very important, because it is the information necessary for calculating the maximum flow that a stream can carry before flooding the surrounding area. The current state of knowledge on this subject is satisfactory for channels with rigid, impermeable boundaries. However, for the case of alluvial channels, things are significantly more difficult. Some of the additional complexities include the presence of a permeable bed, bed forms, channel curvature, varying channel cross-section, sediment transport, vegetation, and overbank flows. An extensive review on channel resistance can be found in Yen (1992).

Uniform, Steady Flow. For the case of a uniform, steady flow in a straight channel with a flat bed, Eq. (24.141) is typically integrated over the entire flow depth to provide the following resistance relation

$$\frac{U_{ave}}{v^*} = \frac{1}{\pi} \ln \left(11 \frac{h}{k_s} \right) \tag{24.143}$$

The shear stress at a point on the channel boundary is given by

$$\tau_0 = \rho \frac{f}{8} U_{ave}^2 \tag{24.144}$$

where U_{ave} is the depth-average velocity and $f = 8[1/\kappa \ln (11 (h/k_s))]^{-2}$ is the Darcy–Weisbach friction factor. For very wide channels, the boundary shear stress, τ_0, does not vary significantly around the channel perimeter, and Eq. (24.144) is generalized to represent cross-sectionally-averaged quantities

$$\tau_0 = \rho g R S_0 = \rho \frac{f}{8} V^2 \tag{24.145}$$

where R is the hydraulic radius of the channel cross section, S_0 is the slope of the channel bed in the downstream direction, and V is the cross sectionally averaged velocity. This extension is based on the results of flow in circular pipes.

   Chezy's and especially Manning's equations are widely used in open channel flows. They relate the resistance of a channel cross section to flow and channel parameters as follows

$$V = C\sqrt{RS_0} \tag{24.146}$$

$$V = \frac{K}{n} R^{2/3} S_0^{1/2} \tag{24.147}$$

where C is Chezy's coefficient and n is Manning's coefficient; $K = 1.0$ in S.I. units and $K = 1.49$ in the English system. The three resistance coefficients are related as follows

$$\sqrt{\frac{8}{f}} = \frac{C}{\sqrt{g}} = \frac{K}{\sqrt{g}} \frac{R^{1/6}}{n} \tag{24.148}$$

Out of the three, f is the only dimensionless coefficient. It is also the only one that changes with the flow depth. The other two coefficients, assume one-dimensional flow and, therefore, lump the resistance of the whole channel section in one number. Strickler expressed n in terms of particle size as [Chang (1988)]

$$n = \frac{D_{50}^{1/6}}{21.1} \tag{24.149}$$

and Meyer–Peter and Müller suggested

$$n = \frac{D_{90}^{1/6}}{26} \qquad (24.150)$$

where both D_{50} and D_{90} are measured in meters.

It is unfortunate that the differences among the shear stress at a point, a cross-section, and a channel reach have been overlooked. The current formulations using the lumped resistance approach are rather simplified. Furthermore, some of the claims made in the literature, such as the constancy of n for given boundary or the use of Moody diagram to determine f, should be viewed cautiously.

It is evident from Eq. (24.148) that all three resistance relations are equivalent for fully rough turbulent flow. Manning's, however, is the most commonly used formula, especially by practitioners. This may be attributed to the long experience with this formula and the availability of an extensive list of tabulated n values for a variety of boundary conditions [Yen (1992) and Chow (1959)].

Nonuniform, Unsteady Flow. Water flow in natural streams is usually nonuniform and unsteady. Flood flows, which have the highest potential for damage, are typically characterized by an initially rapid flow increase followed by a slow recession. To extend the results obtained in the previous section for uniform, steady flow, the 1-D St. Venant equation is used. This is the 1-D shallow-water momentum equation. It can be derived from the Navier–Stokes equations, or by applying Newton's Second Law of motion directly to a control volume in open channel flow [Chaudhry (1993), Mahmood and Yevjevich (1975), and Henderson (1966)]. The basic assumption in deriving the shallow-water equation is that the pressure varies hydrostatically in the vertical direction. This requires that the vertical accelerations are negligible, or that the streamlines do not exhibit strong vertical curvature. Obviously, this equation should not be used for rapidly varying flows. The slope of the channel bed is assumed to be gradual, so that the flow depth normal to the channel bottom can be considered approximately equal to the vertical flow depth. It is also assumed that steady-flow resistance expressions can be used to calculate the energy losses in unsteady flow. Finally, the notion of 1-D flow implies that flow changes are limited to the downstream direction. This approach would not be suitable for flows with strong velocity variability within the channel cross section e.g., flow in compound or curved channels. The St. Venant equation in a form suitable for the present discussion is

$$S_f = S_0 - \frac{\partial}{\partial x}\left(\frac{V^2}{2g} + h\right) - \frac{1}{g}\frac{\partial V}{\partial t} \qquad (24.151)$$

where S_f is the slope of the energy grade line, S_0 is the slope of the channel bed, and x is the coordinate in the downstream flow direction. For uniform, steady flow, $S_f = S_0$. The second term on the right hand side of Eq. (24.151) accounts for flow nonuniformity, while the third term accounts for flow unsteadiness. For most channel flows, the convective and local accelerations are by far the least important terms in Eq. (24.151), therefore they can be dropped. By incorporating a result obtained from the continuity equation for unsteady flow [Henderson (1966)], Eq. (24.151) can be modified to read

$$S_f \cong S_0 + \frac{1}{C'} \frac{\partial h}{\partial t} \qquad (24.152)$$

where C' denotes the kinematic wave speed. By using Manning's formula, the unsteady flow discharge Q_u becomes

$$Q_u = \frac{1}{n} AR^{2/3} S_f^{1/2} \qquad (24.153)$$

In this case, the influence of terms other than S_0 [see Eq. (24.153)] implies that the discharge, Q_u, is not a function of flow depth alone. Furthermore, because $\partial h/\partial t > 0$ during the rising limb and $\partial h/\partial t < 0$ during the falling limb of a flood hydrograph, the stage-discharge relationship for unsteady flows is double-valued. This is known as the *looped rating curve* and is shown in Fig. 24.39. For a given channel cross-section and flow depth, a comparison between the discharges obtained for uniform, steady flow, Q, and unsteady flow conditions, Q_u, is given by

$$\frac{Q_u}{Q} = \left[1 + \frac{1}{C'S_0} \frac{\partial h}{\partial t} \right]^{1/2} \qquad (24.154)$$

The comparison between the steady, uniform and unsteady rating curves is illustrated in Fig. 24.39. During the rising stage, $Q_u > Q$, while during the falling stage,

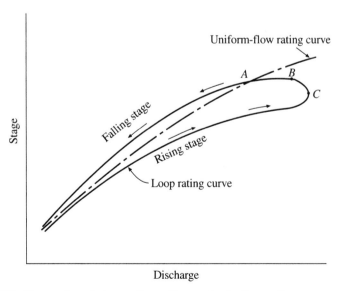

FIGURE 24.39 Comparison between the uniform, steady flow rating curve and the loop rating curve simulating the passage of a flood.

$Q_u < Q$. It is interesting to note that the maximum discharge of a flood (point C in Fig. 24.39) does not result in the maximum flow stage at point B.

Effect of Bed Forms. The flow resistance in natural streams is greatly influenced by the channel bed forms. A list of these bed forms is provided in Sec. 24.4 (Fig. 24.30 and Table 24.3) and in numerous other publications [e.g., Raudkivi (1990) and Vanoni (1975)]. The existence and specific type of bed form depends on several flow, sediment, and fluid parameters that include the dimensionless bottom (*Shields*) *shear stress*, $\tau^* = \tau_0/(\rho_s - \rho)gD_{50}$, the particle or shear Reynolds number, $\mathrm{Re}_P = u_*D_{50}/\nu$, the Froude number $\mathrm{Fr} = V/\sqrt{gh}$, and the relative flow depth h/D_{50} [Vanoni (1974)] where ρ_s is the density of solids. In the presence of bed forms, the channel resistance is typically considered to be made up of two parts—the surface resistance and the form resistance. The first is attributed to grain roughness and coincides with the resistance of a flat bed, while the second is attributed to the pressure difference around the bed form caused by flow separation. The change in water discharge through a stream is usually accompanied by a change in bed form type, which in turn changes the bottom shear or friction factor. This variation in the form, and total, drag is clearly demonstrated in Fig. 24.40 [e.g., Raudkivi (1990)]. This perplexing behavior of rivers may be beneficial to the environment and humans. During periods of low flows, the generation of ripples and dunes results in higher river stages which support flow diversion works and the biotic life within the stream, while during high flow events the bed forms disappear (at near bankfull conditions), thus minimizing channel resistance and flooding. Consequently, the friction factor in a given stream reach can vary by a factor of ten or more, as the bed undergoes

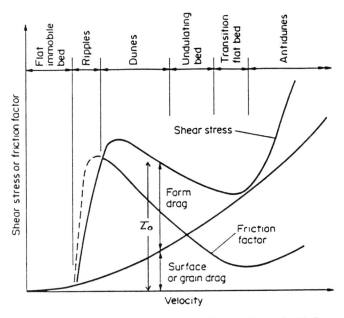

FIGURE 24.40 Variation of total bed stress τ_0 and friction factor f with flow velocity V for a sandy stream. Note also the changes in bed form type with increasing velocity and the attendant changes in the form drag contribution.

all these changes in its configuration. This dramatic variability of the friction factor produces multiple-valued and discontinuous depth-discharge (rating) curves for flows in sand-bed rivers (see Fig. 24.31).

A number of methods have been proposed for determining the friction factor of sandy streams at different flow conditions. The total boundary shear stress is typically partitioned into flat bed and bed form components, as mentioned earlier. A variety of ways have been suggested for calculating each component. Einstein (1950) was probably the first one to devise a complete methodology for obtaining large-discharge predictors by accounting separately for skin friction and form drag. The most popular of the plethora of approaches that have been published since then are summarized in sediment transport books [Simons and Senturk (1993), Raudkivi (1990), Garde and Ranga Raju (1985), and Vanoni (1975)]. The most recent methods for predicting channel resistance include those proposed by Brownlie (1983) and Karim and Kennedy (1990).

The resistance behavior of gravel streams is not as complicated, since the channel bed configuration is weakly formed and does not undergo the drastic changes observed in sandy streams. The gravel bars do not contribute significantly to the overall channel resistance [Parker and Peterson (1980)], which is mainly attributed to grain roughness. For this reason, the resistance relations that have been proposed for gravel streams are similar to those for a fixed-bed, like Eqs. (24.143) and (24.147) [Bray (1979) and Hey (1979)]. However, these relations may not be applicable to streams with high gradient slopes where the grain roughness elements are of the same size with the flow depth.

Finally, the resistance characteristics of compound channels and flow through vegetation constitute two additional and significant difficulties to the overall channel resistance problem. Considering that floodplains are usually vegetated, one needs to determine the combined effect of these two factors. The pronounced nonuniformity of the velocity across the channel cross section makes the use of a lumped resistance factor inappropriate in this case. This is a very important and challenging subject that has received considerable attention during the last decade. Some tentative quantitative results are now available in this area [e.g., Knight *et al.* (1994), Nezu and Nakagawa (1993), Shiono and Knight (1991), and Kadlec (1990)].

24.5.7 River Stability

Rivers are rarely straight for a distance longer than a few channel widths, and they do not have beds that are completely flat. Though a river with a flat bed and straight reach may be in an equilibrium state, it is very unlikely that this condition represents a stable equilibrium. Even if the bed is initially flattened by some artificial means, a turbulence burst, a random pile-up of sediment, or some other irregularity will intervene to create a local disturbance on the bed. This disturbance could subsequently be amplified and become a well-developed bed feature.

The remarkable variety of bed topography and planform geometry observed in natural streams, and partially reproduced in laboratory flumes, is the result of the complex interaction of the fluid flow with the channel boundaries and sediment transport. In their quest to predict the outcome of this interaction, researchers and engineers have resorted to different approaches. The first approach has emphasized the collection of data from field studies and laboratory experiments. Data plots in terms of suitable flow, sediment, and channel parameters are used to delineate the various morphological features of rivers and thus predict their occurrence. Some results for the river planform geometry case are included in Sec. 24.4, and for the bed form case are summarized by Vanoni (1975), (1974), Simons and Senturk (1992), and Raudkivi (1990).

The second approach is based on classical mechanics. Prediction of the conditions for the occurrence and characteristics of the various forms of fluvial morphology is sought by subjecting the governing equations for both fluid flow and sediment motion to a stability analysis. The formulation of the problem for the case of ripples, dunes, and anitdunes as well as the guidelines for their existence obtained from linear stability analysis are described by Richards (1980) and Kennedy (1969).

The problem of identifying the preferred mode of river planform geometry (straight, meandering, braided; see Fig. 24.27) has so far been approached theoretically by two types of stability theory, one for alternate bars and one for bends. Most bar-stability analyses treat shallow flow in a channel with straight, nonerodible, vertical banks and an erodible bed. The appropriate set of equations consists of the 2-D equations of fluid motion and continuity for sediment movement and fluid flow. Constitutive relations for bottom shear stresses and bed load transport in terms of flow parameters are also necessary. The wavelength of the bed disturbances is assumed to be longer than the wavelength of ripples and dunes so that the latter can be considered only as roughness elements. Because of the small growth rate and propagation speed of the bed disturbances, the flow is treated as quasi-steady. A stability analysis is carried out by introducing doubly sinusoidal perturbations for the two velocities, the flow depth, and the elevation of the free-surface into the balance equations. The solution of the resulting eigenvalue problem determines whether a doubly sinusoidal bed perturbation will grow or diminish for given flow conditions. Growth of the alternate bar pattern indicates instability and is considered to be the precursor of incipient meandering or braiding. It has been suggested that the presence of alternate bars may introduce the initial planimetric perturbations that could trigger instabilities due to the bend mechanism. In the bend-stability theories, 2-D flow in a mildly sinuous channel is used. In addition, the banks are allowed to erode, and channel stability is determined in terms of the growth rate of bend amplitude.

Linear and nonlinear stability analyses have been very fruitful areas of research regarding the prediction of river platform geometry. The last two decades have witnessed significant contributions in this area [Tubino and Seminara (1990), Ikeda and Parker (1989), Kuroki and Kishi (1985), Ikeda et al. (1981), Fredsoe (1978), and Parker (1976)]. An important conclusion of the stream stability analyses is that the preferred mode of instability depends on three parameters, dimensionless boundary shear (Shields) stress, τ^*, h/D, B/h, or any other equivalent set. This is in agreement with results obtained from dimensional analysis [Diplas et al. (1988)]. The parameter B/h is probably the most important of the three. Channels are usually stable and remain straight, for $(B/h) \leq 8$. Kuroki and Kishi (1985) chose τ^*, S_0, and B/h as independent parameters and suggested, based on data, that they can be reduced to the following two, τ^* and $(B/h) S_0^{0.2}$. The stability diagram obtained from this theory is tested against field and laboratory data in Fig. 24.41 [Diplas et al. (1988)]. The theory compares favorably with the data. Other stability analyses due to Fredsoe (1978) and Ikeda and Parker (1989) provide equally good agreement. Models de-

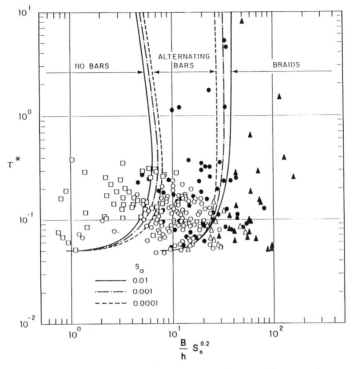

FIGURE 24.41 Straight-Meander-Braided stability diagram. Squares denote ripples and dunes; circles alternate bars, and triangles double-row bars. Solid symbols represent field data and open symbols represent laboratory data. (From Diplas *et al.*, 1988.)

scribing the flow-sediment interaction in weakly and well-developed meander bends are included in Ikeda and Parker (1989) and Falcon (1984). Such models provide estimates of the meander bend evolution and its simultaneous migration in the downstream direction.

24.6 FLUID MECHANICS IN PLANT BIOLOGY
Richard H. Rand and J. Robert Cooke

24.6.1 Introduction

In this section, we describe a variety of phenomena associated with fluid flow in green plants and summarize some attempts to model these effects. We begin by giving a schematic engineering description of the biofluid mechanics of a green plant. The leaves, vascular structure, and roots of a green plant may be thought of as functioning to support the conversion of energy from sunlight into the chemical energy stored in sugars via the process of *photosynthesis*. The chemistry of photosynthesis requires sunlight, CO_2 and water and produces glucose (a simple sugar) and O_2. This process takes place in the leaves, and the manufactured sugars are *translocated* to other parts of the plant via the vascular *phloem* tissue. The water required for photosynthesis is absorbed through the roots, and the necessary CO_2 is found in the ambient atmosphere and enters the leaf by diffusion. Note the contrast with gas exchange in humans, where muscles power breathing, producing a mass flow of fresh air into our lungs. Diffusion is driven by a concentration difference and is effective only over short distances, which is why leaves are relatively thin structures. Accompanying the entry of CO_2 into the leaf is the loss of water vapor by evaporation and diffusion. This loss tends to dessicate the plant, and water lost by evaporation in the leaves is replenished via the vascular *xylem* tissue, resulting in an upward flow called the *transpiration stream*. In contrast to the human circulatory system, which is closed, the vascular system of plants is open. It involves extensive branching at both the leaves and roots. The roots function to absorb the necessary water from the soil, as well as to absorb minerals, store carbohydrates, and anchor the plant in the soil.

A variety of sources are available for an introduction and overview of the engineering aspects of green plants. Nobel (1974, 1991) contains an extensive quantitative introduction and order of magnitude analysis. Merva (1975) presents a shorter quantitative introduction. Meidner and Sheriff (1976) and Milburn (1979) offer introductions which use engineering concepts with a minimum of mathematics. Canny's (1977) brief nonmathematical introduction is aimed at fluid mechanicians. The present authors have attempted several brief summaries. Cooke and Rand (1980) reviewed leaf diffusion models, Cooke (1983) summarized stomatal dynamics and gas exchange, and Rand (1983) presented an overview of fluid dynamics in green plants. Niklas (1992) provides a review of plant biomechanics in relation to plant form and function. The general context for fluid motion in plants is covered in the introductions to plant physiology by Galston *et al.* (1980), Taiz (1991), and Salisbury and Ross (1992). Greater detail is provided in the specialized surveys by Crafts and Crisp (1971), Esau (1977), Jarvis and Mansfield (1981), Zimmerman (1983), Zeiger *et al.* (1987), and Fahn (1990).

24.6.2   Basic Concepts

Water moves in plants as a result of gradients in chemical concentration (cf. *Fick's law*), hydrostatic pressure, and gravitational potential. Plant physiologists have found it convenient to deal with these diverse effects by using a single quantity, a chemical potential called the *water potential*, ψ [Nobel (1974), (1991)]

$$\psi = p - RTc + \rho gz \qquad (24.155)$$

where p is hydrostatic pressure (bar), R is the gas constant $= 83.141$ cm^3-bar/mol K, T is temperature (K), c is concentration of all solutes in an assumed dilute solution (mole/cm^3), ρ is the density of water (g/cm^3), g is the acceleration of gravity $= 980$ cm/sec^2, and z is height (cm). Here, ψ is in bars, a convenient unit commonly used in plant studies for measuring pressure. (One bar equals 10^6 dyne/cm^2 and is approximately equal to one atmosphere.)

An individual plant cell consists of a cell wall surrounding a cell membrane (the *plasmalemma*), inside of which lies the cell protoplasm (see Fig. 24.42). In order for the cell to be in equilibrium with its surrounding medium, the water potential inside the cell must equal the water potential outside the cell. However, since the plasmalemma is able to maintain a concentration difference between the interior and

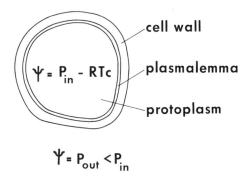

FIGURE 24.42 Schematic diagram of a typical plant cell. For equilibrium, the water potential, ψ, inside the cell must equal that outside the cell.

the exterior of the cell, the hydrostatic pressure inside the cell can be larger than that outside the cell from Eq. (24.155). This situation (of which there is no parallel in the case of animal cells) is resolved by the elastic extension of the plant cell wall, creating a *turgor* (hydrostatic) pressure inside the cell.

An important concept for understanding the flow of water in plants is the distinction between the *symplasm* and the *apoplasm*. The symplasm consists of all the protoplasm (inside the plasmalemma) of all the living cells of the plant, together with the plasmodesmata (thin strands of cytoplasm that go from the interior of a given cell, through the cell wall, and into the interior of a neighboring cell). In terms of *point set topology*, the symplasm is thought to be a *connected set*. The apoplasm consists of those regions of the plant that contain water and are not in the symplasm. In particular, the apoplasm includes the xylem (which consists of dead cells), as well as the fluid in the cell walls of all the cells of the plant. Flow in the symplasm has been estimated to involve a resistance about 50 times as large as that in the apoplasm [Meidner and Sheriff (1976)], although more recent work contests this view [Canny (1993)].

24.6.3 Stomata

A typical leaf of a green plant shown in Fig. 24.43 is filled with *mesophyll cells* which contain the chlorophyll necessary for photosynthesis. These cells are sandwiched between the outer layers of the leaf which consist of epidermal cells covered with a layer of waxy material called *cutin*, which inhibits the loss of water through the surface of the leaf. CO_2 enters the leaf through small holes in the epidermis called *stomata* (or stomates) [Meidner and Mansfield (1968)]. An individual stomate is composed of two specialized cells called *guard cells*. Figure 24.43(c) illustrates the kidney-shaped stomata. A bar–bell shape is common in grasses. Like all cells in a green plant, the guard cell has a cell wall in addition to the cell membrane present in animal cells. This gives the plant cell elastic stiffness, an effect which is enhanced if the cell is *turgid*, i.e., if the cell membrane (or plasmalemma) exerts a hydrostatic pressure on the cell wall due to a solute concentration difference between the interior and the exterior of the cell. In the case of a guard cell, the change in shape upon hydrostatic loading results in a change in the size of the stomatal pore.

When the stomatal pore is open, water vapor diffuses out of the leaf, because the water concentration within the plant is normally higher than in the ambient air. When CO_2 is not needed for photosynthesis, e.g., at night, the stomata can close, acting

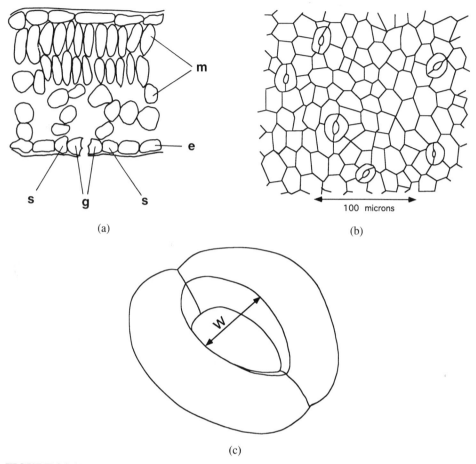

FIGURE 24.43   Typical leaf of a green plant. (a) Schematic diagram of a transverse section of a leaf (after Nobel, 1974). A representative value for leaf thickness is 300 μm. m = mesophyll cell, e = epidermal cell, g = guard cell, and s = subsidiary cell. (b) Stomatal pores on the leaf surface and (c) a pair of guard cells surrounding a stomatal pore; w = pore width.

like valves in a feedback control system to limit water loss. They similarly close when the plant is experiencing *water stress*, i.e., extreme dryness due to lack of rain. For this reason, a great deal of attention has been given to studies of related phenomena, such as diffusive flow through the stomatal pores, the elastostatics of guard cells, and the dynamical behavior of stomata.

How do stomata open and close? Many older plant physiology texts explain this by stating that the walls of the guard cells next to the pore are thicker than the walls on the side away from the pore. They imagine that upon inflation, the outer wall buckles outward, pulling the rest of the guard cell with it and opening the pore. Here is a situation where an engineering approach can clarify things. Cooke *et al.* (1976) have modeled the guard cell as a linear anisotropic thin shell. Using finite element analysis they showed that an increase in hydrostatic pressure in the guard cell (for fixed pressure in neighboring subsidiary cells) tended to open the pore. On the other

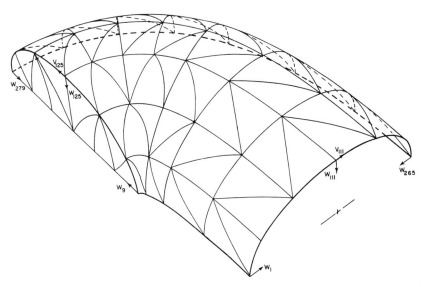

FIGURE 24.44 A finite element discretization of a quadrant of a guard cell. (Reprinted with permission, Cooke, J. R., Debaerdemaker, J. G., Raud, R. H., and Many, H. A., "A Finite Element Shell Analysis of Guard Cells Deformations," *Trans. ASAE*, Vol. 19, pp. 1107–1121, 1976.)

hand, an increase in pressure in neighboring subsidiary cells (for fixed pressure in the guard cell) tended to close the pore (see Fig. 24.44). It was shown that the elliptical shape of the guard cell is critical for opening, and that other features such as wall thickening and radial stiffing (due to the micellae, i.e., radial cellulose microfibrils) could help the opening process, but were not essential for static opening. The effect of the radial stiffing is to permit a smaller pressure fluctuation in the subsidiary cell to counteract a larger pressure fluctuation in the guard cells during the pore opening and closing cycle. The work on the structural analysis was extended by Cooke *et al.* (1977) to include nonlinear effects. When the geometric nonlinearities are included, the pore width, as expected, is shown to plateau as pressure within the guard cell increases.

A problem related to the gaseous fluxes in the leaf concerns the dynamic behavior of the stomatal apparatus. Experimental observations have revealed that the width of the stomatal pore often oscillates, typically with a period ranging from 10 to 50 min. Delwiche and Cooke (1977) modeled this phenomenon by balancing water fluxes between the guard cell, the subsidiary cell, and the rest of the plant. The gaseous flux through the stomatal pore can oscillate periodically under stress conditions—even when all ambient conditions remain constant. The stomatal apparatus acts as a feedback control system. The oscillatory gaseous flux may be described as follows. Water evaporating from the wet mesophyll and subsidiary cell walls diffuses through the stomatal pore to the leaf exterior. This water is replaced both by a flux from the roots via the xylem and by a flux from the guard cells to the subsidiary cells. The resulting decrease in hydrostatic pressure in the guard cells causes the stomatal pore width to decrease [Delwiche and Cooke (1977)]. A smaller pore width slows the rate of evaporation, increasing the water potential in the mesophyll and causing water to accumulate there. In response to this accumulation, water diffuses back to the guard cells, increasing their hydrostatic pressure and increasing the pore width. The model takes the form of an *autonomous* system of two first-order ordinary differential equations for p_g and p_s (the pressures in the guard and subsidiary cells). The resulting flow in the $p_g - p_s$ plane exhibits a *limit cycle* as illustrated in Fig. 24.45. Note that a limit cycle is a strictly nonlinear phenomenon and is due here to the piece-wise linear nature of the relation between pore width w and p_g, p_s. The piece-wise linearity follows from the fact that w can never become negative (even with a linear elastic model of the guard cell).

This work was extended by Rand *et al.* (1981) by embedding the original system of Delwiche and Cooke (1977) into a one-parameter family of systems. It was found that as the parameter (which represents the concentration of the osmotically active solutes in the guard cell) is varied, the dynamical properties exhibited by the system change (see Fig. 24.46). The system was shown to contain a *Hopf bifurcation* [Rand *et al.* (1981) and Rand (1994)] that involved the genesis of an unstable limit cycle. The oscillatory behavior was seen as a kind of dynamical bridge between the open and closed pore equilibrium states. Upadhyaya *et al.* (1980a,b) extended the Delwiche and Cooke (1977) model by including CO_2 feedback effects. This involved modeling the guard cell biochemistry in order to include a CO_2 sensor in the system. The model displayed a limit cycle oscillation, which involved a 2 min CO_2—based oscillation superimposed on the 20 min. water-based oscillation previously discussed (and shown in Fig. 24.45), in agreement with the experimental observations of other investigators.

Oscillatory plant transpiration has been observed in numerous plants and explored experimentally. Gumowski (1981) suggested that the oscillatory behavior avoids prolonged high rates of stomatal transpiration and reduces the internal energy expenditure required for the maintenance of turgor. Upadhyaya *et al* (1981), (1988),

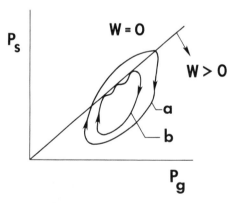

FIGURE 24.45 Limit cycles representing stomatal oscillations in the p_g-p_s plane. Here p_g and p_s represent hydrostatic pressures in guard and subsidiary cells respectively, and w represents pore width. The region above the straight line corresponds to a closed pore ($w = 0$). The arrow perpendicular to the straight line shows direction of increasing pore width. Path a—stomatal oscillation due to hydraulic feedback loop only. (From Delwiche and Cooke, 1977.) Path b—stomatal oscillation due to both hydraulic and CO_2 feedback effects. (From Upadhyaya *et al.*, 1980a,b.)

AMPLITUDE

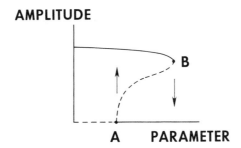

A   PARAMETER

FIGURE 24.46 Changes in the amplitude of the stomatal oscillation of Fig. 24.45 due to changes in a system parameter. (From Rand *et al.*, 1981.) Zero amplitude corresponds to equilibrium behavior. The dashed and solid lines correspond to unstable and stable motions respectively. At point *A* an unstable equilibrium point becomes stable and throws off an unstable limit cycle (a Hopf bifurcation.). At point *B* a stable and an unstable limit cycle coalesce. Arrows represent jump phenomena.

when considering the carbon dioxide feedback loop, independently and simultaneously proposed that the oscillatory behavior reduces water losses while preserving carbon dioxide uptake at minimal levels.

Why do stomata oscillate? That is, in terms of Darwinian evolution, of what advantage to the plant are stomatal oscillations? Upadhyaya *et al.* (1981) investigated this question by comparing gaseous fluxes through a stomatal pore in an open equilibrium state with fluxes through an oscillating pore. For typical values of the system parameters, they found that stomatal oscillations tend to conserve water under relatively dry, i.e., water stressed atmospheric conditions. However, this savings in moisture content occurs at the expense of a reduction in the CO_2 assimilation rate.

Experimental work [Ellenson and Amundson (1982) and Mott *et al.* (1993)] has shown that the process of opening and closing of stomata varies across the leaf surface and involves complicated dynamics taking the form of *waves* of stomatal opening. This organized behavior implies fluid dynamical coupling between the individual stomata. This has been modeled [Rand *et al.* (1982), Rand and Ellenson (1986), and Rand (1987)] by permitting the water potential, ψ, in the leaf to be a field quantity, i.e., $\psi = \psi(x, y, t)$ where x and y measure the position of a point on the leaf surface. ψ satisfies a diffusion equation with a distributed sink term, $Q(w)$, corresponding to the loss of water vapor due to evaporation through open stomatal pores

$$\frac{\partial \psi}{\partial t} = D\left[\frac{\partial^2 \psi}{\partial^2 x} + \frac{\partial^2 \psi}{\partial^2 y}\right] - Q(w) \tag{24.156}$$

where again w is the local pore width. The function $Q(w)$ is essentially nonlinear, since water loss Q is a monotone increasing function of w for $w > 0$, while $Q = 0$ for $w < 0$. The partial differential equation on ψ is accompanied by boundary conditions such as no flux at the leaf edge, or given ψ at the leaf midrib. The dynamics of an individual stomate have been modeled in a variety of ways [Rand *et al.* (1982), Rand and Ellenson (1986), and Rand (1987)], but all involve coupling to other sto-

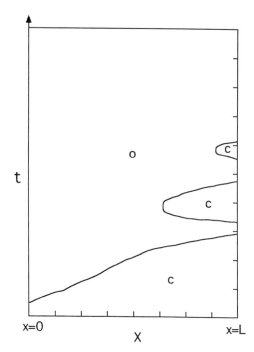

FIGURE 24.47 Pore state displayed as O = open or C = closed in the x–t plane. Here $x = 0$ corresponds to the midrib and $x = L$ to the leaf edge. (From Rand, 1987.)

mata, e.g., by dependence on local water potential, ψ. As an example of the pre-diction of such a model, imagine a leaf which has all its stomata closed, e.g., due to a condition of water stress at the roots, when suddenly the availability of water is increased. This corresponds in the model to a jump in the boundary value of ψ and may result in the eventual opening of all the stomata on the leaf surface. How-ever, the dynamical behavior is wave-like, and may involve *echoes*, as shown in Fig. 24.47 [Rand and Ellenson (1986) and Rand (1987)].

Before leaving this topic, we note that Cooke *et al.* (1988) have provided a com-puter-based introduction to stomatal dynamics.

24.6.4 The Transpiration Stream

Water which enters the plant through the roots rises through the xylem vascular tissue to the leaves, where it evaporates. A natural question concerns how it is pos-sible for the transpiration stream to reach the tree top. Since the pressure exerted by the atmosphere corresponds to a gravitational head of only about ten meters, this process cannot be explained by the creation of a vacuum somewhere in the plant. In fact, the explanation lies in the leaves, where the site of evaporation is the menisci in the mesophyll cell walls. These liquid-air interfaces are bounded by the strands of cellulose that constitute the cell wall. A representative interfibrillar space has a *diameter* of about 0.01 μm [Nobel (1974)]. The pressure difference across a spher-ical meniscus is given by

$$\Delta p = 2\sigma/r \qquad (24.157)$$

where σ is the surface tension coefficient = 73 dyne/cm for an air–water interface at 20°C and r is the radius of curvature of the meniscus (cm). Here, Δp is about

300 bars or about 3000 meters of gravitational head, thus easily accounting for the ascent of water to the tops of the highest trees. Of course, this requires that the continuous fluid column reaching from the roots to the leaves be under considerable tension. Although the theoretical tensile strength of a perfect column of water greatly exceeds 300 bars [Hammel and Scholander (1976) and Nobel (1974)], the presence of small air bubbles and other imperfections reduces the observed tensile strength in laboratory experiments. Nevertheless, the plant is evidently able to maintain a vascular system relatively free from air bubble defects. The phenomenon of xylem cavitation has been studied extensively by Tyree (1991).

The dynamics of a spherical evaporating meniscus has been studied by Rand (1978a). The analysis involved a nonlinear differential–integral equation and predicted damped oscillatory motions for a certain range of parameter values.

The flow of water along the transpiration stream in the leaf proceeds through the branching xylem system to the xylem termini, and it then continues through the apoplastic mesophyll cell walls and symplast to those mesophyll cells near the stomatal pore where evaporation occurs. Stroshine *et al.* (1979, 1985) have studied flow in the leaf and stem. This involves consideration of branching xylem vessels of various sizes as well as a diffusive flow between the xylem and the leaf symplasm (the interior of the mesophyll cells). It was concluded that the large vascular bundles offer relatively little resistance to flow compared with the intermediate and small bundles. This shows the advantage of a branching structure to the vascular system.

24.6.5 Gaseous Diffusion

The problem of gas exchange in leaves was first investigated from a mathematical point of view in a classic paper by Brown and Escombe (1900). By modeling the leaf surface as a plane septum with a circular hole and the pore as a circular cylinder, they explained the experimentally observed relatively large rates of transpiration from leaves (comparable to evaporative fluxes from an equal-sized body of water). Bange (1953) used an approximate analysis in order to consider a realistic geometry for the leaf interior as well as a still air layer outside the leaf. He found that as the wind speed increased, i.e., as the thickness of the still air boundary layer decreased, the stomata played an increasingly important role in controlling gaseous fluxes. Although a wider pore always results in a larger flux, this effect was shown to be negligible for relatively thick boundary layers. Cooke (1967) considered diffusion through an elliptical pore. Using a relationship involving complete elliptic integrals, he showed that a slightly open stomate can permit relatively large diffusion rates. For example, an ellipse with a major to minor axis ratio of 20 has a discharge rate that is 39% of that of a circle of diameter equal to the major axis. Cooke (1969) and Holcomb and Cooke (1977) considered the interaction effects between neighboring stomatal pores. Using separation of variables, they showed that the flux depends on both the spacing between stomata on the leaf surface and the boundary-layer thickness; Fig. 24.48. In particular, the overall diffusion rate from a leaf was shown to not depend linearly on the area of the pores.

Is there an optimal spacing for stomata that will maximize gaseous exchange? Holcomb and Cooke (1977) sought to answer this question by using the analogy between diffusion and the flow of electric current in an aqueous electrolyte solution. They built an electrolytic tank (copper sulphate in a copper and plexiglass container) and used it to study the effects of pore eccentricity, stomatal spacing, boundary-

axis of symmetry

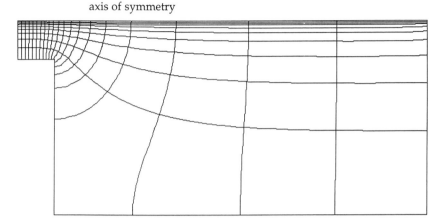

FIGURE 24.48(a)   Diffusion from neighboring stomatal pores. Surfaces of constant partial pressure and lines of diffusion for a stomatal pore modeled as a circular cylinder (at bottom of figure). The atmosphere adjacent to the leaf is also modeled as a circular cylinder (located above the stomatal pore). Concentrations are fixed at the bottom and top surfaces of the figure, and zero flux is assumed at the radial boundaries and at the leaf surface.

layer thickness, and pore depth (see Fig. 24.49). The diffusion rate (per unit leaf area, not per pore) increases monotonically with increasing pore width, with decreasing pore spacing and with decreasing boundary layer thickness. Thus, there is no optimal pore spacing to maximize diffusion rate *per unit leaf area*. In a mathematical sense, as the pore area increases, the passive diffusion rate can approach, but not exceed, that of a free body of water. Note that a relatively thick boundary layer causes the stomate to function as an on–off valve. Parlange and Waggoner (1970) used conformal mapping to study diffusion through a two-dimensional slit. They compared their results with the formula of Brown and Escombe (1900) and found the approximation inherent in the latter to be best for thin, deep slits.

Most studies of gaseous diffusion in leaves have assumed the *cuticle* (i.e., the cutinized epidermal surface) to be completely impermeable. Although diffusion oc-

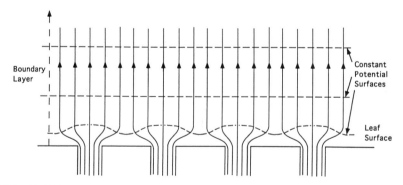

FIGURE 24.48(b)   Sketch of diffusion streamlines away from an array of stomates. (Reprinted with permission, Holcomb, D. P. and Cooke, J. R., "An Electrolytic Tank Analog Determination of Stomatal Diffusion Resistance," ASAE Paper No. 77-5510, 1977.)

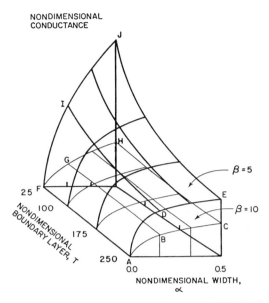

FIGURE 24.49 A nondimensionalized display of stomatal diffusion rate as a function of nondimensional pore width (α), boundary-layer thickness (T), and stomatal spacing, (β). [Reprinted with permission, Cooke, J. R. and Rand, R. H., "Diffusion Resistance Models," *Predicting Photosynthesis for Ecosystem Models*, Hesketh, J. D. and Jones, J. W. (Eds.), Vol. 1, pp. 93–121, CRC Press, 1980.]

curs primarily through the stomata, there are some species, such as ferns (which have a very thin cuticle), in which 30% of the total diffusion rate can occur through the cuticle. Hsu and Ganatos (1983) have modeled this situation as a boundary value problem. They found that cuticular diffusion was decreased as the stomata were spaced closer together.

Current treatments of gaseous diffusion in the leaf [see, for example, Nobel (1991)] utilize a one-dimensional model which, by analogy with Ohm's law, involves a series of resistances, each associated with a portion of the pathway. This is depicted in Fig. 24.50. Parkhurst (1977) compared a three-dimensional field equation approach with the commonly used one-dimensional resistance model and found that the latter involved an error of 44%. Nearly all studies of leaf diffusion have assumed steady-state diffusion. Gross (1981), however, included time-dependent terms in order to estimate the time scale of the gaseous diffusion process. He found equilibrium to be essentially attained in less than one second.

Webster (1981) has applied the concept of the *effectiveness factor* to leaf diffusion in order to gauge the extent to which assimilation is diffusion limited. This factor is defined as the ratio of the actual assimilation rate to the assimilation rate that would occur in the absence of any CO_2 concentration gradients. An effectiveness factor of unity indicates that assimilation is kinetically limited, while a value considerably smaller than unity indicates that losses due to diffusion are significant.

Open stomatal pores allow CO_2 to diffuse into the leaf interior where it is converted into sugars by photosynthesis. At the same time, water vapor may evaporate from the wet mesophyll cell walls and diffuse out of the leaf interior into the neigh-

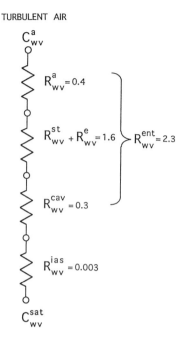

TURBULENT AIR

FIGURE 24.50 Electrical circuit analogy for steady state diffusion of water vapor in a leaf. [Reprinted with permission, Cooke, J. R. and Rand, R. H., "Diffusion Resistance Models," *Predicting Photosynthesis for Ecosystem Models*, Hesketh, J. D. and Jones, J. W. (Eds.), Vol. 1, pp. 93–121, CRC Press, 1980.] C_{wv}^a = concentration of water vapor in the turbulent atmosphere adjacent to the leaf. C_{wv}^{sat} = concentration of water vapor at the mesophyll cell evaporation sites. R_{wv} = resistance values in sec/cm. Superscripts: a = boundary layer adjacent to leaf, st = stomatal pore, e = end effect correction, cav = substomatal cavity, ent = entry region, and ias = intercellular air space.

WET CELL WALLS

boring atmosphere. Although these two diffusion processes appear to be entirely analogous, experimental work [Aston and Jones (1976), Tyree and Yianoulis (1980), and Canny (1993)] has shown that water vapor actually evaporates only from those cell walls which are near the stomatal pore, whereas CO_2 is absorbed into cells throughout the interior of the leaf, including those cell walls which are far from the stomatal pores. Although the diffusion coefficient of water vapor in air is about 1.6 times that of CO_2 in air, this difference (due to the different atomic weights of water vapor and CO_2) is insufficient to account for their difference in performance. Indeed, plant physiologists have been led to conjecture that the mesophyll cell walls far from the stomatal pores are covered with a special layer of cutinized material which inhibits the evaporation of water. No such assumption is needed, however, since the difference in behavior can be explained by the different roles that water vapor and CO_2 play in the physical chemistry of the leaf. Since the liquid in the cell walls of the leaf is a dilute aqueous solution in which water is the solvent and CO_2 the solute, the concentration of water vapor is governed by Raoult's Law, whereas that of CO_2 is governed by Henry's Law. This results in similar differential equations for the gas concentrations c of water vapor and CO_2 [Rand (1977a,b)]

$$\frac{\partial^2 c}{\partial x^2} - \alpha^2 c = 0 \qquad (24.158)$$

where x measures distance along an intercellular pathway and α is a constant that is about 50 times larger for water vapor than for CO_2. For suitable boundary conditions and typical leaf dimensions, this means that most water evaporation occurs within the first 4% of the pathway near the substomatal cavity, whereas 87% of the pathway is needed for comparable CO_2 absorption.

After diffusing as a gas to the mesophyll cell walls, CO_2 continues to diffuse as a solute to the chloroplasts in the cell interior. Sinclair *et al.* (1977) and Sinclair and Rand (1979) have modeled this process by assuming spherical cell geometry and Michaelis–Menten reaction kinetics [Thornley (1976)]. The resulting nonlinear ordinary differential equation for CO_2 concentration as a function of radial position was solved approximately by perturbation methods. Expressions for the rate of CO_2 assimilation by a single cell were obtained in terms of cell size and biochemical parameters. This spherical cell model was incorporated into a more comprehensive model for CO_2 assimilation by Rand and Cooke (1980). The model took account of the gradual absorption of CO_2 into the mesophyll cell walls as CO_2 diffuses inward (i.e., diffusion with a distributed sink), as well as the effects of variation in cell-packing density. An approximate formula for CO_2 flux into the leaf in terms of basic geometrical and biochemical parameters was obtained by perturbations.

24.6.6 Flow in the Xylem and Phloem

The main conduits for fluid flow in the stem of plants consist of the xylem, through which the transpiration stream flows, and the phloem, through which sugars produced in the leaves are translocated to other parts of the plant. The conduits of plants are formed by individual plant cells placed adjacent to one another. During cell differentiation the common walls of two adjacent cells develop holes [called *pits* or *pores*; see Esau (1965)], which permit fluid to pass between them. The xylem contains *tracheids* and *vessel elements* (Fig. 24.51) that die after reaching maturity, while the phloem contains *sieve elements* that remain metabolically active. The Reynolds number for flow in the xylem is about 0.02 [Rand (1983)], which is best modeled by slow viscous (*creeping*) flow [Happel and Brenner (1965)], in which the inertia terms are neglected in the Navier–Stokes equations. The plant physiologist needs to know the pressure drops involved in flow through the vascular tissue. Such questions arise, for example, in the evaluation of various conjectured mechanisms for driving the phloem flow.

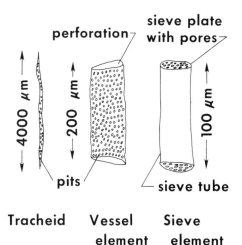

FIGURE 24.51 Fluid-conducting cells in the vascular tissue of plants (after Easu, 1965). Tracheids and vessel elements are found in the xylem, while sieve elements are found in the phloem. Here and in the rest of this paper, the dimensions given are typical but do not represent statistical averages.

The fluid mechanics of phloem flow has been considered by Rand and Cooke (1978) and Rand *et al.* (1980). As shown in Fig. 24.51, this involves flow through a series of cylindrical sieve tubes separated by perforated sieve plates. Due to the mathematical complexities of slow viscous flow, only the relatively unrealistic axi-symmetric case of a single pore has been considered. The boundary value problem involved the field equations

$$\nabla p = \mu \nabla^2 \bar{v} \qquad (24.159)$$

$$\nabla \cdot \bar{v} = 0 \qquad (24.160)$$

with $\bar{v} = 0$ on the boundary, which was modeled as two circular cylinders in series, periodically repeated. The solution involved modified Bessel functions of the first kind, which turned out to be nonorthogonal. The results of the analysis were compared with Poiseuille's Law (which provides the standard approach currently used by plant physiologists). Poiseuille's Law, when applied to the sieve tube and the pore in series, was found to underestimate the exact pressure drop by about a factor of two.

Flow between two neighboring xylem tracheid cells occurs through pits (see Fig. 24.51). A typical bordered pit (in a conifer) consists of a circular border that arches over the pit cavity and contains a closing membrane (Fig. 24.52). The closing membrane is composed of a thick central region, which is relatively impermeable to the flow of fluid and a thin perforated peripheral region through which flow is possible. In nature, the bordered pit is found in both open and closed states. In the open state, flow is possible from one tracheid to another, while in the closed state virtually no flow occurs through the pit. This problem was studied by Chapman *et al.* (1977) by assuming an ideal fluid and using conformal mapping. The thin peripheral region of the closing membrane was modeled as linear springs, and equilibrium for a given flow rate was obtained by balancing the net hydrodynamic force on the central region of the closing membrane with the elastic restoring force of the peripheral region.

(a)

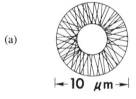

$\vdash$ 10 μm $\dashv$

FIGURE 24.52 Schematic diagram of a bordered pit found in xylem tracheid cells. (a) Top view. The closing membrane is composed of a thick central region and a thin peripheral region. (b) Side view. A circular border arches over the pit cavity and contains the closing membrane. Pit is open. (c) Pit is closed. (d) Two-dimensional hydrodynamical model. (From Chapman *et al.*, 1977.) The dashed and solid lines represent initial and displaced positions respectively.

(b)

(c)

(d)

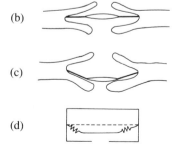

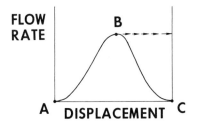

FIGURE 24.53   Results of the analysis of the model in Figure 24.52(d) (from Chapman *et al.*, 1977). Points *A* and *C* correspond to zero and maximum displacement respectively. As the flow rate is increased, the displacement of the membrane is increased until point *B*, after which the pit snaps shut (arrows). The equilibrium states on the curve *BC* are unstable.

Figure 24.53 shows the results of this analysis. It was found that for a given flow rate through the pit there are two equilibrium displacements, one stable and the other unstable. As the flow rate is increased to a value larger than the maximum permissible (see Fig. 24.53), the pit snaps shut. Thus, the pit functions as a valve to limit the flow rate in the xylem pathway.

A problem related to flow in the vascular system concerns observed daily changes in stem diameter accompanying changes in the rate of transpiration. The phenomenon is explained in terms of a decrease in the water content of cells near the xylem tissue resulting from an increase in the rate of transpiration. In order for the transpiration stream to flow, there must be a negative gradient in water potential from the roots to the leaves. This gradient reduces the value of the water potential at all points in the xylem (compared with values corresponding to zero transpiration). This, in turn, causes a decrease in water potential inside a typical cell near the xylem tissue throughout the stem and, accordingly, reduces the cell's turgor pressure and the associated elastic extension of the cell wall. As a result, the size of the cell and the diameter of the stem are decreased. Molz and Klepper (1972) studied this problem by assuming radial diffusion of water potential, a concept first discussed by Philip (1958a,b,c). They obtained good agreement with experimental observations and were able to explain an observed hysteresis loop in the stem diameter–leaf water potential relationship. Their work was extended by Parlange *et al.* (1975), who considered a variable diffusion coefficient and a corresponding nonlinear diffusion equation.

The flow of water in the parallel symplasm and apoplasm pathways has been described by a pair of coupled diffusion equations [Molz (1976), Molz and Ikenberry (1974), and Hornberger (1973)]. The coupling represents the flow between the symplasm and the apoplasm and depends upon various resistances in the model. Molz (1976) has applied these equations to a boundary-value problem representing the immersion of a sheet of tissue initially in equilibrium into a bath of pure water. Aifantis (1977) has decomposed the flow in the apoplasm into two components representing flow in the xylem vessels and flow in the cell walls. His treatment, based on the modern theory of continuum mechanics, neglects viscous effects and results in two coupled diffusion equations. Unger and Aifantis (1979) have applied this theory to a boundary-value problem representing flow in a cylindrical stem. Flow in the plasmodesmata of the symplasm has been studied by Blake (1978). An indi-

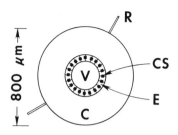

FIGURE 24.54 Schematic diagram of a tranverse section of a root about one cm. from the root tip (after Nobel, 1974). R = root hair, C = cortex, E = endodermis, CS = casparian strip, and V = vascular tissue.

vidual plasmodesma has an internal diameter of about 0.05 μm and a length of about 1 μm. Thus, the scale of this work is much smaller than many of the other problems considered in the biofluid mechanics of plants.

24.6.7 Flow in the Root

Figure 24.54 shows a schematic diagram of a transverse section of a root. Water is absorbed from the soil through the many root hairs (the presence of which greatly increases the absorbing surface area of the root) and flow radially inward across a region of storage tissue called the *cortex*, toward the xylem in the centrally located vascular tissue. Between the cortex and the vascular tissue, however, lies the endodermis, a single layer of cells that are separated from one another by an impermeable barrier called the *casparian strip*. Water must pass through the symplasm of the endodermal cells in order to enter the vascular tissue. Thus, the endodermis and casparian strip locally divide the apoplasm into two disconnected regions. Although the exact role of the endodermis is uncertain, it may function as a filter, selectively absorbing minerals, and it may be the site of observed changes in the plant's resistance to water flux, permitting absorption to occur more readily when the soil is less moist. Once the absorbed water reaches the xylem, it flows axially. See Newman (1976) for a summary of flow in the root.

Unlike the leaf, the root has received relatively little attention from fluid mechanicians. The usual approach has been to use a lumped system resistance–capacitance electric circuit analog [Seaton and Landsberg (1978)]. Although such models yield reasonable estimates for overall plant water fluxes, they do not take account of the geometry of the root. Of greater fluid mechanical interest are the following models, which involve a field-theory approach. Molz (1975) considered radial diffusive flow in a cylindrical root surrounded by a cylindrical region of soil. Continuity of water potential and of water flux were assumed at the soil-root interface. The study indicated that water potential gradients in the soil are small compared with those in the root, except under very dry soil conditions. Landsberg and Fowkes (1978) considered both radial absorption and axial diffusion of water along the length of a root. Their model predicted the value of the plant water potential at the base of the plant necessary to sustain a given flow rate through a root system with given characteristics. An expression was obtained for the optimal root length such that the overall root resistance to water is minimized. It is interesting to note that the mathematical statement of this problem is identical to that used to describe the assimilation of CO_2 in the intercellular air pathway of a leaf [Rand (1977a,b)].

24.6.8   Circumnutation

The phenomenon of *circumnutation* involves the circular motion of the stem of certain vine-like plants, providing them with a way to search for an external support to twine around. Experimental work on this problem goes back to Darwin (1865). Periods of oscillation have been found to range from one to twenty hours. Circumnutation is not a diurnal (daily) phenomenon, and it persists in the dark and in the gravity-free environment of the space shuttle [see Lubkin's (1992) thesis for references to the experimental literature]. Lubkin and Rand (1994) and Lubkin (1992, 1994) developed a mathematical model for circumnutation which involved reaction diffusion equations in the cross-section of the stem modeled as a hollow cylinder. The model supported waves of potassium concentration moving around the circumference of the stem, with water passively following the potassium ions. The side of the stem with high potassium ions would attract more water than the diametrically opposite side, and the corresponding unequal presence of water would cause the stem to bend. The model was used to study the *chirality* of circumnutation, i.e., the observed preference which many plants have to circumnutate in a preferred direction (i.e., clockwise versus counterclockwise).

24.6.9   Conclusions

As in other branches of biomechanics, research work on plants involves greater emphasis on modeling than does work in more traditional areas of mechanics. The researcher is presented with the biological description of the phenomenon to be studied and must invent an appropriate boundary-value problem to represent it. The modeling process has nothing like a unique solution, and the same physical problem can be treated in many different ways, each yielding some additional information. For example, the mesophyll cells in the interior of the leaf have been modeled by spheres [Rand and Cooke (1980) and Sinclair and Rand (1979)], by cylinders [Rand (1978b)], and by the exterior of a cylinder [Rand (1977a,b)]. The strategies encountered in biomechanical modeling involve a great deal of freedom of choice, more akin to sculpture or painting than to the traditional view of science.

Research work on biofluid mechanics of plants is by its nature interdisciplinary. The interaction between mechanics and fields such as agricultural engineering or plant physiology is essential, both to generate the relevant problems and to evaluate the significance of the solutions. In some cases, the use of a mathematical model has been particularly effective in a field where mathematical approaches are less common than experimental or descriptive methods. For example, the question of what makes a stomate open had been incorrectly attributed to differences in guard cell wall thickness prior to biomechanical analysis [Cooke *et al.* (1976)]. Also, the observed absence of evaporation from mesophyll cell walls inside the leaf had led to the conjecture that these walls were covered by waxy cuticle which retarded evaporation, an assumption which was shown to be unnecessary by use of a mathematical model [Rand (1977a,b)].

As we look towards the future of the study of the fluid mechanics of green plants, we hope for applications that permit plant breeders and geneticists to design plants which can withstand unusual stresses, such as frost and dehydrations. The ability to meaningfully incorporate into crop production considerations based on plant physiology will expand rapidly as the utility of mathematical models becomes more apparent to agriculturalists.

24.7 ANIMAL FLUID DYNAMICS
George T. Yates

The dynamics of fluid motion associated with animals encompasses a wide spectrum of flows within and around living creatures. This survey is divided into two major categories—internal flows and external flows. Many important areas of biofluid dynamics, such as the rheology of biological fluids, are difficult to classify into either of these general categories and, consequently, are treated when a particular flow involving them is discussed. Whenever possible, systems utilizing similar principles of operation are considered together. Only topics where direct application of fluid mechanics is made to living animal systems are discussed and the important technologies of biomedical equipment and measurement techniques are not considered. Neither the scope of material nor the references cited are complete, but perhaps this review will serve as a starting point for interested readers to begin their inquiry into the fascinating field of biofluid dynamics.

24.7.1 External Flows

The great variety of animals experiencing the relative motion of fluids over their bodies spans over nineteen orders of magnitude in body mass. These creatures may actively propel themselves through the surrounding fluid, create their own currents, sediment under gravity or other external forces, or experience an externally imposed flow field such as a wind or current. Animal locomotion has received a good deal of attention [Wu *et al.* (1975)] and is discussed first since many classically developed concepts in fluid dynamics are used, and the importance of collaboration between biologists and fluid dynamicists is illustrated.

Swimming at Low Reynolds Number. Motile single cells are typically between 1 and 1,000 μm in length and swim at a Reynolds number, based on body length and swimming speed, less than 0.1 [Brennen and Winet (1977)]. The oscillatory Reynolds number based on the radian beat frequency, ω, and the length of the oscillating organelle, L, can be defined, Re $= \omega L^2/\nu$. Since beat frequencies range from 5 to 30 cycles per second and cilia lengths are usually less than 100 μm, the oscillatory Reynolds number is less than 0.1. At these Reynolds numbers, both the fluid and organism inertia can be neglected, and viscous forces dominate the propulsive mechanism.

Three major categories of microorganism swimmers can be identified. First are flagellated cells or organisms. These include various spermatozoa, flagellated protozoa and bacteria which propel themselves by undulating (or rotating) a long cilium or flagellum or bundles of flagella. Second are ciliated organisms. These are covered with cilia which typically beat in coordinated patterns. The third category includes organisms which move by cytoplasmic flow. This encompasses the amoebic type movement, cytoplasmic streaming and other intracellular transports. Our knowledge of cytoplasmic flows is limited [see review by Stebbings and Hyauis (1979)], thus only the first two categories are considered here.

Slender Body Theory: The flagella of bacteria are about 20 nm in diameter and measure 5 to 15 μm in length, while cilia, including the flagella of eukaryotic cells, are 0.25 μm in diameter and range between 5 to 100 μm in length [Brennen and Winet (1977)]. The long and slender geometry of cilia or flagella is of paramount importance for propulsion at low Reynolds number and also facilitates the theoretical investigations where extensive use of slender body theory and resistive force theory have been made.

At low Reynolds number, the fluid resistance to motion is linearly proportional to the velocity of the body and can be expressed for slender bodies as

$$\Delta F_s = -C_s U_s \quad \text{and} \quad \Delta F_n = -C_n U_n \tag{24.161}$$

where ΔF_s and ΔF_n are the longitudinal and transverse fluid forces per unit length on the body. The constants of proportionality C_s and C_n are generally different and depend on the fluid viscosity and size or shape of the body, but are independent of the tangential and normal velocities U_s and U_n. Using this idea, Gray and Hancock (1955) made a major contribution to the research on microorganisms when they proposed using

$$C_s = \frac{2\pi\mu}{\ln \dfrac{2\lambda}{b} - \dfrac{1}{2}} \tag{24.162}$$

and $\gamma = C_s/C_n = 1/2$ where b is the radius of the flagella cross section, λ is the wave length and μ the fluid viscosity. A better estimate of C_n is obtained from

$$C_n = \frac{4\pi\mu}{\ln \dfrac{2\lambda}{b} + \dfrac{1}{2}} \tag{24.163}$$

Further modifications of the C_s, C_n, and γ have been made for three-dimensional objects [Cox (1970) and Lighthill (1976)], and typically γ ranges between 0.6 and 0.7. An improved slender body theory has been developed by Johnson (1980) which indicates that both force coefficients, Eqs. (24.162) and (24.163), should be increased by about 35% [Johnson and Brokaw (1979)], however the ratio γ remains nearly unchanged. It should be kept in mind that self propulsion is possible as a direct consequence of γ being less than unity.

Hydrodynamic studies of microorganism propulsion have frequently made use of flow singularities. The viscous equations of motion (Stoke's equations) admit a fundamental solution which corresponds to a point force in the fluid, is called a *Stokeslet* and has the velocity field

$$\bar{u}(\bar{x}, \bar{\alpha}) = \frac{\bar{\alpha}}{r} + \frac{(\bar{\alpha} \cdot \bar{x})\bar{x}}{r^3} \tag{24.164}$$

where $8\pi\mu\bar{a}$ gives the strength and direction of the singular force, and $r = |\bar{x}|$. By differentiating the Stokeslet, higher-order singular solutions can be found and include the potential source, Stokes doublet, potential doublet, rotlet, stresslet, and others [Batchelor (1970)]. Solutions for practical application can be found by linear superposition of these point singularities or distributions of singularities and by choosing the singularity strengths to satisfy the appropriate boundary conditions.

The velocity field resulting from a body being forced through a viscous medium penetrates the fluid a great distance, because it decays as the reciprocal of the distance from the object [Eq. (24.164)]. This long range influence of the disturbance adds some difficulty to the study of low Reynolds number flows. For freely swimming organisms in the absence of external forces, however, the long range effects of distributed Stokeslets is modified. When the external force and acceleration or inertia effects are negligibly small, the sum of all the viscous forces must vanish, thus the mean Stokeslet strength must vanish, and, far from the body, the velocity field no longer decays like $1/r$ but rather like a potential doublet, $1/r^3$ [Keller and Wu (1979)].

Flagellar Propulsion: The flagella of bacteria are composed of a helical protein called *flagellin*, have an amorphous core, and are relatively rigid and helically shaped. Propulsion is accomplished by the rotation of the flagella which attaches to the cell body at an ingenious rotary joint [see Routledge (1975) for a review]. Cilia and eukaryotic flagella have a complex internal structure of nine outer pairs of microtubules surrounding a central pair. These microtubules run most of the length of the cilium, slide relative to each other and are joined by cross arms [Satir (1974)].

Both eukaryotic and prokaryotic flagellar configurations and beat patterns are highly varied. For fluid mechanical considerations, however, they are treated together, and two general modes have been identified and analysed: 1) a plane sine wave propagating rearward along the flagella with constant wave speed [Gray and Hancock (1935)], and 2) a helical wave where, in addition to the balance of forces, the torque is also balanced [Chwang and Wu (1971)]. Generally, these works make direct application of slender body theory or resistive force theory. The microscope slide or cover glass and the cell body have a profound influence on the results and need more refined investigation [Lighthill (1976)].

Ciliary Propulsion: Many microorganisms are densely covered with cilia, usually ranging between 5 and 30 μm in length. The distribution and beating pattern of these cilia are highly varied, and several hydrodynamic theories have been developed (all based on low Reynolds number slender body theory).

The *envelope model* assumes the cilia to be sufficiently closely packed so that the individual effects of each cilium can be neglected, and the body surface is considered as an oscillating material surface [Keller and Wu (1977)]. This model is justified for a high concentration of cilia and for wave propagation in the direction of the ciliary power stroke (opposite to the direction of swimming).

The *sublayer model*, first proposed by Blake (1972), treats individual cilia and addresses the flow within the ciliary layer. The force on each cilium is found using slender body theory taking into account the influence of the cell surface and the interaction effects of other cilia. The entire flow field is then computed by adding the contribution from all the cilia. This model is used when the cilia are widely spaced or the interaction between cilia is unimportant.

Swimming at High Reynolds Number. The vast majority of fish inhabiting the oceans, lakes and streams swim at a Reynolds number between ten thousand and five million. In this range of Reynolds numbers and for streamlined bodies, the viscous forces are confined to a boundary layer which is close to the body and which is so thin that the pressure gradient normal to the body surface is small and can be neglected (see Sec. 4.1). Under this assumption, forces calculated by inviscid theory are expected to represent the inertial forces exerted by the fluid on the animal. Since the current understanding of unsteady, three-dimensional boundary layer development is inadequate to accurately specify the boundary layer character (laminar or turbulent) everywhere along the body and the nature of the wake (see Secs. 4.3, 4.8, and 4.12), the calculation of viscous drag force remains a very challenging task. The ability to separate the computation of inertial and resistive forces, however, makes the problem tractable, and the forces acting on the swimmer by the fluid are thus decomposed into thrust, T, calculated from inviscid flow theory, and drag, D, calculated in principle from the boundary layer theory.

The *reduced frequency*, $\sigma = \omega L/U$, is usually based on the radian frequency of the tail beat, ω, fish length, L, and mean swimming speed, U, since these are readily available data. Except for very low speeds, less than one body length per second, σ varies from 8 to 15, and at higher speeds lies between 8 and 10 for a wide range of fish sizes and Reynolds numbers [Yates (1983)]. The reduced frequency, more relevant for evaluating unsteady effects, should be based on the fin dimensions and, thus, should be lowered in most cases by a factor of about ten. An essential consequence for σ in this range is that unsteady effects are important and the much simplified quasi-steady approach must be abandoned. See Chap. 12 for further material on unsteady flows.

Elongated Body Theory: Within the frame work of unsteady, inviscid, potential flow theory, the forces on a swimming fish can be estimated. Using the fact that fish are long and slender, Lighthill (1960) extended classiçal slender body theory to evaluate the propulsive forces for small amplitude lateral motions. He found that the lateral force per unit length of the fish is given by the linearized total (Langrangian) time derivative of the transverse momentum of the fluid per unit length

$$F = -\frac{d}{dt}(mV) = -\left(\frac{\partial}{\partial t} + U\frac{\partial}{\partial x}\right)(mV) \qquad (24.165)$$

where m is the added mass per unit length, and V is the lateral fluid velocity. The presence of dorsal and ventral body fins, which shed vortex sheets, modify Eq. (24.165) by making the longitudinal variation of the added mass unimportant in the lateral force [Wu (1971)]. The lateral force which acts normal to the body surface must be combined with the leading edge *suction force* which acts along the leading edges of the body to give the sectional force. Integrating the component of the sectional force in the direction of swimming over the entire length gives the thrust generated by the fish. This thrust must be balanced by drag which arises from the action of viscosity in a thin boundary layer continguous to the body.

The theory has been extended for finite amplitude motion, vortex shedding from dorsal and ventral fins, and body thickness. Comparison with experimental data on power requirements have been made possible by oxygen consumption measurements made on swimming fish and general agreement with theory is obtained [Yates (1983)].

Propulsion by Fins with High Aspect Ratio: For a wide class of swimmers using lunate tails for propulsion (e.g., tuna, marlin, dolphin, and whales), the elongated body theory must be abandoned, and the propulsive appendage is usually considered independent of the rest of the body. In the simplest case, the local thrust production is estimated using two-dimensional wing theory, where the time dependent forces and moments acting on an infinitely long airfoil can be calculated using unsteady aerodynamics and are given in the context of fish swimming by Wu (1971).

Three-dimensional aspects of oscillating hydrofoils have been investigated with special reference to aquatic swimmers, and methods of evaluating the flow are somewhat varied and lead to differences in thrust and efficiency estimates. Chopra (1974) used the superposition of periodic lifting ribbons of finite span to calculate the performance of rigid rectangular wings undergoing general heave and pitch. Chopra and Kambe (1977) used unsteady lifting surface theory to obtain numerical results for a variety of rigid swept back wings. Using an asymptotic method, Cheng (1975) looked at an unsteady lifting-line theory and identifies various frequency domains for the influence of unsteadiness on the induced downwash. The various theories have been compared by Ahmadi (1980) who employed linearized unsteady lifting-line theory, and by Lan (1979) who used a *quasi-vortex–lattice* method. Ahmadi finds his method predicts a higher thrust than Chopra and Kambe, while Lan finds a 20 percent lower thrust, at high reduced frequencies, than Chopra and Kambe. The efficiency predicted by all three models is in better agreement and varies by only a few percent. Detailed expositions of these methods can be found in the literature and generally require numerical computation.

Other Aquatic Animals: Many other intriguing flows occur in the aquatic environment. The mechanics of jet propulsion has been studied [e.g., Weihs (1977)] where the time rate of change of the fluid momentum expelled from the organism provides a thrust force which is balanced by the viscous drag and the acceleration reaction (the body mass plus apparent mass times the acceleration). Many animals create currents which they filter for feeding (e.g., bivalve mollusks, ascidians and sponges) while others (such as barnacles) remain fixed to the bottom and collect nutrients from the oscillatory tides and waves or from currents [Vogel (1981)].

Flying. Flight is accompanied by a wide variety of birds, insects, mammals, and reptiles. A number of small insects operate at a Reynolds number between 1 and 10 [Weis–Fogh (1975)], while the larger birds can easily exceed a Reynolds number of 10^5. Three basic modes of flight can be identified: 1) gliding or soaring, 2) flapping forward flight, and 3) hovering. The reduced frequency $\sigma = \omega c/U$, where ω is the radian wing beat frequency, c the wing chord and U the forward flight speed, may range from zero for gliding to infinity for hovering and typically has a value between

0.1 and 1 for flapping flight [Lighthill (1975)]. A more detailed discussion on "Animal and Ornithopter Flight" is given in Sec. 24.8.

During gliding flight, the aerodynamic lift and drag force can be calculated from classical aerodynamics, and they are balanced by the component of the animals weight perpendicular to and parallel to the flight path, respectively. The application of large aspect ratio wing theory is straight forward and is reviewed by Lighthill (1977).

For flapping forward flight, the kinematics are more complex, and some excellent observations on the wing motions of birds and insects have been made by Nachtigall (1974). Steady state aerodynamics has been extensively used to estimate lift and thrust forces in flight, and experimental results for the steady lift coefficients of animal wings have been found to balance the weight within the experimental accuracy [Jensen (1956)]. Lift coefficients range from 0.5 to 1.6 [Maxworthy (1981)]. Given the range of the reduced frequencies, these results seem fortuitous, and the effects of unsteady aerodynamics in the flapping flight of a large number of insects has been shown to play a more important role [Maxworthy (1979)]. With the appropriate modifications, the unsteady airfoil theories discussed in the previous section for lunate tail swimmers can be applied to flying.

Hovering flight presents some unique problems in aerodynamics which have been addressed in classical aerodynamics and for which a novel mechanism of lift generation is employed. Although a wide variety of hovering configurations are observed, two major mechanisms are identified: 1) normal *hovering* and 2) *clap-and-fling*. Normal hovering, for which the wing stroke is approximately horizontal and traces out a *Figure-8* pattern, is used by hummingbirds and by most large insects and has been analyzed using steady-state aerodynamics [Weis–Fogh (1973)]. A more detailed examination of the role of unsteady effects in normal hovering seems appropriate in light of more recent studies [see the review by Maxworthy (1981)]. Weis–Fogh (1972) proposed the clap-and-fling mechanism to explain the hovering of the chalcid wasp, *Encarsia formosa*. In this type of hovering, the wings clap together at one extreme of the motion and then fling open to give immediate and large lift. Lighthill (1973) analyzed inviscid source flow into the V formed by the opening wings to predict the generation of large forces and thus explained how these insects may circumvent the gradual lift build-up predicted from the classical Wagner problem.

24.7.2 Internal Flows

Fluids are continuously being transported within animals through a variety of internal tubes. These tubes are generally lined with smooth muscle (however skeletal muscle is sometimes present) and often have smooth or skeletal muscle at their ends which act as flow control valves. The muscle surrounding the tubes is responsible for or aids in the fluid movement by a variety of mechanisms: 1) peristaltic pumping, 2) segmentation, 3) pressure head, and 4) structural support of the walls. Often, several of these mechanisms or others are used in any one tube. Peristalsis and segmentation commonly occur in the gastrointestinal tract (esophagus, stomach, duodenum, small bowel and large bowel), and the urinary tract (ureter and bladder). In addition to muscular contractions and pressure differences, both the male and female reproductive systems, make use of ciliary activity in gamete transport. The flow induced by application of a pressure head occurs in the urethra and in blood vessels where smooth muscle functions as a mechanical wall support [Fung (1981a)]. Since blood flow has been extensively studied and involves a special pumping mechanism it is discussed first.

Hemodynamics. Advances in the understanding of blood flow have been connected with the application of fluid mechanics and have been associated with investigators like Euler, Young, Poiseuille, and others [for a historical review see Skalak *et al.* (1981)]. The study of blood flow in living animals has been the most extensively studied area of biofluid flows and exemplifies the application of analytic methods and the collaboration between engineering and medicine.

Rheology of Blood: Human blood is a suspension of cells in an aqueous solution called *plasma.* The plasma is about 91% water by weight with 7% protein and 2% organic and inorganic substances. The red blood cells, *erythrocytes*, occupy about 45% of the blood volume. White blood cells and the platelets each make up less than 0.2% of the blood. Human red cells are disk shaped and measure 7.6 μm in diameter and 2.8 μm in thickness [Fung (1981a)]. The red blood cells are usually assumed to be fluid filled sacks with constant surface area. The internal fluid (a hemoglobin solution) is Newtonian (Table 24.5), and the enclosing membrane has

TABLE 24.5 Viscosity of Biological Fluids*

Fluid	Viscosity (Nsm^{-2})	Comments
air	1.8×10^{-5}	Newtonian
fresh water	1.00×10^{-3}	Newtonian (20 C)
salt water	1.08×10^{-3}	Newtonian (20 C; 3.5% salinity)
protoplasm and cytoplasm	2×10^{-3}–15×10^{-3}	thixotropic, viscoelastic [Scott Blair (1974)]
blood plasma	1.2×10^{-3}	Newtonian [Fung (1981a)]
hemoglobin solution	6.1×10^{-2}	Newtonian [Evans and Skalak (1979)]
whole blood	3.0×10^{-3}–0.1	viscoelastic; shear thinning [Fung (1981a)]
saliva and sputum	0.5–10	viscoelastic [Fung (1981a)]
periciliary fluid	0.01	Newtonian [Ross and Corrsin (1974)]
mucus	0.1–50	viscoelastic [Litt and Kahn (1976)]
lymph fluid	0.8×10^{-3}–1.2×10^{-3}	Newtonian [Scott Blair (1974)]
semen	3×10^{-3}–3×10^{-2}	shear thinning [Scott Blair (1974)]
synovial fluid	5.0×10^{-3}–10	shear thinning [Fung (1981a)]

*Because most biological fluids have a highly variable composition and a complex rheology (see text), this table serves only to identify approximate valves of viscosity. For any specific conditions, more accurate estimates can be obtained by consulting the appropriate references.

viscoelastic characteristics [Evans and Skalak (1979)]. Under resting conditions, they take on a characteristic biconcave disk shape and deform when subjected to shear flow [Fung (1981a)]. The deformations and membrane properties of red blood cells influence all aspects of whole blood behavior both in the macrocirculation and microcirculation.

Whole blood exhibits considerably higher viscosity than blood plasma, is shear thinning and has viscoelastic properties. The viscosity depends on the shear rates, hematocrit, temperature, clotting, and other conditions. Viscoelastic effects are important during blood coagulation, and the continuum model assumptions may be violated in capillary flow since the red blood cells squeeze through the vessels one by one.

Flow in the Heart: The right half of the heart [Winet (1982)] pumps blood into the pulmonary circulation at pressures which oscillate between 1.2×10^3 and 3.1×10^3 Nm^{-2}, and the left half pumps blood into the systemic circulation at pressures which oscillate between 1.0×10^4 and 1.6×10^4 Nm^{-2}. The latter pressures are the commonly measured *diastolic* and *systolic blood pressure*. Both sides of the heart operate in a similar manner, but the detailed timing of valve opening and muscle contractions and magnitude of the pressures and flow velocities vary somewhat. Investigations of the fluid flow in the heart have centered on the operation of the various valves that enable this magnificent organ to effectively pump blood through the body tissues. These valves permit less than five per cent of the stroke volume to flow back when they close. Flow velocities may exceed 1 ms^{-1}, and Reynolds numbers are as high as 8,000. There appears to be no large scale turbulence in the heart [Taylor and Wade (1973)], however vortex flow is observed [Peskin (1982)]. The presence of a ring vortex in the ventricle during diastole seems important for closing the mitral valve. A similar mechanism is proposed for the aortic valve, but the geometry is more complex and less *in vivo* evidence is available [Peskin (1982)].

Arterial Macrocirculation: The aorta is the largest artery in the body measuring about 2 cm in diameter. Flow velocities entering the aorta from the left ventrical may exceed 1 ms^{-1} and obtain a Reynolds number above 4,500. An estimate of the unsteady viscous boundary layer thickness in the aorta gives approximately 1 mm during systole, and Poiseuille flow is not expected [Jones (1969)]. Mean flow velocity profiles in the aorta of the dog have been measured [Schultz (1972)] and show a gradual development from a plug flow at the aortic valve to a parabolic velocity profile near the distal end.

The oscillatory nature of the flow (1.1 Hz) cannot be neglected and has many important consequences. As the aortic valve opens, blood enters the aorta, the pressure rises, and the aorta stretches. As the cardiac output slows, the aortic pressure drops and the walls contract to their initial position. The fluid inertia causes the fluid to move along the aorta accompanied by a similar expansion and contraction of the vessel walls. Thus, a disturbance is propagated along the arterial system. Using an analogy to sound propagation in an compressible fluid Thomas Young first estimated (in 1809) the wave propagation of an incompressible fluid in an elastic tube and found the wave speed $c = \sqrt{hE/2a\rho}$, where h and E are the wall thickness and Young's modulus of the tube, a the tube radius and ρ the fluid density. The assumptions made in this simplified theory are somewhat restrictive by today's standards. More recent linearized theories for the wave propagation have been developed [Jones (1969)] where, in addition to the compliance of the walls, consideration of fluid resistance and leakage to account for viscosity and flow into side branches have been included. The pressure and flow rate fluctuations decay exponentially with distance along the vessel, and the wave speed is many times the mean fluid velocity.

Nonlinear theories for arterial flows have generally taken mean values over any cross section to reduce the equations of motion and continuity to one-dimensional forms. The method of characteristics has been used to solve the resulting set of ordinary differential equations by numerical methods [Rumberger and Nerem (1977)], and agreement of these theories with experiment has been good in aortic and coronary flows.

Microcirculation: The rheological properties of blood play a more important role in the arterioles, capillaries, and venules than in the macrocirculation where it can be assumed Newtonian. In these smaller vessels (less than about 200 μm in diameter), the Reynolds number is usually less than one, the quasi-steady assumptions are justified and, in the smallest vessels, the blood can no longer be thought of as a homogeneous fluid. In the arterioles and venules, the red blood cells are excluded from the immediate proximity of the wall by geometric constraints and by particle migrations, and the dynamic hematorcit is lower than that of whole blood in a reservoir or in the macrocirculation. Consequently, the small vessels which branch from the larger macrocirculation vessels and in which flow is relatively slow, contain a greater proportion of plasma than the parent tube. In tubes below 300 μm, the reduced hematocrit and exclusion of cells from the tube walls are responsible for reducing the apparent viscosity (the *Fahraeus–Lindquist effect*) by 20–30%. In the capillaries where the Reynolds number is less than 0.01, the red blood cells pass in single file and are greatly deformed. A thin plasma layer between the blood cells and the walls is expected, and lubrication theory can be applied in this narrow gap. The plasma bolus, which fills the spaces between the red blood cells, exhibits a circulating motion [Fung (1981a)].

The capillary flow in the lung is described [Fung and Sobin (1977)] as a sheet flow between two relatively flat walls which are connected by a large number of posts. The pulmonary capillary sheets are compliant and can exhibit Starling resistor type flow behavior [see next section and Fung (1981a)].

Venous Macrocirculation: Veins differ from arteries in that they have much thinner and more flexible walls, have much smaller transmural pressures, may commonly collapse above the heart, have valves to prevent backflow and have confluent rather than divergent flow at their junctions. The Reynolds number in the large veins is usually less than 1,000, and the flow is expected to be fully developed and laminar. At vein junctions, the velocity profile will have two peaks, one arising from the parabolic flow in each vein. Flow oscillations are large, and unsteady effects are important. Skeletal muscle contractions reduce the transmural pressure and, with the help of valves, aid the circulation. The detailed role of the valves as surface roughness, interaction with wave propagation and flow control is not fully understood. Wave propagation occurs in veins, and the wave speed may be comparable to the mean flow velocity. Flow in collapsible tubes has been studied both experimentally and theoretically using a *Starling resistor* [Shapiro (1977)] where the upstream, downstream, and transmural pressure can be adjusted independently. The phenomena of flow limitation due to a reduction in tube cross section and the occurrence of flow and wall oscillations have been observed [Kamm and Shapiro (1979)].

Pulmonary Fluid Dynamics. A wide variety of methods of delivering oxygen from the surrounding environment to the circulatory system and of removing carbon dioxide from the blood are employed by animals. Mammals utilize lungs into which air is inspired and exhaled in a rhythmic manner. Fish and many aquatic creatures make use of a unidirection water flow and counter-current blood flow in gills. Birds have a constant flow lung with an *aerodynamic valve* and a cross-current flow configuration for gas exchange [Lighthill (1975)]. To acquire oxygen, insects show more diverse mechanisms which involve various arrangements of spiracles, tracheae, and tracheoles and which rely mainly on diffusion and body movement [Rockstein (1974)]. This section will concentrate on some aspects of the fluid mechanics of mammalian respiratory systems with particular reference to humans.

Within the lung of mammals there are four major types of fluid flows: 1) inspiration and expiration of air, 2) blood flow in the pulmonary microcirculation, 3) water flow, solute flow, and gas exchange in the alveoli, and 4) mucociliary flow in the bronchi trachea and nasal passages. The blood flow in the pulmonary microcirculation has already been considered.

Air Flow: The human respiratory tract is divided into three regions [Proctor (1964)]: 1) the upper airways (the nose, mouth, sinuses, throat and larynx), 2) the tracheobronchial tree, and 3) the terminal airways and alveoli. The tracheobronchial tree branches from 8 to 25 times before reaching the terminal airways, and the tube diameters decrease from about 1.8 cm in the trachea to about 0.045 cm in the terminal airways. Surprisingly, the ratio of the tube length between branches and the tube diameter remains nearly 3.5, independent of the number of previous branches.

The gas exchange region is distinguished from the conducting airways by the presence of alveoli in the walls of the tubes. The alveoli greatly increase the cross sectional area of the lung, hence the air velocity falls very rapidly in this region. Due to the small diameters and low velocities, the Reynolds number is usually much smaller than one in these regions. Research has concentrated on the air flow in the bronchial tree [see reviews in Fung (1981b) and Pedley (1977)] where the Reynolds number (based on mean flow velocity and tube diameter) is found to vary from 1,000 in the trachea to 0.1 or less in the terminal airways during normal quiet breathing for humans. During rapid breathing, the flow rates, and hence Reynolds numbers, may be elevated by a factor of ten.

The complexity of the flow has forced researchers to construct models and perform detailed flow visualization and velocity measurements using steady flow, and the majority of work has been conducted using single bifurcations. Since the pressure drop over one branch is very small and fluctuations across the channel are of the same order, velocity measurements have been more easily obtainable. The flow patterns in human bronchi are greatly distorted from the Poiseuille predictions due to the intervening high shear boundary layer and in some cases separated flow regions and secondary flow caused by tube curvature. Boundary layer analysis of this type of flow has been applied [Smith (1976)], and computational methods will no doubt play an increasing role in the future. Unsteadiness is important in at least the first few branches of the tracheobronchial tree during rapid breathing and possibly in the terminal airways [Fung (1981b) and Pedley (1977)], however the full importance of the unsteadiness of the flow remains to be clarified [Jaffrin and Hennessey (1972)].

The rate at which pressure does work generally is calculated from the kinetic energy flux and the dissipation function which are in turn obtained from measured velocity profiles. Pedley, Schroter, and Sudlow (1970) and others have calculated the viscous energy dissipation for tubes over a wide range of Reynolds numbers. For all but the lowest flow rates, these calculations show a marked increase in the pressure drop to values greater than the Poiseuille flow estimates. Indications are that the overall airway resistance is dominated by the upper airways, from the mouth to the trachea, and to lesser extent the first several bronchial branches, with the smaller airways making only a minor contribution.

Since all the air in the lung cannot be exhaled in a single expiration, it is important that mixing occur at the next inspiration and thus insure contact of fresh air with the gas exchange regions. Taylor (1954), analyzed a *longitudinal dispersion* model for mixing and found the effective dispersion constant inversely proportional to the coefficient of molecular diffusion. This theory has been applied to airways [Wilson and Lin (1970)].

Mucociliary System: The inside surface of the trachea and bronchi is more or less continuously lined with ciliated epithelial cells. The cilia protrude from the cell surface into a watery periciliary fluid. Overriding the cilia and periciliary fluid is a blanket of viscous non-Newtonian respiratory mucus which is produced by mucus cells also distributed along the epithelium. The mucociliary system functions to trap dust, microorganisms, or other pollutants that are inspired and to remove them from the lung. The respiratory cilia are surprisingly similar for most mammals; they measure about 5–7 μm in length and 0.25 μm in diameter, are very densely distributed, 3–10 cilia per square μm, and beat at 10–25 hertz.

Normal lung mucus is a suspension of glycoproteins (2–4%) and salts (1–2%) in water (95–97%) [Schlesinger (1973)]. The glycoproteins are crosslinked by both secondary bonds and covalent disulfide bonds which give mucus an extended gel network [McIntire (1980)]. This network structure accounts for the viscoelastic and shear thinning properties of mucus [Litt and Kahn (1976)]. Respiratory mucus is usually more than 100 times more viscous than the periciliary fluid, and its viscoelasticity varies widely and depends strongly on glycoprotein composition, and pH and ion concentrations.

The interaction of the cilia with the periciliary fluid and mucus are important for effective fluid transport by this system. Both histological and theoretical evidence indicate that the cilia tips are embedded in the mucus during their forward stroke and totally disengaged from the mucus on their return stroke. However, the versatility of the system and the observed transport in various configurations suggest that effective operation of the system may be more involved and has yet to be adequately described [Brokaw and Verdugo (1982)].

Fluid dynamic models have been based on two Newtonian fluids with large viscosity ratio. Models originally developed for microorganism propulsion have been applied to the mucociliary system and include the *envelope model* [Ross and Corrsin

(1974)] and the *sublayer model* [Blake (1973)]. A model employing the penetration and retraction of the cilia tips into the mucus has been developed [Yates *et al.* (1980)] where effective transport has been shown for the case of Newtonian fluids. The viscoelastic properties of mucus have been considered [Silberberg *et al.* (1976)], but their role in mucociliary transport is yet to be fully realized.

Peristalsis and Segmentation. A variety of tubes transport fluids within the gastrointestinal and urinary tract by contracting the tube walls. Although these tubes have gross similarities, they are very different in their physical size, biological function, and contour. The fluid being transported varies from a simple Newtonian fluid with viscosity near that of water in the ureter, or a very viscous viscoelastic fluid in the bowels to fluid–gas–solid, multi-phase material in the esophagus. They are all discussed together here, since the basic mechanisms of fluid transport appear to be similar and since refined observations on the various flow details are not yet available. Peristalsis has been observed primarily in three biological tubes—the ureter, esophagus, and colon. On occasion, the small bowel also uses peristalsis to purge its contents. Peristaltic pumping is characterized by large amplitude constrictions of the side walls (from 0.2 to 1 times the tube radius) [Jaffrin and Shapiro (1971)]. The wave length of the contractile wave varies from 3 times to 300 times the tube radius and the Reynolds number, based on the tube radius and peristaltic wave speed, ranges from less than 1 to about 100. Shapiro *et al.* (1969) suggest that this Re should be multiplied by the ratio of the tube radius to peristaltic wavelength to indicate the relative importance of inertia and viscosity, thus the viscous terms are expected to play a dominant role in most cases.

Theoretical models have been developed for two-dimensional and axisymmetric geometry and for various ranges of the Reynolds number, tube radius, wavelength, and wave amplitude. Most investigations have concentrated on symmetric contraction and expansion of the tube cross section [Jaffrin and Shapiro (1976)]. The main results are that the viscous dissipation is an essential feature of peristaltic pumping and that fluid motion opposite to the wave motion (*reflux*) sometimes occurs, however many aspects of the fluid flow in these flexible tubes are yet to be fully clarified.

There seems to be no anatomical difference between tubes which propagate peristaltic waves and tubes which exhibit segmentation. Segmentation is characterized by a non-propagating constriction of the lumen and occurs almost exclusively in the digestive tubes but may occur in the oviduct [Verdugo *et al.* (1980)]. The resulting fluid motion is usually a back-and-forth motion or a pendular motion and is commonly observed in the small bowel. Unlike peristaltic pumping, where the main purpose is the transport of fluid along the tube, segmentation functions primarily to mix the nonhomogeneous fluid and to bring fluid into close proximity of the walls where nutrient uptake takes place. There does, however, appear to be a slight bias to move the contents along the bowel. Relatively little theoretical work has been done on this type of fluid mixing or transport beyond stochastic modeling [Bartuzzi (1978)].

Ovum Transport. Often, more than one mechanism is responsible for fluid transport, as in the reproductive tract of mammals where a combination of several mechanisms is utilized. The transport of the ovum from the ovary to the uterus in the oviduct has received the most attention, and transport is accomplished by ciliary activity, muscular activity and by an externally applied pressure difference. Although the cilia appear to be necessary for ovum transport, it seems likely that any one of these mechanisms may produce pro-uterine transport, and the relative importance of all three remains to be fully clarified.

The kinematics of cilia in the oviduct are similar to that in the respiratory system, however the geometry of the oviduct, which is highly convoluted, and the cilia–mucus–periciliary fluid interaction have not been adequately observed and present some challenging problems. Blake *et al.* (1983) has attempted a simplified study, combining the three major mechanisms and using an inner tube surrounded by an outer porous region where the cilia are located. The qualitative importance of pressure difference, segmental muscle contractions and ciliary activity can be compared when this model is combined with more refined physiological observations.

Mechanics of Joints. The synovial joints permit the relative motion of bones, carry large, often rapidly fluctuating, loads (in some joints more than five times the body weight), have low friction and low abrasion and must function for over seventy years. In the proximity of these joints, the bones are covered with a thin layer (less than 2 mm thick), [Mow *et al.* (1980)] of articular cartilage. Ligaments, tendons, and soft tissues surround the joint and insure proper alignment during loading and motion. The articular cartilage and the joint cavity are filled with a small amount of lubricating fluid (the *synovial fluid*); the human knee joint contains about 0.2 ml of fluid [Mow and Lai (1979)].

The synovial fluid is a suspension of large macromolecules (primarily hyaluronic acid protein complexes), is similar to blood plasma, and has viscoelastic properties. Measurements of its storage modulus, G', and loss modulus, G'' (the real and imaginary parts of the complex dynamic shear modulus), in an oscillating Couette rheometer [Balazs and Gibbs (1970)] at various angular velocities, show that G'' is 2 or 3 times larger than G' at low frequencies. As the frequency increased, both G' and G'' increase with the storage modulus increasing more rapidly. At high frequencies, the loss modulus begins to decrease and G' may exceed G'' by a factor of 3 or more.

Articular cartilage is a nonlinear viscoelastic material which undergoes large deformations and is composed of collagen fibers distributed among noncollagenous proteins, proteoglycans, chondrocytes, and some lipid material. Between 75 and 85% of the articular cartilage is fluid [Lipshitz *et al.* (1976)] with the fluid content decreasing with the distance from the surface. This complex network of fluid and proteins results in an anisotropic, inhomogeneous, viscoelastic, fluid filled permeable tissue [Woo *et al.* (1980)] which changes its properties with hydration and age. Under mechanical stress, fluid moves into and out of this porous matrix, and a biphasic model [Mow *et al.* (1980)] has been applied as a rheological model of articular cartilage.

The coefficient of friction of articular cartilage is well below the best man made materials [Malcom (1976)], and ranges from 0.002 to 0.020 for fully hydrated healthy cartilage. This sliding friction is relatively insensitive to the velocity of sliding over a wide range of velocities, but, for sliding velocities below about 1 mm/s, the coefficient of friction increases. Under cyclic dynamic loading, the coefficient of friction has been found to be reduced by about a factor of two from the value for steady motion [Malcom (1976)]. As the articular cartilage dries out and is compressed, the coefficient of friction increases [Malcom (1976)] very dramatically (often several orders of magnitude) and is quickly worn away by abrasion.

The dynamics of synovial joint movement is a complex interaction between a film of highly non-Newtonian fluid and a permeable viscoelastic material, which involves the analysis of squeeze films, lubrication theory, the permeability and deformation of nonlinear biphase materials and fluid structure interactions. This complex system is not yet completely understood, however, some progress has been made [Mow and Lai (1979)].

24.8 ANIMAL AND ORNITHOPTER FLIGHT
Geoffrey R. Spedding and James D. DeLaurier

24.8.1 Introduction

Natural Flight as a Common Phenomenon. Powered animal flight has evolved independently on at least four separate occasions in insects, reptiles (pterosaurs), birds, and bats. Since the majority of animal species living today are insects, and since the great majority of these fly, it may be considered one of the most common forms of transport in nature. This has not gone unnoticed by earthbound humans, and flying creatures of one kind or another feature quite regularly in religion and mythology, even to this day. For reasons that will be outlined in subsequent scaling arguments, *Homo sapiens* (and precursors) is of such a size that dreams of closely imitating flapping wing flight by human-powered machines must remain just that, but the history of *ornithopter* design and development bears many interesting lessons in engineering design. Recent advances in model ornithopter flight embrace leading-edge concepts in composites and flexible structures, together with state-of-the-art, low-speed, unsteady, aerodynamic models. This section discusses some of the basic analytical concepts and methods behind flapping wing flight, both as evolved in nature and as designed by humans.

The basic challenge stems from the fact that, in the absence of a separate power plant enabled by some rotational mechanism, both lift and thrust must be generated by the same device. The unsteady flow due to reciprocal oscillation of beating wings, or wing pairs, significantly complicates the analysis of even simple wing shapes. In analysis of natural flight, the fact that wings do not serve only for flight further complicates the situation.

Scope. To maintain focus, we shall consider only flight domains where natural and ornithopter* flight overlap. The emphasis will be on powered, flapping flight, but partly to establish basic concepts and terms, we will first outline an aerodynamic analysis of gliding flight in animals, treating them as fixed-wing aircraft. More standard texts can be used for further details on fixed-wing aerodynamics and conventional gliders, and discussion of the literature applied to flexible sails and hang-

---

*Since ornithopter flight is itself a product of humankind's natural inquisitiveness, the labelling of non-human flight as 'natural' and presumably therefore, human flight as non-natural might be regarded as an artificial distinction. However, this is more than mere sophistry, as the mechanisms by which one has evolved, while the other has been consciously and purposefully designed, are entirely different.

gliders has also been omitted. Moreover, since here are no hovering, flapping ornithopters, hovering flight, which is of great ecological importance to many flying
insects (and hummingbirds), will not be covered. The very comprehensive analysis
of Ellington (1984) should be consulted on this topic. General reviews and further
references can be found in Norberg (1990) and Spedding (1992), and recent articles
by Lighthill (1990), Spedding (1993), and Rayner (1993) can be used as further
introductions to the literature.

24.8.2 Basic Principles

Re, k Domains: For the purposes of this chapter, the Reynolds number, Re, will
always be defined at the mean chord, c, thus, Re $\equiv u_c c/\nu$, where u_c is the mean
fluid velocity at c, and ν is the kinematic viscosity. Re varies from about 10 for the
smallest of the flying insects to about 10^6 for the larger birds and ornithopter models.
This range is large, but, as Lissaman (1983) notes, the performance of all airfoils
at Re $< 10^6$ is dominated by their comparatively poor resistance to boundary layer
separation, so viscous effects will rarely be completely ignorable in practical applications. Figure 24.55 shows how a threefold increase in Re dramatically affects the
properties of thin and smooth airfoils around Re $\cong 10^5$. Further complications ac-
crue when the mean flow, and even airfoil shape, are allowed to vary with time. A
convenient measure of this is the *reduced frequency*, k, based on the half chord, k
$\equiv \omega c/2U$, where U is the mean forward flight speed, and ω is the flapping frequency. k is a measure of the comparative strength of cross-stream to streamwise
vorticity in an unsteady wake. The wings of an individual flying animal must typically function from values of k close to 0 (gliding) to close to infinity (near-hovering,
take-off and landing), and, much of the time, k must be regarded as an $O(1)$ parameter. Although the implication is that unsteady and viscous effects can only rarely
be ignored completely, remarkable successes have been achieved in essentially in-

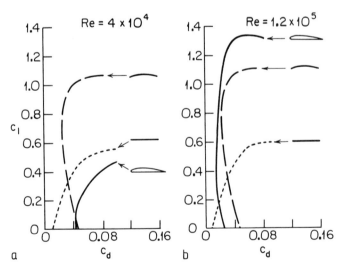

FIGURE 24.55 Lift-Drag polars for smooth airfoils and flat plates at different Re. (After
Jones, 1990.)

viscid, steady or quasi-steady analysis. Following this historical route, this section treats first these classical analyses and then departs from them.

Airfoil Sections. The wing sectional geometry for a bird, a bat, and two insect species, covering a range of Re from 10^5 to 10^2 are shown in Fig. 24.56(a). They represent a similar range to those of Fig. 24.55, with representatives of both flat, cambered plates, and smoother profiles. Moreover, the flight conditions of these particular pigeon and bat examples gave Re $= 2 \times 10^5$, and 3×10^4, respectively, an interesting correspondence with the improvement in properties of the cambered flat plate at lower Re in Fig. 24.55. The performance benefits of increasing the camber of a thin airfoil at moderate Re when there is only partial leading-edge suction has also been discussed by DeLaurier (1983). The pronounced corrugations of the insect sections have been interpreted both as promoting transition, separation, and reattachment of the boundary layer for the dragonfly at Re $\cong 10^4$ and as a structural adaptation for maintaining spanwise stiffness, but with little added wing mass and inertia and without any great aerodynamic performance reduction for the hoverfly (Re $= 4.5 - 9 \times 10^2$).

The *sectional* lift and drag coefficients, c_l and c_d are defined as

$$c_l \equiv \frac{L'}{qc}, \quad c_d \equiv \frac{D'}{qc} \tag{24.166}$$

where the primed quantities are the lift and drag forces per unit length, and $q = 1/2\rho U^2$ is the dynamic pressure. Figure 24.56(b) compares the section polars of $c_l(c_d)$, of a modified high-lift airfoil with profiles based on sections (i) and (iii) for

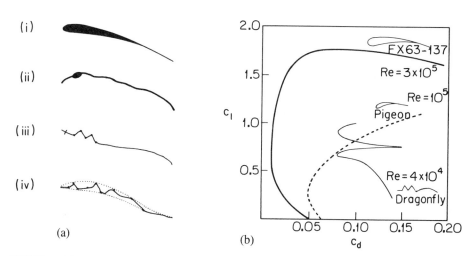

FIGURE 24.56 (a) Natural and man-made airfoil sections. Wing sections of two vertebrate, and two insect species, where $10^5 \geq$ Re $\geq 10^2$. (i) Pigeon, (ii) Dog-faced bat, (iii) Dragonfly, and (iv) Hoverfly. (b) Section lift:drag polars for a modified Wortman FX63-137, and models based on sections (i) and (ii). [Reprinted with permission, Spedding, G. R., "The Aerodynamics of Flight," *Mechanics of Animal Locomotion*, Alexander, RMcN. (Ed.), Springer-Verlag, 1992.]

the pigeon and dragonfly. The lift : drag ratio of the pigeon section is far from the highly specialised FX63-137 curve, having a higher c_d at any given c_l, but it should be noted that this airfoil must also operate in highly unsteady conditions, must be constructed of bones, muscle, connective tissue and feathers, and must be folded away when not in use. The dragonfly wing section apparently results in a highly unusual drag polar allowing a broad range of c_d to be accessed at almost constant c_l and for high angles of attack, α. While the details might be controversial, the general point remains that drag polars of animal wings may have shapes rather different than those to which the aeronautics community is accustomed.

Anticipating somewhat later parts in this section, one may reasonably question the relevance of any such data that assume 2-D, steady, unseparated flow. The magnitude of the frequency parameter k can be used as an indicator of this, and, when $k = O(0.1)$, it will be seen that quantitative agreement has been quite good.

Finite Wings. The dimensionless force coefficients for finite wings are normalized by the wing planform area, S, as

$$C_L \equiv \frac{L}{qS}, \quad C_D \equiv \frac{D}{qS} \qquad (24.167)$$

where L and D are the total lift and drag force on the wing. Figure 24.57 shows pairs of ornithopter, bird, bat, and insect species, respectively. The QN ornithopter design, intended as a reconstruction of a likely pterosaur shape, shows the design influence from the planforms of large sea birds such as the albatross. The aspect ratio, $AR \equiv 2b/c$, planform geometry, and even number of wings per body, change significantly, as does the semispan, b, which varies by more than 3 orders of magnitude. Spanwise twist is allowed/observed in all these wings during flight. Only the albatross-type wing, of those shown here, allows significant variation in span. In this respect, it may be the more typical of a common form of flapping flight, as discussed later. This capacity is limited in the bats due to the requirement of maintaining tension in the elastic wing member. In the swift, the evolution of the lunate planform appears to have been accompanied by a more rigid wing construction, but the wings flex elastically during flight, as do those of the insects. The diversity of forms in Fig. 24.57 hints at the difficulty of constructing any single, reasonable aerodynamic model to cover all cases. Yet, it is still instructive to search for common features and analytical methods.

24.8.3 Gliding Flight

Lifting Line Analysis. A wing moving at mean speed, U, with a bound vortex strength, Γ, experiences a lift per unit span of

$$L' = \rho \Gamma U \qquad (24.168)$$

and the action of a wing with an arbitrary, continuous, spanwise distribution of circulation, $\Gamma(y)$, can be analysed as a *lifting line* composed of an infinite number of coincident vortex filaments of different strengths. The *induced drag* (or drag due to lift) may be calculated, for any U, given only $\Gamma(y)$. Apart from the requirement

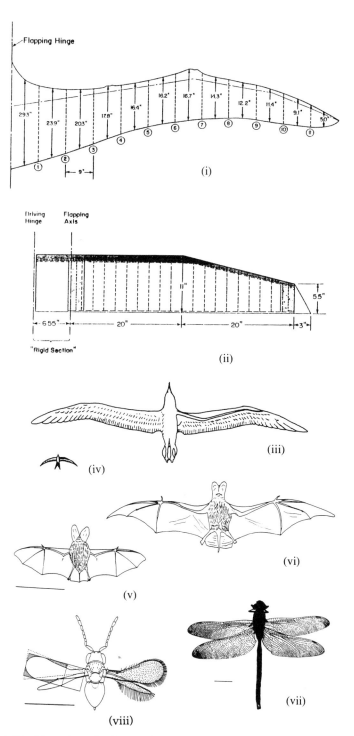

FIGURE 24.57 The wing planforms of certain ornithopters, birds, bats, and insects. [(i) QN, (ii) ComboWing], [(iii) albatross, (iv) Swift], [(v) Long-eared bat, (vi) A free-tailed bat. The scale bar (*sb*) is 0.1 m.] [(vii) Dragonfly. *sb* = 1 cm (viii) Wasp. *sb* = 0.5 mm.] [(iii)–(viii) from Spedding, 1992]. [Reprinted with permission, Spedding, G. R., "The Aerodynamics of Flight," *Mechanics of Animal Locomotion*, Alexander, RMcN. (Ed.), Springer-Verlag, 1992.]

that $\Gamma(-b) = \Gamma(b) = 0$ at the wingtips, $\Gamma(y)$ can be any arbitrary function. A general expression for the induced drag coefficient, C_{D_i}, is

$$C_{D_i} = \frac{C_L^2}{e\pi AR} \tag{24.169}$$

where $e \leq 1$ is the *airfoil efficiency factor*; $e = 1$ when $\Gamma(y)$ is an elliptic function. This is the optimum load distribution for minimum wake energy loss for a given lift on a planar wing in steady motion. It can either be achieved by constructing a wing with an elliptical planform, or the local wing section profile may vary along the span, as may the local wing twist. In practice, small deviations from elliptic loading have a rather small influence on the induced drag, and high AR, rectangular wings work tolerably well, for example. C_{D_i} is proportional to the square of the lift and inversely related to AR, and efficient gliders tend to have very high (> 15) AR wings. Induced drag is not the sole design consideration in practical flying machines. For example, there are conflicting requirements between minimizing induced drag and mean bending moments (long/short span) and between reducing both skin friction drag and maintaining sufficient structural strength (small/large chord). Reviewing these constraints, Jones (1990) reports that the induced drag of a fixed wing may be reduced by about 10% over the elliptic wing, for the same integrated bending moment, for a slightly elongated shape with a more pointed tip.

The total drag force on a finite wing immersed in a viscous fluid can be expressed as a sum of three components, $D_{tot} = D_i + D_f + D_p$, where D_f is the drag component of the surface shear stress, and D_p is the drag component of the pressure due to flow separation. These two terms, which are both due to viscous effects, are frequently added and known as the profile drag, D_{pro}. This can be measured in wind tunnel tests with data such as shown in Figs. 24.55 and 24.56. For a laminar boundary layer over a flat plate, integrating the boundary layer equations for an incompressible flow allows similarity solutions for C_{D_f} as (see Chap. 4)

$$C_{D_f} \approx k_0 \, \mathrm{Re}^{-1/2} \tag{24.170}$$

The proportionality constant, k_0, is approximately 1.33 for a flat plate. Different numbers can be substituted for different geometries. For turbulent boundary layers, the decay is $C_{D_f} \sim \mathrm{Re}^{-1/5}$. Most birds, bats, and ornithopters operate in the range $10^3 < \mathrm{Re} < 10^6$, a domain where estimation of total drag coefficients is fraught with difficulty and uncertainty, even for steady-state conditions in a wind tunnel, and empirical estimates for D_{pro} for both body (referred to as D_{par}) and wings vary widely. Further, they are very sensitive to changes in incidence and surface geometry and roughness, and also to freestream turbulence levels. Issues of unsteady boundary layer separation and reattachment and of time varying geometry can currently be dealt with in only crude approximations, for example with pre- and post-stall coefficients for equations of the form of Eq. (24.170).

Performance Analysis of Gliding in Nature. These simple relationships can be applied to steady gliding flight. From Eq. (24.169), the induced drag can be written

$$D_i = \frac{L^2}{qb^2 e\pi} \tag{24.171}$$

Summing the pressure and frictional drags of Eq. (24.170), the total drag is

$$D_{tot} = qSC_{D_{pro}} + \frac{L^2}{qb^2e\pi}$$ (24.172)

S is the wing surface area, and there will also be a term similar to the first one for the parasite drag on the body, which is commonly asumed to do no useful aerodynamic work. Since the first term increases with q, while the second decreases with q, there exists a value of $U = U_{md}$ at which D_{tot} is minimized. In steady, equilibrium flight, the lift approximately balances the weight (the glide angle, θ, is small), thus $L \cong W$. Ignoring variations in b, S and $C_{D_{pro}}$, U_{md}, the minimum drag speed is,

$$U_{md} = \left(\frac{2L}{\rho b}\right)^{1/2} \left(\frac{1}{SC_{D_{pro}}e\pi}\right)^{1/4}$$ (24.173)

Dimensionally, U_{md} varies with the square root of some characteristic length scale, so geometrically similar gliding animals with linear dimensions different by a factor of 10 will have characteristic gliding speeds different by a factor of around 3. If $L \cong W$, then,

$$\frac{mg}{S} = \frac{1}{2}\rho U^2 C_L$$ (24.174)

and characteristic flight speeds can be expected to vary as the square root of the *wing loading*, $mg/S = Q$. Consequently, the lower the wing loading, the better the ability of a bird to maintain height in only weak updrafts or thermals, and, in locations where the local wind gains strength during the day, one expects that the first gliding birds should be the smaller ones. Manipulation of simple relationships such as these allows useful performance predictions to be generated and tested, and further implications of this cube–square law will be pursued below.

Gliding flight is of great ecological importance to many bird species as the cost of moving unit mass through unit distance is very low. Figure 24.58 compares the glide polars of a bat, a bird, and a motor glider. Characteristic flight speeds increase with size, as expected, and it should be noted that the variable geometry of a bird

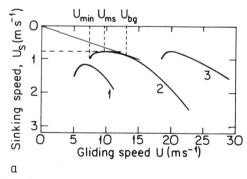

FIGURE 24.58 Glide polars for the fruit-bat (1), the white-backed vulture (2), and the ASK-14 motor glider (3). (After Pennycuick, 1974.)

wing allows not one, but a family of glide polars to be occupied, of which curve 2 is but a single example. In Eq. (24.173), S, b, and probably e and $C_{D_{pro}}$ can be altered by changes in planform and local camber and twist.

24.8.4   Forward Flapping Flight in Birds and Ornithopters

Principles of Lift and Thrust Generation in Flapping Wings. In the absence of any energy input from the environment, the beating wings must provide both lift and thrust required to maintain height and forward speed. Here, we outline the kinematic basis for generating both force components from an oscillating wing section. This is intrinscially a 2-D discussion, and it thus ignores possible variations in span, but such modifications will be discussed later. Some asymmetry must be introduced into the combined pitching and heaving motion of a lifting surface, and it is in fact sufficient just to add a constant angle of attack to the otherwise symmetric motion of a thrust-generating wing. The easiest way to imagine doing this is to tilt the whole body and/or the wings relative to the body, so that, instead of beating vertically up and down with respect to the body, they sweep forwards on the downstroke, making some angle, γ, between the stroke plane and the vertical. Consequently, both the incident velocity, u, and the geometric angle of attack, α, are larger on the down-stroke than on the upstroke, so the magnitude of the resultant force, $\overline{\mathbf{F}}_d$, is greater than that produced during the upstroke, $\overline{\mathbf{F}}_u$. If the time spent on up- and down-strokes is equal (the *downstroke ratio*, $\tau = 0.5$), and the L/D ratio is constant, Fig. 24.59 demonstrates that the mean resultant force, $\overline{\mathbf{F}} = \overline{\mathbf{F}}_d + \overline{\mathbf{F}}_u$, averaged over the wing-beat, can have positive components of lift and thrust. Actually, this figure includes a second strategy for introducing the required asymmetry, as α is even further re-duced during the up-stroke, following a strong pitch-up rotation (or *supination*) at the end of the previous half-stroke. This contributes to the reduction in magnitude of $\overline{\mathbf{F}}_u$. There is considerable scope for exerting fine control over the magnitude and direction of $\overline{\mathbf{F}}$, via changes in γ, $\alpha(t)$ (α can be negative), airfoil section chamber, and τ. Witnn this general framework, ornithopters and animal flappers generate lift and positive net thrust in rather different ways.

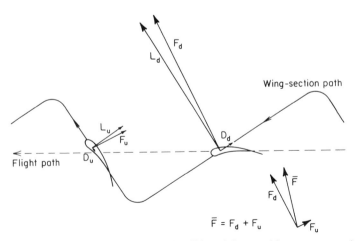

FIGURE 24.59   The production of net positive lift and thrust with an asymmetric wingbeat.

Ornithopter Design: The prototypical, rubber-band powered, model ornithopter was due to Pénaud in 1874 [see Gibbs–Smith (1953)] and is shown in Fig. 24.60(a). It is the basis of the popular toy sold today, and it works essentially by having a large value of γ. This is achieved by adjusting a tailplane so the whole body has a large pitch-up orientation. The wings themselves are composed of a stiff leading-edge spar to which a single sheet of thin flexible material is added. The resulting membranous wing has considerable torsional compliance and almost no leading-edge suction, and thrust is provided by forces normal to the wing chord. As Archer, Sapuppo, and Betteridge (1979) demonstrated, such wings can generate reasonable thrust at low advance ratios, J

$$J \equiv \frac{U}{2\phi nb} \tag{24.175}$$

where n is the flapping frequency, and ϕ is the total angle swept out by the wing of semispan b. However, as evidenced by model aviation records, the efficiency of the Pénaud-type ornithopter is very low. Attempts to improve this have included incorporation of fixed, cambered surfaces to generate the required lift, with outboard flapping elements dedicated to providing thrust in the Lippisch model of Fig. 24.60(b). Alternatively, fixed cambered forward wings can be combined with rearward-mounted flapping surfaces that provide thrust and some stability as shown in Fig. 24.60(c).

Animal Adaptation: Birds, bats, and insects fly with their bodies essentially horizontal and with the stroke plane inclined from the vertical so that $\gamma \cong 20°$. Furthermore, the local angle of incidence is actively changed on down-stroke and up-stroke, as in Fig. 24.59. This occurs through a mix of active twist and passive

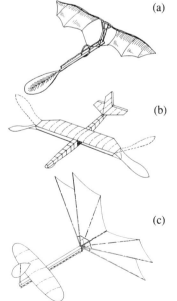

(a)

(b)

(c)

FIGURE 24.60 Time series of evolution in design of three rubber-powered ornithopters. (a) Pénaud, 1874 [see Gibbs–Smith, C. H., "The Eighteen Seventies and Eighties," *A History of Flying*, Batsford, London, 1953], (b) Lippisch, 1938, and (c) Kieser, F., "Canard Ornithopter Construction," *Flapper Facts*, the Ornithopter Modeler Society Newsletter, Vol. 4, No. 4, Autumn 1986.

aeroelastic response to instantaneous forces exerted by the fluid. There is one further modification that has not yet been modeled in ornithopters—the planform geometry and wingspan can be varied. By reducing the span during the up-stroke, the necessary asymmetry in the wingbeat cycle is generated, and a net positive thrust is assured. This strategy of modifying the wing shape during the course of the wingbeat cycle can be quite complicated, involving changes in local camber and incidence, as well as in the total span. In slow flight, many birds reduce the span greatly, *feathering* the folded wing at almost zero local incidence during the up-stroke, which is therefore aerodynamically inactive. In medium speed flight, less drastic span reductions are observed, leading to a different cycle of loading and unloading the outer wing, with the secondary feathers of the inner wing providing positive lift throughout, somewhat analogous to the division of labor in the Lippisch ornithopter. These differing modes of flight lead to the formulation of very different aerodynamic models.

Prevailing aerodynamic analyses and models conveniently divide between those that consider the kinematics and force generation on the flapping wing and those that calculate the net power requirements and energy expenditure by estimating the momentum flux in some kind of vortex wake model. These will be treated in turn.

Flapping Wing Analysis

Quasi-Steady Theories: It is possible, given the instantaneous position of each point on a flapping wing surface, to calculate the total lift and drag by summing all the forces over the span and over the wingstroke cycle, where, at each instant, the lift and drag are calculated from known relationships established in steady flow conditions, so although the time-varying changes in angle of attack are accounted for, the history of the motion is not. The c_l, $c_d(\alpha)$ curves may include nonlinear effects close to stall. The effective angle of attack must be corrected for the induced downwash velocity, w_i, and this can be done with varying degrees of sophistication, the most simple being to use a spanwise-constant estimate from an actuator disk model. Thus, at the nth span location, given the magnitude of the local fluid velocity, $\bar{\mathbf{u}}_n$, and its effective angle of incidence, α_l, on the wing section, the lift per unit span, L', is calculated from

$$L' = \tfrac{1}{2}\rho|\bar{\mathbf{u}}_n|^2(c_l)_n c_n \tag{24.176}$$

where c_n is the local chord. Then, for example, the total lift from both wings is given by the integrated vertical components of these forces over the span and wingbeat cycle

$$L = 2 \int_0^T \int_0^b \{L'_v(r, t) + D'_v(r, t)\}\, dr\, dt \tag{24.177}$$

where r is the spanwise location. If convenient analytical forms for the wingbeat geometry and kinematics are not known, or are not available, then the integrals are replaced by discrete sums. The kinematics and geometry of both live animals and ornithopter model wings can be complicated, and simplifying approximations of some kind are sought.

For animals, the starting point is frequently a detailed description of the wingbeat kinematics and local wing geometry for a particular special. The computed aerodynamic forces are then compared with simultaneous force balance data, or with wind tunnel tests of individual wing sections and checked for self-consistency. The pioneering example of this approach was by Weis–Fogh and Jensen (1956), who took three spanwise locations on the forewing of a locust and computed average lift and drag forces that matched experimental values to within 3–7%. A corollary of this success was the notion that steady-state assumptions were sufficient to account for the aerodynamic performance of the locust, where the reduced frequency, $k =$ 0.15. More recent examples of this approach can be found in Pennycuick (1975), Norberg (1976), and Dudley and Ellington (1990) for birds, bats, and bumblebees, respectively.

From the engineering and design standpoint, if one can construct tractable analytic expressions for the time-dependent, spanwise circulation distribution, then combinations of adjustable parameters can be sought that maximize a propulsive efficiency, η, defined as

$$\eta \equiv U\langle T\rangle/\langle P\rangle \tag{24.178}$$

where $\langle T\rangle$ and $\langle P\rangle$ are the time-averaged thrust and power respectively. Betteridge and Archer (1974), for example, used truncated Fourier series for $\Gamma(y, t)$ in a quasi-steady lifting line model, and found that quite high propulsive efficiencies (compared with conventional propellers) could be attained, but that these declined with increasing advance ratio, J.

Although the remarkable analysis of Jensen and Weis–Fogh argued strongly in favor of the sufficiency of the quasi-steady approach in computing time-averaged force components at low values of $k \approx 0.1$, more recent research indicates that instantaneous forces are poorly resolved, and that even mean values are underpredicted by more than 10% as k rises to 0.4 [cf., Spedding (1992), (1993), Dudley and Ellington (1990)]. More sophisticated analytical models have been proposed that can take into account at least first-order unsteady effects, and they can estimate the magnitude of these corrections to the quasi-steady equations.

Unsteady Analysis: Three contrasting examples are given, where the usual aerodynamic analysis is augmented by addition of unsteady terms that implicitly or explicitly take into account the time-dependent shedding of vorticity into the wake.

Consider first the wing loading and wake geometry estimated from wing kinematics. First-order unsteady corrections to a local strip-wise aerodynamic theory were developed and applied to the forward flight of dragonflies by Azuma et al. (1985) (LCM I) and refined by Azuma and Watanabe (1988) (LCM II). Beginning from detailed measurements of the wingbeat kinematics, the initial $\Gamma(y)$ was approximated by superposing a cascade of elliptic load distributions of diminishing size, each operating in the larger scale shear flow of the previous level. The wake vorticity was assumed to consist of the trailing and transverse (LCM II) vortices fixed by the path of the wing tip and trailing edge, and local circulation values were adjusted iteratively until corrections became small. Nonlinear, empirical $c_l(\alpha)$ curves were used to compute the forces and moments on the wings at each span location. Since the computations were quite lengthy, the wake attenuation coefficients were actually calculated only at $0.75b$. In slow, climbing flight, where k was approximately 1.6, both fore- and hind-wings operated well away from the linear $c_l(\alpha)$ range for a substantial portion of the wingstroke. The computed unsteady terms were shown to be significant improvements over actuator disk-based estimates of constant downwash over the span. In free flight in a wind tunnel ($0.2 \leq k \leq 1.2$), a comparison of LCM II with a constant downwash model at the highest flight speed ($U = 3.2$ ms^{-1}, $k = 0.2$) revealed around 20% differences in the vertical forces and factors of 2–3 difference in the thrust, both for the instantaneous and time-averaged values. Neither of these corrections can be considered small. The estimated power requirements were consistent with the known physio-mechanical properties of insect flight muscle.

Next, consider an unsteady lifting line model for rigid wings. A 3-D unsteady lifting line theory for planar, rigid, unswept wings of high AR was developed by Phlips *et al.* (1981). It is based on a simplified model of the vortex wake, which is divided into near and far wake regions. Transverse vortices are shed at the extremes of the wingstroke, where they roll up and are represented by a single line vortex. Streamwise vortices are shed at the trailing edge and wingtip and are assumed to remain in the path traced by the wings, just as in the previous example. In the far wake, they collect into two trailing vortex lines. This model simplifies the calculation of wake induced velocities on the flapping wing, and the normal forces were calculated assuming a linear $c_l(\alpha)$ relation, effectively restricting the analysis to moderate values of k. When the wingbeat amplitude, ϕ, exceeded approximately 60°, significant departures from the steady state calculations were reported, particularly for $k > 1$. The effect of the induced velocity field in the unsteady calculation was to increase the value of the mean lift coefficient ($\approx 20\%$ for $\phi = \pi/2$, $k = 2$), but the mean thrust coefficients were reduced slightly, so that the efficiency, η, fell with increasing k.

Last, we discuss a strip analysis of a model ornithopter. This represents a point between the previous two examples—the kinematics are completely prescribed, but they are of a real flying machine, so not only is the analysis a little more complicated, but the result must fly in the real world. DeLaurier (1993a) outlined a design-oriented, unsteady, local strip theory for flapping wing flight. At each local span station, unsteady normal force coefficients were taken from a form of modified Theodorsen functions for thin airfoils in potential flow, corrected for finite AR. Although the kinematics allowed for spanwise bending and twist, the downwash, w_i, was calculated for an untwisted elliptical distribution. The local forces were corrected for effects of camber, partial leading edge suction, and partial stall, together with apparent mass effects. These can be thought of as essentially modifications of equations of the form of Eq. (24.176), whose effects are integrated over the wingbeat and over the span as in Eq. (24.177). Very consistent results were obtained for the analysis of two ornithopter models, where $k = 0.11$, and $k = 0.19$. Further performance results will be discussed later in this section.

Unsteady Corrections to Quasi-Steady Models: The results of the models outlined above and their success in apparently accounting for the most important forces and loads on functional flapping wing devices implies: 1) that first-order unsteady terms contribute significantly when $k > 1/4$, and 2) that corrections that account for the nonplanar, unsteady wake terms are quite good at performing a reasonable engineering analysis. Incorporation of spanwise twist allows local stall to be successfully predicted and modeled. In moderately thin airfoils, this can be achieved by passive aeroelastic response of a flexible wing structure. Although the comparative success of these models and design efforts argues indirectly for the *un*importance of three-dimensional and unsteady separated flow effects at low enough k, strong separation and vortex shedding regimes can be observed in cases where flapping amplitudes are high and/or $k > 0.5$. A different class of models can be used in these cases.

Vortex Wake Models

Actuator Disk: The actuator disk is familiar from standard aeronautics texts, and it can be used also to (crudely) model flapping flight. Let the wings be replaced by a conceptual disk with radius, b, and area, $S_d = \pi b^2$. At the disk, downward momentum is imparted to the air, which acquires a vertical velocity, w_i. The final velocity is $2w_i$, so the total rate of change of momentum, which balances the weight in steady level flight is

$$W = \rho U S_d 2w_i \tag{24.179}$$

This can be rearranged to get a simple expression for w_i, which can then be used to estimate a local angle of incidence in a strip theory. The power, P_i, required to overcome induced drag is just $P_i = W w_i$, so the total power requirement for steady flight can be written

$$P_{\text{tot}} = \frac{W^2}{2\rho U S_d} + \frac{1}{2}\rho A U^3 \tag{24.180}$$

The first term is the induced power requirement, and the second is the form that both profile and parasite drags take, where A is a representative surface area. This expression is quite equivalent to the one for the total drag on a gliding wing, Eq. (24.173). This analysis was first applied to animal flight by Pennycuick (1975), who has summarized many of the practical applications in Pennycuick (1989). This is a steady-state model *par excellence*; there is no accounting for time variation of wing loading or span, or, in fact, any wing kinematics at all.

Vortex Rings: Arguing that, in many cases, these assumptions are invalid, Rayner (1979a,b) proposed that when the wings are aerodynamically inactive on the up-stroke, then all bound vorticity must be shed into the wake, and a consideration of Helmholtz's laws suggests that these must eventually form closed vortex loops as in Fig. 24.61(d). If the starting, trailing, and stopping vortices of a single down-stroke do in fact aggregate in a concentrated closed-loop configuration, then one can solve a momentum balance problem by considering the momentum and energy in a chain of discrete vortex rings, whose ellipticity is basically determined by the spanwise circulation distribution and the advance ratio, J, for fixed amplitude ϕ, and down-stroke ratio, τ. In contrast with the actuator disk model, more realistic assumptions can be made concerning these parameters, and the effect of varying them was tested in numerical solutions. Using flow visualization techniques, it was discovered independently by Kokshaysky (1979), and then later by Spedding *et al.* (1984) and

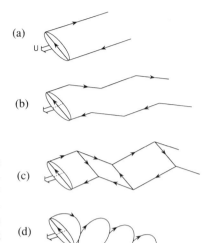

FIGURE 24.61 Topology of vortex lines in 4 wake models, corresponding roughly to domains of increasing k. (a) Steady lifting line. (b) CC/VS—steady lifting line with variable span. (c) Intermediate—unsteady lifting line. (d) Discrete vortex loops. (b)–(d) The wings are imagined to be flapping about the root (denoted by the dashed line), and are in mid-downstroke.

Rayner *et al.* (1986) [see also Brodsky (1991) and Grodnitsky and Morozov (1993) for contrasting views on insect wakes], that wakes of slow-flying birds, bats (and certain insects) are at least qualitatively well-described by a vortex ring-type model. However, if the wing remains even partly loaded during the up-stroke, as appears to be the case in many instances of medium speed or cruising flight of animals and ornithopters both, then the vortex ring wake model cannot be correct.

Constant-Circulation (CC), or Variable-Span (VS) Models: It was noted earlier that the aerodynamic asymmetry required for positive net thrust generation could be generated by corresponding asymmetries in the sectional geometry and kinematics, but it is also implicit in Eq. (24.168) that the *total* lift can be modified not only by changes in Γ, but also by changes in the semispan, b. The latter course of action confers an additional advantage, since it avoids wake energy losses that occur when any $\Delta\Gamma$ on the wings is necessarily accompanied by an equal and opposite increment of wake vorticity. Indeed, Spedding (1987) found that the wake of a kestrel in steady, medium-speed flight was composed not of discrete vortex rings, but rather of two trailing, curved vortices that traced the wingtip path on the down-stroke, but were apparently shed further inboard, from the flexed primary feather at the end of the inner wing, on the up-stroke. There were no detectable transverse vortices in the wake, and their circulation, Γ, was confirmed to be constant along their length. The wake thus takes the form shown in Fig. 24.61(b). As noted above, the absence of transverse wake vortices will coincide with a reduction in wake energy loss, since the induced downwash will be significantly reduced. One might speculate then, that there will be considerable incentive for both animals and ornithopters who care about their efficiency, η, to arrange the local wing twist, camber and span so that such a wake can be produced, as opposed to one where significant transverse vortices appear. This wake model is also attractive to the theoretician, since a very simple formulation that ignores unsteady corrections due to trailing-edge shedding can be constructed. Given the fact that Γ is a constant, then the wake impulse can be computed from the requirement of overcoming the known drag forces on the wings and body, and supporting the weight, and solved for the required Γ and one more un-known geometric parameter that determines the wake shape, such as the ratio of inner to total wing lengths, for example. Example model formulations can be found in Spedding (1987) and Rayner (1986), (1993).

Flapping Flight Performance

Ornithopter Performance: The unsteady strip theory of DeLaurier (1993a) was ap-plied to a numerical analysis of the flapping flight of QN, a 5.5 m span model based on a giant pterosaur [Brooks *et al.* (1985)]. The spanwise twist, $\delta\theta$, was varied linearly with span location, y, and angular position, ϕ, as

$$\delta\theta = -(\beta_0 y) \sin \phi \tag{24.181}$$

The time-averaged thrust, $\langle T \rangle$, and so propulsive efficiency, η, were quite sharply peaked at a particular value of β_0 (twist, in units of radians/unit span). The narrow peak may be due in part to the imposed absence of spanwise bending and the fixed phase angle between pitching and heaving motions. Interestingly enough, the anal-ysis predicted that the outboard wing panels operated under partial stall during much of the down-stroke. Ultimately, QN was not a strong flyer, and increases in drag due to surface roughness of the body and wings could have been sufficient to move off the peak in η. There are also reasons to think that this marginal performance is in fact a realistic result for such a large flapping machine.

The analysis was also applied to the design of a 3 m span model [DeLaurier and Harris (1993)] (see Fig. 24.62) flapping at 3.3 Hz with a mean forward speed of about 15 m/s ($k = 0.19$), which could climb and fly indefinitely depending only on the power plant. In order to retain the high leading-edge suction efficiency of a thick airfoil, and still obtain local twist, torsional compliance was enabled with a novel split trailing edge, and the magnitudes of the dynamic twisting and bending distri-butions were determined by the wing's geometric, inertial, aerodynamic, and elastic characteristics. Such design optimization could be termed *aeroelastic tailoring*. This seems likely to be a crucial ingredient of successful (high η) animal and ornithopter flight.

Scaling Laws in Flapping Flight: A rather basic point about size limitations in flap-ping flight can be made without recourse to any complicated modeling. From an actuator disk theory, one arrives at Eq. (24.180) for the total power requirement; it is a sum of two terms, one of which, the induced power, decreases as U^{-1} and the other, representing frictional and pressure drags on the body and wings, in-creases with U^3. A $P_{tot}(U)$ curve will thus be U-shaped with a minimum, U_{mp}, where $dP/dU = 0$,

$$U_{mp} = \left(\frac{W}{\rho}\right)^{1/2} \left(\frac{1}{3AS_d}\right)^{1/4} \tag{24.182}$$

This is a characteristic flight speed at which energy consumption per unit time is minimized. For geometrically similar animals or ornithopters, $U_{mp} \sim l^{1/2}$. Since $m \sim l^3$, and $P = DU$, then P_{req}, the power required to reach U_{mp}, will scale as P_{req}

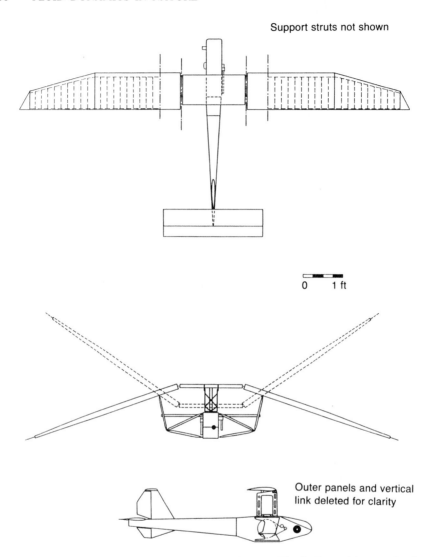

Support struts not shown

Outer panels and vertical
link deleted for clarity

FIGURE 24.62 The 1991 engine-powered ornithopter. The indirect drive mechanism via an articulated fuselage ceiling panel bears a remarkable resemblance to the indirect activation of a resonance-tuned thorax-wing system in flying insects. (Reprinted with permission, DeLaurier, J. D. and Harris, J. M., "A Study of Mechanical Flapping-Wing-Flight," *Aero J.*, Vol. 97, pp. 277–286, 1993.)

$\sim m^{7/6}$. In a biological system, the work, Q, done in one contraction of a flight muscle will scale approximately with its mass, but the contraction frequency, f, scales with l^{-1}, or $m^{-1/3}$. Hence, the power available from the flight muscles, P_{av}, $= Qf$ will scale as $P_{av} \sim m^{2/3}$. If the power available rises less steeply with increasing mass than the power required, then there exists a maximum size beyond which sustained flight is not possible. Although certain assumptions behind this argument may be questioned, the basic result is quite robust to sensible refinements, due, for example, to departures from strict geometric similarity [cf. Ellington (1991), for dissenting arguments, based on observations of *increased* muscle-mass-specific power output with increasing size]. This offers a simple explanation as to why the largest birds walk and run, but do not fly, and also why the flapping flight capabilities of the larger flying birds, such as condors, are rather limited. It is why the original QN is likely to have been primarily a glider, and it is why the possibility of constructing large ornithopters will depend almost entirely on the development of very efficient power plants. Since this is not an option for humans, they are doomed never to fly by flapping wings powered by their own shoulders and arm muscles. This type of scaling behavior actually emerges from any flight model, based as it is on the fundamental way in which the induced drag on one hand, and the pressure and frictional drags on the other hand vary with U, as expressed by the similarity in form of Eqs. (24.172) and (24.180).

Variable-Wing Geometry in Flapping Flight: The CC/VS model can be used to make some observations concerning variations in wing shape. One may consider either the constant-circulation (CC) part of the acronym, or the variable-span (VS) part. For example, in a CC wake, by definition, Γ is constant, so the total lift and thrust are determined by the wake impulse, depending only on the planar areas, $\{A_1, A_2\}$ of the down-stroke and up-stroke wake and the angles $\{\Psi_1, \Psi_2\}$ that they make with the horizontal. It is reasonable to assume that $\Psi_1 = \Psi_2 = \Psi$ and denoting the area ratio, $A_2/A_1 = \zeta$, then, following Pennycuick (1989b), the lift:drag ratio can be expressed as,

$$\frac{L}{D} = \left(\frac{1 + \zeta}{1 - \zeta} \right) \cot \Psi \qquad (24.183)$$

L/D increases with increasing ζ and with decreasing Ψ. On occasions where it is important to maximize this ratio, such as long distance commuting or migrating flight, then a flapper should reduce the amplitude and area ratio as far as possible while still generating sufficient thrust. Long distance commuters should minimize their variations in amplitude and wing shape.

   On the other hand, the VS perspective offers a possible resolution to a curious puzzle raised by the work of DeLaurier (1983), who described how thin, cambered airfoils, such as those possessed by many flying animals, have clear performance optima (in terms of η) at one single value of C_L. How can such a single design-point section operate effectively under the broad range of loads and lift coefficients that one might expect in a flapping cycle? If Γ is constant in the wake and in the bound vortex on the wing, then the airfoil *may* in fact operate, at least in steady flight, in a quite restricted range of C_L. If the animals' spanwise pitching distribution is largely driven by aeroelastic reactions to local forcing from the fluid, then it also may be possible to mechanically replicate this for an ornithopter wing. In fact, this may be the only way to generate efficient flapping flight in small models where airfoils suitable for this Re regime will have low leading-edge suction efficiency. The strategy of CC/VS-type flight in cases of thin, cambered airfoils with no dependence on leading-edge suction contrasts with the alternative for thick, double-surface airfoils at higher Re, where improved leading-edge suction allows for efficient flight with constant semispan. The animal flight world has yet to be researched from this point of view.

24.9 OIL SPREADING ON THE SEA
David P. Hoult

In practice, when oil is spilled on the sea, [Hoult (1972)], we wish to estimate the area of the oil slick and its center of mass in order to then determine where the toxic elements in the oil may affect the environment. We begin by considering an instantaneous release of a volume, V, of oil in a sea with wind and current.

The drift due to wind may be estimated by arguing that the turbulent shear stress law at the water interface is approximately the same in both the air (subscript a) and the water (subscript w). If the wind velocity some distance (usually 10 meters) above the water surface is U_a, then the turbulent stress is

$$\tfrac{1}{2}\rho_a C_{fa} U_a^2 = \tfrac{1}{2}\rho_w C_{fw} U_w^2 \tag{24.184}$$

Our assumption implies that $C_{fa} \cong C_{fw}$. Then it follows that the (drift) velocity in the water is approximately

$$U_w \cong \left[\frac{\rho_a}{\rho_w}\right]^{1/2} U_a \simeq 0.035 \, U_a \tag{24.185}$$

We suppose that placing oil at the air–water interface does not change this result. The drift due to tidal currents is simply taken to be the current velocity. When both wind-driven currents and tidal currents are present, it is supposed that the two vector velocities simply add. The center of mass of the oil is thus supposed to move according to an equation

$$\frac{dX}{dt} = U_c + 0.035 U_a \tag{24.186}$$

where X is the coordinate of the center of mass of the oil, and U_c is current velocity.

Turbulent winds and currents in the oceans disperse as well as convect particles on the surface. It is to be expected that the larger the oil slick, the more rapid this dispersal. In our discussion, this dispersing effect is ignored, because, for a large spill, at least up to times on the order of a few days, the effects of such dispersal are much slower than the tendency for the oil to spread by itself.

We assume a bulk model with constant properties as a first step in developing and understanding the oil spreading process. There are four basic forces that either cause or retard spreading. The oil has a density, ρ_o, only slightly different from water, $\rho_o = \rho_w(1 - \Delta)$, $\Delta \ll 1$. Since the oil is in hydrostatic equilibrium in the vertical, if its thickness is h, it floats a height Δh above the mean-water surface. A simple hydrostatic calculation then shows that there is a pressure acting on the oil in the horizontal direction, of size $(\rho g \Delta h)$. Let the oil pool have a radius, l, that varies with time, t. The force corresponding to this pressure is $(\rho g \Delta h)hl$. If a column of oil is released from rest, inertia effects will tend to retard the motion. The order of magnitude of the inertia term is $(\rho(lt^{-2})hl^2)$.

As the oil slides over the surface of the water there is a viscous drag exerted by the water on the oil. The viscous stress is continuous at the oil–water interface.

However, the oil is much more viscous than the water. Since the oil thickness is smaller than the water boundary layer, we can establish that the velocity gradient in the oil in the vertical direction is negligible compared to that in the water. Thus, there is a *slug flow* in the oil. The retarding force due to viscous drag is $[\mu(lt^{-1})\delta^{-1})]l^2$, where $\delta = (vt)^{1/2}$ is the water boundary-layer thickness, μ is the viscosity of the water, and $v = \mu/\rho$.

Surface tension acts at the edge of the oil slick. The net spreading coefficient is defined as

$$\sigma = \sigma_1 - \sigma_2 - \sigma_3 \qquad (24.187)$$

where σ_1 is the air–water interfacial tension, σ_2 is the oil–water interfacial tension, and σ_3 is the oil–air interfacial tension. The force due to surface tension is σl which may be positive or negative, depending on the sign of σ. For most crude oils, σ is positive; a typical value might be 25 dynes/cm.

Assume that a volume V of oil is released at a point. Then continuity requires

$$hl^2 \sim V \qquad (24.188)$$

If now the various order-of-magnitude estimates are compared, it is clear that for long times, when the oil is very thin and $h \to 0$, the viscous forces must balance the surface-tension forces In this *surface-tension regime* of spreading, we get a spreading law

$$l = 1.33\sigma^{1/2}\left(\frac{\rho}{\mu}\right)^{1/4} t^{3/4} \qquad (24.189)$$

The factor of 1.33 is an empirical constant. It is important to note that in this regime, the rate of oil spread is independent of the volume released.

For very short times, the situation is changed. Of the two forces that tend to retard the oil, inertia is greater, behaving as t^{-2} for fixed dimensions, whereas the viscous drag behaves as $t^{-3/2}$. Provided the oil is thick enough, i.e., if

$$h > \left[\frac{\sigma}{\rho g \Delta}\right]^{1/2} \qquad (24.190)$$

then gravity terms dominate the surface tension driving force. For this *inertial regime* of spreading, we get a second spreading law

$$l \sim (g\Delta V)^{1/4}t^{1/2} \qquad (24.191)$$

which should hold for short times provided Eq. (24.190) is met initially. [The value of $(\sigma/\rho g\Delta)^{1/2}$ is typically 1 cm.]

For intermediate times, when neither Eq. (24.190) (because the time is too long) nor Eq. (24.189) (because time is too short) is valid, there is a third regime that may become important. This regime arises when the viscous retarding force is greater than the interial retarding force, that is, when

$$h < [vt]^{1/2} \qquad (24.192)$$

but when the inequality [Eq. (24.190)] is still met. In this *viscous regime*, the spreading law is derived

$$l \sim \nu^{-1/12}(g\Delta)^{1/6}V^{+1/3}t^{1/4} \tag{24.193}$$

Experiments have shown that the coefficients of the three spreading laws are all near unity for the one dimensional case, however replacing the proportionality sign with a factor near unity does not, in general, improve agreement with field data, due to the large scatter in the latter. In summary then, for a large V, one first gets an inertial spreading phase, with radius increasing as $t^{1/2}$, then a gravity-viscous phase, with width increasing as $t^{1/4}$, and finally a surface tension phase, with width increasing as $t^{3/4}$. For very small spills, only surface tension spreading is present.

For a small, steady leak, on a current with a wind, calculate the drift using Eq. (24.186). Then the width of the plume at a distance X downstream of the source can be obtained by using Eq. (24.189), with t being the time to drift to a distance X. This result also holds a long distance from the site of the spill, provided σ [in Eq. (24.187)] is unchanged.

Of course, the oil properties evolve with time. At mid-latitudes, evaporation of low boiling point compounds to the air takes place in a few hours time. Dissolution of surfactants into the sea water, which can lead to emulsions of water and oil, *chocolate mousse*, takes a few days. Biological degradation of the remaining lighter hydrocarbons takes months, and for the heavier tars, years.

24.10 BUBBLERS FOR ICE MELTING
George D. Ashton

24.10.1 Uses of Bubblers

Use of air bubbler systems to suppress ice formation is one possible means of aiding winter navigation in harbors, ports, and waterways during periods when thick ice would otherwise halt navigation. Bubblers are also used to locally melt the ice in the vicinity of marine structures such as docks and piles with the purpose of minimizing the loads exerted on the structure by ice. The basic concept of operation of a bubbler system is to induce the melting of the ice cover by convecting the warm water beneath the ice against the ice cover as illustrated in Fig. 24.63. A compressor at point A delivers air into a supply line B. It is then released at point C through a small orifice. The rising bubbles then entrain water as they rise (region D) thus forming a rising plume that impinges against the surface and then spreads laterally beneath the ice cover. The heat transfer and consequent melting of the ice occurs

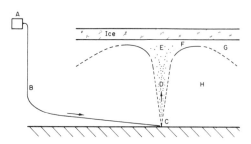

FIGURE 24.63 Schematic drawing of a bubbler system.

primarily by convection at the underside of the ice in regions E, F, and G. Although not treated here, bubblers using high air discharge rates are also used to induce a surface flow of water to flush or move ice or other debris away from such places as lock forebays or gate recesses. To be effective in melting ice, bubblers must have warm water available although surprisingly small temperatures above freezing (on the order of 0.2 C) are often effective. Both point source bubblers, consisting of a single orifice for release of the bubbles, and line source bubblers, consisting of a row of orifices along a distribution line, are used depending on the desired geometry of the area of ice suppression.

24.10.2 Mechanics of Bubbler Flows

The mechanics of bubbler flows may be separated into the mechanics of the bubble formation at the orifice, the mechanics of the plume induced by the rising bubble stream, the mechanics of the impinging and spreading surface water jet and the transport and delivery of thermal energy by these component flows.

The air is delivered to the orifices by a compressor and associated distribution piping. In the case of a line source bubbler, the distribution pipe constitutes a manifold-type flow problem and methods are available for the design by iterative solution [Berlamont and Van der Beken (1973) and Camp and Graber (1968)]. The discharge rate of air at each orifice is determined from the usual discharge equation for an orifice

$$Q_o = C_d(\pi d^2/4) \, (2\Delta p/\rho_a)^{1/2} \tag{24.194}$$

where Q_o is the air discharge from an orifice of diameter d under a differential pressure Δp. C_d is the loss coefficient [Rouse (1946)], and ρ_a is the mass density of air. At low discharge rates, bubbles form individually. As the discharge rate increases, the size and rate of bubble production increases, and at high discharge rates a distribution of bubble sizes results from formation of bubbles by breakup of the air jet stream [Silberman (1957) and van Krevelen and Hoftijzer (1950)]. Bubblers to melt ice generally operate in the latter regime, but the important parameter is the discharge rate of air since it is the buoyancy flux that largely determines the water flow rather than the particular size distribution of the bubbles.

The plume flows induced by the air discharge have been evaluated both analytically and experimentally by Kobus (1968) as jet-type flows with constant buoyancy flux. The mean rising speed of the bubbles, U_b, and the rate of spread, c, of a Gaussian-shaped velocity profile are described as empirical functions of the air discharge rate. For a point source bubbler

$$U_b = C_b Q_o^{0.15} \tag{24.195}$$

$$c = C_c Q_o^{0.15} \tag{24.196}$$

where $C_b = 0.152 \text{ m}^{-0.45} \text{ s}^{0.15}$ and $C_c = 1.83 \text{ m}^{0.55} \text{ s}^{-0.85}$. The induced flow of water at height, H, above the orifice is then

$$Q_w(H) = 2C_c(H + x_o)Q_o^{0.575} \left[\frac{-p_{atm} \, \pi \, \log_e [1 - (H/H^*)]}{\rho_w C_b} \right]^{0.5} \tag{24.197}$$

where x_o is a correction distance ($\doteq$ 0.8 m), p_{atm} is the atmospheric pressure, ρ_w is the water density, and $H^* = H + p_{atm}/\rho_w$ (for sea level conditions $H^* = H + 10.3$ m).

For a line source bubbler with an air discharge rate Q_a per unit length of line ($Q_a = Q_o/s$ where s is the spacing of orifices) the equivalent relationships are

$$U_b = C_b Q_a^{0.15}$$
(24.198)

$$c = C_c Q_a^{0.15}$$
(24.199)

where $C_b = 2.14$ m$^{0.7}$ s$^{-0.85}$ and $C_c = 0.182$ m$^{0.3}$ s$^{0.15}$. The induced flow of water is given by

$$Q_w(H) = \sqrt{2}\pi^{0.25} \left(\frac{p_{atm}C_c}{\rho_w C_b}\right)^{0.5} Q_a^{0.5}(H + x_o)^{0.5}$$
$$\cdot \left[-\log_e\left(1 - \frac{H}{H^*}\right)\right]^{0.5}$$
(24.200)

Figure 24.64 shows the results for the line source case and illustrates the large quantities of water moved by small air discharges. The width of the plume is about 0.1 H over a wide range of air discharges [half-width $b = c(H + x_o)$].

The heat transfer coefficient, h, from the water to the ice undersurface has been analyzed for the point-source case [Ashton (1979a)] and the line source case [Ashton (1978)] using the results [Gardon and Akfirat (1966)] for impinging turbulent jets. The results for the line source case are shown in Fig. 24.65 for the impingement region and are on the order of 10^3 Wm^{-2} C^{-1}. Outward from the impingment region, the heat transfer coefficient rapidly decays (for the line source case $h \sim y^{-0.38}$; for

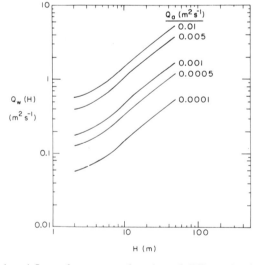

FIGURE 24.64 Induced flow of water as a function of diffuser depth and air discharge for a line source bubbler.

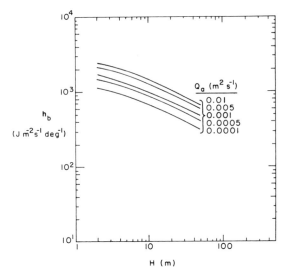

FIGURE 24.65 Heat transfer coefficient in the impingement region as a function of diffuser depth and air discharge for a line source bubbler.

the point source case $h \sim r^{-0.45}$). With the heat transfer coefficient, h, known, the heat flux, q_w, to the ice undersurface is determined from $q_w = hT$ where T is the water temperature in C, and the melting rate is then determined from the energy balance at the ice-water interface

$$q_i - q_w = \rho_i L d\eta/dt \qquad (24.201)$$

where q_i is the conductive flux into the ice cover due to the cold atmosphere above it, ρ_i is the ice density, L is the heat of fusion and $d\eta/dt$ is the rate of thickening of the ice sheet. The temperature of the water can, with good accuracy, be taken as the average over the depth of submergence.

The component mechanics of the bubbler-induced flow and heat transfer lend themselves to quasi-steady numerical simulation using as input data the water temperatures and varying air temperature. Computer codes are available [Ashton (1979b)].

24.10.3 Design of Bubbler Systems

Parameters for which the designer of a bubbler system has a choice, are the compressor output pressure and air discharge rate, the supply line diameter (length is generally dictated by site geography), diffuser line diameter and submergence depth, and orifice diameter and spacing. Each are briefly discussed below, but the designer should realize that the various parameters interact with each other in determining the ultimate performance of a particular system.

Output pressure of the compressor must be sufficient to overcome the hydrostatic pressure at the diffuser line depth, the frictional losses in the supply and diffuser lines and yet provide a pressure differential at the orifices to drive the air out at the desired rate. The air discharge rate is determined by the system geometry (length, pipe diameters, etc.). Typical air discharge rates used in field installations have been on the order of 10^{-4} m^3 s^{-1} per meter of line. A given system is generally either discharge-limited or pressure-limited by the compressor characteristics and system geometry. Ideally, a balanced design would result in discharge pressure output at the peak efficiency point of the compressor performance.

The supply and diffuser line diameters should be chosen sufficiently large that the frictional pressure drop along the line is small, hence a uniform air discharge rate is maintained along the line. Often a small increase in line diameter will result in considerably more uniform air discharge rates and significantly reduced friction losses. Typical line diameters of field installations are 4 to 8 cm.

Submergence depth is governed generally by operational limitations such as depth of water body or required clearance for vessel drafts. The deeper the submergence, the more water will be moved by a given air discharge rate, hence the more suppression effected. In some cases, however, the pressure requirements for very deep depths would make it desirable to suspend the line above the bottom.

Typical orifice diameters are of the order of 1 mm, and typical spacings are of the order of 1/3 the submergence depth. Too large orifice diameters can result in all the air leaving the diffuser line at one end. The pressure-diameter-discharge relationships cannot be easily separated from the total system design.

It is desirable to be able to open the far end of a diffuser line so that water may be easily pumped out after system shutdown. System performance is not affected by type of pipe used (plastic or metal) so the choice should be made on basis of maintenance, reliability, and operational considerations.

REFERENCES

Ahmadi, A. R., "An Asymptotic Unsteady Lifting-Line Theory With Energetics and Optimum Motion of Thrust Producing Lifting Surfaces," Mass. Inst. of Tech., Dep. Aeronautics and Astronautics Fluid Dynamics Res. Lab. Report 80.2, 1980.

Aifantis, E. C., "Mathematical Modelling for Water Flow in Plants," *Proc. 1st Int. Conf. Math. Modeling*, Avula, X. J. R. (Ed.), Rolla, MO, 2, pp. 1083–1090, 1977.

Akins, R. E., Peterka, J. A., and Cermak, J. E., "Mean Force and Moment Coefficients for Buildings in Turbulent Boundary Layers," *J. Industrial Aerodynamics*, Vol. 2, pp. 195–209, 1977.

Andrews, E. D., "Marginal Bed Load Transport in a Gravel Bed Stream, Sagehen Creek, California," *Water Resources Res.*, Vol. 30, No. 7, pp. 2241–2250, 1994.

Archer, R. D., Sapuppo, J., and Betteridge, D. S., "Propulsion Characteristics of Flapping Wings," *Aero. J.*, Vol. 83, pp. 355–371, 1979.

ASCE Standards, "Minimum Design Loads for Buildings and Other Structures," ANSI/ASCE 7-88, American Society of Civil Engineers, New York, 1988.

ASHRAE, "Air Flow Around Buildings," *ASHRAE Handbook, 1993 Fundamentals*, American Society of Heating, Refrigerating and Air-Conditioning Engineers, Inc., Atlanta, GA, 1993.

Ashton, G. D., "Numerical Simulation of Air Bubbler Systems," *Canadian J. Civil Eng.*, Vol. 5, pp. 231–238, 1978.

Ashton, G. D., "Point Source Bubbler Systems to Suppress Ice," *Cold Regions Science and Technology*, Vol. 1, pp. 93–100, 1979a.

Ashton, G. D., "Point Source Bubbler Systems to Suppress Ice," Report 79-12, U.S. Army Cold Regions Research and Engineering Laboratory, Hanover, NH, 13 pp., 1979b.

Aston, M. J. and Jones, M. M., "A Study of the Transpiration Surfaces of *Avena sterilis* L. var. Algerian Leaves Using Monosilicic Acid as a Tracer for Water Movement," *Plant* (Berl.), Vol. 130, pp. 121–129, 1976.

Azad, R., *The Atmospheric Boundary Layer for Engineers*, Kluwer Academic Press, Boston, MA, 1993.

Azuma, A. and Watanabe, T., "Flight Performance of a Dragonfly," *J. Exp. Biol.*, Vol. 137, pp. 221–252, 1988.

Azuma, A., Azuma, S., Watanabe, I., and Furuta, T., "Flight Mechanics of a Dragonfly," *J. Exp. Biol.*, Vol. 116, pp. 79–107, 1985.

Balakrishnan, M. and Dancey, C., "An Investigation of Turbulence in Open Channel Flow via Three-Component Laser Doppler Anemometry," *Proceedings of the Symposium on Fundamentals and Advancements in Hydraulic Measurements and Experimentation*, ASCE, 1994.

Balazs, E. A. and Gibbs, D. A., "The Rheological Properties and Biological Function of Hyaluronic Acid," *Chemistry and Molecular Biology of the Intercellular Matrix*, Balazs, E. A. (Ed.), Academic Press, New York, 1970.

Bange, G. G. J., "On the Quantitative Explanation of Stomatal Transpiration," *Acta Bot. Neerl.*, Vol. 2, pp. 255–297, 1953.

Basdevant, C., Legras, B., Sadourny, R., and Beland, B., "A Study of Barotropic Model Flows: Intermittency, Waves, and Predictability," *J. Atmos. Sci.*, Vol. 38, pp. 2305–2326, 1981.

Batchelor, G. K., "Slender Body Theory for Particles of Arbitrary Cross Section in Stokes Flow," *J. Fluid Mech.*, Vol. 44, pp. 419–440, 1970.

Batts, M. E., Cordes, M. R., Russell, L. R., Shaver, J. R., and Simiu, E., "Hurricane Wind Speeds in the United States," NBS Building Science Series 124, U.S. Government Printing Office, Washington, D.C., 1980.

Berlamont, J. and Van der Beken, A., "Solutions for Lateral Outflow in Perforated Conduits," *J. Hydraulics Div.*, ASCE, Vol. 99, No. HY9, pp. 1531–1549, 1973.

Bertuzzi, A., Mancinelli, R., Ronzoni, G., and Salinari, S., "A Mathematical Model of Intestinal Motor Activity," *J. Biomech.*, Vol. 11, pp. 41–47, 1978.

Betteridge, D. S. and Archer, R. D., "A Study of the Mechanics of Flapping Wings," *Aero. Q.*, Vol. 25, pp. 129–141, 1974.

Bidwell, R. G. S., *Plant Physiology*, Macmillan, New York, 1974.

Blake, J. R., "A Model for Micro-Structure in Ciliated Organisms," *J. Fluid Mech.*, Vol. 55, pp. 1–23, 1972.

Blake, J. R., "Mucus Flows," *Math. Biophys.*, Vol. 17, pp. 301–313, 1973.

Blake, J. R., Vann, G., and Winet, H., "A Model of Ovum Transport," *J. Theoretical Biol.*, Vol. 102, pp. 145–166, 1983.

Blake, J. R., "On the Hydrodynamics of Plasmodesmata," *J. Theor. Biol.*, Vol. 74, pp. 33–47, 1978.

Boggs, D. W. and Peterka, J. A., "Aerodynamic Model Tests of Tall Buildings," *J. of Engineering Mechanics Division*, ASCE, Vol. 115, pp. 618–635, 1989.

Bray, D. I., "Estimating Average Velocity in Gravel-Bed Rivers," *J. Hydraulics Div.*, ASCE, Vol. 105, No. 9, pp. 1103–1122, 1979.

Brennen, C. and Winet, H., "Fluid Mechanics of Propulsion by Cilia and Flagella," *Ann. Rev. Fluid Mech.*, Vol. 9, pp. 339–397, 1977.

Brodsky, A. K., "Vortex Formation in the Tethered Flight of the Peacock Butterfly *Inachis io* L (Lepidoptera, Nymphalidae) and Some Aspects of Insect Flight Evolution," *J. Exp. Biol.*, Vol. 161, pp. 77–95, 1991.

Brokaw, C. J. and Verdugo, P., *Mechanism and Control of Ciliary Movement*, Brokaw, C. J. and Verdugo, P. (Eds.), Alan R. Liss, New York, 1982.

Brooks, A. N., MacCready, P. B., Lissaman, P. B. S., and Morgan, W. R., "Development of a Wing-Flapping Flying Replica of the Largest Pterosaur," AIAA 85-1446, 1985.

Brown, C. B., "Sediment Transportation," *Engineering Hydraulics*, Rouse, H. (Ed.), John Wiley & Sons, New York, 1950.

Brown, H. T. and Escombe, F., "Static Diffusion of Gases and Liquids in Relation to the Assilimilation of Carbon and Translocation in Plants," *Philos. Trans. Roy. Soc. (London) Ser. B*, Vol. 193, pp. 223–291, 1900.

Brownlie, W. R., "Flow Depth in Sand-Bed Channels," *J. Hydraulic Eng.*, ASCE, Vol. 109, No. 7, pp. 959–990, 1983.

Camp. T. R. and Graber, S. D., "Dispersion Conduits," *J. Sanitary Engineering Div.*, ASCE, Vol. 94, No. SA1, pp. 31–39, 1968.

Canny, M. J., "Flow and Transport in Plants," *Ann. Rev. Fluid Mech.*, Vol. 9, pp. 275–296, 1977.

Canny, M. J., "The Transpiration Stream in the Leaf Apoplast Water and Solutes," *Phil. Trans. Roy. Soc. (London) B Biol. Sci.*, Vol. 341, No. 1295, pp. 87–100, 1993.

Carnavale, G. F., McWilliams, J. C., Pomeau, Y., Weiss, J. B., and Young, W. R., "Evolution of Vortex Statistics in Two-Dimensional Turbulence," *Phys. Rev. Lett.*, Vol. 66, pp. 2736–2737, 1991.

Cermak, J. E., "Aerodynamics of Buildings," *Annual Review of Fluid Mechanics*, Vol. 8, pp. 75–106, 1976.

Cermak, J. E., "Applications of Fluid Mechanics to Wind Engineering—A Freeman Scholar Lecture," *J. Fluids Eng.*, Vol. 97, pp. 9–38, 1975.

Cermak, J. E., "Applications of Wind Tunnels to Investigation of Wind-Engineering Problems," *AIAA J.*, Vol. 17, No. 7, pp. 679–690, 1979.

Cermak, J. E., "Laboratory Simulation of the Atmospheric Boundary Layer," *AIAA J.*, Vol. 9, No. 9, Sept., pp. 1746–1754, 1971.

Cermak, J. E., "Physical Modeling of the Atmospheric Boundary Layer (ABL) in Long Boundary-layer Wind Tunnels," *Proceedings of the International Workshop on Wind Tunnel Modeling Criteria and Techniques in Civil Engineering Applications*, Cambridge University Press, 1982.

Cermak, J. E., "Wind Tunnel Design for Physical Modeling of Atmospheric Boundary Layers," *J. Engineering Mechanics Div.*, ASCE, Vol. 102, No. EM3, pp. 623–642, 1981.

Chang, H. H., *Fluvial Processes in River Engineering*, Wiley–Interscience, New York, 1988.

Chapman, D. C., Rand, R. H., and Cooke, J. R., "A Hydrodynamical Model of Bordered Pits in Conifer Tracheids," *J. Theor. Biol.*, Vol. 67, pp. 11–24, 1977.

Charney, J. G., "The Dynamics of Long Waves in a Baroclinic Westerly Current," *J. Meteor.*, Vol. 4, pp. 135–163, 1947.

Chaudhry, M. H., *Open-Channel Flow*, Prentice Hall, Englewood Cliffs, NJ, 1993.

Cheng, H. K., "On Lifting-Line Theory in Unsteady Aerodynamics," *Unsteady Aerodynamics*, Proc. of a Symposium held at the University of Arizona, Mar. 18–20, Kinney, R. B. (Ed.), pp. 719–729, 1975.

Chopra, M. G. and Kambe, T., "Hydromechanics of Lunate-Tail Swimming Propulsion, Part 2," *J. Fluid Mech.*, Vol. 79, pp. 49–60, 1977.

Chopra, M. G., "Hydromechanics of Lunate-Tail Swimming Propulsion," *J. Fluid Mech.*, Vol. 64, pp. 375–391, 1974.

Chow, V. T., *Open-Channel Hydraulics*, McGraw-Hill, New York, 1959.

Church, M. A., McLean, D. G., and Wolcott, J. F., "River Bed Gravels: Sampling and Analysis," in *Sediment Transport in Gravel-Bed Rivers*, Thorne, C. R., Bathurst, J. C., and Hey, R. D. (Eds.), John Wiley & Sons Ltd., London, U.K., 1987.

Chwang, A. T. and Wu, T. Y., "A Note on the Helical Movement of Microorganisms," *Proc. R. Soc. London Ser. B.*, Vol. 178, pp. 327–346, 1971.

Colby, B. R., "Discharge of Sands and Mean-Velocity in Sand-Bed Streams," USGS Prof. Paper 462-A, Washington, D.C., 1964.

Coleman, N. L., "Velocity Profiles with Suspended Sediment," *J. of Hydraulic Research*, IAHR, Vol. 19, No. 3, pp. 211–229, 1981.

Cooke, J. R., Rand, R. H., Mang, H. A., and Debaerdemaeker, J. G., "A Nonlinear Finite Element Analysis of Stomatal Guard Cells," ASAE Paper No. 77-5511, American Society of Agriculture Engineering, 1977.

Cooke, J. R. and Rand, R. H., "Diffusion Resistance Models," *Predicting Photosynthesis for Ecosystem Models*, Hesketh, J. D. and Jones, J. W. (Eds.), Vol. 1, pp. 93–121, CRC Press, Boca Raton, FL, 1980.

Cooke, J. R., Debaerdemaeker, J. G., Rand, R. H., and Mang, H. A., "A Finite Element Shell Analysis of Guard Cell Deformations," *Trans. ASAE*, Vol. 19, pp. 1107–1121, 1976.

Cooke, J. R., Upadhyaya, S. K., Delwiche, M. J., Rand, R. H., Scott, N. S., and Sobel, E. T., *StomateTutor: An Introduction to Stomatal Control of Gas Exchange in Plants*, Cooke Publications, Ithaca, NY (A HyperCard stack for the Macintosh), 1988.

Cooke, J. R., "Some Theoretical Considerations in Stomatal Diffusion: A Field Theory Approach," *Acta Biotheor.*, Vol. 17, pp. 95–124, 1967.

Cooke, J. R., "The Influence of Stomatal Spacing Upon Diffusion Rate," ASAE Paper No. 69-525, American Society of Agricultural Engineering, 1969.

Cooke, J. R., "Water Transport and Balance within the Plant: Stomatal Mechanics and Gas Exchange," *Limitations to Efficient Use of Water in Crop Production*, Sinclair, Tanner, and Jordan (Eds.), American Society of Agronomy, pp. 173–181, 1983.

Cox, R. G., "The Motion of Long Slender Bodies in a Viscous Fluid. Part 1. General Theory," *J. Fluid Mech.*, Vol. 44, pp. 791–810, 1970.

Crafts, A. S. and Crisp, C. E., *Phloem Transport in Plants*, Freeman, W. H. (Ed.), San Francisco, CA, 1971.

Cunge, J. A., Holly, F. M., Jr., and Verwey, A., *Practical Aspects of Computational River Hydraulics*, Pitman, 1980.

Darwin, C., "On the Movements and Habits of Climbing Plants," *J. Linn. Soc. Bot.*, Vol. 9, pp. 1–118, 1865.

Davenport, A. G., "The Application of Statistical Concepts to the Wind Loading of Structures," *Proc. Inst. Civ. Eng. London*, Vol. 19, pp. 449–472, 1961.

Davenport, A. G., "The Relationship of Wind Structure to Wind Loading," *Proceedings of Symposium No. 16—Wind Effects on Buildings and Structures*, National Physical Laboratory, Teddington, England, Her Majesty's Stationery Office, Vol. 1, pp. 54–101, 1965.

DeLaurier, J. D. and Harris, J. M., "A Study of Mechanical Flapping-Wing Flight," *Aero. J.*, Vol. 97, pp. 277–286, 1993.

DeLaurier, J. D., "An Aerodynamic Model for Flapping-Wing Flight," *Aero. J.*, Vol. 97, pp. 125–130, 1993a.

DeLaurier, J. D., "Drag of Wings with Cambered Airfoils and Partial Leading-Edge Suction," *J. Aircraft*, Vol. 20, pp. 882–886, 1983.

DeLaurier, J. D., "The Development of an Efficient Ornithopter Wing," *Aero. J.*, Vol. 97, pp. 153–162, 1993b.

Delwiche, M. J. and Cooke, J. R., "An Analytical Model of the Hydraulic Aspects of Stomatal Dynamics," *J. Theor. Biol.*, Vol. 69, pp. 113–141, 1977.

Diplas, P. and Fripp, J. B., "Properties of Various Sediment Sampling Procedures," *J. Hydraulic Eng.*, ASCE, Vol. 118, No. 7, pp. 955–970, 1992.

Diplas, P., Kennedy, J. F., and Odgaard, A. J., "Stability of Initially Straight Rivers," *River Regime*, White, W. R. (Ed.), John Wiley & Sons, Ltd., 1988.

Diplas, P., "Bedload Transport in Gravel-Bed Streams," *J. Hydraulic Eng.*, ASCE, Vol. 113, No. 3, pp. 277–292, 1987.

Diplas, P., "Characteristics of Self-Formed Straight Channels," *J. Hydraulic Eng.*, ASCE, Vol. 116, No. 5, pp. 707–728, 1990.

Diplas, P., "Modelling of Fine and Coarse Sediment Interaction over Alternate Bars," *J. Hydrology*, Vol. 159, pp. 335–351, 1994.

Drazin, P. G. and Reid, W. H., *Hydrodynamic Stability*, Cambridge University Press, Cambridge, England, 1981.

Dudley, R. and Ellington, C. P., "Mechanics of Forward Flight in Bumblebees. II. Quasi-Steady Lift and Power Requirements," *J. Exp. Biol.*, Vol. 148, pp. 53–88, 1990.

Eady, E. T., "Long Waves and Cyclone Waves," *Tellus*, Vol. 1, pp. 33–52, 1949.

Einstein, H. A., "The Bed Load Function for Sediment Transport in Open Channels," *Technical Bulletin 1026*, U.S. Dept. of Agriculture, 1950.

Ekman, V. W., "On the Influence of the Earth's Rotation on Ocean Currents," *Arkiv för Matematik, Astronomi och Fysik*, Vol. 2, No. 11, p. 52, 1905.

Ellenson, J. L. and Amundson, R. G., "Delayed Light Imaging for the Early Detection of Plant Stress," *Science*, Vol. 215, pp. 1104–1106, 1982.

Ellington, C. P., "The Aerodynamics of Hovering Insect Flight. I. The Quasi-Steady Analysis, II. Morphological Parameters, III. Kinematics, IV. Aerodynamic Mechanisms, V. A Vortex Theory, and VI. Lift and Power Requirements," *Phil. Trans. R. Soc. Lond. B.*, Vol. 305, pp. 1–181, 1984.

EPA, "Guideline for Use of Fluid Modeling to Determine Good Engineering Practice Stack Height," EPA-450/4-79-015, U.S. Environmental Protection Agency, Office of Air Quality Planning and Standards, 1979.

Esau, K., *Anatomy of Seed Plants*, 2nd ed., John Wiley & Sons, New York, 1977.

Esau, K., *Plant Anatomy*, John Wiley & Sons, New York, 1965.

Evans, E. A. and Skalak, R., *Mechanics and Theromdynamics of Biomembranes*, CRC Critical Reviews in Bioengineering, Vol. 3, Issues 3 & 4, CRC Press, Boca Raton, FL, 1979.

Fahn, A., *Plant Anatomy*, 4th ed., Pergamon Press, New York, 1990.

Falcon, M., "Secondary Flow in Curved Open Channels," *Ann. Rev. Fluid Mech.*, Vol. 16, pp. 179–193, 1984.

Ferro, V. and Baiamonte, G., "Velocity Profiles in Gravel-Bed Rivers," *J. Hydraulic Eng.*, ASCE, Vol. 120, No. 1, pp. 60–80, 1994.

Fornberg, B., "A Numerical Study of 2-D Turbulence," *J. Comp. Phys.*, Vol. 25, pp. 1–25, 1977.

Fredsoe, J., "Meandering and Braiding of Rivers," *J. Fluid Mech.*, Vol. 84, pp. 609–624, 1978.

Fripp, J. B. and Diplas, P., "Surface Sampling in Gravel Streams," *J. Hydraulic Eng.*, ASCE, Vol. 119, No. 4, pp. 472–490, 1993.

Fujita, T. T. and Abbey, R. F., Jr., "Tornadoes and High Winds," NUREG/CR-1447, U.S. Nuclear Regulatory Commission, Washington, D.C., 1980.

Fung, Y. C., *Biomechanics: Mechanial Properties of Tissues*, Springer-Verlag, New York, 1981a.

Fung, Y. C., "The Lung—A Perspective of Biomechanics Development," *J. Biomech. Eng.*, Vol. 103, pp. 91–96, 1981b.

Gage, K. S., "Evidence for a $k^{-5/3}$ Law Inertial Range in Mesoscale Two-Dimensional Turbulence," *J. Atmos. Sci.*, Vol. 36, pp. 1950–1954, 1979.

Galston, A. W., Davies, P. J., and Satter, R. L., *The Life of the Green Plant*, 3rd ed., Prentice-Hall, Englewood Cliffs, NJ, 1980.

Garde, R. J. and Ranga Raju, K. G., *Mechanics of Sediment Transportation and Alluvial Stream Problems*, 2nd ed., Wiley Eastern Ltd., New Delhi, 1985.

Gardon, R. and Akfirat, J. C., "Heat Transfer Characteristics of Two-Dimensional Air Jets," *J. Heat Transfer*, Vol. 88, pp. 101–108, 1966.

Gibbs–Smith, C. H., "The Eighteen Seventies and Eighties," *A History of Flying*, Batsford, London, 1953.

Gray, J. and Hancock, G. J., "The Propulsion of Sea-Urchin Spermatozoa," *J. Exp. Biol.*, Vol. 32, pp. 802–814, 1955.

Grodnitsky, D. L. and Morozov, P. P., "Vortex Formation During Tethered Flight of Functionally and Morphologically Two-Winged Insects, Including Evolutionary Considerations on Insect Flight," *J. Exp. Biol.*, Vol. 182, pp. 11–40, 1993.

Gross, L. J., "On the Dynamics of Internal Leaf Carbon Dioxide Uptake," *J. Math. Biol.*, Vol. 11, pp. 181–191, 1981.

Gumowski, I., "Analysis of Oscillatory Plant Transpiration," *J. Interdiscipl. Cycle Research*, Vol. 12, pp. 273–291, 1981.

Hammel, H. T. and Scholander, P. F., *Osmosis and Tensile Solvent*, Springer, New York, 1976.

Hanna, S. R., "Characteristics of Winds and Turbulence in the Planetary Boundary Layer," ESSA Res. Lab. Technical Memorandum ERLTM-ARL8, Air Resources Laboratory, Oak Ridge, TN, 1969.

Happel, J. and Brenner, H., *Low Reynolds Number Hydrodynamics*, Prentice-Hall, Englewood Cliffs, NJ, 1965.

Haugen, D. A. (Ed.), *Workshop on Micrometeorology*, American Meteorological Society, Boston, MA, 1979.

Haurwitz, B., "The Motion of Atmospheric Disturbances on the Spherical Earth," *J. Marine Res.*, Vol. 3, pp. 254–267, 1940.

Helsot, F., Castaing, B., and Libchaber, A., "Transition to Turbulence in Helium Gas," *Phys. Rev.*, Vol. A36, pp. 5870–5873, 1987.

Henderson, F. M., *Open Channel Flow*, Macmillan, New York, 1966.

Herring, J. R., Riley, J. J., Patterson, G. S., and Kraichnan, R. H., "Growth of Uncertainty in Decaying Isotropic Turbulence," *J. Atmos. Sci.*, Vol. 30, pp. 997–1006, 1973.

Herring, J. R., "The Predictability of Quasi-Geostrophic Flows," *Predictability of Fluid Motions*, Holloway and West, B. (Eds.), La Jolla Institute, AIP Conference Proceedings, 1983.

Herring, J. R., "Theory of Two-Dimensional Anisotropic Turbulence," *J. Atmos. Sci.*, Vol. 32, pp. 2254–2271, 1975.

Hey, R. D., "Flow Resistance in Gravel-Bed Rivers," *J. Hydraulic Div.*, ASCE, Vol. 105, No. 4, pp. 365–379, 1979.

Holcomb, D. P. and Cooke, J. R., "An Electrolytic Tank Analog Determination of Stomatal

Diffusion Resistance,'' ASAE Paper No. 77-5510, American Society of Agricultural Engineers, 1977.

Hosker, R. P., ''Flow and Diffusion Near Obstacles,'' *Atmospheric Science and Power Production*, Anderson, D. R. (Ed.), U.S. Department of Energy, Technical Information Center, Oak Ridge, TN, 1983.

Hoskins, B. J., ''The Mathematical Theory of Frontogenesis,'' *Ann. Rev. Fluid Mech.*, Vol. 14, pp. 131-151, 1982.

Hoult, David P., ''Oil Spreading on the Sea,'' *Annual Reviews of Fluid Mechanics*, Annual Reviews, Inc., Palo Alto, CA, Vol. 4, pp. 341-368, 1972.

Hsu, R. S. C. and Ganatos, P., ''A Theoretical Model for the Gaseous Diffusion Across the Surface of Plant Leaves,'' *Advances in Bioengineering*, American Society of Mechanical Engineers, pp. 123-124, 1983.

Hubbell, D. W., ''Bed Load Sampling and Analysis,'' in *Sediment Transport in Gravel-Bed Rivers*, Thorne, C. R., Bathurst J. C., and Hey, R. D. (Eds.), John Wiley & Sons Ltd., Chichester, U.K., 1987.

Ikeda, S. and Parker, G. (Eds.), *River Meandering*, Water Resources Monograph 12, American Geophysical Union, 1989.

Ikeda, S., Parker, G., and Sawai, K., ''Bend Theory of River Meanders. Part 1. Linear Development,'' *J. Fluid Mech.*, Vol. 112, pp. 363-377, 1981.

Jaffrin, M. Y. and Hennessey, T. V., ''Pressure Distribution in a Model of the Central Airways for Sinusoidal Flow,'' *Bull. Physio-Pathol. Respir.*, Vol. 8, pp. 375-390, 1972.

Jaffrin, M. Y. and Shapiro, A. H., ''Peristaltic Pumping,'' *Ann. Rev. Fluid Mech.*, Vol. 3, pp. 13-36, 1971.

Jansen, P. Ph., *Principles of River Engineering*, Pitman, 1979.

Jarvis, P. G. and Mansfield, T. A. (Eds.), *Stomatal Physiology*, Cambridge University Press, New York, 1981.

Jensen, M., ''Biology and Physics of Locust Flight III. The Aerodynamics of Locust Flight,'' *Philos. Trans. R. Soc. London, Ser. B.*, Vol. 239, pp. 511-552, 1956.

Johnson, R. E. and Brokaw, C. J., ''Flagellar Hydrodynamics. A Comparison Between Resistive-Force Theory and Slender-Body Theory,'' *Biophys. J.*, Vol. 25, pp. 113-127, 1979.

Johnson, R. E., ''An Improved Slender-Body Theory for Stokes Flow,'' *J. Fluid Mech.*, Vol. 99, pp. 411-431, 1980.

Jones, R. T., *Wing Theory*, Princeton University Press, Princeton, 1990.

Jones, R. T., ''Blood Flow,'' *Ann. Rev. Fluid Mech.*, Vol. 1, pp. 223-244, 1969.

Kadlec, R. H., ''Overland Flow in Wetlands: Vegetation Resistance,'' *J. Hydraulic Eng.*, ASCE, Vol. 116, No. 5, pp. 691-706, 1990.

Kaimal, J. E., Wyngaard, J. E., Izumi, Y., and Cote, O. R., ''Spectral Characteristics of Surface-layer Turbulence,'' *J. Royal Meteorol. Soc.*, Vol. 98, pp. 563-589, 1972.

Kamm, R. D. and Shapiro, A. H., ''Unsteady Flow in a Collapsible Tube Subjected to External Pressure on Body Forces,'' *J. Fluid Mech.*, Vol. 95, pp. 1-78, 1979.

Karim, M. F. and Kennedy, J. F., ''Menu of Coupled Velocity and Sediment-Discharge Relations for Rivers,'' *J. Hydraulic Eng.*, ASCE, Vol. 116, No. 8, pp. 978-996, 1990.

Kato, T., ''On Classical Solutions of the Two-Dimensional Non-Stationary Euler Equation,'' *Arch. Rat. Mech. and Anal.*, Vol. 25, pp. 188-200, 1967.

Keller, S. R. and Wu, T. Y., ''A Porous Prolate Spheroidal Model for Ciliated Micro-Organisms,'' *J. Fluid Mech.*, Vol. 80, pp. 259-278, 1977.

Kellerhals, R. and Bray, D. I., "Sampling Procedures for Coarse Fluvial Sediments," *J. Hydraulic Eng.*, ASCE, Vol. 103, No. 8, pp. 1165–1180, 1971.

Kennedy, J. F., "Reflections on Rivers, Research, and Rouse," *J. Hydraulic Eng.*, ASCE, Vol. 109, No. 10, pp. 1253–1271, 1983.

Kennedy, J. F., "The Formation of Sediment Ripples, Dunes and Antidunes," *Ann. Rev. Fluid Mech.*, Vol. 1, pp. 147–168, 1969.

Kieser, F., "Canard Ornithopter Construction," *Flapper Facts*, the Ornithopter Modeler Society Newsletter, Vol. 4, No. 4, Autumn 1986.

Knauss, J. A., *Introduction to Physical Oceanography*, Prentice-Hall, Englewood Cliffs, NJ, 1978.

Knight, D. W., Yuen, K. W. H., and Al-Hamid, A. A. I., "Boundary Shear Stress Distributions in Open Channel Flow," in *Mixing and Transport in the Environment*, Beven, K. J., Chatwin, P. C., and Millbank, J. H. (Eds.), John Wiley & Sons Ltd., 1994.

Kobus, H. E., "Analysis of the Flow Induced by Air-Bubble-Systems," Chapter 65 of Part 3, *Coastal Structures*, Vol. II, *Proceedings Conference Coastal Engineering*, 11th ed., pp. 1016–1031, 1968.

Kokshaysky, N. V., "Tracing the Wake of a Flying Bird," *Nature Lond.*, Vol. 279, pp. 146–148, 1979.

Kraichnan, R. H., "Inertial Range in Two-Dimensional Turbulence," *Phys. Fluids*, Vol. 10, pp. 1417–1423, 1967.

Kraichnan, R. H., "Inertial-Range Transfer in Two and Three Dimensional Turbulence," *J. Fluid Mech.*, Vol. 47, pp. 525–535, 1971.

Kuroki, M. and Kishi, T., "Regime Criteria on Bars and Braids," The Research Laboratory of Civil and Environmental Engineering, Hokkaido University, Sapporo, Japan.

Lagasse, P. F., Schall, J. D., Johnson, F., Richardson, E. V., Richardson, J. R., and Chang, F., "Stream Stability at Highway Structures," HEC 20, Pub. No. FHWA-IP-90-014, McLean, VA, 1991.

Lan, C. E., "The Unsteady Quasi-Vortex-Lattice Method With Applications to Animal Propulsion," *J. Fluid Mech.*, Vol. 93, pp. 747–765, 1979.

Landsberg, J. J. and Fowkes, N. D., "Water Movement Through Plant Roots," *Ann. Bot.*, Vol. 42, pp. 493–508, 1978.

Leith, C. E. and Kraichnan, R. H., "Predictability of Turbulent Flows," *J. Atmos. Sci.*, Vol. 29, pp. 1041–1058, 1972.

Leith, C. E., "Atmospheric Predictability and Two-Dimensional Turbulence," *J. Atmos. Sci.*, Vol. 28, pp. 145–161, 1971.

Leith, C. E., "Diffusion Approximation to the Inertial Energy Transfer in Isotropic Turbulence," *Phys. Fluids*, Vol. 10, pp. 1409–1416, 1967.

Leopold, L. B. and Wolman, M. G., "River Channel Patterns: Braided, Meandering and Straight," U.S. Geological Survey, Professional Paper 282-B, 1957.

Leopold, L. B., *A View of the River*, Harvard University Press, Cambridge, MA, 1994.

Lesieur, M., *Turbulence in Fluids*, Kluwer Academic Press, Dordrecht, The Netherlands, 1990.

Lighthill, M. J., *Mathematical Biofluiddynamics*, SIAM, Philadelphia, 1975.

Lighthill, M. J., "Flagellar Hydrodynamics, The John von Neumann Lecture," *SIAM Rev.*, Vol. 18, pp. 161–230, 1976.

Lighthill, M. J., "Introduction to the Scaling of Aerial Locomotion," *Scale Effects in Animal Locomotion*, Pedley, T. J. (Ed.), Academic Press, New York, 1976.

Lighthill, M. J., "Note on the Swimming of Slender Fish," *J. Fluid Mech.*, Vol. 9, pp. 305–317, 1960.

Lighthill, M. J., "On the Weis–Fogh Mechanism of Lift Generation," *J. Fluid Mech.*, Vol. 60, pp. 1–17, 1973.

Lighthill, M. J., "Some Challenging New Applications for Basic Mathematical Methods in the Mechanics of Fluids That Were Originally Pursued With Aeronautical Aims," *Aero. J.*, Vol. 93, pp. 41–52, 1990.

Lilly, D. K., "Numerical Simulation of Developing and Decaying Two-Dimensional Turbulence," *J. Fluid Mech.*, Vol. 45, pp. 395–415, 1971.

Lilly, D. K., "Stratified Turbulence and the Mesoscale Varibility of the Atmosphere," *J. Atmos. Sci.*, Vol. 40, pp. 749–761, 1983.

Linn, F. C., "Lubrication of Animal Joints," *J. Lubr. Technol.*, Vol. 91, pp. 329–341, 1969.

Lippisch, A. M., "Man-Powered Flight in 1929," *J. Roy. Aero. Soc.*, Vol. 64, pp. 395–398, 1960.

Lipshitz, H., Etheridge, R., and Glimcher, M. J., "Changes in the Hexoamine Content and Swelling Ratio of Articular Cartilage as a Function of Depth from the Surface," *J. of Bone and Joint Surgery*, Vol. 58A, pp. 1149–1153, 1976.

Lissaman, P. B. S., "Low-Reynolds-Number Airfoils," *Ann. Rev. Fluid Mech.*, Vol. 15, pp. 223–239, 1983.

Litt, M. and Kahn, M. A., "Mucus Rheology: Relation to Structure and Function," *Biorheology*, Vol. 13, pp. 37–48, 1976.

Lorenz, E. A., "The Predictability of a Flow Which Possesses Many Scales of Motion," *Tellus*, Vol. 21, pp. 289–307, 1969.

Lovera, F. and Kennedy, J. F., "Friction Factor for Flat Bed Flows in Sand Channels," *J. Hydraulics Div.*, ASCE, Vol. 95, No. 4, pp. 1227–1234, 1969.

Lubkin, S. and Rand, R., "Oscillatory Reaction–Diffusion Equations on Rings," *J. Math. Biol.*, Vol. 32, pp. 617–632, 1994.

Lubkin, S. R., "Circumnutation Modeled by Reaction–Diffusion Equations," Ph.D. thesis, Cornell University, Ithaca, NY, 1992.

Lubkin, S. R., "Unidirectional Waves on Rings: Models for Chiral Preference of Circumnutating Plants," *Bulletin Math. Biol.*, Vol. 56, pp. 795–810, 1994.

Lutgen, F. K. and Tarbuck, E. J., *The Atmosphere, an Introduction to Meteorology*, Prentice-Hall, Englewood Cliffs, NJ, p. 478, 1982.

Mahmood, K. and Yevjevich, V. (Eds.), *Unsteady Flow in Open Channels*, Vols. I, II, III, Water Resources Publications, Fort Collins, CO, 1975.

Malcom, L. L., "Frictional and Deformational Responses of Articular Cartilage Interfaces to Static and Dynamic Loading," Ph.D. Thesis, University of California, San Diego, La Jolla, CA, 1976.

Maxworthy, T., "Experiments on the Weis–Fogh Mechanism of Lift Generation by Insects in Hovering Flight, Part I, Dynamics of the Fling," *J. Fluid Mech.*, Vol. 93, pp. 47–63, 1979.

Maxworthy, T., "The Fluid Dynamics of Insect Flight," *Ann. Rev. Fluid Mech.*, Vol. 13, pp. 329–350, 1981.

McIntire, L. V., "Dynamic Materials Testing: Biological and Clinical Applications in Network–forming Systems," *Ann. Rev. Fluid Mech.*, Vol. 12, pp. 159–179, 1980.

McWilliams, J. C., "The Emergence of Isolated Vortices in Turbulent Flows," *J. Fluid Mech.*, Vol. 146, pp. 21–43, 1984.

Mehta, A. J., Hayter, E. J., Parker, W. R., Krone, R. B., and Teeter, A. M., "Cohesive Sediment Transport, I: Process Description," *J. Hydraulic Eng.*, ASCE, Vol. 115, No. 8, pp. 1076–1093, 1989.

Mehta, K. C., Minor, J. E., McDonald, J. R., Manning, B. R., Abernathy, J. J., and Koehler, U. F., "Engineering Aspects of the Tornadoes of April 2–3, 1974," Committee on Natural Disasters, National Academy of Sciences, Washington, D.C., 1975.

Meidner, H. and Mansfield, T. A., *Physiology of Stomata*, McGraw-Hill, New York, 1968.

Meidner, H. and Sheriff, D. W., *Water and Plants*, John Wiley & Sons, New York, 1976.

Mendoza, C. and Zhou, D., "Effects of Porous Bed on Turbulent Stream Flow Above Bed," *J. Hydraulic Eng.*, ASCE, Vol. 118, No. 9, pp. 1222–1240, 1992.

Merva, G. E., *Physioengineering Principles*, AVI Publ., Westport, CT, 1976.

Milburn, J. A., *Water Flow in Plants*, Longmans, New York, 1979.

Miller, D. J., Brinkmann, W. A. R., and Barry, R. G., "Windstorms: A Case Study of Wind Hazard for Boulder, Colorado," *Natural Hazards*, White, G. F. (Ed.), Oxford University Press, Oxford, England, 1974.

Molz, F. J. and Hornberger, G. M., "Water Transport Through Plant Tissues in the Presence of a Diffusable Solute," *Soil Sci. Soc. Am. Proc.*, Vol. 37, pp. 833–837, 1973.

Molz, F. J. and Ikenberry, E., "Water Transport Through Plant Cells and Cell Walls: Theoretical Development," *Soil Sci. Soc. Am. Proc.*, Vol. 38, pp. 699–704, 1974.

Molz, F. J. and Klepper, B., "Radial Propagation of Water Potential in Stems," *Agron. J.*, Vol. 64, pp. 469–473, 1972.

Molz, F. J., "Potential Distributions in the Soil-Root System," *Agron. J.*, Vol. 67, pp. 726–729, 1976.

Molz. F. J., "Water Transport Through Plant Tissue: The Apoplasm and Symplasm Pathways," *J. Theor. Biol.*, Vol. 59, pp. 277–292, 1976.

Monin, A. S. and Obukhov, A. M., "Small Scale Oscillations of the Atmosphere and Adaption of Meteorological Fields," *Izv. Akad. Naut SSSR, Ser Geofiz.*, Vol. 11, pp. 1360–1373, 1958.

Monin, A. S., *Theoretical Geophysical Fluid Dynamics*, Dordrecht/Boston/London, 1990.

Mott, K. A., Cardon, Z. G., and Berry, J. A., "Asymmetric Patchy Stomatal Closure for the Two Surfaces of Xanthium Strumarium Leaves at Low Humidity," *Plant, Cell and Environment*, Vol. 16, pp. 25–34, 1993.

Mow, V. C. and Lai, W. M., "Mechanics of Animal Joints," *Ann. Rev. Fluid Mech.*, Vol. 11, pp. 247–288, 1979.

Mow, V. C., Kuei, S. C., Lai, W. M., and Armstrong, C. G., "Biphase Creep and Stress Relaxation of Articular Cartilage in Compression: Theory and Experiments," *J. Biomech. Eng. Trans.*, Vol. 102, pp. 73–84, 1980.

Mullen, S. L. and Baumhefner, D. P., "Sensitivity of Numerical Simulations of Explosive Cyclogenesis to Changes in Physical Parameterizations," *Monthly Weather Rev.*, pp. 2289–2329, No. 1988.

Munk, W. H., "On the Wind Driven Ocean Circulation," *J. Meteor.*, Vol. 7, No. 2, pp. 79–93, 1950.

Murakami, S. (Ed.), "Computational Wind Engineering 1," Proceedings of the First International Symposium on Computational Wind Engineering, Tokyo, Japan, August 1992, Elsevier Science Publishers, New York, NY, 910 pp.

Nachitgall, W., *Insects in Flight*, McGraw-Hill, New York, 1974.

Nastrom, G. D., Gage, K. S., and Jasperson, W. H., "Atmospheric Kinetic Energy Spectrum, 10^0–10^4 km," *Nature*, Vol. 310, pp. 36–38, 1984.

Neumann, G. and Pierson, W. J., *Principles of Physical Oceanography*, Prentice-Hall, Englewood Cliffs, NJ, 1966.

Newman, E. I., "Water Movement Through Root Systems," *Philos. Trans. Roy. Soc. (London) Ser. B*, Vol. 273, pp. 463–478, 1976.

Nezu, I. and Nakagawa, H., *Turbulence in Open-Channel Flows*, IAHR Monograph, Balkema, Rotterdam, 1993.

Niklas, K. J., *Plant Biomechanics: An Engineering Approach to Plant Form and Function*, The University of Chicago Press, Chicago, IL, 1992.

Nobel, P. S., *Introduction to Biophysical Plant Physiology*, Freeman, San Francisco, CA, 1974.

Nobel, P. S., *Physiochemical and Environmental Plant Physiology*, Academic Press, San Diego, CA, 1991.

Norberg, U. M., "Aerodynamics, Kinematics, and Energetics of Horizontal Flapping Flight in the Long-Eared Bat *Plecotus auritus*," *J. Exp. Biol.*, Vol. 65, pp. 179–212, 1976.

Nordine, C. F., Jr., "Aspects of Flow Resistance and Sediment Transport: Rio Grande nr. Bernalillo, New Mexico," USGS Water Supply Paper 1498-H, Washington, D.C., 1964.

Paola, C., Parker, G., Seal, R., Sinha, S. K., Southard, J. B., and Wilcock, P. R., "Downstream Fining by Selective Deposition in a Laboratory Flume," *Science*, Vol. 258, pp. 1757–1760, 1992.

Parker, G. and Peterson, A. W., "Bar Resistance of Gravel-Bed Streams," *J. Hydraulic Eng.*, ASCE, Vol. 106, No. 10, pp. 1559–1575, 1980.

Parker, G. and Sutherland, A. J., "Fluvial Armor," *J. Hydraulic Res.*, IAHR, Vol. 128, No. 5, pp. 529–544, 1990.

Parker, G., "On the Causes and Characteristic Scales of Meandering and Braiding in Rivers," *J. Fluid Mech.*, Vol. 76, pp. 457–480, 1976.

Parker, G., "Selective Sorting and Abrasion of River Gravel. I: Theory," *J. Hydraulic Eng.*, ASCE, Vol. 117, No. 2, pp. 131–149, 1991.

Parker, G., "Self-Formed Straight Rivers With Equilibrium Banks and Mobile Bed. Part 2. The Gravel River," *J. Fluid Mech.*, Vol. 89, Pt. 1, pp. 127–148, 1978.

Parkhurst, D. F., "A Three-Dimensional Model for CO_2 Uptake by Continuously Distributed Mesophyll in Leaves," *J. Theor. Biol.*, Vol. 67, pp. 471–488, 1977.

Parlange, J. Y. and Waggoner, P. E., "Stomatal Dimensions and Resistance to Diffusion," *Plant Physiol.*, Vol. 46, pp. 337–342, 1970.

Parlange, J. Y., Turner, N. C., and Waggoner, P. E., "Water Uptake, Diameter Change, and Nonlinear Diffusion in Tree Stems," *Plant Physiol.*, Vol. 55, pp. 247–250, 1975.

Pedley, T. J., Schroter, R. C., and Sudlow, M. F., "The Prediction of Pressure Drop and Variation of Resistance Within the Human Bronchial Airways," *Respir. Physiol.*, Vol. 9, pp. 387–405, 1970.

Pedley, T. J., "Pulmonary Fluid Dynamics," *Ann. Rev. Fluid Mech.*, Vol. 9, pp. 229–274, 1977.

Pedlosky, J., *Geophysical Fluid Dynamics*, 2nd ed., Springer-Verlag, New York, 1987.

Pedlosky, J., *Geophysical Fluid Dynamics*, Springer-Verlag, New York, 1979.

Pennycuick, C. J., *Bird Flight Performance. A Practical Calculation Manual*, Oxford University Press, Oxford, England, 1989a.

Pennycuick, C. J., "Mechanics of Flight," *Avian Biology*, Vol. 5, Farner, D. C. and King, J. R. (Eds.), Academic Press, London, pp. 1–75, 1974.

Pennycuick, C. J., "Span-Ratio Analysis Used to Estimate Effective Lift: Drag Ratio in the Double-Crested Cormorant *Phalacrocorax auritus* from Field Observations," *J. Exp Biol.*, Vol. 142, pp. 1–15, 1989b.

Peskin, C. S., "The Fluid Dynamics of Heart Valves: Experimental, Theoretical and Computational Methods," *Ann. Rev. Fluid Mech.*, Vol. 14, pp. 235–259, 1982.

Philip, J. R., "Osmosis and Diffusion in Tissue: Half-Times and Internal Gradients," *Plant Physiol.*, Vol. 33, pp. 275–278, 1958c.

Philip, J. R., "Propagation of Turgor and Other Properties Through Cell Aggregations," *Plant Physiol.*, Vol. 33, pp. 271–274, 1958b.

Philip, J. R., "The Osmotic Cell, Solute Diffusibility, and the Plant Water Economy," *Plant Physiol.*, Vol. 33, pp. 264–271, 1958a.

Phlips, P. J., East, R. A., and Pratt, N. H., "An Unsteady Lifting Line Theory of Flapping Wings With Application to the Forward Flight of Birds," *J. Fluid Mech.*, Vol. 112, pp. 97–125, 1981.

Pond, S. and Pickard, G. L., *Introductory Dynamical Oceanography*, 2nd ed., Pergamon Press, Oxford, England, 1983.

Poreh, M., "Simulation of Plume Rise in Small Wind-Tunnel Models," *J. Wind Engineering and Industrial Aerodynamics*, Vol. 7, No. 1, pp. 1–14, 1981.

Priestley, C. H. B., "Temperature Fluctuations in the Atmospheric Boundary Layer," *J. Fluid Mech.*, Vol. 7, pp. 375–438, 1960.

Proctor, D. F., "Physiology of the Upper Airway," *Handbook of Physiology*, Section 3, *Respiration*, Vol. 1, Fenn, W. O. and Rahn, H. (Eds.), American Physiological Society, Washington, 309–345, 1964.

Rand, R. H., "A Theoretical Analysis of CO_2 Absorption in Sun Versus Shade Leaves," *J. Biomech. Eng.*, Vol. 100, pp. 20–24, 1978b.

Rand, R. H. and Cooke, J. R., "A Comprehensive Model for CO_2 Assimilation in Leaves," *Trans. ASAE*, Vol. 23, pp. 601–607, 1980.

Rand, R. H. and Cooke, J. R., "Fluid Dynamics of Phleom Flow: An Axisymmetric Model," *Trans. ASAE*, Vol. 21, pp. 898–900, 906, 1978.

Rand, R. H. and Ellenson, J. L., "Dynamics of Stomate Fields in Leaves," *Lectures on Mathematics in the Life Sciences, Some Mathematical Questions in Biology—Plant Biology*, Gross, L. (Ed.), Vol. 18, pp. 51–86, American Mathematical Society, 1986.

Rand, R. H., Storti, D. W., Upadhyaya, S. K., and Cooke, J. R., "Dynamics of Coupled Stomatal Oscillators," *J. Math. Biol.*, Vol. 15, pp. 131–149, 1982.

Rand, R. H., Upadhyaya, S. K., and Cooke, J. R., "Fluid Dynamics of Phloem Flow. II. An Approximate Formula," *Trans. ASAE*, Vol. 23, pp. 581–584, 1980.

Rand, R. H., Upadhyaya, S. K., Cooke, J. R., and Storti, D. W., "Hopf Bifurcation in a Stomatal Oscillator," *J. Math. Biol.*, Vol. 12, pp. 1–11, 1981.

Rand, R. H., *Topics in Nonlinear Dynamics with Computer Algebra*, Gordon and Breach, Langhorne, PA, 1994.

Rand, R. H., "Fluid Mechanics of Green Plants," *Ann. Rev. Fluid Mech.*, Vol. 15, pp. 29–45, 1983.

Rand, R. H., "Gaseous Diffusion in the Leaf Interior," *1977 Biomech. Symp.*, Skalak, R. and Schultz, A. B. (Eds.), Vol. 23, pp. 51–53, 1977b.

Rand, R. H., "Gaseous Diffusion in the Leaf Interior," *Trans. ASAE*, Vol. 20, pp. 701–74, 1977a.

Rand, R. H., "Stomatal Dynamics in the Leaves of Green Plants," *Proceedings of the Tenth U.S. National Congress of Applied Mechanics*, Lamb, J. P. (Ed.), American Society of Mechanical Engineers, 1987.

Rand, R. H., "The Dynamics of an Evaporating Meniscus," *Acta Mech.*, Vol. 29, pp. 135–146, 1978a.

Raudkivi, A. J., *Loose Boundary Hydraulics*, 3rd ed., Pergamon Press, 1990.

Rayleigh, Lord, "On the Stability, or Instability, of Certain Fluid Motions, II," *London Math. Soc. Proc.*, Vol. XIX, pp. 67–74, 1887.

Rayleigh, Lord, "On the Stability, or Instability, of Certain Fluid Motions," *London Math. Soc. Proc.*, Vol. XI, pp. 57–70, 1880.

Rayner, J. M. V., Jones, G., and Thomas, A., "Vortex Flow Visualisations Reveal Change in Upstroke Function With Flight Speed in Bats," *Nature, Lond.*, Vol. 321, pp. 162–164, 1986.

Rayner, J. M. V., "A Vortex Theory of Animal Flight. Part 1. The Vortex Wake of a Hovering Animal," *J. Fluid Mech.*, Vol. 91, pp. 697–730, 1979a.

Rayner, J. M. V., "A Vortex Theory of Animal Flight. Part 2. The Forward Flight of Birds," *J. Fluid Mech.*, Vol. 91, pp. 731–763, 1979b.

Rayner, J. M. V., "On Aerodynamics and Energetics of Vertebrate Flapping Flight," *Contemp. Math.*, Vol. 141, pp. 351–400, 1993.

Rayner, J. M. V., "Vertebrate Flapping Flight Mechanics and Aerodynamics, and the Evolution of Flight in Bats," *Biona Report*, Vol. 5, pp. 27–74, 1986.

Reid, I., Layman, J. T., and Frostick, L. E., "The Continuous Measurement of Bedload Discharge," *J. Hydraulic Res.*, Vol. 18, pp. 243–249, 1980.

Richards, K. J., "The Formation of Dunes and Ripples on an Erodible Bed," *J. Fluid Mech.*, Vol. 99, Pt. 3, pp. 597–618, 1980.

Richardson, E. V., Harrison, L. J., Richardson, J. R., and Davis, S. R., "Evaluating Scour at Bridges," HEC 18, Pub. No. FHWA-IP-90-017, McLean, VA, 1993.

Richardson, E. V., Simons, D. B., and Julien, P. Y., "Highways in the River Environment," Pub. No. FHWA-HI-90-016, Washington, D.C., 1990.

Rockstein, M. (Ed.), *The Physiology of Insecta*, Academic Press, New York, 1974.

Ross, S. M. and Corrsin, S., "Results of an Analytical Model of Mucociliary Pumping," *J. Appl. Physiol.*, Vol. 37, pp. 333–340, 1974.

Rossby, C.-G., "The Scientific Basis of Modern Meteorology," *Handbook of Meteorology, Sec. VII*, Berry, G. A., Jr., Bollary, E., and Beers, N. R. (Eds.), McGraw-Hill, New York, 1945.

Rouse, H., *Elementary Mechanics of Fluids*, John Wiley & Sons, New York, 1946.

Routledge, L. M., "Bacterial Flagella: Structure and Function," *Comparative Physiology-Functional Aspects and Structural Materials*, Bolis, L., Maddrell, H. P., Schmidt–Nielsen, K. (Eds.), North–Holland, Amsterdam, 1975.

Salisbury, F. B. and Ross, C. W., *Plant Physiology*, 4th ed., Wadsworth Publishing, Belmont, CA, 1992.

Schlesinger, R. B., "Mucociliary Interaction in the Tracheo-Branchial Tree and Environmental Pollution," *Bioscience*, Vol. 23, pp. 567–573, 1973.

Schlichting, H., *Boundary Layer Theory*, McGraw-Hill, New York, 1980.

Schott, F. and Stommel, H., "Beta-Spirals and Absolute Velocities in Different Oceans," *Deep-Sea Res.*, Vol. 25, pp. 961–1010, 1978.

Schultz, D. L., "Pressure and Flow in Large Arteries," in *Cardiovascular Fluid Dynamics, I*, Bergel, D. H. (Ed.), Academic Press, New York, 1972.

Schumm, S. A. (Ed.), *River Morphology. Benchmark Papers in Geology*, Dowden, Hutchinson & Ross, Inc., Stroudsburg, PA, 1972.

Schumm, S. A., *The Fluvial System*, John Wiley & Sons, New York, 1977.

Scott Blair, G. W., *An Introduction to Biorheology*, Elsevier Scientific, New York, 1974.

Seaton, K. A. and Landsberg, J. J., "Resistance to Water Movement Through Wheat Root Systems," *Aust. J. Agric. Res.*, Vol. 29, pp. 913–924, 1978.

Shapiro, A. H., Jaffrin, M. Y., and Weinberg, S. L., "Peristaltic Pumping with Long Wavelengths at Low Reynolds Number," *J. Fluid Mech.*, Vol. 37, pp. 799–825, 1969.

Shapiro, A. H., "Steady Flow in Collapsible Tubes," *J. Biomech. Eng.*, Vol. 99, pp. 126–147, 1977.

Shaw, J. and Kellerhals, R., "The Composition of Recent Alluvial Gravels in Alberta River Beds," Alberta Research Council, Edmonton, Alberta, Bulletin 41, 1982.

Shiono, K. and Knight, D. W., "Turbulent Open Channel Flows With Variable Depth Across the Channel," *J. Fluid Mech.*, Vol. 222, pp. 617–646, 1991.

Shraiman, B. I. and Siggla, E., "Heat Transport in High Rayleigh Number Convection," *Phys. Rev. A*, Vol. 42, pp. 3650–3653, 1991.

Silberberg, A., Meyer, F. A., Gilboa, A., and Gelman, R. A., "Function and Properties of Epithelial Mucus," *Mucus in Health and Disease*, Elstein, M. and Parke, D. V. (Eds.), Plenum Press, New York, pp. 171–180, 1976.

Silberman, E., "Production of Bubbles by the Disintegration of Gas Jets in Liquid," *Proc. Midwest Conference Fluid Mechanics*, 5th ed., pp. 263–283, 1957.

Simiu, E. and Scanlan, R. H., *Wind Effects on Structures*, John Wiley & Sons, New York, 1978.

Simons, D. B. and Richardson, E. V., "Resistance to Flow in Alluvial Channels," U.S. Geological Survey Prof. Paper No. 422J, Washington, D.C., 1966.

Simons, D. B. and Senturk, F., *Sediment Transport Technology*, Water Res. Pub., Littleton, CO, 1992.

Sinclair, T. R., Gourdriaan, J., and Dewit, C. T., "Mesophyll Resistance and CO_2 Compensation Concentration in Leaf Photosynthesis Models," *Photosynthetica*, Vol. 13, pp. 239–244, 1979.

Skalak, R., Keller, S. R., and Secomb, T. W., "Mechanics of Blood Flow," *J. Biomech. Eng.*, Vol. 103, pp. 102–115, 1981.

Smith, F. T., "Flow Through Constricted or Dilated Pipes and Channels, II," *Q. J. Mech. Appl. Math.*, Vol. 29, pp. 343–376, 1976.

Spedding, G. R., Rayner, J. M. V., and Pennycuick, C. J., "Momentum and Energy in the Wake of a Pigeon (*Columba livia*) in Slow Flight," *J. Exp. Biol.*, Vol. 111, pp. 81–102, 1984.

Spedding, G. R., "On the Significance of Unsteady Effects in the Aerodynamic Performance of Flying Animals," *Contemp. Math.*, Vol. 141, pp. 401–419, 1993.

Spedding, G. R., "The Aerodynamics of Flight," *Mechanics of Animal Locomotion*, Alexander, RMcN (Ed.), *Adv. Comp. Env. Physiol.*, Vol. 11, Springer-Verlag, 1992.

Spedding, G. R., "The Wake of a Kestrel (*Falco tinnunculus*) in Flapping Flight," *J. Exp. Biol.*, Vol. 127, pp. 59–78, 1987.

Stebbings, H. and Hyams, J. S., *Cell Motility*, Longmans, London, 1979.

Stoker, J. J., *Water Waves*, Wiley–Interscience, New York, 1957.

Stommel, H., *A View of the Sea*, Princeton University Press, Princeton, NJ, 1987.

Stommel, H., "The Westward Intensification of Wind-Driven Ocean Currents," *Trans. Amer. Geophys. Un.*, Vol. 29, No. 2, pp. 202–206, 1948.

Stroshine, R. L., Cooke, J. R., Rand, R. H., Cutler, J. M., and Chabot, J. F., "Mathematical Analysis of Pressure Chamber Efflux Curves," ASAE Paper, No. 79-4585, American Society of Agricultural Engineers, 1979.

Stroshine, R. L., Cooke, J. R., Rand, R. H., Cutler, J. M., and Chabot, J. F., "An Analysis of Resistance to Water Flow Through Wheat and Tall Fescue During Pressure Efflux Experiments," *Plant, Cell and Environment*, Vol. 8, pp. 7–18, 1985.

Sverdrup, H. U., "Wind-Driven Currents in a Baroclinic Ocean; With Application to the Equatorial Currents of the Eastern Pacific," *Proc. Nat. Acad. Sci.*, Vol. 33, No. 11, pp. 318–326, 1947.

Tacconi, P. and Billi, P., "Bed Load Transport Measurement by the Vortex-Tube Trap on

Virginio,'' *Sediment Transport in Gravel-Bed Rivers*, Thorne, C. R., Bathurst, J. C., and Hey, R. D. (Eds.), John Wiley & Sons Ltd., Chichester, U.K., 1987.

Taiz, L. and Zeiger, E., *Plant Physiology*, Benjamin/Cummings Publishing, Redwood City, CA, 1991.

Taylor, D. E. M. and Wade, J. D., "Pattern of Blood Flow Within the Heart: A Stable System," *Cardiovasc. Res.*, Vol. 7, pp. 14–21, 1973.

Taylor, G. I., "The Dispersion of Matter in Turbulent Flow Through a Pipe," *Proc. Roy. Soc. London Ser. A*, Vol. 223, pp. 446–468, 1954.

Tennekes, H., "The General Circulation of Two-Dimensional Turbulent Flow on a Beta Plane," *J. Atmos. Sci.*, Vol. 34, pp. 702–712, 1977.

Thornley, J. H. M., *Mathematical Models in Plant Physiology*, Academic Press, New York, 1976.

Tubino, M. and Seminara, G., "Free-Forced Interactions in Developing Meanders and Suppression of Free Bars," *J. Fluid Mech.*, Vol. 214, pp. 131–159, 1990.

Tyree, M. T. and Yianoulis, P., "The Site of Water Evaporation from Sub-Stomatal Cavities, Liquid Path Resistances and Hydroactive Stomatal Closure," *Ann. Bot.*, Vol. 46, pp. 175–193, 1980.

Tyree, M. T., "Tansley Review No. 34: The Hydraulic Architecture of Trees and Other Woody Plants," *New Phytol.*, Cambridge University Press, Cambridge, England, Vol. 119, pp. 345–360, 1991.

U.S. Corps. of Engineers, "Missouri River Channel Regime Studies," MRD Sed. Series No. 13.B., Omaha, NE, 1968.

Unger, D. J. and Aifantis, E. C., "Flow in Stems of Plants," *Proc. 3rd Eng. Mech. Div. Specialty Conf.*, ASCE, 1979.

Upadhyaya, S. K., Rand, R. H., and Cooke, J. R., "A Mathematical Model of the Effects of CO_2 on Stomatal Dynamics," ASAE Paper No. 80-5517, American Society of Agricultural Engineers, 1980b.

Upadhyaya, S. K., Rand, R. H., and Cooke, J. R., "Role of Stomatal Oscillations on Transpiration, Assimilation, and Water Use Efficiency of Plants," *Ecological Modeling*, Vol. 47, pp. 27–40, 1988.

Upadhyaya, S. K., Rand, R. H., and Cooke, J. R., "Stomatal Dynamics," *1980 Advance in Bioengineering*, Mow, V. C. (Ed.), pp. 185–188, ASME, New York, 1980a.

Upadhyaya, S. K., Rand, R. H., and Cooke, J. R., "The Role of Stomatal Oscillations on Plant Productivity and Water Use Efficiency," ASAE Paper No. 81-4017, American Society of Agricultural Engineers, 1981.

van Krevelen, D. W., Hoftijzer, D. W., and Hoftijzer, P. J., "Studies of Gas Bubble Formation—Calculation of Interfacial Area in Bubble Contactors," *Chem. Eng. Progress*, Vol. 46, pp. 29–35, 1950.

Van Zandt, T. E., "A Universal Spectrum of Buoyancy Waves in the Atmosphere," *Geophys. Res. Lett.*, Vol. 9, pp. 575–578, 1982.

Vanoni, V. A. (Ed.), *Sedimentation Engineering*, ASCE Manual and Reports on Engineering Practice No. 54, New York, 1975.

Vanoni, V. A., "Factors Determining Bed Forms of Alluvial Streams," *J. Hydraulic Div.*, ASCE, Vol. 100, No. 3, pp. 363–377, 1974.

Verdugo, P., Lee, W. I., Halbert, S. A., Blandau, R. J., and Tam, P. Y., "A Stochastic Model for Oviductal Egg Transport," *Biophys. J.*, Vol. 29, pp. 257–270, 1980.

Vickery, B. J., "Load Fluctuations on Bluff Shapes in Turbulent Flow," *Eng. Sci. Res. Rept. BLWT-4-67*, University of Western Ontario, 1967.

Vogel, S., *Life in Moving Fluids*, Willard Grant Press, Boston, 1981.

von Schwind, J. J., *Geophysical Fluid Dynamics for Oceanographers*, Prentice-Hall, Englewood Cliffs, NJ, 1980.

Webster, I. A., "The Use of the Effectiveness Factor Concept in CO_2 Diffusion in the Leaf Interior," *Ann. Bot.*, Vol. 48, pp. 757–760, 1981.

Weihs, D., "Periodic Jet Propulsion of Aquatic Creatures," *Bewegungsphysiologie Biomechanik*, Nachtigal, W. (Ed.), Akademie de Wissenschaft, 1977.

Weis–Fogh, T. and Jensen, M., "Biology and Physics of Locust Flight I. Basic Principles in Insect Flight. A Critical Review" (+ 2 following papers), *Phil. Trans. Roy. Soc. London Ser. B*, Vol. 239, pp. 415–458, 1956.

Weis–Fogh, T., "Flapping Flight and Power in Birds and Insects, Conventional and Novel Mechanisms," *Swimming and Flying in Nature*, Vol. 2, Wu, T. Y., Brokaw, C. J. and Brenne, C. (Eds.), Plenum Press, London, 1975.

Weis–Fogh, T., "Quick Estimates of Flight Fitness in Hovering Animals Including Novel Mechanisms for Lift Production," *J. Exp. Biol.*, Vol. 59, pp. 169–230, 1973.

Weiss, J., "The Dynamics of Enstrophy Transfer in Two-Dimensional Hydrodynamics," *La Jolla Institute Report*, La Jolla, CA, 1981; also *Physica D*, Vol. 48, pp. 273–294, 1991.

White, F. M., *Viscous Fluid Flow*, 2nd ed., McGraw-Hill, New York, 1991.

Wilson, T. A. and Lin, K. H., "Convection and Diffusion in the Airways and the Design of the Bronchial Tree," *Airway Dynamics*, Bouhuys, A. (Ed.), Thomas, Springfield, MA, 1970.

Winet, H., "On the Quantitative Analysis of Liquid Flow in Physiological Tubes," MRC Tech. Summary Rep. No. 2456, Math. Res. Center, University of Wisconsin, 1982.

Woo, H. G. C., Peterka, J. A., and Cermak, J. E., "Wind-Tunnel Measurements in the Wakes of Structures," NASA CR-2806, 1977.

Woo, S. L-Y., Simon, B. R., Kuei, S. C., and Akeson, W. H., "Quasi-Linear Viscoelastic Properties of Normal Articular Cartilage," *J. Biomech. Eng.*, Vol. 102, pp. 85–90, 1980.

Wu, T. Y., Brokaw, C. J., and Brenne, C., *Swimming and Flying in Nature*, Vols. 1 and 2, Plenum Press, New York, 1975.

Wu, Y. T., "Hydromechanics of Swimming Propulsion. Parts I, II, III," *J. Fluid Mech.*, (I) Vol. 46, pp. 337–355, (II) Vol. 46, pp. 521–544, (III) Vol. 46, pp. 545–568, 1971.

Yaglom, A. M., "Fluctuation Spectra and Variance in a Convective Atmospheric Surface Layer: A Reevaluation of Old Models," *Phys. Fluids A*, Vol. 6, No. 2, pp. 962–972, 1993.

Yates, G. T., Wu, T. Y., Johnson, R. E., Cheung, A. T. W., and Frand, C. L., "A Theoretical and Experimental Study on Tracheal Muco-Ciliary Transport," *Biorheology*, Vol. 17, pp. 151–162, 1980.

Yates, G. T., "Hydromechanics of Body and Caudal Fin Propulsion," *Fish Biomechanics*, Webb, P. W. and Weihs, D. (Eds.), Praeger, 1983.

Yen, B. C., "Hydraulic Resistance in Open Channels," in *Channel Flow Resistance: Centennial of Manning's Formula*, Yen, B. C. (Ed.), 1992.

Zeiger, E., Farquhar, G. D., and Cowan, I. R. (Eds.), *Stomatal Function*, Stanford University Press, Stanford, CA, 1987.

Zilitinkevich, S., "A Generalized Scaling for Convective Shear Flows," *Boundary-Layer Meteorology*, Vol. 49, pp. 1–4, 1993.

Zimmerman, M. H., *Xylem Structure and the Ascent of Sap*, Springer, New York, 1983.

Zippe, H. J. and Graf, H. G., "Turbulent Boundary-Layer Flow over Permeable and Non-Permeable Rough Surfaces," *J. Hydraulic Research*, IAHR, Vol. 21, No. 1, pp. 51–65, 1983.

25 Static Components of Fluid Machinery

BHARATAN R. PATEL
Fluent, Incorporated
Lebanon, NH

C. SAMUEL MARTIN
Georgia Institute of Technology
Atlanta, GA

JOHN E. MINARDI
University of Dayton
Dayton, OH

PHILIP C. STEIN, JR.
Stein Seal Company
Kulpsville, PA

REINER DECHER
University of Washington
Seattle, WA

STEPHEN HEISTER
Purdue University
West Lafayette, IN

JAMES L. YOUNGHANS
DONALD J. DUSA (Retired)
General Electric Aircraft Engines
Evandale, OH

JAMES L. KEIRSEY (Retired)
WILLIAM B. SHIPPEN (Retired)
EVERETT J. HARDGRAVE, JR. (Retired)
Applied Physics Laboratory,
The Johns Hopkins University
Laurel, MD

Handbook of Fluid Dynamics and Fluid Machinery, Edited by Joseph A. Schetz and Allen E. Fuhs
ISBN 0-471-12598-9 Copyright © 1996 John Wiley & Sons, Inc.

CONTENTS

25.1 DUCT FLOW

Bharatan R. Patel

25.1.1 Introduction

Ducts are commonly used to transport fluid from one point to another. Therefore, determination of the performance characteristics of ducts such as flow rate versus pressure drop is of considerable practical interest. In most cases, the flow in ducts is turbulent and three-dimensional. Hence, a detailed analysis of the flow would require solution of the three-dimensional turbulent flow equations. Computational fluid dynamics has advanced to the stage where this is feasible, and, in most cases, detailed predictions can be made with reasonable accuracies. However, for most practical situations, simple one-dimensional analysis along with empirical performance data will suffice. The purpose of this chapter is to present such simple analysis and empirical data for the prediction of flow characteristics of duct systems.

The flow characteristics addressed in this section will be the relationship between flow rate and pressure loss. At this point, it is useful to define pressure loss. For an incompressible fluid, the energy per unit mass of the fluid at any point in the flow field consists of the internal energy, e, and the mechanical energy, $(p/\rho + U^2/2 + gz)$, where p, ρ, and U are the pressure, density, and velocity, respectively, and z is the elevation above some datum. Total pressure p_0 is then defined as $p_0/\rho = p/\rho + U^2/2 + gz$. Therefore, for an incompressible fluid, the total pressure is a direct measure of the total available mechanical energy. In general, the pressure loss of interest is the loss in total pressure through a duct system. Total pressure loss in a duct occurs due to the irreversible conversion of mechanical energy to heat and, hence, internal energy through the molecular viscosity. This viscous dissipation can occur either in the shear layers at the duct walls or in free shear layers within the main body of the duct flow.

Consider the duct shown in Fig. 25.1. Flow enters the duct at plane 1 and exits at plane 2. These planes are perpendicular to the duct axis. The inlet and exit velocities are u_1 and u_2 and are assumed to be parallel to the duct axis. Taking a control volume shown in dotted lines in Fig. 25.1, for steady flow, mass conservation gives

$$\int_S \rho u_j n_j \, dS = 0 \tag{25.1}$$

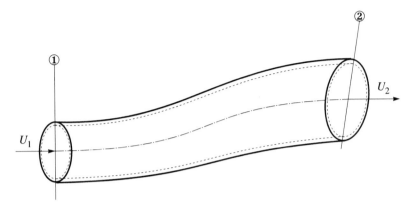

FIGURE 25.1   A general duct.

The unit vector n_j is normal to the control volume surface and points in the outward direction. Since there is no flow normal to the duct walls. Eq. (25.1) becomes

$$\int_{A_1} \rho_1 u_1 \, dA = \int_{A_2} \rho u_2 \, dA = \dot{m} \tag{25.2}$$

where $\dot{m}$ is the mass flow through the duct. For incompressible flow

$$U_1 A_1 = U_2 A_2 = Q \tag{25.3}$$

where U_1 and U_2 are the mean velocities at planes 1 and 2 and Q is the volumetric flow rate through the duct.

Again for steady flow without heat transfer at the walls, the energy conservation equation can be written as

$$\int_S \rho(e + \tfrac{1}{2}u^2) u_j n_j \, dS = -\int_S p u_j n_j \, dS + \int_S \tau_{jk} u_j n_k \, dS + \int_V \rho f_i u_i \, dV \tag{25.4}$$

Briefly, the term on the left-hand side represents the convection of energy in and out of the control volume. The first two terms on the right-hand side represent the work done by the pressure p and shear stresses τ_{ij} at the control volume boundaries. The last term is the work done by the body force per unit mass, f_i (V denotes the volume of the control volume).

Since the boundaries of the duct have fluid velocity zero (due to the no-slip condition imposed by viscosity), the work done by pressure and shear stresses at the duct walls is zero. Further, at the inlet and outlet planes, the viscous shear stresses are generally small hence the second term on the right-hand side can be neglected. Assuming that the body force per unit mass, f_i, is conservative

$$\rho f_i = \frac{\partial(\rho\phi)}{\partial x_i} \tag{25.5}$$

where ϕ is the potential energy per unit mass. With the above considerations, Eq. (25.4) can be simplified to

$$\int_{A_2} \left[e_2 + \frac{1}{2} u_2^2 + \frac{p_2}{\rho_2} + \phi_2 \right] \rho_2 u_2 \, dA$$

$$- \int_{A_1} \left[e_1 + \frac{1}{2} u_1^2 + \frac{p_1}{\rho_1} + \phi_1 \right] \rho_1 u_1 \, dA = 0 \qquad (25.6)$$

For incompressible flow with uniform static pressure at the inlet and outlet planes, the above equation can be further simplified to

$$\left[\frac{p_1}{\rho} + \frac{1}{2} \alpha_1 U_1^2 + \phi_1 \right] - \left[\frac{p_2}{\rho} + \frac{1}{2} \alpha_2 U_2^2 + \phi_2 \right] = (e_2 - e_1) \qquad (25.7)$$

where α_1 and α_2 are the kinetic energy coefficients defined as

$$\alpha = \frac{1}{A} \int \left(\frac{u}{U} \right)^3 dA$$

In general, α is greater than unity. For fully developed turbulent duct flows, α is about 1.2, whereas for laminar fully developed flows, α is 2.

The right-hand side of Eq. (25.7) represents the increase in internal energy due to irreversible viscous dissipation of the flow mechanical energy. It should be noted that although the viscous shear stresses at the wall produce no work at the boundary of the control volume, they cause dissipation of mechanical energy inside the control volume. In this integral analysis, the change in the internal energy cannot be calculated; a more detailed analysis entailing the solution of the entire flow field would be required. Therefore, experimental data are needed to obtain the term on the right-hand side of Eq. (25.7). In practice, Eq. (25.7) is used as the basis for obtaining the change in internal energy through a component by measuring all the quantities on the left-hand side.

The commonly encountered body force is that due to gravity for which $\phi_1 = gz_1$ and $\phi_2 = gz_2$, where z_1 and z_2 are the heights of the inlet and exit planes of the duct measured from some common datum. Equation (25.7) is somewhat difficult to use in practice, since the kinetic energy coefficients α_1 and α_2 are generally not known. Further, their experimental determination requires detailed flow field traverses. This difficulty can be circumvented by rearranging Eq. (25.7) as follows

$$\left[\frac{p_1}{\rho} + \frac{1}{2} U_1^2 + gz_1 \right] - \left[\frac{p_2}{\rho} + \frac{1}{2} U_2^2 + gz_2 \right]$$

$$= (e_2 - e_1) + (1 - \alpha_1) \frac{1}{2} U_1^2 - (1 - \alpha_2) \frac{1}{2} U_2^2 = (e_2' - e_1') \qquad (25.8)$$

In the above form, we have simply consolidated the unknowns into the left-hand side. Now as long as the experimental data are cast in a consistent form, Eq. (25.8) can be used instead of Eq. (25.7), without requiring knowledge of the kinetic energy coefficients. Defining the total pressure as $p_0 = p + \frac{1}{2}\rho U^2 + \rho g z$, Eq. (25.8) can be rearranged to give

$$\frac{p_{01} - p_{02}}{\rho} = \frac{\Delta p_0}{\rho} = (e'_2 - e'_1) \tag{25.9}$$

This expression is used to experimentally evaluate the irreversible loss in total pressure. The loss in total pressure is generally cast into dimensionless *loss coefficients* $K_1 = (p_{01} - p_{02})/\frac{1}{2}\rho U_1^2$ or $K_2 = (p_{01} - p_{02})/\frac{1}{2}\rho U_2^2$ where K_1 and K_2 simply differ in the choice of the velocity (at the inlet or at the exit) used in the definition. In actual experiments to determine the loss coefficient, the piezometric pressure, p', is generally the directly measured quantity: $p' = p + \rho g z$. A pressure coefficient can be constructed based on the difference in the piezometric pressures at the exit and inlet of the duct $C_{p_1} = (p'_2 - p'_1)/\frac{1}{2}\rho U_1^2$ or $C_{p_2} = (p'_2 - p'_1)/\frac{1}{2}\rho U_2^2$. These pressure coefficients can be related to the loss coefficients as follows: $C_{p_1} = 1 - (A_1/A_2)^2 - K_1$ and $C_{p_2} = (A_2/A_1)^2 - 1 - K_2$. For a constant area pipe where from mass conservation, $U_1 = U_2$, $C_{p_1} = C_{p_2} = -K_1 = -K_2$. Using the definitions of the loss coefficients, Eq. (25.8) can be written as:

$$[p_1 + \tfrac{1}{2}\rho U_1^2 + \rho g z_1] = [p_2 + \tfrac{1}{2}\rho U_2^2 + \rho g z_2] + (K_1 \tfrac{1}{2}\rho U_1^2 \text{ or } K_2 \tfrac{1}{2}U_2^2) \tag{25.10a}$$

or

$$\frac{p_1}{\rho g} + \frac{U_1^2}{2g} + z_1 = \frac{p_2}{\rho g} + \frac{U_2^2}{2g} + z_2 + H_L \tag{25.10b}$$

where $H_L = K_1 U_1^2/2g = K_2 U_2^2/2g$. In Eq. (25.10a), the dimension of each term is pressure, whereas in Eq. (25.10b), the dimensions are length or *head*. Either may be used as long as consistency is retained in evaluating all the terms. In summary, Eqs. (25.3) and (25.10) form the basis for obtaining the flow versus pressure drop characteristics of duct systems provided that the appropriate experimental data on loss coefficients are available.

25.1.2 Loss Coefficients for Duct System Components

A duct system is made up of several components such as straight pipes, bends, etc., that are connected in series or parallel. Determination of the overall system flow performance requires knowledge of the loss coefficients of each of the individual components. This section provides experimental data on the loss coefficients for commonly used duct components. Before presenting the data, some background on how these data are obtained is provided to aid in understanding the range of applicability and the appropriate use of these data.

Components are generally tested with sufficient lengths of straight pipes upstream and downstream so that fully developed flow is achieved. A typical set-up is shown in Fig. 25.2 where the component is placed between points B and C. The upstream pipe must be of sufficient length so the influence of the component is not felt at point A, which is the inlet of the pipe, and the flow is fully developed. Similarly, the

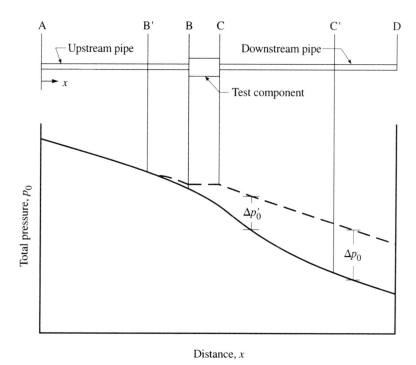

FIGURE 25.2 Test set-up for determination of losses in a duct component.

downstream length of pipe *CD* must be sufficiently long so that fully developed flow exists at the outlet *D*. Figure 25.2 also shows a schematic of the pressure drop characteristic for the component. The dotted lines show the expected pressure drop if the component were absent. The difference between the dotted line and the full line, $\Delta p_0'$, is the additional loss caused by the component. Note that a significant portion of the loss may actually occur upstream and downstream of the component— across the length *B'C'*. Hence, the total pressure loss is a function of distance from the component. At some point *C'* downstream of the component, fully developed flow will be attained, and the loss due to the component will remain constant from that point onwards. The value of this loss is shown as Δp_0 in Fig. 25.2. In most cases, Δp_0 is quoted as the loss for the component. The loss coefficient then becomes $K_1 = \Delta p_0 / \frac{1}{2}\rho U_1^2$ or $K_2 = \Delta p_0 / \frac{1}{2}\rho U_2^2$.

In many practical situations, the length of straight piping between components might not be sufficient to allow establishment of fully developed flow in the connecting pipe. In such cases, the flow fields of the two components will interact with each other, and it is clear that use of K_1 or K_2 derived from the isolated test of a single component will not yield accurate results. In such cases, the choice is to test the actual duct configuration or to accept the estimate provided by use of K_1 or K_2. For more on the subject of component interactions [see Ward–Smith (1976)].

Straight Ducts of Constant Cross Section. A detailed discussion of the various aspects of flow in constant area ducts is given in Chap. 5. In this section, only the information directly relevant to calculating the pressure drops in straight ducts of constant area will be presented. For such ducts, the pressure loss coefficient K, between two points along the duct where fully developed flow exists, can be expressed as $K = (p_{01} - p_{02})/\frac{1}{2}\rho U_2^2 = \lambda L/D$, where, p_{01} and p_{02} are the total pressures at the upstream and downstream points, respectively, L is the duct length between the two points, U is the mean flow velocity, and D is a representative cross-sectional dimension. For circular cross-section ducts, D is the diameter. The coefficient λ is the Darcy friction factor and is a function of the Reynolds number, $\text{Re} = \rho UD/\mu$, and for turbulent flow is a function of the pipe roughness k. For fully developed laminar flow in circular pipes (see Chap. 5), $\lambda = 64/\text{Re}$. For turbulent flows, λ has to be determined from experimental data. Figure 25.3 shows λ as a function of Reynolds number and pipe relative roughness k/D for circular pipes. This figure was developed by Moody (1944) and is, therefore, sometimes referred to as the Moody diagram.

At Reynolds numbers below about 2000, the flow is laminar. Between a Reynolds number of 2000 and 4000, transition from laminar flow to turbulent flow usually occurs, and λ is not well defined in this region. Above a Reynolds number of 4000, the flow is generally turbulent, and λ becomes a function of pipe roughness also. For a smooth pipe, where $k/D = 0$, λ continues to decrease with increasing Reynolds number. For a rough pipe, that is, a nonzero value of k/D, λ decreases up to a certain value and then levels off becoming independent of the Reynolds number. The region where λ is independent of Reynolds number is known as the *fully rough region*. The boundary of the fully rough regime is shown by the dashed line in Fig. 25.3. In the fully rough region, λ increases with increasing pipe roughness, that is, increasing k/D. Typical values of k for various types of pipes is given in Table 25.1.

It should be noted that Table 25.1 provides a general guide, and if high accuracy is desired in the prediction of the pressure drop, actual tests should be conducted on a prototypical pipe specimen to determine the appropriate value of k and/or λ. Also note that k is the *equivalent sand grain* roughness height and does not necessarily coincide with the physical height of the wall roughness element. A more complete discussion of this subject is given in Chap. 5 and by Reynolds (1974). An implicit expression is available (see Chap. 5) for calculating λ directly and can be used instead of Fig. 25.3

$$\frac{1}{\sqrt{\lambda}} = 1.74 - 2.0 \log \left(2\frac{k}{D} + \frac{18.7}{\text{Re}\sqrt{\lambda}} \right) \tag{25.11}$$

For noncircular cross-section ducts, use of a hydraulic diameter provides adequately accurate results for fully developed turbulent flow. The hydraulic diameter is defined as $D_h = 4 \times$ cross-sectional area/wetted perimeter.

A detailed discussion of the hydraulic diameter concept is given in Chap. 5. The concept is that the pressure drop in a noncircular cross-section duct is the same as that for a circular cross-section pipe having a diameter equal to the hydraulic diameter. This approach works well for turbulent flows in ducts having aspect ratios (ratio of the maximum to minimum cross-sectional dimensions) close to unity. However, up to a 20% error can occur in the prediction of turbulent pressure drops in ducts having aspect ratios much greater than unity. The situation is even worse in the case of laminar flow, and correction factors to the hydraulic diameter for common duct shapes are given in Chap. 5.

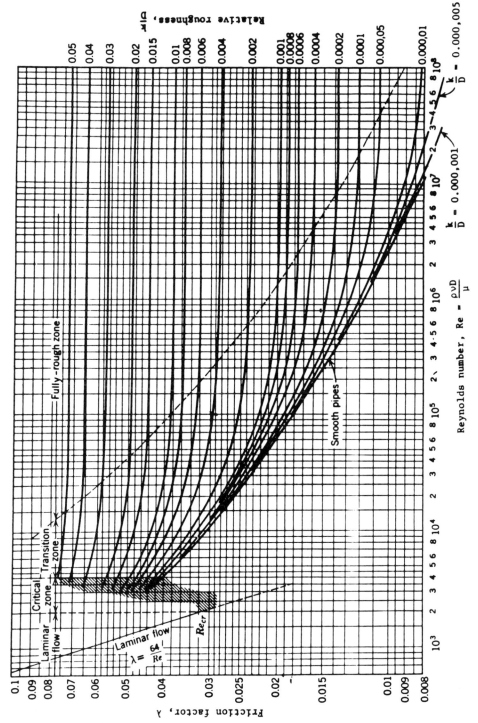

FIGURE 25.3 Friction factor for fully developed flow in circular pipes. (Reprinted with permission, Moody, Lewis F., "Friction Factors for Pipe Flow," *Trans. ASME*, Vol. 66, pp. 671–684, 1944.)

TABLE 25.1   Equivalent Sand Grain Roughness Height for Various Pipes

Type of Pipe (new)	Roughness k, (mm)
Glass, plastic, drawn metal tubing	1.5×10^{-3}
Commercial Steel or wrought iron	4.6×10^{-2}
Ashphalted Cast Iron	0.12
Galvanized Iron	0.15
Cast Iron	0.26
Wood Staves	0.18–0.9
Concrete	0.3–3.0
Riveted Steel	0.9–9.1

(After Moody, 1944).

Entrance and Leaving Losses. The losses at the entrance to a pipe are composed of two components. First, there is the loss K_e, due to the shape of the entrance, and second, there is K_d due to the developing flow in the pipe. The loss coefficient K_e for typical entrance geometries is given in Table 25.2. Note that even a slight rounding produces a considerable reduction in the loss coefficient, K_e.

The second component of the entrance loss, K_d, is due to the acceleration of the flow to the fully developed state. As discussed in Chap. 5, the entrance length, X_e, required to attain fully developed laminar flow is $X_e/D = 0.6 + 0.056$ Re and for turbulent flow, $X_e/D > 40$. In this developing flow region, the pressure drop is higher than that for a fully developed flow over the same length. The incremental loss coefficient K_d for laminar flows is given by Eq. (5.7) in Chap. 5 and can be as high as 1.25 for circular pipes when the entrance length is equal to that required to reach fully developed flow. For turbulent flows, $K_d \sim 0.04$.

The total loss coefficient for the entrance region of a pipe is given by

$$K = K_e + K_d + \lambda L/D \tag{25.12}$$

When a duct discharges into a plenum, all the flow kinetic energy is lost. Therefore, the exit or leaving loss coefficient $K_1 = 1$ for a constant area duct.

Bends. The loss coefficients for bends presented here are for constant area bends where the geometric cross-sectional area through the bend is constant and equal to

TABLE 25.2   Entrance Loss Coefficients K_e

Entrance Type	Diagram	K_e
Re-entrant		0.8–1.0
Square edged		0.5
Rounded edge ($r/R = 0.04$)		0.24
Well rounded ($r/R > 0.15$)		0.04

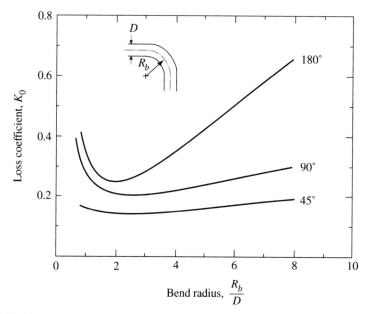

FIGURE 25.4 Loss coefficients for pipe bends for Re $= 10^5$. (Reprinted with permission, Ito, H., "Pressure Losses in Smooth Pipe Bends," *J. Basic Engrg.*, Vol. 82, No. 1, pp. 131–143.)

the area of the pipes attached to it. Figure 25.4 shows the loss coefficients for 45-, 90-, and 180-degree circular pipe bends at a Reynolds number of 10^5 as a function of the radius R_b of the bend. This figure was derived from data of Ito (1960). From Fig. 25.4, it is evident that the minimum loss occurs at R_b/D between 2 and 3. At smaller bend radii, the losses are caused by severe flow distortions in the bend, while at larger bend radii, the frictional losses dominate.

The loss coefficient K_0 shown in Fig. 25.4 is for a Reynolds number of 10^5. Loss coefficients at other Reynolds numbers can be obtained by applying a correction as follows

$$K = \alpha K_0 \qquad (25.13)$$

where α is a correction factor obtained as below.

For $R_b/D = 1$:

$$\alpha = 19.53 \, \mathrm{Re}^{-0.26} \quad \text{for } 2 \times 10^4 \le \mathrm{Re} \le 2 \times 10^5$$

$$\alpha = 0.83 \quad \text{for } \mathrm{Re} \ge 2 \times 10^5 \qquad (25.14a)$$

For $R_b/D \ge 2$:

$$\alpha = 5.080 \, \mathrm{Re}^{-0.141} \quad \text{for } 2 \times 10^4 \le \mathrm{Re} \qquad (25.14b)$$

Losses for rough pipe bends can be obtained by increasing the loss for smooth bends by the ratio of the smooth pipe friction factor λ to the rough pipe friction factor.

Also, note that the loss coefficient is the total loss produced by the bend under the condition that the lengths of the straight pipes at the inlet and exit of the bend are sufficiently long so that fully developed flow is established upstream and downstream of the bend.

Square cross-sectional bend losses are close to those for a circular cross-section bend discussed above. Therefore, the loss coefficients of Fig. 25.4 and the procedure described earlier can be used for square cross-section bends also. For rectangular cross-section bends, the loss coefficient depends on the aspect ratio of the cross-section in addition to the radius of the bend, R_b. The aspect ratio is the ratio of the height of the bend in the plane of the bend to the width. For rectangular cross sections with aspect ratio of 0.5, the loss coefficient can increase by a factor of 2 over that for a circular cross section bend. The loss coefficient for rectangular cross sections with aspect ratio greater than one decreases by up to 30% as compared with that for circular cross-section bends [see Ward–Smith (1980) and Miller (1978)].

The loss coefficients for single miter bends (for Re $= 10^5$) is given by

$$K_b = 2.22 \times 10^{-3}\theta + 4.04 \times 10^{-6}\theta^{2.76} \qquad \text{for } 0 \leq \theta \leq 90 \qquad (25.15)$$

The corresponding loss coefficient for segmented miter bends is shown in Fig. 25.5. The Reynolds number and roughness correction procedure described for smooth bends is applicable to miter bends as well.

Expansions and Contractions. Connection between ducts of differing cross-sectional areas is made using ducts with increasing (expansions) or decreasing (contractions) flow cross-sectional area. When the length of the expansion or contraction is zero, a sudden expansion or contraction is obtained. If the expansion or contraction has a finite length, it is referred to as a *diffuser* or a nozzle (*reducer*), respec-

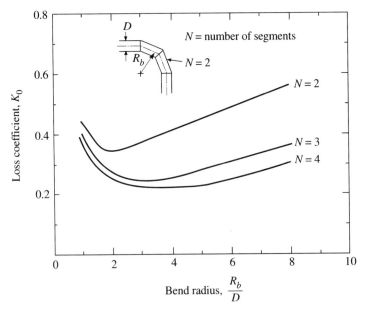

FIGURE 25.5 Loss coefficients for segmented miter bends at Re $= 10^5$.

tively. The details of flow through such ducts are given in Chap. 5. Consider a sudden expansion where the duct area abruptly changes from a smaller upstream value of A_1 to a larger downstream area, A_2. A simple momentum balance across the sudden expansion is used to derive the loss coefficient K, and the expression for incompressible flow is

$$K = \frac{(p_{01} - p_{02})}{\frac{1}{2}\rho U_1^2} = \left(1 - \frac{1}{AR}\right)^2 \tag{25.16a}$$

where AR is the area ratio A_2/A_1. The static pressure rise through the sudden expansion for incompressible flow is given by the pressure recovery coefficient C_p

$$C_p = \frac{(p_1 - p_2)}{\frac{1}{2}\rho U_1^2} = \frac{2}{AR}\left(1 - \frac{1}{AR}\right) \tag{25.16b}$$

The above expressions have been found to agree within 10% with experimental data for turbulent flow. For laminar flow the agreement is quite poor as discussed in Chap. 5. When the increase in area is accomplished over a finite length, a diffuser results. The flow characteristics for *diffusers* is given in Sec. 25.2.

For sudden contractions, simple momentum balance analysis does not yield a satisfactory expression for the loss. The often quoted expression for the turbulent flow loss through a sudden contraction is

$$K = \frac{(p_{01} - p_{02})}{\frac{1}{2}\rho U_2^2} = \frac{1}{2}\left(1 - \frac{1}{AR}\right) \tag{25.17}$$

Note that the loss coefficient is defined using the mean velocity U_2 through the samller cross section. Measurements made by Benedict *et al.* (1966) show that the above expression gives rather poor agreement with data. It is, therefore, suggested that the following correlation be used for turbulent flow

$$K = 0.58(1 - 1/AR)^{0.741} \qquad \text{for } 0 \leq 1/AR \leq 0.5$$

and

$$K = 0.99(1 - 1/AR)^{1.51} \qquad \text{for } 0.5 \leq 1/AR \leq 1.0 \tag{25.18}$$

For gradual contractions with straight converging walls, the following correction is suggested in Crane Co. (1979)

$$C = 1.6 \sin(\theta/2) \qquad \text{for } \theta \leq 45$$

and

$$C = \sqrt{\sin(\theta/2)} \qquad \text{for } 45 \leq \theta \leq 180 \tag{25.19}$$

where θ is the included angle between the converging walls. The loss coefficient for a sudden contraction is multiplied by this correction factor to obtain the loss for straight wall converging sections.

When the objective of the contraction is to produce uniform velocity profile at the exit of the contraction, straight converging walls are inadequate. The contraction wall profile has to be designed with care and such contractions (nozzles) are discussed in details in Chap. 5.

Screens and Perforated Plates. Screens and/or perforated plates are often used to trap debris in pipelines. Another application of screens and perforated plates is to smooth flow distortions. The primary geometric parameter that governs the loss through screens and perforated plates is the fraction of open area, γ. For square mesh screens

$$\gamma = (1 - nd)^2 \tag{25.20}$$

where n is the number of wires per unit length, and d is the diameter of the wire. For rectangular mesh screens

$$\gamma = (1 - n_1 d_1)(1 - n_2 d_2) \tag{25.21}$$

where n_1 and n_2 are the number of wires in the two directions of the weave with diameters d_1 and d_2, respectively. For perforated plates

$$\gamma = \frac{N \pi d^2}{4A} \tag{25.22}$$

where N is the number of holes of diameter d over the cross-sectional area A of the duct.

Ward–Smith (1980) has correlated a large amount of experimental data and arrived at the following correlation for the loss through screens

$$K = K_{in} \cdot C_{Re} \tag{25.23}$$

K_{in} is the loss at $Re_d \geq 1000$ and is given by

$$K_{in} = \left(\frac{1 - \phi\gamma}{\phi\gamma} \right)^2$$

and

$$\phi = 1.09 - 0.35\gamma. \tag{25.24}$$

The Reynolds number correction factor, C_{Re}, is given by

$$\log C_{Re} = 0.182 \, (\log \, (1000/Re_d))^{1.636} \quad \text{for } 10 \leq Re_d \leq 1000 \tag{25.25a}$$

and,

$$C_{Re} = 1 \quad \text{for } Re_d \geq 1000 \tag{25.25b}$$

The Reynolds number, Re_d, is based on the diameter of the screen wire and the approach velocity, that is the mean velocity upstream of the screen.

For perforated plates, the loss is a function of the geometric parameter t/d, where t is the plate thickness and d the hole diameter. For $t/d \leq 0.2$, the flow through the perforated plate can be viewed as being similar to that through a single, sharp-edged orifice with an area equal to the open area of the perforated plate. For turbulent flow through the holes in the perforated plate, the loss for $t/d \leq 0.2$ is given by

$$K = \left(\frac{1 - \phi\gamma}{\phi\gamma} \right)^2 \qquad (25.26a)$$

where ϕ the contraction coefficient is given by

$$\phi = 0.596 + 0.297\gamma^{2.047} \qquad (25.26b)$$

The above expression holds when the Reynolds number based on the hole diameter and the velocity through the hole is greater than 4000.

Ward–Smith (1980) presents the following correlations for losses through perforated plates. For $0.2 \leq t/d < 0.8$ and $0.006 < \gamma < 0.75$,

$$K = \left[\left(0.609\gamma \left\{ 1 - \left(\frac{t}{d} \right)^{3.5} \right\} \{1 - \gamma^{2.6}\} + \gamma^{3.6} \right)^{-1} - 1 \right]^2 \qquad (25.26c)$$

and for $0.8 < t/d < 7.1$ and $0.002 < \gamma < 0.53$,

$$K = \left[\left(\gamma \left\{ 0.872 - 0.0149 \frac{t}{d} - 0.08 \frac{d}{t} \right\} \{1 - \gamma^{3.3}\} \right. \right.$$
$$\left. \left. + \gamma^{4.3} \left\{ 1 + 0.134 \left(\frac{t}{d} \right)^{0.5} \right\}^{-1} \right)^{-1} - 1 \right]^2 \qquad (25.26d)$$

Combining and Dividing Tees. A typical combining or dividing tee used to form junctions in duct systems is shown in Fig. 25.6. The nomenclature for identifying the various legs of the tee is shown in Fig. 25.6. The leg carrying the most flow is labeled as leg 3, and the leg collinear with leg 3 is designated as leg 2. Leg 1 enters at an angle θ with the axis of legs 2 and 3. With this nomenclature, $Q_3 = Q_1 + Q_2$, where Q is the flow rate and subscripts 1, 2, and 3 refer to the three legs of the tee, respectively. The loss coefficient K_{mn} is defined as

$$K_{mn} = (p_{0m} - p_{0n})/\tfrac{1}{2}\rho U_3^2 \qquad (25.27)$$

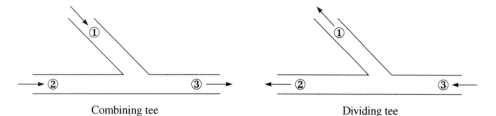

Combining tee   Dividing tee

FIGURE 25.6   Schematic of combining and dividing tees.

where p_{0m} and p_{0n} are the total pressures in legs m and n respectively, and U_3 is the velocity in leg 3.

For combining tees with $\theta = 90°$, $A_1 = A_2 = A_3$, and sharp edges at the junction, Ito and Imai (1973) suggest the following correlations

$$K_{13} = 1.09 - 0.53(1 - q) - 1.48(1 - q)^2$$

$$K_{23} = 0.045 + 1.38q - 0.90q^2 \tag{25.28}$$

where $q = Q_1/Q_3$. The above correlations are for circular cross sections and Reynolds number based on the flow in leg 3 greater than 10^4.

Gardel (1957) tested the effects of varying A_1/A_3, θ, and the radius of the junction edges. In these tests, $A_2 = A_3$, and the flow was turbulent in leg 3. Gardel (1957) suggests the following correlations for combining tees:

$$K_{13} = -0.92(1 - q^2) - q^2 \left[(1.2 - r^{1/2}) \left(\frac{\cos \theta}{a} - 1 \right) \right.$$

$$+ 0.8 \left(1 - \frac{1}{a^2} \right) - \left(\frac{1}{a} - 1 \right) \cos \theta \Bigg] + (2 - a)(1 - q)q$$

$$K_{23} = 0.03(1 - q)^2 - q^2 \left[1 + (1.62 - r^{1/2}) \left(\frac{\cos \theta}{a} - 1 \right) - 0.38(1 - a) \right]$$

$$+ (2 - a)(1 - q)q \tag{25.29}$$

where $r = R_f/D_3$, $a = A_1/A_3$, $q = Q_1/Q_3$, R_f is the fairing radius at the junction with leg 1, and D_3 is the diameter of leg 3. Additional data on combining junctions in turbulent flow are provided by Miller (1978).

Ito and Imai (1973) suggest the following correlations for sharp edged, 90° dividing tees

$$K_{32} = 1.55(0.22 - q)^2 - 0.03, \quad \text{for } 0 \le q \le 0.22$$

$$K_{32} = 0.65(q - 0.22)^2 - 0.03 \quad \text{for } 0.22 \le q \le 1$$

$$K_{31} = 0.99 - 0.82q + 1.02q^2 \quad \text{for } 0 \le q \le 1 \tag{25.30}$$

For dividing tees, Gardell (1957) provides the following correlations that account for variations of θ, fillet radius, and A_1/A_3, with $A_2 = A_3$.

$$K_{32} = 0.03(1 - q)^2 + 0.35q^2 - 0.2q(1 - q)$$

$$K_{31} = 0.95(1 - q)^2 + q^2 \left[\left(1.3 \tan \frac{\phi}{2} - 0.3 + \frac{(0.4 - 0.1a)}{a^2} \right) \right.$$

$$\cdot \left(1 - 0.9 \left(\frac{r}{a} \right)^{1/2} \right) \Bigg] + 0.4q(1 - q) \left(\frac{1 + a}{a} \right) \tan \frac{\phi}{2} \tag{25.31}$$

where $\phi = \pi - \theta$.

TABLE 25.3 Valve Losses

Type of valve (fully open)	Typical loss coefficient K
Gate	0.2
Globe	6
Swing check	1–2
Ball	0.05
Butterfly	0.3–0.7
Plug	0.3

All the correlations presented above can be used if the Reynolds number of the flow in leg 3 is greater than 10^4. Further, these correlations will give reasonable estimates for losses in tees of noncircular cross section.

Valves. Losses through valves depend on the detailed internal geometry of the valve (see Sec. 25.3). Hence, valve loss will vary significantly between two valves of the same general type but manufactured by two different manufacturers. Therefore, it is advised that the loss data for a particular valve be obtained from the actual valve manufacturer. Table 25.3 shows typical valve loss coefficients for different valve types. The information presented in this table is to be used for approximate estimates only.

25.1.3 Duct System Analysis

Duct systems consist of ducts and components placed in series, parallel, or in a network. This section addresses the analysis of flow and pressure drop for ducting systems when the duct system geometry is given. The inverse problem of designing a duct system will be discussed in the next section. The analysis of steady flow through duct systems is based on using three principles. First, mass must be conserved at each point in the duct system. This implies that mass flow remains constant through components connected in series; at junctions where flow is divided or combined, the sum of the mass flows into the junction must be equal to the sum of the mass flows out of the junction. The second principle is that the pressure at any point in the duct system is uniquely determined. Hence, the change in pressure between two points is the same for all flow paths connecting the two points. Stated another way, the net pressure drop around a closed loop is zero. The third principle is that the flow direction must be consistent with the direction of the total pressure gradient across a given component. For example, in duct components where external energy is not added to the flow, the flow must be in the direction of decreasing total pressure.

Duct System With Components in Series. Consider the duct system shown in Fig. 25.7 where all components are connected in series. Components A, B, and C are *discrete* components such as bends, expansions/contractions, valves, etc., connected by constant area ducts ab, cd, etc. Using the first of the three principles discussed earlier, the mass flow through all the components must be the same and equal to the

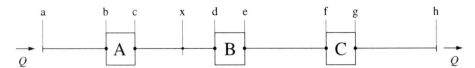

FIGURE 25.7 Duct system with components connected in series.

inlet mass flow. If Q is the volumetric flow input at point a, then for incompressible flow,

$$Q = Q_{ab} = Q_A = Q_{cd} = Q_B = Q_{ef} = Q_C = Q_{gh} \qquad (25.32)$$

The uniqueness of the pressure at any point x in the system implies that,

$$p_{0x} = p_{0a} - \tfrac{1}{2}\rho U_{ab}^2 K_{ab} - \tfrac{1}{2}\rho U_a^2 K_A - \tfrac{1}{2}\rho U_{cd}^2 K_{cx}$$

$$= p_{0h} + \tfrac{1}{2}\rho U_{cd}^2 K_{xd} + \tfrac{1}{2}\rho U_B^2 K_B + \tfrac{1}{2}\rho U_{ef}^2 K_{ef} + \tfrac{1}{2}\rho U_c^2 K_c + \tfrac{1}{2}\rho U_{gh}^2 K_{gh} \qquad (25.33)$$

where K is the loss coefficient for a given component, U is the mean velocity through the component used to define K, p_0 is the total pressure, and subscripts denote the respective components.

The values of K are those given in Sec. 25.1.2, which imply that the losses in the various components are independent of each other, that is there is no interaction between the components. This will be the case if the discrete components are separated by sufficient lengths of straight, constant-area ducts such that fully developed flow is established between the discrete components. If this is not the case, the discrete components will interact with each other, and the use of the loss coefficients given in Sec. 25.1.2 might not be appropriate. For further information on component interactions, the reader is referred to Ward–Smith (1976) and Miller (1978).

Through mass conservation, the velocity $U = Q/A$ and Eq. (25.33) can be written as,

$$p_{0x} = p_{0a} - \sum_a^x CQ^2 = p_{0h} + \sum_x^h CQ^2 \qquad (25.34)$$

where $C = \rho K/2A^2$, and A is the appropriate flow area for a given component. The static pressure p_x at point x can be readily obtained using Eq. (25.9).

From Eq. (25.34), the total pressure loss through the system is,

$$p_{0a} - p_{0h} = \sum_a^h CQ^2 = C_{eq}Q^2 \qquad (25.35)$$

where $C_{eq} = C_{ab} + C_A + C_{cd} + C_B + C_{ef} + C_C + C_{gh}$.

Example 25.1. Calculate the total and static pressure drop through the piping system shown in Fig. 25.8. Water flows through the system at a rate of 471 l/min. The static pressure at the inlet is 200 kPa. Assume that at the exit plane g, the flow goes into a reservoir.

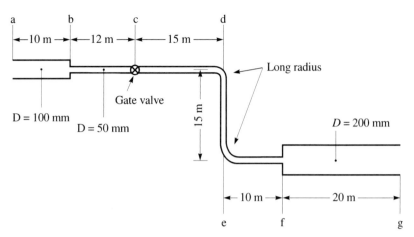

FIGURE 25.8 Duct system for Example 25.1.

The first step is to calculate the flow velocities through various portions of the system using the relation $U = Q/A$. Consider the 100 mm diameter pipe ab,

$$Q = 471 \text{ l/min} = 7.85 \times 10^{-3} \text{ m}^3/\text{sec}$$

and

$$A_{ab} = \frac{\pi}{4} 100^2 \text{ mm}^2 = 7.85 \times 10^3 \text{ mm}^2 = 7.85 \times 10^{-3} \text{ m}^2$$

therefore,

$$U_{ab} = Q/A_{ab} = 1 \text{ m/sec}$$

The next step is to determine the Reynolds number taking $\rho = 1000 \text{ kg/m}^3$ and $\mu = 1 \times 10^{-3}$,

$$Re_{ab} = \frac{\rho D_{ab} U_{ab}}{\mu} = 1000 \frac{100}{1000} \frac{1}{1 \times 10^{-3}} = 10^5$$

Based on this Reynolds number, and assuming that the pipe is smooth, Fig. 25.3 gives a value of 0.018 for the friction factor λ. For the pipe section ab,

$$K_{ab} = \frac{\lambda L_{ab}}{D} = \frac{0.018 \times 10}{100 \times 10^{-3}} = 1.8$$

$$C_{ab} = \frac{\rho K_{ab}}{2A_{ab}^2} = \frac{1000 \times 1.8}{2 \times (7.85 \times 10^{-3})^2} = 1.46 \times 10^7 \text{ kg/m}^7$$

Applying the above procedure to the other sections of straight pipes results in the values of C shown in Table 25.4.

The loss coefficient for the sudden contraction at b is obtained from Sec. 25.1.2.

TABLE 25.4 Calculations for Example 25.1

Pipe Section	Pipe Dia. mm D	Velocity m/sec	Reynolds Number Re	λ	Pipe Section Length, L (m)	$K = \lambda \dfrac{L}{D}$	$C = \dfrac{\rho K}{2A^2}$ (kg/m^7)
bc	50	4	2×10^5	0.016	12	3.84	5×10^8
cd	50	4	2×10^5	0.016	15	4.80	6.25×10^8
de	50	4	2×10^5	0.016	15	4.80	6.25×10^8
ef	50	4	2×10^5	0.016	10	3.20	4.17×10^8
fg	200	0.25	0.5×10^5	0.021	20	2.10	1.06×10^6

The area ratio of the sudden contraction is,

$$AR = A_{ab}/A_{bc} = (100/50)^2 = 4, \quad \text{and } 1/AR = 0.25.$$

Hence, from Eq. (25.15), $K_b = 0.58(1 - 1/AR)^{0.741} = 0.47$, and therefore, $C_b = \rho K_b/2A_{bc} = 4.84 \times 10^8$ kg/m^7. Note that the area of the 50 mm pipe was used in computing C_b.

Next consider the valve at location c. This is a gate valve that is fully open and, therefore, from Table 25.3.

$$K_c = 0.2, \quad \text{and } C_c = \frac{\rho K_c}{2A_{bc}} = 2.6 \times 10^7 \text{ kg/m}^7$$

For the 90° elbow at point d, the procedure described in Sec. 25.1.2 is used. Using Eq. (25.3)

$$K = \alpha K_0, \quad \text{and for } R_b/D \geq 2, \quad \alpha = 5.080 \text{ Re}^{-0.141}$$

therefore, $\alpha = 5.080(2 \times 10^5)^{-0.141} = 0.91$. From Fig. 25.4, $K_0 = 0.21$, and, therefore,

$$K_d = 0.91 \times 0.21 = 0.19$$

and

$$C_d = \frac{\rho K_d}{2A_{de}} = 2.47 \times 10^7 \text{ kg/m}^7$$

Since the second bend at e is identical to the one at point d.

$$C_e = 2.47 \times 10^7 \text{ kg/m}^7$$

The loss for the sudden expansion at point f is given by Eq. (25.16a)

$$K_f = (1 - 1/AR)^2, \quad \text{where } AR = A_{fg}/A_{ef} = (200/50)^2 = 16$$

hence

$$K_f = (1 - 1/16)^2 = 0.88$$

and

$$C_f = \frac{\rho K_f}{2A_{ef}} = 1.04 \times 10^6 \text{ kg/m}^7$$

Finally, since the piping system discharges into a reservoir, all the kinetic energy at point g is lost. Therefore, the leaving loss is,

$$K_g = 1$$

and

$$C_g = \frac{\rho K_g}{2A_{fg}} = 5.07 \times 10^5 \text{ kg/m}^7$$

The overall system loss function C_{eq} is given by,

$$C_{eq} = (C_{ab} + C_b + C_{bc} + C_c + C_{cd} + C_d + C_{de} + C_e$$
$$+ C_{ef} + C_f + C_{fg} + C_g)$$

Substituting the values of the individual component losses,

$$C_{eq} = 27.4 \times 10^8 \text{ kg/m}^7$$

and hence, the loss through the entire system,

$$p_{0a} - p_{0g} = \Sigma CQ^2 = C_{eq}Q^2$$

and

$$p_{0a} - p_{0g} = 27.4 \times 10^8 (7.85 \times 10^{-3})^2 = 169 \text{ kPa.}$$

Next, calculate the total pressure at the inlet

$$p_{0a} = p_a + \tfrac{1}{2}\rho U_a^2 + \rho g z_a$$

Taking the datum elevation to be at the exit $z_a = 15$ m, the gravitational acceleration is 9.81 m/sec, and the static pressure at point a is given as 200 kPa, hence

$$p_{0a} = 200 + \frac{1000 \times (1)^2 \times 10^{-3}}{2} + 1000 \times 9.81 \times 15 \times 10^{-3} = 347.6b \text{ kPa}$$

Note that the factor 10^{-3} is to convert Pa to kPa.

$$p_{0g} = p_{0a} - \text{loss} = 348 - 169 = 179 \text{ kPa}$$

Since the datum elevation was taken at the exit $z_g = 0$, and since the discharge is into a reservoir, all the kinetic energy at point g is lost. Therefore, the static pressure at point g is equal to the total pressure there. Hence,

$$p_g = p_{0g} = 179 \text{ kPa.}$$

Ducting Components in Parallel. Consider the ducting system shown in Fig. 25.9, where the nodes a and b are connected by N parallel ducts. At nodes a and b,

$$Q = Q_1 + Q_2 + \cdots Q_N = \Sigma Q_n \tag{25.36}$$

where Q_1, Q_2, etc. are the flow rates through each of the N parallel duct systems.
  The uniqueness of pressure across the loops requires that,

$$\Delta p_{0ab} = C_1 Q_1^2 = C_2 Q_2^2 = \cdots = C_N Q_N^2 = C_{eq} Q^2 \tag{25.37}$$

where, C_1, C_2, etc., are the appropriate loss coefficients for the N parallel loops, and C_{eq} is the equivalent loss coefficient, that is the loss coefficient for an equivalent single duct connecting the two nodes. From Eq. (25.36)

$$Q_n = Q \sqrt{\frac{C_{eq}}{C_n}} \qquad \text{for } n = 1, N. \tag{25.38}$$

Substituting the above relationship into Eq. (25.36) gives,

$$\frac{1}{\sqrt{C_{eq}}} = \frac{1}{\sqrt{C_1}} + \frac{1}{\sqrt{C_2}} + \cdots + \frac{1}{\sqrt{C_N}} \tag{25.39}$$

  Once C_{eq} is calculated, the pressure drop across the nodes a and b can be determined using Eq. (25.37), and the individual flow rates through the various loops can be calculated from Eq. (25.38). Although this appears to be a straightforward procedure, in practice it is not so, since the loss coefficients C_n are themselves functions of the corresponding flow rates Q_n. Therefore, the solution of parallel duct systems generally involves an iterative procedure where:

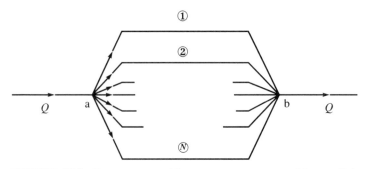

FIGURE 25.9 Duct system with components connected in parallel.

1. The flow rates through the individual loops is guessed such that Eq. (25.36) is satisfied.
2. Based on the assumed values of Q_n, the loss coefficients through the individual loops C_n are determined.
3. The equivalent loss coefficient is calculated using Eq. (25.39).
4. New values of the flow rates through the individual loops are computed using Eq. (25.38), and the procedure is repeated starting from Step 2 using the updated values of Q_n. A final solution is obtained when the difference between the new values of the individual flow rates Q_n in the various parallel paths and the values of Q_n from the previous iteration is small.

From the above procedure, it is clear that even a simple parallel loop analysis can be quite tedious. Hence, such analyses are usually performed using computers.

Duct Networks. A duct network consists of several duct systems connected in series and parallel such as the network shown in Fig. 25.10. The analysis of duct networks is based on applying the three principles discussed at the beginning of this section. For example, consider the network shown in Fig. 25.10. The junctions labeled a, b, c, etc., will be referred to as *nodes*. The ducts connecting the nodes are labeled 1, 2, 3, etc., and the flow rates through the ducts as Q_1, Q_2, and so on. Flows coming into or going out of the network are labeled Q_a, Q_b, etc. For each section of duct between two nodes, the pressure drop is related to the flow rate through the duct. For example, for duct section 1 connecting nodes a and b,

$$p_{0a} - p_{0b} = C_1 Q_1^2 \quad \text{or} \quad p_{0j} - p_{0i} = \sum_j^i C_n Q_n^2 \qquad (25.40)$$

where C_1 is the loss coefficient for duct section 1. If there are M such duct sections, there will be M equations, one for each duct section.

At each node, the sum of the incoming flow rates must be equal to the sum of outgoing flow rates. For example, at node b of the network of Fig. 25.10.

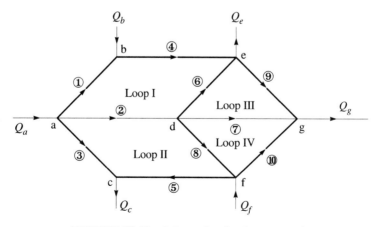

FIGURE 25.10 Schematic of a duct network.

$$Q_b + Q_1 = Q_4 \quad \text{or} \quad \Sigma Q = 0 \tag{25.41}$$

Note that for consistency, a convention has been adopted where flows *entering* a node are considered *positive*, while flows *leaving* a node are *negative*. If the network has N nodes, there will be N such equations. However, only $N - 1$ equations will be independent; the last, i.e., the Nth equation, can always be obtained by some combination of the others. Finally, all the external flow rate entering or leaving the network must satisfy mass conservation, and, therefore, for the network shown in Fig. 25.10.

$$Q_a + Q_b - Q_c - Q_e + Q_f - Q_g = 0 \tag{25.42}$$

There is one such equation. Hence, the total number of independent equations that can be constructed are $(N + M)$. The variables are the N pressures at each of the N nodes, M flow rates Q_1 through Q_m in each of the duct sections connecting the nodes, and I external flows entering or leaving the network. The total number of variables is therefore $(N + M + I)$. Hence, I variables must be specified in order to obtain a unique solution.

Once the adequate number of variables are specified for a given network problem, network analysis reduces to the solution of the system of equations given by Eqs. (25.40), (25.41), and (25.42). Note, however, that Eq. (25.40) is nonlinear due to the term $C_n Q_n^2$; the loss coefficient, C_n, is generally a function of the flow rate, Q_n. Therefore, iterative procedures must be used to derive solutions for network flow problems. Solution of a system of nonlinear equations can be found in textbooks on numerical analysis; solution techniques specific to duct network equations are discussed in depth by Jeppson (1976), and computer programs for network analysis are given by Jeppson (1976), Watters (1984), and Wood and Rayes (1981).

The Hardy–Cross method is commonly used to analyze the network flow when the external flow rates into and out of the network are specified, that is, Q_a, Q_b, etc., for the network of Fig. 25.10 are specified. In the Hardy–Cross method, the pressure drop relations of Eq. (25.40) are combined for duct sections forming a nonoverlapping closed loop to give,

$$\overset{L}{\underset{}{\Sigma}} C_j Q_j^2 = 0 \tag{25.43}$$

where L is the number of duct sections in a given closed loop in the network. For example, for the closed loop I, formed by duct sections 1, 4, 6, and 2 (see Fig. 25.10),

$$C_1 Q_1^2 + C_4 Q_4^2 - C_6 Q_6^2 - C_2 Q_2^2 = 0 \tag{25.43a}$$

The convention used is that the loop is traversed in the clockwise direction and flows in the *clockwise* direction are *positive*, and flows in the *counterclockwise* direction are *negative*. Note that for a unique solution to the flow rates through the network, the number of loop equations needed is equal to $(M - N + 1)$. The procedure then starts by initially providing a guess for the flow rates such that Eq. (25.41) is satisfied at each node. A correction, ΔQ, is sought for each flow loop such that when it is added to the individual flow rates around that loop, Eq. (25.43) will be satisfied

$$\Sigma\, C_j(Q_j + \Delta Q)^2 = 0 \tag{25.44}$$

Neglecting the second-order term ΔQ^2, the above equation can be written as

$$\Delta Q = -\frac{\overset{L}{\Sigma}\, C_j Q_j^2}{2\,\overset{L}{\Sigma}\, C_j Q_j} \tag{25.45}$$

To properly account for the sign of Q_j, the above is written as

$$\Delta Q = -\frac{\overset{L}{\Sigma}\, C_j |Q_j| Q_j}{2\,\overset{L}{\Sigma}\, C_j |Q_j|} \tag{25.46}$$

Once ΔQ is computed for each loop, it is added or subtracted from the individual flow rates in each portion of the loop. Note that since the procedure was initially started with assumed values for Q_j that satisfied Eq. (25.41) at each node, these equations will remain satisfied after the correction ΔQ is applied around each loop. The next step is to use the corrected flow rates to determine new values for C_j (since C_j is generally a function of the flow rate). The updated values of C_j and Q_j are then used to obtain a new value for the correction ΔQ. The entire procedure is repeated until the corrections, ΔQ, become acceptably small. This iterative procedure is necessary because in deriving Eq. (25.46) for ΔQ, the second-order term ΔQ^2 was dropped and also since C_j is a function of Q_j in general.

Example 25.2.   Consider the water pipe network shown in Fig. 25.11, consisting of five sections of 1000 m long, 400 mm I.D., smooth pipes labeled 1 through 5 and connected at four nodes a through d. The flow rates Q_a, Q_b, Q_c, and Q_d, in or out of the network are specified as shown. The lengths and diameters of the various pipe sections are also given in Eq. (25.11). Compute the flow rates through each pipe section using the Hardy–Cross method. Ignore losses at junctions.

Step 1: Guess the flow rate in each pipe section such that continuity is satisfied at each node, i.e., Eq. (25.41) is satisfied at each node. Note that this also requires providing a guessed flow direction in each section of the loop. The initial guesses for the flow rates and direction through each section are shown in Fig. 25.11.

Step 2: Construct the loop equations. There are two nonoverlapping loops, as shown in Fig. 25.11, the corresponding equations for the respective flow corrections ΔQ_I and ΔQ_II are

$$\Delta Q_\text{I} = \frac{-(C_1 |Q_1| Q_1 + C_2 |Q_2| Q_2 + C_4 |Q_4| Q_4)}{2\,(C_1 |Q_1| + C_2 |Q_2| + C_4 |Q_4|)}$$

and

$$\Delta Q_\text{II} = \frac{-(C_1 |Q_1| Q_1 + C_5 |Q_5| Q_5 + C_3 |Q_3| Q_3)}{2\,(C_1 |Q_1| + C_5 |Q_5| + C_3 |Q_3|)}$$

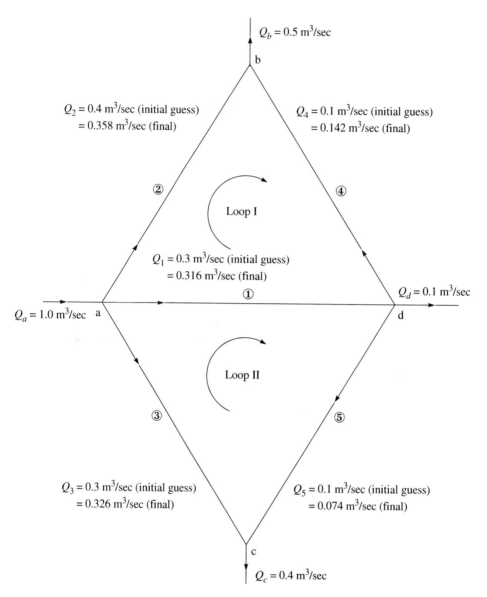

$Q_b = 0.5$ m³/sec

b

$Q_2 = 0.4$ m³/sec (initial guess)
$= 0.358$ m³/sec (final)

$Q_4 = 0.1$ m³/sec (initial guess)
$= 0.142$ m³/sec (final)

②

④

Loop I

$Q_1 = 0.3$ m³/sec (initial guess)
$= 0.316$ m³/sec (final)

①

$Q_d = 0.1$ m³/sec

$Q_a = 1.0$ m³/sec a

d

Loop II

③

⑤

$Q_3 = 0.3$ m³/sec (initial guess)
$= 0.326$ m³/sec (final)

$Q_5 = 0.1$ m³/sec (initial guess)
$= 0.074$ m³/sec (final)

c

$Q_c = 0.4$ m³/sec

FIGURE 25.11   Duct network for Example 25.2.

Note that the previously discussed sign convention will be used. That is, the loop will be traversed in the clockwise direction, and flows in the direction of the traverse will be positive.

   Step 3: Using the values of Q_1, Q_2 etc., determine the loss coefficients, C_1, C_2, etc. This involves calculating the flow velocities, and the Reynolds number for each pipe section. Next, the friction factor is determined. Instead of using the Moody diagram (Fig. 25.3), the following power law expression, which was presented in Chap. 5, will be used to determine λ.

TABLE 25.5 Calculations for Example 25.2

Loop	Pipe	Iteration 1 Q m³/sec	Iteration 1 Re × 10⁻⁵	Iteration 1 λ × 10²	Iteration 1 n kg/m⁷	Iteration 2 Q m³/sec	Iteration 2 Re × 10⁻⁵	Iteration 2 λ × 10²	Iteration 2 C_n kg/m⁷	Iteration 3 Q m³/sec	Iteration 3 Re × 10⁻⁵	Iteration 3 λ × 10²	Iteration 3 C_n kg/m⁷	Iteration 4 Q m³/sec	Iteration 4 Re × 10⁻⁵	Iteration 4 λ × 10²	Iteration 4 C_n kg/m⁷	Final Q m/sec
I	1	−0.3	9.55	1.17	369	−0.322	10.3	1.15	365	−0.315	10	1.16	366	−0.317	10.1	1.16	366	−0.316
	2	0.4	12.7	1.11	351	0.369	11.7	1.12	356	0.364	11.6	1.13	357	0.359	11.4	1.13	358	0.358
	4	−0.1	3.18	1.44	455	−0.131	4.17	1.36	431	−0.136	4.33	1.35	428	−0.141	4.48	1.34	425	0.142
		$\Delta Q_I = -0.0310$				$\Delta Q_I = -0.00529$				$\Delta Q_I = -0.00506$				$\Delta Q_I = -0.0015$				
II	1	0.3	9.55	1.17	369	0.322	10.3	1.15	365	0.315	10	1.16	366	0.317	10.1	1.16	366	0.316
	5	0.1	3.18	1.44	455	0.0914	2.91	1.46	463	0.0792	2.52	1.50	476	0.0761	2.42	1.52	480	0.074
	3	−0.3	9.55	1.17	369	−0.309	9.84	1.16	367	−0.321	10.2	1.15	365	−0.324	10.3	1.15	364	−0.326
		$\Delta Q_{II} = -0.00852$				$\Delta Q_{II} = -0.0122$				$\Delta Q_{II} = -0.00313$				$\Delta Q_{II} = -0.00249$				

2017

$$\lambda = 1.02 \, (\log \, \mathrm{Re})^{-2.5}$$

The loss coefficients C_1, C_2, etc., are then obtained as described in Example 25.1.

Step 4: Calculate ΔQ_I and ΔQ_II using the loop equations developed in Step 2.

Step 5: Correct the flow rates in each loop by adding (or subtracting) the corrections ΔQ_I and ΔQ_II. Note that in this example, the flow rate Q_I in Sec. 1 is corrected by both ΔQ_I and ΔQ_II, since this section is a part of both loops.

Step 6: Return to Step 3 and continue the process until the change in the flow rate between successive calculations is acceptably small.

Table 25.5 shows the calculations for four iterations. At the end of the fourth iteration it was decided that the corrections ΔQ_I and ΔQ_II had reached acceptably low values and therefore the flow rates listed as "Final Q" in Table 25.5 (and Fig. 25.11) constitute the solution.

25.1.4 Design of Ducting Systems

In the previous sections, flow analysis of duct systems was considered where the geometry of the duct system was specified, and the task was to determine the pressure drop versus flow characteristics. The process of design presents the inverse problem in that the geometry of the ducting system must be determined given the flow rate and/or pressure drop. When both flow rate and pressure drop are specified, the selection of duct size is reasonably straightforward for duct systems with all elements in series.

Consider the piping shown in Fig. 25.12, where an incompressible fluid is pumped from point 1 to point 2 at a volumetric flow rate of Q. Further, the pressure at point 1 and 2 are specified, and the pipe has a constant cross-sectional area A. Application of Eq. (25.10a) between the two points gives,

$$p_1 - p_2 = \tfrac{1}{2}\rho U_2^2 - \tfrac{1}{2}\rho U_1^2 + \rho g z_2 - \rho g z_1 + \Delta p_{012} = \rho g H + \Delta p_{012} \quad (25.47)$$

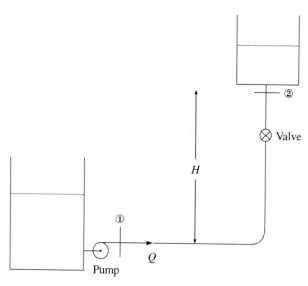

FIGURE 25.12 Schematic of a piping system.

since, for constant cross-section area $U_1 = U_2$ and $z_2 - z_1 = H$ (see Fig. 25.12). The term Δp_{012} is the total pressure loss in the pipe and is given by,

$$\Delta p_{012} = \frac{\rho Q^2}{2A^2} [K_p + K_b + K_v] \tag{25.48}$$

where, K_p, K_b, and K_v are the loss coefficients for the pipe length L, the bend, and the valve, respectively. In general, these loss coefficients are functions of the Reynolds number, and hence, functions of the flow rate and the pipe size. For example, for the pipe loss Δp_{0p},

$$\Delta p_{0p} = \frac{\rho Q^2}{2A^2} \lambda \frac{L}{D} = \frac{8\rho Q^2}{\pi^2} \lambda \frac{L}{D^5} \tag{25.49}$$

where λ is the Darcy friction factor, and D is the pipe diameter. It is clear from Fig. 25.3 that λ is a function of the Reynolds number for smooth pipes and a constant for fully rough pipes. Take the case of a smooth pipe where λ can be approximated by,

$$\lambda = B \, (\text{Re})^{-c} \tag{25.50}$$

where B and c are constants over a given range of Reynolds numbers. For laminar flow, $B = 64$ and $c = 1$, and for turbulent flow in smooth pipes, $B = 0.184$ and $c = 0.2$ for $10^5 \leq \text{Re} \leq 10^6$. Using this functional relationship between λ and Re and noting that $\text{Re} = (4Q\rho)/(\pi \mu D)$,

$$\Delta p_{0p} = \frac{B \, 2^{(3-2c)}}{\pi^{(2-c)}} \mu^c \rho^{(1-c)} Q^{(2-c)} \frac{L}{D^{(5-c)}} \tag{25.51}$$

From this expression it is seen that for laminar flows ($c = 1$), the pressure drop is proportional to D^{-4}, whereas for fully rough turbulent flow ($c = 0$) the pressure drop is proportional to D^{-5}, with the smooth pipe case ($C = 0.2$) falling between the two.

The loss coefficients for the valve and the elbow are virtually independent of the Reynolds number, therefore pressure drops through them can be written as

$$\Delta p_{0v} = \frac{8\rho Q^2}{\pi^2} \frac{K_v}{D^4}; \qquad \Delta p_{0b} = \frac{8\rho Q^2}{\pi^2} \frac{K_b}{D^4} \tag{25.52}$$

Substituting all the individual losses into Eq. (25.47)

$$p_1 - p_2 = \rho g H + \frac{8\rho Q^2}{\pi^2 D^4} (K_v + K_b) + \frac{B2^{(3-2c)}}{\pi^{(2-c)}} \mu^c \rho^{(1-c)} Q^{(2-c)} \frac{L}{D^{(5-c)}} \tag{25.53}$$

Given, p_1, p_2, Q, and H, the required pipe diameter can be obtained from the above expression. Note that since this relation is nonlinear in D, a trial-and-error procedure is required.

The problem of determining the duct size for a network, given the flow rate and the pressure drops to all external nodes, in principle, is solved in a manner similar to that for the single pipe case considered above. However, these computations are quite tedious, therefore the use of a computer is advisable.

The more common design case is when only the flow rate is specified between one or more points, and the designer has to determine the duct system components and sizes. Obtaining an optimal design goes beyond flow aspects, and other considerations such as cost, maintainability etc., must be considered. As an example, consider the case when the flow rate is specified, and the objective is to optimize a simple pipeline with respect to costs. There are two major components of cost that must be considered. The first is the cost of the pipe and fittings. The second is the cost of power to pump the fluid. The cost of the pipe itself increases with increasing pipe diameter. Peters and Timmerhaus (1980) suggest the following relationship between pipe diameter and cost per unit pipe length Z_{pipe}

$$Z_{pipe} = X(D/25.4)^n \tag{25.54}$$

where D is the inside diameter of the pipe (mm), and X is the cost per unit length of a 25.4 mm I.D. pipe. Peters and Timmerhaus (1980) suggest a value of 1 for the exponent n if the pipe diameter is less than 25.4 mm, and $n = 1.5$ for pipe diameters larger than 25.4 mm. Annualized costs of the pipe are proportional to Z_{pipe}, therefore they increase with increasing pipe diameter.

The power required to pump the fluid is $P = Q\Delta p_0$ where Δp_0 is given by Eq. (25.51). Therefore, the pumping power, and hence, annual power cost varies as D^{-4} to D^{-5} and decreases rapidly with pipe diameter. The two cost components are plotted in Fig. 25.13 along with the total annual cost of the system. It is seen that the total cost shows a minimum at some diameter which is, therefore, the optimum diameter. There are, however, many other considerations that must be taken into account when designing duct systems. Peters and Timmerhaus (1980) provide a comprehensive coverage of these as applied to the design of an optimal duct system.

25.1.5 Transients in Ducts

Transients in ducts occur when the flow rate through a duct is increased or decreased as a function of time. This situation commonly occurs when valves are opened or closed, and can cause larger variations in pressures. The extreme case is when a valve is closed suddenly (see Sec. 25.3). A large pressure rise will occur in the duct as the fluid is brought to rest, and this can cause failure of the duct. This phenomenon is generally referred to as *water hammer* due to historic reasons, and it occurs in both compressible and incompressible flows. The subject of transients in ducts is well developed, and for a detailed treatment the reader is referred to books by Wylie and Streeter (1978), Fox (1977), and Watters (1984). This section will focus on common transients in flow of nearly incompressible fluids in straight pipes of constant cross-sectional area.

There are two approaches used in the analysis of transient flow. In the first, the entire fluid column is assumed to move as a rigid body, that is the fluid is assumed to be truly incompressible. Strictly speaking, this *rigid column* theory is an approximation since all fluids are compressible to a certain extent, therefore pressure changes at one end of a fluid column are convected at a finite speed of sound, not

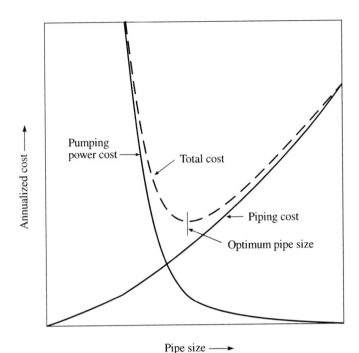

FIGURE 25.13 Trade-off between annualized costs of piping and pumping power.

instantaneously (infinite speed of sound) as assumed in this theory. Nonetheless, the rigid column theory is useful in the analysis of many duct transients of nearly incompressible fluids such as sudden flow establishment and gradual opening or closing of valves.

The second theory takes into account the finite speed of sound, therefore it is well suited to all types of fluid transients including water hammer. Commercial computer programs are available that are capable of solving transients in complex pipe networks.

Sudden Valve Opening. Consider the geometry shown in Fig. 25.14, where a constant area pipe connects two constant pressure plenums at different total pressures p_{01} and p_{02} with $p_{01} \geq p_{02}$. A valve is located near the downstream plenum. The valve, which is initially closed, is opened suddenly to its full open position at time $t = 0$. A transient will ensue, and eventually at some time $t = t'$, steady flow will be established. Assuming that the entire fluid column in the pipe behaves as a rigid body, application of Netwon's second law of motion gives

$$\rho L \frac{dU}{dt} = (p_1 - p_2) + \rho g(z_1 - z_2) - K'_e \frac{1}{2} \rho U^2 \tag{25.55}$$

where p_1 and p_2 are the static pressures, and z_1 and z_2 are the elevations at the inlet plane 1 and at plane 2 just upstream of the valve, respectively (see Fig. 25.14). K'_e is the loss coefficient for all the pipe components between planes 1 and 2. Using Eq. (25.10a) and assuming that the kinetic energy of the fluid exiting plane 3 is lost,

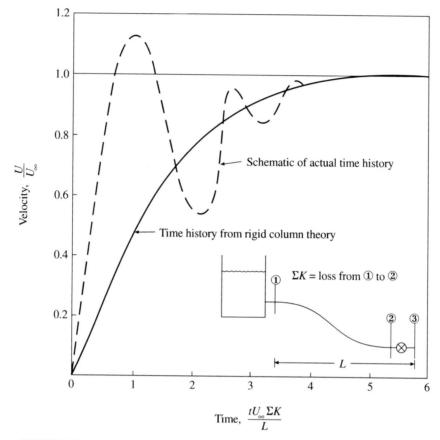

FIGURE 25.14   Schematic of the transient flow due to sudden valve opening.

$$\rho L \frac{dU}{dt} = (p_{01} - p_{03}) - (K'_e + K_v + 1) \frac{1}{2} \rho U^2 \qquad (25.56)$$

where, K_v is the valve loss coefficient for the fully open valve. Equation (25.56) can be integrated giving,

$$t = \frac{L}{(K'_e + K_v + 1) U_\infty} \ln \left(\frac{U_\infty + U}{U_\infty - U} \right) \qquad (25.57)$$

where, U_∞ is the steady-state velocity in the system given by

$$U_\infty = \sqrt{\frac{2(p_{01} - p_{03})}{\rho (K'_e + K_v + 1)}} \qquad (25.58)$$

Equation (25.57) shows that the velocity increases monotonically and reaches the steady state value asymptotically in time. The actual pressure time history following a sudden valve opening would show a damped oscillatory behavior (see Fig. 25.14). This is due to the fact that in all real fluids, pressure is propagated at a finite speed, while the rigid column theory assumes infinite speed of pressure propagation. However, for many engineering design applications, the predictions of Eq. (25.57) are adequate. Defining the time to reach steady-state as t_{99}, when U reaches 99% of U_∞

$$t_{99} = 5.29 \frac{L}{(K'_e + K_v + 1)U_\infty} \tag{25.59}$$

Gradual Valve Opening and Closure. The rigid column theory can be used to determine transients caused by the gradual opening and closing of valves. A gradual transient is one where the duration of the transient is large compared to L/a where a is the speed of sound for the fluid. For such gradual transients, Eq. (25.56) is applicable. However, K_v, the valve loss coefficient, is now a function of time which is determined by the valve closing or opening history. Therefore, in general, Eq. (25.56) cannot be solved analytically, and numerical integration is required with K_v specified as a function of time.

Simple Water Hammer Analysis. Consider the case where steady-state flow in a duct such as the one shown in Fig. 25.14 is decreased suddenly by a sudden closure of a valve. The fluid is decelerated instantaneously at the valve. This decrease in the fluid momentum is translated into a pressure rise at the valve, and the resulting pressure wave is propagated upstream at the speed of sound, a. The fluid upstream of the pressure wave remains at the steady state velocity U, whereas the velocity of the flow downstream of it is at the reduced velocity $(U - \Delta U)$ where ΔU is the velocity change imparted by the sudden valve closure. In a coordinate system moving with the pressure wave, the fluid velocities relative to the pressure wave are shown in Fig. 25.15. Performing a momentum balance across the pressure wave, the pressure rise Δp across the wave is given by,

$$\Delta p = \rho a \Delta U \left(1 + \frac{U}{a} \right) \tag{25.60}$$

If the valve is completely closed, then $\Delta U = U$, and the pressure rise is,

$$\Delta p = \rho a U \left(1 + \frac{U}{a} \right) \approx \rho a U \tag{25.61}$$

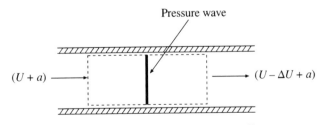

Pressure wave

$(U + a)$ ⟶   ⟶ $(U - \Delta U + a)$

FIGURE 25.15   Flow velocities relative to a traveling pressure wave.

Therefore, given the speed of sound and the velocity change, the resulting pressure rise can be computed. The speed of sound in a fluid is given by

$$a = \sqrt{K/\rho} \tag{25.62}$$

where K is the *bulk modulus of elasticity* (see Sec. 23.14). For pure water, free of any air or vapor bubbles and in a perfectly rigid pipe, the speed of sound is 1440 m/sec. Except for very thick walled pipe, most commonly used pipes are not rigid, and the speed of sound is lower than that given by Eq. (25.62). The speed of sound in a circular pipe of diameter D and wall thickness t is given by

$$a = \frac{\sqrt{K/\rho}}{\left[1 + C\dfrac{K}{E}\dfrac{D}{t}\right]^{1/2}} \tag{25.63}$$

where E is the *modulus of elasticity* for the pipe material, and the constant C is a function of longitudinal restraints on the pipe:

(a) pipe anchored at its upstream end only, $C = 1 - \mu/2$,

(b) pipe anchored against any axial motion, $C = 1 - \mu^2$,

(c) pipe anchored with expansion joints throughout, $C = 1$.

Here, μ is Poisson's ratio for the pipe material.

25.2 DIFFUSERS
B. R. Patel

25.2.1 Introduction

Diffusers are ducts that convert flow kinetic energy to pressure by decelerating the flow. Flow deceleration is affected by varying the cross-sectional area. From mass conservation it follows that diffusers for incompressible fluids, and for subsonic flow as well, are ducts with increasing area along the flow direction; in supersonic diffusers, the area decreases along the flow direction. This section will be limited to subsonic diffusers and will focus on the practical design aspects of subsonic diffusers. The flow physics of subsonic diffusers were detailed in Sec. 5.4.2 and are summarized below.

The flow in the diffuser is governed by the behavior of the boundary layers at the diffuser walls. The deceleration of the flow through the diffuser produces a pressure rise in the streamwise direction. The wall shear layers are therefore subjected to a positive or *adverse* pressure gradient. As is well known, adverse pressure gradients cause the wall boundary layers to thicken and possibly separate from the diffuser walls, forming areas of backflow in the diffuser (see Secs. 4.2, 4.5, and 4.9). The net result of thickening of the wall boundary layers or the formation of regions of backflow, is the *blockage* of flow area which reduces the effective area available to the flow. Reduction in the effective flow area in turn results in a reduced pressure rise through the diffuser. The interactions of the wall shear layers and/or separated zones with the core flow in diffusers is very complicated, therefore diffuser design and performance estimation is largely based on experimental data and empiricism.

25.2.2 Diffuser Geometries

Any duct geometry with an increasing area in the streamwise direction constitutes a subsonic diffuser geometry. Therefore, the number of different diffuser geometries that can be conceived is infinite. However, in practice, adequate design data are available for a limited number of geometries:

rectangular cross section or *planar* diffusers, [Fig. 25.16(a)]
conical diffusers, [Fig. 25.16(b)]
straight walled, annular diffusers, [Fig. 25.16(c)].

The major portion of this section will, therefore, be devoted to these three geometries. Other commonly used diffuser geometries include the radial and axiradial diffuser [see Fig. 25.16(d)] which are used at the exit of radial and axial turbomachines, respectively. Due to limitations of space, these geometries will not be discussed in this section.

The geometric parameters for the three straight wall geometries—planar, conical, and annular are shown in Fig. 25.16. These parameters can be consolidated to a few

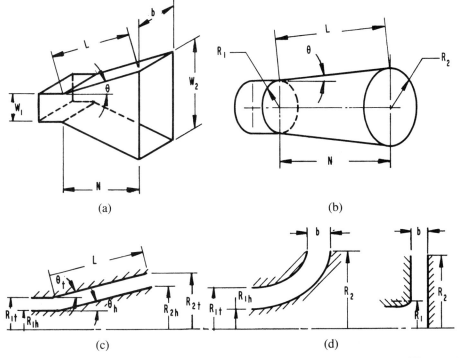

FIGURE 25.16 Common diffuser geometries. (a) Planar diffuser, (b) conical diffuser, (c) straight-wall annular diffuser, and (d) axi-radial and radial diffuser.

dimensionless parameters that are found to be important in terms of diffuser performance. The first is the area ratio, AR, the ratio of the diffuser exit to inlet areas. The area ratio is a measure of the theoretical *diffusion* or pressure recovery expected. The second important parameter is the dimensionless diffuser length defined as: N/W_1 or L/W_1 for planar diffusers, N/R_1 or L/R_1 for conical diffusers, and $L/(R_{1t} - R_{1h})$ for annular diffusers. This dimensionless diffuser length in combination with the area ratio AR is a measure of the overall pressure gradient expected across the diffuser. The third geometric parameter commonly used in displaying diffuser performance is the wall divergence angle—2θ for planar and conical diffusers and θ_h and θ_t, for annular diffusers. The divergence angles, length, and area ratio are related as follows

For planar diffusers: $AR = 1 + 2(N/W_1) \tan \theta$

For conical diffusers: $AR = (1 + (N/R_1) \tan \theta)^2$

For annular diffusers: $AR = \dfrac{(R_{1t} + L \sin \theta_t)^2 - (R_{1h} + L \sin \theta_h)^2}{(R_{1t}^2 - R_{1h}^2)}$ (25.64)

In addition to the three dimensionless geometric parameters discussed above, the aspect ratio $AS = b/W_1$ is an important geometric parameter for planar diffusers.

25.2.3 Diffuser Performance

The performance of diffusers is generally cast in the form of a *pressure recovery coefficient*, C_p

$$C_p = \frac{p_e - p_i}{\frac{1}{2}\rho U_i^2}$$ (25.65)

where p_e and p_i are the diffuser exit and inlet pressures, ρ is the fluid density at the inlet and U_i is the mean velocity at the diffuser inlet. Another form of C_p is

$$C_{pa} = (p_e - p_i)/(P_{oi} - p_i)$$ (25.66)

where P_{oi} is the total pressure in the core region of the flow at the inlet. For incompressible flow, the two definitions are related

$$C_{pa} = (U_i/U_c)^2 C_p$$ (25.67)

where U_c is the maximum velocity at the inlet. Note that these definitions of C_p are unambiguous only when the static pressures are uniform at the inlet and exit planes of the diffuser, which is not the case if there is swirl or significant streamline curvature at these planes. A more general definition based on mass averaged values is

$$C_p = \frac{\bar{p}_e - \bar{p}_i}{\frac{1}{2}\rho U_i^2}$$ (25.68)

where the overbar indicates a mass averaged value of the parameter. The mass value of any parameter ϕ is defined as

$$\overline{\phi} = \frac{\int \phi \rho U \, dA}{\int \rho U \, dA} \tag{25.69}$$

The general definition of C_p degenerates into the simpler forms given previously, only if velocities and pressures are constant at the diffuser inlet and exit planes. Nonetheless, the simpler forms are most commonly used in the literature.

Another parameter of interest is the *ideal pressure recovery*, C_{pi}, which is the pressure recovery coefficient assuming an inviscid flow through the diffuser, which represents the maximum pressure recovery attainable by a given diffuser. For an incompressible fluid, from mass conservation, and Bernoulli's equation

$$C_{pi} = 1 - 1/AR^2 \tag{25.70}$$

A parameter sometimes used as a measure of diffuser performance is the *diffuser effectiveness*, η

$$\eta = C_p/C_{pi} \tag{25.71}$$

The total pressure loss through the diffuser is generally cast in the form of a loss coefficient ζ

$$\zeta = (P_{oi} - P_{oe})/\tfrac{1}{2}\rho U_i^2 = (\overline{u}_i^2 - \overline{u}_e^2)/U_i^2 - C_p$$

$$= (\alpha_i - \alpha_e/AR^2) - C_p \tag{25.72}$$

where P_{oe} is the total pressure in the core region at the exit, the overbar indicates a mass averaged quantity, and α_i and α_e are the kinetic energy parameters at the inlet and exit of the diffuser. The kinetic energy parameter is defined as

$$\alpha = \overline{u}^2/U^2 \tag{25.73}$$

For the case where the velocity profile at the inlet of the diffuser is flat with a thin wall boundary layer, $\alpha_i \approx 1$. However, due to the thickening of the boundary layer through the diffuser, α_e is generally greater than unity. Nonetheless, it is often assumed that the kinetic energy coefficients are equal to unity, then

$$\zeta = C_{pi} - C_p \tag{25.74}$$

Due to the difficulty in calculating the performance of diffusers, performance maps have been developed through model tests. A typical diffuser performance map for a planar diffuser developed by Reneau *et al.* (1966), is shown in Fig. 25.17. This map

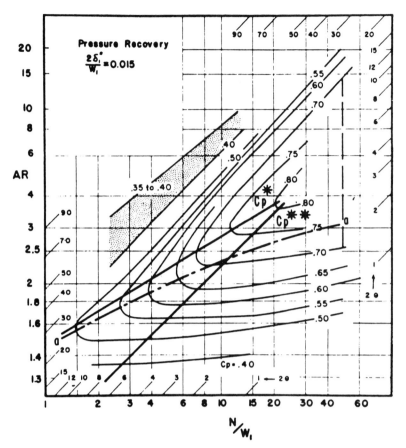

FIGURE 25.17 Performance map for two-dimensional planar diffusers. (Reprinted with permission, Reneau, L. R., Johnston, J. P., and Kline, S. J., "Performance and Design of Straight, Two-Dimensional Diffusers," *J. Basic Engrg.*, Vol. 89, No. 1, p. 141, 1967.)

is for a planar diffuser with a thin inlet boundary layer ($2\delta^*/W_1 = 0.015$) and an aspect ratio $AS > 8$. See Chap. 4 for a discussion of δ^*. The geometric parameter AR and N/W_1 are used as the coordinates, and the 2θ lines run diagonally across. Contours of constant C_p plotted on this map reveal a ridge-like topography with the *peak* in C_p towards the upper right corner. The dashed-dotted line *a-a* on the map represents the line where *stall* (or massive separation) is first observed. Note, therefore, that the peak recovery occurs in the regime of *transitory stall*.

Two lines of performance maxima of interest to designers are the C_p^* and C_p^{**} lines shown on the map. The C_p^* line is the locus of the maximum C_p for a given diffuser length, i.e., the locus of the points where the constant C_p curves are tangent to the constant N/W_1 lines. The C_p^{**} line is the locus of the maximum C_p for a given area ratio, i.e., the locus of points where the constant C_p contours are tangent to the constant AR lines. Note that due to the fact that the constant C_p lines are nearly parallel to the constant AR lines in the lower portion of the map, precise definition of the exact location of the C_p^{**} line is difficult. As drawn in Fig. 25.7, the C_p^{**} line is nearly parallel to the $2\theta = 7°$ line. In contrast, the diffuser angle varies substan-tially along the C_p^* line.

From the diffuser map discussed above, one can construct the map in terms of the diffuser effectiveness in a straightforward manner using Eq. (25.71). The diffuser performance map of Fig. 25.17 recast in terms of the diffuser effectiveness is shown in Fig. 25.18. The C_p^* and C_p^{**} lines are also shown on this figure. It should be noted that the C_p^{**} line nearly coincides with the locus of maximum effectiveness for given area ratios. This means that for a given area ratio, the diffuser length that gives maximum recovery will also have the maximum effectiveness. In constrast, comparing the C_p^* line with the locus of maximum effectiveness for a given diffuser length shows that the maximum pressure recovery and maximum effectiveness for a fixed diffuser length occur at different values of area ratios. The maximum effec-tiveness occurs at a lower value of C_p than C_p^* for this case. The final performance parameter of interest is the *loss coefficient*. Unfortunately loss maps are not generally available. This is primarily due to the fact that construction of a loss coefficient requires careful inlet and exit traverses to determine the inlet and exit kinetic energy parameters, α_i and α_e. In the absence of data for these kinetic energy parameters, often the simplified expression of Eq. (25.74) is used to derive the loss coefficient from the known value of C_p.

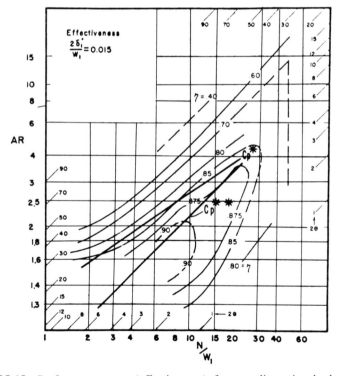

FIGURE 25.18 Performance map (effectiveness) for two-dimensional planar diffusers. (Reprinted with permission, Reneau, L. R., Johnston, J. P., and Kline, S. J., "Performance and Design of Straight, Two-Dimensional Diffusers," *J. Basic Engrg.*, Vol. 89, No. 1, p. 141, 1967.)

The diffuser map dicussed above shows the variation of the C_p with the two important geometric parameters for planar diffusers—AR and N/W_1. The effect of the third geometric parameter, the aspect ratio AS, is not thoroughly investigated at this point. Available data from Dighe (1973) and Runstadler and Dolan (1973) show that the maximum pressure recovery is obtained for diffusers with aspect ratio near 1, as shown in Fig. 25.19. For a given area ratio/length combination, the diffuser recovery drops substantially as the aspect ratio is reduced below 1, especially at low Reynolds number. The decrease in pressure recovery is more gradual as the aspect ratio increases above 1.

Klein (1981a) presents a comprehensive review of the numerous data available on the performance of *conical diffusers*. A typical performance map for conical diffusers for thin inlet boundary layers and plenum discharge conditions is shown in Fig. 25.20. This map was derived by Sovran and Klomp (1967) from the data of Cockrell and Markland (1963). It should be noted that generation of this map required significant extrapolation from the actual data, therefore it should be used with care. A more comprehensive method for the empirical design of conical diffusers will be presented later in this section. Other conical diffuser performance maps are given by Dolan and Runstadler (1973), Miller (1971), and McDonald and Fox

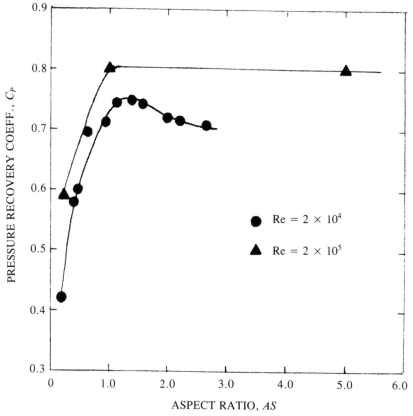

FIGURE 25.19   Effect of aspect ratio on performance of planar diffusers. (Data from Dighe, 1973, and Runstadler and Dolar, 1973.)

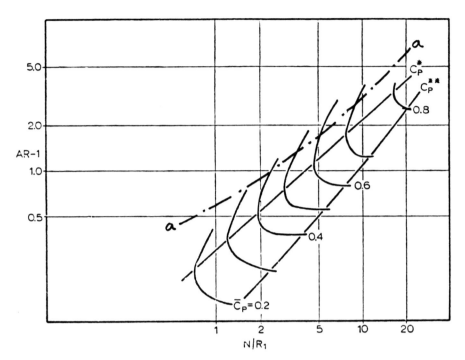

FIGURE 25.20 Performance map for conical diffusers, $B_1 \cong 0.02$. Line a–a is line of appreciable stall. B_1 is blockage at station 1. [Reprinted with permission, Sovran, G. and Klomp, E. D., "Experimentally Determined Optimum Geometries for Rectilinear Diffusers with Rectangular, Conical or Annular Cross-Section," *Fluid Mechanics of Internal Flow*, Sovran, G. (Ed.), Elsevier, New York, 1967.] (Also from data of Cockrell & Markland, 1963.)

(1966). Also shown in Fig. 25.20 is the line of first appreciable stall a–a obtained by McDonald and Fox (1966). It is interesting to note that the C_p^* line falls below line a–a. Therefore, in contrast with planar diffusers, optimum conical diffuser geometries fall in the regime where there is no significant stall in the diffuser.

Extensive performance tests were performed by Sovran and Klomp (1967) on *annular diffuser*. The tests were with thin inlet boundary layers and discharge into a plenum. They tested over 100 geometries having hub to shroud radius ratios of 0.55 and 0.70. Various combinations of inner and outer cone angles were tested. They concluded that the two variables of primary importance were the overall diffuser area ratio and the diffuser length made dimensionless by the inlet passage height. The other geometric parameters had a much smaller effect on diffuser performance. The annular diffuser performance map they derived is shown in Fig. 25.21.

Effect of Inlet Conditions on Diffuser Performance. As discussed earlier, diffuser performance is determined primarily by the behavior of the wall boundary layers. Therefore, it is not surprising that the inlet flow characteristics that have the most significant effect on diffuser performance are those that affect the health of the wall boundary layers as they grow in the diffuser. Thin energetic turbulent boundary

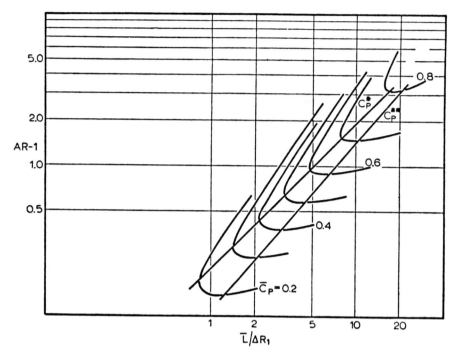

FIGURE 25.21   Annular-diffuser performance map, $B_1 = 0.02$. B_1 is blockage at station 1. [Reprinted with permission, Sovran, G. and Klomp, E. D., "Experimentally Determined Optimum Geometries for Rectilinear Diffusers with Rectangular, Conical or Annular Cross-Section," *Fluid Mechanics of Internal Flow*, Sovran, G. (Ed.), Elsevier, New York, 1967.]

layers at the inlet generally result in better diffuser performance, since such boundary layers can sustain an adverse pressure gradient longer and hence a higher diffuser pressure rise. Further, the presence of any mechanism such as free stream turbulence, that helps energize the wall boundary layers, will improve diffuser performance. Keeping this in mind will help understand the effects of inlet flow characteristics on diffuser performance.

The parameters that characterize the inlet flow to a diffuser are numerous. The most commonly used *inlet flow parameters* are Reynolds number, boundary layer properties, free stream turbulence level, velocity profile, swirl, and Mach number.

The diffuser inlet is generally taken to be the plane at which area expansion starts, and this plane is often referred to as the *diffuser throat*. The inlet Reynolds number, Re, is based on the flow properties and the hydraulic diameter at the throat. In most practical applications, $Re > 10^4$, therefore one can conclude that the inlet flow is turbulent based on the usual rule of thumb for duct flow. However, in most applications, the length of the duct leading up to the diffuser throat is shorter than that required to achieve fully developed turbulent flow. Therefore, depending on the shape of the duct leading to the diffuser throat, the boundary layers at the diffuser inlet can be laminar or transitional even for $Re > 10^5$ as shown by Klein (1981b). As stated earlier, inlet boundary layer characteristics are important in determining the diffuser performance. Therefore, inlet Reynolds number by itself is not sufficient for correlating the available diffuser data. In fact, as will become apparent in the following discussions, the widely used rule of thumb that diffuser performance is insensitive to inlet Reynolds number if Re $> 10^5$ must be applied with caution.

Inlet boundary layer characteristics have the most significant effect on diffuser performance. A boundary layer can be characterized by many parameters such as the displacement and momentum thicknesses, shape factor, etc. In most diffuser tests to date, the type of inlet boundary layer examined is the kind that develops in a constant area duct. For such an *approach length* type turbulent boundary layer, Sovran and Klomp (1967) found that a single parameter, blockage, was sufficient to correlate the effect of inlet boundary layer characteristics on the diffuser performance. The *blockage parameter B*, is related to the displacement thickness and is defined as

$$B = \frac{1}{A} \int \left(1 - \frac{u}{U_c} \right) dA = 1 - U/U_c \qquad (25.75)$$

where A is the cross-sectional area, and U_c and U are the maximum and average velocities at that cross section. Notice that the determination of the blockage simply requires the knowledge of the flow rate through the diffuser and the maximum velocity at a given cross section; the detailed velocity profile is not needed. Blockage for a fully developed turbulent pipe flow is approximately 0.15 for Re $\approx 5 \times 10^5$.

The effect of increasing approach length blockage (blockage generated by an upstream length of constant area duct) is to reduce diffuser performance. This is illustrated in Fig. 25.22 which shows the effect of approach length blockage on diffuser performance for a conical diffuser of area ratio 5. The decrease in C_p with initial increase in blockage is most dramatic for short diffusers that operate in the stalled

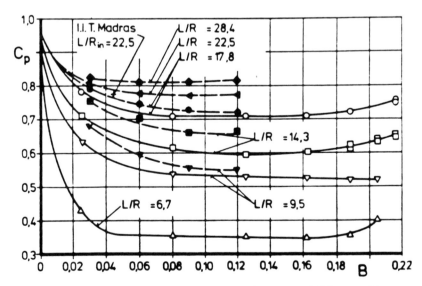

FIGURE 25.22 Effects of inlet blockage on the performance of diffusers with an area ratio $AR = 5$. (From Klein, 1981a.) Open symbols: Cockrell and Markland, 1963, Re$_D$ = 18.10^5; Full symbols: Dolan and Runstadler, 1973, Re$_D$ = 1.2 · 10^5; Slashed symbols: entry flow classified as laminar in Dolan and Runstadler 1973.

region. A diffuser length of between 20 and $24L/R_i$ corresponds to the C_p^* line at this area ratio of 5. For these longer diffusers, the effect of blockage is quite small. Also note that some data in Fig. 25.22 show that the diffuser performance increases with blockage for blockages above 0.16, i.e., blockages greater than those with a fully developed turbulent flow at the diffuser inlet. Klein (1981) found that these higher blockages occurred when the upstream duct was about 40 throat diameters long. In developing flows, the wall boundary layers reach the duct centerline at this duct length, and the flow has a higher turbulence level (see Chap. 5 for a more complete discussion). Klein (1981b) attributes the increase in C_p at the higher blockages to this increased turbulence level.

Based on data of Reneau *et al.* (1967) it appears that approach length type blockage has no effect on the location of the C_p^* line for planar diffusers. That is, the optimum area ratio for a given diffuser length remains unaffected by a change in inlet blockage of this type. However, the diffuser pressure recovery decreases with increasing inlet blockage similar to that discussed above for conical diffusers. The C_p^* line for conical diffusers does show some sensitivity to inlet blockage. Based on available data, Klein (1981a) concludes that the slope of the C_p^* line decreases with increasing inlet blockage. That is, the optimum area ratio for a given diffuser length decreases as inlet blockage increases. However, this trend is not clear enough to be conclusive.

It should be noted that for blockages other than the approach length type blockage discussed above, the blockage parameter alone is not sufficient to predict diffuser performance. For example, Klein (1981a) in his conical diffuser data review shows several cases where, for the same value of blockage, markedly different diffuser performance is obtained. This occurs when the boundary layer is laminar or transitional at the diffuser inlet or when the boundary layer has been energized (or de-energized) by artificial means. At $Re > 10^5$, a laminar or transitional boundary layer at the diffuser inlet produces higher pressure recovery than a turbulent boundary layer. Similarly, boundary layers in which turbulence levels have been increased by devices such as *vortex generators* produce an increased diffuser pressure recovery. About the only consistent trend observed for the various types of nonapproach length type boundary layers is that increasing boundary layer shape factor leads to decreasing diffuser performance. Indeed, much needs to be learned before all the effects of various boundary layer parameters on diffuser performance are clearly understood.

Increased free-stream turbulence levels at the diffuser inlet almost always improve diffuser performance. High free-stream turbulence pumps energy into the boundary layer and hence delays its separation. This results in higher pressure recoveries. The magnitude of the increase in pressure recovery with free-stream turbulence levels is not well quantified, however pressure recovery improvement of up to 20% has been reported for high free-stream turbulence ($u'/U > 3\%$) at the diffuser entrance compared with the low turbulence case. In their review of inlet effects on diffuser performance, Cutler and Johnston (1981) conclude that the length scale of the free-stream turbulence is an important parameter in addition to the turbulence intensity. They surmise that turbulence length scales of the order of the boundary layer displacement thickness are the most beneficial.

In many practical applications, diffusers operate downstream of fluid elements that produce distortions in the free-stream velocity profiles at the diffuser inlet. As discussed in Chap. 5, the diffuser acts as a *distortion amplifier*, and, when the free-stream turbulence is low, nonuniformities in the mean velocity at the inlet will be amplified as the flow proceeds through the diffuser. This in turn leads to a lower diffuser performance than for the case with a uniform inlet velocity. High free-stream turbulence will mitigate this amplification of inlet flow nonuniformities and therefore helps in reducing the performance degradation due to nonuniform inlet velocity profiles. A review of investigations performed on the effects of inlet velocity distortions can be found in Cutler and Johnston (1981), and the following rules of thumb are presented. Uniform or step shear type profiles almost always degrade diffuser performance. Mild wake-type profiles (velocity deficit at the diffuser centerline) produce a slight improvement over the uniform inlet velocity case. Strong wake type profiles degrade diffuser performance. Jet type profiles (pronounced velocity peak at the diffuser centerline) produce severe degradation in diffuser performance. Again, available data are insufficient to quantify the effect of inlet velocity profile distortions on diffuser performance. However, for diffusers operating in the unstalled regime with distorted inlet flows, reasonable performance predictions can be obtained using computer programs that solve the Navier–Stokes equations with second order turbulence closure models such as the k-ϵ model.

Effects of Swirl on Performance of Conical Diffusers. In the discussions above, we have considered the effects of inlet boundary layer thickness and distortions of the axial velocity profiles at the inlet to the diffuser. The other important class of inlet *distortion* is the presence of *swirl* at the inlet of the diffuser. Inlet swirl is encountered in many applications of conical and annular diffusers operating downstream of turbomachines. The presence of swirl produces a centripetal acceleration that tends to push the flow to the outer wall with the result that a radial pressure gradient is produced with the pressure increasing in the radial direction. Intuitively, therefore, it is expected that swirl will tend to reduce or even eliminate flow separation at the diffuser walls and hence improve diffuser performance. In fact, as discussed below, available data support this intuition and significant improvement in diffuser performance due to swirl at the diffuser inlet has been observed.

Swirl produces a variation in the static pressure across the diffuser inlet. Therefore, the simple definition of diffuser pressure recovery given earlier in Eq. (25.65) no longer holds, and the more general definition given in Eq. (25.68) must be used. Note that this more general definition involves the mass averaged values of the pressures and the inlet kinetic energy that have to be obtained from detailed flow field traverses which are both difficult and tedious. It is not surprising, therefore, that the data available are not extensive and in many cases qualitative.

A comprehensive investigation of the effect of inlet swirl on conical diffuser performance was carried out by McDonald *et al.* (1971). They used solid body type swirl where the swirl velocity varies linearly with radius. This solid body swirl was generated by a rotating honey-comb placed upstream of the diffuser inlet. The axial velocity profile was uniform (with the usual boundary layer profile at the walls). The effect of swirl on diffuser performance was found to be a function of diffuser geometry. For diffuser geometries that were unstalled under purely axial inlet flow conditions (that is, diffuser geometries lying below the C_p^* line), swirl had little or no effect on diffuser C_p. Large improvements in performance (up to 40%) were observed for diffuser geometries above the C_p^* line, i.e., diffuser geometries that would have significant stall under purely axial inlet flow conditions. In effect, the

C_p^* line moves up and to the left with increasing swirl. Therefore, at a given diffuser length, the optimum area ratio (and hence C_p) is significantly greater than for the case with axial inlet flow. McDonald *et al.* (1971) also show that the diffuser C_p increases with increasing swirl up to some value of the inlet swirl, after which C_p decreases. This optimum value of swirl at which maximum C_p occurs is a function of the diffuser geometries and generally increases with increasing diffuser wall divergence angle.

For subsonic up to sonic flow at the diffuser throat, inlet Mach number has no significant effect on diffuser performance for diffuser geometries that lie on or below the C_p^* line. As pointed out by Runstadler and Dean (1969), the notion that diffusers choke at throat Mach numbers below sonic is a myth fostered by misinterpretation of data.

Effect of Outlet Conditions on Diffuser Performance. The performance maps discussed previously are for *plenum discharge conditions*, that is the exhaust of the diffuser discharges into a large plenum. Other outlet conditions such as discharge into a tail-pipe, can significantly affect diffuser performance. Klenhofer and Derrick (1970) provide data on the effect of a *tail-pipe* on the performance of planar diffusers. Diffuser C_p is based on the pressure rise to the exit of the tail-pipe. They tested AS $= 1$ diffusers with varying area ratios. Tail-pipes had the following effects: 1) tail pipes did not have any significant influence on the location of the C_p^{**} line, 2) tail pipes increased the overall C_p significantly for diffusers operating above the C_p^* line especially for short diffusers with $L/W_1 < 15$, and 3) overall C_p increased with increasing tail pipe length; it is expected that the C_p will level-off at sufficiently long tail pipe length. It should be noted, however, that for a given area ratio, a diffuser tail pipe combination has a slightly lower overall C_p than a diffuser of the same overall length.

The effect of tail pipes on conical diffuser performance has been investigated by Miller (1971). The effects of adding a tail pipe to a conical diffuser are qualitatively the same as those discussed above for planar diffuser, i.e., tail pipes significantly improve the performance of diffuser geometries above the C_p^{**} line. The performance improvement for diffuser geometries below the C_p^{**} line is slight. Miller (1971) also found that the benefits in terms of performance improvement were achieved at a tail-pipe length of $4L/D_2$, where D_2 is the diffuser exit diameter.

25.2.4 Diffuser Design and Performance Predictions

In the previous sections, the relevant background on diffuser performance and the factors affecting diffuser performance were discussed. From these discussions, it should be evident that the diffuser designer has to rely heavily on empiricism. Even then, reasonable performance predictions can be made only for diffusers with rather simple inlet flow profiles. This section is intended to serve as a guide in the *design* and *performance predictions* of diffusers. Before proceeding with the design of a diffuser, the designer must clearly define the design constraints. Typical design constraints are diffuser length, area ratio, or desired pressure recovery level. The most common of these being the length constraint, where maximum pressure recovery is required for a given diffuser length. In addition to geometric constraints, the condition of the flow exiting the diffuser might also be an important consideration, especially if the diffuser interfaces with a downstream flow element. For example, if the degree of flow nonuniformity and/or unsteadiness is to be minimized, the diffuser design must be such that it lies well below the line of first appreciable stall.

As discussed earlier, diffuser performance is strongly affected by diffuser inlet and outlet conditions. The inlet conditions that are of primary importance are those that affect the health of the boundary layers at the diffuser inlet. However, only the effects of the blockage parameter have been investigated to the point where quantitative predictions can be made with any degree of certainty. Effects of other parameters such as turbulence, mean velocity profile, and swirl can only be treated qualitatively. Therefore, in this section, the only inlet parameter that will be considered in the diffuser design procedure will be the inlet blockage. Estimation of the inlet blockage is the first step in the design process. Obviously, the inlet blockage is a function of the flow elements upstream of the diffuser. For the case where the diffuser is preceded by a section of constant area duct, the inlet blockage can be determined from data on developing flow in constant area ducts (see Chap. 5). Klein (1981b) presents a comprehensive review of developing circular pipe turbulent flows and shows that the development of inlet blockage is very sensitive to small perturbations at the pipe inlet such that there is significant scatter between different data sets examined. Figure 25.23 shows the blockage as a function of the distance x from the entrance of a circular pipe of diameter D for the case where the boundary layer is turbulent at the pipe inlet and is close to the mean of all the data analyzed by Klein (1981b). In most practical situations, the diffuser is downstream of another flow element such as a turbomachine. In such cases, the determination of inlet blockage can become quite difficult. Nonetheless, an estimate of the inlet blockage is

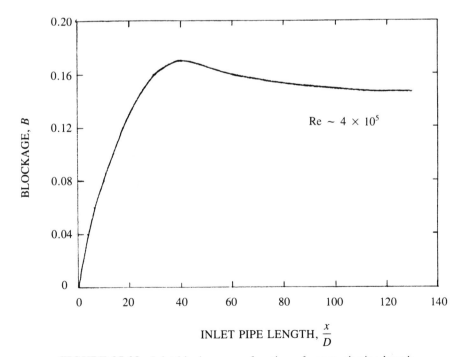

FIGURE 25.23 Inlet blockage as a function of approach pipe length.

essential to estimating the diffuser performance, therefore even a rough estimate is better than none at all.

Diffuser exit conditions commonly encountered are for a discharge into a plenum or a tail pipe. These two conditions will be treated in the following sections. The effect of other outlet conditions are not sufficiently quantified to justify inclusion in the design procedure given here.

Optimum Diffuser Design for a Given Area Ratio. The first design situation is where an optimum design for a given area ratio is required. These diffuser geometries lie on the C_p^{**} line. For planar diffusers, the C_p^{**} line lies along the line of constant diffuser angle of $2\theta \cong 7°$. Utilizing the expression relating the area ratio of a planar diffuser to the diffuser length and setting the diffuser angle $2\theta = 7°$ gives

$$N/W_1 = 8.17(AR - 1) \tag{25.76}$$

Reference to Fig. 25.17 shows that the C_p^{**} line lies below the line of first stall *a-a* for $AR < 2.4$. For higher area ratios, the C_p^{**} planar diffuser geometries will be in the transitory stall regime, therefore unsteady flow at the diffuser exit will occur. Further, from Fig. 25.17 it can be noted that the C_p contours below the C_p^{**} line are quite flat, therefore if steady flow conditions are desired at the diffuser exit for $AR > 2.4$, the diffuser angle can be reduced so that the diffuser geometry falls below the line *a-a*, without a large penalty in pressure recovery. The performance of C_p^{**} planar diffuser geometries can be estimated using Fig. 25.24 which shows the effectiveness as a function of the inlet blockage for diffuser $AR < 2.4$.

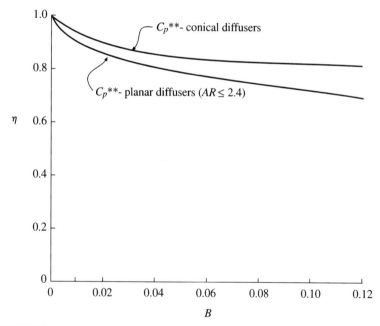

FIGURE 25.24 Effectiveness of C_p^{**} diffuser geometries as a function of inlet blockage.

For conical diffusers, the C_p^{**} line is very nearly coincident with the line for constant diffuser angle of 5°. From geometrical relations for conical diffusers and using $2\theta = 5°$, the diffuser length N/R_1 for the optimum diffuser geometry is

$$N/R_1 = 22.9(\sqrt{AR} - 1) \tag{25.77}$$

As shown in Fig. 25.20, the line of first appreciable stall for conical diffusers lies above the C_p^{**} for all practical diffuser lengths. Therefore, optimum diffusers for a given area ratio will operate in the unstalled region with generally steady exit flow. The performance of C_p^{**} conical diffuser geometries can be estimated using the diffuser effectiveness plotted in Fig. 25.24 as a function of inlet blockage.

The optimum geometry for a given area ratio annular diffuser lies on the C_p^{**} line shown in Fig. 25.21. Unfortunately, definitive data are not available to define the line a-a, the line of no appreciable stall. Further, data are not available to allow estimation of the performance of the C_p^{**} geometries as a function of inlet blockage. Reasonable estimates are probably obtained by using the effectiveness versus blockage curve given in Fig. 25.24 for planar diffusers.

Example 25.3: Design an optimum conical diffuser for an area ratio of 2.5. The diffuser inlet diameter is 10 cm and the diffuser is preceded by a straight circular duct of diameter 10 cm and length of 80 cm. The inlet Reynolds number is approximately 10^5, and the flow is turbulent at the diffuser inlet.

Since the area ratio is given, the optimum diffuser geometry lies on the C_p^{**} line. The diffuser angle $2\theta = 5°$. The diffuser length is given by Eq. (25.77)

$$N/R_1 = 22.9(\sqrt{2.5} - 1) = 13.3,$$

Therefore, $N = 13.3 \times 10/2 = 66.5$ cm.

The inlet pipe length $x/D = 80/10 = 8$. From Fig. 25.23, the estimated inlet blockage is 0.065. From Fig. 25.24, the diffuser effectiveness, η, for an inlet blockage of 0.065 is 0.84. The ideal pressure recovery coefficient C_{pi} from Eq. (25.70) is,

$$C_{pi} = (1 - 1/AR^2) = (1 - 1/2.5^2) = 0.84$$

Therefore, the estimated diffuser C_p is,

$$C_p = C_{pi} \times \eta = 0.84 \times 0.84 = 0.71$$

Design of Optimum Diffusers for a Given Length. In most practical applications, the diffuser length is given, and the task is to design the optimum diffuser within this constraint. As described earlier, the optimum diffuser geometries for a given diffuser length fall on the C_p^* line. For planar diffusers, Reneau et al. (1967) obtained the C_p^* line for a range of inlet blockages and found that all the C_p^* geometries lie on the same line on the diffuser map irrespective of inlet blockage as shown in Fig. 25.25. Therefore, given a diffuser length, the area ratio for the optimum diffuser can

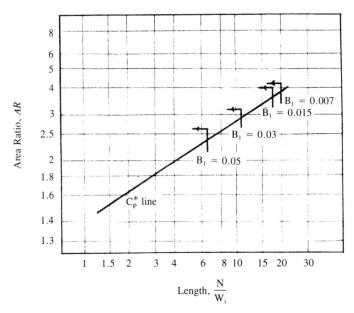

FIGURE 25.25   Optimum planar diffuser geometries for a given diffuser length. (Reprinted with permission, Reneau, L. R., Johnston, J. P., and Kline, S. J., "Performance and Design of Straight, Two-Dimensional Diffusers," *J. Basic Engrg.*, Vol. 89, No. 1, p. 141, 1967.)

be determined from this figure. There are two important points to keep in mind when using this figure. First, although the optimum area ratio at a given length is independent of inlet blockage, the actual value of the pressure recovery coefficient, C_p, is a function of the inlet blockage, and it decreases as the inlet blockage increases. Second, as the inlet blockage increases, the maximum usable diffuser length decreases as shown by the short vertical lines on the C_p^* line in Fig. 25.25. For example, at inlet blockage of 0.05, the maximum diffuser length, N/W_1, is limited to around 6.5. At this blockage, use of this figure for longer diffuser lengths will result in geometries that have severe unsteady flow and degraded performance. On the other hand, for inlet blockage of 0.007, Fig. 25.25 can be used for diffuser lengths N/W_1 up to about 17. Finally, note that this figure was developed from data for planar diffusers with large aspect ratios and should, therefore, be used only when the aspect ratio is greater than 1. There is no corresponding curve for small aspect ratios, and the design of small aspect ratio diffusers has to be derived from actual performance maps such as those of Runstadler and Dolan (1973).

As shown on the planar diffuser performance map of Fig. 25.17, the C_p^* line lies mostly above the line *a-a*, therefore C_p^* planar diffuser geometries operate in the transitory stall region. The diffuser exit flow for such geometries will be unsteady and nonuniform. The performance of diffuser geometries on or near the C_p^* line can be estimated using the procedure developed by Sovran and Klomp (1967). From one-dimensional analysis of diffuser flows where the inlet flow consists of an inviscid core with thin wall boundary layers, diffuser effectiveness can be related to the inlet and exit blockages and the area ratio as follows

$$\eta = \frac{1}{E_1^2}\left[\frac{1 - (E_1/E_2)^2/AR^2}{1 - (1/AR^2)}\right] \tag{25.78}$$

where E_1 and E_2 are the effective diffuser areas at the inlet and outlet respectively, which are related to the blockage by

$$E = 1 - B \qquad (25.79)$$

Sovran and Klomp (1967) found that E_1, E_2, and AR could be correlated as shown in Fig. 25.26 for all diffuser geometries that are near or on the C_p^* line. Hence, using Fig. 25.26 and Eq. (25.78), the performance of optimum diffuser geometries at a given length can be determined.

Example 25.4:   Design an optimum planar diffuser with an inlet width of 15 cm and length of 60 cm. The diffuser aspect ratio is 5, and the inlet blockage is 0.02.

Since the diffuser aspect ratio is greater than 1, the C_p^* line of Fig. 25.25 can be used to determine the optimum area ratio.

$$\text{Dimensionless diffuser length } N/W_1 = 60/15 = 4$$

From Fig. 25.25, this diffuser length is well below the maximum usable diffuser length at inlet blockage of 0.03, therefore the C_p^* line of Fig. 25.25 can be used to determine the optimum area ratio for this diffuser. From Fig. 25.25 for $N/W_1 = 4$, the area ratio $AR = 2$.

Using this area ratio and the inlet blockage value, $AR(100B_1)^{0.25} = 2.38$.

From Fig. 25.26, $E_2 = 0.75$

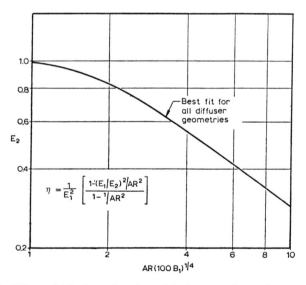

$$\eta = \frac{1}{E_1^2}\left[\frac{1-(E_1/E_2)^2/AR^2}{1-1/AR^2}\right]$$

Best fit for all diffuser geometries

E_2

$AR(100\,B_1)^{1/4}$

FIGURE 25.26   Effect of inlet-boundary-layer blockage on the performance of diffusers on C_p^* and near C_p^*. [Reprinted with permission, Sovran, G. and Klomp, E. D., "Experimentally Determined Optimum Geometries for Rectilinear Diffusers with Rectangular, Conical or Annular Cross-Section," *Fluid Mechanics of Internal Flow*, Sovran, G. (Ed.), Elsevier, New York, 1967.]

By definition, $E_1 = 1 - B_1 = 1 - 0.02 = 0.98$. Using Eq. (25.78)

$$\eta = 0.8 = C_p/C_{pi}, \quad \text{and } C_{pi} = (1 - 1/AR^2) = 0.75$$

Therefore, the estimated pressure recovery coefficient for this optimum diffuser geometry is

$$C_p = \eta \times C_{pi} = 0.6$$

For conical diffusers, based on a review of available data, Klein (1981a) found that the optimum diffuser geometries for a given length were dependent on the inlet blockage. That is, the C_p^* line for conical diffusers depends on the inlet blockage. Figure 25.27 shows C_p^* lines for inlet blockages of 0.02 and 0.12, respectively. Note that the C_p^* line for $B_1 \geq 0.12$ in this figure is close to that derived by Sovran and Klomp (1967) and shown in Fig. 25.20. Klein (1981a) recommends the use of the C_p^* line for 0.02 for inlet blockages less than 0.02. For inlet blockages between 0.02 and 0.12, the appropriate C_p^* line can be obtained by linear interpolation between the two lines shown in Fig. 25.27. For inlet blockages greater than 0.12, the C_p^* line for inlet blockage of 0.12 should be used. For conical diffuser, the C_p^* line lies mostly below the line a-a (see Fig. 25.20), hence the diffuser exit flow will be reasonably steady. Usable estimates for conical diffuser geometries on or near the C_p^* line can be obtained using Fig. 25.26 and the procedure developed by Sovran and Klomp (1967), discussed above for planar diffusers.

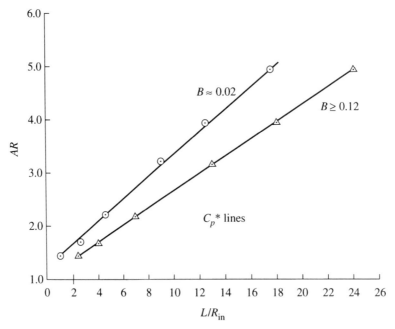

FIGURE 25.27 Optimum conical diffuser geometries for a given diffuser length. (Reprinted with permission, Klein, A., "Review: Effects of Inlet Conditions on Conical Diffuser Performance," *J. Fluids Engrg.*, Vol. 103, No. 2, pp. 250–257, 1981.)

The optimum annular diffuser geometries for a given length can be estimated from the C_p^* line developed by Sovran and Klomp (1967), shown in Fig. 25.21. There are no data for how the C_p^* line is affected by inlet blockage. Therefore in practice, the C_p^* line of Fig. 21 is used irrespective of the inlet blockage. Finally, the performance of annular geometries on or near the C_p^* line can also be estimated using the procedure of Sovran and Klomp (1967) described above.

Estimation of Effect of a Tail Pipe on Diffuser Performance. As discussed earlier, the addition of tail pipes improves the performance of diffuser geometries that lie above the C_p^{**} line. For diffuser geometries below the C_p^{**} line, tail pipes have little or no effect. Further, tail pipe lengths of about four times the diffuser exit width or diameter are needed to achieve the full benefit of pressure recovery in the tail pipe.

The effect of tail pipes on diffuser performance can be estimated by assuming that the addition of the tail pipe does not affect the flow in the diffuser itself, but only serves to mix the nonuniform velocity profile at the diffuser exit, thereby achieving additional pressure recovery. There is some justification to this view in that Klenhofer and Derrick (1970) found that addition of a tail pipe did not change the location or size of the stall in the diffuser itself. Consider the case where the flow at the diffuser inlet consists of a thin boundary layer and a potential core with uniform velocity U_1. First, consider the diffuser alone discharging into a plenum. The C_p can be obtained from diffuser maps or using the method outlined earlier. Based on this C_p, the effective area fraction at the diffuser exit can be obtained from

$$E_2 = \frac{(1 - B_1)}{AR\sqrt{(1 - C_p)}} \tag{25.80}$$

Making a simple momentum balance between the diffuser exit and the exit of the tail pipe, where the flow velocity is assumed to be uniform, and neglecting friction losses, the tail pipe recovery coefficient C_{ptp} can be obtained from

$$C_{ptp} = \frac{2}{AR^2}\frac{(1 - E_2)}{E_2} \tag{25.81}$$

The overall pressure recovery in the diffuser and tail pipe is then given by

$$C_{ptotal} = C_p + \eta_t * C_{ptp} \tag{25.82}$$

Comparison with experimental data shows that the tail pipe efficiency, η_t, varies between 0.5 and 0.7. For diffuser geometries above the C_p^{**} line, a value of η_t of 0.6 gives results that are generally within 10% of the data.

25.3 HYDRAULICS OF VALVES
C. Samuel Martin

25.3.1 Introduction

Valves are integral elements of any piping system used for the handling and transport of liquids. Their primary purposes are flow control, energy dissipation, and isolation of portions of the piping system for maintenance. It is important for the purposes of design and final operation to understand the hydraulic characteristics of valves under both steady and unsteady flow conditions. Examples of dynamic conditions are servovalves in a hydraulic system, product control valves in a petrochemical plant, direct opening or closing of valves by a motor, and the response of a swing check valve under unsteady conditions.

The hydraulic characteristics of valves under either noncavitating or cavitating conditions vary considerably from one type of valve design to another. Moreover, valve characteristics also depend upon the particular valve design for a special function, upon absolute size, on manufacturer as well as the type of pipe fitting employed. In this section, the fundamentals of valve hydraulics are presented in terms of flow rate versus pressure drop characteristics. Typical flow characteristics of selected valve types—gate, ball and spherical, butterfly, globe, and swing check—are presented. For rotary valves, torque characteristics are discussed with relationship to flow quantities. Finally, the effect of *cavitation* on valve characteristics is addressed.

25.3.2   Valve Identification

Valves used for the control of liquid flow vary widely in size, shape, and overall design due to vast differences in application. They can vary in size from a few millimeters as needle valves in small tubing to many meters in hydroelectric installations, for which spherical and butterfly valves of very special design are built. The hydraulic characteristics of all types of valves, albeit different in design and size, can always be reduced to the same basic coefficients, notwithstanding fluid effects such as viscosity and cavitation.

Because of space limitations the hydraulic characteristics of only a few types of valves—gate, ball and spherical, butterfly, globe, and swing check—will be presented here. For other types of valves, the reader will need to refer to the references and other sources. For control valves in general and specially designed valves in particular, for which the hydraulic characteristics have been modified on purpose, a good source of information is the ISA Handbook of Control Valves [Hutchinson (1976)]. This source includes information and references for characterized ball valves, plug valves, various designs of globe valves, V-orifice gate valves, diaphragm valves, and variations of butterfly valves. Valve types such as angle, Y, hollow jet, pressure relief, and multisleeve also are not covered herein. Figure 25.28 shows photographs of the valve types to be discussed in relation to hydraulic performance.

25.3.3   Geometric Characteristics of Selected Valves

The hydraulic performance of a valve depends upon the flow passage through the valve opening and the subsequent recovery of pressure. The valve geometry, expressed in terms of cross-sectional area at any opening, sharpness of edges, type of passage, and valve shape, has a considerable influence on the eventual hydraulic characteristics. Except to a limited extent for particular shapes, as shown by Sarpkaya (1961) for butterfly valves, hydrodynamic theory falls short in the prediction of valve characteristics in general. Indeed, the hydraulic characteristics of valves must be determined from experiments. In order to understand better the hydraulic

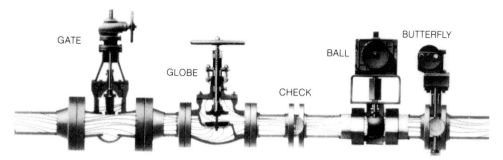

FIGURE 25.28 View of postulated flow paths through selected valves. (Courtesy of Mark Controls Corporation.)

characteristics of valves, it is useful, however, to express the projected area of the valve in terms of geometric quantities. With reference to Fig. 25.29, the ratio of the projected open area of the valve, A_v, to the full open valve, A_{vo}, which in many instances is equal to the pipe area A, has been formulated by Wood and Jones (1973), as follows:

Ball Valve

$$\frac{A_v}{A_{vo}} = \{1 + \cos \theta\} \left\{ \frac{1}{2} - \frac{1}{\pi} \left\{ \sin^{-1}(x) + \frac{1}{2} \sin [2 \sin^{-1}(x)] \right\} \right\} \quad (25.83)$$

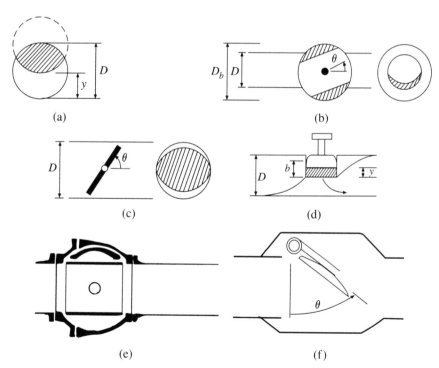

(a)

(b)

(c)

(d)

(e)

(f)

FIGURE 25.29 Definition sketches of selected valves. (a) Circular gate valve, (b) ball valve, (c) bufferfly valve, (d) globe valve, (e) spherical valve, and (f) tilting-disk check valve.

for which

$$x = \frac{\sin\theta \sqrt{\left(\dfrac{D_b}{D}\right)^2 - 1}}{1 + \cos\theta} \tag{25.84}$$

Butterfly Valve

$$\frac{A_v}{A_{vo}} = 1 - \cos\left(\frac{\pi}{2} - \theta\right) \tag{25.85}$$

Gate Valve (Circular Gate)

$$\frac{A_v}{A_{vo}} = 1 - \frac{2}{\pi}\left[\cos^{-1}\left(\frac{y}{D}\right) - \frac{y}{D}\sqrt{1 - \left(\frac{y}{D}\right)^2}\right] \tag{25.86}$$

Globe Valve

$$\frac{A_v}{A_{vo}} = \frac{y}{b} \tag{25.87}$$

in which y is the opening for stem valves, θ the angle for rotary valves, D is the diameter of the valve for ball, butterfly, and gate valves, and b is the full opening for a globe valve. In this section all valves of type other than globe will have the same diameter D as that of the pipe itself. In practice, there are reduced area port ball valves and installations with valves having smaller diameters than that of the pipe.

The projected area ratios for these four widely used valves are shown in Fig. 25.30 as a function of percentage open. The globe valve shown in Fig. 25.29 yields a linear relationship, while the other three have either increasing or decreasing rates of area change as the valve is closed. Although Fig. 25.30 suggests how the area change can affect hydraulic performance in terms of what range a valve would be most effective in restricting the flow, the actual performance can only be ascertained in terms of hydraulic characteristics, which must be determined experimentally. Because of a double-orifice effect, the hydraulic characteristics of a ball valve would certainly be expected to be different from that of the other types.

The curve for the ball valve is for a given value of $D/D_b = 0.629$ for which the angle corresponding to closure is 78°. For a ball valve of the full ball variety, the angle at which the valve becomes closed is given by

$$\theta_c = 2\tan^{-1}\frac{D/D_b}{\sqrt{1 - (D/D_b)^2}} \tag{25.88}$$

Geometric relationships for valves of the type reported here and others are available in Pickett *et al.* (1971).

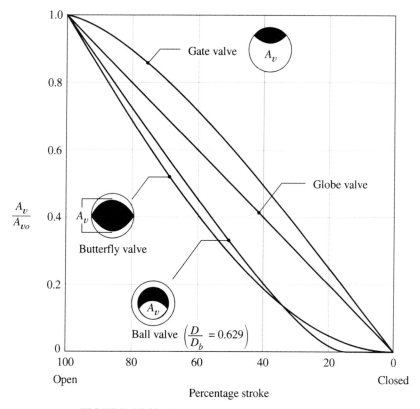

FIGURE 25.30 Projected area of selected valves.

25.3.4 Flow Characteristics

The hydraulic characteristics of a valve under partial to fully opened conditions typically relate the *volumetric flowrate* to a characteristic *valve area* and the *head drop* across the valve. The principal fluid properties that can affect the flow characteristics are fluid density, ρ, fluid viscosity, μ, and liquid vapor pressure, p_v, if cavitation occurs. Except for very small valves and/or quite viscous liquids, Reynolds number effects are not important. In any event, Reynolds number effects will not be considered in this treatment.

A valve in a pipeline acts as an obstruction, disturbs the flow, and in general causes a loss in total pressure (or *head*), as well as affecting the pressure distribution both upstream and downstream. The characteristics are expressed either in terms of (a) flow capacity as a function of a defined pressure drop or (b) energy dissipation (head loss) as a function of pipe velocity. In both instances, the pressure or head drop is usually the difference in total head caused by the presence of the valve itself, minus any loss due to regular pipe friction between measuring stations. Definition of *discharge coefficients*, which vary from user to user, are also influenced by downstream conditions, which may consist of a long continuous pipe of the same diameter, or submerged discharge into a reservoir or vessel, or even free discharge into the atmosphere or into a pressurized tank.

For a valve located in the interior of a long continuous pipe, as shown in Fig.

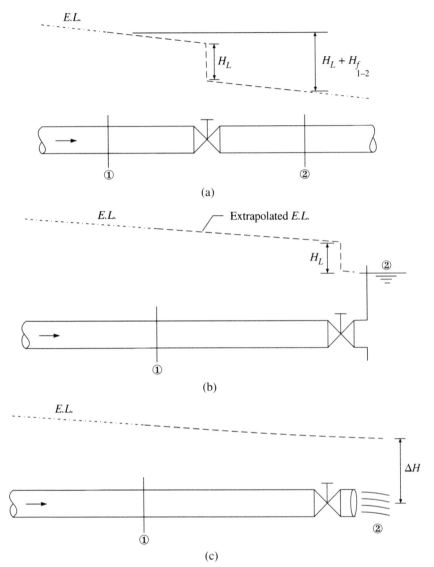

FIGURE 25.31   Definition of head on valve. (a) Continuous pipe of uniform area, (b) reservoir at end of pipe, and (c) free discharge.

25.31(a), the presence of the valve disturbs the flow both upstream and downstream of the obstruction as reflected by the velocity distribution and the pressure variation which will be nonhydrostatic in the regions of nonuniform flow. In terms of the *hydraulic grade line* (HGL) or *piezometric head*

$$h = \frac{p}{\rho g} + z \tag{25.89}$$

in which p is the pressure, g is the gravitational acceleration, and z is the elevation, the total head equations between uniform flow sections ① and ② can be expressed as

$$h_1 + \frac{V_1^2}{2g} = h_2 + \frac{V_2^2}{2g} + H_{f_{1-2}} + H_L \tag{25.90}$$

where $H_{f_{1-2}}$ is the uniform flow pipe friction. For equal pipe diameters $D_1 = D_2$, the cross-sectional average velocities

$$V_1 = V_2 = V = \frac{Q}{A} \tag{25.91}$$

where Q is the volumetric flow and A the cross-sectional area of the pipe. For this case of a valve in a continuous pipe, the valve head loss becomes

$$H_L = h_1 - h_2 - H_{f_{1-2}} \tag{25.92}$$

The proper manner in determining H_L experimentally is to measure the HGL far enough both upstream and downstream of the valve so that uniform flow sections to the left of ① and to the right of ② can be established, allowing for the extrapolation of the energy lines (EL) to the plane of the valve. Otherwise, the valve head loss is not properly defined. It is common to express the hydraulic characteristics either in terms of a *head loss coefficient*, K_L

$$H_L = K_L \frac{V^2}{2g} \tag{25.93}$$

or as a *discharge coefficient* C_d

$$Q = C_d A_v \sqrt{2g\Delta H} \tag{25.94}$$

where A_v is the area of the valve at any opening, and ΔH is the head drop defined for the valve which, in this case, is H_L. Frequently, a discharge coefficient is defined in terms of the fully-open valve area

$$Q = C_f A_{vo} \sqrt{2g\Delta H} \tag{25.95}$$

The relationship between geometry, discharge coefficients, and loss coefficients from Eqs. (25.93)–(25.95) becomes

$$\tau = \frac{C_d A_v}{C_{do} A_{vo}} = \sqrt{\frac{K_{Lo}}{K_L}} = \frac{C_f}{C_{fo}} \tag{25.96}$$

Equation (25.96) through τ embodies not only the geometric features of the valve through A_v/A_{vo} but also the hydraulic characteristics as influenced by C_d/C_{do}. This parameter τ has been effectively employed in the calculation of hydraulic transients associated with valve operation. The relationship between the head-loss coefficient and discharge coefficient becomes

$$K_L = \frac{1}{C_d^2} \left(\frac{A}{A_v} \right)^2 \tag{25.97}$$

If the fully-open valve area A_{vo} is used in the definition of the discharge coefficient, Eq. (25.95), and $A_{vo} = A$, then

$$K_L = \frac{1}{C_f^2} \tag{25.98}$$

For a valve discharging into a reservoir at the termination of a pipe, Fig. 25.31(b), the *total head equation* is

$$h_1 + \frac{V_1^2}{2g} = h_2 + H_L + H_{f_{1-2}} \tag{25.99}$$

yielding

$$\Delta H = H_L = h_1 + \frac{V^2}{2g} - z_2 - H_{f_{1-2}} = H_1 - z_2 - H_{f_{1-2}} \tag{25.100}$$

Because of the lack of complete pressure recovery for this case of submerged discharge, the discharge coefficient C_d differs from that of continuous pipe flow as will be discussed later. Furthermore, some authors have used another definition of head drop for the continuous pipe flow case rather than Eq. (25.95), for which ΔH, the energy loss across the valve, is also equal to the drop in piezometric head, Δh, caused by the valve. Such a definition, as employed by Grein and Pirchl (1970), Guins (1968), Strohmer (1977), and Csemniczky (1972) is

$$Q = C_Q A_{vo} \sqrt{2gH} \tag{25.101}$$

for which

$$H = h_1 + \frac{V^2}{2g} - h_2 - H_{f_{1-2}} = \Delta H + \frac{V^2}{2g} \tag{25.102}$$

the flow coefficient, C_f, and the loss coefficient, K_L, are then related to C_Q by

$$C_f = \frac{C_Q}{\sqrt{1 - C_Q^2}} \tag{25.103}$$

and

$$K_L = \frac{1}{C_Q^2} - 1 \tag{25.104}$$

For the case of free discharge of a valve at the end of a pipe, Fig. 25.31(c), the head difference across the valve is assumed equal to the kinetic energy remaining in the jet, or

$$\Delta H = H_1 - H_{f_{1-2}} = \frac{V_2^2}{2g} \qquad (25.105)$$

The user must be careful as to the conditions under which the valve was tested and the definition of the flow coefficient, whether C_f [Eq. (25.95)] or C_Q [Eq. (25.101)], which are both dimensionless, or the U.S. definition

$$Q(gpm) = C_v \sqrt{\Delta p \text{ (psi)}} \qquad (25.106)$$

for which C_v is defined in terms of water for that particular valve. The flow coefficient, C_f, for any liquid flow for which there are no Reynolds number effects can be determined from C_v values from manufacturers by

$$C_f = \frac{0.0335 C_v}{D^2} \qquad (25.107)$$

In Eq. (25.107), D is in inches.

The use of the definition of H rather than ΔH for the head difference in a continuous pipe case has probably been employed by investigators in the hydroelectric field to spherical and butterfly valves, because the value of C_Q does not vary as much from full open to closed as does C_f. In fact, for a valve without losses, $C_Q = 1$ and $C_f = \infty$.

Unless uniform flow is established far upstream and downstream of a valve in a pipeline, the value of any of the coefficients can be affected by nonuniform flow. It is not unusual for investigators to use only two pressure taps—one upstream and one downstream, frequently one and ten diameters, respectively. Careful measurements of the pressure distribution on the walls of a channel were conducted by Jung (1969) for a butterfly valve in a rectangular conduit (see Fig. 25.32). The valve leaf, which closed at $\theta = 75°$, was positioned at $\theta = 45°$. These results illustrate the extent of

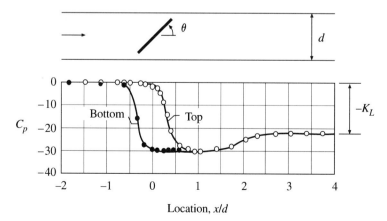

FIGURE 25.32 Pressure distribution resulting from butterfly valve in rectangular channel.

the nonuniform flow conditions across the channel as well as the effect upstream and the number of channel widths necessary to recover pressure downstream for the particular condition. Additional results showing pressure distributions are available from Grein and Osterwalder (1978).

The flow characteristics of valves in terms of pressure drop or head loss have been determined for numerous valves by many investigators and countless manufacturers. Only a few sets of data and typical curves will be presented here for ball, spherical, butterfly, gate, globe, and tilting-disc check valves. The characteristics will be expressed either in terms of C_f or K_L. For plots of raw data, good sources are McPherson *et al.* (1957) and Nece and Dubois (1955). Miller (1978) summarizes head-loss data for a number of commercial valves.

Ball Valves and Spherical Valves. The hydraulic characteristics of ball valves (full ball) and spherical valves are somewhat similar even though the latter can have considerable flow around its body, whereas the former may not. For small sizes, the loss coefficient, K_L, at nearly full open to full open position ($\theta = 0°$) is greater than for larger valves due to more resistance in the port as well as the result of a greater influence of the pipe fitting. In Fig. 25.33 are plotted the loss coefficient, K_L, versus θ for 16 mm and 100 mm full port ball valves tested at Georgia Tech and at Delft University [Safwat (1973)], respectively. The value of θ_c corresponding to initial closure is 78° for the smaller valve and 82° for the larger valve. The same results are plotted in Fig. 25.34 in terms of C_f versus θ. The difficulty in measuring the head loss at $\theta = 0°$ leads to uncertainty in producing accurate values of C_f at that fully open position.

Flow coefficient curves are plotted in Fig. 25.35 from tests on spherical valves conducted in air by Guins (1968) and in water by Strohmer (1977). These larger valves are more efficient at the fully-open position, yielding values of C_f approaching 5, compared to the range of 2 to 2.5 for the smaller ball valves. In fact, a spherical valve is more efficient hydraulically over the entire range of operation than a ball valve because of its greater size and the flow around the body itself.

Butterfly Valves. The leaf design of butterfly valves varies considerably with application depending upon size, loss considerations, strength, and torque requirements. Wafer and lenticular leaves are built with various thickness ratios, defined by the maximum thickness, s, to the leaf diameter, D, or s/D. Other variations in geometry are the through flow or lattice-type body, the use of guide vanes, and the eccentric location of the center of the valve stem for purposes of improvement in valve actuation. All of these factors affect the flow and torque characteristics of the valve.

The effect of the thickness ratio, s/D, is most pronounced at the fully-open position, $\theta = 0°$. Ball and Tullis (1973) investigated the effect of leaf thickness, leaf location, and leaf eccentricity for disc-type bodies. The results of those measurements and those of Harrison and Schweiger (1963) for commercial valves are plotted in Fig. 25.36. In Fig. 25.37, results for the loss coefficient, K_L, versus angle of rotation, θ, are given for four large butterfly valves, tested by Jung (1969) (disc), Guins (1968) (wafer), Grein and Pirchl (1970) (eccentric disc), and Csemniczky (1972). Csemniczky (1972) found that the flow coefficients, C_f, and K_L, were not too sensitive to leaf location, to the relative axis of rotation and flow direction, or

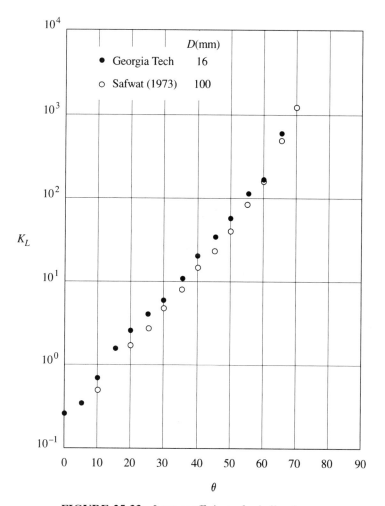

FIGURE 25.33 Loss coefficients for ball valves.

to eccentricity. Hence, all of his data are plotted as one curve. Manufacturers' data for commercial butterfly valves yield higher values of K_L for $0 < \theta < 20°$, but sometimes lower values for $45° < \theta < 90°$.

Gate Valves. Frequently, as reported by Pickett *et al.* (1971) and by Nece and Dubois (1955), the loss or discharge characteristics of gate valves are presented in terms of the area ratio, A_v/A_{vo}, instead of the ratio y/D. The two curves on Fig. 25.38 for K_L versus y/D span the range of performance of typical gate valves of the circular type. The effect of valve size on K_L is given by Miller (1978) for commercial gate valves fully open.

Globe Valves. Flow or loss coefficients for globe valves, for which there are numerous designs, are not widely available for partial openings. Figure 25.39 is a single plot from Miller (1978) of K_L versus y/b for a globe valve. Further informa-

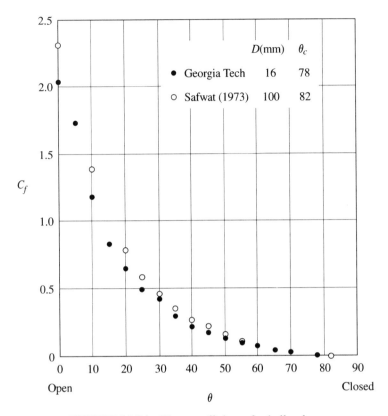

FIGURE 25.34 Flow coefficients for ball valves.

tion concerning the effect of the ratio of valve area to pipe area and the size of the valve are available from this same source.

Check Valves. As with many types of valves, the design of check valves can vary considerably from manufacturer to manufacturer. The same as for globe valves, there is much more information for the hydraulic characteristics of check valves fully or nearly fully open than partially open. From two sources, Nece and Dubois (1955) and Kane and Cho (1976), values of K_L versus the disc angle α are given in Fig. 25.40 for tilting-disc swing-check valves. For dual wafer type check valves, loss coefficients are given by Collier and Hoerner (1981).

25.3.5 Torque Characteristics

For rotary valves such as butterfly and ball types, the fluid torque is important from the standpoint of the selection of actuators. For swing-check valves, the dynamic response of the door depends upon the torque characteristics of the valve as well as the head-loss characteristics. For the actuator-controlled butterfly and ball or spherical valves, the fluid torque produces a moment in the closing direction, unless the valve design is modified from the standard shape. For butterfly valves, the fluid

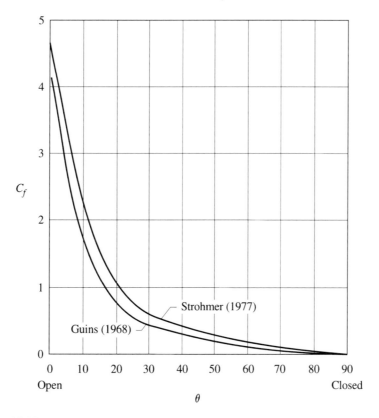

FIGURE 25.35 Flow coefficients for spherical valves. (Reprinted with permission, Guins, V. G., "Flow Characteristics of Butterfly and Spherical Valves," *J. of Hydraulics Div.*, ASCE, Vol. 94, HY3, pp. 675–690, 1968.)

torque can be significantly altered by eccentrically locating the axis of valve rotation relative to the pipe centerline. The fluid torque characteristics can also be modified by changing the valve body shape; for example, modifications include using lattice- or through-flow leaves for butterfly valves [Strohmer (1977)], or so-called characteristic or *modified ball valves* [Hutchison (1976)].

Dynamic similitude considerations would yield a torque coefficient which, in the absence of viscous (Reynolds number) and cavitation effects, depended solely upon valve geometry and position. As with the correlation of pressure-drop and discharge characteristics of valves, there are several definitions in the literature for *torque coefficients* as well. If T is the fluid or hydrodynamic torque on a valve body, being positive in the closing direction, the definition used here is

$$T = C_T \rho g D^3 \Delta H \tag{25.108}$$

in which C_T is the dimensionless torque coefficient. Other definitions used are

$$T = C_M \rho D^3 V^2 \tag{25.109}$$

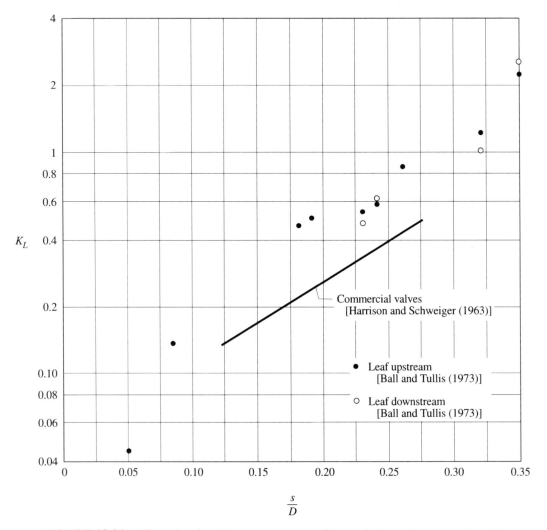

FIGURE 25.36   Effect of wafer thickness on loss coefficients for butterfly valves ($\theta = 0$).
(Reprinted with permission, Harrison, P. and Schweiger, F., "Head-Loss Characteristics of
Butterfly Values," IAHR Congress, London, paper 4.15. 1963.) (Also from Ball, J. W. and
Tullis, J. P., "Cavitation in Butterfly Values," *J. of Hydraulics Div.*, ASCE, Vol. 99, HY9,
pp. 1303–1318, 1973.)

and

$$T = K_T \rho g D^3 H \tag{25.110}$$

Using the definitions of K_L, C_f, and C_Q employed for the hydraulic characteristics,
the various torque coefficients are related by

$$C_T = \frac{2C_M}{K_L} = \frac{1 + K_L}{K_L} K_T \tag{25.111}$$

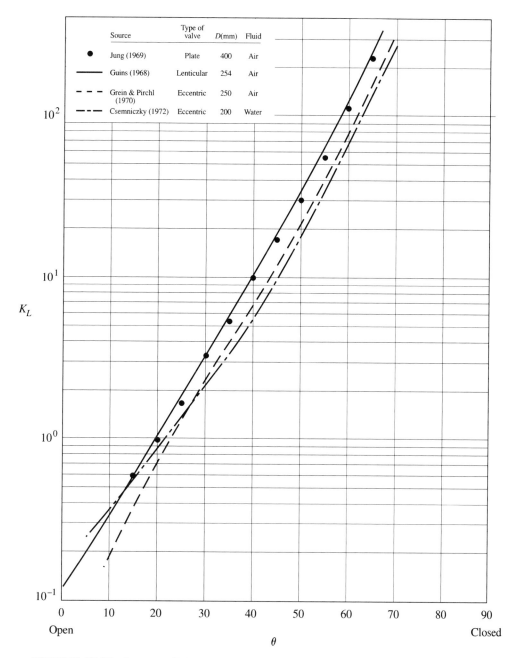

FIGURE 25.37 Loss coefficients for butterfly valves. (Reprinted with permission, Jung, R., "Beiträge angewandter Strömungsforschung zur Entwicklung der Kohlenstaubfeuerung," *VDI-Forschungsheft*, Vol. 532, pp. 1–40, 1969.)

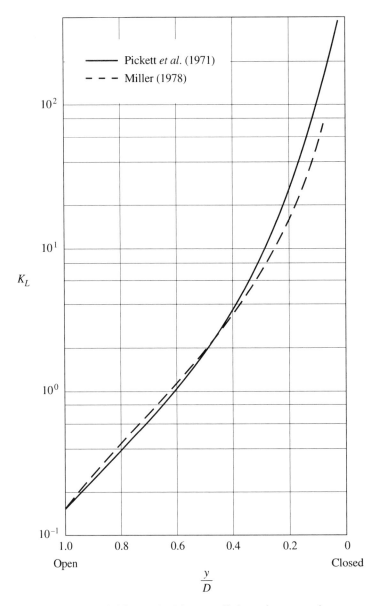

FIGURE 25.38   Typical loss coefficients for gate valves.

For butterfly valves of the disc and lenticular shape, the variation of the torque coefficient C_T with the angle θ measured from the fully-open position is plotted in Fig. 25.41. As noted, all of the test data were taken with air as the fluid except for the tests by Csemniczky (1972), who investigated the effect of valve eccentricity and leaf location relative to the valve axis using water. The valve leaf used by Jung (1969) was a flat disc, concentrically located on the pipe axis. As shown by the curve for the valve tested by Grein and Pirchl (1970), eccentricity can increase the closing moment for small angles near the fully open position, yet produce negative

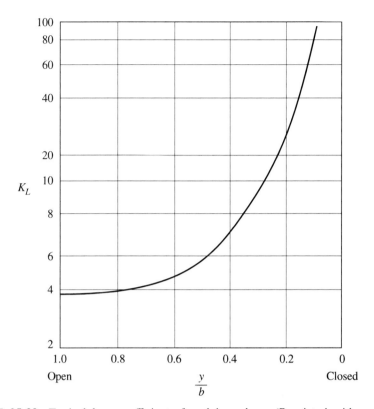

FIGURE 25.39 Typical loss coefficients for globe valves. (Reprinted with permission, Miller, D. S., *Internal Flow Systems*, BHRA Fluid Engineering, Cranfield, 1978.)

or opening torques as the valve is further closed; in this instance, C_T changes sign for $\theta = 27°$. As determined by Csemniczky (1972), more extreme eccentricities can either give greater closing torques throughout the range of θ, or yield nearly only opening torques, or negative values of C_T, as shown by the lowest curve on Fig. 25.41. Obviously, great care should be exercised in the choice of valve eccentricity to produce desired values of C_T over the range of operation.

For standard ball and spherical valves, the fluid torque will also tend to close the valve over the entire range of θ. The variation of C_T versus the angle θ from full open position is given in Fig. 25.42 for two model spherical valves, tested in air by Guins (1968) and in water by Strohmer (1977). The characteristics of these two model valves appear to be quite similar. Torque data for ball valves are available from Ytzen (1965).

The calculation of the time-varying position of a check valve during transient condition can only be accomplished if the hydraulic characteristics of the valve, namely loss and torque coefficients, are known. This information coupled with knowledge regarding such mechanical features of the valve as moment of inertia, added mass, damping, spring constant, etc., is needed for transient analyses such as *water hammer*. Unfortunately, only limited information is available in the literature. Figure 25.43 is a plot of data from Nece and Dubois (1955) and Kane and Cho (1976) for tilting-disc check valves over a range of angles of the leaf. For wafer

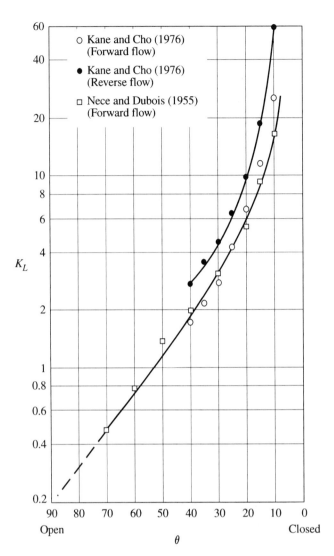

FIGURE 25.40 Loss coefficients for tilting-disc check valves. (Reprinted with permission, Kane, R. S. and Cho, S. M., "Hydraulic Performance of Tilting-Disc Check Valves," *J. of the Hydraulics Division*, ASCE, Vol. 102, HYI, 57-72, 1976; also from Nece, R. E. and Dubois, R. E., "Hydraulic Performance of Check and Control Valves," *J. of the Boston Society of Civil Engineers*, Vol. 42, pp. 263–286, 1955.)

valves of the dual-leaf variety, torque coefficients are also available from Collier and Hoerner (1981).

25.3.6 Effect of Cavitation

Cavitation in valves manifests itself in three important ways—degradation of hydraulic performance, noise, and material damage through erosion (see Sec. 23.10). Although all of these are significant, only results of the first, which is the effect of

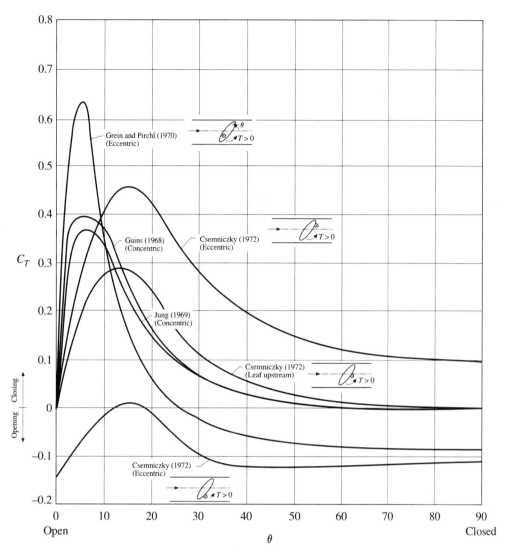

FIGURE 25.41 Torque coefficients for lenticular butterfly valves. (Reprinted with permission, Csemniczky, J., "Hydraulic Investigations of Check Valves and Butterfly Valves," *Proc. of the 4th Conference on Fluid Machinery*, pp. 293–307, 1972.)

cavitation on flow and torque coefficients, will be presented here. Frequently the *cavitation index* for valves is defined as

$$\sigma = \frac{h_d - h_v}{h_u - h_d} \tag{25.112}$$

where h_u and h_d are the upstream and downstream pressure heads, respectively, and h_v is the vapor pressure head. As with pumps, hydrofoils, and other devices, there is a value of σ above which there is absolutely no cavitation. Investigations con-

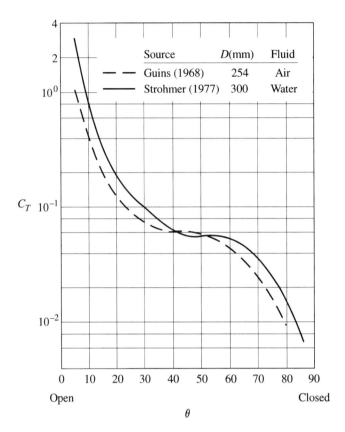

FIGURE 25.42 Torque coefficients for spherical valves. (Reprinted with permission, Strohmer, F., "Investigating the Characteristics of Shutoff Valves by Model Tests," *Water Power*, Vol. 29, pp. 41–46, 1977.)

ducted at Colorado State University [Ball and Tullis (1973) and Tullis (1971)] have resulted in a definition of values of σ at cavitation inception, so-called *critical cavitation*, and *super-cavitating*, or choked, conditions. Martin *et al.* (1981) used spectral analysis to ascertain incipient cavitation of spool valves. Although material damage may occur to valve bodies and the pipe for values of σ less than the value at inception, studies by Tullis (1971) and by Rao and Martin (1980) have shown that cavitation must be quite severe before hydraulic performance in terms of a flow coefficient is affected. As with orifices, the effect of cavitation on the flow coefficient depends strongly upon the location of the pressure taps in terms of the number of pipe diameters upstream and downstream of the obstruction. Moreover, the torque characteristics of valves are influenced more by cavitation than the direct flow characteristic.

Figure 25.44 is a plot of the flow coefficient, C_f, as a function of the cavitation number σ for a 254 mm butterfly valve [Tullis (1971)]. The valve is at two different openings, corresponding to $K_L = 15$ and 330. Although Fig. 25.44 does not clearly indicate the value of σ_{ch} at which the hydraulic performance really begins to dete-

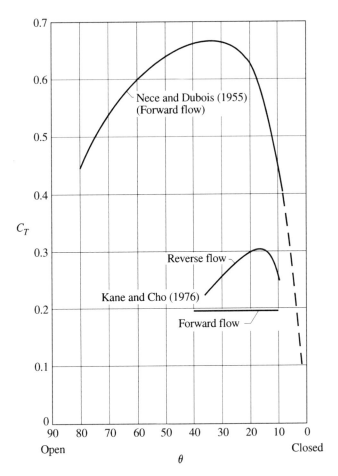

FIGURE 25.43 Torque coefficients for tilting-disc check valves. (Reprinted with permission, Kane, R. S. and Cho, S. M., "Hydraulic Performance of Tilting-Disc Check Valves," *J. of the Hydraulics Division*, ASCE, Vol. 102, HYI, 57–72, 1976; also from Nece, R. E. and Dubois, R. E., "Hydraulic Performance of Check and Control," *J. of the Boston Society of Civil Engineers*, Vol. 42, pp. 263–286, 1955.)

riorate, as a general rule the value of the cavitation index at choked conditions gradually decreases as the valve is closed further. Additional results similar to Fig. 25.44 are available in Tullis (1971) for gate, ball, and globe valves of various sizes. For spherical and lattice-type butterfly valves Strohmer (1977) shows the effect of cavitation on the flow coefficient C_Q for various angles θ. Tullis (1971) has correlated the cavitation index at choked conditions with the noncavitating flow coefficient, C_f, for several types of valves. In terms of the coefficients preferred here, his results are plotted in Fig. 25.45.

The torque characteristics of rotary valves are more sensitive to cavitation than the flow coefficient, C_f, as shown by Strohmer (1977) for spherical valves in Fig. 25.46 for various angles of the rotor.

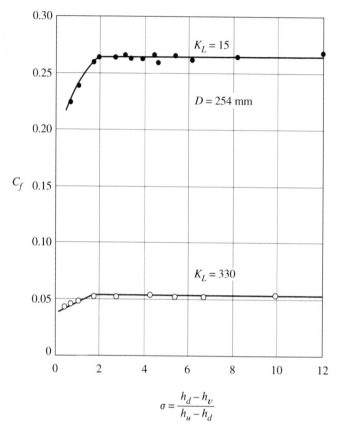

$$\sigma = \frac{h_d - h_v}{h_u - h_d}$$

FIGURE 25.44   Effect of cavitation on flow coefficient of butterfly valve. [Reprinted with permission, Tullis, J. P., "Choking and Supercavitating Valves," *J. of the Hydraulics Division, ASCE*, Vol. 97, HY12, pp. 1931–1945, 1971; also from "Prototype Experience with Large Gates and Valves," in *Control of Flow in Closed Conduits*, Tullis, J. P. (Ed.), pp. 510–511, 1971.]

25.4   EJECTORS
John E. Minardi

25.4.1   Introduction

An ejector, shown schematically in Fig. 25.47, is a device in which a high-velocity (*primary*) fluid stream entrains and compresses a low-pressure (*secondary*) fluid stream, with further compression of the mixture taking place in the diffuser (Sec. 25.2 discusses diffusers) and discharging to a back pressure which is usually higher than the total pressure of the secondary fluid. The chief components of the ejector are the primary nozzles (or single nozzle), the mixing section, and the diffuser. The multiple nozzles shown in Fig. 25.47 promote more rapid mixing, thereby permitting a shorter mixing section. The shorter section reduces the *parasitic drag* and permits the ejector to discharge into a higher back pressure. The shorter mixing section also reduces the length and weight of the ejector which may be important

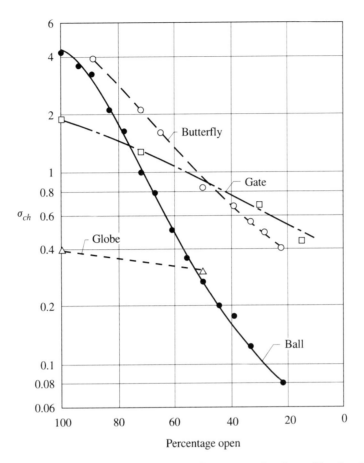

FIGURE 25.45 Choking cavitation index for four types of valves. [Reprinted with permission, Tullis, J. P., "Choking and Supercavitating Valves," *J. of the Hydraulics Division, ASCE*, Vol. 97, HY12, pp. 1931–1945, 1971; also from "Prototype Experience with Large Gates and Valves," in *Control of Flow in Closed Conduits*, Tullis, J. P. (Ed.), pp. 510–511, 1971.]

for flight applications. Of course, an ejector is not as efficient as a turbofan or a turbopump, but its simple construction, lack of moving parts, lightweight, compact design, and low cost make it desirable for many applications.

For many years, ejectors have found wide applications in jet pumps, steam-jet refrigeration, vacuum pumps, and mercury diffusion pumps. More recently, ejectors have been investigated for flight applications, especially as a method of thrust augmentation, and their potential usefulness has been demonstrated in experimental aircraft.

The topic of ejectors has been extensively studied over the years—a publication by Porter and Squyers (1981) lists over 1,600 references concerning ejector systems and related topics. Analyses of the mixing problem have been divided into two general types: 1) detailed mixing models using the Navier–Stokes equations, or 2) the control volume approaches which use integrated forms of the conservation

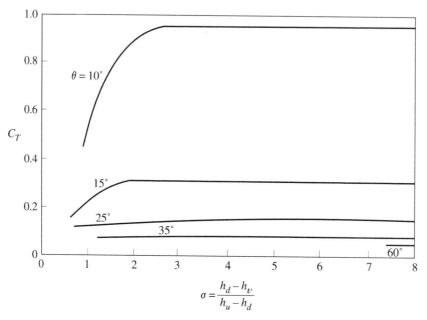

$$\sigma = \frac{h_d - h_v}{h_u - h_d}$$

FIGURE 25.46 Effect of cavitation on torque coefficient for spherical valve. (Reprinted with permission, Strohmer, F., "Investigating the Characteristics of Shutoff Valves by Model Tests," *Water Power*, Vol. 29, pp. 41–46, 1977.)

equations of mass, momentum, and energy. The one-dimensional control volume approach, using a compressible fluid, was chosen for this discussion, since it affords the best vehicle for the parametric studies required to understand the potential of ejectors for a given application.

It has been known for some time [see Kennedy (1955), Fabri and Siestrunck (1958), Hoge (1959), and Kisela (1949)] that analyses of an ejector with a constant

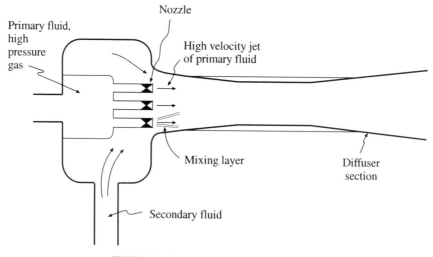

FIGURE 25.47 Schematic of an ejector.

area mixing section lead to a double valued solution. One solution occurs where the mixed flow is subsonic and the other where the mixed flow is supersonic. These two solutions are related by the normal shock relations.

Recently, Alperin and Wu (1979) have pointed out the potential advantages of the *supersonic branch* for applications to thrust augmentation. Minardi *et al.* (1982), studying two fluid ejectors for applications in turbines, also have found better results for their applications using the supersonic branch.

The use of ejector-refrigeration systems has found wide application using water as the working fluid [Jordan and Priester (1956)]. More recently, Balasubramaniam *et al.* (1976) and Hamner (1981) have suggested the use of ejector-refrigeration systems for automotive air conditioning. Again, results for this application can be improved by utilizing the supersonic branch solution [Minardi (1982a)].

Therefore, for completeness, results will be presented for both the supersonic mixed flow branch and the more commonly used subsonic mixed flow branch for constant area ejectors.

25.4.2 Compressible Flow Ejector Analysis for Ideal Gases

In this section a compressible flow analysis of an ejector for both constant area and constant pressure operation is presented. Both problems can be solved in closed form if the fluids are assumed to be ideal gases. However, the solutions are somewhat complex, therefore computer programs should be developed to obtain solutions over a wide range of parameters for a number of combinations of fluids.

The constant area analysis is presented first, followed by the constant pressure analysis.

Constant Area Analysis. A schematic of a constant area ejector is shown in Fig. 25.48. Also shown is the *control volume* used in the analysis. At station 1, the flows are completely unmixed, and each flow is assumed to be uniform and parallel. Station 1 is located at the exit plane of the primary nozzle. The exit area of the nozzle

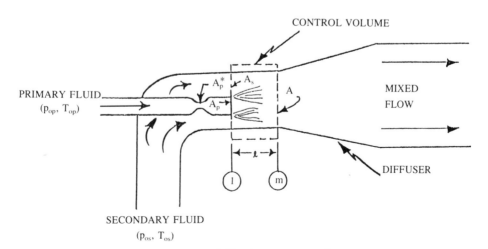

FIGURE 25.48 Constant area ejector.

is A_p, and the area occupied by the secondary flow at station 1 is A_s. The exit area, A, at station m is the sum of the areas

$$A = A_p + A_s \tag{25.113}$$

Station m is assumed to be located far enough downstream from station 1 so that complete mixing has taken place, and the flow is uniform and parallel.

For the control volume shown, the continuity, momentum, and energy equations are

$$\dot{m}_p + \dot{m}_s = \dot{m}_m \tag{25.114}$$

$$p_m A - p_{1p} A_p - p_{1s} A_s + 2\pi r l \tau = \dot{m}_p (V_p - V_m) + \dot{m}_s (V_s - V_m) \tag{25.115}$$

$$\dot{m}_p h_{op} + \dot{m}_s h_{os} = \dot{m}_m \left(h_m + \frac{V_m^2}{2} \right) \tag{25.116}$$

In general, the pressure p_{1p} does not have to be equal to p_{1s}, since the primary nozzle is a supersonic nozzle in the applications of interest for compressible flows.

Algebraic manipulation of the basic equations presented above results in a quadratic equation

$$(C2 - C6)\bar{V}_m^2 + (C3 + C4 + C5)\bar{V}_m - C1 = 0 \tag{25.117}$$

where $\bar{V} = V/V_s$. The values of the C's can be determined from the known information and the following equations

$$C1 = \frac{\Gamma \bar{m} \cdot TR + \bar{W}}{\Gamma \bar{m} + \bar{W}} \left(1 + \frac{\gamma_s - 1}{2} M_s^2 \right) \tag{25.118}$$

$$C2 = \frac{\bar{W}(\bar{m} + 1)}{\Gamma \bar{m} + \bar{W}} \frac{\gamma_s - 1}{2} M_s^2 \tag{25.119}$$

$$C3 = \left(\bar{p}_1 \bar{A} + 1 - \frac{2\tau l}{P_{1s} r} (1 + \bar{A}) \right) \frac{\bar{W}}{\bar{m} + \bar{W}} \tag{25.120}$$

$$C4 = \frac{\bar{m}\,\bar{W}\,\bar{V}}{\bar{m} + \bar{W}} \gamma_s M_s^2 \tag{25.121}$$

$$C5 = \frac{\bar{W}}{\bar{m} + \bar{W}} \gamma_s M_s^2 \tag{25.122}$$

$$C6 = \frac{\bar{W}(\bar{m} + 1)}{\bar{W} + \bar{m}} \gamma_s M_s^2 \tag{25.123}$$

Here, $\bar{m} = \dot{m}_p/\dot{m}_s$, $\bar{W} = W_p/W_s$ the ratio of molecular weights, l is the length from station 1 to m, $A = A_p/A_s$, τ is the wall shear, r is the radius, $TR = T_{op}/T_{os}$ and $\bar{p}_1 = p_{1p}/p_{1s}$. A complete derivation is presented in Minardi *et al.* (1982) and will not be repeated here, however Table 25.6 presents definitions and formulas helpful in the derivation and use in any computer programs based on these results.

TABLE 25.6   Useful Equations for Flow of an Ideal Gas

Isentropic Flow

$$\frac{T_0}{T} = 1 + \frac{\gamma - 1}{2} M^2 \tag{T1}$$

$$\frac{p_0}{p} = \left(1 + \frac{\gamma - 1}{2} M^2\right)^{\gamma/(\gamma - 1)} \tag{T2}$$

$$\frac{\rho_0}{\rho} = \left(1 + \frac{\gamma - 1}{2} M^2\right)^{1/(\gamma - 1)} \tag{T3}$$

$$\frac{A}{A^*} = \frac{1}{M}\left[\frac{2}{\gamma + 1}\left(1 + \frac{\gamma - 1}{2} M^2\right)\right]^{(\gamma + 1)/2(\gamma - 1)} \tag{T4}$$

$$\frac{T_0}{T}\frac{u^2}{2h_0} = \frac{\gamma - 1}{2} M^2 \tag{T5}$$

$$\frac{\dot{m}}{A} = \sqrt{\frac{\gamma W}{\overline{R}}}\frac{p_0}{\sqrt{T}}\frac{M}{\left(1 + \dfrac{\gamma - 1}{2} M^2\right)^{(\gamma + 1)/(2(\gamma - 1))}} \tag{T6}$$

Mixing Formula

$$C_{pm}\,\dot{m}_m = C_{pp}\,\dot{m}_p + C_{ps}\,\dot{m}_s \tag{T7}$$

$$C_{vm}\,\dot{m}_m = C_{vp}\,\dot{m}_p + C_{vs}\,\dot{m}_s \tag{T8}$$

$$R_m\,\dot{m}_m = R_p\,\dot{m}_p + R_s\,\dot{m}_s \tag{T9}$$

$$\gamma_m = \gamma_p \frac{1 + \dfrac{W_p}{W_s}\dfrac{\dot{m}_s}{\dot{m}_p}\dfrac{\gamma_p - 1}{\gamma_p}\cdot\dfrac{\gamma_s}{\gamma_s - 1}}{1 + \dfrac{W_p}{W_s}\dfrac{\dot{m}_s}{\dot{m}_p}\cdot\dfrac{\gamma_p - 1}{\gamma_s - 1}} \tag{T10}$$

$$\Gamma \equiv [\gamma_p/(\gamma_p - 1)]/[\gamma_s/(\gamma_s - 1)]$$

$$C_{pp}/C_{ps} = \Gamma/\overline{W} \tag{T11}$$

$$C_{pm}/C_{ps} = (\Gamma\overline{m} + \overline{W})/[\overline{W}(\overline{m} + 1)]$$

$$R_m/R_s = (\overline{m} + \overline{W})/[\overline{W}(\overline{m} + 1)] \tag{T12}$$

$$C_{pp}\dot{m}_p/C_{pm}\dot{m}_m = \Gamma\overline{m}/(\Gamma\overline{m} + \overline{W}) \tag{T13}$$

$$R_p\dot{m}_p/R_m\dot{m}_m = \overline{m}/(\overline{m} + \overline{W}) \tag{T14}$$

Where $\overline{W} = W_p/W_s$ and $\overline{m} = \dot{m}_p/\dot{m}_s$ \tag{T15}

If Eq. T6 of Table 25.6 is evaluated for both the secondary flow rate and the primary flow rate, then it can be shown that

$$\frac{\overline{m}}{\overline{A}} = \sqrt{\overline{\gamma}}\,\overline{W} \left(\frac{PR}{\sqrt{TR}}\right) \left(\frac{2}{\gamma_p + 1}\right)^{(\gamma_p + 1)/2(\gamma_p - 1)} \left(\frac{A_p^*}{A_p}\right) \frac{\left(1 + \dfrac{\gamma_s - 1}{2} M_s^2\right)^{(\gamma_s + 1)/2(\gamma_s - 1)}}{M_s}$$

(25.124)

where $PR = p_{op}/p_{os}$. In Eq. (25.124), it is assumed that the primary nozzle is choked. The value of M_s can be controlled in an operating ejector by adjusting the back pressure on the ejector at station m for a given value of p_{os}. Thus, it is convenient to use M_s as an independent variable in computer calculations. A geometry, pair of fluids, and stagnation conditions are chosen, and Eq. (25.124) is solved for $\overline{m}$ (or hold $\overline{m}$ fixed and calculate $\overline{A}$) for various values of M_s. Equations (25.118)–(25.124) can then be solved for each value of M_s.

Once V_m is determined, the static temperature and pressure are obtained by the following equations

$$\overline{T}_m = \frac{\overline{W}(\overline{m} + 1)}{\Gamma \overline{m} + \overline{W}} \left[\frac{\Gamma \overline{m} TR + \overline{W}}{\overline{W}(\overline{m} + 1)} - (1 - \overline{T}_{1s})\overline{V}_m^2\right] \qquad (25.125)$$

$$\overline{p}_m = \frac{\overline{W} + \overline{m}}{\overline{W}(1 + \overline{A})} \cdot \frac{\overline{T}_m}{\overline{T}_{1s}} \cdot \frac{\overline{p}_{1s}}{\overline{V}_m} \qquad (25.126)$$

The ratios $\overline{T}_{1s} = T_{1s}/T_{os}$ and $\overline{p}_{1s} = p_{1s}/p_{os}$ can be found from the isentropic relations for the secondary gas from M_s (Eqs. T1 and T2 from Table 25.6). Consequently, the final state of the gas is completely determined since the velocity, pressure, and temperature are determined at the exit from the mixing section.

Ejector Efficiency. Many definitions of *ejector efficiency* can be made and found to be useful. Some of them, however, can be greater than unity under certain conditions of operation, therefore they are not true efficiencies but rather pseudo efficiencies. For purposes herein only one efficiency is used. It is based on the *thermodynamic availability*, ψ, of the primary fluid and the availability of the mixed primary and secondary fluids leaving the ejector. The availability of a fluid in steady flow, neglecting potential energy terms, is

$$\psi = h_0 - h_R - T_R(s - s_R) \qquad (25.127)$$

Ordinarily, the reference values would be evaluated at atmospheric pressure and temperature; however, here the reference values are evaluated at the stagnation temperature and pressure of the secondary gas. This ensures that the secondary gas will have zero availability before mixing and that all the availability results only from the primary flow.

It is shown in Minardi *et al.* (1982) that the efficiency based on availability, η_{av}, is

$$\eta_{av} = \frac{\dot{m}_m C_{pm}}{\dot{m}_p C_{pp}} \frac{\dfrac{T_{om}}{T_{os}} - 1 - \ln\left[\dfrac{T_{om}}{T_{os}}\left(\dfrac{p_{os}}{p_{om}}\right)^{(\gamma_m - 1)/\gamma_m}\right]}{\dfrac{T_{op}}{T_{os}} - 1 - \ln\left[\dfrac{T_{op}}{T_{os}}\left(\dfrac{p_{os}}{p_{op}}\right)^{(\gamma_p - 1)/\gamma_p}\right]} \tag{25.128}$$

Equation (25.128) can be solved for η_{av}, since all the quantities are known from the previous calculations. It is a true efficiency for the ejector if the primary and secondary gases are the same, but only a pseudo efficiency if not, since the availability due to dissimilar gases is neglected in Eq. (25.128).

Constant Pressure Analysis. For the constant pressure case, the momentum equation takes a simple form if shear stresses are neglected, and it can be immediately solved for V_m. Using the same notation as for the constant area case the result is

$$\overline{V}_m = \frac{\overline{m}\,\overline{V}_p + 1}{\overline{m} + 1} \tag{25.129}$$

The temperature can be found from Eq. (25.125) which is also valid for the constant pressure case. The pressure is, of course, equal to the inlet value, therefore Eq. (25.126) can be solved for the exit area. (The term $(1 + A)$ can be replaced by A_m/A_{1s}.) Thus, a complete solution can be found for constant pressure which is simpler than the constant area case. The efficiency definition can be used for either the constant pressure or constant area case.

25.4.3 Computer Results of Ejector Studies

For any given pair of gases, three approaches to obtaining solutions to the equations presented in the previous sections readily come to mind in view of Eq. (25.124), which relates the mass flow ratio and area ratios to the fluid properties and M_s. In all of them, M_s is considered the independent variable, and the total pressure and temperature ratios are given. Two of them represent design approaches, and the third represents an operating characteristic of a given design or geometry. The design approaches will be discussed first.

In either design approach, it is assumed that at the inlet of the ejector (station 1 of Fig. 25.48) the static pressures of the two streams are equal (i.e., $p_{1p} = p_{1s}$). For each value of M_s, value of p_{1s}/p_{os} is determined from the isentropic pressure relation, Eq. T2 in Table 25.6. Since $p_{1p} = p_{1s}$, the value of p_{1p}/p_{op} is determined from the given total pressure ratio and the relation $p_{1p}/p_{op} = (p_{1s}/p_{os})(p_{os}/p_{op})$. Hence, M_p and A_p/A_p^* can be determined from the isentropic pressure relation and area relation, Eqs. T2 and T4 in Table 25.6. Then, one of the two approaches is chosen by assuming: 1) that mass flow $\overline{m} = \dot{m}_p/\dot{m}_s$ is given, or 2) the area ratio A/A_p^* is given.

If condition (1) is chosen, then Eq. (25.124) is solved for the only unknown, A, and all other equations can then be solved as discussed previously [where Eq. (25.124) was presented]. On the other hand, if condition (2) is chosen, then the constant area relations [Eq. (25.113)] and the calculated values of A_p/A_p^* (as before from M_s) are used to determine $\bar{A}$ ($= A_p/A_s$). Equation (25.124) is next used to determine the remaining unknown, $\bar{m}$, and all other values can then be determined.

Thus, for either procedure the geometry of the ejector is different at each value of M_s, and the curves generated are considered as design curves.

In the third approach, a design is selected, then the geometry is fixed, and Eq. (25.124) is used to calculate the mass flow ratio, $\bar{m}$, for each choice of M_s. However, it is now clear that at station 1, the static pressure will not be equal except at the design condition (if it was chosen from one of the design curves). These curves represent an *operating characteristic* for the given ejector. Various conditions on these curves are achieved by adjusting the back pressure into which the ejector is exhausting.

Figure 25.49 presents the results without friction for both branches of the solution of the equations for a constant area ejector with a constant mass flow ratio and equal static pressures at the inlet. The two solution branches are marked subsonic and supersonic on Fig. 25.49. The branch with supersonic mixed flow is sometimes referred to as the *second solution*. Also shown on Fig. 25.49 are the results for a constant pressure solution. The solutions presented are for air driving air with a pressure ratio of 6 and a temperature ratio of 3.7. The *bypass ratio* or *entrainment ratio* $(\dot{m}_s/\dot{m}_p)$ was taken as 10. The efficiency based on availability is plotted as a function of the secondary inlet Mach number, M_s. The primary inlet Mach number, M_p, is adjusted to match the pressure at the inlet (i.e., $p_{1p} = p_{1s}$).

As seen in Fig. 25.49, the value of the efficiency, η, exceeds one on the supersonic branch at subsonic inlet Mach numbers, M_s. Also, note that a choking of the flow takes place in the constant area case, and no real solution exists for a range of secondary inlet Mach numbers (this condition disappears at lower bypass ratios or lower temperature ratios).

It is still important to understand the physical significance of the extremely high efficiencies apparent on the supersonic branch as indicated on Fig. 25.49. In Minardi (1982b)], it is shown that the constant area geometry is a sufficient condition to derive the set of equations used for its analysis, but it is not a necessary condition.

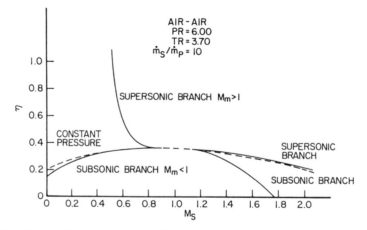

FIGURE 25.49 Efficiency based on availability versus secondary inlet Mach number for a bypass ratio of 10.

Since it is not a necessary condition, some of the solutions to the equations may not be possible with an ejector. On the other hand, since it is a sufficient condition, all of the solutions possible with an ejector will be found using the equations.

The constant area ejector with a fixed geometry and a diffuser operates on the subsonic branch in response to high enough back pressures. However, if the back pressure is dropped low enough, the ejector will transition to the supersonic branch and operate at one single operating point for any further reductions in the back pressure. Thus, not all of the solutions are available to the ejector, but only those on the subsonic branch between zero and the value of the secondary Mach number where transition occurs to the supersonic branch, and only the one single operating point is available on the supersonic branch. The transition point from the subsonic branch to the supersonic branch is not determined by either the back pressure or the exit diffuser (or nozzle) design. It is, in fact, uniquely determined by the inlet conditions as briefly discussed in the next section.

Determination of the Supersonic Operating Point. Minardi (1982b), (1982c) following Fabri and Siestrunck (1958) derived procedures for determining the transition point from the subsonic branch to the supersonic branch for an ejector of fixed geometry and a given pressure ratio.

An important design result for the studies presented is that the most efficient operation of an ejector when operating with a given mass flow ratio, occurs near the point where $M_s = 1$ and the primary nozzle has an area ratio (A_p/A_p^*) sized to match the pressure of the secondary flow at the inlet. This procedure determines nearly the best operating point for either the subsonic or supersonic branch. Thus, the design total pressure ratio satisfies the following equation

$$p_{op}/p_{os} = \left(\frac{2}{\gamma_s + 1}\right)^{\gamma_s/(\gamma_s - 1)} \left(1 + \frac{\gamma_p - 1}{2} M_p^2\right)^{\gamma_p/(\gamma_p - 1)} \tag{25.130}$$

The area ratio, A_s/A_p^*, is then chosen to produce the desired mass flow ratio. If the pressure ratio is known, Eq. (25.130) is used to find M_p and then A_p/A_p^*.

If the ejector operates at pressure ratios less than that given by Eq. (25.130), the secondary inlet Mach number is still one. However, the inlet pressures will not match, and shocks will be required in the primary flow. The efficiency of the ejector will be reduced and the entrainment, $\dot{m}_s/\dot{m}_p$, will be increased, since $\dot{m}_p$ is decreased (assuming a constant p_{os}). This regime is referred to by Fabri and Siestrunck (1958) as the *supersonic-saturated regime*.

If the ejector operates at a higher pressure ratio than the value given by Eq. (25.130), the efficiency will increase and the entrainment ratio will decrease (again assuming a constant p_{os}). However, the previous studies [Minardi (1982b)] indicated that it would be possible to redesign the ejector to operate at this pressure ratio and mass flow ratio with a higher efficiency. This regime is referred to by Fabri and Siestrunck (1958) as the *supersonic regime*. In this regime, as the entrainment ratio approaches zero, boundary layer effects along the wall of the ejector become significant and cannot be neglected.

Procedures for determination of the operating point of a fixed-geometry, constant area ejector when Eq. (25.130) is not satisfied are presented in Fabri and Siestrunck (1958), Minardi (1982b), and Addy *et al.* (1982). These approaches lead to the following pair of equations for an ejector using a single species (e.g., air driving air).

$$\frac{1}{M_{ep}^*}\left(\frac{\gamma+1}{2}-\frac{\gamma-1}{2}M_{ep}^{*2}\right)^{-1/(\gamma-1)}$$

$$= \frac{A}{A_p^*}-\frac{A_s}{A_p^*}M_s^*\left(\frac{\gamma+1}{2}-\frac{\gamma-1}{2}M_s^{*2}\right)^{1/(\gamma-1)} \qquad (25.131)$$

$$\frac{1+M_{ep}^{*2}}{M_{ep}^*}=\frac{1+M_p^{*2}}{M_p^*}+\frac{A_s}{A_p^*}\frac{p_{os}}{p_{op}}$$

$$\cdot\left(\frac{\gamma+1}{2}-\frac{\gamma-1}{2}M_s^{*2}\right)^{1/(\gamma-1)}(1-M_s^*)^2 \qquad (25.132)$$

In Eqs. (25.131) and (25.132), the dimensionless ratio M^* was used in place of the Mach number. The value of M_{ep}^* is the Mach number within the mixing tube where the secondary system reaches Mach one. It is assumed that mixing is negligible up to that point.

A computer program can be written to solve these equations for a given geometry including the primary nozzle, which fixed M_p^*. The continuity equation, Eq. (25.131) can be solved for M_{ep}^* for a given choice of M_s^*. These values can then be used in Eq. (25.132) to determine the value of p_{op}/p_{os} that is consistent with the choice of M_s^*. In this way, a curve of p_{op}/p_{os} versus M_s^* can be constructed which completely determines the operational characteristics of the particular ejector when coupled with the results from the supersonic-saturated regime where $M_s^* = 1$, but p_{op}/p_{os} can vary.

Addy (1972) presents an approach using the method of characteristics that is applicable to ejectors of quite general shape. However, the results cannot be reduced to a simple set of equations such as Eqs. (25.131) and (25.132).

The simple theory presented here predicts the mass flow ratios very well as shown on Fig. 25.50. These results are from Minardi et al. (1982) and are for carbon dioxide driving air in a constant area ejector driven by eight primary nozzles. The secondary enters the mixing tube through a bellmouth. The relative location of the primary nozzles in the bellmouth could be adjusted. The simple theory is only valid for the nozzles inserted, and the comparison to the data shown on Fig. 25.50 is very good. The more sophisticated method of Addy (1972) is required for comparison to the data in the case where the nozzles were not fully inserted. Using the Fabri and Siestrunck inlet conditions given by Eqs. (25.131) and (25.132), a good agreement between theory and experiment is shown in Fabri and Siestrunck (1958), Addy et al. (1982), and Addy (1972).

However, it is more difficult to accurately predict efficiency or total pressures into which the ejector can discharge. Figure 25.51, taken from Minardi et al. (1982), shows a comparison of theoretical and experimental efficiencies for a number of configurations. Predicted efficiencies for both the supersonic and subsonic branch are shown with and without friction. The efficiency used is based on kinetic energy and is a pseudo-efficiency and can, therefore, exceed once as it did on the supersonic

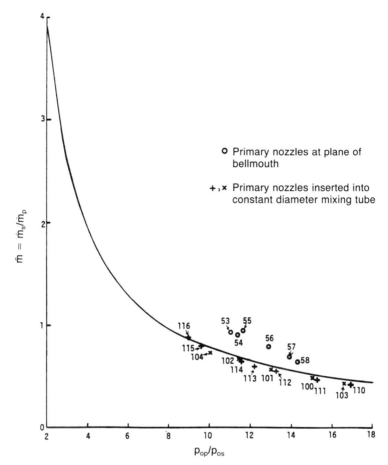

FIGURE 25.50 Comparison of mass flow ratio, $\dot{m}_s/\dot{m}_p$, predicted by theory and measured. The theory should only be correct for nozzles inserted.

branch. The addition of frictional effects brings the data and experiment closer together but does not fit the shape of the data which is somewhat irregular. It is of interest to note that the experimental efficiency at some of the points is higher than that predicted for the subsonic branch without friction. Clearly, the ejector is operating on the supersonic branch at such points.

In the next section some of the factors that influence performance and design are discussed.

25.4.4 Discussion of Factors Influencing Performance and Design

In an ejector, there are intrinsic losses resulting from differences at the entrance in velocity, properties (such as pressure and temperature), and in constituents (two different fluids). These losses are automatically accounted for in the solution of the control volume equations on the subsonic branch when the inlet pressures are assumed to match or results are determined for a fixed geometry ejector.

However, not all of the results achieved on the supersonic branch can be obtained

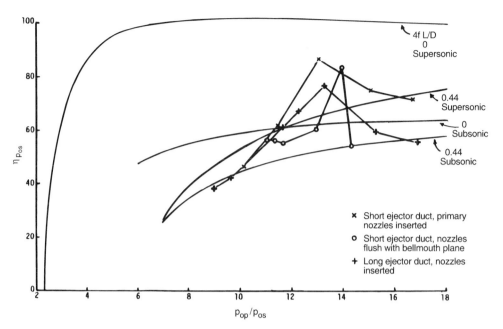

FIGURE 25.51 Ejector efficiency versus ratio primary total pressure to secondary total pressure.

from an ejector. Rather, the Fabri and Siestrunck (1958) inlet conditions must be used to determine where on the supersonic branch the ejector will actually operate when the inlet secondary Mach number is subsonic. If a throat is upstream in the secondary flow, the ejector can be forced to operate on the supersonic branch (i.e., the mixed Mach number is supersonic) at the appropriate value of M_s as determined by the area ratio.

It is also possible that the ejector is choked because of thermal effects or species differences. Hoge (1959) presents a thorough treatment of this phenomena. Figure 25.49 shows an example of this phenomena where the supersonic and the subsonic branches join (the mixed Mach number is one at this point), and there is no solution for a range of M_s about one.

In addition to the *intrinsic losses* resulting from the *mixing processes*, there are additional *parasitic losses* resulting from wall friction, diffuser losses, and inlet losses. *Friction losses* can be reduced by reducing the length of the mixing tube while still maintaining adequate mixing. The best approach for reducing the ejector length while maintaining mixing, is to use multiple primary nozzles that provide the same mass flow rate as the single nozzle which is replaced. An accelerated rate of entrainment from a single nozzle is referred to as hypermixing. Bevilaqua (1974) evaluated a hypermixing nozzle for thrust augmenting ejectors. A slot nozzle with a segmented exit plane that alternately directs the jet slightly up or down generates streamwise vortices which serve to entrain additional fluid to promote mixing. By completing the mixing in a shorter distance, the flow could be diffused more efficiently, and the *thrust augmentation* increased. Other approaches attempt to use transient phenomena to reduce the mixing length, some of which are discussed in Braden *et al.* (1982).

Diffusers will, of course, induce additional losses and reduce the final value of the mixed total pressure. Diffuser design is discussed in Sec. 25.2 and will not be repeated here. However, if one wants to achieve the improved efficiencies obtainable from the supersonic branch, then a supersonic diffuser will be required if the flow must be brought to its total pressure before it can be used. If a normal shock occurs in the constant area section of the ejector, or even at a higher Mach number or in the expanding area of a diffuser designed for subsonic flow, then the final total pressure would be equal to or less than that achieved on the subsonic branch. In many discussions of ejector design [e.g., see Uebelhack (1972) and Defrate and Haer (1982)], it is assumed that either the subsonic solution is achieved or a normal shock occurs within the duct. (Actually, because of shock boundary layer interactions, a series of shocks occur in a straight tube which produce nearly the same results as a normal shock.) If no special care is taken to design a supersonic diffuser, then that is a reasonable assumption and is quite satisfactory for most applications. With that approach, the optimum design should be chosen from the results on the subsonic branch as discussed earlier.

For thrust augmentation in flight, the effect of the forward motion is to increase the secondary total pressure above ambient, therefore a diffuser may not be required, but further expansion of the mixed flow may be needed. However, even for performance of the supersonic branch, the thrust augmentation decreases for increasing Mach number and falls below one near Mach one [Minardi and von Ohain (1983)]. The effect of temperature ratio (high temperature primary) is just the opposite at supersonic speeds and subsonic speeds. The thrust augmentation falls at subsonic speeds with increasing temperature, and the thrust augmentation increases at supersonic speed with increasing temperature [Minardi and von Ohain (1983)]. The improvement in the latter case is more akin to a ramjet effect than that of an ejector nozzle, and inlet losses are quite small except in flight applications of ejectors where the inlet might substantially reduce the secondary total pressure as compared to the free stream total pressure.

If calculations are made neglecting friction, the mass flow ratios are determined with good accuracy if inlet secondary Mach number is greater than 0.3 [see DeFrate and Hoerl (1982)]. They suggest that the final total pressure, however, should be reduced to 90% of the frictionless value.

25.5 SEALS
Philip C. Stein, Jr.

25.5.1 Principles of Sealing

It is well to dispel the common notion that seal design is a secretive, occult art. The design and understanding of sealing devices is rather straightforward, requiring only the ordinary acumen of a good conventional engineer and a recognition that a number of interrelated parameters are involved. In this respect, seal problems resemble most other problems. The only essential difference pertains to the magnitude of the significant dimensions. In sealing devices, behavior is governed by certain small dimensions defining the interface separation of the leakage paths. Therefore, all influences affecting this interface distance become of large importance and all deviations from geometry, such as structural and thermal deformations, must be considered rather exactly. This means that, whereas in conventional engineering, deflections in millimeters have significance, with seals, deformations of micrometers may be important. Consequently, discrimination is required in evolving a design so that adequate precision is engineered into those dimensions which may affect the interface dimensions and the thickness of the leakage film.

A complete treatment of a problem in any one field must draw on the experience and knowledge of many other fields. This is particularly true in this field where structural theory, heat flow, lubrication theory, fluid mechanics, thermodynamics, chemistry, metallurgy, and other fields of knowledge all make a contribution. Seal problems consist of a superposition of interrelated effects. Usually, each can be analyzed by itself and all resynthesized to evaluate the total content. It can be seen that, by and large, seal problems are fine grained, and various small effects associated with thermal gradients, deformations, pressure gradients, friction, and inertia forces and similar causative or resultant parameters require detailed consideration. Apart from this change in scale, the engineering of sealing devices does not depart in substance from any other type of engineering.

Flow of leakage through a seal is controlled by a combination of two factors. The leakage is controlled by the size of the gap between the seal and the mating part. Flow is also controlled by the length that the fluid must travel across the gap. The effects of these two factors will vary with the seal design. A *labyrinth seal*, (see Fig. 25.52) depends almost entirely on the size of the gap to limit flow through velocity dissipation, while a *bushing seal* (see Fig. 25.53) inhibits leakage mainly through fluid friction from the boundary layer drag of the seal surfaces.

Flow through a single-tooth labyrinth seal for an incompressible fluid, ignoring entrance losses can be calculated using the formula

$$\text{FLOW} = A \cdot \sqrt{2pg\rho} \qquad (25.133)$$

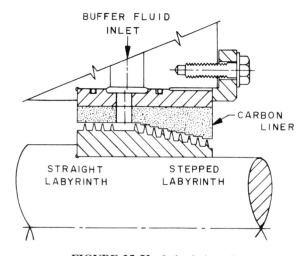

FIGURE 25.52   Labyrinth seals.

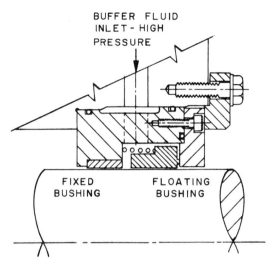

FIGURE 25.53 Bushing seal with both fixed and floating sleeves or bushings.

For laminar incompressible flow through a *bushing*, *face seal*, or *circumferential seal*, the flow can be expressed by

$$\text{FLOW} = \frac{\pi D h^3 \Delta p}{12 \mu L} \qquad (25.134)$$

where A is the area, h is separation distance of the walls, L is the length of the sealing dam, μ is absolute viscosity, D is shaft diameter, ρ is the density, and p is pressure. It can be seen from the above formula that the leakage is dependent on face opening, pressure across the seal, viscosity, and geometric parameters, including waviness and coning of the seal face. The drag and, therefore, the face opening are speed dependent and the viscosity, which is susceptible to variation with temperature due to frictional heating, is also speed dependent [O'Conner (1968)]. Thus, speed affects leakage in a multitude of interrelated ways. The interested reader should study Sec. 23.1 on Lubrication which discusses similar features.

The power consumption of a seal [Stein (1961)] is a function of both the surface speed of the rotating member and the friction forces between the rotating and stationary parts. The following formula approximates the horsepower consumed by a seal.

$$HP = \frac{FU}{550 \times 12} \qquad (25.135)$$

where F is the friction force in pounds, and U is surface speed in inches/second. The frictional force may be calculated by the following

$$F = \frac{2 \pi R U L \mu}{h} \qquad (25.136)$$

where R is the shaft radius in inches, L is the width of the seal dam in inches, μ is the viscosity in lb-sec/in.^{2}, and h is the clearance in inches. Using metric units, Eq. (25.135) becomes for power, P, watts

$$P = FU \tag{25.137}$$

Equation (25.136) applies with units as follows: R, m; L, m; μ, Ns/m^2; U, m/s; and h, m. See Chap. 2 for values of μ for various fluids.

The accuracy of this formula with respect to actual performance depends on the assumptions made with regard to seal clearance and fluid viscosity.

25.5.2 Purpose of Sealing

Nearly every rotating machine uses some sort of sealing device. Seals in fluid machinery should be designed for more than just good leakage. Other considerations should include power consumed, life, ease of installation, and the possibilities of sudden failure.

All seal systems take power from the machine to overcome friction during rotation. Contact seals at start-up require additional power to overcome the static friction forces. Power consumption of some types of seals, such as packings and long bushings, can be substantially higher than other types of seals for the same leakage. Sophisticated seal systems cost significantly more than *packings* and *bushings*, but the reduced power consumption of the seals and the improved efficiency of the machine justify their use solely on a cost saving basis. Low leakage, longer periods between routine maintenance and low fluid loss may also justify the initial seal costs.

Seals are necessary to prevent the intermixing of fluids in machine components. The leakage of *bearing oil* into *process fluid* can contaminate product mixtures. With the increased attention to contamination of consumer products, absolute contaminate-free products become more and more critical. Likewise, process fluid leakage into a bearing cavity can lead to oil deterioration and resulting premature bearing failure.

A seal system may be necessary for reasons of safety or pollution. This has prompted the use of complex seal systems to eliminate environmental emissions. These systems may use a gas or liquid buffering chamber to prevent seal leakage. These buffering chambers, in which inert gas is bled between two seals, can be used when sealing flammable fluids as well, since the inert gas hampers the initiation of fire or explosion (refer to Sec. 23.11 on Combustion).

The most obvious use of seals is to prevent the loss of fluid to the atmosphere. Large leakage losses are usually enough to justify the expense of a seal system which eliminates the leakage and also increases the machine efficiency. Additionally, the elimination of unsightly leakage to the machine surroundings is generally desirable.

25.5.3 Information Required for Seal Design

There are a myriad of types and combinations of seals which can be used in any particular seal cavity [Dobek (1973), Stein (1979), Zuk et al. (1970), Zuk (1976), and Mayer (1973)]. The seal designer must analyze the machine operating parameters and leakage requirements to determine what particular design is the most inexpensive yet reliable enough for the application. It is advisable for the seal to be designed at the same time as the machine. If the seal designer is not brought in until after the machine is under construction, then the choices of seal design are limited.

There are many factors which affect seal operation, and each one may need to be examined in detail. The amount of space available, as well as the method of component attachment, is critical in the choice of a seal design. The method of *clamping* the seal and its mating parts has a direct effect on the performance of the seal. Shaft and housing seal seat surfaces for high performance seals should be smooth and flat. In many cases, provisions should be made for *lapping* the seating lands.

The fluid characteristics must be obtained. The viscosity of the fluid affects the leakage rate, the seal drag, and the leakpath openings. The effect of the fluid on the seal materials must be evaluated with respect to material decomposition.

Pressure differentials across the seal determine the leak rate in an obvious way, but the less apparent influence on structural change to seal geometry must be analyzed. Deformation of seal hardware can have leakage effects greatly in excess of those calculated due to pressure changes.

The seal designer must be informed of the temperature of the seal environment. Standard elastomeric secondaries fail due to embrittlement at temperatures over 120 C. Thermal expansion of seal parts occurs under even moderate temperature changes, and if there are temperature gradients or dissimilar materials involved, the relative clearances of the seal parts can change, resulting in higher leakage, or at worst, catastrophic failure.

The *surface velocity* at the seal interface affects the seal in a number of ways. Parts of the seal that are subjected to substantial rotational speeds may tend to grow and distort due to centrifugal loadings. The drag and, therefore, the face openings are speed dependent, and the viscosity, which varies with temperature due to frictional heating, is also speed dependent. Rubbing seals will wear in proportion to speed, unless hydrodynamic lifting is introduced.

A properly designed seal will maintain contact with the mating part despite dynamic eccentricity of the seal rotating member. With large *runouts* on the mating ring of a face seal, springs must be able to accelerate the face seal into the mating ring faster than the mating ring rotates away from the seal. In addition, the primary face seal must be guided on its holder at the secondary seal member with enough freedom to allow it to adjust to the tilt of the mating ring without binding.

If the seal application is of such a nature that no outside leakage of sealed fluid or gas is permitted, then it may be necessary to mount two seal elements back to back and to inject a *buffering medium* in the space created between the seals. Externally pressurized buffers should be regulated to pressures specified by the seal designer.

25.5.4 Types of Seals

Labyrinth Seals. The labyrinth [O'Connor (1968)] is the simplest of sealing devices, consisting of a series of circumferential strips of metal extending from the shaft or from the bore of a housing about the shaft, to form a restricted clearance between the rotating and the static parts. Two types of labyrinths are shown in Fig. 25.52. The strips of the labyrinth are made thin enough to wear away or to *mushroom* if they should rub. The labyrinth strips restrict leakage by reducing the flow area, each restriction having an area equal to the product of the radial clearance and the circumference. Labyrinths are suitable for the most severe systems involving high temperature, pressure, and speed but where relatively high leak rates are acceptable.

Labyrinth clearance must allow for bearing clearances, shaft vibrations, deflections, thermal growth of the shaft, and similar factors. When leakage requirements prohibit adequate running clearances, tighter labyrinths can be devised with shaft knives inside closely fitted sleeves which melt or wear readily, such as of babbitt, soft carbon, phenolic composition, etc. With these, grooves will be cut in the sleeves without serious wear or *mushrooming* of knives. The V-shaped cut in the sleeve does not greatly increase the effective clearance area of the labyrinth. Obviously, damage will result if the labyrinth strips are imbedded in the grooves coincidental with large axial shaft travel.

Labyrinths do not have pressure restrictions and, since the choice of materials is almost boundless, temperature limits are defined by those of adjacent parts. Because labyrinths are clearance devices, shaft speeds are not limiting, once rubs are eliminated.

Various aspirating, pressurizing, and barrier systems can be used with labyrinth structures. Fluid inlets and outlets can be located between the strip locations. By such means, labyrinths can prevent entry of air into low pressure processes as well as outleakage of noxious fluids when processes are above atmospheric pressure.

Labyrinths can also be used in conjunction with rubbing and other types of positive seals. The labyrinth sometimes is used as an internal breakdown seal at the high pressure end of a compressor, upstream of a face seal, with a vent line connecting this space with the suction side of the machine. This pressure breakdown arrangement allows a positive final closure to be employed in a situation where it ordinarily cannot be used.

Bushings. The bushing seal (see Fig. 25.53) is a sleeve which forms a close clearance around a shaft. The fluid flow through the seal is restricted by the space between the shaft and the sleeve due to the flow resistance from the wall friction and the velocity head loss.

Leakage through a bushing seal is of the order of one tenth that of a labyrinth seal of equal axial length. Most bushing seals are made with a carbon, babbitt, or bronze bore to minimize serious damage due to rubbing. The use of carbon limits the temperature range of the bushing to about 300 C, but with the use of other materials, the temperature limitations can be those of the surrounding material. Bushings can be used on applications from low pressure to pressures in excess of 300 kg/cm^2.

Bushing seals can be designed with either *fixed* or *floating sleeves*. A typical application using both a fixed and a floating bushing is shown in Fig. 25.53. If the bushing is fixed to the machine housing, the clearance must be at least as great as the combination of the shaft housing misalignment, vibration, bow and deflection of the shaft, and bushing. With inadequate clearance, rubbing of the shaft on the bushing can occur inducing excessive bearing loads, serious vibration, and possible *seizure* of the shaft in the bushing due to *thermal growth* of the shaft.

If the desired leak rate of a bushing seal cannot be obtained using a fixed bushing seal, the designer may use a floating bushing. Floating bushing seals are sleeves which are mounted in a chamber and held in an axial position by springs but are allowed freedom to move in a radial direction. Radial positioning around the shaft by the bushing is aided by hydrodynamic forces generated by rotation. If the bushing is not centered, the circumferential fluid velocity created by the shaft rotation will form a pressure wedge at the closest clearance point, helping to center it. *Diametral clearances* for floating bushings average about 0.1 mm/cm of shaft diameter.

High pressure bushings have the additional problems of diametral deformation, and higher seating loads on the bushing face due to pressure. Such factors counteract the bushing hydrodynamic centering forces. Under high pressure, bushings also cone to form a converging flow path from the high pressure area to the lower pressure area. This converging flow path will tend to center the bushing about the shaft. However, the contraction in the bore, especially near the low pressure end of the bushing, necessitates the bushing clearance with the shaft to be increased to compensate for the decrease in bushing diameter at high pressure.

Power consumption of long bushings at high speeds is usually higher than that of other seal types for a given leakage. The fluid drag between the stationary bushing and the rotating shaft occurs over a longer area than a face seal, for example. The drag can be minimized by making the bushing width smaller and decreasing the clearance a proportional amount. A 20% reduction in the bushing clearance allows the bushing length to be half the previous length for the same leakage and reduces the bushing power consumption by 38%.

Both gases and liquids can be sealed with bushing seals. The sealing of gas under high pressure applications presents a multitude of design problems, because its compressible nature makes the effects of hydrodynamic stability much less certain and the cooling effects of a gas are much less than of a liquid (see Sec. 23.1).

Windbacks. When oil at a low rate enters a confined interspace between a shaft with moderately high surface speed and a housing bore, it is carried continuously around the bore by the *windage* of the shaft. This effect is utilized by a screw thread device which winds the oil into an internal drain for return to the system. A typical *windback* appears in Fig. 25.54. Windbacks can be very effective in restraining the leakage of oil splash from unpressurized sumps and gear cases where there must be no outward blow-through of air and gas.

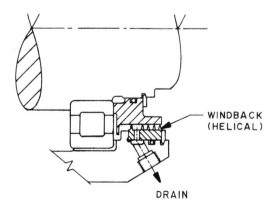

FIGURE 25.54 Windback to return fluid to a system via a drain.

Windbacks are also used as adjuncts to other types of seals. With circumferential seals, windbacks can be used to keep oil splash from reaching the seal carbons when *coking* problems exist. In oil-buffered seals for gas pipe line compressors, windbacks are used to direct the small internal leakage into a pressurized drain, thereby achieving practically complete recovery of the leakage.

Face Seals. *Face seals* (see Fig. 25.55) are widely used throughout industry on all types of machinery [Zuk *et al.* (1970)]. Face seals, whether sealing liquids or gases, are usually classified as *rubbing* or *contacting* types as opposed to *noncontacting* or *film riding* devices.

Essentially, the face seal consists of two radially flat surfaced rings, one rotating with the shaft and the other nonrotating and carried by the machine housing. One of the rings should be designed with enough freedom of movement to enable it to maintain continuous positioning with respect to the other ring, despite relative axial shaft and housing movement, lack of squareness between the two rings, and machine vibration. The axially free member, which is designated the face seal, requires a force to overcome system friction, inertia, and other restraints in order to maintain contact with the fixed or mating ring. This force is commonly supplied by mechanical springs. A static or secondary seal is required to prevent leakage between the free ring and its holder; in Fig. 25.55, an "O" ring is used.

At low shaft speeds, either the static or the rotating ring may have axial freedom. With high speeds and large diameters, centrifugal effects on the rotating member introduce forces which inhibit the freedom of movement of a freely suspended ring, and the seal cannot function properly. This makes it mandatory for the shaft member to be rigidly attached, and the axially free member to be nonrotating.

Rubbing Face Seals: Rubbing face seals are the most common type of face seal. A rubbing face seal is illustrated in Fig. 25.55. These seals are usually made of carbon

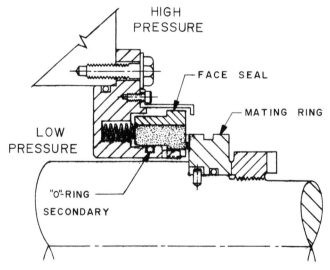

FIGURE 25.55 Rubbing face seal with an "O" ring secondary seal. Mating ring rotates while face seal is stationary.

graphite which rubs on a mating ring made of appropriate wearing material, such as hard cast irons, tool steels, carbides, ceramics, as well as a variety of hard surface coatings overlayed onto ductile metals. Flatness and surface finish of the mating surfaces are critical, and in many applications, flatness must be held to 1 or 2 wavelengths of visible radiation (1 helium wavelength = 0.3 micron).

Long wear life is achieved by keeping the rubbing load between the face seal and the mating ring minimal, yet adequate to overbalance axial friction, inertia, and film pressure forces. Face seals with moderate and high pressure differentials must be carefully designed to take into account the effects of the pressure on the rubbing load at the interface. Pressure control of face loading is achieved by balancing the pressure load on the back of the seal with pressure loading at the front of the seal. Face loadings at the seal interface should be high enough to allow the seal to maintain proper contact while avoiding the problems associated with overloading. Face loading values vary widely, depending upon the seal application. Substantial rubbing loads can readily be tolerated when high viscosity fluids are being sealed at low surface speeds, but minimum loads are needed for dry gas seal applications where the surface speeds can be in the order of 150 meters per second.

Excess loadings, especially in fluids which have relatively low heat transfer coefficients, will not reduce leak rates significantly and can result in high thermal gradients in the seal and mating ring resulting in excessive leak rates and, in severe cases, in *cracking* and *pitting* in the mating ring. This is especially true if it has been plated or hard surfaced with a flame spray.

Gas face seals operating at low temperature, speed, and pressure conditions can be designed with relative ease, but, under more severe conditions, the seal must be designed to prevent excessive face loading. Heat generation from rubbing between the static and rotating members is difficult to dissipate in a gas environment due to the low heat transfer properties of gases. The pressure loadings between the face and mating ring must be kept as low as possible but must be sufficient to overcome friction from the seal secondary which is designed with low friction as a prime consideration. Where there is no lubrication in a dry gas environment, the increase in the secondary seal friction coefficient causes additional problems.

A grooved face design can be used to minimize the change in interface pressure caused by thermal and structural deflections. The grooved face permits pressure balance to be maintained despite deflections, since it reduces the degree of convergence or divergence across the dam when expressed in terms of the ratio of inlet to outlet opening. The seal face has a narrow sealing dam with inner and outer circumferential support pads of the same height on either side of the sealing dam. These allow the seating forces needed to insure contact with the mating ring to be spread over a large area, therefore the specific loading on the seal is lower than with a single dam design. Because of the low positive rubbing load which this design allows, these seals can operate dry in the regime of gliding wear at high speeds and at significant pressure differentials. The grooved face designs are viable solutions to many gas and vapor sealing problems where low rates of leakage are specified. Depending on operating conditions, leakage rates to the atmosphere of 150 to 600 cc per minute of gas (at STP) per cm of diameter per bar are obtained for outleakage and less, of course, for inleakage to subatmospheric applications.

Unlike single nose face designs, in which pressure loading is highly sensitive to deflections, the grooved face performs most satisfactorily when completely unlubricated. Where moderately low gas leakage to the environment can be tolerated, buffering and leak-off systems can be avoided. Therefore, dry gas seals can reduce auxiliary equipment, conserve shaft length, and result in simplification and cost reduction of the machine.

Nonrubbing Face Seals: Under severe conditions, it is advantageous to avoid rubbing between the face seal and mating ring. In these cases, the seal face is designed to take advantage of pressure generated between the seal and mating ring to maintain a small clearance of the order of 3 μm between the seal and mating ring. There are two main types of nonrubbing face seals: 1) those that use the pressure distribution across the seal interface to provide a *lifting force* and 2) those that utilize hydrodynamic principles to develop low grade gas bearings on the seal face thereby creating a dynamic separating force between the seal and its mating ring.

Hydrostatically lifted face seals are sometimes used in high pressure applications where the larger amount of leakage due to the interface clearance can be tolerated. The design of the seal face is such that any excursion of the seal face from its equilibrium position will change the interface pressures and generate a restoring force.

Hydrostatically supported seals may be divided into two groupings, those that are externally pressurized and those that are self-energized. The externally pressurized group utilizes orifices or other constricting conduits from an external high pressure source to pressurize a groove in the seal dam. This pressure is channeled to the seal through the seal housing and enters the seal through a chamber formed by two secondary seals. The *self-energized hydrostatic seal* depends on the ambient pressure of the system to provide the forces necessary to maintain a limited clearance. These devices differ from externally pressurized designs by having only a single secondary seal and are classified as stepped face, tapered face, and orifice compensated types. Figure 25.56 shows a tapered face hydrostatically lifted seal.

The stepped face or tapered face seal relies on face geometry to provide a pressure profile which changes with variation in the face opening. When the seal approaches the closed position, the upstream portion of the seal face is subjected to full pressure, but as the seal lifts off, the pressure profile changes so that pressure diminishes across the face.

The externally pressurized type can provide supporting forces at times when the ambient pressure is very low, whereas the self-energized seal must depend on the ambient pressure for support. Externally pressurized face seals depend on the flatness of the face dam for uniform pressure loading, any twist or coning in the face can cause pressure losses on the seal face which alters the face radically. Stepped or tapered face seals are subjected to this pressure loss also but are much less sensitive to the amount of coning. Externally energized face seals are vulnerable to loss of pressure from their external source which can cause deterioration of the seal in a short time.

Hydrodynamic Face Seals: Face seals can be made to operate by employing a variety of lift-bearing geometries at the primary seal interface. Lift augmenting devices having plane wedge, taper-flat, taper step, or spiral groove geometry are configured onto the seal face upstream of the sealing dam. These devices can be incorporated onto the face of the static or rotating ring and, where possible, have side rails to restrict side leakage. A hydrodynamically lifted face seal is shown in Fig. 25.57.

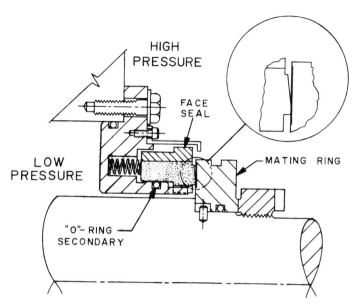

FIGURE 25.56 A tapered-face hydrostatically lifted seal. Note regions of high and low pressure.

Dynamically supported floating seals depend upon the velocity of the rotating member to drag fluid into the bearing pad areas, causing bearing pressures to build up to sufficient magnitude to reduce or totally support the spring load required to compensate for the inertia and friction requirements of the seal ring. Hydrodynamic seals are relatively insensitive to large pressure variations. These seals are not as

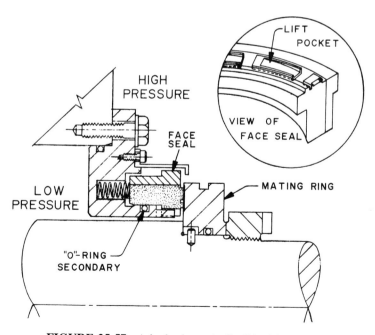

FIGURE 25.57 A hydrodynamically lifted face seal.

severely affected by the excessive face loading which the hydrostatic type must survive in the event that pressure or thermal distortion alter the geometry of coacting surfaces.

Hydrodynamic type seals have the additional advantage of being independent of the machine pressure in achieving lift off from rubbing contact. Hydrodynamic face seals remain in rubbing contact until the surface speed is sufficient to cause separation, however these lower *pre-separation* speeds are usually of such short duration as to be tolerable.

Hydrodynamic lifting devices develop relatively low lifting forces compared to hydrostatic seals. Thus, it is more difficult to cope with excursions of the seal away from its mating surface.

With the taper-flat design, a moderate amount of wear does not greatly damage the lift characteristics, thus this type is well suited to applications involving frequent starts and stops or prolonged running at low speeds. In comparison, a pocketed step bearing pad has better lift capacity, but any decrease in step depth, due to wear, changes the lift and causes a shift of the seal operating clearance.

Spiral groove seals are designed with both outward and inward pumping grooving. With the outward pumping sole acting geometry, the direction of rotation moves the fluid along the spiral groove channel so that the groove pressure peaks radially outward. The groove configuration for the inward pumping spiral groove is designed, in contrast, to force the fluid radially inward.

Secondary Seals. Face seals must be capable of maintaining position with their mating rings despite axial excursions due to shaft travel, lack of squareness of the shaft and housing shoulder, rotational restraints, vibration of the shaft, etc. To insure adequate response to such changes in axial location, the face seal must be capable of shifting its position relative to the position of the seal holder. This need for low friction relative movement between the seal and its holder creates a potential leak path which must be sealed.

The various devices used to restrict this secondary leak path are commonly termed *secondary* seals in comparison with the leakage path across the face or *primary* seal path. Piston ring types, elastomeric rings, bellows, and diaphragms of various sorts are all utilized as leakage inhibitors. Three standard types of secondary seals are shown in Fig. 25.58.

The selection of a secondary seal depends on the fluid to be sealed, the operating conditions in terms of temperature, pressure, and speed, as well as other characteristics which affect the secondary seal element.

Elastomeric Rings for Secondary Seals: The most common secondary seals are made of *elastomeric compounds*. An elastomeric compound can usually be found that is inert to the sealed fluid. Standard elastomeric seals of *Buna-N* will function in temperature ranges of −60 to 120 C. Silicone or fluorocarbon seals can operate up to 240 C, and *Kalrez* can be used in 280 C environments.

Friction from elastomeric secondary seals should be a minimum. Face load between the seal and the mating ring from the springs and pressure bias must be sufficient to overcome face seal inertia, anti-rotation lock, and secondary seal friction. Particularly in gas applications where face loading must be minimal to avoid rapid wear, the secondary seal plays a critical part in the success of the seal system. For

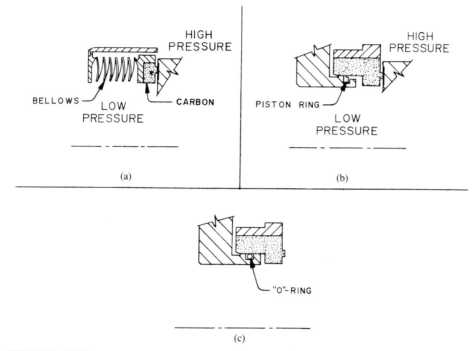

FIGURE 25.58 Three standard types of secondary seals. (a) Bellows secondary, (b) piston ring, and (c) elastomeric seal.

both liquid and gas seals, the elastomeric seal *squeeze* at the secondary should be reduced as far as practical to reduce friction while still insuring sealing. For some applications, the secondary surface is treated with a low friction coating, or a teflon sheath covers the elastomeric seal.

Piston Ring Secondary Seals: Piston ring secondaries are used in applications where temperature or low friction considerations preclude the use of elastomeric seals. Proper material selection allows use of piston ring secondaries over a wide temperature spectrum. Secondaries of this type can be constructed of metal, carbon, teflon, ceramic, and other materials.

The seal ring can be made as either a single gapped ring or from multiple segments. Leakage across the piston ring is due to the lack of conformability of the ring to its mating surface or leakage through ring gaps. Multiple segment rings reduce leakage by virtue of better conformity with their mating surfaces, but the additional segment gaps must be sealed to prevent their increasing leakage.

If leakage considerations are secondary, a simple straight gap may be used. In more sophisticated designs, the gaps are sealed by overlapping the leading edge of each segment with the trailing edge of the next. The more effective joint designs are those that have maintained contact across the sealing surface of both the circumference and the transverse sealing faces of the piston ring. This type of gap joint is similar to that on circumferential seal joints as shown in Fig. 25.59.

The initial preload of piston rings of one-piece design sealing on their outer circumference results from compressive forces built into the ring when it is radially

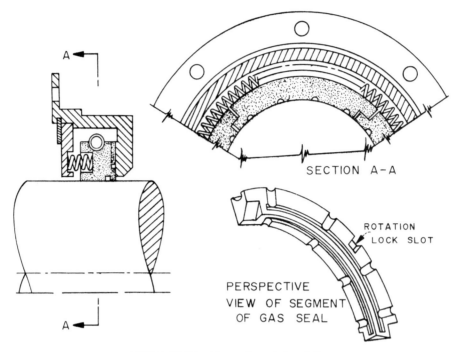

SECTION A-A

ROTATION
LOCK SLOT

PERSPECTIVE
VIEW OF SEGMENT
OF GAS SEAL

FIGURE 25.59   Circumferential seal for gas.

restrained. The ring can be manufactured so that, when it is radially restrained in operation, the radial force exerted by the ring on the bore of the mating liner is equal around the circumference. In operation, the piston ring is subjected to the added force from the pressure differential acting on the piston sealing surfaces. In seal applications requiring very low secondary seal friction, the piston ring is usually substantially pressure relieved, as described in the circumferential seal section below, thereby reducing the load transmitted from the secondary to the primary seal member.

Bellows Secondary Seals: The use of a bellows secondary seal (see Fig. 25.58) affords a five-fold benefit: 1) serves as the holder for the face seal ring, 2) provides a frictionless secondary, 3) furnishes the rotational restraint, 4) supplies axial springing, and 5) avoids secondary seal leakage. The elimination of secondary sliding friction inherent in other types of secondary seals, as well as the elimination of friction from anti-rotation devices, gives bellows seals a marked advantage for certain applications. Metallic bellows are also particularly useful where high temperature or corrosive fluids preclude the use of elastomeric secondaries.

The number, thickness, and shape of *bellows convolutions* determine the spring constant of the bellows. Under pressure, a bellows with a long axial working length may become laterally unstable. Bellows seals will oscillate if subjected to nonuniform seal face loads. *Bellows oscillation*, which may lead to premature fatigue failure, can be controlled through the use of vibration dampeners.

Circumferential Seals. The circumferential seal which is illustrated in Fig. 25.59 is a nonrotating, shaft-rubbing, multisegmented ring made from a suitable carbon grade [Stein (1979)]. The matched segments are individually pin or key locked against rotation and use an extension spring to keep the seal in contact with the rotating shaft. Sealing along the transverse static face is accomplished by using springs which keep the segments in contact with the metal housing face.

In many instances, circumferential seals are operated on replaceable shaft sleeves with carefully finished hard, wear-resistant surfaces. The *runner* (see Fig. 25.59) is an integral part of the seal, and it must be given careful consideration particularly its thermal properties. Seal runners operated at elevated pressures and temperatures must be made from high thermal conductivity materials. Compatability of thermal expansion between runner and shaft should also be considered.

Because circumferential seals are shaft contacting devices, their leakage is a fraction of the flow through shaft clearance seal hardware, such as bushings or labyrinths. The anticipated flow of 100 C air across a single tooth labyrinth seal subjected to a three bar (0.987 atmosphere or 10^5 N/m^2) differential pressure to atmosphere with an operating clearance to its liner of one millimeter per centimeter of sealing diameter, would be 100 times that for a bore-rubbing carbon circumferential seal. Metal type clearance seals have an advantage due to their ability to tolerate high gas temperatures and surface speeds, however carbon circumferential seals are routinely expected to perform in 300 C ambients.

Segmented circumferential seals, because pressure cannot be balanced but only pressure relieved, have a practical pressure limit which varies with the severity of other operating parameters and the degree of heat dissipation. In general, differential gas pressures in excess of four bars are avoided if possible. Liquid circumferential seals are sometimes operated with pressure differentials as high as 10 bar.

Circumferential seals have several advantages. These seals are capable of accepting unlimited axial shaft travel, and, with careful design, can perform despite large variation in sealing diameter. The segmented design allows use in applications where a split seal assembly is essential. These seals are also well suited as very large diameter seals. Segmentation permits the seal to better conform to the required flexibility inherent in large hardware. Because circumferential seals have short axial length and radial height, the seals require relatively little space and are less costly in comparison with more complex face seals.

Circumferential Seals for Liquids: Standard circumferential seals, when used in a liquid environment at moderate and high speeds, form a fluid film across the dynamic sealing surface. A *hydraulic pressure wedge* develops and lifts the segments off the mating part. Increased circumferential springing is usually insufficient to create enough force to overcome the hydraulic wedge, and, in applications in which the seal is subjected to both gas and fluid, the increased circumferential loading in the gas environment causes increased wear.

The hydraulic wedge problem is solvable through the use of pocket bearings in the carbon segment bore which are structured to create a negative pressure differential between the seal bore and the rotating shaft. Refer to Fig. 25.60 for the geometry of this type seal. This negative bearing concept offers a simple method of producing negative rather than positive lift when the seal interface area is subjected to fluid.

Liquid circumferential seals show excellent wear life when sealing oil and can be made with fairly shallow pocket depths without fear of rapid pad wear destroying the negative lift concept. The design of liquid circumferential seals is such that these

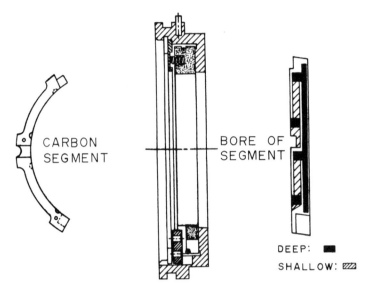

FIGURE 25.60 Circumferential seal for liquids with pocket bearings to create negative pressure differential.

seals can perform when used in applications where periods of totally dry running can occur.

25.5.5 Sealing Systems

Uncomplicated sealing applications usually require the use of only a single seal. Difficult applications involving the sealing of *toxic* or contaminating substances where external leakage is not permissible or which involve stringent leakage requirements when high pressure, temperature, and speed conditions are present may dictate the use of multiple seal system designs. Such systems not only provide positive sealing capability, but often extend reliability to regions well beyond the physical limitations of the sealing devices alone. When the requirements are especially severe, design of the seal system is as important as design of the seals themselves.

Controlling toxic leakage usually necessitates the use of a *buffer fluid* between the toxic substance and the atmosphere. Usually a chamber is created through the use of *back-to-back* seals, and an external buffer fluid pressure, higher than that of the toxic fluid, is maintained in this chamber. Although the buffering fluid may be allowed to leak into the toxic fluid, the toxic fluid itself is excluded from the surrounding atmosphere.

Tandem or back-to-back seals are also frequently used when little or no leakage to the outside is permissible. It is simple to collect the leakage across the inboard high differential seal in a drain space formed between it and an outboard atmospheric side closure. The drainage can then readily be accumulated and returned to the process.

Where the sealed pressure is excessive and it is convenient to employ a particular seal type which has a defined maximum working pressure, a straight forward solution is to *stage* the pressure over multiple seals set in series.

Seal systems which must contend with elevated temperature generally aim at ameliorating the environment by introducing external cooling around the seal or by use of heat shielding, baffling, and insulation blankets. Use of cooling jets which impinge moderate temperature gas or liquid directly onto the seal to remove heat is used when feasible, as is the introduction of an external cooling buffer or flush between two seals. On occasion, the entire seal area is surrounded by a cooling jacket, through which a suitable heat exchange medium is circulated.

25.6 NOZZLES
Reiner Decher

25.6.1 Introduction

A nozzle is a device that converts thermal energy to directed kinetic energy. The high-momentum jet thus created generates a force that may in turn be used to accelerate the vehicle from which the jet is issued. The following paragraphs will describe the thrust performance of two types of nozzles: convergent and convergent-divergent.

Figure 25.61 shows a convergent-divergent nozzle on a propulsion engine. On a convergent nozzle, the area A_e is equal to A_t.

The approach we will take to discuss nozzle performance is known as *gas generator* methodology, which relies on the measurement or knowledge of nozzle-entrance conditions and on the calibration characteristics of the nozzle.

Figure 25.61 is a schematic of either an airbreathing engine or a rocket engine generating supply gas for a nozzle. We will focus our attention on the exhaust flow from the gas generator and seek to determine its thrust given knowledge of the total pressure, p_0, the total temperature, T_0, and the environmental or ambient pressure, p. Because it is important to determine accurately the momentum entering the engine, we will also be interested in the equation for determining exhaust gas mass flow rate.

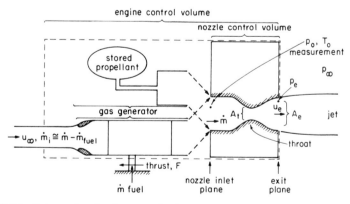

FIGURE 25.61 Schematic of a thrust producing gas generator and its nozzle. Gas generator carrying its stored propellant is a rocket while generator with incoming flow is an air breathing engine. Note control volume definitions.

The parameter that describes the gas dynamics is the *nozzle pressure ratio* (NPR), defined as

$$\text{NPR} = \frac{\text{stagnation pressure of fluid entering nozzle}}{\text{ambient pressure of nozzle discharge environment}} = \frac{p_0}{p_\infty} \quad (25.138)$$

For diatomic gases flowing through the nozzle, an NPR of approximately 2 results in sonic jet velocity. A lower NPR yields subsonic flow, and a higher NPR, supersonic flow. Considerations of conversion efficiency dictate that convergent nozzles are operated at a low pressure ratio, typically less than 3, unless such other advantages as weight or cost are valued, and con-di (con-di is a commonly used term for *convergent-divergent* nozzles) nozzles are operated at a relatively high NPR (3 or more). While it is possible to operate nozzles outside their ideal pressure ratios, it is not common practice and will not be discussed in detail here. Suffice it to say that convergent nozzles operated at a high pressure ratio fail to convert all the available thermal energy to directed kinetic energy, and con-di nozzles operated at a low NPR are accompanied by complex shock systems in the nozzle. The reader is encouraged to refer to analyses available on the behavior of nozzle flows that are describable as piece-wise isentropic and have normal shocks [see Shapiro (1953) and Chap. 8]. Such analyses are usually overly simplistic for describing real nozzle flows (the shocks are usually not planar), but they do illuminate the need to consider shocks in order to properly describe the flow field. Whatever the details of the shocks in the nozzle, their presence is the source of conversion inefficiency. Therefore, because we are primarily interested in practical nozzle designs that operate efficiently, we consider only shockless designs for which total pressure and total temperature are constant. This is, however, an approximation to the extent that there are losses at the nozzle entrance due to drag-producing walls, probes, or flameholders.

Newton's second law of motion applied to a control volume that processes fluid in a steady fashion, i.e., the force acting on the control volume is [Shapiro (1953), Crocco (1958), and Hill and Peterson (1992)]

$$F = \dot{m}u_e + (p_e - p_\infty)A_e \quad (25.139)$$

where $\dot{m}$ = mass flow rate leaving the control volume and mass flow rate entering the control volume, u_e = exit velocity, p_e = exit plane static pressure, p_∞ = external pressure, and A_e = jet cross-sectional area over which p_e acts.

In the discussion to follow, we adopt the view that the nozzle's exit jet is the only momentum-carrying flow. That is, ignore the momentum of the nozzle's incoming flow, which is of no direct interest because it is canceled by an identical term in the momentum balance on the forward portion of the engine. Alternatively, one could design the entrance flow area sufficiently large so that the velocity (and momentum) is negligibly small. With such an accounting scheme we obtain the so-called *nozzle gross thrust*, F. The net thrust on the complete engine is obtained from [Shapiro (1953) and Crocco (1958)]

$$F_\text{net} = F - \dot{m}u_\infty \quad (25.140)$$

where u_∞ is the velocity of the incoming airstream. Here, we neglect the small fuel flow contribution, which, in actuality, distinguishes $\dot{m}$ in Eqs. (25.139) and (25.140) (see Fig. 25.61).

The dominant, and therefore most interesting, term in the gross thrust expression is $\dot{m}u_e$. This is true in the case of subsonic jet flow where $p_e = p_\infty$ and in supersonic jet flow where we will see that the design that gives $p_e = p_\infty$ is advantageous.

25.6.2 Flow Equations

Here, we summarize a few key equations and results from compressible flow for later application to nozzle flows. See Chaps. 1 and 8 for more details.

The *momentum equation* applied to a steady, compressible, inviscid flow can be combined with the definition of a streamline to read (see Sec. 1.2)

$$d\frac{u^2}{2} + \frac{dp}{\rho} = 0 \text{ along a streamline} \qquad (25.141)$$

Because the flow is assumed isentropic (i.e., adiabatic reversible) this may be integrated using $s = \text{constant}$, giving (see Sec. 1.5)

$$d(p\rho^{-\gamma}) = 0 \quad \text{or} \quad p \cdot T^{-\gamma/(\gamma-1)} = \text{constant} \qquad (25.142)$$

Here, the additional approximations of ideal gas ($R = \text{constant}$) and thermally perfect gas ($c_p = \text{constant}$) have been used.

The *energy equation*, for adiabatic flow is

$$h_0 = h + \tfrac{1}{2}u^2 = c_p T_0 = \text{constant} \qquad (25.143)$$

Here, h is the enthalpy of the gas, and the subscript, $_0$, denotes the hypothetical state that could be attained by isentropically bringing the flow to rest, i.e., the *stagnation state*.

From Eq. (25.143) it is apparent that for a given energy content of the gas entering the nozzle (i.e., T_0 or h_0), a large u is obtained for h or T small in the jet. This occurs when the flow is discharged into a low-pressure environment. Recall that the one-dimensional jet exit velocity, u_e, reached is related to the jet exit Mach number, M_e, as

$$u_e = a_e M_e = \sqrt{(\gamma - 1)c_p T_0} \cdot \frac{M_e}{\sqrt{1 + \dfrac{\gamma - 1}{2} M_e^2}} \qquad (25.144)$$

Figure 25.62 shows a plot of $u(M)$; u is proportional to M at low M and asymptotic to $\sqrt{2C_p T_0}$ at large M. The isentropic relation [Eq. (25.142)] and the energy equation [Eq. (25.143)] can be manipulated to yield

$$\left(\frac{p_0}{p_e}\right)^{(\gamma-1)/\gamma} = \frac{T_0}{T_e} = 1 + \frac{\gamma - 1}{2} M_e^2 \qquad (25.145)$$

also shown in Fig. 25.62.

We note that the expansion pressure at the nozzle exit, p_e, is controlled by the

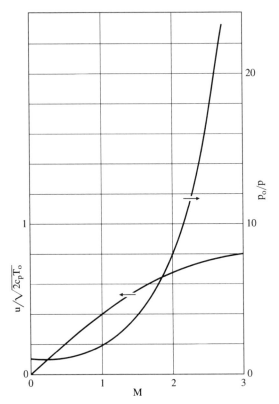

FIGURE 25.62 Adiabatic flow expansion. Relation between velocity and Mach number. Stagnation to static pressure ratio is that required for isentropic flow. Note a Mach 3 jet will realize conversion of 64% of the medium's thermal energy into jet kinetic energy ($\gamma = 1.4$).

flow area of the nozzle and is not necessarily equal to the pressure of the ambient environment into which the jet issues. Thus, p_0/p_e is not the NPR mentioned earlier.

We summarize here with the calculation procedure for thrust, given knowledge of p_∞ and $\dot{m}$.

Known: p_0, p_e, p_∞, T_0, A_e, $\dot{m}$ as well as γ, c_p

$$M_e = \left(\frac{2}{\gamma - 1} \left(\frac{p_0}{p_e} \right)^{(\gamma - 1)/\gamma} - 1 \right)^{1/2} \tag{25.146}$$

$$u_e = \sqrt{2C_p T_0} \sqrt{\frac{\gamma - 1}{2}} M_e \left(1 + \frac{\gamma - 1}{2} M_e^2 \right)^{-1/2} \tag{25.147}$$

$$F = \dot{m} u_e + A_e(p_e - p_\infty) \tag{25.139}$$

The quantities p_0, T_0, p_e, A_e, and $\dot{m}$ are not independent. The mass flow rate, $\dot{m}$, may be written in terms of A_e and thermodynamic variables. Furthermore, the conservation-of-mass statement applied to the nozzle-control volume eliminates the need for another of these variables. Counting p_∞ (and not γ or c_p), four independent measurements are required to determine thrust.

In addition to satisfying the momentum and energy equations, the flow must obey the mass conservation or continuity equation.

The continuity equation for a nozzle (or any duct) reads

$$\dot{m} \equiv \rho u A = \text{constant} \tag{25.148}$$

Chapter 8 showed that for a 1-D flow with area change but without heat addition and friction, the differential change in velocity is given by

$$\frac{du}{u} = \frac{dA}{A} \frac{1}{M^2 - 1} \tag{25.149}$$

It is readily seen that M equals unity at the place in the flow where A is a minimum. This *nozzle throat* is a logical reference area for defining the geometry. There, p and T are known from the isentropic relations with given p_0 and T_0. Defining A^* as the area where $M = 1$ ($= A_t$ in Fig. 25.61, if $M \geq 1$ anywhere in the nozzle) we may write Eq. (25.148) as (see Sec. 4.4)

$$\frac{A}{A^*} = \frac{1}{M}\left\{\frac{2}{\gamma + 1}\left(1 + \frac{\gamma - 1}{2} M^2\right)\right\}^{(\gamma + 1)/(2(\gamma - 1))} \tag{25.150}$$

The relation between p/p_0 ($p_0 =$ fixed) and the local flow M is given by Eq. 25.146. The combination of these equations is plotted in Fig. 25.63. Consider a process where we take the flow entering the nozzle at chosen M with known and constant

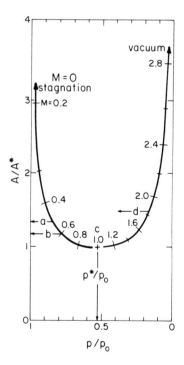

FIGURE 25.63 Mach number-flow area relationship for 1-D isentropic flow.

p_0, T_0 (point "a"). This fixes the mass flow rate, $\dot{m}$. We may accelerate the flow by reducing the pressure of the environment into which the flow discharges. For the incoming flow to be unaltered we require a reduced flow area corresponding to that pressure according to Eqs. (25.145) and (25.150), and shown in Fig. 25.63 as point "b." Thus, an area decrease from "a" to "b" is required to increase M from 0.5 to 0.6, which also requires the ambient pressure to be lower as indicated. When the pressure is lowered to the point where $p = p^* = p_0 \cdot ((\gamma + 1)/2)^{-\gamma/(\gamma - 1)}$ and the area corresponds to A^*, point "c", then the local flow is sonic, $M = 1$. Under these conditions, the area behind the throat and the pressure in the duct, p, can be varied independently, because signals that in subsonic flow could travel upstream and change the mass flow rate are no longer able to do so. This results in a difference between p and p_∞. The difference is supported by an oblique shock wave or a Prandtl–Meyer expansion fan outside the nozzle. Specifically, we have the opportunity of increasing the flow Mach number to supersonic values by increasing the flow area, point "d" in Fig. 25.63. The exit Mach number and corresponding pressure, p_e, can be chosen freely, provided p_∞ is low enough to achieve $M = 1$ at the throat. We speak of flow with sonic velocity at the minimum area as *choked*, while the alternative, *unchoked* flow will have subsonic flow at the minimum area, as well as subsonic flow throughout the nozzle. Thus, a nozzle flow with subsonic starting conditions and a convergent-divergent area variation will follow a state path "a-b-a" if unchoked (relatively high exit pressure), or "a-b-c-d" if choked.

The important consequence of choking the flow is that when $M = 1$ at the throat, the maximum mass flow rate per unit area, ρu, passes through it. This can be seen from the fact that A experiences a minimum with $\rho u A$ constant.

For unchoked, 1-D flow the specified stagnation and ambient pressures, i.e., the NPR, determines the exit flow Mach number. Under such conditions, the exhaust velocity is given by Eq. (25.144). With the density obtained from the isentropic relations [Eq. (25.142)], we may write, for unchoked flow

$$\text{NPR} = \frac{p_0}{p_\infty} = \frac{p_0}{p_e} = \left(1 + \frac{\gamma - 1}{2} M_e^2\right)^{\gamma/(\gamma - 1)}$$

$$\dot{m} = \rho_e u_e A_e = \frac{p_0 A_e}{\sqrt{RT_0}} \sqrt{\gamma}\, M_e \left(1 + \frac{\gamma - 1}{2} M_e^2\right)^{-(\gamma + 1)/(2(\gamma - 1))}$$

or

$$\frac{\dot{m}\sqrt{RT_0}}{p_0 A^*} = \frac{A_e}{A^*} \sqrt{\frac{2\gamma}{\gamma - 1}}\, \text{NPR}^{-(\gamma + 1)/2\gamma} (\text{NPR}^{(\gamma - 1)/\gamma} - 1)^{1/2} \qquad (25.151)$$

For choked flow, the mass flow rate is easily related to the throat conditions where $M = 1$, i.e., $u^* = a^*$.

$$\text{NPR} = p_0/p_\infty \geq p_0/p_e$$

$$\dot{m} = \rho^* u^* A^* = \frac{p_0 A^*}{\sqrt{RT_0}} f(\gamma)$$

or

$$\frac{\dot{m}\sqrt{RT_0}}{p_0 A^*} = f(\gamma) \equiv \sqrt{\gamma}\left(\frac{\gamma + 1}{2}\right)^{-(\gamma + 1)/(2(\gamma - 1))} \tag{25.152}$$

25.6.3 Nozzle Performance Indices

It is a relatively straightforward task to develop expressions giving mass flow rate and thrust for a chosen nozzle geometry. A number of factors, including the presence of nonuniformities (such as arise from nonuniform work or heat input, boundary layers, or wakes), unsteadiness, and variability in fluid properties, among others, contribute to the failure to realize ideal performance. By *ideal performance* we mean that which is calculated by a relatively simple procedure using several idealizing assumptions. We assume a steady, one-dimensional isentropic flow with an ideal, calorically perfect gas; and the geometric flow area is assumed to be the effective flow area. (This last assumption ignores the presence of boundary layers and the fact that constant property surfaces normal to the velocity vector may not be planar.) After deriving the expressions for the ideal mass flow, jet velocity, and thrust for convergent and convergent-divergent nozzles, the actual performance of the nozzle may be compared with the calculated ideal performance.

Mass Flow. A nondimensional measure of mass flow may be defined as

$$\text{Discharge Coefficient, } C_D = \frac{\text{actual mass flow}}{\text{ideal mass flow}} \tag{25.153}$$

The ideal mass flow for the definition of C_D [Eq. (25.153)] is given by Eqs. (25.151) and (25.152).

The *choked convergent nozzle* operates with a one-dimensional Mach number equal to unity at the throat. Thus, a flow-area convergence sufficient to bring the flow conditions to sonic is a means for determining mass flow rate. The details of the nozzle design determine the value of the discharge coefficient, C_D. Figure 25.64 shows two types of nozzles (a) is a standardized and calibrated design that produces nearly 1-D flow, but because of the long contraction, it suffers the presence of boundary layers on the walls. These (usually thin) boundary layers cause a blockage, and C_D is therefore close to unity but Reynolds number dependent. The (throat diameter) Reynolds number where the boundary layer influence is most pronounced ranges between 30 and 3000 [Kuluva and Hosack (1971), Tang and Fenn (1978)]. Design (b) produces such a rapid acceleration that boundary layer thickness plays little part. The dominant determinant of C_D is the radially inward momentum, causing effective area contraction downstream of the geometric throat. In this case, C_D is relatively far from unity, insensitive to Reynolds number and capable of being calibrated [ASME (1959)].

Figure 25.64 also shows the instrumentation for a mass-flow-measuring nozzle. Assuming the chemical composition is known so that R and γ are calculable, one needs total temperature and pressure probes (for T_0 and p_0) as well as a static pressure measurement downstream (p_D) to determine that the NPR is sufficiently large

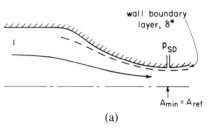

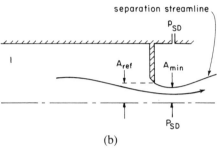

FIGURE 25.64 Nozzles for mass flow measurement. Smoothly convergent nozzle has one-dimensional flow but boundary layer whose displacement thickness δ^* reduces flow area. Sharp edged orifice nozzle performance is less Reynolds number dependent but requires calibration. (a) Smoothly convergent nozzle and (b) sharp edged orifice nozzle.

for choked flow (NPR > 2). The calibration C_D then gives the mass flow as $\dot{m} = C_D \cdot \dot{m}_{\text{ideal}}$ with $\dot{m}_{\text{ideal}}$ given by Eq. (25.152).

Thrust. A *nozzle thrust coefficient* is theoretically defined as the ratio of actual to ideal thrust. In practice, it is often more convenient to define this coefficient to reflect the measurements for thrust. Thrust may be determined from measurements of pressures and temperature alone or from a separate, and possibly more accurate, measurement of mass flow rate and pressures.

Functional expressions for thrust that are dependent only on pressure ratios (and γ) are the following

$$\frac{F}{\dot{m}\sqrt{RT_0}} \quad \text{and} \quad \frac{F}{p_\infty A_e} \tag{25.154}$$

These ratios are referred to as *thrust per unit mass flow* and *thrust per unit area* functions. The thrust per unit mass flow function would be used if one wanted to characterize the performance of a nozzle and use the data on a nozzle of a larger or smaller scale, knowing the mass flow rate in both applications. By contrast, the thrust per unit area function would be used if an accurate mass flow rate were not available so that measurement of pressure, temperature, and area would be required. The nondimensional *thrust coefficient* is defined as

$$C_{FG,W} \equiv (F/\dot{m}\sqrt{RT_0})/(F/\dot{m}\sqrt{RT_0})_{\text{ideal}} \tag{25.155}$$

and

$$C_{FG,A} \equiv (F/p_\infty A_e)/(F/p_\infty A_e)_{\text{ideal}} \tag{25.156}$$

Here, the ideal value of thrust F is given by Eq. (25.139).

We note that the definition of the ideal thrust is arbitrary and depends on the needs of the user. One could define the ideal fictitiously, as a nozzle with a flexible divergent section that always adjusts A_e so that $p_e = p_\infty$. In that case

$$F_{\text{ideal}} = \dot{m}u_e \tag{25.157}$$

The reader is cautioned to note definitions of ideal when interpreting nozzle coefficient data. In this section, we choose to define the ideal as a fixed geometry nozzle with $1-D$, inviscid, adiabatic flow of an ideal, calorically perfect gas.

A nondimensional measure of nozzle-thrust performance often encountered is the so-called *velocity coefficient*, defined as

$$C_v = \frac{\text{Nozzle gross thrust}}{(\dot{m}_{\text{actual}})\,(u_{e,\text{ideal}};\,p_e = p_\infty)} \tag{25.158}$$

and related to the thrust coefficient $C_{FG,W}$ where F_{ideal} is given by Eq. (25.157) (variable A_e nozzle).

25.6.4 Ideal Nozzle Thrust Performance

For simplicity we drop the *ideal* subscript on F for this section.

Convergent Nozzles. The convergent nozzle shown in Fig. 25.65 has the advantages of simplicity and light weight. Its performance is good in the pressure ratio range of 1.0 to 5.0.

Unchoked Convergent Nozzle: The performance of the unchoked convergent nozzle is determined as follows: Given p_0 and the environmental pressure at the exit, one

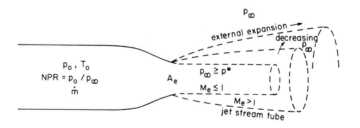

Convergent Nozzle

criterion	unchoked (subcrit) $p_\infty = p_e$			choked: critical $p_\infty = p^*,\ A_e = A^*$	:super critical $p_\infty < p^*;\ A_e = A^*$
M_e	< 1				1.0
$\dfrac{\dot{m}\sqrt{RT_0}}{p_0 A_c}$	$\sqrt{\dfrac{2\gamma}{\gamma-1}}\ \text{NPR}^{-\frac{\gamma+1}{2\gamma}}\,(\text{NPR}^{\frac{\gamma-1}{\gamma}}-1)^{\frac{1}{2}}$				$\sqrt{\gamma}\left(\dfrac{\gamma+1}{2}\right)^{-\frac{\gamma+1}{2(\gamma-1)}}$
$\dfrac{F}{\dot{m}\sqrt{RT_c}}$	$\dfrac{\sqrt{\gamma}\,M_e}{\sqrt{1+(\gamma-1)M_e^2/2}}$		$\sqrt{\dfrac{2\gamma}{\gamma+1}}$		$\sqrt{\dfrac{2}{\gamma+1}}\left[\sqrt{\gamma}+(1-\frac{1}{\pi})\right],\ \pi=\left(\dfrac{\gamma+1}{2}\right)^{\frac{\gamma}{\gamma-1}}\cdot\text{NPR}$
$\dfrac{F}{p_\infty A_e}$	γM_e^2			γ	$(\gamma+1)\pi-1$

FIGURE 25.65 Convergent nozzle and underexpanded external flow stream tubes.

obtains p_0/p, which must be less than $((\gamma + 1)/2)^{\gamma/(\gamma - 1)}$. This subcritical pressure ratio gives M_e by means of Eq. (25.145) or (25.146).

The equation for thrust (with $p_e = p_\infty$) is $F = \dot{m}u_e$, where u_e and $\dot{m}$ are given by Eqs. (25.144) and (25.151), respectively. Thus, for $M_e < 1$

$$\frac{F}{\dot{m}\sqrt{RT_o}} = \sqrt{\gamma}\,\frac{M_e}{\sqrt{1 + \dfrac{\gamma - 1}{2}M_e^2}}, \qquad \left(= \sqrt{\frac{2\gamma}{\gamma + 1}} \text{ if critical, } M_e = 1\right)$$

and

$$\frac{F}{p_\infty A_e} = \gamma M_e^2 \qquad (= \gamma \text{ if critical, } M_e = 1) \tag{25.159}$$

Choked Convergent Nozzle: When the pressure ratio is larger than the critical value, the flow seeks to attain supersonic velocities [see Eq. (25.144) or Fig. 25.62]. At the nozzle exit, however, the flow reaches the sonic condition, and, when the ambient pressure is equal to or lower than p^*, the fact that the flow in the throat is at the local signal-communication speed means that further pressure decreases cannot be communicated upstream. The exit then becomes the area minimum with sonic flow, and further expansion occurs outside the nozzle.

Just at the point where $p_\infty = p^*$, i.e., $M = 1$ at the nozzle exit, the unchoked flow equations can be used to determine thrust. The simple expressions obtained are given parenthetically in Eq. (25.159).

For supercritical flow, $p_\infty < p^*$, the mass flow rate equals the critical value, given by Eq. (25.152). Although further expansion occurs beyond the nozzle exit, u_e is fixed and given by

$$u_e = a^* = \sqrt{\frac{2(\gamma - 1)}{\gamma + 1}}\,(c_p T_o) = \sqrt{\frac{2\gamma}{\gamma + 1}}\,(RT_o) \tag{25.160}$$

from Eq. (25.144). The thrust functions are [from Eqs. (25.139), (25.152), and (25.160)]

$$\frac{F}{\dot{m}\sqrt{RT_o}} = \sqrt{\frac{2}{\gamma + 1}}\left[\sqrt{\gamma} + \left(1 - \frac{1}{\pi}\right)\right] \qquad \pi \geq 1$$

and

$$\frac{F}{p_\infty A_e} = \pi(\gamma + 1) - 1 \qquad\qquad \pi \geq 1 \tag{25.161}$$

Here π is shorthand for the algebraic combination

$$\pi = \left(\frac{\gamma + 1}{2}\right)^{-\gamma/(\gamma - 1)} \cdot \text{NPR} \tag{25.162}$$

which may be interpreted as the actual-to-critical nozzle pressure ratio.

While the subcritical expressions for the thrust functions [Eq. (25.159)] could also be written in terms of π, their algebraic complexity does not justify it, since M_e is more convenient. The table in Fig. 25.65 summarizes these expressions for mass flow and thrust.

Convergent-Divergent Nozzles. If the nozzle pressure ratio, i.e., p_0/p_∞, is greater than that required to reach sonic conditions ($\pi > 1$) at the exit of a convergent nozzle, then expansion to supersonic velocities is possible. In Fig. 25.64 it is apparent that the nozzle area must increase beyond the sonic flow plane. The size of the throat is dictated by the requirement that the incoming flow rate, $\dot{m}$, must be accommodated with its total temperature and pressure as required by Eq. (25.152).

The flow, in following an ever increasing area, must fall in pressure as indicated in the figure. The designer chooses the final area of the flow as it exits the nozzle and thus, in effect, specifies the pressure to which the flow is expanded. If the nozzle exit pressure is greater than the ambient pressure ($p_e > p_\infty$), then one speaks of the nozzle as being *under-expanded*. (The designer has chosen not to add the additional nozzle hardware, perhaps for weight reasons, that would completely expand the flow.) On the other hand, if the pressure at the nozzle exit plane is lower than ambient ($p_e < p_\infty$), the nozzle is *overexpanded*.

The thrust functions [Eq. (25.156)] for the convergent-divergent nozzle are

$$\frac{F}{\dot{m}\sqrt{RT_o}} = \sqrt{\gamma}\, M_e \left(1 + \frac{\gamma - 1}{2} M_e^2\right)^{-1/2} \left\{1 + \frac{1}{\gamma M_e^2}\left(1 - \frac{p_\infty}{p_e}\right)\right\} \quad (25.163a)$$

and

$$\frac{F}{p_\infty A_e} = (1 + \gamma M_e^2)\frac{p_e}{p_\infty} - 1 \quad (25.163b)$$

which equal those obtained for the convergent nozzle in the limit $p_\infty = p_e$ [see Eq. (25.159)]. Here M_e is determined from

$$M_e^2 = \frac{2}{\gamma - 1}\left(\left(\frac{p_0}{p_\infty}\frac{p_\infty}{p_e}\right)^{(\gamma - 1)/\gamma} - 1\right) \quad (25.164)$$

One can show that a nozzle is optimally expanded when $p_e = p_\infty$. The *underexpanded* nozzle fails to realize all the conversion of energy to thrust-producing momentum, while the *overexpanded* nozzle experiences a pressure force, or drag, on the last section, because $p_e < p_\infty$. This may be shown algebraically using Eq. (25.139). For these purposes, we fix the upstream conditions, namely p_0, T_0, and the mass flow rate (equivalently A^*), and vary only p_e relative to p_∞. Because A_e varies with the changing p_e, the appropriate thrust function to use is the thrust-per-unit-mass flow. Eliminating M_e^2 between Eqs. (25.163b) and (25.164), one obtains

$$\frac{F}{\dot{m}\sqrt{RT_o}} = \sqrt{\frac{2\gamma}{\gamma - 1}}\left\{\sqrt{1 - \text{NTR}} + \frac{\gamma - 1}{2\gamma}\frac{\text{NTR}(1 - p_\infty/p_e)}{\sqrt{1 - \text{NTR}}}\right\} \quad (25.165a)$$

where $\text{NTR} = [(1/\text{NPR})\,(p_e/p_\infty)]^{(\gamma - 1)/\gamma}$ which may be interpreted as a nozzle temperature ratio. [Note that a thrust coefficient also used in practice and defined as

F/p_0A^* is directly related to this thrust-per-unit-mass flow function through Eq. (25.152).]

Following Eq. (25.165a), and if the nozzle is properly expanded, $p_e = p_\infty$, the thrust becomes

$$\left(\frac{F}{\dot{m}\sqrt{RT_o}}\right)_{\text{max}} = \sqrt{\frac{2\gamma}{\gamma - 1}}\,[1 - (NPR)^{-(\gamma - 1)/\gamma}]^{1/2} \qquad (25.165b)$$

This result represents a maximum as can be shown by plotting or differentiating Eq. (25.165a). The ratio F/F_{max} gives a measure of the nozzle's thrust efficiency when improperly expanded and may be interpreted as a velocity coefficient for expansion effects. This parameter is plotted in Fig. 25.66.

Figure 25.67 shows the key physical features of nozzle flows with over- and under-expansion. Note particularly the change in flow direction at the nozzle exit. Ultimately, the bounding streamlines of these three flows will be parallel to the jet axis and nearly equal in size. The difference in size is traceable to the shock wave irreversibilities and viscous mixing with the external flow.

The performance equations for the supersonic convergent-divergent nozzle are valid near the design point where $p_e = p_\infty$. The most important limitation encountered in departing from this regime is where the flow is severely overexpanded. Under such circumstances, the compression required to meet ambient pressure is so

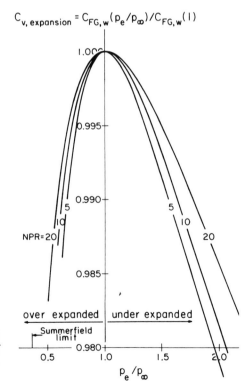

FIGURE 25.66 Under and overexpansion losses in a convergent-divergent nozzle of fixed geometry. Design value of nozzle pressure ratio is indicated ($\gamma = 1.4$).

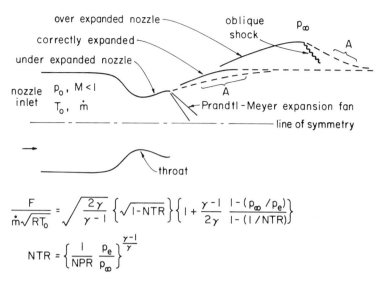

$$\frac{F}{\dot{m}\sqrt{RT_0}} = \sqrt{\frac{2\gamma}{\gamma-1}} \left\{\sqrt{1-NTR}\right\}\left\{1 + \frac{\gamma-1}{2\gamma}\frac{1-(p_\infty/p_e)}{1-(1/NTR)}\right\}$$

$$NTR = \left\{\frac{1}{NPR}\frac{p_e}{p_\infty}\right\}^{\frac{\gamma-1}{\gamma}}$$

FIGURE 25.67 Convergent-divergent nozzle with choked flow. For the cases of under- and over-expansion, the regions (noted "A") of streamline curvature are due to shock and expansion interactions with the jet boundary.

great that the relatively strong oblique shock may separate the boundary layer on the nozzle wall. A detailed numerical analysis may be carried out to determine the behavior of this boundary layer, but often a sufficiently accurate design method involves the use of the *Summerfield criterion* [Summerfield (1959)], which states that the boundary layer will separate at the location where

$$\frac{p_s}{p_\infty} \sim \frac{1}{2.8} \tag{25.166}$$

The flow outside the bounding streamline will recirculate at nearly ambient pressure (see Fig. 25.68). The net result is that the nozzle performs as if the duct beyond the separation point did not exist. This means that the effective exit area, A_e, is the flow area at the separation point.

25.6.5 Nonideal Flow Effects

Boundary Layers. The boundary layer on the external nozzle walls grows due to the frictional retarding force, which may be large because of the roughness of walls resulting from acoustic treatment and because of steps and gaps resulting from assembly. For acoustic liners, experimentally determined values of the friction coefficient are generally used. Steps and gaps are handled as described in Hoerner (1965) if such accuracy is warranted.

Boundary layer build up on the inside of the nozzle reduces the internal flow area and brings about a thrust penalty due to skin friction. The loss of thrust may be calculated from the boundary layer momentum loss as follows. Consider the axisymmetric convergent nozzle shown in Fig. 25.69, and assume fully expanded flow.

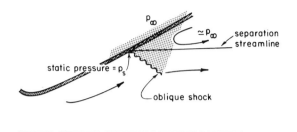

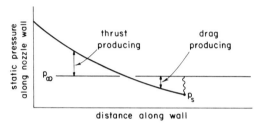

FIGURE 25.68 Pressures on the wall of an overexpanded converging-diverging nozzle with internal flow separation.

The *gross thrust* is the integral of the exit momentum, viz

$$F = 2\pi \int_0^R \rho u^2 r \, dr \simeq \pi \rho_\infty u_\infty^2 (R - \delta)^2 + 2\pi R \int_0^\delta \rho u^2 \, dy \quad (25.167)$$

where the approximation applies for $\delta \ll R$ with the definitions of *displacement thickness*

$$\delta^* = \int_0^\delta \left(1 - \frac{\rho u}{\rho_\infty u_\infty}\right) dy \quad (25.168a)$$

and *momentum thickness* (see Chap. 4)

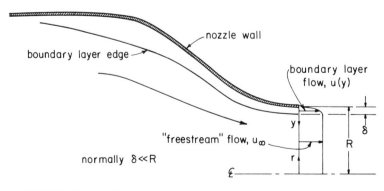

FIGURE 25.69 Nozzle flow with internal boundary layer buildup.

$$\theta = \int_0^\delta \frac{\rho u}{\rho_\infty u_\infty} \left(1 - \frac{u}{u_\infty} \right) dy \qquad (25.168b)$$

we have for F

$$F = \pi R^2 \rho_\infty u_\infty^2 \left(1 - 2\frac{\delta^*}{R} - 2\frac{\theta}{R} \right) \qquad (25.169)$$

The air mass flow rate is

$$\dot{m} = 2\pi \int_0^R \rho u r \, dr \simeq \pi \rho_\infty u_\infty (R - \delta)^2 + 2\pi R \int_0^\delta \rho u \, dy \qquad (25.170)$$

or

$$\dot{m} = (\pi R^2) \rho_\infty u_\infty (1 - 2\delta^*/R) \qquad (25.170a)$$

Then, defining C_D as before [see Eq. (25.153)], we obtain

$$C_D = \frac{\dot{m}}{\dot{m}(\delta = 0)} = 1 - 2\delta^*/R \qquad (25.171)$$

which is the reduced discharge coefficient due to the presence of the boundary layer.

In addition, C_D is found to be sensitive to the flow pressure gradient in the throat as well as the *throat radius of curvature* [Tang and Fenn (1978)]. In the limit of large pressure gradient, it is found that the ratio of boundary layer displacement thickness to throat radius (and therefore discharge coefficient) [see Eq. (25.171)] is well approximated by a functional dependence on a modified Reynolds number defined as $\overline{\mathrm{Re}_D^*} = \mathrm{Re}_D^* \cdot \sqrt{r_t/r_c}$, where subscripts t and c refer to *t*hroat and *c*urvature and * refers to sonic conditions, thus

$$C_D = 1 - 2A(\gamma) \, (\overline{\mathrm{Re}_D^*})^{-1/2} \qquad (25.172)$$

is found to agree well with experimental observation. See Tang and Fenn (1978) and Emanuel (1979) for a discussion of the magnitude of A as well as the role of specific heat ratio, γ.

Similarly, the velocity coefficient [Eq. (25.158)] is

$$C_{V,BL} = \frac{F}{F(\delta = 0)} = 1 - \frac{2\theta}{R} \qquad (25.173)$$

Boundary layer calculations may be useful for estimating δ^* and θ in model nozzles with large wetted areas. However, in real propulsion nozzles, internal steps, leaking, and secondary and cooling flow preclude the application of elementary boundary layer theory.

Flow Spreading at Nozzle Exit. Convergent-divergent nozzles are usually terminated with a diverging cone section. This means that the flow near the wall will have a momentum vector that parallels the wall, with the result that the axial momentum is reduced from the ideal value. A correction for this effect is calculated by determining a ρu^2 weighted average of the cosine of the angle between the streamline and the axial flow direction. Such a flow angularity correction factor requires knowledge (or an estimate) of the local value of ρu^2 as a function of r. Then

$$C_{V,\theta} = \frac{1}{\dot{m}u_{e,u}} \int_0^R \cos\theta \rho u^2 2\pi r \, dr \qquad (25.174)$$

may be determined. Simplifying approximations of ρu^2 constant and a simple function $\theta(r)$ may be adequate for some purposes.

Sonic Line Location and Shape. For reasons of compactness, many nozzle designs operate with chemically reacting, nonadiabatic flows, and all flows are subject to viscous friction at the boundary. If friction is viewed as a volumetric effect [see Shapiro (1952) and Chap. 4 in this handbook], then the velocity-area relationship [Eq. (25.149)], without body forces or heat interaction, becomes

$$\frac{du}{u} = \frac{1}{M^2 - 1}\left(\frac{dA}{A} - \frac{q}{c_p T} - \frac{\gamma \tau A_w}{\rho a^2 A}\right) \qquad (25.175)$$

Here, q is the volumetric heat addition, τ is the wall shear, and A_w is the flow wetted area. This equation implies that a region of positive dA/A (divergent section), where the term in parentheses is zero, is the location where $M = 1$. Thus, in real nozzle flows the sonic line is slightly downstream of the throat.

Most practical nozzles are designed to be short for lessened frictional effect, for light weight, and for minimizing cooling requirements in hot nozzle flows. This results, for a chosen area contraction in the convergent section, in a rapid area variation and to some degree a violation of the assumption of one-dimensionality. The rapid turning around the throat is accompanied by radial pressure gradients and thus nonuniform conditions across the throat plane. The static pressure is, therefore, higher at the nozzle center than near the wall. The flow at the center must expand to lower pressure (further downstream) before $M = 1$ is reached locally. Figure 25.70 shows the sonic line shape for two nozzles that differ in radial pressure gradient at the exit plane. The reference nozzle is clearly more uniform and thus has its sonic line closer to the throat. The sonic line of the 15° conical converging nozzle is behind the geometrical throat, a location that is pressure-ratio dependent near choking.

Experimentally determined sonic line locations for a typical convergent nozzle are shown in Fig. 25.70. These data taken from Thornock (1968) show the variation of sonic line shape and location with nozzle pressure ratio. The nozzle discharge coefficient curves in Fig. 25.71 show that the discharge coefficient becomes constant at the same pressure ratio at which the sonic line position becomes fixed. Note that this pressure ratio for hard choking (i.e., mass flow independent of pressure ratio) does not correspond to the critical pressure ratio where a one-dimensional sonic line spans the nozzle exit. This results from the two-dimensional nature of the flow field. It will be evident from the following discussion that the pressure ratio for choked

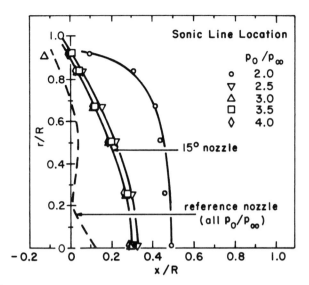

FIGURE 25.70 Sonic line location for two nozzles sketched in Fig. 25.71. Note dependence on nozzle pressure ratio even above the critical value where 1-D theory gives $M = 1$.

flow allows a Prandtl–Meyer expansion angle equal to approximately the nozzle half angle.

Consider the nozzle in Fig. 25.72(a) showing a two-dimensional supercritical, but unchoked, nozzle of half angle α. The sonic line and flow streamlines are shown. From point Q, a centered Prandtl–Meyer fan expands the flow to p_∞. If p_∞ is not low enough (i.e., if the Mach number behind the expansion is not high enough), characteristics from such points on the boundary as A, C, and E intersect the sonic

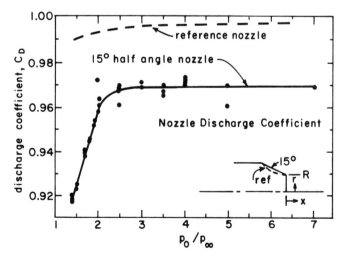

FIGURE 25.71 Mass flow characteristics of two nozzle designs. Hard choking is not achieved until NPR $\simeq$ 3. Data points attest to the achievable repeatability of such tests.

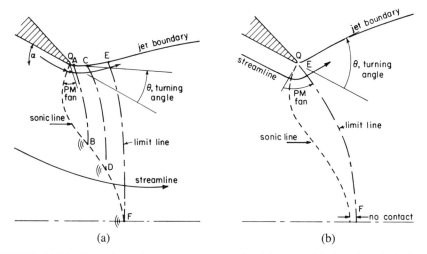

(a)   (b)

FIGURE 25.72   Flow through a convergent nozzle: (a) supercritical and unchoked and (b) supercritical and choked. Limit lines shown are the possible signal communication paths between the external pressure and the flow field ahead of the sonic line. In (b) no pressure changes can be felt by the subsonic flow to alter the mass flow rate.

line at B, D, and F. Thus, the pressure p_∞ is felt upstream of the nozzle, affecting the nozzle mass flow rate. See Chap. 4 for a discussion of Prandtl–Meyer fans and characteristics. The point E is the origin of the last characteristic that intersects the sonic line (at F) and is, therefore, the *limit line*. When this limit line originates from the nozzle lip and fails to intersect the sonic line, the nozzle is choked. This flow configuration is shown in Fig. 25.72(b). Note that the flow area is a minimum at the nozzle exit plane under the choked condition, and the average Mach number at the exit plane is less than unity.

In the unchoked flow case, the identified points are located in the *hodograph plane* in Fig. 25.73(a). The inviscid flow stream line along the wall is initially inclined at an angle α. Around the point Q, from Q to Q', the flow turns through the

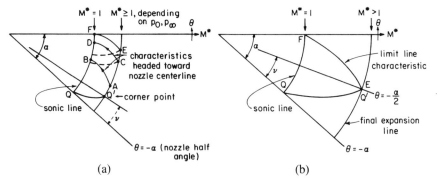

(a)   (b)

FIGURE 25.73   Hodograph representation of the flow fields in Fig. 25.72: (a) supercritical and unchoked and (b) supercritical and choked. Limit lines shown are the possible signal communication paths between the external pressure and the flow field ahead of the sonic line. In (b) no pressure changes can be felt by the subsonic flow to alter the mass flow rate.

Prandtl–Meyer angle, ν, consistent with the pressure ratio. Because the static pressure on the jet boundary is constant and equal to ambient pressure, the velocity on the jet boundary (assuming inviscid flow) is everywhere equal to that at point Q'. Thus, the jet boundary is represented on the hodograph diagram by a circle of radius M^*. The sonic line is represented by the circle $M^* = 1$ in the hodograph plane.

The hodograph plane is shown for the *just choked* flow situation in Fig. 25.73(b). In this case, the limit line EF originates from the point Q, i.e., points E and Q' on the hodograph plane coincide. Thus, the Prandtl–Meyer turning angle, ν, equals $\alpha/2$ as shown, because, for two-dimensional flow, the characteristics in the hodograph plane are independent of the geometry in the physical plane and are identical to or reflections of one another. In axisymmetric flow, the shape of the characteristics depends on the geometry of the flow; the characteristics, therefore, are not symmetrical. Consequently, the choked pressure ratio for an axisymmetric nozzle is slightly higher than that for a two-dimensional nozzle of the same half angle.

Sonic line shape and location for convergent nozzles may be computed for pressure ratios above 2. Using this as a starting line, a supersonic two-dimensional flow analysis may be carried out to give the inviscid nozzle discharge and velocity coefficients by integration of flow properties over the exit plane. Such inviscid discharge and velocity coefficients are shown in Figs. 25.74 and 25.75.

For smooth-walled nozzles, boundary layer contributions to discharge and velocity coefficient losses may be predicted by the above methods. Comparison with experimental data shows that discharge and velocity coefficient predictions are accurate to within 0.5 and 0.2% of measured values, respectively.

Nozzle Flow with Nonuniform Fluid Medium. In a number of applications, for example, aircraft propulsion nozzles where afterburning occurs, the fluid processed by the nozzle is nonuniform in total temperature or total pressure or both. Nozzles are usually short enough that one may consider the adjacent streamtubes of fluid to expand through the nozzle without mixing or viscous interaction. Because the streams are and remain parallel, cross-sectional flow areas are planes of uniform static pressure. This fact allows one to determine the appropriate average total pressure and total temperature so that nozzle thrust and mass flow rate may be calculated.

The key to calculating the choked nozzle performance is the compound nozzle-

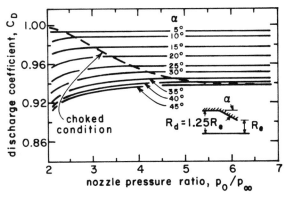

FIGURE 25.74 Variation of choking nozzle pressure ratio for nozzle designs of varying degrees of two-dimensional flow. Influence on nozzle C_D can also be seen.

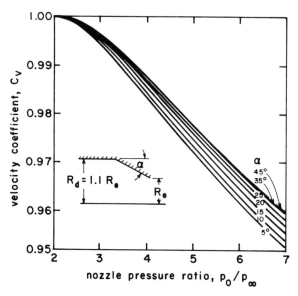

FIGURE 25.75   Calculated velocity coefficients of convergent, conical nozzles without viscous effects.

flow choking criterion given by Decher (1978, 1994, 1995)

$$\int_{A*} \left(\frac{1}{M^{*2}} - 1 \right) \frac{dA*}{\gamma} = 0 \tag{25.176}$$

where M^{*2} is given by

$$\frac{2}{\gamma - 1} \left[\left(\frac{p_0(A)}{p*} \right)^{\gamma - 1/\gamma} - 1 \right] \tag{25.177}$$

Here, $p*$ is the throat static pressure, and the nonuniform total pressure $p_0(A)$ is known at some upstream condition. The continuity condition, Eq. (25.150), is required to relate A to $A*$ and thus combine these two statements. Bernstein *et al.* (1967) develop this condition, as well as an expression for nozzle thrust, given nonuniformities in total pressure, total temperature, and static exit pressure.

Nozzle Flows with Nonperfect Gases.   The temperature change undergone by a typical fluid element in a nozzle flow is often very slight so that specific heat may be assumed constant. When the expansion is more extreme, such as in rocket engines or when the end state in the nozzle is close to the saturation line of the gas, one or both of the assumptions of perfect (c_p = constant) or ideal (R = constant) gas may break down. The problem is then to find the exit velocity from a nozzle without resorting to the use of $\gamma = c_p/(c_p - R)$. The utilization of γ is relevant to a flow problem only when it is constant and the isentropic relation

$$\frac{p_2}{p_1} = \left(\frac{T_2}{T_1} \right)^{\gamma/\gamma - 1} \tag{25.178}$$

can be obtained by straightforward integration of the combined first and second laws of thermodynamics. Prior to the introduction of γ, the combination of these laws reads

$$ds = c_p \frac{dT}{T} - R \frac{dp}{p} \tag{25.179}$$

At a temperature sufficiently low that chemical reactions (such as dissociation) are negligible, a gas remains constant in composition even though the temperature may vary. Thus, the molecular weight and the gas constant, R, also remain constant. For many nozzle flows, it is indeed realistic to assume R constant even though C_p may vary considerably.

For a single component gas, Eq. (25.179) may be integrated to arrive at an isentropic process of such a gas from state 1 to state 2

$$\ln \frac{p_2}{p_1} = \frac{1}{R} \int_1^2 c_p \frac{dT}{T} \tag{25.180}$$

The integral from state $T = 0$ to T is listed in thermochemical tables (such as JANNAF) as $S^0(T)$, and Eq. (25.180) allows the determination of T_2 given p_2/p_1, T_1, and the specific heat variation of the gas.

Example 25.5: Consider CO_2 at stagnation conditions of 1000 K and 10 atm and assume the gas is expanded by a convergent-divergent nozzle to 1 atm. We might first ask what are the jet velocities obtained using perfect gas and real gas assumptions? For the perfect gas, we need guess an appropriate γ. Say we choose a value of 1.181 [the value at 1000 K from tables in Stull and Prophet (1971)], we obtain using Eq. (25.178)

$$T_e = 1000 \cdot 10^{-(0.181/1.181)} = 703 \text{ K}$$

Using the value of γ at 700 K, we would have obtained 678 K. The correct temperature is obtained from Eq. (25.180).

$$S^0(T) = S^0(1000) - R \ln (p_1/p_2)$$

$$S^0(T) = 64.344 - 1.988 \ln 10 = 59.776$$

Interpolating linearly in the tables between 600 and 700 K gives $T_e = 692$ K.

For this nonperfect gas, the exhaust velocity reached is given by Eq. (25.143)

$$u^2 = 2[h(1000) - h(692)]$$

with h given in energy per unit mass

$$u = \sqrt{2(44.011)^{-1} [7.984 - 4.152]1000(4184)} = 854 \text{ m/sec. (exact)}$$

For the case of the assumed constant c_p, i.e., γ, there are many ways of getting the approximate answer. For example, from Eq. (25.145) we obtain

$$M_e = \sqrt{\frac{2}{\gamma - 1}} \left\{ \left(\frac{p_0}{p}\right)^{\gamma - 1/\gamma} - 1 \right\}^{1/2} = 2.162$$

$$a_e = \sqrt{\gamma R T_e} = \sqrt{\frac{1.181 \cdot 1.988 \cdot 4184 \cdot 703}{44.011}} = 396.1 \text{ m/sec}$$

$$u_e = M_e a_e = 856.4 \text{ m/sec}$$

In this example, we obtained the temperature to a 1% accuracy and the exhaust velocity to within 0.25%. These are not, however, the accuracy levels for such calculations in general, rather for this simple example the constant c_p and γ assumptions are fairly good.

The method for determining A_e/A^* should also be mentioned. It is quite clear that because $M_e > 1$ in the example, a convergent-divergent nozzle is required. The throat is that point where $M = 1$. In the preceding paragraphs we outlined a method to determine u, so that all that is required is a way of finding the value of the local speed of sound. Equation (25.142) can be manipulated together with the state equation to give $a^2 = (\partial p/\partial \rho)_s = (c_p/(c_p - R))RT$ which is valid for nonconstant c_p (but $R = $ constant). A numerical procedure for determining the area $- M$ relation for an imperfect gas described by thermochemical tables is summarized below:

1) Given: p_0, T_0
2) assume: p
3) determine   a) p_0/p
                b) T using $s^0(T) - R \ln p_0/p$ (entropy chart)
                c) $u = \sqrt{2[h(T_0) - h(T)]}$ (enthalpy chart)
                d) $a = \sqrt{[c_p/(c_p - R)]RT}$ (c_p chart)
                e) $M = u/a$
                f) $\rho = p/RT$
                g) $A = 1/\rho u$
4) when $M = 1$; $A = A^*$
5) and, finally, form A/A^* vs M

Analysis of Flows With Chemical Reaction. Here, we examine the salient features of flows through nozzles where simultaneous chemical composition changes are important by virtue of the temperature changes encountered. If these temperature changes are sizable, then we are most probably concerned with convergent-divergent nozzles operating at relatively large nozzle-pressure ratios (rocket engines, for example).

Typically, the gas generation process is the high-temperature oxidation of a fuel. We assume that we know the conditions of the fluid entering the nozzle and proceed to calculate the exhaust velocity u_e given the NPR, which is a measure of thrust assuming ideal expansion. Hill and Peterson (1992) discuss a procedure to calculate stagnation temperature given reactants and their entry conditions. We assume T_0 is known. We assume further that p_0 is known, because it can be fixed by controlling propellant feed rate and nozzle throat area [see Eq. (25.152)].

The determination of exhaust velocity requires calculation of the change in enthalpy per unit mass of the gas [viz. Eq. (25.143)]. The enthalpy of a gas mixture is obtained from

$$\sum_i [H_i(T) - H_{i,298}]x_i = [H_m(T) - H_{m,298}] \tag{25.181}$$

where $H_i(T) - H_{i,298}$ is the tabulated enthalpy per mole referenced to $T = 298$ K for component, i. The mole fraction of component i is indicated by X_i, and the subscript m refers to the mixture. The molecular weight of the mixture, W_m is obtained from

$$W_m = \sum X_i W_i \tag{25.182}$$

From Eq. (25.143), it is apparent that h must be known up- and down-stream of the nozzle and the difference is $u_e^2/2$, assuming the velocity upstream (in the plenum) is negligibly small. Generally, one can calculate the state of the gas in the plenum assuming local chemical equilibrium (see Chap. 7).

The falling temperature and pressure of each fluid element as it proceeds to the nozzle exit dictate that the composition must change, provided that the interacting atoms meet often enough. Initially, when the flow is subsonic, the gas density is usually high enough for this to be true. As the flow proceeds past the throat to supersonic velocities, the density falls and the mean random speed falls so that the collisional communication between reacting molecules becomes more difficult. At sufficiently high Mach number, the reaction rate becomes so slow compared to the time scale of interest (the nozzle fluid transit time), that composition is effectively constant. Such flow is said to be (chemically) *frozen*. Alternatively, a continuously reacting fluid is referred to as being an *equilibrium* flow (shifting equilibrium). Real nozzle flows originate as equilibrium flows but eventually freeze if the expansion is to a sufficiently low pressure.

Both *frozen-* and *equilibrium-flow* limits may be treated as isentropic flow. The first, because of no heat flow from the chemical energy reservoir is involved, and the second, because the heat transfer is reversible. Thus these two cases are easily calculated as performance limits for nozzle flows.

Frozen Flow: The case of frozen flow for a gas mixture is a straightforward extension of the discussion on the calorically imperfect, ideal, single-component gas described earlier.

Given entrance conditions p, T, composition

1) calculate mixture enthalpy per mole from Eq. (25.181)

$$(H - H_{298})_m = \sum_i X_i(H - H_{298})_i$$

2) molecular weight from Eq. (25.182)

$$W_m = \sum X_i W_i$$

3) mixture enthalpy per mass from $(h - h_{298})_m = (H - H_{298})_m / W_m$

4) mixture entropy per mole from $S_m^o = \Sigma X_i S_i^o$

5) mixture entropy per mass from $s_m^o = S_m^o / W_m$

6) gas constant of the mixture from $R = R_u / W_m$

Equation (25.180) can be rewritten to state that for an isentropic process between states 1 and 2

$$s_1^o - s_2^o = R \ln p_2/p_1 \qquad (25.183)$$

where all but s_2^o are known. (Note that division of entropy and gas constant by W_m is superfluous for the frozen flow case, but maintained for consistency with the equilibrium case that follows.) Solving for s_2^o gives the expansion temperature T_2 and the mixture enthalpy per mass at state two. The exhaust velocity follows from Eq. (25.143).

Equilibrium Flow: The equilibrium flow calculation is more complex, because the composition is not known at T_2. The procedure is therefore iterative. Given the same information and the results of calculations as above leading to s_2^o, we assume T_2 and know p_2; the laws of mass action and species conservation then give x_i. This allows the calculation of a trial value for s_2^o, which will not equal that required by Eq. (25.183). A second trial establishes the sensitivity to choice of guessed T_2, and rapid convergence follows. The nozzle exhaust velocity is then obtained by the procedure described above for frozen flow.

Note that the conservation-of-energy statement, Eq. (25.143), that gives u_e involves the enthalpy per unit mass, not per mole (although the latter is more convenient in formulating thermochemical tables), hence the importance of calculating molecular weight, particularly for the equilibrium flow calculation. In addition, it is worthwhile to remember that entropy per unit mass (not per mole) is conserved in an isentropic process, because the number of moles in a unit mass may change.

The *CRC Handbook of Tables for Applied Engineering* (1970) provides sample calculation results that illustrate the impact of chemical reactions on thrust performance of a rocket nozzle. Various combinations of fuel and oxidizer are assumed to react, and the results are presented. Numerical calculations are given for cases of both frozen flow and equilibrium flow. The resulting exhaust velocities differ from 1 to 5%.

Many descriptions of nozzle flow with finite-rate chemical process are available. Discussion of that work is beyond the scope of this article.

Flow codes which embody the processes discussed in this article as well as extensions thereof are available from *COSMIC* (1980).

25.7   ROCKET PROPULSION SYSTEMS
Stephen Heister

Rocket propulsion systems are devices which must provide high thrust-to-weight ratios in order to overcome the acceleration due to gravity or to insure high agility. It is not uncommon for rocket propulsion systems to develop thrust levels of 50–75 times the weight of the engine itself. Rockets utilize some of the highest combustion intensities and mass fluxes of any fluid machinery device made by man.

In this section, we will first describe the most common types of rocket propulsion systems in use today. We will then investigate the theory of operation of these devices under ideal conditions followed by a brief description of various types of rocket nozzles. Deviations from ideal performance are then discussed. The next part gives a brief overview on rocket propellants. The section ends with a discussion of performance prediction methodologies.

25.7.1 Classification of Rocket Propulsion Systems

There are two fundamental parameters by which we characterize the relative merits of various rocket propulsion systems. Probably the best known parameter used to describe the performance of rocket propulsion systems is its *specific impulse* or *Isp*. A rocket's specific impulse is simply the current thrust, F, of the rocket, divided by the massflow (generally regarded as a weightflow) of propellant gases currently exiting the nozzle

$$I_{sp} = \frac{F}{\dot{m}g} \tag{25.184}$$

where g is the acceleration due to gravity which is required to convert massflow, $\dot{m}$, to weight-flow. Using this definition, *Isp* values are typically reported in units of seconds. Systems with high specific impulse are efficient, because only a small massflow is required to obtain the desired thrust level.

Recognizing the importance of mass characteristics in aerospace propulsion systems, a measure of structural efficiency is also useful to consider. This parameter, called the *propellant mass fraction*, λ, reflects the mass of useful propellant, m_p, as a percentage (or fraction) of the overall propulsion system, m_o, mass

$$\lambda = m_p/m_o \tag{25.185}$$

where m_o is the overall mass of the loaded propulsion system. We desire high λ values so as to minimize the mass of structural components within the system.

Figure 25.76 highlights one of the more common types of rocket propulsion systems in use today—the solid rocket motor or SRM [Anon. (1971), (1972), (1987),

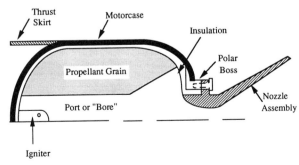

FIGURE 25.76 Major components of a typical solid rocket motor (SRM).

(1991)]. A SRM is conceptually a very simple device with few moving parts. The entire propellant charge, known as a *propellant grain* is contained within the pressure vessel generally called the *motor case*. The case is protected from hot combustion gases by internal insulation. The nozzle serves to convert the thermal energy of the combustion gases to thrust. The nozzle may in some cases be vectorable in order to provide vehicle control. The *polar boss* provides a means of connecting the nozzle assembly to the motor case, while the *thrust skirt* provides a mechanism to transmit thrust loads to an upper stage or payload.

The performance of current solid propellant/motor combinations leads to vaccuum *Isp* values between 150 and 300 seconds with values approaching 350 seconds for exotic fuels. While this *Isp* does not compare favorably with liquid systems, the high mass fraction of SRM systems ($\lambda > 0.9$ in many systems), due to simplicity and high fuel density, does make these devices attractive for many applications.

Figure 25.77 highlights the features of a typical liquid rocket engine (LRE) system using a *gas generator* as a means to provide hot gases to a turbine which drives both fuel and oxidizer pumps. Note that there are several other types of power generation schemes which have been utilized in liquid engines of this type [Sutton (1993) and Huzel and Huang (1971)]. In the gas generator cycle shown in Fig. 25.77, the

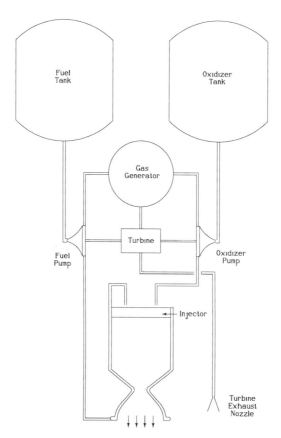

FIGURE 25.77 Components of a liquid rocket engine (LRE) using a gas generator power cycle.

turbine exhaust pressure is not high enough to permit expansion of these gases through the main nozzle. For this reason, a separate turbine exhaust nozzle is required. Another alternative would involve dumping the turbine exhaust into the main nozzle at some point downstream of the throat.

Liquid propellant combinations requiring fuel and oxidizer stored in separate tanks are called *bipropellant systems* as in Fig. 25.77. Some propellants (such as hydrazine, N_2H_4) can decompose to high temperature gases upon contact with a suitable catalyst material. These types of propellants are termed *monopropellants* [Altman *et al.* (1960)]. Bipropellant engines require an *injector* which is made up of a series of fuel and oxidizer orifices responsible for metering and atomizing the propellants as they enter the chamber. We also require a *feed system* to force the liquid propellants into the chamber. In Fig. 25.77, a pump fed system is depicted; another alternative would involve using an inert gas stored in a separate bottle to pressurize propellant tanks and force propellants into the chamber. This *pressure fed* arrangement is generally restricted to smaller systems, since an enormous amount of pressurant gas would be required for systems with large propellant tanks.

Bipropellant systems typically perform in the range $300 < Isp < 460$ sec, while monopropellants generally operate in the $150 < Isp < 220$ sec range. The relatively simple monopropellant systems have applications in small LREs used as satellite propulsion, while the larger LREs are generally bipropellant systems. Finally, we note that while the performance of bipropellant liquid systems is significantly higher than solid rocket motors, the propellant mass fraction is usually markedly lower ($\lambda < 0.9$ in many cases), since these devices are more complex than SRM's.

Figure 25.78 highlights the features of a rocket which combines the SRM and LRE ideas into a single system, the hybrid rocket. A gas generator using a separate tank of liquid fuel is assumed to provide power to the turbine in this popular design concept. The fuel is cast in solid form into a chamber which will serve as the combustion chamber. The liquid oxidizer is then injected at either the head or aft end of the chamber and reacted with the fuel. Since the engine thrust is proportional to the oxidizer flowrate, thrust can be modulated using a valve on the oxidizer. The hybrid engine is inherently safer than a SRM, since the oxidizer and fuel are stored separately until engine operation. The device is simpler than bipropellant liquid systems, since all the fuel is stored in the chamber. Performance and propellant mass fractions for hybrid systems lie somewhere between the SRM and LRE extremes. Typical specific impulse values lie near 300 seconds, while mass fractions are higher than most liquid systems and lower than most solid systems.

Nuclear rockets can be separated into two general categories. In a *nuclear thermal rocket*, energy from a nuclear process is used to heat a working fluid (usually hydrogen) to high temperatures, and the hot gases are then expanded through a typical Delaval nozzle. A schematic of this type of rocket is shown in Fig. 25.79. The *nuclear electric rocket*, utilizes electricity generated from a nuclear powerplant to generate thrust through electric rocket principles as discussed below. Nuclear thermal rockets have demonstrated *Isp* values in excess of 900 sec, which is much higher than either solid or liquid systems [Anon. (1990)]. However, the mass fraction of the nuclear devices can be quite low, since heavy shielding is required to avoid high amounts of radiation from the reactor core. In spite of this, nuclear thermal rockets are attractive alternatives for interplanetary missions where high *Isp* is a necessity. While nuclear rockets have been ground-tested, none have ever flown on vehicles in actual missions.

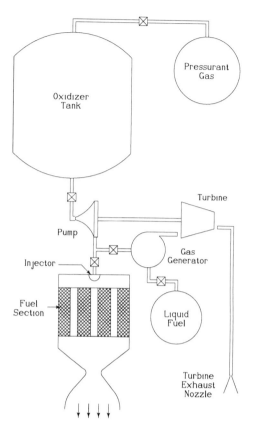

FIGURE 25.78 Typical design of a hybrid rocket engine.

Figure 25.80 highlights the two most common electric propulsion concepts. In an *arcjet* or *electrothermal thruster*, an electric potential is *arced* across a gap between two electrodes much like the operation of a spark plug in your car. Fluid passing through the arc is superheated to extremely high temperatures and expanded through a nozzle. Typical *Isp* values for the electrothermal devices lie between 400 and 1500 seconds. Hydrogen, ammonia, and helium have all been used successfully as working fluids for these devices. Currently, arcjet thrusters are being designed for use aboard communication satellites.

In *electrostatic*, or *ion thrusters*, heavy atoms (usually mercury or xenon) are ionized via electric heating coils, and the resulting positively charged ions are accelerated using an electric potential to very high velocities. The ions are neutralized by recombination with an electron beam at the exit of the thruster. Specific impulse values for these engines are measured in the thousands of seconds, with values of 10,000 seconds attainable under certain conditions.

While the specific impulse of the electric devices is certainly attractive, these devices utilize large amounts of electric power to generate very small thrust values. For example, thrust levels of these devices are usually measured in millipounds, with power requirements in the kilowatt to megawatt range. In spite of the low thrust, these devices are attractive for satellite maneuvering and possibly interplan-

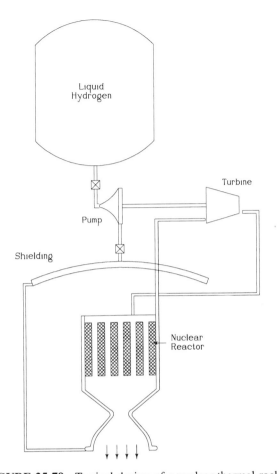

FIGURE 25.79   Typical design of a nuclear thermal rocket.

etary travel applications. Use of a nuclear generator has been proposed to supply the large amounts of power required by these types of thrusters.

25.7.2 Theory of Operation of Ideal Rockets

An upper bound on the performance of a rocket propulsion system can be obtained by assuming steady, one-dimensional isentropic flow through the nozzle. Further, we will assume that the gases traveling through the nozzle are of a fixed composition, i.e., chemical reactions are neglected. Under this assumption, the gas molecular weight, W, and ratio of specific heats, γ, remain constant throughout the nozzle. Since rocket systems typically operate at high chamber pressures, p_c, the Delaval nozzle used is choked except during the early stages of ignition and the latter stages of engine shutdown. Finally, since the gas velocities in the chamber are typically small, we generally assume that chamber conditions (p_c, T_c) are equivalent to stagnation conditions.

Consider a rocket engine mounted on a test stand as indicated in Fig. 25.81. We can use a simple, one-dimensional momentum balance on a control volume which

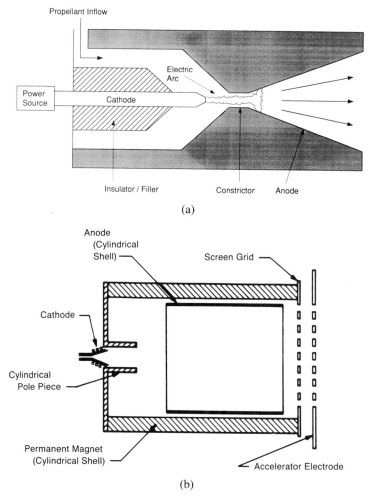

FIGURE 25.80   Designs of arcjet (a) and ion-type (b) electric rockets.

cuts through the test stand and through the exit plane of the nozzle to determine the rocket's thrust. Under this assumption, the only external forces on the control volume are the negative of the thrust (the restraining force from the stand) and the net pressure force acting over the nozzle exit area, A_e. In addition, the only momentum flow, $\dot{m}v_e$, through the selected control surfaces occurs at the nozzle exit plane. For a steady flow, the momentum must balance the sum of the forces on the control volume. Therefore, the thrust of the engine may be written

$$F = \dot{m}u_e + (p_e - p_a)A_e \qquad (25.186)$$

where u_e is the nozzle exit velocity, and p_a and p_e are the ambient and nozzle exit pressures, respectively. The first term on the right-hand side of this equation is the *jet thrust*, while the second term is the *pressure thrust*. Thrust is dependent on altitude due to the presence of p_a, but the relation indicates that thrust is independent

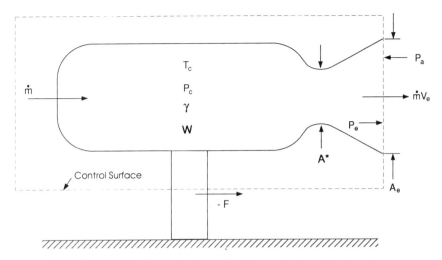

FIGURE 25.81 Control volume for determination of rocket thrust.

of flight speed. For launch vehicles, the pressure thrust can change a large amount (from negative to positive in sign) over the firing duration due to large atmospheric pressure changes. (Note that the thrust is a maximum in a vacuum, where $p_a = 0$.)

If we divide Eq. (25.186) by $\dot{m}g$, we can write the Isp as

$$Isp = \frac{F}{\dot{m}g} = u_e/g + \frac{A_e}{\dot{m}g}(p_e - p_a) \tag{25.187}$$

so we can see that the Isp is related to the exhaust velocity of the engine and that high exhaust velocities imply high performance.

To express the nozzle exit velocity in terms of chamber conditions, begin with the first law of thermodynamics assuming an adiabatic flow with only $p - dV$ work. In this case, the internal energy and $p - dV$ work are measured by the specific enthalpy h

$$dh + u \, du = 0 \tag{25.188}$$

This equation can be integrated between states c and e in Fig. 25.81 to give

$$h_e - h_c = \tfrac{1}{2}(u_c^2 - u_e^2) \tag{25.189}$$

The flow within the chamber is at very low Mach numbers so that p_c and T_c can be assumed to be stagnation conditions, and $u_c \approx 0$. Under these conditions

$$u_e = \sqrt{2(h_c - h_e)} \tag{25.190}$$

Now if assuming perfect gases, we can write $h = c_p T = 2\gamma R T/(\gamma - 1)$, so that

$$u_e^2 = \frac{2\gamma R_u T_c}{W(\gamma - 1)}(1 - T_e/T_c) \tag{25.191}$$

where R_u is the universal gas constant and W is the molecular weight of the mixture of gases within the chamber. And finally, if we write the temperature ratio in terms of the pressure, ratio using isentropic flow relations

$$u_e^2 = \frac{2\gamma R_u T_c}{W(\gamma - 1)} [1 - (p_e/p_c)^{(\gamma - 1)/\gamma}] \tag{25.192}$$

which gives the ideal exit velocity for a rocket nozzle.

The quantity p_e/p_c is the ratio of the static pressure to stagnation pressure at the nozzle exit. Isentropic flow relations give

$$p_e/p_c = \left(1 + \frac{\gamma - 1}{2} M_e^2\right)^{-\gamma/(\gamma - 1)} \tag{25.193}$$

where M_e is the Mach number at the exit plane. However, this quantity is an implicit function of the *nozzle expansion area ratio*, $\epsilon = A_e/A^*$. By presuming a choked flow and that massflow is constant through the nozzle, this relationship is given by

$$A/A^* = \epsilon = \frac{1}{M_e} \left[\frac{2 + (\gamma - 1)M_e^2}{(\gamma + 1)}\right]^{(\gamma + 1)/(2(\gamma - 1))} \tag{25.194}$$

which sets the nozzle exit Mach number, M_e, for a given expansion ratio. Therefore, we can presume that the quantity p_e/p_c is a function of ϵ alone for a given choked nozzle flow.

For steady flow, the nozzle massflow rate can be written

$$\dot{m} = \rho u A = \frac{\gamma p M A}{a} \tag{25.195}$$

where a is the local speed of sound in the flow. Concentrate on the massflow through the throat. Since the massflow is constant at all points in the nozzle and since the Mach number is unity for choked flow we may write

$$\dot{m} = \frac{\gamma p_0 A^*}{a_t} \tag{25.196}$$

Now, we can write the throat pressure and sonic velocity

$$p_c/p_0 = \left(1 + \frac{\gamma - 1}{2} 1^2\right)^{\gamma/(\gamma - 1)} = \left[\frac{\gamma + 1}{2}\right]^{\gamma/(\gamma - 1)} \tag{25.197}$$

$$(a_c/a_t)^2 = T_c/T_0 = \frac{\gamma + 1}{2} \tag{25.198}$$

Substituting Eqs. (25.197), (25.198), and (25.196) to write the massflow in terms of chamber conditions and nozzle throat area

$$\dot{m} = p_c A^* \sqrt{\frac{\gamma W}{R_u T_c}} \, [2/(\gamma + 1)]^{(\gamma + 1)/(2(\gamma - 1))} \qquad (25.199)$$

The quantity involving γ and variables under the radical is like the inverse of a velocity. Define this quantity as c^*, and we may write

$$c^* = \frac{p_c A^*}{\dot{m}} = \sqrt{\frac{R_u T_c}{\gamma W}} \, [2/(\gamma + 1)]^{-(\gamma + 1)/(2(\gamma - 1))} \qquad (25.200)$$

This quantity is known as the *characteristic velocity* for a rocket. Note that $c^* = c^*(\gamma, T_c, W)$, so that the characteristic velocity is a function of the propellant combination only. Therefore, c^* is a measure of the energy available in the propellant combination. From the definition of c^* we can write the massflow

$$\dot{m} = p_c A^*/c^* \qquad (25.201)$$

so that high c^* value reduces the massflow required to sustain a pressure p_c with the throat area A^*. For this reason, propellants with high characteristic velocity are desirable.

Finally, note from Eq. (25.200) that c^* is independent of pressure. In many analyses, we will assume this is true to permit integration of the governing equations. In actuality, there is a weak pressure dependence, since the chamber composition (and hence W, γ and T_c) do vary a small amount with pressure due to recombination of radical species at higher pressures. Therefore, c^* will tend to increase slightly with pressure. Typical c^* values are in the range of 4800–5000 f/s for solid propellants, while most liquid propellants lie in the range of 6000–8000 f/s.

Thrust Coefficient. Substituting Eqs. (25.192) and (25.199) into Eq. (25.186), we obtain

$$F = p_c A^* \left[\frac{2\gamma^2}{\gamma - 1} \left(\frac{2}{\gamma + 1} \right)^{(\gamma + 1)/(\gamma - 1)} (1 - (p_e/p_c)^{(\gamma - 1)/\gamma}) \right]^{1/2} + (p_e - p_a)A_e$$

$$(25.202)$$

which indicates that $F \propto p_c A^*$. We can introduce a dimensionless parameter called *thrust coefficient*, C_f, by dividing Eq. (25.202) by $p_c A^*$

$$C_f = \frac{F}{p_c A^*} = \left[\frac{2\gamma^2}{\gamma - 1} \left(\frac{2}{\gamma + 1} \right)^{(\gamma + 1)/(\gamma - 1)} (1 - (p_e/p_c)^{(\gamma - 1)/\gamma}) \right]^{1/2}$$

$$+ (p_e/p_c - p_a/p_c)\epsilon \qquad (25.203)$$

We can see from Eq. (25.203) that $C_f = C_f(\gamma, \epsilon, p_e/p_c, p_a/p_c)$, and, since we had noted that p_e/p_c was a function of ϵ, C_f is really a measure of nozzle performance for a given ambient condition.

1. $p_e > p_a$; This is the case for an *underexpanded* nozzle, where we obtain positive pressure thrust.

2. $p_e < p_a$; This is the case for an *overexpanded* nozzle, where we obtain negative pressure thrust.

3. $p_e = p_a$; This is the case for a *perfectly expanded* nozzle where we obtain no pressure thrust.

All three of these situations can occur in rocket propulsion nozzles due to the fact that the ambient pressure and the nozzle exit pressure may both vary appreciably over the course of the firing. If the flow is highly overexpanded, the potential exists for boundary layer separation within the nozzle. The separation can occur over just a portion of the circumference (asymmetric separation), or uniformly around the entire circumference. If the flow is mildly overexpanded, the fluid has significant energy to overcome the adverse pressure gradient, and the nozzle flows full. For launch vehicle applications, nozzles are typically designed to be near the separation point at ignition so that improved performance is obtained at higher altitudes where ambient pressure is lower.

Predicting the separation limit *a priori* is a difficult task, since this analysis requires knowledge of the behavior of turbulent boundary layers in adverse pressure gradients. For this reason, nozzle designers typically resort to empirical correlations [Kalt and Bendall (1965), Morisette and Goldberg (1978), and Schmucker (1984)] to predict separation pressures. Assuming the separation pressure is roughly 30% of the ambient pressure is a reasonable estimate for preliminary calculations. For this condition, we predict separation will occur at nozzle exit pressures below about 5 psi at sea level.

We can plot the behavior of the nozzle thrust coefficient by selecting a γ value and a given p_c/p_a ratio. A typical thrust coefficient map derived using this approach is shown in Fig. 25.82 for $\gamma = 1.2$ (a typical value for rocket nozzles). One can

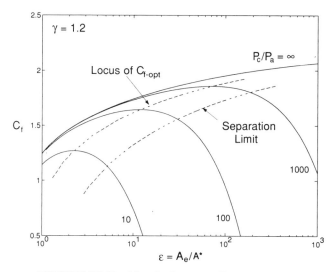

FIGURE 25.82   Nozzle thrust coefficient behavior.

prove that C_f is maximized when the nozzle is perfectly expanded; the locus of C_{f-max} values corresponds to this condition. Assuming the nozzle will separate at pressures below 30% of ambient leads to the nozzle separation line shown in the figure. Finally, if we operate in a vacuum, $p_c/p_a = \infty$ which is the uppermost line of constant p_c/p_a in the figure. If we increase the nozzle size to infinite expansion ratio under these conditions, we can find the ultimate achievable value for C_f. This C_{f-ult} is a function of γ alone and has a value of 2.246 for $\gamma = 1.2$.

If a rocket is designed to operate at a fixed altitude (fixed p_a) with a constant chamber pressure, it is possible to design a nozzle for optimal performance under these conditions. Since thrust coefficient is maximized when the nozzle is perfectly expanded ($p_a = p_e$), the optimal value from Eq. (25.203) is

$$C_{f-opt} = \left[\frac{2\gamma^2}{\gamma - 1} \left(\frac{2}{\gamma + 1} \right)^{(\gamma + 1)/(\gamma - 1)} (1 - (p_e/p_c)^{(\gamma - 1)/\gamma}) \right]^{1/2} \quad (25.204)$$

the unique expansion ratio corresponding to this result can be obtained by solving Eqs. (25.193) and (25.194) for M_e and ϵ, respectively.

Specific Impulse. Having defined the new variables c^* and C_f, which describe propellant energy and nozzle performance, we can now relate Isp to these fundamental quantities. From Eqs. (25.184), (25.201), and (25.203), we can write

$$Isp = \frac{F}{\dot{m}g} = \frac{C_f p_c A^*}{g p_c A^*/c^*} = C_f c^*/g \quad (25.205)$$

Therefore, Isp is a direct function of both propellant energy and nozzle performance. We can now see why Isp is the most fundamental performance parameter, since it factors in both propellant and nozzle attributes. If we substitute in for c^* and C_f into Eq. (25.205), we can write

$$Isp \cdot g = \sqrt{\frac{2\gamma R_u T_c}{(\gamma - 1)W} (1 - (p_e/p_c)^{(\gamma - 1)/\gamma})} \left[1 + \frac{(p_e/p_c - p_a/p_c)(p_c/p_e)^{1/\gamma}}{\frac{2\gamma}{\gamma - 1}(1 - (p_e/p_c)^{(\gamma - 1)/\gamma})} \right]$$

$$(25.206)$$

While this relation is rarely used to calculate Isp [Eq. (25.205) is much simpler], it does demonstrate the very important result that

$$Isp \propto \sqrt{T_c/W} \quad (25.207)$$

which indicates one desires high chamber temperature and low molecular weight of chamber gases to maximize performance. This is the fundamental reason that the Liquid Oxygen/Liquid Hydrogen (LOX/LH2) propellant system provides some of the highest Isp values for chemical propulsion systems. The presence of large amounts of hydrogen lowers the molecular weight substantially and thereby increases *Isp*, even though the flame temperature of this combination is similar to many other propellants.

Other rocket propulsion systems which make use of a *working fluid* (rather than combustion gases) to achieve high specific impulse will invariably choose hydrogen as this fluid. Systems of this type include nuclear thermal rockets and laser propelled rockets in which nuclear or laser energy is used to heat hydrogen to high temperatures in the chamber. Electrothermal thrusters (arcjets) can also operate using hydrogen as fuel, but oftentimes, less energetic fuels (such as ammonia or helium) are selected based on thermal considerations.

Nozzle Designs. An efficient nozzle contour will enable optimal expansion of the high pressure chamber gases to supersonic conditions without any shocks. Most rocket engines make use of a conventional Delaval nozzle with circular cross-section to accomplish this task. Some of the parameters which describe this contour are given in Fig. 25.83. The nozzle entry region is shown as a simple conical surface at *convergence angle* of θ_i. Designs of this region vary considerably depending on volume constraints. For example, a steeper inlet angle can be used to effectively shorten the nozzle and therefore minimize engine length.

The throat region is usually formed by circular arcs through the angles θ_i and θ_d for the inlet and exhaust sides of the nozzle. While θ_i values can vary considerably, θ_d is generally restricted to angles between 10 and 25 degrees. The radii of curvature defining upstream and downstream circular arcs are denoted R_{wtu} and R_{wtd}, respectively. In general, the two radii of curvature, which are nondimensionalized with respect to the throat radius, are not equal. Typical values lie in the range $0.2 < R_{wtu}, R_{wtd} < 4.0$, in many current nozzle designs.

The exit cone geometry can be specified as a simple cone at angle θ_d, or as a series of points defining a contour. *Contoured nozzles* can generally be made shorter (and lighter) than a *conical nozzle* with the same expansion ratio. In addition, a contoured design often can give a lower nozzle exit angle, θ_e, which we will be advantageous in reducing 2-D flow losses. Method of characteristics or CFD techniques are often utilized to optimize this contour and insure that no shocks are formed. The contoured nozzle has the disadvantage of being more difficult to fabricate, although the production complications are minor compared to a conical design.

Since most solid propellant combustion products contain particles, erosion of in-

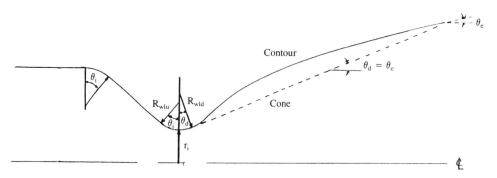

FIGURE 25.83   Contour definition for a conventional DeLaval nozzle.

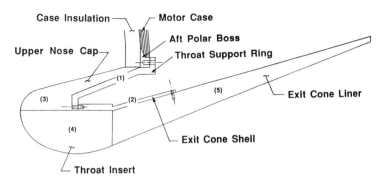

FIGURE 25.84 Major components of a submerged nozzle design frequently used in solid rocket motors.

sulation and nozzle surfaces can be a concern. Particles tend to impact in the converging portion of the nozzle, leading to high insulation erosion in this region. One technique used to alleviate aft dome erosion problems is to move the nozzle assembly forward, thus *submerging* the nozzle within the aft dome of the motor case. A schematic of such a design is shown in Fig. 25.84. The *upper nose cap*, *throat insert*, and *exit cone liner* are typically made from fibrous materials cured in a high strength resin to enhance insulatory performance in areas exposed to flame. The nozzle support structure [parts (1) and (2) in Fig. 25.84] is generally steel or aluminum.

Submergence of the nozzle reduces erosion of the insulation in the reentrant region above the upper nosecap and also reduces the overall length of the motor. This reduction in length can be quite beneficial for *upper stage* motors which must be stored within the fairing of a launch vehicle or within the bay of the Space Shuttle. Systems of this type are said to be *volume limited* (or volume constrained), and it is desirable to package the propulsion system efficiently for these applications. Submerged designs are not normally considered for cooled nozzles, since the surface area exposed to the flame is much greater than in a conventional design. Effects of submergence on aerodynamic efficiency are quite minor. Results in the literature give conflicting reports; Landsbaum, *et al.* (1980) found that nozzle submergence improved efficiency, while Kordig and Fuller (1967) indicate that nozzle efficiency drops with increasing submergence.

For vehicles operating within the earth's atmosphere over a range of altitudes, the ambient pressure can vary considerably during the course of engine operation. To avoid low altitude performance losses associated with this phenomenon, *altitude compensating* nozzles have been devised. The *aerospike* nozzle shown in Fig. 25.85 exemplifies this feature. The nozzle can be either axisymmetric or two-dimensional, and the base region may or may not have exhaust gases exiting through its surface. As the external pressure changes, the outer flow boundary automatically adjusts and flow remains attached to the inner wall at all times. The aerospike nozzle also enjoys the advantage of being shorter in length (and possibly lighter in weight).

The difficulties associated with these nonconventional nozzles generally involve the throat region. Manufacturing of an annular or linear throat to exacting tolerances is difficult for small engines, and next to impossible for larger engines, because the

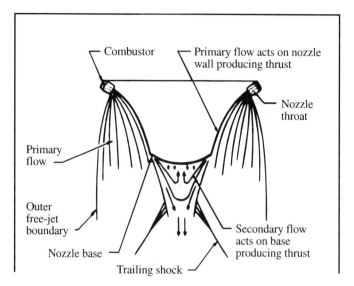

FIGURE 25.85   Plug (aerospike) nozzle design. (Courtesy of Rocketdyne Corporation.)

throat region is defined as the difference between two large radii. In addition, cooling of the throat (and the base of the plug) can be a challenge which requires a substantial weight penalty as compared to conventional designs. For this reason, aerospike engines which have multiple chambers feeding the nozzle have been considered. A view of the linear engine is shown in Fig. 25.86.

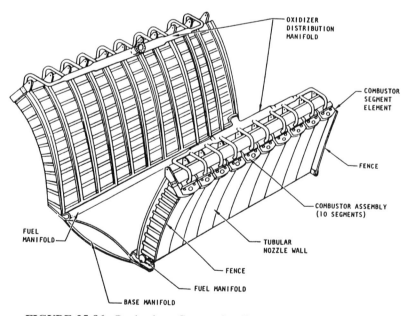

FIGURE 25.86   Rocketdyne Corporation linear aerospike engine design.

25.7.3 Deviations from Ideal Performance

Current state-of-the-art performance calculations begin by allowing chemical reactions to occur as the gases expand through the nozzle. An industry-standard computer code [Gordon and McBride (1971)] is readily available and is commonly used for this purpose. Recombinations of chemical species within the exit cone is almost always important in performance calculations due to the very high levels of dissociation within the chamber. The Gordon/McBride code is typically used in a one-dimensional, equilibrium calculation which assumes that kinetic processes are infinitely fast and that all recombinations possible do actually occur at a given nozzle station. This calculation represents an upper bound on the *Isp* for a given engine.

Starting with this *Isp* value, various loss mechanisms are accounted for by applying *efficiency factors* estimated from more detailed calculations. Two-dimensional flow effects cause the flow exiting the nozzle to have a small radial component (especially near the wall). The radial momentum amounts to a loss in thrust as compared to the idealized one-dimensional situation in which all velocity vectors are oriented parallel with the nozzle centerline. To minimize this effect, the designer desires a low θ_e value (see Fig. 25.84). However, decreasing this angle usually implies increasing nozzle length (and weight). These two factors must both be considered in optimizing the nozzle design. Overall, two-dimensional flow effects are not a major loss mechanism; this loss generally is less than 1% in thrust or *Isp*.

Viscous effects due to the presence of a wall boundary layer are also taken into account. In general, the Reynolds number associated with moderate to large engines is very high which implies that viscous effects will be restricted to a very small region near the wall. Overall, the boundary layer losses amount to less than 1% in larger engines.

Chemical kinetics losses are also taken into account by considering the reaction rates of various recombinations at the local pressure and temperature of the mixture. The kinetics loss measures the extent to which recombination reactions actually occur within the abreviated time the gases spend in the nozzle. Larger motors and engines typically have low kinetics loss (less than 0.5%) which indicates that the equilibrium assumption is typically quite good.

Since many solid rockets utilize metallized propellants which lead to droplets/particles in the exhaust, two-phase flow losses can be an important performance consideration. The two-phase flow loss stems from the fact that particles in the exhaust do not accelerate to the same velocities as the gas, thus yielding an equivalent loss in momentum and thrust. This factor can be the dominant loss mechanism for solid motors, yielding thrust losses of up to 10%.

In addition, solid motors are subject to erosion of the throat (which leads to a degradation in average expansion ratio). This effect is more pronounced on motors with smaller throat diameters, since a given erosion will change the expansion ratio more dramatically in this case. Typical losses for moderate to large motors are in the 1–2% range. Overall efficiencies, which incorporate all these effects, vary between 88 and 96% depending on motor size and propellant composition.

Liquid engines are subject to additional loss mechanisms as well. Typically, the heat transfer to the wall leads to a loss in available energy of the combustion gases. In addition, a vaporization loss is included to account for a small fraction of larger drops which fail to evaporate and combust prior to entry into the nozzle. Maldistri-

butions in local mixture ratio due to the detailed injector configuration are also estimated. These three effects are lumped into an *energy release efficiency* which has
values around 99% for large liquid engines. For large engines, kinetics losses and
boundary losses generally lie in the 1–2% range. Typical overall efficiencies, which
take all factors into account, are in the 85–96% range depending on engine size.

The Two-Dimensional Kinetics (TDK) program [Nickerson *et al.* (1988)] serves
as an industry-standard tool for analysis of liquid engine performance. The analog
in the solid motor community is referred to as the Solid Propellant Rocket Motor
Performance Prediction Computer Program (SPP) [Nickerson *et al.* (1987)]. These
codes were developed through United States Air Force and NASA sponsorship, and
are utilized by virtually all manufacturers and consultants involved in the rocket
propulsion community.

25.7.4 Propellants

Solid Rocket Propellants. The propellant in a SRM must not only burn at the proper
rate so as to maintain the desired chamber pressure, but it must also have enough
structural continuity to withstand pressure and acceleration loads introduced during
flight. Most physical characteristics (including burning rate) cannot be predicted
analytically, so heavy use of test data is normally required [Kuo and Summerfield
(1984)]. The high density of solid propellants (near that of cement) provides a large
impulse capability in a limited volume.

Solid propellants fall into two general categories. In *double base* propellants, fuel
and oxidizer are mixed on a molecular level. These propellants generally contain a
nitrocellulose type of gun powder dissolved in nitroglycerine with additional amounts
of minor additives. Both of these primary ingredients are explosives containing both
fuel and oxidizer (oxygen) within their molecular structures. Double-base propellants have been used most frequently in military applications (primarily in ballistic
missiles), but their use is currently declining due to an increased emphasis on safety
of munitions devices by the military.

The other general type of solid propellant is called a *composite propellant* to
highlight the fact that this is a heterogeneous mixture of fuel, oxidizer, and binder
ingredients. Fuels for composite propellants are generally metallic powders, with
aluminum being the choice in most instances. Oxidizers are generally crystalline
materials; ammonium perchlorate (NH_4ClO_4), also know as AP is the most popular
choice in today's motors. The binder is a rubber-based material which serves as a
means to hold the powder/crystal mixture together in a cohesive grain. Current bind-
ers include hydroxyl-terminated polybutadiene (HTPB), and polybutadiene nitrile
(PBAN). Occasionally, small amounts of very energetic ingredients such as Her
Majesty's Explosive (HMX) are included to improve performance.

Solid propellant burning rates have been found (both theoretically as well as experimentally) to be a function of the chamber pressure. Most manufacturers assuming a burning rate dependence according to St. Robert's law

$$r_b = ap_c^n \tag{25.208}$$

where r_b is the burning rate (in inches or cm per second), a is the burning rate
coefficient, and n is the burning rate exponent. The constants a and n are determined

TABLE 25.7   Characteristics of Composite Propellants used in Solid Rocket Motors

Item	ρ_p kg/m^3	$c*$ m/s	T_c °K	Pressure Range, MPa	$r = aP_c^n$, cm/s a	n
A	1840	1541	3636	2.75–7.0	0.415	0.31
B	1800	1527	3396	2.75–5.0	0.399	0.30
C	1760	1568	3392	2.75–7.0	0.561	0.35

A: TP-H-1202, 21% Al, 57% AP, 12% HMX, 10% HTPB; Star 63D
B: TP-H-3340, 18% Al, 71% AP, 11% HTPB; Star 48, Star 37
C: TP-H-1148, 16% Al, 70% AP, 14% PBAN; RSRM, Titan IV

from test data. Table 25.7 summarizes some state-of-the-art propellant formulations used in current motors.

Liquid and Hybrid Propellant Combinations.   Performance of liquid and hybrid systems can be characterized in terms of the *mixture ratio, r*. This quantity is expressed as a ratio of oxidizer and fuel flow rates

$$r = \dot{m}_{ox}/\dot{m}_f \qquad (25.209)$$

Typical engine performance is characterized as shown in Fig. 25.87 for the liquid oxygen (LOX)/liquid hydrogen (LH2) cryogenic propellant combination used in the Space Shuttle Main Engine (SSME). The chamber temperature, T_c, is maximized at the stoichiometric mixture ratio. However, the molecular weight increases steadily with mixture ratio, since the oxidizer typically has a higher molecular weight than the fuel. For this reason, the characteristic velocity is maximized at a slightly fuel rich condition. (Remember we want to maximize $\sqrt{T_c/W}$ to get highest *Isp*.)

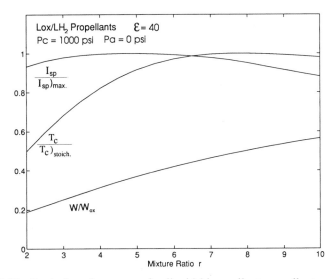

FIGURE 25.87   Typical performance of a liquid bipropellant propellant as a function of mixture ratio.

For example, the LOX/LH2 propellant combination has a stoichiometric mixture ratio of 8, while actual LOX/LH2 engines (such as the SSME) operate at r values between 5 and 6.

If we choose a mixture ratio, the propellant characteristic velocity can be determined. This information, along with the associated nozzle geometry, permits calculation of engine specific impulse. If the desired thrust level is known, the total propellant flowrate can be calculated. We can express fuel and oxidizer flowrates in terms of this massflow and the selected mixture ratio

$$\dot{m}_f = \frac{\dot{m}}{1 + r} \qquad (25.210)$$

$$\dot{m}_{ox} = \frac{r\,\dot{m}}{1 + r} \qquad (25.211)$$

Another important variable is the propellant bulk density, ρ_b, which can be written in terms of fuel and oxidizer ratios, as well as the mixture ratio

$$\rho_b = \frac{1 + r}{1/\rho_f + r/\rho_{ox}} \qquad (25.212)$$

By knowing this parameter, one can determine the total volume of tankage required. High bulk density is desirable, since a smaller volume of tankage will be required. This is the primary problem with the use of LH2 as a fuel. While LOX has relatively high density (71 lb/ft^3), LH2 density is only 4 lb/ft^3. At a mixture ratio of 6, this cryogenic propellant combination has a bulk density of 21 lb/ft^3 which is only about 1/3 that of water. Therefore, large tanks are required to house this propellant combination.

The most common oxidizer in use today is liquid oxygen. LOX is a cryogenic liquid which does present unique problems, but its high density, low cost, and high performance make it a good alternative for many systems. If storability is desirable, nitrogen tetroxide (N_2O_4) is an often-used alternative. Its high density and hypergolic tendencies (ignites spontaneously when combined with fuel) are desirable, but it is also hygroscopic, difficult to handle safely, and quite expensive. The fluorine oxidizers (liquid fluorine and chlorine pentafluoride) are the most energetic available due to the low molecular weight of this element. Fluorine is very reactive and difficult to handle which has precluded its use in most applications. The ClF$_5$ oxidizer is currently under development for use in space-based kinetic energy weapons.

As far as fuels go, liquid hydrogen gives the highest performance. Negative aspects associated with hydrogen include the fact that it is cryogenic, very low in density, and quite expensive. On the positive side, hydrogen has excellent heat transfer characteristics and makes an excellent coolant. The most common storable (non-cryogenic) fuels include RP-1 (kerosene) and the hydrazine family of hydrazine (N_2H_4), monomethyl hydrazine (MMH), and unsymmetrical dimethyl hydrazine (UDMH).

While hybrid propellant combinations also exhibit an optimal mixture ratio as shown in Fig. 25.87, it is more difficult to achieve this value in practice due to the fact that there is no explicit control of the fuel flowrate. Optimization of the fuel grain design and empirical observation of fuel regression rate is required to tailor the mixture ratio to the desired value. The combustion of a hybrid propellant is limited by oxidizer diffusion through the turbulent boundary layers formed near the fuel surface. Correlations of experimental data [Estey *et al.* (1991)] indicate the regression rate, r_b, is influenced most directly by the mass flux, G, through the local port cross-sectional area as well as the length of the port, L

$$r_b = aG^n/L^m \tag{25.213}$$

where a, n, and m are empirical constants. Limited theory on the behavior of turbulent boundary layers suggests that $n = 0.8$ and $m = 0.2$. Actual correlations reveal exponents near these values for several propellant combinations. Experimental results indicate that hybrid regression rates are roughly an order of magnitude lower than those for a solid propellant (in the range of 0.05–0.25 cm/s) which implies that hybrid motors will require several ports (exposing additional surface area) to generate mass flows and thrust levels equivalent to a single port solid rocket design.

Liquid oxygen (LOX) is the most popular oxidizer in hybrid propellant combinations. The most common fuel is HTPB, although numerous plastics and rubber fuels can be utilized. The fuel can be loaded with powdered metal (usually aluminum) to increase performance. Recently, some researchers have made measurements with nitrous oxide and high-concentration hydrogen peroxide (H_2O_2) [Wernimont and Heister (1995)] in aqueous solution as potential storable (noncryogenic) oxidizers. While these oxidizers offer lower specific impulse than LOX, they are higher in density and offer some systems advantages associated with non-cryogenic operations.

Table 25.8 summarizes some state-of-the-art liquid and hybrid propellant combinations including optimal mixture ratio, and characteristic velocity values. These data were generated using the Gordon–McBride computer code assuming a chamber pressure of 1000 psi (6.9 Mpa) and a nozzle expansion ratio of 40. The optimal mixture ratio (and other performance values) does change slightly with chamber pressure due to recombination effects within the nozzle.

TABLE 25.8   Characteristics of Propellant Combinations used in Current Liquid Rocket Engines

Propellant Combination	Optimal Mixture Ratio	ρ_b, kg/m^3	c^*, m/s	T_c, °K	Vacuum Isp, sec
LOX/LH$_2$	4.83	320	2385	3250	455.3
LOX/RP1	2.77	1030	1780	3700	358.2
N$_2$O$_4$/MMH	2.37	1200	1720	4100	341.5
N$_2$O$_4$/N$_2$H$_4$	1.42	1220	1770	3260	343.8
ClF$_5$/MMH	2.83	1400	1890	4100	375.0
LOX/HTPB	2.5	1061	1780	3770	357.0
90% H$_2$O$_2$/HTPB	7.5	1320	1610	2830	316.1
N$_2$O/HTPB	8.5	800	1600	3450	315.3

25.7.5 Engine/Motor Internal Ballistics

The term *internal ballistics* is used to describe the methodology employed in calculating the motor/engine chamber pressure and thrust for a given chamber design and propellant flowrate. Actually, the term evolved from early artillery performance analysis and is most often applied to solid rocket motor calculations. This case is actually the most interesting, since the propellant massflow into the chamber is controlled by the propellant burning rate, the exposed surface area, A_b, and the propellant density, ρ_p

$$\dot{m}_{in} = \rho_p r_b A_b \tag{25.214}$$

We have already shown that the massflow out of the nozzle depends on c^* [see Eq. (25.201)] and that the propellant burning rate depends on the chamber pressure [see Eq. (25.208)]. Under the assumption of steady state, the two massflows must balance exactly. Making use of this requirement and Eq. (25.208), we can solve for the chamber pressure in the motor

$$p_c = \left[\frac{a\rho_p A_b c^*}{A_t} \right]^{1/(1-n)} \tag{25.215}$$

which defines the motor chamber pressure for given motor geometry and propellant characteristics. Note that the pressure will essentially follow the trend the burn surface history $A_b(t)$ provides. Since thrust is proportional to p_c, the thrust will also mimic the burn surface history trend.

Also note from Eq. (25.215) that the chamber pressure gets large as $n \rightarrow 1$. In fact, we can show that we must have $n < 1$ for the equilibrium condition to be attained. If $n > 1$ a perturbation in the chamber pressure will cause this quantity to grow without bound. This behavior is consistent with what we call *explosives*. Fortunately, typical n values lie in the range $0.2 < n < 0.6$, so we need not be concerned with this point on most motors.

For hybrid motors, we control the oxidizer massflow, but the fuel flowrate depends on the regression rate, fuel geometry, and fuel density, ρ_f. Since regression rates are much lower than solid propellants, we typically require multiple ports to achieve desired thrust levels. In this case, the exposed fuel surface area can be written

$$A_f = N \cdot Per \cdot L \tag{25.216}$$

where we have assumed N identical ports of perimeter Per and length L. Under this assumption, the fuel flowrate becomes

$$\dot{m}_f = \rho_f r_b N \cdot Per \cdot L \tag{25.217}$$

Combining this result with a known oxidizer flowrate, $\dot{m}_o$, and assuming a regression behavior as in Eq. (25.213) under a steady-state assumption gives

$$p_c = \frac{c^*}{A_t} (\dot{m}_o + N \cdot Per \cdot \rho_f a L^{1-m} (\dot{m}_o/A_p)^n) \tag{25.218}$$

where we have used $G_o = \dot{m}_o/A_p$ with A_p representing the current port cross-sectional area. Remember, that c^* is a function of the current mixture ratio which will in general be a function of the current fuel port geometry. Equation 25.218 indicates that we can achieve a constant chamber pressure (with $\dot{m}_o$ constant) only if the ratio Per/A_p^n remains fixed as the ports burn out. In general, this is not possible, and some shift in mixture ratio and chamber pressure will occur during the firing.

The ballistics of the bipropellant liquid engine are the simple to describe, because we have control over both fuel and oxidizer massflows. The applicable form of Eq. (25.201) gives

$$P_c = \frac{(\dot{m}_o + \dot{m}_f)c^*}{A_t} \tag{25.219}$$

where once again c^* is a function of the overall mixture ratio implied by the known flowrates. By maintaining constant flowrates, it is possible to operate liquid engines at constant pressure conditions at the desired optimal mixture ratio. However, during throttling excursions, mixture ratio shifts may occur, because the pressure drop characteristics of the fuel system are not identical to those of the oxidizer. While it is possible to control both flowrates in theory, the simplest and most efficient throttling schemes typically tolerate minor shifts in mixture ratio for throttling events.

Given the chamber pressure prediction, it is simple to calculate engine thrust from $F = C_f p_c A_t$; i.e., the thrust coefficient can be calculated once the chamber pressure is known.

25.8 AIR INTAKES AND DIFFUSERS FOR AIRCRAFT ENGINES
James L. Younghans

25.8.1 The Diffusion Process

The primary purpose of air intakes and diffusers associated with typical aircraft gas turbine engines is to efficiently convert the available kinetic energy of the free-stream, air into a static pressure rise in the free stream fluid. The amount of static pressure rise typically required is consistent with a diffuser exit Mach number in the range of 0.5 for conventional axial flow turbomachinery.

This diffusion process is illustrated in Fig. 25.88 where a typical inlet/diffuser process is indicated by the path 0–2. Entropy is increased as a result of diffuser friction and, in the case of supersonic inlets, the presence of shocks in the kinetic energy conversion process. The overall diffusion process can be represented in general as

$$p_{02} = p_{0\infty} - \text{pressure losses} \tag{25.220}$$

where

$$p_{0\infty} = p_\infty + \frac{\varrho V_\infty^2}{2g} \quad \text{for incompressible flow} \tag{25.221a}$$

$$p_{0\infty} = p_\infty \left(1 + \frac{\gamma - 1}{2} M_\infty^2\right)^{\gamma/(\gamma - 1)} \quad \text{for compressible flow} \tag{25.221b}$$

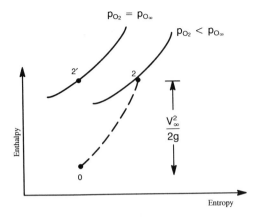

FIGURE 25.88   Diffuser process.

25.8.2   Efficiency Measure for Inlets/Diffusers

A measure of combined efficiency for the inlet and diffuser is typically referred to as *inlet recovery*, defined as

$$\text{Inlet Recovery} = \eta_R = \frac{p_{02}}{p_{0\infty}} \tag{25.222}$$

Inlet recovery characteristics are dependent on the free stream velocity the inlet is operating at as well as the amount of flow passing through the inlet.

25.8.3   Subsonic Inlet and Diffuser Design Considerations

Figure 25.89 illustrates an air induction system for a high bypass ratio aircraft gas turbine installation. The inlet and diffuser combination is designed to operate over

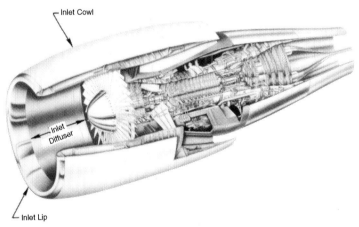

FIGURE 25.89   Subsonic inlet and diffuser.

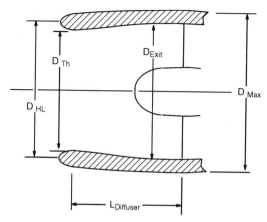

FIGURE 25.90 Inlet geometric design variables.

a wide range of airflows and aircraft operating velocities from static up to a cruise Mach number of 0.85. The relevant design variables are shown in Fig. 25.90. Table 25.9 presents the ratios of these design variables and their sizing conditions. Table 25.10 portrays representative ranges of design parameters for a transport aircraft gas turbine installation.

The spatial and temporal uniformity of the diffuser exit flow is critical to the satisfactory operation of the downstream turbomachinery. An example of spatial pressure nonuniformity, generally termed *distortion*, is depicted in Fig. 25.91 where the lower portion of the annulus exhibits asymmetric losses. A typical measure for pressure distortion would be

TABLE 25.9 Inlet Design Parameters

Design Parameter	Physical Significance	Design Conditions
D_{HL}/D_{max}	Aerodynamic fineness ratio of external cowl contour	• High speed cruise • Climb and cruise with unpowered engine
D_{HL}/D_{TH}	Aerodynamic fineness ratio of internal lip contour	• Low speed angle of attack operation for takeoff and landing • Crosswind takeoff
$(D_{exit}/D_{TH})^2$	Diffuser area ratio from inlet throat to diffuser exit	• Low speed angle of attack operation for takeoff and landing
$L_{Diffuser}/D_{exit}$	Subsonic diffuser length	• Low speed angle of attack operation for takeoff and landing
M_{TH}	Throat one-dimensional Mach number	• Maximum airflow requirements

TABLE 25.10 Typical Ranges of Key Inlet Design Parameters

Design Parameter	Values	
	Maximum	Minimum
D_{HL}/D_{max}	0.840 Top 0.885 Side 0.860 Bottom	0.815 Top 0.820 Side 0.732 Bottom
D_{HL}/D_{TH}	1.13 Top 1.16 Side 1.20 Bottom	1.11 Top 1.12 Side 1.12 Bottom
$(D_{exit}/D_{TH})^2$	1.24	1.16
$L_{Diffuser}/D_{exit}$	0.60	0.30
$M_{TH}(@ \text{ max W2R})$	0.78	0.65

$$\left(\frac{\Delta p_0}{p_0}\right)_{max} = \frac{p_{0max} - p_{0min}}{\bar{p}_0} \qquad (25.223)$$

where p_{0max} and p_{0min} are the respective maximum and minimum diffuser exit pressures. The consequences of diffuser exit pressure distortion for the turbomachinery can be:

i) Compression system flow instability
ii) Aeromechanical excitations of blading
iii) Hot streaks in the combustor and turbine
iv) Losses in thrust and component efficiency
v) Increased specific fuel consumption

Designers would typically be required to keep $(\Delta p_0/p_0)_{max} < 0.20$ for a viable design.

Air intakes for transport aircraft experience supersonic velocities even while op-

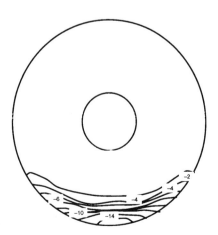

FIGURE 25.91 Pressure isobars at diffuser exit (% from average exit pressure).

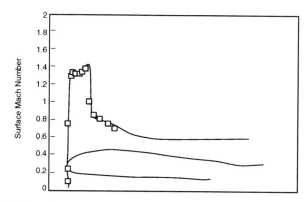

FIGURE 25.92   Inlet diffuser wall Mach number distribution.

erating at relatively modest flight velocity conditions. For an inlet operating on an aircraft that is in the takeoff mode with a free stream Mach of 0.3, the high airflow rate of the engine coupled with aircraft altitude and local wall curvature effects produces local supersonic velocities near the wall that are on the order of $M = 1.5$ (Fig. 25.92). The diffusion of the flow from $M = 1.5$ to $M = 0.5$ at the fan entrance must be accomplished with great design and analysis fidelity so that the flow does not separate even in a transient sense. Flow separation along the diffuser can lead to unacceptable engine thrust or engine surge due to the low inlet pressures associated with diffuser separation.

The relatively high transonic wall velocities around the inlet lip and along the diffuser wall necessitate sophisticated analysis techniques for acceptable levels of numerical fidelity. Three-dimensional, compressible flow calculations with wall boundary layer modeling are typically required for design purposes. A number of Navier–Stokes codes presently exist which are capable of analyzing most of the critical flight conditions. Final design verification of a selected geometry is usually accomplished with wind tunnel testing. Reynolds number effects are a testing problem if laminar separations are present and would require full scale simulation of Reynolds numbers via tunnel pressurization rather than sub scale Reynolds number simulations.

25.8.4 Supersonic Inlet and Diffuser Design Considerations

The design of supersonic inlets and diffusers addresses issues in addition to those discussed in the subsonic inlet and diffuser section, such as the need for diffuser boundary layer bleed and the need for variable geometry to control the inlet shock system. An illustration of the geometric and aerodynamic classification of supersonic inlets is contained in Fig. 25.93. The aerodynamic classification deals with where the shock system compresses the flow, i.e., *mixed compression inlets* have a shock system which is both external and internal to the inlet. The inlets themselves are not typically pod mounted as with the subsonic high bypass applications and are highly integrated with the aircraft forebody, fuselage, and wing as depicted in Fig. 25.94.

Details of a typical supersonic inlet geometry are represented in Fig. 25.95 where variable compression ramps as well as a variable diffuser ramp and bleed/bypass exit

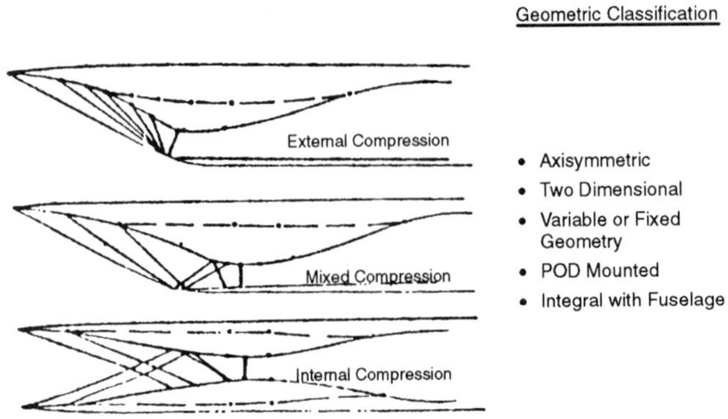

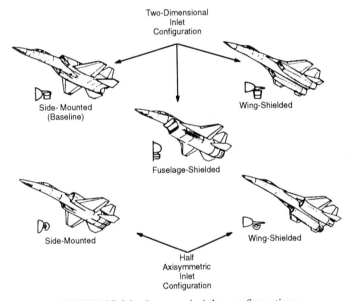

Geometric Classification

External Compression

Mixed Compression

Internal Compression

- Axisymmetric
- Two Dimensional
- Variable or Fixed Geometry
- POD Mounted
- Integral with Fuselage

FIGURE 25.93   Types of supersonic inlets.

Two-Dimensional
Inlet
Configuration

Side- Mounted
(Baseline)

Wing-Shielded

Fuselage-Shielded

Side-Mounted

Wing-Shielded

Half
Axisymmetric
Inlet
Configuration

FIGURE 25.94   Supersonic inlet configurations.

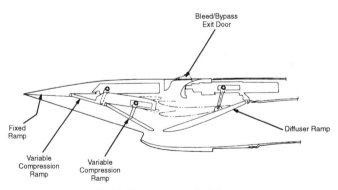

Bleed/Bypass
Exit Door

Fixed
Ramp

Variable
Compression
Ramp

Variable
Compression
Ramp

Diffuser Ramp

FIGURE 25.95   Supersonic inlet geometry.

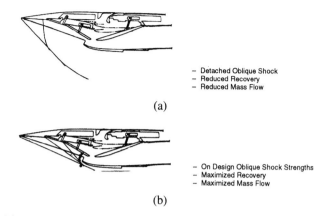

- Detached Oblique Shock
- Reduced Recovery
- Reduced Mass Flow

(a)

- On Design Oblique Shock Strengths
- Maximized Recovery
- Maximized Mass Flow

(b)

FIGURE 25.96   Inlet operation with fixed geometry. (a) $M_\infty = 1.3$ and (b) $M_\infty = 2.5$.

door are depicted. The ramps are variable so that inlet recovery can be maximized over the transonic and supersonic operating region. Ramp angles which would provide optimal recovery at supersonic cruise Mach numbers are inconsistent with maximum transonic recovery. Fig. 25.96(a) illustrates the effect of fixed geometry operation (ramp geometry set for $M_o = 2.5$ flight Mach number) when operated at a lower Mach number. Variable geometry allows the inlet to be operated *on design* over the entire Mach number range rather than just as the sizing point depicted in Fig. 25.96(b).

Figure 25.97 contains representative inlet recovery characteristics for supersonic inlets of varying weight and complexity. The *normal shock inlet* has the least weight and complexity but also provides the lowest inlet recovery above $M = 1.2$. Utilization of *external compression inlets* provides increased recovery relative to the normal shock or Pitot inlet but typically requires variable ramps and boundary layer bleed capability. This increases inlet length, weight, and cost. Mixed compression inlets extend acceptable levels of inlet recovery to even higher Mach numbers but

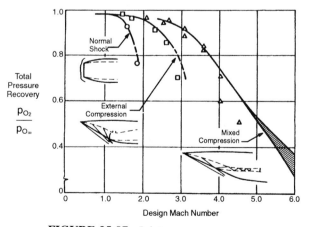

FIGURE 25.97   Inlet pressure recovery.

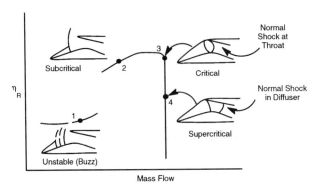

FIGURE 25.98   Inlet operational modes for external compression inlet.

at the expense of weight, length, cost, and complexity relative to the external compression inlet.

Supersonic inlets must be designed to provide acceptable levels of distortion, just as subsonic inlets must. However, another phenomenon must be addressed during supersonic inlet design, and it is referred to as *inlet buzz*. Many theories exist relative to the physics behind inlet buzz. The phenomenon occurs at low inlet mass flow and is characterized by a periodic fluctuation in inlet pressure and mass flow. Figure 25.98 characterizes the inlet operational modes possible with an external compression inlet. Operating at point 4 is undesirable because of the high levels of inlet pressure distortion that occur with the normal shock positioned in the diffuser. Point 1 is an undesirable operating condition, because the large pressure and mass flow fluctuations associated with inlet buzz can cause engine surge as well as create undesirable mechanical pressure loads.

To accommodate varying engine airflow demands around the aircraft supersonic flight placard while avoiding the operational conditions of points 1 and 4 in Fig. 25.98, inlets usually contain some sort of flow bypass system to match inlet supplied airflow with the airflow required by the engine. Excess air can be bypassed through a controlled area louver system such as depicted in Fig. 25.99.

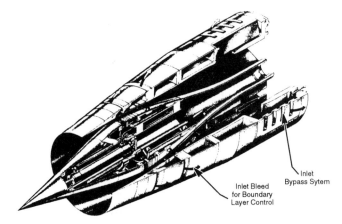

FIGURE 25.99   Inlet bypass system for controlling inlet supplied airflow.

25.9   NACELLES AND EXHAUST SYSTEMS FOR AIRCRAFT ENGINES
Donald J. Dusa

25.9.1   Introduction

This section covers the practical application of the principles of fluid dynamics relative to the aero design of nacelles and exhaust systems for aircraft engines. Inlets and exhaust systems for aircraft engine applications can basically be separated into two categories, 1) subsonic aircraft, e.g., commercial transports, and 2) multimission aircraft, e.g., military fighters and bombers. The material in this section covers the nozzle design for both applications, i.e., nacelles/exhaust systems for high bypass turbofan engines and exhaust systems for afterburning turbofan engines. In addition, because these components impact the performance of both the propulsion system and aircraft system, installed aero design procedures are also covered. The details of the flow inside nozzles are covered in Sec. 25.6.

25.9.2   High Bypass Turbofan Nacelles for Subsonic Transports

Nacelle Design Fundamentals.   A typical example of a high bypass turbofan engine on a subsonic transport is a Boeing 747. It is important to note that the design and development of the nacelle must take into account the integration of the nacelle with the wing. To the layman, a nacelle is usually looked upon as a piece of sheet metal that simply covers the turbomachinery and provides a duct for the air/gases to flow through. To the contrary, it is a multifunctional system that must act as, 1) an inlet, delivering air to the fan both efficiently and at low distortion levels, 2) a nozzle to efficiently exhaust the gases that pass through the turbomachinery, 3) a thrust reverser, to assist in slowing down the aircraft on landing, and 4) a sound suppressor, to maintain acceptable levels of fan and jet noise. In addition, the nacelle must integrate well (low installed drag) with the airplane, and be as lightweight as possible.

There are a number of factors that influence the overall design of the nacelle. These include, 1) specific nacelle aero design criteria, 2) turbomachinery integration requirements, 3) performance requirements, 4) acoustic requirements, and 5) aircraft integration and installed performance. The successful achievement of these requirements and design considerations, by necessity, involves an integrated, interactive process between those individuals involved with the nacelle aero and mechanical design, turbomachinery design, engine operability, acoustics, and the aircraft design. An example of the nacelle design requirements and considerations is illustrated in Fig. 25.100.

Nacelle Types.   Fundamentally, there are two types of nacelles for high bypass turbofan engines—a separate flow nacelle, and a long duct nacelle, as shown in Fig. 25.101. The separate flow nacelle has separate exhaust nozzles for the fan and core gases. The long duct nacelle has one exhaust nozzle and normally employs a forced mixer to gain the cycle performance advantage for the thermodynamic mixing of the two gas streams. Most of the current commercial transports which utilize wing mounted, high bypass turbofan engines, employ separate flow nacelles. On a relative basis, even though the Long Duct Mixed Flow (LDMF) nacelle has significantly

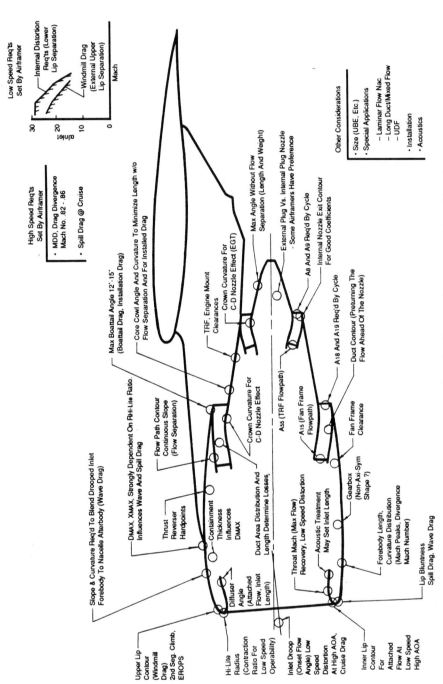

FIGURE 25.100 Nacelle design: Design requirements and considerations.

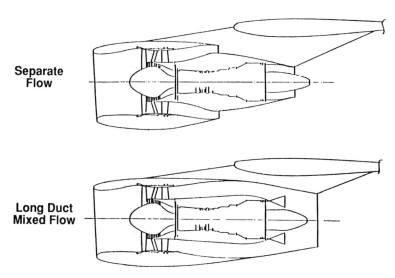

FIGURE 25.101 High bypass turbofan nacelle types.

better (lower) uninstalled specific fuel consumption (SFC), its added weight and higher installed drag have generally precluded its use on the past generation of engines and airplanes (1970s and 1980s). However, with advances in engine and aircraft technology, materials, and analytical methods, the LDMF nacelle is making strides in the marketplace (1990s and beyond). In addition, the LDMF nacelle also has advantages over the separate flow nacelle in that it can provide more reverse thrust and jet noise suppression. The practical application of LDMF nacelles, for high bypass turbofan engines, is normally for bypass ratios of less than 10. This is because weight and drag become too prohibitive for higher bypass ratio engines. Both of these type nacelles are discussed later in the exhaust system part of this section. The inlet is essentially the same for both nacelles.

Performance Perspective. In order to get a better appreciation for the impact of the nacelle on total aircraft performance, a simple performance perspective is shown in Table 25.11 for a separate flow nacelle. This table lists the derivatives and status of four of the nacelle performance elements—inlet recovery, nozzle efficiency, isolated

TABLE 25.11 Performance Perspective Separate Flow Nacelle Components Mach 0.8 Cruise

	Derivative	Status Levels
Inlet Recovery	1% Rec ≈ 0.7% SFC	$\eta_{REC} \approx .997$
Nozzle Efficiency	1% Thrust ≈ 2.5 to 3.0% SFC	$C_{FG} \approx .970$ to .983
Cowl Drag	1% Drag ≈ 1% SFC	2 to 3%
Interference Drag	1% Drag ≈ 1% SFC	1 to 4%
Fuel Burned Incentive	1% SFC ≈ 130,000 gals/yr/747	

cowl drag (uninstalled), and interference drag (increase in total aircraft drag for installing the propulsion system). Even though the status performance of these elements is high (inlet, nozzle efficiency) and the drags are relatively low, the importance of making small improvements, not necessarily performance, becomes obvious when considering the fuel burned incentive. For example, improvements in the aircraft system won't come from inlet performance gains but from shorter, lightweight inlets with the same performance and distortion levels. On the other hand, the exhaust system has a large impact on propulsion system performance and small changes in efficiency have a significant impact on the propulsion system performance (SFC). It is important to be sure that changes in propulsion system design, operation, or installation do not adversely affect the nozzle performance. To reduce nacelle cowl drag, laminar flow control is being investigated. Improvements in interference drag are expected to be realized through the use of improved CFD methods and the design of the wing and nacelle as an integrated entity as opposed to designing them separately. Each of these four elements is discussed in more detail later on in this section.

Installed Thrust and Net Propulsive Force. The definition of engine installed thrust, F_N, and net propulsive force, NPF, is shown in Eqs. (25.224)–(25.227).

$$F_N = F_{GAct} - D_R = \text{engine installed thrust (quoted by engine co.)} \quad (25.224)$$

where

$$D_R = (W_\infty/g)V_\infty = \text{ram drag}$$

$$W_\infty = \text{inlet airflow}$$

$$V_\infty = \text{aircraft velocity}$$

$$F_{GAct} = \text{engine gross thrust} \quad (25.225)$$

$$NPF = F_N - D_{\text{Fan Cowl}} - \Delta D_{\text{Inlet}} - \Delta D_{\text{Inter}} = \text{net propulsive force} \quad (25.226)$$

where

$$D_{\text{Fan Cowl}} = \text{cowl drag}$$

$$\Delta D_{\text{Inlet}} = \text{spillage drag} \quad \left.\right\} \quad \text{determined by scale model wind tunnel tests}$$

$$\Delta D_{\text{Inter}} = \text{interference drag}$$

$$(25.227)$$

The engine company normally provides the aircraft company with the installed net thrust, F_N, which is the gross thrust, F_G, (determined by the thermodynamic properties of the exhaust gases and efficiency of the nozzle) less the ram drag, D_R, of the air entering the inlet. In order to determine the net propulsive force NPF, adjustments are made to F_N to account for inlet spillage drag, fan cowl drag, and interference drag. Normally, the fan cowl drag and the reference spillage and interference drags are accounted for in the aircraft polar for a reference condition. Because the spillage and interference drags are throttle related (i.e., they change with changes in inlet flow via throttle position), they have to be accounted for on a delta basis. All of the above performance elements are discussed later on in this section.

Exhaust System Design Procedures and Performance. The definition of nozzle efficiency and gross thrust is shown in Fig. 25.102 for both LDMF and separate flow nacelles. The nozzle efficiency, C_{FG}, is defined as the gross thrust coefficient. The sketches illustrate the engine stations and relevant thermodynamic parameters used in determining the ideal gross thrust. The thrust coefficient can be estimated by analytically determining the pressure losses from the reference station to the nozzle exit (due to friction, protuberances, steps, and gaps, etc.) plus any external losses due to the exhaust gases scrubbing the core, plug, and pylon (separate flow only). The C_{FG} is normally verified through scale model static performance testing. The C_{FG}, along with the thermodynamic properties at the reference station (for determining the ideal gross thrust) are used to calculate the actual gross thrust, F_{GACT}, as described in Fig. 25.102.

Separate Flow Exhaust System: A schematic and performance summary of a separate flow exhaust system are presented in Fig. 25.103. The performance loss elements are identified, and a range of losses for each element is shown (items 1, 2, and 3). The largest portion of the loss is due to the fundamental duct friction, and the core, pylon, and plug scrubbing drag. Losses for steps and gaps, and protuberances, etc., must also be accounted for in the entire exhaust system flow field (internal and external). The aerodynamic losses associated with expansion, flow angularity, unguided turning, and pylon interference are the items the nacelle designer has to work with regarding performance improvements. The aerodynamic losses for separate flow nacelles (late 1980's and early 1990's generation aircraft engines) are normally $<0.5\%$ ΔC_{FG}, and it is obvious that a plateau is being approached regarding potential performance improvements for these types of exhaust systems. However, it is important to note that small changes in C_{FG}, e.g., 0.3%, can result

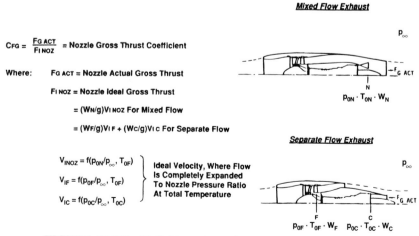

FIGURE 25.102 Definition of nozzle efficiency and gross thrust.

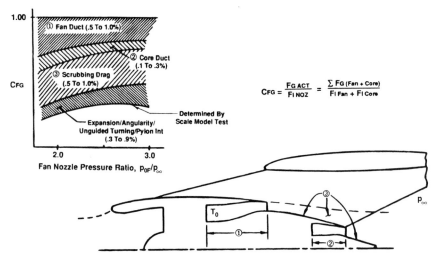

FIGURE 25.103 Separate flow exhaust system loss elements.

in a significant change in SFC ($> 1.0\%$), because the sensitivity relative to nozzle performance is so high (see Table 25.11).

The nozzle coefficients can be estimated analytically by calculating all of the individual loss elements, however, in practice, the C_{FG} is normally determined by scale model test. The values in Fig. 25.103 are good estimates for conceptual and preliminary design studies, however, for engine performance guarantees, nozzle scale model performance validation tests are always conducted.

Long Duct Mixed Flow Exhaust Systems: A schematic and trimetric cut-away view of an engine with a long duct mixed flow nacelle are shown in Figs. 25.104(a) and (b). A forced (lobed) mixer [see Fig. 25.104(b)] is normally employed to maximize the thermodynamic performance advantage for mixing the two streams. In reality, a pressure loss occurs in the mixing process. This is due to the inherent loss associated with the mixing of two streams and the loss due to the presence of the mixer hardware. Although the total energy is still the same, the pressure loss results in an increase in entropy resulting in a decrease in available energy and an attendant loss in thrust. This is illustrated in Fig. 25.105 and indicates the importance of designing a low pressure loss mixer. For example, a mixer could be designed to produce 100% mixing, but if it results in high pressure losses it would negate the advantage for mixing. Essentially, the mixed flow point in Fig. 25.105 would move farther to the right, at constant total enthalpy, H, and the Δh_{Ma} would get smaller. A practical example of uninstalled mixer performance gains (not accounting for the additional external drag or added weight) as a function of mixer aerodynamic pressure loss and mixing effectiveness is shown in Fig. 25.106. In practice, about 80% of mixing effectiveness can be achieved with a 0.6% Δp_0 loss.

Figure 25.107 shows a performance summary for a long duct nacelle with both a free and forced mixer at a typical subsonic cruise operating condition. The fundamental losses for the mixer are included in elements 1 and 2. The tailpipe loss (the same for both examples) is element 3. The thrust gain due to thermodynamic mixing

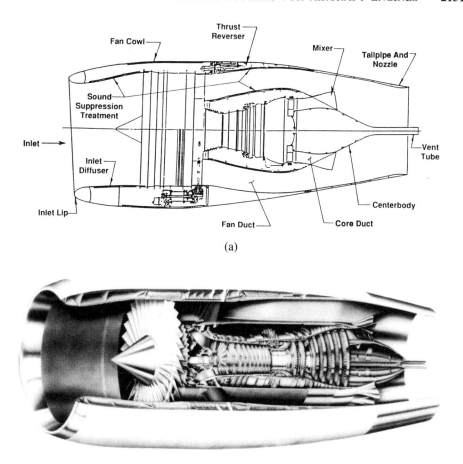

(a)

(b)

FIGURE 25.104   (a) Cross section long duct mixed flow nacelle components and (b) trimetric view.

is shown as element 4. The forced mixer provides about 0.5% higher C_{FG} which is equivalent to ~1.3 to 1.5% higher net thrust. The weight and cost penalty for the forced mixer relative to the free mixer is insignificant compared to this performance advantage. Also shown in Fig. 25.107 is a 1.5% ΔC_{FG} (~3.5% higher net thrust) internal performance improvement relative to a separate flow exhaust system. This is an internal performance comparison and it should be noted that the improvement would be about 1% net thrust lower when the additional external scrubbing drag is taken into account for the long duct nacelle. Another factor, which has negated long duct nacelles from applications, is the interference drag which offsets the internal performance gains. However, with improved analytical tools and advancements in technology, long duct nacelles show promise in the future.

In addition to performance, there are other advantages for long duct nacelles. These advantages are in the form of lower jet noise (lower exit velocities and potential for acoustic treatment) and higher reverse thrust. These advantages must be weighed against higher installed drag, a heavier nacelle, and poorer accessibility.

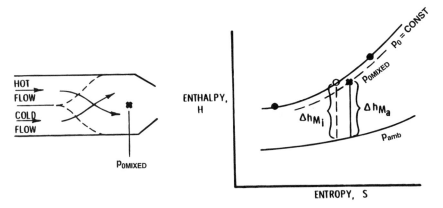

- $P_{OMIXED} < P_0$ AND $\Delta h_{Ma} < \Delta h_{Mi}$

    - THERMODYNAMIC MIXING p_0 LOSS

    - MIXER HARDWARE AERODYNAMIC p_0 LOSS

- PRESSURE LOSS IN MIXER DESIGN IMPORTANT

FIGURE 25.105   Mixed flow nozzle thrust.

Thrust Reversers. Thrust reversers, although not used for aircraft certification, are an important aspect of the nacelle design process and they do reduce wear on tires and brakes and, obviously, are useful in case of emergencies. Figs. 25.108(a) and (b) show a separate flow nacelle and long duct nacelle in both the forward and reverse thrust modes. For the separate flow nacelle, the net reverse thrust effectiveness at static conditions (Mach 0.0) is only 8.0%. For the long duct nacelle at static

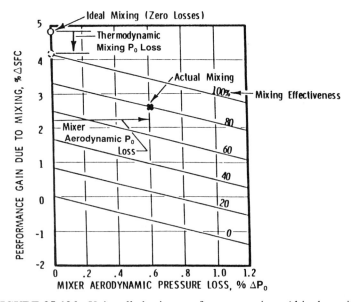

FIGURE 25.106   Uninstalled mixer performance gains—Altitude cruise.

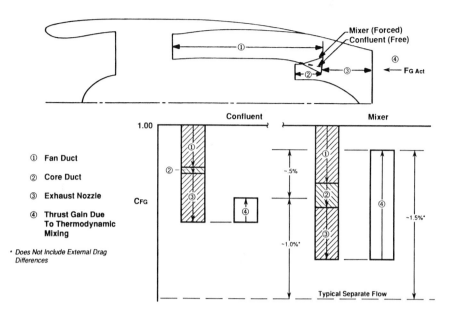

FIGURE 25.107 Mixed flow exhaust system—Altitude cruise performance (uninstalled).

conditions, the reverse thrust effectiveness is 35.0%. This is primarily due to the fact that the core thrust is spoiled for the long duct nacelle. In actuality, the reverse thrust on landing is significantly higher for both nacelles due to the ram drag at the landing velocities (see Fig. 25.109). It is very important for reverser efflux to be tailored in order to avoid engine reingestion and adverse aircraft flow fields. Integration with the airframe companies is extremely important here.

Installed Performance. A chart illustrating aircraft/nacelle installed performance is presented in Fig. 25.110. Specifically, the chart identifies the total nacelle drag and quantifies the two elements making up that drag (isolated nacelle drag and interference drag). The isolated nacelle drag is the uninstalled (isolated) external friction and pressure drag. The interference drag is the difference between the total nacelle drag and the isolated nacelle drag and is essentially the drag increase due to installing the nacelle on the airplane. Fundamentally, the interference drag is a result of increased drag on both the nacelle and airplane and is illustrated by the schematics on Fig. 25.111. As a result of integrating the nacelle with the wing, the nacelle pressure distributions, relative to isolated, become more negative on the core cowl, thus increasing the nacelle drag. In addition, the wing pressure in the vicinity of the nacelle becomes more negative thus reducing the lift. The reduced lift requires a higher angle-of-attack and thus higher aircraft drag. It is important for the nacelle designer to work closely with the propulsion system and airplane designers in order to optimize overall aircraft system performance.

25.9.3 Exhaust Systems for Multimission Applications

Nozzle Description. Figure 25.112 illustrates a typical propulsion system installation for a multimission fighter type aircraft, and Fig. 25.113 shows a schematic of

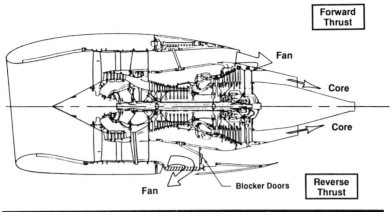

Forward Thrust	Reverse Thrust
Fan ~ 77% Of Total Thrust Core ~ 23%	Fan ~ 40% Reverse Thrust Blocker Leakage ~ 9% Forward Thrust Core ~ 23% Forward Thrust Net Reverser Effectiveness ~ 8% Reverse Thrust

(a)

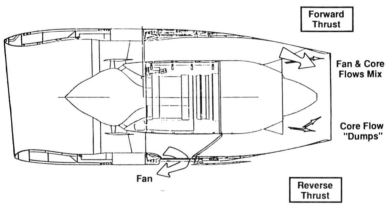

Fan Flow ~ 45% Reverse Thrust

Core Flow ~ 10% Forward Thrust (Core Thrust Spoiled)

Net Reverser Effectiveness = 35% Reverse Thrust

(b)

FIGURE 25.108 Thrust reverser arrangements. (a) Separate flow turbofan thrust reverser and (b) mixed flow turbofan thrust reverser.

the various turbomachinery components. Multimission applications normally utilize an afterburning turbofan engine, and the nozzle is cooled by air from the fan bypass duct. The air is directed through a liner in the tailpipe and exits the liner at the primary (convergent) nozzle hinge location. The air provides a cooling film for both the convergent and divergent nozzle during afterburning operation. Figure 25.113 shows the nozzle in both the dry (dashed line) and afterburning operation. Because

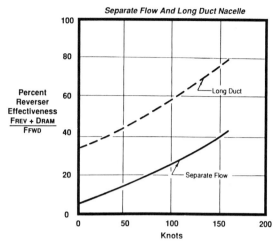

FIGURE 25.109   Engine reverse thrust effectiveness.

the engines for these types of applications are primarily afterburning engines the nozzles are variable geometry.

Performance Perspective. Table 25.12 presents typical nozzle performance, C_{FG}, and performance sensitivity, ΔF_N and ΔSFC, to changes in thrust coefficient at several operating conditions. In addition, external drag, leakage, and cooling performance values are presented.

Nozzle Aero Design Performance Elements. The definition of nozzle efficiency and the design and performance parameters are given in Eq. (25.228) with Fig. 25.114.

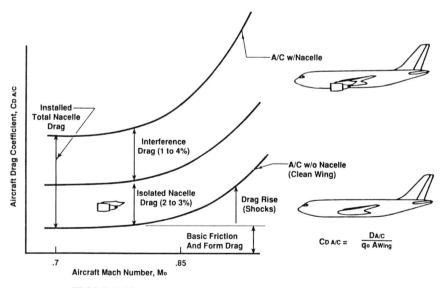

FIGURE 25.110   Aircraft/nacelle installed performance.

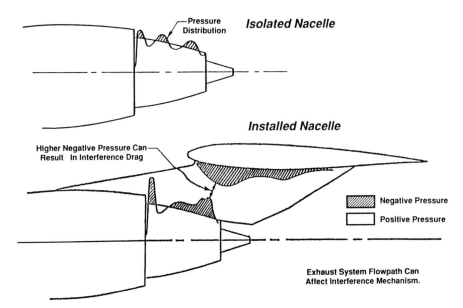

FIGURE 25.111   Nacelle/wing interference drag mechanism.

$$C_{FG} \text{ (gross thrust coefficient)} = \frac{F_{GACT}}{F_{INOZ}} = \frac{C_{FGPeak} \ (W_7/g)V_9 + (p_9 - p_\infty)A_9}{F_{iNOZ}}$$

(25.228)

where

$$F_{GACT} = \text{nozzle actual gross thrust}$$

$$F_{iNOZ} = (W_7/g)V_{iNOZ} = \text{nozzle ideal gross thrust}$$

$$W_7 = \text{nozzle actual mass flow}$$

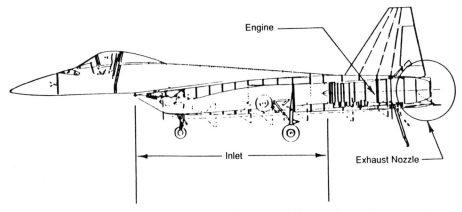

FIGURE 25.112   Airplane schematic illustrating inlet, engine, exhaust nozzle.

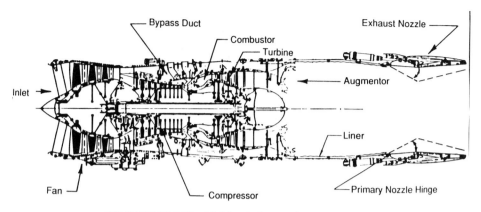

FIGURE 25.113 Multi-mission engine components.

$$V_{iNOZ} = f(p_{07}/p_\infty, \gamma_7) = \text{ideal velocity}$$

$$C_{FGPeak} = \text{peak thrust coefficient, } f(A_9/A_8, \alpha_7), \text{ where } p_9 = p_\infty$$

$$V_9 = \text{nozzle ideal exit velocity, } f(A_9/A_8, \gamma_7)$$

$$p_9 = \text{nozzle exit static pressure, } F(A_9/A_8, \gamma_7)$$

$$p_\infty = \text{ambient pressure}$$

p_{07} nozzle total pressure
T_{07} nozzle total temperature
γ_7 ratio of specific heats for nozzle airflow
$\overline{m_8}$ nozzle throat flow function $= f(\gamma_7, M_8) = \dfrac{W_7\sqrt{T_{07}}}{p_{07}\,A_{e8}}$
M_8 nozzle throat mach number

TABLE 25.12 Exhaust Nozzle Performance Sensitivities

Operating Condition	%ΔC_{FG}	%ΔF_N	%ΔSFC	Status (C_{FG})
Take Off	+1.0	+1.0		0.985
Subsonic Cr. (Mach .85)	+1.0	+1.5	−1.5	0.980–0.985
Transonic Accel (Mach 1.2)	+1.0	+1.5	−1.5	0.985
Supersonic Cr. (Mach 3.0)	+1.0	+2.0	−2.0	0.980–0.985
Hypersonic (Mach 5.0)	+1.0	+8.0	−8.0	0.975

- Nozzle external drag (boattail or aftbody drag)
  - —Can be as much as 35% of airplane drag
  - —Normally in the order of 4% to 10%
- Leakage: 1% $W_8 \approx 0.75\% \ \Delta C_{FG}$ dry and 0.5% ΔC_{FG} A/B
- Cooling impacts engine sizing/TOGW:
  - —1% engine air to cool nozzle $\approx$ −0.5% unrealized gross thrust during A/B
  - —Ejectors for cooling result in system performance reduction, and added weight and complexity, but, may be necessary to solve system problem

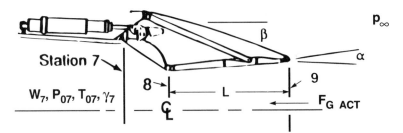

FIGURE 25.114   Nozzle nomenclature.

A_{e8} nozzle throat effective area, station 8
A_8 nozzle physical area at station 8
A_9 nozzle exit area, size based on performance requirements, station 9

C_{F8} nozzle flow coefficient $= \dfrac{A_{e8}}{A_8}$

The loss elements are, 1) leakage, 2) friction, 3) angularity, and 4) expansion. For variable geometry nozzles, leakage occurs along hinge lines and between flaps and seals. The friction loss is simply the result of the exhaust gases scrubbing the nozzle surface. This can be as much as a 0.5% to 1% of the nozzle gross thrust coefficient depending on the size and operating condition of the nozzle. The angularity loss is a result of the flow not exiting the nozzle parallel to the engine centerline. Both the friction and angularity loss coefficients are normally derived from scale model tests, however they can also be calculated analytically. Expansion losses occur when the nozzle exit pressure is not equal to ambient (see Sec. 25.6).

The nozzle throat (station 8) is an engine control parameter, and as such, it is important to know the effective area. In order to determine the effective area, it is necessary to know the flow coefficient, which is normally determined empirically.

The previous discussion only addresses internal (uninstalled) nozzle performance. In order to optimize the nozzle performance, for a variable geometry exhaust system, the afterbody drag (a.k.a. *boattail drag*) must also be considered. This will be discussed in more detail in the *Installed Performance* part of this section.

Nozzle Internal Performance Optimization. The chart of Fig. 25.115 shows a plot of the nozzle thrust coefficient (efficiency) as a function of nozzle pressure ratio. This illustrates that, for a given nozzle geometry, the peak thrust coefficient occurs at a pressure ratio where the expansion loss is zero. The peak thrust coefficient is not the highest nozzle efficiency that can be obtained with a given nozzle at a given pressure ratio. This will be discussed in more detail later on in this section. It should be noted that the leakage losses, because they are essentially constant, are not included on this notational chart. Leakage losses of $.0075 \Delta C_{FG}$ for dry operation and $.005 \Delta C_{FG}$ for afterburning operation are typical for variable geometry axisymmetric exhaust systems.

A theoretical angularity loss coefficient is shown in Eq. (25.229) for both axisymmetric convergent-divergent and two

$$C_A, \, Axi_{\text{Nozzle}} = \frac{1 + \cos \alpha}{2}$$

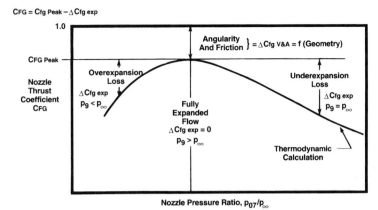

FIGURE 25.115 Nozzle thrust coefficient curve performance loss elements. For simplicity this definition does not include leakage which is approximately a constant value; e.g., for axisymmetric exhaust nozzle systems $= 0.0075\Delta Cfg$ dry and $.005\Delta CfgA/B$.

$$C_A, 2DCD_{\text{Nozzle}} = \frac{\sin \alpha}{\alpha \text{ (rad)}}$$

$$\Delta C_{fgA} = 1 - C_A \tag{25.229}$$

dimensional convergent-divergent nozzles. The theoretical angularity coefficients are simply derived from an integration of the velocity vectors emanating from a point source in the spherical sector (or cylindrical sector for $2DCD$ nozzles) of the nozzle exit. An analytical procedure for calculating the friction loss is shown in Eq. (25.230).

$$\Delta C_{fgv} = \frac{\int C_{f,l} q dS}{(W_7/g)V_i} \tag{25.230}$$

For a given nozzle, the combination of friction and angularity is defined as the

$$C_{FGPeak} = 1.0 - \Delta C_{fgv} - \Delta C_{fgA} \tag{25.231}$$

peak thrust coefficient in Eq. (25.231). The peak thrust coefficient can be defined analytically, or by scale model tests. Once the peak thrust coefficient is defined, the expansion losses are determined thermodynamically.

The effect of area ratio variation on thrust coefficient as a function of nozzle pressure ratio, for a given nozzle and gas properties is shown in Fig. 25.116. The peak thrust coefficient for each area ratio curve is shown to occur at a higher pressure ratio as nozzle area ratio increases. Also, the peak coefficient decreases as the area ratio increases because, for a nozzle with a fixed flap length, the angularity loss increases (α increases) as the area ratio increases. The shape of the curve on each side of the peak is a function of the under or over expansion loss.

By cross plotting the thrust coefficient of Fig. 25.116 at a constant pressure ratio (operating condition), the nozzle thrust coefficient as a function of nozzle area ratio

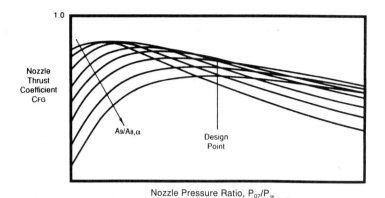

FIGURE 25.116   Nozzle area ratio effects.

can be derived as shown in Fig. 25.117. By definition, the peak thrust coefficient occurs at zero expansion loss and is strictly a function of nozzle geometry (angularity and friction). As can be seen on Fig. 25.117, the optimum nozzle performance for a given pressure ratio occurs at an area ratio smaller than that defined for the geometric peak thrust coefficient. The reason for this is that as the nozzle area ratio is reduced the initial rate of increase in underexpansion loss is less than the rate of decrease in angularity loss, which results in a nozzle thrust coefficient higher than the peak thrust coefficient.

Installed Performance.   The analysis discussed relative to Fig. 25.117, which identifies maximum internal performance, can be used to assess the performance loss associated with changes in area ratio relative to the maximum internal performance area ratio. In Fig. 25.118, for example, the area ratio reduction for 0.0025 and 0.005 loss in C_{FG} is also illustrated. Similarly, losses in C_{FG} for increases in area ratio can be determined. This can be done for a wide range of C_{FG} losses, for both higher and lower area ratios, and the information can be used to provide nozzle internal performance sensitivity to area ratio as shown in Fig. 25.118. This information is used by the nozzle designer/aircraft designer to trade off nozzle internal performance ver-

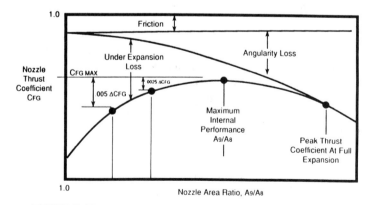

FIGURE 25.117   Nozzle internal performance optimization.

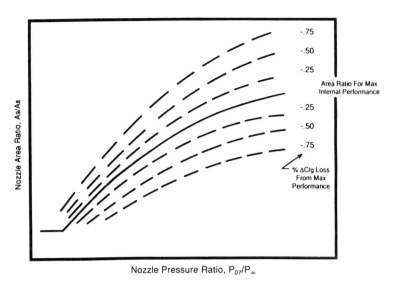

FIGURE 25.118 Nozzle internal performance sensitivity to A_9/A_8.

sus external flap (boattail) drag as a function of A9. This is a key study in order to arrive at the best installed area ratio schedule which maximizes internal thrust minus external drag.

Figure 25.119 is a chart showing nozzle boattail drag versus aircraft Mach number for various values of nozzle exit area to fuselage projected area (A9/A10). From this type of information, a change in thrust due to boattail drag can be calculated as a function of A_9 for a particular operating condition. This information is then added to the internal performance to determine optimized installed performance as shown in Fig. 25.120. Again it is seen that the maximum installed performance occurs at

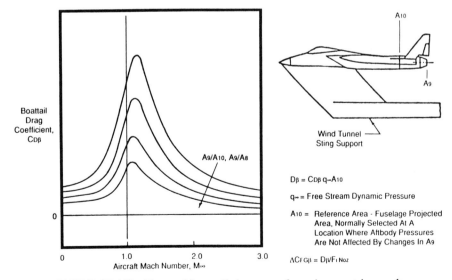

FIGURE 25.119 External/boattail drag map for axisymmetric nozzle.

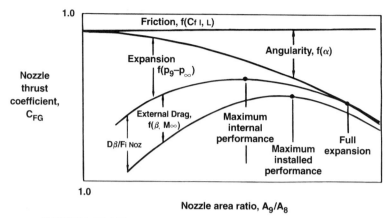

FIGURE 25.120   Installed nozzle performance optimization.

yet a different A_9/A_8. In order to optimize the overall aircraft performance, it is important for the nozzle designer to understand all of the loss elements.

25.10   RAMJETS
James L. Keirsey, William B. Shippen, and Everett J. Hardgrave, Jr.

25.10.1   Ramjet Engine Description and Analysis

The ramjet is functionally simple (see Fig. 25.121), with no moving parts except for fuel supply and control, but it does require a booster for acceleration to sufficient speed for operation with thrust in excess of drag. The ramjet is similar to the turbojet, but it does not require the complex rotating compressor and turbine.

The supersonic inlet is designed to catch and compress air efficiently and with minimum external drag. The condition of high pressure and low internal velocity tends to minimize flow pressure loss. The subsonic diffuser and through-duct carry air to the combustor, and an *aerogrid* makes the flow uniform. The combustor burns the fuel/air mixture to supply energy in the form of high temperature gas. It operates at the lowest practical velocity and minimum blockage necessary to sustain combustion and to minimize pressure loss across the flameholder and loss due to heat addition in a moving stream. The convergent-divergent nozzle transforms the potential energy of the high temperature and high pressure gas to kinetic energy in the exhaust jet. The maximum jet reaction is obtained by expanding to ambient pressure at the highest possible velocity. This is achieved by maximum exhaust gas temperature and maximum pressure ratio.

For the ideal case, air enters the engine at flight velocity and ambient pressure and is exhausted at maximum jet velocity and ambient pressure. Thrust is equal to change in momentum, i.e.,

$$\text{Thrust} = (\dot{w}_{ex}V_{ex} - \dot{w}_{air}V_\infty)/g \qquad (25.232)$$

In actual practice, pressure-area forces must be considered at the inlet and exit. This is done through the use of *stream thrust* [Rudnick (1947)]. Referring to Fig.

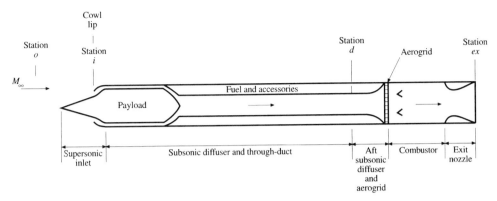

FIGURE 25.121 Ramjet engine schematic.

25.122 we can write

$$\text{Reaction on duct} = p_2 A_2 - p_1 A_1 + (\dot{w}/g)(V_2 - V_1)$$

$$= \int p_D \, dA - \text{wall friction force}$$

$$= (p_2 A_2 + \dot{w} V_2 / g) - (p_1 A_1 + \dot{w} V_1 / g)$$

$$= F_2 - F_1 \tag{25.233}$$

where

$$F = pA + \dot{w} V / g$$

$$f = F/A = p + \rho V^2 = p(1 + \gamma M^2) \tag{25.234}$$

Weight flow rate is $\dot{w}$. *Stream thrust* per unit area, f, has the dimensions of a pressure and is treated similarly. Stream thrust acts parallel to the flow, and reacts on the duct towards the section of duct under consideration. The reaction of the gas stream on the duct is equal to the difference in stream thrust of the entering and leaving streams.

Referring again to the engine schematic (Fig. 25.121) overall gross thrust is the difference in the inlet and exit stream thrust. Thus, thrust is found by establishing inlet and exit flow conditions only. A detailed study would show thrust equal to the summation of pressure and frictional forces on all interior surfaces.

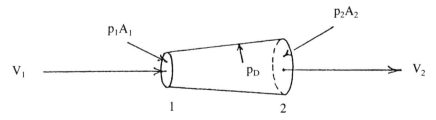

FIGURE 25.122 Duct flow schematic.

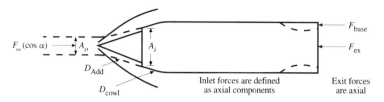

FIGURE 25.123 Engine force balance.

In most practical cases, expansion to ambient pressure would require exit areas greater than the vehicle body area otherwise necessary. Exit area can be limited to body area to minimize external drag and body weight with only a slight loss of jet thrust.

Engine analytical procedures used by others in the ramjet field may vary in detail from that described herein, which was developed at the Johns Hopkins University, Applied Physics Laboratory [Hardgrave (1977)]. Refer to Fig. 25.123, engine force balance, and we can say

$$F_\infty \; (\cos \alpha) = \text{Free stream stream thrust} = A_\infty p_\infty \; (1 + \gamma M_\infty^2) \; (\cos \alpha)$$

$$D_{\text{Add}} = \text{Additive drag} = \int p \; dA \; \text{on streamline between } A_\infty \text{ and } A_i$$

$$D_{\text{cowl}} = \text{Cowl wave drag} = \int p \; dA \; \text{on cowl projected area}$$

$$F_{\text{ex}} = \text{Exit stream thrust} = A_{\text{ex}} p_{\text{ex}} (1 + \gamma M_{\text{ex}}^2)$$

$$F_{\text{base}} = \text{Base force} = \text{pressure} \times \text{base area}$$

$$T_{\text{Gross}} = F_{\text{ex}} - F_\infty$$

$$T_{\text{Net}} = (F_{\text{ex}} + F_{\text{base}}) - (F_\infty + D_{\text{Add}} + D_{\text{cowl}}) \tag{25.235}$$

Here, M_∞ is the free stream Mach number. The forces itemized above are on an *absolute* force basis. For conversion to coefficients, all forces are adjusted to a *gage* basis by subtracting $(p_\infty \times \text{Area})$ from each. All force components and summations are then converted to dimensionless coefficients through division by $q_\infty A_R$, thus normalizing for pressure level and body size, e.g., $C_{TN} = T_{\text{Net gage}} / q_\infty A_R$, for the net gage *thrust coefficient*.

It is common to make the following simplifying assumptions in ramjet analysis:

i) Perfect gas, $p/\rho = RT$

ii) Specific heat ratio, γ, is constant, $7/5 =$ for air and $9/7 =$ for combustion products

iii) Steady flow

iv) Unidimensional flow (uniform, parallel); using *average* values of properties at given station, except for design of supersonic inlets

v) Adiabatic (no heat transferred, except for combustion)

vi) Nonisentropic, i.e., shock, turbulent and friction losses are accounted for

Finally, we need to define some performance parameters starting with *air specific impulse*

$$S_a \equiv F*/\dot{w}_a \qquad (25.236)$$

where * refers to sonic conditions and $\dot{w}_a$ is the air weight flow ($\dot{w}_f$ below is the fuel weight flow). Developing the relation of S_a to total temperature, note

$$\dot{w} = \dot{w}_a + \dot{w}_f \qquad (25.237a)$$

$$\dot{w}_a = \dot{w}/(1 + \dot{w}_f/\dot{w}_a) \qquad (25.237b)$$

so

$$S_a \equiv F*/\dot{w}_a = F*(1 + \dot{w}_f/\dot{w}_a)/\dot{w} = \left[\frac{Ap(1 + \gamma M^2)}{Ap\dot{m}/\sqrt{T_0}}\right]^* (1 + \dot{w}_f/\dot{w}_a) \qquad (25.238)$$

Substituting the definition of the *weight flow* function

$$\dot{m}(\text{weight flow function}) = g\sqrt{\gamma/R}\, M\left[1 + \frac{\gamma - 1}{2}M^2\right]^{1/2}$$

$$= \frac{\dot{w}\sqrt{T_0}}{Ap} \qquad (25.239)$$

with $M = 1$

$$S_a = \frac{(1 + \dot{w}_f/\dot{w}_a)}{g}\left[2\frac{(\gamma + 1)}{\gamma}RT_0\right]^{1/2} \qquad (25.240)$$

S_a is the stream thrust per unit of air flow per second at the sonic condition and is substantially proportional to the square root of gas total temperature, T_0. Also,

$$S_a/\sqrt{T_{0\infty}} = K(1 + \dot{w}_f/\dot{w}_a)\sqrt{T_0/T_{0\infty}} \qquad (25.241)$$

where K is a constant depending on the value of γ. The *ideal* S_a (Fig. 25.124) is calculated from initial air temperature and fuel–air ratio for a specific fuel, assuming complete chemical equilibrium (perfect combustion with dissociation).

Ideal
S_a

$T_{0\infty}$

Equivalence Ratio, E.R.

FIGURE 25.124 Ideal air specific impulse. (From Billig *et al.*, 1974.)

Next, we define *combustion efficiency*

$$\eta_c = \frac{E.R._{\text{ideal, for measured } S_a}}{E.R._{\text{actual}}} \times 100 \tag{25.242}$$

where the *equivalence ratio* is

$$E.R. = \frac{\text{actual fuel/air ratio}}{\text{stoichiometric fuel/air ratio}} \tag{25.243}$$

Performance Analysis of Ramjet Components. Engine performance is calculated using data correlations from individual components (air inlet, combustor, and exit nozzle).

Inlet Diffuser: The characteristics of the diffuser relating to engine analyses are *air capture*, *pressure recovery*, and *inlet drag*.

Air capture is the quantity of air entering the diffuser measured in terms of free stream area, A_o, of the stream tube captured by the inlet, in ratio to the area defined by the cowl lip, A_i. Refer to the engine force balance diagram (Fig. 25.123). The *supercritical air capture ratio*, $A_{o\max}/A_i$, represents the maximum amount of air that can be intercepted by the diffuser at a given Mach number and angle of attack. The *subcritical air capture ratio*, $A_o/A_{o\max}$, represents the fraction of this maximum which is actually accepted, when subsonic spillover occurs. Total air capture is the product of these two ratios.

Air capture of a diffuser is governed by the size of the inlet, the compression surface shape, and the relative position of the nose and cowl lip. These factors are selected to give air capture compatible with overall engine requirements for the flight Mach number range and thrust variation desired. A typical relation of $A_{o\max}/A_i$ is shown in Fig. 25.125.

Pressure recovery is the efficiency of the diffuser in converting the kinetic energy of motion into pressure measured as *total pressure recovery*. This is the ratio of the total pressure actually achieved at the exit of the diffuser to the isentropic free stream total pressure. Typical pressure recovery data correlations are given in Figs. 25.126 and 25.127.

Since the frontal drag of a ramjet is primarily a function of the air capture and pressure recovery characteristics of the inlet as selected for propulsive purposes, it is customary to charge this drag against the thrust summation as *inlet drag*. Inlet drag coefficient, C_{AE}, includes additive drag and cowl drag. The sum of these depends on Mach number and air capture as indicated in Fig. 25.128.

Combustor: Combustor characteristics pertinent to overall engine performance are *combustion efficiency*, η_c, *combustion stability*, and *combustor pressure recovery*.

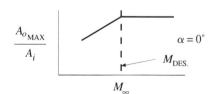

FIGURE 25.125   Typical maximum capture area variation with M_∞.

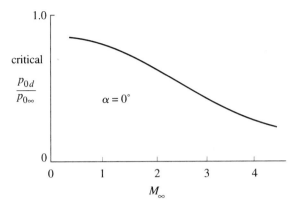

FIGURE 25.126 Typical variation of inlet critical total pressure recovery with M_∞.

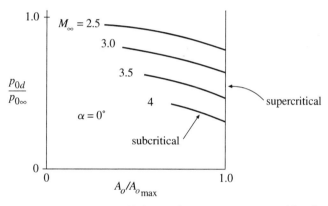

FIGURE 25.127 Typical variation of inlet total pressure recovery with subcritical spillage and M_∞.

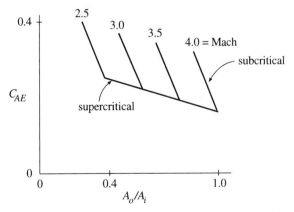

FIGURE 25.128 Typical variation of inlet drag coefficient.

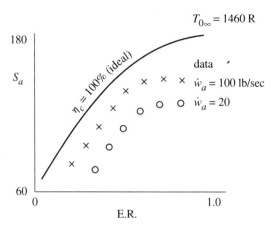

FIGURE 25.129   Typical measured air specific impulse variation with equivalence ratio.

Combustion efficiency depends on fuel type, hardware design, and the primary operating variables, fuel–air equivalence ratio, E.R., and air temperature and pressure. Although combustion efficiency is a scale for measuring the completeness of chemical reaction, in ramjet practice it is a correlating factor between the above operating variables and the thrust potential of the burned gas mixture. Experimentally, a thrust stand is used to measure the sonic exit stream thrust. This quantity is divided by the weight flow rate of air to give S_a and then converted to combustion efficiency in relation to the ideal S_a described earlier. In engine analysis, combustion efficiency is used to calculate S_a. A typical relation is shown in Figs. 25.129 and 25.130.

Combustion Stability: Combustion limits depend on the same variables as combustion efficiency. It is necessary to define the range of operating variables for which stable combustion is possible in the engine.

Combustor Pressure Recovery: Loss of total pressure across the combustor occurs in two ways, aerodynamic drag of the flameholder and the loss due to heat addition.

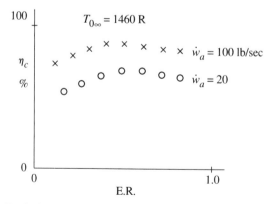

FIGURE 25.130   Typical measured combustion efficiency variation with equivalence ratio.

Calculation of these losses would be feasible only by assuming each loss to occur separately, and by assuming each loss to take place in a constant area duct.

Aerodynamic drag of the flameholder alone may be found by conventional wind tunnel techniques. The following calculation may be used to find the loss of total pressure due to heat addition for subsonic flow in a constant area duct. Since there is no change in stream thrust in a constant area duct, and neglecting wall friction,

$$F_1 = F_2, \qquad \dot{w}_1 = \dot{w}_2 \tag{25.244}$$

so

$$S_{a1}(F/F^*)_1 = S_{a2}(F/F^*)_2 \tag{25.245}$$

and

$$Ap_{01}(f/p_0)_1 = Ap_{02}(f/p_0)_2 \tag{25.246}$$

where

$$f/p_0 = (1 + \gamma M^2) \cdot (p/p_0) \tag{25.247}$$

Thus

$$p_{02}/p_{01} = (f/p_0)_1/(f/p_0)_2 \tag{25.248}$$

For a sonic exit, $(F/F^*)_1 = S_{a2}/S_{a1}$. We can find M_1 from the relation

$$F/F^* \text{ (stream thrust function)} = (1 + \gamma M^2)/M \sqrt{2(\gamma + 1)\left(1 + \frac{\gamma - 1}{2} M^2\right)} \tag{25.249}$$

Results are seen in Fig. 25.131 for $M_2 = 1$. If the Mach number at station 2 is reduced by constriction in an exit nozzle, the total pressure loss is reduced.

In practice, the drag loss and the loss due to heat addition are not measured separately, but are jointly accounted for in a correlation experimentally determined between $S_a/\sqrt{T_{0\infty}}$ and combustor inlet Mach number, M_d, as illustrated in Fig. 25.132. This correlation is suggested by the fact that, for a particular geometrical configuration between the matching station and the exit nozzle throat, the only substantial operating variable affecting the matching station Mach number, M_d, is the heat release parameter, S_{a2}/S_{a1}, or its equivalent $Sa/\sqrt{T_{0\infty}}$. Variations of specific heat ratio, γ, or heat addition other than in the constant area combustor section have some effect on the correlation, but these effects are not sufficient to displace the curve in Fig. 25.132 significantly. A single curve is found for the complete spread of combustor operating conditions.

Exit Nozzle: Exit nozzle parameters are *contraction ratio*, *expansion ratio*, and *expansion angle*.

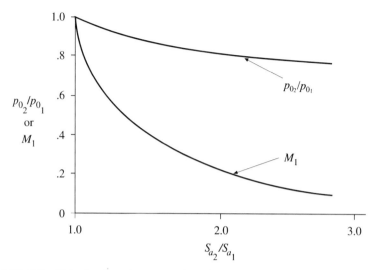

FIGURE 25.131   Calculated total pressure loss due to heat addition in a constant area duct (with no friction) for a sonic condition at the exit (station 2).

The *contraction ratio* (together with heat addition in the combustor and total pressure losses) controls the duct Mach number at the combustor entrance, M_d, as well as elsewhere in the duct. The *expansion ratio* (on the divergent side downstream of the throat) establishes the pressure ratio, or Mach number, and hence the velocity to which the gases can expand. The *expansion angle* largely determines nozzle efficiency, since radial velocity has no value in producing axial thrust. For a 15° half-angle nozzle, the thrust efficiency is approximately 97%.

Overall Engine Performance.   For a typical supersonic ramjet, the ratio of internal to ambient pressure is always sufficient to produce a sonic condition at the throat of the exit nozzle. From the experimental correlation, the Mach number at the combustor/inlet matching station, d, is then a function only of $S_a/\sqrt{T_{0\infty}}$, the heat addition parameter. For high values of $Sa/\sqrt{T_{0\infty}}$, the low diffuser exit Mach number may lead to inlet bow shock displacement forward of the cowl lip and *subcritical operation*.

For some intermediate value of $S_a/\sqrt{T_{0\infty}}$, the stream Mach number established at the cowl lip by the combustor is just equal to the Mach number behind the bow shock, resulting in *critical inlet operation*. For lower values of $S_a/\sqrt{T_{0\infty}}$, the higher

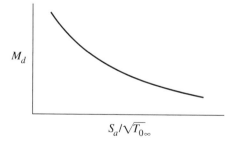

FIGURE 25.132   Typical correlation of heat addition parameter, $S_a/\sqrt{T_{0\infty}}$ with inlet/combustor matching station Mach number, M_d, for a specific combustor/exit nozzle configuration.

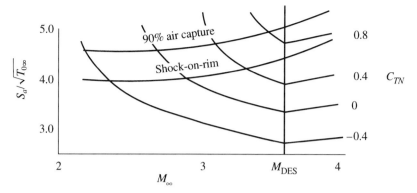

FIGURE 25.133 Typical engine operating map, showing the variation of thrust coefficient and inlet air spillage with heat addition parameter and flight Mach number.

duct Mach number permits entry of the bow shock into the annulus aft of the cowl lip, and *supercritical operation* exists. For subcritical operation, the pressure developed inside the engine is limited by the pressure recovery characteristic of the diffuser.

Engine thrust coefficient at a given Mach number varies linearly with $S_a/\sqrt{T_{0\infty}}$ for supercritical operation, but increases more slowly with $S_a/\sqrt{T_{0\infty}}$ in the subcritical regime due to inlet drag and reduced air capture. Flight Mach number, $S_a/\sqrt{T_{0\infty}}$, air capture, and thrust coefficient relations are seen in Fig. 25.133.

Figures 25.134–25.137 show some pertinent ramjet engine performance characteristics. Figure 25.134 shows the effect of altitude and Mach number on maximum C_T. As altitude increases, ambient temperature falls more rapidly than S_a. For a given Mach number, C_T is directly related to $S_a/\sqrt{T_{0\infty}}$, and C_T increases with altitude until ambient temperature stabilizes in the constant temperature region of the

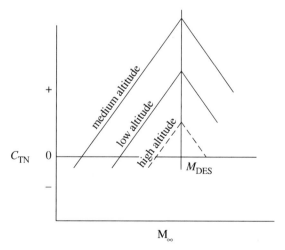

FIGURE 25.134 Typical variation of thrust coefficient with flight Mach number and altitude.

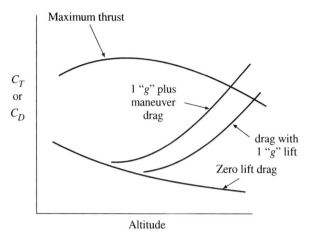

FIGURE 25.135   Variation of drag and thrust coefficients with altitude for a typical ramjet powered vehicle.

atmosphere. As altitude further increases, air flow decrease causes S_a and C_T to decrease due to a pressure effect on combustion efficiency. The Mach number effects on C_T are due to the inlet performance. At the *inlet design Mach number*, shocks and compression waves are focused at the cowl lip, permitting near maximum pressure recovery with zero flow spillage. Below the inlet design Mach number, the inlet is operating at less than full capture. Hence, as Mach number increases up to the design Mach number, air capture also increases. Once full air capture is reached, continued increase in Mach number results in lower overall engine pressure recovery and a decrease in C_T.

Figure 25.135 shows the variation of drag and thrust coefficients of a typical vehicle as a function of altitude. It is assumed that Mach number increases with altitude. The maximum cruise altitude for constant speed is indicated by the intersection of the maximum thrust line with the drag line for 1g lift. As drag is added by additional maneuver g's, the altitude ceiling without slowdown decreases.

The virtues of the ramjet are its fuel specific impulse at high Mach numbers and its ability to control thrust and speed to suit the environment. Figure 25.136 is a comparison of fuel specific impulse for various propulsion systems from Billig (1983). Figure 25.137 shows a typical ramjet (subsonic combustion) operating regime.

25.10.2   Ramjet Component Design Characteristics

Air Intakes. Inlets for ramjets have much in common with those for supersonic aircraft (see Sec. 25.8). General references on ramjet inlets are Faro and Keirsey (1967), Mahoney (1991), and McLafferty *et al.* (1955).

Figure 25.138 illustrates a typical operating characteristic of pressure recovery vs. air capture ratio. The diffuser stability limit is indicated on the figure. Operation of the inlet past this limit results in flow instability (*inlet buzz*), which is characterized by an oscillating normal shock, which produces significant internal pressure, and airflow oscillations.

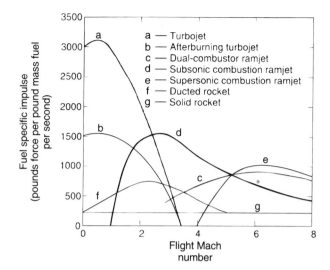

FIGURE 25.136 Fuel-specific impulse for various missile propulsion systems using dense hydrocarbon fuels. (Reprinted with permission from *Johns Hopkins APL Technical Digest*, Billig, F. S., "Tactical Missile Design Concepts," Vol. 4, No. 3, p. 144, Johns Hopkins University, Applied Physics Laboratory, Baltimore, MD, 1983.)

The minimum inlet operating Mach number determines the limit of external compression and sets the maximum turning angle on the innerbody compression surface. This condition also sets the local flow condition at the cowl lip, and the cowl internal angle is defined to allow an attached oblique shock. This minimizes cowl drag and produces acceptable pressure recovery.

A nose mounted inlet operates at near zero angle of attack at ramjet take-over (low Mach number and altitude). At a cruise condition (high Mach number and altitude), it operates at a significant but acceptable angle of attack, typically 6°–8°. Much higher angles of attack can be required during a brief terminal flight phase,

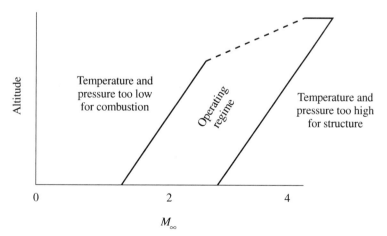

FIGURE 25.137 Typical ramjet engine operating regime.

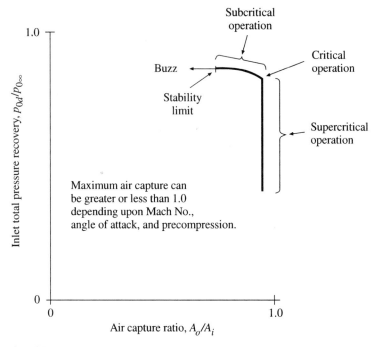

FIGURE 25.138 Typical inlet total pressure recovery characteristic at specific M_∞ and angle of attack.

calling for special fuel control and terminal shock limiting provisions. For some vehicle configurations, the air inlet(s) may be positioned under a wing or forebody to provide some pre-compression ahead of the inlet. For such designs, the inlet will provide peak performance at high angles of attack ($5°-10°$), and reduced performance at ramjet takeover (low Mach number and angle of attack) due to lack of pre-compression.

Aft of the cowl lip, a near-constant area throat section is provided to stabilize the normal shock in the presence of the boundary layer. The shock pressure rise interacts with the boundary layer to produce a *shock train* of weaker oblique shocks, which spread the pressure rise over a length necessary to sustain the pressure gradient without separation. The throat length required is several hydraulic diameters and is a function of boundary layer thickness Reynolds number [McLafferty *et al.* (1955)]. For some inlet designs, boundary layer *bleed* orifices and forebody boundary layer diverters have been applied to minimize boundary layer effects.

Combustion Chambers

Liquid Fueled Combustors: The design objectives for liquid fueled combustion chambers are, 1) uniform air velocity profiles at the entrance and exit, 2) fuel injection allowing complete mixing and combustion in minimum length and with uniform exit temperature distribution, 3) a flameholder producing low velocity recirculation zones with fuel/air ratios within combustion limits, 4) a supersonic exit nozzle of minimum throat area, and 5) chamber and exit nozzle wall temperatures within material structural limits [Petrein *et al.* (1955) and Bunt *et al.* (1972)].

The degree of difficulty in meeting these design objectives is dependent on free stream Mach number and altitude. At low Mach numbers, the engine pressure ratio is low, requiring a large exit nozzle throat area and high internal velocities. Internal air temperatures are low, reducing combustion reaction rates and requiring complex, high-drag flame holders. Low pressure at high altitude also reduces combustion reaction rates.

Figure 25.139 illustrates a combustion chamber designed for a flight Mach number of 1.8 at sea level to 2.3 at 70,000 feet altitude, with a minimum combustion chamber pressure of 5 psia at an air inlet temperature of 200°F. Relatively smooth velocity and temperature profiles result from the high blockage of the *can-type* flameholder. The piloting section is an annular can with small holes, which operates at nearly constant fuel/air ratio at very low velocity and is optimized for combustion stability. The main burner is a central can with larger holes where the fuel/air ratio varies with thrust demands. Holes in the central cone produce vortex wakes of relatively low speeds for flame holding action. Air from the outer periphery of the duct is diverted to a louvered cooling shroud for air–film cooling of the chamber wall, the exit nozzle, and the shroud itself.

Figure 25.140 shows a chamber designed for Mach number 2.8 at sea level and 4.0 at 70,000 feet altitude and above, with a minimum combustion chamber pressure of 14.7 psia. These flight Mach numbers give higher inlet air temperature and higher fuel vaporization and combustion reaction rates. Uniform profiles are provided by an aerogrid at the chamber inlet, and the favorable combustion conditions permit a low drag flameholder with recirculation zones in the triangular (*gutter*) base. The fuel is injected close to the flameholder to produce a range of local fuel–air ratio around the periphery, and some areas within combustion limits at all times. A separate piloting fuel control is not required. Cooling of the combustion chamber and exit nozzle is critical at high Mach numbers due to high temperature and high heat transfer rates at high pressure. Here also, the chamber wall and exit nozzle are air film-cooled. Air for the cooling shroud comes through the outer periphery of the aerogrid, and the shroud is extended into the nozzle entrance for more positive cooling of the nozzle throat. Both of these first two combustion chamber illustrations are for engines with an axisymmetric air inlet.

Figure 25.141 shows an arrangement with side-mounted inlet ducts and an Integral Rocket-Ramjet (IRR) combustion chamber. In the ramjet operating mode, fuel is injected from the walls of the inlet ducts, a few inches upstream of the dump ports in the chamber dome. A flameholding recirculation zone is provided by the head end of the chamber, and the long chamber required for the rocket grain assures nearly complete mixing and combustion. It was designed for approximately the same flight condition as the previous gutter type combustor. The combustion chamber wall is protected by an ablative coating of silicone rubber.

Solid Fueled Combustors: Ramjets utilizing solid, rocket-type fuel grains with low oxidizer content pose different design requirements [Wanstall (1984)]. Possible configurations are indicated in Figs. 25.142 and 25.143.

The ducted-rocket type, shown in Fig. 25.142(a) can have a fixed or variable gas generator nozzle with an end burning or radial burning grain. Programmed fuel flow

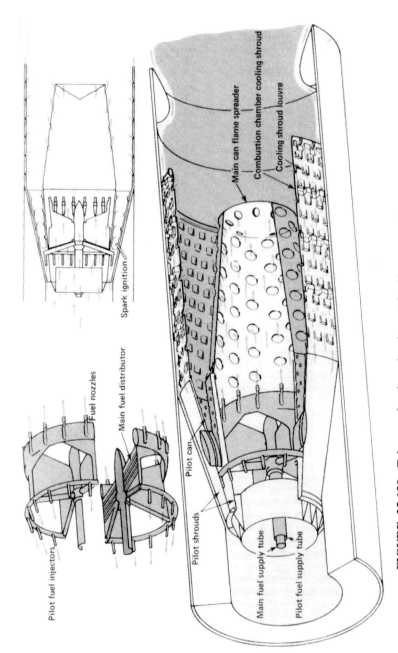

FIGURE 25.139 Talos combustion chamber. ''Can'' type combustor permits efficient combustion at low pressure and temperature corresponding to high altitude flight. (Reprinted with permission from *Johns Hopkins APL Technical Digest*, Shippen, W. B., Berl, W. G., Garten, W., Jr., and Hardgrave, E. J., Jr., ''The Talos Propulsion System,'' Vol. 3, No. 2, p. 133, Johns Hopkins University, Applied Physics Laboratory, Baltimore, MD, 1982.)

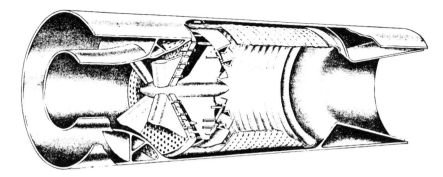

FIGURE 25.140 Long range typhon combustion chamber. High flight Mach number produces high inlet air temperature which permits use of *Gutter* type flameholders. (Reprinted with permission from *Johns Hopkins APL Technical Digest*, Keirsey, J. L. "Airbreathing Propulsion for Defense of the Surface Fleet," Vol. 13, No. 1, p. 62, Johns Hopkins University, Applied Physics Laboratory, Baltimore, MD, 1992.)

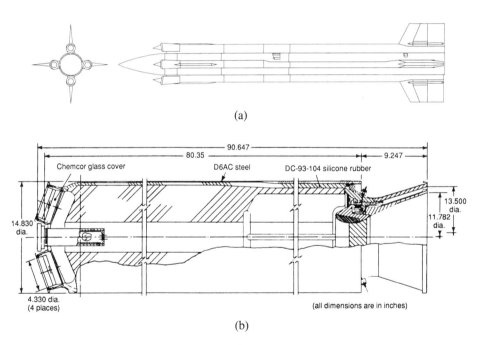

FIGURE 25.141 Integral rocket-ramjet. (a) Outboard profile showing four axisymmetric air inlets. (b) Combustion chamber showing solid propellant booster rocket configuration. The forward end of the motor case has four air inlet *stacks* for the air inlet admission ports, which are sealed with frangible glass port covers and are retained in tapered sleeves. The glass port covers are fragmented by explosive charges, which opens the ports for ramjet operation. The motor throat insert needed during rocket booster operation is also separated at pressure tailoff, leaving the properly sized exit nozzle for ramjet operation. (Reprinted with permission from *Johns Hopkins APL Technical Digest*, Keirsey, J. L., "Airbreathing Propulsion for Defense of the Surface Fleet," Vol. 13, No. 1, p. 66, Johns Hopkins University, Applied Physics Laboratory, Baltimore, MD, 1992.)

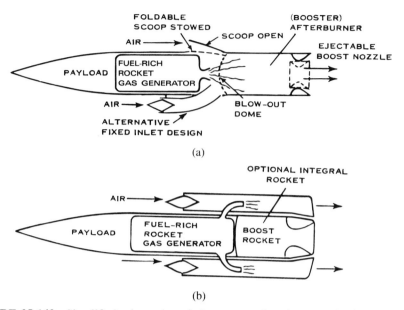

(a)

(b)

FIGURE 25.142   Simplified schematics of air-augmented rockets. (a) Design with in-line (common) afterburner and (b) design with parallel outboard engines. (Reprinted with permission from *Johns Hopkins APL Technical Digest*, Keirsey, J. L., "Airbreathing Propulsion for Defense of the Surface Fleet," Vol. 13, No. 1, p. 67, Johns Hopkins University, Applied Physics Laboratory, Baltimore, MD, 1992.)

rates can be provided by tailoring either the grain composition for variable regression rates and/or the grain shape for variable surface area. A controllable fuel flow rate can be achieved with a variable pintle nozzle controlling the regression rate by way of chamber pressure, but response time with a variable chamber volume must be carefully considered.

Figure 25.142(b) illustrates another ducted rocket configuration. The separate combustor sections are fed fuel by a single propellant chamber, and hot fuel is injected from a circular *blast-tube* aligned with each combustor centerline. The combustors are cylindrical tubes.

Solid-fueled ramjet (SFRJ) configurations similar to Fig. 25.143 with a centrally ported fuel grain have also been tested in experimental programs. Here, the chamber pressure and the regression rate of the fuel grain and the airflow are primarily a

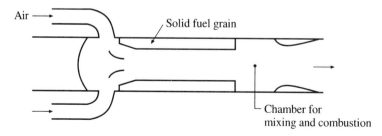

FIGURE 25.143   Solid fueled combustion chamber conceptual design.

function of inlet air pressure and velocity, with a resulting tendency toward limited range of fuel–air ratio. Fuel–air ratio varies with pressure, inlet air temperature, grain surface area, erosion rate, and local heat transfer rate. Additional chamber length is required at the aft end of the fuel grain for mixing and completion of combustion. Predictability of in-flight fuel–air ratios and inlet matching are significant problems.

25.10.3 Applied Engine Designs

The earliest operational ramjets were Talos [Goss (1982), Shippen *et al.* (1982)] for the U.S. Navy and BOMARC [Mahoney (1991)] for the U.S. Air Force. These pioneering designs for Mach 1.8–2.0 operation at high altitude, posed difficult conditions for combustion. The Navy's Long Range Typhon [Cronvich (1965) and Keirsey (1992)] was a Talos follow-on design and was developed through the flight demonstration phase. It was smaller, operated in excess of Mach 4, and had longer range. The Sea-Dart was a British design similar to Typhon-LR. A number of ramjet designs by the Marquardt Company have also been flight tested [Wanstall (1984) and Mahoney (1991)].

The Air Force ASALM (Advanced Strategic Air Launched Missile) [Webster (1978) and Wanstall (1984)] was also developed through flight demonstration. It was a Mach 4, integral rocket-ramjet (IRR). The U.S. Navy also has conducted flights of IRR engines in the LVRJ (Low Volume Ramjet) and SLAT (Supersonic Low Altitude Target) programs [Mahoney (1991) and Wanstall (1984)].

Wanstall (1984) gives information on Russian IRR engines using liquid and solid fuels, French IRR engines by Aerospatiale and others, and a German program by MBB (Messerschmidt–Blokow–Blohm, now Deutsche Aerospace).

25.10.4 Higher Speed Potential Ramjets (Scramjets)

Ramjets with supersonic combustion have been the subject of considerable investigation for operation at *hypersonic* speeds [Gilreath (1990), Waltrup *et al.* (1976), and Waltrup (1987)]. Early theoretical investigations (late 1950s and 60s) of these engines (called *Scramjets*) indicated potential superiority over conventional ramjet performance for operation above about Mach 5. Figure 25.144(a) is an engine schematic. The principal differences between a scramjet and a ramjet are that the scramjet inlet decelerates the air to a supersonic rather than a subsonic speed, combustion occurs in supersonic rather than subsonic flow, and the scramjet does not require an exit nozzle throat. The superior potential performance of the scramjet results from the lower combined inlet and combustor total pressure losses at higher flight speeds. For the scramjet, however, key objectives are, 1) to add heat supersonically at as low a Mach number as possible to avoid excessive pressure losses in the combustion process, and 2) to utilize fuel capable of producing high performance, while having acceptable physical properties of practical applications.

Early laboratory experiments with scramjet combustors were based on highly reactive fuels such as triethyl-aluminum (TEA) and HiCal 3-D (principally ethyldecarborane), for which high combustion efficiency was found with short combustor lengths, while avoiding large pressure losses due to wall friction and shocks [Waltrup *et al.* (1976)]. While these experiments produced useful data on combustor

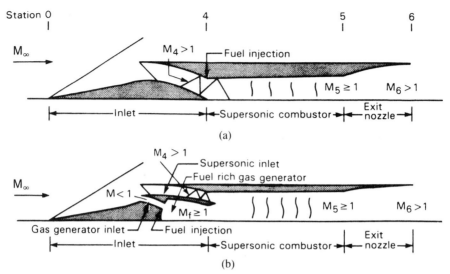

FIGURE 25.144 Schematics of supersonic combustion ramjet engines. (a) Supersonic combustion ramjet and (b) dual combustor ramjet. (Reprinted with permission from *Johns Hopkins APL Digest*, Waltrup, P. J., "Hypersonic Airbreathing Propulsion: Evolution and Opportunity," AGARD/PEP Symposium, Paper No. 12, April 1987.)

performance, the fuels were not regarded as acceptable for practical applications. Most applications of interest required storable fuels.

Emphasis shifted to the use of heavy hydrocarbon fuels. A dual-combustor ramjet [Keirsey (1992), Billig *et al.* (1980), and Keirsey and Cusick (1983)] was invented in 1977 to address the problem of supersonic combustion with these hydrocarbon fuels. The engine is a hybrid cycle, with a small subsonic combustion dump-type combustor as a pre-burner for preparation of the fuel, in tandem with a supersonic combustor [see Fig. 25.144(b)]. A small portion of the air captured by the inlet and all of the fuel is injected in the dump preburner which operates fuel-rich as a hot gas generator. The remainder of the air is ducted supersonically to the supersonic combustor. Heavyweight combustor hardware was used with hydrocarbon fuels, JP-5 and RJ-5 (Shelldyne-H). High combustion efficiency was measured in these experiments with supersonic combustor lengths of only 3 to 5 ft [Keirsey (1992)].

For very high Mach number applications interest shifted to the use of pure scramjets with cryogenic fuels, mainly hydrogen, for other applications. Much effort has been directed to demonstration testing and showing the potential of such fuels in the pure scramjet combustor. Major objectives of this work are for the National AeroSpace Plane (NASP) program, where design aspects are those of a manned hypersonic aircraft.

REFERENCES

Addy, A. L., "The Analysis of Supersonic Ejector Systems," *AGARDograph No. 163 on Supersonic Ejectors*, Ginoux, J. J. (Ed.), AGARD (NTIS, 5285 Port Royal Road, Springfield, VA), 1972.

Addy, A. L., Dutton, J. C., and Mikkelsen, C. C., "Supersonic Ejector Diffuser Theory and Experiments," *Proceedings: Ejector Workshop for Aerospace Applications*, AF-WAL-TR-82-3059, Flight Dynamics Laboratory, Air Force Wright Aeronautical Laboratories, Wright-Patterson Air Force Base, OH, 1982.

Alperin, M. and Wu, J., *High Speed Ejectors*, AFFDL-TR-79-3048, Flight Dynamics Laboratory, Wright–Patterson Air Force Base, OH, 1979.

Altman, D., Carter, J. M., Penner, S. S., and Summerfield, M., *Liquid Propellant Rockets*, Princeton Aeronautical Paperbacks, Princeton University Press, Princeton, New Jersey, 1960.

Anon., *Design Methods in Solid Rocket Motors*, AGARD LS-150, 1987.

Anon., *Nuclear Thermal Propulsion, A Joint NASA/DOE/DOD Workshop*, NASA Conference Publication 10079, Cleveland, OH, 1990.

Anon., *Solid Propellant Grain Design and Internal Ballistics*, NASA SP-8076, 1972.

Anon., *Solid Propellant Selection and Characterization*, NASA SP-8064, 1971.

Anon., *Combustion of Solid Propellants*, AGARD LS-180, 1991.

Anon., *Solid Rocket Motor Performance Analysis and Prediction*, NASA SP-8039, 1971.

Balasubramaniam, M., Lowi, A., Schrenk, G. L., and Denton, J. C., "Fuel Economy Potential of a Combined Engine Cooling and Waste Heat Driven Automotive Air-Conditioning System," *11th IECEC Meeting*, State Line, NV, September 1976.

Ball, J. W. and Tullis, J. P., "Cavitation in Butterfly Valves," *J. Hydraulics Division, ASCE*, Vol. 99, HY9, pp. 1303–1318, 1973.

Benedict, R. P., Carlucci, N. A., and Swetz, S. D., "Flow Losses in Abrupt Enlargements and Contractions," *Trans. ASME, Ser A*, Vol. 88, p. 73, 1966.

Bernstein, A., Heiser, W., and Hevenor, C., "Compound Compressible Nozzle Flow," *J. Appl. Mech.*, vol. 34, pp. 548–554, 1967.

Bevilaqua, P. M., "Evaluation of Hypermixing for Thrust Augmenting Ejectors," *J. Aircraft*, Vol. 11, No. 6, pp. 348–354, 1974.

Billig, F. S., "Tactical Missile Design Concepts," *Johns Hopkins APL Technical Digest*, Vol. 4, No. 3, 1983.

Billig, F. S., Waltrup, P. J., and Stockbridge, R. D., "The Integral-Rocket, Dual-Combustion Ramjet: A New Propulsion Concept," *J. of Spacecraft and Rockets*, Vol. 17, No. 5, 1980.

Braden, R. F., Nagaraja, K. S., and von Ohain, H. J. P. (Eds.), *Proceedings: Ejector Workshop for Aerospace Applications*, AFWAL-TR-82-3059, Flight Dynamics Laboratory, Air Force Wright Aeronautical Laboratories, Wright–Patterson Air Force Base, OH, June 1982.

Bunt, E. A., McMurray, G. S., and Dugger, G. L., "Ramjet Technology—Chapter 9, Combustor Design," TG 610-9, Johns Hopkins University, Applied Physics Laboratory, Baltimore, MD, 1972.

Cockrell, D. J. and Markland, E., "A Review of Incompressible Diffuser Flow," *Aircraft Eng.*, Vol. 35, No. 10, p. 286, 1963.

Collier, S. L. and Hoerner, C. C., "Development of Affinity Relations for Modeling Characteristics of Check Valves," *J. Energy Resources Technology*, Vol. 103, pp. 196–200, 1981.

COSMIC, A Catalog of Selected Computer Programs, NASA CR 163728 (N81-11690), University of Georgia, Athens, 1980.

Crane Co., *Flow of Fluids through Valves, Fittings, and Pipe*, Tech. Paper No. 410, Crane, New York, 1979.

Crocco, L., "One-Dimensional Treatment of Steady Gas Dynamics," *Fundamentals of Gas*

Dynamics, High Speed Aerodynamics and Jet Propulsion Series, Vol. III, Princeton University Press, Princeton, NJ, 1958.

Cronvich, L. L. (Ed.), "Summary Report on Airframe, Aerodynamics and Propulsion Design of Typhon LR Missiles, S/N 1 to 9," TG 679, Johns Hopkins University, Applied Physics Laboratory, Baltimore, MD, 1965.

Csemniczky, J., "Hydraulic Investigations of Check Valves and Butterfly Valves," *Proceedings of the 4th Conference on Fluid Machinery*, Budapest, pp. 293–307, 1972.

Cutler, A. D. and Johnston, J. P., "The Effects of Inlet Conditions on the Performance of Straight-walled Diffusers at Low Subsonic Mach Numbers—A Review," Thermosciences Div., Report PD-26, Stanford University, Stanford, CA, 1981.

Decher, R., "Nonuniform Flow through Nozzles," *AIAA J. Aircraft*, Vol. 15, No. 7, pp. 416–421, 1978.

Decher, R. "Mass Flow and Thrust Performance of Nozzles with Mixed and Unmixed Non-Uniform Flow," *J. Fluids Eng.*, Sept. 1995.

Decher, R., *Energy Conversion*, Oxford University Press, NY, 1994.

DeFrate, L. A. and Hoerl, A. E., "Optimum Design of Ejectors Using Digital Computers," *Chemical Engineering Progress Symposium Series*, No. 21, Vol. 55, 1982.

Dighe, A. S., "Effects of Aspect Ratio and Low Reynolds Number on the Pressure Recovery and Flow Regime Behavior of Plane Wall Diffusers," Ph.D. thesis, School of Mechanical Engineering, Purdue University, Lafayette, IN, May 1973.

Dobek, L. J., "Labyrinth Seal Testing for Lift Fan Engines," NASA Contract #NAS3-14409, 1973.

Dolan, F. X. and Runstadler, P. W., Jr., "Pressure Recovery Performance of Conical Diffusers at High Subsonic Mach Numbers," NASA CR-2299, 1973.

Emanuel, G., "Comment on 'Experimental Determination of Discharge Coefficient for Critical Flow through an Axisymmetric Nozzle,'" *AIAA J.*, Vol. 17, No. 1, p. 126, 1979. "Flow Measurement," ASME Test Code 19.5.4, 1959.

Estey, P., Altman, D., and McFarlane, J., "An Evaluation of Scaling Effects for Hybrid Rocket Motors," AIAA-91-2517, AIAA 27th Joint Propulsion Conference, 1991.

Fabri, J. and Siestrunck, R., "Supersonic Air Ejectors," *Advances in Applied Mechanics*, Vol. 5, Academic Press, New York, 1958.

Faro, I. D. V. and Keirsey, J. L., "Ramjet Technology, Supersonic Inlets," TG 610-3B, Johns Hopkins University, Applied Physics Laboratory, Baltimore, MD, 1967.

"Flow Measurement," ASME Test Code 19.5.4, 1959.

Fox, J. A., *Hydraulic Analysis of Unsteady Flow in Pipe Networks*, John Wiley & Sons, New York, 1977.

Gardel, A., "Pressure Drops in Flows through T-Shaped Pipe Fittings," *Bull. Techn. de la Suisse Romande*, Vol. 83, Nos. 9 & 10, pp. 123, 143, 1957.

Gilreath, H. E., "The Beginning of Hypersonic Ramjet Research at APL," *Johns Hopkins APL Technical Digest*, Vol. 11, Nos. 3 and 4, 1990.

Gordon, S. and McBride, B. J., "Computer Program for Calculation of Complex Chemical Equilibrium Composition, Rocket Performance, Incident and Reflected Shocks, and Chapman-Jouget Detonations," NASA SP-273, NASA Lewis Research Center, 1971.

Goss, W. H., "Talos in Retrospect," *Johns Hopkins APL Technical Digest*, Vol. 3, No. 2, 1982.

Grein, H. and Osterwalder, J., "Hydraulic Forces Acting on the Upstream and Downstream Pipelines and the Casing of Butterfly Valves During Closing," *Escher Wyss News*, No. 1, pp. 27–32, 1978.

Grein, H. and Pirchl, H., "The Butterfly Valve Used as Damper in Pumped Storage Tail-water Systems," Paper G1, IAHR Symposium, Stockholm, 1970.

Guins, V. G., "Flow Characteristics of Butterfly and Spherical Valves," *J. Hydraulics Division*, ASCE, Vol. 94, HY3, pp. 675–690, 1968.

Hamner, R. M., "The Use of Waste Heat for Automotive Air Conditioning," SAE Technical Paper Series, Detroit, MI, 1981.

Hardgrave, E. J., Jr., "Ramjet Description and Analysis," Lectures at the Johns Hopkins University, Applied Physics Laboratory, unpublished, 1977 (est.).

Harrison, P. and Schweiger, F., "Head-Loss Characteristics of Butterfly Valves," *Proceedings, Congress of the International Association for Hydraulic Research*, London, Paper 4.15, 1963.

Hill, P. G. and Peterson, C. R., *Mechanics and Thermodynamics of Propulsion*, 2nd ed., Addison-Wesley, Reading, MA, 1992.

Hoerner, S. F., *Fluid Dynamic Drag*, Hoerner, 1965.

Hoge, H. J., *On the Theory of Mixing of Fluid Streams*, Quartermaster Research and Engineering Center, Pioneering Research Division, Technical Report PR-2, 1959.

Hutchison, J. W., *ISA Handbook of Control Valves*, 2nd Ed., Instrument Society of America, Pittsburgh, PA, 1976.

Huzel, D. K. and Huang, D. H., "Design of Liquid Propellant Rocket Engines," 2nd ed., NASA SP-125, 1971.

Ito, H., "Pressure Losses in Smooth Pipe Bends," *Trans. ASME, Ser. D*, Vol. 82, No. 1, p. 131, 1960.

Ito, H. and Imai, K., "Energy Losses at 90 Degree Pipe Junctions," *Proc. ASCE, J. Hydraulics Division*, Vol. 99, p. 1353, 1973.

Jeppson, R. W., *Analysis of Flow through Pipe Networks*, Ann Arbor Science, Ann Arbor, MI, 1976.

Jordan, R. C. and Priester, G. B., *Refrigeration and Air Conditioning*, 2nd ed., Prentice-Hall, Englewood Cliffs, NJ, 1956.

Jung, R., "Beiträge angewandter Strömungsforschung zur Entwicklung der Kohlenstaubfeuerung," *VDI-Forschungsheft* 532, pp. 1–40, 1969.

Kalt, S. and Bendall, D., "Conical Rocket Performance Under Flow Separated Conditions," *J. of Spacecraft and Rockets*, Vol. 2, No. 3, pp. 447–449, 1965.

Kane, R. S. and Cho, S. M., "Hydraulic Performance of Tilting-Disc Check Valves," *J. Hydraulics Division, ASCE*, Vol. 102, HY1, pp. 57–72, 1976.

Keirsey, J. L. and Cusick, R. T., "Automated Airbreathing Propulsion Test Facilities," *Johns Hopkins APL Technical Digest*, Vol. 4, No. 3, 1983.

Keirsey, J. L., "Airbreathing Propulsion for Defense of the Surface Fleet," *Johns Hopkins APL Technical Digest*, Vol. 13, No. 1, 1992.

Kennedy, E. D., "The Ejector Flow Process," M.S. Thesis, University of Minnesota, Minneapolis, MN, 1955.

Kerrebrock, J. L., *Aircraft Engines and Gas Turbines*, MIT Press, Cambridge, MA, 1977.

Kiselev, B., *Calculation on One-Dimensional Gas Flows*, USAF Technical Report No. F-TS-1209-1A, 1949 (translated from Russian by Brown University).

Klein, A., "Review: Effects of Inlet Conditions on Conical-Diffuser Performance," *J. Fluids Eng.*, Vol. 103, No. 2, p. 250, 1981a.

Klein, A., "Review: Turbulent Developing Pipe Flow," *J. Fluids Eng.*, Vol. 103, No. 2, p. 243, 1981b.

Klenhofer, W. J. and Derrick, C. T., "Tailpipe Effects on Gas Turbine Diffuser Performance with Fully Developed Inlet Conditions," ASME Paper No. 70-GT-86, 1970.

Kording, J. W. and Fuller, G. H., "Correlation of Nozzle Submergence Losses in Solid Rocket Motors," *AIAA J.*, Vol. 5, pp. 175–177, 1967.

Kuluva, N. M. and Hosack, G. A., "Supersonic Nozzle Discharge Coefficients at Low Reynolds Numbers," *AIAA J.*, Vol. 9, No. 9, p. 1876, 1971.

Kuo, K. K. and Summerfield, M. (Eds.), *Fundamentals of Solid Propellant Combustion*, Progress in Astronautics and Aeronautics, Vol. 90, AIAA, New York, 1984.

Landsbaum, E. M., Salinas, M. P., and Leary, J. P., "Specific Impulse Predictions of Solid-Propellant Motors," *J. Spacecraft and Rockets*, Vol. 17, No. 5, pp. 400–406, 1980.

Liepmann, H. W. and Roshko, A., *Elements of Gas Dynamics*, John Wiley & Sons, New York, 1957.

"Liquid Propellant Theoretical Performance," in *CRC Handbook of Tables for Applied Engineering*, Chemical Rubber Co., Cleveland, OH, 1970.

Mahoney, J. J., "Inlets for Supersonic Missiles," AIAA Education Series, AIAA, Washington, D.C., 1991.

Martin, C. S., Medlarz, H., Wiggert, D. C., and Brennen, C., "Cavitation Inception in Spool Valves," *J. of Fluids Engineering*, Vol. 103, pp. 564–576, 1981.

Mayer, E. (Nau, B. S., translator): *Mechanical Seals*, 2nd ed., Elsevier, New York, 1973.

McDonald, A. T. and Fox, R. W., "An Experimental Investigation of Incompressible Flow in Conical Diffusers," *Intern. J. Mech. Sci.*, Vol. 8, No. 2, p. 125, 1966.

McDonald, A. T., Fox, R. W., and Van DeWostine, R. V., "Effects of Swirling Inlet Flow on Pressure Recovery in Conical Diffusers," *AIAA J.*, Vol. 9, No. 10, p. 2014, 1971.

McLafferty, G. H., Krasnoff, E. L., Ranard, E. D., Rose, W. G., and Vergara, R. D., "Investigation of Turbojet Inlet Design Parameters," United Aircraft Corp. Rept. R-0790-13, 1955.

McMahon, P. J., *Aircraft Propulsion*, Harper & Row, New York, 1971.

McPherson, M. B., Strausser, H. S., and Williams, J. C., "Butterfly Valve Flow Characteristics," *J. Hydraulics Division, ASCE*, Vol. 83, HY1, Paper 1167, 1957.

Miller, D., "Performance of Straight Diffusers," Part II of *Internal Flow: A Guide to Losses in Pipe and Duct Systems, BHRA Fluid Eng.*, Cranfield/Bedford, UK, 1971.

Miller, D. S., *Internal Flow Systems*, BHRA Fluid Engineering Series, Vol. 5, 1978.

Minardi, J. E. and von Ohain, H. P., *Thrust Augmentation Study of High Performance Ejectors*, AFWAL-TR-83-3087, Flight Dynamics Laboratory, Air Force Wright Aeronautical Laboratories, Wright-Patterson Air Force Base, OH, 1983.

Minardi, J. E., Lawson, M. O., Krolack, R. B., Newman, R. K., von Ohain, H. P., Salyer, I. O., Watterndorf, F. L., Boehman, L. I., and Braden, R. P., *Ejector-Turbine Studies and Experimental Data*, DOE/ER/10509-1, 1982.

Minardi, J. E., "Compressible Flow Ejector Analysis with Application to Energy Conversion and Thrust Augmentation," AIAA-82-0133, 1982b.

Minardi, J. E., *Characteristics of High Performance Ejectors*, AFWAL-TR-81-3170, Flight Dynamics Laboratory, Air Force Wright Aeronautical Laboratories, Wright–Patterson Air Force Base, OH, 1982b.

Minardi, J. E., "Ejector-Refrigeration for Automotive Application," UDRTR-82-20, Harrison Radiator, Division of General Motors Corporation, Dayton, OH, 1982a.

Moody, L. F., "Friction Factors for Pipe Flow," *Trans. ASME*, Vol. 66, No. 8, p. 671, 1944.

Morisette, E. L. and Goldberg, T. J., "Turbulent Flow Separation Criteria for Overexpanded Supersonic Nozzles," NASA Technical Report 1207, 1978.

Nece, R. E. and Dubois, R. E., "Hydraulic Performance of Check and Control Valves," *J. Boston Society of Civil Engineers*, Vol. 42, pp. 263–286, 1955.

Nickerson, G. R., Coats, D. E., Dang, A. L., Dunn, S. S., Berker, D. R., Hermsen, R. L., and Lamberty, J. T., *The Solid Propellant Rocket Motor Performance Prediction Computer Program (SPP) Version 6.0*, Air Force Astronautics Laboratory, AFAL-TR-87-078, 1987.

Nickerson, G. R., Coats, D. E., Dang, A. L., Dunn, S. S., and Kehtarnavaz, H., *Two-Dimensional Kinetics (TDK) Nozzle Performance Computer*, NAS8-36863, February 1988.

Oates, G. C. (Ed.), "The Aerothermodynamics of Aircraft Gas Turbine Engines" AFAPL-TR-78-52, Wright Patterson Air Force Base, OH, 1978.

O'Connor, J. J., *Standard Handbook of Lubrication Engineering*, McGraw-Hill, New York, 1968.

Peters, M. S. and Timmerhaus, K. D., *Plant Design and Economics for Chemical Engineers*, 3rd ed., McGraw-Hill, New York, 1980.

Petrein, R. J., Longwell, J. P., and Weiss, M. A., "Flame Spreading from Baffles," Esso Research and Engineering Co., Linden, NJ, 1955.

Pickett, E. B., Hart, E. D., and Neilson, F. M., "Prototype Experiences with Large Gates and Valves," *Control of Flow in Closed Conduits*, Colorado State University, Fort Collins, CO, 1971.

Porter, J. L. and Squyers, R. A., *A Summary/Overview of Ejector Augmentor Theory and Performance*, USAF Technical Report No. R-91100-9CR-47, Vols. I and II, 1981.

Rao, P. V. and Martin, C. S., "Choking in a Cavitating Spool Valve," *ASME Cavitation and Polyphase Flow Forum*, New Orleans, LA, pp. 21–24, 1980.

Reneau, L. R., Johnston, J. P., and Kline, S. J., "Performance and Design of Straight, Two-Dimensional Diffusers," *J. Basic Engrg.*, Vol. 89, No. 1, p. 141, 1967.

Reynolds, A. J., *Turbulent Flows in Engineering*, John Wiley & Sons, New York, 1974.

Rudnick, P., "Momentum Relation in Propulsive Ducts," *J. Aero. Sci.*, Vol. 14, p. 540, 1947.

Runstadler, P. W., Jr. and Dean, R. C., Jr., "Straight Channel Diffuser Performance at High Inlet Mach Numbers," *J. Basic Eng.*, Vol. 91, No. 3, p. 397, 1969.

Runstadler, P. W., Jr. and Dolan, F. X., "Further Data on the Pressure Recovery Performance of Straight-Channel, Plane-Divergence Diffusers at High Subsonic Mach Numbers," *J. Fluids Eng.*, Vol. 95, No. 3, p. 373, 1973.

Safwat, H. H., "Experimental and Analytic Data Correlation Study of Water Column Separation," *J. Fluids Eng.*, Vol. 95, pp. 91–97, 1973.

Sarpkaya, T., "Torque and Cavitation Characteristics of Butterfly Valves," *J. Appl. Mech.*, Vol. 28, pp. 511–518, 1961.

Schmucker, R. H., "Flow Processes in Overexpanded Chemical Rocket Nozzles. Part 1: Flow Separation," NASA TM-77396, 55 pp., 1984.

Shapiro, A. H., *The Dynamics and Thermodynamics of Compressible Fluid Flow*, Vol. I, Ronald Press, 1953.

Shippen, W. B., Berl, W. G., Garten, W., Jr., and Hardgrave, E. J., Jr., "The Talos Propulsion System," *Johns Hopkins APL Technical Digest*, Vol. 3, No. 2, 1982.

Sovran, G. and Klomp, E. D., "Experimentally Determined Optimum Geometries for Rectilinear Diffusers with Rectangular, Conical or Annular Cross Section," *Fluid Mechanics of Internal Flow*, Sovran, G. (Ed.), Elsevier, New York, p. 270, 1967.

Stein, P. C., "A Discussion of The Theory of Sealing Devices," Presented to The Society of Automotive Engineers, 1961.

Stein, P. C., "Circumferential Seals for Use as Oil Seals," American Society of Lubrication Engineers Paper #79062, 1979.

Strohmer, F., "Investigating the Characteristics of Shutoff Valves by Model Tests," *Water Power*, Vol. 29, pp. 41–46, 1977.

Stull, D. R. and Prophet, H., *JANNAF Thermochemical Tables*, 2nd ed., Office of Standard Reference Data, National Bureau of Standards, Washington, D.C., 1971.

Summerfield, M., "The Liquid Propellant Rocket Engine," *High Speed Aerodynamics and Jet Propulsion*, Vol. III, Princeton University Press, Princeton, NJ, 1959.

Sutton, G. P., *Rocket Propulsion Elements, Sixth Edition*, John Wiley & Sons, New York, 1993.

Tang, S. P. and Fenn, J. B., "Experimental Determinations of the Discharge Coefficient for Critical Flow through an Axisymmetric Nozzle," *AIAA J.*, Vol. 16, No. 1, p. 41, 1978.

Thornock, R. L., "Experimental Investigation of the Flow through Convergent Conical Nozzles," The Boeing Company, D6-20375, 1968, NTIS AD 841-418.

Tullis, J. P., "Choking and Supercavitating Valves," *J. Hydraulics Division, ASCE*, Vol. 97, HY12, pp. 1931–1945, 1971.

Uebelhack, H. T., "One-Dimensional Inviscid Analysis of Supersonic Ejectors," *AGAR-Dograph No. 163 on Supersonic Ejectors*, Ginoux, J. J. (Ed.), AGARD (NTIS, 5285 Port Royal Road, Springfield, VA), 1972.

Waltrup, P. J., "Hypersonic Airbreathing Propulsion: Evolution and Opportunities," *Proc. AGARD Conference, Aerodynamics of Hypersonic Lifting Vehicles*, NATO, Bristol, England, 1987.

Waltrup, P. J., Anderson, G. Y., and Stull, F. D., "Supersonic Combustion Ramjet (Scramjet) Engine Development in the United States," *Proc. 3rd International Symposium on Air Breathing Engines*, Munich, Germany, 1976.

Wanstall, B., "Integral Rocket Ramjets for Long Legs at Supersonic Speeds," *Interavia*, Vol. 12, pp. 1331–1334, 1984.

Ward–Smith, A. J., "Component Interactions and Their Influence on the Pressure Losses in Internal Flow Systems," *Proc. Inst. Mech. Engrs.*, Vol. 190, p. 349, 1976.

Ward–Smith, A. J., *Internal Fluid Flow*, Clarendon Press, Oxford, 1980.

Watters, G. Z., *Analysis and Control of Unsteady Flow in Pipelines*, Butterworths, Boston, MA, 1984.

Webster, F. F., "Liquid Fueled Integral Rocket/Ramjet Technology Review," AIAA paper 78-1108, 14th Joint Propulsion Conference, Liquid Fuel IRR Technology, Martin–Marietta, Orlando, FL, 1978.

Wernimont, E. J. and Heister, S. D., "Performance Characterization of Hybrid Rockets using Hydrogen Peroxide Oxidizer," AIAA 95-3084, 1995.

Wood, D. J. and Jones, S. E., "Waterhammer Charts for Various Types of Valves," *J. Hydraulics Division, ASCE*, Vol. 99, HY1, pp. 167–178, 1973.

Wood, D. J. and Rayes, A. G., "Reliability of Algorithms for Pipe Network Analysis," *Jour. Hydraulics Div.*, ASCE, Vol. 107, No. 10, 1981.

Wylie, E. B. and Streeter, V. L., *Fluid Transients*, McGraw-Hill, New York, 1978.

Ytzen, G. R., "Ball Valves for Throttling Control," *ISA Journal*, Vol. 12, pp. 78–80, 1965.

Zuk, J., "Fundamentals of Fluid Sealing," NASA TN D-8151, 1976.

Zuk, J., Ludwik, L., and Johnson, R., "Design Study of Shaft Face Seal with Self-Acting Lift Augmentation," NASA TN D-5744, 1970.

26 Positive Displacement Compressors, Pumps, and Motors

RICHARD NEERKEN (Retired)
The Ralph M. Parsons Company
Pasadena, CA

CONTENTS

26.1 INTRODUCTION

No discussion of fluid handling machinery can be complete without considering positive displacement machinery in addition to turbomachinery. Most of the earliest fluid handling machines were positive displacement types, and modern forms of these are still in wide use for many application areas. This chapter will provide a review of types, areas of usage, and fundamentals of selection and application.

Handbook of Fluid Dynamics and Fluid Machinery, Edited by Joseph A. Schetz and Allen E. Fuhs
ISBN 0-471-12598-9 Copyright © 1996 John Wiley & Sons, Inc.

In Positive Displacement (PD) machines, energy is imparted to a liquid or gas contained in a fixed displacement volume such as a casing or a cylinder, by the rotary motion of mechanisms such as gears, screws or vanes, or by reciprocating pistons or plungers. Table 26.1 illustrates an overall view of types currently in use today.

PD machines have several distinct advantages over turbomachines for certain application areas, such as the following:

- Low volume flows where a suitable turbomachine cannot be designed
- Intermediate flow ranges where the PD machine will have higher efficiency than the turbomachine
- High pressure, low flow applications
- Gas compression of low molecular weight gases at relatively high pressure ratios, where the head required for a turbo-compressor may be excessively high
- Pumping of viscous materials, where the effect of viscosity causes centrifugal pump efficiency to be lower than a PD pump
- Pumping non-Newtonian liquids
- Slurry pumping in certain higher pressure applications
- Fluids of widely varying composition, or in systems where the flows or operating pressures may vary widely

Reference to specfic speed charts for turbomachinery published by Balje (1962) will show clearly where the PD machine should be applied from a flow standpoint. Although high speed turbomachines are also available for certain low flow, high head applications, a study of both types may show in many cases that the PD machine is best suited for the particular job.

TABLE 26.1   Types of Positive Displacement Machinery*

Reciprocating			Rotary		
Compressors	Pumps	Expansion Engines	Compressors	Pumps	Hydraulic Motors
Piston	Piston	Steam engines	Lobe	Gear	Gear
Plunger	Plunger	Gas expander	Vane	Internal	Screw
Air-cooled	Direct-acting	engines	Dry-screw	External	Vane
cylinder	Power driven	(cryogenic)	Oil-flooded	Lobe	
Water-cooled	Controlled-	Air motors	screw	Vane	
cylinder	volume		Liquid ring	Screw type	
Diaphragm	Diaphragm			Single	
Non-lubricated				Double	
cylinder				Triple	
				Flexible-	
				member	
				Cam-and-	
				piston	
				Circumferential	
				piston	

*Excluding Combustion Engines.

PD machines have the theoretical capability of producing whatever pressure rise is required for a particular application. Of course, limitations apply for a real machine. For example, although a fixed dimension cylinder and piston combination will produce pressure beyond the original design condition, the power input will rise until the driver is overloaded or stalls, the pressure will rise until the cylinder maximum working pressure is exceeded, or the discharge temperature may rise until it exceeds what the fluid can stand for safety reasons (e.g., the discharge temperature of air in an oil lubricated air compressor must not rise above a limit of about 200°C, or an explosive mixtue of lubricating oil and air may enter the discharge system). Since the PD machine produces pressure, not head, it is to some extent unaffected by changes in fluid characteristics such as gas molecular weight or liquid density. This is a useful feature in certain applications where operating conditions are subject to wide variation.

26.2 POSITIVE DISPLACEMENT COMPRESSOR TYPES

26.2.1 Reciprocating Compressors

This type of compressor makes use of the motion of pistons or plungers in one or more cylinders, connected through piston rods, crossheads, and connecting rods to a crankshaft which transmits power from a prime mover. High pressures and relativly low volume flows usually require reciprocating machines. The number of stages or cylinders must be selected with consideration for stage discharge temperature, available cylinder size, and compressor power frame capability. Pistons with metal or nonmetal piston rings are used for lower pressures, to about 7000 kPa, depending on size of piston and type of duty. For almost all higher pressure applications, plungers are required. A multistage compressor may have both piston type cylinders in the lower stages, and plunger type in the higher stages. Two cylinders may be arranged in tandem on one piston rod/crosshead assembly to save space. Figure 26.1 shows a typical multistage compressor for process duty with several sizes and types of cylinder construction and arrangement.

Smaller machines using air-cooled cylinders are applied mainly as air compressors. These usually have trunk-type pistons similar to an automobile engine with the inherent feature of oil from the crankcase mixing with the air being compressed. Such construction is satisfactory for many industrial applications, but if no oil carryover can be tolerated in the compressed air such as in an instrument air supply, nonlubricated compressor cylinders are required. Figure 26.2 illustrates a small two-stage air compressor with air-cooled cylinders.

Diaphragm type cylinders are used on small capacity units for special applications. Here, the reciprocating motion is transmitted to a flexible diaphragm which prevents leakage of the compressed gas to atmosphere, a most important factor where the gas is toxic, flammable, or very costly. Figure 26.3 illustrates this type.

26.2.2 Rotary Compressors

Rotary compressors, blowers, and vacuum pumps are also PD machines, in which a rotating mechanism displaces a fixed volume during each revolution. These types have no pistons, piston rods, or crankshafts. Also, no valve motion is required to

FIGURE 26.1 Typical multistage reciprocating compressor. (Courtesy of Ingersoll-Rand Co.)

admit and discharge gas or air during each revolution. Current types of rotary compressors available today can be grouped into four general categories (Table 26.1). The oldest, and perhaps the best-known, is the *lobe-type*, where two figure eight shaped rotors mesh together, driven through timing gears attached to each shaft (Fig. 26.4). Sizes range from about 1 dm^3/s to 10,000 dm^3/s. Pressures available with

FIGURE 26.2 Single-acting air-cooled air compressor. (Courtesy of Ingersoll-Rand Co.)

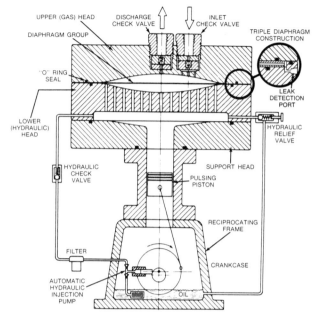

TYPICAL ASSEMBLY OF A
DIAPHRAGM COMPRESSOR

FIGURE 26.3 Section of Diaphragm type compressor. (Courtesy of Pressure Products Ind., Duriron Company.)

this type are low, up to about 300 kPa maximum. The lobe type machine is also widely used as a *vacuum pump*, which is the name given to a compressor which takes suction at less than atmospheric pressure and discharges at or slightly above atmospheric.

A second style utilizes sliding vanes which move radially within a rotor, trapping air or gas and gradually reducing its volume with each revolution. This type can produce up to about 700 kPa per stage and can be arranged in two stages. Its main uses are in compressing air and as a vacuum pump.

The most versatile type of rotary compressor is the *screw-type*, which utilizes male and female rotors meshing together within a cylindrical or slightly oval-shaped casing. These rotors use a somewhat complex form, either symmetrical or asymmetrical, generated on specialized milling machines and carefully fitted to provide the correct clearance between the mating parts. The rotor form was conceived in

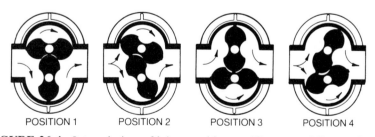

POSITION 1 POSITION 2 POSITION 3 POSITION 4

FIGURE 26.4 Internal view of lobe type blower. (Courtesy of Dresser Roots.)

FIGURE 26.5   Rotary screw-type compressor. (Courtesy of Atlas Copco Comptec Inc.)

Europe before World War II and is sometimes named the *Lysholm rotor* after its inventor, A. J. R. Lysholm (Fig. 26.5). The screw compressor is built in both oil flooded (oil cooled) and dry-screw types. The oil flooded machines have found wide acceptance for services such as plant air, service air, portable air compression, and as refrigeration compressors. Rotating speed is limited due to the oil-flooded design, where the oil performs the function of removing the heat of compression as well as lubricating the rotors. Because of this cooling, a one-stage machine can produce a considerably higher ratio of compression than any other type of rotary machine.

For most chemical process applications handling gases where oil contamination is not permissible, the dry-screw type will be required. This type operates at speeds up to 12000 rev/min in smaller sizes and has been used for discharge pressures up to 3000 kPa, with pressure rise per stage limited to about 500 kPa. Another variation of the screw-type, not related to the Lysholm rotor design, is the spiral-axial rotor type, following the idea of male and female helical rotors meshing together but not reaching pressures as high as the helical-screw type.

The fourth type of rotary is the *liquid-ring* type, used either as a compressor or as a vacuum pump (Fig. 26.6). A circular or elliptical vaned-type rotor moves in a circular or oval-shaped casing where water or other sealing liquid is present. Centrifugal force causes the liquid to form a ring around the periphery of the casing when in operation. The air or gas being compressed flows inward towards the center of the rotor gradually decreasing in volume and increasing in pressure until passing discharge ports where it leaves the casing. The liquid carried into the air or gas is

FIGURE 26.6 Liquid ring type rotary compressor or vacuum pump. (Courtesy of Nash Engineering Co.)

separated and then cooled and recirculated, or discharged in a once-through system. Pressure limitations of the liquid-ring type are low, limited to about 700 kPa per casing. This type finds its greatest use as a vacuum pump in such applications as in a power generation plant to exhaust air and other noncondensible gases from a turbine steam condenser. Stainless steel construction is also available, which is useful in certain chemical process applications handling difficult, toxic or corrosive gases at low pressures.

26.3 POSITIVE DISPLACEMENT COMPRESSOR PERFORMANCE

26.3.1. Reciprocating Compressor Performance Calculations

Reference to any basic thermodynamic text will give fundamentals for the compression of air or gas in PD machines (See Chaps. 7 and 8). Compression processes are shown as isothermal, isentropic, and polytropic. Experience with real PD machines indicates that seldom, if ever, is isothermal compression achieved. A very small diameter piston in a well-designed, water-jacketed cylinder may approach isothermal conditions, but for an estimate of compressor performance, it is preferable to use isentropic or polytropic formulae. Assume isentropic compression and recall that the work, W, performed is

$$W = \frac{p_1 V_1}{[(\gamma - 1)/\gamma]} \left[\left(\frac{p_2}{p_1} \right)^{(\gamma - 1)/\gamma} - 1 \right] \tag{26.1}$$

or

$$W = \frac{\dot{m} z_1 R T_1}{[(\gamma - 1)/\gamma]} \left[\left(\frac{p_2}{p_1} \right)^{(\gamma - 1)/\gamma} - 1 \right] \tag{26.2}$$

and discharge temperature

$$T_2 = T_1[p_2/p_1]^{(\gamma - 1)/\gamma} \qquad (26.3)$$

This work, divided by an assumed or measured overall efficiency, will give an estimate of the power input required. Note that absolute values of pressure and temperature must be used. The same formulae may be used for polytropic compression, $pV^n = C$, where the specific heat ratio γ is replaced with the polytropic exponent n. Polytropic compression in a PD machine occurs when the cylinder cooling causes the compression to deviate from isentropic. Heat may be added in the process, in which case $n > \gamma$, or, in some well-cooled cylinders, the compression is actually between isentropic and isothermal, with $n < \gamma$. However, for preliminary estimates of real PD compressor performance, the use of the isentropic relation is sufficiently accurate. More empirical data regarding the effect of cylinder cooling is needed to accurately predict how much the compression will deviate from isentropic [see Faires and Simmang (1978) for further information on the thermodynamics of compression].

For more advanced problems, especially with gases at high pressures or where gas properties deviate significantly from an ideal gas, it is more accurate to solve power requirements based on computations using equations of state to find the actual gas properties at the inlet and outlet conditions. Due to the scope and complexity of the computations, large computers are frequently used. It is important to observe that the power requirement must be calculated per stage of compression. Otherwise, the compression process deviates too far from isentropic. If the ratio of compression is too high for one stage (about 3 to 3.5 maximum) as determined from temperature rise or from compressor frame rating, then the square root of the total ratio will equal the ratio per stage for two stages, or the nth root for n stages. The actual ratio per stage must be adjusted when interstage losses due to intercoolers, separators, interstage piping, and pulsation suppression devices have been included.

The compressor industry has long recognized the convenience of pre-calculated values of power as functions of gas quantity, compression ratio per stage, and ratio of specific heats of the gas being compressed. Such data is available in chart form for reciprocating compressors [NGPSA (1972)]. The charts are accurate only for air or gases with molecular weights near that of air. Gases significantly heavier or lighter than air must utilize correction factors against these chart values or deviation factors in the basic efficiency. Also, the effect of compressibility of the gas mixture, or deviation from perfect gas relationships, must be included in the overall determination of power requirements. The compression efficiency in reciprocating machines is very much influenced by the losses through the inlet and discharge valves and the sizing of the valves in proportion to the sizing of the cylinder displacement (see Sec. 25.3). Thus, it can readily be seen that a compressor handling a much lighter gas such as hydrogen will probably have somewhat lower valve losses than if handling air, and although the ratio of specific heats for hydrogen is almost the same as for air, the power required for a given volume flow at a given pressure ratio is somewhat less with hydrogen. (Important note: Volume flow rates must be used with these chart methods of calculation, not weight or mass flow.)

Rather than using the chart method, a first approximation of power can be obtained by using the basic textbook relations for isentropic compression together with

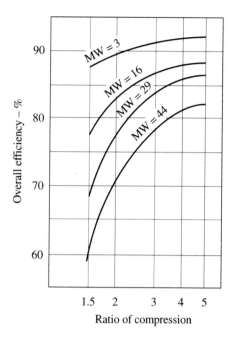

FIGURE   26.7 Reciprocating   compressor overall efficiency.

an assumed overall isentropic efficiency. This efficiency may be expressed as a function of gas properties and the ratio of compression (Fig. 26.7). These curves are typical for modern, slow to medium speed (277 to 720 rev/min), heavy duty type machines. Values of efficiency for smaller or higher speed compressors will be somewhat less, due to differences in valve and mechanical friction losses.

Sizes of reciprocating compressor cylinders are determined by calculating the expected volumetric efficiency of the compression stage. Because the piston or plunger does not sweep the entire volume of the cylinder due to operating clearances and allowance for valve passageways, the actual capacity of a given cylinder is less than its displacement volume. Cylinder displacement is found from speed, cylinder, and frame dimensions as

$$\text{Displ., } m^3/s = (\text{Total cyl. area, } m^2) \ (\text{stroke, } m) \ (\text{rev/min})/60 \qquad (26.4)$$

Volumetric efficiency is calculated and used with the required actual capacity to determine the displacement required. Faires and Simmang (1978) show the classic definitions of cylinder clearance and volumetric efficiency. For real machines, an empirical relation such as the following will be more accurate

$$\text{Volumetric efficiency} = 97 - (\% \text{ clearance}) \left[\left(\frac{p_2}{p_1} \right)^{1/\gamma} - 1 \right] \qquad (26.5)$$

$$(\text{Displ., } m^3/s) \ (\text{volumetric efficiency}) = \text{actual capacity, } m^3/s \qquad (26.6)$$

When Eq. (26.4) is applied to a compressor with a double acting piston, such as illustrated in Fig. 26.8, the following formulae should be used for cylinder area

$$\text{Total cylinder area} = \text{area of head end} + \text{area of crank end} \qquad (26.7)$$

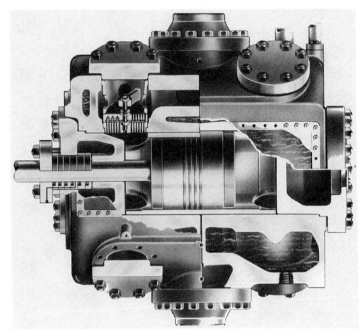

FIGURE 26.8   Section of double acting compressor cylinder. (Courtesy of Cooper Energy Services.)

or

$$\text{Total cylinder area} = (2)(\text{area head end}) - (\text{area of piston rod}) \quad (26.8)$$

Frame load or piston rod load must be considered, as every compressor frame design will have a limit on the loads that can be applied. For a double-acting cylinder when the piston is moving inward toward the crankshaft, the parts are in compression and the frame load is

$$F_c = (p_2 A_h) - (p_1 A_c) \quad (26.9)$$

Similarly, when the piston is moving outward away from the crankshaft, the parts are in tension, and the frame load is

$$F_t = (p_1 A_h) - (p_2 A_c) \quad (26.10)$$

These values will be the maximum during each reversal of the piston as it moves back and forth during one complete revolution of the crankshaft. The limits for a given compressor frame are usually available from manufacturer's published data and should not be exceeded, or mechanical problems will result, such as bending in piston rods or excessive wear or failure of crosshead pins or bushings. Generally, it is desirable to rate a machine for a given installation at not over 50 to 75% of the manufacturer's published frame rating. Note that the above simplified formulae do not take into account inertia loads from the mass and motion of the moving parts.

Compressor speed is limited by the average piston speed which has been proven to give acceptable wear life. This value for process type compressors should be held to the range 3 to 5 m/s as determined by the following relation

$$\text{Ave. piston speed, m/s} = \left[\text{crankshaft speed,} \frac{(\text{rev/min})}{60} \right] (\text{stroke, m}) \ (2)$$

$$(26.11)$$

26.3.2 Control of Reciprocating Compressors

Since a reciprocating compressor is a fixed volume machine, once the cylinder and piston dimensions have been selected, the capacity of the machine will vary only slightly as the ratio of compression changes [see Eq. (26.5)]. Assuming the compressor will run at a constant speed, three alternatives appear as means of controlling the output capacity. It is apparent that by increasing the cylinder clearance, the volumetric efficiency of a given size cylinder will be reduced, resulting in a lower output. *Clearance pockets* may be built into the cylinders or bolted on to special openings on the cylinders (Fig. 26.1). The pockets may be actuated by hand (manual clearance pockets) or automatically actuated by a pneumatic or electric control signal. Clearance pockets may be of fixed-volume design, which gives a definite *step* of capacity reduction based on the amount of clearance added. Variable volume pockets are also available, which are designed to give variable increases in clearance. Note that volumetric efficiency is also a function of compression ratio; thus, it can be seen that on low-ratio applications, clearance pockets may not be feasible. Generally, if the compression ratio is 2 or less, the size of the pocket may be too large to be practical.

Cylinder unloaders are devices which physically hold open the suction valves or remove them from action on one or both ends of a cylinder. Thus, the compressor does no work on that portion of the stroke (head end or crank end) when the unloader is activated. The capacity of a cylinder can be reduced first to approximately 50% of design, then to 0% by the second set of unloaders. Unloaders are usually also applied as a means for unloading the cylinders during startup, to avoid the need for excessively high starting torque in the prime mover. Like clearance pockets, unloaders may be manually or automatically actuated.

The third, and least efficient but most commonly used, method of capacity control for fixed speed compressors, is *external bypass control*. In this case, all the air or gas will have been compressed, then the excess amount not required for the process is bypassed around the cylinder, back to the suction source, sometimes requiring a bypass cooler. There is no saving in power when using the bypass, but in most actual process systems, due to limitations on the sizing of clearance pockets and unloaders, some bypass control is required to give steady control of the net forward flow. On plant or instrument air applications, the bypass is seldom needed. Excess air capacity is stored in an air receiver or reservoir (Fig. 26.2), from which system demand can be drawn at a variable rate, while the compressor cycles back and forth between 50 and 100% capacity, for example, with compressor control accomplished by unloaders and/or clearance pockets.

Variable speed drivers may be applied to reciprocating compressors. The gas

FIGURE 26.9   Gas engine driven reciprocating compressor. (Courtesy of Cooper Energy Services.)

engine driven compressor has been used extensively in the gas processing and pipeline transmission industries (Fig. 26.9). The steam turbine driver is sometimes used in process plant applications and will require a double-reduction gearbox to reduce the relatively high turbine speed to the low reciprocating compressor speed. Torsional problems may arise with a variable speed design, which must be carefully checked for each application. The reciprocating steam engine once was a popular prime mover for reciprocating compressors, but it is obsolete today.

Example 26.1:   Consider a requirement to compress 56710 *normal* m³/h* of low molecular weight gas such as a hydrogen/hydrocarbon mixture in a petroleum refinery process application. Inlet pressure is 1460 kPa, temperature 40 C, and discharge pressure is 15,000 kPa. Molecular weight is given (or calculated) as 3.28, ratio of specific heats at average temperature during compression is 1.38, and compressibility factors are 1.01 at inlet and 1.09 at discharge. [See Scheel (1972), Gibbs (1956), and Faires and Simmang (1978) for information on how to obtain these gas mixture properties.]

Looking first at the ratio of compression shows a total overall ratio of 15000/1460 or 10.274, which is too high for one stage (discharge temperature would be 322°C). For a first approximation, three stages might appear to be suitable, with a

---

Normal or *standard* volume expressed at 15 C and 101.325 kPa.

ratio per stage of $\sqrt[3]{10.274}$ or 2.1738. Flow at suction conditions to the first stage would be 1.2 m³/s from the relationship $p_1V_1/z_1T_1 = p_2V_2/z_2T_2$. A first estimate of the work required, using isentropic compression per stage is

$$\frac{kJ}{s} = \frac{(1.2 \text{ m}^3\text{/s}) \ (1460 \text{ kPa})}{0.275} (2.1738^{0.275} - 1) \qquad (26.12)$$

or 1523 kJ/s for the first stage. (Note: $(\gamma - 1)/\gamma = 0.275$.) Taking an approximate efficiency from Fig. 26.7 of 89% on this low molecular weight gas, the power required for the first stage is approximately 1523/0.89 or 1711 kW, or for three equal stages, about 5133 kW. With a compression ratio of 2.1738, the isentropic discharge temperature will be 115 C. The discharge gas temperature must be within the range of 100 to 150 C for a safe application on a gas mixture such as this.

Next, an allowance must be made for interstage pressure losses due to interstage cooling, piping, and pulsation dampers. Assume that the gas is cooled to 40 C between each successive stage of compression. Experience shows that reasonable values for interstage pressure losses can be expected to be from 3 to 5% of the absolute discharge pressure. Using this for a first estimate, the stage-by-stage performance is tabulated as shown in Table 26.2. The actual compressor pressures, flows, and estimated power requirements of each stage are shown, then the summation which gives the total power required for the three stage machine. If an electric motor were to be used as the driver for this compressor, the next largest commercially available size (at least 5 to 10% above the calculated kW) would be chosen, in this case probably a 6000 kW driver. Alternate operating conditions might make it desirable to oversize the driver so that the compressor would still perform even if, for example, the suction pressure were reduced for an alternate process condition.

If desired, additional estimating could be performed to arrive at cylinder sizes and actual machine preliminary outline dimensions and weights. Table 26.3 shows how a commercially available frame size with given stroke, speed, and frame rating might be selected from a specific manufacturer's data. Assuming some typical values for cylinder clearance (or taking these from manufacturer's data, if available), it would then be possible to estimate volumetric efficiencies, cylinder displacements, cylinder diameters, and actual frame loads. Having confirmed that cylinders of such dimensions are available with the required pressure ratings, a preliminary outline could be derived from the same manufacturer's published information, which would be useful in making an overall estimate of the total installed cost of such a compressor. After such preliminary estimates and studies have been completed, purchase specifications could be prepared, inquiries made, and machines offered from which an optimum choice could be made. Then, much additional information and data about the actual machine would be available. However, by the use of the simple fundamentals given here, a preliminary solution can be found for any reciprocating compressor problem.

Note in this example why a PD compressor is the only commercially practical machine which could be used. Assuming three centrifugal compressor sections with two intercoolers, just as with the PD compressor, the adiabatic head per section would exceed 700 kJ/kg, probably requiring at least 20 wheels in each casing, totally unfeasible for commercial machinery. Due to the large reduction in volume through

TABLE 26.2   Calculation Example for Reciprocating Compressor

Service: Hydrogen Feed Compressor

Flow, normal m^3/h	56,710
Molecular weight	3.28
Ratio Specific Heats, $c_p/c_v = \gamma$	1.38
Inlet Pressure, kPa	1460
Inlet Temperature, C	40
Discharge Pressure, kPa	15,000
Overall Compression Ratio	10.274
Number of Compression Stages	3

Stage	1	2	3
Inlet volume flow rate, m^3/s	1.2	0.5575	0.2627
Inlet pressure, kPa	1460	3174	6900
Inlet temperature, C	40	40	40
Compressibility at inlet, z_s	1.01	1.02	1.045
Discharge pressure, kPa	3174	6900	15,000
Compressibility at discharge, z_d	1.015	1.04	1.09
Interstage pressure loss, 5%	158	345	—
Actual discharge pressure at cylinder, kPa	3332	7245	15,000
Actual ratio of compression	2.282	2.282	2.174
Adiabatic discharge temperature, C	120	120	115
Adiabatic work, kJ/s	1622.9	1639	1569.5
*Adiabatic efficiency, percent	89	89	89
Power required per stage, kW	1823	1842	1763
Total power required, kW		5428	

*From Fig. 26.7.

this compression cycle, it is likely that all centrifugal casings would not run at the same speed, hence a long, five or six casing machine with at least two speed-changing gearboxes would be required. Industry would insist on a PD compressor for an application such as this.

Although it might seem that a rotary screw type compressor could also be used in this example, with two or three casing connected together in series, actual investigation with manufacturers of this type would show that casings have not been designed for pressures such as these. Unless new designs were developed, the rotary type could not be used.

26.3.3   Rotary Compressor Performance Calculations

Calculation methods for rotary type compressors, blowers, and vacuum pumps are not as widely known in the industry as for reciprocating types. Adiabatic head, weight flow, and inlet capacity calculations may be applied using the same basic thermodynamic relations discussed earlier. However, there is no well recognized source of information regarding efficiencies for these various types of machines, and most preliminary sizing must be done on the basis of manufacturer's catalog data. One original, independent method for sizing and selecting rotary compressors is

TABLE 26.3 Calculation Example for Reciprocating Compressor

Service: Hydrogen feed compressor

Frame rating
(from manufacturer's data)

Stroke, mm	450
Operating Speed, rev/min	277
Piston Speed, m/s	4.155
Piston Rod Diameter, mm	125
Maximum Frame Load, Compression, N	586
Maximum Frame Load, Tension, N	586

Stage	1	2	3
Cylinder clearance (assume)	16%	16%	18%
Calculated volumetric efficiency (eq. 26.5)	84%	84%	83.4%
Numer of cylinders (assume)	2	2	2
Cylinder type		Double-acting	
Displacement required/cylinder, m^3/s (eq. 26.6)	0.71429	0.33185	0.15749
Area of piston rod, m^2	0.01227	0.01227	0.01227
Total cylinder area required, m^2 (eq. 26.7)	0.34289	0.15879	0.07821
Cylinder diameter, mm	476	331	237
Frame load—in compression, N	349.65	387.14	451.10
in tension, N	−290.85	−259.34	−182.39
*Maximum cylinder operating pressure, kPa	3832	8332	17,250
Manufacturer's listed maximum cylinder working presure, kPa	4200	10,000	20,000

*Discharge pressure × 115% for relief valve setting.

presented in Scheel (1972) if the reader is interested in further analysis of this type PD machine.

26.4 POSITIVE DISPLACEMENT PUMP TYPES

As with compressors, rotary and reciprocating pumps are the two major types to be considered. Variations of the reciprocating type include controlled-volume (metering) pumps and diaphragm pumps for certain specialized applications.

Rotary pumps cover a field of application from the very smallest flows, such as might be required in a laboratory application, to medium flows up to about 250 $dm^3/$ s. Above this flow, either multiple rotary pumps will be required or the application will be satisfactory for centrifugal pumps. Reference to specific speed charts [Balje (1962)] will show the application area for this type.

Rotary pumps include many types or varieties of the same basic pump principle, a definition of which is given very clearly in Hydraulic Institute Standards [HIS (1983)] as ''. . . a positive displacement pump consisting of a chamber containing gears, cams, screws, vanes, plungers, or similar elements actuated by relative ro-

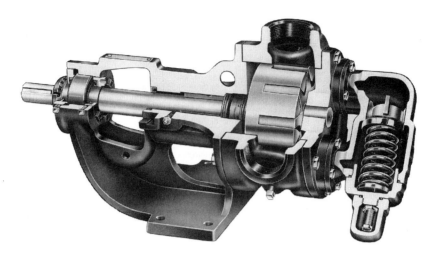

FIGURE 26.10 Cut-away view of internal gear pump. (Courtesy of Viking Pump Division.)

tation of the drive shaft and casing, and which has no separate inlet and outlet valves.'' That reference also shows schematic views of most of the major types of pumps presently in use. The most common type is the *gear pump*, where two helical or spur gears mesh, causing liquid to flow through a casing (Fig. 26.10). Pumps with helical screws meshing together, with or without timing gears, are another widely used type (Fig. 26.11). Pumps with a single helical rotor (known as the progressing cavity type), pumps with twin lobes, sliding vanes, circumferential piston, cam-and-piston, pumps using flexible tubes or flexible cylindrical liners, all have found good acceptance for many special applications.

Reciprocating pumps have widespread application for comparatively low flows and for high pressures to 70,000 kPa or higher. This type is also widely used in pumping slurries. For pressures above about 7000 kPa, the reciprocating pump is often the only choice, unless the flow is high enough to allow a multistage centrifugal pump to be used. Reciprocating types include *power pumps*, where a prime mover transmits power through a crankshaft to one or more pistons or plungers having either horizontal or vertical plunger orientation (Fig. 26.12). Another type

FIGURE 26.11 Sectional view of rotary screw pump. (Courtesy of Allweiler Pump Inc.)

FIGURE 26.12   Triplex plunger pump with electric motor driver and gear speed reducer. (Courtesy Wilson-Snyder Pumps, Division of U.S. Steel.)

is the *direct-acting pump*, where the driving medium is steam, air, or gas expanding in one or two cylinders, directly connected through a drive rod to one or two liquid cylinders containing the pumped liquid. Valve motion for the driving end is actuated by a linkage from the moving rods causing the motive fluid to change from one side of the piston to the other as the piston ends its stroke (Fig. 26.13). The crankshaft drive type may also be connected to a diaphragm type liquid end using an intermediate hydraulic fluid, thus isolating the pumped fluid from any leakage to atmosphere. Such pumps are only available in relatively small sizes, to about 10 dm³/s flow.

Either the plunger type or the diaphragm type is also used as a *controlled-volume pump*, where the pump output capacity can be varied accurately throughout a given flow range by means of stroke adjustment. Such adjustment can be accomplished manually, or more commonly, by an automatic mechanism receiving a pneumatic or electronic signal. The controlled-volume pump has the important feature of repeatability, that is, the ability to reduce the flow by stroke adjustment and then return to the original stroke setting where the pumping rate accuracy is repeated. Such

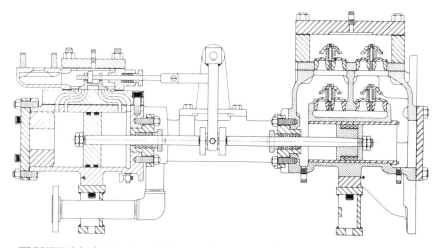

FIGURE 26.13   Section of direct-acting pump. (Courtesy of Union Pump Co.)

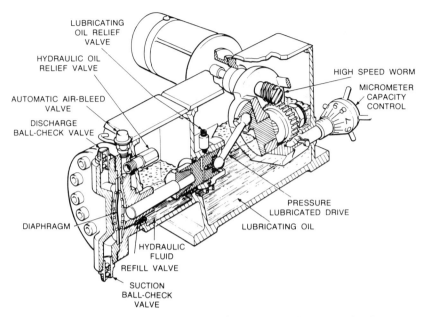

FIGURE 26.14   Diaphragm type controlled-volume pump. (Courtesy of Milton Roy Co.)

pump types have their primary applications in systems where precise accuracy of the pumped fluid delivery is required. Figure 26.14 shows a typical diaphragm type controlled-volume pump.

Numerous articles and books on positive displacement pumping are available, and the reader is referred to HIS (1983), Karassik *et al.* (1985), Henshaw (1981), Bristol (1981), Poynton (1983), and Neerken (1980) for more details.

26.5 POSITIVE DISPLACEMENT PUMP PERFORMANCE

For PD pumps, the net forward flow is less than the displacement of the elements within the casing or cylinder. The difference is referred to as *slip* (in a rotary pump), or the effect is measured in terms of the volumetric efficiency of the pump, thus

$$\text{net flow} = \text{displacement} - \text{slip} \tag{26.13}$$

$$\text{volumetric efficiency} = \frac{\text{net flow}}{\text{displacement}} \tag{26.14}$$

The internal clearances in a PD pump thus become very important in determining how much net flow a given pump will produce in a certain application. Figure 26.15 shows the expected performance of a certain size external gear type rotary pump with variations in clearance and discharge pressure. Similarly, clearance between the end of the piston or plunger stroke and the end of the cylinder, plus the volume contained in valve ports leading to and from the cylinder will determine the volumetric efficiency of a reciprocating pump.

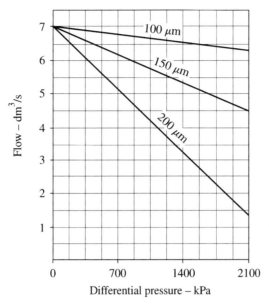

FIGURE 26.15 Rotary pump curve showing effect of clearance.

Pump performance can be estimated by using data from pump manufacturers' charts or tables, size by size, showing displacements, speed, actual net flows, and power requirements for liquids of different viscosities and at different pressures. Table 26.4 shows a typical manufacturer's chart for a certain size and type of rotary pump. Power requirements can be estimated using the formula for pump power

$$kW = \frac{\text{flow, dm}^3/\text{s} \times \Delta p, \text{kPa}}{1000 \times \text{efficiency (decimal)}} \qquad (26.15)$$

Figure 26.16 shows performance for a reciprocating pump with varying size plungers at different speeds. PD pump performance is seldom shown as a function of efficiency; instead, the charts are likely to indicate the power required for a given duty from which overall efficiency could be figured if required. Power requirements for a given size and type PD pump will obviously increase as the viscosity increases.

TABLE 26.4 Rating Chart for Rotary Pump

100 mm Twin-Screw Pump at 870 rev/m	Viscosity, cP							
	50		100		200		1000	
Differential Pressure, kPa	dm³/s	kW	dm³/s	kW	dm³/s	kW	dm³/s	kW
350	19.6	10.5	19.9	12.8	20.2	14.5	20.4	20.6
700	18.5	19.5	19.0	20.6	19.4	22.0	19.7	28.0
1050	17.5	26.8	18.3	28	18.8	29.5	19.2	35.5
1400	16.6	34.8	17.5	35.6	18.2	37.5	18.7	43.3

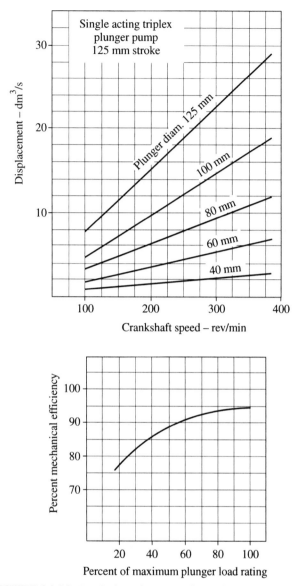

FIGURE 26.16   Typical performance for reciprocating pump.

If needed, performance curves can be constructed for pumps including the variables of flow, pressure, efficiency, viscosity, and power at a particular speed (Fig. 26.17) or at a variety of speeds (Fig. 26.18). In Fig. 26.17, notice that the flow is constant over the range of pressure shown for 100 cP viscosity, which confirms the fact that the slip inside the pump on viscous liquids is the same within these pressure ranges. With lower viscosity liquids this is not the case. Another way to express the rotary pump performance is shown in Fig. 26.19, where a fixed pressure is assumed, and the capacity, power, and efficiency are plotted as a function of viscosity.

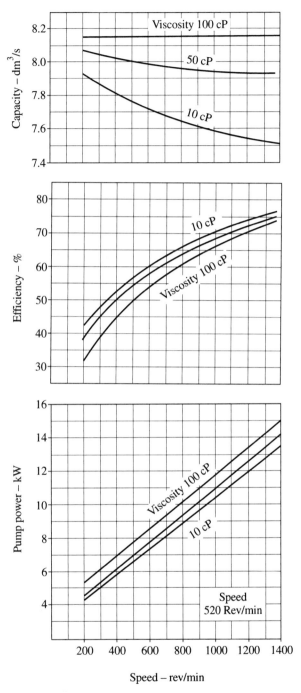

FIGURE 26.17 Performance for circumferential piston pump.

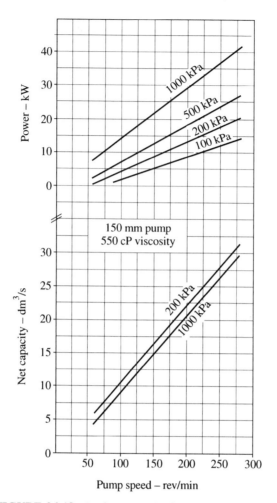

FIGURE 26.18 Performance for internal gear pump.

26.5.1 Pumps for Viscous Liquids

The viscosity of a liquid to be pumped influences the selection of pump type in any system. Centrifugal pumps are derated from their water performance by various methods, including the widely accepted method shown in the Hydraulic Institute Standards (1983). Reciprocating and rotary pumps are well-suited to pumping viscous liquids due to their inherent positive displacement, low-velocity principle. These types may be preferred, or even mandatory, for many high viscosity liquids which could not be pumped efficiently or at all, in a centrifugal machine. Because of parts such as meshing gears or screws which require lubrication, most rotary pumps perform better on viscous liquids than on thin liquids such as water or gasoline. Certain types of PD pumps are well-suited to pumping non-Newtonian liquids due to the fact that velocity changes in the pumped fluid are quite low in these pumps causing little or no change in the flowing liquid as it passes through the pump. See Chapter

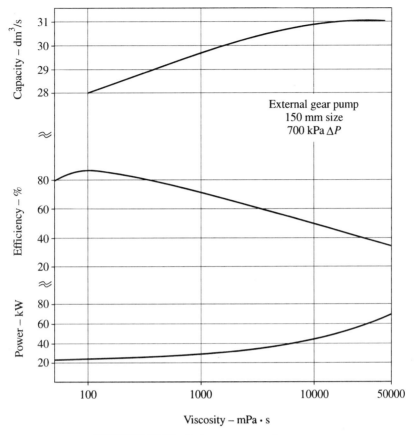

FIGURE 26.19   Rotary pump performance.

3 for a discussion on non-Newtonian liquids. See Skelland (1967) and Walters (1975) for information on pumping such materials.

26.5.2   Pumps for High Pressures

PD pumps are often preferred when the duty requirements are for a comparatively low capacity at medium to high pressures. Rotary pumps are regularly applied for applications requiring pressures to 20,000 kPa, and sometimes higher. Reciprocating plunger pumps may also be used for such pressures, and they will often be required for pressures higher than a rotary, to 70,000 kPa or higher.

Example 26.2:   A pump selection is required for 2 dm^3/s of water at ambient temperature, suction pressure 350 kPa, discharge pressure 35,000 kPa. Ample Net Positive Suction Head (NPSH) can be provided in the design of the pumping system. Since the pressure is high, the liquid viscosity low, the reciprocating pump appears to be the best choice. (Few rotary pump casings are available for a discharge pressure this high.) Reference to a manufacturer's published performance curve such as

Fig. 26.16 would show that a 40 mm × 125 mm *triplex* pump running at 260 rev/min would do the job. Efficiency would be estimated at 93% and power requirement at 74.5 kW. Since the speed is low, a motor with a speed reducer gear is probably the preferred drive arrangement. For certain applications, multiple v-belt drives may also be considered.

On a somewhat lower pressure, say 12,000 kPa, and for a fluid with higher viscosity than water, such as seal oil as used by a turbocompressor oil film seal system, the rotary screw type pump would be the preferred choice, partly because of smaller size, but mainly to avoid the pulsating flow of a reciprocating pump.

26.5.3   Controlled-Volume Pumping

The controlled-volume or metering type pump is mandatory when the requirement for a very accurate flow is present. Most such applications are for comparatively low flows, up to about 0.2 dm^3/s and to whatever pressure the system demands. Again, catalog information from reputable manufacturers represents the quickest and most reliable way to size and price such a pump. Special applications may require variations of manufacturers' standard designs, but the engineer is well advised to check the standard designs first.

Example 26.3:   A dosing pump is required to inject chemicals into a boiler feed water system. To maintain the proper water chemistry, the injection rate must be accurately controlled. Typical flow might be 0.05 dm^3/s at 7000 kPa. Either a plunger type or a diaphragm type pump would be suitable. Rating charts would show this to be an easy selection, for example a 30 mm × 80 mm plunger pump (with one plunger), having automatic stroke adjustment, suitable metallurgy, and flange-mounted 2 kW motor, pumping at a speed of something less than 70 strokes/min. Table 26.5 shows such a rating chart.

26.5.4   NPSH and Acceleration Head

Net Positive Suction Head (NPSH) has long been recognized as a factor in correctly applying a centrifugal pump in a given pumping system. NPSH is also required for PD pumps, which must fill completely with each stroke or revolution if they are to pump the rated capacity for which they were selected. A rotary or reciprocating pump will pump a mixture of liquid and air or gas, and does not exhibit the phenomenon of cavitation as does a centrifugal pump. However, the PD pump capacity will be reduced if the casing or cylinder does not fill completely on suction, and other harmful effects such as short-stroking (in direct-acting pumps), valve problems in reciprocating pumps, and excessive wear in certain rotary pumps, make such operation undesirable for long periods of time. NPSH for PD pumps is sometimes termed Net Inlet Pressure (NIP), but the majority of pump manufacturers seem to have accepted the term NPSH in an effort to avoid further confusion on a subject which has not been altogether clear to many users of pumps. NPSH for PD pumps is usually expressed in pressure, not head.

TABLE 26.5 Controlled-Volume Pump Rating Chart

Pump Type C, Single Plunger, 0 to 80 mm Stroke

Plunger Diameter mm	Strokes per Minute	Motor Speed rev/min	Maximum Capacity dm^3/h	Maximum Discharge Pressure, kPa	
				Packed Plunger	Disc Diaphragm
20	46	1140	61		
	70	1750	92		
	92	1140	122	28,000	24,000
	140	1750	184		
30	46	1140	140		
	70	1750	213		
	92	1140	280	16,600	16,500
	140	1750	426		
40	46	1140	249		
	70	1750	379		
	92	1140	498	9400	9300
	140	1750	758		
50	46	1140	389		
	70	1750	592		
	92	1140	778	5100	5000
	140	1750	1184		
60	46	1140	560		
	70	1750	853	3200	2000
	92	1140	1120		
70	46	1140	763		
	70	1750	1160	2000	N/A
	92	1140	1525		

For PD pumps, NPSH *required* by a pump is a function of speed, viscosity of liquid, size of pump, configuration of inlet casing or cylinder passages, and for reciprocating types, valve type and size. The objective is to have the pump properly sized and running at a proper speed so that the chamber will fill completely without formation of vapor. A viscous liquid will take longer to flow into the pump. Hence, NPSH *required* on viscous liquids will be higher than for the same pump at the same speed on a liquid with low viscosity. A pump of given dimensions running at a speed slower than the maximum rated speed will have more time for the liquid to flow in, hence the NPSH requirement is lower at less than full speed. Figure 26.20 shows how manufacturers display information about NPSH requirements for individual pumps.

The user of the pump or the designer of the pumping system must insist that enough NPSH has been made *available* in the system design for a reasonable pump selection to be made. *Available* NPSH is the difference between the suction pressure at the source and the liquid vapor pressure, plus the static pressure caused by elevation of the liquid above the pump, (or minus the static pressure if a suction lift is

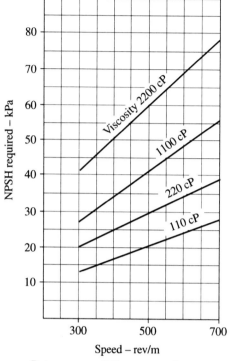

Rotary pump – 100 mm external gear type

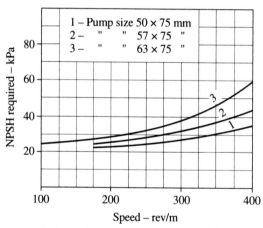

Reciprocating pump – single acting triplex type

FIGURE 26.20   NPSH curves for rotary and reciprocating pumps.

required), minus the friction loss in the suction piping system. Increasing the size of suction piping, raising the height of a suction vessel, and moving the pump closer to the suction source are ways of making more NPSH available. In some systems, where the pumped liquid is at a pressure significantly above its vapor pressure (e.g., pumping fuel oil from a tank at atmospheric pressure), the PD pump will successfully handle a suction lift without the need to be primed before operation begins. If the pump is in a process application where the liquid is at its vapor pressure in the suction vessel, then it is readily seen that the only ways to make more NPSH available to the pump are to raise the static elevation of the source liquid higher above the pump or decrease the friction losses in the piping leading to the pump. If this cannot be done, a pump requiring less NPSH must be chosen. Several rotary pump designs are made in a vertical configuration, allowing the pump to be installed below grade in a pit or a tank, sometimes to solve the NPSH problem.

In addition to NPSH for reciprocating pumps, the acceleration head in the pumping system must be considered to be sure that the pump speed is not too high or that the suction lines are too small or too long to cause trouble from the acceleration and deceleration of the pumped liquid on each stroke of the pump. Several methods have been suggested for determining acceleration head for pumps and, if carefully followed, should result in a trouble-free system design [HIS (1983), Karassik *et al.* (1985), Henshaw (1981), and Miller (1966)].

26.6 PULSATING FLOW FROM POSITIVE DISPLACEMENT MACHINES

The reciprocating machine, either compressor or pump, has the inherent disadvantage of causing a pulsating flow in the fluid leading to or from the unit. In the single cylinder compressor or pump, with one piston or plunger moving back and forth through one 360 degree revolution of the crankshaft, the effect of pulsation will be the highest. Where two or more cylinders are connected to the same crankshaft and the flow is manifolded together, as in a three or five cylinder pump or a two cylinder single stage compressor, some of the pulsation effects of each plunger will cancel each other, making the net pulsating effect lower. Figure 26.21 shows a typical pressure pulsation diagram for a three cylinder plunger pump to illustrate this effect.

Applications such as plant and instrument air compressors, which take suction from the atmosphere, and some comparatively simple reciprocating pump circuits, may not require special apparatus to reduce or dampen the effect of pulsation. On

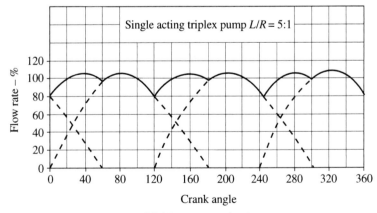

FIGURE 26.21 Pump pulsation curve.

most process applications, especially on higher pressures or larger size machines, some device will generally be required to reduce pulsation. On compressors, volume bottles (pressure vessels with no internals, also called *surge bottles*, *surge drums*, or *pulsation bottles*) may be attached directly either before or after each cylinder or group of cylinders on a particular stage. This provides surge capacity which reduces the pulsating flow effect downstream of the machine and insures that the suction stroke will receive a relatively pulsation free supply. As pressures increase, it is common practice to substitute *pulsation dampers* for volume bottles. These are specially designed vessels containing orifices, choke tubes, baffles, or other internal geometrics causing the pulsating flow to be cancelled or reduced to acceptable levels (Fig. 26.22). On large sophisticated systems, especially for multistage compressors and for high pressures, it is customary to conduct an analog or digital simulation of the pressure pulsations by means of an electrical model, study the results, and make needed changes during the study which will result in a system with known, predictable pulsation levels. Such studies are available from most compressor manufacturers, manufacturers of the various pulsation damping devices, and several well-known research organizations specializing in this subject. API (1984) has a section on this subject, with guidelines given for how to choose the type of pulsation suppression devices needed for a given application. Sparks (1970) gives more information and describes the analog study approach (see also Chap. 28).

For pumps, similar apparatus can be supplied which has been correctly sized for

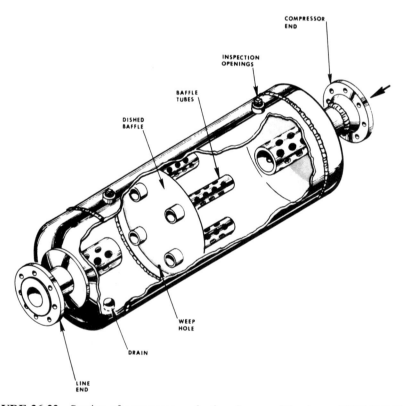

FIGURE 26.22   Section of compressor pulsation damper. (Courtesy of VANEC Corp.)

liquids rather than gases. Accumulators with nonmetallic bladders are often used for pump applications and have proven to be successful if sized correctly. Large complex pump systems may also warrant the added expense of an analog study of the piping and the machine, although this is less common for pumps than for compressors. Use of a suction bottle or accumulator is often a mandatory requirement in a pump system to avoid problems with acceleration head (see Sec. 26.5.4).

Rotary machines can also generate pulsating flow by the nature of the interaction of the rotating elements. Many units are so small, or operate at such low pressures, that the pulsating effect can be ignored. However, a lobe type blower, for example, even on a low pressure rise application, will require large suction and discharge volume bottles to reduce noise levels caused by pulsing flow. Such devices are not normally required on rotary pumps except for the most difficult, high pressure applications.

26.7 POSITIVE DISPLACEMENT MACHINES AS PRIME MOVERS

It should also be recognized that either the reciprocating or rotary machine can be used as an expansion machine in which gas or liquid enters at a higher pressure, expands to a lower pressure and produces work. The most common examples of these devices in use today are the rotary air motor and the hydraulic motor.

The vane type rotary air motor (Fig. 26.23) is the standard of the portable tool industry, and it is used for pneumatic tools throughout the construction, mining, and manufacturing industries (see Chap. 29). Even a very small air-powered wrench or drill will use a small rotary air motor. Larger, slower speed loads such as a pneumatically operated hoist or crane may use reciprocating type air motors, which are multicylinder machines with two or more pistons arranged around one crankshaft. Reciprocating machines have also been used in process applications as low temperature expanders but are gradually disappearing in favor of turboexpanders.

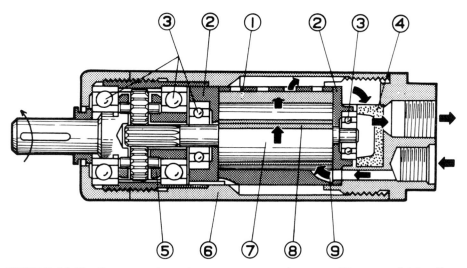

FIGURE 26.23 Cut-away view of rotary vane-type air motor. (Courtesy of Atlas Copco Comptec Inc.)

Rotary or reciprocating motors are also used with liquids to provide the motive power for hydraulic systems. Similar principles of application apply as for pumps, i.e., for small to medium flows at inlet pressures up to about 20,000 kPa, a rotary motor will probably be used, or for pressures higher than that, a reciprocating machine.

A quick estimate of power available can be made using similar formulae as for compression or pumping but with the efficiency in the numerator.

Example: How much power is obtained from a rotary, vane-type air motor for use as a power tool driver? Assume a flow of 50 dm³/s from 700 to 100 kPa, 60% efficiency, air with $\gamma = 1.4$ [$(\gamma - 1)/\gamma = 0.286$]

$$\text{Power} = \frac{p_1 V_1}{0.286} \left(1 - \frac{p_2}{p_1} 0.286 \right) \text{(efficiency)}$$

$$= \frac{(700)\,(50)}{0.286} \left(1 - \frac{100}{700} 0.286 \right) (0.60) = 31 \text{ kW} \qquad (26.16)$$

Example: Assume a rotary type hydraulic motor with a flow of 2 dm³/s, p of 10,000 kPa, and efficiency of 70%. How much power is produced?

$$\text{Power} = \frac{(\text{flow, dm}^3/\text{s})\,(p, \text{kPa})\,(\text{efficiency})}{1000}$$

$$= \frac{(2)\,(10,000)\,(0.70)}{1000} \text{ or } 14 \text{ kW} \qquad (26.17)$$

Karassik *et al.* (1985) and Atlas Copco (1978) give more information about air motors and hydraulic motors. More data is available from manufacturers' literature. More sophisticated applications may require further study and design developments.

Although a discussion of combustion engines as prime movers is not included in this chapter, it should be observed that most combustion engines used in automotive, industrial, commercial, and portable services are PD machines. Basic thermodynamic texts give fundamental principles for such engines, and many references are available to provide some application guidance.

REFERENCES

Atlas Copco Manual, 3rd ed., Atlas Copco AB, Stockholm, Sweden, 1978.

Balje, O. E., "A Study on Design Criteria and Matching of Turbomachines—Part B," *Transactions of ASME, J. Eng. Power*, 1962.

Bristol, J. M., "Diaphragm Metering Pumps," *Chem. Eng.*, p. 124, 1981.

Engineering Data Book, 9th ed., Natural Gas Processors Suppliers Association, Tulsa, OK, 1972.

Faires, V. M. and Simmang, C. M., *Thermodynamics*, 6th ed., Macmillan, New York, 1978.

Gibbs, C. W. (Ed.), *Compressed Air and Gas Data*, Ingersoll-Rand Co., New York, 1969.

Henshaw, T. L., "Reciprocating Pumps," *Chem. Eng.*, p. 105, 1981.

Hydraulic Institute Standards, 14th ed., Hydraulic Institute, Cleveland, OH, 1983.

Karassik, I. J., Krutzsch, W. C., Fraser, W. H., and Messina, J., *Pump Handbook*, 2nd ed., McGraw-Hill, New York, 1985.

Miller, J. E., "Effect of Valve Design on Plunger Pump Net Positive Suction Head Requirements," ASME Petroleum Engineering Conference, New Orleans, LA, Sept. 18–21, 1966.

Neerken, R. F., "How to Select and Apply Positive-Displacement Rotary Pumps," *Chem. Eng.*, p. 76, 1980.

Poynton, J. P., *Metering Pumps—Selection and Application*, Marcel Dekker, New York, 1983.

Reciprocating Compressors for General Refinery Services, 3rd ed., American Petroleum Institute Standard 618, 1984.

Scheel, L. F., *Gas Machinery*, Gulf, Houston, TX, 1972.

Skelland, A. H. P., *Non-Newtonian Flow and Heat Transfer*, John Wiley & Sons, New York, 1967.

Sparks, C. R., "Analog Simulation of Compressible Pipe Flow," ASME Paper 70-Pet-33, American Society of Mechanical Engineers, 1970.

Walters, K., *Rheometry*, John Wiley & Sons, New York, 1975.

27 Turbomachinery

DAVID JAPIKSE
Concepts ETI, Incorporated
Wilder, VT

WALTER S. GEARHART (Retired)
ROBERT E. HENDERSON
Pennsylvania State University
State College, PA

J. GORDON LEISHMAN
ALFRED GESSOW
University of Maryland
College Park, MD

NICHOLAS A. CUMPSTY
University of Cambridge
Cambridge, UK

COLIN RODGERS
Los Angeles, CA

TERRY WRIGHT
University of Alabama at Birmingham
Birmingham, AL

MICHAEL W. VOLK
Oakland, CA

EDWARD M. GREITZER
Massachusetts Institute of Technology
Cambridge, MA

MICHAEL V. CASEY
HELMUT KECK
Sulzer Innotec/Sulzer Hydro
Winterthur/Zurich, Switzerland

WILLIAM G. STELTZ (Retired)
Westinghouse Electric Company
Orlando, FL

Handbook of Fluid Dynamics and Fluid Machinery, Edited by Joseph A. Schetz and Allen E. Fuhs
ISBN 0-471-12598-9 Copyright © 1996 John Wiley & Sons, Inc.

TSUKASA YOSHINAKA
Concepts ETI, Incorporated
Wilder, VT

P. SAMPATH
HANY MOUSTAPHA
Pratt and Whitney Canada
Montreal, Canada

ERNEST W. UPTON
Bloomfield Hills, MI

LOUIS V. DIVONE
DANIEL F. ANCONA
U.S. Department of Energy
Washington, D.C.

CONTENTS

27.1   INTRODUCTION TO TURBOMACHINERY
David Japikse

27.1.1   Defining Concepts

Virtually all of the power generated worldwide is obtained by extracting energy from a moving stream of water, air, steam, or some other gas or liquid. On the other end of the energy cycle, a significant portion of all energy consumed is dissipated in propelling other streams of fluids by the use of pumps, compressors, or blowers. With such a profound impact on the flow of energy through society, it is altogether appropriate to ask: What is a piece of turbomachinery? The answer is simple: Any machine is a turbomachine if it follows the Euler turbomachinery equation. The Euler turbomachinery equation is derived from Newton's Second Law of Motion for a system operating in a rotational frame. It gives the following

$$W_x = U_2 C_{\theta 2} - U_1 C_{\theta 1} \tag{27.1}$$

where W_x = specific work, U = wheel speed, C_θ = tangential component of absolute velocity. Any fluid machine whose work input or work extraction is governed by this equation is a piece of turbomachinery. It tells us that the work is transferred through exchanges of kinetic energy which may involve changes in radius (U_2 and U_1 need not be equal) and changes in flow direction. But, it is important to remember that a turbomachine is a kinetic energy transfer device which follows Euler's basic principle given above.

FIGURE 27.1 Industrial axial fan.

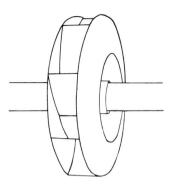

FIGURE 27.2 Industrial centrifugal fan.

The first distinction to be drawn while considering the fluid moving devices (*stages*) is the contrast between axial and centrifugal, or radial, stages. One of each is shown in Figs. 27.1 and 27.2, respectively. There are important reasons for the differences in the blading. The axial blower is configured to provide a high volume, low head characteristic whereas the radial blower is configured to provide a comparatively lower volume and a higher head characteristic. Parts of these characteristics can be immediately appreciated from the most essential equation of turbomachinery, given above as Eq. (27.1). This equation relates the work delivered to the flow (which in turn gives a pressure and temperature rise) to the change in important velocities leaving (state 2) and entering (state 1) the stage. For the axial machine, the inlet and outlet radii are approximately constant, therefore $U_2 = U_1$ as a very good approximation. Hence, for the axial machine, the work input is affected principally by changes in flow angle. However, for the radial machine, the work input and, in turn, the pressure rise, or *head*, is affected not only by angle changes but also by changes in radii. Clearly, the inlet average radius for the radial blower shown in Fig. 27.2 is far less than the exit radius. Hence, the radial machine is preferred for high head, or high exit pressure applications, whereas the axial machine will give larger flow rates for a given pressure rise. The axial machine is also limited to a much lower pressure rise per stage than is the radial.

The variations in flow and head (or pressure ratio) for various machines can be put into a simple and logical order if appropriate dimensionless parameters are employed. It has been found, based on experience and the appropriate use of similitude principles, that a proper relationship between a flow coefficient and a head coefficient gives the *specific speed* as follows

$$N_S = \phi^{1/2}/\psi^{3/4} = NQ^{1/2}/\Delta h^{3/4} \qquad (27.2)$$

where $\phi = Q/ND^3$, $\psi = \Delta h/N^2 D^2$, Q is the volumetric flow rate, N is the rotational speed, D is the diameter, and Δh is the pressure change. In this relationship, we have two dimensionless parameters, the *flow coefficient* and the *head coefficient* which, by using the proper powers, form the *specific speed*. The specific speed has only one unique characteristic: it is independent of length scale. Therefore, we can have a given specific speed for a small stage and the same specific speed for a very large stage which has a comparable flow and head coefficient. The utility of this relationship can be appreciated by considering Fig. 27.3 which shows a wide range of work input and work extraction devices arranged in the order of decreasing specific speed, from top to bottom in the chart. The domain of operation of these devices is shown in Fig. 27.4(a) and (b) for work input and work extraction devices, respectively. These charts were first presented by Turner (1959).

In addition to the specific speed, an additional dimensionless parameter is occasionally used to categorize different types of machinery. This is the *specific diameter* $D_s = D(gH)^{1/4}/\sqrt{Q}$, where H is the stage head. It has its appropriateness when one wishes to systematically correlate performance for a limited set of machinery which represents their own product line and the serial relationship of various elements in the product line. See Fig. 27.5 for a carpet plot of D_s vs. N_s.

The specific speed parameter has been used effectively to correlate efficiencies, in a broad and general sense, for a given class of machinery. Two examples are shown in Fig. 27.6. Figure 27.6 shows a set of calculations of compressor efficiency for various different applications as a function of specific speed, taken from a study by Japikse (1976). A similar analytical study was carried out for a specific radial turbine application by Rohlik (1968) as shown in Fig. 27.7. In each case, a possible optimum specific speed for the particular machine can be detected. It should be emphasized that the choice of optimum specific speed will depend on many other parameters and is not unique for all machines of a given type. For example, the notion that $N_S = 80$ (U.S. Customary Units, as per Fig. 27.6) might be optimum for centrifugal compressors is false. It may be the correct optimum value, from a design standpoint, for one particular type of centrifugal compressor but may not be the optimum for another (designed, for example, with different range and mechanical requirements). To be sure, a true design optimum depends on many fluid dynamic and structural considerations before the best possible machine can be defined. These charts give examples which are reasonably illustrative of the application.

Another important lesson can be realized from N_s consideration. All turbomachines are kinetic energy and work exchange devices. However, the significance of this statement varies depending on the level of specific speed. For example, at very low specific speeds, the relative kinetic energy approaching the eye of a centrifugal compressor or pump may amount to only 5 to 15% of the total work input. In this case, one need not be greatly concerned with the precise detailed design of the blading in this region. However, at high specific speeds, the inlet relative kinetic energy can amount to about 40% of the total power being delivered to the fluid. In this case, one must be careful to design for good diffusion (in the relative coordinate system). A similar point can be made in Fig. 27.7 for the radial turbine. It is clear that the exit kinetic energy is not important at low specific speeds, but at high specific speed its impact is dramatic. Thus, it follows that for all turbomachinery, the kinetic energy effects are very significant at high specific speed and less so at lower specific speeds. Frictional effects are very important at low specific speed in contrast to the high specific speed concern with kinetic energy (diffusion) problems. In the intermediate zone, both effects have a balanced impact.

To appreciate the processes in both radial and axial machinery more deeply, a sample radial flow compressor or pump, a radial flow turbine, an axial compressor, and an axial turbine is considered in the following four subsections.

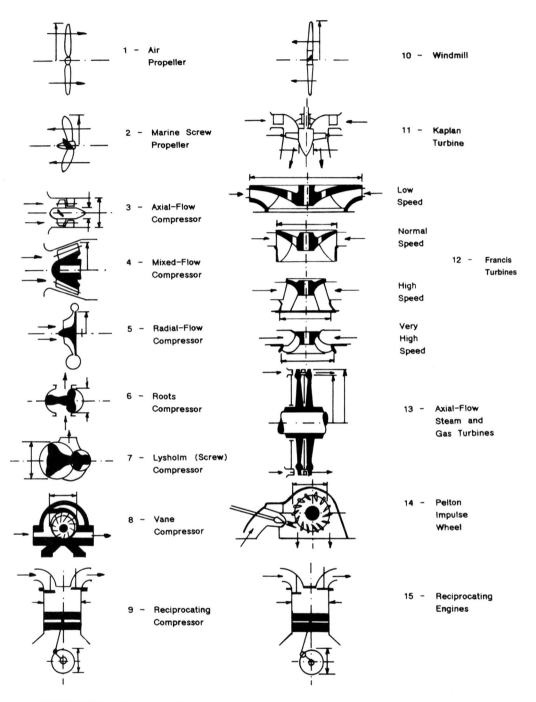

FIGURE 27.3 Schematic illustrations of various classes of industrial turbomachinery. (From Turner, 1959.)

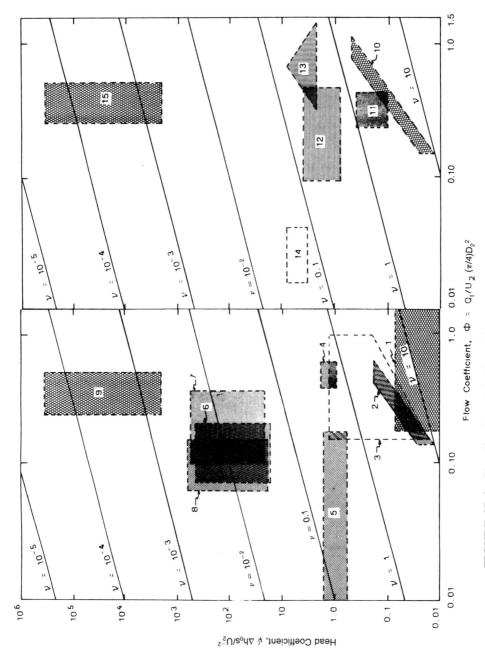

FIGURE 27.4 Classification of fluid machinery. Graph on the left-hand side is for fluid machinery with work input whereas the graph on the right-hand side is for work extraction. (From Turner, 1959.)

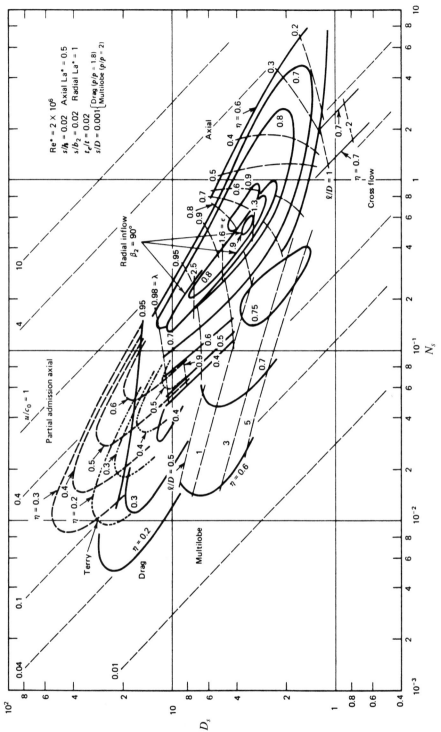

FIGURE 27.5 Carpet plot of specific diameter versus specific speed showing efficiencies for various types of turbomachinery.

2227

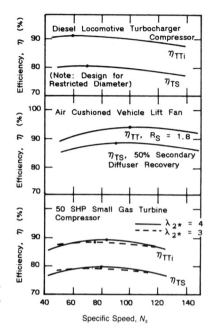

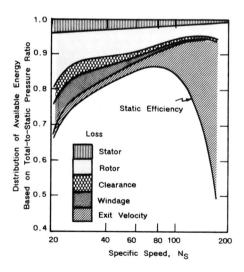

FIGURE 27.6 Examples of optimum performance, with respect to specific speed, for various industrial applications.

27.1.2 Centrifugal Pumps and Compressors

In a typical centrifugal compressor or pump stage, the flow approaches the impeller through an inlet duct or an inlet *bellmouth* in an axial inward direction, unless *preswirl* has been used to establish a nonzero value of $C_{\theta 1}$. Frequently, preswirl is not employed; in other cases, it is used for load control. One of the functions of the impeller is to create a low pressure region at the inlet face of the impeller in order that the flow does proceed through the inlet duct and into the impeller. The impeller admits this flow from the rotating frame, hence the approaching fluid enters the

FIGURE 27.7 Breakdown of losses within a radial inflow turbine stage. (From Rohlik, 1966.)

blading with some incidence angle. In order to set this incidence angle intelligently, one must bend the blade in a preferred direction to minimize and control the incidence. The flow then enters into the rotating impeller and is drawn through the impeller. The kinetic energy associated with the inlet *relative* flow is frequently equal to 20 to 40% of the total work input.

A special consideration must be made for the inlet portion of a centrifugal pump. As the flow enters the eye of the wheel, it is accelerated in order to enter the eye and give the low static pressure required to draw the fluid into the impeller. It will also be further accelerated as it enters into the passages between adjacent blades. This is brought about by the finite blade blockage. In addition, there will be even further acceleration over either the suction or pressure sides in the immediate vicinity of the blade leading edge, depending on the level of incidence. Thus, there is a significant reduction in static pressure as the flow enters the inlet of the pump, and, in fact, it may fall low enough that it drops below the vapor pressure of the liquid being pumped. In this case, vapor bubbles begin to form and a process known as *cavitation* ensues (see Sec. 23.11). To avoid pump cavitation, it is necessary, first of all, to use appropriate blading for the particular application at hand. Secondly, it requires the use of sufficient hydrostatic head at the inlet to avoid cavitation. Frequently, the *net positive suction head* (NPSH) is specified to avoid cavitation. The NPSH is given as follows

$$\text{NPSH} = h_0 - h_v \tag{27.3}$$

where

h_0 = absolute stagnation head
h_v = vapor pressure head

The flow then is propelled through the impeller with work being continuously transferred to the flow as it transits through the impeller passages. As the flow bends from the axial to the radial direction, very complex forces come to act on the flow field including centrifugal and Coriolis forces. Complicated flow fields can be established in each passage. Near the impeller discharge, flow leaves the impeller according to the blade exit angle. This, of course, is in the relative frame, that is, with respect to an observer rotating with the impeller. From an absolute frame, the flow field is in fact thrown off nearly in a tangential direction of rotation because the rotational velocity component is very large compared with the radial velocity component, with which the fluid is conveyed through the impeller. Consequently, if a channel diffuser is to be employed, then the vanes must be set at a proper angle in order that incidence can again be controlled. The level of kinetic energy exiting from the impeller, using the absolute velocity, is frequently equal to 20 to 50% of the total work input. Therefore, it is important that care be taken to diffuse this flow efficiently in order that the maximum pressure rise can be achieved. Many designers have tried diverse diffusers in order to achieve high efficiency.

Upon exit from the stage, the flow must be collected and delivered to the required downstream elements. Downstream elements include heat exchangers, combustors, or just flow networks to chemical reactors or other processes. The flow may be

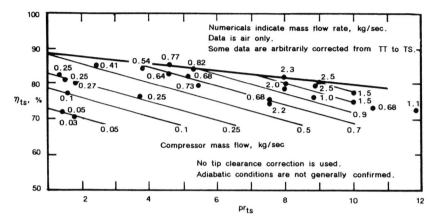

FIGURE 27.8. Illustration of various compressor performance levels as a function of flow rate (and therefore size) and stage pressure ratio.

collected in a volute, in a collector, or taken through a return bend and entered into the eye of a subsequent stage.

Centrifugal compressor stages can be used for a very wide variety of pressure ratio and flow conditions and frequently with substantial operating range. Figures 27.8 and 27.9 show typical efficiency levels as a function of stage pressure ratio and mass flow for centrifugal compressors as well as typical flow range versus efficiency. Centrifugal stages have been made in quite a variety of sizes ranging from less than 1 cm to greater than 2 meters, and they may operate at speeds from a few hundred rpm to in excess of 500,000 rpm. A typical centrifugal compressor performance map is shown in Fig. 27.10.

A detailed discussion of centrifugal pumps can be found in Sec. 27.7, and more information on centrifugal compressors is in Sec. 27.5.

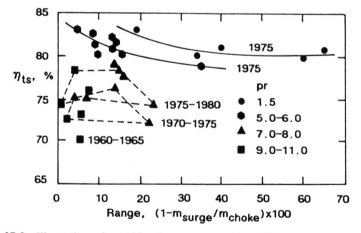

FIGURE 27.9 Illustration of centrifugal compressor stage efficiency versus range for various operating pressure levels.

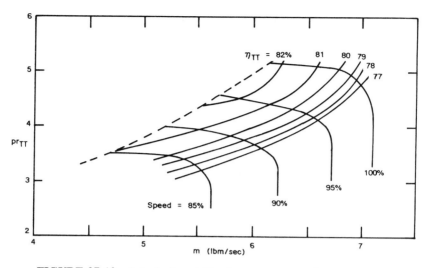

FIGURE 27.10 A typical centrifugal compressor performance map.

27.1.3 Radial Inflow Gas and Hydraulic Tubines

The radial inflow turbine is similar in appearance to the centrifugal compressor stage. However, the flow is in the opposite sense, and the Euler turbine pump equation (for a turbine $W_x = U_2 C_{\theta 2} - U_3 C_{\theta 3}$) is brought into action by creating a very high degree of initial swirl ($C_{\theta 2}$) which serves as the source of kinetic energy from which the stage work is extracted. If the stage is designed for zero exit swirl ($C_{\theta 3} = 0$), then the work input is given simply by $U_2 C_{\theta 2}$. However, residual exit swirl is frequently encountered. Thus, the stage is broken into 2 or 3 major components including a swirl generation section which can be either a swirl volute or a nozzle ring. It is followed then by the impeller. The final element is frequently an exhaust diffuser.

The nozzle ring is designed to create the required level of swirl in order that the work requirement of the stage is met. The flow angles must be adjusted by control of nozzle setting angle and passage depth so that the impeller has a reasonable inlet relative angle and a reasonable velocity ratio across the stage. The nozzle ring is one common method for achieving the desired level of swirl but similar levels of swirl can also be achieved with the nozzleless volute. In fact, the nozzleless volute can operate with higher angles of swirl, but the nozzle ring gives, under most design conditions, lower losses and a circumferentially more uniform flow field. The impeller is designed so that the inlet matches with the nozzle ring to obtain reasonable inlet velocity triangles and with controlled velocity ratio across the stage. In addition, it is important to control the exit blade angle of the stage in order that the final $C_{\theta 3}$ meets the required level and, hence, gives the required stage work. The exhaust diffuser is configured to accept the particular flow conditions leaving the rotor, thus giving the best possible diffuser performance under these particular exit conditions. The flow is then decelerated in the exhaust diffuser so that the static pressure is forced to rise. The diffuser discharge is to the atmosphere. Alternatively, the process can be explained by indicating that the diffuser discharge pressure is fixed at atmospheric, and the rotor exit pressure is, therefore, decreased, hence increasing the

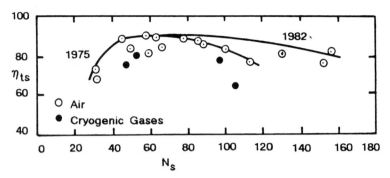

FIGURE 27.11 Illustration of peak radial turbine performance as a function of specific speed.

rotor pressure ratio and the level of work that can be extracted. A plot of radial turbine efficiency vs. specific speed is shown in Fig. 27.11. In addition, a representative radial turbine performance map is shown in Fig. 27.12.

Hydraulic turbines are discussed in some detail in Sec. 27.9 of this handbook.

27.1.4 Axial Turbines

We now turn to an axial turbine stage. Here, the objective is to exploit the Euler turbomachinery equation by changes principally in flow angle, not with changes in

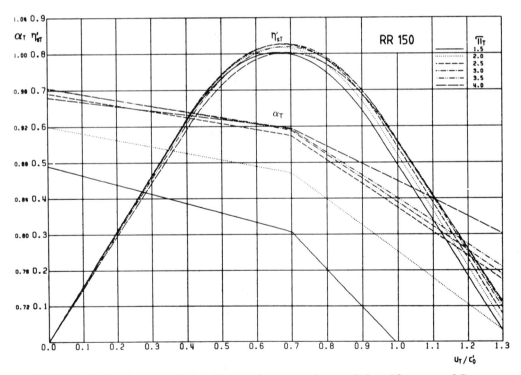

FIGURE 27.12 Typical radial turbine performance characteristics. [Courtesy of Brown Boveri Corporation (now Asea Brown Boveri).]

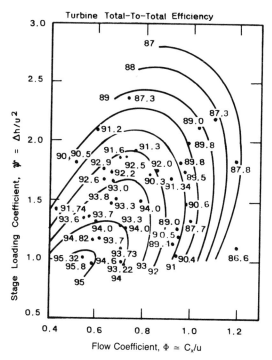

FIGURE 27.13 Contour plot of axial turbine efficiencies in terms of stage loading coefficient and flow coefficient. (Reprinted with permission, Smith, S. F., "A Simple Correlation of Turbine Efficiency," *J. Royal Aero. Soc.*, Vol. 69, pp. 467–470, 1965.)

radius. This provides the capability of handling high levels of work with high mass flows but with a greater restriction on the level of pressure ratio that can be handled in a given stage. The principal flow elements in the axial turbine stage are a nozzle ring, a rotor, and frequently, an exhaust diffuser. These three components serve the same function as their corresponding elements in the radial turbine stage, except that the mean line of these elements is typically a constant radius surface. Hence, the nozzle ring is designed to accept flow coming to it (with whatever level of preswirl it might have, usually zero) and then turn it in a direction such that the required level of rotor preswirl is established. The rotor, in turn, is configured to accept this level of swirl with reasonable, controlled levels of incidence and then, by turning the flow, extract work and achieve a desired exit state. Sometimes an exit guide vane is employed in order that greater design freedom may be achieved at the rotor exit. Subsequently, the exhaust diffuser is designed to serve the same function as discussed above. Typical examples of single-stage turbine efficiency are shown in Fig. 27.13, and a typical axial turbine map is shown in Fig. 27.14.

More information on axial turbines is available in Secs. 27.10, 27.11, and 27.13.

27.1.5 Axial Compressors

Finally, the axial compressor is considered. The axial compressor stage usually consists of a rotor and a subsequent stator. Sometimes, an upstream stator is used to establish a desired level of preswirl. The axial compressor is occasionally used as a

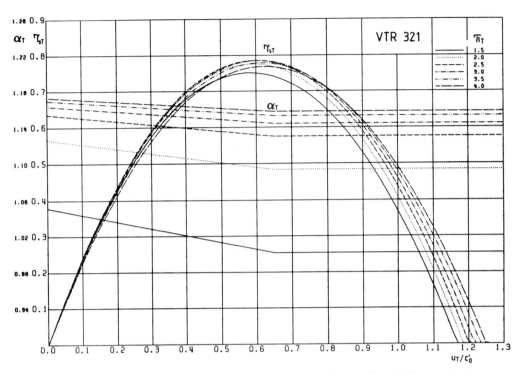

FIGURE 27.14 Axial turbine characteristics. [Courtesy of Brown Boveri Corporation (now Asea Brown Boveri).]

single-stage fan, but most frequently it is used in a multi-stage axial compression machine. As with the axial turbines, these stages are designed for large mass flows and hence very high power levels. On a single stage basis, they are limited in that much higher pressure ratios can be achieved in a single-stage centrifugal compressor than in a single-stage axial compressor. Whereas pressure ratios of 5 to 10 are frequently achieved in centrifugal compressors, industrial axial compressor stages are usually under a pressure ratio of 2. The purpose of the rotor is to impart work to the fluid and raise the pressure by work input, through flow angle changes (consider the Euler turbomachinery equation) and by diffusion with the stage. Diffusion is an essential process and an inherent risk for good efficiency. As the flow leaves the rotor, the exit flow angles are usually such that an exit guide vane is required to achieve the required level of exit swirl or to prepare the flow for a subsequent stage. A typical axial compressor map is shown in Fig. 27.15, and sample levels of compressor efficiency are shown in Fig. 27.16.

Section 27.4 contains a more detailed discussion of axial flow compressors.

27.1.6 Process Turbomachinery

Radial flow compressors and blowers are frequently employed in process applications. Within the family of radial machinery, there is a strong variation in types of stage. Figures 27.2 and 27.16 contrast two important types. The radial blower shown in Fig. 27.2 is a comparatively simple construction with blading principally in the

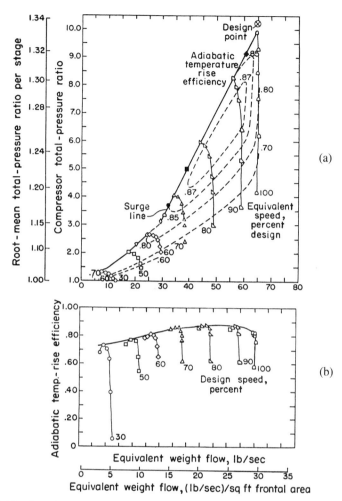

FIGURE 27.15 Sample axial compressor performance map. (a) Pressure ratio and flow and (b) efficiency versus flow.

radial section only. By contrast, the radial impellers shown in Fig. 27.16 are more complicated with the blading starting near the axial region and then turning for a radial discharge. The first impeller (see Fig. 27.2) should be familiar to everyone: it is identical in type to the impellers used in the common household vacuum cleaner. Clearly, cost of manufacture would be a difference between the two design types and also the need to be concerned with inlet flow state would clearly be a difference. Note that the stage shown in Fig. 27.16 is typical of moderate to high specific speed designs whereas the stage shown in Fig. 27.2 is characteristic of low specific speed designs.

Figure 27.16 introduces the major components of a radial stage as might be found amongst many stages of a process compressor. The first component in the system, working from left to right as the flow proceeds, is the IGV's (*inlet guide vanes*) or PRV's (*pre-rotation vanes*) as they are sometimes called. The function of the inlet guide vanes is to set the inlet swirl angle either with or against the direction of

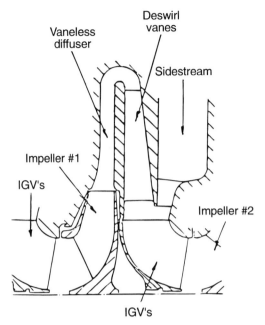

FIGURE 27.16   Illustration of principal elements of a typical industrial inline centrifugal compressor stage.

rotation in order to create a nonzero $C_{\theta 1}$ component in the Euler equation. Therefore, a designer can derive greater or less work from a stage (and therefore more or less pressure rise) by this inlet control. However, the slope and stability of the stage will also be affected. The second component is the impeller, which is responsible for the work input. According to the Euler equation, the impeller is responsible for the work input and pressure rise subject to the given rotational speed. As the flow leaves the impeller, it moves in a highly tangential, *not radial*, direction, with a comparatively small radial component. The kinetic energy level is very high. The high kinetic energy level is a fundamental requirement of the Euler equation if any reasonable pressure rise or work input is to be accomplished. Approximately two-thirds of the pressure rise occurs in the impeller. An additional one-third occurs in the subsequent diffuser. Depending on the nature of the design, 50 to 70% of the kinetic energy leaving the impeller will be recovered as a static pressure rise in the diffuser in the regions of best performance. At extreme off-design points, poor recovery will result.

As the flow leaves the vaneless diffuser, it enters the return bend at the top of the system. The purpose of the return bend is to guide the flow from the diffuser to the inlet of the return channel, or deswirl vanes. The function of the return channel is to remove the very high level of swirl that exists at the diffuser exit, prior to introduction at the eye of the second stage. The swirl at the inlet to the return channel can typically be 50 to 85 degrees from the radial, or 5 to 40 degrees from the tangential direction. This flow angle must be changed to a nearly straight axial flow prior to the eye of the second impeller. As the flow leaves the deswirl vanes, it may be mixed with a sidestream and then some additional turning may be required in order to establish a particular flow angle prior to the introduction of the flow to the subsequent impeller. An additional set of guide vanes may be required to do this. It should be understood that the flow state coming from the deswirl vanes and the sidestream is a very chaotic flow with strong variations in angle and total pressure across the flow field. It makes the precise design and development of the machinery extremely difficult.

In some machines, both axial and centrifugal stages are used, and respectable levels of efficiency can be obtained. Figure 27.17 serves to illustrate this point. In this figure, a plot based on efficiency vs. specific speed is presented. Multi-stage axial compressors give comparatively high efficiencies but at the expense of numerous stages and comparatively low head relative to the volume flow provided. Centrifugal stages, by contrast, provide much higher head relative to the flow and still decent stage efficiencies. It is clear that improvements in technology have benefited the radial flow machinery more than improvements in the axial in recent years. For higher heads and lower flow rates, positive displacement equipment is required. However, it must be understood that Fig. 27.17 is incomplete and only serves to illustrate a facet of machinery efficiency characteristics.

A more comprehensive (but still incomplete) representation was observed in Figs. 27.3 and 27.4. In these figures, a wide representation of machinery types is given for both work input and work extraction devices. The head coefficient and flow coefficient are used as coordinates with all machines mapped into this domain with specific speed as a parameter. More information, but still incomplete, was shown in Figs. 27.8 and 27.9. In Fig. 27.8, efficiency characteristics vs. pressure ratio with flow as a coordinate was shown. This representation and Fig. 27.17 are not equiv-

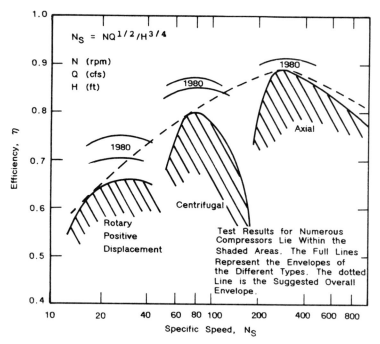

FIGURE 27.17 Survey of anticipated performance levels for radial and axial turbomachinery.

alent. For a true mapping of the characteristics, a composite of the two would be required. Further, separate mappings would be required for each sub-class of machinery type. For example, if Fig. 27.9 is considered, it is clear that different characteristics are obtained in terms of efficiency and range at different levels of design pressure ratio. Process machinery is usually interested in the low pressure ratio range which involves vaneless diffusers. Hence a reasonable mapping could be made for vaneless diffuser (only) stages with plots of the type shown as Figs. 27.8 and 27.9. However, if channel diffusers are introduced, then less range and higher efficiencies would be expected, and separate maps should be used. These figures serve to illustrate the major aspects of stage performance representation and some of the variants that can be found. Specific plots cannot be used in general to consider process machines. Indeed, it is recommended that the only characteristics that should be considered for evaluating process machinery are characteristics that are provided by a manufacturer for a series of machines that they are producing. Such information, when interpreted with experience, will provide the best guidance.

27.1.7 Turbochargers

Many types of internal combustion engines utilize turbochargers ranging from very small motorcycle engines to extremely large ship diesel engines. Typical examples of turbochargers are shown in Figs. 27.18 and 27.19. In the first figure, an automotive size turbocharger is shown using only radial components. In the second figure, a larger turbocharger, of the class used for locomotive engines and large marine

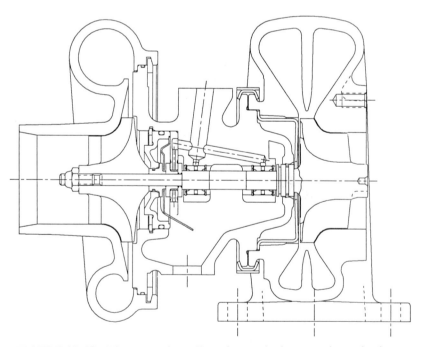

FIGURE 27.18 Diagrammatic outline of a standard automotive turbocharger.

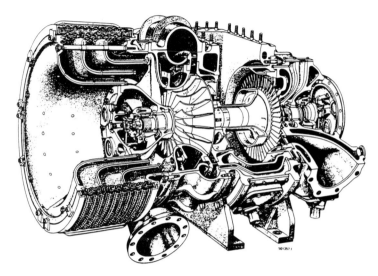

FIGURE 27.19 Cross section of a typical large diesel turbocharger. [Courtesy of Brown Boveri Corporation (now Asea Brown Boveri).]

engines, is illustrated. These larger turbochargers are a hybrid of centrifugal compressor and axial turbine staging. Some of the details of these configurations should be examined carefully to further appreciate the design problems. Consider first Fig. 27.18. Air is drawn to the centrifugal compressor impeller due to the low pressure created in front of the impeller by the action of the impeller itself. Flow then proceeds from the axial to the radial direction in the compressor impeller with a very significant increase in tangential momentum (as dictated by the Euler turbomachinery equation). As the flow enters the impeller, it attains a relative velocity of such a magnitude that the kinetic energy of the relative flow is equivalent to approximately 30–40% of work input. Clearly, such a flow must be dealt with very carefully or losses will result. As the flow proceeds through the impeller, it is turned in direction and has had significant work done upon it. It may be necessary to diffuse the flow substantially in the impeller, or very large exit velocities will result. As the flow discharges from the impeller, it is largely tangential in direction (typically less than 20° from the tangential direction as it enters the vaneless diffuser) and will have an absolute kinetic energy entering the diffuser equal to approximately 40% of the work input. Again, the flow must be carefully diffused. The flow is then permitted to migrate tangentially (and slightly radially) out through the vaneless diffuser where an attempt is made to reduce the velocity and increase the static pressure. Finally, the flow enters into a scroll or collector in order that it may be collected and delivered to the engine cylinders.

After the flow is expanded in the engine, it is discharged into the turbine, which is on the right hand side of Fig. 27.18. In this particular case, flow from two different cylinder banks is brought to the turbine separately, and half is permitted to enter into each half of the turbine volute. The flow moves rapidly in a circumferential direction around the turbine volute and moves radially inward achieving a very high swirl angle as it approaches the turbine rotor. By this means, the kinetic energy of the flow delivered to the turbine is increased and is oriented in a highly tangential direction to give the swirl velocity necessary to extract work (via the Euler turbomachinery equation). The flow then passes through the rotor with substantial changes in radius and direction until it exits from the blading. Ideally, a designer attempts to match the components so that, in the regions of desired best efficiency, the flow is exiting from the turbine rotor in a nearly axial direction. If possible, an exhaust diffuser would be added to this configuration, but this is often prohibited due to installation requirements. Also, circumferential pressure distortions caused by the volute may cripple a diffuser.

Between the compressor and turbine is the bearing system which is frequently comprised of two radial or journal bearings and one axial or thrust bearing. In addition, various seals are included to keep oil from the turbine rotor and also away from the compressor rotor. Further, a heat shield is located immediately aft of the turbine rotor to minimize the flow of heat into the bearing housing, but this flux still remains substantial. Approximately 3–10% of the power extracted by the turbine is absorbed by bearing and seal friction. In order to design a successful turbocharger, it is necessary to match together all the individual elements shown in order to meet the pressure and flow characteristics of the engine.

The second figure, Fig. 27.19, depicts an even more complicated turbocharger for the larger engine applications. On the left is the centrifugal compressor with an axial turbine on the right. This system has outboard bearings of a rolling element type. Lubrication is provided through an oil sump and a slinger in the two extreme cavities. The compressor and turbine aerodynamic components are similar to the above case (notwithstanding the axial vs. radial turbine configuration) with several notable exceptions. The compressor inlet is now far more complicated and contains a very large silencer which is designed to prohibit compressor noise from propagating upstream and into the environment. This noise is produced principally by the interaction of the rotor and diffuser vane pressure fields. At the discharge of the turbine is a very short diffuser and then a single exit collector.

27.1.8 Gas Turbines

The gas turbine market, particularly the aircraft gas turbine, demands extreme reliability and very high efficiency and minimal weight in order to offer attractive economics. An example of a gas turbine system is shown in Fig. 27.20. This particular system is an early schematic for a cruise missile, and it is interesting in that it uses many different types of turbomachinery. Beginning on the left is a two-stage axial fan followed by a bypass duct and a low pressure ratio axial compressor. The axial compressor is then followed by a centrifugal compressor which has been employed to achieve high pressure ratio in the smallest possible space. Due to the diameter restriction, the channel diffuser which follows the compressor impeller is quite limited in radial extent. The discharge from this compressor then feeds the burner which in turn delivers high pressure, high temperature gas to the high pressure turbine. Two stages are used for the high pressure turbine which drives the centrifugal compressor and part of the axial compressor. Subsequently, there is a transition duct, which involves some diffusion of swirling flow through that passage, and which is followed, in turn, by the low pressure turbine. The low pressure turbine stage ex-

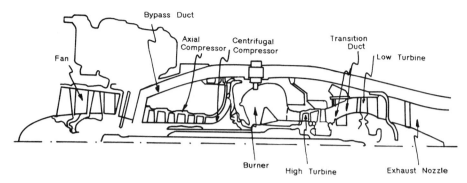

FIGURE 27.20   Cross section of a small aviation gas turbine.

tracts sufficient power to drive the inlet fan system. Since this is an aircraft propulsion system, an exhaust diffuser is not employed but instead an exhaust nozzle is included in order to obtain maximum thrust.

Although the gas turbine system is an extremely complicated system to design and develop, each of the basic rotating components is governed by the Euler turbomachinery equation. Each of the elements, whether they are an axial or centrifugal stage, follows the basic principles presented previously. A more comprehensive discussion is given in Sec. 27.11.

Industrial Gas Turbines.   Industrial gas turbines have the same variety of flow problems as shown in the preceding illustration with one fundamental difference: the machines may be vastly larger. This means that the blades tend to be very long and subject to vibratory stress. Although secondary flows are still important in the large industrial gas turbines, the flow field is dominated more by the potential core flows due to the very large blade span characteristic of the industrial gas turbine. By contrast, the small gas turbine shown in Fig. 27.20 frequently has highly viscous flow throughout the entire machine with end wall boundary layers merging at the passage mid-section due to the very small span.

27.1.9 Steam Turbines

The industrial steam turbine and the industrial gas turbine are similar in appearance, both having a large size due to the high power levels involved. The steam turbine is invariably used as a drive, either for a generator or for process compression equipment. The gas turbine, or course, is a complete gas turbine cycle system and, therefore, includes compression, combustion, and expansion. There are, however, two specific characteristics of steam expansion which distinguish the steam turbine from the axial turbine portion of the gas turbine. The first important characteristic is that the steam turbine is frequently exhausted to a condenser whereas the gas turbine is invariably exhausted to atmosphere. The condensation pressure of steam is such that the back pressure on the steam turbine is therefore invariably very low (on the order of 1 psia). Therefore, one can expand over a much greater pressure ratio through a typical steam turbine than through a typical gas turbine. As a consequence, this higher available expansion ratio is exploited in steam turbine design and each individual blade row in a steam turbine has a higher expansion ratio than one typically finds in a gas turbine. Thus the exit Mach numbers from a steam turbine blade row may easily reach a level of 1.5 or on occasion 2.0. By contrast, the gas turbine has exit Mach numbers from each blade row in the range of 0.8 to 1.2, with the latter being quite high. The other basic characteristic difference is the formation of condensate in the steam turbine. If this condensate is not removed from the gas path, the performance of the turbine will be degraded gas dynamically, and the blading will be eroded yielding early structural failure. Thus blade shapes must be adapted to direct the condensate out of the turbine into the shroud area. Consequently, the shroud regions must be designed so as to permit the continuous removal of condensate.

See Sec. 27.10 for a more detailed coverage of steam turbines.

27.1.10 Conclusion

In concluding the Introduction, it may be noted that all of these turbomachines are quite similar in conceptual form, when considered from the perspective of the Euler turbomachinery equation, and yet can differ substantially in geometric form. As mentioned before, all work exchange in a turbomachine is done with changes in kinetic energy. The Euler turbomachinery equation suggests that these are symmetric, continuous processes. Therefore, discrete repetitive processes *can* be avoided if the designer so desires, whereas for positive displacement equipment, discrete repetitive processes are absolutely essential (see Chap. 26). Hence, there is a fundamental difference in the character of noise generation between positive displacement and rotating turbomachinery. The positive displacement equipment must have discrete (repetitive) tones due to the very basic nature of the machinery. Rotating machinery, however, can avoid (repetitive) discrete tones. On the other hand, rotating machinery effects its work exchange by changes in kinetic energy, which therefore can become powerful sources to drive noise emitting processes. Indeed, many turbomachinery stages introduce discrete items (nozzle rings, volute tongues, diffuser leading edges, etc.) which then interact with a very high kinetic energy level and emit substantial noise. However, if the designer wishes to design for low noise conditions, then turbomachinery can offer a good foundation for such design. There is, however, a trade-off between efficiency and noise in most applications.

27.2 SHIP AND AIRCRAFT PROPELLERS
Walter S. Gearhart and Robert E. Henderson

27.2.1 Introduction

The propeller represents the earliest form of turbomachinery employed to propel marine vehicles and aircraft. As with any known method for propelling a vehicle, the propeller thrust is the reaction to the fluid force created by increasing the momentum of the fluid passing through the propeller in the direction of vehicle motion. Similarly, the change in angular momentum of the fluid passing through the rotating propeller produces a reaction moment, torque, about the propeller axis of rotation which is related to the energy placed in the fluid. A propeller can also be analyzed by considering the forces acting on its blades.

In many writings, the subjects of marine and aircraft propellers are presented separately. The presentation here will discuss the characteristics common to both applications and then discuss the differences. These differences are primarily the occurrence of cavitation on marine propellers and the effects of compressibility in certain aircraft propellers. The photographs of a modern marine propeller and a modern aircraft propeller shown in Fig. 27.21 demonstrate the similarities between the two.

27.2.2 Design and Analysis Methods

Actuator Disk Model. The classical way to model momentum changes through a propeller is to represent the propeller as an *actuator disk* [Durand (1963)]. This representation reduces the propeller to a solid circular area with infinite number of blades and an infinitesimal length in the direction parallel to the axis of shaft rotation.

The actuator disc model shown in Fig. 27.22 can be used to estimate the overall performance of the propeller by considering flow through the control volumes shown. A simplified actuator disk model employs values of velocity and pressure averaged over the entering and leaving surfaces of the control volume. For example, the flow entering a propeller mounted on the stern of a vehicle (a *pusher* propeller) is partially or entirely due to a boundary layer flow over the stern and, therefore, varies in the r-direction. If $w_i(r)$ is the local velocity at i, then the average velocity, $\overline{w}_i$, is defined as

$$\overline{w}_i = 2 \int_0^{R_i} \frac{w_i(r)}{R_i^2} r \, dr \tag{27.4}$$

The averaged fluid velocities at other stations through the control volume are related by the Law of Conservation of Mass.

Thrust and Energy Transfer: The net thrust, F_n, produced by the propeller on the fluid is related to the flow entering and leaving the control volume *abcd* [Fig. 27.22(a)] by the Law of Conservation of Linear Momentum. Assuming the fluid density to be constant, $\rho = \rho_D = \rho_i = \rho_j$, and $\overline{p}_\infty$ acting along surfaces *ab*, *ad*, and *dc*, we get

$$F_n = \rho \pi R_p^2 \overline{w}_D (\overline{w}_j - \overline{w}_i) - (\overline{p}_\infty - \overline{p}_j) A_j \tag{27.5}$$

For the propeller-alone configuration shown in Fig. 27.22, F_n is equal to the thrust on the propeller shaft, F_s. In many applications, the propeller is operated near the surface of the vehicle, and the accelerated flow due to the propeller increases the drag of the vehicle when compared with the vehicle-alone configuration. In this case, $F_n < F_s$, since the flow induces by the propeller changes the pressure distribution on the vehicle which increases its drag and increases the required shaft thrust.

From the Law of Conservation of Energy, the energy (power) added to the fluid by the propeller is

$$\text{energy added} = \rho \pi R_p^2 \overline{w}_D \left[\frac{\overline{p}_j}{\rho} + \frac{\overline{V}_j^2}{2} - \frac{\overline{p}_\infty}{\rho} - \frac{\overline{V}_i^2}{2} \right] \tag{27.6}$$

(a)

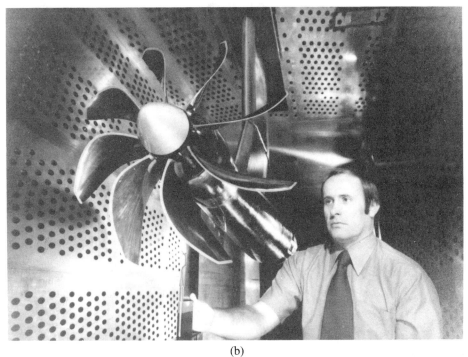

(b)

FIGURE 27.21 Comparison of advanced marine and aircraft geometries. (a) Marine propeller (courtesy of David Taylor Ship Research and Development Center) and (b) aircraft propeller (courtesy of NASA).

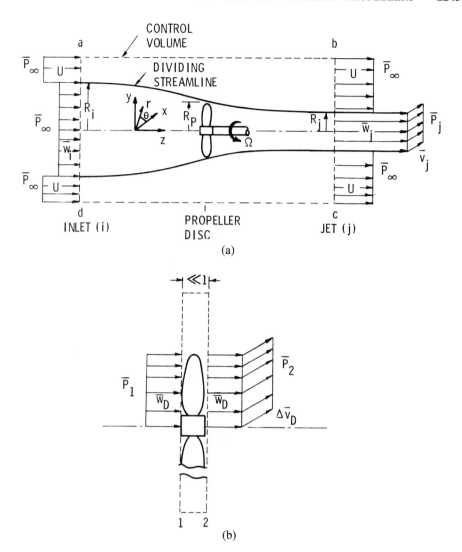

FIGURE 27.22 Actuator disk control volume. (a) Propeller slipstream and (b) propeller control volume.

The Law of the Conservation of Angular Momentum allows the energy added to the fluid to be related to the torque on the propeller shaft, T.

$$T = \rho \pi R_p^2 \bar{w}_D [\bar{r} \Delta \bar{v}_D] \tag{27.7}$$

where $\bar{r}$ is a mean radius of the disc, approximately $0.7R_p$, and $\Delta \bar{v}_D$ the change in tangential velocity through the propeller disk.

The thrust on the propeller shaft can be determined by considering a control volume between 1 and 2 as in Fig. 27.22(b).

$$F_S = \pi R_p^2 (\bar{p}_2 - \bar{p}_1) \tag{27.8}$$

The Law of the Conservation of Energy for this control volume states

$$\bar{p}_1 + \tfrac{1}{2}\rho\bar{V}_1^2 + \rho g H = \bar{p}_2 + \tfrac{1}{2}\rho\bar{V}_2^2 \tag{27.9}$$

where $gH = 2\eta_h\Omega\bar{r}\Delta\bar{v}_D$ is the fluid total head rise added by the propeller, η_h is the hydraulic efficiency, and $\bar{V}$ is the averaged resultant velocity. Thus,

$$
\begin{aligned}
F_S &= \frac{\rho\pi R_p^2}{2} [\bar{V}_1^2 - \bar{V}_2^2 + 4\eta_h\Omega\bar{r}\Delta\bar{v}_D] \\[2mm]
&= \frac{\rho\pi R_p^2}{2} \Delta\bar{v}_D[-\Delta\bar{v}_D + 4\eta_h\Omega\bar{r}]
\end{aligned}
\tag{27.10}
$$

The overall efficiency of the propeller is then

$$
\begin{aligned}
\eta_0 &= \eta_m\eta_p\eta_h \\[2mm]
&= \eta_m U \left\{ \frac{\rho\pi R_p^2\bar{w}_D(\bar{w}_j - \bar{w}_i) - (p_\infty - \bar{p}_j)A_j}{\rho\pi R_p^2\bar{w}_D \left[\dfrac{\bar{p}_j}{\rho} + \dfrac{\bar{V}_j^2}{2} - \dfrac{\bar{p}_i}{\rho} - \dfrac{\bar{V}_i^2}{2} \right]} \right\} \eta_h
\end{aligned}
\tag{27.11}
$$

where η_m = mechanical efficiency of the shaft bearings and seals, η_p = propulsive efficiency the effectiveness of the propeller in converting fluid energy to thrust), and η_h = hydraulic efficiency (effectiveness of the propeller in converting available shaft power to fluid energy). Typically, $\eta_m \cong 0.95$ and $\eta_h \cong 0.88$. The value of η_p is dependent upon the type of propeller employed and the flow field in which it operates.

The thrust and torque on the propeller are usually expressed in a nondimensional form as

$$K_F = \frac{\text{THRUST}}{\rho n^2 D_p^4} = \eta_p \frac{\pi}{8} \left(\frac{\bar{w}_D}{U} \right) \left(\frac{U}{nD_p} \right)^2 \psi \tag{27.12}$$

$$K_T = \frac{\text{TORQUE}}{\rho n^2 D_p^5} = \frac{\pi}{16} \left(\frac{\bar{w}_D}{U} \right) \left(\frac{U}{nD_p} \right)^3 \psi \tag{27.13}$$

where

$$\eta_p = \frac{1}{2\pi} \left(\frac{K_F}{K_T} \right) \left(\frac{U}{nD_p} \right) \tag{27.14}$$

and

$$\psi \equiv \frac{2\Omega\bar{r}\Delta\bar{v}_D}{U^2} = \frac{1}{\eta_h} \left[\frac{\bar{p}_j - \bar{p}_\infty}{1/2\rho U^2} + \frac{\bar{V}_j^2}{U^2} - \frac{\bar{V}_i^2}{U^2} \right] \tag{27.15}$$

Propulsive Efficiency: The definition of η_p can be simplified by assuming the inlet velocity is axial and has no swirl, $\bar{w}_i \equiv \bar{V}_i$. Then

$$\eta_p = \frac{U(\bar{w}_j - \bar{V}_i) + \dfrac{(\bar{p}_j - \bar{p}_\infty)}{\rho}\left(\dfrac{D_j}{D_p}\right)^2\left(\dfrac{U}{\bar{w}_D}\right)}{\dfrac{(\bar{p}_j - \bar{p}_\infty)}{\rho} + \dfrac{1}{2}\left[(\bar{w}_j^2 - \bar{V}_i^2) + \bar{v}_j^2\right]} \qquad (27.16)$$

This expression shows that the existence of swirl in the jet (*slipstream*) $\bar{v}_j$, acts to increase the energy which must be added to the fluid to produce a given thrust. The term $(\bar{p}_j - \bar{p}_\infty)$ is < 0 and represents a negative thrust or drag. Slipstream swirl reduces efficiency as illustrated in Fig. 27.23.

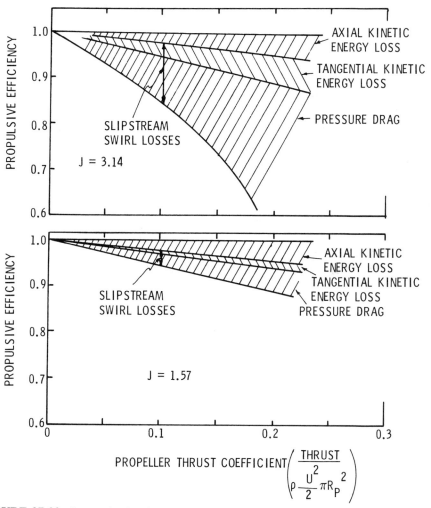

FIGURE 27.23 Losses in the slipstream as a function of thrust coefficient for two values of the advance ratio, J.

In many cases, it is appropriate to neglect the swirl kinetic energy term, i.e., $\bar{v}_j \cong 0$ and $(\bar{p}_j - \bar{p}_\infty) \cong 0$, without affecting the predicted performance. These cases include single propellers for which $\psi(U/nD_p)^2 < 0.340$ and counterrotating propellers or a propeller with stationary blades which cancel the swirl. In these cases,

$$\eta_p = \cfrac{1}{\cfrac{\bar{V}_i}{U} + \cfrac{1}{2}\left(\cfrac{\bar{w}_j - \bar{V}_i}{U}\right)} = \cfrac{1}{\cfrac{\bar{V}_i}{U} + \cfrac{1}{2}\cfrac{F}{\rho \bar{w}_D A_D}} \qquad (27.17)$$

These expressions of η_p demonstrate the effect of several propeller characteristics on η_p: 1) operation of a propeller in a boundary layer flow, $\bar{V}_i/U < 1$, results in a higher value of η_p than operation in a free stream, $\bar{V}_i = U$, 2) as the thrust loading of a propeller, (F/A_D), decreases, η_p increases, and 3) for a constant thrust, η_p increases as the mass flow rate of fluid, $\rho \bar{w}_D A_D$, or the size of the propeller increases. For cases where $\bar{v}_j \neq 0$ and $(\bar{p}_j - \bar{p}_\infty) \neq 0$ these same effects of the propeller characteristics on η_p will exist.

Blade Element Model. The momentum actuator disk model does not provide a means to relate the total thrust and torque produced by the propeller to the geometry of the propeller blades. This relation is accomplished by considering a finite number of blades which are composed of a series of elements of spanwise length, dr, as in Fig. 27.24(a).

Figure 27.24(b) depicts the blade element at a radius, r, the forces on the element, and its inlet and exit velocity diagrams when viewed looking along a radial line. The blade appears as an airfoil section, with increments of the total propeller thrust and torque, dF and dT, respectively, acting at its center. The total force on the blade element can also be represented as the lift, dL, and drag, dD, components of force on the element which act normal and parallel to the relative velocity W_m.

Many times, propellers are categorized by their *geometric pitch* $P = 2\pi r \tan \beta$, the distance an ideal propeller would move forward in one revolution as a *screw*. The *effective pitch*, $P_e = 2\pi r \tan \phi$, is the actual distance the propeller moves forward through the fluid. The term *slip* refers to the difference between the ideal and actual motion divided by the ideal forward motion.

Estimation of Forces: Referring to Fig. 27.24(d) the contribution of one blade element to the total propeller thrust and torque is

$$dF = dL \cos (\phi + \alpha_i) - dD \sin (\phi + \alpha_i) \qquad (27.18)$$

$$dT = r[dL \sin (\phi + \alpha_i) + dD \cos (\phi + \alpha_i)] \qquad (27.19)$$

The values of dL and dD can be expressed in terms of section lift and drag-coefficients, C_l and C_D, respectively, as

$$dL = \tfrac{1}{2}\rho W_m^2 C_l c \, dr \qquad (27.20)$$

$$dD = \tfrac{1}{2}\rho W_m^2 C_D c \, dr \qquad (27.21)$$

The section lift coefficient, C_l, is a function of the airfoil section geometry, the section angle of attack α, the proximity or spacing of the other blades on the pro-

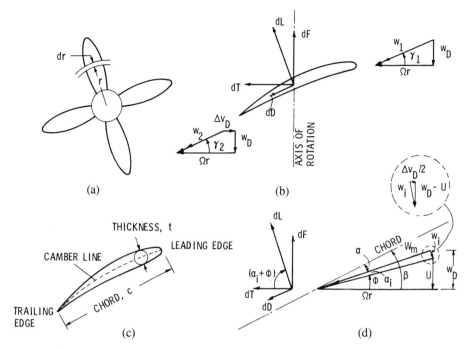

FIGURE 27.24 Propeller blade element. (a) Full propeller, (b) blade element, (c) airfoil notation, and (d) force and velocity components.

peller, and the Mach number of the flow relative to the blade. The value of C_D is primarily a function of C_l and the local Reynolds and Mach numbers.

The value of C_l can be determined for a particular airfoil section shape from experimental data such as presented in Abbott and von Deonhoff (1959). These data are for isolated airfoil sections and do not include the interference effects from adjacent blades. The interference effect can be estimated from Fig. 27.25. The ordinate of this figure is the ratio of the C_l for a finite spacing, s, to the C_l for an isolated airfoil $s = \infty$. The abscissa is the space to chord ratio, s/c, of the propeller blades at a constant radius

$$\frac{s}{c} = \frac{2\pi r}{Bc} \tag{27.22}$$

where B = number of blades.

If experimental data are not available, the value of C_l can be estimated from the analysis of the previous section as

$$C_l = \frac{L}{\frac{1}{2}\rho W_m^2 c} = \frac{\rho W_m \Gamma}{\frac{1}{2}\rho W_m^2 c} \tag{27.23}$$

$$= \frac{4\pi r}{Bc} \frac{\Delta \bar{v}_D}{W_m} \tag{27.24}$$

where Γ is the blade circulation.

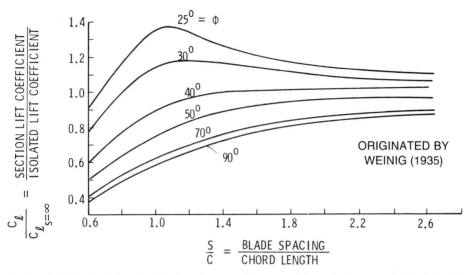

FIGURE 27.25 Blade-to-blade interference effect coefficient. (Courtesy of G. F. Wislicenus.)

If the flow Mach number relative to the blade, $M_r \equiv W_m/a$, is $0.3 < M_r < 0.99$, compressible flow effects on C_l must be included. In the absence of reliable experimental data, the Prandtl–Glauert compressibility correction is used. It can be shown by comparison with experimental data [McCormick (1979)] that this correction must be applied with some caution, particularly as M_r approaches 1.0. Daley and Dick (1956) and USAF (1975) present collections of experimental data which demonstrate the effects of compressibility on airfoil section data.

Maximum Blade Element Efficiency: Equations (27.18) and (27.19) can be rewritten as

$$dF = \frac{1}{2} \rho \bar{w}_D^2 c \, dr C_l \frac{\cos (\phi + \epsilon)}{\cos \epsilon \sin^2 \phi} \qquad (27.25)$$

$$dT = \frac{1}{2} \rho \bar{w}_D^2 cr \, dr C_l \frac{\sin (\phi + \epsilon)}{\cos \epsilon \sin^2 \phi} \qquad (27.26)$$

where $\tan \epsilon \equiv C_D/C_l$, and it is assumed that $\alpha_i \ll \phi$. The *blade element efficiency* is defined as

$$\eta_b = \frac{U \, dF}{\Omega \, dT} = \frac{\cot \epsilon - \tan \phi}{\cot \epsilon + \cot \phi} = \frac{C_l/C_D - \tan \phi}{C_l/C_D + \cot \phi} \qquad (27.27)$$

The value of ϕ that makes η_b a maximum is termed the *optimum pitch angle*, ϕ_{opt}. Thus,

$$\phi_{\text{opt}} = \frac{\pi}{4} - \frac{\epsilon}{2} = 45° - \frac{57.3}{2(C_l/C_D)} \qquad (27.28)$$

and

$$(\eta_b)_{max} = \frac{2(C_l/C_D) - 1}{2(C_l/C_D) + 1} \tag{27.29}$$

Propeller Efficiency: The *shaft power coefficient* of a propeller, C_p, less mechanical power losses, can be approximated as a function of the *propeller shaft thrust coefficient*, C_F

$$C_P = \frac{\text{POWER}}{\frac{1}{2}\rho U^3 \pi R_p^2} = C_F + \frac{C_F}{2}[\sqrt{1 + C_F}] - 1 + \frac{Bc}{\pi R_p}\frac{C_D}{4}\left(\frac{\pi}{J}\right)^3\left[1 + \frac{J^2}{\pi}\right]^{3/2}$$

$$\tag{27.30}$$

where J is the *advance* ratio

$$J \equiv U/2nR_p \tag{27.31}$$

and

$$C_F \equiv (\text{SHAFT THRUST})/\tfrac{1}{2}\rho U^2 \pi R_p^2 = 8K_F/\pi J^2 \tag{27.32}$$

In this expression, c and C_D are average values of chord length and section drag coefficient along the blade span. The propeller efficiency is then

$$\eta = \frac{FU}{\text{POWER}} = 2\left\{1 + \sqrt{1 - C_F} + \frac{Bc}{\pi R_p}\frac{C_D}{2C_F}\left(\frac{\pi}{J}\right)^3\left[1 + \frac{J^2}{\pi}\right]^{3/2}\right\}^{-1}$$

$$\tag{27.33}$$

For a given *solidity*, $Bc/\pi R_p$, this equation predicts an optimum value of C_F. If $C_D \equiv 0$, the ideal efficiency increases as C_F decreases.

Lifting-Line and Lifting-Surface Models. With the advent of high-speed digital computer methods, potential flow models, Sec. 1.3, have been developed for the design and analysis of propellers. The use of these methods results in a complex mathematical formulation and very large computer codes [see Kerwin (1973), Cox and Morgan (1972), Chaussee (1979), and Greeley and Kerwin (1982)]. As a result, these flow models will not be discussed here.

27.2.3 Propeller Performance Characteristics

There are applications which require high levels of propeller *thrust loading* (thrust per unit disk area) that cannot be efficiently attained with the open propeller. In these cases, some form of jet propulsion or ducted propeller is required. The discussion here will be directed mainly toward ducted propellers for marine application.

Open and Ducted Propellers. The primary difference between open and ducted propellers involves the shape of the slipstream that passes through the propulsor. For

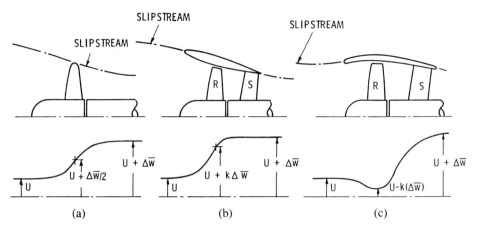

FIGURE 27.26 Schematic of ducted and open propellers. (a) Open propeller, (b) accelerating shroud arrangement, and (c) decelerating shroud arrangement.

the ducted propeller, the shape of the slipstream can be controlled by proper shaping of the duct which surrounds the blade rows. In contrast, the slipstream through an open propeller is always contracting and the fluid accelerates as it approaches the propeller disk. The ducted propeller provides the option of either accelerating or decelerating the flow before it enters the rotating blade row. A schematic of the open propeller, and two ducted propellers employing counterswirl vanes is shown in Fig. 27.26. The axial velocity distribution in the slipstream at various stations through each propulsor is also indicated.

Assuming the pressure in the propeller jet $p_j = p_\infty$, the thrust generated by a ducted or open propeller can be expressed using Eq. (27.6) as

$$C_F = \frac{\text{THRUST}}{\rho \dfrac{U^2}{2} \pi R_p^2} = \frac{U \rho \pi R_P^2 \Delta \overline{w}}{\rho \dfrac{U^2}{2} \pi R_P^2} = 2 \frac{\overline{w}_D}{U} \frac{\Delta \overline{w}}{U} \tag{27.34}$$

where Δw is the increase in axial velocity. It is shown in Greeley and Kerwin (1982) that $\Delta \overline{w}/U$ values for open propellers typically have a maximum value of 0.2, but normally they are on the order 0.1 or lower. This implies that only a small component of tangential velocity is placed in the flow which makes the kinetic energy losses in the slipstream very small. Small diameter propellers with low shaft speeds tend to be inefficient, since the swirl losses in the slipstream become relatively high. On this basis, it can be shown that open propeller advance coefficients, J, higher than about 1.80 are undesirable [McCormick (1979)] because of penalties in efficiency due to slipstream losses.

As indicated in Eq. (27.11), the propulsive or propeller efficiency for an open propeller is defined as the ratio of the thrust power to the power added to the fluid. In the case of ducted propellers, the duct inlet loss must also be included in defining propulsive efficiency [Wislicenus (1973)]. If a ducted propeller is considered which employs counterswirl vanes upstream or downstream of the rotor to eliminate swirl in the slipstream, the propulsive efficiency can be defined as

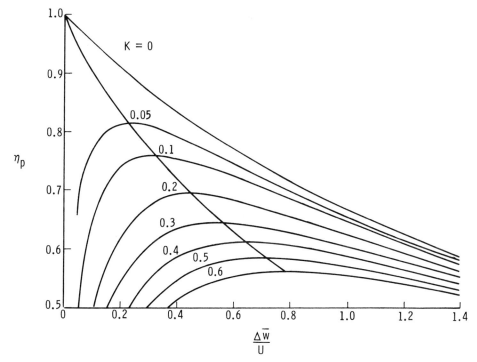

FIGURE 27.27 Propulsive efficiency as a function of inlet loss.

$$\eta_p = \frac{1}{1 + \dfrac{\Delta\overline{w}}{2U} + \dfrac{K}{2}\dfrac{U}{\Delta\overline{w}}}$$ (27.35)

where K is the inlet loss expressed as a fraction of the dynamic head, $U^2/2g$. The results of this relation are plotted in Fig. 27.27.

If the inlet loss to the duct is as low as 5% of the dynamic head, the value of $\Delta\overline{w}/U$ required to obtain a maximum efficiency is 0.22. This is more than twice the value of $\Delta\overline{w}/U$ typically imparted to the fluid by an open propeller and implies, for a given thrust, that the mass flowrate through a ducted propeller is normally about one half that through an open propeller. Therefore, the diameter of a ducted propeller, when designed to produce the same thrust, will be smaller than that of the open propeller.

Cavitation and Blade Sonic Velocities. If aircraft and marine propellers are compared, it is apparent the marine propeller blades fill a much larger portion of the disk area. This increased solidity or wetted surface area of the marine propeller for a given disk diameter is required to reduce thrust loading per unit of blade surface area. In this way, the pressure on the surface of the blades is controlled to prevent excessive *cavitation* which could result in thrust breakdown. Cavitation is discussed in Sec. 23.11.

The occurrence of cavitation can be understood by considering the velocity W_1 relative to a blade (Fig. 27.24). The blade relative velocity is highest at the propeller tip, and cavitation will usually occur first in this region, as will the sonic velocities associated with aircraft propellers.

The change in static pressure from station (1), just upstream of the blade, to the point of minimum pressure on the blade surface is a function of the relative velocity squared and the geometric shape of the blade section. The minimum pressure coefficient of the section can be defined as

$$c_b = \frac{p_1 - p_{min}}{\rho \dfrac{W_1^2}{2}} = \left(\frac{W_{min}}{W_1}\right)^2 - 1 \tag{27.36}$$

Values of $-c_b$ are given in Abbott and von Doenhoff (1959) for different blade sections. By setting p_{min} equal to the liquid vapor pressure, this expression can be rearranged to give, for a given ship speed, the depth below the surface of the liquid at which cavitation will first occur.

$$h_{sub} = c_b \frac{U^2}{2g}\left[1 + \left(\frac{\Omega R_P}{U}\right)^2\right] - 9.75 \tag{27.37}$$

Unsteady Forces and Vibration. The vibration of a propeller originates from its operation in a distorted inflow and from vortex shedding at the blade trailing edges. The variations of the inflow velocity can occur in both the radial and circumferential directions. The circumferential velocity variations at a given radius result in a time-dependent variation of the blade inlet flow angle and, subsequently, propeller vibration and unsteady blade forces.

The frequency of the periodic force fluctuations due to a propeller operating in a circumferentially distorted flow field will be equal to the product of shaft rps and the number of blades, and multiples thereof. The harmonic content of the inflow to the propeller should be considered in selecting the number of blades on the propeller as described [Tsakomas *et al.* (1967), McCarthy (1961), and Thompson (1976)]. Vortex shedding at the blade trailing edge will occur at a discrete frequency which depends upon the blade thickness at the trailing edge and the velocity relative to the blade. To *detune* this frequency and the blade structural response, the trailing edge can be shaped to change the shedding frequencies [Thompson (1976)].

The presence of cavitation on the blades also gives rise to unsteady forces and vibration since cavitation itself is an unsteady process. The nonradial or *skewed* blades shown in Fig. 27.21 are used to increase the propeller forward velocity before cavitation inception, or sonic velocities, occur and to reduce the unsteady blade response.

Selection of Design Parameters; Performance Tradeoffs. The selection of the propeller diameter is normally based on attaining a maximum propulsive efficiency. In most cases, gearbox and powerplant constraints limit the shaft speed to some specified value at a given forward speed. With this information, Eq. (27.30) provides an approximate relation which, for a given solidity, indicates an optimum value of C_F. Knowing the forward speed and thrust required, the propeller tip diameter can thereby be found for maximum efficiency.

In similar fashion the ducted propeller can be sized by considering Fig. 27.27.

The duct-loss coefficient is estimated based on the type of duct to be used, either accelerating or decelerating. The value of $\Delta\overline{w}/U$ is selected to give a maximum efficiency for the estimated value of duct-loss. The impulse–momentum relationship, Eq. (27.34), will then provide the volumetric flow rate, Q, required to produce a given thrust with a specified value of, $\Delta\overline{w}$.

With the flow rate, Q, known, the requirement to maintain maximum cavitation or sonic velocity resistance requires limiting the relative velocity at the blade tip. This can be achieved by expressing the relative velocity, W_{TIP}, in terms of the tip radius

$$W^2_{\text{TIP}} = \overline{w}^2_D + (\Omega R_p)^2 \tag{27.38}$$

where

$$\overline{w}_D = \frac{Q}{\pi(R^2_P - R^2_{\text{HUB}})} \tag{27.39}$$

Combining these relations, the relative velocity at blade tip becomes

$$W_{\text{TIP}} = \left\{\left[\frac{Q/\pi}{R^2_P - R^2_{\text{HUB}}}\right]^2 + (\Omega R_P)^2\right\}^{1/2} \tag{27.40}$$

With the hub diameter, flowrate, and shaft speed fixed, Eq. (27.40) can be used to obtain the value of R_P which gives minimum, W_{TIP}.

Constraints on Propeller Design. The expanded area ratio of a marine propeller, the ratio of blade wetted area to propeller disk area, must be large to prevent thrust breakdown due to excessive cavitation. The blade area required can be selected by the method of Burrill and Emerson (1963).

Due to inflow distortions and variations in flow incidence to the blade as it rotates, it is difficult to avoid local areas of cavitation and sonic velocity. This is shown in Fig. 27.28 where the relative blade pressure coefficient defined in Eq. (27.36) is depicted as a function of flow incidence angle and thickness to chord ratio, τ, for a typical marine propeller blade section. The shaded zones for a given τ are regions of cavitation-free operation. Outside of these shaded regions, cavitation will occur on the suction face of the blade at positive angles of incidence and on the pressure face for negative angles. These curves indicate the necessity for a trade-off. With the selection of a blade section having a given τ, usually dictated by blade strength requirements, it is possible to achieve cavitation free operation at the leading edge of the blade over a wide range of flow incidence angles. However in so doing, the cavitation resistance aft of the leading edge is compromised. Similar effects are experienced with local sonic velocities on aircraft propeller blades. Strength requirements which result in large values of τ near the propeller hub can also lead to either cavitation or sonic velocities in the hub region.

27.2.4 Advanced Propeller Concepts

The elimination of swirl in the slipstream results in efficiency gains by reducing the kinetic energy losses and pressure drag associated with a swirling slipstream dis-

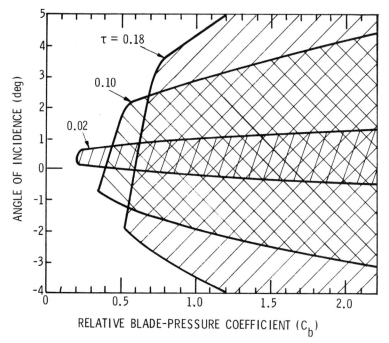

FIGURE 27.28 Minimum pressure coefficient of airfoil sections of varying thickness with flow incidence angle.

cussed in Sec. 27.2.2. In Takekuma (1980), this concept has been applied to large tankers which incorporate a set of stationary counterswirl vanes forward of the propeller. Efficiency gains as high as 12% were recorded when compared to a propeller alone.

Counterrotating propellers provide similar gains in efficiency. The counterrotating arrangement presents a mechanically more complex system than the single-shafted propeller. However, the shaft power and loading can be divided between the two propellers, and that will result in improved cavitation and sonic velocity resistance.

Mitchell and Mikkelson (1982) and Rohrbach *et al.* (1982) describe the development of aircraft having a forward speed at 80% of the speed of sound and propelled by a turboprop. The rotational blade velocity at the tip also approaches this value. The velocity relative to the blade tip will, therefore, exceed the speed of sound by about 13%. The operation of a propeller with transonic velocities requires a large number of blades and a high solidity [see Fig. 23.21(b)]. The swept or skewed blade shapes reduce the blade unsteady pressure fields, noise, and vibration. It was previously noted that a penalty in efficiency occurs if a propeller is designed to operate above an advance ratio of about 1.80. This propeller operates at an advance ratio approaching 3.0 and does experience an efficiency penalty. However, at this cruise Mach number, the efficiency of this propeller is significantly higher than that of existing designs or turbofans.

Efforts to reduce propeller induced vibrations caused by the propeller operating in a nonuniform flow field behind a surface ship have resulted in propellers with swept blading. The design philosophy and experimental evaluation of these highly swept or skewed propellers are reported in Valentine and Dashaw (1975), Hammer and McGinn (1978), and Wilson *et al.* (1982).

27.3 HELICOPTER ROTOR AERODYNAMICS
J. Gordon Leishman and Alfred Gessow

27.3.1 Introduction

Unlike a fixed wing aircraft, the helicopter is not dependent upon forward speed to generate lift and control. It can hover, fly forward or backwards, sideways and vertically, because lift, control, and propulsion are generated by one or more turning rotors that operate across the entire speed range. These functions are basically separated in the fixed-wing airplane. The unique hover and vertical flight capability of the helicopter comes, however, at a price: aerodynamic, dynamic, and structural complexity, limits on maximum forward speed, and larger power requirements for the same gross weight compared to its fixed-wing counterpart.

Although the helicopter concept can be traced back to early Chinese *tops* and to Leonardo de Vinci's aerial-screw machine of 1483, its inherent mechanical and aerodynamic complexities resisted serious efforts to bring it to fruition as a practical aircraft. The first successful approach to a helicopter was not, in fact, this breed of machine but an evolutionary precursor that depended on the propulsion and control systems of a conventional airplane. In 1923, Spanish engineer Juan de la Cierva built and flew an aircraft that was driven by a tractor propeller without lifting fixed wings, instead an arrangement of four blades that could be turned on a vertical shaft provided the lift. This rotor was not powered directly, but was turned by the action of the airflow on the blades as the aircraft moved forward. Under these conditions, the power to drive the rotor and produce lift comes from the relative velocity through the rotor, so-called *autorotation*.

As a stall-proof, safe aircraft that could land almost anywhere, de la Cierva's *autogiro* proved to be a convenient and practical innovation. Improved versions were developed, including direct control and jump take-off, but it became evident that the autogiro was only a makeshift hybrid, and the logical development of the machine was to directly power the rotor. By so doing, it would be possible to dispense with the propeller, and give the vehicle the ability to take off and climb vertically, hover, and perform desirable maneuvers. Several inventors went on to this decisive step, and by the late 1930's the first true helicopters were flying, lagging the development of fixed-wing aircraft by some thirty years. During World War II, successful models were developed and employed. Since then, helicopters have advanced substantially in capability and versatility, with scores of military and commercial uses. Making a virtue out of necessity, modern helicopters equal or exceed fixed wing aircraft in making maximum use of leading-edge technology, such as advanced composite structures, fly-by-wire automatic control systems, active vibration control, innovative high-torque transmissions, and high performance gas turbines.

From an aerodynamics perspective, the flow field of a helicopter is much more complex compared to a fixed wing aircraft. This complexity stems from an aerodynamic environment that is unique to rotors. As compared to a fixed wing, the lifting rotor has a velocity that varies along the span of each blade. This results in a high concentration of aerodynamic forces over the outer portions of the blade and the formation of strong vortices that trail from each of the blade tips. Unlike a fixed-wing aircraft, these tip vortices remain close to the rotor and produce a complicated three-dimensional induced velocity field. This, in turn, strongly affects the aerodynamic loads, performance, aeroelasticity, and noise generation of the entire rotor. Some idea as to the complexity of the rotor flow field can be seen in the experimental works of Larin (1970), (1973), Lehman (1968), and Landgrebe and Bellinger (1971).

All helicopters spend some time in hovering flight, a regime in which they are operationally efficient. However, helicopters are also required to travel in forward flight, and under these conditions the rotor moves edgewise through the air. As a consequence, the blades encounter an asymmetric velocity field biased toward the advancing blade, and the blade airloads become periodic at multiples of the rotational speed of the rotor. The flow field at the rotor, consequently, becomes much more unsteady and three-dimensional, and the path of the rotor tip vortices trailed into the wake becomes considerably more complicated. In addition, the blades can locally interact with these vortices to produce strong time-dependent airloads. From an acoustics perspective, this interaction is a source of obtrusive low and high frequency rotor noise.

Furthermore, in high speed forward flight (i.e., between 150 and 200 kts.), the blade tips on the advancing side of the rotor disk can penetrate into transonic flow regimes, with the associated formation of compressibility zones and shock waves. Besides the occurrence of wave drag and the possibilities of shock induced flow separation, the periodic formation of shock waves is also a source of noise. On the retreating side of the disk, the relative velocity is quite low, and the blades are required to operate at very high angles of attack to maintain lift, often into partially separated or stalled flow regimes. Due to the inherent time-dependent nature of the rotor environment in forward flight, this stall process is unsteady and is referred to as *dynamic stall*. Some idea of the complexity of the helicopter flow field can be gained from Fig. 27.29.

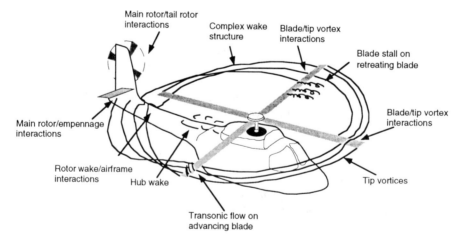

FIGURE 27.29 Basic flow structure generated by a helicopter in forward flight, and some problem areas.

27.3.2 Axial Momentum Theory

The helicopter or other rotary wing aircraft may operate in a variety of flight regimes, including hover, climb or descent, sideward or forward flight, and various maneuvers. In hover or axial (climb or descent) flight, the flow is essentially axisymmetric throught the rotor disk(s), either upward or downward. Although the actual physical flow through and about the rotor will include discrete vortices trailed from the blade tips, the basic performance of the rotor can be analyzed by an idealized approach called *momentum theory*. This theory allows us to derive a physical understanding of helicopter rotor performance, and it also forms a foundation for more elaborate treatments of helicopter characteristics. In the *momentum* approach, the three basic conservation laws (conservation of mass, momentum, and energy) are applied in a quasi-one-dimensional integral form to the rotor and its flow field. This is analogous to the actuator disk approach discussed for propellers in Sec. 27.2.

Momentum Theory in Hovering Flight. Hover is a unique flight condition wherein the rotor has zero forward (edgewise) speed and zero vertical speed (no climb or descent). Consider the rotor and its flow field, as shown in Fig. 27.30, and a control volume v, with surface area S. Let $d\overline{S}$ be the unit normal area vector, which by convention always points out of the control volume. The rotor disk area is A.

It is assumed that the fluid velocity increases smoothly as it is entrained into and through the rotor disk; there is no jump in velocity across the disk, although there is a jump in pressure. Further, it is normally assumed that there is no induced rotational motion in the slipstream, and this is found justified for the low rpm's of helicopter rotors. Denote the section far upstream of the rotor as section 0, just above the rotor by section 1, just below the rotor by section 2, and in the *far wake* (well downstream of the rotor) by section 3. Far upstream of the rotor, the velocity is zero. At the plane of the rotor (sections 1 and 2), the induced velocity is v_i. In the far wake, the velocity is increased over that at the plane of the rotor to a velocity, w.

By the principle of conservation of mass, the mass flow rate, $\dot{m}$, is constant within the boundaries of the rotor wake. The general equation for the mass flow rate through the rotor disk is

$$\dot{m} = \iint_3 \rho \overline{V} \cdot d\overline{S} = \iint_2 \rho \overline{V} \cdot d\overline{S} \tag{27.41}$$

where ρ is the fluid density. For this one-dimensional situation it follows that

$$\dot{m} = \rho A_3 w = \rho A v_i \tag{27.42}$$

The principle of conservation of momentum can be used to equate the rotor thrust to the net time rate of change of momentum out of the control volume (Newton's second law). More specifically, we find the force on the fluid; therefore, by Newton's third law, the force on the rotor (the thrust, T) is equal and opposite to the force on the fluid. Therefore,

$$\overline{F} = T = \iint_3 \rho(\overline{V} \cdot d\overline{S})\overline{V} - \iint_0 \rho(\overline{V} \cdot d\overline{S})\overline{V} \tag{27.43}$$

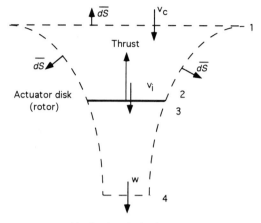

Idealized control volume

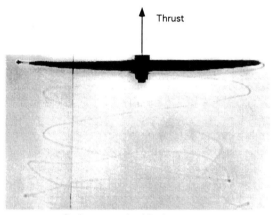

Real vortex wake (shadowgram
of two-bladed rotor)

FIGURE 27.30   Actual vortex flow from rotor and idealized control volume surrounding
rotor and its flow field in axial flight state.

Since in hover the velocity far upstream of the rotor is zero, the second term on
the right-hand side of the above equation is also zero. Therefore, for hovering flight,
the rotor thrust is

$$T = \left| \iint_3 \rho(\overline{V} \cdot d\overline{S})\overline{V} \right| = \dot{m}w \qquad (27.44)$$

From the principle of the conservation of energy, the work done on the rotor is
equal to the gain in energy of the fluid per unit time. The work done by the rotor is
Tv_i, so we have

$$Tv_i = \iint_3 \tfrac{1}{2}\rho(\overline{V} \cdot d\overline{S})|\overline{V}|^2 - \iint_0 \tfrac{1}{2}\rho(\overline{V} \cdot d\overline{S})|\overline{V}|^2 = \tfrac{1}{2}\dot{m}w^2 \qquad (27.45)$$

Therefore, eliminating $T/\dot{m}$ from Eqs. (27.44) and (27.45) gives $w = 2v_i$, which simply shows that the slipstream velocity is twice the induced velocity in the plane of the rotor.

Wake Contraction: Because the flow velocity increases in the rotor wake from continuity considerations, the area of the slipstream must decrease. It follows from Eq. (27.42) that the ratio of the area of the far wake to the area of the disk is one-half. Alternatively, considering the diameter of the rotor wake relative to the rotor we find that the wake *contraction ratio* is $1/\sqrt{2}$. In practice, it has been found experimentally that the wake contraction ratio is a little less than the theoretical value of 0.707 (typically 0.75 to 0.8), mainly as a consequence of viscosity which tends to entrain fluid outside of the ideal control volume. Also, there is a small but significant swirl component of velocity in the wake induced by the spinning rotor, which tends to reduce the net change of the fluid momentum in the vertical direction. Such effects tend to decrease the rotor thrust for a given shaft torque (power supplied) or increase the power required to produce a given thrust.

Pressure Variation: The pressure variation through the rotor flow field can be found by the application of Bernoulli's equation to a streamline above and below the rotor disk. Between stations 0 and 1

$$p_0 = p_1 = \tfrac{1}{2}\rho v_i^2 \tag{27.46}$$

and between stations 2 and 3

$$p_2 + \tfrac{1}{2}\rho v_i^2 = p_0 + \tfrac{1}{2}\rho w^2 \tag{27.47}$$

Because the jump in pressure Δp across the disk is uniform across the disk, it must be equal to the disk loading, T/A, so we can write

$$\frac{T}{A} = p_2 - p_1 = \left(p_0 + \frac{1}{2}\rho w^2 - \frac{1}{2}\rho v_i^2\right) - \left(p_0 - \frac{1}{2}\rho v_i^2\right) = \frac{1}{2}\rho w^2 \tag{27.48}$$

This simply proves that the rotor disk loading is equal to the dynamic pressure of the rotor slipstream. We also find that just above the disk the pressure is reduced to $p_1 = p_0 - \tfrac{1}{4}(T/A)$, and just below the disk the pressure is increased to $p_2 = p_0 + \tfrac{3}{4}(T/A)$.

Induced Velocity: Rearranging Eq. (27.44) and using the result that $w = 2v_i$ gives the induced velocity at the plane of the rotor disk in hover, v_h, as

$$v_i = v_h = \sqrt{\frac{T}{2\rho A}} = \sqrt{\left(\frac{T}{A}\right)\frac{1}{2\rho}} \tag{27.49}$$

(Note that v_h is used to represent the induced velocity in hover, and is later used as a reference when considering the climb, descent, and forward flight cases.) Therefore, the power required to hover will be

$$P_i = Tv_i = Tv_h = T\sqrt{\frac{T}{2\rho A}} = \frac{T^{3/2}}{\sqrt{2\rho A}} = \frac{W^{3/2}}{\sqrt{2\rho A}} \qquad (27.50)$$

where W is the weight of the helicopter. This power is entirely induced in origin since viscous effects have not been considered thus far. This is the minimum power required to hover. From this equation it is clear that to make a helicopter hover at a given thrust (and density altitude) with minimum induced power, the induced velocity at the disk must be small. Therefore, the mass flow through the disk must be large, and this consequently requires a large rotor disk area.

Nondimensional Coefficients: As in the case of fixed wing aircraft, nondimensional coefficients are normally employed in helicopter rotor analysis, but with some differences. The *rotor thrust coefficient* is defined as

$$C_T = \frac{T}{\rho A (\Omega R)^2} \qquad (27.51)$$

Velocities are nondimensionalized by the hover tip speed, ΩR, so that the *inflow ratio*, $\lambda_i = v_i/\Omega R$, is related to the thrust coefficient in hover by

$$\lambda_i = \lambda_h = \frac{1}{\Omega R}\sqrt{\frac{T}{2\pi A}} = \sqrt{\frac{C_T}{2}} \qquad (27.52)$$

and this, of course, is based on the assumption of a uniform inflow over the disk. The *rotor power coefficient* is defined as

$$C_P = \frac{P}{\rho A (\Omega R)^3} \qquad (27.53)$$

so that for the hovering rotor, the induced power coefficient is

$$C_{P_i} = C_T \lambda_i = \frac{C_T^{3/2}}{\sqrt{2}} \qquad (27.54)$$

which is known as the *ideal power coefficient.*

Note that the definition of the thrust and power coefficients is a little different in some parts of the world, including Europe. In some countries, a factor of $\frac{1}{2}$ is used in the denominator.

Figure of Merit: An important parameter that is used frequently in helicopter rotor design for hovering flight is called the *figure of merit*, and is calculated using momentum theory as a basis. The figure of merit is defined as the ratio of the ideal power required to hover to the actual power required, i.e.,

$$FM = \frac{\text{Ideal power required to hover}}{\text{Actual power required to hover}} \qquad (27.55)$$

Since the application of momentum theory assumes an ideal fluid, uniform inflow, and zero swirl, the ideal power is entirely induced in origin. Therefore, the figure of merit can be used as a measure of the efficiency of the rotor in hover relative to the ideal. The most useful purpose of figure of merit is for making a comparative study of various rotors.

Since the momentum theory gives the ideal or minimum induced power loss as

$$P_i = P_{\text{ideal}} = Tv_h = T\sqrt{\frac{T}{2\rho A}} \qquad (27.56)$$

the figure of merit, *FM*, is given by

$$FM = \frac{T\sqrt{T/2\rho A}}{P_{\text{actual}}} = \frac{C_T^{3/2}}{\sqrt{2}C_P} \qquad (27.57)$$

Since the actual power consumed by a rotor will involve both induced drag and profile drag parts, the figure of merit can also be written as

$$FM = \frac{P_{\text{ideal}}}{P_{\text{actual}}} = \frac{P_{\text{ideal}}}{P_i + P_0} = \frac{Tv_i}{\kappa Tv_i + P_0} \qquad (27.58)$$

where P_0 is called the *profile power*, and is the power contribution due to the viscous shear and viscous effects on the pressure drag forces on the blades. As will be explained shortly, the factor κ can be used to account for *nonideal effects* on the induced drag part of the power. In rotor tests, both induced and profile power contributions are present and cannot be measured separately, so the figure of merit must be computed from the measured or calculated thrust and total power coefficients using Eq. (27.57).

Note that a figure of merit of unity represents the unattainable ideal. In practice, *FM* values greater than 0.7 represent a good hovering performance for a helicopter, and anything less than 0.6 represents a comparatively poor performance. A typical figure of merit plot versus blade loading (C_T/σ) is shown in Fig. 27.31 for different values of *rotor solidity*, σ. Here, σ is simply the ratio of the total blade area to the area of the disk, i.e., $\sigma = N_b cR/\pi R^2 = N_b c/\pi R$ where N_b is the number of blades. Also, as will be shown, the quantity C_T/σ reflects the mean lift coefficient of the rotor. It can be seen from this figure that the *FM* reaches a maximum and then remains constant. For some rotors, especially those with older and less efficient airfoils, the curve will exhibit a peak *FM*, followed by a drop. This is because of the effects of flow separation and blade stall over parts of the rotor blades, which produces both a higher profile drag coefficient along with some loss of lift relative to the ideal.

Momentum Theory in Vertical Climb and Descent. Basic momentum theory can also be used to examine rotor performance in climb or descent. The climb case is a simple extension of hover, but with a nonzero axial (far-upstream) velocity, v_c. If the hover induced velocity is used as a reference, then the solution for the induced velocity is

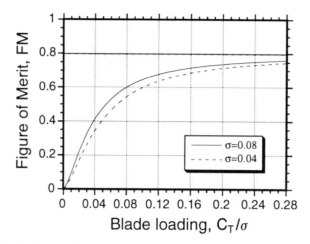

FIGURE 27.31   Typical figure of merit variation versus thrust for a helicopter rotor.

$$\frac{v_i}{v_h} = -\left(\frac{v_c}{2v_h}\right) + \sqrt{\left(\frac{v_c}{2v_h}\right)^2 + 1} \qquad (27.59)$$

which is valid for $v_c/v_h \leq -2$. Therefore, as the descent velocity increases, the induced velocity also decreases and asymptotes smoothly to zero at high descent rates. For the descent case, the corresponding power ratio is

$$\frac{P}{P_h} = \frac{v_c}{2v_h} + \sqrt{\left(\frac{v_c}{2v_h}\right)^2 + 1} \qquad (27.60)$$

Note that although there is a decrease in induced power, the total power required to climb is always greater than the power required to hover, because power is required to increase the potential energy of the aircraft. However, at the same time, the induced power becomes a progressively smaller percentage of the total power required to climb.

For the descent case, the climb model used previously cannot be used for all $v_c < 0$, because, in a descent, v_c is directed upward with the slipstream above the rotor. This is the case whenever v_c is at least twice the averaged induced velocity at the disk. For cases where the descent velocity is less than this, the flow velocity at the plane of the rotor can take on two possible directions. Under these circumstances momentum theory cannot be used. However, if $|v_c| > 2v_h$, so that a well defined slipstream will exist above the rotor, we find that

$$\frac{v_i}{v_h} = -\left(\frac{v_c}{2v_h}\right) - \sqrt{\left(\frac{v_c}{2v_h}\right)^2 - 1} \qquad (27.61)$$

which is valid for $v_c/v_h \leq -2$. Therefore, as the descent velocity increases, the induced velocity also decreases and asymptotes smoothly to zero at high descent rates. For the descent case, the corresponding power ratio is

$$\frac{P}{P_h} = \frac{v_c}{2v_h} - \sqrt{\left(\frac{v_c}{2v_h}\right)^2 - 1} \tag{27.62}$$

Thus, the rotor extracts power from the air in a descent and for $v_c < 2v_h$ it is always less than the power required to hover. This is called the *windmill brake state*. For $-2 \leq v_c/v_h < 0$, the induced velocity and power are estimated by experimental means, and the results approximated empirically.

Nonideal Effects on Rotor Performance. We find that, in practice, the actual induced rotor power is always higher than that given by simple momentum theory considerations. Although these effects can be calculated by more advanced treatments, for engineering use they can be empirically described in the induced power equation by modifying Eq. (27.54) to read

$$C_{P_i} = \frac{\kappa C_T^{3/2}}{\sqrt{2}} \tag{27.63}$$

where κ is a coefficient derived from flight test results. This coefficient is effectively a catchall parameter that can be used to encompass a number of nonideal effects such as nonuniform inflow, tip losses, root cut-out, wake swirl, less than ideal wake contraction, etc. For example, the formation of a trailed vortex at the tip of each blade produces a high local inflow over the tip region and reduces the lifting capability of each blade and, therefore, the thrust of the rotor. This is often referred to as *tip loss*. A *tip loss factor*, B, is used to account for this effect such that the product BR corresponds to an effective blade radius and, therefore, a reduction in rotor disk area by a factor $1/B^2$. This will result in a higher effective disk loading and, therefore, an increase in induced velocity and a corresponding increase in induced power. In practice, typical values of B range from about 0.95 to 0.98. One equation for the tip loss factor is due to Prandtl and is frequently employed in engineering studies. For low inflow it is

$$B = 1 - \frac{\sqrt{C_T/2}}{N_b} \tag{27.64}$$

Several other forms of losses can also be combined into the value of κ. For example, in practice the distribution of inflow across a rotor disk is biased toward the tips, and this produces another source of induced power that is somewhat greater than the ideal case. Swirl and interference effects may also be accommodated, albeit very approximately. However, usually κ is not much more than 1.15; anything more means a poorly designed rotor.

Forward Flight Performance. The momentum theory can also be extended to rotors operating in forward flight on the basis of certain assumptions. This treatment of rotor performance in forward flight was first derived by Glauert (1928). By the application of the conservation equations to a control volume surrounding the rotor and its wake, we find the rotor thrust in forward flight is given by

$$T = 2\dot{m}v_i = 2\rho A v_i \sqrt{V_\infty^2 + 2V_\infty v_i \sin \alpha + v_i^2} \tag{27.65}$$

where α is the disk angle of attack.

Note that for hovering flight, $V_\infty = 0$, so that Eq. (27.65) reduces to the result for hover as derived previously. In high speed forward flight, $V_\infty \gg v_i$, so we see that Eq. (27.65) reduces to

$$T \approx 2\rho A v_i V_\infty \tag{27.66}$$

This result is especially interesting, since this is exactly equal to the lift on an elliptically loaded fixed wing with a circular planform of radius R. This can be easily proved by using classical fixed wing theory. There is also considerable experimental evidence to support the conclusion that a rotor acts aerodynamically very much like a fixed wing when operating at high forward speeds [Heyson and Katzoff (1957) and Leishman and Bi (1990)].

Induced Velocity in Forward Flight: In forward flight, the induced velocity can be shown to be

$$v_i = \frac{v_h^2}{\sqrt{(V_\infty \cos \alpha)^2 + (V_\infty \sin \alpha + v_i)^2}} \tag{27.67}$$

Also, since the *advance ratio* is defined as $\mu = V_\infty \cos \alpha / \Omega R$, and the *inflow ratio* is defined as $\lambda = (V_\infty \sin \alpha + v_i)/\Omega R$, we find

$$\lambda = \frac{V_\infty \sin \alpha}{\Omega R} + \frac{v_i}{\Omega R} \tag{27.68}$$

$$= \mu \tan \alpha + \lambda_i \tag{27.69}$$

and

$$\lambda_i = \frac{\lambda_h^2}{\sqrt{\mu^2 + \lambda^2}} = \frac{C_T}{2\sqrt{\mu^2 + \lambda^2}} \tag{27.70}$$

Finally, we can write

$$\lambda = \mu \tan \alpha + \frac{C_T}{2\sqrt{\mu^2 + \lambda^2}} \tag{27.71}$$

Because the inflow ratio λ appears on both sides of the above equation, for the general case an iterative procedure must be used to solve for λ. Note that for a constant rotor thrust, the induced part of the total inflow decreases with increasing advance ratio, μ, and the inflow becomes dominated by the free stream component $\mu \tan \alpha$ term at higher advance ratios.

Power in Forward Flight: In an ideal fluid, the rotor power in forward flight is given by

$$P = T(V_\infty \sin \alpha + v_i) = TV_\infty \sin \alpha + Tv_i \tag{27.72}$$

The first term on the right-hand side is the power required to propel the rotor forward, and the second term is the induced power. As for the axial flight case, we may reference the rotor power in forward flight to the hover result so we can write

$$\frac{P}{P_h} = \frac{P}{Tv_h} = \frac{T(V_\infty \sin \alpha + v_i)}{Tv_h} = \frac{V_\infty \sin \alpha + v_i}{v_h} = \frac{\lambda}{\lambda_h} \tag{27.73}$$

Therefore,

$$\frac{\lambda}{\lambda_h} = \frac{P}{P_h} = \frac{\mu}{\lambda_h} \tan \alpha + \frac{\lambda_h}{\sqrt{\mu^2 + \lambda^2}} \tag{27.74}$$

In a real fluid, the total power, P, required in forward flight can be written as

$$P = P_i + P_0 + P_p + P_c \tag{27.75}$$

where P_i is the induced power, P_0 is the profile power required to overcome viscous losses, P_p is the *parasite power* required to overcome the drag of the helicopter, and P_c is the *climb power* required to increase the gravitational potential of the helicopter.

Consider now the equilibrium of forces on the helicopter. If θ_{FP} is the flight path angle, then the climb velocity is $V_\infty \theta_{FP}$. For small angles, vertical equilibrium gives

$$T = W \tag{27.76}$$

where W is the weight of the helicopter. Horizontal equilibrium gives

$$T(\alpha - \theta_{FP}) = D \tag{27.77}$$

where α is the angle of attack of the rotor, and D is the drag of the helicopter. Solving for α gives

$$\alpha = \theta_{FP} + \frac{D}{T} \tag{27.78}$$

Considering the power to climb and propel the rotor forward, we have

$$TV_\infty \sin \alpha \approx TV_\infty \alpha = TV_\infty \left(\theta_{FP} + \frac{D}{T} \right)$$
$$= TV_\infty \theta_{FP} + DV_\infty$$
$$= Tv_c + DV_\infty \tag{27.79}$$

The term Tv_c is known as the climb power, P_c, and $D_p V_\infty$ is known as the *parasite power*, P_p.

If the forward velocity is high enough, we can approximate the induced velocity by the result given in Eq. (27.67). Therefore, the power equation can be written as

$$P = P_0 + Tv_c + P_p + \frac{\kappa T^2}{2\rho A V_\infty} \tag{27.80}$$

This equation can be used to solve for the climb velocity, v_c, giving

$$v_c = \frac{P - \left(P_0 + P_c + \frac{\kappa T^2}{2\rho A V_\infty}\right)}{T} \tag{27.81}$$

If we assume that P_0 and the aircraft drag, D_p, are not influenced by the climb (or descent) velocity, we have

$$v_c = \frac{P - P_{\text{level}}}{T} = \frac{\Delta P}{T} \tag{27.82}$$

where P_{level} is the power required for level flight at the same forward speed. There-fore, we see that just like a fixed wing aircraft, the climb (or descent) velocity of a helicopter is determined simply by the excess (or decrease) in power required, ΔP, relative to level flight conditions at the same forward speed.

We already know from momentum theory that the induced power of the rotor can be written in coefficient form as

$$C_{P_i} = \frac{\kappa C_T^2}{2\sqrt{\lambda^2 + \mu^2}} \approx \frac{\kappa C_T^2}{2\mu} \text{ (for larger } \mu) \tag{27.83}$$

where as described previously, κ is an empirical correction to account for a multitude of second-order aerodynamic phenomena. The profile power coefficient of the rotor can be estimated from blade element considerations [Gessow and Myers (1952)] as

$$C_{P_0} = C_{Q_0} = \frac{\sigma C_{D_0}}{2} \int_0^1 \left(\frac{U_T^2}{\Omega R}\right)^3 dr$$

$$= \frac{\sigma C_{D_0}}{8} (1 + 3\mu^3) \tag{27.84}$$

Furthermore, the parasite power is expressed as

$$C_{P_p} = \frac{1}{2} \mu^3 \left(\frac{f}{A}\right) \tag{27.85}$$

where f is known as the *equivalent flat plate area* of the hub, fuselage, landing gear etc., and A is the rotor disk area. Typically, values of f/A range from about 0.03 to 0.05 on practical helicopter designs. Obviously, the addition of external stores on military helicopters tends to dramatically increase the equivalent flat plate area, thereby significantly reducing maximum speed capability.

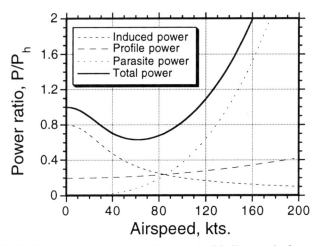

FIGURE 27.32 Power break-down for a typical helicopter in forward flight.

The climb power is equal to the time rate of increase of potential energy $= Tv_c$ $\approx Wv_c$. This climb power coefficient can be written as

$$C_{P_c} = \lambda_c C_T \tag{27.86}$$

For larger values of μ, $\lambda \ll \mu$ so that the final power equation can be simplified to

$$C_P = C_Q = \frac{\kappa C_T^2}{2\mu} + \frac{\sigma C_{D_0}}{8}(1 + 3\mu^2) + \frac{1}{2}\left(\frac{f}{A}\right)\mu^3 + \lambda_c C_T \tag{27.87}$$

A typical example of the power breakdown for a typical helicopter is shown in Fig. 27.32. The general performance characteristics of the helicopter in forward flight may be obtained from the power available relative to that required for straight and level flight at the same speed. The maximum rate of climb is obtained at the speed for minimum power required in level flight. Also, this will be the optimum speed for *minimum autorotative rate of descent*, since the power required at the rotor is a minimum. Thus, in the event of engine failure in hover, it is beneficial for the pilot to translate into forward flight, since the autorotative rate of descent under these conditions will be lower. Also, it can be seen that the power required in high speed forward flight increases dramatically. Since the airframe drag is the major contributor to the power in high speed flight, much can be done to expand the flight envelope by a general drag clean-up of the airframe.

27.3.3 Blade Element Theory

To calculate the distribution of aerodynamic forces on the rotor blades, the *blade element theory* can be used. Blade element theory was originally devised by Drzewiecki, [see Durand (1963)], and forms the basis of all modern analyses of rotor aerodynamics because it provides details on the distribution of blade loading and, therefore, rotor performance in terms of such design parameters as blade twist, planform, chord, airfoil shape, etc.

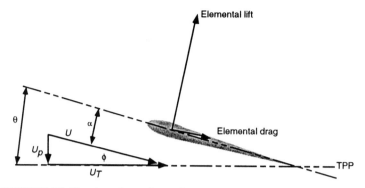

FIGURE 27.33 Aerodynamic environment at a typical blade element.

Induced Velocity Considerations. Blade element theory assumes that each rotor blade section acts as a quasi-two-dimensional airfoil to produce aerodynamic forces. The influence of the rotor wake, etc., is contained in an *induced angle of attack* component at each blade section. The induced angle of attack arises because of the induced velocities from the rotor and its wake, therefore the induced velocity modifies the angle of attack at each blade element (see Fig. 27.33). The blade element method, therefore, requires some estimate of the induced velocity through the rotor. By implication, this will usually include the self-induced velocity field from the wake trailed from each blade, as well as the influence of all the blades and other airframe components.

The velocity at a blade element with a pitch angle, θ, has an out-of-plane component, U_P, and an in-plane component, U_T, relative to the rotor tip-path-plane (TPP) plane (see Fig. 27.33). The resultant velocity is

$$U = \sqrt{U_T^2 + U_P^2} \approx U_T \qquad (27.88)$$

since U_P is generally small relative to U_T over most of the blade. The relative inflow angle (or induced angle of attack), ϕ, at the blade section is

$$\phi = \tan^{-1}\left(\frac{U_P}{U_T}\right) \approx \frac{U_P}{U_T} \text{ for small angles} \qquad (27.89)$$

Therefore, the aerodynamic angle of attack at a blade element is given by $\alpha = \theta - \phi$. Resolving the resultant lift and drag forces perpendicular and parallel to the rotor disk gives

$$dF_z = dL \cos \phi - dD \sin \phi \approx dL$$

$$= \frac{1}{2} \rho c C_{L_\alpha}(\theta U_T^2 - U_P U_T) \, dy \qquad (27.90)$$

$$dF_x = dL \sin \phi + dD \cos \phi \approx \phi \, dL + dD$$

$$= \frac{1}{2} \rho c C_{L_\alpha}\left(\theta U_P U_T - U_P^2 + \frac{C_D}{C_{L_\alpha}} U_T^2\right) dy \qquad (27.91)$$

Hover and Vertical Flight. Induced velocity can be determined for hover or axial flight by a blade element/momentum analysis. However, one assumption made in the simple momentum theory is that the induced velocity is uniformly distributed along the blade span and across the disk. In reality, this is often not the case. One way of determining this nonuniformity and examining the effects on the rotor performance is to use a method comprising both the blade element and momentum approaches.

In this method, we still apply the conservation laws, but now to an annulus of the rotor disk. The annulus is at a distance y from the rotational axis, and has radial thickness dy. The area of this annulus is therefore $dA = 2\pi y \, dy$. We may calculate the incremental thrust, dT, on this annulus on the basis of simple momentum theory, with the implicit assumption that successive rotor annuli have no mutual effect.

On the basis of simple one-dimensional momentum theory, the incremental thrust on the rotor annulus is the product of the mass flow rate across the annulus and twice the induced velocity at the section. It is convenient to work in nondimensional quantities with $r = y/R$, and it can be shown that

$$dC_T = 4\lambda\lambda_i r \, dr = 4\lambda(\lambda - \lambda_c)r \, dr \tag{27.92}$$

since $\lambda_i = \lambda - \lambda_c$. From simple blade element theory, it can be shown that the incremental rotor thrust coefficient on the sections of the blades that sweep out the same area as the annulus is

$$dC_T = \frac{\sigma C_{L\alpha}}{2} (\theta r^2 - \lambda r) \, dr \tag{27.93}$$

where $C_{L\alpha}$ is the lift curve slope of the airfoils comprising the rotor and θ is the local blade pitch angle. Equating the incremental thrust coefficients from the momentum and blade element theories, Eqs. (27.92) and (27.93), we get

$$\frac{\sigma C_{L\alpha}}{2} (\theta r^2 - \lambda r) = 4\lambda(\lambda - \lambda_c)r \tag{27.94}$$

this gives

$$\lambda^2 + \left(\frac{\sigma C_{L\alpha}}{8} - \lambda_c\right)\lambda - \frac{\sigma C_{L\alpha}}{8}\theta r = 0 \tag{27.95}$$

which has the solution

$$\lambda(r, \lambda_c) = \sqrt{\left(\frac{\sigma C_{L\alpha}}{16} - \frac{\lambda_c}{2}\right)^2 + \frac{\sigma C_{L\alpha}}{8}\theta r} - \left(\frac{\sigma C_{L\alpha}}{16} - \frac{\lambda_c}{2}\right) \tag{27.96}$$

Therefore, at a given climb velocity and for any given blade pitch, twist, chord, and airfoil section, the inflow over the disk (blade) may be calculated as a function of r. For hovering conditions, the climb velocity is zero. Therefore, with $\lambda_c = 0$, the above equation simplifies to

$$\lambda(r) = \frac{\sigma C_{L_\alpha}}{16}\left[\sqrt{1 + \frac{32}{\sigma C_{L_\alpha}}\theta r} - 1\right] \tag{27.97}$$

The rotor thrust may then be found by integration across the rotor disk using

$$C_T = \frac{\sigma C_{L_\alpha}}{2}\int_0^1 (\theta r^2 - \lambda r)\, dr \tag{27.98}$$

and a corresponding equation can be obtained for the power.

For an untwisted blade of constant chord and uniform airfoil section, the distribution of inflow as predicted by Eq. (27.97) is shown in Fig. 27.34, and is compared with uniform inflow at the same thrust coefficient. It is clear that the distribution of inflow (and, therefore, of lift) is concentrated toward the blade tips. Obviously, this combination of rotor geometric parameters is nonideal, since the induced power will be higher than the minimum possible that results from uniform inflow.

This analysis can now be used to show how the blade geometric properties can be adjusted to give a more uniform inflow and, therefore, minimum induced power. Returning to Eq. (27.97), there is a special solution that gives $\lambda(r) = $ constant, that is $\theta r = $ constant $= \theta_{tip}$. Thus, the blade pitch angle required to give uniform inflow is given by $\theta(r) = \theta_{tip}/r$. This *ideal twist distribution* is realizable everywhere except at the root since $\theta \to \infty$ as $r \to 0$.

Another interesting output from this type of analysis is the distribution of local lift coefficient along the blade, as shown in Fig. 27.35. Although the ideal twist produces uniform inflow and minimum induced power, the corresponding distribution of C_L is far from optimal. At the root end of the blade, the lift coefficients become increasingly large, and some portions of the blade will ultimately stall.

Consider a rotor with constant blade chord and the ideal twist distribution, $\theta = \theta_{tip}/r$. In this case, λ is a constant, so the blade loading is triangular.* For the ideal

*It is interesting to show that the ideal twist distribution also gives constant bound circulation and uniform disk loading.

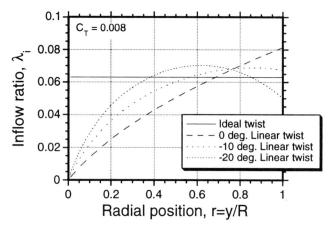

FIGURE 27.34 Distribution of inflow for blades of various twists predicted by the blade element/momentum theory.

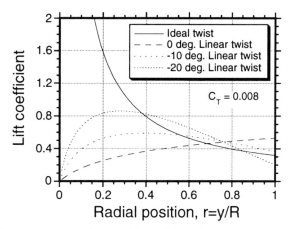

FIGURE 27.35   Distribution of blade lift coefficient for different values of blade twist.

rotor (constant chord, ideal twist, uniform inflow) we find that the local blade lift coefficient is

$$C_L(r) = C_{L_\alpha}\alpha = \frac{4C_T}{B^2\sigma}\left(\frac{1}{r}\right)$$
(27.99)

Consider the behavior of the lift coefficient near the root end of the blade, i.e., as $r \to 0$. Clearly, the values of C_L become increasingly large when moving inboard from the tip, and ultimately there is some point where the attainable section C_L will be limited by the onset of blade stall. This is accompanied by high drag coefficients and a corresponding increase in profile power. Therefore, there will only be one blade station operating at its best lift-to-drag ratio and this compromises the performance of the rotor. The best C_L/C_D for a rotor airfoil is a function of airfoil shape and Mach number. However, assuming for the moment that this angle is independent of Mach number, consider a rotor designed to give both minimum induced power and minimum profile power. Minimum induced power requires uniform inflow and the angle of attack is $\alpha(r) = \alpha_{tip}/r$. Minimum profile power requires that each blade station operate at the best C_L/C_D, i.e., at the condition $\alpha = \alpha_{opt}$. We can show that if $\alpha_{opt} = $ constant, then $\sigma r = $ constant or $c(r) = c_{tip}/r$; that is, to provide the optimum lift-to-drag ratio at each blade station, the local blade chord must vary inversely with span. In a practical sense, the result means that the use of planform taper has a beneficial effect on rotor performance. Therefore, maximum hovering performance considerations require that helicopter rotor blades incorporate both twist and taper.

   Manufacturing considerations often dictate variations that are different from the optimum defined on the basis of the blade element/momentum theory, and usually linear twist and taper variations are employed on a real rotor. This is found to be sufficient for several reasons. First, a real rotor blade has a significant root cut-out to accommodate blade hinges and structural attachments. Therefore, the aerodynamically active portion of the blade may be over the outboard 70% of the blade radius. The optimum twist and taper over this region can be very closely approximated by linear variations. Secondly, the finite root cut-out generates trailed vorticity, which produces a higher induced velocity than the ideal in this inboard part of the blade. Combined with the trailed vorticity from the blade tip, this tends to produce a fairly uniform lift coefficient distribution over the blade, regardless of the actual taper. The use of significant taper may also compromise the rotor performance in forward flight, so moderate amounts of taper are used in practice.

Now we see that on the basis of a blade element treatment of profile power, the rotor figure of merit can be written as

$$FM = \frac{C_{P_{\text{ideal}}}}{C_{P_i} + C_{P_0}} = \frac{\dfrac{C_T^{3/2}}{\sqrt{2}}}{\dfrac{\kappa C_T^{3/2}}{\sqrt{2}} + \dfrac{\sigma C_{do}}{8}} = f(C_T/\sigma) \qquad (27.100)$$

when using a constant section drag coefficient approximation. Clearly, to maximize the figure of merit, the profile part of the power must be kept as low as possible. We see that the profile power is affected by the rotor solidity, σ. Therefore, one way to minimize profile power is to keep the rotor solidity as low as possible. However, this must be done with some caution, because decreasing the solidity reduces the blade lifting area and elevates the local blade lift coefficients for a given rotor thrust. Therefore, decreasing the solidity decreases the *stall margin*.

Consider the mean rotor lift coefficient, which can be defined as that to give the same thrust coefficient as

$$C_T = \tfrac{1}{2} \int_0^1 \sigma r^2 C_L \, dr \qquad (27.101)$$

when the entire blade is assumed to be working at the same local lift coefficient as $\overline{C}_L$, i.e., an optimum rotor. Thus,

$$C_T = \tfrac{1}{2} \int_0^1 \sigma r^2 \overline{C}_L \, dr = \tfrac{1}{2} \overline{C}_L \int_0^1 \sigma r^2 \, dr = \tfrac{1}{6} \sigma \overline{C}_L \qquad (27.102)$$

Therefore, the mean lift coefficient is

$$\overline{C}_L = 6 \left(\frac{C_T}{\sigma} \right) \qquad (27.103)$$

Typically, $\overline{C}_L$ is in the range 0.5 to 0.8 for helicopter rotors, and in practice, it is found to be a good overall indicator as to the working state of the blades. Since the onset of stall sets the performance limits of a conventional rotor, it is very important to leave sufficient stall margin in the rotor design to allow for maneuvers and gusts.

Another way to help minimize profile power is to consider the use of a low drag airfoil. The pressure drag can be minimized by using relatively thin airfoil sections, although the use of thin airfoils has some attendant disadvantages as well. The viscous drag component, C_{do}, can be minimized by carefully controlling the airfoil pressure distribution, and maximizing the run of laminar flow in the boundary layer. This produces a so-called *laminar flow* airfoil. Unfortunately, there are other aspects of airfoil performance such as a need to maintain a low pitching moment that usually precludes the use of special *low drag* or *laminar flow* airfoil sections on rotor blades.

Furthermore, environmental factors tends to produce blade erosion at the blade leading edge, which causes premature boundary layer transition and reduces the run of laminar flow in any case.

Forward Flight. For a helicopter, the rotor is unique, since, unlike a propeller, it must provide both a lifting force and a propulsive force in forward flight. In forward flight, the rotor moves edgewise through the air and blade sections encounter a periodic variation in local velocity (and hence dynamic pressure and Mach number). The inplane, U_T, component now consists of a rotational part and a free-stream (translational) part such that

$$U_T(y, \psi) = \Omega y + V_\infty \sin \psi = \Omega y + \mu \Omega R \sin \psi \qquad (27.104)$$

assuming that no rotor shaft tilt is imposed. The out-of-plane component consists of three parts, comprising the inflow velocity, a perturbation velocity due to the blade flapping velocity ($\dot{\beta}$ term) and another perturbation velocity due to the free-stream influence with blade coning (β term). Therefore, the velocity perpendicular to the disk can be written in nondimensional form as

$$\frac{U_T}{\Omega R} = \left(\frac{y}{R} + \mu \sin \psi \right) = (r + \mu \sin \psi)$$

$$\frac{U_P}{\Omega R} = \left(\lambda + \frac{y\dot{\beta}}{\Omega R} + \mu \beta \cos \psi \right) = \left(\lambda + \frac{r\dot{\beta}}{\Omega} + \mu \beta \cos \psi \right)$$

$$\frac{U_R}{\Omega R} = \mu \cos \psi \qquad (27.105)$$

In the blade element theory, the aerodynamic effects due to the radial velocity component, U_R, are neglected.

Blade Motion: In hover as each blade rotates, it encounters an identical aerodynamic environment, and the blades reach an equilibrium position under the action of the aerodynamic and centrifugal forces. In forward flight, the aerodynamic forces become functions of blade azimuth position due to the nonuniform velocities and angles of attack produced along the blades. *Articulated rotors* use a flapping hinge incorporated at the root of each blade, and this allows each blade to freely flap with respect to the plane of rotation under the action of the azimuthally varying aerodynamic loads. This hinge also eliminates cyclic bending loads at the blade root. *Hingeless* and *bearingless rotors* use flexures or virtual hinges instead of mechanical hinges, but the objective is the same. The rotor blades reach an equilibrium when the local changes in angle of attack due to blade flapping are sufficient to compensate for local changes in dynamic pressure. Another hinge (actual or virtual) allows each blade to lead and lag in the plane of rotation to alleviate in-plane bending loads due to Coriolis effects. Control of the rotor orientation is obtained by means of cyclic pitch variations that are imposed through a mechanism known as a *swashplate*.

In hover, the blade pitch is constant, and, since the blades see an identical aerodynamic environment, the flapping angle is constant. However, in forward flight the blade pitch angle becomes periodic and can be described by the first harmonic Fourier series

$$\theta(r, \psi) = \theta_0 + \theta_{tw}r + \theta_{1c} \cos \psi + \theta_{1s} \sin \psi \qquad (27.106)$$

where θ_{tw} is a linear twist rate. Similarly, the blade undergoes a flapping motion in response to the aerodynamic forces. Blade flapping can be described by the first harmonic motion

$$\beta(\psi) = \beta_0 + \beta_{1c} \cos \psi + \beta_{1s} \sin \psi \qquad (27.107)$$

Further details on blade flapping are beyond the scope of this section, and we will simply accept these equations without further explanation.

Rotor Forces, Moments, and Torque: The rotor thrust is simply the average of the blade lift during one revolution multiplied by the number of blades.

$$T = \frac{N_b}{2\pi} \int_0^{2\pi} \int_0^R dF_Z \, d\psi \qquad (27.108)$$

The thrust coefficient can therefore be written as

$$C_T = \frac{\sigma C_{L_\alpha}}{2} \frac{1}{2\pi} \int_0^{2\pi} \int_0^1 \left[\left(\frac{U_T}{\Omega R}\right)^2 \theta - \left(\frac{U_P}{\Omega R}\right) \left(\frac{U_T}{\Omega R}\right) \right] dr \, d\psi \qquad (27.109)$$

Because of the complexity of the expressions for U_P, U_T, and θ, this equation is usually solved numerically. However, by assuming uniform inflow, that is $\lambda(r, \psi) = \lambda = $ constant, $C_D = C_{do}$, and $c = $ constant, then the result for the rotor thrust coefficient can be obtained analytically as

$$C_T = \frac{\sigma C_{L_\alpha}}{2} \left[\frac{\theta_0}{3} \left(1 + \frac{3}{2} \mu^2\right) + \frac{\theta_{tw}}{4} (1 + \mu^2) + \frac{\mu}{2} \theta_{1s} - \frac{\lambda}{2} \right] \qquad (27.110)$$

In addition, the rotor torque, side force, drag force, and moments about the respective axes can be analytically computed by a similar process [Gessow and Myers (1952)].

Induced Velocity Variation in Forward Flight: Note that the inflow velocity is required for the integration of the above equations. The calculation of this quantity can be accomplished approximately by using momentum theory, but the complex nature of the inflow velocity across the disk is properly predicted only by using vortex theory. Nevertheless, inflow models are widely used. These are basically variations of the simple momentum theory, and they ignore the true vortical nature of the rotor wake.

The simplest inflow models assume uniform or linearly varying inflow over the rotor disk. In forward flight, Glauert (1926), (1928) suggested the linear inflow model

$$\lambda_i = \lambda_0(1 + k_x r \cos \psi) \qquad (27.111)$$

where λ_0 is the induced velocity at the center of the rotor as given by simple momentum theory. Note that with a positive value of k_x, the inflow is a minimum at the leading edge of the disk and a maximum at the trailing edge. Note also that $\lambda_i = \lambda_0$ at the center of the disk. As shown in Johnson (1980), a local momentum analysis can be used to find an approximation for k_x, namely

$$k_x = \frac{1}{\sqrt{1 - r^2 \sin^2 \psi}} \qquad (27.112)$$

As a second approximation to the nonuniform inflow distribution at the rotor in forward flight, consider both a longitudinal and lateral linear variation of the form

$$\lambda_i = \lambda_0(1 + k_x r \cos \psi + k_y r \sin \psi) \qquad (27.113)$$

The weighting factors, k_x and k_y, represent the distortion of the inflow from the uniform inflow value predicted by the simple momentum theory. The sign of the coefficients is such that the maximum inflow is biased toward the rear and the retreating side of the disk. Also, note that in hover, k_x and k_y must both be zero, so that k_x and k_y must be written as a function of advance ratio. One approximation for k_x is $k_x = \tan(\chi/2)$, where χ is the *wake skew angle* given by $\chi = \tan^{-1}(\mu/\lambda_i)$.

Another simple linear inflow model that is frequently employed in rotor analyses is due to Drees (1949). In this model the coefficients of the linear part of the inflow are obtained in terms of rotor geometric and operational factors. Like all linear inflow models, the Drees model is very easy to implement in numerical models and gives a reasonably good description of the rotor inflow for advance ratios upwards from about 0.1. Another inflow model that has found use in Europe is due to Mangler and Squire (1950), where the inflow is represented by a series of shape functions.

27.3.4 Vortex Models

As shown in Fig. 27.29, the helicopter rotor wake consists of strong vortices trailed from the tip of each blade. These vortices are laid down by the blade as it rotates, and then the vortices are convected to new positions that depend on the local flow velocity. The vortices may also interact with other vortices (or with other rotor blades) resulting in considerable distortion to the trajectories of the tip vortex filaments. The accurate prediction of rotor performance, blade loads, and acoustics requires a detailed mathematical model that faithfully represents this continuously evolving wake geometry. This wake model must also be numerically coupled to a representation of the blade aerodynamics.

The rotor wake may be modeled explicitly or implicitly. Implicit or *inflow* models have been discussed previously. Explicit wake models employ vortex filaments positioned in the flow field, and the induced velocity at any point in the flow can be calculated by the application of the Biot–Savart Law (see Chap. 1 and Sec. 13.1). However, accurate prediction of both the position and strengths of the vortex filaments relative to the rotor is essential. This requires the consideration of both convective and dissipative flow processes. Up until about 1980, vortex wake models used in helicopter analyses were of a semi-empirical nature, so-called *prescribed wakes.* Since then, increasingly sophisticated *free wake models* have become available. In the following sections, we will compare and contrast some of the capabilities and limitations of these vortex models.

To reduce the magnitude of the problem, the complexity of the real rotor wake model is usually reduced to a simplified model. A common approximation is to treat only the trailed vortex filaments, that is the filaments that are trailed perpendicular to the blade. These arise because of the spatial distribution of circulation loading on the blades. The temporal loading results in *shed* wake vorticity, and this can be either neglected completely, or handled by means of one of the unsteady aerodynamic models. Even with the trailed wake system alone, the problem of calculating the induced velocity field is still one of formidable complexity, however experiments have shown that the trailed wake sheet from each blade is unstable and rolls-up into a single dominant vortex that is effectively trailed from the blade tips. Therefore, many vortex wake models consider just these *tip vortices* in the problem.

The Vortex Ring Wake Model. One simple way to approximate the trailed wake vorticity from the rotor is by a series of vortex rings. The advantage of using vortex rings is that an exact solution for the induced velocity can be obtained in terms of elliptic integrals. In the vortex ring model, one ring represents the tailed wake system generated by one blade during one rotor revolution. The positioning of the rings is defined on the basis of simple momentum theory. The induced velocity from one ring of strength Γ can be obtained by using the stream function for a vortex ring [Payne (1959)]. From this basic result, the net induced velocity at any point in the flow field can be obtained by summing the effects of all the rings. It has been found from the vortex ring model that the coefficient k_x in the linear inflow model (discussed above) can be approximated using $k_x = \tan(\chi/2)$.

Discrete Vortex Models. In discrete vortex models, a general equation describing the geometry of the wake can be derived by assuming that at each point in the flow field the local vorticity is convected at the local velocity. Considering a trailed vortex filament, the equation describing the filament can be expressed as

$$\frac{d\bar{\mathbf{r}}}{dt} = \bar{\mathbf{V}}_{loc}(\bar{\mathbf{r}}(\psi_w, \psi_b), t) \tag{27.114}$$

where $\bar{\mathbf{r}}(\psi_w, \psi_b)$ is the location of a point on the filament at the azimuth angle ψ_w that was trailed from the blade at an azimuth angle ψ_b. The derivative on the left hand side can be expanded by the use of the Chain Rule to give

$$\frac{d\bar{\mathbf{r}}(\psi_w, \psi_b)}{dt} = \frac{\partial\bar{\mathbf{r}}(\psi_w, \psi_b)}{\partial\psi_w} \cdot \frac{\partial\psi_w}{\partial t} + \frac{\partial\bar{\mathbf{r}}(\psi_w, \psi_b)}{\partial\psi_b} \cdot \frac{\partial\psi_b}{\partial t} \tag{27.115}$$

This equation can be simplified by noting that

$$\frac{\partial\psi_w}{\partial t} = \frac{\partial\psi_b}{\partial t} = \Omega \tag{27.116}$$

Therefore, the equation of the wake geometry can be written as

$$\frac{d\bar{\mathbf{r}}(\psi_w, \psi_b)}{d\psi_w} + \frac{\partial\bar{\mathbf{r}}(\psi_w, \psi_b)}{\partial\psi_b} = \frac{1}{\Omega}\bar{\mathbf{V}}_{loc}(\bar{\mathbf{r}}(\psi_w, \psi_b), t) \qquad (27.117)$$

This is a first-order, quasi-linear, partial differential equation. The homogeneous portion of the equation (the left-hand side) is the linear wave equation. However, the right-hand side will be, in general, a nonlinear function of the complete wake geometry. This is the portion of the equation that complicates the solution of the helicopter rotor wake problem. The geometry of each discretized trailed vortex filament from each blade is governed by one of these equations, and the velocity term on the right-hand side couples the equations representing each of the trailed filaments. This means that the equations for all of the filaments must be solved simultaneously at considerable computational expense.

The primary difference between the various discretized rotor wake models is the method employed for solution of the system of differential equations. The *prescribed wake method* is the lowest in the hierarchy, and the *free wake method* is currently the most sophisticated. In prescribed-wake models, the difficulties inherent in trying to solve the differential equations for the wake geometry are avoided by either specifying the location of the wake filaments directly, or by using an approximate velocity field. With the latter method, a very simple velocity field is usually prescribed that decouples the equations and allows an analytical solution to the wake geometry. When specifying the wake geometry directly, the geometry can be derived from a combination of momentum theory and flow visualization experiments.

Rigid Wake. The classical rigid vortex wake is one where the rotor wake is represented by skewed helical vortex filaments. The position of the vortex filaments is defined geometrically based on the flight conditions and momentum theory considerations. There is no self or mutual interaction between vortex filaments, so the local velocity at any vortex filament is simply

$$\bar{\mathbf{V}}_{loc} = \Omega R(\mu\bar{\mathbf{i}} - \lambda\bar{\mathbf{k}}) \qquad (27.118)$$

The differential equations of the wake can then be solved exactly leading to a skewed helical wake geometry. In the classic problem we assume only tip vortices, and so the solution of the tip vortex geometry can be described by the equations

$$x_{tip}/R = \cos(\psi_w - \psi_b) + \mu\psi_w$$

$$y_{tip}/R = -\sin(\psi_w - \psi_b)$$

$$z_{tip}/R = -\mu\psi_w \tan\chi_{TPP} = \lambda\psi_w \qquad (27.119)$$

where ψ_w is defined as the *wake age*, and ψ_b is the azimuth angle of the blade from which the vortex filament was generated. There is no contraction assumed in the classical vortex wake, however the tip vortices can be assumed to originate at some point radially inboard to take account of wake contraction, if this is justified from experiments. The first two of the above equations are easily modified to take this into account.

The wake skew angle, χ_{TPP}, in the last equation is given by $\chi_{TPP} = \tan^{-1}(-\lambda/\mu)$ where

$$\lambda = \mu \tan \alpha_{\text{TPP}} + \frac{C_T}{2\sqrt{\mu^2 + \lambda^2}} \qquad (27.120)$$

Note that as $\mu \to 0$, the wake geometry reduces to a helix. In actuality, the wake geometry of a helicopter rotor in hover is not exactly a helix, and the assumption that it is can introduce significant errors in the prediction of blade loads.

Undistorted Wake. As an extension to the classical rigid wake model, the skewed helical wake may be coupled to a *near wake model* of the blade and its distribution of trailed circulation [Piziali (1966)]. There may also be some model of the roll-up of the tip vortex filaments into a single tip vortex, which also occurs within some prescribed small azimuth angle of the blade; one approach that has been used is based on a *center of vorticity concept* [Rossow (1973)]. However, the inner trailed vortex filaments are still assumed to follow undistorted trajectories, that is with the same vertical trajectories, but trailed from radial station r, and they are prescribed according to the equations

$$x/R = r \cos (\psi_w - \psi_b) + \mu \psi_w$$

$$y/R = -r \sin (\psi_w - \psi_b) \qquad (27.121)$$

Note that if additional inner trailed wake filaments are included in the problem, the size and computational time of the induced velocity calculation increases enormously.

Prescribed Wake in Hover. To remove the limitations of the classical rigid wake models, various prescribed wake models have been developed. Many of these models have also been generalized in terms of rotor geometric and operational parameters. These models prescribe the locations of the rotor tip vortices (and sometimes also the inner vortex sheet) by means of functions that are expressed in terms of wake age, ψ_w. In the generalized wake models, the coefficients of these functions are determined from a systematic series of rotor experiments, where the effects of the rotor geometric and operational parameters on the rotor wake geometry are determined.

Such a study was performed by Landgrebe (1969), (1971). On the basis of these experiments, the tip vortex geometry can be modeled by the following relations

$$z_{\text{tip}}/R = \begin{cases} k_1 \psi_w & \text{for } 0 \leq \psi_w \leq 2\pi/N_b \\ (z_{\text{tip}}/R)_{\psi_w = 2\pi/N_b} + k_2 \left(\psi_w - \dfrac{2\pi}{N_b} \right) & \text{for } \psi_w > 2\pi/N_b \end{cases} \qquad (27.122)$$

and

$$\frac{y_{\text{tip}}}{R} = A + (1 - A) \exp (-\lambda \psi_w) \qquad (27.123)$$

Note that the vertical displacements are linear with respect to wake age. At the first blade passage when $\psi_w = 2\pi/N_b$, there is a sudden change in the tip vortex convection velocity due to the downwash produced by the blade and its own tip vortex. This is reflected in the change of the coefficient from k_1 to k_2 in the equations for the axial displacements. Note, however, that the radial contraction is smooth and asymptotic.

The coefficients of the equations are expressed in terms of the rotor thrust and blade twist using

$$k_1 = -0.25(C_T/\sigma + 0.001\theta_t)$$

$$k_2 = -(1.41 + 0.0141\theta_t)\sqrt{C_T/2} \approx -(1 + 0.01\theta_t)\sqrt{C_T} \qquad (27.124)$$

where the blade twist, θ_t, is measured in degrees. The coefficients for the radial contraction are given by $A = 0.78$, and $\lambda = 0.145 + 27C_T$. Note that while the theoretical contraction ratio of a rotor wake in hover is 0.707, the contraction ratio is modified here to 0.78 to take account of the nonideal effects discussed previously. The inner vortex *sheet* can also be modeled by means of a similar set of equations [Landgrebe (1971)].

A typical experimental comparison with the prescribed wake model is shown in Fig. 27.36. The experimental data were obtained from a different experiment [Bagai and Leishman (1992)] from that used to generate the original prescribed wake model, and it can be seen that the agreement is very good.

Prescribed Wake for Forward Flight. Several types of prescribed wake models for use in forward flight have also been derived. It has been observed from experiments that the longitudinal and lateral distortions of the wake are relatively small compared to the vertical distortions. This means that the self-induced velocities in a plane parallel to the rotor are small. Therefore, forward flight prescribed wake models normally use the undistorted tip vortex equations

$$x_{tip}/R = \cos(\psi_w - \psi_b) + \mu\psi_w$$

$$y_{tip}/R = -\sin(\psi_w - \psi_b) \qquad (27.125)$$

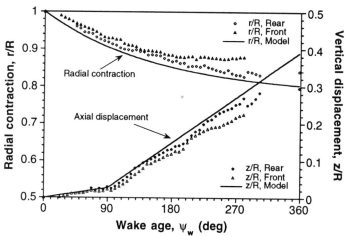

FIGURE 27.36 Comparison of prescribed wake model with experimental data, hovering flight.

The vertical displacements in the prescribed distorted wake are defined according to some weighting functions. One such approach has been used in Egolf and Landgrebe (1983) where the vertical displacement equation is written as

$$z_{\text{tip}}/R = \lambda \psi_w + E\,G \qquad (27.126)$$

It will be seen that the second term is simply a modification to the undistorted rigid wake. The E term is called an *envelope function* and is effectively an amplitude term, and the G term is a *shape* or *geometric function*.

Like the hover case, the coefficients of this model have been deduced from experimental studies of the wake geometry. However, forward flight tests are enormously more difficult to undertake, so the amount of data available is much more limited. Nevertheless, the agreement of the model with measured wake geometries has been shown to be good, and there is usually a significant improvement in the prediction of blade loads compared to those obtained using a rigid wake. There are other prescribed wake models in use for engineering studies that use different weighting functions applied to the vertical displacements [Beddoes (1985)].

As an alternative to the preceding, the form of the inflow can be specified and then used to numerically solve the differential equations that govern the wake geometry. Since the velocity field, and therefore the right-hand side of the differential equations, can be specified *a priori*, the system of equations are linearly decoupled and it can be solved without much difficulty. One major advantage with this approach to the prescribed wake problem is that other sources of wake distortion can be included, such as the local effects of airframe components. For example, we may write

$$\overline{\mathbf{V}}_{\text{loc}}(\overline{\mathbf{r}}) = \overline{\mathbf{V}}_\infty + V_i \overline{\mathbf{k}} + \overline{\mathbf{V}}_{\text{airframe}}(\overline{\mathbf{r}}) \qquad (27.127)$$

where $\overline{\mathbf{V}}_{\text{airframe}}$ is a perturbation velocity that can be used to account for the perturbation due to the airframe [Crouse and Leishman (1992)].

Free-Wake Analysis. While a prescribed-wake analysis can predict the inflow velocities to a much higher level of detail than a linear inflow model or an undistorted vortex wake, the prescribed-wake analysis does not predict the wake geometry as part of the solution. It will be clear, therefore, that the wake geometry in a prescribed-wake analyses is not self-consistent, since the velocity field used for the construction of the wake geometry is different from the velocity field that will be computed using this prescribed geometry. These and other deficiencies can be avoided by using the next level of sophistication in rotor wake modeling, the so-called *free-wake analysis*.

A free-wake analysis numerically solves the differential equations governing the wake geometry. The wake vortices are allowed to convect through the flow field under the action of the local velocity field to force-free positions. The resulting wake geometry is found by integrating the equations with the local velocity used at each point. These local velocities are calculated by integrating the influence of all the rotor blades, all the wake vortex filaments, and any additional perturbations that may be present in the flow, say due to the airframe.

Two approaches are commonly used for solving the free-wake problem: 1) the time-stepping approach, and 2) the relaxation approach. In the time-stepping approach, the first boundary condition that is used is the condition that the trailed vortices meet the blades at $\psi_b = 0$. An initial condition must also be specified, and this is typically done using an initial wake geometry specified for some azimuth angle.

An alternative approach to the boundary conditions is the relaxation method. Here, the periodicity of the wake is enforced as a boundary condition. The initial condition for this method is that at $\psi_b = 0$, the wake filaments are attached to the blades. For this method, in addition to the specified boundary and initial conditions, an approximate wake geometry is specified to start the calculations. Then, based on this geometry, a new wake is generated in an iterative manner.

Some typical results for rotor wake geometries obtained using a modern free-wake model are shown in Figs. 27.37 and 27.38. The first set of figures show the development of the rotor wake during the transition from hover through low speed forward flight. Note how the wake is progressively skewed back behind the rotor with increasing forward speed and also how the individual tip vortices roll-up into two main vortex bundles. This can also be seen in Fig. 27.38, where the tip vortex trajectory for a single blade of a four-bladed rotor is shown in comparison with experimental measurements.

27.3.5 Unsteady Aerodynamics

Besides the need to predict the spatial variations in inflow across the rotor disk due to the wake, the inherent difficulty with the prediction of airloads on rotors is to fully account for unsteady aerodynamic effects. Blade airloads are greatly influenced by the unsteady variations in angle of attack induced by blade motion, such as pitch

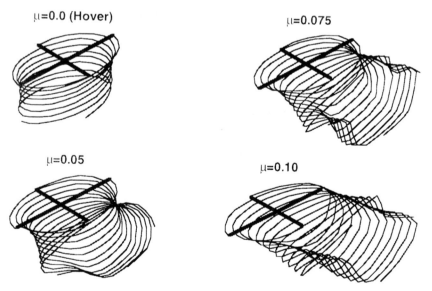

FIGURE 27.37   A series of isometric views of predicted tip vortex geometries in forward flight as generated by a single rotor helicopter (free-stream from left).

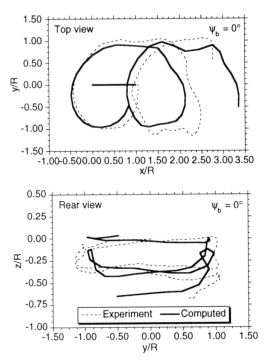

FIGURE 27.38 Typical predictions of the rotor tip vortex geometries from a free-wake model and comparison with experimentally obtained data.

due to control inputs and elastic response, and particularly blade flapping. Also, the discrete encounters of the blades with tip vortices give rise to large unsteady effects that may not just be local. The resulting unsteady effects on the rotor manifest themselves as significant amplitude and phase variations between the blade motion and the aerodynamic loads such as the lift, drag, and pitching moment. Unless these unsteady effects are properly accounted for in any realistic analysis of the rotor system, erroneous predictions of the blade loads and aeroelastic response will be obtained.

When considering unsteady effects, the interrelated effects of flow compressibility cannot be overlooked. In forward flight, helicopter rotor blades encounter a significant range of Mach numbers from very low subsonic on the retreating blade to transonic conditions (which require a nonlinear theory) on the advancing blade tip. If the mach number is sufficiently high, the occurrence of transonic flow may result in further nonlinear effects due to shock induced flow separation. Nevertheless, most of the blade sections operate at subcritical Mach numbers where compressibility effects can be modeled quite adequately using a linear theory. Furthermore, if the aerodynamic perturbations at transonic speeds are small, then the transonic flow problem can also be modeled within the bounds of a linear theory.

Even though the basic attached flow problem can be linearized with some level of confidence, it should be remembered that since the flow is also unsteady, compressibility effects are more difficult to handle. For the unsteady case, compressibility effects cannot be accounted for by a simple application of the Prandtl–Glauert correction to the incompressible loads, as can be done for the linearized quasi-steady case. As a rule of thumb, the flow field can only be considered as truly incompressible if both the Mach number is low and the degree of unsteadiness (often characterized in terms of reduced frequency) is small.

Blade Element Approaches. Nearly all of the blade element based unsteady aerodynamic models used for helicopter work are variations or extensions of the corresponding fixed wing 2-D unsteady section models (see Chap. 12). When the flow remains attached to the blades, the airloads can be adequately represented using a linear model, and, although the nature of the helicopter rotor wake is different, the exact solutions that can be found based on simplified wake assumptions are often justified for rotors as well. Extensions of these approaches to rotary-wing problems have also been attempted. One is Loewy's theory, [Loewy (1957)] which is basically Theodorsen's frequency domain theory extended to account for the returning nature of the shed wake vorticity below the rotor. Theodorsen's lift deficiency function (and frequency domain theories in general) have found great use in fixed wing aeroelasticity. However, for rotor work, Theodorsen's theory is much less useful, since in a rotor environment the argument k (the reduced frequency) is an ambiguous parameter.

For rotor work, an unsteady theory formulated in the time domain is more useful and more powerful. Wagner has obtained a time domain solution for the indicial lift response to a step change in angle of attack, α. In general, the section lift coefficient for a step input in α_0 can be written as

$$C_L(t) = 2\pi\alpha_0\phi_w(s) \qquad (27.128)$$

where $2\pi\alpha_0$ is just the steady state lift coefficient. The nondimensional time, s, represents the distance traveled by the airfoil in semi-chords, and is proportional to the area under the velocity versus time curve. In the Wagner problem, the lift at $s = 0$ is infinite, and it then builds asymptotically from half of the steady state lift. This initial lift pulse is due to the acceleration terms (apparent mass). Also, it is known that the aerodynamic center always acts at the 1/4-chord for all s. Furthermore, it can be shown that Wagner's result (time domain) and Theodorsen's theory (frequency domain) are related through a Fourier transform pair.

For helicopter rotor analyses, the indicial response is a useful starting point in the development of a general time domain theory, since, if the indicial response is known, then the unsteady loads due to arbitrary changes in angle of attack can be readily obtained in terms of Duhamel's integral. For example, for incompressible flow, we can write the lift coefficient in terms of the Wagner function and the Duhamel integral as

$$C_L(t) = 2\pi\left(\alpha(0)\phi_w(s) + \int_0^s \frac{d\alpha(\sigma)}{dt}\phi_w(s - \sigma)\,d\sigma\right) = 2\pi\alpha_{eff}(t) \quad (27.129)$$

and the appropriate apparent mass terms must be added to get the total lift. Note that the term α_{eff} can be considered as the *effective circulatory angle of attack* of the airfoil, which contains within it all the time history effects on the lift due to the shed wake vorticity. For incompressible flow, the apparent mass terms are proportional to the instantaneous motion; they appear outside the Duhamel integral. However, for compressible flow, this is not the case, and the problem becomes somewhat more complicated [Bisplinghoff *et al.* (1955), Leishman (1990)].

Dynamic Stall. When flow separation occurs, the airloads become nonlinear functions of angle of attack, and linear methods cannot be used without appropriate corrections or modification. A severe case of separation that results in highly non-linear airloads is classified as *dynamic stall.* Since the flow environment on the rotor is intrinsically unsteady, rotor stall is always dynamic in nature. The dynamic stall characteristics of airfoils is known to be fundamentally different from the quasi-steady stall that would be found on fixed wing aircraft. Under dynamic conditions, it is known that substantial lift *overshoots* occur relative to the quasi-steady case, and this is accompanied by large pitching moments. There is also considerable hysteresis in the airloads that can lead to reduced or negative pitch damping which can result in a form of aeroelastic instability known as *stall flutter.*

It is known that dynamic stall may occur on the rotor under many different flight conditions; however, it is also known that minor separation effects may still occur on the rotor in the absence of any significant stall. This is particularly so on the advancing blade tip at high advance ratios (shock induced trailing edge separation), and on the retreating blade where, by virtue of the high angle of attack, limited trailing edge flow separation may occur even at quite low rotor thrust levels. Stall effects are too complicated to be predicted by means of first principle methods, and they are usually represented in helicopter analysis by means of semi-empirical models [Reddy and Kaza (1987), Leishman and Beddoes (1989)].

Dynamic Inflow. Many rotor aeroelasticity analyses are so complex that blade-element based unsteady aerodynamics are impractical. In these cases, inflow models based on simple momentum theory are used. However, under time dependent conditions, the rotor wake will also be unsteady and will result in an unsteady induced inflow called *dynamic inflow.* This is an effect that will also be produced with the vortex wake models, given the correct boundary conditions. These time history effects on the inflow are known to be a significant factor in the calculation of the aeroelastic behavior of the rotor.

It has been shown previously that simple momentum theory gives the relationship between the average induced inflow and the rotor thrust using Eq. (27.52). A simple dynamic inflow model for hover is

$$\tau \dot{\lambda}_i + \lambda_i = \kappa \sqrt{\frac{C_T}{2}} \tag{27.130}$$

or

$$\tau \Delta \dot{\lambda}_i + \Delta \lambda_i = \kappa^2 \frac{\Delta C_T}{4\lambda_0} \tag{27.131}$$

where τ is an unsteady aerodynamic time lag. One approximation that has been used is $\tau = 0.85/4\lambda_0 \Omega$.

Dynamic inflow models in forward flight are somewhat more complex, but they retain overall simplicity so that they can be used in complex aeroelasticity analyses. A simple model is to assume a linear perturbation to the induced inflow, much along the lines of Eq. (27.113). Further details on such models are given in Pitt and Peters (1981).

27.3.6 Computational Fluid Dynamics

Modern supercomputers have enabled new advances to be made in the field of helicopter aerodynamics using computational fluid dynamics (CFD). Here, finite-difference approximations to the governing flow equations are used to solve for the complex flow field about the rotor. For the rotor problem, documented approaches have ranged from the transonic small disturbance equations, through full-potential to Euler and the Navier–Stokes equations.

The computer resources required with CFD methods have limited their use in helicopter work to mostly as a research tool. Unlike a fixed wing aircraft, a helicopter rotor induces significant velocities at large distances from the rotor. Therefore, the computational boundaries, which are typically small for economy, need to be larger than for corresponding fixed wing problems to avoid the possibilities of artifical flow recirculation within the computational domain.

While the rotor blades themselves have received the most attention from CFD methodologies, there has also been some progress in modeling the flow field around the entire rotorcraft. A summary of rotorcraft CFD prior to 1985 is given in McCroskey and Baeder (1985). At present, the blade (rotor) and wake are usually dealt with using different techniques. For example, a CFD based solution scheme for the blade aerodynamics can be coupled to such integral wake models as the vortex models discussed previously. The inflow from the rotor wake then forms a boundary condition to the computational domain. Full-potential methods [Strawn and Cardoma (1987)] have been used for this purpose, and have shown reasonable agreement with experimental measurements of the blade pressure distributions, including the critical area near the tip.

The treatment of the rotor wake itself has proved a more daunting task for CFD methods. Euler methods have been used [Agarwal and Deese (1987)], but since the vortex formation near a blade tip is a result of complex three-dimensional separated flow, it is not clear whether Euler based methods can adequately capture the physics of the tip vortex formation. Nevertheless, like full-potential methods, Euler methods have also been used to successfully predict the chordwise pressure distributions and spanwise blade loading. The increasing use of swept tips on helicopter rotors may mean that the Euler based methods will become the lowest order set of equations that will enable accurate flow field predictions.

There are increasing numbers of techniques that deal with the entire solution process for the rotor and its wake. One approach, in particular, is called the transonic rotor Navier–Stokes (TURNS) method [Srinivasan *et al.* (1992)]. The blades and computational domain are enveloped in body conforming C–H grids. This approach is based on the use of Roe's upwind differencing scheme in all three spatial directions. The method allows the rotor wake to be computed well enough to approximately simulate the correct inflow through the rotor, thus obviating the need for a separate vortex wake model.

27.4 AXIAL FLOW COMPRESSORS
Nicholas A. Cumpsty

27.4.1 Introduction

Axial compressor aerodynamics has occupied the attention of very many capable people for 50 years or more, and to summarize this in one section is difficult. More detail is given in a book by Cumpsty (1989), with a large list of references. A very interesting description is given in lecture notes by Wisler (1988), and these are highly recommended. The background is given in the famous NACA reports "The Aerodynamic Design of Axial Flow Compressors," originally written in 1956 and republished by NASA in 1965 as SP 36.

The design of an axial flow compressor is literally something of an uphill struggle. The flow is progressing against the axial pressure gradient, and, given the opportunity, it will reverse. The reversal leads to rotating stall or to surge, and these represent effective limits on what can be achieved. It follows that the single most important design decision is the selection of realistic goals. If the aims are too ambitious, the consequences may be more than just a drop in operating efficiency (as it might be for a turbine, for example) but a complete failure to operate unstalled or an inadequate range of flow between stalling and choking.

Given the difficulty that the axial compressor causes to the designer, it is reasonable to ask why they are used and why in some applications they are the normal choice. The alternative is the radial compressor, for which much of the pressure rise is produced by the centrifugal effect, as opposed to being produced by decelerating the flow in the blades of the axial compressor. The radial compressor is able to produce much larger pressure rises in a single stage, often giving stage pressure ratios in excess of 5, whereas for an axial compressor the stage pressure ratio is normally less than 1.5. The axial compressor remains attractive when the inlet area is an important consideration, and for some applications it is possible to design the inlet to pass a flow as high as 85% of that needed to choke the empty (i.e., unbladed) annulus. As obvious example for which inlet area is important is the jet engine, since a large frontal area would be incompatible with low drag; for the jet engine, axial compressors are very nearly universal. For industrial compressors where the volume flows is very large it is also common to use the axial type; at large flow rates the overall diameter of a radial machine would be large enough to be very inconvenient.

There are several ways of selecting the parameters to describe the operation of a compressor. One can use the overall pressure ratio, usually using the inlet and outlet stagnation pressures and carrying out a mass-average at each station. The overall pressure ratio gives a measure of the difficulty of the task. This is mainly because it becomes progressively more difficult to match the front and back stages as the density difference (which is related to pressure ratio) gets bigger. The problem of matching the front and back is more acute when variation in speed has to be allowed for. One way of ameliorating this is to allow the stators to be adjustable in stagger as speed changes, another is to have the ability to bleed off some of the flow part way along the compressor at reduced speeds. Many high pressure ratio machines will have both of these features for off-design operation.

One can also use the nondimensional pressure rise of a stage to describe the performance; again, the stagnation pressure rise is normally used, nondimensionalized by U, the rotor blade speed (usually the mid-span speed) to give $\Delta p_0/\rho U^2$. Sometimes the rise in stagnation enthalpy, $\Delta h_0/U^2$, is used instead of the pressure, this having the advantage that it depends only on the velocity triangles and not on the efficiency. As discussed below, the value of $\Delta p_0/\rho U^2$ which can be obtained depends on the blading and on the value of the flow coefficient, V_x/U, but designers know approximately what is acceptable and for a particular pressure ratio would choose the number of stages accordingly.

The choice of V_x/U has a large effect on the geometry of the compressor as well as its performance. Figure 27.39 is a schematic showing the rotor blades and velocity triangles for two different possible arrangements, one with $V_x/U = 0.75$, the other with $V_x/U = 0.5$. It is clear that the shapes of the velocity triangles are very different in each case; with low V_x/U the blades are highly staggered and have little camber, while with high V_x/U the stagger is much smaller and the camber higher. As will be seen, the nondimensional work input $\Delta h_0/U^2 = (V_{\theta 2} - V_{\theta 1})/U$ is normally greater when the stagger is low. The velocity triangles have been drawn in Fig. 27.39 so that the axial velocity is equal into and out of the blade row. This, often referred to as a repeating stage, is a common design choice; similar levels of axial velocity are normally maintained through most of a multistage compressor by contracting the annulus (reducing the blade span) to compensate for the increase in density. The current trend is to go for relatively low values of V_x/U, typically around 0.5–0.6, mainly for reasons of stability which will be discussed briefly later.

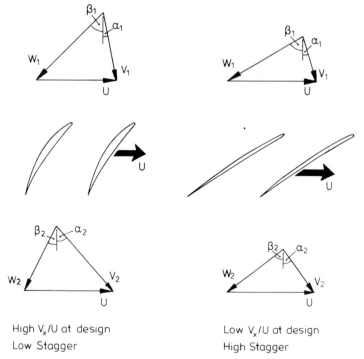

High V_x/U at design
Low Stagger

Low V_x/U at design
High Stagger

FIGURE 27.39 Schematic rotor blades and velocity triangles for two stages with markedly different V_x/U at design. (Reprinted with permission from Longmans Group UK, Cumpsty, N. A., *Compressor Aerodynamics*.)

Choosing the V_x/U is not normally separate from decisions regarding the mass flow capacity of the compressor, since the overall size and allowable blade speed are constrained. Typically, one designs for V_x as high as possible. Because the front stages may be quite close to choking at design speed, it is common to consider the mass flow in terms of the compressible flow variables. One way of doing this is to use corrected mass flow or a flow function such as $\dot{m}_{air}\sqrt{c_p T_{01}}/A_{p01}$ based on inlet stagnation conditions. The overall performance of compressors is usually expressed in terms of curves of pressure ratio versus $\dot{m}_{air}\sqrt{c_p T_{01}}/A_{p01}$ for given nondimensional rotation speed $N/\sqrt{c_p T_{01}}$; the latter usually simplified to $N/\sqrt{T_{01}}$. (Particularly for fans for aeroengine application the mass flow is normally very close to that necessary to choke the annulus.) It is easy to show that for constant rotational speed (more precisely, constant $N/\sqrt{c_p T_{01}}$) the choice of $\dot{m}_{air}\sqrt{c_p T_{01}}/A_{p01} = $ constant is equivalent to maintaining $V_{x1}/U = $ constant.

It is never possible to design a compressor without serious external constraints. Some are obvious, like the allowable tip speed which is determined by material strength and cost. The size of the annulus for a given mass flow rate fixes V_x. The allowable axial length fixes either the number of stages or the aspect ratio of the blades (the spanwise length divided by the chord). If high aspect ratio is chosen, it is possible to have more stages in a given axial length, and there is a necessary trade between the number of stages and the aspect ratio. The current trend is to go for lower aspect ratio, with values between about 1.0 and 2.0 being typical for a multistage compressor. One reason for this trend is that an important part of the cost is related to the number of blades used; low aspect ratio with fewer stages means many fewer blades.

The level of difficulty depends not only on the aerodynamic specifications of flow rate and pressure rise, but also on the geometric constraints. This is illustrated in Fig. 27.40 which shows three schematic meridional sections through a compressor rotor. It is necessary to reduce the area as the pressure and density rise, and, depending on how this is done, the aerodynamic duty alters. If the radius can increase for a streamline as it goes through the rotor, there is an additional work input. This extra work input carries with it no losses and is not limited by boundary layer type effects in the way that work input by turning is. (It is the same mechanism which is responsible for most of the pressure rise in radial compressors.) It can be understood by recalling that the work input is given by $\Delta h_0 = \omega(r_2 V_{\theta 2} - r_1 V_{\theta 1})$ and the radius change is a bonus quite separate from the change in V_θ.

The effect of radius change through a rotor can also be understood by considering the *rothalpy*

$$I = h + (W^2 - U^2)/2, \; = h_{0rel} - U^2/2 \qquad (27.132)$$

which is the relative stagnation enthalpy minus half the square of the local blade speed. Rothalpy is a quantity which is conserved along streamlines across rotor rows, and in this it is analogous to stagnation enthalpy in stator blades. (This is only true when there is no heat transfer between streamlines, but this is generally a good approximation.) Suppose at entry to the rotor the rothalpy on a particular streamline is I_1 and at outlet the rothalpy is I_2; by equating these it follows that $h_{0rel2} - h_{0rel1} = (U_2^2 - U_1^2)/2$. Assuming a perfect gas it follows that the isentropic relative stagnation pressure at outlet is given by

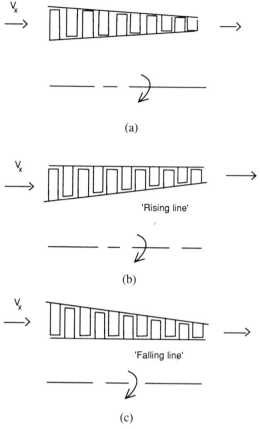

FIGURE 27.40 Different multistage compressor layouts. (Reprinted with permission from Longmans Group UK, Cumpsty, N. A., *Compressor Aerodynamics*.)

$$p_{02s}/p_{01} = (T_{02}/T_{01})^{\gamma/(\gamma - 1)} = (h_{0rel2}/h_{0rel1})^{\gamma/(\gamma - 1)} \qquad (27.133)$$

where the subscript "rel" is dropped from the pressure and temperature for the sake of brevity. The actual stagnation pressure may be obtained from

$$p_{02} = p_{02s} - \Delta p_0 \qquad (27.134)$$

where Δp_0 is the loss incurred. It is normal to assume that the losses are the same in rotating and stationary blades, at least in regions away from the endwalls.

By increasing U across a rotor, as in Fig. 27.40(b), the static enthalpy is increased for the same magnitude of change in the relative velocity; in Fig. 27.40(a) there is much less radius increase near the hub, and more of the pressure increase would need to come from deceleration of the flow. Deceleration of the flow is always a potential source of trouble, being limited in its maximum extent by separation and being a source of increased loss. The reason that not all compressors are like the example in Fig. 27.40(b) is that the inlet and outlet radii are usually fixed by other constraints.

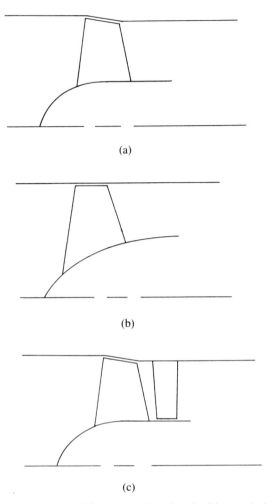

(a)

(b)

(c)

FIGURE 27.41   Schematic layout of fan rotors. (Reprinted with permission from Longmans Group UK, Cumpsty, N. A., *Compressor Aerodynamics.*)

Figure 27.41 shows schematic meridional sections through a fan. For the reasons discussed in connection with Fig. 27.40, it is much easier to achieve high work inputs as pressure rises for case (b) than case (a), since there is a marked increase in streamline radius across the rotor near the hub. The reason that all fans are not like (b) is that consideration must be given to the downstream components; for an aeroengine, an increase in the hub radius through the fan would exacerbate the conditions in the following compressor.

Figure 27.41(c) shows the rotor of Fig. 27.41(a) with a stator blade row behind, as would be normal in practice. This introduces another problem associated with the variation in static pressure in the radial direction. Downstream of the rotor the whirl (tangential velocity, V_θ) is substantial and gives rise to a radial pressure gradient; as an approximation this pressure gradient is given by

$$dp/dr = \rho V_\theta^2/r \qquad (27.135)$$

Downstream of the rotor the static pressure is, therefore, invariably lower near the hub than near the casing, and this has the effect of making things easier across the rotor hub than it would otherwise have been without the radial pressure gradient and correspondingly harder near the casing. Near the casing, the rotor speed is normally sufficiently high that this can be accommodated. The problem arises across the stator. Downstream of the stator the whirl is lower, or even zero, and the static pressure gradient in the radial direction is generally negligible. As a result, the static pressure across the stator near the hub has to rise markedly from the low value downstream of the rotor. This makes flow separation a serious problem in the stator hub region and would not be expected based on a simple two-dimensional treatment of the flow.

27.4.2 Parametric Study for Low Mach Number Stages

Although it is clearly a rather sweeping assumption to ignore compressibility, much can be learned about the way blades determine the overall performance by treating the flow in that very simplified manner. It is useful to restrict attention to the mid-span regions of the blades, where the flow is relatively well understood and the loss is predominantly that of the blade profile which can be determined from cascade tests, or from correlations produced from such tests. It should be pointed out that, in fact, most of the loss and the crucial conditions which determine the limit on stage pressure rise are in the endwall regions. Near the endwall the flow is very complicated, and at the present time it is not possible to give simple relations to describe any aspect of the behavior very satisfactorily. Although the endwall region is very important for creating loss and crucial for determining the conditions at which stall occurs, near optimum operating conditions for a good design the overall performance is determined by the flow around mid-span. In other words, it is the mid-span region which has the overriding influence on the pressure rise and mass flow rate at design point.

One can examine stages using the velocity triangles as shown in Fig. 27.39, to estimate the blade row pressure rise and efficiency. Figure 27.39 shows only the rotor row, but the stator must also be included; since the repeating stage condition is assumed, the velocity magnitude and direction out of the stator (V_3 and α_3) are equal to those into the rotor (V_1 and α_1). It also follows that the axial velocity V_x is taken to be uniform, i.e., $V_{x1} = V_{x2} = V_{x3}$. In conducting any parametric study to consider efficiency, it is necessary to make some assumptions for losses and, with attention here restricted to around midspan, just profile losses are included. Then, following Lieblein (1956), who showed that the nondimensional loss was only a very weak function of the loading of the blades, it suffices for the present purpose of examining design conditions to specify the loss by

$$\omega = \frac{p_{01} - p_{02}}{p_{01} - p_1} = \frac{0.014\sigma}{\cos \beta_2} \tag{27.136}$$

where $\sigma = c/s$ is *solidity* (c is the blade chord and s is the blade pitch) and β_2 is the flow angle out of the blade row (in this example a rotor). If the compressor were ill-designed, or were operated well away from its design point, this simple expression for loss would not be adequate; loss rises rapidly when large separations develop on the blade suction surface giving rise to wide and deep wakes. For incompressible or low Mach number flow, it is possible to write the dynamic pressure in the form

$$p_{01} - p_1 = \tfrac{1}{2}\rho V_1^2 \tag{27.137}$$

In calculating the stage loss, the individual contributions of rotor and stator are added so that the efficiency, η, can be written in terms of the loss in stagnation pressure in both rows as

$$\eta = \frac{\text{useful work}}{\text{work input}} = \frac{\text{work input} - \text{losses}}{\text{work input}} = 1 - \frac{(\Delta p_{0\text{rotor}} + \Delta p_{0\text{stator}})}{\rho \Delta h_0} \tag{27.138}$$

where Δh_0 is the stage enthalpy rise and Δp_0 here represents just the losses. Because the efficiency η is often nearly equal to unity, it is more helpful to consider

$$1 - \eta = \frac{\Delta p_{0\text{stator}} + \Delta p_{0\text{rotor}}}{\rho \Delta h_0} \tag{27.139}$$

The relative velocity into the rotor is given by

$$W_1^2 = V_x^2 + W_{\theta 1}^2 \tag{27.140}$$

so that

$$(W_1/U)^2 = \phi^2 + (W_{\theta 1}/U)^2 \tag{27.141}$$

where $\phi = V_x/U$.

Since a repeating stage (V_x unchanged) is assumed, the *static* enthalpy change for the stage is also the *stagnation* enthalpy change. Thus, the *degree of reaction, R*, the ratio of static enthalpy rise in the rotor to that for the whole stage, can be written

$$R = \frac{W_1^2 - W_2^2}{2\Delta h_0} = \frac{W_1^2 - W_2^2}{2U(W_{\theta 1} - W_{\theta 2})} \tag{27.142}$$

Since $V_{x1} = V_{x2}$ it follows that

$$W_1^2 - W_2^2 = W_{\theta 1}^2 - W_{\theta 2}^2 \tag{27.143}$$

and the degree of reaction is given by

$$R = (W_{\theta 1} + W_{\theta 2})/2U \tag{27.144}$$

$$2RU = W_{\theta 1} + W_{\theta 2} \tag{27.144a}$$

while from the definition for the *stage loading*, $\Psi = \Delta h_0/U^2$,

$$\Psi U = W_{\theta 1} - W_{\theta 2} \tag{27.145}$$

On rearranging

$$W_{\theta 1} = (2R + \Psi)U/2 \text{ and } W_{\theta 2} = (2R - \Psi)U/2 \qquad (27.146)$$

The relative velocity into the rotor can then be written, using this equation and Eq. (17.141), as

$$(W_1/U)^2 = \phi^2 + (\Psi/2 + R)^2 \qquad (27.147)$$

Similarly for the absolute velocity into the stator

$$V_2^2 = V_{x2}^2 + V_{\theta 2}^2 \qquad (27.148)$$

leading to

$$(V_2/U)^2 = \phi^2 + (\Psi/2 + 1 - R)^2 \qquad (27.149)$$

The expressions for V_2 and W_1 are needed in the calculation of loss. The flow angles are needed in order to find loss and diffusion factor. With the repeating stage condition, $V_x = $ constant, it is easy to show that

$$\cos \beta_2 = \frac{\phi}{[\phi^2 + (R - \psi/2)^2]^{1/2}}$$

$$\cos \beta_1 = \frac{\phi}{[\phi^2 + (R + \Psi/2)^2]^{1/2}} \qquad (27.150)$$

For the parametric study to be useful, it is necessary to have some way of putting in the limits on blade performance. The most common way of doing this is to use the diffusion factor, originally proposed by Lieblein (1956). The *diffusion factor*, DF, is given in terms of the velocities (in the example below for the rotor) as

$$DF = \left(1 - \frac{W_2}{W_1}\right) + \frac{\Delta W_\theta}{2\sigma W_1} \qquad (27.151)$$

which for uniform axial velocity reduces to

$$DF = \left(1 - \frac{\cos \beta_1}{\cos \beta_2}\right) + \frac{\cos \beta_1}{2\sigma} (\tan \beta_1 - \tan \beta_2) \qquad (27.152)$$

(For a stator the expressions are the same, but absolute velocity magnitude and direction, V and α, replace the corresponding relative quantities, W and β.) The first term in the diffusion factor is simply the one-dimensional deceleration produced by the increase in area consequent on the turning of the flow; the second term takes account of the cross-passage pressure field necessary to turn the flow and, therefore, contains a dependence on solidity $\sigma = c/s$. Typical design values of diffusion factor are $DF = 0.45$, though these are creeping up, partly as a result of the use of low aspect ratio blading.

An alternative way of looking at what limits the possible pressure rise that the blade row can stand is to use the de Haller (1953) criterion: $W_2 > 0.75W_1$ for rotors and $V_3 > 0.75V_2$ for stators. The de Haller criterion is equivalent to Δp across the row being not greater than about 0.44 times the inlet dynamic pressure to that row. This criterion, which is too simplistic to really account for all the complexities and subtleties, does at least allow the preponderance of the endwall region in limiting the pressure rise to be recognized. In the assessment of designs, it quite common to make sure that both the diffusion factor and the de Haller number are below the normal limit of acceptable performance; as discussed below there is a newer and better method for assessing performance due to Koch (1981), though one which is less convenient for simple parametric studies.

Figure 27.42 shows a result for a parametric study of a 50% reaction stage, one with equal static enthalpy and pressure rises for the rotor and stator. For this example, the flow coefficient $\phi = V_x/U = 0.6$. Figure 27.42(a) shows how the efficiency would vary (the loss can be thought of as the term $1 - \eta$) with loading for a given level of diffusion factor DF $= 0.45$. Figure 27.42(a) demonstrates the expected trend, that efficiency tends to decrease as machines are designed with higher loading, and one way of increasing efficiency is to lower the pressure rise per stage. The reason for this is that by holding diffusion factor constant the greater loading requires higher solidity blading, see Fig. 27.42(c), and in consequence the loss rises. (In crude terms the loss has gone up because the wetted area has increased.) The nondimensional static pressure rise is presented in Fig. 27.42(b), and this shows that the de Haller criterion is exceeded for stage loading above about 0.37, near the value normally regarded as an upper limit. Figure 27.42(d) shows how the diffusion factor varies with loading and solidity; for a given loading, one can reduce the diffusion factor by increasing solidity—in simple terms, by putting more blades in or increasing the chord length of the existing number of blades. A similar parametric study, again with reaction held constant, would show that by increasing the flow coefficient one can obtain much higher loading for the same level of diffusion factor. For reasons discussed below, it is nevertheless normally found better to operate with flow coefficients near to about 0.5.

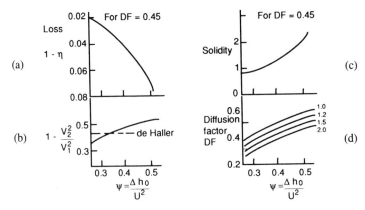

FIGURE 27.42   Calculated results for a parametric study of a two-dimensional stage. Degree of reaction 50 percent. Flow coefficient, $\phi = V_x/U = 0.6$. (Reprinted with permission from Longmans Group UK, Cumpsty, N. A., *Compressor Aerodynamics*.)

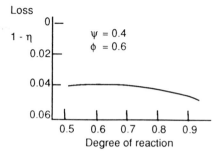

FIGURE 27.43 The calculated variation in efficiency for a two-dimensional stage, considering only profile loss, as a function of degree of reaction. Diffusion factor = 0.45. Profile loss calculated from $(\omega \cos \alpha_{out})/2\sigma = 0.007$. (Reprinted with permission from Longmans Group UK, Cumpsty, N. A., *Compressor Aerodynamics.*)

Figure 27.43 shows the variation in efficiency or loss with degree of reaction. The relation from the parametric study is symmetric about a reaction of 0.5, so between reactions of about 0.2 and 0.8 there is almost no variation in loss with reaction. (This dispels any idea that there is a natural advantage to 50%.) Some compressors, such as those on the RB211, engine have axial absolute flow into each rotor, and, for these, the reaction is 80% or more. Figure 27.43 shows why this is no disadvantage.

It may have been noted that the efficiencies shown in Fig. 27.43 are around 96%, but it is generally regarded at the present time that a compressor which achieves 92% efficiency is a very good one. In other words, there is another source of loss at least as important as the profile loss which was included in the parametric study. This loss is associated with the endwall regions, and a large part of it is specifically related to the tip clearance flow. In terms of clear evidence there is still nothing more definite than the plot of Howell (1945), shown here as Fig. 27.44. Nowadays, one would probably not use his classification for the endwall loss, but the breakdown, in which less than half the loss comes from the profile, is still widely accepted. It might also be noticed that the design flow coefficient in Fig. 27.44 is 1.0,

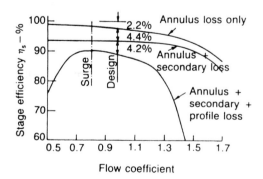

FIGURE 27.44 Howell's breakdown of loss for an axial stage. (Reprinted with permission from Howell, A. R., "Fluid Dynamics of Axial Compressors," *Proc. I. Mech. E.*, Vol. 153, pp. 441–482, 1945.)

much higher than would now be used. The high values of ϕ were used in the early days, because mechanical considerations limited the blade speed and only with high ϕ was it possible to get adequate stage pressure rise.

It has been said that compressors are now normally designed with flow coefficient $\phi = V_x/U$ nearer to 0.5 than to 1.0, even though this necessarily leads to a much smaller stage pressure rise. This is essentially because compressors designed in this way have been found to work better and to be more stable. The trend to low flow coefficient can be rationalized in more than one way. For example the stage enthalpy rise can be written

$$\Delta h_0 = U(V_{\theta 2} - V_{\theta 1}), \qquad (27.153)$$

or, in terms of the tangential velocities leaving each blade row ($V_{\theta 1} = V_{\theta 3}$ in the absolute frame of reference from the stator, $W_{\theta 2}$ in the relative frame from the rotor)

$$\Delta h_0 = U\{(U - W_{\theta 2}) - V_{\theta 1}\} \qquad (27.153a)$$

Then, using the assumption that the axial velocity does not alter across the stage

$$\Delta h_0 = U\{(U - V_x \tan \beta_2) - V_x \tan \alpha_1\} \qquad (27.153b)$$

so that

$$\Delta h_0/U^2 = 1 - \phi (\tan \beta_2 + \tan \alpha_1) \qquad (27.153c)$$

where $\alpha_1 = \alpha_3$ gives the direction of the flow out of the stator measured from axial in the absolute frame of reference and β_2 is the flow direction out of the rotor in the relative fame. The advantage of the form of Eq. (27.153c) for the enthalpy rise is that as a reasonable approximation well away from stall, the *outlet* flow angles do not change substantially with incidence, i.e., with respect to V_x/U. The outlet flow directions can be written $\alpha_1 = \chi_1 + \delta_1$ for the stator and $\beta_2 = \chi_2 + \delta_2$ for the rotor, where χ and δ are the blade outlet angles and the deviations, respectively. (It is the deviation which is almost constant, being largely determined by the blade geometry. This is discussed further below.)

Equation (27.153c) shows that the gradient of $\Delta h_0/U^2$ with respect to ϕ is equal to ($\tan \beta_2 + \tan \alpha_1$), and the relation is sketched in Fig. 27.45. If for any reason the value of V_x is locally reduced, the compressor blading is such that the work input increases in proportion to ($\tan \beta_2 + \tan \alpha_1$). The greater the magnitude of this quantity, the greater is the stability of the flow. By reference to the velocity triangles in Fig. 27.39, it can be seen that large β_2 and α_1 correspond to designs with low values of flow coefficient V_x/U. In practice, the flow is always reduced near to the endwalls, and the stability of the flow in these regions is very important.

An alternative view, which is important for the consideration of the endwall region, is to consider the velocity triangles of Fig. 27.46. As drawn, they may be imagined to be the mid-span triangles at the design point. Suppose now that the axial velocity is reduced, as it might be in the endwall region, without change in direction from the upstream blades. For the high flow coefficient design [for which ($\tan \beta_2 + \tan \alpha_1$) is small] the magnitude of relative velocity (i.e., the speed) into each blade

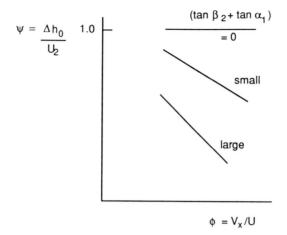

FIGURE 27.45 Idealized compressor performance. (Reprinted with permission from Long-mans Group UK, Cumpsty, N. A., *Compressor Aerodynamics.*)

row is reduced when V_x is reduced. For the low flow coefficient design [for which $(\tan \beta_2 + \tan \alpha_1)$ is large] the speed into each row may be little changed, and possibly increased, when V_x is reduced. In other words, the drop in speed in the boundary layer near the endwall is compensated for in designs with low V_x/U by the change in frame of reference. Since the conditions near the endwall are almost always what limits pressure rise and create a large part of the loss, ameliorating conditions there is very beneficial.

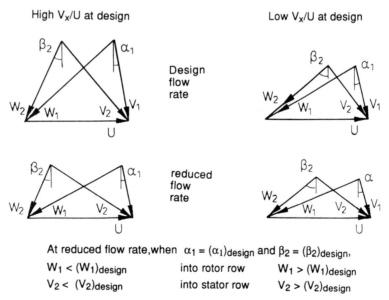

FIGURE 27.46 The effect of flow rate reduction for stages with different V_x/U at design. (Reprinted with permission from Longmans Group UK, Cumpsty, N. A., *Compressor Aerodynamics.*)

27.4.3 The Initial Design Process

The introduction and the section on parametric studies have given some idea of the way that the preliminary design of compressors is approached, but this is far too simple to be of direct application.

Any design will be affected by overall constraints on, for example, size, speed, inlet and outlet radii, and area. Then, at a very early stage of the design, it is necessary to specify the annulus shape on the basis of estimates of likely efficiency, thereby allowing sufficient reduction in area to compensate for the density rise and providing means for operating at reduced speed. It should be emphasized that decisions taken at the early stage, such as the goals and the overall geometry, can have an overwhelming impact on the difficulty in achieving the goals and on the level of final performance achieved. Indeed, some of these early decisions can be far more influential than, for example, the exact specification of blade shape.

Pitchline. Although every design is going to be slightly different, since the constraints and goals will differ, it is possible to outline the steps which must be gone through and the normal order in which they are taken. Sometimes steps will be repeated, and it is in the nature of design that it is iterative. Design usually begins with a specification along the mid-span, sometimes called the *pitchline*. This allows the choice of the number of stages, the decisions of rising or falling line and other specifications or requirements to be tested. It is common to choose that the stage loading is not equal throughout the machine; often the rear stages are unloaded and the axial velocity deliberately reduced so that there is less exit loss. A very interesting account of the preliminary design for the General Electric E^3 core compressor is given by Wisler, Koch, and Smith (1977). The overall performance characteristics of the compressor which was finally built are shown in Fig. 27.47. A meridional section through a compressor from the same company, possibly the very same compressor, is shown in Fig. 27.48. The general shape for this multistage compressor is not far from what one might describe as optimum. In contrast, the meridional section through the fan shown in Fig. 27.49 is a case for which the geometric constrains have strongly affected the layout in ways which are not likely to be wholly desirable from an aerodynamic point of view.

The mid-span (pitchline) analysis referred to above can contain varying levels of sophistication. Possible choking can be examined when estimates are available for blade thickness and flow velocity. The loading of the blades can be assessed with diffusion factor, and the blade row pressure rise can be assessed by the de Haller criterion, as discussed briefly in Sec. 27.4.2. Neither of these loading criteria is wholly satisfactory, and a better system is due to Koch (1981), who considered each blade passage like a diffuser. For a diffuser of given area ratio, a crucial geometric feature is the axial length. Koch used the ratio of the blade length along the camber line to the width of the passage at outlet to characterize this. The resulting data correlation is shown in Fig. 27.50, after including corrections for tip clearance, Reynolds number, axial gap between blade rows and mean stagger. (The mean stagger is equivalent to including the effect of design V_x/U, since a high value of flow coefficient occurs with a low value of stagger. It will be recalled that when the design value V_x/U is low the effect of the change of frame of reference is to make low velocity flow in one blade row appear to have high velocity in the next, thereby appearing to reengerize itself.)

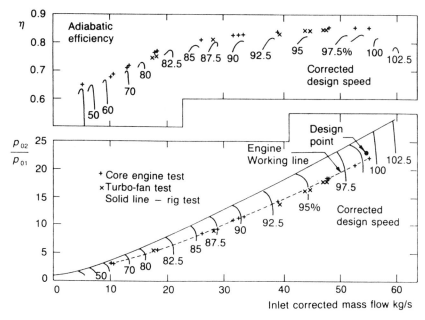

FIGURE 27.47 Measured pressure rise and efficiency of General Electric E^3 compressor. (Reprinted with permission of General Electric.)

Koch's method was devised for *stages* and not for individual blade rows. When the Koch correlation has been applied to various high-speed compressors for which test results exist, it has been found that the stages which it highlights as being overloaded are the ones shown by measurement to be in difficulty. It has been found to give a much better indication of the likelihood of major breakdown of the flow in a stage (often referred to as stall, though distinct from the rotating stall pattern often observed) than any other method. Designers usually choose a pressure rise something like 80% of the value given by the curve in Fig. 27.50.

Spanwise Specification. When a satisfactory design has been arrived at using the pitchline analysis, it is time to move on to the next phase of the design. First, de-

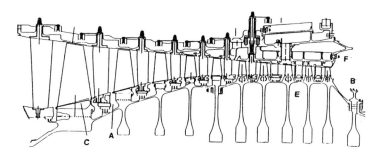

FIGURE 27.48 Meridional section through a multistage compressor. (Reprinted with permission of General Electric Co., Wisler, D. C., *Advanced Compressor and Fan Systems*, 1988.)

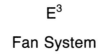

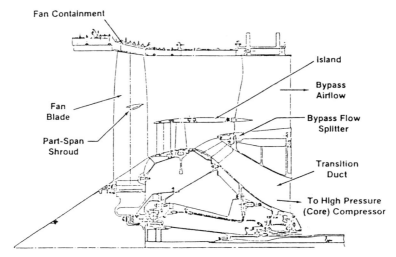

FIGURE 27.49 Meridional section through a fan. (Reprinted with permission of General Electric Co., Wisler, D. C., *Advanced Compressor and Fan Systems*, 1988.)

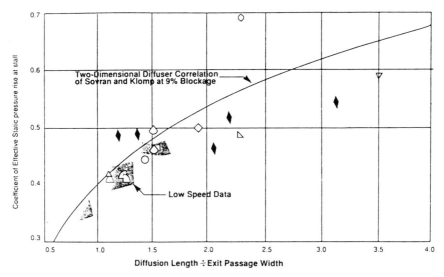

FIGURE 27.50 Correlation of stalling effective static pressure rise coefficient for high speed compressor stages. Diffusion length is length of blade camber line. (Reprinted with permission of ASME, Koch, C. C., "Stalling Pressure Rise Capability of Axial Flow Compressor Stages," *J. of Engineering for Power*, Vol. 103, pp. 645–656, 1981.)

cisions need to be made about the variation in loading with span. For most stages in the multistage axial compressor the aim is to obtain uniform rise in stagnation pressure with respect to span. Then, after making allowance for the losses, the required radial variation in work input and stagnation temperature rise is fixed. In many cases the requirement in terms of stagnation temperature rise is not far from uniform, and this means that the change in tangential velocity across each rotor is inversely proportional to radius, $r\Delta V_\theta$ = constant. In some important cases, like fans with low hub-casing radius ratios, it is impossible to maintain $r\Delta V_\theta$ = constant down to the hub, and this condition must be relaxed because ΔV_θ becomes unacceptably large towards the hub. Indeed, for the outer part of some aircraft engine fans the condition $\Delta V_\theta/r$ = constant is more nearly true. Although the work distribution fixed the *change* in tangential velocity across the rotor, the distribution of tangential velocity at each station (sometimes called the *vortex distribution*) must be chosen separately. The distribution chosen is often an approximation to a free-vortex distribution, rV_θ = constant, since this leads to uniform axial velocity in the ideal case. The actual V_θ distribution tends to emerge from other requirements, for example uniform axial velocity (giving the free vortex) or a wish to maintain degree of reaction uniform along the span.

With the spanwise specification in loading and velocity distribution made, simple methods can be used to estimate loss variations in the spanwise direction and checks can be made on the possibility of blades being able to achieve the turning necessary. With the blades specified in a preliminary and simple manner, and with spanwise estimates for loss, it is possible to move to the next phase of the convential design process, which is the calculation of the meridional flow.

27.4.4 Meridional Flow

The flow through a turbomachine is really three-dimensional. Consider a particular blade row as in Fig. 27.51(a), and imagine streamsurfaces through it. One set of streamsurfaces originate well upstream of the blades as a surface of revolution (and these are known conventionally as the S1 surfaces). The other set of streamsurfaces originate well upstream as meridional planes (and these are known as the S2 surfaces). In passing through the blade row, these would twist and warp so that they would lose their simple shape, as is shown in Fig. 27.51(a). To take account of this streamsurface distortion is really quite difficult, and three-dimensional calculations which are performed now are not based on streamsurfaces.

Although calculations can now be performed of the three-dimensional flow, including such effects as tip clearance, viscous, and compressibility, most design is still carried out in a way which is essentially two-dimensional. More precisely, it is two coupled two-dimensional calculations based on simple surfaces, as sketched in Fig. 27.51(b). The blade-to-blade surface is a simple surface of revolution, while the other surface is the meridional plane, i.e., the plane in which the radial and axial directions lie. These are not normally streamsurfaces and the conventional treatment based upon their use is approximate. (For historical reasons, however the blade-to-blade surface is sometimes called the S1 surface and the meridional plane the S2 surface.) The calculation on the meridional plane is often referred to as a *through-flow calculation*. So long as the design does not go wrong (that is to say, so long as

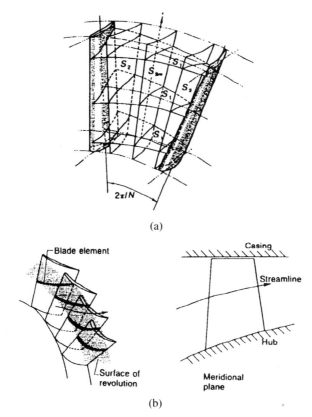

(a)

(b)

FIGURE 27.51 Streamsurfaces in a blade row: (a) intersecting S1 and S2 streamsurfaces in a blade row and (b) conventional description of flow in compressors in a surface of revolution (the blade-to-blade surface) and on a meridional place. (Reprinted with permission from Longmans Group UK, Cumpsty, N. A., *Compressor Aerodynamics.*)

there are not massive separations in parts of the flow) it appears that the approximation involved in using the blade-to-blade surface and meridional plane is a small one, much less than the errors associated with inaccuracy in estimation of boundary layer effects, for example. At off-design conditions the two surfaces may not provide a satisfactory description of the flow.

This is not the place to discuss in detail the solution of flow on the meridional surface. The most common method of solution is known as *streamline curvature* and is based on the use of streamlines in the meridional surface. A method widely used is described by Denton (1978), and many large companies have methods developed inside their own organizations. The meridional streamlines are, or course, the circumferential average of the streamlines at a particular radius. The solution also uses lines approximately normal to the streamlines, referred to as *quasi-orthogonals*. The quasi-orthogonals are usually nearly radial in an axial machine and are often placed along the leading and trailing edges of blades. Sometimes quasi-orthogonals are placed inside the blades as well, but problems due to insability can occur, and it gets more problematic as the aspect ratio of the blades gets higher. The calculation of meridional throughflow is now a very standard procedure, and many methods are available for doing it. All are iterative, so that the streamline shape is altered as the calculation proceeds, but by modern standards the methods are extremely fast. A flow chart for one possible scheme is shown in Table 27.1.

A meridional section through a three-stage compressor is shown in Fig. 27.52 on which the streamlines and the quasi-orthogonals are drawn; the quasi-orthogonals here are at the leading and trailing edges, specified at the start of the calculation, whereas the streamlines shown have been drawn after calculating their position with the streamline curvature method. The machine shown in Fig. 27.52 is one with low hub-casing radius ratio and, therefore, one in which the variations in the spanwise direction would be large.

It turns out that the most important contribution to the acceleration of the flow in the radial direction (and therefore in the quasi-orthogonal direction for axial compressors) is due to the swirl velocity. Expressed as a radial pressure gradient this is

$$\frac{\partial p}{\partial r} = \frac{\rho V_\theta^2}{r} \tag{27.154}$$

The analysis based on only this component in the radial direction is known as *simple radial equilibrium*. As noted in the *Introduction*, the radial pressure gradient is such that the static pressure is always lower at the hub than at the casing. This has important implications for blades, for it means that downstream of rotors, where the swirl velocity is usually high in compressors, there will be significantly lower static pressure at the hub than the casing. Downstream of the stator, the swirl velocity is normally low, so the static pressure is required to rise much more across the stator hub than the stator casing. Put another way, the degree of reaction is lower at the hub than at the casing. As a result, the hub region of stators is one of the areas where flow breakdown is likely to occur; the other is at the rotor tips.

In Fig. 27.52 the broken lines show the streamlines which were calculated when the only radial acceleration term retained in the streamline curvature equation was $\rho V_\theta^2/r$. In most regions the meridional streamlines are very similar to the predictions of the full streamline curvature equation, shown as solid lines, and the meridional velocities V_m do not differ very much. This is shown in Fig. 27.53 where the meridional velocities calculated by the streamline curvature method and by simple radial equilibrium are compared at stations in the compressor shown in Fig. 27.52. As already noted, the compressor used for Figs. 27.52 and 27.53 is one in which the spanwise variations and the effect of streamline curvature in the meridional plane are comparatively large. In machines in which the hub-casing radius ratio is larger, so blade span is less in proportion to mean radius, the effects of streamline curvature are likely to be still less, and the simple radial equilibrium approximation even better.

At the outset to this section it was stressed that the manner in which a surface of revolution is used for the blade-to-blade flow in conjunction with the meridional plane is an approximation, the real streamsurfaces twist. The reason that this does not normally give rise to concern is that there are more serious causes of error, most notably the losses and the blockage caused by the boundary layer developing on the endwalls. Together, the loss and blockage exert a very large effect on the matching of the compressor, since the axial velocity through each stage depends on this. The blockage is vitally important for each blade row, because it determines the effective

TABLE 27.1   Streamline Curvature Method Flow Chart for Throughflow on Hub-Casing Surface

1

Choose positions for quasi-orthogonals (q-os), for example at blade leading and trailing edges

2

Guess positions of streamlines in meridional plane and then evaluate the streamline curvature and streamtube contraction at intersections with q-os

3

Guess meridional velocity V_m at each intersection of q-o and streamline, guess T_0, entropy etc., along first q-o.

4

Use blade-to-blade calculation or correlation in conjunction with specified geometry and estimates for V_m, T, etc., to calculate flow outlet direction and loss. (Streamline convergence gives important contribution to AVDR which strongly affects blade performance.) Hence V_θ and p_0 along the q-o. Note that stagnation enthalpy (or rothalpy) is conserved along meridional streamlines and losses are convected along meridional streamlines.

5

Evaluate terms in the equation of radial equilibrium along q-o, beginning with first, using current estimate for shape of meridional streamlines.

6

Integrate dV_m^2/dq along q-o to get V_m with an arbitrary or guessed constant

7

Calculate overall mass flow and adjust constant in predicted V_m distribution to get prescribed overall mass flow. Return to 6 unless no adjustment needed, in which case go to 8

8

Integrate V_m to find new locations of meridional streamlines along q-o for correct mass

flow between them and store this information

9

Move to next q-o and repeat steps 4 to 8; after last q-o go to 10

10

Allow intersection of streamlines with q-os to move towards new position stored in step 8 but use relaxation factor (so that movement is a small fraction of difference between current and stored positions) to ensure stability. This gives new streamline shape and thence curvature

11

Go to 5 unless movement required of streamlines is less than a convergence threshold, i.e., meridional solution is converged, in which case go to 12

12

Print out results

As an alternative to going from step 11 to 12 the calculation could return to step 4 and recalculate the blade-to-blade flow in the light of the improved estimate for the meridional flow. The meridional flow could then be recalculated in steps 5 to 11. This refinement seems to be rarely performed.

The streamline curvature method is extremely sensitive to the shape of the hub and casing. Great care needs to be taken to make the surfaces used in the calculation smoothly curved in the meridional plane, even if the actual compressor has significant discontinuities of radius or curvature. (The real discontinuities are smoothed out by the boundary layer.) Because this requires some smoothing, for example the fitting of a low-order polynomial to describe the endwall, there can be problems when the meridional curvature of the endwalls is large.

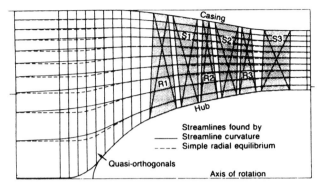

FIGURE 27.52 Meridional section through a three-stage compressor showing quasi-orthogonals and streamlines. Streamlines calculated by streamline curvature method (solid lines) and by simple radial equilibrium (broken lines).

flow area and this then determines the level of flow velocity over most of the span. The current methods for estimating blockage are crude in the extreme, a typical starting estimate would be 0.5% for each blade row until 4% is reached and then remaining constant. More refined estimates can be obtained from correlations of data or from tests on similar machines.

There are many sources of losses, but at least half of the loss is produced near the endwalls, and there is currently no satisfactory method of predicting it. Designs are generally based on estimates for loss from previous tests. Since so much of the loss is generated near the endwalls, there must be a greater work input per unit mass of flow there to give the same rise in static pressure. As a result there is a greater temperature rise near the endwalls. In the conventional throughflow calculation methods the temperature rise and stagnation pressure on one quasi-orthogonal is

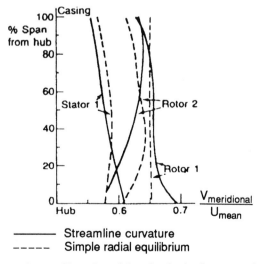

FIGURE 27.53 Spanwise profiles of meridional velocity into rotor 1, stator 1, and rotor 2 of the three compressors shown in Fig. 27.52.

passed along the meridional streamlines to the next quasi-orthogonal. The cumulative effect of the lower velocity and higher work input near the endwalls is either to make the calculation go unstable or else to give far too high a temperature near the endwall and too low a temperature near mid-span at compressor outlet. The problem arises because conventional meridional calculation methods have approximated the flow as inviscid and nonconducting; they do not have the diffusive properties of the real highly turbulent flow, which is able to spread temperature and velocity. The *traditional* way of getting around this has been to put higher losses into the calculation near mid-span and lower losses near the endwalls than are plausible, in other words to smear out the losses while retaining the same overall loss. A more recent method, [Adkins and Smith (1982), Gallimore and Cumpsty (1986), and Gallimore (1986)] is to include spanwise mixing. This is based more soundly on the physics of the flow and takes account of the process whereby the high loss, high temperature fluid is redistributed. It turns out that the calculated results are fairly insensitive to the magnitude of the mixing, though some mixing appears to be essential.

The output of the meridional throughflow calculation is, in addition to the streamline shape, the velocities at several spanwise stations along each blade row at both the leading and trailing edge and perhaps at more stations as well. This makes it possible to carry out calculations of the blade-to-blade flow with much better defined input data. For example, at a particular spanwise position the following would now be known: inlet velocity (and hence Mach number and Reynolds number), the incidence, the radial shift in the streamlines in going through the blade row, and the convergence or divergence of the meridional streamlines. A radial shift outwards, gives a marked lessening of the effective loading, for reasons discussed in the *Introduction*, while a convergence of the meridional streamlines effectively unloads the blades by reducing the overall deceleration of the flow in them.

27.4.5   Blade-to-Blade Flow

This is the most studied aspect of axial compressor flow, partly because it lends itself to study in two-dimensional geometries, experimentally in cascades or theoretically using two-dimensional methods which have generally been very much easier to implement than three-dimensional ones. A sketch of a stator blade row is shown in Fig. 27.54, defining the geometry and flow direction. Here, Station 1 is that into the stator and Station 2 out of it. The flow angles into and out of the row are denoted by α_1 and α_2, and the blade angles by χ_1 and χ_2.

The blade-to-blade consideration is expected to give many things including:

   i) The turning of blades of given geometry, or the geometry to give the required turning (this really means fixing the deviation $\delta_S = \beta_2 - \chi_2$ of Fig. 27.54)

  ii) fix the optimum blade geometry, including the type of shape (if it is a simple type of blade family, to fix the stagger, camber and solidity)

 iii) indicate whether or not the chosen turning and deceleration is possible or prudent,

 iv) indicate whether the blades can pass the required mass flow (i.e., are they choked)

  v) predict the losses.

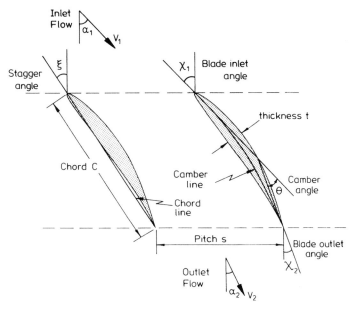

FIGURE 27.54 Blade geometry and notation.

The consideration is very different in character for blades with subsonic inlet flow from those with supersonic entry. Since the supersonic blade is so specialized, it is better to omit much consideration of this here. Supersonic blades are discussed by Cumpsty (1989).

Deviation. Determining the flow turning is one of the most important aspects of the study of blade-to-blade flow. In the simplest possible treatment of flow through blades, the flow would leave the blade passage in the direction of the blade at the trailing edge. This is too simple to be useful in most cases; the flow is turned less than if it were to leave the blade in the trailing edge direction. The difference between the blade trailing edge direction and the flow direction is referred to as the *deviation*. The most common basis of methods currently in use is Carter's (1950) rule

$$\delta = m\theta \left(\frac{s}{c} \right)^{n} \tag{27.155}$$

where θ is the blade camber. On the basis of experience, the index, n, is taken as 0.5 for decelerating blades and 1.0 for accelerating blades such as inlet guide vanes. The parameter, m, is obtained from the diagram given by Carter, shown here as Fig. 27.55. It is commonly found that this underestimates the deviation, and many empirical corrections exist, one of the simplest is to add 2°. Now it is possible to calculate the deviation reasonably accurately, though a company with a large data base may still achieve more accurate results using their own correlations. The prediction of deviation is only reliable, however, when boundary layer separation in the trailing edge region can be avoided. It is also unreliable in the endwall regions for a number of different reasons.

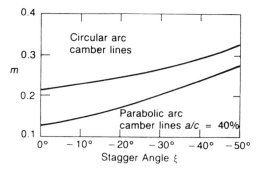

FIGURE 27.55 Carter's rule: *m* is the empirical parameter in the deviation rule. (Reprinted with permission from Aeronautical Research Council CP 29, Carter, A. D. S., "The Low Speed Performance of Related Aerofoils in Cascade," 1950.)

The deviation has often been ascribed to viscous effects, but these definitely produce only a small part of it unless the flow is heavily separated. The deviation arises because the camber line on conventional blades continues curving right up to the trailing edge. If the flow is to follow the camber line, it too must follow a curved path, and to achieve this there must be a pressure gradient normal to the streamlines. The Kutta–Joukowsky condition, however, requires that the pressure difference across the blades goes to zero at the trailing edge, and, as a result, the pressure gradient that would curve the flow no longer exists. Instead of following the blades, the majority of the flow reverts back to traveling in a straight line, and the direction therefore deviates from that of the blade trailing edge. By reducing the amount of camber towards the rear of the blades the amount of deviation is reduced. This is the case for the parabolic arc blades in Fig. 27.54 and is even more pronounced for the modern blades with *supercritical* type shapes.

Profile Optimization and Selection. In the early days of compressor design, the blade profiles were wing sections, perhaps with some modification. This served the industry well, and these profiles still find wide acceptance. The best known is the NACA-65 profile, though this is almost never used in quite the way it was designed [see Cumpsty (1989)]. Another well known set of profiles is the C series, of which the best known is the C4. These have generally less desirable aerodynamic features than the NACA-65, at all but the lowest Mach numbers, but they are mechanically more robust because of the thick leading and trailing edges. The double circular arc blade (DCA) has the simplest possible geometry and would, at first sight, not seem a suitable shape. In fact, it works very well, particularly at high subsonic Mach numbers where it was used in preference to the other types of blades until recently. In terms of loss and of turning, there is very little to choose between the NACA-65, C4, and DCA profiles until the Mach number is high enough for the shocks to separate the boundary layers, first on the C4 and then the NACA-65 blades. Examples of the various profiles, including a more modern *supercritical* blade for a similar duty are shown in Fig. 27.56.

The most widely used design system with conventional blades is to choose the stagger and camber so that the outlet flow is in the desired direction, after allowing for the deviation, and the inlet flow is such as to put the stagnation point near the

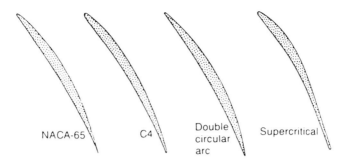

FIGURE 27.56   Four blades of nominally identical turning.

leading edge. The stagnation point on the leading edge gives the most uniform velocity distribution and, therefore, the least boundary layer or shock problems. Putting the stagnation point on the leading edge does not in general occur at zero incidence. For a given specified inlet flow, increasing the blade camber moves the stagnation point towards the pressure surface (thereby putting a velocity spike on the suction surface), while increasing blade thickness moves the stagnation point onto the suction surface. These two effects, therefore, tend to cancel, as described by Lieblein (1956). Lieblein also gives a convenient way of estimating the minimum loss incidence; the correct incidence normally differs from zero by no more than about 3° and is usually negative. The common assumption that a good choice is zero incidence based on the mean upstream conditions is, therefore, usually not far wrong.

It is not always the case that a blade row will be designed to operate at the *optimum* or minimum-loss incidence at the design condition. It is common to select a slightly different incidence so as to anticipate the problems at off-design conditions. The front stages are thus set at slightly negative values of incidence relative to the *optimum*, since at reduced speed V_x/U tends to fall and these stages are inclined to stall, whereas the rear stages are designed at slight positive incidence, since these tend to choke at reduced speed.

During the 1980's, the power of two-dimensional calculation methods was used to design blade profiles to give the desired or prescribed velocity distributions. This began with the aim of designing shock-free blades for high transonic operation, and that gave rise to the name *supercritical*. In fact, it turns out that the flow is not made shock free, but the shock is weak and unsteady. It also turns out that this type of velocity distribution gives improved performance at fully subsonic conditions as well, and the name *controlled diffusion* is also used. The characteristics of the desirable Mach number distribution for transonic operation, as set out by Hobbs and Weingold (1984) are shown in Fig. 27.57. The reduction in the velocity gradient towards the rear of the profile leads to an unloading near the rear, which is reflected in the blade shape by a straightening near the trailing edge as can be seen in Fig. 27.56. For the reasons given above, this leads to much smaller (and more predictable) deviation with this type of profile. By choosing the velocity distribution such that the boundary layer does not separate, it is possible to design for higher blade loading, equivalent to lower solidity (higher pitch-chord ratio). As a result, the profile losses can be lower, although as tests show they can also be higher if the design

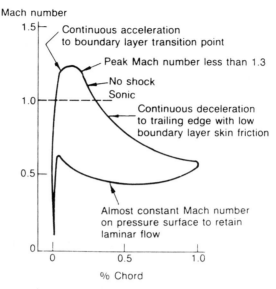

FIGURE 27.57 A schematic representation of the design Mach number distribution for a supercritical (controlled diffusion) type blade. (Reprinted with permission from ASME, Hobbs, D. E. and Weingold, H. D., "Development of Controlled Diffusion Aerofoils for Multistage Compressor Applications," *J. of Engineering for Gas Turbines and Power*, Vol. 106, pp. 171–178, 1984.)

goes wrong. It also needs to be remembered that the largest source of loss is near the endwalls, where the considerations giving rise to the profile loss do not apply.

Early designs of *supercritical* or *controlled diffusion* airfoils used *inverse* methods; the desired velocity distribution is specified and the calculation produces the blade shape which should produce this. Once the characteristic shape of blades to give the type of velocity distribution desired had been recognized, it became possible to use the more common (and easier to use) *direct* methods. The blade shape is specified and the velocity distribution is then calculated. The direct calculations are sufficiently rapid that iterations on shape can be performed to get an adequate approximation to the desired velocity distribution. The limit on accuracy hangs mainly on boundary layer laminar-turbulent transition and the effects of possible weak shocks on the boundary layer. In addition, three-dimensional effects, such as separations spreading along the span from the endwalls, may completely invalidate the basis of the optimization procedure.

Limits on Blade Loading. The limit on the amount of blade loading, expressed in terms of flow turning, is usually given by the diffusion factor given by Lieblein (1956)

$$DF = \left(1 - \frac{V_2}{V_1}\right) + \frac{\Delta V_\theta}{2\sigma V_1} \qquad (27.156)$$

expressed here in terms of velocities for the stator. Lieblein chose this form of the expression to give a reasonable fit to measured data in terms of a real diffusion ratio (called the *local diffusion factor*)

$$D_{local} = \frac{V_{max} - V_2}{V_{max}} \qquad (27.157)$$

The fit between Lieblein's DF and the measured D_{local} was made for the NACA-65 blades at the incidence for minimum loss. In fact, the diffusion factor, DF, has been extensively used at a wide range of incidences other than minimum loss and found to be fairly reliable. This is illustrated and discussed at some length by Cumpsty (1989). Typical design values for DF are about 0.45, though some successful compressors have diffusion factors at the design point of around 0.55; since the real limit on compressor operation comes from the endwall region, careful design with high solidity and low V_x/U probably makes this possible.

An earlier criterion was proposed by Howell (1945) which, because of its simplicity, has some advantages. Howell specified the *nominal* conditions to be those to give 80% of the flow deflection for stall of the blade row, the extra 20% being some leeway for error and some stall margin. The simple correlation in terms of outlet flow angle and solidity is given here in Fig. 27.58 and it is useful as a quick and simple check on the appropriateness of a design choice.

Whether Howell's approach or the more usual diffusion factor method is used, it is important to take account of the effect of Mach number, which exacerbates the deceleration for a given turning. In addition, some allowance must be made for the change of radius between inlet and outlet along the blade-to-blade surface and the variation in the axial velocity across the blade row. The criteria for loading normally assume constant radius and uniform axial velocity.

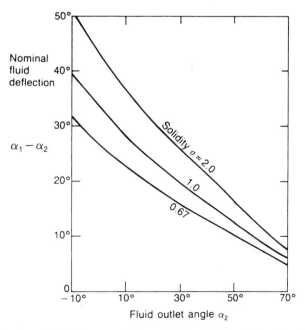

FIGURE 27.58 Cascade loading correlation of Howell. Nominal deflection is defined as 80% of deflection at stall. (Reprinted with permission from Howell, A. R., "Fluid Dynamics of Axial Compressors," *Proc. I. Mech. E.*, Vol. 153, pp. 441–482, 1945.)

Compressibility and Choking. As the inlet Mach number to a blade row increases, the effect of camber and thickness increase, and it becomes more important to operate at an incidence close to that at which the stagnation point is on the leading edge. Eventually, a Mach number is reached at which the flow is just sonic on the suction surface. This has no deleterious effects on the performance. The problems only start when the shock wave is strong enough to have a significant effect on the boundary layer, perhaps even separating it. Figure 27.59 shows schlieren photographs taken by Hoheisel of a cascade of double circular arc blades at two different Mach numbers, but the same inlet flow direction. At Mach 0.80, there is evidence of some compressibility effects near the leading edge, but at 0.85 there is a shock right across the passage. The corresponding losses are shown in Fig. 27.60 for a range of incidences. The schlieren pictures are for an incidence of 2°, and in Fig. 27.60 it is evident that there is a steep rise in loss between $M_1 = 0.80$ and 0.85. (It may also be seen that the loss is barely greater at 0.80 than at 0.30.)

Eventually, the Mach number is high enough for the flow to choke, and this represents a real and insurmountable barrier. The choking condition can be predicted by reference to Fig. 27.61, where for simplicity the flow is treated as uniform across the streamtube bounded by the two blades. If the flow is sonic across the throat, width b in the diagram, the variation in premissible inlet Mach number as a function of $(b/s \cos \alpha_1)$ is shown in Fig. 27.62. All the measured data lies below the calcu-

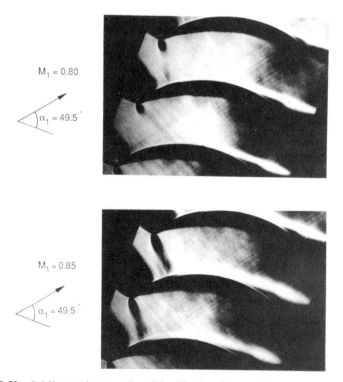

FIGURE 27.59   Schlieren photographs of double circular arc blades in cascade at two inlet Mach numbers, both at 1.9° incidence. (Pictures by Hoheisel; reproduced by permission of Rolls Royce plc.)

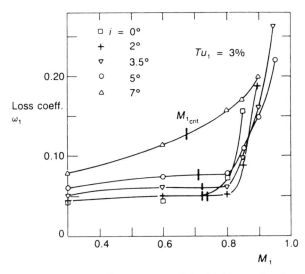

FIGURE 27.60 Measured loss coefficients versus inlet Mach number for a cascade of double circular arc blades for various incidences. Stagger 19.2°, camber 56.8°, and thickness to chord ratio 2.2. (Measurements by Hoheisel; published by permission of Rolls Royce plc.)

lated line, because the flow is nonuniform across the throat. The very simple analysis which led to Fig. 27.62 can be extended so that for a given blade row (one for which b/s is known) the inlet flow angle for choking is plotted against the inlet Mach number as in Fig. 27.63. Assuming a constant outlet flow angle, one can also draw on a line of constant diffusion factor as it varies with inlet flow direction and Mach number. (In Fig. 27.63 the line for constant diffusion factor falls as Mach number increases to take account of the exacerbation of effects at the higher speeds.) Two choking lines are shown on Fig. 27.63 for two different blade thickness-chord ratios. The effect of changing thickness is small at low Mach numbers, but, at the higher Mach numbers, the reduction in thickness gives a very marked proportional improvement in operating incidence range because of closeness of the choke line to the upper limit on α_1 provided by the diffusion factor.

The example in Fig. 27.63 was for a double circular arc blade. It is in the nature of the shape of these blades that each surface has a large radius of curvature, and

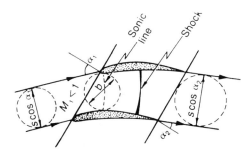

FIGURE 27.61 Schematic of choked flow in a blade passage with subsonic inlet and outlet flow.

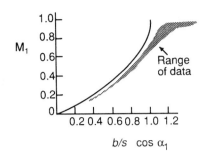

FIGURE 27.62 The theoretical choke line with a range of measured data; b denotes the blade throat width. (Reprinted with permission, Scholz, N., "Aerodynamics of Cascades," AGARDograph No. 220, 1977.)

the degree of nonuniformity across the passage is correspondingly small. For such profiles as the C4 or NACA-65 blades, the experimentally determined choke flow would be significantly less than that for the double circular arc, and, for a given inlet Mach number the minimum inlet flow angle for choking would be greater.

Profile Loss. The loss associated with the blade-to-blade flow is the profile loss, and for wholly subsonic blades Lieblein (1956) showed that this is virtually independent of loading (i.e., independent of diffusion factor) so that it can be approximated by

$$\omega = \frac{p_{02} - p_{01}}{p_{01} - p_1} = \frac{0.014\sigma}{\cos \alpha_2} \tag{27.158}$$

This expression is far from exact, but since the profile losses are normally only a fraction of the total losses, it suffices. If the local Mach number on the blades becomes high enough for a shock on a blade surface to separate the boundary layer, or if the blade chokes, there is a steep rise in loss. Many aspects of the loss on profiles can be estimated by two-dimensional methods. There are residual problems,

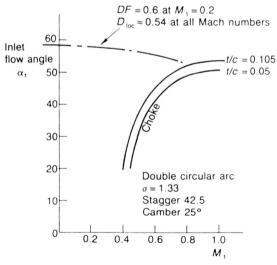

FIGURE 27.63 Inlet flow angle versus inlet Mach number for cascades of DCA blades. Lines of choke at different thickness-chord ratio. Line for constant local diffusion factor for the thicker blade.

however, associated with the behavior of the boundary layers. There is not sufficient information on the nature of transition in the presence of strong pressure gradients (much higher than are common in aeronautical applications), high levels of free-stream turbulence and high levels of unsteadiness. In addition, it has become apparent that the leading edge region can have a determining influence on the behavior of the boundary layer over the entire suction surface. Compressor blades are commonly so small, with a chord of order an inch (25 mm) that it is normally impossible to manufacture the leading edges to the precision needed to ensure unseparated laminar flow or even to be able to specify the geometry with sufficient accuracy to be able to calculate the flow adequately.

For blades with supersonic inlet flow, the treatment is quite different [see Freeman and Cumpsty (1992)], and the loss can be quite a large fraction of the inlet dynamic pressure.

27.4.6 Conclusions and Overview

Inevitably, a short description of design is an oversimplification. Compressor design is an art, and much of this is in the intuitive grasp and experience of the designers. It can be an enormously costly process if the design does not work as planned, and here the selection of realistic goals is very important. Since so much of the important input, such as the loss and blockage from the endwalls, is known only empirically, the organization which can base a new design on a similar previous one has an enormous advantage. It seems very likely that the availability of three-dimensional Navier–Stokes solvers will change this situation in the next few years, but this is only beginning.

The description here has been traditional in the way it has portrayed the design process and in the emphasis it has given to the topics. In practice, one of the most important problems is the matching of the stages on-design and off-design. The problem stems from the way in which uncertainties and errors from one stage are passed to the next, to give very large cumulative errors at the rear; if this leads to choking of the rear stages, the effect is felt powerfully at the front by a restriction in the mass flow. Catastrophic performance is much more likely to arise from inaccurate stage matching than from details of blade design, though clearly knowledge of blade performance is necessary in order to carry out the matching. Some consideration of off-design operation cannot be avoided because of the need to start the machine. Again, there is a great advantage in having similar machines on which to base the new design. There is also a great advantage in designing in a large number of variable or adjustable stators so that the prototype can be adjusted to run satisfactorily even when some of the estimates on which the design was based turn out to have been inaccurate.

27.5 CENTRIFUGAL AND MIXED FLOW COMPRESSORS
Colin Rodgers

27.5.1 Introduction

The centrifugal compressor, and pump, is probably the most predominant type of basic turbomachine encompassing an extremely wide and diverse application field from household washing machines, refrigeration equipment, piston engine turbochargers, to gas turbine engines. The current centrifugal compressor technology level is discussed in this section together with sufficient information to enable at least generating, or corroborating, a preliminary compressor design for a specific application requirement.

27.5.2 Basic Performance Parameters

The relevant flowpath stations and symbols used in this section are illustrated in Fig. 27.64.

Work and Flow Factors. The Euler equation for the impeller equates the specific energy (work/mass flow) to the product of the stagnation temperature rise and specific heat at constant pressure, i.e.,

$$\text{Energy/Mass flow} = \Delta T c_p = U_2 C_{u2} - U_1 C_{u1} \tag{27.159}$$

Figure 27.64 depicts a typical centrifugal compressor flowpath and defines the nomenclature, where U is the blade speed, W the gas velocity relative to the impeller, and C is the absolute velocity to the observer. Inlet guide vanes (IGV's) may be used to impart prewhirl to the impeller entry (*Inducer*), thereby tending to increase flow stability and/or decrease inlet relative Mach number as will be addressed later.

It is possible to amplify the Euler equation as

$$C_{u2} = U_2 - \tan \beta_2 C_{r2}/U_2 \tag{27.160}$$

$$C_{u1} = C_{r1} \tan \theta_1 \phi \tag{27.161}$$

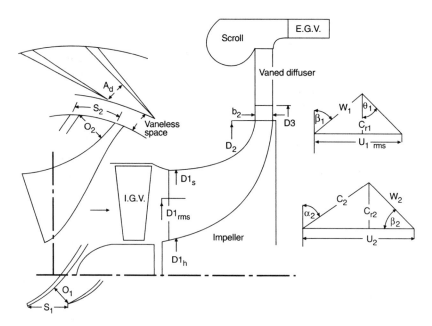

FIGURE 27.64   Centrifugal compressor flowpath and nomenclature.

The impeller exit gas angle, β_2, is composed of two components: 1) the blade exit metal angle, β_{2m}, (radial blade = 0), and 2) the *slip* or *deviation* of the flow from the blade exit angle. Note that for relatively thick blades at the impeller, $\beta_{2m} = \cos^{-1}(O_2/S_2)$. Various relationships exist for defining the slip factor or deviation. The Weisner (1966) relationship is used herein, which for typical diameter ratios $D_2/D_{1s} > 1.2$ becomes

$$\text{Slip} = 1 - (\sqrt{(\cos \beta_{2m})}/Z^{0.7} = C_{u2}/(U_2 - C_{r2} \tan \beta_{2m}) \qquad (27.162)$$

where Z = Blade number.

The two primary performance parameters for centrifugal impellers are the *work* and *flow factors* defined here as

$$\text{Work factor, } q = (U_2 C_{u2} - U_1 U_{u1})/U_2 = gJc_p\Delta T_{1\text{-}2}/U_2$$

$$\text{Flow factor, } \phi = C_{r1}/U_2 \qquad (27.163)$$

These two parameters in combination with stage efficiency define the basic compressor stage characteristic. The work factor can be expressed as

$$q = \text{Slip} (1 - \tan \beta_{2m} C_{r2}/U_2) - \phi \tan \theta_1 D_{1\text{rms}}/D_2 \qquad (27.164)$$

For nearly constant meridional velocity ($C_{r1} \cong C_{r2}$), this equation can be written as

$$q = \text{Slip} (1 - \phi \tan \beta_{2m}) - \phi \tan \theta_1 D_{1\text{rms}}/D_2 \qquad (27.164a)$$

In multistage compressors, the stage characteristic may be defined with the through-flow velocity based upon the projected area at the impeller tip, or impeller exit flow conditions, such that

$$\phi_{\text{inlet}} = \frac{\text{Inlet volume flow}}{U_2\pi D_2^2/4}, \quad \phi_{\text{exit}} = \frac{\text{Exit volume flow}}{U_2\pi D_2^2/4} \qquad (27.165)$$

It is important to note that as the flow reduces (decreasing ϕ), the work factor increases as dependent upon the magnitude of the impeller backsweep angle, β_{2m}, and the IGV prewhirl, θ_1. Therefore, increasing backsweep and prewhirl both tend to increase the work factor. As flow is reduced even further towards shut-off, stagnation temperature rise increases as a result of flow recirculation back to the inlet and disk windage effects. Without ventilation complete shut off would probably result in permanent damage. The effect of recirculation and windage on the work factor can be approximated by

$$q = \text{Slip} (1 - \phi \tan \beta_{2m}) - \phi \tan \theta_1 D_{1\text{rms}}/D_2 + q_w/\phi \qquad (27.166)$$

where q_w is a recirculation and windage factor of the order 0.01.

The work factor, Eq. (27.166) is shown plotted on Fig. 27.65 as a function of the inlet flow factor with blade backsweep angle and inlet prewhirl as parameters for a conventional impeller with 20 blades, using the Weisner slip equation. Work

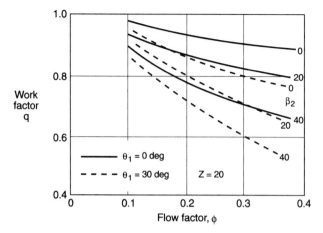

FIGURE 27.65   Work factor versus flow factor.

factors of 0.9 are shown for radial impellers ($\beta_{2m} = 0$ deg), decreasing to 0.7 for $\beta_{2m} = 40$ deg, at a flow factor of 0.3. Since impeller flow stability is primarily influenced by the slope of work factor with flow factor, increased blade backsweep and prewhirl promote extended impeller flow range between maximum (choke) flow and minimum (surge) flow limits as will be discussed later. If the mass flow through a stage is steadily increased, then the flow velocity across the whole of the passage will at some point reach the speed of sound. No further increase in mass flow is then possible, and the stage is said to be *choked*.

Conversely, if the mass flow is steadily reduced, the flow will break down at some point and become unstable. Two different types of instability can be observed, the most serious of which is *surge* characterized by violent low frequency fluctuations in mass flow, coupled with increased noise and vibration. The second type of instability is known as *rotating stall*, characterized by small pressure fluctuations and a steady mass flow through the compressor. Although the two phenomena are different, they are not completely unrelated, since the stall of a component can trigger the onset of surge.

Compressor Efficiency. The *isentropic compressor efficiency*, also referred to as the *adiabatic efficiency* is defined by the ratio between the isentropic temperature rise and actual temperature rise, i.e.,

$$\eta_{c,\text{AD}} = \Delta T_{\text{AD}}/\Delta T_{\text{ACTUAL}} \tag{27.167}$$

Some compressor manufacturers prefer to use the *small stage* or *polytropic efficiency* which can be computed from

$$\eta_{c,\text{POLY}} = \ln \, (\text{pressure ratio})^{k-1/k}/\ln \, (\text{temperature ratio}) \tag{27.168}$$

On single stage or multistage compressors of low pressure ratio, there is hardly any difference between the numerical values corresponding to both definitions. On multistage and high pressure ratio compressors, the difference can be significant, with the overall adiabatic efficiency being the lower. The ASME Power Test Code No.

10 uses the polytropic efficiency for comparison purposes between the design point in the field and the equivalent design point determined during test in a closed loop using a different gas. Code 10 considers that, although the adiabatic efficiency during test could be different, the polytropic efficiency should remain unchanged. In essence, it does not make any difference what definition a manufacturer uses, since one definition of efficiency can be obtained from the other using the following relationship;

$$\eta_{c,\text{AD}} = \frac{(\text{pressure ratio})^{(k-1/k)} - 1}{(\text{pressure ratio})^{(k-1/k\eta_{c,\text{POLY}})} - 1} \qquad (27.169)$$

The state of the art for single stage centrifugal compressor efficiencies is discussed in Sec. 27.5, as being essentially influenced by specific speed, size, and Mach number.

Specific Speed. The importance of the specific speed/efficiency relationship and the design parameter selection that enables optimum performance to be attained along the peak efficiency locus is paramount technology for successful competitive designs. The derivation and use of specific speed, N_s, has been extensively treated by many authors [Rodgers (1991)]. Further treatment would be superfluous, except to reiterate that N_s combines two primary nondimensional parameters (q_{ad} and ϕ) with a constant in the following relationship;

$$\text{Specific Speed}, N_s = \text{constant} \cdot \phi^{0.50}/q_{\text{ad}}^{0.75} \qquad (27.170)$$

The most common forms for N_s used in the United States are

$$N_s(\text{pump}) = N \cdot \text{GPM}^{0.5}/H_{\text{ad}}^{0.75} \qquad (27.171)$$

where $H_{\text{ad}} = q_{\text{ad}}U_2^2$

$$N_s(\text{compressor}) = N \cdot \text{CFS}^{0.5}/H_{\text{ad}}^{0.75} \qquad (27.172)$$

$$\text{Nondimensional } N_s = \omega \cdot \text{CFS}^{0.5}/(gH_{\text{ad}})^{0.75} \qquad (27.173)$$

$$\text{Nondimensional } N_s = N_s(\text{pump})/2736 = N_s(\text{compressor})/129 \qquad (27.174)$$

To aid in comparison between U.S. and Metric designed pumps and compressors, the nondimensional form of N_s is adapted herein.

Commercial pump design practice in the U.S. is to essentially rely upon a single N_s curve, often known as *the chart*, as a yardstick with which designers may judge the merit of their preliminary designs before commencing detailed design, or to assess potential efficiency level. Rodgers (1980) recommends the following optimum flow factor choice

$$\phi = 0.2 + 0.125N_s \qquad (27.175)$$

For low Mach numbers, it is possible to compare centrifugal pumps and compressors at the same Reynolds number on the basis of N_s, since the density ratio is small. Tests on centrifugal pumps using both air and water are commonly conducted.

Mach Number. The previous description of the compressor stage characteristic was in terms of a unique relationship between the flow factor, work factor, and efficiency. However, this is strictly true only at low speeds and with low molecular weight gases, where the density variations inside the impeller are small. At higher tip speeds, compressibility effects become significant as a consequence of increasing Mach number. Two definitions of Mach number are often used in centrifugal compressor design practice,

$$\text{Mach number, } M = \text{velocity/speed of sound} = \text{velocity}/(gkRt_1)^{0.5} \quad (27.176)$$

where R = gas constant and k = specific heat ratio

$$\text{Stage Mach number (\textit{De Laval No.}), } Mu = U_2/(gkRT_1)^{0.5} \quad (27.177)$$

where U_2 is the tip speed. The *stage Mach number* does not agree with the true definition of Mach number, because it relates the impeller tip speed (not gas velocity) to the speed of sound at compressor inlet stagnation conditions. Stage Mach number is readily computed from stagnation conditions and provides an index of Mach number effects upon the compressor. Mach number effects upon compressor flow range and efficiency become noticeable as $Mu > 0.6$, and significant at $Mu > 1.0$. The typical effect of Mu on stage performance is shown on Fig. 27.66. Although stage performance representation using the Mu parameter is convenient, detailed internal component design necessitates flow examination on the basis of the local (true) Mach numbers.

Reynolds Number. Reynolds number also effects the compressor stage performance, and thus, the stage characteristic. Figure 27.66 only applies for one Reynolds number.

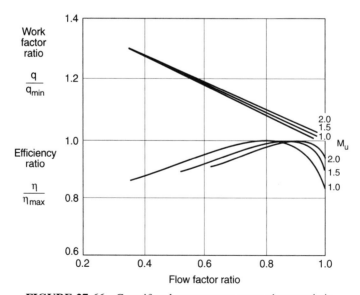

FIGURE 27.66   Centrifugal compressor stage characteristic.

The compressor Reynolds number, also often called the *Stage* or *Machine Reynolds number* is a reference value used as an index of the true Reynolds numbers through the compressor,

$$\text{Re} = U_2 \times \text{characteristic dimension/kinematic viscosity} \quad (27.178)$$

For convenience, the kinematic viscosity at inlet stagnation conditions is used. The characteristic dimension according to ASME PTC-10 definition corresponds to the impeller tip width. In some Reynolds number test comparisons, e.g., Skoch and Moore (1987), the impeller tip diameter has been used. A thorough treatise on Re effects is presented by Casey (1984), recommending revision of ASME PTC-10 with improved Re corrections for efficiency, work factor, and flow factor, based on both kinematic viscosity change and relative surface roughness difference. For simplified analysis (assuming aerodynamically smooth surfaces) the following Reynolds number corrections are suggested

$$\text{efficiency:} \quad \left(\frac{1 - \eta}{1 - \eta_{\text{REF}}}\right) = \left(\frac{\text{Re}}{\text{Re}_{\text{REF}}}\right)^n \quad (27.179)$$

where the exponent $n \cong 0.1$ to 0.2

$$\text{work factor:} \quad q/q_{\text{REF}} = 0.5 + 0.5(\eta/\eta_{\text{REF}}) \quad (27.180)$$

$$\text{flow factor:} \quad \phi/\phi_{\text{REF}} = (q/q_{\text{REF}})^{0.5} \quad (27.181)$$

where η_{REF}, q_{REF}, and ϕ_{REF} are all at Re_{REF}.

27.5.3 Component Characteristics

The typical centrifugal compressor of Fig. 23.64 comprises the following components: inlet or suction pipe, inlet guide vanes (of the axial type), inducer, impeller, vaneless space, vaned diffuser, and discharge scroll or exit guide vanes. The geometric features and performance characteristics of each of these components will be discussed in order to provide specific component design information and to enable a meanline, or two-dimensional, model for overall compressor performance analysis and component matching optimization.

Inlet or Suction Pipe. Centrifugal compressor inlets are typically of either the axial or radial single entry type. In some instances, the installation may require dual, or bifurcated, inlets which may impose flow distribution problems and or additional pressure losses dependent upon specific geometric features. Maximum performance is normally obtained with an axial bellmouth type inlet. In some cases, the radial inlet with circumferentially accelerating flow around a 90° bend towards the inducer eye is often selected, such as in gas turbine engine and multistage centrifugal compressor configurations. Rodgers (1988) describes the results of tests concerning radial inlet geometric features and, in particular, the influence of axial length extension upstream of the inducer leading edge. The test data indicated that relatively short axial extensions could be used at inducer tip relative Mach numbers up to 1.15

without significant performance penalties providing: 1) the ratio of the axial extension (hub contour)/$D_{1s} > 0.4$, 2) the mean streamline velocity maintains a positive acceleration gradient, especially in the last third of the flow path up to the inducer leading edge, and 3) a general acceleration ratio of 2.0 (exit/entry) is maintained.

Inlet Guide Vanes (IGV's). As previously discussed the use of IGV's changes the slope of work factor versus flow factor according to the simplified relationship of Eq. (27.166). Positive prewhirl (in the direction of rotation) reduces the work factor and increases the slope of work factor versus flow factor, adding impeller flow stability. Negative prewhirl has the opposite effect, increasing work and reducing stability.

IGV's are most effective in combination with a vaneless diffuser, since the impeller tends to control both maximum and minimum flow limits (choke and surge). IGV's are somewhat less effective (dependent upon $D_2/D_{1\text{rms}}$ and N_s) in shifting the surge line for a vaned diffuser compressor configuration.

The typical effect of IGV's on the maximum flow factor and peak efficiency of a single stage centrifugal compressor are shown on Fig. 27.67. Negative (against rotation) prewhirl tends to be less effective than positive prewirl and may cause premature impeller choking with reduced efficiency. Examination of Eq. (27.166) shows that the effectiveness of IGV's on the impeller is also dependent upon the diameter ratio D_2/D_{2s}; low N_s impellers with high diameter ratios responding less to IGV modulation, and vice versa.

Inducer. The main function of the impeller is to provide the required pressure and flow levels with minimum energy input and with stable flow operation. The desired pressure level is idealistically attained merely by providing sufficient tip speed, with the product of rotational speed and tip diameter, whereas the required flow and stability is mostly dependent upon the correct shaping and sizing of the inducer section. The inducer is the critical section of the impeller, especially for efficient operation at high inlet relative Mach numbers. Analysis of inducer experimental studies revealed the similarity of inducer operating characteristics to those of the axial compressor rotor stage, even though followed by a flow turn from the axial to the radial direction. It was subsequently possile to correlate inducer characteristics with extensive existing cascade data of axial compressors as discussed by Rodgers (1961).

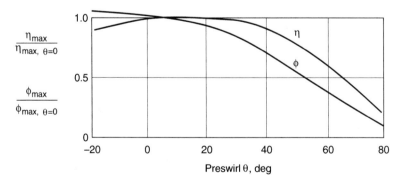

FIGURE 27.67 Typical effect of IGV's on efficiency and and flow factor.

For zero prewhirl, the inducer tip relative velocity is given by

$$W_{1s}^2 = U_{1s}^2 + [\text{Mass flow} \cdot W/\rho_1\pi/4(D_{1s}^2 - D_{1h}^2)]^2 \qquad (27.182)$$

where ρ_1 = density. Examination of Eq. (27.182) shows that for constant massflow and density that the right-hand term increases with increasing diameter, while the left-hand term decreases with increasing diameter. The relative velocity, W_{1s}, therefore, has a minimum which for incompressible flow is at a tip relative flow angle β_{1s} of 55 deg. For compressible flow, W_{1s} is minimum with β_{1s} between 55° to 65°, depending upon the inducer hub to tip ratio. Choice of D_{1s} providing this blade angle may however, on high Ns designs, result in excessive shroud curvature ($D_{1s} \rightarrow D_{1h}$). Thus, selection of a lower inducer tip flow angle may be considered.

Having made a preliminary selection of the inducer hub and tip diameters, attention is focused towards sizing the inducer throat to pass the desired flow, of sufficient area to avoid choking (or cavitation in pumps). A choke flow margin of 10% is typical for $Mu > 1.5$ compressors, the appropriate throat of which, must be compatible with a selected blade entry angle and possess adequate stall margin. The following conditions should therefore be satisfied:

i) Smallest possible inducer hub diameter consistent with blade manufacture and shaft dynamic concerns

ii) Optimum inducer tip diameter for minimum tip relative Mach number and tolerable shroud curvature

iii) Computation of corresponding hub, RMS, and tip velocity triangles

iv) An inducer choke margin of at least 10%, with minimal blade turning between the leading edge and the throat

v) Choice of a radial distribution of blade incidence giving adequate stall margin

Two axial compressor blade design procedures adopted in the design of inducers for centrifugal compressors were the correlation of maximum inlet relative Mach number versus blade throat opening to pitch ratio, and blade stalling incidence as a function of turning and solidity. Inducer choke flow data is shown in Fig. 27.68, and it parallels the theoretical choke line up to near the sonic limit, after which maximum flows are highly dependent upon such factors as blade thickness, leading edge radius, and blade turning between the leading edge and throat. The iteration of the throat opening the inlet blade angle to provide adequate choke and stall margins with, as is often, the desire to have blade elements stacked radially is resolved computationally. As a design guideline, the stalling incidences shown in Fig. 27.69 from Rodgers (1991) can be used. In many instances, alternate inducer blade cutbacks are preferred to maximize the throat area for a given blade angle and tip diameter or simply for ease of manufacture, especially in small turbocharger type impellers. Alternate blade cutbacks or splittered inducers provide a larger throat area for a given blade angle and, therefore, tend to exhibit a wider flow range between impeller stall and choke, at the expense of slightly lower impeller efficiency (of the order 1% point) as compared to fully bladed inducers.

Inducer stalling has been delayed in some cases by the use of inducer shroud bleed. Circumferential bleed hole ports located adjacent to the inducer throat permit

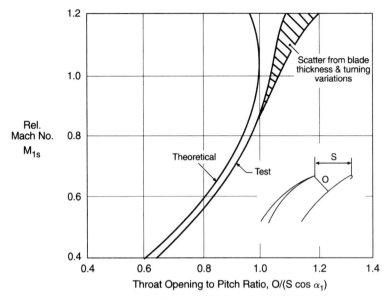

FIGURE 27.68   Inducer choke flow correlation.

flow to be either blown out, or sucked in, as dependent upon the port internal static pressure, and thus, throttle position. In effect, active flow in, or out of, the port is regulating the effective inducer throat area.

Impeller. With the preliminary design of the impeller inlet (inducer) defined, attention may now be focused upon selection of the two important impeller tip design features, impeller backsweep angle, β_{2m}, and tip width, b_2. These two features, in

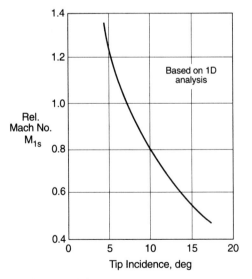

FIGURE 27.69   Inducer stalling incidences.

combination with the rotational speed vector, set the impeller exit flow to provide the required angular momentum change and thus work factor, which together with the impeller efficiency governs the impeller stagnation pressure radio according to

$$p_{02}/p_{01} = (1 + \eta_{1-2}\Delta T_{1-2}/T_1)^{(k/k-1)} \tag{27.183}$$

The sweep back angle and the tip width, together with the inducer diameters and entry angle, are the primary corner points of the impeller, as connected by the meridional and blade surface contours. If there are no blade stress concerns, a backsweep angle between 30° and 50° is usual, as compared to centrifugal pumps where the *chart* optimum is 68°. Tip speeds exceeding 600 m/sec (2000 ft/sec) with 40° backsweep are close to current acceptable lift limits for titanium. The incentive to use high backsweep is illustrated in Fig. 27.70 from Kenny (1984), where stage efficiency gains of up to 5% points are shown increasing backsweep from zero to 45°, as well as reducing the diffuser inlet Mach number with higher reaction.

Representation of the impeller tip flow conditions by either blockage, jet and wake momentum deficiency, or profile corrections becomes a matter of experience of the individual designer. Whichever analysis is used for arriving with a mixed out impeller efficiency and work factor, the impeller stagnation pressure ratio becomes defined, which is the predominant parameter separating the impeller from the downstream diffusion system. It is the impeller total pressure and vaneless space losses which eventually size the vaned diffuser throat or scroll tongue areas, independent of the flow angle. As will be shown later, the inducer and diffusion system throat areas are the two dominant features molding the shape of the compressor map.

Rodgers (1980) used an impeller average specific speed for correlation of impeller polytropic efficiency defined as

$$\text{Impeller average specific speed, } N_{s,\text{ave}} = N_s(0.5 + 0.5\rho_1/\rho_2)^{0.5} \tag{27.184}$$

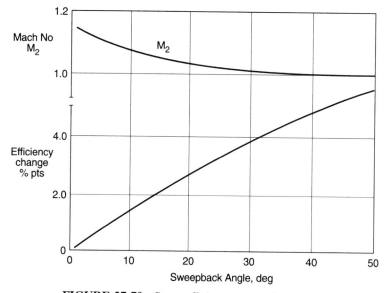

FIGURE 27.70 Stage efficiency versus backsweep.

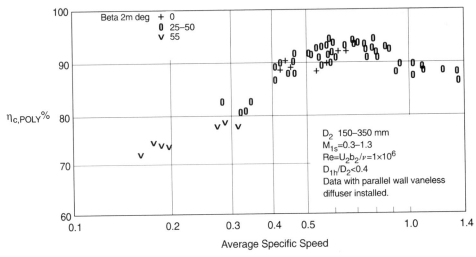

FIGURE 27.71   Impeller efficiency.

It is to be anticipated that a higher density ratio design will result in lower polytropic efficiency (apart from clearance effects), since the impeller passage friction losses increase with reduced hydraulic diameter. The correlation of impeller polytropic efficiency versus average specific speed is shown on Fig. 27.71.

The basis of compressor design is to define a geometry which will exhibit a pressure, flow, efficiency, and stability margin characteristic optimum for the intended application. Prior to the advent of computational fluid dynamics (CFD), and still in most instances today, compressor characteristics were computed on the basis of a *meanline flow model*, or simply by scaling. In the case of the centrifugal compressor, meanline modeling encompasses at least the following flow stations: 1) inducer entry RMS section (1), 2) mixed outflow at the impeller tip (2), 3) diffuser throat (3), and 4) compressor exit (e). For the purposes of abridged analysis in this section, it is elected to discuss the simplistic blockage model as a means of setting a mixed out impeller tip velocity triangle. The tip annulus area blockage may either be selected as a constant of the order 0.7 to 0.9, or a function of the relative velocity diffusion ratio across the impeller, W_1/W_2.

As a design guideline, impeller flow instability may be encountered when the diffusion ratio, W_1/W_2, exceeds 1.6 [Rodgers (1978)]. This is predated by design guidelines for refrigeration centrifugal compressors which established allowable impeller deceleration limits based upon the ratio of C_{r1}/C_{r2} not greater than unity.

A recommended impeller tip blockage correlation is

$$\text{Tip annulus area blockage factor, } \lambda_2 = 0.95 - 0.035(W_1/W_2)^3 \quad (27.185)$$

At this point, with prescription of the stage pressure ratio, stage η, Mu, and $\beta 2m$, a first iteration at the tip triangle can be computed by selecting an initial tip width, b_2, resulting in a diffusion ratio less than 1.6. This first iteration may not, however, be consistent with other criteria, e.g., factors influencing the impeller efficiency, thus iterative computations are normally necessary for closure at all the meanline flow stations. Connecting the impeller inlet and exit corner points by the blade surfaces is a more complex task most tractable by prepackaged blade geometry computational programs, combined with streamline curvature or CFD solutions of the blade surface velocities. Utilization of such programs presents a rapid means of searching for an optimum blade geometry within the following general criteria: 1) smooth suction and pressure surface velocity distributions with no regions of negative velocities, and 2) blade to blade loadings not to exceed Kenny (1984) limits of Δ (relative velocity)/mean velocity = 0.6.

With the impeller geometer defined, the impeller efficiency becomes dictated by specific speed, inducer incidence, or choke flow factor ratio, ϕ/ϕ_{choke}, inducer, tip relative Mach number, shroud clearance, and Reynolds number. A normalized impeller efficiency characteristic is shown in Fig. 27.72, where impeller efficiency ratio is plotted versus choke flow factor ratio. Plots of this type are used together with the downstream diffuser characteristic for component matching optimization at design and off design conditions.

Vaneless Space. The term, *zone of rapid flow adjustment*, is appropriately descriptive of the vaneless space between the impeller tip and vaned diffuser or scroll tongue. In multistage centrifugal gas compressors, the vaneless space may extend around the stage crossover to the deswirl vanes of the next stage. The static pressure recovery characteristics of radial vaneless diffusers are of particular concern to high pressure industrial compressor, turbocharger compressor, and very high Mach number centrifugal compressor designers. The concern relates a so-called *critical* value of the mass-averaged impeller tip flow angle, α_2, (approximately 80°) at which either rotating stall, or the slope of the vaneless diffuser static pressure recovery with flow becomes positive, potentially precipitating undesirable compressor mechanical operating conditions.

Even though operating on a rising pressure characteristic (negative slope), a high pressure industrial compressor can experience harmful sub-synchronous shaft vibrations triggered by rotating stall cell pressure instabilities or localized reverse flows penetrating the impeller from a downstream vaneless diffuser. Further increases in back pressure may eventually precipitate a rapid reduction in vaneless diffuser static

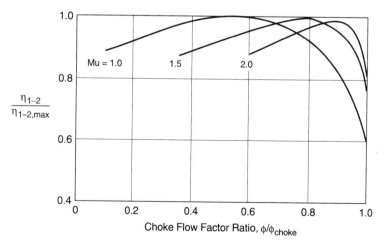

FIGURE 27.72 Normalized impeller characteristic.

pressure recovery and, thereby, cause compression system surge adjacent to the delivery pressure inflection point.

At lower suction pressure near ambient conditions, the impeller aerodynamic loads are considerably reduced, thus actual compressor surge rather than rotating stall or reverse flows is the primary concern. It is important to distinguish between the static pressure recovery of a vaneless diffuser alone and the overall static pressure recovery of a vaneless diffuser followed by a vaned or scroll diffusion system. Typical static pressure recovery trends for each of these systems are shown in Fig. 27.73. It is observed that increased peak static presure recovery via the use of either a collector or vaned diffuser row normally leads to reduced flow range.

The static pressure recovery for the vaneless diffuser may be defined by

$$C_{p,2-3} = (p_3 - p_2)/(p_{02} - p_2) = [1 - (D_2/D_3)^2] [1 - \Delta p_0/(p_{02} - p_2)]$$

(27.186)

The *total pressure loss ratio* or *loss coefficient*, $\Delta p_0/(p_{02} - p_2)$, can be expressed in terms of an *equivalent skin friction factor* CFE (of the order 0.005) by

$$C_{p,2-3} = [1 - (D_2/D_3)^2] - (CFE)D_2[1 - D_2/D_3]/(\cos \alpha_2 \cdot b_2) \quad (27.187)$$

Rodgers (1984) shows that this CFE is apparently more influenced by impeller tip width to diameter ratio, b_2/D_2, than by Reynolds number, as a consequence of non-uniform impeller tip flows with increasing tip widths (increasing specific speed impellers). It was hypothesized that the equivalent CFE could be separated into a wall friction component and a dissipation component to account for energy losses other than wall friction, i.e., mixing losses between jet and wake flows. These equivalent

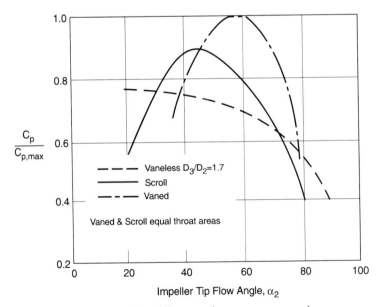

FIGURE 27.73   Diffuser static pressure recoveries.

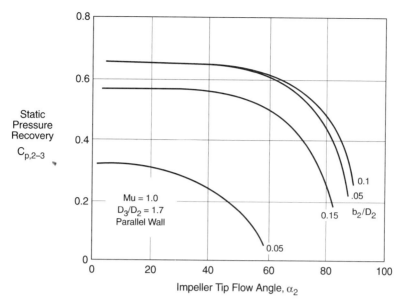

FIGURE 27.74 Vaneless diffuser static pressure recoveries.

skin friction factors were used to predict vaneles diffuser static pressure characteristics shown in Fig. 27.74, at *Mu* = 1.0, with a vaneless diffuser diameter ratio of 1.7. Low static pressure recoveries with width to diameter ratios less than 0.015, typical of low specific speed impellers, is a consequence of smaller equivalent hydraulic diameters.

Vaned Diffuser. The stationary vaned diffuser tends to be the flow controlling component, in that its overall Mach number level and inlet blockage are higher than those of the inducer, which operates with a radial variation of Mach number from hub to shroud. The diffuser must also accept an already diffused flow from the impeller with resulting nonuniform entrance conditions, further aggravating its stalling sensitivity. Attainment of a large flow range requires that the impeller and the diffuser must be capable of extended operation into their stalled or positive incidence regions to a flow where the static pressure rise plateaus, and compressor surge is eventually triggered. Stage surge is believed to stem from operation on an unstable (positive slope) portion of the overall compressor characteristic, where the static pressure ratio decreases with decreasing flow. One effective method of increasing compressor operating range is to provide sufficient impeller stability such that the downstream diffuser can tolerate operation into its stalled regime (and *vice versa*). The limiting impeller diffusion concepts described in the *Impeller* section above suggested that such a simple stall parameter might also be formulated for vaned diffusers.

Diffusion limitations for straight channel diffusers are discussed by Reneau *et al.* (1967) where it was shown that an important performance parameter is the channel diffuser included angle, which should be of the order 12°. The major diffuser design concern is, however, sizing of the diffuser throat to provide an adequate choke margin. Fortunately, with accelerating flow towards the throat, as choke is approaches, the blockage diminishes to a minimal level. Indeed, blockage is usually assumed to be zero at choke.

The *choke flow parameter* or *choke ratio* can be expressed as

$$\text{RWD} = W\sqrt{T_3}/p_{03}A_DQ_{3,\text{crit}} \tag{27.188}$$

where A_D = throat area, and $Q_{3,\text{crit}}$ = *Sonic Flow Function*. Compressor test data typically shows this parameter to be close to unity at diffuser choke.

For centrifugal compressor meanline flow analysis, the flow may be simplistically isolated at three stations-impeller tip, downstream diffuser throat, and final compressor discharge. The static pressure recoveries from impeller tip to diffuser throat and throat to final exit may be defined as for tip to throat

$$C_{p,2-3} = p_3 - p_2/p_{02} - p_2 \tag{27.189}$$

and for throat to exit

$$C_{p,3-e} = p_e - p_3/p_{03} - p_3 \tag{27.190}$$

Since $C_{p,2-3} = 1 - (C_3/C_2)^2$, and for tip to exit

$$C_{p,2-e} = [1 - (C_3/C_2)^2] + (C_3/C_2)^2 Cp_{3-e} \tag{27.191}$$

The overall static pressure rise from impeller tip to final exit is clearly dominated by the diffusion ratio, C_2/C_3, in the uncovered tip to throat region rather than the covered channel, as can be seen from Fig. 27.75. Nevertheless, maximization of covered channel static pressure recovery is warranted to achieve maximum overall stage efficiency.

Examination of a family of centrifugal compressor vaned diffusers with back-swept impellers is presented by Rodgers (1982), where it is revealed that the stalling

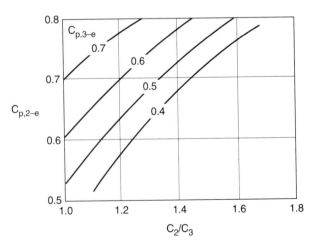

FIGURE 27.75   Influence of vaneless space and covered channel on $C_{p,2-e}$.

value of C_2/C_3 lies close to 2.0, based on an unblocked (blockage = zero) throat. Note that the anticipated stall point can be assessed on the basis of at least three parameters:

i) Diffusion ratio C_2/C_3, blocked throat, close to 1.7
ii) Diffusion ratio C_2/C_3, unblocked throat, close to 2.0
iii) Choke flow ratio, RWD = function M_2

Many other vaned diffuser stall parameters can be found in the open literature, among which one described by Elder and Gill (1984), covers the influence of diffuser vane number. A normalized diffuser static pressure recovery characteristic suitable for centrifugal compressor matching analysis is shown in Fig. 27.76, where overall static pressure recovery ratio is plotted versus the choke flow parameter, RWD, with machine Mach number, Mu, as a parameter. Increasing Mu, diminishes the usable flow range between choke and peak recovery near stall. A compressor matching analysis to be discussed later will show that discrete definition of the actual diffuser stall point may be secondary to the predicted overall pressure ratio inflection point on the compressor map as being more indicative of the surge position. The inflection point tends to be more dependent upon the relative alignment of both the diffuser and impeller capacities (their respective throat sizes) than the individual stall points unless, coincidentally, both stall simultaneously at the unique incidence crossover point.

The importance of the throat areas cannot be overemphasized, rather than the particular incidences, i.e., selection of blade and vane angles, and, as a result, the design of channel-type diffusers do not require extensive analysis. The simple geometry of the straight wall channel diffuser with the throat area and vane number selection plus covered channel diffusion limits just about constrain everything, as is apparent from Fig. 27.77. The simple virtues of the channel type, especially in a pinned separate vane configuration make it amenable to quick modifications, thereby aiding rapid compressor development. The influence of channel diffuser geometry

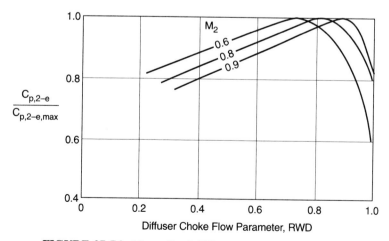

FIGURE 27.76 Normalized diffuser recovery characteristic.

FIGURE 27.77   Channel diffuser geometry.

on the flow range and efficiency of a centrifugal compressor is further elaborated by Clements and Artt (1987), (1989).

In summary, therefore, the key points in vaned diffuser design for centrifugal compressors are:

i) Throat area sizing is recommended to set the vane position rather than choice of a particular incidence

ii) The vaneless space diameter ratio should be from 5 to 20% as first determined by impeller tip high cycle fatigue margin and, second, flow range considerations

iii) Higher static pressure recoveries are attained with high vane numbers, in the 20 to 30 range. Lower vane numbers in the 10 to 20 range generally exhibit increased range with slightly lower recoveries

iv) Diffuser stall is primarily dominated by the static pressure rise, (diffusion ratio, C_2/C_3) between the impeller tip and throat. Covered channel or passage diffusion should be within open literature limits for two-dimensional diffusers

Interstage Crossover Duct. The purpose of the crossover duct, or return channel, in multistage centrifugal compressors is to transfer fluid from one stage exit to the inlet of the following stage. The crossover duct is the least understood component of the centrifugal compressor, and it has been afforded limited research and analysis. Aungier (1993) presents a meanline aerodynamic design methodology for vaneless diffusers and return channels, supported by test data from several low *Mu* industrial type centrifugal gas compressors.

The design intent of the crossover duct is to feed the next stage, removing most of the swirl with minimum pressure drop, maximum static pressure rise, and with a tolerable flow profile, after negotiating 270° meridional turning in limited axial space. As mentioned previously, axial space becomes a premium in very high suction pressure multistage centrifugal compressors as a consequence of high shaft stiffness necessities. A sharp 180° crossover bend is, therefore, the general practice in such applications. Total pressure loss trends for typical multistage crossover ducts are shown on Fig. 27.78.

Discharge Casing. Discharge casings for centrifugal compressors can be classified into three basic types: 1) volute or exit scroll, 2) exit guide vane type, or 3) plenum discharge port. Exit guide vanes (EGV's) or deswirl cascades are used where performance considerations require the recovery of residual angular momentum prior to discharging into a downstream plenum, typical of most gas turbine designs.

The exterior dimensions of centrifugal compressors for small turbochargers and aircraft gas turbines are envelope and cost constrained, and, as a consequence, the exit Mach number from the vaneless or vaned diffuser can exceed 0.3, with some 7% residual kinetic energy. Recovery of this energy is critical to attaining maximum compressor efficiency, at least in terms of final discharge static pressure.

Exit scroll design is particularly important for small, relatively high specific speed turbocharger centrifugal compressors, where tight engine compartment envelopes impose scroll dimensions. The work of Mishina and Gyobu (1978) is most inform-

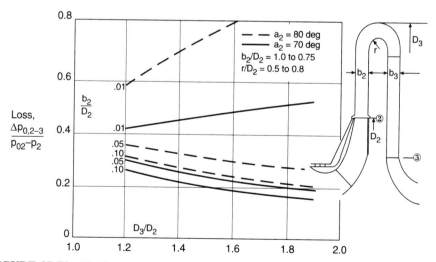

FIGURE 27.78 Multistage crossover duct pressure losses. (Reprinted with permission of ASME, Mishina, H. and Gyobu, I., "Performance Investigations of Large Capacity Centrifugal Compressors," Figure 16, ASME 78-GT-3, 1978.)

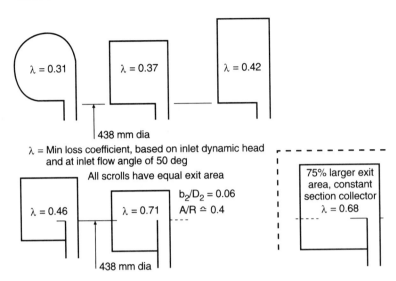

FIGURE 27.79   Scroll configurations and losses.

ative in providing relevant information concerning the effect of scroll geometry on exit pressure losses, some of which is reproduced on Fig. 27.79. The *folded back* scroll configurations where the centroid initially decreases below the discharge diameter and the tangential component of the flow has to reaccelerate resulting in higher losses. The scroll tongue area is analogous to the vaned diffuser throat area, and indeed the scroll may be visualized as a single vane diffuser wrapping around to form a throat on itself at the full 360°. Discharge casings configured with scrolls are almost universally adopted where the flow is to be delivered via a pipe, or conduit, to the process component or reservoir. In some installations, the flow is required to be de-swirled prior to discharge in a plenum, or in gas turbine applications, the downstream combustor.

27.5.4 Compressor Matching

The merit of a compressor design is not only assessed in its ability to attain the required adiabatic efficiency, pressure ratio, and flow, but also in the ability of the compressor to perform completely stable and efficiently within the total operating envelope of the system to which it is subordinate. The shape of the compressor map, the broadness of the efficiency islands, and the pressure rise towards surger are fundamentally controlled by the flow matching of the respective components, be they the individual stages of a multistage compressor, or the impeller and diffuser of an individual stage. Both aspects of compressor matching are discussed below.

Impeller/Diffuser Matching. The overall performance of a single stage compressor is composed from the characteristics of the impeller and diffuser, which can be matched together on the basis of the normalized flow at the impeller tip on the assumption that component interaction effects are negligible. Interaction effects do occur in some instances adjacent to choke and stall or with the approach of supersonic conditions at the impeller exit.

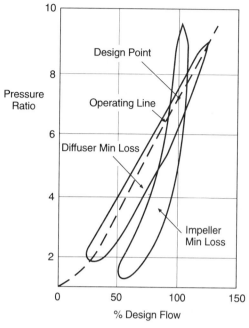

FIGURE 27.80 Impeller and diffuser minimum loss regimes.

Component matching optimization is of crucial importance in high pressure ratio, single stage centrifugal compressors. High stage pressure ratios cause conflicting demands between the impeller and diffuser if both are to operate in their respective regimes of minimum loss over the complete compressor operating range. Figure 27.80 illustrates the relative positioning of these minimum loss zones with the design point at which both impeller and diffuser minimum loss coincides. At pressure ratios other than design, the minimum loss regimes of the impeller and diffuser depart, the departure increasing with change in pressure ratio from the design condition. The higher the design pressure ratio, the greater the departure at part load conditions. The relative displacement of the minimum loss regimes results in reduced overall compressor efficiency and, moreover, irregular surge lines. Selection of optimum impeller and diffuser matching at the design pressure ratio can, consequently, induce stalled impeller operation at part load.

Since component matching is more crucial with high pressure ratio compressors, the component matching procedure is best illustrated by selecting a specific example of a small single stage high pressure ratio centrifugal compressor reported by Rodgers (1977). The impeller and diffuser matching of the compressor is shown on the left-hand side of Fig. 27.81, depicting the impeller stagnation pressure ratio and diffuser pressure loss ratio versus the normalized flow at the impeller tip. The slopes of both the impeller and diffuser characteristics commence to inflex adjacent to surge, which was coincident with an impeller diffusion ratio, $W_1/W_2 = 1.6$, and diffuser ratio, $C_2/C_3 = 1.65$. The steeper positive slope of the impeller characteristic at 100% speed is indicative of imminent inducer choke, thus operation at higher speeds would be expected to both reduce efficiency and the rate of massflow increase. The difficulties in single stage matching at higher stage pressure ratios are mitigated

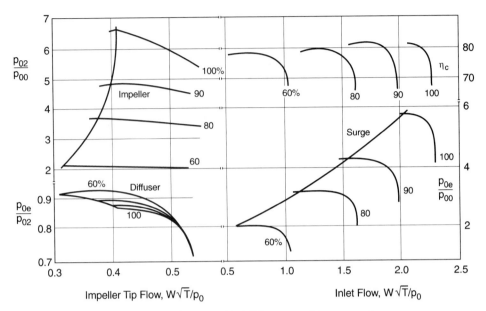

FIGURE 27.81   Impeller/diffuser matching example.

partially by the use of bleed and or variable geometry, but become even more amplified in multistage applications. As a consequence, stage pressure ratios of less
than 2.0, are generally adopted for multistage compressors, with the exception of
advanced high pressure ratio aviation two stage machines similar to that described
by Rodgers and Langworthy (1974).

Multistage Centrifugal Compressor Matching. The performance of a multistage
centrifugal compressor is generally computed from the individual stage characteristics using the *stage stacking* procedure. The computation of the performance of
each successive stage determines the inlet conditions and the operating point (ϕ, q,
η, Mu) of the next stage. The overall compressor performance is then determined
by the cumulative performance of the individual stages.

In multistage compressors, the maximum flow is normally determined by the
choke of the first stage, while the overall surge may be controlled by the stage that
stalls first. The optimum matching producing the maximum flow range and efficiency is the one where all stages are at their peak efficiency and where all stages
reach choke and surge simultaneously.

27.5.5   State-of-the-Art Efficiencies

The attainable total to static efficiency levels of high pressure ratio single stage centrifugal compressors with ambient sea level (air), conditions is largely dependent
upon five parameters: 1) inlet specific speed, N_s, 2) impeller tip backsweep, 3) impeller tip diameter, 4) inducer tip Mach number, and 5) exit Mach number.

Representative impeller geometry covering a wide range of specific speeds are
shown in Fig. 27.82. At low specific speeds, the equivalent hydraulic radius of the
blade passage is small; this precipitates high frictional losses. At high specific speeds,

FIGURE 27.82   Impellers of varying specific speed.

the tighter passage curvatures incur increased losses and the susceptibility of shroud separation.

27.5.6 Mixed Flow Compressors

A generalized distinction between axial and centrifugal or mixed flow compressors is that static pressure rise in axial compressors is accomplished essentially by diffusion of the throughflow relative velocity, whereas both diffusion of the relative velocity and change in radius produce static pressure in centrifugal and mixed flow compressors. Furthermore, centrifugal and mixed flow compressors are differentiated by the blade exit angle relative to the meridional plane, mixed flow impellers being characterized by a generally diagonal flowpath and discharge. Accordingly, centrifugal impellers have constant tip diameter, and mixed flow impellers have increasing tip diameter from hub to shroud.

Mixed flow compressors provide a natural geometric transition between the centrifugal and axial compressor, and they attain peak performance in the specific speed range spanning the two predominant compressor types, mainly $N_s = 1.0$ to 3.0. Centrifugal impellers with specific speeds higher than unity typically feature tight shroud curvature precipitous to separation and loss generation. The diagonal discharge of the mixed flow impeller permits relaxation of the shroud curvature with reduced losses. Mixed flow compressors have a rather modest history, early development work at NASA in the 1950's having been hampered by structural constraints and diagonal flowpath diffuser mismatching. In the same time period, a limited number of mixed flow compressors penetrated the gas compressor market as low pressure ratio, high flow, gas booster compressors. Increased power density demands in both small turbochargers and expendable turbojets focused towards minimizing frontal area have fostered renewed interest in the mixed flow compressor. Frontal area can

FIGURE 27.83   Mixed flow compressor.

be further reduced with mixed flowpaths by adoption of a short vaneless space turning from the diagonal to the axial direction, followed by an axial diffusing cascade, as illustrated in Fig. 27.83.

27.6 LOW PRESSURE AXIAL FANS
Terry Wright

In the low pressure, low speed range of turbomachinery, relatively simple axial flow machines dominate. They may range from simple propeller-type fans, for providing a breeze in a workspace, to an axial-flow blower with two or more stages comprised by axial rotating impellers and stationary vane rows. This section is intended to give a summary of the characteristics of these fans and to provide means for evaluation, sizing, and preliminary layout of such fans for a specified performance requirement stated as pressure rise, volume flow rate, and fluid density, with efficiency and noise constraints. Figure 27.84 shows typical configurations for these fans. To characterize these fans, performance coefficients [Shepherd (1956)] are defined as

$$C_Q \text{ (flow coefficient)} = Q/nD^3 \qquad (27.192)$$

$$C_p \text{ (pressure coefficient)} = \Delta p_T/\rho n^2 D^2 \qquad (27.193)$$

$$C_P \text{ (power coefficient)} = P/\rho n^3 D^5 \qquad (27.194)$$

$$\eta_T \text{ (total efficiency)} = C_Q C_p/C_P \qquad (27.195)$$

Q is the volume flow rate, Δp_T is the total to static pressure rise $(\Delta p_s + \frac{1}{2}\rho C_x^2)$, $C_x = Q/(\pi D^2/4)$, P is the shaft power, n is rotational speed, and D is the fan diameter.

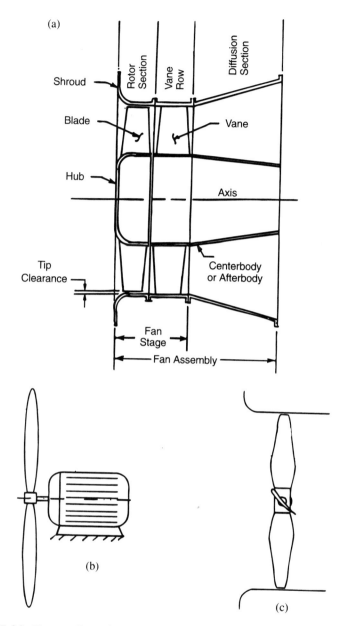

FIGURE 27.84 Fan configurations and nomenclature. (a) Full stage vane type, (b) propeller type, and (c) tube type.

Fundamental units for Q, Δp_T, P, ρ, n, and D are used as (m³/s, N/m², N · m/s, Kg/m³, radians/s, m) for SI calculation or (ft³/s, lbf/ft², lbf · ft/s, slugs/ft³, radians/s, ft) for English unit calculations. So defined, the performance coefficients are dimensionless numbers. Compared to centrifugal fans, pumps, or compressors, the fans of interest here will be typified by relatively large values of C_Q and small values of C_p. That is, for a given size and speed, low pressure and high flow are characteristic of low-speed, axial fans. The characteristic performance curves using these coefficients are illustrated in Fig. 27.85.

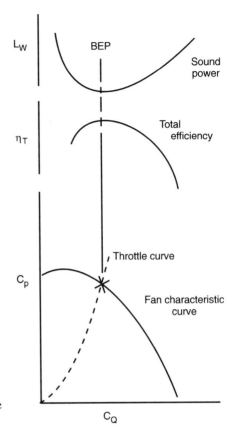

FIGURE 27.85 Characteristic performance curves.

The use of these performance coefficients implies the process of scaling the fan size and speed (or fluid density) from one configuration to another and predicting reasonably accurately the changes in performance. If changes in Reynolds Number or Mach Number (not really a concern in low-speed, low-pressure fans) are moderate, the performance coefficients are invariant. Therefore, C_Q, for example, can be written as

$$C_Q = \frac{Q_m}{n_m D_m^3} = \frac{Q_p}{n_p D_p^3} \tag{27.196}$$

So that the change in flow from model $(\ )_m$ to prototype $(\ )_p$ is

$$Q_p = Q_m \left(\frac{n_p}{n_m}\right) \left(\frac{D_p}{D_m}\right)^3 \tag{27.197}$$

Similarly, if C_P and C_p are invariant

$$\Delta p_{T,p} = \Delta p_{T,m} \left(\frac{\rho_p}{\rho_m}\right) \left(\frac{n_p}{n_m}\right)^2 \left(\frac{D_p}{D_m}\right)^2 \tag{27.198}$$

$$P_p = P_m \left(\frac{\rho_p}{\rho_m}\right) \left(\frac{n_p}{n_m}\right)^3 \left(\frac{D_p}{D_m}\right)^5 \qquad (27.199)$$

and implicitly

$$\eta_{T,p} = \eta_{T,m} \qquad (27.200)$$

These scaling rules (or *fan laws*) are based on the assumption of complete geometric similarity, right down to surface relative roughness, and negligible influence of Reynolds and Mach numbers. In fact, roughness and Reynolds number (ϵ/D and Re) frequently lie outside a reasonable range of variation, and how to account for their influence has been the subject of some controversy for many years [see AMCA (1985), ASME (1965), Pampreen (1973), Casey (1985), Simon and Buelskaemper (1984), Strub *et al.* (1987), and Wright (1989)].

The AMCA rules state simply that no accounting for the influence of *scale*, *size-effect*, or Reynolds number is allowed when scaling, but a minimum model test size and speed assure values of Re = $ND^2/\nu > 3 \times 10^6$. Size ratio must be at least 1/5 of full scale (or diameter at least 35 inches). These model restrictions preclude strong influences of roughness and Reynolds number, so that, for low speed fans, the AMCA standard of no adjustment to efficiency, flow, or pressure rise due to *size-effect* is a reasonable compromise. If extreme ranges are encountered, the use of some form of the methods of Casey (1985) or Strub *et al.* (1987) appear to represent the best choice available. The form for ratio of losses

$$(1 - \eta_{T_p})/(1 - \eta_{T_a})$$
$$= [a + (1 - a)(f_p/f_{fr,p})]/[a + (1 - a)(f_m/f_{fr,m})] \qquad (27.201)$$

gives conservative, reliable results for centrifugal compressors, but appears to be equally useful for small axial flow machines using $f = f(\epsilon/D, \text{Re})$ and $a = 0.3$ [Wright (1988)]. This result allows scaling of efficiency, while pressure and flow may be adjusted according to

$$C_{pp}/C_{pm} = 0.5 + 0.5(\eta_{Tp}/\eta_{Tm}) \qquad (27.202)$$
$$C_{Qp}/C_{Qm} = (C_{pp}/C_{pm})^{1/2} \qquad (27.203)$$

Then

$$C_{P_p} = \eta_{T,p}^{-1} Q_{Q_p} C_{p,p} \qquad (27.204)$$

It is important to remember that such scaling of efficiency and performance is not fully agreed upon in the international turbomachinery community.

These performance coefficients can be used to form dimensionless values of *specific speed* and *specific diameter* according to

$$N_s \text{ (specific speed)} = C_Q^{1/2}/C_p^{3/4} = \frac{Q^{1/2}}{(\Delta p_T/\rho)^{3/4}} n \qquad (27.205)$$

$$D_s \text{ (specific diameter)} = C_p^{3/4}/C_Q^{1/2} = \frac{(\Delta p_T/\rho)^{1/4}}{Q^{1/2}} D \qquad (27.206)$$

These *specific* variables serve to characterize the performance and geometric parameters of the machines in a unique way. Cordier (1955), Balje (1962 and 1981), and Csanady (1964) show the separability and means of classification of turbomachines with N_s and D_s when attention is directed to machines representing the *best* of a given type. For example, taking *best* to mean a propeller fan whose efficiency is high as compared to others of the type, would allow grouping of excellent propellers in a zone of $N_s - D_s$ values. This concept, usually referred to as a *Cordier line* or the *Balje approach*, is illustrated in Fig. 27.86 for low-pressure axial fans. The zones "a," "b," and "c" roughly define the proper range of specific diameter for the three types of axial fans considered here as propeller, tube axial, and vane axial fans.

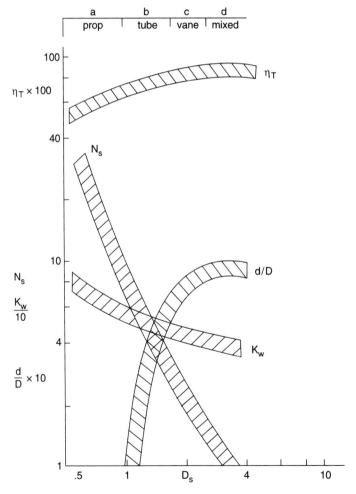

FIGURE 27.86   Concept of the Cordier diagram shows *zones* of data for *best* equipment.

The range of efficiencies related to the values of D_s are shown in Fig. 27.86 as well. Note that, for a given value of D_s, the values of N_s and η_T for a good fan are more or less uniquely defined. By restricting attention to good fans or best designs, the locus of $N_s - D_s$ and $\eta_T - D_s$ are identified with design points or *best efficiency points* (BEP) of operation of the machines involved. This BEP is illustrated in Fig. 27.85. Available performance to the left and right of C_{QBEP} are the less desirable values in terms of the efficiency or relative amount of power required.

An approximate quantitative treatment of the Cordier diagram for low-speed axial fans is shown in Fig. 27.87, where the basic concept of the Graham (1991) treatment of specific noise is incorporated into the diagram as $K_w - Vs - D_s$. The *zones* shown can be approximated by the following equations.

$$N_s \doteq 9.0D_s^{-2.1}, \; D_s \doteq 2.8N_s^{-.48} \qquad (27.207)$$

$$\eta_T \doteq .664D_s^{.425}, \; \eta_T \le .86 \qquad (27.208)$$

$$K_w \doteq 72D_s^{-0.8} \qquad (27.209)$$

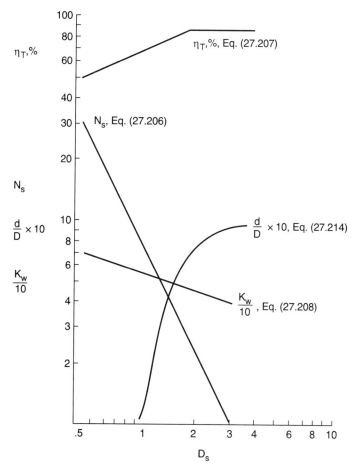

FIGURE 27.87 Approximate characterization of parameters for low-pressure axial fans.

For total sound power level, L_w (dB), is given by Graham as

$$L_w = K_w + 10 \log (Q) + 20 \log (\Delta p) \qquad (27.210)$$

where the simple form shown here for K_w as a function of D_s is meant to be a coarse approximation to Graham's original detailed treatment of the noise prediction. (Note that Graham's Eq. (27.210) requires Q as ft^3/min and Δp as in. W.G.) These rough equations suffice to provide estimates of the performance levels to be expected from low-pressure fans and may be used to initiate preliminary equipment selection studies or preliminary design.

Example 27.1:   A fan is required to supply air at standard density ($\rho g = 0.075$ lbf/ft^3) at 10,000 ft^3/min of flow with a total pressure rise of 2 in W.G. (note that US industrial units have been used). If a choice of fan is to be made from vendor's catalogs or a new design attempted, then the first decision to be made is—what type of fan is a reasonable choice? Converting the requirements to fundamental units yields: $Q = 167$ ft^3/s, $\Delta p_T = 10.4$ lbf/ft^2, $\rho = 0.00233$ slugs/ft^3. The specific speed and diameter may then be formed as $N_s = 0.0273n$ and $D_s = 0.633D$. An initial attempt could be to choose $n = 183$s$^{-1} \doteq 1,750$ rpm, then $N_s = 4.34$. The corresponding specific diameter is $D_s = 1.41$. The choice of $n = 183$s^{-1} or 1750 rpm forces a fan diameter $D = (1.41/.633)$ ft $= 2.23$ ft $\cong 27$ inches. Cordier's concept also indicates that the fan will be tube axial with $\eta_T \cong 0.65$, with a noise sound power level of $L_w = 55 + 40 + 4 = 95$ dB. Choosing other speeds of course changes all of these parameters as seen in Table 27.2. Considering such a range of speeds covers virtually the entire range of low-pressure axial fans. In fact, the selection at 600 rpm is inadmissible as an axial fan, and would require a mixed-flow configuration (see Sec. 27.5), unless a two-stage design is allowed. Clearly, the advantage of lower speed is lowered noise and increased efficiency (for lowered motor size and kilowatt cost). The offsetting penalty lies in rapidly increasing fan size. Between the speeds of 3600 rpm and 900 rpm, the diameter doubles, while power is reduced by 30% and noise is reduced by 20 dB. Identification of the best choice depends on the criteria for selection. For example, if low cost and small size are particularly important, then the 2400 rpm fan is a good selection, if 104 dB sound power is acceptable. The ability to examine an optimal pool of candidates serves both the fan designer and the engineer selecting a fan from available equipment.

TABLE 27.2   Candidate Fans for Example 27.1

n (rpm)	N_s	D_s	D (in.)	η_T	L_w (dB)	P (hp)	Fan Type
600	1.49	2.31	43.8	0.86	83	3.67	Mixed
900	2.17	1.93	36.6	0.86	88	3.67	Vane
1200	2.90	1.68	31.8	0.83	93	3.78	Vane
1800	4.34	1.41	27.8	0.77	99	4.10	Tube
2400	5.80	1.20	22.8	0.72	104	4.39	Tube
3000	7.23	1.08	20.5	0.69	106	4.57	Tube
3600	8.93	0.98	18.6	0.66	102	4.80	Prop

If, as is often the case, the specifications are written in terms of static pressure rise, where $\Delta p_s = \Delta p_T - \frac{1}{2}\rho C_x^2$, then the above selection would have to be iterated to assure the required level of Δp_s. If for this case $\Delta p_s = 2$ in W.G., then $C_x = Q/A = Q/(\pi D^2/4) = (167 \text{ ft}^3/\text{s})/(2.84 \text{ ft}^2) = 59$ ft/s $\Delta p_T = \Delta p_s + \frac{1}{2}\rho C_x^2 = (10.4 + 4.0)$ lbf/ft $= 14.4$ lbf/ft^2. Using $n = 251$ s^{-1}, gives $N_s = 4.56$ instead of 5.80. Correspondingly, $D_s = 1.34$ yields $D = 1.95$ ft $= 23.4$ in. The resulting efficiency and power are $\eta_T = 0.75$ and $P = 4.21$ hp. Since D changed, C_x is recalculated as 56 ft/s and $\Delta p_T = 1.41$ lbf/ft^2. Again, using $n = 251$ s^{-1} gives $D = 1.94$ ft $= 23.3$ in. No significant change occurs with the second iteration so that the optimal solution or selection is not greatly changed by the use of static pressure rise. The first iteration, however, should be made and the other entries in Table 27.2 should be adjusted for a new pool of candidates.

At a more fundamental level, the performance of a fan and its required geometry are governed by the transfer of angular momemtum to the fluid by the rotating blade row. The relationship is known as the *Euler equation* of turbomachinery which for low pressure fans is [White (1994)]

$$(\Delta p_T/\rho)_{\text{ideal}} = U_2 C_{\theta 2} - U_1 C_{\theta 1} \tag{27.211}$$

Here, U is the rotational velocity of a blade element located at a radial distance, r, from the fan axis, and turning at n radians per second. That is, $U = nr$, and the 1 and 2 subscripts indicate upstream and downstream locations relative to the blade. C_θ is the value of the tangential velocity component in the absolute reference. Figures 27.88 and 27.89 illustrate the absolute blade-relative velocity vectors and the concept of a cascade of blade elements. Considering a simplified case where $U_1 = U_2$ and $C_{\theta 1} = 0$ (pure axial entry flow) with $C_x = C_{x1} = C_{x2} = Q/A_{\text{fan}}(A_{\text{fan}} = (\pi_1/4)(D^2 - d^2))$, the Euler equation reduces to

$$\Delta p_T = \rho \eta_T nr C_{\theta 2} \tag{27.211a}$$

where η_T accounts for real viscous flow effects and Δp_T is no longer an ideal value. The blade relative inlet and outlet velocities are calculated as

$$W_1 = (C_x^2 + U^2)^{1/2} \tag{27.212}$$

$$W_2 = (C_x^2 + (U - C_{\theta 2})^2)^{1/2} \tag{27.213}$$

so that $W_2 \le W_1$, if positive pressure rise is to occur. Equivalently $W_2/W_1 \le 1$. This behavior is analogous to flow in a diffuser and, as in all diffusion flows, the amount of reduction in velocity is limited by the process of wall flow separation or diffuser stall. W_2/W_1 is often given the name *de Haller ratio* [Wilson (1984) and de Haller (1953)]. The accepted guideline for preliminary design layout or flow studies is $(W_2/W_1) > 0.7$.

The concept of blade *solidity* is introduced as $\sigma = c/s$ where c is blade chord and s is circumferential spacing, blade to blade, as shown in Fig. 27.90. The ratio c/s is analogous to the planar diffuser parameter, L/w_1, so that for values of L/w_1 in the

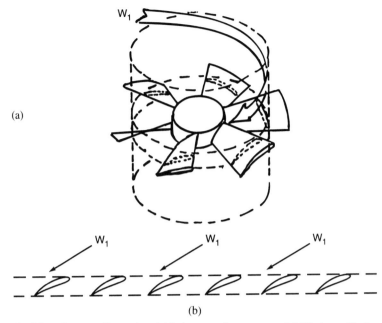

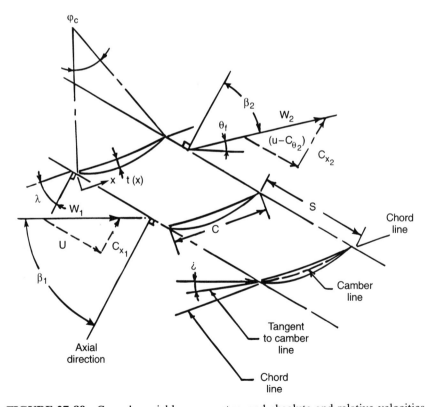

FIGURE 27.88   The two-dimensional blade cascade concept. (a) Blade-to-blade proximinity in an axial fan and (b) two-dimensional analog showing relative inflow.

FIGURE 27.89   Cascade variables, geometry, and absolute and relative velocities.

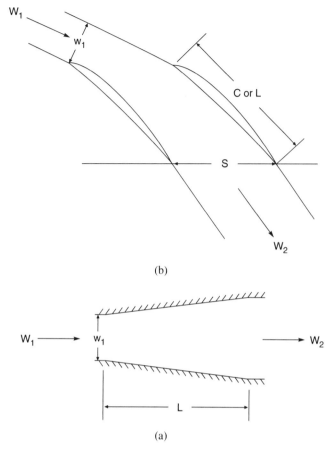

(b)

(a)

FIGURE 27.90 Analogy between geometry and performance of a planar diffuser and a blade row cascade. (a) Plane diffuser and (b) typical low-pressure fan cascade.

range of 4 to 5, pressure recovery or $C_d = 1 - (W_2/W_1)^2$ is limited to perhaps 0.5 or $(W_2/W_1) = 0.71$. The indication is that the value of σ may be increased somewhat to control the allowable diffusion or blade row loading.

Example 27.2: A vane axial fan must generate 5,000 cfm with $\Delta p_T = 3$ in W.G. ($\rho g = 0.075$ lbf/ft^3). A good value for N_s is around 2.3 from the Cordier diagram with a corresponding D_s of 1.88 ($\eta_T = 0.86$), resulting in $n \doteq 1800$ rpm and $D \doteq 1.89$ ft. If the hub diameter, d, is chosen as $d = (3/4)D = 1.42$ ft, then the mean radial station of the blade row (at $r = 0.83$ ft) can be examined in terms of W_2/W_1 with $nr = U = 155$ ft/s, $C_\theta = \Delta p_T/(\rho \eta_T U) = 50$ ft/s, and $C_x = 58$ ft/s. Then, $W_1 = 155$ ft/s, $W_2 = 120$ ft/s, and $(W_2/W_1) = 0.72$. This is a conservative level of blade loading and may indicate an unnecessarily large hub. Table 27.3 summarizes the de Haller calculations at the mean radial station for both blade row and vane row with broad variation of the hub-tip ratio d/D.

The values for both blade and vane similar are similar and show the very favorable influence of increasing hub size on the diffusion. Below $d/D = 0.65$, the

TABLE 27.3 Mean Station Data for Example 27.3

d/D	Um	C_x	C_θ	W_1	β_1	W_e	β_2	DH_B	C_2	DH_v	DPR	θv
0.20	106.5	31.11	85.46	111.0	73.73	37.61	34.19	0.34	90.94	0.34	0.93	70.00
0.25	111.0	31.85	82.04	115.5	73.99	43.07	42.31	0.37	88.01	0.36	0.92	68.78
0.30	115.4	32.82	78.88	120.0	74.14	49.15	48.11	0.41	85.44	0.38	0.92	67.41
0.35	119.9	34.04	75.96	124.6	74.16	55.59	52.25	0.45	83.24	0.41	0.91	65.87
0.40	124.3	35.55	73.25	129.3	74.05	62.26	55.18	0.48	81.42	0.44	0.91	64.11
0.45	128.8	37.45	70.72	134.1	73.79	69.10	57.19	0.52	80.03	0.47	0.90	62.10
0.50	133.2	39.82	68.37	139.0	73.36	76.12	58.46	0.55	79.12	0.50	0.88	59.78
0.55	137.6	42.81	66.16	144.1	72.73	83.35	59.09	0.58	78.81	0.54	0.86	57.09
0.60	142.1	46.66	64.09	149.5	71.82	90.92	59.12	0.61	79.28	0.59	0.84	53.94
0.65	146.5	51.71	62.15	155.4	70.57	98.99	58.51	0.64	80.85	0.64	0.80	50.24
0.70	151.0	58.55	60.32	161.9	68.81	107.94	57.15	0.67	84.07	0.70	0.74	45.85
0.75	155.4	68.26	58.60	169.7	66.29	118.48	54.82	0.70	89.96	0.76	0.65	40.65
0.80	159.8	82.95	56.97	180.1	62.58	132.18	51.13	0.73	100.6	0.82	0.49	34.48
0.85	164.3	107.61	55.43	196.4	56.78	153.10	45.34	0.78	121.0	0.89	0.14	27.25
0.90	168.7	157.17	53.97	230.6	47.04	194.63	36.14	0.84	166.1	0.95	—	18.95
0.95	173.2	306.29	52.59	351.8	29.49	329.18	21.50	0.94	310.7	0.99	—	9.75

de Haller ratios are clearly unacceptable for blade and vane. For values above about $d/D = 0.70$, the loading or diffusion is well controlled and the design sizing acceptable. However, as seen in the table, the decreasing fan annulus area is steadily increasing the velocity pressure (or dynamic pressure) at the expense of static pressure rise (DPR $= \Delta p_S/\Delta p_T$). At $d/D = 0.8$, the need for a very effective downstream annular diffuser is clear and achievement of acceptable static pressure and efficiency is doubtful.

The mean radial station considered above is characteristic of the overall behavior of the blade and vane row, but the blade and vane diffusion at the radial stations nearest the hub generally provide the critical limit in minimum speed, through-flow velocity or hub size for a given performance requirement. If radially uniform work and total pressure rise are to be maintained across the blade from hub to tip, then the reduction of $U = rn$ with decreasing r leads to the requirement of larger and larger values of $C_\theta = \Delta p_T/(\rho \eta_T nr)$, varying as r^{-1}. The fan example is extended to show the hub station diffusion behavior as seen in Table 27.4, for the same range of d/D. The results for the smaller hubs are clearly unacceptable with absolute fluid turning angles θ_v in the 70° to 100° range, and acceptable blade performance is not reached until the hub-tip ratio exceeds about 0.70. Even for large solidity blades and vanes in the hub region, the requirement for uniform work forces a large hub, small annulus, and low static efficiency. One alternative path involves choosing a higher specific speed which may also result in a reduction of total and static efficiency. The η_T and η_S values for $d/D = 0.75$ are 0.86 and 0.42, respectively.

If the specific speed range is explored using fan speeds from 900 to 4200 rpm, and acceptable values of de Haller ratio are used to establish minimum hub sizes (d/D values), Table 27.5 can be constructed. If static efficiency were the criterion for choice, then a speed of about 2400 rpm looks best.

The systematic variation of the acceptable value of d/D with speed is a characteristic behavior for the axial fans considered here. A more thorough investigation with N_s varied from about 1.0 to values above 6.0 is summarized in Fig. 27.91. The data appear to be self limiting at the extremes of the range and a heuristic curve fit is shown with the calculated points. Approximately and conservatively chosen, d/D may be estimated by

$$d/D = (1/2)\{1 - (2/\pi) \tan^{-1} [(2/\pi)(N_s - 3.8)]\} \qquad (27.214)$$

These conservative hub-size values are based on the restriction of uniform total pressure rise, or uniform work, across the blade from hub-to-tip.

As seen above, this restriction forces the behavior of $C_\theta \sim r^{-1}$ or the *free-vortex* blade loading condition. Smaller hubs and higher static efficiencies can also be sought by relaxing the free-vortex constraint so that near-hub C_θ values may be reduced to more acceptable levels [Kahane (1948)]. This shifting of blade loading from the near-hub blade region outward results in a distortion of the axial component of velocity, reducing C_x near the hub and increasing it near the tip. Figure 27.92 illustrates this behavior for a heavily loaded axial fan with $N_s = 1.3$. The results are shown as $C_{\theta 2}/U_T$, value at the blade tip and C_{x2}/U_T along the blade span ($x = 2r/D$). Here, $d/D = 0.5$. Case 1 is the free vortex limit, and the progressive devia-

TABLE 27.4 Hub Station and Tip Station Data for Example 27.3

d/D	Um	C_x	C_θ	W_1	β_1	W_e	β_2	DH_B	C_2	DH_v	DPR	θv
0.20	35.5	31.11	256.3	47.22	48.80	223.0	−81.98	4.72	258.2	0.12	0.93	83.08
0.25	44.4	31.85	205.1	54.66	54.35	163.8	−78.79	3.00	207.5	0.15	0.92	81.17
0.30	53.3	32.82	170.9	62.59	58.38	122.1	−74.41	1.95	174.0	0.19	0.92	79.13
0.35	62.1	34.03	146.5	70.88	61.31	90.93	−68.02	1.28	150.4	0.23	0.91	76.92
0.40	71.0	35.55	128.1	79.46	63.42	67.29	−58.10	0.85	133.0	0.27	0.91	74.50
0.45	79.9	37.45	133.9	88.28	64.90	50.58	−42.24	0.57	119.9	0.31	0.90	71.81
0.50	88.8	39.82	102.5	97.34	65.86	42.12	−19.02	0.43	110.0	0.36	0.88	68.78
0.55	97.7	42.81	93.23	106.6	66.34	43.05	5.98	0.40	102.5	0.42	0.86	65.33
0.60	106.5	46.66	85.46	116.3	66.36	51.22	24.37	0.44	97.37	0.48	0.84	61.37
0.65	115.4	51.71	78.88	126.5	65.88	63.35	35.28	0.50	94.32	0.55	0.80	56.75
0.70	124.3	58.55	73.25	137.4	64.79	77.72	41.11	0.57	93.78	0.62	0.74	51.36
0.75	133.2	68.26	68.37	149.7	62.87	94.17	43.54	0.63	96.61	0.71	0.65	45.05
0.80	142.1	82.95	64.09	164.5	67.73	113.8	43.25	0.69	104.3	0.79	0.49	37.69
0.85	151.0	107.6	60.32	185.4	54.52	140.7	40.12	0.76	123.3	0.87	0.14	29.27
0.90	159.8	157.1	56.97	224.2	45.49	187.8	33.22	0.84	167.1	0.94	—	19.92
0.95	168.7	306.2	53.97	349.7	28.86	327.0	20.55	0.94	311.0	0.98	—	9.99
						TIP Station Data						
0.20	177.6	31.11	51.27	180.3	80.07	130.1	76.17	0.72	59.97	0.52	0.93	58.76
0.25	177.6	31.85	51.27	180.4	79.83	130.3	75.85	0.72	60.36	0.53	0.92	58.15
0.30	177.6	32.82	51.27	180.6	79.53	130.5	75.44	0.72	60.88	0.54	0.92	57.38
0.35	177.6	34.03	51.27	180.8	79.16	130.8	74.93	0.72	61.54	0.55	0.91	56.43
0.40	177.6	35.55	51.27	181.1	78.68	131.2	74.29	0.72	62.39	0.57	0.91	55.26
0.45	177.6	37.45	51.27	181.5	78.10	131.8	73.50	0.73	63.49	0.59	0.90	53.86
0.50	177.6	39.82	51.27	182.0	77.37	132.5	72.51	0.73	64.92	0.61	0.88	52.17
0.55	177.6	42.81	51.27	182.7	76.45	133.4	71.28	0.73	66.80	0.64	0.86	50.14
0.60	177.6	46.66	51.27	183.6	75.28	134.7	69.73	0.73	69.33	0.67	0.84	47.70
0.65	177.6	51.71	51.27	185.0	73.77	136.5	67.75	0.74	72.82	0.71	0.80	44.76
0.70	177.6	58.55	51.27	187.0	71.76	139.2	65.14	0.74	77.83	0.75	0.74	41.21
0.75	177.6	68.26	51.27	190.3	68.98	143.6	61.63	0.75	85.37	0.80	0.65	36.91
0.80	177.6	82.95	51.27	196.0	64.97	151.1	56.72	0.77	97.52	0.85	0.49	31.72
0.85	177.6	107.6	51.27	207.7	58.79	165.9	49.58	0.80	119.2	0.90	0.14	25.48
0.90	177.6	157.1	51.27	237.2	48.50	201.6	38.80	0.85	165.3	0.95	—	18.07
0.95	177.6	306.2	51.27	354.0	30.11	331.3	22.42	0.94	310.5	0.99	—	9.50

TABLE 27.5 d/D Values for Example 27.3

N (rpm)	D (ft)	N_s	D_s	d/D	DPR	η_T	η_s	L_w (dB)
900	2.63	1.16	2.61	0.93	0.04	0.86	0.04	88
1200	2.29	1.55	2.27	0.87	0.41	0.86	0.41	90
1800	1.88	2.32	1.87	0.75	0.65	0.86	0.56	92
2400	1.64	3.10	1.63	0.60	0.72	0.82	0.59	93.5
3000	1.48	3.87	1.46	0.48	0.70	0.78	0.55	95
3600	1.35	4.65	1.34	0.36	0.67	0.75	0.50	96
4200	1.26	5.42	1.26	0.25	0.61	0.73	0.45	99.5

tion from free vortex includes cases 2, 3, and 4. For the free vortex case, $C_x/U_{tip} = 0.25$ and $C_{\theta m}/U_T = 0.25$ so that $W_2/W_1 = (W_2/U_T)/(W_1/U_T) = 0.71$ at the mean station with $d/D = 0.5$. The approximate throughflow analysis yields, for this case, a hub-station de Haller ratio of $(W_2/W_1)_{hub} = 0.5$ with the $d/D = 0.50$ value. Rather than increase the hub size, we can shift the blade loading as shown in Fig. 27.92 and modify the hub-station de Haller ratio as shown in Table 27.6.

Skewing the load and the axial velocity, C_{x2}, away from the hub can substantially relieve diffusion requirements at the hub. (C_{x1}/C_{x2}) varies from 1.0 for free vortex to 0.76 with the *solid body* swirl of case 4 so that severe skewing of the axial profile is not caused by this load shift. Excessively low values of C_{x2}/C_{x1} (below 0.5) will limit the advantages of shifting the work outboard.

Figure 27.93 shows the extension of this behavior, in terms of $(C_{x2}/C_{x1})_{hub} = \hat{C}_x$ (at the hub station), as a function d/D for values of C_p with the $\hat{C}_x$ ratio as a

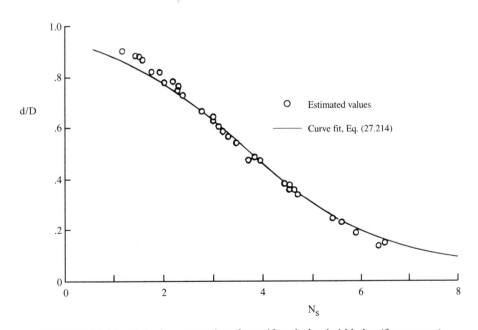

FIGURE 27.91   Hub-size constraints for uniformly loaded blades (free-vortex).

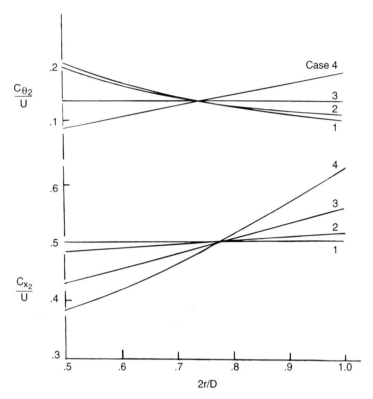

FIGURE 27.92 Nonuniform loading and velocity skewing. (Reprinted with permission from ASME, Wright, T., "A Closed Form Algebraic Approximation to Quasi-Three-Dimensional Flow in Axial Fans," ASME Paper No. 88-GT-15, 1988.)

parameter. The set of curves allows for skewing of the load and gives an estimated minimum hub size (to avoid excess diffusion) as a function of C_P. The values are based on an assumption of relatively low values of C_Q ($C_Q^2 \ll 1$), so the curves are inherently somewhat conservative. For example, if C_{x2} must be at least 40% of C_{x1}, then at $C_P = 0.05$, d/D should be at least 0.55. For no skewing, d/D would have to be 0.66. For the fan used in Example 27.1, with $Q = 10{,}000$ ft^3/min, $\Delta p_T = 2$ in W.G., $\rho g = 0.075$ lbf/ft^2, $n = 183$ s^{-1}, we found $N_s = 4.3$ and $C_p = \Delta p_T/(\rho n^2 D^2) = 0.027$ for which Fig. 27.93 gives $d/D \geq 0.50$ at $\hat{C}_x = 1.0$. According to Eq. (27.214), with $N_s = 4.3$, $d/D \geq 0.4$. Figure 27.90 would indeed appear to

TABLE 27.6 Hub Station Properties

Case	C_{x2}/U_T	$C_{\theta 2}/U_T$	W_1/U_T	W_2/U_T	DH_{hub}
1	0.26	0.413	0.560	0.274	0.490
2	0.25	0.375	0.560	0.280	0.500
3	0.24	0.338	0.560	0.290	0.520
4	0.215	0.248	0.560	0.331	0.590
5	0.190	0.167	0.560	0.380	0.680

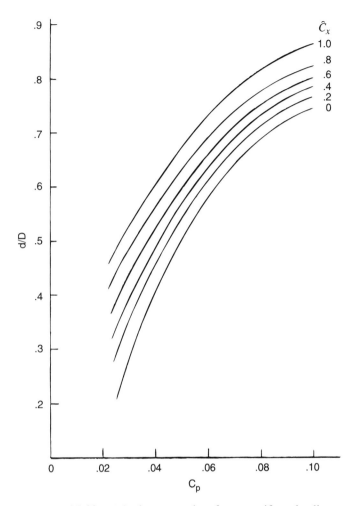

FIGURE 27.93 Hub-size constraints for nonuniform loading.

be conservative, but it offers a good starting point for exploration. With 60% skew on the axial throughflow ($\hat{C}_x = 0.4$ at the hub), the figure indicates $d/D \geq 0.35$. For the example with $Q = 5,000$ cfm and $\Delta p_T = 3$ in W.G., ($N_s = 2.3$ and $C_P = 0.053$), Eq. (27.214) yields $d/D \geq 0.74$, and Fig. 27.93 gives $d/D \geq 0.68$. Agreement appears better for more heavily loaded fans. Again, 60% skewing ($\hat{C}_x = 0.4$ at the hub) would allow d/D around 0.59 for a 20% reduction in the velocity pressure.

The preceding treatment of low speed, low pressure axial fans is both approximate and terse. Completely missing is any predictive process for losses due to viscous flow, leakage, tip-clearance flow, hub windage, or any mechanistic treatment for the estimation of efficiency. The use of Fig. 27.87 or Eqs. (27.207) through (27.209) leads to estimates of *typical* or characteristic values. The reader is referred to the literature for detailed loss estimation [Wilson (1984) or Koch and Smith (1976)]. Likewise, the detailed blading parameters such as blade camber, ϕ_c, blade stagger, λ, blade thickness, t/c, and further details for blade layout are referred to the literature on fans and axial compressors [Bullock and Johnson (1965) or Wright and Ralston (1987)].

27.7 CENTRIFUGAL PUMPS*
Michael W. Volk

27.7.1 Introduction

Centrifugal pumps are among the most widely used machines, playing important roles in industrial, commercial, and residential settings. This section summarizes the significant aspects of centrifugal pump hydraulic principles, operating characteristics, and recent technological advances. For a more in-depth examination of centrifugal pumps, together with the other major pump type, positive displacement [see Volk (1996)].

Simply stated, a pump is a machine used to move liquid through a piping network and to raise the pressure of the liquid. A centrifugal pump (Fig. 27.94) consists of an impeller, attached to and rotating with the shaft, and a casing which encloses the impeller. Liquid is forced into the inlet of the pump by atmospheric pressure or some upstream pressure. With the exception of a particular type of centrifugal pump called a *self-priming centrifugal*, centrifugal pumps are not inherently self-priming. In other words, the suction piping and inlet side of the pump must be filled with liquid and vented of air before the pump can be started.

Once the liquid entering the pump reaches the rotating impeller, it begins to move along the curved impeller vanes, increasing in velocity and kinetic energy. The vanes on a centrifugal pump impeller are curved backwards to the direction of rotation, this being the best method to achieve the maximum fluid velocity.

When the liquid leaves the impeller vane tip, it is at its maximum velocity. Then it enters the casing, where an increase in the cross-sectional area of the flow path occurs. Since the cross-sectional area is increasing as the liquid moves through the casing, the velocity decreases. The Bernoulli equation demonstrates that the reduced

*Excerpted from *Pump Characteristics and Applications*, by Michael W. Volk, P.E., to be published by Marcel Dekker, Inc., New York, 1996.

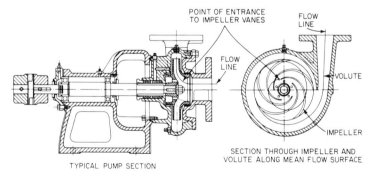

FIGURE 27.94 Centrifugal pump with single volute casing. (From *Pump Handbook*, Karassik, I. J. *et al.*, 1986. Reproduced with permission of McGraw-Hill, New York.)

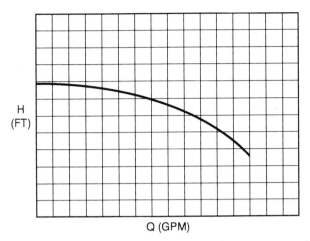

FIGURE 27.95 Characteristic head-capacity (*H–Q*) performance curve for a centrifugal pump operating with fixed speed and impeller diameter.

kinetic energy has to go somewhere, so it is transformed into increased potential energy. This causes the pressure of the liquid to increase as the velocity decreases. Note that there is some increase in the fluid pressure as the liquid moves along the impeller, but this is mainly due to centrifugal force acting on the liquid. The primary pressure increase occurs after the liquid leaves the impeller, in the pump casing.

A centrifugal pump operating at a fixed speed and with a fixed impeller diameter produces a pressure, or *head*, which varies with the flow rate delivered by the pump, as illustrated by the characteristic *head-capacity* (*H-Q*) curve shown in Fig. 27.95. As the *head* produced by the pump (usually expressed in feet of water, also called total *dynamic head*, or *TDH*, or simply *H*) decreases, the *capacity* produced (usually expressed in gallons per minute, or *gpm*) increases and, as the head increases, the flow produced decreases.

The centrifugal pump casing is one of several types. A *single volute* casing is illustrated in Fig. 27.94. Note that the single volute casing has a single *cutwater* where the flow is separated. As the flow moves around the volute casing, the pressure increases as the flow cross-sectional area expands, and this produces a radial force at each point on the periphery of the impeller. Summing all of these radial forces produces a net radial force, which must be carried by the radial bearing in the pump. The radial bearing also supports the weight of the shaft and impeller.

The radial force generated by a centrifugal pump also varies as the pump operates at different points on the *H-Q* curve, with the minimum radial forces being developed at the *best efficiency point* (BEP) of the pump (see Sec. 27.7.3 for discussion of the pump BEP). Operation at points to the right or to the left of the BEP produce higher radial loads. This is especially true for pumps with single volute casings.

A *diffuser casing* (Fig. 27.96) is a more complex casing arrangement, consisting of multiple flow paths around the periphery of the impeller. The liquid leaving the impeller vanes, rather than wrapping around the casing as it does with the single volute, merely enters the nearest flow path in the diffuser casing. The diffuser has multiple cutwaters, evenly spaced around the impeller, versus the single volute casing's one cutwater. The main advantage of the diffuser design is that this results in

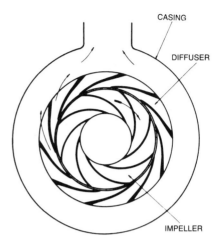

FIGURE 27.96 Diffuser casing design minimizes radial loads in a centrifugal pump. (From *Pump Handbook*, Karassik, I. J. *et al.*, 1986. Reproduced with permission of McGraw-Hill, New York.)

a near balancing of radial forces, eliminating the need for a heavy duty radial bearing. The dead weight of the rotating element must still be carried by the radial bearing, but overall the diffuser design minimizes radial bearing loads.

A hybrid between a single volute casing and a diffuser casing is a *double volute casing* (Fig. 27.97). This design has a splitter located in the casing which creates a second cutwater, located 180° from the first cutwater, an arrangement which results in much lower radial loads than single volute designs. Double volute casings are usually used by pump designers for larger, higher flow pumps (usually for flows higher than about 1500 to 2000 gpm) to allow the use of smaller radial bearings.

27.7.2 Capacity and Total Dynamic Head

In order to choose the size of a centrifugal pump for a particular application, the first necessary operation is to select the pump capacity and the total dynamic head, TDH.

Capacity of the pump is normally dictated by the requirements of the system in which the pump is located. A process system is designed for a particular throughput, or a vessel must be emptied in a certain amount of time. An air conditioning system requires a particular flow of chilled water in order to function properly, etc. Regardless of the pump system being designed, it is usually possible to arrive at a

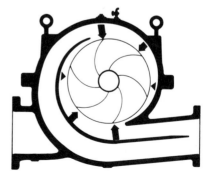

FIGURE 27.97 Double volute casing design has two cutwaters to reduce radial bearing loads below that developed in single volute designs. (Courtesy of Goulds Pumps, Inc., Seneca Falls, NY.)

design capacity for the pump. Sometimes there will be a planned duty cycle for the pump that will require it to operate at only a percentage of the full design capacity during certain periods of time. Examples of this include a process plant with a variable capacity, or a chilled water system designed to meet a variable air conditioning load. The operating duty cycle of the pump should be estimated, using the best available process and operations estimates, as this will help choose the best type of pump and control system.

In sizing a centrifugal pump, all of the components of system head must be added up to determine the pump TDH. The four separate components of system head are *Static Head*, *Friction Head*, *Pressure Head*, and *Velocity Head*. Each of these must be considered for the system in which the pump is to be operated, and the sum of these is the pump TDH.

Static Head. *Static Head* is the total elevation change which the liquid must undergo. It is measured from the surface of the liquid in the suction vessel to the surface of the liquid in the delivery vessel. The static head is measured from suction vessel surface to delivery vessel surface regardless of whether the pump is located above the liquid level in the suction vessel (an arrangement which is referred to as a *suction lift*), or below the liquid level in the suction vessel (called a *suction head*). For a pump in a closed loop system, the static head is zero.

Friction Head. *Friction Head* is the head necessary to overcome the frictional losses in the piping, fittings, and valves of the system in which the pump operates. Generally speaking, frictional losses in a piping system must be calculated by means of empirical formulae, taking into account such factors as liquid density and viscosity, and pipe diameter and material. Fortunately, these formulae have been reduced to tables and charts [Volk (1996)] which, though somewhat tedious, are nevertheless not too complex.

Pressure Head. *Pressure Head* is the head required to overcome a pressure or vacuum in the system upstream or downstream of the pump. If the pressure in the suction vessel and the pressure in the delivery vessel are identical (e.g., if both are atmospheric tanks) then there is no required pressure adjustment to TDH. There is also no pressure adjustment to TDH for a closed loop system.

If the suction vessel is under a vacuum, or under a pressure different from the delivery vessel, then there is a pressure head adjustment to TDH. In that case, the pressure or vacuum must be converted to feet of water. Pressure in units of pounds per square inch (psi) is converted to units of feet of liquid using the following formula

$$\text{Lb/sq. in. (psi)} = \frac{\text{Ft} \times \text{s.g.}}{2.31} \tag{27.215}$$

where s.g. = specific gravity. Vacuum, usually expressed in inches of mercury (" Hg), is converted to feet by the formula

$$\text{Vac. (Ft)} = \frac{\text{Vac. (" Hg)} \times 1.133}{\text{s.g.}} \tag{27.216}$$

Velocity Head. *Velocity Head* is the energy of a liquid as a result of its motion at some velocity, V. The formula for velocity head is

$$H_v = \frac{V^2}{2g} \qquad (27.217)$$

This value is usually found in the friction tables discussed above and is expressed in feet of head. The value of velocity head will be different at the suction and discharge of the pump, since the size of the pump suction flange is usually larger than the size of the discharge flange.

Note that the normal procedure in sizing centrifugal pumps measures the required pressure head at the surface of the suction and delivery vessel, as well as establishing static head values from these levels. In this situation, since velocity is zero at the surface of the supply and delivery vessels, velocity head at these points is zero. Velocity head is only included in the calculation of required pump total head when the pressure head requirements are given as gauge readings at the pump suction and discharge flanges.

In order to determine the velocity head component of TDH in those situations where it is appropriate, it is necessary to calculate the change in velocity head from pump suction to discharge. Since the net change in velocity from the pump suction to discharge is relatively small, the velocity head component of TDH is quite small, usually on the order of no more than one or two feet of water. With the TDH of many pumps being several hundred feet of water or more, many pump selectors choose to totally ignore the effect of velocity head, since the change of velocity head is often less than one percent of TDH. Ignoring velocity head is usually a safe assumption, given the other inaccuracies in sizing a pump; the one situation when velocity head cannot be ignored is when sizing a very low head pump, where velocity head may represent a more significant component of TDH.

Once the pump rating (capacity and TDH) have been determined, the next step in the selection process is to decide which pump speeds will be considered. It is quite often the case that two or more operating speeds may be commercially available for a particular pump rating and configuration. Each of these offered speeds will result in a different sized pump, each having different first costs, operating costs, and maintenance costs.

The head-capacity relationship for a constant speed centrifugal pump was depicted in Fig. 27.95, for a fixed impeller diameter. Most centrifugal pumps, however, have the capability to operate over an extended range of head and flow, by trimming, or cutting the impeller diameter from its maximum size down to some pre-determined minimum size. Thus, for a given pump speed, a centrifugal pump produces an envelope of head-capacity performance, as illustrated by the family of curves in Fig. 27.98. The upper and lower *H-Q* curves are based on the maximum and minimum impeller sizes available for that particular pump. The right and left boundaries of the performance envelope are the maximum and minimum flows recommended for each impeller diameter.

The convention for designating the size of a particular centrifugal pump is (*Discharge Size*) × (*Suction Size*) − *Maximum Impeller Diameter*, with all terms expressed in inches. So, a centrifugal pump with a three inch discharge flange, a four inch suction flange, and a maximum impeller diameter of 12 inches would be given

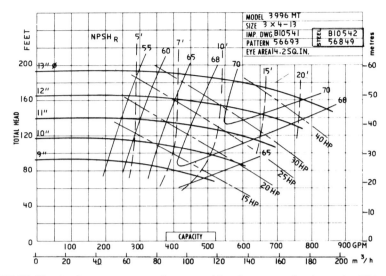

FIGURE 27.98 Performance curves for a centrifugal pump at fixed speed (1780 rpm) over its range of impeller diameters. (Courtesy of Goulds Pumps, Inc., Seneca Falls, NY.)

a size designation of 3 × 4-12 (which is read "three by four, twelve"). Note that the maximum impeller number is nominal. The actual maximum impeller for a pump designated a 3 × 4-12 may be 12-1/8″.

In addition to radial loads, centrifugal pumps also develop axial loads, referred to as *axial thrust*. These loads, caused by the differential pressure acting against the front and back faces of the impeller, must be carried by the pump's thrust bearing. Axial thrust varies with pump TDH, so is at a maximum when the pump is operating at its minimum flow.

27.7.3 Power and Efficiency

Power refers to the amount of power which must be supplied to operate a pump. There are two commonly designated expressions for horsepower. *Hydraulic Horsepower*, or WHP, refers to the output of the pump handling a liquid of a given specific gravity with a given capacity and TDH. The formula for WHP is

$$\text{WHP} = \frac{Q \times H \times \text{s.g.}}{3960} \tag{27.218}$$

where Q = flow rate in gpm, and H = TDH in feet of water.

Brake Horsepower, or BHP, is the actual amount of power which must be supplied to the pump from the motor in order to obtain the desired capacity and TDH. The formula for BHP is

$$\text{BHP} = \frac{Q \times H \times \text{s.g.}}{3960 \times \eta} \tag{27.219}$$

where η = pump efficiency. BHP is greater than WHP because of the fact that a pump is not a perfectly efficient machine, due to internal frictional losses and recirculation of a small portion of pumped fluid. The pump efficiency is expressed as a decimal number less than one, for example 0.75 for 75% efficiency. Best efficiencies for various types and sizes of centrifugal pumps can vary widely, over a range from less than 30% to over 90%.

Figure 27.98 shows typical centrifugal pump performance curves of BHP and efficiency as a function of pump capacity. Since BHP and efficiency vary with the impeller diameter as well as with the capacity of the pump, these curves are shown as lines of constant BHP (*iso-horsepower*) and efficiency. These curves must be interpolated (e.g., for choosing the required motor size) once the pump rating and impeller diameter are chosen.

The iso-horsepower data in Fig. 27.98 shows that with a typical centrifugal pump at a fixed impeller diameter and speed, BHP is rising as flow increases, although Sec. 27.7.5 below will illustrate that this isn't always the case. BHP data is based on a specific gravity of 1.0 unless otherwise indicated, and it must be adjusted if the liquid has a specific gravity other than 1.0. The pump efficiency curves in Fig. 27.98 show that efficiency varies with flow, rising to a peak value, known as the *Best Efficiency Point* (BEP). This is the ideal point for the pump to operate.

27.7.4 NPSH and Cavitation

The one set of curves in Fig. 27.98 which have not yet been discussed are the NPSH$_r$ curves. *Net Positive Suction Head* (NPSH) must be examined when using centrifugal pumps in order to predict the possibility of *cavitation*, a phenomenon which has both hydraulic and sometimes destructive mechanical effects on pumps. Cavitation is caused by the formation and subsequent collapse of vapor bubbles at a particular point along the flow path in a pump. For a more complete discussion of the causes and effects of cavitation in pumps, see Volk (1996). Also see Sec. 23.11 for coverage of cavitation in general.

Cavitation in a centrifugal pump produces several symptoms. First, the collapsing bubbles make a distinctive noise, which has been described as a sound like the pump is pumping gravel. This can be a nuisance, or even a health hazard in an extreme situation where a cavitating pump is operating where people are working. This physical symptom is usually the least area of concern with cavitation, however. Of far greater concern is the effect on the pump's hydraulic performance, and perhaps even the mechanical integrity of the pump. The hydraulic effect of a cavitating pump is that the pump performance drops off of its expected performance curve, producing a lower than expected head and flow.

An even more serious effect of cavitation is the mechanical damage which can occur due to the shock wave that is created when the vapor bubbles collapse. This shock wave can physically damage the impeller, causing the removal of material from the surface of the impeller. This has the effect of upsetting the dynamic balance of this rotating component thus causing the pump to vibrate excessively, which in turn causes a likely failure of the pump seal and/or bearings. Excessive vibration from cavitation can occur even without the removal of material from the impeller, but rather from the uneven loading of the impeller by the cavitation shock waves. Vibration is the most likely failure mode of a cavitating pump and the reason why

NPSH and cavitation must be properly understood by the system designer and pump user.

In analyzing a pump operating in a system to determine if cavitation is likely, there are two aspects of NPSH which must be considered, $NPSH_a$ and $NPSH_r$, which will be defined below. In order for a pump to be operating free of cavitation, $NPSH_a$ must be greater than $NPSH_r$. So, in determining the acceptability of a particular pump operating in a particular system with regard to NPSH, the $NPSH_a$ for the system must be calculated by the system designer or pump user at the planned pump capacity. This calculated value must then be compared with the $NPSH_r$ for the pump to be used, at the same capacity value, by looking at this information on the pump performance curve (Fig. 27.98).

The accepted convention is that both $NPSH_a$ and $NPSH_r$ are given at the centerline of the pump impeller. To be conservative, the minimum value of $NPSH_a$ should be calculated.

NPSH_a. *Net Positive Suction Head Available* is the suction head present at the pump suction flange over and above the vapor pressure of the liquid. $NPSH_a$ is a function of the suction system and is independent of the type of pump in the system. The formula for calculating $NPSH_a$ is

$$NPSH_a = P \pm H - H_f - V_{vp} \qquad (27.220)$$

where:

P = the absolute pressure on the surface of the fluid in the suction vessel, expressed in feet of fluid.

H = the static distance from the surface of the fluid in the supply vessel to the centerline of the pump impeller, expressed in feet. (The term is positive if the pump has a static suction head, negative if the pump has a static suction lift.)

H_f = the friction loss in the suction line, including all piping, valves, fittings, filters, etc., expressed in feet of fluid. (Note that this term varies with flow, so $NPSH_a$ must be based on a particular flow rate, usually the maximum flow), and

H_{vp} = vapor pressure of the fluid at the pumping temperature, expressed in feet of fluid.

NPSH_r. *Net Positive Suction Head Required* is the suction head required at the pump suction flange over and above the vapor pressure of the fluid. $NPSH_r$ is strictly a function of the pump inlet design and is independent of the suction piping system. The pump requires a pressure at the suction flange greater than the vapor pressure of the fluid, because merely getting the fluid to the pump suction flange in a liquid state isn't enough. The fluid experiences pressure losses when it first enters the pump, before it gets to the point on the impeller vane when pressure begins to increase. These losses are caused by frictional effects as the fluid passes through the pump suction nozzle, across the impeller inlet, and changes direction to begin to flow along the impeller vanes. The $NPSH_r$ for a particular pump is shown on the pump performance curve (Fig. 27.98) using lines of constant $NPSH_r$, similar to the way that BHP is shown.

27.7.5 Specific Speed

Specific speed, N_s, is a dimensionless index, primarily used by pump designers, to describe the geometry of pump impellers and to classify them by type. The formula for specific speed is

$$N_s = \frac{N \times \sqrt{Q}}{H^{3/4}}\tag{27.221}$$

where:

  N = pump speed in rpm
  Q = capacity at BEP, full diameter, in gpm, and
  H = pump total head per stage at BEP, full diameter, in feet.

Figure 27.99 illustrates the range of specific speeds commonly found with centrifugal pumps. Low specific speed impellers (N_s in the range of about 300 to 2,000) are called *radial flow* impellers. These impeller types produce relatively low flows and high heads, relatively flat *H-Q* curves, and horsepower curves which rise as capacity increases. This is the most common specific speed range for process pumps. High specific speed impellers (N_s between roughly 8,000 and 20,000) are called *axial flow* or *propeller* pumps. These impeller types produce relatively high flows and low heads, very steep *H-Q* curves, and horsepower curves which decrease as capacity increases. Impellers in between these two ranges are called *mixed flow impellers.* They have performance characteristics and curve shapes between the other two types.

27.7.6 Affinity Laws

The *pump affinity laws* are rules which govern the performance of a centrifugal pump when speed or impeller diameter are changed. The basis for the derivation of the affinity laws is that a pump's specific speed, once calculated, does not change. So, if the performance of a pump at one speed and impeller diameter are known, it is possible to predict the performance of the same pump if the pump's speed or impeller diameter are changed. There are two sets of pump affinity laws. With constant impeller diameter, D, the first set of laws is

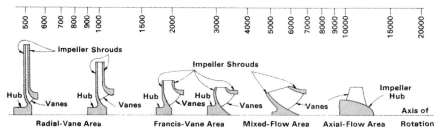

FIGURE 27.99 Specific speed (N_s) range for centrifugal pumps. (Courtesy of the Hydraulic Institute.)

$$\frac{Q_1}{Q_2} = \frac{N_1}{N_2} \tag{27.222}$$

$$\frac{H_1}{H_2} = \left(\frac{N_1}{N_2}\right)^2 \tag{27.223}$$

$$\frac{BHP_1}{BHP_2} = \left(\frac{N_1}{N_2}\right)^3 \tag{27.224}$$

with constant speed, N, the second set of laws is

$$\frac{Q_1}{Q_2} = \frac{D_1}{D_2} \tag{27.225}$$

$$\frac{H_1}{H_2} = \left(\frac{D_1}{D_2}\right)^2 \tag{27.226}$$

$$\frac{BHP_1}{BHP_2} = \left(\frac{D_1}{D_2}\right)^3 \tag{27.227}$$

27.7.7 System Head Curves

The system head curve is a plot of the head requirement of a system as it varies with flow. Figure 27.100 illustrates a system head curve plotted on the same scale as the H-Q curve for the pump operating in the system. The system head curve is developed by computing the total head required by the system at several arbitrarily selected values of flow. The total head required by the system is composed of static components (e.g., elevation changes, pressure in delivery vessels), and frictional components which vary with flow through the system (e.g., friction losses through pipe

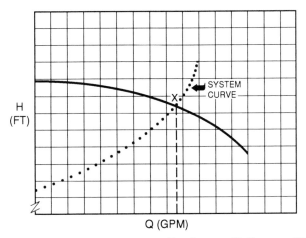

FIGURE 27.100 System head curve and centrifugal pump H-Q curve. (From Volk, 1996, courtesy of Marcel Dekker, New York.)

and fittings). The pump seeks the point on its head-capacity curve where it delivers the head required by the system. So, the pump operates at the point of intersection of the pump H-Q curve and the system head curve [see Volk (1996)].

The system head curve describing a particular fluid piping arrangement may change shape over time due to a variety of circumstances. Changes in the static component of the system head curve might occur due to changes in the level of either the supply vessel or the delivery vessel, or due to variations in pressure in either vessel. Changes in the frictional component of the system head curve might be due to several causes, including the build-up of solids on a filter in the system, the throttling of a valve in the system, or the gradual build-up of corrosion products on the inside of system piping.

It is important for the system designer to produce and evaluate system head curves, particularly if the system is complex or has many expected variations in flow or fluid properties over time. This is necessary in order to establish the proper settings required on any system control valves, to insure that the pump is always operating at a point on its H-Q curve that is healthy for the pump, and to operate where $NPSH_a$ and motor horsepower are adequate.

27.7.8 Multiple Pump Operation

Multiple pumps may be installed in a single fluid piping system, arranged either in parallel or series. When parallel or series pumps are being considered for a system design, the pumps must be carefully matched to each other and to the system to insure that the pumps are always operating at a healthy point on their H-Q curves, and to insure that the system is such that true benefits are achieved from the parallel or series pumping arrangement. This is not always the case, depending on the shape of the system head curve. For a more detailed discussion of parallel and series pumping arrangements [see Volk (1996)].

Parallel operation of pumps is illustrated in Fig. 27.101, where one pump, some of the pumps, or all of the pumps may be operated at the same time. The primary purpose of operating pumps in parallel is to allow a wider range of flow than would be possible with a single, fixed-speed pump, for systems with widely varying flow demand. This may also allow the use of more commercially available and/or more economical pump selections. Examples of applications for parallel pumping include municipal water supply and wastewater pumps, HVAC system chilled water pumps,

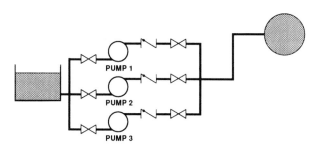

FIGURE 27.101   Parallel pump operation. (From Volk, 1996, courtesy of Marcel Dekker, New York.)

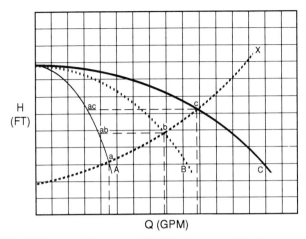

Q (GPM)

FIGURE 27.102 *H-Q* curve for a centrifugal pump, combined curves for two or three identical such pumps operating in parallel, and system head curve. (From Volk, 1996, courtesy of Marcel Dekker, New York.)

and main process pumps in a variable capacity process plant. Usually there are no more than three or four pumps operating in parallel.

In order to analyze the parallel pumping arrangement, it is necessary to construct the system head curve, as explained in Sec. 27.7.7 above. Then, a combined pump curve must be developed depicting the head-flow relationship for the pumps while pumping in parallel. The combined pump curve for parallel operation of two or three identical pumps is shown in Fig. 27.102, along with the curve for single pump operation and the system head curve. The combined curve for two pump operation is developed by doubling the capacity at arbitrary values of head. (For nonidentical pumps, the procedure involves adding together the capacity of the two pumps at the same head.) Once these curves are constructed, the same rule for total system flow as discussed in Sec. 27.7.7 applies, i.e., that the total flow through the system is represented by the intersection of the system head curve with the combined pump curve.

Series operation of pumps is illustrated in Fig. 27.103. In series operation, the discharge of one pump feeds the suction of a second pump. When two or more pumps are operated in series, the flow through all of the pumps is equal, since

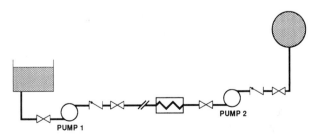

FIGURE 27.103 Series pump operation. (From Volk, 1996, courtesy of Marcel Dekker, New York.)

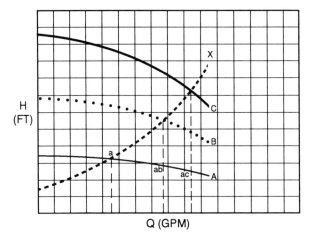

Q (GPM)

FIGURE 27.104 *H-Q* curve for a centrifugal pump, combined curves for two or three identical such pumps operating in series, and system head curve. (From Volk, 1996, courtesy of Marcel Dekker, New York.)

whatever flows though one pump has to flow through the next pump in series. Series operation of pumps has several purposes. These include insurance that commercially available equipment can be used in a particular system, reduction of overall system costs, reduction of the required design pressure for the system piping, valves and instrumentation, allowance of a variation of flow through the system by varying the number of pumps which are pumping at any one time, and insurance of adequate available NPSH for the second pump.

As with parallel pumping arrangements, series pumping arrangements are analyzed by overlaying the system head curve with a combined curve for the two pumps. Figure 27.104 illustrates a pump *H-Q* curve and a combined curve for two or three identical such pumps in series. The combined curve for two pump operation is developed by doubling the head at arbitrary values of flow (or adding together the heads of two nonidentical pumps at the same value of flow).

27.7.9 Pump Configuration Types

Figure 27.94 illustrates the most common type of centrifugal pump configuration, the *end suction* configuration. With this pump type, the impeller is cantilevered (i.e., supported by bearings on only one side of the impeller). Smaller sizes of this pump type, in a version known as *close-coupled*, may have the impeller and casing directly coupled to a C-Face motor, with no separate pump bearing system or coupling. In the *frame mounted* arrangement, the pump has an integral bearing frame with a radial and thrust bearing, is connected to a foot mounted motor by a coupling, and is mounted with the motor on a bedplate.

Besides end suction pumps, there are numerous other pump configurations, each with distinct characteristics as to capacity and head range, temperature limits and liquid types which can be accommodated, suction limitations, commercially available materials of construction, installed cost of the equipment, and costs of operation and maintenance. Some of these configurations include *inline, self-priming, split case double suction, multistage, submersible,* and *vertical turbine* pump types. Details of these and other configuration types can be found in Volk (1996).

27.7.10 Centrifugal Pump Technology Trends

For the most part, the fundamentals of centrifugal pump design have remained unchanged for the past 50 years. The basic designs of pump impellers, volutes, and diffusers, the shapes of pump performance curves, and the characteristics of pumps operating in systems are little different from these same parameters as they appeared in pumps and systems half a century ago. This is not meant to imply that nothing new has happened in pump technology in the past fifty years. There have been a great many achievements in manufacturing techniques resulting in higher quality and more durable pumps. Improvements in these areas include: better metal casting techniques such as investment casting, producing cast parts of greater integrity and requiring less machining and repair, and more precise machine tools and techniques, producing more accurate fits and closer tolerances. Other relatively recent advancements in pump technology, described in more detail in Volk (1996), are discussed below.

Materials of Construction. Many improvements in pump reliability and ability to handle highly corrosive and abrasive fluids have been achieved through the development of superior materials, including both metal alloys and nonmetallic materials. Engineered plastics such as PVC, CPVC, polypropylene, PVDF, and PTFE are being used more and more in pump designs for their economy and performance.

Sealing Systems and Sealless Pumps. Many new mechanical seal configurations, materials, and features have been developed in recent years. Advancements in seal design allow higher pressures and temperature limits, eliminate or greatly reduce product leakage through the use of multiple sealing systems, extend seal life, reduce maintenance costs, and allow monitoring and control of seal operating conditions.

The need for sealless pumping continues to grow as more and more liquids are put into the category for which zero leakage is accepted, and as the search for greater pump reliability attempts to eliminate one of the leading contributors to pump downtime, the pump sealing system. The two major types of sealless centrifugal pumps, *magnetic drive* and *canned motor* pumps, are examined in detail in Volk (1996). Both of these design types completely eliminate the shaft penetration of the casing.

Neither the magnetic drive nor the canned motor design type is a panacea. The major *Achilles heel* for both design types is the fact that radial and thrust loads generated by the pump must be accommodated by sleeve bearings and thrust plates which are ordinarily exposed to the liquid being pumped. Other shortcomings include upper limits of viscosity, concerns for inadvertent dry running (especially for magnetic drive pumps), special motors required for canned motor pumps, complication of design compared to sealed pumps, and the resulting high level of maintenance and reduced reliability that these limitations suggest. New designs and materials are being developed to address these limitations.

Variable Speed Operation. Operating a centrifugal pump at varying speeds to achieve a range of operating conditions has been shown to be a highly effective way to reduce total pumping costs for systems which require a wide range of flow, as an alternative to throttling of a single pump or the use of multiple pumps in the system. The technology for achieving variable speed operation of pumps has evolved in the past ten years. The use of *variable frequency drives* (*VFD*'s) to achieve this end is considered by many to be the best alternative for the broadest range of pump applications. The advantages of using VFD's to achieve variable speed pumping include energy savings, the capability to retrofit existing equipment without buying new motors, the high turndown or speed change achievable without serious energy loss consequences, the lower bearing loads and overall better health of the pump due to operating at or near the pump's BEP, and, finally, the ability to start the pump at reduced speed to achieve even more savings in power costs.

Computer Software Analysis of Piping Networks. For complicated systems, manual computation of piping losses and pump TDH can be tedious and time consuming. This is particularly true if the system contains multiple flow branches or if multiple pumps are used in the system. Furthermore, if the pumped liquid is something other than water, tables to determine friction losses through the pipes, valves, and fittings may not be readily available. The same dilemma may exist if the system piping is something other than standard pipe size and material.

Because of these limitations, many system designers and engineers must make their best approximations of pump head in the situations described above. This usually results in pumps being oversized, with the final adjustment to make the pump operate at the desired flow rate being done with control valves or orifices. Oversizing pumps results in a more expensive pump, a sizeable waste of energy, and an increase in operating costs. Fortunately, technology has provided a solution to this dilemma. There are several computer software packages available which greatly reduce the time it takes to design a piping system, and improve the capability to optimize the design of the system and to select the right pump. The software packages currently available on the market [Volk (1996)] vary widely as to their ease of use, their capability to model complex systems, their flexibility to handle a wide range of fluids and pipe materials, and the ease with which they allow the designer to modify and optimize the system design and pump selection.

Precision Alignment Techniques. Lack of proper alignment may very well be the single most important cause of premature pump failure. Like many other activities, the degree of pump alignment is not a specific point, but rather a spectrum, with more precise alignment generally being reserved for larger, higher speed, more expensive, and more critical equipment. To achieve the full range of the possible spectrum of alignment accuracies, many approaches can be used, ranging from very simple and rudimentary alignment using straight edges and feeler gauges, to the use of recently developed laser alignment equipment at the other end of the spectrum. In between these two extremes of the spectrum are other alignment techniques, such as single dial indicator (rim and face) alignment, reverse indicator alignment, and alignment using electronic gauges.

27.8 STABILITY IN PUMPS AND COMPRESSORS
Edward M. Greitzer

27.8.1 Introduction

This section discusses the instabilities that can occur in systems in which one moves a fluid through pipes, ducts, etc., by means of some type of turbomachine. The term *stability* can be defined in a general manner with respect to the equilibrium operating point of such a system. An operating point is stable if, when the system is disturbed slightly, it tends to return to the equilibrium point or at least does not keep moving further away from it. For the most part, only overall guidelines can be given as to when instability is to be expected as well as for methods to enhance stability, since, in many cases, predictive methods have not yet been developed for the complex situations typical of engineering practice.

To illustrate the basic ideas, consider the simple pumping system shown in Fig. 27.105. An incompressible fluid is pumped from a large, constant pressure reservoir through a closed tank which contains a compressible gas (air) and then through a throttle valve into another large reservoir. The second reservoir does not necessarily have to be at the same pressure level as the first, however, in this example it is taken as such. If end effects are neglected, the inlet of the pump (the *Inlet* station) and the exit of the throttle (the *Exit* station) can also be assumed to be at this pressure, so that only the parts of the system within the dotted control surface need be discussed. The essential elements of the system are therefore the pump, the *compliance* (or mass storage capability) of the closed volume, the throttle (which controls the system flow rate), and the *inertance* of the fluid in the inlet and exit lines.

The steady-state system operating point is set by two conditions, namely that the flow through the pump and the flow through the throttle are the same and that the pressure rise through the pump is equal to the pressure drop due to the system resistance. These conditions imply that the steady-state operating point is at the intersection of the pumping characteristic and the throttle (or system resistance) curves.

Consider the stability of an arbitrary steady-state operating point. With reference

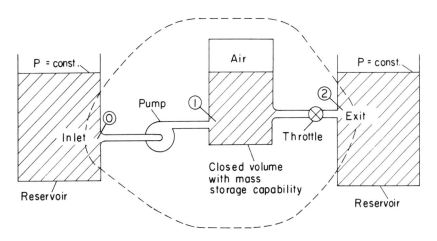

FIGURE 27.105   A basic pumping system.

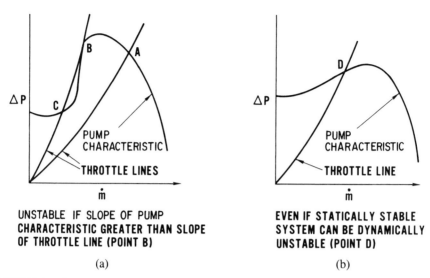

UNSTABLE IF SLOPE OF PUMP
CHARACTERISTIC GREATER THAN SLOPE
OF THROTTLE LINE (POINT B)

(a)

EVEN IF STATICALLY STABLE
SYSTEM CAN BE DYNAMICALLY
UNSTABLE (POINT D)

(b)

FIGURE 27.106   Static and dynamic system instabilities. (a) Static instability and (b) dynamic instability.

to the left-hand side of Fig. 27.106, the effect of a small perturbation in mass flow (a decrease, say) at operating point A is to cause the pressure drop across the throttle to be smaller than that produced by the pump. The resulting pressure imbalance causes fluid accelerations that return the system to operation at the initial point, so that point A is stable. This is true for all points to the right of point A or for points between A and B. At point B, however, where the throttle line is tangent to the pumping characteristic, the pressure forces that arise due to a small decrease in mass flow will cause the system to depart further from the initial operating point, so point B is an unstable operating point. This is the most basic of pumping system stability criteria. The system will become unstable if the slope of the pump (or compressor) pressure rise curve is steeper than the slope of the throttle curve.

This criterion is, however, too simple to describe many of the real phenomena that are observed in pumping systems, since it only considers the *static stability* of the system (static instability is associated with pure divergence from the initial point). It is often the criteria for *dynamic stability* (dynamic instability leads to growing *oscillatory* motion about the initial point) which are violated first and which are most important in practice.

Static instabilities can be inferred from viewing the transient performance of the system as a sequence of quasi-steady states. Hence, knowledge of the steady state pump characteristics and throttle lines is enough to define the stability. In the prediction of dynamic instability, on the other hand, parameters such as system inertances and capacitances must be included, since they play an essential role in determining the transient response of the system to disturbances. Thus, knowledge of steady state performance curves alone is not sufficient for prediction of dynamic instability, and additional information about quantities such as volumes, duct lengths, etc., must also be included. As shown on the right-hand side of Fig. 27.106, the point to be emphasized is that a pumping system can be statically stable and still exhibit (dynamic) instability. The terms *dynamic* and *static instability* can be made more quantitative by the following illustration. Consider a simple, second-order system described by the equation

$$\frac{d^2x}{dt^2} + 2\alpha \frac{dx}{dt} + \beta x = 0 \tag{27.228}$$

where α and β are constants of the system. The transient response of the system to an initial perturbation is given by

$$x = A \exp\left(\{-\alpha + \sqrt{\alpha^2 - \beta}\}t\right) + B \exp\left(\{-\alpha - \sqrt{\alpha^2 - \beta}\}t\right) \tag{27.229}$$

where the constants A and B are determined by the initial conditions. If $\beta > \alpha^2$, the condition for instability is simply $\alpha < 0$, which corresponds to oscillations of exponentially growing amplitude. Instability will also occur if $\beta < 0$, independent of the value of α, however, in this case, the exponential growth is nonoscillatory. It is usual to denote these two types of instability as dynamic and static, respectively. Static stability is a necessary but not sufficient condition for dynamic stability.

The dynamic stability of the system shown in Fig. 27.105 can be analyzed using a lumped parameter model [Greitzer (1981)]. In this, all the kinetic energy of the unsteady flow in the system is associated with the flow in the pump and throttle lines, and all the potential energy associated with the system transients is taken to arise from the expansion and compression of the fluid in the storage volume. The system is thus viewed essentially as a *Helmholtz Resonator*. The mass-spring-damper mechanical analogue of such a model is illustrated schematically in Fig. 27.107, where the essential components of the simple pumping system are shown. As is indicated, a key feature due to the pump is the ability to provide a negative damping (i.e., a net input of mechanical energy) to the system transients.

Examining the response of this system to small perturbations about a given operating point, one finds that there are two conditions for *stability* [Greitzer (1981)]

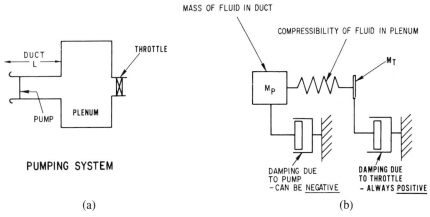

FIGURE 27.107 Mechanical analogue of simple pumping system. (a) Pumping system and (b) mass/spring/damper system.

$$\left(\overline{\frac{d\Delta p_p}{d\dot{m}}}\right) < \frac{\gamma p_1}{\rho V_a} \frac{L}{A} \frac{1}{\left(\overline{\frac{d\Delta p_T}{d\dot{m}}}\right)} \tag{27.230a}$$

or

$$\left(\overline{\frac{d\Delta p_p}{d\dot{m}}}\right) < \left(\overline{\frac{d\Delta p_T}{d\dot{m}}}\right) \tag{27.230b}$$

where γ is the ratio of specific heats, ρ is the density, V_a is the air volume, L is the length, and A is the flow-through area. In this expression, the quantities $(\overline{d\Delta p_p/d\dot{m}})$ and $(\overline{d\Delta p_T/d\dot{m}})$ are the slopes of the compressor and throttle (system resistance) curves, evaluated at the mean given operating point. In general the condition for dynamic stability, Eq. (27.230a), is the more critical, and, since throttle slopes are often steep and volumes large, it may occur very near the peak of the pump characteristic curve.

A similar analysis can also be carried out for a compressor (with pressure rise small compared to ambient) operating in a gas. In this case, the capacity for mass storage in the volume is due to the compressibility of the flowing medium, but the analysis is not changed materially. The condition for static stability, Eq. (27.230b), is unchanged, however the condition for dynamic stability, Eq. (27.230a), is now

$$\left(\overline{\frac{d\Delta p_p}{d\dot{m}}}\right) < \frac{a^2}{V_a} \frac{L}{A} \frac{1}{\left(\overline{\frac{d\Delta p_T}{d\dot{m}}}\right)} \tag{27.231}$$

where a is the sound speed. Note also that for small oscillations, the *natural frequency* of such a system will scale with the Helmholtz Resonator frequency, i.e.,

$$\omega = a \sqrt{\frac{A}{V_a L}} \tag{27.232}$$

27.8.2 Physical Mechanism for Dynamic Instability

The physical mechanism associated with dynamic instability can be understood by considering the system undergoing oscillations about a mean operating point. Since the flow through the throttle is dissipative, there must be energy put into the system to maintain the oscillation (or to increase its amplitude in the case of instability). The only source of this energy is the pump. The mass flow and pressure rise perturbations through the pump are shown in Fig. 27.103 which presents the *perturbations* in mass flow, $\delta\dot{m}$, and pressure rise through the pump, $\delta\Delta p$, plotted versus time over a period of one cycle (assuming that the pump responds quasi-steadily to fluctuations in mass flow), as well as the product of the two, $(\delta\dot{m}) \times (\delta\Delta p)$, whose integral over a cycle is equal to the net *excess* (over the steady-state value) of the rate of production of mechanical energy.

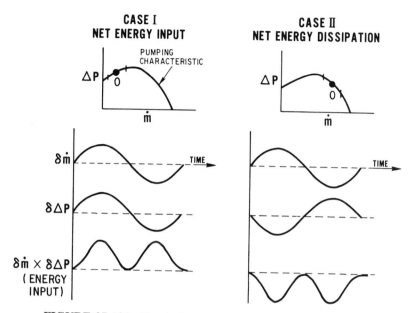

FIGURE 27.108 Physical mechanism for dynamic instability.

In the case of a positive slope, it is seen that a favorable condition for energy addition occurs, since high mass flow rate and high rate of mechanical energy addition (in the form of pressure rise) go together. The net amount of mechanical energy that the pump puts into the flow will thus be higher than if the system were in steady operation at the mean flow rate. In a similar fashion the net dissipation due to the throttle will also be higher than if the system were in steady operation. When the net energy input over a cycle balances the dissipation, a periodic oscillation can be maintained. This corresponds to the boundary between stability and instability. For an operating point on the negatively sloped region, as shown in the right-hand side of Fig. 27.108, the pump adds less mechanical energy over a cycle than in steady operation, because high mass flow is associated with low energy input. Perturbations will therefore decay and the operating point will be stable. Dynamic instability for this system can thus occur when the mechanical energy input from the pump is greater than during a mean (steady) flow (i.e., when there is the required amount of negative damping), and this will be true only if the pump characteristic is positively sloped so that high mass flow and high mechanical energy input per unit mass flow go together.

27.8.3 Rotating Stall

The system instabilities that have been described will occur when the slope of the compressor or pump characteristic becomes positive to some degree. In the systems that were discussed above, this peaking of the characteristic is due in general to the presence of stall. Looked at from the point of view of the individual diffusing passages in the compressor, stall generally implies separation of the flow from one or more of the passage walls. However, compressor blade rows consist of many of these diffusing passages in parallel, so that phenomena can occur which do not happen with a single airfoil or diffusing passage.

One of the most notable of these is *rotating stall*. This is a flow regime in which one or more *stall cells* propagate around the circumference of the compressor with a constant rotational speed, which is generally between fifteen and seventy percent of the rotor speed. Rotating stall can occur in axial and centrifugal compressors and pumps. In the cells, the blades are very severely stalled; typically, there is negligible net through-flow, with areas of local reverse flow in these regions. The cells can range from covering only part of the span (either at the root or at the tip) and being only a few blades in angular width, to covering the full span and extending over more than 180 degrees of the compressor annulus. This latter situation commonly occurs in multistage axial compressors at speeds near design [Cumpsty (1989)].

A basic explanation of the mechanism associated with the onset of stall propagation, originally given by Emmons, can be summarized as follows. Consider a row of axial compressor blades operating at a high angle of attack, such as shown on the left-hand side of Fig. 27.109. Suppose that there is a nonuniformity in the inlet flow such that a locally higher angle of attack is produced on blade *B* which is enough to stall it. If this happens, the flow can separate from the suction surface of the blade so that a substantial flow blockage occurs in the channel between *B* and *C*. This blockage causes a diversion of the inlet flow away from blade *B* and towards *C* and *A* to occur (as shown by the arrows), resulting in an increased angle of attack on *C* and a reduced angle of attack on *A*. Since *C* was on the verge of stall before, it will now tend to stall, whereas the reduced angle of attack on *A* will inhibit its tendency to stall. The stall will thus propagate along the blade row in the direction shown, and, under suitable conditions, it can grow to a fully developed cell covering half of the flow annulus or more, as shown on the right-hand side of Fig. 27.109. In this fully developed regime, the flow at any local position is quite unsteady, although the annulus averaged mass flow is steady with the stall cells serving only to redistribute this flow.

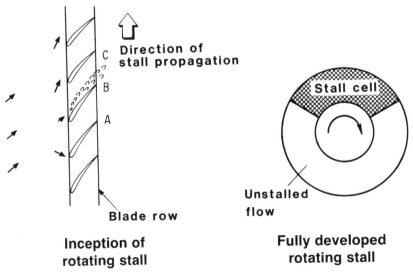

FIGURE 27.109   Rotating stall.

The onset of rotating stall is associated with an instability which arises due to the stall of the compressor blade passages. (The onset of stall can, however, be affected by other closely coupled components in the system as described below.) As far as the overall system is concerned, this can be regarded as a localized instability. However, there is also a more global system instability [as described in Greitzer (1981) or Cumpsty (1989)] which leads to *surge*. In contrast to the behavior during rotating stall, the annulus averaged mass flow and system pressure rise undergo large amplitude oscillations during surge. The frequencies of these oscillations are generally at least an order of mgnitude below those associated with passage of a rotating stall cell and depend on the system parameters. During the surge cycles, the instantaneous mass flow through the compressor changes from values at which (in steady state operation) the compressor would be free from stall, to values at which one would find rotating stall or totally reversed flow. Because of the low frequency of the oscillations, the compressor may pass in and out of these flow regimes in an approximately quasi-steady manner.

The two phenomena (surge and rotating stall) are thus quite distinct. However, they are not unrelated, since often the occurrence of the local instability (associated with the onset of rotating stall) can *trigger* the more global type of system instability leading to surge. It is therefore necessary to consider the possibility for both types of instability and develop methods for their prediction.

27.8.4 Prediction of Stall Onset in Axial Compressors

The prediction of the point at which stall occurs has been attacked by many investigators at quite different levels of approach. The most empirical are the correlations that have been developed for stall onset. The basic concept is to find a parameter (or parameters) which correlates the *stall point* (defined here as the condition at which the steady, axisymmetric flow becomes unstable) for a number of different blade geometries, compressor designs, etc. In a design procedure for a low hub tip ratio fan, for example, the parameter could be applied at different span locations along the blading, using the local flow conditions generated by use of one of the many axisymmetric compressor flow field calculations, to see whether any section would be operating under too adverse a condition, while for a multistage compressor the parameter might be applied only on a *meanline* or averaged basis.

One well-known example of this type of approach, which is still much in use, is the work of Lieblein (1965). He developed a parameter which he called the *diffusion factor* (or D-factor) defined as $D = 1 - W_2/W_1 + \Delta W_\theta/2\sigma W_1$, where W_1 is the inlet relative velocity, W_2 is the exit relative velocity, ΔW_θ is the change in circumferential velocity component, and σ is the solidity. This parameter, D, is related to the adverse pressure gradient on the suction surface of the airfoils.

It is found that the total pressure loss correlates quite well with D, and, based on Lieblein's cascade results, one can see a rather sharp rise in loss occur as D is increased past a value of roughly 0.6. This can, therefore, be taken as an approximate criterion for the onset of *airfoil* stall in a cascade. Although much of the work done by Lieblein was based on two-dimensional cascades, the use of the D-factor has been carried over to axial as well as to centrifugal compressors [Rodgers (1978)]. Features such as the differences between the flows in a cascade and the flows at the tips of axial compressor rotors, for example, are recognized by noting that different limiting values of the D-factor are used for the rotor tips than for other sections.

The diffusion factor, however, is really a measure of the tendency of the boundary layers on the compressor *airfoil* to separate and does not account for the fact that it is very often the flow in the annulus *endwall* regions which is the central cause of rotating stall onset. In order to approach this, it has been found useful to view the blade passages in another way, namely as a diffuser. A method of doing this is described by Koch (1981) who has examined data from over fifty compressor builds to obtain a correlation for the peak (stalling) pressure rise capability of axial compressors.

The results of the correlation are presented as an adjusted stage average peak pressure rise coefficient, $\Delta P/(1/2)\rho W_1^2$, versus the nondimensional passage length, L/g_2. This latter parameter, g_2, is essentially the chord length divided by the *staggered pitch* (the pitch $\times$ cos (β_2), where β_2 is the exit metal angle of the blade). Adjustments in the pressure rise coefficient are made due to Reynolds number effects, tip clearance effects, blade stagger angle (essentially inlet boundary layer skew), and axial spacing. To account for the third of these, an effective dynamic pressure, $\mathcal{F}_{ef}$, at blade inlet is defined. This is shown in Fig. 27.110 which is for a *stator*. Similar considerations are used for rotors. Reynolds number effects on the peak pressure rise are illustrated in Fig. 27.111.

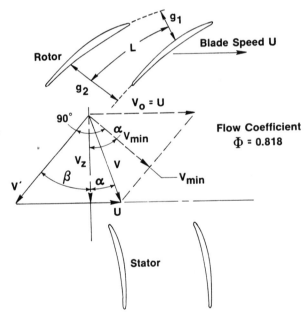

For a Stator:

$$\mathcal{F}_{ef} = \frac{V^2_{ef}}{V^2} = \left(V^2 + 2.5 \quad V^2_{min} + 0.5 \quad V_o{}^2\right)\Big/ 4.0\, V^2$$

$$\frac{V^2_{min}}{V^2} = \sin^2\left(\alpha + \beta\right), \text{ if } \left(\alpha + \beta\right) \leq 90° \text{ and } \beta \geq 0°$$

FIGURE 27.110 Effective dynamic pressure. (Reprinted with permission of ASME, Koch, C. C., "Stalling Pressure Rise Capability of Axial Flow Compressor Stages," *ASME J. of Engineering for Power*, Vol. 103, pp. 645–656, Oct. 1981.)

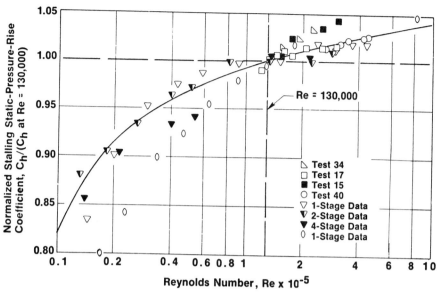

FIGURE 27.111   Effect of Reynolds number on stalling pressure rise coefficient. (Reprinted with permission of ASME, Koch, C. C., "Stalling Pressure Rise Capability of Axial Flow Compressor Stages," *ASME J. of Engineering for Power*, Vol. 103, pp. 645–656, Oct. 1981; reference numbers shown in the figure are those of the original publication.)

Although this correlation has been derived from multistage machines with near *equilibrium* endwall layers, it has also been applied with good success to single stages. The basic correlation is shown in Fig. 27.112 which presents the adjusted peak pressure rise achievable in an axial compressor row plotted versus the nondimensional length parameter, L/g_2. Also shown is a curve from a two dimensional diffuser correlation by Sovran and Klomp (1967). All data has been corrected to a Reynolds number of 130,000, a tip clearance/staggered gap ratio of 5.5%, and an axial gap of 38% of stator chord. The main point is that it is generally unlikely that a pressure rise above the curve can be achieved in a fan or compressor. For further details on the basis of the correlation (e.g., the influence of tip clearance on peak pressure rise) and its use, one can refer to Koch (1981).

27.8.5   Analyses of Compressor/Pump Instability Onset

Stability analyses have also been carried out for prediction of compressor rotating stall [Longley (1993)], but in general these have not been as quantitatively precise as desired. As an example, let us consider one of the best known of these criteria which states that rotating stall inception will occur at the *zero slope point* of the exit static pressure minus inlet total pressure compressor characteristic. This criterion has been applied with some success and does appear to furnish a rough *rule of thumb*. However, counter examples in which it does not hold can also readily be found. An illustration of this, Fig. 27.113, shows data from a sample of low-speed, multistage compressors. In this figure, the horizontal axis is axial velocity parameter, $\phi = C_x/U$, (C_x is the axial velocity) and the vertical axis is ψ_{TS}, the nondimensional total-to-static pressure rise, $\psi_{TS} = (p_{exit} - p_{0, inlet})/(\rho U^2)$.

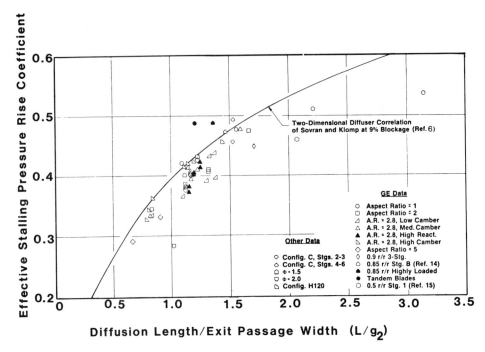

FIGURE 27.112 Correlation of stalling effective static pressure rise coefficient for low speed stages. (Reprinted with permission of ASME, Koch, C. C., "Stalling Pressure Rise Capability of Axial Flow Compressor Stages," *ASME J. of Engineering for Power*, Vol. 103, pp. 645–656, Oct. 1981; references numbers are those of original publication.)

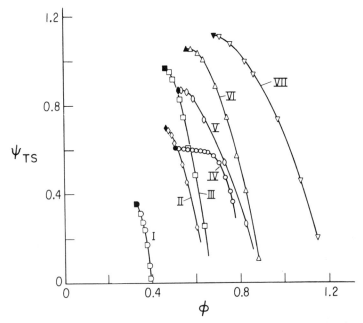

FIGURE 27.113 Multistage compressor performance (low speed rigs).

Curves V, VI, and VII do appear to show approximately zero slope (within the accuracy of the data), but curves I–IV have a negative slope right up to the stall points. This situation, where there does not appear to be a zero slope region of the compressor characteristic, is even more apparent in high-speed, multistage compressor data. The use of this criteria in any quantitative manner thus should be treated with some caution.

Analytical methods also exist for predicting the onset of the more global type of instability leading to surge. This is a system phenomenon and one in which, in contrast to rotating stall, the compressor appears to participate in a roughly one-dimensional manner. The calculation of the system stability point is based on using analyses of the type described in Sec. 27.8.1 and leads to a situation where the critical parameters are the slopes of the steady-state (uniform flow) compressor speed lines at the stall point. In general, however, these are not known *a priori* with the desired accuracy.

The relation between the two types of instability can be viewed as follows. First, system instability is a basically one-dimensional phenomenon, involving an overall, annulus averaged (in some sense) compressor performance curve. However, the flow through the compressor can also be locally unstable and exhibit rotating stall. The onset of rotating stall is often associated with a precipitous drop in the overall pressure-rise-mass-flow curve of compressor performance and can thus lead to a situation where the instantaneous compressor operating point is on a steeply positively sloped part of the characteristic. In this sense, the onset of the local compressor instability can trigger the more global compression system instability.

27.8.6 Inlet Distortion Effects on Axial Compressor Stability

The stall point of the turbomachine has been regarded until now as being set by the geometrical parameters of the machine. In practice, other factors can also affect the point at which the onset of instability occurs. An important one of these is *inlet distortion*, which is a term used to describe a situation in which substantial total pressure, velocity, and/or flow angle variations exist at the compressor inlet face. Inlet distortion problems can occur in aircraft turbine engines due to changes in aircraft attitude, as well as in industrial installations where poorly designed bends have been installed upstream of the compressor. In these situations, some portion of the blading is likely to be operating under more unfavorable conditions than would occur with a uniform flow at the same mass flow rate, leading to a decrease in the useful range of operation at the machine.

Because of the widespread occurrence of inlet distortion and its adverse effects on system stability, there has been a large amount of work on the problem of predicting compressor response to flow distortion ranging from correlations to more basic analyses [Longley and Greitzer (1992)]. In these investigations, the flow nonuniformities are commonly divided into radially varying steady state, circumferentially varying steady state, and unsteady distortions. In reality, the distortions encountered are combinations of two or possibly all three of these types, but significant progress has been made using the above simplifications.

One important result is that for transient distortions, or for a steady, circumferentially nonuniform flow (which the moving compressor rotor sees as an unsteady flow), the effect of unsteady blade row response is to mitigate the degradation of stall point by the distortion. This effect increases as the reduced frequency (based on blade length, through-flow velocity, and a frequency which characterizes the time scale of the distortion) increases. Hence, for the circumferential distortions, it is the distortions with low harmonic content, i.e., a small number of *lobes* around the circumference, that are most serious, and distortions with numbers of lobes greater than two or three or of small circumferential extent, will not affect the stall point to an appreciable extent.

27.8.7 Effect of Downstream Components on Compressor Stability

The stability boundary for a compressor, i.e., the inception point of rotating stall, can also be affected by other components in a compression system. This should be apparent for components upstream of the compressor, since they affect the compressor inlet conditions. Although perhaps not so apparent, however, the stability boundary can also be influenced by components downstream of the compressor, since they alter the downstream boundary conditions on flow perturbations. The extent of the change in stability boundary is dependent on not only the nature of the downstream component but on the circumferential length scale of the flow perturbation, since this determines how close the coupling is between compressor and downstream component. For many situations of practical interest, the predominant mode of instability occurs with a one-lobed (single cell rotating stall) type of disturbance, so that the relevant axial distance within which there can be a strong interaction is on the order of the machine diameter. This generally means that in terms of the stability boundary the compressor is not isolated from the influence of downstream components [Greitzer (1980)].

It can be shown that, relative to the situation with a constant area exit annulus, an exit (annular) diffuser should be destabilizing (i.e., the onset of rotating stall should occur at a higher flow rate), whereas an exit (annular) nozzle should be stabilizing. As an example, Fig. 27.114 shows data from a three-stage compressor

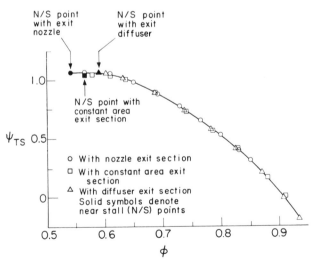

FIGURE 27.114 Effect of exit conditions on axial compressor stability, *N/S* is *Near Stall*.

run with three different downstream components. The stability limits are marked for the three conditions. It can be seen that even with these relatively passive devices, a shift in the stall point of roughly 10% in flow occurs. For a situation with a downstream turbomachine, therefore, even larger changes in stability limits may occur.

27.8.8 Stability Enhancement with *Casing Treatment*

Another area of interest concerns techniques for enhancing the stability margin of a turbomachine. The most obvious of these is to achieve the needed stability margin by matching the compression system below its peak efficiency point (in effect setting the match point so that the compressor blading has incidences and pressure rises far below the maximum). Although this provides an increase in airfoil incidence range between the operating line and the stability limit, it leads to decreased efficiency on the (down-rated) operating line, and this is generally unacceptable.

Another solution to this problem is the use of so-called *rotor casing treatment* to improve the stability of compressors. This casing treatment consists of grooves or perforations, over the tips of the rotors in an axial compressor, or located on the (outer) shroud in a centrifugal machine. Numerous investigations of these types of configuration have been carried out under widely varying flow conditions, and these have demonstrated that the range of usefulness of these casing configurations extends from compressor operation in basically incompressible flow (relative Mach numbers of ~0.15) to the supersonic flow regime (relative Mach numbers ~1.5). A sketch of one of the more successful of these casing configurations, known as *axial skewed grooves*, is shown in Fig. 27.115. A typical improvement in stall line brought about by use of these grooves is shown in Fig. 27.116 for a transonic axial fan. It is to be noted that far larger improvements have been seen. Casing treatment using axial skewed grooves has also been used to inhibit instability in centrifugal compressors

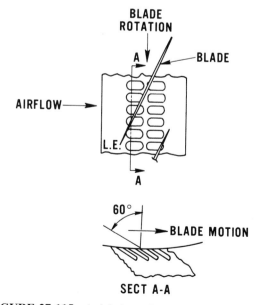

FIGURE 27.115 Axial skewed groove casing treatment.

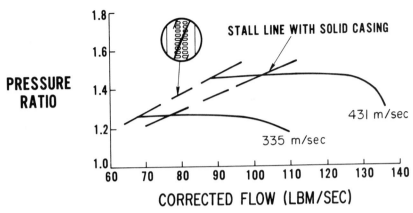

FIGURE 27.116 Axial compressor stall margin increase due to casing treatment.

(as will be discussed in a subsequent section), and stator *hub* treatment has also been used to increase stall range.

The basic mechanism of operation of casing treatment has not yet been fully elucidated, but some important points for its use have emerged. First, it is clear that for casing treatment to work it must be the rotor tip section that is setting the stall point. Second, it appears that the type of stall must be an endwall related stall rather than an airfoil stall. Thus, although casing treatment does have the potential for large increases in stable flow range, one should be sure that its application is warranted before it is used.

Finally, it should be noted that although casing treatment is a potent remedy in increasing stall range, it is not a *panacea*, since there can be some penalty in efficiency for treatments which have the most success in improving stall range.

27.8.9 Stability of Centrifugal Compressors and Pumps: Stage Stall in a Centrifugal Compressor

Although many of the same phenomena occur with centrifugal compressors as with axial turbomachines (especially when viewed on an overall system basis) there seem to be, especially in high pressure ratio centrifugal compressors, differences in the fluid mechanic features associated with the onset of instability. At the present level of understanding of compression system instability, one can say that the basic ideas concerning overall system behavior apply to centrifugal as well as axial compression systems. However, when one examines the specifics of the local phenomena that can *trigger* this overall instability, there do appear to be differences between the two types of machine. Because of this, it is useful to discuss the centrifugal compressor separately and to focus on the phenomena that are associated with the stall of the centrifugal compressor *stage*, since this lies at the root of the system instability.

Centrifugal compressor stages consist of an impeller and a diffuser, with the latter of either vaned or vaneless design. The type of diffuser used can have a considerable effect on the stable flow range of the stage. As an illustration of this, Table 27.7, Dean (1974) divides diffusers into two types according to whether the principal objective is high efficiency or wide flow range, with the choice of design depending

TABLE 27.7 Diffuser Types

Wide Range	High Efficiency
Dump Collector	Vaneless plus cascade (1–3 row)
Vaneless diffuser	Thin vane:
with:	single row
dump collector or	multiple row
scroll or	contoured vane
scroll plus discharge	single row
diffuser	multiple row
	Vane-island with:
	curved
	straight centerline passages
	Pipe diffuser (UACL patent) with:
	circular
	noncircular "pipe" cross
	section
	Rotating-wall vaneless plus vaned
	diffuser

From Dean, 1974.

strongly on the application. Process compressors, which can require very broad range, may use vaneless diffusers. Centrifugal compressor stages in aircraft gas turbine engines, where efficiency is very important, will tend to have vaned diffusers.

As with axial compressors, there are several levels of approach to finding the most important contributor to stage stall. The most empirical are based on correlations of impeller or diffuser flow range as functions of various flow parameters. However, from the diversity of the correlative procedures, one can infer that none has provided a criterion that applies in all cases of interest, and, in general, designers appear to rely heavily on data correlations from similar geometries that have been run previously. References to these are given by Cumpsty (1989) and by Greitzer (1981).

A somewhat different approach has been based on the idea that, as discussed previously, the slope of the overall total-to-static stage pressure ratio is an indication of the onset of instability. The overall ratio can be separated into the product of separate ratios for inlet, impeller, impeller exit to vaned diffuser throat, and channel diffuser to diffuser exit. The first of these is a total-to-static pressure ratio, and the rest are all static-to-static pressure ratios. Writing this overall ratio as a product, an expression can be obtained for the normalized slope of the overall stage pressure ratio, $PR_{overall}$, in terms of the different element pressure ratios

$$\left(\frac{1}{PR_{overall}}\right) \frac{\partial (PR_{overall})}{\partial \dot{m}} = \sum_i \frac{1}{(PR)_i} \frac{\partial (PR_i)}{\partial \dot{m}} \qquad (27.233)$$

where the sum over i indicates the different elements of the stage. This normalized slope may be regarded as a *stability slope parameter*. An illustration of the behavior of this parameter is plotted in Fig. 27.117 taken from Dean (1974). The values of the stability slope parameter are presented for the individual elements as well as for

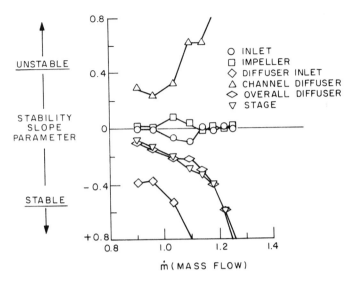

FIGURE 27.117   Influence of stage elements on centrifugal compressor stage stability. (Reprinted with permission, Dean, R. C., Jr., "The Fluid Dynamic Design of Advanced Centrifugal Compressors," Creare Technical Note TN-185, 1974.)

the overall stage plotted as functions of the mass flow rate. Positive values indicate a tendency toward instability (positive slopes), negative values indicate stability. It can be seen that, in this example at least, the channel diffuser appears to be the element that has the largest destabilizing effect and that, as the flow is reduced, the counteracting influence of other elements decreases so that the overall value moves towards positive.

The conclusion from Dean (1974) appears to be that in medium to high pressure ratio centrifugal compressors it is very often the vaned diffuser that sets the stall point, although this is not true in all cases. For example, a variable geometry diffuser, which could be set close to choke over a wide flow range, could well be used as the stabilizing element in some stages. (Rotating vaned diffusers have also been used as a stabilizing influence to increase flow range compared to the situation with a stationary, vaneless diffuser.) It should also be emphasized that, as with the axial compressor, the usefulness of the overall total to static pressure ratio is primarily as a guide to the initiation of stage stall, rather than as a direct quantitative criterion.

The situation is, thus, that the stalling element cannot always be regarded as being the diffuser but must be found from examination of the aerodynamics of each component. In addition, the choice of parameters to correlate the onset of stall is still under considerable debate as are the unsteady physical phenomena that characterize the stage stall process. At present, therefore, one can say only that stage stall in a moderate to high pressure ratio centrifugal compressor with vaned diffuser is very often, but not always, set by the diffuser at speeds near design, that there can be substantial fluctuations in mass flow rate and pressure before the onset of *surge* (reverse flow), and that the presence of rotating stall in the diffuser may not be necessary for the initiation of overall system instability.

27.8.10 Stability Enhancement in Centrifugal Compressors

There are many applications of centrifugal compressors in which considerable flow range is required, even if some efficiency must be sacrificed. As with the axial compressor, there are several techniques that have been developed to do this. One of these is to use inlet guide vanes to impart a *prewhirl* to the inducer, shifting the pumping characteristics on the map and altering the flow rate at which one encounters instability. The effects of using this technique are well documented in the literature, and a discussion of this method, with reference, can be found in Dean (1974). Other means for increasing stability are the use of backward leaning impeller blades to create a negatively sloped pumping curves and/or the use of vaneless diffusers (as mentioned above) rather than vaned.

Apart from these methods, however, there are other approaches that are perhaps less well known. The first is the use of a closely coupled system resistance to extend the stable operating range. The basic idea can be illustrated as in Fig. 27.118. If operation on the positively sloped characteristic leads to instability, we change the characteristic slope by *closely coupling* another element to the stage to make the combined slope negative. Thus, curve C is the pumping characteristic, and curves A and B are two examples of resistance curves (exit pressure dropping as flow is increased). If operation at a low flow rate such as A' is needed, then resistance A can be closely coupled to the pump; if further reduction to point B' is needed, then resistance B can be used. The combined curves in the two cases are shown as having basically zero slope at the desired operating points; negative slopes could be obtained with larger resistances.

This remedy has been applied to a compressor that had exhibited large amplitude surge cycles (including severe reverse flow) as a result of system instabilities [Dussourd *et al.* (1977)]. The closely coupled downstream throttling was achieved by means of overlapping plates with slots cut in them so that the open area, i.e., the

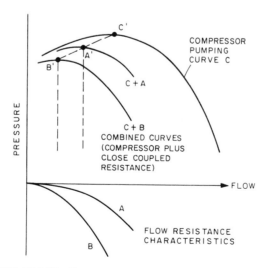

FIGURE 27.118 Surge control using close coupled resistances.

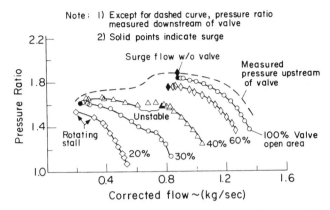

FIGURE 27.119   Centrifugal compressor performance with close coupled surge control. (Reprinted with permission of ASME, Dussourd, J. L., Pfannebecker, G. W., and Singhania, S. K., "An Experimental Investigation of the Control of Surge in Radial Compressors Using Close Coupled Resistance," *ASME J. of Fluids Engineering*, Vol. 99, pp. 64–76, 1977.)

downstream resistance curve, could be varied. The results are shown in Fig. 27.119, which gives the measured compressor characteristics for one configuration tested. The data shown are from the baseline (no downstream resistance) run and from several other runs with varying resistance. The possibility of extending the stable flow range is definitely exhibited.

Note that the instability in the region of steep positive slope was not suppressed with the more open configurations, although it was much less severe than with no resistance. However, this instability was suppressed with the more closed (higher resistance) configurations. The downstream resistance did not completely inhibit the onset of rotating stall, although this was associated with a gradual decrease in performance so that the overall *stage* curve was still rising and the system behavior was stable.

A related use of this technique has been to reduce pressure pulsations encountered in a boiler feedwater system (due to the multistage centrifugal feed pumps) by insertion of an orifice in the pump delivery line close to the pump [see Greitzer (1981)].

A second approach is the use of casing treatment. If the stage stall line is controlled by the inducer (impeller) at low speed and by the vaned diffuser at high speed, one might expect to improve stability over a considerable speed range by the use of treatment on both the impeller and the diffuser. This has been done using an essentially axial, skewed, groove over the tip of the impeller at inlet, as well as with the impeller hub wall extended under the inlet of the vaned diffuser so that a treated (grooved) wall moved under the diffuser inlet [Jansen *et al.* (1980)]. The locations of the casing treatment are indicated in Fig. 27.120. The (rotating) treatment used under the diffuser inlet was a radial skewed groove. Tests performed with various centrifugal compressor stages showed significant improvements in stall range with grooves at the impeller inlet for speeds from 0.70 to 1.05 of design speed although there was some small loss in efficiency. The *hub treatment* under the vaned diffuser also gave an improvement in stall flow margin as well as an increase in choke flow.

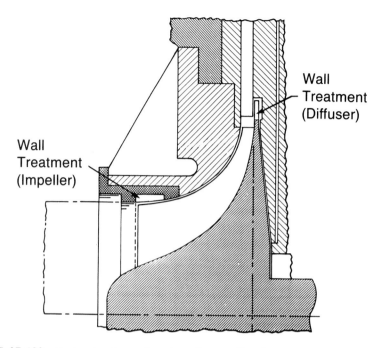

Wall
Treatment
(Diffuser)

Wall
Treatment
(Impeller)

FIGURE 27.120 Casing treatment locations in centrifugal compressor stage. (Reprinted with permission, Jansen, W., Carter, A. F., and Swarden, M. C., "Improvements in Surge Margin for Centrifugal Compressors," *Centrifugal Compressors, Flow Phenomena, and Performance*, AGARD CP-282, 1980.)

27.8.11 Instability in Radial Vaneless Diffusers

Vaneless diffusers (as well as vaned diffusers) can also exhibit a type of propagating disturbance at high inlet swirl angles. This has also come to be called *rotating stall*, although there does not appear to be as direct a connection with stall as there is in the case of the bladed cascades.

Experiments have shown that there can be (at least) two different types of oscillatory behavior, one occurring at a speed of roughly one-fourth to one-third of the impeller velocity and one which occurs at less than 10%. These oscillations are local, in the sense that they merely redistribute the flow around the diffuser and that the overall mass flow is not affected, and they are not truly a system phenomena, since the *system* behavior is *stable*. This phenomenon has been studied, and references can be found in Greitzer (1981).

27.8.12 Oscillations in Systems with Cavitating Turbopumps

The instabilities occurring in the systems discussed above have been associated with stall. However, there are other classes of instabilities in pumping systems in which stall is not the primary cause, and these can occur even when the turbomachine is operating at its design flow rate. Such types of instability occur in systems with cavitating turbopumps or in pumps with two-phase flow. See Secs. 27.7 and 23.11 for more details on cavitation phenomena.

The instabilities associated with cavitating turbopumps fall into two general categories. One of these affects only the local flow in the inducer and inlet, and is known as *rotating cavitation*. It is manifested as an unsteady cavitation pattern at the inducer inlet that rotates with respect to the inducer blades, and it does not appear to involve fluctuations in the overall (annulus averaged) mass flowrate through the pump. The other type of instability encountered with cavitating pumps (or inducers) is associated with overall mass flow oscillations through the entire hydraulic system and is known as *auto-oscillation* or *surge*.

For pumps in which there is the possibility of cavitation, another parameter, in addition to flowrate, is necessary to characterize the steady state performance. A suitable nondimensional parameter which is often used is the cavitation number, defined by $\sigma = (p_1 - p_v)/(1/2)\rho U^2$ where p_1 = pump inlet pressure, p_v = vapor pressure of the liquid, ρ = liquid density, and U = the rotor (tip) speed. This parameter indicates, for a given geometry, the extent of the cavitation which will occur. A decrease in cavitation number is associated with an increase in cavitation extent.

The performance of the pump can therefore be expressed as $\psi_p = f(\phi, \sigma)$ where ψ_p is a pressure rise coefficient ($\psi_p = (p_2 - p_1)/\rho U^2/2$), p_2 is the pump discharge pressure, and ϕ is a flow coefficient or nondimensional mass flow ($\phi = C_x/U$ where C_x is the axial velocity based on the volumetric flow rate and the inlet area). Under noncavitating conditions, the performance is only dependent on ϕ. As cavitation becomes important, however, the pressure rise becomes a function of cavitation number as well. Representative performance curves for a rocket pump impeller are given in Fig. 27.121 for a range of cavitation numbers. In this figure, the horizontal axis is cavitation number, and the vertical axis is the pressure rise coefficient, ψ_p, with the different curves corresponding to different nondimensional mass flows. The design value of ϕ is 0.07. Indicated on the figure by the stars are the points at which the instability known as auto-oscillation was encountered with this system. It can be seen that this instability occurred at design (and higher) flow on the negatively sloped part of the performance curve. This is in direct contrast to the behavior that is found in the single phase systems, and it is evident that stall is not involved in this instance. It is apparent, however, that cavitation is connected with the onset of instability.

A *simplified* analysis that shows the qualitative features of these instabilities can be performed by modelling the pump performance as a quasi-steady function of two variables, the inlet pressure and mass flow. For small perturbations, therefore

$$\delta \Delta p_p = \delta p_2 - \delta p_1 = \left(\overline{\frac{\partial \Delta p_p}{\partial p_1}}\right)\delta p_1 + \left(\overline{\frac{\partial \Delta p_p}{\partial \dot{m}_1}}\right)\delta \dot{m}_1 \qquad (27.234)$$

$$\delta \dot{m}_2 - \delta \dot{m}_1 = \rho\left(\left(\overline{\frac{\partial V_c}{\partial p}}\right)\frac{d\delta p_1}{dt} + \left(\overline{\frac{\partial V_c}{\partial \dot{m}}}\right)\frac{d\delta \dot{m}_1}{dt}\right) \qquad (27.235)$$

In these relations, Δp_p, is the pump pressure rise, $\dot{m}$ is the mass flow, V_c is now the volume occupied *by the cavitation*, and the subscripts 1 and 2 denote pump inlet and outlet.

Using this representation of the transient pump behavior, if one analyzes the sta-

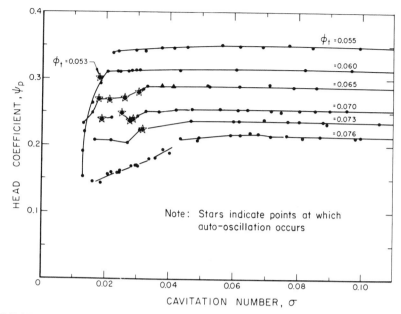

FIGURE 27.121 Cavitation performance of impeller. (Reprinted with permission of ASME, Braisted, D. M. and Brennen, C. E., "Auto-Oscillation of Cavitating Inducers," *Proc. of Conference on Polyphase Flows*, ASME, July 1980.)

bility of a system with a cavitating turbopump, it is found that for pumps operating near the design value of ϕ, the system will become unstable when $(\partial \overline{V}_c / \partial \dot{m})$, which is referred to as the *mass flow gain factor*, becomes sufficiently negative. A further result that can also be extracted from the stability analysis based on this simple model is the determination of the natural frequency of the system. The result is that for a given cavitation number, the square of the natural frequency is proportional to $1/[\partial \overline{V}_c / \partial P]$. If this is written in terms of cavitation number and pump rotor speed, the frequency ω is seen to scale as $[U/\partial \overline{V}_c / \partial \sigma]$. In other words, for a given system at specified mean values of ϕ and cavitation number, the natural frequency for small oscillations scales linearly with speed. This can be contrasted with the behavior in the single phase systems discussed previously, where the natural frequency of small amplitude oscillations was independent of rotor speed, being set only by the physical *system* parameters. The reason for this is that the *pressure compliance* of the system, $\partial V_c / \partial P$, which plays a role analogous to the spring constant, decreases as the blade speed increases, since a fixed increment in inlet absolute pressure changes the cavitation number less and less as the speed increases.

27.8.13 Dynamic Performance of Cavitating Pumps

Using the simple model one can discuss qualitatively some of the relevant features of the basic fluid mechanics of the instabilities that occur in cavitating turbopumps. Quantitative prediction of these oscillations, however, requires a more precise description of the pump dynamic performance. A useful approach to developing such a description [Ng and Brennan (1978) and Braisted and Brennan (1980)] is based on a *transfer matrix* representation of the pump performance to relate the quantities at the inlet and exit of the pump. Although the approach assumes that the system transients are small enough in amplitude so that a linearized description of the motion can be adopted, it has proved of value in clarifying the nature of dynamic performance of cavitating pumps.

The forms of these transfer matrices have been examined theoretically and experimentally for a cavitating pump (a rocket pump impeller) and have been used in a detailed system stability calculation. The approach is to focus on the net flux of mechanical energy out of the various elements in the system. This can be found for the pump by using the experimentally measured transfer matrices, and Fig. 27.122(a) presents (a cross plot of) a typical result which shows the *activity parameter* (essentially the nondimensional flux of mechanical energy out of the pump), as a function of cavitation number. The flow rate is $\phi = 0.07$, the design value. The vertical scale in this figure indicates the nondimensionalized value of this net flux. Positive values indicates a dynamically active element, i.e., one that is feeding energy into the oscillations. The activity parameter is strongly dependent on oscillation fre-

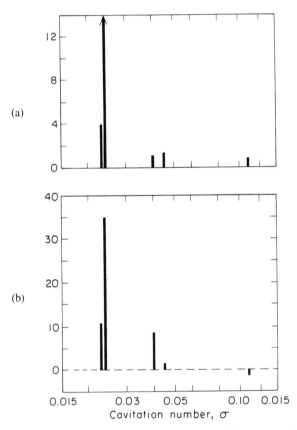

FIGURE 27.122 Effect of cavitation number on pump activity parameter and net system energy flux. (a) Pump activity parameter and (b) net system energy flux. (Reprinted with permission of ASME, Braisted, D. M. and Brennen, C. E., "Auto-Oscillation of Cavitating Inducers," *Proc. of Conference on Polyphase Flows*, ASME, July 1980.)

quency as well as cavitation number and mass flow. The quantity shown in the figure is the maximum occurring at a particular cavitation number, which would be the frequency that characterizes the largest outflux of mechanical energy.

A positive value of the dynamic activity of the pump is a necessary, but not sufficient condition for system instability, since there can be dissipation occurring in the rest of the system to offset the pump behavior. To find the overall stability of the system, one must couple the dynamic pump behavior to the rest of the system dynamics. This can be done using the transfer matrix techniques. Figure 27.122(b) shows the results of this overall energy analysis and gives net flux of mechanical energy as a function of cavitation number. As before, the convention used is that a positive number indicates a trend toward instability. Note that it is the value of this quantity, rather than that shown in Fig. 27.122(a) that is directly tied to whether the system is unstable or not, and it is again pointed out that the bars indicate the maximum value (as a function of frequency) for the particular cavitation numbers shown.

The important point is that there are large differences in the values of these activity parameters as the cavitation number is reduced. These differences appear to be strongly associated with the mass flow gain factor which becomes large in magnitude over certain ranges in cavitation number. To analyze the system stability, it is thus important to be able to understand how the mass flow gain factor behaves over the different regimes of pump performance.

27.8.14 Cavitation Induced Instabilities at Off-Design Conditions

Although the discussion has been focused so far on the flow regimes near design, instability is also seen during operation of cavitating pumps at *throttled*, i.e., low flow, operation. In many instances, it appears that system oscillations can in fact become more severe as the flow is reduced from the design point. The occurrence of severe system oscillations due to cavitation is not limited to axial pumps described above, and they can also occur with centrifugal pumps, for example, boiler feed pumps (see Sec. 27.7).

Observations of the flow fields in both types of pumps (axial and centrifugal) reveal that during operation at low flows there is a strong backflow in the tip region of the pump. This can give rise to a strong tangential velocity component in the fluid upstream of the inducer. The inlet flow can also contain a significant cavitation volume *upstream* of the pump and can thus contribute to the mass flow gain factor. Flow visualization has shown that the extent of the upstream prerotation region and the cavitation volume can vary substantially during the surge cycle. In this connection, it has been found that augmentation of the *prewhirl* upstream of the pump, by injecting high pressure fluid from the pump discharge in a tangential direction, can have a significant stabilizing effect.

27.8.15 Pump Surge Due to Two-Phase Flow

Although not directly tied to cavitation, a related type of instability has also been encountered in two-phase flow systems. This has been termed *pump surge* in view of the apparent similarity to the types of instability observed with single phase flow in pumps and compressors. Large amplitude oscillations have been found in two-phase flow systems with both centrifugal and axial pumps.

At present no in-depth quantitative analysis of this phenomenon appears to have been conducted, although it seems that the techniques mentioned in the previous two sections could be usefully applied here as well. However, some qualitative observations relating to the system stability can be made. The steady state behavior of the headrise (pressure rise) coefficient of a centrifugal pump in a two-phase flow as the inlet *void fraction* (the volumetric concentration of the gas phase) is increased as shown in Fig. 27.123(a). The ordinate is the head coefficient (pump head rise

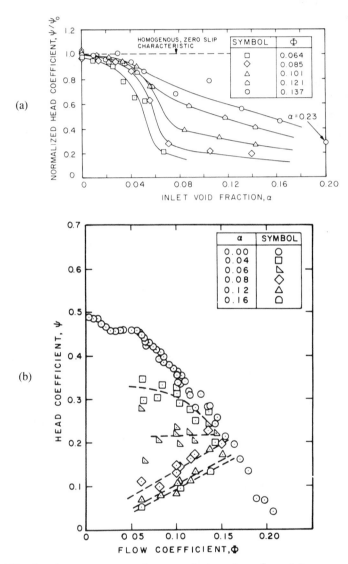

FIGURE 27.123   Steady state pump performance in two-phase flow: (a) pressure rise coefficient versus inlet void fraction for different flows and (b) pressure rise coefficient versus flow for different inlet void fraction (crossplot of *a*). (Reprinted with permission of ASME, Rothe, P. H., Runstadler, P. W., Jr., and Dolan, F. X., "Pump Surge Due to Two-Phase Flow," *Polyphase Flow in Turbomachinery*, ASME, Winter 1978.)

divided by impeller tip wheel speed squared) normalized by the head coefficient for single phase flow, and the abscissa is the inlet void fraction. The different symbols represent different total flow coefficients (defined as total volumetric flow at inlet divided by the product of impeller tip wheel speed and impeller discharge area). As the void fraction is increased, a significant falloff in head coefficient can be seen. In addition, at constant void fraction, the falloff in head coefficient increases as the flow coefficient is decreased.

With this as background, it is pertinent to plot this data in the usual headrise versus flow coefficient format as is done in Fig. 27.123(b). Whereas the single phase curve (the circles) is negatively sloped except for a small region at quite low flow rates, the curves for higher inlet void fractions have a considerable positive slope over much of the range shown. Thus, if one considered small perturbations which occurred in a system with a pumping characteristic of this type, there would be a definite possibility for the promotion of an instability as described previously.

For the general case of an arbitrary perturbation in system operating point, there is no need for the inlet void fraction to remain constant, and even if one considers the pump transient performance to be approximately quasi-steady, the headrise will then be a function of both the instantaneous inlet void fraction and the flow coefficient. The pump operating point could thus traverse a path that cuts across the curves in Fig. 27.123(b), and the effective slope could be more or less steeply sloped than shown. Nevertheless, it does appear that the falloff in head coefficient as the result of a two phase inlet flow can be a potentially important factor in promoting instability, in addition to the effects associated with the mass flow gain factor, for these types of pumps operating in two phase flow.

27.8.16 Self-Excited Oscillations in Hydraulic Systems

There can be many other instabilities encountered in the general area of hydraulic pumping systems [Greitzer (1981) and Wylie and Streeter (1993)] although space precludes a detailed description here (see Chap. 28). Severe pulsations have been known to arise, due not only to system oscillations, but also to local instability (i.e., rotating stall) in the pumps. For example, self-excited oscillations of check valves with spring dampers can also occur. A simplified model of this phenomenon has been studied experimentally to clarify the cause of instability which was associated with high rates of change of discharge in the last few degrees of closing. Instabilities have been encountered with hydraulic gate seals and hydraulic turbine penstock valves, and these have also been analyzed in terms of a negative damping which was responsible for the growth of the oscillations.

27.8.17 Flow Transients and Instabilities in Compound Pumping Systems (Systems with Pumps in Parallel)

In many practical situations, two or more pumping devices operate in parallel. Under these conditions, not only are there possibilities for instabilities of the types that we have already discussed, but new problems, associated with the compound system, can also appear. In particular, there can be difficulties associated with transients, such as startups.

To examine some of these problems, consider the model pumping system pictured

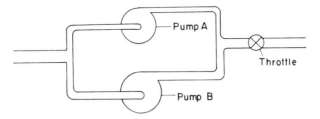

FIGURE 27.124 Compound pumping system with pumps in parallel.

in Fig. 27.124. Suppose that the two pumps have different pressure rise and mass flow capabilities. The steady rate operating points of such a configuration might be as given in Fig. 27.125. Curves A and B are the individual curves for each machine. The dash-dot curve C is for the combined system. For system pressure requirement curve 1, the two machines do better than one, while for the system resistance curve 3, the capacity of the two machines is actually smaller than that of the larger machine by itself. As it is drawn, the smaller machine would be operating with reverse flow. The overall system operation, however, is stable in that the pump performance changes smoothly and continuously as the throttle area is decreased.

An often encountered situation is that one or the other of the pumps has been shut off and is brought on line. To be specific, let us suppose that pump A is operating with pump B off, and B is now brought on line. If the throttle curve is considered fixed at 3, A's operating point would undergo a large transient from a to a', with operation of B at point b'. If the demand (flow volume) is considered to be fixed, then the transient would be to a". In either case, the result could be that a pump that was running near peak efficiency undergoes a transient to a region of much less benign operation.

The difficulties can be compounded if the pump curves have regions with positive slope, as one can see using arguments similar to those presented in Sec. 27.8.1. In general, it is found that for pumps or fans to operate satisfactorily in parallel they

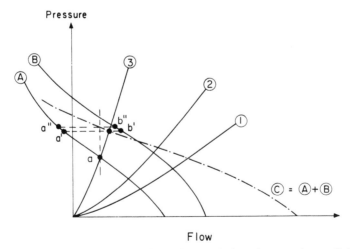

FIGURE 27.125 Pressure-flow characteristics of pumps in parallel.

should have negatively sloped pressure rise versus flow curves, and they should have roughly the same percentage reduction in capacity over the operating range. Some design rules for different specific combinations of fans and compressors have been developed by Eck (1973).

The transients just described can be large and can lead to considerable difficulty. In addition to this excursion behavior, there are also situations in which compound systems give rise to an oscillatory transient. One simple example of this is the system associated with a branched diffuser configuration, a situation encountered in gas turbine annular combustors. Each branch of the system has a diffuser, a volume with a mass storage capability, and a downstream resistance, similar to the configuration shown in Fig. 27.105 for the simple system. Analysis of this system shows there is a range of operating points for which a large amplitude oscillation would be set up. For symmetric diffusers, this results in the flow through the combustor cells, i.e., the downstream resistance, varying periodically in an approximately triangular fashion, while the system pressure drop varies periodically (approximately as a sawtooth wave) with twice the frequency of the other parameters [Ehrich (1970)].

A further example of flow transients that can arise in pumping systems in parallel has been seen in dual entry compressors. In one situation, a large amplitude, sudden flow shift occurred, accompanied by a substantial pressure change [Dussourd and Putnam (1960)]. The cure for the instability was to reduce the pressure rise in the *stronger* side by using an inlet screen so that the two sides were matched. Although this was tried successfully, one must be careful to tailor the screens properly, since too much blockage can cause a mismatch of the two sides in the opposite direction. The best *remedy* is to match the two inlets carefully.

Although there does not appear to be an extensive literature covering specific instabilities that occur in complex pumping systems with parallel pumping elements, there are many instances in which these can be important. To analyze such systems, one must dynamically couple the various components. For small oscillatory perturbations, one method of doing this is to use the transfer matrix approach that was described in the section on cavitating inducer stability. This technique has been used to analyze hydraulic transients in complex piping systems [Wylie and Streeter (1993)]. For predicting large amplitude transients, one must integrate the equations of motion for the flow in the system numerically. However, if the stability boundary is what is desired, a linear analysis will suffice. As far as overall results are concerned, one cannot make a general statement about the stability of the complicated pumping systems that exist in practice, except to note that a very conservative criterion would be that if none of the components in the system are dynamically active, i.e., if none of them have a net outflux of mechanical energy over a complete period, then the system will be dynamically stable.

27.9 HYDRAULIC TURBINES
Michael V. Casey and Helmut Keck

27.9.1 Introduction

The use of hydraulic machines for the generation of power has a very strong historical tradition [see Rouse and Ince (1957)]. In the middle ages, various types of water wheels were used for grinding grain, sawing wood, and to power bellows for furnaces. The industrial revolution in the early nineteenth century increased the need for motive power for driving cotton mills and other machinery and promoted the development of turbines to replace water wheels. Since the turn of the century, most new applications of hydropower have been for the generation of electricity.

The first hydraulic turbine worthy of the name, the Fourneyron turbine, was developed in France around 1830 and predates by more than half a century the invention of the first practical thermal turbomachine, the Parsons steam turbine. The Fourneyron turbine made use of a radial outward flow through a runner with stationary guide vanes in the central portion. The first truly effective inward flow reaction turbine was developed and tested by Francis and his collaborators around 1850 in Lowell, MA [see Francis (1909)]. Modern Francis turbines have developed into very different forms from the original, but they retain the concept of radial inward flow.

The modern impulse turbine was also developed in the USA and takes its name from Pelton, a mining engineer from California, who invented the split bucket with a central ridge around 1880. The modern Pelton turbine owes its current form, with a double elliptical bucket including a notch for the jet and a needle control for the nozzle, to Doble who used these techniques for the first time around 1900. The axial flow propeller turbine with adjustable blades was developed by the Austrian engineer Kaplan from 1910 to 1924.

During their long history, there has been continuous development of the design of hydraulic turbines, particularly with regard to improvements in efficiency, size, power output, and the head of water being exploited. Most of this development has been highly empirical with little attention paid to detailed fluid dynamic analysis of the flowfield. Recently, the use of modern techniques of computational fluid dynamics (CFD) for predicting the flow in these machines has brought further substantial improvements in their hydraulic design, in the detailed understanding of the flow and its influence on turbine performance, and in the prediction and prevention of cavitation inception. The efficient application of advanced CFD tools is of great practical importance as the design of hydraulic turbines is custom-tailored for each project. Standardized components are only used for units with small power outputs (below 5 MW).

The hydraulic turbine itself usually contributes only between 5 and 10% of the investment costs of a large hydropowerplant. The high civil engineering costs determine that the choice of turbine type is strongly influenced by features of its design that lead to lower capital costs for the complete plant. For example, the type, diameter, and orientation of the runner strongly influence the size of excavation needed for the powerhouse. Nevertheless, the value of the extra power generated by a small improvement in efficiency forces the hydraulic designer to give a great deal of attention to the sources of loss in the flow, both in the stationary and the rotating components. This has promoted the development of highly efficient units, with typical peak hydraulic efficiency levels being around 96% for the best Francis units.

Hydraulic turbomachines are not only used to convert hydraulic energy into electricity, making use of a completely renewable source of energy (the global H_2O cycle triggered by solar power). They are also used in pumped-storage schemes, which are the most efficient large-scale technology available for the storage of electrical energy. Separate pumps and turbines or reversible machines, so called *pump turbines*, are used in such schemes, the largest of which have a capacity above 2000 MW.

This short review of hydraulic turbines describes the applications and the different types of modern turbines of most economic importance, with emphasis on the fundamental fluid mechanic features that determine their losses and performance. Special designs, such as crossflow machines are omitted. Further details on industrial aspects of large hydraulic machines can be found in specialist texts, such as Höller and Grein (1984), Raabe (1989), and Henry (1992).

27.9.2 Basic Fluid Mechanics of Hydraulic Turbines

Euler Equation and Velocity Triangles. The torque on any turbomachinery rotor can be estimated from the inlet and outlet velocity triangles by application of Newton's second law of motion applied to the moments of forces in a rotating coordinate system. For a hydraulic turbine, the resulting equation is known as *Euler's turbine equation* and gives the specific energy transferred by the runner as

$$E = \frac{T\omega}{\rho Q} = U_1 Cu_1 - U_2 Cu_2 \qquad (27.236)$$

Here, E is the energy per unit mass, T is torque, ω is rotational speed, ρ is density, Q is flow rate, U is blade speed, and Cu is the circumferential component of velocity. The associated velocity triangles can be seen in Fig. 27.126.

Degree of Reaction. For hydraulic turbines, the *degree of reaction* is classically defined as the ratio of the static pressure drop across the runner to the static pressure drop across the stage. An *impulse stage* (such as a Pelton turbine) has zero reaction with all the pressure drop occurring across the stationary components, and no pressure drop across the runner. *Reaction stages* (such as Kaplan and Francis turbines) have a proportion of the pressure drop occurring in the rotor and a proportion in the stator. Typically, at their design points, a Kaplan turbine has a reaction around 90%, a Francis turbine around 75%, and a pump-turbine around 50%. At off-design operating points, these values change.

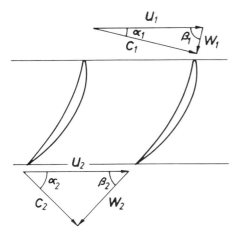

FIGURE 27.126 Velocity triangles for an axial turbine runner.

Efficiency Definitions. The *overall efficiency*, η_o, of a turbine is defined as the ratio of the power delivered to the shaft to that available in the water entering the turbine,

$$\eta_o = \frac{T\omega}{\rho g Q H} \tag{27.237}$$

where the net head, H, is the difference between the total pressure at turbine inlet and turbine outlet. Losses upstream of the turbine inlet (penstock, trash rack, etc.) and downstream of turbine outlet (Carnot loss at draft tube outlet, tailrace tunnel) are considered as *plant losses*, i.e., as the difference between gross head and net head. The overall efficiency can be broken down into three components, the *hydraulic efficiency*, the *volumetric efficiency*, and the *mechanical efficiency*.

The hydraulic efficiency of a turbine is the ratio of the actual power transferred to the rotor to that available in the water flowing through the machine,

$$\eta_h = \frac{H - H_f}{H} \tag{27.238}$$

where H_f is the fluid friction head loss in the machine. The volumetric efficiency, η_v, takes account of the small loss of efficiency due to leakage around the outside of the turbine runner. If the leakage flow around the runner is denoted by Q_L, while Q is the total flow entering the turbine, then $Q - Q_L$ is the flow that effectively passes through the runner blades, and the volumetric efficiency is given by

$$\eta_v = \frac{Q - Q_L}{Q} \tag{27.239}$$

Normally, the design of the machine includes appropriate sealing arrangements to keep the leakage flow around the turbine as small as possible, but under some conditions, such as erosion of the labyrinth seals by sand particles, this may become important.

The mechanical efficiency of the turbine is the ratio of the power available at the shaft to that exerted by the fluid on the runner. The difference between these two is the power needed to overcome mechanical friction and includes friction in the bearings, the stuffing boxes, and the disk friction on the sides of the rotor caused by the fluid in the gap between the rotor and the casing. It includes the windage and splashing losses of a Pelton runner.

Performance Parameter Definitions. For pumps and pump turbines, the *flow coefficient* and *head coefficient* (parameters ϕ and ψ) are generally nondimensionalized with the rotor blade speed, in a similar manner to other turbomachines, as

$$\phi = \frac{Q}{UD^2(\pi/4)}$$

$$\psi = \frac{H}{(U^2/2g)} \tag{27.240}$$

Otherwise, the performance parameters used for hydraulic turbines generally differ from those used for other turbomachinery applications.

In hydraulic turbines, the usual performance parameters, such as the flow velocity and the circumferential blade speed, C and U, are made dimensionless with respect to $\sqrt{2gH}$

$$Kc = \frac{C}{\sqrt{2gH}}$$

$$Ku = \frac{U}{\sqrt{2gH}} \qquad (27.241)$$

The meridional and circumferential components of Kc are denoted by Kcm and Kcu. The term $\sqrt{2gH}$ can be associated with the spouting velocity, C_o, created when the entire head of water is converted into kinetic energy through a jet. These nondimensional parameters relate the most relevant variables and are popular in turbines for the following reasons. First, for turbines, the available head of water, H, is imposed externally and can be considered as the input parameter, whereas the flow-rate and speed, Q and ω, are internal variables determined by the mode of operation of the turbine. It should be noted, however, that turbines normally drive synchronous generators with fixed speed, i.e., the turbine speed has to be controlled by the turbine governor to yield $\omega = $ const. For comparison, in pumps, the rotational speed is imposed externally by the motor, and the flowrate and head are inner variables determined by the characteristics of the pump and the system. Second, the nondimensional parameters Kcm and Ku can be used directly to define dimensionless velocity triangles. The Euler equation in dimensionless form reads

$$\eta_o = \frac{E}{gH} = 2(Ku_1 Kcu_1 - Ku_2 Kcu_2) \qquad (27.242)$$

or

$$\eta_o = 2(Ku_1 Kcm_1 \cot \alpha_1 - Ku_2 Kcm_2 \cot \alpha_2) \qquad (27.242a)$$

Last, during transient modes (start-up, shut-down, 4-quadrant-characteristics, etc.) it is more convenient not to have the blade speed as the denominator of the dimensionless groups, especially if values $\omega = 0$ and $\omega < 0$ occur.

Specific Speed. For both pumps and turbines, a useful parameter is the *dimensionless specific speed* defined as

$$\nu_s = \frac{\omega}{\pi^{1/2}} \frac{Q^{1/2}}{(2gH)^{3/4}} = \frac{\phi^{1/2}}{\psi^{3/4}} \qquad (27.243)$$

Model Test Conditions and Similarity Parameters. A key instrument for the verification of predicted performance characteristics of hydraulic turbines is the model test. Typically, closed loop test stands for models with runner diameters between 300 and 400 mm and test heads between 10 and 100 m are used. The laws of similarity are respected by running the tests at the same values of the dimensionless performance parameters for model and prototype. However, it is not possible to respect both kinematic similarity (*Euler similarity*) together with similarity with respect to Reynolds number and Froude number. The lower values of Reynolds number in the model compared to the full-scale prototype lead to higher friction losses (so-called *scale effects*). Various correction formulae have been established to convert measured efficiencies from model to prototype, and these provide typical efficiency increments of between 1 and 3% for reaction turbines [see IEC code 995 (1991)]. The correction for Pelton turbines depends only weakly on the Reynolds number, but in addition the Froude and Weber numbers need to be considered [see Grein (1986)]. Respecting the Froude number similarity has priority for Pelton turbine model tests.

27.9.3   Types of Hydraulic Turbomachines

Modern hydraulic turbomachines are generally classified into three categories:

i) Impulse turbines operating with a degree of reaction of zero. These are generally used for low flow volumes and high heads of water, such as the Pelton turbine and various forms of cross-flow turbine.

ii) Reaction turbines operating with a pressure drop across both the runner and the stator vanes, which are generally used for lower heads and higher flow volumes, such as the Francis and Kaplan turbines.

iii) Pump turbines (reversible machines for both modes of operation). The impellers are generally designed to be similar to a pump, whereas the stationary cascades (mostly comprising two blade rows, one of which is adjustable) resemble the design of Francis turbines.

Each type of turbine design can be further classified according to such criteria as

shaft orientation	horizontal axis or vertical axis
specific speed	high, medium, or low specific speed
operating head	high pressure $200 < H < 2000$ m
	medium pressure $20 < H < 200$ m
	low pressure $H < 20$ m
type of regulation	single (variable stator vanes, e.g., Francis)
	double (variable runner and stator vanes, e.g., Kaplan, or variable needle stroke and variable numbers of jets for Pelton)
design concepts	single stage or multistage
	single volute or double volute
	single jet or multijet

Large variations in size and design are possible with each particular type in order to adapt them to the local hydraulic conditions. Small machines may develop typically only 100 kW whereas the following examples of large hydraulic turbines demonstrate their potential.

TABLE 27.8   Characteristics of Some Large Hydraulic Turbines

Plant	Type	Power (MW)	Diameter (m)	Head (m)
Bieudron	Pelton (5 jets)	416	4.6	1874
Itaipu	Francis	740	8.1	126
Gezhouba	Kaplan	176	11.3	27
Racine	Bulb	25	7.7	7
Edolo	Pump-turbine (5 stages)	130	2.2	1287
Bath County	Pump-turbine (1 stage)	420	6.4	329

More details of the fields of application of the different turbines can be found in Fig. 27.127. This figure shows which type of turbine is used for different volume flows and head rise and also gives the absolute power output generated for each category. A further diagram showing the different impeller forms and the head for a variation of the specific speed is given in Fig. 27.128. In general, the number of runner vanes (or buckets) decreases with increasing specific speed, from about 26 to 18 in a Pelton wheel, to 19 to 11 in a Francis turbine, and 7 to 3 in an axial turbine.

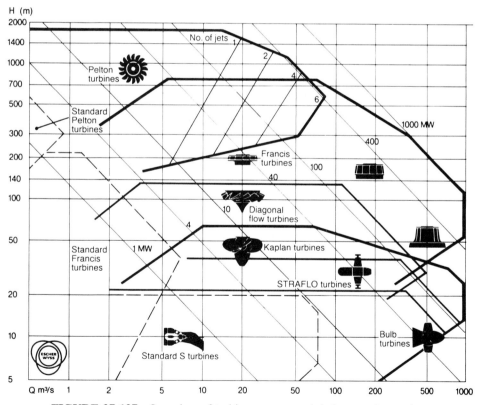

FIGURE 27.127   Overview of turbine runners and their operating regimes.

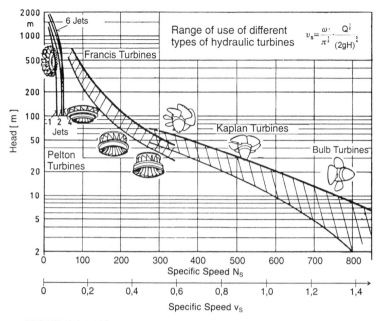

FIGURE 27.128   Selection diagram for different types of turbine.

27.9.4   Pelton Turbines

Introduction.   The impulse turbine extracts energy from the water by first converting the available head into kinetic energy in the form of a high-speed jet discharged from a nozzle. All the pressure drop occurs across the nozzle, and the runner operates at constant static pressure (giving zero degree of reaction). The jet is directed onto buckets fixed around the rim of a runner, and these are designed to remove the maximum energy from the water. The power of a given runner may be increased by using more than one jet, the two jet arrangement being common for horizontal shafts and up to six jets being used for vertical shafts. A cross section through a typical vertical unit with six jets is shown in Fig. 27.129, showing the inlet piping arrangement, the nozzles with needle valves, the deflectors, and the runner with buckets.

   Up to about 1960, the horizontal axis Pelton turbine, with one or two nozzles per runner and with one or two runners on the generator shaft, dominated the market. Since then, vertical axis multinozzle units such as in Fig. 27.129 have been built in increasing numbers in response to the demand for higher speeds and power per unit, and the development of cavern-type powerhouses where the reduced overall dimensions of multinozzle units is of importance [see Brekke (1994)].

Elementary Theory of the Pelton Turbine.   The hydrodynamics of the Pelton turbine are simple to understand, and a one-dimensional steady analysis based on the velocity triangles provides much insight into the design principles. The Pelton turbine remains, however, the most complicated of all types of turbomachinery with respect to detailed flow analysis, as it involves a highly three-dimensional, viscous, unsteady flow with a free surface on a moving boundary, and this is just on the limit of the capability of the most advanced CFD methods.

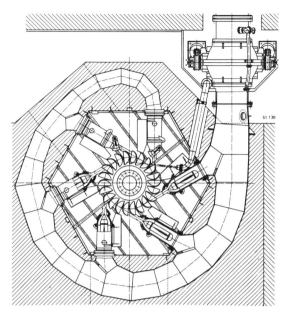

FIGURE 27.129 Cross section through a horizontal Pelton turbine with six jets.

By application of the momentum theorem to an ideal bucket section, the force, F, and power, P, delivered to a Pelton wheel can be derived to be

$$P = Fu = \rho Qu(c_o - u)(1 - \cos \beta_2) \qquad (27.244)$$

where β_2 is the angle turned by the bucket. A bucket turning angle of $\beta_2 = 180°$ gives maximum power but is physically impractical because of the presence of adjacent buckets. In practice, a typical bucket angle of $\beta_2 = 170°$ is used (giving $1 - \cos \beta_2 \approx 1.985$), which is only 0.75% less than maximum power.

The theoretical power of a Pelton turbine is parabolic in bucket speed and is a maximum at $u = c_o/2$. For a perfect nozzle, the entire head of water would be converted into kinetic energy in the jet $c_o = \sqrt{2gH}$, but in practice, the losses in the nozzle are between 2% (fully open needle valve) and 8% (partly closed needle valve), such that

$$c_o = \kappa\sqrt{2gH}, \text{ where } 0.92 < \kappa < 0.98 \qquad (27.245)$$

These losses are not uniformly distributed through the jet but are mainly concentrated in the edge of the jet where the velocity is lower.

Available Head. Usually, the pressure within the case of an impulse turbine is atmospheric, and this determines the pressure at which the jet is discharged from the nozzle. The height of the Pelton runner above the tailwater in such an installation is unavailable head for power generation. If the tailwater level varies by too much, this unavailable head can become the source of a power loss of undesirable proportions. In general, the Pelton runner will be installed to be as near the surface of the tailwater as possible, taking account of the risk of submergence at high tailwater levels. If the tailwater rises above a critical level, η drops dramatically due to interference of splashing water with the runner.

In order to increase the available head, various Pelton turbines have, in recent years, been installed with back pressure. In these cases, the turbine is installed at a lower level than the tailwater and the water level in the turbine housing is lowered by means of compressed air at a pressure given by the backpressure of the tailwater.

Nozzle and Deflector Design. The quantity of water flowing through each jet is controlled primarily by a needle valve sliding inside the nozzle. It is not possible to rapidly reduce the water quantity flowing through the nozzle using the needle, as this would cause large pressure surges in the penstock. If the discharge through the turbine needs to be reduced rapidly to avoid speed rise up to *runaway speed* at load rejection, a second safety device, a deflector, is moved into the stream of the jet to deflect it from the buckets. Two types of safety devices are possible, a *jet deflector* which deflects the whole jet or a *jet cutter* which cuts out only part of the jet depending on its position. The latter can also be used as an additional control device to control the output other than by needles.

Flow Pattern in the Buckets. Figure 27.130 shows an instantaneous snapshot of the flow interaction between the jet and the buckets of a Pelton runner [see Bachmann *et al.* (1990)]. It can be seen that the jet acts on several buckets at the same time. The uppermost bucket is just about to cut the jet. The second bucket almost fully intercepts the jet, but very little flow is yet leaving this bucket. The jet entering the third bucket is about to be fully cut off by the second bucket. The surface of the water in the bucket can clearly be seen, and it is noteworthy that the jet has spread out very flat on the surface of the bucket. The exit angle of the flow from the bucket can also be seen to be around 170°.

Much empirical development work has been carried out to determine the shape of the bucket for optimum performance. In general, standard bucket shapes from a family of profiles can be modified for a wide range of applications.

Performance Characteristics. For Pelton turbines, the dimensionless flow coefficients can be defined on the basis of either the nozzle diameter, D_0, or the bucket width, B_2, as

$$Kc_0 = \frac{C_0}{\sqrt{2gH}} = \frac{Q}{(D_0^2\pi/4)\sqrt{2gH}}$$

$$Kc_2 = \frac{C_2}{\sqrt{2gH}} = \frac{Q}{(B_2^2\pi/4)\sqrt{2gH}} \qquad (27.246)$$

The distinction between Kc_0 and Kc_2 is only relevant if the ratio B_2/D_0 is changed from one design to another, which is only the case within relatively narrow limits $(2 < B_2/D_0 < 3)$. *Hill charts* are usually plotted with Ku_1 and Kc_2 as coordinates (see Fig. 27.131). It is evident that the discharge is a function of the needle stroke, s/s_{max}, alone and is independent of Ku_1.

FIGURE 27.130 Flow visualization oɪ the flow in a Pelton runner. (Reprinted with permission, Bachman, P., Schärer, Ch., Staubli, T., and Vullioud, G., "Experimental Flow Studies in a 1-Jet Model Pelton Turbine," IAHR Symposium, 1990.)

To explain the effect of head and flow on the performance curves of Pelton turbines, it is useful to examine two different velocity triangles, as shown in Figs. 27.132 and 27.133. Figure 27.133 shows the velocity triangle at the bucket tip as it first enters the jet. The lip of the bucket is designed to accept flow at a certain relative inlet angle. At lower available head, $H < H_{\text{opt}}$ [or from Eq. (27.241) $Ku_1 > Ku_{1\,\text{opt}}$], the inlet angle β_1 becomes larger than optimal. The negative incidence angle leads to an efficiency drop as the relative direction of the jet acts to decelerate the buckets. The poor incidence may also cause cavitation erosion within the bucket. Values of $Ku_1 > 0.52$ are usually avoided in the lay-out of Pelton buckets to minimize these problems.

Figure 27.132 shows the velocity triangles of the water entering and leaving the bucket in the middle section of its passage through the jet. At a single value of the jet velocity, or available head, that is for $U_2 < U_{2\,\text{opt}}$ and $H = H_{\text{opt}}$, the optimum exit angle of $\alpha_2 = 90°$ is attained with minimum kinetic energy in the flow leaving the bucket. For $H < H_{\text{opt}}$, the relative velocity of the flow leaving the bucket, W_2, is reduced and the outlet water contains a rotating component in the same direction as the runner. For $H > H_{\text{opt}}$ the outlet water rotates against the rotational direction

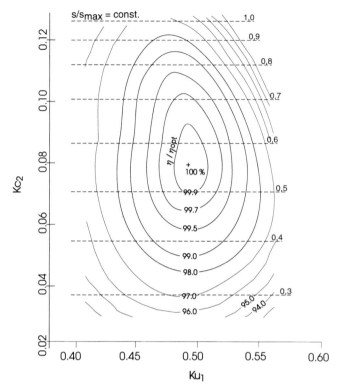

FIGURE 27.131   Hill chart for a Pelton turbine.

of the runner. The outlet velocity triangle explains two phenomena. First, B_2/D_1 is proportional to the specific speed in Pelton turbines. If B_2/D_1 is small, then the blade speed where the jet leaves the bucket, U_2, varies only little over the whole bucket and α_2 deviates only little from 90° for the different fluid particles leaving the bucket. With increasing ratio, B_2/D_1, the variation of U_2, and thus of α_2, increases. For this reason, the efficiency of Pelton turbines decreases with increasing specific speed (see

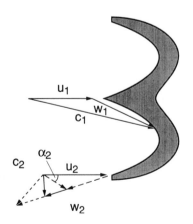

FIGURE 27.132   Velocity triangles for analysis of a Pelton bucket.

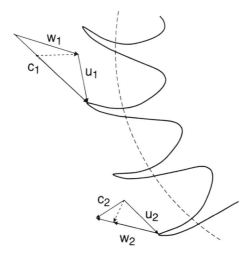

FIGURE 27.133 Velocity triangles of flow entering and leaving Pelton bucket.

Fig. 27.134). Second, operation with $H < H_{opt}$ [or from Eq. (27.241) $Ku_1 > Ku_{1\,opt}$], is dangerous for multijet Pelton turbines. The lower values of relative velocity mean that the final water droplets from the preceding jet might not have left the bucket before the water of the next jet is already entering the bucket. Also, the positive circumferential component of outlet velocity might cause the outlet water to impinge on the following jet. This leads to efficiency drop and can occasionally lead to cavitation erosion from the droplets of the disturbed jet.

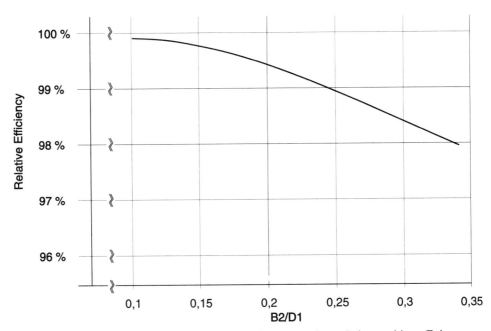

FIGURE 27.134 Effect of bucket width to diameter ratio on Pelton turbine efficiency.

27.9.5 Francis Turbines

Introduction. The original Francis turbine utilized a radial flow inwards through both the stator and the rotor, and the leading and trailing edges of both the stator and the rotor blades were parallel to the axis of rotation. The modern Francis turbine has retained the purely radial inlet flow through stationary guide vanes (*wicket gates*), but the runners are mixed-flow devices with a component of the flow in the axial direction. The trend from a purely radial-flow device through mixed-flow devices to near axial-flow devices increases as the specific speed is increased.

The Components of a Francis Turbine. The flow channel of a modern Francis turbine with a vertical axis is shown in Fig. 27.135 comprising a spiral inlet case, stay vanes, wicket gates, runner, and draft tube.

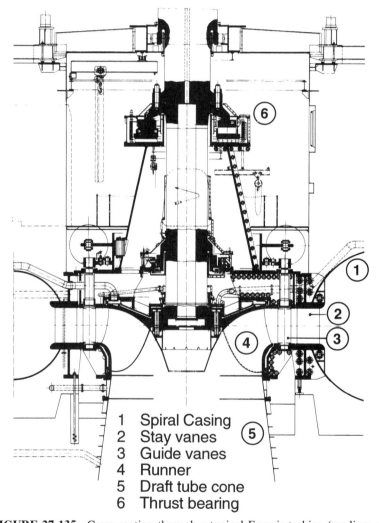

1 Spiral Casing
2 Stay vanes
3 Guide vanes
4 Runner
5 Draft tube cone
6 Thrust bearing

FIGURE 27.135 Cross-section through a typical Francis turbine (medium ν_s).

The spiral case of a Francis turbine is designed such that the velocity distribution in the circumferential direction at the inlet to the stay vanes is uniform and the incidence angle over the height of the stay vanes varies only little. As a consequence of the need for cost reduction in manufacturing, the second requirement is partly sacrificed in some modern stay ring designs with parallel instead of curved side walls. The main function of the stay vanes is to carry the pressure loads in the spiral case and turbine head cover. Their secondary purpose is to direct the flow towards the adjustable guide vanes with an optimal incidence angle. Improper shaping of the trailing edges of stay vanes has caused severe stay vane cracks due to excitation from the von Karman vortex street shed from the vane trailing edges.

The adjustable guide vanes (or wicket gates) are the only device available to control the flow, and thus the power output, of a Francis turbine. The vane setting is usually controlled by servomotors located on top of the turbine head cover. Leakage flow through the gaps between the guide vane tips and facing plates causes efficiency losses (especially for low specific speed turbines) and can cause local erosion in water flows with a high sand content.

The runner consists of a crown and band supporting highly curved, three-dimensional sculpted blades. Some aspects of blade design are described below. To reduce the leakage flow between the runner and the casing, labyrinth seals at the crown and band are provided. Their design and location also controls the pressure distribution on the runner outside of the main water passage and thus has a strong impact on the axial thrust of the runner. For low specific speed designs, the high pressure difference between inlet and outlet of the runner often calls for a multistep labyrinth seal.

The choice of the number of runner blades, z_R, is generally not independent of the number of guide vanes, z_0. In order to avoid self-induced oscillations, one should especially avoid $z_R = z_0$ and $z_R = z_0 - 1$. The condition $z_R = z_0 - 1$ leads to a cumulation of pressure waves in the spiral casing and oscillations of the penstock (a phenomenon called *phase resonance*) [Dörfler (1984)]. For low specific speeds, runners with splitter blades are sometimes used, i.e., there is an alternating sequence of long and short blades, as in some radial compressor impeller configurations.

The diffuser downstream of the runner is usually an elbow-type draft tube similar to that of Kaplan turbines. The vortices in the draft tube at off-design conditions often give rise to severe oscillations.

Loss Distribution Within a Francis Turbine. Figure 27.136 shows a breakdown of the losses in Francis turbines at the design point as a function of the design specific speed, v_s. The following aspects are of interest. First, the highest efficiencies occur at medium v_s. Second, with decreasing v_s, the losses due to disc friction (of crown and band) and the labyrinth losses increase. Third, with increasing v_s the losses in the draft tube increase, whereas the friction losses in the other components (spiral casing, guide vanes) can be kept unchanged if the main dimensions of these components are also increased as v_s is increased. Last, the losses inside the runner at design point are relatively small. At off-design conditions, the losses attributable to the runner (such as incidence losses, friction inside blade passages, and exit swirl losses) strongly increase.

Performance Characteristics. The performance characteristics of Francis turbines with adjustable guide vanes are represented by *hill charts* with lines of constant

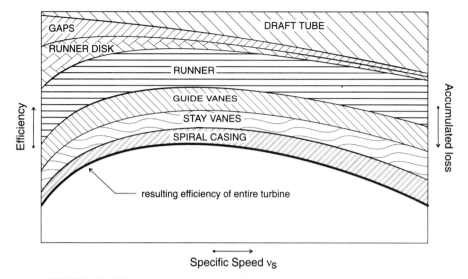

FIGURE 27.136   Loss breakdown of Francis turbines over a range of ν_s.

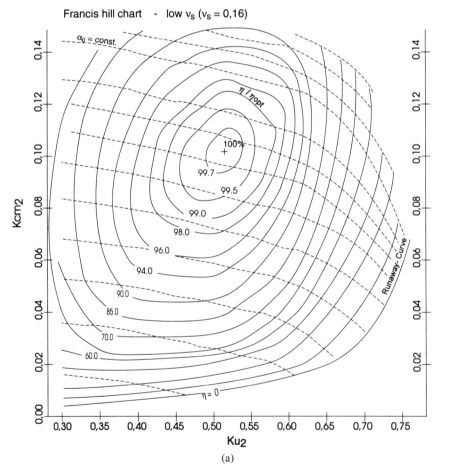

FIGURE 27.137   Hill charts for Francis turbines. (a) Low ν_s and (b) high ν_s.

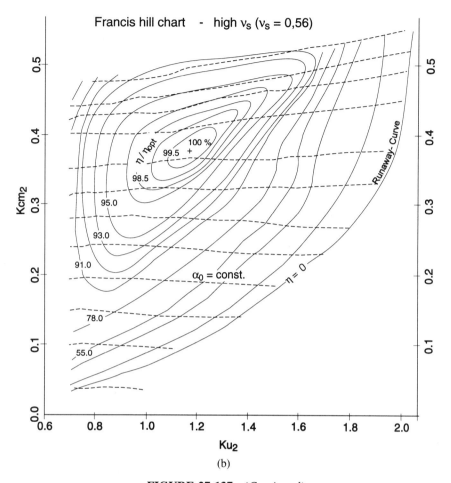

FIGURE 27.137   (*Continued*)

efficiency and lines of constant guide vane angle, α_0 [see Fig. 27.137(a) and (b)]. The dimensionless coefficients used in these diagrams, Ku_2 and Kcm_2, are defined by using the runner outlet diameter, D_2, as the characteristic length. Significantly different shapes of hill charts are observed for low and high v_s turbines, both with respect to the position of the lines $\eta =$ const. as well as of the lines $\alpha_0 =$ const.

Typical velocity triangles for a mean streamline for low and high v_s turbines are shown in Fig. 27.138(a) and (b). Note that the values of Ku_1 are nearly the same. For very high v_s, the flow through the runner is nearly axial giving $Ku_1 \approx Ku_2$. With decreasing v_s, the flow becomes increasingly radial with larger differences between inlet and outlet diameters, D_1 and D_2, and hence between Ku_1 and Ku_2. The increasing difference between inlet and outlet velocity, U_1 and U_2, explains the increasing head as v_s decreases.

At the best efficiency point (BEP) we have two conditions leading to high efficiency, small mean swirl in the outlet, $\alpha_2 \approx 90°$ and correct incidence angles at runner inlet ($\beta_{1\,\text{flow}} \approx \beta_{1\,\text{blade}}$). At other operating points, there are zones where either of these conditions are not satisfied (see Fig. 27.139). Outside the zone $\alpha_2 \approx 90°$, we get strong swirl at runner outlet with consequential risk of oscillations. Outside

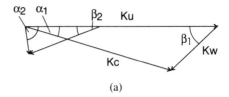

velocity triangles
low v_s (v_s= 0.1878)

vectors shown in the middle
between band and crown

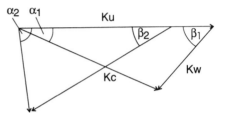

(a)

velocity triangles
high v_s (v_s= 0.5588)

vectors shown in the middle
between band and crown

FIGURE 27.138 Velocity triangles for Francis turbines. (a) Low v_s and (b) high v_s.

(b)

the zone where $\beta_{1\,\text{flow}} \approx \beta_{1\,\text{blade}}$, we get high incidence at runner inlet. Large positive incidence angles (low Ku_2) lead to cavitation at the runner leading edge, and large negative incidence angles (high Ku_2) lead to flow separation at the pressure side of the leading edge.

Use of CFD in Francis Turbine Design. The design of a Francis runner is characterized by careful optimization of the hydrodynamic form to tailor the geometry to

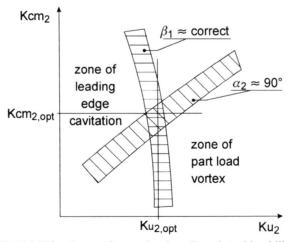

FIGURE 27.139   Zones of operation in a Francis turbine hill chart.

the local flow and head available. Unlike Kaplan and Pelton turbines, it is not satisfactory to use blades or geometries that belong to a standard profile family. Until the mid-eighties the design process for the runner geometry was essentially based on semi-empirical data, and the subsequent optimization was carried out through experimental development on a model runner in a test rig. The semi-empirical approach is now used only to provide the preliminary basis of a design that can be analyzed and refined with CFD computations. With a numerically designed runner, the models tests can begin with a runner that is much closer to the performance guarantees than with empirically designed runners. The CFD computations can also be used to help with the fine tuning of the geometry following the model test.

It is important to realize that the flow in a Francis runner is a strongly three-dimensional rotational flow and that the assumption of potential flow is a strong simplification. The close proximity of the wicket gate to the highly curved meridional flow channel in the runner leads to a nonuniform meridional velocity at the outlet of the stator. This gives rise to a strongly rotational flow at the outlet of the stator vane and a severe 3-D flow pattern inside the rotor. Methods of runner analysis that do not take the rotationality into account (such as 3-D potential flow methods) or that assume meridional stream sheets (such as Q3D Euler throughflow methods) are not as effective as fully 3-D Euler methods for the analysis of the design. Currently, finite volume, 3-D Euler methods, which neglect viscous effects, are used for the design of turbines [see Goede and Rhyming (1987) and Keck *et al.* (1990)]. These are quicker and require less investment in computational power than the alternative 3-D Navier–Stokes methods which would take into account the viscous effects. The neglect of the viscous effects in the calculations is not crucial, as the viscous losses and boundary layer blockage in an accelerating turbine flow are low.

The hydraulic parameters H, Q, and ω (head, flow, and speed) determine the specific speed of the turbine which represents the starting point of any design. Based on previous experience, and taking account of any special geometrical constraints and particular manufacturing requirements, the specific speed defines a preliminary meridional contour for the runner including the position of the leading and trailing edges. An elementary meridional throughflow calculation allows a first estimate of the meridional streamlines and flow angles to be determined, from which the blading geometry can be estimated. Due consideration needs to be taken of the flow incidence at the leading edges, the deviation at the trailing edges, the vent area between the blades, the blade loading distribution, and the number of blades.

The CFD analysis for hydraulic turbines, like other turbomachinery, breaks down into several stages: 1) pre-processing, which involves definition of the geometrical boundaries of the solution domain, generation of a suitable mesh for the numerical computation and specification of the boundary conditions and initial solution, 2) solution process, which for 3-D Euler solvers requires a few thousand iteration steps for the time-marching algorithms in use, and 3) post-processing, which involves preparation of the visual data in the form of 2-D and 3-D plots from the numerical results. These methods have brought substantial improvements in the performance of Francis runners, and particularly in the achievement of design objectives for new designs with fewer model tests. Two features of the CFD computation are of particular value to the designer. First, the Euler computation, in combination with a blade geometry generator, allows fine control during the design process of the static pressure distribution on the blades. By achieving a uniform static pressure along the blade suction side, local low pressure areas with early cavitation onset can be avoided. Secondly, the CFD codes provide better estimates of the flow angles in the highly nonuniform flow downstream of the runner than earlier empirical methods. This means that the designer can avoid unwanted swirl in the exit flow and has better control over the location of the peak efficiency point in the characteristic curves.

27.9.6   Axial Turbines

Mode of Operation. Axial turbines are used in river power stations where the discharge is large and available head rather low. Whereas Pelton and Francis turbines may be used in *peaking mode* (operating only a few hours a day at best efficiency and retaining the water in the upper reservoir during the rest of the day), axial turbines are mostly operated in the run-off-river mode. River water is passed through the turbines under all head and flow conditions with only little regard to the electrical demand. The ratios H_{max}/H_{min} and Q_{max}/Q_{min} under which axial turbines have to be operated may go up to about five and ten, respectively. The following concepts are used to adapt the turbine to the variable conditions: 1) double-regulated, with adjustable stator and rotor blades, also called the full-Kaplan concept, 2) fixed blade propeller, with adjustable stator blades, but fixed rotor blades, or 3) the semi-Kaplan (or *Kapeller*) concept, with fixed stator blades and adjustable rotor blades. Double-regulated turbines are best suited for large H and Q variations but are expensive. Propeller turbines are cheaper per unit, but in order to cope with large variations of river flows, one needs a large number of units which can be turned on and off depending on the flow. Semi-Kaplan turbines are well suited for a medium degree of Q and H variation. The cost reduction over the full-Kaplan turbine, due to the cheaper stator design, is partially offset by the requirement of an additional downstream gate for emergency closure and synchronizing of the units.

Types of Axial Turbines. An overview of axial turbine designs is given in Figs. 27.140 through 27.143, and explained below.

Vertical Kaplan (or Propeller) Turbines: The classical Kaplan turbine has either a steel scroll-case (like a Francis unit) for heads between 30 and 60 cm or a concrete semi-spiral casing for heads between 10 and 40 m. The largest Kaplan turbines have runner diameters up to 10 m and may have an output up to 250 MW. In comparison to Francis turbines, Kaplan turbines allow much larger Q and H variations, have higher specific speed and higher specific discharge but require a deeper setting level relative to the tailwater level due to inferior cavitation behavior.

Horizontal Bulb (or Pit) Turbines: The design with a horizontal axis has the advantage of a more or less straight flow path through the intake and draft tube. The friction losses are considerably lower in these components than in the spiral casing and elbow draft tube of the vertical Kaplan turbine. The classical bulb turbine has a direct driven, low-speed generator in a bulb type casing within the water passage of the intake. The generator has limited this concept to sizes below 60 MW, whereby runner diameters up to 7.7 m have been built.

For very low heads and speeds it is advantageous to provide a high-speed generator connected to the turbine rotor either by a planetary gear or a bevel gear. For

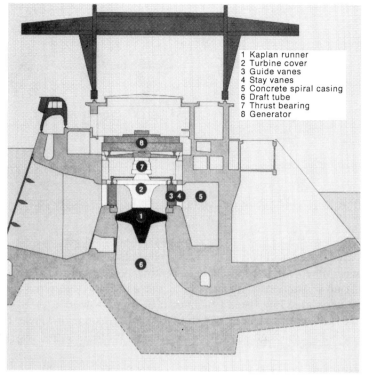

1 Kaplan runner
2 Turbine cover
3 Guide vanes
4 Stay vanes
5 Concrete spiral casing
6 Draft tube
7 Thrust bearing
8 Generator

FIGURE 27.140 Cross section of a vertical Kaplan turbine.

the first type, the so called pit turbine, units up to 20 MW and runner diameters up to 8.3 m have been built. The bevel gear turbine is typically used for small standardized units up to 3 MW with runner diameters up to 2.5 m.

Straflo Turbines: In the Straflo turbine design (*Straflo = straight flow*), the turbine and generator form an integral unit without a driving shaft. The turbine rotor blades are connected to an outer ring which directly carries the generator rotor poles. Straflo turbines are horizontal units mostly with fixed blades and have the following advan-

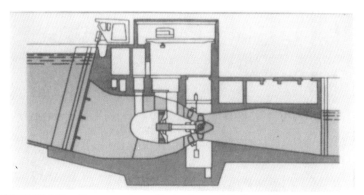

FIGURE 27.141 Cross section of a horizontal Bulb turbine with direct drive.

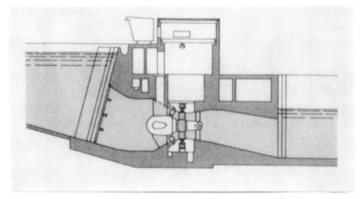

FIGURE 27.142   Cross section of a Straflo turbine with rim generator.

tages compared to classical bulb turbines: 1) reduced power house dimensions, making them well suited for integration into existing power plants and for large tidal schemes, 2) no restrictions with respect to the generator (the problems with the cooling of a very large low speed generator inside a narrow bulb are avoided), and 3) the intake structure is slender and built in solid concrete, whereas the hollow casing of the bulb turbine gives increased blockage to the flow and less stiffness to the structure.

The disadvantages are the small additional losses due to friction of the outer rim, the rim sealing, and the proper functioning of the rim sealing and its durability in cases with sand laden water.

S-Turbines: S-turbines derive their name from the S-shaped draft tube, which has an advantage for small standardized units, as it allows the generator to be mounted outside of the water passage. The S-shaped draft tube has lower efficiencies than straight draft tubes and requires some care in its design, as the flow curvature limits the amount of diffusion that is possible [Keck (1984)].

Blading Design.   The design of the rotor blades is mainly influenced by the following design features:

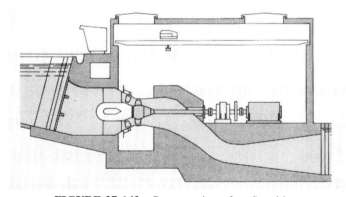

FIGURE 27.143   Cross section of an S-turbine.

Hub-to-Tip Ratio: This ranges from 0.35 for 3-blade runners up to 0.65 for 7-blade runners. Because of the large u variation over the blade span, the velocity triangles differ strongly between the root and tip section (see Fig. 27.144). The profiles at the root section have large camber and thickness (high turning of the flow and high stresses), the profiles at the tip section have small camber and thickness (low turning of the flow and low stresses).

Inflow Conditions: The stator vanes are usually designed such that they close off the flow completely at their minimum vane angle. This leads to a constant profile across the relatively large span in a Kaplan turbine and to conical vanes of similar profiles in a bulb turbine. The swirl at the stator outlet is no longer of the irrotational, potential-vortex type. As in Francis turbines, a complete flow analysis with 3-D Euler methods is therefore superior to a 3-D potential flow analysis. However, secondary flows are less pronounced than in Francis turbines and, because of this, axial runner design has for many decades been successfully based on 2-D profile theory.

Outflow Conditions: The swirl at the runner outlet strongly influences the performance of the draft tube, which is the largest contributor to the losses of low head turbines. In cases where the draft tube is designed to be close to the stability limit (onset of diffuser stall), the swirl at runner outlet, especially at the tip section, has to be carefully controlled.

Performance Curves. Typical performance curves of axial turbines are shown in Fig. 27.145. The efficiency curves at constant rotor blade angles are called propeller curves and the envelope over all the propeller curves is the so-called Kaplan curve. Based on these curves the propeller- and Kaplan hill charts are generated. To operate

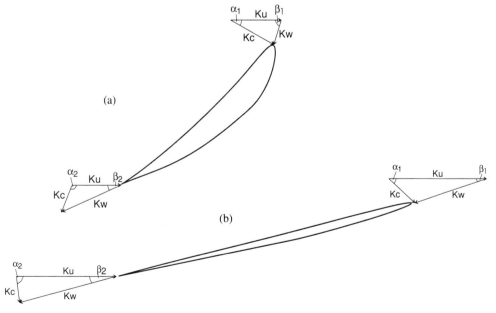

FIGURE 27.144   Velocity triangles over the span of a Kaplan turbine blade: (a) root and (b) tip.

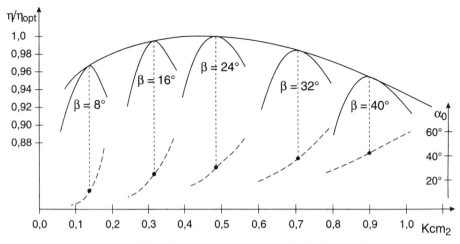

FIGURE 27.145   Performance curve of a Kaplan turbine.

the units along the peak of the Kaplan hill chart, the correct relationship between stator and rotor blade angles has to be provided, and this is done by means of a digital governor.

27.9.7 Pump Turbines

In the area of large energy storage, the most attractive and interesting technique in areas where there are suitable storage reservoirs at different altitudes is hydro-electric pump storage. Such systems use the excess energy available from thermal, nuclear, or river power plants during a period of low demand to pump water from a low level to a higher storage reservoir. At periods of peak electrical demand, or in the case of failure of electrical power plants, the stored water is allowed to flow back to the low level and used to drive a turbine to generate electricity. The overall efficiency of the energy recovery from such schemes is typically between 70 and 80%.

Pumped-storage schemes can make use of separate pumps and turbines, both of which are designed for their optimal performance at the heads and flows concerned. These can be on separate shafts with their own motor and generator, but they are usually mounted on the same shaft and coupled to the same motor-generator so that less investment in electrical equipment is needed. A further reduction in the investment costs is obtained when the storage scheme is operated with a reversible pump-turbine unit, which operates both as a pump and as a turbine with a change in the sense of rotation and flow direction. The much lower investment costs of a reversible pump-turbine have promoted the development of highly efficient pump-turbine units in which the hydraulic efficiency of the pump-turbine is well above 90% in both modes of operation. For high head applications ($H > 700$ m), a multistage pump turbine is used. For medium heads ($400 < H < 600$) a single stage pump turbine of the Francis type with adjustable wicket gates is used. For low head applications, a mixed flow pump turbine with adjustable runner blades (*Deriaz turbine*) or a Francis type device with variable speed is used.

The hydrodynamic design of a reversible pump turbine is strongly constrained by the need to avoid cavitation and to obtain high efficiency and stable flow in the pump mode. To achieve high efficiency and stable flow for the whole head range while operating as a pump, the deceleration of the flow in the impeller has to be controlled carefully. This, together with the need to avoid cavitation at the leading edge, determines the selection of blade angles and flow channel areas.

As a consequence of the blade angles being determined by the pump mode, the unit does not operate at its best efficiency point when operating as a turbine; it operates at a lower head than that at peak efficiency. To shift the peak efficiency for the turbine mode to lower heads would require much higher blade loading in the pump mode, with severe consequences on its performance and stability. The accelerating flow in the turbine mode is much less sensitive to deviations in the flow angles from the optimal values, so it has been usual practice in the past to design a pump-turbine impeller as a pump. Its performance as a turbine is then simply accepted. (It is interesting to note that thanks to the accelerating flow, the efficiency in the turbine mode is still generally higher than that in the pump mode.) The use of modern 3-D simulation methods for flowfield calculations has contributed to improved understanding of possible design compromises for the operation of pumps as turbines [Goede *et al.* (1989)].

27.9.8 Special Fluid Dynamic Problems in Hydraulic Turbines

Cavitation. Cavitation occurs in flows of water when, due to regions of high flow velocity, the local static pressure decreases below the vapor pressure and the water effectively begins to boil. There are three main areas where this can occur in hydraulic turbines: 1) attached to the blade on the blade suction surface in areas of low local static pressure, 2) at the runner leading edge when operating with variable head (off-design) (this can occur on both sides of the blade, at low head operation it tends to be on the pressure side), and 3) inside the vortices associated with the tip leakage vortex on axial runners, the draft tube vortex in Francis machines, and also in vortices occurring inside the blade passages of Francis turbines when operating at extreme off-design operating points.

The effects of cavitation are harmful, both on performance and on the erosion of materials. Erosion is caused by the extremely high pressure peaks (*microjets* or shock waves) which occur during the implosion of the cavitation bubbles in the vicinity of a material surface.

The design attempts to avoid cavitation by optimizing the pressure distribution on the blades to avoid areas of high local velocity. In addition, the setting level of the turbine relative to the tailwater can be reduced (with higher civil engineering costs), the turbine can be designed for larger diameter and lower flow velocities (with higher turbine costs) or special cavitation resistant steel or welding overlays can be used. Typical industrial design practice is to design for bubble free operation in turbines operating with $H > 200$ m, otherwise the pitting rate would be unacceptably high. For machines with heads in the range $20 > H > 200$ m, the turbine is designed to be bubble free at the design point, but small areas of cavitation tend to be accepted at off-design points as a trade-off between investment costs and the costs of cavitation repair. For low heads $H < 20$ m, there is no erosion on modern stainless steels, and cavitation is a concern in the design only in terms of its effect on the efficiency and power output.

Sand Erosion. A further problem that can also lead to material erosion in hydraulic turbines is abrasion by sand particles carried by the flow. This affects the stator vanes, runner vanes, and the labyrinth seal flow channels in reaction machines and the buckets and nozzles in Pelton turbines. Turbines installed in areas where the water is heavily sand-laden require frequent revision and repair.

Two types of abrasion can be distinguished, namely, that caused by direct bombardment of the sand particles almost at right angles to the surface and that where the surface experiences a glancing blow from sand particles at low angles of impingement. The rates of material erosion for both types are linearly dependent on the hardness and concentration of the sand particles in the flow, and on the third power of the flow velocity [see Krause and Grein (1993)]. Recently, the use of ceramic coatings for the components most prone to wear has proved successful in increasing the interval between major revisions. CFD computations of the sand particle paths and their interaction with the turbine components are useful for predicting erosion rates.

Dynamic Behavior. Turbines may be subject to relatively fast changes of their operating conditions, with typical time intervals of a few seconds for load rejection or one second per period for pressure surges. The dynamic response of turbines to the unsteady pressure fields caused by these changes is an active area of research [see Dörfler (1982)].

An important pressure pulsation and power variation excited by Francis turbines at off-design conditions is known as *draft tube surging*. Two flow phenomena are generally responsible for this. One is the distortion of the radial flow profile downstream of the turbine at part load, which can give rise to a central back-flow region and a nonsymmetric flow pattern with one or more strong vortex filaments that precess around the draft tube axis (vortex breakdown of the spiral type). The second important feature is cavitation along the axis of the vortex in the draft tube. This produces a free internal surface and a zone of highly compressible flow that changes the dynamic characteristics of the turbine system. Traditional remedies for these pressure pulsations consist either in acting directly on the swirl by mounting two to four axial fins on the draft tube wall or by changing the compressibility by admission of air into the vortex core.

27.10   STEAM TURBINES
William G. Steltz

27.10.1   Historical Background

The process of generating power depends on several energy conversion processes starting with the chemical energy in fossil fuels or the nuclear energy within the atom. This energy is converted to thermal energy which is then transferred to the working fluid, in our case, steam. This thermal energy is converted to mechanical energy with the help of a high speed turbine rotor and a final conversion to electrical energy is made by means of an electrical generator in the electrical power generation application. The presentation in this section focuses on the electrical power application, but is also relevant to other applications such as ship propulsion.

Throughout the world, the power generation industry relies primarily on the steam turbine for the production of electrical energy. In the USA, approximately 77% of installed power generating capacity is steam turbine driven. Of the remaining 23%, hydroelectric installations contribute 13%, gas turbines account for 9% and the remaining 1% is split between geothermal, diesel, and solar power sources. In effect, over 99% of electric power generated in the United States is developed by turbomachinery of one design or another, with steam turbines carrying by far the greatest share of the burden.

Steam turbines have had a long and eventful life since their initial practical development in the late 19th century due primarily to efforts led by C. A. Parsons and G. deLaval. Significant developments came quite rapidly in those early days in the fields of ship propulsion and later in the power generation industry. Steam conditions at the throttle progressively climbed, contributing to increases in power production and thermal efficiency. The recent advent of nuclear energy as a heat source for power production had an opposite effect in the late 1950's. Steam conditions tumbled to accommodate reactor designs, and unit heat rates underwent a step change increase. By this time, fossil unit throttle steam conditions had essentially settled out at 2400 psi and 1000 F with single reheat to 1000 F. Further advances in steam power plants were achieved by the use of once-through boilers delivering supercritical pressure steam at 3500 to 4500 psi. A unique steam plant utilizing advanced steam conditions is Eddystone No. 1, designed to deliver steam at 5000 psi and 1200 F to the throttle, with reheat to 1050 F and second reheat also to 1050 F.

Unit sizes increased rapidly in the period from 1950 to 1970; the maximum unit size increased from 200 MW to 1200 MW (a six-fold increase) in this span of 20 years. In the 1970's, unit sizes stabilized with new units generally rated at substantially less than the maximum size. At the present time, however, the expected size of new units is considerably less, appearing to be in the range of 350 to 500 MW.

In terms of heat rate (or thermal efficiency) the changes have not been so dramatic. A general trend showing the reduction of power station heat rate over an 80 year period is presented in Fig. 27.146. The advent of regenerative feedwater heating in the 1920's brought about a step change reduction in heat rate. A further reduction was brought about by the introduction of steam reheating. Gradual improvements continued in steam systems and were recently supplemented by the technology of the combined cycle, the gas turbine/steam turbine system (see Fig. 27.147). In the same period of time that unit sizes changed by a factor or six (1950 to 1970), heat rate diminished by less than 20%, a change which includes the combined cycle. In reality, the improvement is even less as environmental regulations and the energy required to satisfy them can consume up to 6% or so of a unit's generated power.

The rate of improvement of turbine cycle heat rate is obviously decreasing. Power plant and machinery designers are working hard to achieve small improvements both in new designs and in retrofit and repowering programs tailored to existing units. Considering the worth of energy, what then are our options leading to thermal performance improvements and the management of our energy and financial resources? Exotic energy conversion processes are a possibility: MHD, solar power, the breeder reactor, and fusion are some of the longer range possibilities. A more near-term possibility is through the improvement (increase) of steam conditions. The effect of improved steam conditions on turbine cycle heat rate is shown in Fig. 27.148, where

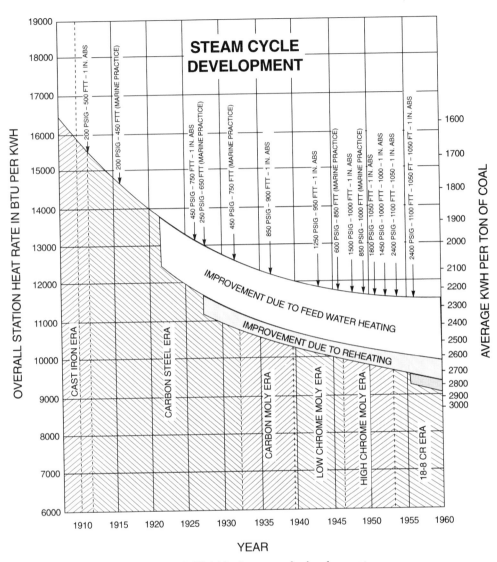

FIGURE 27.146   Steam cycle development.

heat rate is plotted as a function of throttle pressure with parameters of steam temperature level. The plus mark indicates the placement of the Eddystone unit previously mentioned.

27.10.2   The Heat Engine and Energy Conversion Processes

The mechanism for conversion of thermal energy is the *heat engine*, a thermodynamic concept, defined and sketched out by Carnot, and applied by many, the power generation industry in particular. The heat engine is a device which accepts thermal energy (heat) as input and converts this energy to useful work. In the process, it rejects a portion of this supplied heat as unusable by the work production process.

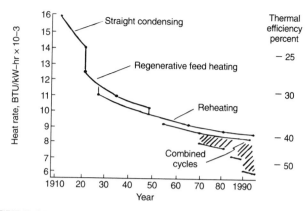

FIGURE 27.147 Fossil fueled unit heat rate as a function of time.

The efficiency of the ideal conversion process is known as the *Carnot efficiency*. It serves as a guide to the practitioner and as a limit for which no practical process can exceed. The Carnot efficiency is defined in terms of the absolute temperatures of the heat source, T_{hot}, and the heat sink, T_{cold}, as follows

$$\text{Carnot efficiency} = \frac{T_{hot} - T_{cold}}{T_{hot}} \tag{27.247}$$

Consider Fig. 27.149 which depicts a heat engine in fundamental terms consisting of a quantity of heat supplied, heat added, a quantity of heat rejected, heat rejected, and an amount of useful work done, work done. The *thermal efficiency* of this basic engine can be defined as

$$\text{efficiency} = \frac{\text{work done}}{\text{heat added}} \tag{27.248}$$

This thermal efficiency is fundamental to any heat engine and is, in effect, a measure of the heat rate of any turbine-generator unit of interest. Figure 27.150 is

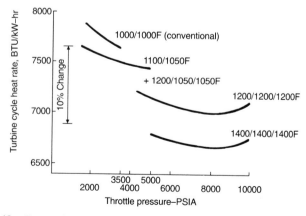

FIGURE 27.148 Comparison of turbine cycle heat rate as a function of steam conditions.

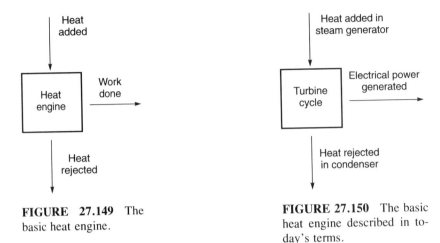

FIGURE 27.149 The
basic heat engine.

FIGURE 27.150 The basic
heat engine described in today's terms.

the same basic heat engine redefined in terms of turbine cycle terminology, i.e., heat
added is the heat input to the steam generator, heat rejected is the heat removed by
the condenser, and the difference is the work done (power) produced by the turbine
cycle. Figure 27.151 is a depiction of a simple turbine cycle showing the same
parameters, but described in conventional terms. *Heat rate* is now defined as the
quantity of heat input required to generate a unit of electrical power (kW).

$$\text{heat rate} = \frac{\text{heat added}}{\text{work done}} \qquad (27.249)$$

The units of heat rate are usually in terms of Btu/kW-hr.
    Further definition of the turbine cycle is presented in Fig. 27.152 which shows
the simple turbine cycle with pumps and a feedwater heater included (of the open

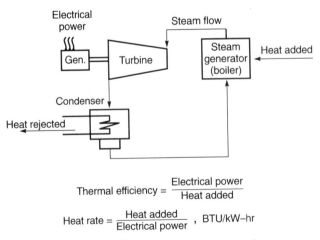

$$\text{Thermal efficiency} = \frac{\text{Electrical power}}{\text{Heat added}}$$

$$\text{Heat rate} = \frac{\text{Heat added}}{\text{Electrical power}} \text{ , BTU/kW–hr}$$

FIGURE 27.151   A simple turbine cycle.

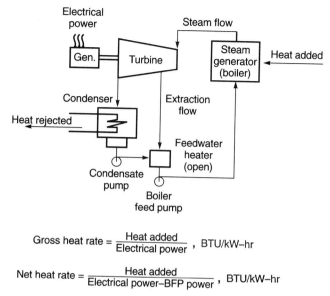

$$\text{Gross heat rate} = \frac{\text{Heat added}}{\text{Electrical power}}, \text{ BTU/kW–hr}$$

$$\text{Net heat rate} = \frac{\text{Heat added}}{\text{Electrical power–BFP power}}, \text{ BTU/kW–hr}$$

FIGURE 27.152   A simple turbine cycle with an open heater and a boiler feed pump.

type). In this instance, two types of heat rate are identified: 1) a *gross heat rate*, in which the turbine-generator set's natural output (i.e., gross electrical power) is the denominator of the heat rate expression, and 2) a *net heat rate*, in which the gross power output has been debited by the power requirement of the boiler feed pump resulting in a larger numeric value of heat rate. This procedure is conventional in the power generation industry, as it accounts for the inner requirements of the cycle needed to make it operate. In other more complex cycles, the boiler feed pump power might be supplied by a steam turbine driven feed pump, these effects are then included in the heat balance describing the unit's performance.

The same accounting procedures are true for all cycles, regardless of their complexity. A typical 450 MW fossil unit turbine cycle heat balance is presented in Fig. 27.153. Steam conditions are 2415 psia/1000 F/1000 F/2.5 inHga, and the cycle features seven feedwater heaters and a motor driven boiler feed pump. Only pertinent flow and steam property parameters have been shown in order to avoid confusion and to support the conceptual simplicity of heat rate. As shown in the two heat rate expressions, only two flow rates, four enthalpies, and two kW values are required to determine the gross and net heat rates of 8044 and 8272 Btu/kW-hr, respectively.

To supplement the fossil unit of Fig. 27.153, Fig. 27.154 presents a typical nuclear unit of 1000 MW capability. Again, only the pertinent parameters are included in this sketch for simplicity. Steam conditions at the throttle are 690 psi with $\frac{1}{4}$% moisture, and the condenser pressure is 3.0 inHga. The cycle features six feedwater heaters, a steam turbine driven feed pump, and a moisture separator reheater (MSR). The reheater portion of the MSR takes throttle steam to heat the low pressure (LP) flow to 473 F from 369 F (saturation at 164 psia). In this cycle, the feed pump is turbine driven by steam taken from the MSR exit, hence only one heat rate is shown, the net heat rate, 10,516 Btu/kW-hr. This heat rate comprises only four numbers,

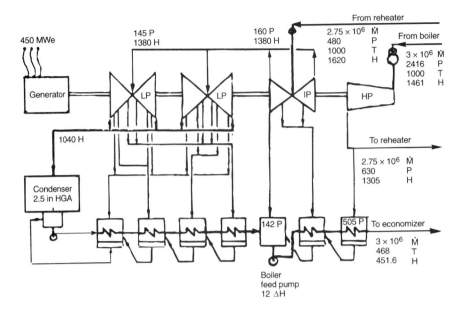

$$\text{Gross heat rate} = \frac{3{,}000{,}000 \, (1461 - 451.6) + 2{,}760{,}000 \, (1520 - 1305)}{450{,}000} = 8044 \text{ BTU/kW–hr}$$

$$\text{Net heat rate} = \frac{3{,}000{,}000 \, (1461 - 451.6) + 2{,}760{,}000 \, (1520 - 1305)}{450{,}000 - 12{,}400} = 8272 \text{ BTU/kW–hr}$$

FIGURE 27.153 Typical fossil unit turbine cycle heat balance.

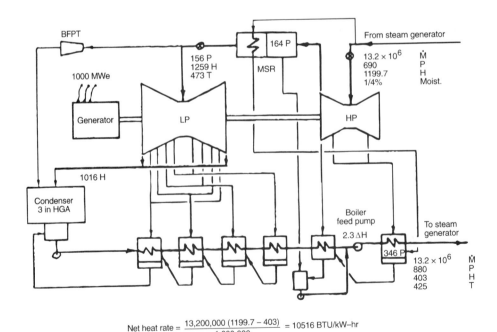

$$\text{Net heat rate} = \frac{13{,}200{,}000 \, (1199.7 - 403)}{1{,}000{,}000} = 10516 \text{ BTU/kW–hr}$$

FIGURE 27.154 Typical nuclear unit turbine cycle heat balance.

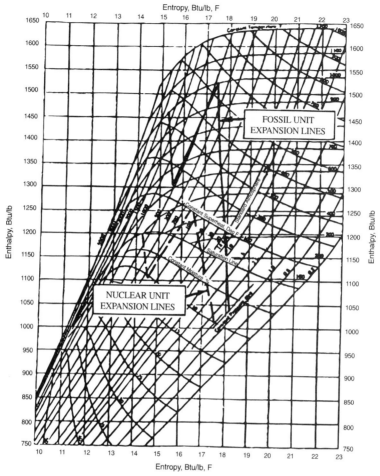

FIGURE 27.155 Fossil and nuclear unit turbine expansion lines superimposed on the Mollier diagram.

the throttle mass flow rate, the throttle enthalpy, the final feedwater enthalpy, and the net power output of the cycle.

For comparative purposes the expansion lines of the fossil and nuclear units of Figs. 27.153 and 27.154 have been superimposed on the *Mollier diagram* of Fig. 27.155. It is easy to see the great difference in steam conditions encompassed by the two designs and to relate the ratio of cold to hot temperatures to their Carnot efficiencies. In the terms of Carnot, the maximum fossil unit thermal efficiency would be 61%, and the maximum nuclear unit thermal efficiency would be 40%. The ratio of these two Carnot efficiencies (1.53) compares somewhat favorably with the ratio of their net heat rates (1.27).

To this point, emphasis has been placed on the *conventional* steam turbine cycle, where conventional implies the central station power generating unit whose energy source is either a fossil fuel (coal, oil, gas) or a fissionable nuclear fuel. Figure 27.147 has shown a significant improvement in heat rate attributable to *combined cycle* technology, i.e., the marriage of the gas turbine used as a *topping unit* and the steam turbine used as a *bottoming unit*. The cycle efficiency benefits come from the high firing temperature level of the gas turbine, current units in service are operating at 2300 F, and the utilization of its waste heat to generate steam in a heat recovery steam generator (HRSG). Figure 27.156 is a heat balance diagram of a simplified combined cycle showing a two pressure level HRSG. The purpose of the two pressure level (or even three pressure level) HRSG is the minimization of the temperature differences existing between the gas turbine exhaust and the evaporating water/steam mixture. Second Law analyses (commonly termed *availability* or *exergy analyses*) result in improved cycle thermal efficiency when integrated average values of the various heat exchanger temperature differences are small. The smaller, the better from an efficiency viewpoint, however the smaller the temperature difference, the larger the required physical heat transfer area. These Second Law results are then reflected by the cycle heat balance which is basically a consequence of the First Law of thermodynamics (conservation of energy) and the conservation of mass. As implied by Fig. 27.156, a typical combined cycle schematic, the heat rate is about 6300 Btu/kW-hr, and the corresponding cycle thermal efficiency is about 54%, about ten points better than a conventional stand-alone fossil steam turbine cycle.

A major concept of the Federal Energy Policy of 1992 is the attainment of an Advanced Turbine System (ATS) thermal efficiency of 60% by the year 2000. Needless to say, significant innovative approaches will be required in order to achieve this ambitious level. The several approaches to this end include the increase of gas turbine inlet temperature and probably pressure ratio, reduction of cooling flow requirements, and generic reduction of blade path aerodynamic losses. On the steam turbine side, reduction of blade path aerodynamic losses and most likely increased inlet steam temperatures to be compatible with the gas turbine exhaust temperature are required.

A possibility, one that is undergoing active development, is the use of an ammonia/water mixture as the working fluid of the gas turbine's bottoming cycle in

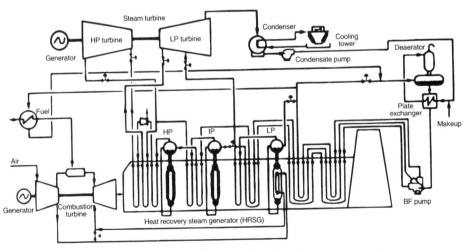

FIGURE 27.156   A typical combined cycle plant schematic.

place of pure water. This concept known as the *Kalina Cycle* [Kalina (1993)] prom-
ises a significant improvement to cycle thermal efficiency primarily by means of the
reduction of losses in system *availability*. Physically, a practical ammonia/water
system requires a number of heat exchangers, pumps and piping, and a turbine which
is smaller than its steam counterpart due to the higher pressure levels that are a
consequence of the ammonia/water working fluid.

27.10.3 Selected Steam Thermodynamic Properties

Steam has had a long history of research applied to the determination of its thermodynamic and transport properties. The currently accepted description of steam's
thermodynamic properties is the ASME Properties of Steam publication [ASME
(1967)]. The Mollier diagram, the plot of enthalpy versus entropy, is the single most
significant and useful steam property relationship applicable to the steam turbine
machinery and cycle designer/analyst (see Fig. 27.157).

There are, however, several other parameters which are just as important and
which require special attention. Although not a perfect gas, steam may be treated as
such provided the appropriate perfect gas parameters are used for the conditions of
interest. The cornerstone of perfect gas analysis is the requirement that $pv = RT$.
For nonperfect gasses a factor Z may be defined such that $pv = RZT$ where the
product RZ in effect replaces the particular gas constant R. For steam this relationship is described in Fig. 27.158 where RZ has been divided by J, Joule's constant.

A second parameter pertaining to perfect gas analysis is the isentropic expansion
exponent given in Fig. 27.159. (The definition of the exponent is given in the caption on the figure.) Note that the value of γ well represents the properties of steam
for a *short* isentropic expansion. It is the author's experience that accurate results
are achievable at least over a $2:1$ pressure ratio using an average value of the exponent.

The first of the derived quantities relates the critical flow rate of steam [Steltz
(1961)] to the flow system's inlet pressure and enthalpy as in Fig. 27.160. The
critical (maximum) mass flow rate, $\dot{M}$, assuming an isentropic expansion process
and equilibrium steam properties, is obtained by multiplying the ordinate value, K,
by the inlet pressure, p_1, in psia and the passage throat area, A, in square inches

$$\dot{M}_{\text{critical}} = p_1 K A \tag{27.250}$$

The actual steam flow rate can then be determined as a function of actual operating
conditions and geometry.

The corresponding choking velocity (acoustic velocity in the superheated steam
region) is shown in Figs. 27.161 and 27.162 for superheated steam and wet steam,
respectively. The range of Mach numbers experienced in steam turbines can be put
in terms of the *wheel speed Mach number*, i.e., the rotor tangential velocity divided
by the local acoustic velocity. In the HP turbine, wheel speed is on the order of 600
ft/sec, while the acoustic velocity at 2000 psia and 975 F is about 2140 ft/sec, hence
the wheel speed Mach number is 0.28. For the last rotating blade of the LP turbine,
its tip wheel speed could be as high as 2050 ft/sec. At a pressure level of 1.0 psia
and an enthalpy of 1050 Btu/lb, the choking velocity is 1275 ft/sec, hence the wheel
speed Mach number is 1.60. As Mach numbers relative to either the stationary or

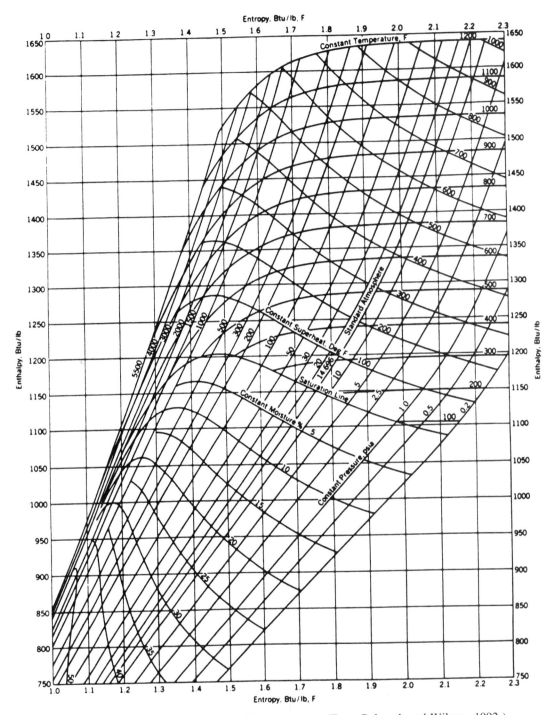

FIGURE 27.157 Mollier diagram (*h-s*) for steam. (From Babcock and Wilcox, 1992.)

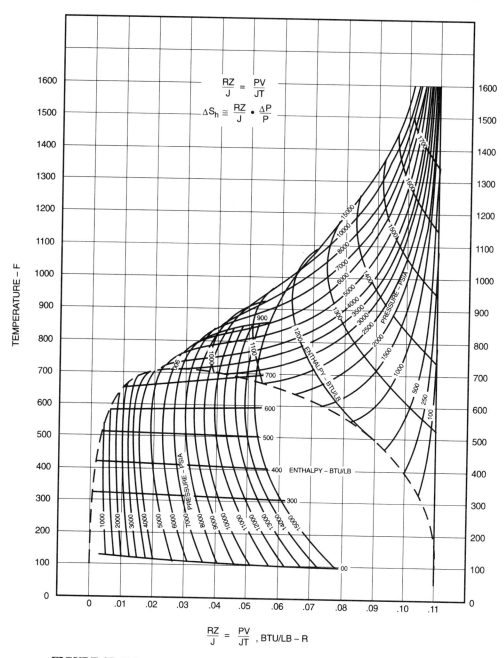

FIGURE 27.158 (RZ/J) for steam and water. (From Westinghouse, 1969.)

rotating blading are approximately comparable, the steam turbine designer must negotiate flow regimes from incompressible flow, low subsonic Mach number of 0.3, to supersonic Mach numbers on the order of 1.6.

Another quite useful characteristic of steam is the product of pressure and specific volume plotted versus enthalpy in Figs. 27.163 and 27.164 for low temperature/wet

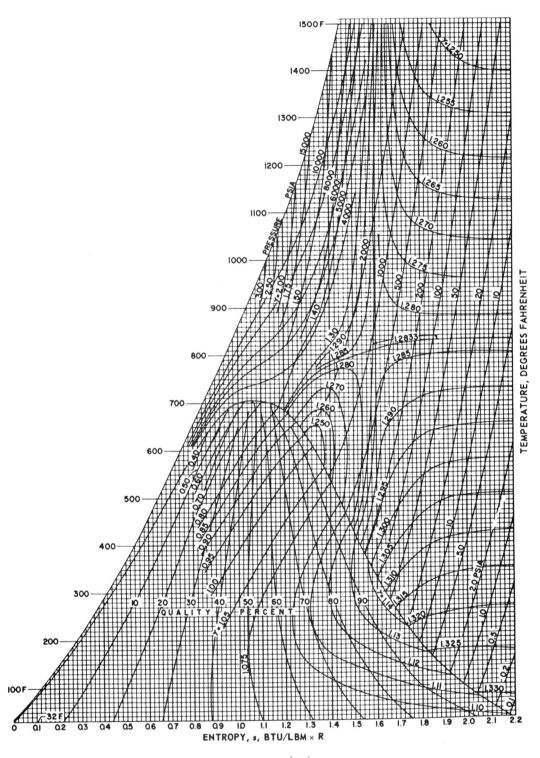

FIGURE 27.159 Isentropic exponent, $\gamma = -\dfrac{v}{p}\left(\dfrac{\partial p}{\partial v}\right)_s$. $pv^{\gamma} = $ constant for a short expansion. (From ASME, 1967.)

2434

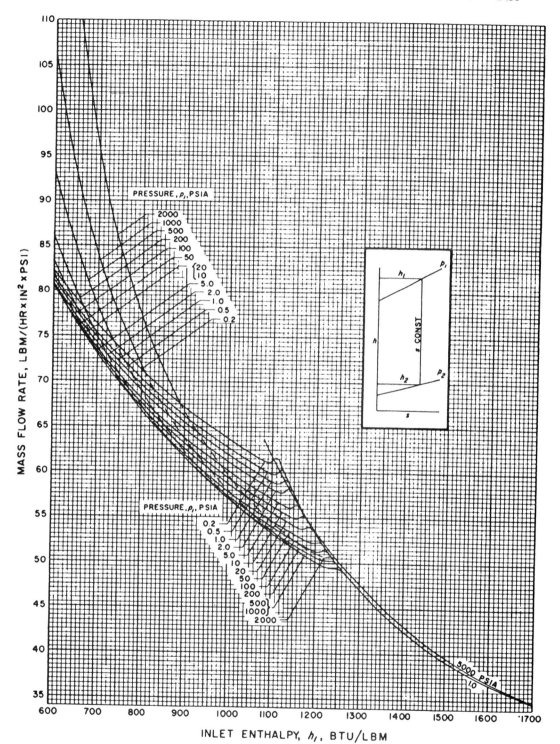

FIGURE 27.160 Critical (choking) mass flow rate for isentropic process and equilibrium conditions. (From ASME, 1967.)

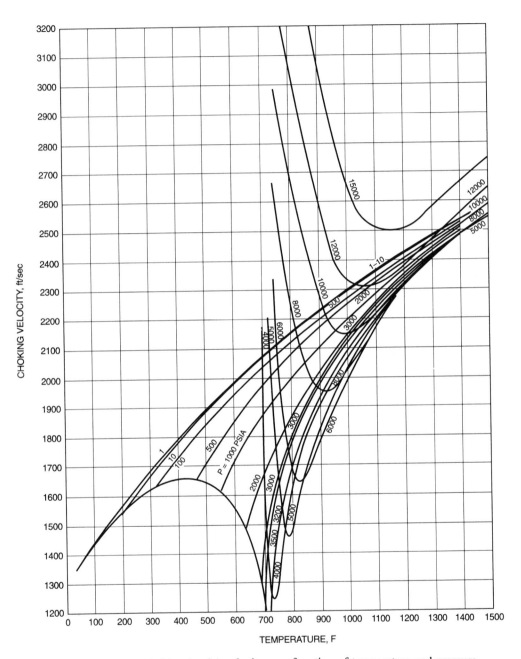

FIGURE 27.161   Choking (sonic) velocity as a function of temperature and pressure.

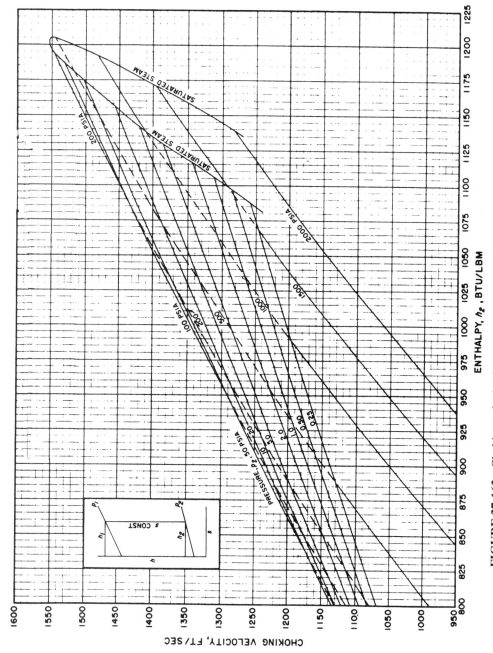

FIGURE 27.162 Choking velocity for water-steam mixture for isentropic process and equilibrium conditions. (From ASME, 1967.)

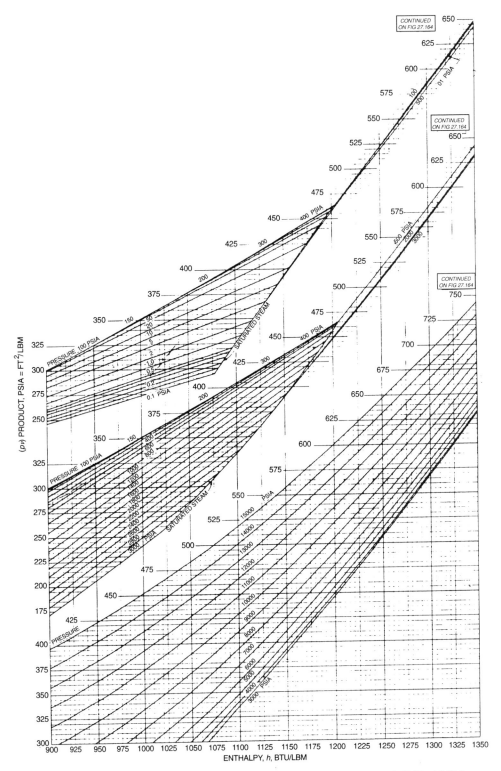

FIGURE 27.163   (ρv) product for low temperature steam. (From ASME, 1967.)

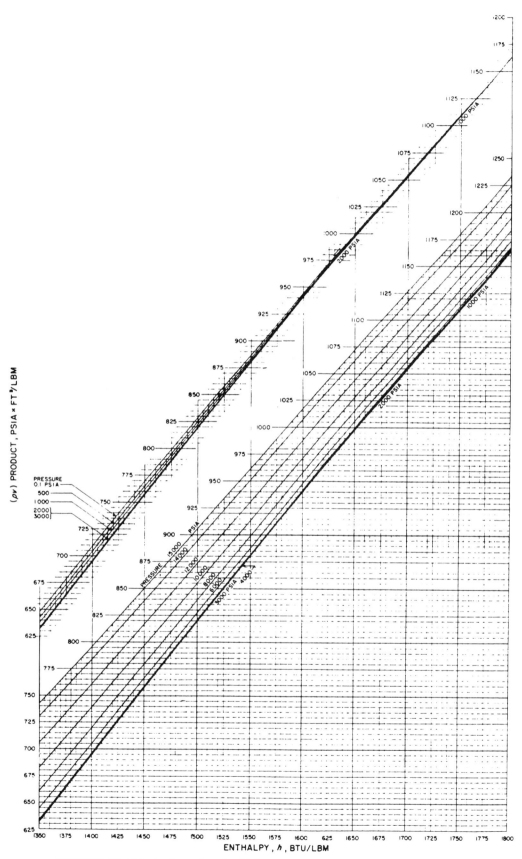

FIGURE 27.164 (ρv) product for high temperature steam. (From ASME, 1967.)

steam and superheated steam, respectively. If the fluid were a perfect gas, this plot would be a straight line. In reality, it is a series of nearly straight lines, with pressure as a parameter. A significant change occurs in the wet steam region where the pressure parameters spread out at a slope different from that of the superheated region. These plots are quite accurate for determining specific volume and for computing the often used *flow number*

$$\frac{\dot{M}\sqrt{pv}}{p} \tag{27.251}$$

A direct application of the above mentioned approximations is the treatment by perfect gas analysis techniques of applications where the working fluid is a mixture of air and a significant amount of steam (*significant* implies greater than 2 to 4%). Not to limit the application to air and steam, the working fluid could be the products of combustion and steam, or other arbitrary gases and steam.

27.10.4 Blade Path Design

The accomplishment of the thermal to mechanical energy conversion process in a steam turbine is, in general, achieved by successive expansion processes in alternate stationary and rotating blade rows. The turbine is a heat engine, working between the thermodynamic boundaries of maximum and minimum temperature levels, and as such is subject to the laws of thermodynamics prohibiting the achievement of engine efficiencies greater than that of a Carnot cycle. The turbine is also a dynamic machine in that the thermal to mechanical energy conversion process depends on blading forces, traveling at rotor velocities, developed by the change of momentum of the fluid passing through a blade passage. The laws of nature as expressed by Carnot and Newton govern the turbine designer's efforts and provide common boundaries for his achievements.

The purpose of this section is to present the considerations involved in the design of steam turbine blade paths and to indicate means by which these concerns have been resolved. The means of design resolution are not unique. An infinite number of possibilities exist to achieve the specific goals, such as efficiency, reliability, cost, and time. The real problem is the achievement of all these goals simultaneously in an optimum manner, and even then, there are many approaches and many very similar solutions.

Thermal to Mechanical Energy Conversion. The purpose of turbomachinery blading is the implementation of the conversion of thermal energy to mechanical energy. The process of conversion is by means of curved airfoil sections which accept incoming steam flow and redirect it in order to develop blading forces and resultant mechanical work. The force developed on the blading is equal and opposite to the force imposed on the steam flow, i.e., the force is proportional to the change of fluid momentum passing through the turbine blade row. The magnitude of the force developed is determined by application of Newton's laws to the quantity of fluid occupying the flow passage between adjacent blades, a space which is in effect, a *streamtube*. The assumptions are made that the flow process is steady and that flow conditions are uniform upstream and downstream of the flow passage.

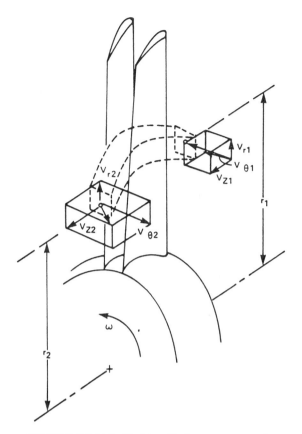

FIGURE 27.165 Turbine blade row streamtube.

Figure 27.165 presents an isometric view of a turbine blade row defining the components of steam velocity at the inlet and exit of the blade passage streamtube. The tangential force impressed on the fluid is equal to the change of fluid momentum in the tangential direction by direct application of Newton's laws.

$$F_{\theta \text{ on fluid}} = \int_{A_2} V_\theta \, d\dot{M} - \int_{A_1} V_\theta \, d\dot{M} \qquad (27.252)$$

$$F_{\theta \text{ on fluid}} = \int_{A_2} \rho_2 V_{Z2} V_{\theta_2} \, dA - \int_{A_1} \rho_1 V_{Z1} V_{\theta_1} \, dA \qquad (27.252a)$$

Since the flow is assumed steady and the velocity components are assumed constant within the streamtube upstream and downstream of the blade row, the force on the fluid is

$$F_{\theta \text{ on fluid}} = \frac{\dot{M}}{g} (V_{\theta_2} - V_{\theta_1}) \qquad (27.252b)$$

If the entering and leaving streamtube radii are identical, the power, P, developed by the moving blade row is

$$P = \frac{\dot{M}}{g} r\omega(V_{\theta 1} - V_{\theta 2}) \tag{27.253}$$

where the algebraic sign is changed to determine the fluid force on the blade. With unequal radii, implying unequal tangential blade velocities, the change in angular fluid momentum requires the power relation to be

$$P = \frac{\dot{M}}{g} \omega(r_1 V_{\theta 1} - r_2 V_{\theta 2}) \tag{27.253a}$$

Consideration of the general energy relationship (the First Law) expressed as

$$\delta Q - \delta W = du + d(pv) + \frac{V dV}{g} + dz \tag{27.254}$$

(Q is heat, W is work, and u is internal energy) and applied to the inlet and exit of the streamtube yields

$$\delta W = dh + \frac{V dV}{g} = dh_0 \tag{27.255}$$

where h is static enthalpy and h_0 is stagnation enthalpy. The change in total energy content of the fluid must equal that amount absorbed by the moving rotor blade in the form of mechanical energy

$$h_{01} - h_{02} = \frac{\omega}{g} (r_1 V_{\theta 1} - r_2 V_{\theta 2}) \tag{27.256}$$

or

$$h_1 + \frac{V_1^2}{2g} - h_2 - \frac{V_2^2}{2g} = \frac{\omega}{g} (r_1 V_{\theta 1} - r_2 V_{\theta 2}) \tag{27.256a}$$

Both of these may be recognized as alternate forms of *Euler's turbine equation.*

In the event the radii r_1 and r_2 are the same, Euler's equation may be expressed in terms of the velocity components relative to the rotor blade as

$$h_{01} - h_{02} = \frac{U}{g} (W_{\theta 1} - W_{\theta 2}) \tag{27.256b}$$

or

$$h_1 + \frac{V_1^2}{2g} - h_2 - \frac{V_2^2}{2g} = \frac{U}{g} (W_{\theta 1} - W_{\theta 2}) \tag{27.256c}$$

Turbine Stage Designs. The means of achieving the change in tangential momentum, and hence blade force and work, are many and result in varying turbine stage designs. Stage design and construction falls generally into two broad categories, *impulse* and *reaction*. The former implies that the stage pressure drop is taken entirely in the stationary blade passage, and that the relative entering and leaving rotor blade steam velocities are equal in magnitude. Work is achieved by the redirection of flow through the blade without incurring additional pressure decrease. At the other end of the scale, one could infer that for the reaction concept, all the stage pressure drop is taken across the rotor blade, while the stator blade merely redefines steam direction. The modern connotation of reaction is that of *50% reaction*, wherein half the pressure drop is accommodated in both stator and rotor. Reaction can be defined in two ways, on an enthalpy drop or a pressure drop basis. Both definitions ratio the change occurring in the rotor blade to the stage change. The pressure definition is more commonly encountered in practice.

Figure 27.166 presents a schematic representation of impulse and reaction stage designs. The impulse design requires substantial stationary diaphragms to withstand the stage pressure drop and tight sealing at the inner diameter to minimize leakage. The reaction design (50%) can accept somewhat greater seal clearances under the stator and over the rotor blade and still achieve the same stage efficiency.

The work per stage potential of the impulse design is substantially greater than that of the reaction design. A comparison is presented in Fig. 27.167 of various types of impulse designs with the basic 50% reaction design. A typical impulse stage (*Rateau stage*) can develop 1.67 times the work output of the reaction stage; for a given energy availability, this results in 40% fewer stages.

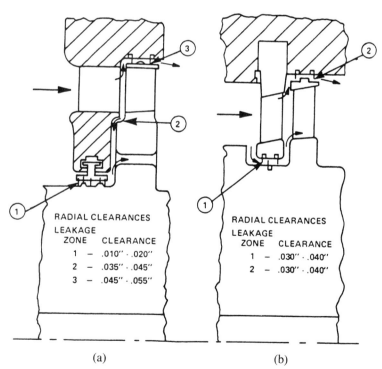

RADIAL CLEARANCES
LEAKAGE
ZONE   CLEARANCE
1 — .010" - .020"
2 — .035" - .045"
3 — .045" - .055"

RADIAL CLEARANCES
LEAKAGE
ZONE   CLEARANCE
1 — .030" - .040"
2 — .030" - .040"

(a)   (b)

FIGURE 27.166   Comparison of impulse and reaction stage geometries. (a) Impulse construction and (b) reaction construction.

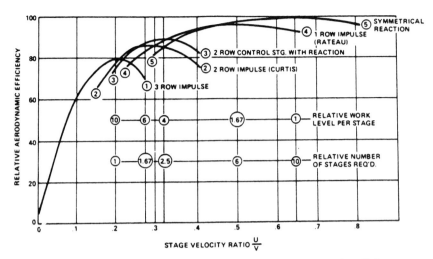

FIGURE 27.167   The effect of stage design on aerodynamic efficiency.

Performance characteristics of these designs are fundamentally described by the variation of stage aerodynamic efficiency as a function of velocity ratio. The *velocity ratio* can be defined many ways dependent on the reference used to define the steam speed, V. The velocity ratio in general is

$$\nu = \frac{U}{V} \qquad (27.257)$$

where U is the blade tangential velocity.

The relative level of stage aerodynamic efficiency is indicated in Fig. 27.167 for these various geometries ranging from the symmetrical reaction design (50% reaction) to a three rotating row impulse design. As the work level per design increases, the aerodynamic efficiency decreases. Some qualifying practicalities prevent this from occurring precisely this way in practice. That is, the impulse concept can be utilized in the high pressure region of the turbine where densities are high and blade heights are short. When blade heights necessarily increase in order to pass the required mass flow, significant radial changes occur in the flow conditions resulting in radial variations in pressure and, hence, reaction. All impulse designs have some amount of reaction dependent on the ratio of blade hub to tip diameters. Figure 27.168 presents the variation of reaction in a typical impulse stage as a function of the blade diameter/base diameter ratio.

Further complications arise when comparing the impulse to the reaction stage design. The rotor blade turning is necessarily much greater in the impulse design, the work done can be some 67% greater, and the attendant blade section aerodynamic losses tend to be greater. As an extreme case, Fig. 27.169 presents blade losses as a function of turning angle, indicating much greater losses incurred by the impulse rotor blade design.

Test data is, of course, the best source of information defining these performance relationships. Overall turbine test data provides the stage characteristic which includes all losses incurred. The fundamental blade section losses are compounded by

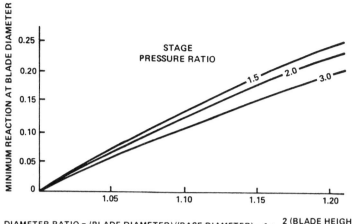

FIGURE 27.168 Blade reaction required to prevent negative base reaction.

leakages around blade ends, three dimensional losses due to finite length blades and their end walls, moisture losses if applicable, disk and shroud friction, and blading incidence losses induced by the variation in the velocity ratio itself. A detailed description of losses has been presented by Craig and Cox (1970) and more recently by others including Kacker and Okapuu (1982).

Stage Performance Characteristics. Referring again to Fig. 27.165, the velocities relative to the rotor blade are related to the stator velocities and to the wheel speed by means of the appropriate velocity triangles as in Fig. 27.170. Absolute velocities are denoted by V and relative velocities by W. The blade sections schematically shown in this figure are representative of reaction blading. The concepts and relationships are representative of all blading.

The transposition of these velocities and accompanying steam conditions are presented on a Mollier diagram for superheated steam in Fig. 27.171. The stage work

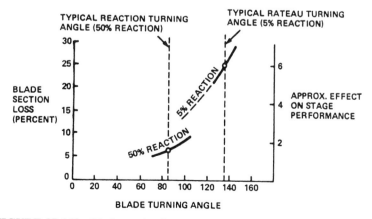

FIGURE 27.169 Blade section losses as a function of turning and reaction.

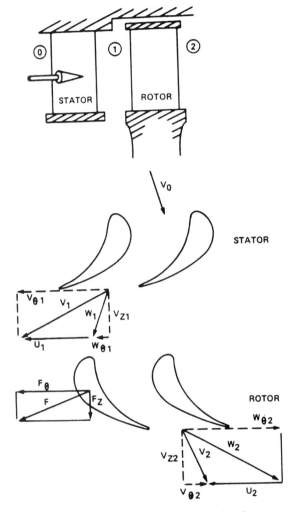

FIGURE 27.170   Stage velocity triangles.

will be compatible with Euler's turbine equation shown in Eqs. (27.256). The *stage work* is the change in total enthalpy ($\Delta h_w = h_{00} - h_{02}$) of the fluid passing through the stage. The total pressures and temperatures before and after the stage, the local thermodynamic conditions, and total pressure and total temperature relative to the rotor blade (p_{0r} and T_{0r}) are all indicated in Fig. 27.171.

This ideal description of the steam expansion process defines the local conditions of a particular streamtube located at a certain blade height. For analysis calculations and the prediction of turbine performance, the mean diameter is usually chosen as representative of the stage performance. This procedure is, in effect, a one-dimensional analysis, and it represents the turbine performance well if appropriate corrections are made for three-dimensional effects, leakage, and moisture. In other words, the blade row, or stage efficiency must be known or characterized as a function of operating parameters. The velocity ratio for example, is one parameter of great significance. Dimensional analysis of the individual variables bearing on turboma-

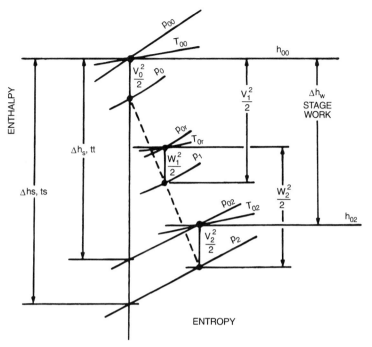

FIGURE 27.171 Stage thermodynamic conditions on a Mollier diagram.

chinery performance has resulted in several commonly used dimensionless quantities

$$\nu = \text{velocity ratio } (U/V) \tag{27.258}$$

$$\rho = \text{pressure ratio } (p_0/p) \tag{27.259}$$

$$\frac{\dot{M}\sqrt{\theta}}{\delta} = \text{referred flow rate} \tag{27.260}$$

$$\frac{N}{\sqrt{\theta}} = \text{referred speed} \tag{27.261}$$

$$\eta = \text{efficiency} \tag{27.262}$$

where $\theta = T_0/T_{\text{ref}}$ and $\delta = p_0/p_{\text{ref}}$.

The *referred flow rate* is a function of pressure ratio. In gas dynamics, the referred flow rate through a converging nozzle is primarily a function of the pressure ratio across the nozzle (Fig. 27.172). When the pressure ratio, p_0/p, is approximately 2 (depending on the ratio of specific heats) the referred mass flow maximizes, that is,

$$\frac{\dot{M}\sqrt{T_0}}{p_0} = \text{constant} \tag{27.263}$$

This referred flow rate is also termed a *flow number* and takes the following form which is more appropriate for steam turbine usage.

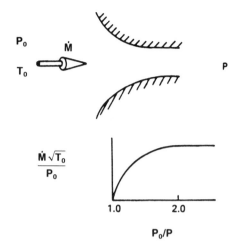

$$\frac{\dot{M}\sqrt{T_0}}{P_0}$$

1.0   2.0

FIGURE 27.172   Gas dynamic flow relationships.

P_0/P

$$\frac{\dot{M}\sqrt{p_0 v_0}}{p_0} = \text{flow number} \tag{27.264}$$

The turbine behaves in a manner similar to a nozzle, but it is also influenced by the wheel speed Mach number which is represented by $N/\sqrt{T_0}$. In effect, a similar flow rate–pressure ratio relationship is experienced as shown schematically in Fig. 27.173.

A combined plot describing turbine performance as a function of all these dimensionless quantities is shown schematically in Fig. 27.174. Power producing steam turbines, however, run at constant speed and at essentially constant flow number, pressure ratio, and velocity ratio, and in effect operate at nearly a fixed point on this

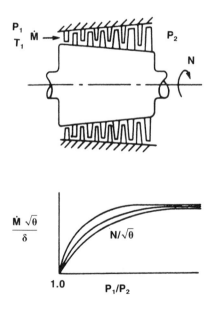

FIGURE 27.173   Turbine flow relationships.

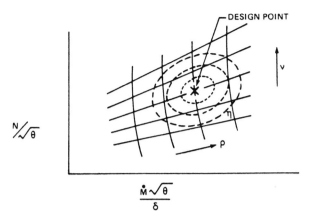

FIGURE 27.174 Turbine performance map.

performance map. Exceptions to this are the control stage, if the unit has one, and the last few stages of the machine. The control stage normally experiences a wide range of pressure ratios in the situation where throttle pressure is maintained at a constant value and a series of governing valves admits steam to the control stage blading through succeeding active nozzle arcs.

The last stage of the low pressure end exhausts to an essentially constant pressure zone maintained by the condenser. Some variation occurs as the condenser heat load changes with unit flow rate and load. At a point several stages upstream where the pressure ratio to the condenser is sufficiently high, the flow number is maintained at a constant value. As unit load reduces, the pressure level and mass flow rate decrease simultaneously, and the pressure ratio across the last few stages reduces in value. This change in pressure ratio changes the stage velocity ratio, and the performance level of these last few stages change. Figure 27.175 indicates the trend of operation of these stages as a function of load change superimposed on the dimensionless turbine performance map. It is these last few stages, say two or three, which confront the steam turbine designer with the greatest challenge.

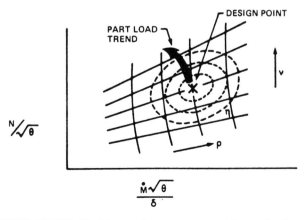

FIGURE 27.175 Turbine performance map with part load trend.

Low Pressure Turbine Design. The low pressure element of the steam turbine is normally a self-contained design in the sense that it comprises a rotor carrying several stages, a cylinder with piping connections, and bearings supporting the rotor. Complete power generating units may use from one to three low pressure elements in combination with high pressure and intermediate pressure elements depending on the particular application.

Symmetric double flow designs utilize from 5 to 10 stages on each end of the low pressure element. The initial stage accepts flow from a centrally located plenum fed by large diameter, low pressure steam piping. In general, the upstream stages feature constant cross-section blading, while the latter stages require blades of varying cross-section. Varying design requirements and criteria result in these latter twisted and tapered low pressure blades comprising from 40 to 60% of the stages in the elements.

The particular design criteria used significantly affects the end product. For example, the low pressure end (one-half of the element) of a nuclear unit might contain 10 stages and rotate at 1800 RPM and generate 100 MW, while the fossil machine would spin at 3600 RPM, carry five or six stages, and develop 50 MW. The physical size could double and the weight increase by a factor of 8.

Further differences in design are immediately implied by the steam conditions to be accepted by the element. The nuclear unit inlet temperature would be approximately 550 F (288 C) as compared to the fossil condition of 750 F (399 C). The nuclear unit must also be designed to effectively remove blade path moisture in order to achieve high thermal performance levels (heat rate). Moisture removal devices in both the high pressure and low pressure elements are normally incorporated for this purpose. Blading physical erosion is also an important consideration. Material removal by the high speed impingement of water droplets can be minimized by proper design. Moore and Sieverding (1976) describe the phenomena controlling wet steam flow in turbines.

Flow Field Considerations: Substantial advances have been achieved in the aerothermodynamic processes of low pressure turbines. A significant factor is the availability of high speed digital computers in the solution of the system of turbomachinery aerothermodynamic equations. The coupling of the solution of this complex equation system with the necessary verification as derived from experimental testing programs is the subject of this section.

The development of the system of aerothermodynamic equations is traced from basic considerations of the several conservation equations to working equations which are then generally solved by numerical techniques. Pioneering work in the development of this type of equation systems was performed by Wu (1952). Subsequent refinement and adaptation of his fundamental approach has been made by several workers in recent years.

The general three dimensional representation of the conservation equations of mass, momentum, and energy can be written for this adiabatic system

$$\frac{\partial \rho}{\partial t} + \nabla \cdot \rho \overline{V} = 0 \tag{27.265}$$

$$\frac{\partial \overline{V}}{\partial t} + \overline{V} \cdot \nabla \overline{V} = -\frac{1}{\rho} \nabla p \tag{27.266}$$

$$\nabla h = T \nabla s + \frac{1}{\rho} \nabla p \qquad (27.267)$$

Assuming the flow to be steady and axisymmetric, the conservation equations can be expanded in cylindrical coordinates as

$$\frac{\partial}{\partial r}(\rho r b V_r) + \frac{\partial}{\partial z}(\rho r b V_z) = 0 \qquad (27.268)$$

$$V_r \frac{\partial}{\partial r}(r V_\theta) + V_z \frac{\partial}{\partial z}(r V_\theta) = 0 \qquad (27.269)$$

$$V_r \frac{\partial V_z}{\partial r} + V_z \frac{\partial V_z}{\partial z} = -\frac{1}{\rho} \frac{\partial p}{\partial z} \qquad (27.270)$$

$$V_r \frac{\partial V_r}{\partial r} + V_z \frac{\partial V_r}{\partial z} - \frac{V_\theta^2}{r} = -\frac{1}{\rho} \frac{\partial p}{\partial r} \qquad (27.271)$$

$$\frac{1}{\rho} dp = dh_0 - T ds - \frac{1}{2} d(V_\theta^2 + V_z^2 + V_r^2) \qquad (27.272)$$

where

$$h_0 = h + \tfrac{1}{2}(V_\theta^2 + V_z^2 + V_r^2) \qquad (27.273)$$

and b has been introduced to account for *blade blockage*. The properties of steam may be described by equations of state, such that $p = p(\rho, T)$ and $h = h(p, T)$. These functions are available in the form of the steam tables [ASME (1967)].

As the increase in system entropy is a function of the several internal loss mechanisms inherent in the turbomachine, it is necessary that definitive known (or assumed) relationships be employed for its evaluation. In the design process, this need can usually be met, while the performance analysis of a given geometry usually introduces loss considerations more difficult to completely evaluate. The matter of loss relationship has been discussed with regard to stage design and will be further reviewed in a later section. In general, however, the entropy increase can be determined as a function of aerodynamic design parameters, i.e.,

$$\Delta s = f(V_i, W_i, \text{Mach number}) \qquad (27.274)$$

Equations (27.268) through (27.274) form a system of nine equations in nine unknowns, V_r, V_θ, V_z, ρ, p, T, h, h_0, and s, the solution of which defines the low pressure turbine flow field.

The *meridional plane*, defined as that plane passing through the turbomachine axis and containing the radial and axial coordinates, can be used to describe an additional representation of the flow process. The velocity, V_m, (see Fig. 27.176) represents the streamtube meridional plane velocity with direction proportional to the velocity components V_r and V_z. If the changes in entropy and total enthalpy along the streamlines are known, or specified, it can be shown that Eqs. (27.270) and (27.271) are equivalent and that Eq. (27.269) is also satisfied. Application of

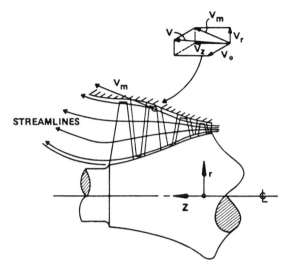

FIGURE 27.176   Turbine blade path flow field.

Eq. (27.269) to the rotating blade in effect describes Euler's turbine equation when equated to the blade force producing useful work. In this situation, it is most convenient to choose Eq. (27.271), the commonly known *radial equilibrium equation*, as the relationship for continued evaluation.

Flow Field Solution Techniques. Two commonly encountered techniques have been employed to solve this set of fundamental differential equations and relationships. They are usually referred to as the *streamline curvature* and *matrix solution* techniques. The streamline curvature technique is structured to evaluate meridional velocities and to trace streamlines in the flow field allowing the calculation of the streamline curvature itself. The streamline curvature technique has been developed to a high degree of sophistication and has been applied to axial flow compressor design as well as to axial flow turbine problems.

The matrix approach utilizes a stream function satisfying the equation of continuity. The calculation procedure then determines the stream function throughout the flow field. This technique has also been developed satisfactorily with particular application to low pressure steam turbines. The matrix approach asserts the existence of a stream function which is defined such that

$$V_r = -\frac{1}{\rho r b}\frac{\partial \psi}{\partial z} \tag{27.275}$$

and

$$V_z = \frac{1}{\rho r b}\frac{\partial \psi}{\partial r} \tag{27.276}$$

With this definition, ψ, V_r, and V_z identically satisfy Eq. (27.268). The combination of the equations of momentum, energy, and continuity, Eqs. (27.271), (27.272), (27.275), (27.276), and the definition of total enthalpy, Eq. (27.273) results in

$$\frac{\partial}{\partial r}\left(\frac{1}{\rho r b}\frac{\partial \psi}{\partial r}\right) + \frac{\partial}{\partial z}\left(\frac{1}{\rho r b}\frac{\partial \psi}{\partial z}\right) = \frac{1}{V_z}\left[\frac{\partial h_t}{\partial r} - \frac{V_\theta}{r}\frac{\partial(rV_\theta)}{\partial r} - T\frac{\partial s}{\partial r}\right] \quad (27.277)$$

This is the basic flow field equation which can then be solved by numerical techniques. This equation is an elliptic differential equation provided the meridional Mach number is less than unity. This is probably the case in all large steam turbines. The absolute Mach number can of course exceed unity and usually does in the last few blade rows of the low pressure turbine.

The streamline curvature approach satisfies the same governing equations but solves for the meridional velocity, V_m, rather than the stream function, ψ, as in the matrix approach. Equations (27.271) and (27.272) can be combined and expressed in terms of directions along the blade row leading or trailing edges. An equation used to describe the variation of meridional velocity is

$$\frac{dV_m^2}{dl} + A(l)V_m^2 = B(l) \quad (27.278)$$

where l is the coordinate along the blade edge and the coefficients A and B depend primarily on the slopes of the streamlines in the meridional plane. An iterative process is then employed to satisfy the governing equations.

Field Test Verification of Flow Field Design. The most conclusive verification of a design concept is the in-service evaluation of the product with respect to its design parameters. Application of a matrix type flow field design program has been made to the design of the last three stages of a 3600 RPM low pressure end. Figure 27.177 is indicative of the general layout of this high speed fossil turbine.

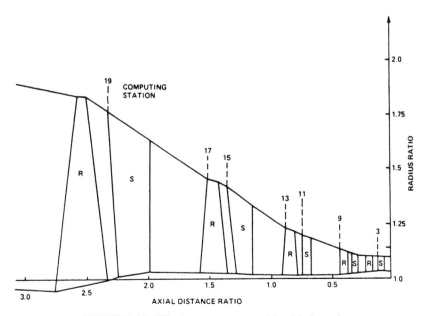

FIGURE 27.177 Low pressure turbine blade path.

The aerodynamic design process is to a great degree an iterative process, i.e., the specification of the design parameters (for example, work per stage and radial work distribution) are continuously adjusted in order to optimize the flow field design. Just as important, feedback from mechanical analyses must be accommodated in order to achieve reliable long low pressure blades.

A key design criterion is the requirement that low pressure blades must be vibrationally tuned in that the lowest several natural modes of vibration must be sufficiently removed from harmonics of running speed during operation. This tuning process is best represented by the *Campbell diagram* of Fig. 27.178. The first four modes of the longest blade are represented as a function of running speed in RPM. At the design speed, the intersection of the mode lines and their band widths indicate nonresonant operation. A comparison is also shown between full-size laboratory test results and shop test data. The laboratory results, [Steltz *et al.* (1977)], were obtained by means of strain gage signals transmitted by radio telemetry techniques. In the shop tests, the blades were excited once per revolution by a specially designed steam jet. The strain gage signals were delivered to recording equipment by means of slip rings.

If it turns out that the mode lines intersect a harmonic line at running speed, a resonant condition exists which could lead to a fatigue failure of the blade. The iterative process occurs prior to laboratory and shop verification. Analytic studies and guidance from experience gained from previous blade design programs, enables the mechanical designer to determine blade shape changes which will eliminate the resonance problem. This information is then incorporated by the aerodynamicist into the flow field design. As this process continues, manufacturing considerations are incorporated to ensure the design can be produced satisfactorily.

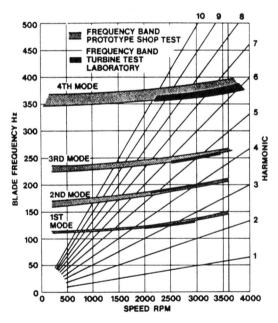

FIGURE 27.178 Campbell diagram of low pressure blade.

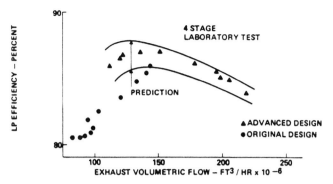

FIGURE 27.179 Low pressure turbine laboratory test results.

Construction and testing then followed; test data was obtained from several sources, a field test and an in-house test program [Steltz *et al.* (1977)]. Figure 27.179 displays low pressure turbine efficiency versus exhaust volumetric flow for the advanced design turbine, designed by means of the matrix type flow field design process, and original design turbine. Although the same range of exhaust volumetric flow was not achievable in both test series, the improvement is apparent and is further indicated by the extended performance line of the original design as determined by predictive techniques.

Field test results were obtained in which the original design and the advanced design were evaluated by procedures defined by the ASME Performance Test Code. Figure 27.180 presents these results in the form of low pressure turbine efficiency, again as a function of exhaust volumetric flow. The improvement in turbine efficiency is equivalent to 70 Btu/kWh in turbine cycle heat rate (18 kcal/kWh).

Aerodynamic losses in the blade path are an integral part of performance determination as they directly affect the overall efficiency of the turbine. The losses at part load, or off-design operation, are additive to those existing at the design point and may contribute significantly to a deterioration of performance. Many factors influence these incremental losses, but they are due primarily to blading flow conditions which are different from those at the optimum operating condition. For example, lower (or higher) mass flow rates through the machine have been shown to

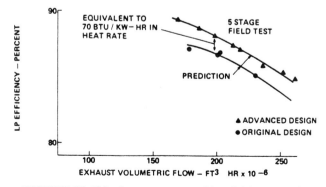

FIGURE 27.180 Low pressure turbine field test results.

change the pressure levels, temperatures, velocities, flow angles, moisture content, etc., within the blading. These changes induce flow conditions relative to the blading which create additional losses. As different radial locations on the same blade are affected to a different degree, the overall effect must be considered as a summation of all individual effects for that blade. In fact, different blade rows, either stationary or rotating, are affected in a like manner, and the complete effect would then be the summation of the individual effects of all the blade rows.

Detailed verification of the design process has been obtained from the analysis and comparison of these internal flow characteristics to expected conditions as determined by off-design calculations employing the same principles as incorporated in the design procedure. From measurements of total and static pressure and a knowledge of the enthalpy level at the point of interest, it is possible to determine the steam velocity, Mach number, and local mass flow rate. The flow angle is measured simultaneously. Traverse data for the advanced design turbine is presented in Figs. 27.181 and 27.182 for the last stationary blade inlet. Flow incidence is presented as a function of blade height for high and low values of specific mass flow. Good agreement is indicated between prediction and the traverse data. As flow rate (and load) decrease, the correspondence between test and calculation becomes less convincing, indicating that loss mechanisms existing under part-load operation are not as well defined as at high load near the design point.

By calculation, utilizing the rotor blade downstream traverse data, it is possible to determine the operating conditions relative to the rotor blade. That is, the conditions as seen by a traveler on the rotor blade can be determined. Figure 27.183 presents the Mach number leaving the last rotor blade (i.e., relative to it) as a function of blade height compared to the expected variation. Very high Mach numbers are experienced in this design and are a natural consequence of the high rotor blade tip speed which approximates 2010 ft/sec which converts to a wheel speed Mach number of 1.61.

In a manner similar to that applied at the inlet of the last stationary blade, the characteristics at the exit of this blade can be determined. Results of this traverse

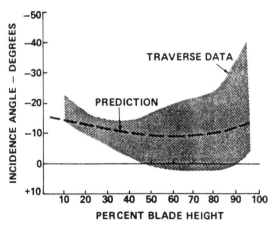

FIGURE 27.181   Last stationary blade incidence angle at high end loading—12,000 lb/hr-ft².

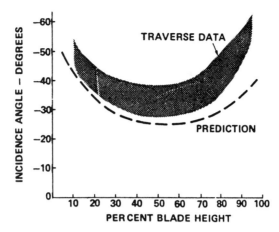

FIGURE 27.182   Last stationary blade incidence angle at low end loading—6,000 lb/hr-ft^2.

are shown in Fig. 27.184 where the effects of the stator wakes can be identified. The high Mach numbers which are experienced at the stationary blade hub make this particular measurement extremely difficult. The comparison shows good agreement with expectations for this difficult measurement location.

The combination of these stationary blade exit traverses with velocity triangle calculations enables the flow conditions at the inlet of the rotor blade to be determined. Flow incidence as a function of blade height (diameter) is presented in Fig. 27.185 for the high end load condition of Fig. 27.181. Significant flow losses can be incurred by off-design operation at high incidence levels. The hub and tip are particularly sensitive, and care must be taken in the design process to avoid this situation.

Further confirmation of the design process can be achieved by evaluation of the kinetic energy leaving the last stage. This energy is lost to the turbine and represents

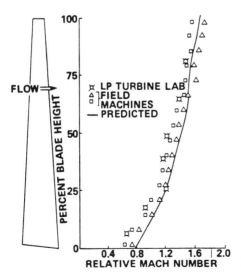

FIGURE 27.183   Mach number relative to last rotating blade.

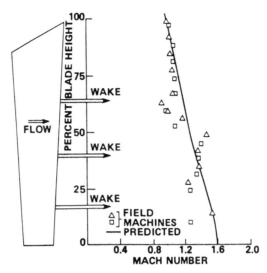

FIGURE 27.184 Mach number at exit of last stationary blade.

a significant power output if it were possible to convert it to useful work. This kinetic energy, or *leaving loss* is defined as

$$\overline{V}^2 = \int_{\text{Hub}}^{\text{Tip}} (\rho V) \frac{V^2 \, dA}{\dot{M}} \tag{27.279}$$

and plotted in Fig. 27.186 as a function of exhaust volumetric flow. Excellent agreement has been achieved between the test data and prediction.

The foregoing represents conventional practice in the design of low pressure blading as well as the design and analysis of upstream blading located in the HP and IP cylinders. The capability of solving the Navier–Stokes equations is now available.

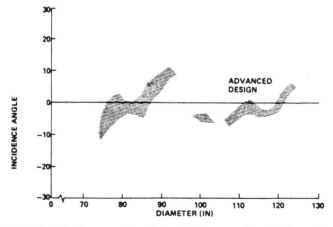

FIGURE 27.185 Last rotating blade incidence angle at high end loading.

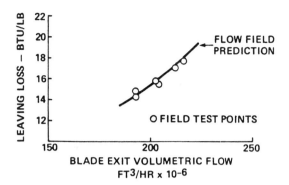

FIGURE 27.186 Leaving loss as a function of exhaust volumetric flow for an advanced design.

Drawbacks to this approach are the complexities of constructing a three-dimensional model of the subject of interest and the computational time required. The current philosophy regarding the use of Navier–Stokes solvers is that their practical application is to new design concept developments in order to determine valid and optimum resolutions of these concepts. In this process, design guidelines and cause and effect relationships are developed for subsequent application to production machinery. As of this writing, their day-to-day use is prohibited by manpower and computer time requirements. Nevertheless, significant improvements in, and changes to, the conventional design process have been made by the Navier–Stokes solvers. For the steam turbine aerodynamicist, the most influential approaches are those developed by Denton (1992), a program *not* a Navier–Stokes solver but having three-dimensional inviscid capability, and that by Dawes (1992), a truly three-dimensional Navier–Stokes solver.

Blade to Blade Flow Analysis. The blade to blade flow field problem confronts the turbine designer with a variety of situations, some design, and some analysis of existing designs. The flow regime can vary from low subsonic (incompressible) to transonic with exit Mach numbers approaching 1.8. The questions to be resolved in blade design are what section will satisfy the flow field design, and how efficiently will it operate?

From the mechanical viewpoint, will the blade section, or combination of sections be strong enough to withstand the steady forces required of it, centrifugal and steam loading, and will it be able to withstand the unknown unsteady forces impressed upon it? The structural strength of the blade can be readily evaluated as a function of its geometry, material characteristics, and steady stresses induced by steady loads. The response of the blade to unsteady forces, generally of an unknown nature, is a much more formidable problem. Tuning of the long, low pressure blade is a common occurrence, but as blades become shorter, in the high and intermediate pressure elements, and in the initial stages of the low pressure element, tuning becomes a difficult task. Blades must be designed with sufficient strength, or margin, to withstand these unsteady loads and resonance conditions at high harmonics of running speed.

The aerodynamic design of the blade depends on its duty requirements, i.e., what work is the blade expected to produce? The fluid turning is defined by the flow field design which determines the radial distribution of inlet and exit angles. Inherent in the flow field design is the expectation of certain levels of efficiency (or losses) which have defined the angles themselves. The detailed blade design must then be accomplished in such a manner as to satisfy, or better, these design expectations. Losses then, are a fundamental concern to the blade designer.

Blade Aerodynamic Considerations. Losses in blade sections, and blade rows, are a prime concern to the manufacturer and are usually considered proprietary information, as these control and establish its product's performance level in the market place. These data and correlative information provide the basic guidelines affecting the design and application of turbine-generator units.

Techniques involved in identifying and analyzing blade sections are more fundamentally scientific and have been the subject of development and refinement for many years. Current processes utilize the digital computer and depend on numerical techniques for their solution.

Early developments in the analysis of blade to blade flow fields, [Benedict and Meyer (1957)] utilized analogies existing between fluid flow fields and electrical fields. A specific application can be found in Steltz *et al.* (1976).

Transonic Blade Flow: The most difficult problem has been the solution of the transonic flow field in the passageway between two blades. The governing differential equations change from the elliptic type in the subsonic flow regime to the hyperbolic type in the supersonic regime, making a uniform approach to the problem solution mathematically quite difficult. In lieu of, and also in support of, analytic procedures to define the flow conditions within a transonic passage, experimental data have been heavily relied upon. These test programs also define the blade section losses, which is invaluable information in itself. Testing and evaluating blade section performance is an art (and a science) unto itself. Of the several approaches to this type of testing, the air cascade is by far the most common and manageable. Pressure distributions (and hence, Mach number) can be determined as a function of operating conditions such as inlet flow direction (incidence), Reynolds number, and overall pressure ratio. Losses also are determinable. Optical techniques are also useful in the evaluation of flow in transonic passages. Constant density parameters from interferometric photos can be translated into pressure and Mach number distributions.

A less common testing technique is the use of a free surface water table wherein local water depths are measured and converted to local Mach number [Benedict (1963)]. The analogy between the water table and gas flow is valid for a gas with a ratio of specific heats equal to 2, an approximation to be sure. A direct comparison of test results for the same blade cascade is presented in Fig. 27.187. The air test data were taken from interferometric photos and converted to parameters of constant Mach number shown in heavy solid lines. Superimposed on this figure are test data from a free surface water table in lighter lines. A very good comparison is noted for these two sets of data from two completely different blade testing concepts. The air test is more useful in that the section losses can be determined at the same time the optical studies are being done. The water table however, produces reasonable results, comparable to the air cascade as far as blade surface distributions are concerned, and is an inexpensive and rapid means of obtaining good qualitative information, and with sufficient care in the experiment, good quantitative data.

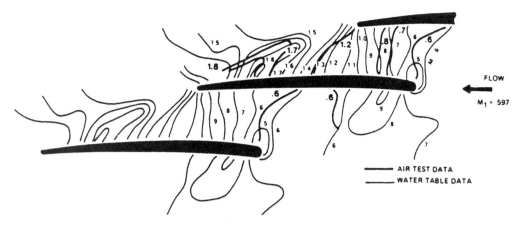

FIGURE 27.187 Comparison of water table test results with interferometry air test data for Mach number distribution in transonic blade.

Analytical Techniques: Various analytic means have been employed for the determination of the velocity and pressure distributions in transonic blade passages. A particular technique, termed *time marching*, solves the unsteady compressible flow equations by means of successive calculations in time utilizing a flow field comprised of finite area, or volume, elements. Figure 27.188 presents the general calculation region describing the flow passage between two blades. A result of the time marching method is shown in Fig. 27.189 which presents a plot of local Mach number along both the suction and pressure sides of a transonic blade. Parameters of exit isentropic Mach number show a significant variation in local surface conditions as the blade's overall total to static pressure ratio is varied. These data were obtained in air from a transonic cascade facility. In summary, sophisticated numerical processes are available and in common use for the determination of blade surface pressure and velocity distributions. These predictions have been verified by experimental programs.

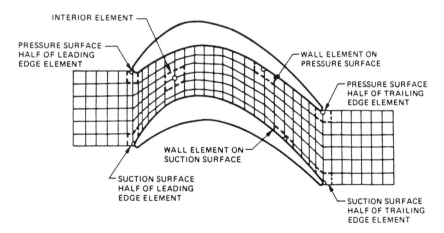

FIGURE 27.188 Blade to blade calculational region.

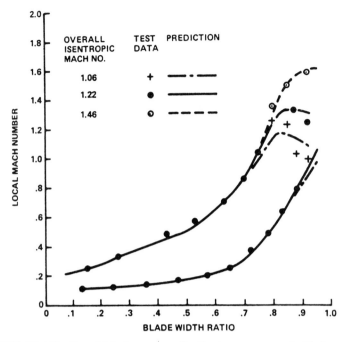

FIGURE 27.189   Local Mach number distribution on transonic blade surfaces.

27.11   GAS TURBINES

27.11.1   Cycles
T. Yoshinaka

Introduction. The gas turbine is a power plant which is thermodynamically based on the Joule–Brayton cycle. It consists of, in its simplest form, an adiabatic compression, an isobaric heat addition (or combustion), an adiabatic expansion and an isobaric heat rejection process. It has a closed circuit as well as an open circuit configuration, with the latter being used predominantly, today. Here, the atmosphere is considered as an enormous heat sink as well as a gas (air) reservoir, so that the flow in an open circuit gas turbine can be treated as being cyclic. Its application extends to include electricity generation, oil pumping, and propulsion of air, land, and sea transport.

The following parameters are used to evaluate plant performance

$$\eta \text{ (cycle efficiency)} = \frac{P \text{ (power output)}}{E_i \text{ (energy input)}} \tag{27.280}$$

where for internal combustion

$$E_i = \dot{m}_f \, (CV)_0 \tag{27.281a}$$

and for combustion external to gas turbine

$$E_i = \dot{m}_f \, (CV)_0 \eta_B \tag{27.281b}$$

Also, $\dot{m}_f$ is the mass flow rate of fuel, $(CV)_0$ is the heating value of the fuel at T_0 and η_B is the combustor efficiency.

$$\hat{P} \text{ (specific power)} = \frac{P \text{ (power output)}}{\dot{m}_1 c_{p1} T_{01} \text{ (Total energy of incoming flow)}} \qquad (27.282)$$

where subscript 1 denotes compressor inlet, c_p is the specific heat at constant pressure and T_0 is total temperature.

sfc (specific fuel consumption based on shaft power)

$$= \frac{\dot{m}_f \times 3600}{P \text{ (power)}} \text{ [kg/kw/hr, (lb/hp/hr)]} \qquad (27.283)$$

(for turboshaft engines)

tsfc (specific fuel consumption based on net thrust)

$$= \frac{\dot{m} \times 3600}{F_n \text{ (net thrust)}} \text{ [kg/KN/hr, (lb/lb}_f\text{/hr)]} \qquad (27.284)$$

(for turbojet or turbofan engines)

esfc (specific fuel consumption based on equivalent power)

$$= \frac{\dot{m}_f \times 3600}{P + F_n V_a} \text{ [kg/kW/hr, (lb/hp/hr)]} \qquad (27.285)$$

(for turboprop engines with exhaust jet thrust)

where V_a is aircraft speed

$$HR \text{ (heat rate)} = \frac{\dot{m}_f (CV)_0 \times 3600}{P} \text{ [kJ/kW/hr, (Btu/hp/hr)]} \qquad (27.286)$$

Performance of gas turbine engines for aircraft propulsion is evaluated in terms of *specific fuel consumption*, sfc, as opposed to *heat rate* for industrial gas turbines. For turboshaft engines, sfc is defined in Eq. (27.283) which is related to the cycle efficiency [Eq. (27.280)], as follows

$$(\text{sfc}) \cdot \eta = \frac{3600}{(CV)_0} = \text{Constant} \qquad (27.287)$$

See Fig. 27.190.

The sfc of a turbojet or turbofan engines is given in Eq. (27.284), which can be rewritten as

$$\text{tsfc} = \frac{\dot{m}_f \times 3600}{F_n} = \frac{\dot{m}_f \times 3600}{(\dot{m}_f + \dot{m}_c)V_j - \dot{m}_c V_a} \qquad (27.284a)$$

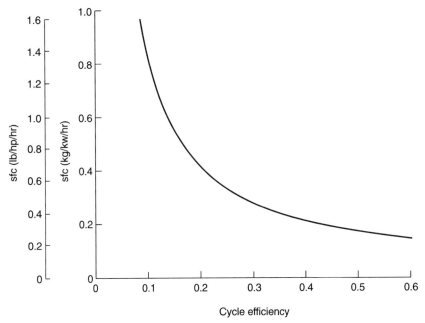

FIGURE 27.190   Cycle efficiency vs. specific fuel consumption (based on shaft power).

where V_j is the exhaust jet velocity. Since power output for the turbojet or turbofan engines is equivalent to the increase of kinetic energy of the flow through the engine,

$$\eta \text{ (cycle efficiency)} = \frac{\left|\dfrac{(\dot{m}_f + \dot{m}_c)}{2} V_j^2 - \dfrac{\dot{m}_c}{2} V_a^2\right|}{\dot{m}_f (CV)_0} \tag{27.288}$$

By introducing *propulsive efficiency*, η_{pr}, which is a performance index of the kinetic energy being effectively used to propel the aircraft,

$$\eta_{pr} = \frac{[(\dot{m}_f + \dot{m}_c)V_j - \dot{m}_c V_a]V_a}{\left[\dfrac{(\dot{m}_f + \dot{m}_c)}{2} V_j^2 - \dfrac{\dot{m}_c}{2} V_a^2\right]} \tag{27.289}$$

tsfc may be coupled with η as follows

$$(\text{tsfc}) \cdot \eta \cdot \eta_{pr} = \frac{3600}{(CV)_0} V_a \tag{27.290}$$

See Fig. 27.191. The esfc for turboprop engines with useful jet thrust may be related to the cycle efficiency in a similar manner to sfc for turboshaft engines.

$$\text{esfc} \cdot \eta = \frac{3600}{(CV)_0} = \text{Constant} \tag{27.291}$$

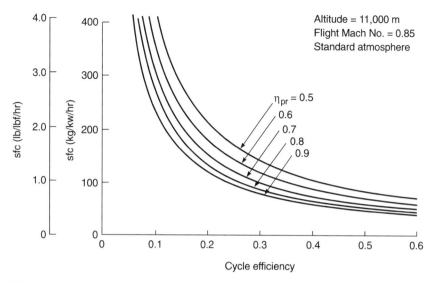

FIGURE 27.191 Cycle efficiency vs. specific fuel consumption (based on net thrust).

Note that Eqs. (27.287) and (27.291) are identical. However, if a turboshaft engine and a turboprop engine with usable jet thrust have the same value of shaft power for the same fuel consumption, the latter has a higher *equivalent shaft power* than the former due to the available jet thrust. Therefore, the cycle efficiency of the turboprop engine becomes greater than that of the turboshaft, or esfc < sfc.

There are two types of power turbine configurations, a fixed-shaft power turbine in which the turbine drives both the compressor and load on a single shaft [see Fig. 27.192(a)], and a free power turbine in which the compressor and load are not mechanically coupled and are driven by two separate turbines [see Fig. 27.192(b)]. For a load with a fixed rotational speed, the load control of the fixed-shaft power turbine type is achieved by varying the compressor operating point at the fixed rotational speed line [see Fig. 27.193(a)]. On the other hand, a change of the load on the free power turbine type results in varying the compressor rotational speed while maintaining its steady state operating line [see Fig. 27.193(b)]. These changes of the

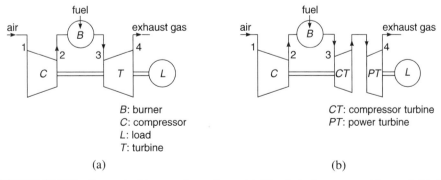

FIGURE 27.192 Power turbine configurations. (a) Fixed-shaft power turbine and (b) free power turbine.

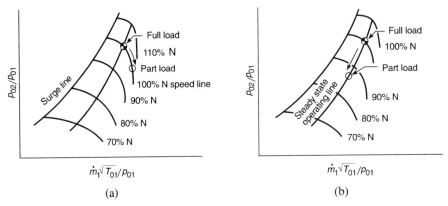

FIGURE 27.193 Shift of compressor operating point. (See Fig. 27.192 for subscripts 1 and 2.) (a) Fixed-shaft power turbine and (b) free power turbine.

compressor operating point are automatically obtained when the fuel flow rate is controlled to maintain the constant power turbine rotational speed as the magnitude of the load varies.

Cycle efficiency and specific power for various gas turbine cycles are defined in the next three sections. In these definitions, the following assumptions are made: 1) parasitic losses such as those of bearings, gears, and windage are neglected for simplicity (although this is not realistic), and 2) the mass flow rate in the turbine is the addition of that at compressor inlet and fuel flow rate.

Open Cycles. Simple (CBT) Cycle: Figure 27.194 depicts the h-s (enthalpy-entropy) diagram of a simple [compressor, burner, turbine (CBT)] cycle. A majority of gas turbines used today, including those for aircraft propulsion, fall into this cycle category. The turbojet or turbofan engines (see Fig. 27.195) do not produce *shaft power*, thus they have no need for power turbines. Instead, they emit exhaust gas

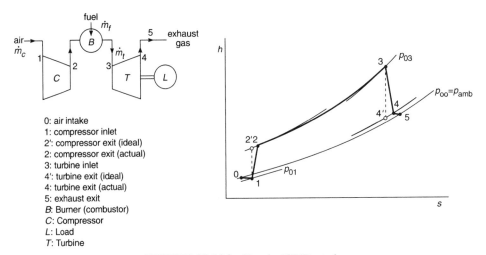

0: air intake
1: compressor inlet
2': compressor exit (ideal)
2: compressor exit (actual)
3: turbine inlet
4': turbine exit (ideal)
4: turbine exit (actual)
5: exhaust exit
B: Burner (combustor)
C: Compressor
L: Load
T: Turbine

FIGURE 27.194 Simple (CBT) cycle.

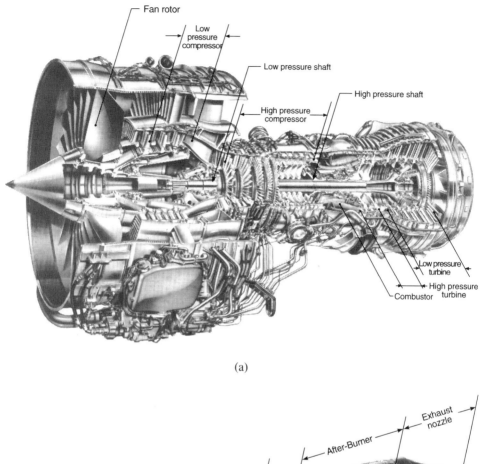

(a)

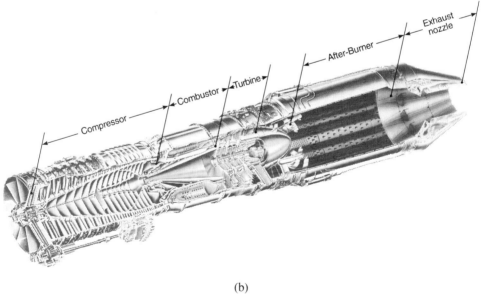

(b)

FIGURE 27.195 Aircraft jet engines: (a) VT2500-A1 turbofan engine (courtesy of International Aero Engines) and (b) J79 turbojet engine (courtesy of General Electric).

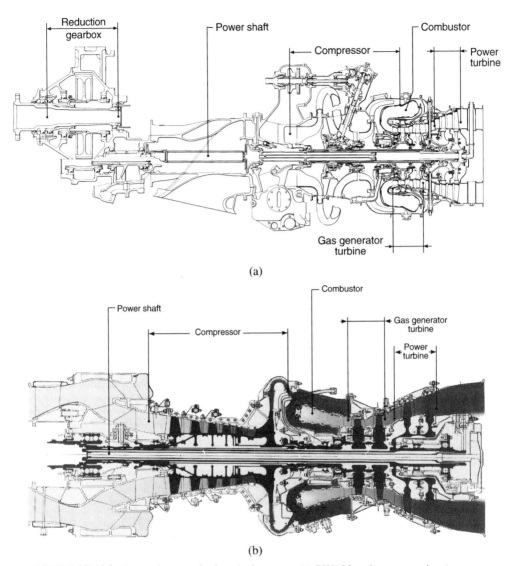

FIGURE 27.196   Jet engines producing shaft power. (a) PW100 turboprop engine (courtesy of Pratt & Whitney Canada) and (b) T700 turboshaft engine (courtesy of General Electric).

with high kinetic energy generating thrust. In the case of turboprop or turboshaft engines (see Fig. 27.196), shaft power is produced by the turbine. The load is a propeller (for a turboprop) or a helicopter main rotor (for a turboshaft).

The *cycle efficiency*, η and *specific power*, $\hat{P}$, of the simple cycle gas turbine is defined

$$
\eta = \frac{\dot{m}_t(h_{03} - h_{04}) - \dot{m}_c(h_{02} - h_{01})}{\dot{m}_t h_{03} - \dot{m}_c h_{02}} = \frac{\left(\dfrac{\dot{m}_t}{\dot{m}_c}\right)\left(\dfrac{c_{p3}}{c_{p1}}\right) t(1 - r_t^{-\alpha}) - (r_c^{\beta} - 1)}{\left(\dfrac{\dot{m}_t}{\dot{m}_c}\right)\left(\dfrac{c_{p3}}{c_{p1}}\right) t - r_c^{\beta}}
$$

$$(27.292)$$

where h_0 is the stagnation enthalpy, η_{pt} is the turbine polytropic efficiency, η_{pc} is the compressor polytropic efficiency, $\dot{m}_c$ and $\dot{m}_t$ are the compressor and turbine mass flows and

$$t = \frac{T_{03}}{T_{01}}, \quad r_c = \frac{P_{02}}{P_{01}}, \quad r_t = \frac{P_{03}}{P_{04}}, \quad \alpha = \frac{(\gamma_t - 1)}{\gamma_t}\eta_{pt}, \quad \beta = \frac{(\gamma_c - 1)}{\gamma_c \eta_{pc}}$$

$$\hat{P} = \left(\frac{\dot{m}_t}{\dot{m}_c}\right)\left(\frac{c_{p3}}{c_{p1}}\right) t(1 - r_t^{-\alpha}) - (r_c^\beta - 1) \tag{27.293}$$

Note that

$$r_t = k_{0-1} \cdot k_{2-3} \cdot k_{4-5} \cdot r_c \tag{27.294}$$

where

$$k_{i-j} = \left(1 - \frac{\Delta P_{0i-j}}{P_{0i}}\right)$$

and ΔP_{0i} are total pressure losses. As shown in Fig. 27.197, the cycle efficiency is a strong function of compressor pressure ratio, r_c, and a weak function of cycle temperature ratio, t. On the other hand, specific power is more significantly influenced by cycle temperature ratio than compressor pressure ratio except for very low compressor pressure ratio conditions (see Fig. 27.198).

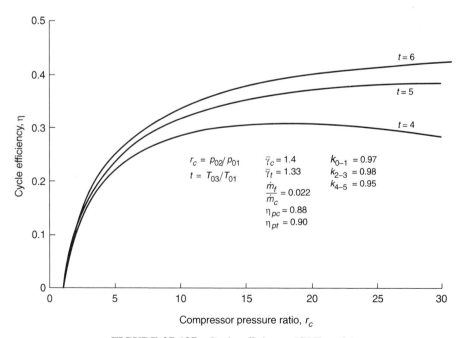

FIGURE 27.197 Cycle efficiency (CBT cycle).

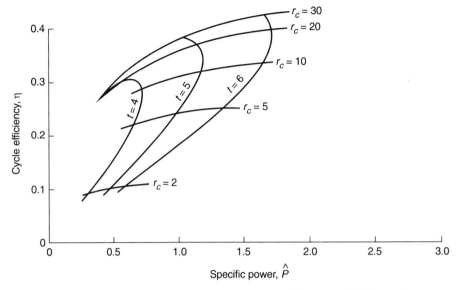

FIGURE 27.198   Cycle efficiency versus specific power (CBT cycle).

Heat-Exchange (CBTX) Cycle: The *h-s* diagram of the *heat-exchange* [compressor, burner, turbine, exchanger (CBTX)] cycle is shown in Fig. 27.199. The advantage of this cycle over the simple cycle is the pre-heating of compressor delivery air by the exhaust gas, thus improving the cycle efficiency. The improvement is particularly significant at low compressor pressure ratios. However, this advantage diminishes rapidly with increasing pressure ratio for a fixed temperature ratio, $t = T_{04}/T_{01}$, since the turbine exit temperature, T_{05}, decreases.

The cycle efficiency and specific power of the heat-exchange cycle are

$$\eta = \frac{\dot{m}_t(h_{04} - h_{05}) - \dot{m}_c(h_{02} - h_{01})}{\dot{m}_t h_{04} - \dot{m}_c h_{03}}$$

$$= \frac{\left(\dfrac{\dot{m}_t}{\dot{m}_c}\right)\left(\dfrac{c_{p4}}{c_{p1}}\right) t(1 - r_t^{-\alpha}) - (r_c^\beta - 1)}{\left(\dfrac{\dot{m}_t}{\dot{m}_c}\right)t\left[\left(\dfrac{c_{p4}}{c_{p1}}\right) - \epsilon_{ex} r_t^{-\alpha}\right] - (1 - \epsilon_{ex})\, r_c^\beta + \dfrac{\dot{m}_f}{\dot{m}_c}\,\epsilon_{ex} t r_t^{-\alpha}} \qquad (27.295)$$

where

$$t = \frac{T_{04}}{T_{01}}, \qquad r_c = \frac{p_{02}}{p_{01}}, \qquad r_t = \frac{p_{04}}{p_{05}}$$

$$\alpha = \frac{\gamma_t - 1}{\gamma_t}\,\eta_{pt}, \qquad \beta = \frac{\gamma_c - 1}{\gamma_c \eta_{pc}}$$

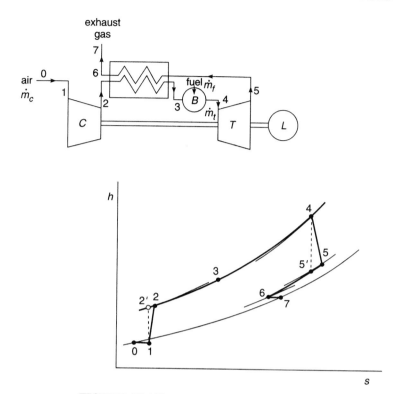

FIGURE 27.199 Heat exchange (CBTX) cycle.

and

$$\epsilon_{ex} \text{ (exchanger effectiveness)} = \frac{(T_{03} - T_{02})}{(T_{05} - T_{02})} \qquad (27.296)$$

$$\hat{P} = \left(\frac{\dot{m}_t}{\dot{m}_c}\right)\left(\frac{c_{p4}}{c_{p1}}\right) t(1 - r_t^{-\alpha}) - (r_c^{\beta} - 1) \qquad (27.297)$$

r_t is a function of r_c and aerodynamic losses of components other than the compressor and turbine described by

$$r_t = k_{0-1} \cdot k_{2-3} \cdot k_{3-4} \cdot k_{5-6} \cdot k_{6-7} \cdot r_c$$

where

$$k_{i-j} = \left(1 - \frac{\Delta p_{0i-j}}{p_{0i}}\right) \qquad (27.298)$$

Some numerical examples of the heat-exchange cycle are shown in Figs. 27.200 and 27.201. The cycle efficiency of the simple cycle is also shown for comparison. Exchanger effectiveness strongly influences the cycle efficiency of the heat-

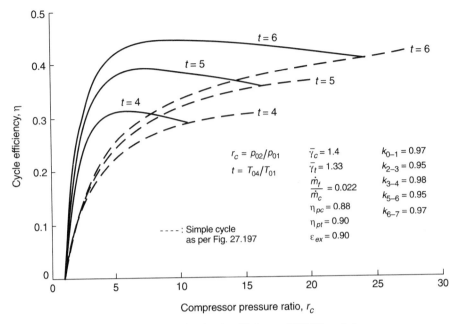

FIGURE 27.200   Cycle efficiency (CBTX cycle).

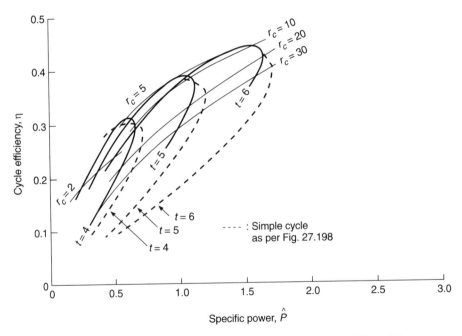

FIGURE 27.201   Cycle efficiency vs. specific power (CBTX cycle).

exchanger cycle, while specific power is lower than that of the simple cycle due to the added pressure loss in the heat exchanger.

Intercooled (CICBT) Cycle: The *intercooled cycle* whose h-s diagram is shown in Fig. 27.202 attempts to improve specific power by reducing the power required by the compressor. The power reduction is achieved by adding *intercooling* to the compression cycle. The cycle efficiency of the intercooled cycle may be higher or lower than that of the simple cycle, depending upon a trade-off between the reduction of power required by the compressors and pressure losses in the intercooler. Numerical examples shown in Figs. 27.203 and 27.204 indicate that there is a general trend in which the former becomes competitive with or superior to the latter at high compressor pressure ratio values.

The performance parameters of this cycle are

$$\eta = \frac{\dot{m}_t(h_{06} - h_{07}) - \dot{m}_c(h_{02} - h_{01}) - \dot{m}_c(h_{04} - h_{03})}{(\dot{m}_c + \dot{m}_f)h_{05} - \dot{m}_c h_{04}}$$

$$= \frac{\left(\dfrac{\dot{m}_t}{\dot{m}_c}\right)\left(\dfrac{c_{p5}}{c_{p1}}\right) t(1 - r_t^{-\alpha}) - (r_{c1}^{\beta_1} - 1) - \left(\dfrac{T_{03}}{T_{01}}\right)(r_{c2}^{\beta_2} - 1)}{\left(\dfrac{\dot{m}_t}{\dot{m}_c}\right)\left(\dfrac{c_{p5}}{c_{p1}}\right) t - \left(\dfrac{T_{03}}{T_{01}}\right) r_{c2}^{\beta_2}} \qquad (27.299)$$

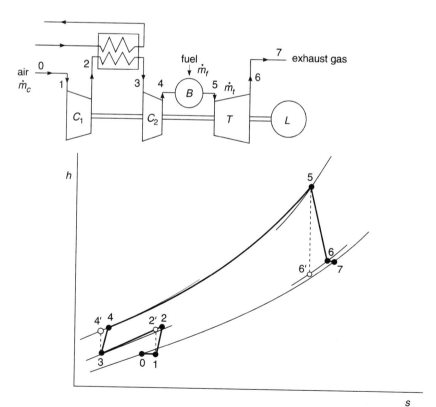

FIGURE 27.202 Intercooled (CICBT) cycle.

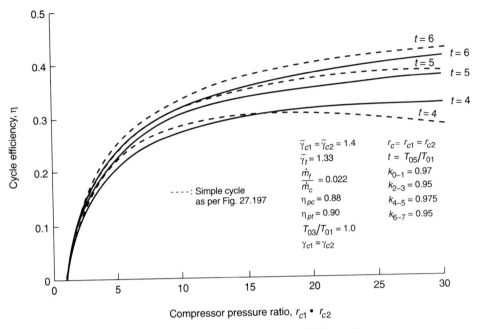

$\bar\gamma_{c1} = \bar\gamma_{c2} = 1.4$ $r_c = r_{c1} = r_{c2}$
$\bar\gamma_t = 1.33$ $t = T_{05}/T_{01}$
$\dfrac{\dot m_f}{\dot m_c} = 0.022$ $k_{0-1} = 0.97$
$k_{2-3} = 0.95$
----: Simple cycle $\eta_{pc} = 0.88$ $k_{4-5} = 0.975$
as per Fig. 27.197 $\eta_{pt} = 0.90$ $k_{6-7} = 0.95$
$T_{03}/T_{01} = 1.0$
$\gamma_{c1} = \gamma_{c2}$

FIGURE 27.203   Cycle efficiency (CICBT cycle).

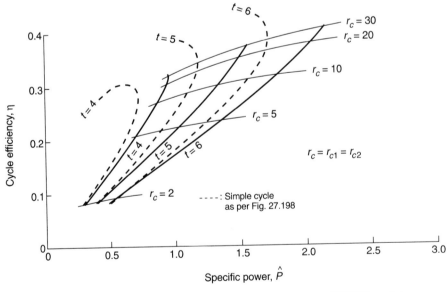

FIGURE 27.204   Cycle efficiency vs. specific power (CICBT cycle).

where

$$t = \frac{T_{05}}{T_{01}}, \quad r_{c1} = \frac{p_{02}}{p_{01}}, \quad r_{c2} = \frac{p_{04}}{p_{03}}, \quad r_t = \frac{p_{05}}{p_{06}},$$

$$\alpha = \frac{\gamma_t - 1}{\gamma_t} \eta_{pt}, \quad \beta_1 = \frac{\gamma_{c1} - 1}{\gamma_{c1}\eta_{pc1}}, \quad \beta_2 = \frac{\gamma_{c2} - 1}{\gamma_{c2}\eta_{pc2}}$$

$$\hat{P} = \left(\frac{\dot{m}_t}{\dot{m}_c}\right)\left(\frac{c_{p5}}{c_{p1}}\right) t(1 - r_t^{-\alpha}) - (r_{c1}^{\beta_1} - 1) - \left(\frac{T_{03}}{T_{01}}\right)(r_{c2}^{\beta_2} - 1) \quad (27.300)$$

$$r_t = k_{0-1} \cdot k_{2-3} \cdot k_{4-5} \cdot k_{6-7} \cdot r_{c1} \cdot r_{c2} \quad (27.301)$$

where

$$k_{i-j} = \left(1 - \frac{\Delta p_{0i-j}}{p_{0i}}\right)$$

It should be noted that the performance related equations described above do not include any performance penalty which may be associated with the cooling flow cycle.

Intercooled, Heat Exchange (CICBTX) Cycle: The low cycle efficiency of the intercooled cycle in the low compressor pressure ratio range may be vastly improved, if a heat exchanger is added to the intercooled cycle. The h-s diagram of the CICBTX cycle is depicted in Fig. 27.205, while Figs. 27.206 and 27.207 show some numerical examples of its performance characteristics. The performance parameters are defined

$$\eta = \frac{\dot{m}_t(h_{06} - h_{07}) - \dot{m}_c(h_{02} - h_{01}) - \dot{m}_c(h_{04} - h_{03})}{(\dot{m}_c + \dot{m}_f)h_{06} - \dot{m}_c h_{05}}$$

$$= \frac{\left(\frac{\dot{m}_t}{\dot{m}_c}\right)\left(\frac{c_{p0}}{c_{p1}}\right) t(1 - r_t^{-\alpha}) - (r_{c1}^{\beta_1} - 1) - \left(\frac{T_{03}}{T_{01}}\right)(r_{c2}^{\beta_2} - 1)}{\left(\frac{\dot{m}_t}{\dot{m}_c}\right) t\left[\left(\frac{c_{p6}}{c_{p1}}\right) - \epsilon_{ex}r_t^{-\alpha}\right] - (1 - \epsilon_{ex})\left(\frac{T_{03}}{T_{01}}\right) r_{c2}^{\beta_2} + \left(\frac{\dot{m}_f}{\dot{m}_c}\right)\epsilon_{ex}tr_t^{-\alpha}} \quad (27.302)$$

where

$$t = \frac{T_{06}}{T_{01}}, \quad r_{c1} = \frac{p_{02}}{p_{01}}, \quad r_{c2} = \frac{p_{04}}{p_{03}}, \quad r_t = \frac{p_{06}}{p_{07}}$$

$$\alpha = \frac{\gamma_t - 1}{\gamma_t} \eta_{pt}, \quad \beta_1 = \frac{\gamma_{c1} - 1}{\gamma_{c1}\eta_{pc1}}, \quad \beta_2 = \frac{\gamma_{c2} - 1}{\gamma_{c2}\eta_{pc2}}$$

$$\epsilon_{ex} = \frac{(T_{05} - T_{04})}{(T_{07} - T_{04})}$$

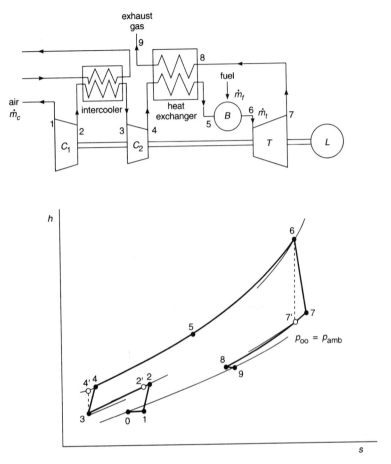

FIGURE 27.205   Intercooled, heat exchange (CICBTX) cycle.

$$\hat{P} = \left(\frac{\dot{m}_t}{\dot{m}_c}\right)\left(\frac{c_{p6}}{c_{p1}}\right) t(1 - r_t^{-\alpha}) - (r_{c1}^{\beta_1} - 1) - \left(\frac{T_{03}}{T_{01}}\right)(r_{c2}^{\beta_2} - 1) \quad (27.303)$$

$$r_t = k_{0-1} \cdot k_{2-3} \cdot k_{4-5} \cdot k_{5-6} \cdot k_{7-8} \cdot k_{8-9} \cdot r_{c1} \cdot r_{c2} \quad (27.304)$$

where

$$k_{i-j} = \left(1 - \frac{\Delta p_{0i-j}}{p_{0i}}\right)$$

Reheat (CBTBT) Cycle: The simple cycle indicates that specific power increases significantly as the temperature ratio increases. However, this advantage may diminish or even become negative as the mass flow rate of turbine cooling air increases. However, if the fuel is burnt in two stages between which one part of the turbine is located, the temperature ratio would be kept much lower than that of the simple cycle which burns the same amount of fuel. This implies that it would pro-

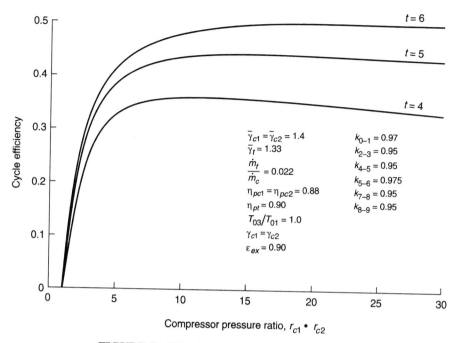

FIGURE 27.206 Cycle efficiency (CICBTX cycle).

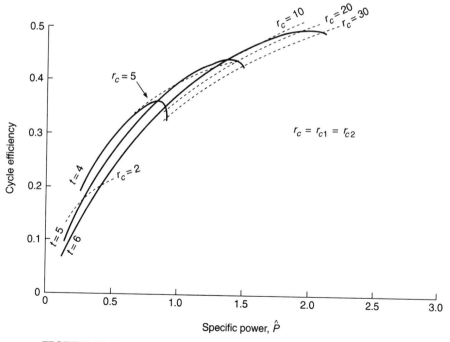

FIGURE 27.207 Cycle efficiency vs. specific power (CICBTX cycle).

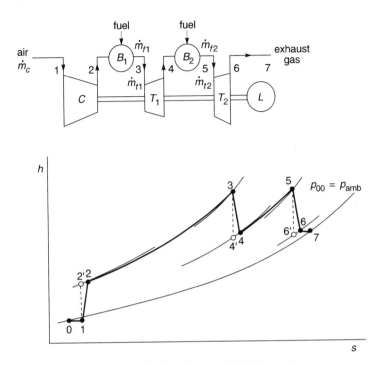

FIGURE 27.208   Reheat (CBTBT) cycle.

duce an increased specific power in comparison with a simple cycle with the same temperature ratio. This type of cycle is called a *reheat* cycle, and Fig. 27.208 shows its h-s diagram. The cycle efficiency and specific power cycle are

$$\eta = \frac{\dot{m}_{t1}(h_{03} - h_{04}) + \dot{m}_{t2}(h_{05} - h_{06}) - \dot{m}_c(h_{02} - h_{01})}{(\dot{m}_c + \dot{m}_{f1})h_{03} - \dot{m}_c h_{02} + (\dot{m}_c + \dot{m}_{f1} + \dot{m}_{f2})h_{05} - (\dot{m}_c + \dot{m}_{f1})h_{04}}$$

$$= \frac{\left(\dfrac{\dot{m}_{t1}}{\dot{m}_c}\right)\left(\dfrac{c_{p3}}{c_{p1}}\right) t_1(1 - r_{t1}^{-\alpha_1}) + \left(\dfrac{\dot{m}_{t2}}{\dot{m}_c}\right)\left(\dfrac{c_{p5}}{c_{p1}}\right) t_1\left(\dfrac{T_{05}}{T_{03}}\right)(1 - r_{t2}^{-\alpha_2}) - (r_c^\beta - 1)}{\left(\dfrac{\dot{m}_{t1}}{\dot{m}_c}\right)\left(\dfrac{c_{p3}}{c_{p1}}\right) t_1(1 - r_{t1}^{-\alpha_1}) + \left(\dfrac{\dot{m}_{t2}}{\dot{m}_c}\right)\left(\dfrac{c_{p5}}{c_{p1}}\right) t_1\left(\dfrac{T_{05}}{T_{03}}\right) - r_c^\beta}$$

(27.305)

where

$$t_1 = \left(\frac{T_{03}}{T_{01}}\right), \quad \dot{m}_{t1} = \dot{m}_c + \dot{m}_{f1}, \quad \dot{m}_{t2} = \dot{m}_{t1} + \dot{m}_{f2},$$

$$r_{t1} = \frac{p_{03}}{p_{04}}, \quad r_{t2} = \frac{p_{05}}{p_{06}}, \quad r_c = \frac{p_{02}}{p_{01}}$$

$$\alpha_1 = \frac{\gamma_{t1} - 1}{\gamma_{t1}} \eta_{pt1}, \quad \alpha_2 = \frac{\gamma_{t2} - 1}{\gamma_{t2}} \eta_{pt2}, \quad \beta = \frac{\gamma_c - 1}{\gamma_c \eta_{pc}}$$

$$\hat{P} = \left(\frac{\dot{m}_{t1}}{\dot{m}_c}\right)\left(\frac{c_{p3}}{c_{p1}}\right) t_1(1 - r_{t1}^{-\alpha_1}) + \left(\frac{\dot{m}_{t2}}{\dot{m}_c}\right)\left(\frac{c_{p5}}{c_{p1}}\right) t_1 \left(\frac{T_{05}}{T_{03}}\right)(1 - r_{t2}^{-\alpha_2}) - (r_c^\beta - 1)$$

(27.306)

$$r_{t1} \cdot r_{t2} = k_{0-1} \cdot k_{2-3} \cdot k_{4-5} \cdot k_{6-7} \cdot r_c$$

(27.307)

where

$$k_{i-j} = \left(1 - \frac{\Delta p_{0i-j}}{p_{0i}}\right)$$

Some numerical examples of the performance of this cycle are shown in Figs. 27.209 and 27.210. The following assumptions are used for the two turbines in these examples: $r_{t1} = r_{t2}$ and $T_{03} = T_{05}$. The cycle efficiency of the simple cycle is also plotted in Fig. 27.209 for comparison.

Reheat, Heat Exchange (CBTBTX) Cycle: As was seen in the heat exchange cycle, heating the compressor delivery air with exhaust gas thermal energy significantly improves cycle efficiency at low compressor pressure ratios. This preheating is particularly suitable for the reheat cycle, since the exhaust thermal energy level is high. Then, the combination of reheat and heat exchange (see Fig. 27.211) should provide the desirable combination of high cycle efficiency and high specific power, particularly in the low compressor pressure ratio range.

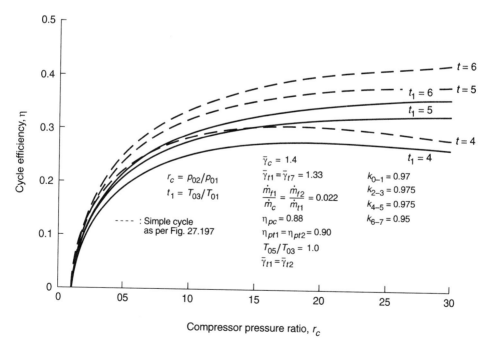

FIGURE 27.209 Cycle efficiency (CBTBT cycle).

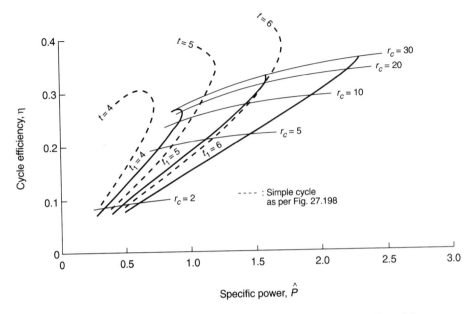

FIGURE 27.210   Cycle efficiency vs specific power (CBTBT cycle).

The performance parameters of the CBTBTX cycle are

η (cycle efficiency)

$$= \frac{\dot{m}_{t1}(h_{04} - h_{05}) + \dot{m}_{t2}(h_{06} - h_{07}) - \dot{m}_c(h_{02} - h_{01})}{(\dot{m}_{c1} + \dot{m}_{f1})h_{04} - \dot{m}_c h_{03} + (\dot{m}_c + \dot{m}_{f1} + \dot{m}_{t2})h_{06} - (\dot{m}_c + \dot{m}_{f1})h_{05}}$$

$$= \frac{\left(\dfrac{\dot{m}_{t1}}{\dot{m}_c}\right)\left(\dfrac{c_{p4}}{c_{p1}}\right) t_1 (1 - r_{t1}^{-\alpha_1}) + \left(\dfrac{\dot{m}_{t2}}{\dot{m}_c}\right)\left(\dfrac{c_{p6}}{c_{p1}}\right)\left(\dfrac{T_{06}}{T_{04}}\right) t_1 (1 - r_{t2}^{-\alpha_2}) - (r_c^\beta - 1)}{\left(\dfrac{\dot{m}_{t1}}{\dot{m}_c}\right)\left(\dfrac{c_{p4}}{c_{p1}}\right) t_1 \left[1 - r_{t1}^{-\alpha_1} + \left(\dfrac{\dot{m}_{t2}}{\dot{m}_{t1}}\right)\left(\dfrac{c_{p6}}{c_{p4}}\right)\left(\dfrac{T_{06}}{T_{04}}\right)\right]} \\ - \epsilon_{ex} t_1 \left(\dfrac{T_{06}}{T_{04}}\right) r_{t2}^{\alpha_2} - (1 - \epsilon_{ex}) r^{\beta_c}$$

$$\tag{27.308}$$

where

$$t_1 = \left(\frac{T_{04}}{T_{01}}\right), \quad r_c = \frac{p_{02}}{p_{01}}, \quad r_{t1} = \frac{p_{04}}{p_{05}}, \quad r_{t2} = \frac{p_{06}}{p_{07}}, \quad \epsilon_{ex} = \frac{(T_{03} - T_{02})}{(T_{07} - T_{02})}$$

$$\alpha_1 = \frac{\gamma_{t1} - 1}{\gamma_{t1}} \eta_{pt1}, \quad \alpha_2 = \frac{\gamma_{t2} - 1}{\gamma_{t2}} \eta_{pt2}, \quad \beta = \frac{\gamma_c - 1}{\gamma_c \eta_{pc}}$$

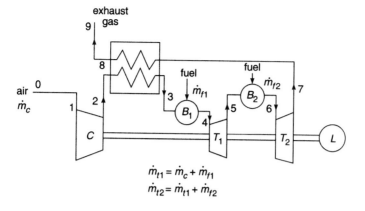

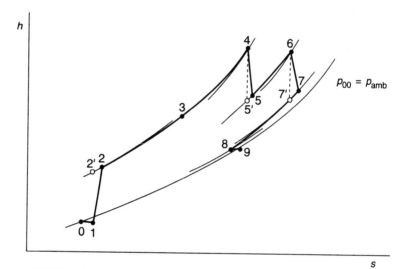

FIGURE 27.211 Reheat, heat exchanger (CBTBTX) cycle.

$$\hat{P} \text{ (specific power)} = \left(\frac{\dot{m}_{t1}}{\dot{m}_c}\right)\left(\frac{c_{p4}}{c_{p1}}\right) t_1(1 - r_{t1}^{-\alpha_1})$$

$$+ \left(\frac{\dot{m}_{t2}}{\dot{m}_c}\right)\left(\frac{c_{p6}}{c_{p1}}\right)\left(\frac{T_{06}}{T_{04}}\right) t_1(1 - r_{t2}^{-\alpha_2}) - (r_c^{\beta} - 1) \quad (27.309)$$

$$r_{t1} \cdot r_{t2} = k_{0-1} \cdot k_{2-3} \cdot k_{3-4} \cdot k_{5-6} \cdot k_{7-8} \cdot k_{8-9} \cdot r_c \quad (27.310)$$

where

$$k_{i-j} = \left(1 - \frac{\Delta p_{0i-j}}{p_{0i}}\right)$$

Figures 27.212 and 27.213 depict performance characteristics of the CBTBTX cycle.

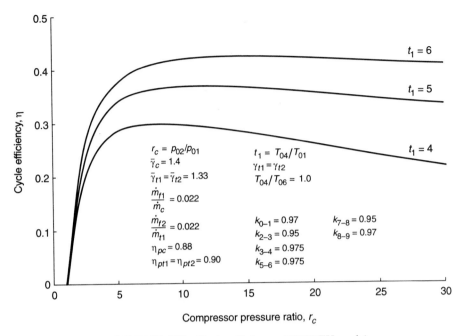

FIGURE 27.212   Cycle efficiency (CBTBTX cycle).

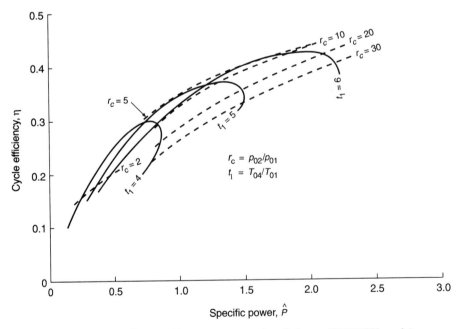

FIGURE 27.213   Specific power vs. cycle efficiency (CBTBTX cycle).

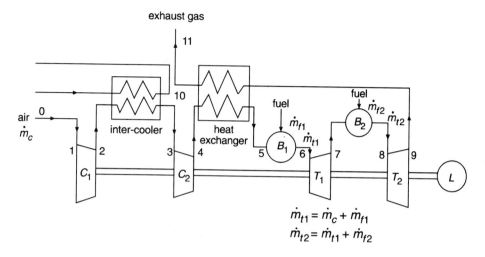

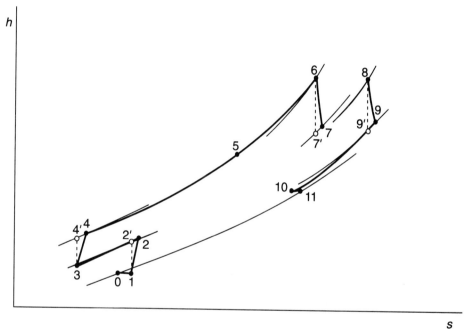

FIGURE 27.214 Reheat, intercooled, heat exchange (CICBTBTX) cycle.

Reheat, Intercooled, Heat Exchange (CICBTBTX) Cycle: As seen above, the cycle efficiency of the intercooled (CICBT) cycle peaks at high compressor pressure ratio conditions. Then, the addition of intercooling to the CBTBTX cycle should provide a combination of high cycle efficiency and high specific power and also should rectify the weakness of the CBTBTX cycle (low cycle efficiency at high compressor pressure ratio). The h-s diagram of this cycle is shown on Fig. 27.214.

The performance parameters are defined as follows

η (cycle efficiency)

$$= \frac{\dot{m}_{t1}(h_{06} - h_{07}) + \dot{m}_{t2}(h_{08} - h_{09}) - \dot{m}_c(h_{02} - h_{01}) - \dot{m}_c(h_{04} - h_{03})}{(\dot{m}_c + \dot{m}_{f1})h_{06} - \dot{m}_c h_{05} + (\dot{m}_{t1} + \dot{m}_{f2})h_{08} - \dot{m}_{t1}h_{07}}$$

$$= \frac{\left(\dfrac{\dot{m}_{t1}}{\dot{m}_c}\right)\left(\dfrac{c_{p6}}{c_{p1}}\right) t_1(1 - r_{t1}^{-\alpha_1}) + \left(\dfrac{\dot{m}_{t2}}{\dot{m}_c}\right)\left(\dfrac{c_{p8}}{c_{p1}}\right) t_1(1 - r_{t2}^{-\alpha_2})}{\begin{array}{c} - (r_{c1}^{\beta_1} - 1) - \left(\dfrac{T_{03}}{T_{01}}\right)(r_{c2}^{\beta_2} - 1) \\[2mm] \hline \left(\dfrac{\dot{m}_{t1}}{\dot{m}_c}\right)\left(\dfrac{c_{p6}}{c_{p1}}\right) t_1 \left[1 - r_{t1}^{-\alpha_1} + \left(\dfrac{\dot{m}_{t2}}{\dot{m}_{t1}}\right)\left(\dfrac{c_{p8}}{c_{p6}}\right)\left(\dfrac{T_{08}}{T_{06}}\right)\right] \end{array}}$$

$$\qquad\qquad - (1 - \epsilon_{ex})\left(\dfrac{T_{03}}{T_{01}}\right) r_{c2}^{\beta_1} - \epsilon_{ex}\left(\dfrac{T_{08}}{T_{06}}\right) t_1 r_{t2}^{-\alpha_2}$$

$$\tag{27.311}$$

where

$$t_1 = \left(\frac{T_{06}}{T_{01}}\right), \qquad r_{c1} = \frac{p_{02}}{p_{01}}, \qquad r_{c1} = \frac{p_{04}}{p_{03}},$$

$$r_{t1} = \frac{p_{06}}{p_{07}}, \qquad r_{t2} = \frac{p_{08}}{p_{09}}, \qquad \epsilon_{ex} = \frac{(T_{05} - T_{04})}{(T_{09} - T_{04})}$$

$$\alpha_1 = \frac{\gamma_{t1} - 1}{\gamma_{t1}}\eta_{pt1}, \qquad \alpha_2 = \frac{\gamma_{t2} - 1}{\gamma_{t2}}\eta_{pt2},$$

$$\beta_1 = \frac{\gamma_{c1} - 1}{\gamma_{c1}\eta_{pc1}}, \qquad \beta_2 = \frac{\gamma_{c2} - 1}{\gamma_{c2}\eta_{pc2}}$$

$$\hat{P} \text{ (specific work)} = \left(\frac{\dot{m}_{t1}}{\dot{m}_c}\right)\left(\frac{c_{p6}}{c_{p1}}\right) t_1(1 - r_{t1}^{-\alpha_1})$$

$$+ \left(\frac{\dot{m}_{t2}}{\dot{m}_c}\right)\left(\frac{c_{p8}}{c_{p1}}\right)\left(\frac{T_{08}}{T_{06}}\right) t_1(1 - r_{t2}^{-\alpha_2}) - (r_{c1}^{\beta_1} - 1)$$

$$- \left(\frac{T_{03}}{T_{01}}\right)(r_{c2}^{\beta_2} - 1) \tag{27.312}$$

$$r_{t1} \cdot r_{t2} = k_{0-1} \cdot k_{2-3} \cdot k_{4-5} \cdot k_{5-6} \cdot k_{7-8} \cdot k_{9-10} \cdot k_{10-11} \cdot r_{c1} \cdot r_{c2}$$

$$\tag{27.313}$$

where

$$k_{i-j} = \left(1 - \frac{\Delta p_{0i-j}}{p_{0i}}\right)$$

Figures 27.215 and 27.216 show cycle efficiency and its relationship with specific power, respectively.

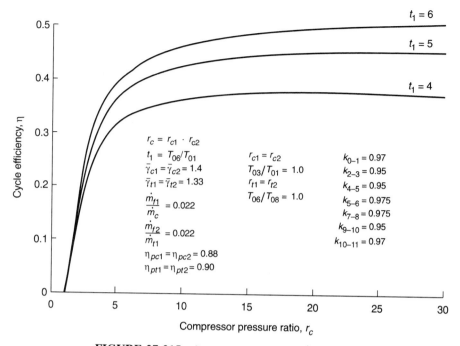

$$r_c = r_{c1} \cdot r_{c2}$$
$$t_1 = T_{06}/T_{01}$$
$$\bar{\gamma}_{c1} = \bar{\gamma}_{c2} = 1.4$$
$$\bar{\gamma}_{t1} = \bar{\gamma}_{t2} = 1.33$$
$$\frac{\dot{m}_{f1}}{\dot{m}_c} = 0.022$$
$$\frac{\dot{m}_{f2}}{\dot{m}_{t1}} = 0.022$$
$$\eta_{pc1} = \eta_{pc2} = 0.88$$
$$\eta_{pt1} = \eta_{pt2} = 0.90$$

$$r_{c1} = r_{c2}$$
$$T_{03}/T_{01} = 1.0$$
$$r_{t1} = r_{t2}$$
$$T_{06}/T_{08} = 1.0$$

$$k_{0-1} = 0.97$$
$$k_{2-3} = 0.95$$
$$k_{4-5} = 0.95$$
$$k_{5-6} = 0.975$$
$$k_{7-8} = 0.975$$
$$k_{9-10} = 0.95$$
$$k_{10-11} = 0.97$$

FIGURE 27.215 Cycle efficiency (CTCBTTX cycle).

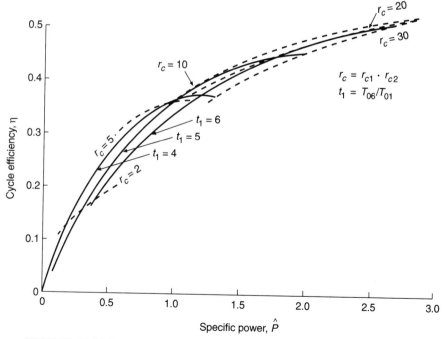

$$r_c = r_{c1} \cdot r_{c2}$$
$$t_1 = T_{06}/T_{01}$$

FIGURE 27.216 Cycle efficiency vs. specific power, $\hat{P}$ (CICBTBTX cycle).

Closed Cycle: Figure 27.217 shows the h-s diagram of a simple, closed cycle. It has a gas heater instead of a combustor used on the simple, open cycle, and a pre-cooler in order to decrease the turbine exit gas temperature to repeat the cycle. The cycle efficiency and specific power of this cycle are readily derivable from Eqs. (27.292) and (27.293). Note that γ (and therefore c_p) is assumed to be constant throughout the cycle in the derivation.

$$\eta \text{ (cycle efficiency)} = 1 - r_t^{-(\gamma-1)\eta_{pt}/\gamma} \cdot \frac{[t - r_t^{(\gamma-1)\eta_{pt}/\gamma}]}{[t - r_c^{-(\gamma-1)/\gamma\eta_{pc}}]} \quad (27.314)$$

where

$$t = \frac{T_{03}}{T_{01}}, \qquad r_c = \frac{p_{02}}{p_{01}}, \qquad r_t = \frac{p_{03}}{p_{04}}$$

$$\hat{P} \text{ (specific power)} = [t - r_c^{-(\gamma-1)/\gamma\eta_{pc}}] - r_t^{-(\gamma-1)\eta_{pt}/\gamma} \cdot [t - r_t^{(\gamma-1)\eta_{pt}/\gamma}] \quad (27.315)$$

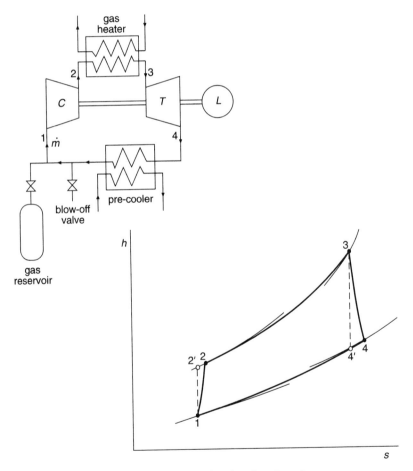

FIGURE 27.217 Simple, closed cycle.

$$r_t = k_{2-3} \cdot k_{4-1} r_c \qquad (27.316)$$

where

$$k_{i-j} = \left(1 - \frac{\Delta p_{0i-j}}{p_{0i}}\right)$$

Some numerical examples are shown in Figs. 27.218 and 27.219 in a comparative manner with the simple, open cycle. In these numerical examples, $\gamma = 1.4$ is assumed for the gas in the closed cycle.

The advantages of the closed cycle over the open cycle are as:

i) Since the cycle is not open to the atmosphere, gas pressure may be varied as a function of load. Thus, the optimum pressure ratio may be maintained for a wide range of load.

ii) High density gas may be used as the working fluid, which results in a smaller unit than the open cycle for the same magnitude of output.

iii) Gases other than air ($\gamma = 1.4$) may be used. The use of gases such as Argon or Helium ($\gamma = 1.67$) would provide higher cycle efficiencies than the open cycle at low compressor pressure ratios (see Figs. 27.220 and 27.221).

On the other hand, the closed cycle needs auxiliary equipment for gas heating and pre-cooling.

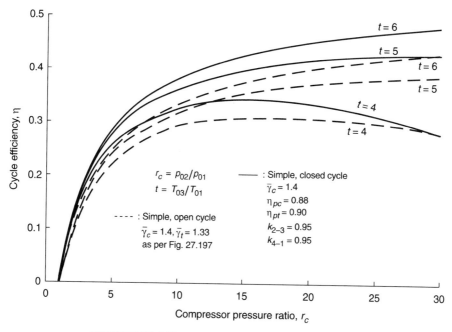

FIGURE 27.218 Cycle efficiency (closed CBT cycle).

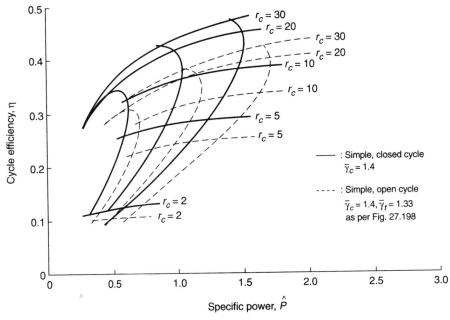

FIGURE 27.219   Specific power vs. cycle efficiency (closed CBT cycle).

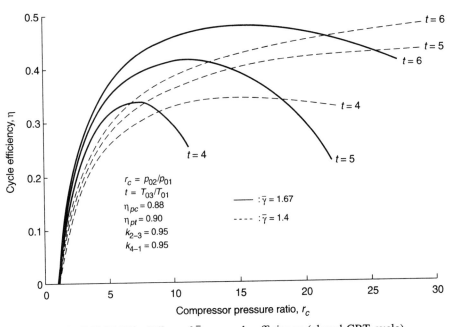

FIGURE 27.220   Effect of $\bar{r}_c$ on cycle efficiency (closed CBT cycle).

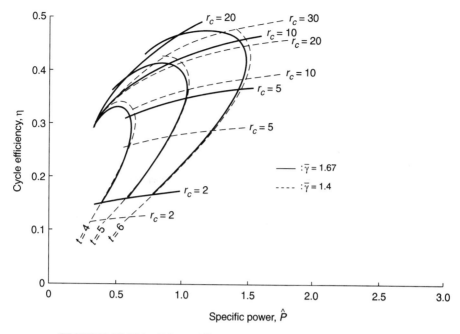

FIGURE 27.221 Effect of $\bar{r}_c$ on specific power vs. cycle efficiency.

Combined Cycles: Horlock (1992) classifies *combined cycle* into four categories which produce power only, and not a combination of heat and power (co-generation): 1) *double cycles*, with two closed cycles which employ a separate working fluid, 2) *binary cycles*, with two closed cycles which employ a separate working fluid, 3) *open cycle/closed cycle* combination, and 4) *double, open cycles*. Among these four categories, the combined cycles commonly used today are those in Categories 3) and 4).

Open Cycle/Closed Cycle Combination: Figure 27.222 shows the simplest configuration of an open cycle/closed cycle combination in which a gas turbine is used as a higher temperature cycle, while a steam turbine is used as a lower temperature cycle. The overall thermal efficiency of the combined cycle is defined as

$$\eta_0 A = \frac{p \text{ (total power output)}}{E_i \text{ (energy input)}} = \frac{W_H + W_L}{\dot{m}_f (CV)_0} \qquad (27.317)$$

where

$$W_H = \dot{m}_t(h_{03} - h_{04}) - \dot{m}_c(h_{02} - h_{01})$$

$$W_L = \dot{m}_{st}(h_{oc} - h_{od})$$

Since,

$$\eta_H \text{ (higher temperature cycle efficiency)} = \frac{W_H}{\dot{m}_f (CV)_0} \qquad (27.318)$$

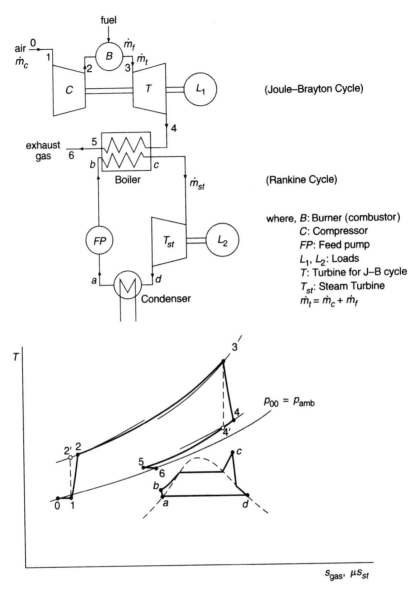

FIGURE 27.222 T-s diagram of a combined (Joule–Brayton/Rankine) cycle. [Reprinted with permission, Horlock, J. H., *Combined Power Plant*, Elsevier Service Ltd. (Pergamon Press), Oxford, 1992.] μ is a scale factor on steam entropy.

and

$$\eta_L \text{ (lower temperature cycle efficiency)} = \frac{W_L}{\dot{m}_t(h_{04} - h_{05})} \qquad (27.319)$$

therefore the overall efficiency becomes

$$\eta_{OA} = \eta_H + \frac{\eta_L[(\dot{m}_t h_{03} - \dot{m}_c h_{02}) - W_H - (\dot{m}_t h_{05} - \dot{m}_c h_{01})]}{\dot{m}_f(CV)_0}$$

$$= \eta_H + \eta_L(1 - \eta_H - v) = \eta_H + \eta_L - \eta_H\eta_L - v\eta_L \qquad (27.320)$$

where

$$v \text{ (nondimensional heat loss)} = \frac{(\dot{m}_c + \dot{m}_f)(h_{05} - h_{01})}{\dot{m}_f(CV)_0} \qquad (27.321)$$

Figure 27.223 clearly indicates the high efficiency potential of the combined cycle, while the effect of the heat loss on the overall thermal efficiency is shown on Fig. 27.224.

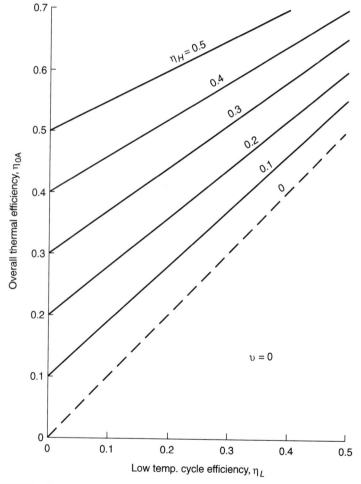

FIGURE 27.223 Overall thermal efficiency without heat loss. [Reprinted with permission, Horlock, J. H., *Combined Power Plant*, Elsevier Service Ltd. (Pergamon Press), Oxford, 1992.]

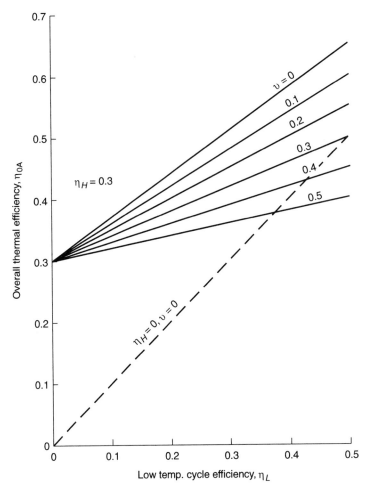

FIGURE 27.224   Effect of heat loss on overall thermal efficiency. [Reprinted with permission, Horlock, J. H., *Combined Power Plant*, Elsevier Service Ltd. (Pergamon Press), Oxford, 1992.]

Supplementary Firing: Figure 27.225 depicts the simplest configuration of an open cycle/closed cycle combination as shown on Fig. 27.222 except for a second combustor inserted between the higher temperature cycle turbine and the boiler. The advantage of this configuration is higher specific power capability. The overall thermal efficiency of this cycle is defined as

$$\eta_{0A} = \frac{\text{(total power output)}}{\text{(total energy input)}} = \frac{W_H + W_L}{(\dot{m}_{f1} + \dot{m}_{f2})(CV)_0} \qquad (27.322)$$

where

$$W_H = \dot{m}_t(h_{03} - h_{04}) - \dot{m}_c(h_{02} - h_{01})$$

$$W_L = \dot{m}_{st}(h_{oc} - h_{od})$$

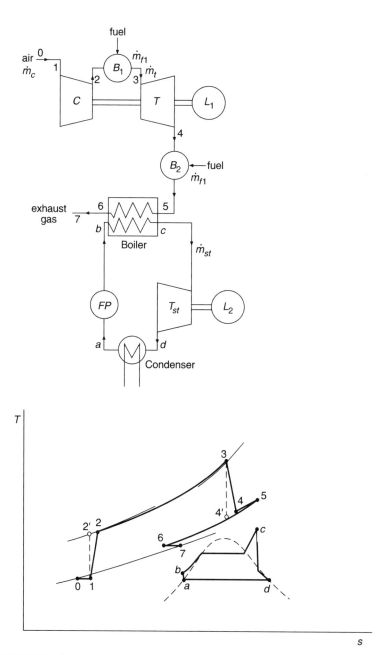

FIGURE 27.225 T-s diagram of a combined (Joule-Brayton/Rankine) cycle with supplementary firing. [Reprinted with permission, Horlock, J. H., *Combined Power Plant*, Elsevier Service Ltd. (Pergamon Press), Oxford, 1992.]

Since

$$\eta_H = \frac{W_H}{\dot{m}_{f1}(CV)_0}$$

(27.323)

and

$$\eta_L = \frac{W_L}{\dot{m}_t h_{04} + \dot{m}_{f2}(CV)_0 - (\dot{m}_t + \dot{m}_{f2})h_{06}}$$

(27.324)

therefore,

$$\eta_{OA} = \eta_L + \frac{\dot{m}_{f1}}{(\dot{m}_{f1} + \dot{m}_{f2})}\eta_H(1 - \eta_L) - \upsilon\eta_L$$

(27.325)

where

$$\upsilon = \frac{(\dot{m}_c + \dot{m}_{f1} + \dot{m}_{f2})(h_{06} - h_{01})}{(\dot{m}_{f1} + \dot{m}_{f2})(CV)_0}$$

(27.326)

Steam Injection Gas Turbine (STIG) Cycle: Figure 27.226 depicts the flow diagram of a basic steam injection gas turbine cycle. This is considered as the simplest form of a double, open combined cycle [Horlock (1992)]. Water pressurized by a pump is delivered to a boiler (heat exchanger) in which it becomes steam and is then introduced to an open cycle gas turbine at its combustor inlet. Both the compressed air and steam are further heated up in the combustor prior to entering the turbine. The turbine, therefore, become a gas turbine as well as a steam turbine. The exhaust becomes a heat source for the heat exchanger.

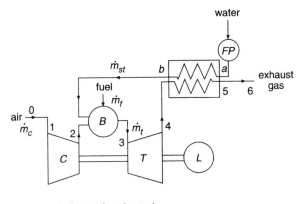

FIGURE 27.226 Basic steam injection gas turbine (STIG) cycle.

The high c_p value of steam makes a power plant with this cycle relatively compact and simple. The use of the heat exchange improves cycle efficiency as well. The cycle efficiency and specific power may be defined as

$$\eta = \frac{\dot{m}_t(h_{03} - h_{04}) - \dot{m}_c(h_{02} - h_{01})}{\dot{m}_t h_{03} - (\dot{m}_{st} h_{0b} + \dot{m}_c h_{02})}$$

$$= \frac{\left(1 + \dfrac{\dot{m}_{st}}{\dot{m}_c} + \dfrac{\dot{m}_f}{\dot{m}_c}\right) \dfrac{C_{p3}}{C_{p1}} t(1 - r_t^{-\alpha}) - (r_c^{\beta} - 1)}{\left(1 + \dfrac{\dot{m}_{st}}{\dot{m}_c} + \dfrac{\dot{m}_f}{\dot{m}_c}\right) \dfrac{C_{p3}}{C_{p1}} t - \left(\dfrac{\dot{m}_{st}}{\dot{m}_c} \dfrac{C_{pstb}}{C_{p1}} \dfrac{T_{0st2}}{T_{02}} + 1\right) r_c^{\beta}} \tag{27.327}$$

where

$$t = \frac{T_{03}}{T_{01}}, \qquad r_c = \frac{p_{02}}{p_{01}}, \qquad r_t = \frac{p_{04}}{p_{03}}, \qquad \alpha = \frac{(\gamma_t - 1)}{\gamma_t} \eta_{pt}, \qquad \beta = \frac{(\gamma_c - 1)}{\gamma_c \eta_{pc}}$$

Here, c_p is the average c_p of air and steam based on mass flow ratio [Fraize and Kinney (1979)].

$$\hat{P} = \left(1 + \frac{\dot{m}_{st}}{\dot{m}_c} + \frac{\dot{m}_f}{\dot{m}_c}\right) \frac{C_{p3}}{C_{p1}} t(1 - r_t^{-\alpha}) - (r_c^{\beta} - 1) \tag{27.328}$$

Note that the specific power defined above is based on the energy level of air flow at the compressor entry, m_c.

$$r_t = k_{0-1} \cdot k_{2-3} \cdot k_{4-5} \cdot k_{5-6} \cdot r_c \tag{27.329}$$

where

$$k_{i-j} = \left(1 - \frac{\Delta p_{0i-j}}{p_{0i}}\right)$$

Fraize and Kinney (1979) point out some heat recovery limitations existing in the cycle- the *pinch point limit* and the *minimum exhaust stack temperature* limit. The *pinch point* is defined as the minimum difference between the water/steam and exhaust temperatures in the heat exchanger (see Fig. 27.227). The efficiency and specific power of this cycle are calculated as a function of steam flow rate and compressor pressure ratio, and the results are shown in Figs. 27.228 and 27.229.

27.11.2 Compressors
T. Yoshinaka

Type. The gas turbine, which requires high mass flow rate, demands the exclusive use of a turbocompressor. These are axial, centrifugal, or mixed (axial/centrifugal)

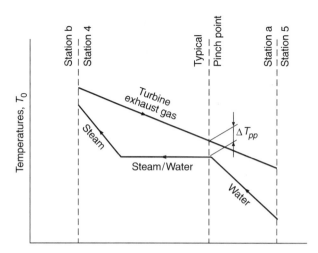

FIGURE 27.227 Pinch point in boiler. (Reprinted with permission of ASME, Fraize, W. E. and Kinney, C., "Effects of Steam Injection on the Performance of Gas Turbine Power Cycles," *J. of Engrg. for Power*, Vol. 101, No. 2, pp. 217–227, 1979.)

flow types. The choice of the compressor type largely depends on performance, hence the mass flow rate as depicted in Fig. 27.230.

The axial compressor is superior in performance to the centrifugal compressor at high mass flow rate, while the opposite is true in small machines. The mixed flow compressor having its optimum specific speed between that of the axial and the centrifugal compressor [Baljé (1981)] may be usable in the gas turbine from a performance perspective. Weight and space penalties appear to be the reasons for the rare use of this type of compressors. See Secs. 27.4 and 27.5 for a more detailed coverage of axial, centrifugal, and mixed flow compressors.

Pressure Ratio Capability. Figure 27.230 also indicates a general trend towards higher pressure ratio on larger compressors. Large power plants which require high initial cost are typically used for high running hour duties. Therefore, high cycle efficiency (or low fuel consumption) is a must. This requirement drives compressor pressure ratio to increase, along with an elevated level of turbine inlet temperature. Today, the number of stages of a multistage axial compressor required to produce pressure ratios higher than $3:1$ is approximately as large as the value of the pressure ratio itself (see Fig. 27.231). This limited capability can be enhanced by 2 to 2.5 times by applying a multispool configuration in which the rear part of a multistage axial compressor is mechanically separated from the front part as in Fig. 27.232 and driven by its own turbine faster than the front part.

Small power plants are typically intended for low time usage, such as emergency power, peak load, or helicopter propulsion. Therefore, low initial and maintenance costs become of prime importance. These requirements are addressed by the choice of a simple construction and the use of the minimum number of compressor stages. Figure 27.233 depicts this technology trend for small gas turbine engines for aircraft propulsion (turboprop and turboshaft engines).

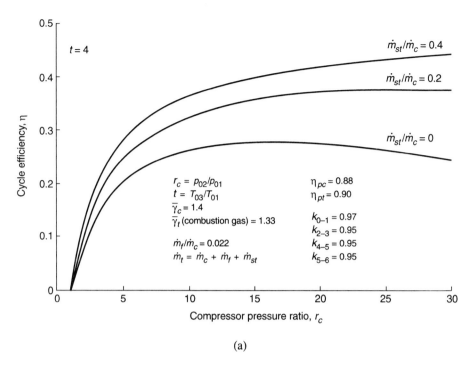

(a)

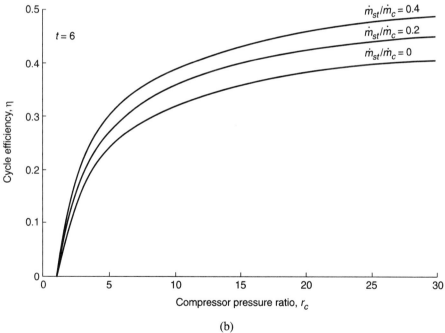

(b)

FIGURE 27.228   Cycle efficiency (STIG cycle). (a) $t = 4$ and (b) $t = 6$.

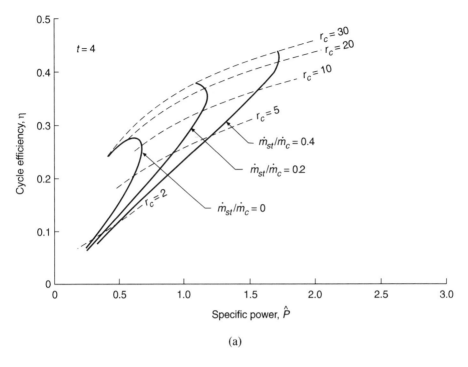

(a)

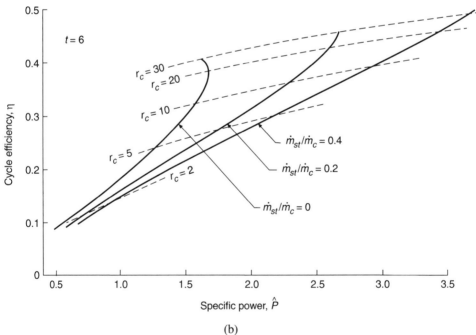

(b)

FIGURE 27.229 Cycle efficiency vs. specific power (STIG cycle). (a) $t = 4$ and (b) $t = 6$.

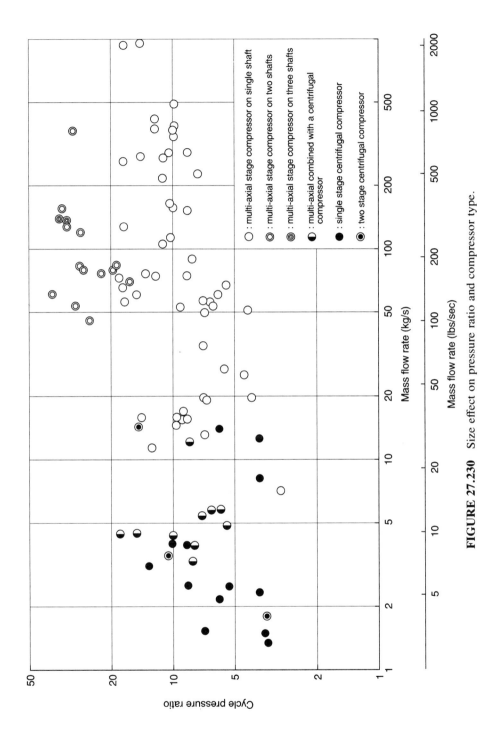

FIGURE 27.230 Size effect on pressure ratio and compressor type.

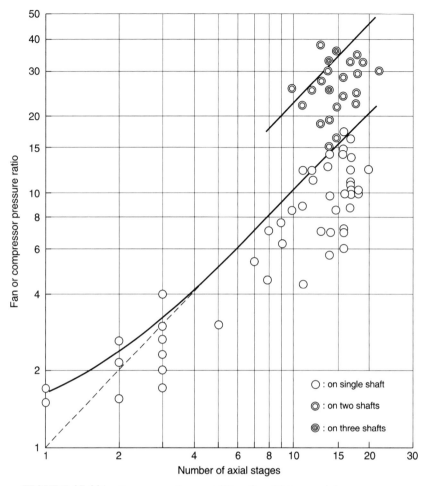

FIGURE 27.231   Pressure ratio capability of multistage axial compressors.

27.11.3   Combustors
P. Sampath

Basic Requirements. The combustion system of a gas turbine must satisfy a wide range of requirements: 1) raise the air temperature, achieving high levels of combustion efficiency, 2) operate with low system pressure loss, 3) provide exit temperature uniformity and satisfy radial temperature requirements of the turbines, 4) have stable operation over a wide operating range, including transients, 5) obtain reliable ignition at very low temperatures and high altitudes, 6) produce low exhaust emissions and operate without forming carbon, and 7) have mechanical reliability and long life. In the case of aircraft gas turbines, these standards of performance must be achieved with minimum weight and bulk. Industrial gas turbines must also achieve very high levels of reliability and low exhaust pollution; they are required to operate with a wide range of liquid and gaseous fuels.

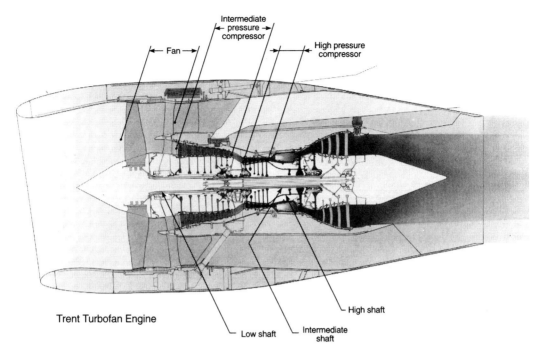

FIGURE 27.232 Three shaft configurations. (Courtesy of Rolls-Royce.)

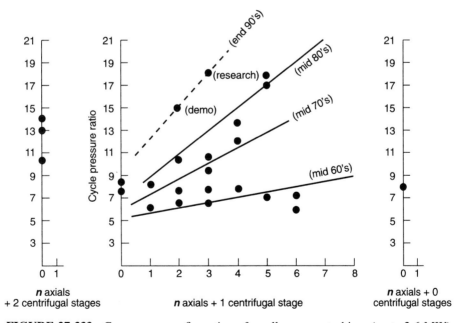

FIGURE 27.233 Compressor configuration of small aero gas turbines (up to 2.6 MW).

Aircraft Gas Turbine Combustors

Combustor Types: The commonly used combustor types (see Fig. 27.234) are tubular, annular, tubo-annular, reverse-flow annular, and radial-flow annular. The
choice of combustion system is influenced by the engine layout, operating conditions, and engine size. The size and shape of the space into which the combustor
must fit are usually dictated by the dimensions and orientation of the surrounding
turbomachinery. The diameter is normally influenced by the size of the final stage
of the compressor and the length by the desirability of minimizing the axial distance
between the compressor and the turbine. Tubular and reverse-flow annular combustors are common in small engines, while annular and tubo-annular combustors are
commonly used by large engines. Table 27.9 shows the relative merits of the various
combustor types.

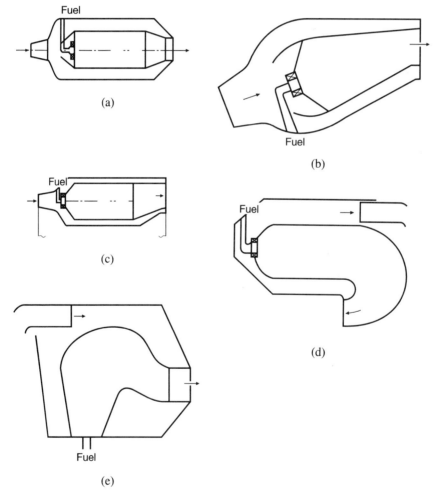

FIGURE 27.234 Aircraft combustor types. (a) Tubular combustor, (b) annular combustor,
(c) tubo-annular combustor, (d) reverse-flow annular combustor, and (e) radial-flow combustor.

TABLE 27.9 Comparison of Combustor Types

Type	Benefits	Drawbacks
Tubular	• Simple Construction • Symmetrical Air & Fuel Flow Fields • Performance/Emission Optimization Relatively Easy	• Ducting Losses Due to Compressor & Turbine Transition Ducts
Tubo-Annular	• Good Matching of Air & Fuel Flow Fields in Combustion Zone • Front End Optimization Easier to Achieve	• Interconnectors Required for Light Around • Cooling of Metal in Transition Areas may be Difficult
Annular	• Compact Geometry Achievable • Configuration Matched to Compressor & Turbine Geometries • Low Pressure Loss with Matched Diffuser • Cooling Flow More Easily Optimized	• Flow Fields Typically 3-Dimensional • Rig Development More Difficult Than Tubular
Reverse-Flow Annular	• Compact for Engines With Centrifugal Compressors • Reduced Engine Length • Low Pressure Loss	• High Cooling Flows Required Due to Transition Duct • High Surface to Volume Ratio can Affect Idle η • Pattern Factor Affected by Relatively Large Nozzle Pitching
Radial-Flow Annular	• Compact Construction With Reduced Engine Length • Adaptable to "Slinger" Fuel Injection	• Rig Development Difficult • Accelerating/Decelerating Flows can Create Mixing or Separation Problems

Combustor Aerothermodynamics: Combustion and mixing processes occur in three regions of the combustor (see Fig. 27.235): 1) *primary zone* for producing burning mixture conditions and stable flame front location during combustion, 2) *intermediate zone* for completion of combustion and recovery of dissociated combustion products, and 3) *dilution zone* for mixing of combustion products with the balance of air to produce uniform entry conditions to the turbine. A combination of jets and wall flows are typically used in all three zones. Typical sizes of these regions for various types of combustors are shown in Table 27.10.

The mixture conditions in the primary zone are normally close to the theoretical or stoichiometric air to fuel ratios at the high end of the engine operating range (typically 14.7 to 1), although recent low emission combustor designs have resorted to both rich and lean primary zone concepts. The flow paths in the primary zone typically comprise of swirler or wall flow generated recirculations to achieve good stability, as shown in Fig. 27.235.

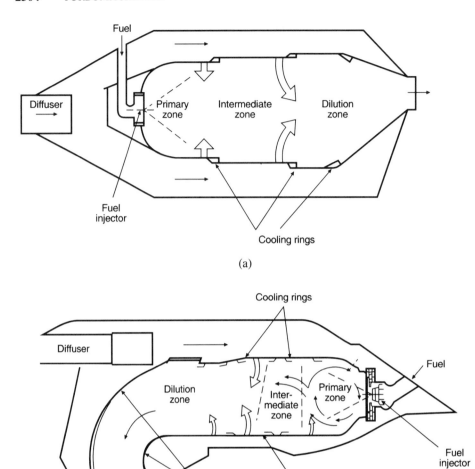

FIGURE 27.235 Combustor flow partitioning. (a) Tubo-annular combustor and (b) re-verse-flow annular combustor.

TABLE 27.10 Typical Combustor Size

	Primary Zone (L/D)	Intermediate Zone (L/D)	Dilution Zone (L/D)
Tubular	0.5	0.75	>1.5
Annular	0.5	0.75	>1.5
Rev-Flow Annular	1.0	0.5	>1.5

The total pressure loss through a combustion system, Δp_{3-4}, is composed of three parts.

$$\frac{\Delta p_{3-4}}{q_{REF}} = \frac{\Delta p_{0,\text{diff}}}{q_{REF}} + \frac{\Delta p_{0,L}}{q_{REF}} + \frac{\Delta p_{0,\text{HOT}}}{q_{REF}} \tag{27.330}$$

$$\frac{\Delta p_{0,L}}{q_{REF}} = \left[\frac{A_{REF}}{A_H C_d}\right]^2 \tag{27.330a}$$

$$\frac{\Delta p_{0,\text{HOT}}}{q_{REF}} = \frac{T_4}{T_3} - 1 \tag{27.330b}$$

$$q_{REF} = \frac{W_3^2}{2\rho_3 A_{REF}^2} \tag{27.330c}$$

where q is dynamic pressure, $\Delta p_{0,\text{DIFF}}$ is diffuser pressure loss, and $\Delta p_{0,L}$ is linear pressure loss, and W_3 is combustor inlet flow. Thus, the overall pressure loss is dictated by the casing size, which is normally defined as A_{REF}.

The combustion chamber typically includes a large number of air holes strategically located along the liner skin to achieve good combustion, mixing, and wall cooling. The flow through combustion orifices is given by

$$\dot{m}_H = A_H C_d (2 g \rho_3 (p_{0,3} - p_4))^{1/2} \tag{27.331}$$

The coefficient of discharge, C_d, is affected by the type and shape of the orifices, pressure drop factor $(\Delta p_0/q)$ and local flow condition. Lefebvre (1983) gives C_d data for plain and plunged orifices typically used in gas turbine combustor designs. Figure 27.236 shows common types of combustor air holes.

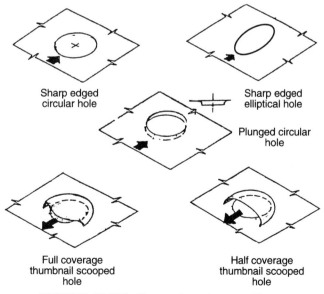

Sharp edged
circular hole

Sharp edged
elliptical hole

Plunged circular
hole

Full coverage
thumbnail scooped
hole

Half coverage
thumbnail scooped
hole

FIGURE 27.236   Types of combustor orifices.

Efficient combustion of fuel in the combustor results in heat release close to the theoretical value available to raise the temperature of the air. The temperature rise across the combustor is a function of mixture fuel–air ratio, *far*, and the heating value of the fuel. Figure 27.237 shows the theoretical temperature rise for Jet A-1 fuel as a function of overall fuel–air ratio and combustor inlet temperature [Jones *et al.* (1984)]. The actual temperature rise would depend on, η_c, the *efficiency of combustion* thus

$$\Delta T_{\text{ACT}} = \Delta T_{\text{THEO}} \eta_c \qquad (27.332)$$

The combustion process is extremely complex involving 3-dimensional, 2-phase flow fields influenced by atomization, evaporation, mixing, and reaction. Combustion efficiency is influenced by operating variables through the *combustor loading parameter*, θ, [Lefebvre (1983)], i.e., $\eta_c = f(\theta)$, where

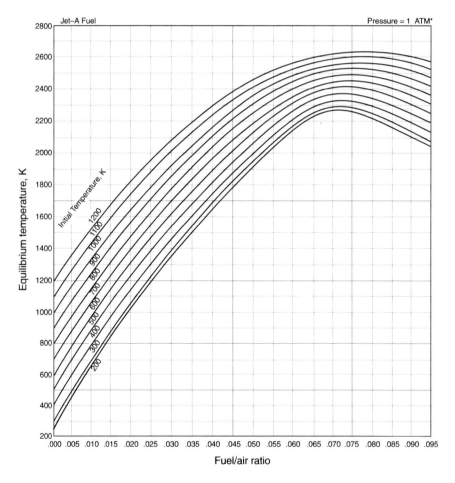

FIGURE 27.237 Equilibrium temperature as function of initial temperature and fuel-air ratio at pressure of 1 atmosphere. (From Jones *et al.*, 1984.)

$$\theta = \frac{(p_{0,3})^{1.75} A_{REF} D_{REF}^{0.75} e^{T3/540} (\Delta p_{0,3-4}/q_{REF})^{0.4}}{W_3} \qquad (27.333)$$

Other commonly used *loading factors* are

$$V_L \text{ (velocity loading)} = \frac{W_3 \sqrt{T_3}}{p_{0,3}} \qquad (27.334a)$$

$$\Omega \text{ (Longwell air loading)} = \frac{W_3}{p_{0,3}^{1.8} T_\Omega V_c}; \; T_\Omega = f(e^{T3/540}) \qquad (27.334b)$$

$$I \text{ (volumetric heat loading)} = \frac{W_3 \cdot LHV}{p_{0,3} V_c} \qquad (27.335)$$

$$t_{RES} \text{ (combustor residence time)} = \frac{L_C A_C \rho_3 10^3}{W_3} \qquad (27.336)$$

The air loading parameter, Ω, represents a rough approximation of the extent to which the fuel combustion should have proceeded. Specifically

$$\Omega = \frac{1}{(\text{reaction rate}) \times (\text{residence time})} \qquad (27.337)$$

The efficiencies of combustors can be compared usefully using the above loading parameters. Figure 27.238 shows typical values of combustion efficiency with the θ factor.

Combustor Stability: Combustors must be designed to operate stably over the entire engine operating range including transients. Figure 27.239 shows the *rich* and *weak extinction limits* of typical gas turbine combustors, and this is called the *Stability Loop*. A characteristic parameter commonly used for stability is the lean limit fuel–

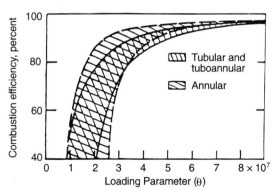

FIGURE 27.238 Combustion efficiency variation with loading parameter. (From *Gas Turbine Combustion*, Lefebvre, A. H., Taylor and Francis, 1983. Reproduced with permission. All rights reserved.)

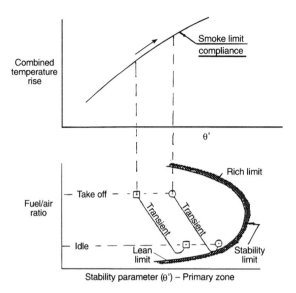

FIGURE 27.239   Effect of operating parameters on combustor stability.

air ratio influenced by loading parameter (θ'); thus the lean limit fuel–air ratio, $LL\ far = f(\theta')$ where

$$\theta' = \frac{1}{V_C p_{0.3}^{1.8}} \tag{27.338}$$

where V_c is the combustor velocity. Generally, weak extinction presents the most challenges to operability during transients, especially when the combustor smoke and emission needs require lean mixture ratios at high power conditions (see Fig. 27.239).

The combustor must also be stable during aircraft operation under conditions conducive to water, ice, or hail ingestion, which can occur on the ground or at altitude. The reduction in *flameout margin* is typically the result of heat absorption by water under marginal conditions. A way to reduce the occurrence of flameout is to operate the combustor at higher primary zone fuel to air ratios at the marginal condition and reduce entry of water/ice into the primary zone around the fuel nozzles.

Combustor Lighting: Aircraft gas turbines must be capable of reliable lightoffs over a wide operating envelope, which normally covers ambient temperatures from $-65°F$ to $+135°F$ and ambient pressures from sea level to 20,000 feet. In addition, the engine must be capable of reliable relights following incidences of flameouts, which can occur over any portion of the flight envelope, typically up to 35,000 feet. Reliable ignition is ensured by: 1) good atomization of fuel at lightoff flows, 2) adequate lighting sources, called *igniters*, situated in a relatively fuel rich zone of the combustor, and 3) mixture ratios which are within the ignitable range as defined by the *Ignition Loop* of the combustor (see Fig. 27.240). Cold ignition is influenced by atomization quality, which is also dependent upon fuel properties such as vis-

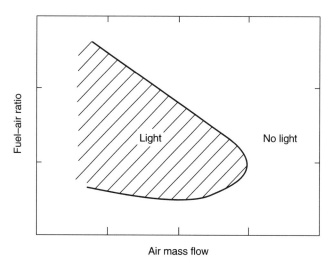

FIGURE 27.240 Typical aircraft combustor ignition envelope.

cosity, surface tension and density, and by the type of atomizer. Lightoff and acceleration to idle are also dependent on the torque supplied by the starter motor, which may be electrically or pneumatically powered. Starting performance is usually expressed in terms of time-to-light and time-to-idle of the engine.

Altitude relight capability is affected by combustion reaction rate and residence time, expressed in terms of the loading parameter. The bounds of the relight maps are influenced by combustion efficiency variation with the loading parameter (Fig. 27.238), lean blowout characteristic (Fig. 27.239), and the transient temperature limits of the engine. For very high altitude application, a quenching limit criteria may apply, where the flame in the combustor may be quenched by the combustor walls, which is influenced by physical size, air pressure, temperature, and loading. Figure 27.241 shows a typical relight envelope, with satisfactory relights possible in the shaded area bounded by the above mentioned factors.

Fuel Preparation: Atomizers, vaporizers, and slingers are used to prepare the fuel for combustion. Table 27.11 (ν is the kinematic viscosity, σ' is the surface tension, ρ is the density, and μ is the viscosity) summarizes fuel preparation methods and influencing factors. Hybrid airblast nozzles combine pressure atomizing starter nozzles with airblast main nozzles. Vaporizers and slingers typically need starter nozzles to achieve reliable ignition.

Fuel atomization quality is influenced by fuel properties such as viscosity, density, and surface tension. Atomization, therefore, tends to be poor at low temperatures where the fuel viscosity is high, and this may cause low temperature ignition problems. Grouping or staging of fuel nozzles is a common practice to ensure adequate atomization for ignition at low temperature. Fuel atomization quality can also influence gaseous exhaust emissions and smoke from gas turbine combustors (see Fig. 27.242).

The optimum number of fuel injectors for a gas turbine combustion system is a function of combustor type, fuel pump pressure limit, minimum passage size of the atomizer, and the fuel flow range (max. to min.) required. A characterizing param-

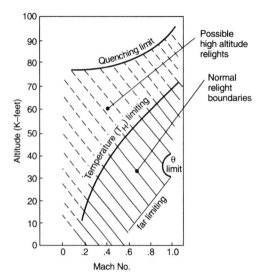

FIGURE 27.241 Typical aircraft combustor relight envelope.

eter determining nozzle capacity is the *Flow Number* defined as $FN = \dot{w}_F/(\Delta p_{0F})^{1/2}$. The total system Flow Number must satisfy the flow needs of the combustor within the pressure capability of the fuel delivery system. The number of fuel injectors, N, is also typically matched to be symmetrical with the number of compressor diffusers and/or turbine nozzles. In some cases, engine dynamics considerations may preclude achieving such symmetry. The number of fuel injectors also influences combustor exit temperature quality, or *Pattern Factor (PF)* defined as

$$PF = \frac{T_{4,\text{MAX}} - T_{4,\text{Mean}}}{T_{4,\text{Mean}} - T_3} = f(N, p_{0,3}, \Delta p_{3\text{-}4}, L, D, T_3, T_4) \qquad (27.339)$$

Figure 27.243 shows a typical circumferential and radial temperature distribution at the exit plane of an annular combustor. The radial temperature distribution is expressed as the *Radial Profile Factor* defined as

$$RPF = T_4/T_{4,\text{Mean}} = f(T_3, T_4, p_{0,3}, \Delta p_{3\text{-}4}, W_{\text{dil}}/W_3, L, D) \qquad (27.340)$$

Combustor Durability: Combustors in aircraft turbine engines are expected to operate reliably without distress over several overhaul periods. Combinations of high metal temperatures and thermal gradients, however, often initiate cracking and buckling of combustor walls and these can be controlled by: 1) ensuring good temperature distribution of combustion products, and 2) good cooling of combustor walls. Efficient wall cooling is also important to minimize quenching of combustion reactions at low power, which can result in poor efficiency and high exhaust pollutants.

Combustor liners are subjected to convection, radiation, and conduction heat loads as indicated in Fig. 27.244. Since the axial conduction is usually small, the heat transfer relationship is

TABLE 27.11 Comparison of Fuel Preparation Methods

Type	Benefits	Drawbacks	Performance Influencing Factors
Pressure Atomizer	• Low Cost • Good Atomization at All Conditions • Duel Orifice for Large Turn-Down • Good Stability	• Fuel Passages Prone to Coking • No Air Mixing • High Press Fuel System • Smoke/Emissions • Atomization Influenced Primarily By Fuel Properties	Droplet Diameter $$SMD^* = K\,\frac{\dot{W}_F \nu_F \rho'_F}{\Delta p_F^{0.4}}$$
Airblast Atomizer	• Good Atomization During Normal Operation • Low Smoke/Emissions • Large Passage Sizes • Longer Life	• Atomization At Light-Off Condition Can Be Marginal • Higher Cost • Stability Can Be a Concern • Air Temp Can Affect Coking	Droplet Diameter $$SMD = K_{11}\,\frac{\sigma'\rho_F}{\rho_a V_a}\left(1 + \frac{\dot{W}_F}{\dot{W}_a}\right)$$ $$+ K_{22}\left(\frac{\mu_F^2}{\sigma'\rho_a}\right)\left(1 + \frac{\dot{W}_F}{\dot{W}_a}\right)^2$$
Vaporizer	• Relatively Insensitive to Fuel Type • Good Combustion η • Low Fuel Pressures • Reduced Radiation to Burner Walls • Simple Fuel Injector	• Starter Pressure Atomizers Required for Light-Off • Durability of Vaporizer Tubes • Carbon Build Ups • Relatively Poor Stability	Combustion Performance Affected by Varpoizer Air-Fuel Ratio, Turbulence in Vaporizer Tubes, Surface Area of Tube, Air Temp
Slinger	• Simple Construction • Relatively Insensitive to Fuel Type • Good Pattern Factor	• Atomization During Windmilling Condition • Altitude Relights	Droplet Diameter, SMD, Inversely Proportional to Shaft Speed

*Sauter Mean Diameter.

2511

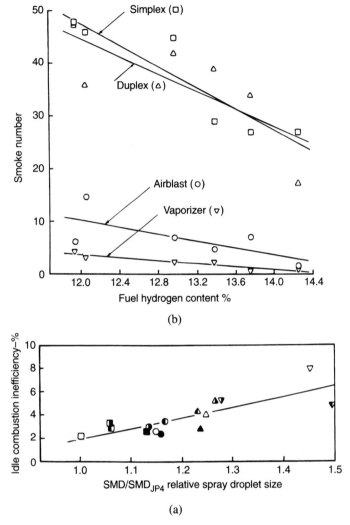

FIGURE 27.242 Effects of atomization quality and fuel properties on combustion η and smoke emissions. (From Gratton *et al.*, 1983–1984.) (a) Efficiency and (b) smoke.

$$R_1 + C_1 = R_2 + C_2 = H_{CR} \qquad (27.341)$$

$$R_1 = \tfrac{1}{2}\sigma(1 + \alpha_w)\,(\epsilon_g T_g^4 - \epsilon_{wg} T_{wg}^4) \qquad (27.342)$$

α_w is the absorptivity of the wall and ϵ_g is the emissivity of hot gases which is influenced by temperature T_g, gas pressure, beam length, and gas composition. Radiation is also influenced by the presence of soot particles in the gas stream. Internal convection, C_1, is given by [Claus *et al.* (1979)]

$$C_1 = h_1(T_f - T_{wg}), \qquad h_1 = 0.023\,\frac{K_f}{D_C}\,(\mathrm{Re}_C)^{0.8}(\mathrm{Pr}_C)^{0.4} \qquad (27.343)$$

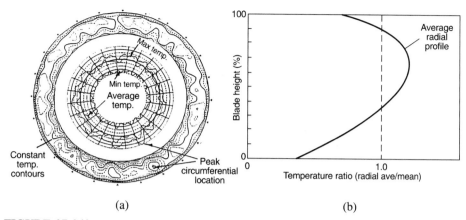

(a)   (b)

FIGURE 27.243 Typical annular combustor exit temperature distribution. (a) Circumferential and (b) radial.

(T_{wg} is the wall temperature on the hot gas side and the subscript f denotes "film". Also, subscript C denotes "combustor".) External convection is given by

$$C_2 = h_2(T_w - T_3)$$

$$h_2 = 0.023 \frac{K_c}{D_{an}} (Re_{an})^{0.8} (Pr_{an})^{0.33} \qquad (27.344)$$

where the subscript (an) denotes flow in the annulus. The external radiation is given as

$$R_2 = \frac{\epsilon_w \epsilon_c}{\epsilon_c + \epsilon_w(1 - \epsilon_c)(A_w/A_c)} (T_w^4 - T_c^4) \qquad (27.345)$$

where

$$Re = \frac{\rho VD}{\mu} \quad \text{and} \quad Pr = \frac{\mu c_p}{K}$$

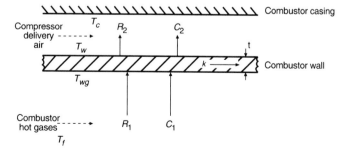

FIGURE 27.244 Combustor heat transfer schematic.

Radial conduction through the combustor wall is given by

$$H_{CR} = \frac{K}{t} (T_{wg} - T_w) \qquad (27.346)$$

Equations (27.341) to (27.346) can be used to calculate combustor wall temperatures, knowing hot gas and air stream temperatures. They can also be extended for cases involving multilayered combustor walls and thermal barrier coatings.

The most common method of cooling combustor walls is by film cooling, which consists of a film of cold air being introduced at regular intervals along the combustor walls. Figure 27.245 shows some common methods of film cooling. The *effectiveness* of film cooling is defined as

$$\eta_f = (T_g - T_w)/(T_g - T_c) \qquad (27.347)$$

where subscript c denotes "coolant." The effectiveness is highest at the beginning of the film or cooling louvre and decays as the coolant interacts with hot gases. Figure 27.246 shows some alternative methods of wall cooling and their relative efficiencies. As the temperature in the combustors increase, it is common to use several cooling schemes concurrently.

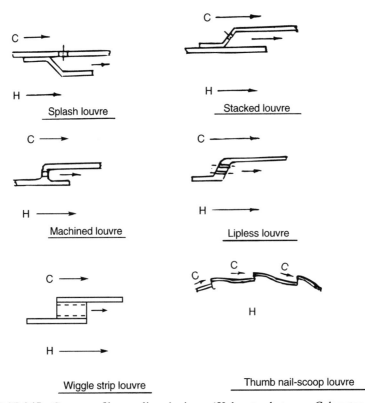

FIGURE 27.245 Common film cooling devices. (*H* denotes hot gas, *C* denotes cold gas.)

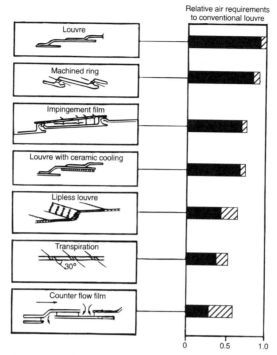

FIGURE 27.246 Relative effectiveness of various cooling schemes.

The commonly used materials for combustor walls are nickel and cobalt alloys. Depending on composition, their temperature capability varies between 1500 and 2000 F. Higher temperature materials actively being researched include ceramics and matrix alloys. Thermal barrier coatings such as Yttria stabilized zirconia ceramic are commonly applied to combustor walls to increase temperature capability and reduce thermal gradients.

Exhaust Emissions: The principal pollutants from gas turbines are unburned hydrocarbons (HC), carbon monoxide (CO), oxides of nitrogen (NO_x), and oxides of sulphur (SO_x). Since aviation fuels have little sulphur concentration (< 0.003), SO_x emissions are not a concern. HC and CO emissions are the result of inefficient combustion and are mainly produced at low power such as ground idle. NO_x emissions consisting of NO and NO_2 are dependent on gas temperatures and residence times and are substantial at high power levels. Figure 27.247 shows the trends in exhaust emissions as a function of engine power.

CO and HC emissions are influenced by primary zone atomization and mixing as well as quenching effects of air jets and cool combustor walls. The air loading parameter, Ω, can be used to correlate and compare CO, HC, emissions of combustors; NO_x formation on the other hand, is influenced primarily by thermal gradients and by the presence of nitrogen and oxygen reactants. Controlling CO and HC emissions from gas turbines will require optimizing burning mixture ratios, achieving good fuel atomization quality, providing sufficient gas phase residence times, and operations at optimum combustor wall temperatures. Table 27.12 shows various means

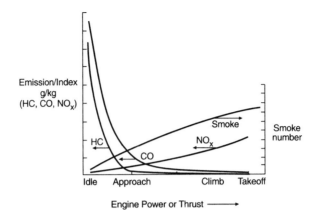

FIGURE 27.247 Exhaust emission variation as a function of power.

TABLE 27.12   Techniques for Controlling CO, HC Emissions

Technique	Mechanism of Control of HC & CO	Possible Disadvantages
Richer primary zone Fuel–air ratio at idle	Increased reaction zone temperatures	Increased NO_x emissions Increased smoke Impaired durability and carbon
Delayed quench of intermediate zone reactions	Increased residence time in reaction and intermediate zones	Increased NO_x emission Worse pattern factor
Hotter walls	Reduced wall quenching	Impaired durability Increased NO_x and carbon
Increased pressure drop Increased inlet air-swirl	Improved mixing	Increased NO_x Higher SFC
Optimized flow pattern in combustor	Improved mixing Increased residence time	Increased NO_x Impaired stability
Improved atomization and air flow distribution to suit	Improved mixing	Increased NO_x
Fuel scheduling	Improved atomization at low power	Starting problems, higher fuel pressure requirements at low power
Increased compressor Bleed at idle Increased idle setting	Increased fuel–air ratio at idle, improved fuel atomization, reduced wall quenching	Increased fuel consumption Increased noise
Variable geometry	Fuel–air ratio control over entire operating range	Complicated and expensive Unknown reliability

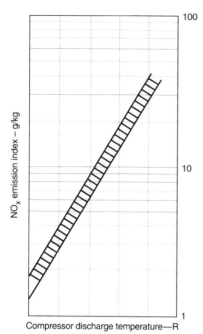

100

10

NO_x emission index – g/kg

1

Compressor discharge temperature—R

FIGURE 27.248 Trend of NO_x emissions with combustor inlet temperature.

of reducing CO and HC emissions in aircraft gas turbines. One of the concerns with some of the techniques is the impact on NO_x emissions. Combustors that have not been optimized for NO_x typically correlate well with combustor inlet temperature, as shown in Fig. 27.248. The other correlating parameter is the primary zone equivalence ratio, ϕ_{pz}, with maximum formation rate corresponding to stoichiometric ($\phi = 1$) combusting conditions as in Fig. 27.249. A method of reducing NO_x is to set the primary zone mixture conditions on the rich ($\phi > 1$) or the lean ($\phi < 1$) side of the stoichiometric operating condition [Sturgess (1992)]. Table 27.13 shows common NO_x control techniques and some of the tradeoffs. Concepts such as rich burn-quick quench and rich-lean burning are aimed at satisfying the conflicting requirements for the control of CO, HC at low power and NO_x, smoke at high power.

Industrial Gas Turbine Combustors: All modern aircraft engines are designed with compact combustors with the aim of keeping the length and the weight of the engine to a minimum. The combustion chambers are usually optimized for kerosene and wide-cut type jet fuels, and combustion intensities are high, in the range of 5–10 $\times$ 10^6 Btu/hr/atm ft^3. Industrial gas turbines, on the other hand, are required to operate with many different types of liquid and gaseous fuels achieving high levels of durability and reliability, with low exhaust emissions. Some of the industrial gas turbines are derivatives of aeroengines with relatively highly loaded combustors, but optimized to burn liquid and gaseous fuels. These tend to maintain a distinct weight advantage over the second category where the engine and combustor depart radically from aircraft practice and are designed with clear objectives of long life, low emissions and multifuel capability. Loading intensities of these combustors tend to be lower than aero-derived industrial combustors. Some manufacturers prefer tubular combustors, which can be designed for good exit temperature quality, with long life

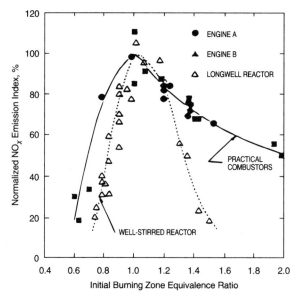

FIGURE 27.249 Variation of NO_x emissions with fuel–air ratio. (Reprinted with permission, Sturgess, G. J., McKinney, R., and Harford, S., "Modification of Combustor Stoichiometry Distribution for Reduced NO_x Emissions from Aircraft Engines," ASME 92-GT-108, 1992.)

and ease of inspection and service. Figure 27.250 shows an industrial gas turbine combustor utilizing the tubular design. Figure 27.251 shows an industrial derivative of a large turbofan aircraft engine using tubo-annular design, and Fig. 27.252 shows an industrial derivative of a small turboprop aircraft engine using a reverse-flow annular combustor.

A recent major concern of industrial gas turbine installations has been NO_x emissions. While lean-burn combustion similar to aircraft engines can be used to lower NO_x, this is usually inadequate to achieve the very low standard established in many countries. A powerful method of NO_x control is water or steam injection which lowers the temperatures in the combustor primary zone, thus drastically reducing

TABLE 27.13 Techniques for Controlling NO_x Emissions

Technique	Mechanism of Control of NO_x	Possible Disadvantages
Rich burn	Reduce reaction zone temperature	• Increased smoke, carbon • Pattern factor degradation
Lean burn	Reduce reaction zone temperature	• Increased CO, HC and low power • Stability degradation
Quick quench	Reduce residence time	Increased CO emissions
Water injection	Reduce reaction zone temperature	Need demineralized water supply, not practical for aircraft engines

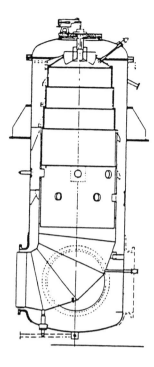

FIGURE 27.250 Tubular industrial gas turbine combustor. (Courtesy of Brown Boveri Co.)

NO_x formation rates. However, demineralized water is required for this application. Other advanced methods of NO_x control have been developed, including multistage combustion using partially mixed or fully premixed reaction zones and optimized fuel air ratios in the combustion zones. Some of these techniques are more effective with gaseous fuels than with liquid fuels.

A further contribution to NO_x formation is given by nitrogenous species, like

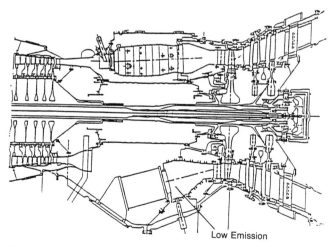

Low Emission

FIGURE 27.251 Tubo-annular industrial gas turbine combustor. (Courtesy of Pratt & Whitney Aircraft TPM.)

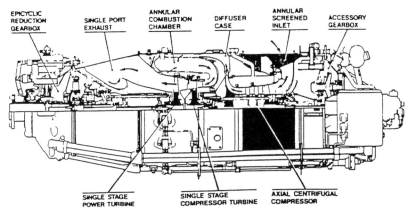

FIGURE 27.252   Reverse flow-annular industrial gas turbine combustor. (Courtesy of Pratt & Whitney Canada.)

ammonia, contained in fuel which undergo direct conversion to nitrogen oxides. The contribution of fuel NO_x is more significant for fuels such as low heating value coal-derived gaseous fuels, which can be optimized to generate low thermal NO_x. The relatively fewer constraints for industrial gas turbines compared to aero engines, generally give more freedom to the combustor designer, to satisfy performance, life, and emission requirements.

27.11.4   Turbines
H. Moustapha

Turbines can be generally classified into radial, mixed-flow, and axial types. The radial and mixed-flow turbines are used for compact and high efficiency power sources such as automotive and diesel engine turbochargers, aircraft auxiliary power units, and space power systems. The main applications of the axial turbine are in electric power generation, aircraft and naval propulsion, and pumping drives for gas and liquid pipelines. The turboprop multistage axial turbine [Fig. 27.196(a)] is divided into the high (compressor) turbine and the low (power) turbine. The low turbine could be rotating in the same direction as the high spool or contrarotating in order to reduce the turning and hence the loss in the first vane. Depending on the inlet temperature and pressure ratio, turbines could be either uncooled or cooled, unshrouded or shrouded, and subsonic or supersonic.

Turbine Parameters

Thermodynamic and Velocity Triangles: Figure 27.253 shows the velocity triangles across a single stage axial turbine (gas, steam, or hydraulic). The stage comprises a row of stationary vanes (also called *stator* or *nozzle*) and a row of rotating blades (also called *rotor*). All angles are measured from the axial direction. The rotor relative velocities are found by vectorially substracting the blade speed from the absolute velocities. Figure 27.254 shows the corresponding enthalpy–entropy diagram, including the effects of irreversibility.

The *specific work*, w, done by the rotor, equal to the stagnation enthalpy drop

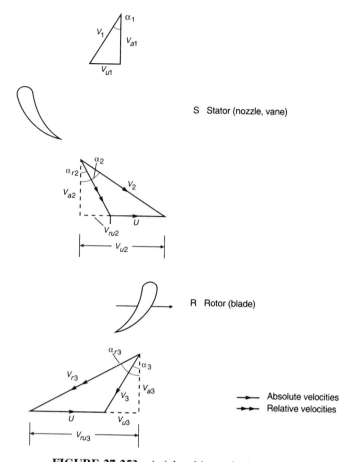

FIGURE 27.253 Axial turbine velocity triangles.

across the stage, can be obtained from the Euler turbine equation

$$W = h_{01} - h_{03} = U_2 V_{u2} - U_3 V_{u3} = \tfrac{1}{2}[(U_2^2 - U_3^2) - (V_{r2}^2 - V_{r3}^2) + (V_2^2 - V_3^2)]$$

(27.348)

For a nozzle with no cooling, the stagnation enthalpy across it remains constant, $h_{01} = h_{02}$, and

$$h_1 - h_2 = \tfrac{1}{2}(V_2^2 - V_1^2)$$

(27.349)

For a blade with no change in radius, the relative stagnation enthalpy is conserved, $h_{0r2} = h_{0r3}$, and

$$h_2 - h_3 = \tfrac{1}{2}(V_{r2}^2 - V_{r3}^2)$$

(27.350)

Degree of Reaction: The stage reaction is a measure of the degree of acceleration in the rotor passage with respect to the stage. The pressure based reaction is given

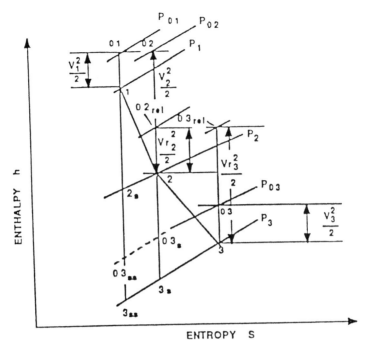

FIGURE 27.254   Axial turbine enthalpy–entropy diagram.

as the ratio of the static pressure drop in the rotor to the static pressure drop in the stage. The widely used temperature based reaction is defined as the static enthalpy drop in the rotor to the static enthalpy drop across the stage

$$R_d = (h_2 - h_3)/(h_1 - h_3) \tag{27.351}$$

Blading Terminology: Figure 27.255 shows the airfoil geometric parameters and terminology used in the gas turbine industry. Flow turning and aspect ratio could be as low as 50° and 0.4 and as high as 130° and 5.0, respectively. The incidence angle at design conditions is caused by the fact that the minimum loss occurs at a certain flow angle which is not equal to the metal angle. Off-design incidences are caused by the mismatch of the inlet velocity vectors and the blade leading edge angles. Exit angle deviation is a result of difference in boundary layer thicknesses on both airfoil surfaces and three-dimensional flow overturning and underturning due to vortices.

Loss Coefficients and Efficiency: The irreversibilities in the flow may be expressed in terms of increase in entropy, enthalpy loss coefficient, stagnation pressure loss coefficient, velocity coefficient, or drag coefficient. The relation between these parameters can be obtained as function of the velocity triangles and Mach numbers.

The loss parameter widely used is the *total pressure loss coefficient* defined as

$$Y_N = (p_{01} - p_{02})/(p_{02} - p_2) \text{ for the nozzle} \tag{27.352}$$

$$Y_R = (p_{0r2} - p_{0r3})/(p_{0r3} - p_3) \text{ for the rotor} \tag{27.353}$$

Leading Edge	(L.E.)
Trailing Edge	(T.E.)
Pressure Surface	(PS)
Suction Surface	(SS)
Chord	(C)
Axial Chord	(Ca)
Stagger Angle	(γ)
Pitch or Spacing	(S)
Height or Span	(h)
Throat	(O)
Blade or Metal Angles	(β_1, β_2)
Flow or Gas Angles	(α_1, α_2)
Incidence Angle	$I = \alpha_1 - \beta_1$
Deviation Angle	$\delta = \alpha_2 - \beta_2$
Camber Angle	$\theta = \beta_1 - \beta_2$
Flow Deflection	$\epsilon = \alpha_1 - \alpha_2$
Aspect Ratio	AR = h/c
Solidity	σ = C/S
Pitch Chord Ratio	S/C

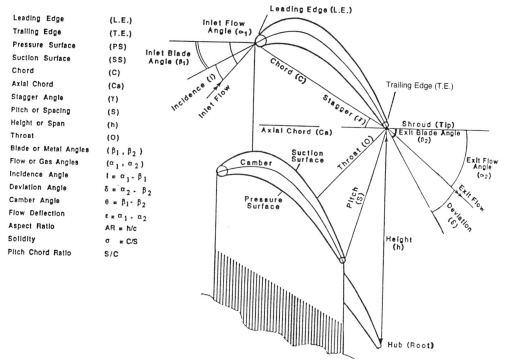

FIGURE 27.255 Turbine airfoil geometric parameters.

The *turbine efficiency* is generally expressed as a total-to-total, η_{tt}, and a total-to-static, η_{ts}, efficiency (see Fig. 27.254)

$$\eta_{tt} = (h_{01} - h_{03})/(h_{01} - h_{03ss}) \tag{27.354}$$

$$\eta_{ts} = (h_{01} - h_{03})/(h_{01} - h_{3ss}) \tag{27.355}$$

The overall (multistage) efficiency of a turbine is higher than the stage efficiency due to the reheat effect.

Optimum Airfoil Loading—Zweifel Coefficient: The optimum pitch to chord ratio is a trade-off between the high friction losses resulting from high airfoil count and the losses resulting from flow separation for low airfoil count. Zweifel (1945) defined a *blade loading coefficient* as the ratio of the actual to an ideal tangential force acting on a blade

$$\Psi = 2S/C_a(\tan \alpha_1 + \tan \alpha_2) \cos \alpha_2^2 \tag{27.356}$$

Based on a number of experiments carried out on turbine cascades, Zweifel found that Ψ was approximately equal to 0.8 for minimum losses.

Stage Loading: The *stage loading* and *flow factor* for a turbine are defined as

$$\varphi = \Delta h_0/U^2 \tag{27.357}$$

$$\phi = V_a/U \tag{27.358}$$

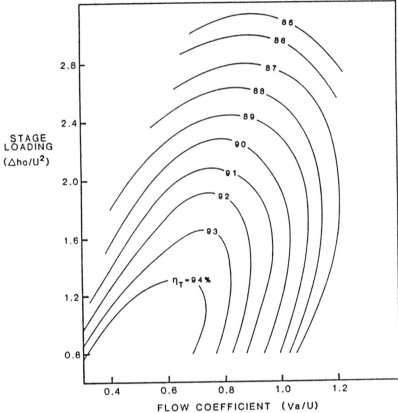

FIGURE 27.256   Turbine stage loading diagram. (Reprinted with permission, Smith, S. F., "A Simple Correlation of Turbine Efficiency," *J. Roy. Soc.*, Vol. 69, pp. 457–470, 1965.)

Figure 27.256 shows the stage loading or the *Smith chart* [Smith (1965)] for zero tip clearance and 50% reaction. Typical loading and flow coefficients are between 0.7 to 2.2 and 0.4 and 0.8, respectively. Increasing the stage loading and flow coefficient (by reducing speed for example) generally results in lower efficiency due to the higher turning and Mach number required to achieve the same work. Figure 27.256 is useful only to show general trends, particularly at the early design stage, and no reliance should be placed on the absolute efficiency levels due to the differing details of each turbine.

Performance Characteristics: Figure 27.257 shows an example of axial flow turbine characteristics. Increasing the pressure ratio beyond the design value at constant speed will eventually result in choking in the turbine. Choking may occur in the stator and/or rotor depending on the design. The drop in efficiency is mainly due to positive blade incidences for increased pressure ratio and reduced speed and negative incidences for reduced pressure ratio and increased speed. Because of the acceler-

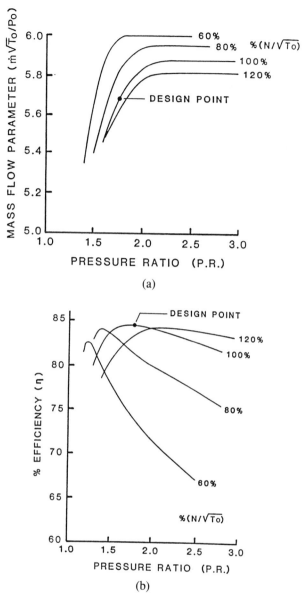

FIGURE 27.257 Turbine performance characteristics. (a) Mass flow and (b) efficiency.

ating nature of the flow, large leading edge diameters and lower inlet Mach number, turbines are tolerant to off-design operation, in general.

Aerodynamic Losses. Figure 27.258 shows the different loss components for a single stage high pressure ratio turbine. The *profile losses* are due to the skin friction and the boundary layer build up on both surfaces of the airfoil. They are a function of the blade geometry and the flow parameters such as Mach and Reynolds numbers, turbulence levels, and flow turning. The losses caused by the *trailing edge wakes*

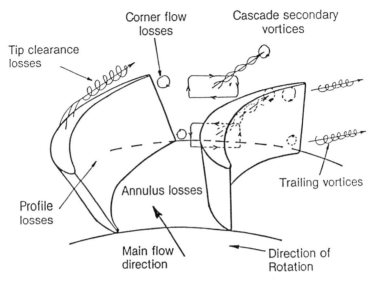

FIGURE 27.258 Turbine aerodynamic losses.

are generally small, except where both airfoils are cooled when the value increases due to the thick trailing edges needed to accommodate the internal cooling passages (see Fig. 27.259). The basis for the profile and trailing edge losses are due to the correlations of Ainley and Mathieson (1951).

The *secondary losses* amount to about 40 to 50% of the total losses. These losses are generated by the interaction between the endwall boundary layer and the blade to blade pressure gradient (see Fig. 27.260). Strong passage vortices are formed [Sieverding (1985)] and combined with the horseshoe vortex created by the inter-action between the oncoming endwall boundary layer and the airfoil leading edge, resulting in high pressure losses and regions of flow overturning and underturning as shown in Fig. 27.261.

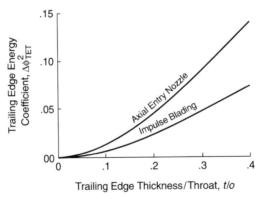

FIGURE 27.259 Trailing edge losses. (Reprinted with permission, Kacker, S. C. and Okapuu, U., "A Mean Line Prediction Method for Axial Flow Turbine Efficiency," *ASME J. of Engineering for Power*, Vol. 104, pp. 111–119, 1982.)

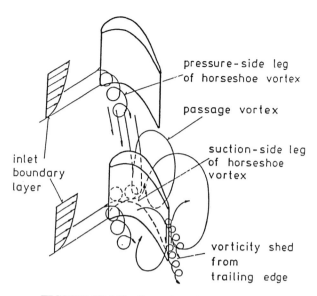

FIGURE 27.260 Secondary flow vortices.

For a single stage unshrouded turbine, *tip leakage losses* amount to about one third of the total stage losses. Rotor tip leakage flow is mainly driven by the difference in pressure between the pressure and suction sides of the airfoil. The tip pressure difference is a function of the stage work, pressure ratio, and the blade aerodynamic loading. The leakage flow emerging from the gap mixes with the main flow

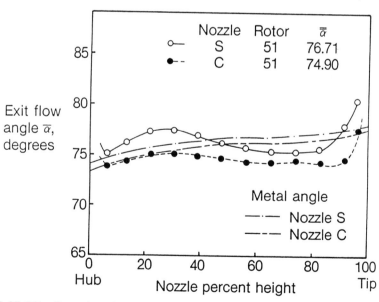

FIGURE 27.261 Secondary flow overturning and underturning. (Reprinted with permission, Moustapha, S. H., Okapuu, U., and Williamson, R. G., "Influence of Rotor Blade Aerodynamic Loading on the Performance of a Highly Loaded Turbine Stage," *ASME J. of Turbomachinery*, Vol. 109, pp. 155–162, 1987.)

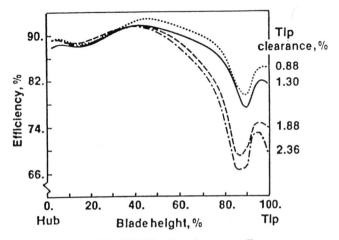

FIGURE 27.262   Tip clearance effects.

in the adjacent passage to produce a leakage vortex. That vortex then diffuses downstream causing losses and flow underturning which affect the performance of the downstream component. The extent of the tip leakage effect could cover up to 50% of the blade span (see Fig. 27.262). The efficiency gradient, defined as the percentage drop in efficiency for a one percent increase in clearance to blade span ratio, is about 2% and 1% for unshrouded and shrouded turbines [Kacker and Okapuu (1982)] respectively (see Fig. 27.263).

The Mach and Reynolds numbers have a strong effect on the above losses. The effect of Mach number is manifested in the trailing edge shock illustrated in Fig. 27.264 impinging on the suction surface of the adjacent airfoil, disturbing the boundary layer, and resulting in the additional losses as given in Fig. 27.265. According to Ainley and Mathieson (1951), below a Reynolds number of 2×10^5 based on mean chord and exit conditions, a correction should be made to the efficiency. For

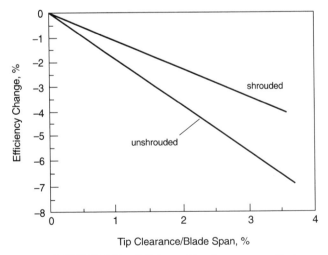

FIGURE 27.263   Efficiency/tip clearance gradient.

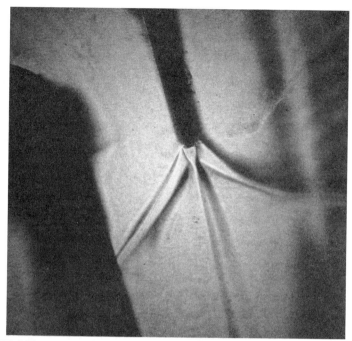

FIGURE 27.264 Mach number effects: trailing edge shock pattern. (Reprinted with permission, Moustapha, S. H., Carscallen, W. E., and McGeachy, J. D., "Aerodynamic Performance of a Transonic Low Aspect Ratio Turbine Nozzle," *ASME J. of Turbomachinery*, Vol. 115, pp. 400–408, 1993.)

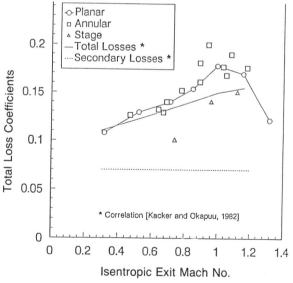

FIGURE 27.265 Effect of Mach number on losses. (Reprinted with permission, Moustapha, S. H., Carscallen, W. E., and McGeachy, J. D., "Aerodynamic Performance of a Transonic Low Aspect Ratio Turbine Nozzle," *ASME J. of Turbomachinery*, Vol. 115, pp. 400–408, 1993.)

cooled turbines, the airfoil cooling flows, platform and disk cooling, and secondary air entering the turbine gas path have a detrimental effect on turbine performance. Figure 27.266 shows the efficiency penalty due to the *mixing losses* between the mainflow and cooling air for turbine airfoils.

All of the above losses are at design conditions or at zero incidence. As a result of variation in power and speed, the inlet velocity vectors become mismatched with the leading edge angles of the blades causing additional *incidence losses*. The off-design performance of a turbine could also be influenced by a number of additional factors. For multistage turbines, the losses and exit conditions in an upstream component may cause the downstream stage to operate at off-design condition. Altitude effects for aircraft turbines due to a drop in Reynolds number, variation in cooling

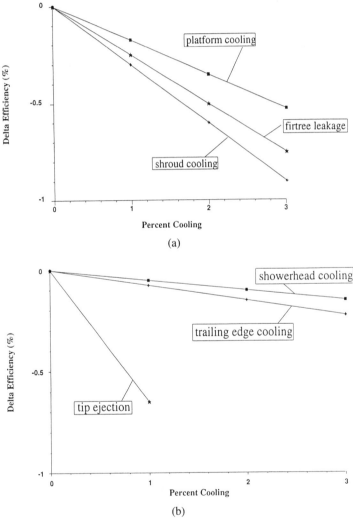

FIGURE 27.266 Effect of cooling on efficiency. (a) Cooling schemes and (b) further cooling schemes.

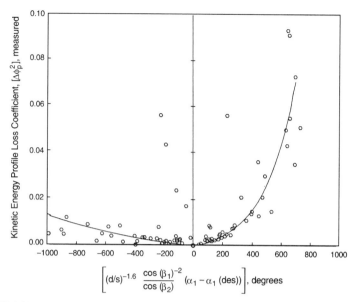

FIGURE 27.267 Effect of incidence on profile losses. (Reprinted with permission, Moustapha, S. H., Kacker, S. C., and Tremblay, B., "An Improved Incidence Losses Prediction Method for Turbine Airfoils," *ASME J. of Turbomachinery*, Vol. 112, pp. 267–276, 1990.)

air, and running tip clearances will also affect the design point performance. Figure 27.267 shows the effect of incidence on the profile losses from Moustapha *et al.* (1990).

Turbine Design. Figure 27.268 shows the gas turbine aerodynamic design procedure. The iterative process starts with the preliminary (also called *meanline*) design where given parameters such as inlet total pressure, temperature, mass flow, work, speed, and geometric limitations are used to define the turbine gas path, airfoil geometric parameters, meanline velocity triangles, and achievable efficiency. The objective of the through flow design is to determine the velocity triangles from hub to shroud and from leading to trailing edges in order to achieve an optimum radial distribution of exit angle, reaction, and work. The airfoils are then designed at different sections, radially stacked, and their performance is assessed using 3-D flow solvers.

Selection of Speed: Factors which affect the selection of speed for turbines include stage loading, allowable stress levels, noise and gearbox limitations, and compressor characteristics. As shown in Fig. 27.256, higher efficiencies are obtained at lower stage loadings through increasing the speed. The limitation on speed is defined by the flow area speed square stress parameter, AN^2, which should be a minimum for best blade life. Hence, the optimum aerodynamic design is a trade-off between low flow area which leads to small aspect ratio airfoils and high Mach numbers and low speed which results in high turning and stage loading. The AN^2 ranges from about 3×10^{10} to 9×10^{10}, with the low values for high turbines and the high values for low turbines.

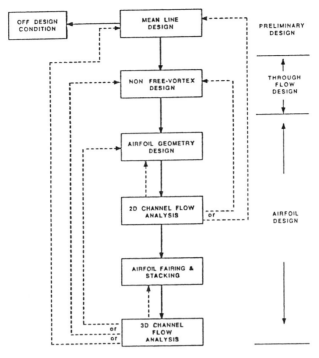

FIGURE 27.268   Gas turbine design procedure.

Selection of Reaction: Factors which enter into the selection of mean stage reaction are pressure ratio and efficiency, blade temperature and life, exit swirl and duct loss, blade spacing and platform fit, root stagger and fixing stress, cooling supply pressure, nozzle design, and bearing loads. The final selected reaction is not necessarily the optimum aerodynamic value which is mainly a function of pressure ratio and stator/rotor design. Figure 27.269 gives a parametric study showing the effect of blade number and reaction on blade loading (Zweifel coefficient), blade root stagger angle, and relative temperature. Reducing the reaction is beneficial to lower the blade temperature (increased life) and to reduce the root stagger (lower fixing stress), however at the expense of increased blade loading (lower efficiency).

Selection of Airfoil Number: The airfoil number is dictated by factors such as optimum loading, root and shroud stresses, platform fit, rotor dynamics, weight, and cost. The optimum number is based on a trade-off between trailing edge blockage, blade loading, and aspect ratio. For a given trailing edge thickness, set by the minimum manufacturing and cooling requirements, the trailing edge loss (see Fig. 27.259) is minimized by increasing the throat and, hence, the pitch. This results in lower airfoil count and possibly a higher than optimum blade loading. To reduce the loading by increasing the chord will result in lower aspect ratio and, hence, a higher secondary loss. Figure 27.269 showed that lowering the airfoil count, though beneficial for reducing the disc stresses, resulted in higher root stagger and, hence, higher fixing stress. The effect of reducing the blade number on turbine efficiency and exit angle is shown in Fig. 27.270.

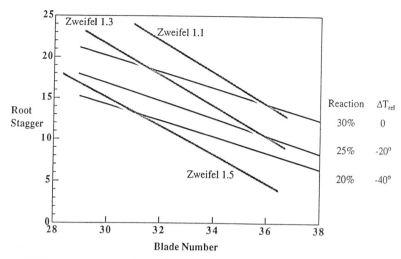

FIGURE 27.269 Effect of blade number and reaction on Zweifel coefficient and root stagger.

Subsonic versus Transonic Design: Table 27.14 shows a comparison between the stage and the airfoil parameters for a subsonic and a transonic turbine design. The effects of higher stage loading, trailing edge blockage, turning, and exit Mach number are apparent in the lower efficiency attained by the transonic turbine.

Airfoil Geometry Design: The profile of the vanes and blades, at different design sections, is determined using geometry generator packages (see Fig. 27.271). The

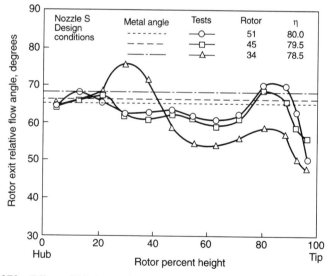

FIGURE 27.270 Effect of blade number on exit angle and efficiency. (Reprinted with permission, Moustapha, S. H., Okapuu, U., and Williamson, R. G., "Influence of Rotor Blade Aerodynamic Loading on the Performance of a Highly Loaded Turbine Stage," *ASME J. of Turbomachinery,* Vol. 109, pp. 155–162, 1987.)

TABLE 27.14   Turbine Design Parameters

(a) Subsonic

Design Point Parameters

Pressure Ratio	1.97
Stage Loading	1.31
Flow Coefficient	0.47
Mean Reaction	50%
Mean Exit Swirl	22°
Efficiency	88.0%

Airfoil Parameters	Stator	Rotor
Inlet Mach Number	0.14	0.23
Exit Mach Number	0.67	0.82
Flow Turning	60°	78°
Aspect Ratio	0.71	1.25
Zweifel Loading	0.74	0.88
Trailing Edge/Throat	12%	9%
Tip Clearance/Height	0.0	1.5%

Aerodynamic Losses		Stator	Rotor	Total
Profile	Y_P	0.024 (19%)	0.020 (12%)	0.044 (15%)
Trailing Edge	Y_{TET}	0.023 (18%)	0.012 (7%)	0.035 (12%)
Secondary	Y_S	0.080 (63%)	0.063 (36%)	0.143 (47%)
Tip Clearance	Y_{TC}	0.000 (0%)	0.079 (45%)	0.079 (26%)
Total	Y_T	0.127 (100%)	0.174 (100%)	0.300 (100%)

(b) Transonic

Design Point Parameters

Pressure Ratio	3.76
Stage Loading	2.47
Flow Coefficient	0.64
Mean Reaction	30%
Mean Exit Swirl	37°
Efficiency	83.5%

Airfoil Parameters	Stator	Rotor
Inlet Mach Number	0.1	0.55
Exit Mach Number	1.1	1.14
Flow Turning	76°	124°
Aspect Ratio	0.7	1.44
Zweifel Loading	0.84	0.76
Trailing Edge/Throat	17%	10%
Tip Clearance/Height	0.0	1.5%

Aerodynamic Losses		Stator	Rotor	Total
Profile	Y_P	0.029 (20%)	0.058 (16%)	0.087 (17%)
Trailing Edge	Y_{TET}	0.046 (31%)	0.011 (12%)	0.057 (12%)
Secondary	Y_S	0.071 (49%)	0.180 (51%)	0.251 (50%)
Tip Clearance	Y_{TC}	0.000 (0%)	0.110 (30%)	0.110 (21%)
Total	Y_T	0.147 (100%)	0.356 (100%)	0.500 (100%)

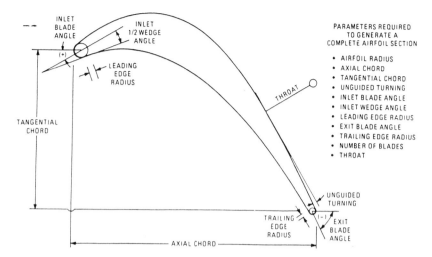

FIGURE 27.271 Airfoil geometry generator. (Reprinted with permission, Pritchard, L. H., "An Eleven Parameter Axial Turbine Airfoil Geometry Model," ASME Paper No. 85-GT-219, 1985.)

input used is the inlet/exit velocity triangles and geometric parameters such as leading/trailing edge diameters and wedge angles, incidence and deviation, uncovered turning, etc. Mechanical and structural requirements, as well as blade to blade Mach number distribution, are monitored during the airfoil definition. After the airfoil is stacked, a 3-D flow solver [e.g., Ni and Bogian (1982)] is used to finally assess the Mach number distribution in the blade passage as in Fig. 27.272. Stacking through meridional sweep, bowing, and tangential lean (see Fig. 27.273) and gas path contouring are used to optimize the velocity distributions.

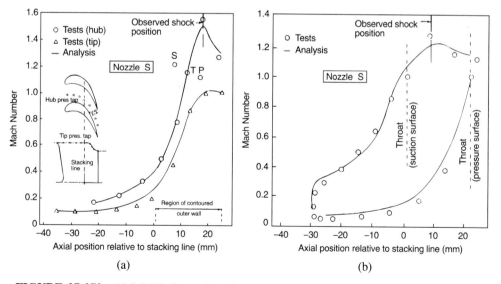

FIGURE 27.272 Airfoil Mach number distributions. (a) Endwall Mach number distributions and (b) mid-span surface Mach number distributions.

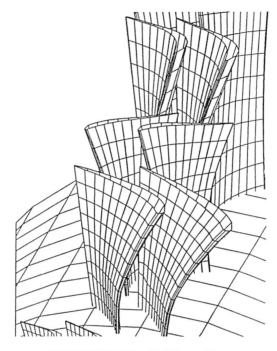

FIGURE 27.273   Airfoil stacking.

Cooling Design and Lifing: The benefit in reduced specific fuel consumption and increased specific power through increasing the turbine inlet temperature has led to the necessity of cooling first stages, airfoils, and endwalls. Air bled from the compressor is routed through internal passages in the airfoils for cooling and then is dumped into the main gas stream at discrete locations around the airfoils. Typical cooling flow mass flow per airfoil is about 1 to 5% of the main flow. The discharged air results in additional turbine loss as well as a drop of the overall cycle efficiency. Depending on the inlet temperature, various cooling schemes are used such as convection, impingement and film-cooling (see Fig. 27.274). The cooling air is ejected near the trailing edge or through the blade tip. The cooling design consists of optimizing the internal cooling passages through the use of heat augmentation devices and multipass channels with minimum pressure loss. The cooling scheme is also designed so that the ejected flow causes minimum disturbance to the main flow.

27.12 FLUID COUPLINGS AND TORQUE CONVERTERS
Ernest W. Upton

These devices are not new. Development of the torque converter as a marine drive reduction gear is credited to Dr. Herrman Fottinger near the beginning of this century. A few years later, Dr. Bauer, also of the Vulcan Ship Yards of Hamburg, Germany, built a fluid coupling similar to the torque converter but without the stationary reaction blades. Fluid couplings and torque converters are applied in drive lines where it is desirable to avoid a positive mechanical connection between the power source and driven load.

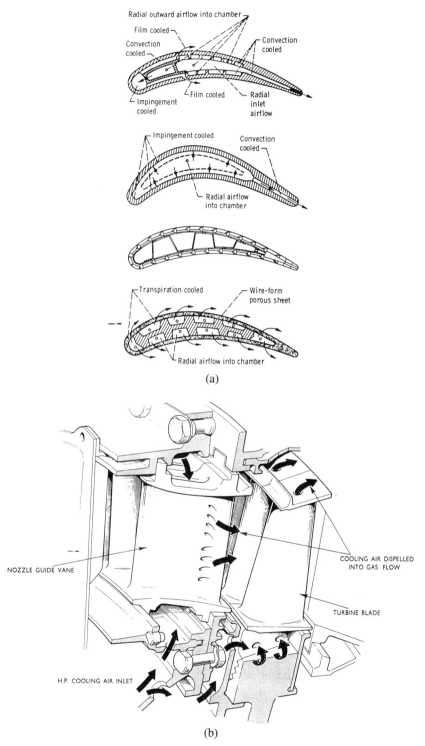

FIGURE 27.274 Airfoil cooling schemes: (a) cross-section views (Glassman, A. J., "Turbine Design and Applications," NASA SP-290, Vols. 1, 2, 3, 1975) and (b) overall view. (Courtesy of Rolls Royce Publication, 1965.)

In this section, the terminology and basic equations governing the design and application of these drives are presented. Examples demonstrate the process by which a reader can determine the operating relationship between a specific combination of power plant, hydrodynamic device, and driven load. Some understanding is provided about where and why these transmissions are used and how they are constructed. References provide more extensive information on design details, varieties of design, and applications.

27.12.1 Function and Terminology

The *fluid coupling* and *fluid torque converter* are hydrodynamic drive devices that transfer rotary power through a working fluid by the action of the fluid flowing through the bladed passages in the driving and driven members arranged in a closed loop circuit. Both devices allow a smooth, continuously changing speed relationship between the driving member and the driven member. The driving member functions as a centrifugal pump. In the most common couplings and converters, the driven member functions as a radially inward flowing turbine. The torque converter includes a set of fixed reaction blades redirecting the flow from the turbine into the pump in the direction of pump rotation.

The *fluid coupling* transfers power without changing torque. It includes only two members, a *pump*, and a *turbine* as shown in Fig. 27.275. This figure is a cross section in the axial plane, i.e., any plane containing the axis of the device.

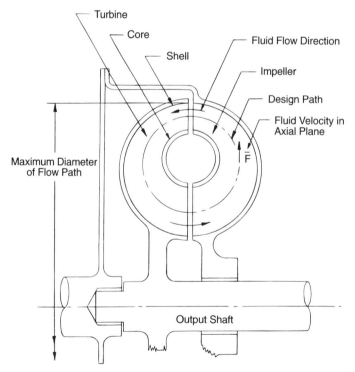

FIGURE 27.275 Fluid coupling. (Reprinted with permission from SAE, *SAE Handbook*, Section 29, "Fluid Coupling," 1992.)

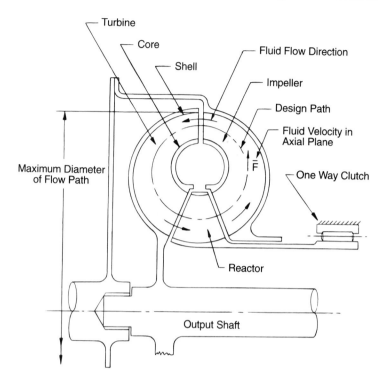

FIGURE 27.276 Fluid torque converter using three elements. The one-way clutch permits first phase operation as a torque converter and second phase operation at high speed ratios as a fluid coupling. (Reprinted with permission from SAE, *SAE Handbook*, Section 29, "Fluid Coupling," 1992.)

The *fluid torque converter*, Fig. 27.276, provides continuously changing torque and speed while transmitting power. The torque change requires a third set of blades which are stationary and provide the reaction point in what can be thought of as a fluid lever system. This reactive *stator (reactor)* redirects the flow from the turbine into the pump inlet in the direction of rotation. For most automotive applications, the stator (reactor) is mounted on a *one-way clutch* which allows it to rotate in the direction of the pump at high speed ratio conditions. Such a unit combines the characteristics of the torque converter for acceleration and the fluid coupling for cruising.

Components of the Flow Circuit. The *torus* is the closed loop circuit that confines the operating fluid, directing the flow in sequence through the main members, pump, turbine, and stator. The torus and other key components of typical fluid couplings and torque converters are identified in Figs. 27.275 and 27.276. The *shell*, or outer shroud, is the outer boundary of the torus, and the *core*, or inner shroud, is the interior boundary of the torus. The *maximum diameter* of the toric fluid is commonly used to define the *size* of the hydrodynamic drive. The *blades*, or vanes, control the tangential components of the flow and thus are the surfaces where the exchanges between mechanical and fluid power occur. The design path is the line within the torus where the mass of the working fluid is considered to act for design purposes.

Blade Angle System. To obtain a set of compatible equations suitable for system-ized computations, it is mandatory to establish a logical and consistent system for defining blade angles. The convention adopted here is described in Fig. 27.277. The base vector for the angle is aligned with the toric velocity, $\overline{F}$, and thus lies in the axial plane. A positive angle produces a tangent vector in the direction of pump rotation, and a negative angle produces a tangent vector opposite to pump rotation. Angles range from $0°$ to $90°$ and $0°$ to $-90°$. All cosine values are positive and tangent values carry the sign of the angle. This convention, system A in the SAE Handbook [SAE (1992, vol. 4)], can be readily understood by nontechnical people. For this reason, it is preferred over other systems. Flat radial blades do not alter the influence of rotation. Positive blade angles increase fluid whirl in the direction of pump rotation and negative blade angles reduce whirl relative to pump rotation. At this point, it is important to note that the blades shown in Fig. 27.277 do not show the radius change typical of the general case which is clarified by the blade tip diagram in Fig. 27.278.

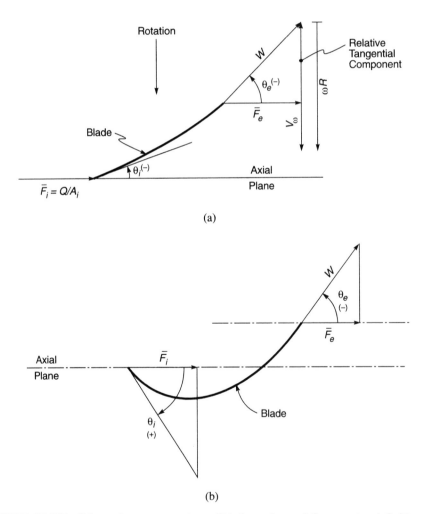

FIGURE 27.277   Schematic representation of blade angles and flow vector definition. (a) θ_i negative and (b) θ_i positive.

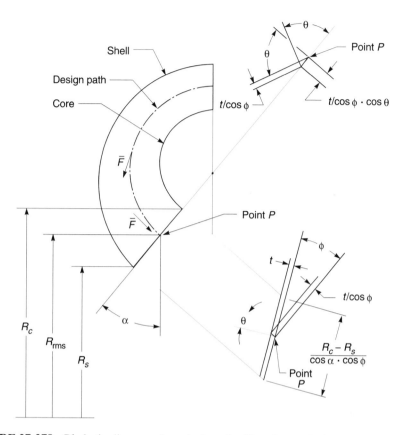

FIGURE 27.278 Blade tip diagram. A conical section line element is shown passing through point A on the blade tip. Bias angle ϕ increases the length and width of the tip in the conical section. Blade angle ϕ further increases the tip width in this section. These effects on net flow area are included in Eq. (27.364).

The equations relating the components of the flow vectors as defined are

$$W = \frac{\overline{F}}{\cos \theta} \tag{27.359}$$

$$V_\omega = \overline{F} \tan \theta + \omega R \tag{27.360}$$

$$\overline{F} = Q/A \tag{27.361}$$

where

W = relative flow velocity in the direction of the blade passage, m/s
V_ω = absolute tangential flow component, m/s
R = radius at point of calculation, m
ω = angular velocity of blade (positive in direction of pump rotation), rad/s
$\overline{F}$ = meridional (toric) flow velocity, m/s (subscript i for inlet and e for exit)
Q = volume flow rate of fluid, m³/s

TABLE 27.15 Blade Angle Relationships for Three Systems

System	Blade Angles Degrees							Reference
Preferred	90	60	30	0	−30	−60	−90	SAE A
Alternate 1	0	30	60	90	120	150	180	SAE B
Alternate 2	180	150	120	90	60	30	0	—

A = blade passage area, m^2
θ = flow angle measured from axis to relative flow velocity vector, degrees

Two other blade angle systems are found in textbooks and other references. These both use the plane tangent to the conical surface cross section of the torus as the zero angle reference. Angles vary from 0° to 180°. The first of these systems, system B in the SAE Handbook, sets zero angles in the direction of rotation where angle values increase opposite to rotation. The second of these systems sets zero opposite to rotation where angle values increase in the direction of rotation. All three systems allow correct solutions if correctly applied. Table 27.15 demonstrates the relationship between angles in the three systems.

27.12.2 Principles of Operation

Both fluid couplings and torque converters depend upon the centrifugal action of the pump to drive the fluid around the torus and a turbine to absorb the fluid power and deliver mechanical power to the load [SAE (1992), Waclawek (1973), and Upton (1973)]. In the common form of fluid coupling (Fig. 27.275) the torus is divided radially into two equal sections, with each section containing a set of flat radial vanes. Physically, this coupling is similar in appearance to a grapefruit sliced through at its equator.

In the most common form of torque converter (Fig. 27.276) the stator is inserted between the pump and turbine, where it provides the reaction necessary to obtain greater torque on the turbine than that driving the pump. To perform this function, the reactor blades must increase the moment of momentum rate in the direction of pump rotation relative to the value at the stator entrance. Therefore, the external torque on the stator blades is in the direction of rotation.

All concepts developed here apply equally to fluid couplings and torque converters. The equations are developed for the torque converter; equations for the fluid coupling omit terms related to the stator. The flow within the toric system is extremely complex. The toric shape alone is equivalent to a 360° pipe bend. Superimposed upon this are the blade curvatures in directions normal to the toric plane. The blades are typically a combination of long thin vanes and varying thickness hydrofoils. The entire system is subjected to and interacts with centrifugal pressure systems acting within and between the various rotating parts of the assembly. Careful application of fluid mechanics principles has resulted in practical methods for predicting, designing, and applying hydrodynamic drives. Four concepts are critical to understanding the operation of hydrodynamic drives: 1) continuity of flow, 2) conservation of energy, 3) moment of momentum, and 4) hydrodynamic similarity. These are discussed in the context of steady state operation. A brief discussion of transient operation is found in Sec. 27.12.3.

Continuity of Flow. The continuity concept, expressed in Eq. (27.362), provides the basis for a mathematical description of the flow around the toric circuit. The volume flow rate, Q, is constant and, with a physical description of the passages, allows calculation of the fluid velocity vectors. All other flows such as those into and out of the converter housing for cooling purposes are considered incidental.

$$Q = \bar{F}Q = \bar{F}_1 A_1 = \bar{F}_2 A_2 = \bar{F}_3 A_3 \qquad (27.362)$$

where A is the conical section area normal to the toric path, m^2, and subscripts 1, 2, and 3 identify specific toric sections where calculations are made.

The nominal toric velocity, $\bar{F}$, implies a constant velocity profile across the torus. The center of fluid mass at this section is then the circle at the center of the conical area. Integration shows this to be located at the square root of the mean of the squares of R_s and R_c in Fig. 27.278 and Eq. (27.363).

$$R_{\text{rms}} = \left(\frac{R_c^2 + R_s^2}{2} \right)^{1/2} \qquad (27.363)$$

The net conical area at this section includes adjustment for blade thickness

$$A = \left[\frac{\pi(R_c^2 - R_s^2)}{\cos \alpha} \right] - \left[\frac{B \cdot t(R_c - R_s)}{\cos \alpha \cdot \cos^2 \phi \cdot \cos \theta} \right] \qquad (27.364)$$

where

$$
\begin{aligned}
R_{\text{rms}} &= \text{radius on design path, m} \\
R_c, R_s &= \text{radius on core, shell, m} \\
t &= \text{blade thickness, m} \\
B &= \text{number blades} \\
\alpha &= \text{toric cross-section angle relative to a plane perpendicular to the axis of rotation, degrees} \\
\theta &= \text{blade angle, degrees} \\
\phi &= \text{blade bias angle (angle blade is tipped out of the axial plane), degrees}
\end{aligned}
$$

Refer to Fig. 27.278 which illustrates the geometry and the symbols used above.

Conservation of Energy. Conservation of energy leads to the mathematical expressions defining system equilibrium. The appropriate expression of conservation of energy is

$$\text{pump power} = \text{turbine power} + \text{loss power} \qquad (27.365)$$

where

$$\text{pump power} = T_p \omega_p = H_p \rho g Q \qquad (27.366)$$

and

$$\text{turbine power} = T_T \omega_T = H_T \rho g Q \qquad (27.367)$$

where T_P and T_T are the torques on the pump and turbine.

By substituting and expressing the energy balance in terms of hydraulic head per unit mass flow, the losses in the circuit are defined as

$$\left(\frac{T_p \omega_p}{\rho g Q}\right) - \left(\frac{T_T \omega_T}{\rho g Q}\right) = H_{\text{loss}} \tag{27.368}$$

with

$$H_{\text{loss}} = H_f + H_j \tag{27.369}$$

where T = torque, N · m, H = fluid head, m, ρ = fluid mass per unit volume, kg/m³, and g = gravitation acceleration. Subscripts P, T, and S designate pump, turbine, and stator, while f and j designate frictional losses and losses attributed to the gaps between the bladed members. Losses in the toric circuit are considered to be of two distinct types: 1) *friction losses*, H_f, attributed to flow in the blade passages, and 2) *junction losses*, H_j, commonly called *shock losses* attributed to the transition from one blade row to the next.

Friction Losses: Blade passage losses occur because of friction at the walls and turbulence caused by the curvature and forces of the torus and the blades. Included are effects of the preferred orderly flow in the direction of the passages and secondary flows which may be very strong. Equation (27.370) is a summation of terms for each blade element in the system. The calculation makes an approximation of the length of each passage and the velocities relative to the surfaces. The equation for friction losses is

$$H_f = \frac{1}{2gD} \sum_f \left[\frac{C_f L \overline{F}_{ei}^2}{\langle \cos \theta \rangle^3} \right]_{P,T,S} \tag{27.370}$$

The symbols used are defined as: C_f = friction flow coefficient, dimensionless, L = length of the toric segment along the design path within the specified member, m, $\langle \cos \theta \rangle$ = effective average value of cosine of the blade angle, dimensionless, and $\overline{F}_{ei}$ = blade passage average toric flow velocity for the specific member, m/s.

The length, L, in combination with the averaged velocity and cosine values, $\overline{F}_{ei}$ and $\langle \cos \theta \rangle$, provide an approximate alternative to the integration calculation for the true passage length and relative flow velocities.

$$\overline{F}_{ei} = \frac{\overline{F}_e + \overline{F}_i}{2} = \frac{Q}{2}\left(\frac{1}{A_e} + \frac{1}{A_i}\right) \tag{27.371}$$

where e = exit blade tip location and i = inlet blade tip location.

When averaging the effect of angle along the blade length, it is essential to recognize whether the blade changes direction, in which case, the angle decreases in magnitude from the inlet to zero and increases to the exit, or whether the angles are in the same direction (refer to Fig. 27.277). For the case of angles in one direction as shown in Fig. 27.277(a)

$$\langle \cos \theta \rangle = \frac{(\cos \theta_e + \cos \theta_i)}{2} \tag{27.372}$$

For the case of the blade changing direction as shown in Fig. 27.277(b)

$$\langle \cos \theta \rangle = \frac{[\theta_e \, (1 \, + \, \cos \theta_e) \, + \, \theta_i \, (1 \, + \, \cos \theta_i)]}{2(\theta_e \, + \, \theta_i)} \tag{27.373}$$

where θ = magnitude of the angle in degrees.

Junction Losses: Junction losses account for the sudden velocity changes which occur when the velocity vector of the flow approaching a blade row is not equal to the ideal vector for entrance to the blade row. Torque transfer does not occur across the gap between blade rows. Using the *moment of momentum* concept (see below) the exit velocity vector is converted to the equivalent entering vector at the inlet radius and area. The difference in tangential velocities between this entering vector and the ideal entering vector is the change the fluid must accommodate. Distortion of the ideal flow and the related turbulent losses extend into the passage. These losses are approximated using the following equation for H_j

$$H_j = \frac{1}{2g} \sum_m [C_j (\Delta V \omega)_i^2]_m \tag{27.374}$$

where the summation index m refers to three junctions as follows: pump/turbine, turbine/stator, stator/pump, and C_j = junction loss coefficient.

$$(\Delta V_\omega)_i = \left[\frac{(\omega_e R_e^2 \, + \, \overline{F}_e \, \tan \theta_e R_e)}{R_i} - (\omega_i R_i \, + \, \overline{F}_i \, \tan \theta_i) \right] \tag{27.375}$$

For blades with a thin leading edge, $C_j \approx 1.0$. For properly designed hydrofoil blades which are frequently used in stators, $C_j < 1.0$.

Moment of Momentum. Moment of momentum provides the basis for writing equations defining torque in terms of the operating conditions and flow conditions. Analysis of moments for the entire system leads to the equation for system equilibrium. Using the preferred system of blade angles

$$\Sigma T = T_P + T_T + T_S = 0 \tag{27.376}$$

$$T = (\dot{m} V_\omega R)_e - (\dot{m} V_\omega R)_i \tag{27.377}$$

$$T_P = \dot{m}\{(\overline{F}_{Pe} \, \tan \theta_{Pe} + \omega_P R_{Pe})R_{Pe} - (\overline{F}_{Se} \, \tan \theta_{Se} + \omega_S R_{Se})R_{Se}\} \tag{27.378}$$

where $\dot{m} = \rho Q$ = mass flow rate, kg/s, and T_s is the torque of the stator.

Equation (27.377) states that the torque resultant of the fluid flow acting against the blades of a specified member is equal to the time rate of outflow of moment of momentum. This equation accounts for blade forces acting at different radii all resulting from concurrent changes in blade angle, toric direction, and flow magnitude. Of prime importance is the recognition of three facts: 1) the specified number is responsible for all changes in moment of momentum rate from the exit of the previous blade row to its own exit, 2) regardless of the complexity of the path the total change is the difference between these terminal values, and 3) the total moment of momentum rate change affecting torque is dependent only on the tangential velocities. Changes in the axial momentum rate results only in axial thrusts. The application of this concept and Eq. (27.377) to the pump provides Eq. (27.378), which is an example for a specific member in a torque converter represented by Fig. 27.276. Equations (27.360) to (27.361) are used in obtaining correct system operating solutions for flow velocities and rotational speeds. For a fluid coupling, Eq. (27.378) must be revised substituting turbine exit values for stator exit values.

Hydrodynamic Similarity. Hydrodynamic similarity applied to hydrodynamic drives results in two equations, one equation for torque and another equation for power. The variables for the equations are related to the speed of a member, the characteristic size of the drive, and the operating speed ratio.

$$T = C\omega^2 D^5 \tag{27.379}$$

$$P = T \cdot \omega = \overline{C}\omega^3 D^5 \tag{27.380}$$

where P = power, W, J/s, C = capacity coefficient, N $\cdot$ m/(rad/s)^{2}m^5, and $\overline{C}$ = power coefficient, W/(rad/s)^{3}m^5. *Capacity coefficient, C,* and *power coefficient, $\overline{C}$,* are characteristic of the geometric design and are unique for each *speed ratio* condition.

$$N_r = \text{speed ratio} = \frac{\omega_T}{\omega_P} \tag{27.381}$$

These equations provide the basis for adapting a given design to a specific application through conversion of performance data for a specific design to performance of the same design at different speeds, torques, powers, and sizes in any combination. The values substituted in the equation may represent pump, turbine, or stator conditions or a mix of these. The results of a conversion calculation represent the original members at new conditions.

Example of a Conversion Calculation:

Initial Conditions = 1   Desired Conditions = 2

$T_P = 100$ N $\cdot$ m $T_P = 200$ N $\cdot$ m

$n_T = 1000$ r/min $n_T = 1200$ r/min

$D = 230$ mm D = new size

$N_r = 0.5$ $N_r = 0.5$

where n_T is the shaft speed. Using Eq. (20.379) and assuming $C_1 = C_2$

$$D_2 = d_1 \left[\left(\frac{T_{P_2}}{T_{P_1}} \right) \left(\frac{n_{T_1}}{n_{T_2}} \right)^2 \right]^{1/5}$$

$$D_2 = 245.6 \text{ mm}$$

27.12.3 Operating Characteristics

The basic parameters which define the operating characteristics of hydrodynamic drives are the efficiency, E; speed ratio, N_r; torque ratio, $T_r = T_T/T_p$; input speed, ω_p; and output speed, ω_T. The expression for efficiency, E, establishes the relationship between these parameters. Efficiency is the ratio of power out to power in

$$E = \frac{T_T \omega_T}{T_p \omega_p} = T_r N_r \qquad (27.382)$$

The primary graphical presentation of these parameters for a torque converter is shown in Fig. 27.279. The input speed is dependent upon input torque and the size of the unit according to Eq. (27.379). Both must be stated as part of the data. Two additional input speed characteristics are derived from Eq. (27.379). Both must be stated as part of the data. Two additional input speed characteristics are derived from Eq. (27.379).

$$K \text{ (capacity factor)} = \left(\frac{\omega}{\sqrt{T}}\right)_P = \left(\frac{1}{CD^5}\right)^{1/2} \qquad (27.383)$$

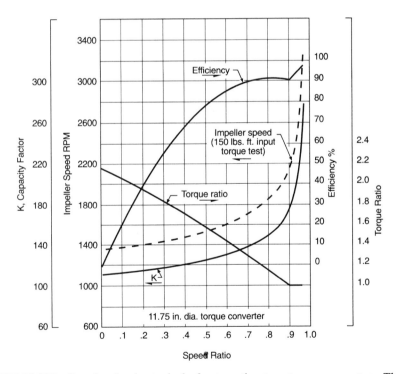

FIGURE 27.279 Speed ratio plot typical of automotive type torque converters. The sharp breaks in the efficiency and torque ratio curves occur at the coupling point above which the device functions as a fluid coupling as noted in Sec. 27.12.1. (Reprinted with permission from SAE, *SAE Handbook*, Section 29, "Fluid Coupling," 1992.)

$$\overline{K} \text{ (unity capacity factor)} = \left[\omega\left(\frac{D^5}{T}\right)^{1/2}\right]_P = \left(\frac{1}{C}\right)^{1/2} \qquad (27.384)$$

The *capacity factor*, K, represents the input speed of a specific size drive at unity input torque. The *unity capacity factor*, $\overline{K}$, represents the input speed of a unit diameter hydrodynamic drive at unity input torque. Both factors depict the characteristic shape of the input speed curve at constant input torque. Their usefulness is in selecting an appropriate hydrodynamic drive for a particular application. Values for C, K, and $\overline{K}$ are calculated from performance data which has been obtained through design prediction methods based upon equations in Sec. 27.12.2 or through actual tests of a specific unit. If data at constant input speed is desired, an input factor can be calculated using Eq. (27.379). A second basic presentation of performance characteristics presents the key parameters against turbine speed as the independent variable as in Fig. 27.280. Turbine speed and torque reflect the load operating requirements. The torque ratio curve presents the direct torque relationship between turbine and pump torques.

A third type of graph is often referred to as an *absorption plot*. Figure 27.281 shows output torque versus output speed on logarithmic scales with each line representing a specific speed ratio. The graph is based on a rearrangement of Eq. (27.379).

$$\log T = 2 \log \omega + 5 \log D + \text{constant} \qquad (27.379a)$$

If test data for a specific hydrodynamic drive is plotted at several input torque levels the actual exponent for the speed, ω, is represented by the slope of the line on the absorption plot. The exponent so determined usually ranges from 1.82 to 2.0, compared to the theoretical value of 2.0.

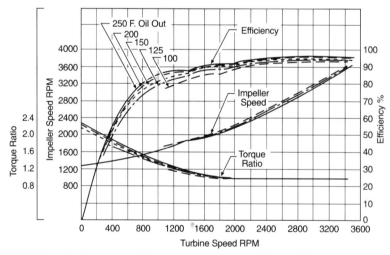

FIGURE 27.280   Performance of fluid torque converter. Effect of fluid temperature on converter performance. (170 lb-ft input torque test.) (Reprinted with permission from SAE, Upton, E. W., "Design Practices—Passenger Car Auto. Trans.," 1994.)

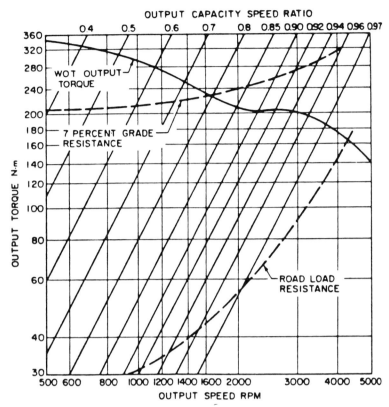

FIGURE 27.281 Typical output capacity plot used to determine the operating conditions of the automobile and hydrodynamic unit. (Reprinted with permission from SAE, Upton, E. W., "Design Practices—Passenger Car Auto. Trans.," 1994.)

27.13 WIND TURBINES
Louis V. Divone and Daniel F. Ancona

27.13.1 History and Background

The first practical use of wind as a stationary power source is believed to have occurred in Persia between the first and third century A.D. Over the next millennium, its use spread to the Mediterranean and then to western Europe, where the technology evolved to the familiar Dutch windmill. This type was not effective in the New World due principally to its labor-intensive nature and the competition from water-power. However, Halladay and Wheeler's development of the automatic, self-regulating, multibladed, water-pumping windmill in the mid-1800's helped lead to the opening of the arid West. Brush, in Cleveland, in the 1890's pioneered the adaptation of the windmill to electricity generation. This was further developed by LaCour in Denmark and Jacobs in the United States, so that by the 1920's small kilowatt size, direct current (DC), wind electric generators were highly successful. In Europe, larger scale systems were attempted during and after both World Wars. In the United States, the 53-meter diameter, 1.25-megawatt (MW) Smith Putnam turbine

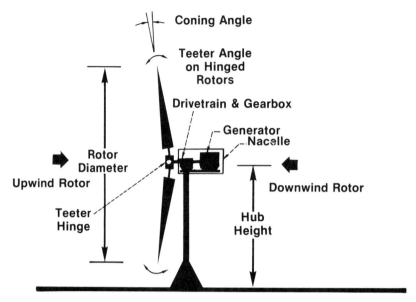

FIGURE 27.282   Wind turbine configurations.

was built at Grandpa's Knob, Vermont, in 1940 but later suffered a blade failure. Interest in wind power ceased in the 1950's until the oil shortages and rising prices in the mid-1970's again led to major research and development activities worldwide. Typical modern electric generating wind turbines and their major features are shown in Fig. 27.282 [ASME (1994)].

Essentially, any device which can produce a resultant asymmetric force from the wind can be made to produce power, and the generic term for such devices is *wind energy conversion system* (WECS). The modern electric generating systems are termed *wind turbines* or *wind turbine generators*. The many possible concepts and types have usually been characterized by several different taxonomies as shown in Table 27.16.

The conventional horizontal axis, propeller-type, electric generating wind turbine is usually further characterized by number of blades, rotor location (upwind or downwind of the tower), electrical application (e.g., stand-alone, battery charging, engine-generator backup, or grid intertie), and aerodynamic control technique (e.g., fixed pitch, full span variable pitch, tip control, ailerons, or spoilers).

An additional category of innovative concepts is utilized to include major variants such as elevated systems (including balloon or kite-suspended turbines), augmenting devices (e.g., diffusers, ducts, and tip vanes), and other unconventional approaches.

27.13.2   Power Extraction

The kinetic energy in a streamtube of uniform flow is given by $E_k = \frac{1}{2}mV_\infty^2$ where m is the mass and V_∞ is the free stream velocity of the moving fluid. The mass flow rate is $\rho A V_\infty$, where ρ is the air density and A is the area of the stream tube. The power in a streamtube of air is $P_\infty = \frac{1}{2}\rho A V_\infty^3$. The existence of a device to extract energy will decelerate the windstream and, from continuity and Newton's laws, will

TABLE 27.16 Wind Turbine Taxonomies

General Classification	Specific Classification	Typical Characteristics
Basic Configuration	Type of Motion	Rotating, translating, oscillating
	Primary Force	Lifting airfoil, semilift, drag
	Axis of Rotation	Horizontal, vertical, crosswind, oblique
Power Output	Electrical	DC, DC with inverter, synchronous AC, induction AC, variable–frequency AC, other
	Mechanical	Water and air pumping, direct heat (viscous churns), heat pumps, other
Size	Very Small <2 kW and/or <3 m dia.	Small battery chargers and water pumping
	Small 2–100 kW and up to 25 m dia.	Residential, agricultural, village, and industrial power, "wind farms" (clusters of turbines)
	Large 100–1000 kW and up to 60 m dia. and larger	Isolated or small utilities and "wind farms" or wind power plants

cause the streamtube to expand, with a portion of the flow passing outside the rotor disk (or swept) area, A_d, as in Fig. 27.283. An axial *induction (interference) factor*, a, is defined as the fractional decrease in average from stream velocity across the rotor disk, $a = 1 - V_d/V_\infty$, where V_d is the average fluid velocity at the plane of the rotor disk. Neglecting drag and rotational losses and assuming two-dimensional flow and isentropic pressure drop, the power extracted by decelerating the air is

$$P_0 = \tfrac{1}{2}\rho A_d V_\infty^3 (4a(1 - a)^2) \tag{27.385}$$

The *power coefficient*, C_p, is defined as the ratio of the power extracted by the wind turbine, P_0, to the power in an undisturbed windstream tube of the same diameter as the rotor disk

$$C_p = \frac{P_0}{\tfrac{1}{2}\rho A_d V_\infty^3} \tag{27.386}$$

In 1926 using momentum considerations, Betz [Glauert (1948)] determined the maximum theoretical C_p was $16/27 = 0.593$, and it occurred when the axial induction factor, $a = 1/3$, which also leads to a downstream wake velocity decrement of 2/3, i.e., $V_2 = V_\infty/3$. The Betz limit is, to first order, independent of blade number, area, or design. It should be noted that C_p is not efficiency, which is

$$\xi = \frac{P_0}{\tfrac{1}{2}\rho A_d V_\infty^2 V_d} \tag{27.387}$$

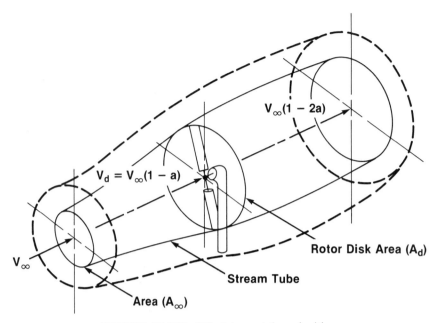

FIGURE 27.283   Windstream tube velocities.

$\xi = 0.88$ at $C_{p\text{max}}$ when $a = 0.32$ and $\xi_{\text{max}} = 1.0$ at $a = 0.5$. C_p, however, being related to free stream (undisturbed site wind) velocity, V_∞, and rotor area (and hence cost) is the accepted measure of wind turbine performance.

The thrust, T, is the axial downwind (overturning) force

$$T = \tfrac{1}{2}\rho A_d V_\infty^2(4a(1 - a)) \qquad (27.388)$$

and the *thrust coefficient* is defined as

$$C_t = \frac{T}{\tfrac{1}{2}\rho A_d V_\infty^2} \qquad (27.389)$$

$C_t = 8/9$ at $C_{p\text{max}}$ when $a = 1/3$ and $C_{t\text{max}} = 1$ at max ξ when $a = 1/2$ for inviscid flow.

The flow conditions as shown in Fig. 27.284 and the primary coefficients are functions of the induced velocity factor, a, which is in turn a function of the design (e.g., number and area of blades) and operating conditions (e.g., blade pitch angle, and rotational speed) of the rotor.

Glauert (1948) further developed this approach using two-dimensional, inviscid flow assumptions for an operating wind turbine rotor. Momentum considerations cause the introduction of a rotational velocity into the fluid and the resulting rotational energy into the wake. Thus, a rotational induction factor, $a' = V_t/r\Omega$, must be considered, where V_t is the tangential wind velocity component at a given radial station along the blade span, r, and Ω, is the rotational speed of the rotor blade.

The power coefficient, C_p, is generally plotted as a function of *tip speed ratio*,

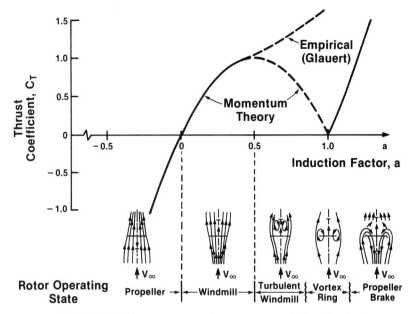

FIGURE 27.284 Rotor operating state versus induced velocity.

λ, the ratio of the tangential velocity at the blade tip, V_T, compared to free stream wind velocity.

$$\lambda = \frac{V_T}{V_\infty} = \frac{R\Omega}{V_\infty} \qquad (27.390)$$

where R is the blade radius. Typical tubine rotor performance is given for various types of rotors in Fig. 27.285. It can be shown that for a rotor with optimum blade twist and solidity, $a' = (1 - 3a)/(4a - 1)$. *Solidity*, σ, is defined as the ratio of total blade area, BA_b, to disk area, A_d, where B = number of blades and A_b = individual blade area projected on the plane of rotation. Machines of high solidity with large area blades and/or large numbers of blades produce high torque, Q_t, thus putting large rotational energy into the wake and decreasing the energy extracted by the rotor. Since power, $P_0 = Q_t\Omega$, high-solidity machines operate at low λ and Ω and are material intensive and limited in the peak power coefficient they can achieve. At high λ, peak theoretical C_p approaches the Betz limit. Such machines have low solidity with typically two or three slender blades, lower torque, and higher rpm, thereby reducing drive-train and gearbox torque and size. The optimum λ for maximum C_p is not highly sensitive to rotor diameter, thus tip speeds tend to be similar for a wide range of machine sizes, and larger machines, therefore, operate at lower Ω.

27.13.3 Rotor Aerodynamics

The flow velocity vectors and primary rotor forces are shown in Fig. 27.286(a) and (b), the 2/3 blade radius station being traditionally used for general reference. The

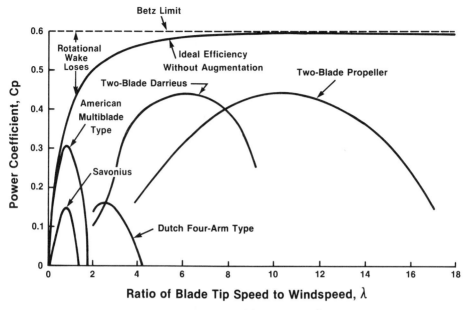

FIGURE 27.285   Wind turbine rotor performance.

relative or *apparent wind velocity* as seen by the blade, V_R, is the vector sum of the free stream wind velocity, V_∞, and the blade velocity, $V_t = r\Omega$ as modified by the axial, a, and rotational, a', induced-velocity factors. The lift, dL, and drag, dD, incremental forces for a radial blade element, following aeronautical conventions, act perpendicular and parallel to V_R, and are given by

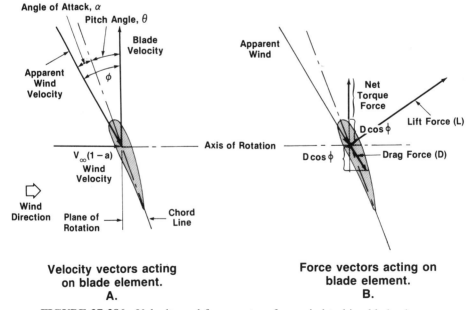

FIGURE 27.286   Velocity and force vectors for a wind turbine blade element.

$$dL = C_l(\tfrac{1}{2})\rho V_R^2 c \; dr \tag{27.391}$$

$$dD = C_d(\tfrac{1}{2})\rho V_R^2 c \; dr \tag{27.392}$$

where c is the airfoil chord length, and C_l and C_d are the blade airfoil section lift-and-drag coefficients which are functions of the local blade shape and angle of attack, α. The angle θ represents the local blade pitch angle setting.

The torque force is given by

$$dF = dL \sin \phi - dD \cos \phi \tag{27.393}$$

and the thrust force by

$$dT = dL \cos \phi + dD \sin \phi \tag{27.394}$$

The importance of a high airfoil lift-to-drag ratio can be seen.

Strip element theory may be used to estimate torque, Q_t, power, P_0, and axial thrust, T. For each value of r, V_∞, and Ω, and for the geometry of Fig. 27.286, the induced velocities a and a' may be assumed, Q_t, T, and P_0 calculated, and a and a' recomputed until the calculation converges. This process is performed for a series of stations and integrated along the blade from hub to tip. This technique is based on momentum theory, and the assumptions are only valid for turbines operating under conditions where the induced velocity factor is $0 < a < 0.5$ (Fig. 27.284). For other regimes, and for better accuracy, more complex vortex line or vortex sheet theory must be utilized. The torque of the electrical or mechanical load, Q_1, will also vary as a function of Ω, and the operating rpm point is determined by the intersection of Q_t and Q_1 versus Ω curves for each V_∞. Computer models for these techniques have been developed [Wilson and Lissaman (1976)].

A constant rpm wind turbine with fixed-pitch blades will operate off the peak C_p point as the wind varies and changes the angle of attack from that for maximum lift-to-drag ratio, L/D_{max}. Variable-pitch blades allow broadening of the C_p versus λ curve and thus improve performance. Further improvement in performance is possible by operating at constant λ, however the resulting variable Ω then involves added structural and electrical complexity.

27.13.4 Wind Characteristics

Windspeed and direction usually vary greatly over short time periods (i.e., minutes and seconds) as well as through diurnal and seasonal cycles. Windspeed and direction can also vary significantly over relatively short distances, particularly in rough terrain. The general pattern of average wind power density for the United States is shown in Fig. 27.287. The Midwest and Great Plains, portions of California, the Pacific Northwest and New England, and most coastal, mountain-ridge, and island locales have the higher average windspeeds. Multiyear wind resource data have been compiled and summarized in Table 27.17. Wind power classes have been defined, as shown in Table 27.18, to provide a generalized description of wind power potential. Sites with wind power classes 1 and 2 are considered poor; 3 and 4 good; and 5 and above, of excellent potential. However, wind characteristics should be mea-

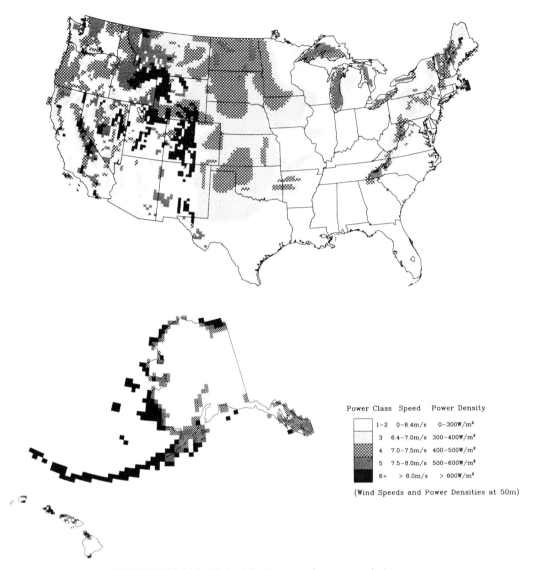

Power Class Speed   Power Density

	1-2	0-6.4m/s	0-300W/m²
	3	6.4-7.0m/s	300-400W/m²
	4	7.0-7.5m/s	400-500W/m²
	5	7.5-8.0m/s	500-600W/m²
	6+	> 8.0m/s	> 600W/m²

(Wind Speeds and Power Densities at 50m)

FIGURE 27.287   United States annual average wind power.

sured for each individual site, since they are highly sensitive to local terrain and climatological effects. Wind velocity duration and energy distribution curves (Figs. 27.288 and 27.289) are generally based on hourly average measurements over a one-year period. A yearly average is considered to be adequate with uncertainty varying as a function of the frequency and duration of wind records.

Windspeed and power characteristics can be obtained for many United States sites from the National Climatic Data Center in Ashville, North Carolina and the National Renewable Energy Laboratory in Golden, Colorado. Extrapolation of wind characteristics from a measured site to a site of interest must be performed with great care, since the increase in uncertainty with distance is a function of numerous and

TABLE 27.17 Wind Resource Summary

| | | | | | Measurement Height | | | | | |
| | | | | | 10 Meter | | | 50 Meter | | |
State	Name	Lat	Long	Elev (m)	Wind Speed (m/s)	Power (w/m²)	C	Wind Speed (m/s)	Power (w/m²)	C
AK	COLD BAY-DOE	55.20	162.72	29	6.6	371.8	6	8.2	719.4	6
CA	PITTSBURG PGE TWR	38.03	121.92	6	4.6	113.7	2	5.9	241.6	2
CA	POINT ARENA-DOE	38.93	123.72	21	4.8	136.4	2	6.67	335.7	3
CA	PT CONCPTION SCE TWR	34.45	120.42	30	5.9	237.8	4	6.3	289.3	2
CA	ROMERO OVERLOOK-DOE	37.07	121.22	458	4.9	132.3	2	6.5	277.5	2
CA	SAN GORGONIO-DOE	33.95	116.58	344	6.3	366.6	6	7.8	731.0	6
HI	ILIO POINT-DOE	21.22	157.25	61	7.2	367.1	6	11.0	1077.1	7
HI	KAHUA RANCH-DOE	20.12	155.78	1030	9.0	768.4	7	12.5	1809.0	7
HI	KAHUKU-DOE	21.70	158.00	108	7.4	452.8	7	8.1	590.0	5
KS	MEADE-DOE	37.30	100.32	756	5.8	215.6	4	8.4	486.5	4
KS	PLAINVILLE	39.23	99.30	655	5.2	153.0	3	7.1	392.4	3
KS	RUSSELL-DOE	38.83	98.85	564	5.4	181.2	3	7.4	388.7	3
KS	TUTTLE CREEK	39.25	96.60	384	6.1	258.8	5	7.7	510.1	5
KS	WRIGHT	37.62	99.88	792	6.2	254.2	5	8.1	580.8	5
MA	HOLYOKE-DOE	42.30	72.58	372	4.3	92.0	1	7.0	406.4	4
MA	NANTUCKET ISLAND-DOE	41.25	70.00	12	5.9	293.7	5	9.1	670.4	6
MA	PROVINCETOWN-DOE	42.05	70.20	10	6.3	247.5	4	9.8	815.5	7
MI	BIG SABLE PT-DOE	44.05	86.52	179	5.8	218.7	4	8.4	547.0	5
MI	LUDINGTON-DOE	43.88	86.43	213	4.7	139.9	2	7.6	482.3	4
MT	LIVINGSTON-DOE	45.67	110.50	1420	6.8	477.0	7	8.5	826.3	7
NC	BOONE-DOE	36.23	81.68	1347	5.6	243.5	4	7.4	469.7	4
ND	FINLEY-DOE	47.52	97.87	472	6.2	245.8	4	9.2	771.3	6
ND	MINOT-DOE	48.00	101.30	675	6.6	283.3	5	8.5	555.1	5
NE	KINGSLEY-DOE	41.20	101.67	1024	5.4	167.7	3	6.6	297.3	2
NM	CLAYTON-DOE	36.45	103.17	1536	5.5	169.7	3	7.4	348.2	3
NM	LOVINGTON	33.00	103.32	1192	4.9	122.5	2	7.8	492.1	4
NM	SAN AUGUSTIN PASS	32.42	106.57	1859	7.8	558.2	7	8.8	597.7	5
NM	TUCUMCARI-DOE	35.17	103.83	1354	6.3	232.2	4	8.7	536.3	5
NV	WELLS-DOE	41.03	114.57	2268	6.9	317.6	6	7.3	326.1	3

TABLE 27.17 (Continued)

					Measurement Height					
					10 Meter			50 Meter		
State	Name	Lat	Long	Elev (m)	Wind Speed (m/s)	Power (w/m²)	C	Wind Speed (m/s)	Power (w/m²)	C
NY	DUNKIRK	42.47	79.33	192	4.5	104.6	2	5.7	211.7	2
NY	MONTAUK-DOE	41.05	71.97	2	5.6	239.6	4	7.3	454.2	4
NY	NINE MILE POINT	43.47	76.50	80	5.5	239.0	4	6.6	570.0	5
OK	FORT SILL-DOE	34.65	98.45	366	5.6	215.6	4	9.1	661.1	6
OR	BOARDMAN-DOE	45.68	119.83	212	3.8	107.3	2	5.2	241.3	2
OR	CAPE BLANCO-DOE	42.83	124.52	30	4.8	133.3	2	8.0	468.8	4
OR	PEBBLE SPRINGS	45.70	120.14	220	4.2	205.0	4	5.3	360.0	3
OR	SEVENMILE TOWER	45.65	121.27	573	6.3	279.3	5	7.6	493.2	4
PR	CULEBRA IS.-DOE	18.33	65.32	80	6.2	217.8	4	7.1	298.4	2
RI	BLOCK ISLAND-DOE	41.17	71.57	14	5.1	138.5	2	7.5	424.1	4
SD	HURON-DOE	44.40	98.13	396	4.8	136.4	2	6.9	345.1	3
TX	AMARILLO	35.22	101.55	1067	5.8	219.0	4	7.8	456.0	4
TX	AMARILLO-DOE	35.28	101.75	1091	6.4	238.5	4	8.2	483.3	4
TX	CHILDRESS	34.43	100.23	609	4.2	94.0	1	6.1	228.0	2
TX	DALHART	36.10	102.50	1219	5.0	181.0	3	7.0	395.0	3
TX	FORT WORTH	32.70	97.37	243	4.9	133.0	2	6.1	265.0	2
TX	MULESHOE	34.23	102.72	1158	4.5	118.0	2	5.9	236.0	2
TX	PERRYTON	36.43	100.80	899	6.0	231.0	4	7.5	401.0	4
TX	STANTON	32.10	101.75	807	4.7	133.0	2	6.6	349.0	3
TX	WICHITA FALLS	33.90	98.58	283	4.2	82.8	1	5.9	232.8	2
VT	STRATTON MT.-DOE	43.08	72.83	1183	6.7	259.9	5	12.7	1647.8	7
WA	AUGSBURGER MT.-DOE	45.75	121.68	853	6.9	320.7	6	8.3	569.6	5
WA	CRESTON	47.80	118.50	762	5.6	193.5	3	6.3	276.6	2
WA	DIABLO DAM-DOE	48.50	121.12	500	3.3	51.1	1	5.9	206.8	2
WA	GOODNOE HILLS-DOE	45.75	120.50	805	5.6	180.4	3	6.7	343.6	3
WY	BRIDGER BUTTE-DOE	41.28	110.48	2292	7.1	387.4	6	8.5	612.2	6
WY	MEDICINE BOW	41.51	106.15	2060	6.8	300.2	6	7.9	463.9	4

TABLE 27.18 Classes of Wind Power Density at 10 m and 50 m[a]

Height Above Ground	10 m (33 ft)		50 m (164 ft)	
Wind Power Class	Wind Power Density Watts/m^2	Speed,[b] m/s (mph)	Wind Power Density Watts/m^2	Speed,[b] m/s (mph)
	0	0	0	0
1	100	4.4(9.8)	200	5.6(12.5)
2	150	5.1(11.5)	300	6.4(14.3)
3	200	5.6(12.5)	400	7.0(15.7)
4	250	6.0(13.4)	500	7.5(16.8)
5	300	6.4(14.3)	600	8.0(17.9)
6	400	7.0(15.7)	800	8.8(19.7)
7	1000	9.4(21.1)	2000	11.9(26.6)

[a]Vertical extrapolation of windspeed based on the 1/7 power law.
[b]Mean windspeed is based on Rayleigh speed distribution of equivalent mean wind power density. Windspeed is for standard sea-level conditions. To maintain the same power density, speed increases 5%/5000 ft (3%/1000 m) of elevation.

difficult to quantify terrain and climatic factors. Siting handbooks provide a systematic approach to selecting appropriate sites [Hiester and Pennell (1981)].

For typical sites without excessive terrain roughness or unique conditions, a *Rayleigh distribution* may be used to represent the wind duration curve for initial design or siting estimates

$$F(V) = \frac{\pi V}{2\overline{V}^2} \exp \frac{-\pi}{4} \left(\frac{V}{\overline{\overline{V}}}\right)^2 \qquad (27.395)$$

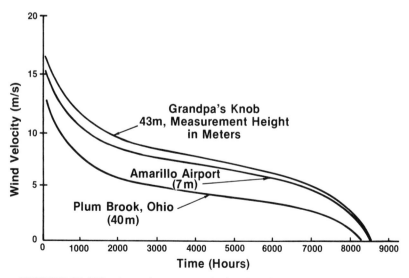

FIGURE 27.288 Annual average velocity duration curves for three sites.

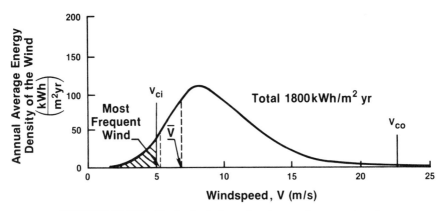

FIGURE 27.289   Typical windspeed and annual energy density.

where $F(V)$ is the *probability density function* of windspeed over $0 < V < \infty$ and $\bar{V}$ is the mean windspeed. For more complex cases, a *Weibull distribution* may be used.

Wind velocity generally increases with height while the earth's boundary layer reduces windspeed near the surface. The wind velaocity at the height of the turbine rotor hub may be estimated by the equation

$$V = V_{\text{ref}} \left(\frac{H}{H_{\text{ref}}} \right)^N \tag{27.396}$$

where V_{ref} and H_{ref} are the velocity and height of the referenced measurement station, typically 10 meters above ground level. The exponent, N, varies considerably depending on both site surface roughness and current climatological conditions. $N = 1/7$ is generally used as an overall average for relatively flat terrain and neutrally stable atmospheric conditions. N increases with increasing surface roughness to $\frac{1}{4}$ for forested hills and up to $\frac{1}{2}$ in very rough terrain or urban areas.

Turbulence levels and wind direction rate of change also vary widely. The angle of attack of a wind turbine blade will, thus, vary widely as a function of blade rotational position, and that can introduce yaw motion as well as undesirable loads. Unsteady flow effects on airfoil-section properties may also become important.

27.13.5   System Design and Performance

Wind turbine system designs depend on a complex set of variables and tradeoffs. The bulk of the annual wind energy at a typical size occurs when $0.5\bar{V} < V < 2.0\bar{V}$, where $\bar{V}$ is the mean wind velocity. There is generally little energy to be captured when $V < 0.3\bar{V}$ or $V > 3.0\bar{V}$. Therefore, a limitation on power output is selected based on an economic tradeoff between the marginal difference in energy capture versus the marginal cost of using a larger or smaller generator. The selected point is called the *rated power*, P_r, of the wind turbine which occurs at the *rated windspeed*, V_r, (Fig. 27.290). P_r is typically selected with V_r in the range of $1.5\bar{V} < V_r < 2.5\bar{V}$. The additional power available in the wind at velocities greater than

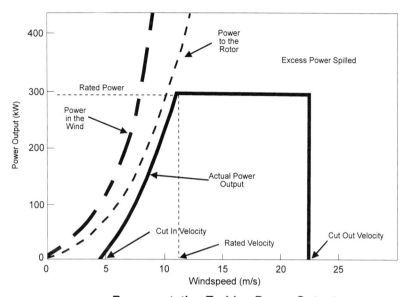

Representative Turbine Power Output

FIGURE 27.290 Wind turbine power output curves.

V_r, may be *spilled* by utilizing variable-pitch blades or other aerodynamic techniques.

Similar engineering-economics tradeoffs lead to the selection of *cut-out windspeed*, V_{co}. Operation above V_{co} does not provide enough added energy to offset the added high wind loads and wear. *Cut-in windspeed*, V_{ci}, is the point below which the low available power (and hence energy) to be captured also does not warrant the added hours of wear. There is, in practice, some hysteresis in V_{ci}, V_r, and V_{co} between the wind-increasing and wind-decreasing situations. Control logic usually incorporates windspeed-time-delay relationships to minimize repeated start-stop cycles during rapidly varying wind conditions.

Average power, $\overline{P}$, can be estimated from Fig. 27.291 for given wind turbine characteristics and site mean windspeeds where a Rayleigh distribution and $V_{ci} = 0$ are assumed. Annual energy capture in kWh/yr is stated as $E_a = 8,760\overline{P}$. This annual energy would be reduced by mechanical and electrical losses as well as the effects of control operation, V_{ci} and startup cycle times, yaw angles, and maintenance needs. The net annual energy capture and rated powers of various typical modern wind turbines are presented in Table 27.19.

It should be emphasized that wind turbines are in reality energy- rather than power-oriented machines (Fig. 27.292) and that the rated power can be traded off with the *capacity factor* (the ratio of annual energy produced to the annual energy which would be produced if the system operated at rated power the full 8,760 hours in the year).

27.13.6 Vertical Axis Systems

Vertical axis systems have the advantage that they can accept the wind from any direction thus requiring no yaw control. The high *L/D* Darrieus, gyromill, and vari-

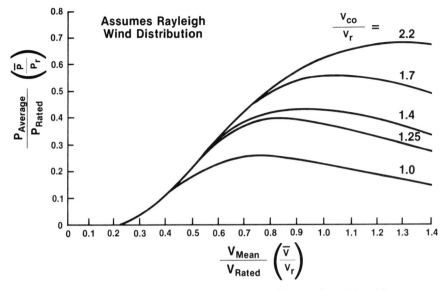

FIGURE 27.291 Average expected output of a generic wind turbine.

able-geometry machine (Fig. 27.293) concepts have peak C_p's equivalent to propeller-type systems, although usually at slightly lower blade-to-wind speed ratios.

The velocity and force vectors for the Darrieus are shown in Fig. 27.294. In addition to the omission of yaw control requirements, other advantages of the Darrieus include the placement of drive and control equipment near the ground for ease of maintenance and the structural advantages of simplicity and double-ended, simply-supported, troposkein shaped, fixed-pitch blades. Conversely, as can be seen from Fig. 27.293, the Darrieus requires a significant tangential velocity, V_t (as well as high L/D airfoils) to allow for the generation of positive torque. Thus, the Darrieus is not normally self-starting but requires supplemental electrical or aerodynamic techniques for that purpose. The Darrieus, since it is a fixed-pitch machine

TABLE 27.19 Approximate Net Annual Energy Output (MWh/yr)

				Site Characteristics			
				Average Annual Windspeed m/s (MPH)			
	Machine Characteristics			Measured at 10 m above ground			
Rotor Diameter (m)	Rated Power (kW)	Rated Windspeed (m/s)	Rotor C_p @ V_r	5.4 (12)	6.3 (14)	7.2 (16)	8.0 (18)
3	4	13.4	0.35	5	6	8	10
9	30	13.4	0.35	30	54	75	100
18	140	13.4	0.35	135	210	300	360
38	200	9.4	0.36	640	825	980	1100
40	500	12	0.37	1240	1600	1900	2160
91	2500	12.1	0.37	4800	6400	7600	8400

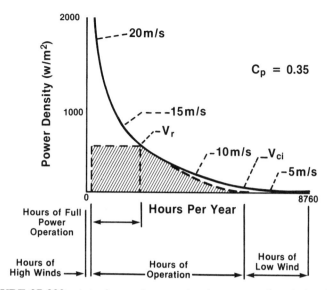

FIGURE 27.292 Actual annual power density output of a wind turbine.

and relies on generator load to control rotational speed, will tend to have a higher rated power and rated velocity than an equivalent propeller-type system with the same swept area and annual energy capture. In addition, the Darrieus has larger tower (diameter rather than radius plus ground clearance) and longer blades (circumference rather than diameter) and more severe materials requirements than propeller systems. At the present stage of technology, the Darrieus and propeller types appear approximately competitive.

The *gyromill* concepts use variable-pitch blades to increase $C_{p_{max}}$, broaden the C_p versus λ curve, and reduce peak loads. Yaw control, in terms of varying the pitch angle as a function of wind direction is, however, required as are straight blades.

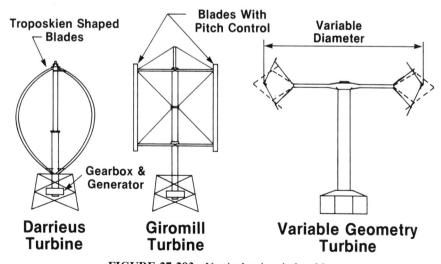

FIGURE 27.293 Vertical axis wind turbines.

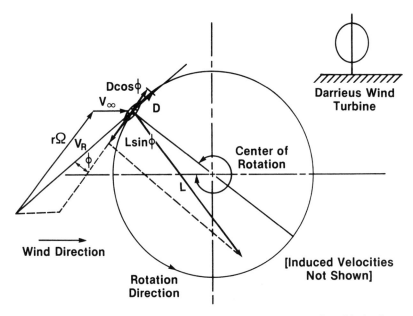

FIGURE 27.294   Simplified velocity and force vectors on a Darrieus blade element.

These requirements mean the gyromill sacrifices the structural and control simplicity of the Darrieus. Thus, it has seldom been adopted commercially.

The *Savonius rotor* [Savonius (1931)], while a simple drag or semi-lift device, has only been used for a few applications due to its low C_p and Ω and its large material usage and high aerodynamic load.

27.13.7   Structural Dynamics

In large rotating objects such as wind turbines, structural vibrations are unavoidable and must be considered along with normal static loads. The primary forcing functions are gravity and the aerodynamic loads induced by variations in the wind as a function of time and space across the rotor disc. Factors to consider in vibration analysis are: 1) rotor blade aeroelastic response including the effects of variations in lift, drag, stall, turbulence, and irregularities of the blades, 2) the blade pitch-control subsystem, 3) drive-train dynamics including the drive-shaft flexibility, hydraulic couplings, gearbox dynamics, and torsional dampers, 4) generator and power subsystem stiffness and electrical control subsystem parameters, and 5) tower stiffness, and foundation and soil properties. All of these system characteristics interact, with many degrees of freedom, to determine the overall structural response of the wind turbine. Resonant vibration must be prevented, and the forced response of the turbine must be controlled to prevent excessive fatigue damage.

The primary forcing functions have both static and dynamic components. Gravity induces edgewise bending loads in the blades that reverse once every rotation (Fig. 27.295) with more than 4×10^8 cycles over the normal machine life of 30 years. Flapwise blade loads (perpendicular to the chord line) result from the horizontal wind velocity producing thrust, bending due to blade pitch control, rotor coning

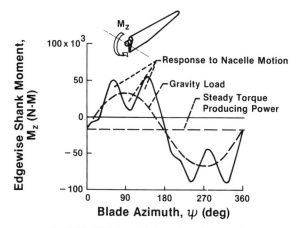

FIGURE 27.295 Blade edgewise bending.

angle, and nacelle motion. On downwind machines (where the rotor operates downwind of the tower), additional loads result from the *wind shadow* behind the tower. Flapwise loads at the rotor hub can be significantly reduced by hinging the rotor or allowing it to teeter (Fig. 27.296).

Wind velocity variations in both direction and magnitude are significant factors in determining the forced response of wind turbines. As atmospheric turbulence increases, these effects can induce thrust and torque changes, but more importantly, natural frequencies of wind turbine systems can be excited. Power spectral density of lower frequency components, in the wind-gust velocity frequencies, have sufficient energy content to excite turbine component natural frequencies.

Structural properties of wind turbines are significantly different from most other electric power generating equipment. Wind turbine rotors have a very high polar

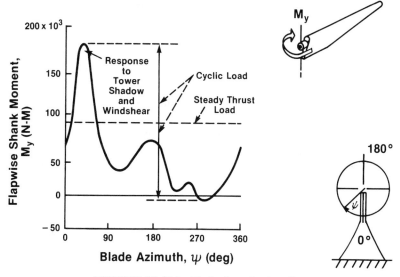

FIGURE 27.296 Blade flapwise bending.

moment of inertia relative to the generator unlike hydro and steam turbines where the turbine wheels and generators have roughly similar inertias. In addition, wind turbines typically have very flexible, lightly damped shafts connecting the rotor to the step-up gearbox. These dynamic characteristics affect the design as well as the control of wind turbines and mean drive trains are more likely to experience load and power fluctuations. A number of structural dynamic computer models have been developed to predict natural frequencies and structural loads for both horizontal and vertical axis wind turbines [Thresher (1982)].

27.13.8 Economics and Environmental Matters

The large-scale use of wind power has fluctuated through several cost-effectiveness cycles during its history. Attractiveness depends primarily on the cost of building and maintaining the wind turbine relative to the competing cost of equivalent conventional energy. The final measure of effectiveness is the energy output value versus its cost, and the primary determinants are the site wind resource, energy capture, material, fabrication, interconnection and land costs, operations and maintenance cost, and the effect on value of the transient and intermittent nature of the wind system output in the particular application.

Regardless of the application, the wind resource is critically important in determining the value of a wind system. Typically, a 6 to 7 m/s average annual wind resource is needed for most applications, and such winds are available over large land areas and in shallow water near the shore in the United States and most countries in temperate or higher latitudes. Seasonal variations and daily patterns are also important and must be considered because of the effects of matching electrical demand to wind availability.

Environmental issues must also be considered in the siting and economics of wind turbines. Ecological impacts have been found to be minimal to nonexistent. The rotating blades, depending on size, material, and sitting geometry, can cause interference with electromagnetic signals, particularly in the upper-VHF and lower-UHF television and microwave bands [Senior (1978)]. The rotating blades and their interaction with the flow field can also cause acoustic noise [Hubbard 1983)]. Sound pressure levels can vary widely depending on detailed rotor and system design, although most wind turbines cannot be heard above the ambient wind noise several rotor diameters away. Bird strikes are also a concern at some sites especially where protected species are present.

Electric utility uses represent the largest potential application for wind systems in the United States. Approximately one-third of the energy used in the United States goes to produce electricity, and fossil fuel is the chief energy source. Wind turbines are fundamentally different from conventional electric power sources, since wind as a *fuel supply* is free, albeit variable, and somewhat unpredictable. Consequently, wind turbines are normally connected to the existing power system network and operated as *fuel-savers*, generally displacing higher priced fuels such as oil or gas and reducing atmospheric pollution. The value of wind energy is based on the cost of fuel saved, which is typically one-third of the busbar cost of conventional power (which includes plant and equipment, real estate, transmission, and operating and maintenance costs).

In addition to fuel cost savings, there is a capacity credit associated with the installation of wind systems based on the statistical availability of the wind resource, which would thus reduce somewhat the conventional power plant size requirement. The effect of this capacity credit on the economics of wind power is, however, relatively small compared with the fuel savings.

Rural residential and agricultural uses are also attractive applications for wind-generated electricity. Wind turbines can be used to produce heat for direct use or for DC electricity which can be stored in batteries and, if necessary, inverted back to alternating current (AC) power. The most common application, however, is to generate AC power by means of an induction or synchronous generator in the wind turbine. The generator is connected directly to the utility grid such that power which exceeds local requirements can be fed back to the grid and utility power used during times of low wind [Johansson (1992)].

REFERENCES

Abbott, I. H. and Von Doenhoff, A. E., *Theory of Wing Section (Including a Summary of Airfoil Data)*, Dover Publications, New York, 1959.

Adkins, G. G. and Smith, L. H., "Spanwise Mixing in Axial Flow Turbomachines," *J. of Engineering for Power*, Vol. 104, pp. 97–110, 1982.

Agarwal, R. K. and Deese, J. E., "Euler Calculations for a Flowfield of a Helicopter Rotor in Hover," *J. Aircraft*, Vol. 24, No. 4, 1987.

Ainley, D. G. and Mathieson, G. C. R., "A Method of Performance Estimation for Axial Flow Turbines," Aero. Res. Council, R&M 2974, 1951.

AMCA, Bulletin 210-85, "Laboratory Methods of Testing Fans for Rating," *ANSI/AMCA/ASHRAE Standards*, 1985.

ASME, *Power Test Codes*, Test Code for Compressors and Exhausters (PTC-10), 1985.

ASME, *The 1967 ASME Steam Tables*, ASME, New York.

ASME, "Wind Turbine Technology, Fundamental Concepts of Wind Turbine Engineering," ASME, New York, 1994.

Aungier, R. H., "Aerodynamic Design and Analysis of Vaneless Diffuser and Return Channels," ASME 93-GT-103, 1993.

Babcock and Wilcox, *Steam, Its Generation and Use*, Babcock and Wilcox, 1992.

Bachman, P., Schärer, Ch., Staubli, T., and Vullioud, G., "Experimental Flow Studies in a 1-Jet Model Pelton Turbine," IAHR Symposium, Belgrade, Sept. 1990.

Bagai, A. and Leishman, J. G., "A Study of Rotor Wake Developments and Wake/Body Interactions in Hover," *J. American Helicopter Soc.*, Vol. 37, No. 4, pp. 48–57, 1992.

Baljé, O. E., "A Study on Design Criteria and Matching of Turbomachines," *J. Eng. for Power*, Vol. 84, No. 1, p. 83, 1962.

Baljé, O. E., *Turbomachines—A Guide to Design, Selection and Theory*, John Wiley & Sons, New York, 1981.

Benedict, R. P. and Meyer, C. A., "Electrolytic Tank Analog for Studying Fluid Flow Fields within Turbomachinery," ASME Paper 57-A-120, 1957.

Benedict, R. P. and Steltz, W. G., "A Generalized Approach to One-Dimensional Gas Dynamics," *J. Engrg. for Power*, Vol. 84, Series A, No. 1, 1962.

Benedict, R. P. and Steltz, W. G., *Handbook of Generalized Gas Dynamics*, Plenum Press, New York, 1966.

Benedict, R. P., "Analog Simulation," *Electro-Technology*, Dec. 1963.

Bisplinghoff, R. L., Ashley, H., and Halfman, R. L., *Aeroelasticity*, Addison–Wesley Publishers, 1955.

Braisted, D. M. and Brennen, C. E., "Auto-Oscillation of Cavitating Inducers," *Proc. of Conference on Polyphase Flows*, ASME, July 1980.

Brekke, H., "The State of the Art in Pelton Turbine Design," *Int. J. Hydropower and Dams*, Vol. 1, No. 2, pp. 21–28, 1994.

Bullock, R. O. and Johnson, I. A., *Aerodynamic Design of Axial-Flow Compressors*, NASA SP-36, 1976.

Burrill, L. C. and Emerson, A., "Propeller Cavitation: Further Tests on 16 In. Propeller Models in the Kings College Cavitation Tunnel," *Trans. North East Coast Institution of Engineers and Shipbuilders*, Vol. 79, pp. 295–320, 1963.

Carter, A. D. S., "The Low Speed Performance of Related Aerofoils in Cascade," Aeronautical Research Council CP29, 1950.

Casey, M. V., "The Effects of Reynolds Number on the Efficiency of Centrifugal Compressor Stages," ASME 84-GT-247, 1984.

Casey, M. V., "The Effects of Reynolds Number on the Efficiency of Centrifugal Compressor Stages," *J. Eng. for Gas Turbines and Power*, Vol. 107, No. 2, pp. 541–548, 1985.

Chaussee, D. S., "Computational of Three Dimensional Flow Through Prop Fans," Nielsen Engineering and Research, Inc., Mountain View, CA, NEAR-TR 199, 1979.

Claus, R. W., Wear, J. D., and Liebert, C. H., "Ceramic Coating Effect on Liner Metal Temperatures of Film Cooled Annular Combustor," NASA Technical Paper 1323, 1979.

Clements, W. E. and Artt, D. W., "The Influence of Diffuser Leading Edge Geometry on the Performance of a Centrifugal Compressor," ASME 89-GT-163, 1989.

Clements, W. W. and Artt, D. W., "The Influence of Channel Geometry on the Flow Range and Efficiency of a Centrifugal Compressor," *Proc. Inst. Mech. Engrs.*, Vol. 201, 1987.

Cohen, H., Rogers, C. F. C., and Saravanamuttoo, H. I. H., *Gas Turbine Theory*, 2nd ed., Longmans, London, 1972.

Cordier, O., "Similarity Considerations in Turbomachines," *VDI Reports*, Vol. 3, p. 85, 1955.

Cox, G. G. and Morgan, W. B., "The Use of Theory in Propeller Design," *Marine Technology*, Vol. 9, No. 4, 1972.

Craig, H. R. M. and Cox, H. J. A., "Performance Estimation of Axial Flow Turbines," *Proc. Inst. of Mech. Engs.*, Vol. 185, No. 32/71, 1970–1971.

Crouse, G. L. and Leishman, J. G., "Interaction Aerodynamic Effects on Rotor Performance in Hover and Forward Flight," *Proc. 48th Annual Forum of the American Helicopter Soc.*, Washington, D.C., 1992.

Csanady, G. T., *Theory of Turbomachines*, McGraw-Hill, New York, 1964.

Cumming, R. A. *et al.*, "Highly Skewed Propellers," *Trans. Soc. Naval Architects and Marine Engineers*, Vol. 80, 1972.

Cumpsty, N. A., *Compressor Aerodynamics*, Longmans, Harlow, England, 1989.

Daley, B. N. and Dick, R. S., "Effect of Thickness, Camber, and Thickness Distribution on Airfoil Characteristics at Mach Numbers up to 1.0," NACA TN 3607, 1956.

Dawes, W. N., "Toward Improved Throughflow Capability: The Use of Three-Dimensional Viscous Flow Solvers in a Multistage Environment," *J. of Turbomachinery*, Vol. 114, 1992.

Dean, R. C., Jr., "The Fluid Dynamic Design of Advanced Centrifugal Compressor," Creare Technical Note TN-185 1974.

de Haller, P., "Das Verhalten von Tragflugelgittern in Axialverdichtern und in Windkanal," *Brennstoff-Warmer-Kraft*, Vol. 5, Heft 10, 1953.

Denton, J. D., "The Calculation of Three-Dimensional Viscous Flow Through Multistage Turbines," *J. of Turbomachinery*, Vol. 114, 1992.

Denton, J. D., "Throughflow Calculation for Transonic Axial Flow Turbines," *J. of Engineering for Power*, Vol. 100, pp. 212–218, 1978.

Dörfler, P., "On the Role of Phase Resonance in Vibrations Caused by Blade Passage in Radial Hydraulic Turbines," IAHR Symposium, p. 228, 1984.

Dörfler, P., "System Dynamics of the Francis Turbine Hard Flow Surge," IAHR Symposium, Amsterdam, September, 1982.

Drees, J. M., "A Theory of Airflow Through Rotors and its Application to Some Helicopter Problems," *J. Helicopter Association of Great Britain*, Vol. 3, No. 2, 1949.

Durand, W. F., *Aerodynamic Theory*, Vol. IV, Dover Publications, New York, 1963.

Dussourd, J. L. and Putnam, W. C., "Instability and Surge in Dual Entry Centrifugal Compressors," *Symposium on Compressor Stall, Surge, and System Response*, ASME, 1960.

Dussourd, J. L., Pfannebecker, G. W., and Singhania, S. K., "An Experimental Investigation of the Control of Surge in Radial Compressors Using Close Coupled Resistances," *J. of Fluids Engineering*, Vol. 99, pp. 64–76, 1977.

Eck, Bo, *Fans, Design and Operation of Centrifugal, Axial-Flow, and Cross-Flow Fans*, Pergamon Press, 1973.

Egolf, T. A. and Landgrebe, A. J., "Helicopter Rotor Wake Geometry and its Influence in Forward Flight, Vol. I—Generalized Wake Geometry and Wake Effect on Rotor Airloads and Performance," NASA CR 3726, 1983.

Ehrich, F. E., "Aerodynamic Stability of Branched Diffusers," *J. of Engineering for Power*, Vol. 92, pp. 330–334, 1970.

Elder, R. L. and Gill, M. E., "A Discussion of the Factors Affecting Surge in Centrifugal Compressors," ASME 84-GT-194, 1984.

Fraize, W. E. and Kinney, C., "Effects of Steam Injection on the Performance of Gas Turbine Power Cycles," *Engineering for Power*, Vol. 101, pp. 217–227, 1979.

Francis, J. B., *Lowell Hydraulic Experiments*, 5th ed., Van Nostrand, Princeton, NJ, 1909.

Freeman, C. and Cumpsty, N. A., "Method for the Prediction of Supersonic Blade Performance," *J. Propulsion*, Vol. 8, No. 1, pp. 199–208, 1992.

Gallimore, S. J. and Cumpsty, N. A., "Spanwise Mixing in Multistage Compressors: Part 1, Experimental Investigation," *J. of Turbomachinery*, Vol. 108, pp. 2–9, 1986.

Gallimore, S. J. "Spanwise Mixing in Multistage Compressors: Part II Throughflow Calculations Including Mixing," *J. of Turbomachinery*, Vol. 108, pp. 2–6.

Gessow, A. and Meyers, G. C., *Aerodynamics of the Helicopter*, College Park Press, College Park, MD, 1952.

Glassman, A. J., "Turbine Design and Applications," NASA SP-290, Vols. 1, 2, and 3, 1975.

Glauert, H., "A General Theory of the Autogyro," Aero. Res. Council, Rep. Memo. 111, 1926.

Glauert, H., "Aerodynamic Theory—Airplane Propellers," Vol. 4, Division L, *Aerodynamic Theory*, Durand, W. F. (Ed.), Dover, New York, 1963.

Glauert, H., "On the Horizontal Flight of a Helicopter," British Aero. Res. Council R&M 1157, 1928.

Glauert, H., *The Elements of Aerofoil and Airscrew Theory*, 2nd ed., Cambridge University Press, Cambridge, U.K., 1948.

Goede, E. and Rhyming, I. L., "3-D Computation of the Flow in a Francis Runner," Sulzer Technical Review, 1987.

Goede, E., Cuenod, R. *et al.*, "3-Dimensional Flow Simulation in a Pump Turbine," ASME FED, Vol. 86, *Industrial and Agricultural Uses of Fluid Mechanics*, Ed Morrow, Sherif and Marshall, 1989.

Graham, J. B., "Prediction of Fan Sound Power," *ASHRAE Handbook HVAC Applications*, American Soc. of Heating, Refrigerating and Air-Conditioning Eng., Inc., pp. 42.7–42.9, 1991.

Gratton, M., Critchley, I. L., and Sampath, P., "Alternate Fuels Combustion Research," AFWAL TR-84-2042, 1984.

Greeley, D. S. and Kerwin, J. E., "Numerical Methods for Propeller Design and Analysis in Steady Flow," *Trans. Soc. Naval Architects and Marine Engrs.*, Vol. 90, 1982.

Grein, H., Meier, J., and Klicov, D., "Efficiency Scale Effects in Pelton Turbines," IAHR Symposium, Montreal, 1986.

Greitzer, E. M., "The Stability of Pumping Systems—The 1980 Freeman Scholar Lecture," *J. of Fluids Engineering*, Vol. 103, pp. 193–242, 1981.

Greitzer, E. M., "Review—Axial Compressor Stall Phenomena," *J. of Fluids Engineering*, Vol. 102, pp. 134–151, 1980.

Hammer, N. O. and McGinn, R. F., "Highly Skewed Propellers—Full-Scale Vibration Test Results and Economic Considerations," *Proceedings of Ship Vibration, Symposium*, Soc. Naval Architects and Marine Engineers Publication SY-8, 1978 (discussion by R. J. Boswell).

Henry, P., *Turbomachines Hydrauliques*, Presse Polytechniques et Universitaires Romandes, Lausanne, Switzerland, 1992.

Heyson, H. H. and Katzoff, S., "Induced Velocities Near a Lifting Rotor with Nonuniform Disk Loading," NACA TR 1319, 1957.

Hiester, T. R. and Pennell, W. T., *The Meteorological Aspects of Siting Large Wind Turbines*, Battelle Pacific Northwest Lab., National Technical Information Service, Springfield, VA, PNL-2522, 1981.

Hobbs, D. E. and Weingold, H. D., "Development of Controlled Diffusion Aerofoils for Multistage Compressor Applications," *J. of Engineering for Gas Turbines and Power*, Vol. 106, pp. 271–278, 1984.

Höller, H. K. and Grein, H., *Utilisation of Water Power by Means of Hydraulic Machines*, Escher Wyss Publication, 1984.

Horlock, J. H., *Axial Flow Turbines*, Krieger, New York, 1973.

Horlock, J. H., *Combined Power Plant*, Elsevier Service Ltd. (Pergamon Press), Oxford, 1992.

Howell, A. R., "Fluid Dynamics of Axial Compressors," *Proc. I. Mech. E.*, Vol. 153, pp. 441–482, 1945.

Hubbard, Harvey, H., "Noise Characteristics of Large Wind Turbine Generators," *Noise Control Engineering Journal*, Vol. 21, No. 1, July–Aug. 1983.

IEC Code 995, "Determination of the Prototype Performance from Model Acceptance Tests of Hydraulic Machines with Consideration of Scale Effects," International Electrotechnical Commission, 1991.

Jansen, W., Carter, A. F., and Swarden, M. C., "Improvements in Surge Margin for Centrifugal Compressors," AGARD CP-282, *Centrifugal Compressors, Flow Phenomena and Performance*, AGARD, Brussels, 1980.

Japikse, D., "Design Optimization and Performance Map Prediction for Centrifugal Compressors and Radial Inflow Turbines,'' AGARD Lecture Series No. 83 on *Modern Prediction Methods for Turbomachine Performance*, Munich, Germany, 1976.

Johansson, T. B. *et al.*, *Renewable Energy: Source for Fuel and Electricity*, Island Press, Washington, DC, 1992.

Johnson, I. A. and Bullock, R. O., ''Aerodynamic Design of Axial Flow Compressors—Revised,'' NASA SP-36, 1965.

Johnson, W., *Helicopter Theory*, Princeton University Press, Princeton, NJ, 1980.

Jones, R. E., Trout, A. M., Wear, J. D., and McBride, B. J., ''Combustion Gas Properties,'' NASA Technical Paper 2359, Oct. 1984.

JSME, *The Mechanical Engineering Handbook—Hydraulic and Pneumatic Machines*, Volume 9, 5th ed., Japanese Soc. of Mechanical Engineers, Tokyo, 1972.

Kacker, S. C. and Okapuu, U., ''A Mean Line Prediction Method for Axial Flow Turbine Efficiency,'' *J. Eng. for Power*, Vol. 104, pp. 111–119, 1982.

Kahane, A., ''Investigation of Axial-Flow Fan and Compressor Rotor Designed for Three-Dimensional Flow,'' NACA TN-1652, 1948.

Kalina, A. I., ''Recent Improvements in Kalina Cycles: Rationale and Methodology,'' *American Power Conf.*, Vol. 55, 1993.

Keck, H., ''Entwicklung und Projektierung von Saugrohren für Wasserturbinen grosser Schnellaüfigkeit,'' 3rd International Seminar on ''Wasserkraftanlagen,'' Technical University, Vienna, Nov. 1984.

Keck, H., Goede, E., and Pestalozzi, J., ''Experience with 3D Euler Flow Analysis as a Practical Design Tool,'' IAHR Symposium, Belgrade, Sept. 1990.

Kenny, D. P., ''The History and Future of the Centrifugal Compressor in Aviation Gas Turbines,'' SAE SP-602, 1984.

Kerwin, J. E., ''Computer Techniques for Propeller Blade Section Design,'' *Int. Shipbuilding Progress*, Vol. 20, No. 227, 1973.

Koch, C. C. and Smith, L. H., Jr., ''Loss Sources and Magnitudes in Axial-Flow Compressors,'' *J. Eng. for Power*, Vol. 98, pp. 411–424, 1976.

Koch, C. C., ''Stalling Pressure Rise Capability of Axial Flow Compressor Stages,'' *J. of Engineering for Power*, Vol. 103, pp. 645–665, 1981.

Krause, M. and Grein, H., ''Abrasion Research and Prevention,'' Sulzer Technical Review, Feb. 1993.

Landgrebe, A. J. and Bellinger, E. D., ''An Investigation of the Quantitative Applicability of Model Helicopter Rotor Wake Patterns Obtained from a Water Tunnel,'' UARL K910917-23, 1971.

Landgrebe, A. J., ''An Analytical and Experimental Investigation of Helicopter Rotor Performance and Wake Geometry Characteristics,'' USAAMRDL TR 71-24, June 1971.

Landgrebe, A. J., ''An Analytical Method for Predicting Rotor Wake Geometry,'' *J. American Helicopter Soc.*, Vol. 14, No. 4, pp. 20–32, 1969.

Larin, A. V., ''Vortex Formation in Oblique Flow Around a Helicopter Rotor,'' *Uchenyye Zapiski TSAGI*, Vol. 1, No. 3, pp. 115–122, 1970. (Avail. NTIS as translation AD781245.)

Larin, A. V., ''Vortex Wake Behind a Helicopter,'' *Aviatsiya i kosmonavtika*, No. 3, pp. 32–33, 1973. (Avail. NTIS as translation ADA005479.)

Lefebvre, A. W., *Gas Turbine Combustion*, Taylor and Francis, Washington, D.C., 1983.

Lehman, A. F., ''Model Studies of Helicopter Rotor Patterns,'' *Proceedings of the 24th Annual Forum of the American Helicopter Society*, 1968.

Leishman, J. G. and Beddoes, T. S., "A Semi-Empirical Model for Dynamic Stall," *J. of the American Helicopter Society*, Vol. 34, No. 3, pp. 3–17, 1989.

Leishman, J. G., "Modeling of Subsonic Unsteady Aerodynamics for Rotary Wing Applications," *J. of the American Helicopter Society*, Vol. 35, No. 1, pp. 29–38, 1990.

Leishman, J. G. and Bi, Nai-pei, "Measurements of a Rotor Flowfield and the Effects of a Fuselage in Forward Flight," *Vertica*, Vol. 14, No. 3, 1990.

Lieblein, S., "Experimental Flow in 2D Cascades," Chapter VI of *The Aerodynamic Design of Axial Flow Compressors*, reprinted as NASA SP36 in 1965 (originally NACA RME 56BO3), 1956.

Lieblein, S., "Experimental Flow in Two Dimensional Cascades," Chapter VI in *Aerodynamic Design of Axial Flow Compressors*, NASA SP-36, 1965.

Lindsay, W. T., Jr., "Behavior of Impurities in Steam Turbines," *Power Engineering*, Vol. 83, No. 5, 1979.

Loewy, R. G., "A Two-Dimensional Approximation to the Unsteady Aerodynamics of Rotary Wings," *J. Aero. Sciences*, Vol. 24, No. 2, 1957.

Longley, J. P. and Greitzer, E. M., "Inlet Distortion Effects in Aircraft Propulsion System Integration," *Steady and Transient Performance Prediction of Gas Turbine Engines*, AGARD Lecture Series LS-183, 1992.

Longley, J. P., "A Review of Non-Steady Flow Models for Compressor Stability," ASME Paper 93-GT-17, 1993 to be published in *J. Turbomachinery*.

Mangler, K. W. and Squire, H. B., "The Induced Velocity Field of a Lifting Rotor," British Aero. Res. Council, Rep. Memo. 2642, 1950.

McCarthy, J. H., "On the Calculations of Thrust and Torque Fluctuations of Propellers in Nonuniform Flow," David Taylor Ship R&D Center Report 1533, 1961.

McCormick, B. W., *Aerodynamics, Aeronautics and Flight Mechanics*, John Wiley & Sons, New York, 1979.

McCroskey, W. J. and Baeder, J. D., "Some Recent Advances in Computational Aerodynamics for Helicopter Applications," NASA TM 86777, 1985.

Mishina, H. and Gyobu, I., "Performance Investigations of Large Capacity Centrifugal Compressors," ASME 78-GT-3, 1978.

Mitchell, G. A. and Mikkelson, D. C., "Summary and Recent Results from the NASA Advanced High-Speed Propeller Research Program," NASA TM 82891, 1982.

Mizuno, Y. (Ed.), *The World Aircraft Year-Book 1992*, Kantosha, Tokyo, 1993.

Moore, M. J. and Sieverding, C. H., *Two Phase Steam Flow in Turbines and Separators*, McGraw-Hill, New York, 1976.

Moustapha, S. H., Carscallen, W. B., and McGeahy, J. D., "Aerodynamic Performance of a Transonic Low Aspect Ratio Turbine Nozzle," *J. Turbomachinery*, Vol. 115, pp. 400–408, 1993.

Moustapha, S. H., Kacker, S. C., and Tremblay, B., "An Improved Incidence Losses Prediction Method for Turbine Airfoils," *J. Turbomachinery*, Vol. 112, pp. 267–276, 1990.

Moustapha, S. H., Okapuu, U., and Williamson, R. G., "Influence of Rotor Blade Aerodynamic Loading on the Performance of a Highly Loaded Turbine Stage," *J. Turbomachinery*, Vol. 109, pp. 155–162, 1987.

Ng, S. L. and Brennen, C. E., "Experiments on the Dynamic Behavior of Cavitating Pumps," *J. of Fluids Engineering*, Vol. 100, pp. 166–176, 1978.

Ni, R. H. and Bogoian, J., "Prediction of 3D Multi-Stage Turbine Flow Field Using a Multiple-Grid Euler Solver," AIAA 89-0203, 1989.

Pampreen, R. C., "Small Turbomachinery and Fan Aerodynamics," ASME Paper No. 73-GT-6, 1973.

Payne, P. R., *Helicopter Dynamics and Aerodynamics*, Macmillan, New York, 1959.

Pitt, D. M. and Peters, D. A., "Theoretical Prediction of Dynamic Inflow Derivatives," *Vertica*, Vol. 5, 1981.

Piziali, R. A., "A Method for the Solution of Aeroelastic Response of Rotating Wings," *J. Sound and Vibration*, Vol. 4, No. 3, 1966.

Raabe, J., *Hydraulische Maschinen und Anlagen*, VDI-Verlag, Düsseldorf, 1989.

Reddy, T. S. R. and Kaza, K. R. V., "A Comparative Study of Some Dynamic Stall Models," NASA TM-88917, 1987.

Reneau, L. R., Johnston, J. P., and Kline, S. J., "Performance and Design of Two-Dimensional Diffusers," *J. Basic Engr.*, Vol. 89, 1967.

Rodgers, C., "A Diffusion Correlation for Centrifugal Compressor Impeller Stalling," ASME 78-GT-61, 1978.

Rodgers, C., "A Diffusion Factor Correlation for Centrifugal Impeller Stalling," *J. of Engineering Power*, Vol. 100, pp. 592–603, 1978.

Rodgers, C., "Efficiency of Centrifugal Compressor Impellers," *Performance and Prediction of Centrifugal Pumps and Compressors*, ASME, 1980.

Rodgers, C., "Impeller Stalling as Influenced by Diffusion Limitations," *J. Fluids Engr.*, Vol. 99, 1977.

Rodgers, C., "Influence of Impeller and Diffuser Characteristics and Matching on Radial Compressor Performance," *SAE Centrifugal Compressors*, 1961.

Rodgers, C., "Static Pressure Recovery Characteristics of Some Radial Vaneless Diffusers," *CASI Proceedings*, Vol. 30, 1984.

Rodgers, C., "Effect of Inlet Geometry on the Performance of Small Centrifugal Compressors," AIAA 88-2812, 1988.

Rodgers, C., "The Efficiencies of Single Stage Centrifugal Compressors for Aircraft Applications," ASME 91-GT-77, 1991.

Rodgers, C., "The Performance of Centrifugal Compressor Channel Diffusers," ASME 82-GT-10, 1982.

Rodgers, C. and Langworthy, R. A., "Design and Test of a Two Stage High Pressure Ratio Centrifugal Compressor," ASME 74-GT-137, 1974.

Rohlik, H. E., "Analytical Determination of Radial Inflow Turbine Design for Maximum Efficiency," NASA TN D-4384, 1966.

Rohrbach, C., Metzger, F. B., Black, D. M., and Ladden, R. M., "Evaluation of Wind Tunnel Performance Testings of an Advanced 45° Swept Eight-Bladed at Mach Numbers from 0.45 to 0.85," NASA CR 3505, 1982.

Ross, D., *Mechanics of Underwater Noise*, Pergamon Press, New York, 1976.

Rossow, V. J., "On the Inviscid Rolled-Up Structure of Lift Generated Vortices," *J. of Aircraft*, Vol. 10, No. 11, pp. 647–650, 1973.

Rothe, P. H., Runstadler, P. W., Jr., and Dolan, F. X., "Pump Surge Due to Two-Phase Flow," *Polyphase Flow in Turbomachinery*, ASME, 1978.

Rouse, H. and Ince, S., *History of Hydraulics*, Dover, New York, 1963.

SAE, *SAE Handbook*, Section 28, Society of Automotive Engineers, 1987.

Saintsbury, J. A. and Sampath, P., "A Review of Small Gas Turbine Combustion System Development," ASME 79-GT-136, 1979.

Savonius, S. G., "The S-Rotor and Its Applications," *Mechanical Engineering*, Vol. 53, pp. 333–338, 1931.

Sawyer, J. W. and Japikse, D. (Eds.), *Sawyer's Gas Turbine Engineering Handbook*, Vol. 1, 3rd ed., Turbomachinery International Publications, Norwalk, CT, 1985.

Sawyer, R. T. *et al.* (Eds.), *Sawyer's Gas Turbine Catalog 1975*, Volume 13, Gas Turbine Publications, Stamford, CT, 1975.

Scholz, N., "Aerodynamics of Cascades," AGARDograph No. 220, 1977. (Translated and revised Klein, A., from "Aerodynamic der Schaufelgitter," Verlag Braun, 1965.)

Senior, T., *Wind Turbine Siting and TV Reception Handbook*, Univ. of Michigan, National Technical Information Service, Springfield, VA, C00-2846-1, 1978.

Sheperd, D. G., *Principles of Turbomachinery*, Macmillan, New York, 1956.

Sieverding, C., "Secondary Flows in Straight and Annular Turbine Cascades," *Thermodynamics and Fluid Mechanics of Turbomachines*, Vol. 2, NATO ASI Series, 1985.

Simon, H. and Buelskaemper, A., "On the Evaluation of Reynolds Number and Surface Roughness Based on Systematic Experimental Investigations," *J. Eng. for Gas Turbines and Power*, Vol. 106, No. 2, pp. 489–501, 1984.

Skoch, G. J. and Moore, R. D., "Performance of Two 10 lb/sec Centrifugal Compressors with Different Blade and Shroud Thicknesses Operating Over a Range of Reynolds Numbers," NASA TM 10015, 1987.

Smith, S. F., "A Simple Correlation of Turbine Efficiency," *J. Royal Aero. Soc.*, Vol. 69, pp. 467–470, 1965.

Sovran, G. and Klomp, E. D., "Experimentally Determined Optimum Geometries for Rectilinear Diffusers with Rectangular Conical, or Annular Cross-Section," *Fluid Mechanics of Internal Flow*, Sovran, G. (Ed.), Elsevier Publishing, 1967.

Srinivasan, G. R., Baeder, J. D., Obayashi, S., and McCroskey, W. J., "Flowfield of a Lifting Rotor—A Navier–Stokes Simulation," *AIAA J.*, Vol. 30, No. 10, 1992.

Steltz, W. G., "The Critical and Two Phase Flow of Steam," *J. Engrg. for Power*, Vol. 83, Series A, No. 2, 1961.

Steltz, W. G., Evans, D. H., and Stahl, W. F., "The Aerodynamic Design of High Performance Low Pressure Steam Turbines," *Transactions of the Institute of Fluid Flow Machinery*, Polish Academy of Sciences, Vols. 70–72, 1976.

Steltz, W. G., Lindsay, W. T., Jr., and Lee, P. K., "The Verification of Concentrated Impurities in Low Pressure Steam Turbines," *J. Engrg. for Power*, Vol. 105, No. 1, 1983.

Steltz, W. G., Rosard, D. D., Maedel, Jr., P. H., and Bannister, R. L., "Large Scale Testing for Improved Reliability," American Power Conference, 1977.

Strawn, R. C. and Caradonna, F. X., "Conservative Full-Potential Model for Unsteady Transoic Flows," *AIAA J.*, Vol. 25, No. 2, 1987.

Strub, R. A., Bonciani, L., Borer, C. J., Casey, M. V., Cole, S. L., Cook, B. B., Kotzur, J., Simon, H., and Strite, M. A., "Influence of the Reynolds Number on the Performance of Centrifugal Compressors," *J. of Turbomachinery*, Vol. 109, No. 4, pp. 541–544, 1987.

Sturgess, G. J., McKinney, R., and Harford, S., "Modification of Combustor Stoichiometry Distribution for Reduced NO_x Emissions from Aircraft Engines," ASME 92-GT-108, 1992.

Takekuma, K., "Evaluation of Various Types of Nozzle Propellers and Reaction Fin as the Device for the Improvement of Propulsive Performance at High Block Coefficient Ships," *Proceedings of Shipboard Energy Conservation Symposium*, Soc. Naval Architects and Marine Engineers Publication SY-12, 1980.

Taylor, J. W. R., *Jane's All the World's Aircraft, 1987–1988*, Jane's Publishing Co., Ltd., London, 1987.

Thompson, D. E., "Propeller Time-Dependent Forces Due to Nonuniform Inflow," Ph.D. thesis, Pennsylvania State University, University Park, 1976.

Thresher, R. W., "Structural Dynamic Analysis of Wind Turbine Systems," *J. Solar Energy Engineering*, May, 1982.

Tsakonas, S., Breslin, J., and Miller, M., "Correlation and Application of an Unsteady Flow Theory for Propeller Forces," *Trans. Soc. Naval Architects and Marine Engrs.*, Vol. 75, pp. 158–193, 1967.

Turner, J., Lectures, Lehigh University, Bethlehem, PA, February–May, 1959.

United Technologies Corporation, *The Aircraft Gas Turbine Engines and Its Operation*, P&W Operation Instruction 200, Hartford, CT, 1988.

Upton, E. W., "Design Practices—Passenger Car Automatic Transmissions," Sec. II, Soc. of Automotive Engineers, 1994.

USAF, *USAF Stability and Control Datcom*, Flight Control Division, Air Force Flight Dynamics Laboratory, Wright–Patterson Air Force Base, OH, 1960, revised 1975.

Valentine, D. T. and Dashaw, F. J., "Highly Skewed Propeller for San Clemente Class Ore/Bulk/Oil Carrier Design Considerations, Model, and Full-Scale Evaluation," *Proceedings of First Ship Technology and Research Symposium*, Washington, D.C., 1975.

Volk, M. W., *Pump Characteristics and Applications*, Marcel Dekker, New York, 1996.

Waclawek, M. J., "Altering Hydrodynamic Torque Converter Performance," SAE Paper 730001, 1973.

Weinig, F., *Die Stromung um die Schaufeln von Turbomachinen*, Joh Ambr. Barth, Leipzig, 1935.

Weisner, F. R., "A Review of Slip Factors for Centrifugal Impellers," ASME 66-WA/FE-18, 1966.

Westinghouse, *Westinghouse Steam Charts*, Westinghouse Electric Co., 1969.

White, F. M., *Fluid Mechanics*, McGraw-Hill, New York, 1994.

Wilson, D. G., *The Design of High Efficiency Turbomachinery and Gas Turbine Engines*, MIT Press, Cambridge, 1984.

Wilson, M. B., McCallum, D. N., Boswell, R. J., Bernhard, D. D., and Chase, A. B., "Causes and Corrections for Propeller-Excited Airborne Noise on a Naval Auxiliary Oiler," *Trans. Soc. of Naval Architects and Marine Engineers*, Vol. 90, 1982.

Wilson, R. E. and Lissaman, P. B. S., *Aerodynamic Performance of Wind Turbines*, Oregon State Univ., 1976.

Wisler, D. C., Koch, C. C., and Smith, L. H., "Preliminary Design Study of Advanced Multistage Axial Flow Core Compressor," NASA CR 135133, 1977.

Wisler, D. C., *Advanced Compressor and Fan Systems*, GE Aircraft Engines, Cincinnati Ohio, Copyright by General Electric Co. USA (also 1986 lecture to ASME Turbomachinery Institute, Ames, IA) 1988.

Wislicenus, G. F., "Hydrodynamic Design Principles of Pumps and Ducting for Waterjet Propulsion," Publication #3990, David W. Taylor Ship R and D Center, Bethesda, MD, 1973.

Wright, T. and Ralston, S. A., "Computer-Aided Design of Axial Flow Fans Using Small Computers," *ASHRAE Transactions*, Vol. 93, Part 2, No. 3072, 1987.

Wright, T., "A Closed Form Algebraic Approximation to Quasi-Three-Dimensional Flow in Axial Fans," ASME Paper No. 88-GT-15, 1988.

Wright, T., "Comments on Compressor Efficiency Scaling with Reynolds Number and Relative Roughness," ASME Paper 89-GT-31, 1989.

Wu, C. H., "A General Theory of Three-Dimensional Flow in Subsonic and Supersonic Turbomachines of Axial-, Radial-, and Mixed-Flow Types," NASA TN 2604, 1952.

Wylie, E. B. and Streeter, V. L., *Fluid Transients in Systems*, Prentice Hall, Englewood Cliffs, NJ, 1993.

Zweifel, O., "The Spacing of Turbomachine Blading Especially with Large Angular Deflection," *Brown Boveri Rev.*, Vol. 32, pp. 436–444, 1945.

28 Hydraulic Systems

HUGH R. MARTIN
University of Waterloo
Waterloo, Ontario, Canada

CONTENTS

28.1 HYDRAULIC FLUIDS

One of the results of the study of fluid mechanics has been the development of the use of hydraulic oil, a so called incompressible fluid, for performing useful work. The use of a fluid to transmit power has been utilized for many centuries, where the most available fluid was water. While this fluid is cheap and usually readily available, it does have the distinct disadvantages of promoting rusting, of freezing to a solid, and of having relatively poor lubrication properties.

The introduction of mineral oils have provided for superior properties. Much of the success of modern hydraulic oils is due to the relative ease with which their properties can be altered by the use of additives, for example rust and foam inhibitors, without significantly changing fluid characteristics.

Handbook of Fluid Dynamics and Fluid Machinery, Edited by Joseph A. Schetz and Allen E. Fuhs
ISBN 0-471-12598-9 Copyright © 1996 John Wiley & Sons, Inc.

Although hydraulic oil is used mainly to transmit fluid power, it must also: 1) provide lubrication for moving parts, such as spool valves, 2) absorb and transfer heat generated within the system, and 3) remain stable, both in storage and in use, over a wide range of possible physical and chemical changes.

It is estimated that 75% of all hydraulic equipment problems are directly related to the improper use of oil in the system. Contamination control in the system is a very important aspect of circuit design.

In certain industries such as mining, and nuclear power, it is critically important to control the potential for fire hazards. Hence, fire resistant fluids have been playing an ever increasing role in these types of industry. The higher pressure levels in modern fluid power circuits have made fire hazards more serious when petroleum oil is used, since a fractured component or line will result in a fine mist of oil which can travel as far as 40 ft. and is readily ignited. The term fire resistant fluid (FRF) generally relates to those liquids which fall into two broad classes: a) those where water provides the fire resistance, and b) those where a fire retardant is inherent in the chemical structure [Stanley (1971), Dalibert (1971), Kelly (1971), and Louie *et al.* (1981)]. Fluids in the first group are water/glycol mixtures, water in oil emulsions (40 to 50% water) and oil in water emulsions (5 to 15% water). The second group are synthetic materials, in particular chlorinated hydrocarbons and phosphate esters.

A disadvantage with water based fluids is that they are limited to approximately 50–60 C operating temperature because of evaporation. The high vapor pressure indicates this group is more prone to cavitation than mineral oils. Synthetic fluids such as the phosphate esters do not have this problem and also have far superior lubrication properties. Some typical characteristics of these various types of fluids are shown in Table 28.1.

Of all the physical properties that can be listed for hydraulic fluids, the essential characteristics of immediate interest to a designer are: 1) bulk modulus, to assess system rigidity and natural frequency, 2) viscosity, to assess pipe work and component pressure losses, 3) density, to measure flow and pressure drop calculations, and 4) lubricity, to determine threshold and control accuracy assessments. The first three items are discussed in separate sections as they relate directly to circuit design. *Lubricity*, the final item, is difficult to define as it is very much a qualitative judgment. Lubricity affects the performance of a system, since it is a major factor in determining the level of damping in the system, that is *viscous* or velocity dependent

TABLE 28.1   Comparison of Some Hydraulic Fluids

Property	Units	FRF (Ester)	Mineral Oil	Water in Oil
Density (38 C)	kg m^{-3}	1136.0	858.2	980.0
Viscosity (38 C)	m^2 s^{-1}	4.6×10^{-6}	4.0×10^{-5}	0.15×10^{-5}
(99 C)		4.9×10^{-6}	5.8×10^{-6}	
Bulk modulus (38 C and 34.5 MPa)	N m^{-2}	2.25×10^9	1.38×10^9	2.18×10^9
Vapor pressure	kPa (abs)	6×10^{-5}	6×10^{-5}	1.0

damping. It also affects the accuracy of operation of a system because of its influence on the other type of friction, *coulomb friction*, which is velocity independent.

Oil film strength is often referred to as the *anti-wear value* of a lubricant which is the ability of the fluid to maintain a film between moving parts and thus prevent metal to metal contact. These characteristics are important for the moving parts in valves, cylinders, and pumps [Anon., *Hydraulics & Pneumatics* (1980)].

28.2 CONTAMINATION CONTROL

There is little doubt that component failure or damage due to fluid contamination is an area of major concern to both the designer and user of fluid power equipment. Sources of contamination in fluid power equipment are many. Although oil is refined and blended under relatively clean conditions, it does accumulate small particles of debris during storage and transportation. It is not unusual for hydraulic oil circulating in a well maintained hydraulic circuit to be cleaner than that from a newly purchased drum. New components and equipment invariably have a certain amount of debris left from the manufacturing process, in spite of rigorous post production flushing of the unit.

The contaminant level in a system can be increased internally due to local burning (oxidation) of oil to create sludges. This can be a result of running the oil temperature too high, normally 40 to 60 C is recommended, or due to local cavitation in the fluid.

The trend towards the use of higher system pressures in hydraulics generally results in narrower clearances between mating components. Under such design conditions, quite small particles in the range 2 to 20 micron can block moving surfaces.

Extensive work on contamination classification has been carried out by Fitch and his co-workers [Fitch *et al.* (1978)].

To take a specific example, consider the piston pump shown in Fig. 28.1. Component parts of the pump are loaded towards each other by forces generated by the pressure, and this same pressure always tends to force oil through the adjacent clearance. The life of the pump is related to the rate at which a relatively small amount of material is being worn away from a few critical surfaces. It is logical to assume therefore, if the fluid in a clearance is contaminated with particles, rapid degradation and eventual failure can occur.

Although the geometric clearances are fixed, the actual clearances vary with eccentricity due to load and viscosity variations. Some typical clearances between moving parts are shown in Table 28.2.

Contamination control is the job of filtration. System reliability and life are related not only to the contamination level but also to contaminant size ranges. To maintain contaminant levels at a magnitude compatible with component reliability requires both the correct filter specification and suitable placement in the circuit. Filters can be placed in the suction line, pressure line, return line, or in a partial flow mode. To use a broad approach of just inserting a filter with a very low rating is unsatisfactory from both the aspects of cost and high pressure loss. The optimization of choice can be approached using simple computer modeling as described by Foord (1978).

Dirt in hydraulic systems consists of many different types of material ranging in

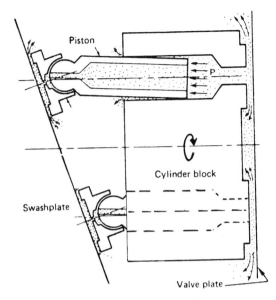

FIGURE 28.1 Piston pump clearances.

size from less than 1 micron to greater than 100 microns. Most general industrial hydraulics operating below 14 MPa are able to tolerate particles up to 25 microns, therefore a 25 micron-rated filter is satisfactory. Equipment operating at pressures in the 14–21 MPa range should have 10–15 micron-rated filters, while high pressure pumps and precision servo valves need 5 micron-rated filtration. A good practical reference for filter selection has been written by Spencer (1980).

The size distribution of particles is of course random, and, generally speaking, the smaller the size range the greater the number of particles per 100 ml of fluid. Filters are not capable of removing all the contaminants, but for example, a 10 micron filter is one capable of removing about 98% of all particles exceeding 10 microns of a standard contaminant in a given concentration of prepared solution.

28.3 POSITIVE ASPECTS OF CONTAMINATION

Contamination build up in a system can be used as a diagnostic tool. Regular sampling of the oil and examination of the particles can often give a clue to potential failure of components. In other words, this is a preventive maintenance tool. Many methods can be used for this type of examination such as spectrochemical [Forgeron

TABLE 28.2 Typical Clearances in Pumps

Component	Clearance Range (micron)
Spool to sleeve in valve	1–10 diametrical
Gear pump tip to casing	0.5–5
Piston to bore	5–40
Valve plate to body of pump	0.5–5

TABLE 28.3 Some Typical Normal Contaminant Levels

Material	Source in System	Max Level (ppm)
Iron	Bearings, gears, or pipe rust. Pistons and valve wear.	20
Chromium	Alloyed with bearing steel	4
Aluminum	Air cooler equipment	10
Copper	Bronze or brass in bearings. Connectors. Oil temperature sensor bulb. Cooler core tubes	30
Lead	Usually alloyed with copper or tin. Bearing cage metal	20
Tin	Bearing cages and retainers	15
Silver	Cooling tube solder	3
Nickel	Bearing steel alloy	4
Silicon	Seals; dust and sand from poor filter or air leak	9
Sodium	Possible coolant leak into hydraulic oil	50

and McCormack (1978)] or Ferrographic [Tessman (1978)] methods. Sampling of the oil can be taken at any time and does not interfere with the operation of the equipment.

Table 28.3 shows the normally expected contaminant levels in parts per million (ppm); levels rising above these values and particularly rates of change of levels are indicative of potential failures.

The Ferrographic technique allows the separation of wear debris and contaminants from the fluid and allows arrangement as a transparent substrate for examination. When wear particles are precipitated magnetically, virtually all nonmagnetic debris is eliminated. The deposited particles deposit according to size and may be individually examined. By this method it is possible to differentiate cutting wear, rubbing wear, erosion, and scuffing by the size and geometry of the particles. However, the Ferrographic method is expensive compared to other methods of analysis [Collacott (1977)].

28.4 DESIGN EQUATIONS—ORIFICES AND VALVES

The main controlling element in any hydraulic circuit is the orifice. It is the fluid equivalent of the electrical resistance and can be fixed in size or can be variable in the case of a spool valve. The orifice in its various configurations is also the main source of heat generation, resulting in the need for cooling techniques and a major source of noise.

The orifice equation is developed from Bernoulli's energy balance approach, which results in the following relationship [Martin and McCloy (1980)].

$$Q = \frac{C_c C_v A_o}{\sqrt{1 - \left(\frac{C_c A_o}{A_u}\right)^2}} \sqrt{\frac{2(p_u - p_{vc})}{\rho}} \qquad (28.1)$$

where

Q = volume flow rate, m³/s
A_o = orifice area, m²
A_u = upstream area, m²
p_u = upstream static pressure, Pa
p_{vc} = static pressure at *Vena contracta*, Pa
C_c = contraction coefficient
C_v = velocity coefficient
ρ = mass density of hydraulic fluid, kg/m³

These parameters are shown in Fig. 28.2, together with the static pressure distribution on either side of a sharp edged orifice. Experimental measurements show that
the actual flow is about 60% of that given by Bernoulli's equation. Hence, the need

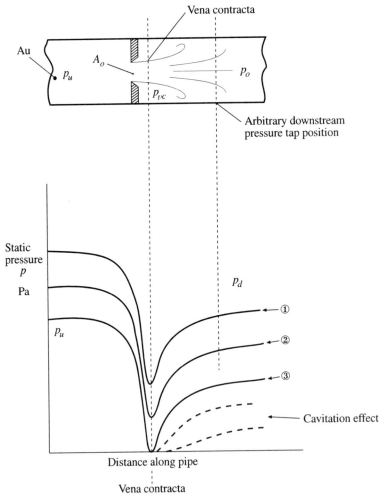

FIGURE 28.2   Static pressure distribution.

TABLE 28.4 Comparing Experimental Data to Predictions of Eq. (28.2)

Supply Pressure = 13.78 MPa   Valve overlap = ±0.0127 mm

Valve Displacement (mm)	Calculated Flow (ml/sec)	Measured Flow (ml/sec)
+0.3810	104.140	—
+0.3048	—	79.540
+0.2540	56.088	59.860
+0.2032	40.180	45.264
+0.1270	20.172	24.272
+0.0508	4.920	6.560
+0.0254	1.968	0.820
Center	0	0
−0.0254	1.968	0.820
−0.0508	4.920	4.264
−0.1270	20.172	20.992
−0.2032	40.180	41.000
−0.2540	56.088	58.384
−0.3048	—	76.588
−0.3810	104.14	—

for the contraction and velocity coefficients. This results in the practical form of Eq. (28.1) for typical industrial hydraulic oil

$$Q = 3.12 \times 10^{-2} A_o \sqrt{p_u - p_d} \ \mathrm{m^3 s^{-1}} \tag{28.2}$$

The symbols have the same definition as those for Eq. (28.1). The adequacy of Eq. (28.2) is demonstrated in Table 28.4.

In the case of a variable orifice, such as that found in a spool type valve, the orifice area is a variable. In fact, it can be seen from Fig. 28.3 that the exposed area available for oil flow is part of a circle. If the orifice, in this case called a control orifice or port, is of radius r and the spool displacement from the closed position is x, then the uncovered area is

$$A_o = [\theta - \cos(\theta/2)] \left(\frac{r^2}{2}\right) \tag{28.3}$$

$$\theta = 2 \cos^{-1}[1 - (x/r)] \tag{28.4}$$

The area displacement characteristic plotted in Fig. 28.3 shows the nonlinear nature of the curve.

One of the significant differences between the theoretical valve and the practical valve is the *lap*. It is not economical to produce zero lapped valves, so that only at the center position is the flow through the valve zero. Normally, the valve is either overlapped or underlapped as shown in Fig. 28.4. An overlapped valve saves fluid loss when the spool is central. This is fine for directional control valves, but it produces both accuracy and stability problems if the valve is a precision control valve within a closed loop configuration.

An underlapped valve gives much better control and stability, at the expense of

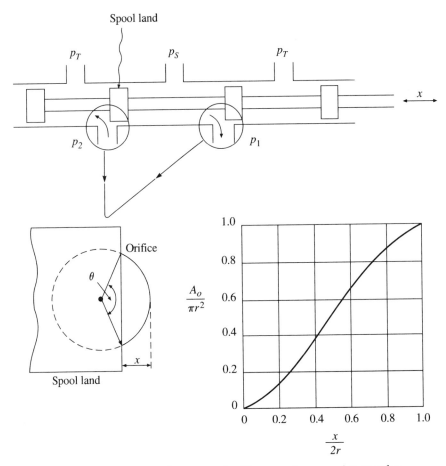

FIGURE 28.3 Effective exposed orifice area for a spool-type valve.

a higher leakage rate (power loss). Many more details of valve design can be found in Martin and McCloy (1980).

28.5 DESIGN EQUATIONS—PIPES AND FITTINGS

While orifices serve the important function of controlling flow in the system, pipes and fittings are necessary to transmit fluid power from the input (usually a pump) to the output (usually a ram or motor). It is important to minimize losses through these conductors as well as through other components so that the maximum power is available for useful work at the circuit output. It is equally important to minimize component and piping cost. In some applications, it is also important to minimize weight and bulk size.

Pipe sizes are specified by nominal diameters, and the wall thickness by schedule number. The three schedules (or wall thicknesses) used in hydraulic piping are 40, 80, and 160 corresponding to *standard* pipe, *extra heavy*, and a little less than *double extra heavy*. The metric system of units has helped to complicate things for the

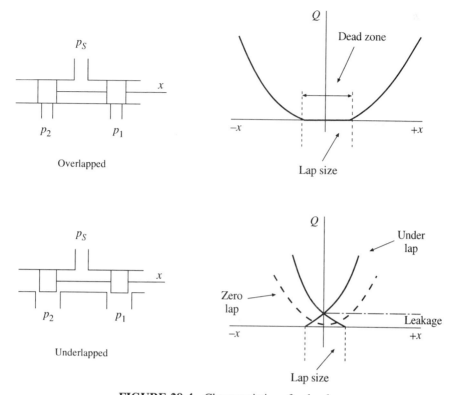

FIGURE 28.4 Characteristics of valve lap.

designer during this transition period, more details can be found in Martin and McCloy (1980).

In the selection of piping for hydraulic circuits the following are suggested: a) suction lines to pumps should not carry fluid at velocities in excess of 1.5 m/s in order to reduce the possibility of cavitation at the pump inlet, b) delivery lines should not carry fluid at velocities in excess of 4.5 m/s in order to prevent excessive shock loads in the pipework due to valve closure. Pressure loss due to friction in pipes should be limited to approximately 5% of the supply pressure, and the recommendation also keeps heat generation to a reasonable level, and c) return lines should be of larger diameter than delivery lines to avoid back pressure build-up.

For typical industrial hydraulic oil, we can write

$$\Delta p = K_L/K_1 Q^2 \tag{28.5}$$

where

Δp = pressure drop along a straight pipe (kPa)
K_L = loss coefficient = $f\ell/d$
  f = friction factor (see Chapter 1)
  ℓ = pipe length (m)
  d = internal pipe diameter (m)

TABLE 28.5   Coefficients for Eq. 28.5 and 28.6

Nominal Bore		S.I. System		Old System		Pipe Area	
mm	in.	K_1	K_2	K_1	K_2	m^2	$in.^2$
8	$\frac{1}{4}$	1.027×10^{-11}	138	12.64	10379.12	6.64×10^{-5}	0.1041
10	$\frac{3}{8}$	3.506×10^{-11}	102	42.26	7663.28	12.27×10^{-5}	0.1909
15	$\frac{1}{2}$	8.949×10^{-11}	81	108.08	6073.93	19.60×10^{-5}	0.3039
20	$\frac{3}{4}$	2.740×10^{-10}	61	333.0	4584.95	34.30×10^{-5}	0.5333
25	1	7.195×10^{-10}	47	874.66	3601.52	55.57×10^{-5}	0.8643
32	$1\frac{1}{4}$	2.181×10^{-9}	36	2620.49	2737.68	96.76×10^{-5}	1.496
40	$1\frac{1}{2}$	4.014×10^{-9}	31	4854.14	2340.66	13.13×10^{-4}	2.036
50	2	1.090×10^{-8}	24	13180.81	1823.64	21.65×10^{-4}	3.355

K_1 = see Table 28.5
Q = flow rate (m³/s)

The friction factor f has been shown experimentally to be a function of Reynolds number (Re) and of pipe roughness (see Chap. 1). The Reynolds number for industrial hydraulic calculations can be calculated from

$$Re = \frac{K_2}{\nu} Q \tag{28.6}$$

where the kinematic viscosity, ν, has a typical value of 4.0×10^{-5} m²/s, and K_2 can be found in Table 28.5.

Given the flow through a section of straight pipe the procedure to calculate the pressure loss is simple. Using Eq. 28.6 and ν given above calculate the Reynolds number. Using the Reynolds number to calculate the appropriate value for f, calculate a value for K_L. Referring to Table 28.5 for K_1, the pressure drop can be calculated using Eq. 28.5.

Unfortunately, not all piping is in straight runs so when a bend occurs the loss of pressure will be greater. The effective bend loss can be estimated from Eq. 28.7 and Figs. 28.5 and 28.6. These results are from AFRPL (1964)

$$\Delta p = (K + cK_B)Q^2/K_1 \tag{28.7}$$

where

c = correction factor for bend angle (Fig. 28.5)
K_B = resistance coefficient for 90° bends (Fig. 28.6)

Further useful information about circuit design can be found in Keller (1974).

28.6   HYDROSTATIC PUMPS AND MOTORS

The source of power in a hydraulic circuit is the result of *hydrostatic* flow under pressure with the energy being transmitted by static pressure. Another type of fluid power is termed *hydrokinetic* in which the transmission of energy is related to the

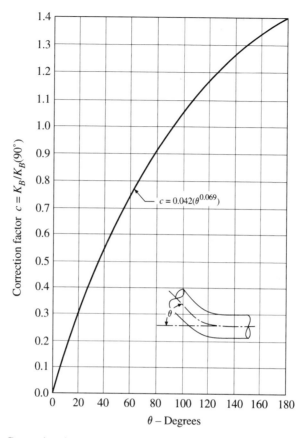

FIGURE 28.5 Correction factor c. (Reproduced from AF Rocket Propulsion Lab., 1964.)

change in velocity of the hydraulic fluid. While hydrostatic systems use positive displacement pumps, hydrokinetic systems use centrifugal pumps [Thomas (1970)].

Positive displacement machines have been in existence for many years (see Chap. 26). The concept is simply a variable displacement volume which can take the form of a piston in a cylinder, gear teeth engaging, or the sweeping action of a vane with eccentric axis placement. All these configurations are positive displacement in the sense that for each revolution of the pump shaft, a nearly constant quantity of fluid is delivered. In addition, there is some form of valving which either takes the form of nonreturn valves or a porting arrangement on a valve plate.

Examples of different types of pumps are shown in Figs. 28.7, 28.8, and 28.9. While torque and speed are the input variables to a pump, the output variables are pressure and flow. The product of these variables will give the input and output power. The difference between these values is a measure of the fluid and mechanical losses through the machine. These factors should be taken into account even for a simple analysis.

The torque required to drive the pump at constant speed can be divided into 5 components

$$T_p = T_i + T_v + T_f + T_c \tag{28.8}$$

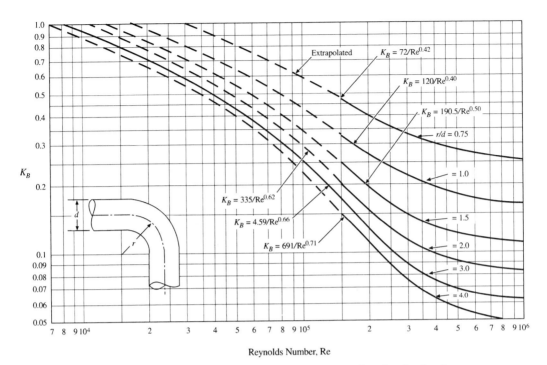

FIGURE 28.6   Correction factor K_B for pressure loss in pipe bends.

Where

T_p = actual required input torque (Nm)
T_i = ideal torque due to pressure differential and physical dimensions only
T_v = resisting torque due to viscous shearing of the fluid between stationary
        and moving parts of the pump, that is, viscous friction

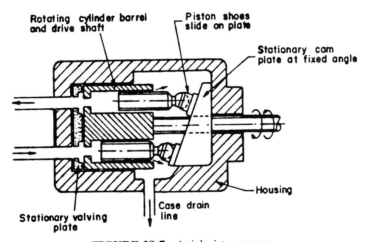

FIGURE 28.7   Axial piston pump.

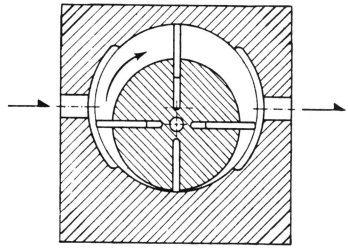

FIGURE 28.8 Schematic cross section through a vane pump. Reprinted with permission, Thoma, J., *Modern Oil Hydraulic Engineering*, Technical and Trade Press, 1970.

T_f = resisting torque due to pressure and speed dependent friction sources such as bearings and seals

T_c = remaining dry friction effects due to rubbing

The delivery from the pump can be expressed in a similar manner

$$Q_p = Q_i - Q_l - Q_r \tag{28.9}$$

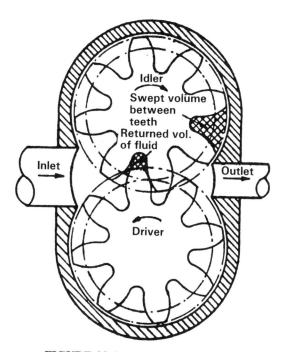

FIGURE 28.9 Gear pump construction.

where

Q_p = actual pump delivery (ml/sec)
Q_i = ideal delivery of a pump due to geometric shape only.
Q_l = viscous leakage flow
Q_r = loss in delivery due to inlet restriction [Wilson (1950)]

If the pump is well designed and operating under its working specification, the loss represented by Q_r should not occur.

For a hydraulic motor, the procedure is reversed in the sense that flow and pressure are the input variables, and torque with angular velocity appears at the output. The corresponding equations are therefore

$$T_p = T_i - T_v - T_f - T_c \tag{28.10}$$

$$Q_p = Q_i + Q_l \tag{28.11}$$

Q_r is not a factor in motor performance.

The ideal positive displacement machine displaces a given volume of fluid for every revolution of the input shaft. This value is given the name *displacement* of the pump or motor and is extensively used by manufacturers to label the pump size. Some typical characteristics for a hydraulic radial piston motor are shown in Fig. 28.10 and Table 28.6.

If the pump or motor rotates at N rpm, then

$$Q_i = D_p N \tag{28.12}$$

where

D_p = swept volume per revolution = nV
V = swept volume per cylinder per revolution
n = number of cylinders in the pump or motor

The leakage term Q_l can be expressed in terms of a leakage coefficient C_s which is sometimes called the slip coefficient

$$Q_l = \frac{\Delta p D_p C_s}{\mu} \tag{28.13}$$

For most designs, the slip coefficient is proportional to the cube of typical clearances within the machine [Turnball (1976)].

While the volume of fluid theoretically pumped per revolution can be calculated from the geometry of the design, in practice, a pump does not deliver that amount. The *volumetric efficient*, η_V, is used to assess this characteristic and is essentially a measure of the quality of machining or of wear in a pump.

$$\eta_V = \frac{Q_p}{Q_i} = \frac{Q_i - Q_l}{Q_i}$$

$$= 1 - \frac{Q_l}{Q_i} = 1 - \frac{\Delta p}{\mu N} C_s \tag{28.14}$$

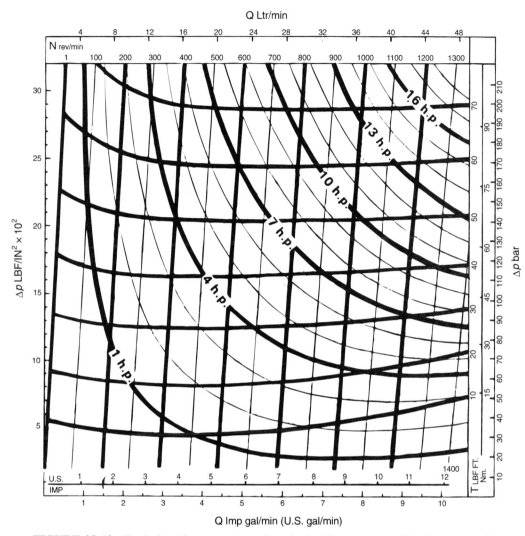

FIGURE 28.10 Typical performance range for a hydraulic motor, specifications appear in Table 28.6. (Courtesy of Kontak Manufacturing, Lincolnshire, England.)

The value of η_V can vary from about 75% up to 97%. In general, the cheaper the pump, the lower the volumetric efficiency.

The theoretical applied torque to a pump is given by

$$T_i = \Delta p D_p / 2\pi \qquad (28.15)$$

In this case, the losses are assessed by the viscous drag coefficient, C_d, which is inversely proportional to the typical pump clearances, and by the drag coefficient C_f, which is proportional to the size of the pump. Referring to Eq. 28.10, T_c in a well designed pump is normally small enough to ignore.

$$T_v = C_d D_p \mu N \qquad (28.16)$$

TABLE 28.6   Typical Performance Specification for a Radial Piston Motor

Displacement	in.³/rev (cm³/rev)	1.99 (32.6)
Max. torque	lbf · ft (Nm)	74 (100)
Max. recommend speed	rev/min	1360
Max. output	h.p.	18
Max. recommended flow rate		10 imp.gal/min (45 ltr/min)
		(12 U.S.gal/min)
Max. pressure		3000 lbf/in.² (207 bar)
Approx. overall efficiency		85%
Max. back pressure (reversible)		3000 lbf/in.² (207 bar)
Max. drain line pressure (reversible)		50 lbf/in.² (3.5 bar)
Max. back pressure (uni-directional)		50 lbf/in.² (3.5 bar)
Weight		15 lb (6.8 kg)
Max. permissible shaft end load		2000 lbf (907 kgf)
Max. permissible shaft side load (3/4 in. (19 mm) from shaft end)		2000 lbf (907 kgf)

Acknowledgments to Kontak Manufacturing, Lincolnshire, England.

$$T_f = C_f \frac{\Delta_p D_p}{2\pi} \tag{28.17}$$

Wilson (1950) gives guidance as to the magnitude of these coefficients, and his figures are given in Table 28.7.

The *mechanical efficiency*, η_m, of the device is a measure of the power wasted in friction. A reduction in the mechanical efficiency could, for example, be an indicator of bearing failure due to lack of lubrication.

$$\eta_m = \frac{T_i}{T_p} = \frac{T_i}{T_i + T_v + T_f}$$

$$\eta_m = \frac{1}{1 + \dfrac{2\pi C_d \mu N}{\Delta p} + C_f} \tag{28.18}$$

Finally, the *overall efficiency* of a pump or motor is the product of the volumetric and mechanical efficiencies. In general, gear pumps are suitable for pressures up to 17 MPa and have overall efficiencies of approximately 80%. A good quality piston pump has an overall efficiency of 95% and is capable of operating with pressures up to 68.9 MPa.

TABLE 28.7   Typical Hydraulic Pump Coefficients

Pump Type	D_p, in.³/rev	C_d	C_s	C_f
Piston	3.600	16.8×10^4	0.15×10^{-7}	0.045
Vane	2.865	7.3×10^4	0.477×10^{-7}	0.212
Spur gear	2.965	10.25×10^4	0.48×10^{-7}	0.179
Internal gear	2.965	9.77×10^4	1.02×10^{-7}	0.045

28.7 STIFFNESS IN HYDRAULIC SYSTEMS

One important and often neglected aspect of hydraulic circuit design is the fact that in practice hydraulic oil is compressible. So far, only steady flow through the circuit has been discussed. However, when a demand for flow is changed or a valve is shut, flow and pressure in the system became subject to the rates of change. Under these conditions, natural modes of resonance can be excited which can result in seemingly endless problems ranging from excessive noise to fatigue failures.

Referring to Fig. 28.11, an increase in the applied force F to the piston will cause the volume of trapped oil to compress according to the relationship

$$-\Delta V = \frac{(p_2 - p_1)V_0}{\beta} \tag{28.19}$$

where

$p_1 = F1/A =$ steady initial pressure
$p_2 = F2/A =$ steady final pressure
$V_0 =$ original volume
$\beta =$ bulk modulus of oil (see Table 2.1)

The negative sign is to indicate that the oil volume reduces as the applied pressure increases. It is assumed that the walls of the container are rigid.

Although the change in volume is small with a value of about 0.5% per 7 MPa applied pressure, it does result is high transient flow rates. As a comparison, air compresses about 50% for a pressure change of 0.1 MPa (1 atmosphere). The transient flow rates due to oil compressibility effects can be estimated from the first derivative of Eq. 28.19

$$Q_c = \frac{V_0}{\beta} \frac{d\Delta p}{dt} \tag{28.20}$$

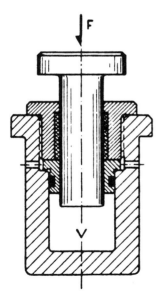

FIGURE 28.11 Pressure chamber for measuring compressibility. Reprinted with permission, Thoma, J., *Modern Oil Hydraulic Engineering*, Technical and Trade Press, 1970.

The actual value of oil bulk modulus is strongly dependent on the amount of air present in the form of bubbles. In practice, it is impossible not to have some level of air entrainment. The effective bulk modulus can then be estimated using

$$\beta_e = \frac{1}{\left[\dfrac{1}{\beta_0} + \dfrac{\alpha}{p} \right]} \qquad (28.21)$$

where

β_0 = oil bulk modulus with no air present
α = ratio of air volume to oil volume (typically 0.5%)
p = operating oil pressure.

These effects are illustrated in Fig. 28.12. The bulk modulus of the oil and the entrained air contribute to the effective spring a hydraulic system exhibits. For example, the hydraulic braking system of a vehicle feels spongy if there is air in the brake fluid, as a result of the circuit not being *bled* correctly.

The third factor in the system *stiffness* is the contribution from the containment vessel which in this case is the steel pipework or reinforced rubber hose [Martin (1981)]. For a thin walled metal pipe the effective bulk modulus is estimated from [Merritt (1967)]

$$\beta_c = \frac{TE}{D} \qquad (28.22)$$

where

T = wall thickness, m
E = modulus of elasticity, Pa
D = pipe diameter, m

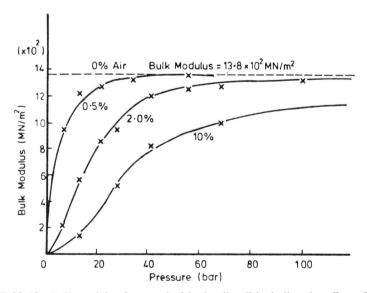

FIGURE 28.12   Bulk modulus for a typical hydraulic oil including the effect of free air.

When the pipeline is a hydraulic hose, there is some difficulty in obtaining design information. Values for β_c in the range 6.8×10^7 to 7.7×10^8 Pa have been quoted. Some further guidance is given in Martin (1981).

The total effective system bulk modulus taking all these effects into account can be calculated from:

$$\beta_e = \left[\frac{1}{\beta_0} + \frac{\alpha}{p} + \frac{D}{TE} \right]^{-1} \tag{28.23}$$

The effective stiffness of a system is important when the designer is concerned with reverse loading. For example, consider a hydraulic ram controlling a metal cutting tool. The loads on the tool can vary as it cuts through metal. If the hydraulic system is not very stiff, the tool will move about giving a poor finish. Obviously there will be some movement, as it is not possible to design an infinitely stiff system. However, a high stiffness will make any such tendency to move very little.

The second problem due to system stiffness is related to dynamic behavior. Since hydraulic machines have moving parts, there will be masses and inertias to accelerate. The interaction of mass with stiffness results in natural resonant modes. These natural frequencies are normally passive, but, if excited by a power source of comparable frequency, the result can be significant noise and vibration, or in the extreme, structural failure. It is very important, therefore, for the designer to estimate these passive modes at the design stage.

Consider the case of a simple ram shown in Fig. 28.13, which is used to position a mass M. It can be shown in Martin and McCloy (1980) that the flow into the ram is

$$Q_1 = \frac{V_1}{\beta_1} \frac{dp_1}{dt} + A \frac{dx_0}{dt} \tag{28.24}$$

In other words, the first term on the right-hand side of Eq. (28.24) represents the contribution of compressibility to the total oil flow. If the flow is steady, this term disappears. The second term is the more commonly recognized flow into the ram as the piston bore volume geometry changes.

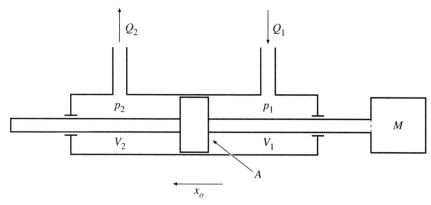

FIGURE 28.13 Compressibility effects in a cylinder.

A similar argument can be applied to the left-hand side of the ram where the oil is being pushed out

$$Q_2 = -\frac{V_2}{\beta_2}\frac{dp_2}{dt} + A\frac{dx_0}{dt} \qquad (28.25)$$

The sign change is to differentiate between oil that is being compressed and oil that is expanding. The average flow through the ram can now be estimated by combining Eqs. (28.24) and (28.25). In practice it is unlikely that there are two different fluids in the ram unless it's an air–oil system. Therefore, $\beta_1 = \beta_2 = \beta$. If V is the swept volume of the ram, then $V_1 = V_2 = V/2$ for the piston control. This results in the load flow equation

$$Q = \frac{V}{4\beta}\frac{d(p_1 - p_2)}{dt} + A\frac{dx_0}{dt} \qquad (28.26)$$

Now, the pressure drop across the piston $(p_1 - p_2)$ is, in this case, used to accelerate the mass attached to this piston rod.

$$p_1 - p_2 = \frac{M}{A}\frac{d^2x_0}{dt^2} \qquad (28.27)$$

Combining Eqs. (28.26) and (28.27) gives

$$Q = \frac{VM}{4\beta A}\frac{d^3x_0}{dt^3} + A\frac{dx_0}{dt}$$

Operating on both sides with the Laplace operator yields

$$\int_0^\infty Q(t)e^{-st}\,dt = \frac{VM}{4\beta A}\int_0^\infty e^{-st}\frac{d^3x_0}{dt}\,dt + A\int_0^\infty \frac{dx_0}{dt}e^{-st}\,dt$$

With the appropriate initial conditions, namely $x_0 = dx_0/dt = 0$ at $t = 0$, the result is

$$\overline{Q}(s) = A\left[\frac{VM}{4\beta A^2}s^2 + 1\right]S\overline{x}_0(s) \qquad (28.28)$$

The bar over the symbol denotes the Laplace transform of the function, and s is the Laplace transform independent variable. The term inside the square brackets of Eq. (21.28) can be compared directly with the general equation for a second order system with zero damping ratio. Hence, the term $VM/4\beta A^2$ is the reciprocal of the system natural frequency squared.

$$f_n = \frac{1}{2\pi}\sqrt{\frac{\text{Stiffness}}{\text{mass}}} = \frac{1}{2\pi}\sqrt{\frac{4\beta A^2}{V}\cdot\frac{1}{m}} \qquad (28.29)$$

If this system is excited, say by an impact on the mass, it will oscillate at the frequency defined by Eq. (28.29). Since there is no damping term, the mass would oscillate continuously. In practice there will always be some damping, however small, from seal rubbing and oil film shear. The difficulty for the designer is to make some meaningful guesses as to what value to use. For example, one might use 15% of the stall load on the ram divided by the maximum velocity of the piston as a first guess.

28.8 SYSTEM CLASSIFICATIONS

Most hydraulic circuits, regardless of the application, fall into one or two general classifications [*Fluid Power Designers Handbook*]. The two major divisions are constant flow and constant pressure depending on whether the output is mainly a function of flow, (i.e., velocity, displacement, or acceleration) or mainly related to pressure, (i.e., force or torque).

The simplest hydraulic circuits fall into the constant flow classification with open center valving. A simple example of this is shown in Fig. 28.14(a). It is open center in the sense that when the control valve is centered the fluid is circulated directly back to tank. This method ensures minimum power loss and fluid heating in the quiescent periods. Compare this with Fig. 28.14(b) which is a simple constant flow circuit in which oil is dumped through the relief valve at the end of the ram stroke. The restriction of the orifice in the relief valve generates high levels of heat and noise thereby wasting power.

Figure 28.14(a) introduces the use of standard symbols for circuit design. The pressure relief valve is represented by a square with an offset arrow indicating that this valve is normally closed until sufficient pressure is developed in the pilot line (dotted) to push the valve open against the mechanical spring. This symbol is quite different from the physical drawing of a relief valve as shown in Fig. 28.15, but it certainly conveys to the reader how the device is expected to operate [Esposito (1980)].

In the case of a directional control valve, it is always shown in its normally closed position. The reader is expected to visualize what happens when the valve is moved into its other two operating positions, as shown in Fig. 28.14(c). In Europe, these symbols are standardized by the International Organization of Standards (ISO) [ISO (1976)] and also by the British Standards Institution (BSI) [BSI (1977)]. In the USA, the American National Standards Institute develops the standards for the fluid power industry [ANSI (1967)].

The simplest form of flow control uses an adjustable orifice. These circuit configurations are shown in Fig. 28.16. When the orifice control is placed in the supply line upstream of the hydraulic cylinder, the system is said to be under *meter-in* flow control. Flow control is operative when flow is directed to the large-area side of the piston in Fig. 28.16(a), while in the reverse direction, flow is dumped freely through the check valve. This type of control is best suited for resistive type loads that the piston rod pushes against and not for overrunning type loads.

When the orifice control is placed in the return line, Fig. 28.16(b), the system is said to be under *meter-out* flow control. This type of control is best suited over running loads which are moving in the same direction as the motion of the actuator.

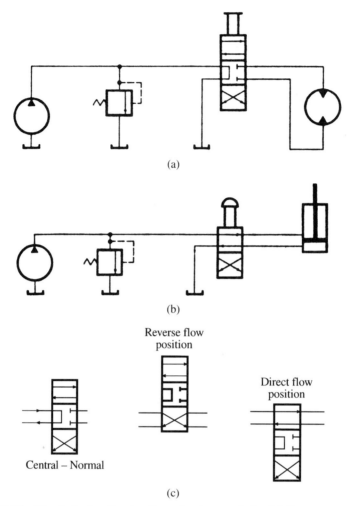

FIGURE 28.14 Simple hydraulic circuits and valve symbols. (a) Open-center valve control, (b) simple constant flow circuit, and (c) valve symbol meanings.

In *bleed-off* control, Fig. 28.16(c), the flow control parallels the cylinder feed line. This approach can be used to adjust the cylinder speed over a range that is less than the maximum speed available. It has the advantage of using a small control valve, only large enough to handle the bleed flow and not the total flow. It also does not introduce a pressure drop in the main delivery line to the ram.

There are many pitfalls in designing fluid power circuits for the inexperienced, some of which are due to lack of design information. Some of these problems are reviewed in an excellent article by Achariga (1982).

28.9 PUMP SETS AND ACCUMULATORS

Any hydraulic circuit is useless unless there is a unit to provide the fluid power. Hence, the pump set is an important component of the system. Depending on the

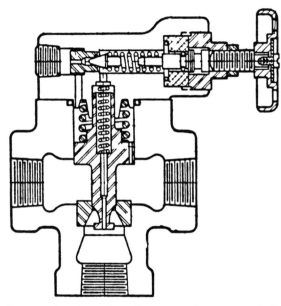

FIGURE 28.15 Cross section through a pilot operated pressure relief valve. (Courtesy of Vickers.)

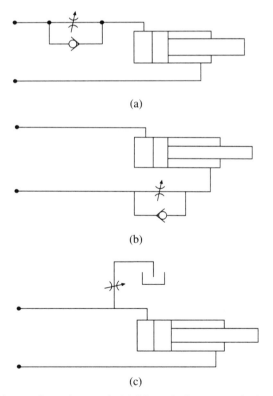

FIGURE 28.16 Means of speed control. (a) Meter-in flow control, (b) meter-out flow control, and (c) bleed-off.

application, this can be designed and constructed by the user or the decision may be and to purchase one of the many commercially available complete packages.

A typical design [Martin and McCloy (1980) and Martin (1995)] is shown in Fig. 28.17. The motor driving the pump is usually electric for industrial application and runs at 1740 rpm. Other prime movers, such as diesel engines could also be used. The outlet from the fixed displacement pump is piped to a relief valve. This allows the working pressure to be selected. The nonreturn valve, N, prevents flow being forced back into the pump and also helps to stiffen the hydraulic circuit. An accu-

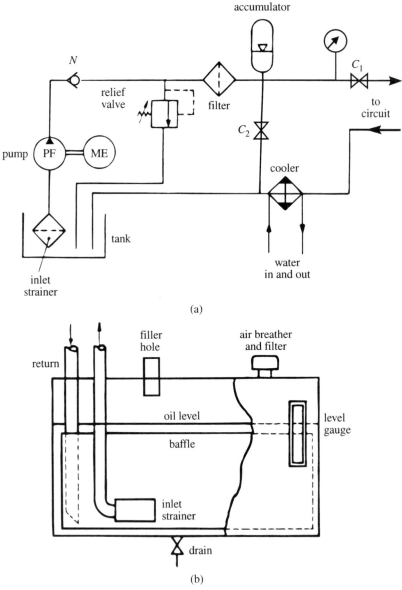

FIGURE 28.17 Pump set design. (a) The basic pump circuit and (b) layout of a typical hydraulic reservoir.

mulator is included to smooth the pressure pulses developed in the pump. It will also provide additional flow for short time, high transient demands.

It is wise to include a shut off cock, C1, to prevent oil spillage when the delivery line is disconnected and also to include shut off cock, C2, so that the accumulator can be discharged safely.

Good reservoir design as in Fig. 2.17(b) is probably the most important aspect of preventive maintenance. It fulfills many functions besides containing sufficient oil to meet the demands of the complete system. It acts as a cooler and allows time for contaminants such as foam and dirt to settle out. The tank capacity is usually at least three times the maximum delivery of the pump in one minute and may be as large as six times if there are numerous valves in the circuit to generate heat. The inlet strainer removes larger debris, but care must be taken to ensure that it or any other component does not create significant pressure loss in the inlet (suction) side of the pump.

In starting a new hydraulic circuit, the following steps should be followed: 1) make certain the pump is being driven in the correct direction for its design, 2) make sure all nonreturn valves are located in the correct flow direction, 3) jog the pump drive motor two or three times watching for leaks, and note system pressure for an indication of pump priming, 4) check oil level in the tank and top off if a new circuit has used a significant amount of oil in filling, 5) check that the accumulator is charged correctly, and 6) after 5–10 hrs. of running, check the filter and strainer for debris left in the new system from parts manufacture.

The accumulator is an energy storage element into which hydraulic oil is pumped by the system to compress the contained gas. The accumulator can be used as a low pass filter to smooth out pump delivery fluctuations, or it can be used to supply small amounts of additional power for transient demands. It is sometimes cost effective to use a small pump in a circuit where the duty cycle calls for low power requirements most of the time and large demands for relatively short periods. In this case, a larger accumulator can be used to store energy during the low level part of the duty cycle and release it when high demand is required. A third application is to provide a stand-by, short-term power source in case of failure of the pump set. This is particularly important in aerospace applications.

Accumulation size estimates are normally based on Boyle's law and assume that the charging gas temperature remains constant. The gas precharge in the accumulator is selected at about 1/3 of the final maximum pressure required in the oil. It is advisable, of course, to use nitrogen as the gas. The usual stages in the operation of an accumulator are shown in Fig. 28.18, where,

p_1, V_1 = gas precharge pressure and volume.
p_2, V_2 = gas charge pressure when the pump is turned on and will correspond to the system maximum pressure in the oil.
p_3, V_3 = minimum pressure required in the circuit.

If the volume of oil delivered from the accumulator is,

$$V_0 = V_3 - V_2 \tag{28.30}$$

then for gas compressed isothermally,

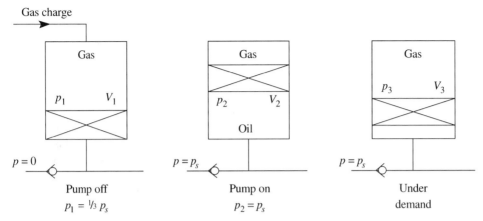

FIGURE 28.18   Accumulator operation.

$$V_1 = \frac{p_2/p_1 V_0}{(p_2/p_3 - 1)} \tag{28.31}$$

If there is shock loading present, the process is closer to adiabatic, and Eq. 28.31 is modified to

$$V_1 = \frac{(p_2/p_1)^{1/\gamma} V_0}{[(p_2/p_3)^{1/\gamma} - 1]} \tag{28.32}$$

where γ is the ratio of specific heats for the gas, which normally has a value $\gamma = 1.4$.

28.10   HYDROSTATIC TRANSMISSIONS

The hydrostatic transmission is usually a closed circuit system in which the pump output flow is sent directly to a hydraulic motor. This package finds extensive use in many industries, for example steel and paper manufacturing. The electro mechanical version is the Ward–Leonard drive which consists of a d.c. generator supplying current to a d.c. motor, often used in locomotives. However, this units tends to be very bulky in comparison to the hydraulic transmission.

Because of the ease of control of speed, torque, and direction of rotation, the hydrostatic transmission has become a popular choice in industrial equipment design. As can be seen from Fig. 28.19, there are several combinations whereby a positive displacement pump can be used to drive a motor. The choice will depend on the application.

The combination of a fixed displacement motor with a fixed displacement pump, Fig. 28.19(a) gives a fixed drive ratio. This is the simplest configuration where the output speed can only be controlled by altering the speed of the prime mover. For speed control, an alternative approach would be to include a bleed flow control valve from the main delivery line [Yeaples (1966) and Kern (1969)].

When the motor is variable displacement, Fig. 28.19(b), a fixed ratio drive at any given motor setting is obtained. The torque decreases with speed increase making

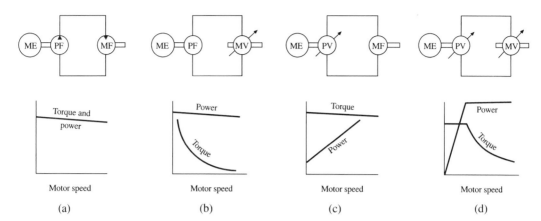

FIGURE 28.19 Pump and motor combinations. (a) Fixed displacement pump and fixed displacement motor, (b) variable displacement motor and fixed displacement pump, (c) variable displacement pump and fixed displacement motor, and (d) variable displacement pump and variable displacement motor.

the characteristics compatible with winding machines. As a roll diameter increases, the rotational speed must decrease to hold a constant linear velocity of the material being wound while at the same time mass is increasing requiring more drive torque.

By making the pump variable displacement instead of the motor as in Fig. 28.19(c), the torque remains constant over the speed range. With the pump at zero output, an idle condition is produced which is similar to a disengaged clutch.

The final configuration, shown in Fig. 28.19(d) introduces variability for both the pump and the motor. This combination can produce either a constant power or a constant torque drive. The combination has great flexibility—at a cost of course— and can have both pump and motor adjusted together or separately. For example, where two separated parts of the same machine are to be driven at different speeds, a variable pump with two variable displacement motors can be used.

28.11 CONCEPT OF FEEDBACK CONTROL IN HYDRAULICS

A simple open loop hydraulic servomechanism is shown in Fig. 28.20(a). It consists of a spool valve, moving in a sleeve, so as to uncover two sets of control orifices and, therefore, allowing flow to and from the ram. Details of the valve design are discussed in Sec. 28.4.

A displacement of the spool, x_v, to the right will allow the supply pressure, p_s, and flow Q, to pass into the ram causing the piston to move to the left. Consequently, the oil flow, $Q2$, in the left-hand ram chamber will be exhausted through one set of control orifices to the tank. Three important facts should be noted about this device. First, a displacement of the spool causes the piston to move at a constant velocity. Hence, a simple hydraulic servomechanism has the characteristics of an integrator

$$y = K \int x_v \, dt \qquad (28.33)$$

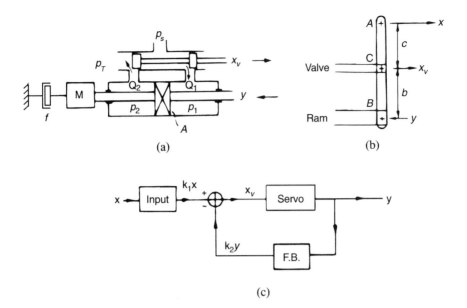

FIGURE 28.20 Simple open loop hydraulic servomechanism. (a) Open loop configuration, (b) feedback linkage, and (c) closed loop block diagram.

Second, it only requires a relatively small effort in displacing the spool valve to make considerable force available at the output of the ram. Third, the system is nonlinear and difficult to analyze accurately, except under certain simple loading conditions. This is mainly due to the fact that the valve flow is a fraction formed from two variables, orifice area, and pressure drop [see Eq. (28.1)].

It is shown in Martin and McCloy (1980) that if the load pressure is defined as $p_L = p_1 - p_2$, then the load flow is given by,

$$Q_L = C_d A_0 \sqrt{\frac{2}{\rho}} \sqrt{\frac{p_s - p_L}{2}} \tag{28.34}$$

If the mass M and damping f are very small, see Fig. 28.20(a), then $p_1 = p_2 \approx \frac{1}{2}p_s$, and Eq. (28.34) becomes

$$Q_L = K_1 x_v \sqrt{\frac{p_s}{2}} = A \frac{dy}{dt} \tag{28.35}$$

where

$$K_1 x_v = C_d A_0 \sqrt{\frac{2}{\rho}}$$

and finally,

$$y = \frac{K_1}{A}\sqrt{\frac{p_s}{2}}\int x_v\, dt \tag{28.36}$$

Consider now the lever system shown in Fig. 28.20(b). When this follow-up mechanism is attached to the valve and ram rod, a closed loop configuration results. A displacement of the input, x, causes a movement of the valve, x_v, also to the right from a central closed position, since initially the lever $(c + b)$ pivots about B. The contribution to the valve displacement is now initially,

$$x_v = \frac{b}{c + b}x = k_1x \tag{28.37}$$

Once the spool valve opens, flow passes to the ram, and the pivot B starts to move to the left. The top pivot point A is now held fixed by the input, hence the original valve displacement, Eq. 28.37, is now closed. The control equation is,

$$x_v = \frac{b}{c + b}\cdot x - \frac{c}{c + b}\cdot y = k_1x - k_2y \tag{28.38}$$

The block diagram in Fig. 28.20(c) should clarify the arrangement.

Combining Eq. (28.35) with Eq. (28.38) results in the closed loop transfer function for this configuration

$$y(s)/x = \frac{b/c}{(1 + Ts)} \tag{28.39}$$

where the gain equals b/c, and the time constant, T, is,

$$T = \frac{c + b}{c}\frac{A}{K_1\sqrt{p_s/2}} \tag{28.40}$$

Equation (28.39) means that the closed loop servo operates as a simple exponential type lag, instead of an integrator as in the open loop configuration.

The performance of the unit can be assessed in several ways, depending on the type of information needed and the test equipment available. The transient response is a plot of the output movement, y against time, for a defined magnitude of step input. Mathematically the solution to Eq. (28.39) is

$$y = \frac{b}{c}(1 - e^{-t/T})\, x \tag{28.41}$$

The plot of this equation is shown in Fig. 28.21(a), where the time constant can be found from 63.2% of the final steady state point. In theory, the steady state given by

$$y = bx/c \tag{28.42}$$

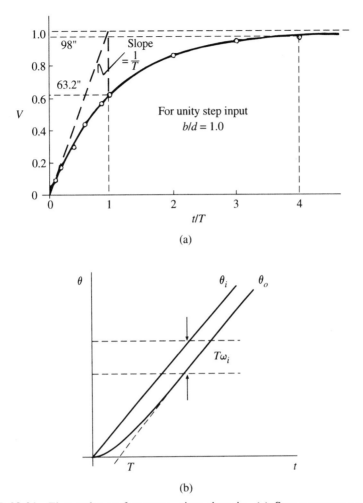

FIGURE 28.21 First order performance—time domain. (a) Step response and (b) ramp input; $\theta = \omega_i t$ where t is time.

will only be reached when t reaches infinity, however it is normal in practice to use $4T = 98\%$ of final steady state as the practical steady state value. Transient response testing is less costly and easy to perform. However, the information is limited in that spectral information is difficult to interpret especially phase shift between the input and output. Some nonlinearities such as dead zone, clipping, and small amplitude parasitic oscillations can be identified.

A simple production test for assessing the level of friction in the moving parts of the servomechanism is to apply the ramp test depicted in Fig. 28.21(b). In this case, the input is suddenly subjected to a constant velocity of ω_i rad/sec. The output tries to follow but lags by a steady state error of $T\omega_i$ where T is the system time constant. If the system were modeled by Eq. (28.39), this error could be reduced by changing the lever ratio $(c + b)/c$ for example.

28.12 IMPROVED MODEL

Experimental testing of an actual hydraulic servo system immediately reveals that the first order model discussed in the previous section does not adequately represent the actual performance. This is especially true at the prototype testing phase where the system response has not been optimized. Instead of the smooth exponential type behavior shown in Fig. 28.21(a), the response is likely to be quite oscillatory. The main reason for this is the fact that compressibility of the oil and the mass of the moving parts were considered negligible in the simple first order model.

In addition, if mass is to be included, the load pressure p_L, cannot be ignored as it was in Eq. (28.35). This means that the valve equation becomes a function of two variables: pressure drop and valve displacement.

Early attempts to solve this problem are recorded in a classic paper by Harper (1953) using a small perturbation method. This had the disadvantage that the valve characteristics at any instant in the motion of the servo were defined by the instantaneous values of the slopes of two nonlinear curves. In order to improve this situation, an alternative approach was suggested [Keating and Martin (1973)] which minimizes the average error and to a large extent overcomes this problem. Referring once more to Fig. 28.20, the flow into the actuator is given by,

$$Q_1 = KA_{01}\sqrt{p_s - p_1} \tag{28.43}$$

and the flow out is given by

$$Q_2 = KA_{02}\sqrt{p_2 - p_T} \tag{28.44}$$

where $K = C_d\sqrt{2/\phi}$ and the valve is symmetrical $A_{01} = A_{02} = A_0$. It can be shown [see Martin and McCloy (1980)] that the load flow through the valve is

$$Q_L = KA_0\sqrt{\frac{p_s - p_L}{2}} \tag{28.45}$$

If compressibility of the oil is now included, see Eq. (28.26) of Sec. 28.7, then the load flow can be equated to the flow through the actuator so that;

Q_L = (flow due to piston movement) + (flow due to compressibility)

$$= A\frac{dy}{dt} + \frac{V}{4\beta}\frac{d}{dt}(p_L) \tag{28.46}$$

By equating Eqs. (28.45) and (28.46), the relationship between valve displacement x_v and output movement y can be obtained, and it is much more complex than in the previous model shown in Eq. (28.36).

$$K_v x_v\sqrt{\frac{p_s - p_L}{2}} = A\frac{dy}{dt} + \frac{V}{4\beta}\frac{d(p_L)}{dt} \tag{28.47}$$

where a linear relationship between valve displacement and uncovered orifice area has been assumed.

The worst case for loading a system is when the output load is pure mass. It is the load pressure p_L, that is used to accelerate the mass.

$$p_L = \frac{M}{A} \frac{d^2 y}{dt^2} \tag{28.48}$$

Introduce this into Eq. (28.47)

$$\frac{K_v x_v}{A} \sqrt{\frac{p_s}{2}} \left(1 - \frac{M}{p_s A} \frac{d^2 y}{dt^2} \right) = \frac{dy}{dt} + \frac{VM}{4\beta A^2} \frac{d^3 y}{dt^3} \tag{28.49}$$

If the constants are lumped together so that $K_r = K_v/A\sqrt{p_s/2}$, and if the square root term is expanded by the binomial theorem, neglecting terms greater than first order, Eq. (28.49) becomes

$$K_r x_v \left[1 - \frac{1}{2} \left(\frac{M}{p_s A} \right) \frac{d^2 y}{dt^2} \right] = \frac{dy}{dt} + \frac{VM}{4\beta A^2} \frac{d^3 y}{dt^3} \tag{28.50}$$

Rearranging and transforming to Laplace domain as was done previously with Eq. (21.28)

$$\left[\frac{VM}{4\beta A^2} s^2 + \frac{x_v}{2} \left(\frac{MK_r}{p_s A} \right) s + 1 \right] s = K_r x_v \tag{28.51}$$

The equation within the square brackets is equivalent to the general equation for a spring-mass-damper system, hence Eq. (28.51) can be written,

$$\frac{1}{\omega_n^2} s^2 + \frac{2\xi}{\omega_n} s + s = K_r x_v \tag{28.52}$$

It can now be seen that the damping contributed from the valve is partially determined by the valve displacement; the symbol ξ is the damping ratio. This explains why in practical test results, the frequency response curves of a hydraulic servo change depending on the size of the input amplitude. Keating and Martin (1973) discuss a further refinement to this model which results in an even better estimate of the dynamic response of this type of system.

28.13 ELECTROHYDRAULIC SYSTEMS—ANALOG

In the arrangements discussed in the previous section, signal transfer for feedback was done using mechanical linkages. It is much more convenient to employ electrical means to achieve these loops. However, this does require the use of transducers to convert mechanical and fluid signals into an electrical form.

Typical electro hydraulic servo valves use an electrical torque motor to move the spool arrangement. The most famous of these types of valves is the Moog 1500 series two stage valve shown in Fig. 28.22(a). It is also possible to have a single stage spool as shown in Fig. 28.22(b). In the example shown, the first stage is a double nozzle pilot valve controlling a second stage spool valve. The torque motor is really a limited movement electric motor, arranged so that the flapper extends between two nozzles. This allows differential pressure to be applied across the sliding valve, whose movement in term meters fluid out of the valve.

While the flexibility of combining electrics with hydraulics is a major advantage, it should be noted that the torque motor does introduce an additional transfer function into the system. This in itself need not be a problem, provided that care is taken in

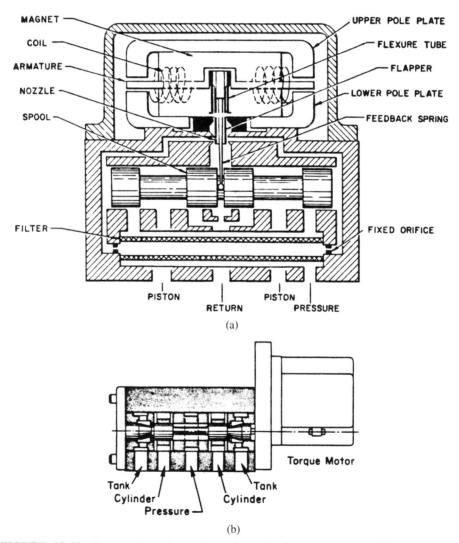

FIGURE 28.22 Typical electrohydraulic valves. (a) Two stage valve (Moog Inc., East Aurora, NY) and (b) single stage valve.

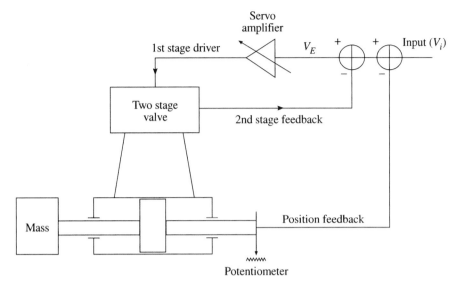

FIGURE 28.23 Two stage valve–electrohydraulic servo system.

the design process that the overall phase shift is not increased. Like any series type system, the overall performance is determined by the component with the poorest dynamic characteristics. It is important, therefore, to ensure that the torque motor valve assembly is not the weak link in the chain.

Feedback can now be achieved with the use of electrical transducers such as precision potentiometer, pressure devices, and accelerometers. A typical arrangement is shown in Fig. 28.23. For this particular arrangement, the minor loop is closed with position of the second stage spool valve, and the major loop is closed with position of the driving ram which controls the inertia type load. All the signals around the servo are now electrical and easy to adjust. The amplifier is used both as an easy method of gain adjustment and as a summing junction.

Potentiometers are the cheapest and simplest devices for converting linear and angular displacement information to electrical signals, but unless their performance, especially when coupled to other parts of the circuit are understood, then there can be real practical problems in getting an electro hydraulic servo to function correctly. The selection of a suitable instrument potentiometer requires the following specification items to be addressed [Dummer (1963), Davis and Ledgerwood (1961), and Neubert (1963)]. The most commonly occurring terms are:

1. *Linearity* The deviation of the output voltage from a potentiometer from a linear law related to shaft rotation, Fig. 28.24(a).

2. *Conformity* Similar to linearity, but used in relation to potentiometers designed to follow a nonlinear law, Fig. 28.24(b).

3. *Deviation* Some suppliers of potentiometers quote deviation instead of linearity. The deviation is defined as the maximum permissable offset from the best straight line which can be drawn through the experimental points of measured resistance against rotation of the potentiometer. This tolerance is expressed as a percentage of the total resistance.

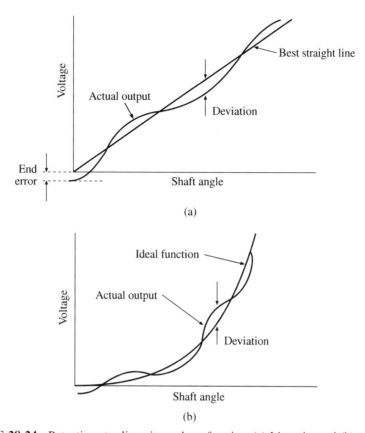

FIGURE 28.24 Potentiometer linearity and conformity. (a) Linearity and (b) conformity.

4. *Resolution* The incremental rotation of the shaft necessary to produce the smallest incremental change of output voltage.

5. *Power rating* The maximum continuous power that can be safely dissipated by the resistance at a specific temperature.

6. *Operating Torque* This is the torque required to start the wiper moving from rest. For small, general-purpose, wire-wound potentiometers this is in the range of 1–5 oz/in., while special purpose units can be obtained having torques as low as 0.005 oz in.

7. *Electrical and Mechanical Angle* The mechanical angle is the angle through which the potentiometer shaft can be rotated freely, while the electrical angle is that angle of rotation from which a voltage can be recorded.

A wide range of potentiometers are available, e.g., using wire wound cores, carbon tracks, and film deposits. Their mode of operation can be linear or nonlinear. Multiturn potentiometers are formed using a helix configuration and give much better resolution and accuracy than single turn units.

Wire wound potentiometers have the disadvantage that, when used in high gain systems, the spacing between the wires produces a staircase effect in the output voltage. This results in a distinct roughness of motion in the servo output shaft. This problem can be resolved by using the more expensive film potentiometer. Carbon track versions tend to leave a deposit of carbon particles as the potentiometer wears in use. This also makes the servo output motion rough. These units need to be cleaned regularly. Operating torques, especially the starting torque for a potentiometer, become important when the unit is being driven from a low power source, for example, the first or second stage of a spool valve. Another critical problem in electrohydraulic applications is to minimize loading errors on potentiometers, or in fact, any other transducer associated with the cricuit.

Consider the arrangement shown in Fig. 28.25. In this circuit, a potentiometer is loaded by a circuit of resistance, R_L, which represents the input impedance of the next stage. For the input circuit, Fig. 28.25(a), $V_i = RI$.
For the output circuit with $R_L = \infty$, the ideal case, $V_o = kRI$.

Since the current, I, in the circuit is common $V_i/V_o = 1/k$ which satisfies the conditions: $V_o = 0$ when wiper is at B, $k = 0$ and $V_o = V_i$ when wiper is at A, $k = 1$.

If now the practical case is considered, where R_L is finite in value, the circuit can be interpreted as shown in Fig. 28.25(b). The parallel resistances can be replaced by

$$\frac{1}{R_T} = \frac{1}{kR} + \frac{1}{R_L}$$

$$R_T = \frac{kRR_L}{R_L + kR} \tag{28.53}$$

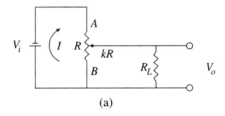

(a)

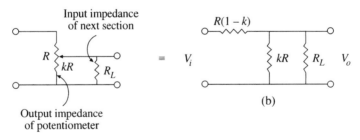

(b)

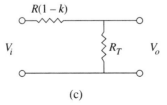

(c)

FIGURE 28.25 Potentiometer loading. (a) Actual circuit, (b) equivalent circuit, and (c) least complex equivalent circuit.

as shown in Fig. 28.25(c). The total resistance as seen by V_i is now

$$R_i = R(1 - k) + \frac{kRR_L}{R_L + kR}$$

The current in the circuit must now be,

$$I = \left(\frac{R_L + kR}{kR^2(1 - k) + RR_L} \right) V_i$$

But

$$V_o = R_T I = \frac{kRR_L}{R_L + kR} I$$

hence

$$\frac{V_o}{V_i} = \frac{k}{k(1 - k) \dfrac{R}{R_L} + 1} \tag{28.54}$$

The error in measurement due to potentiometer loading can be expressed as,

$$\text{error } \% = \frac{V_o \text{ (ideal)} - V_o \text{ (actual)}}{V_i} \times 100$$

$$= \frac{k^2(1 - k)}{k(1 - k) + \dfrac{R_L}{R}} \times 100 \tag{28.55}$$

This equation shows that the error is variable over the range of potentiometer shaft angle positions. The shaft angle yielding maximum error can be found from $\partial(\text{error})/\partial k = 0$ and is a function of R_L/R. The shaft angle with maximum error occurs near $k^* = 2/3$ for $0.2 < R_L/R$. As R_L/R becomes large, k^* becomes precisely 2/3. If the error is to be less than 2% at this position, then $R_L/R \geq 7.2$.

The term *noise* in electrohydraulic circuits refers to any undesirable electrical circuit signal which is superimposed on the desired command signals. The noise signal will cause roughness in actuator movement or can cause mechanical parts to *buzz*. Such signals are often random in nature. Sources of circuit noise are many, including poorly soldered connections (dry joints), voltaic effects arising from an electrolyte in the presence of two dissimilar metals, unshielded wiring and resistances of high impedance. Probably the major offender is the potentiometer. Noise due to vibration occurs when the wiper jumps away from the track; this can be controlled by careful adjustment of the contact arm pressure. Another cause of noise can be excessive rotational speed of the potentiometer shaft causing the wiper to bounce along the track. Noise will also be generated from dirt on the track and, in more unusual circumstances, by chemical action due to moisture, oil, or other liquids which may have accidentally penetrated the equipment. Amplification of such noise signals in the servo amplifier will result in transient spikes which can cause the servo amplifier or another instrument amplifier to saturate.

28.14 ELECTROHYDRAULIC SYSTEMS—DIGITAL

The electrohydraulic stepping motor is used as a high torque, high speed drive whose output motion is precise and repeatable. Some of the unique advantages are summarised in Benson (1982) as: 1) position feedback is not needed for positional control, unlike the electrohydraulic servo, 2) the operation of the unit is such that a microprocessor can often be interfaced directly (hence, A/D and D/A converters are not needed), and 3) electrical tuning is not required other than input of the command profile.

The electrohydraulic stepping motor has three main components, an electric stepping motor, a servo valve, and a rotary or linear actuator. A typical arrangement is shown in Fig. 28.26. The block diagram describes the function of the unit. The electric pulse motor controls the rate of flow and direction of flow of oil to the hydraulic motor [Samar (1970)].

Command pulses are directed to the drive circuit amplifier. The pulses are fed in a phased sequence to the electric pulse motor, which, in turn, is connected through gears to a four-way spool valve. The gear on the stepping motor shaft has wide teeth so that the gear on the spool valve can move axially without disengaging. The other end of the spool has a lead screw, which is engaged with a nut on the hydraulic motor shaft. Thus, rotary motion of the stepping motor is transformed into axial motion of the spool which in turn opens the four-way valve. This allows pressurized oil to flow to the hydraulic motor and cause it to rotate. As the hydraulic motor rotates, the nut on the motor shaft rotates and moves in a direction opposite to the

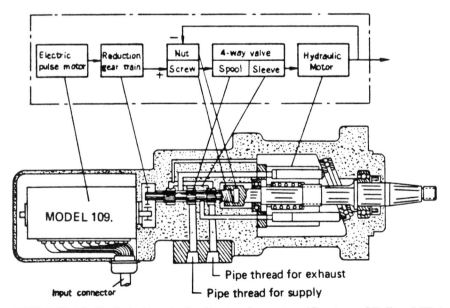

FIGURE 28.26 Typical electrohydraulic stepping motor. (Courtesy of Fujitsu LTD.)

motion of the spool returning the spool to its original position. Hence, the mechanical coupling of the spool valve and the hydraulic motor through the lead screw and nut form a negative feedback loop.

The electrical stepping motor is a fractional horespower device which can be based on the concepts of variable reluctance, permanent magnet, rotating disc, or flexspline for the method of operation. In the case of the permanent magnet type, the magnetic rotor aligns itself with the magnetic orientation of the stator. When the stepping motor windings are correctly energized, the rotor will rotate one unit of angular displacement, typically $2\pi/200$ degrees/revolution and then stop. Thus, any magnitude of motor rotation can be equated into the summation of some number of steps, as a result of a string of pulses from a microcomputer, applied at the input. Since the output position is not verified, the accuracy is solely a function of the ability of the motor to step through the exact number of steps commanded by the input. Hence, one important characteristic of a stepping motor is the maximum rate of the input pulses that can be followed.

The quantization of the motion into discrete steps is particularly well suited for a digital control device such as a microprocessor. The stepping motor drive (translator) accepts position and velocity profile commands in the form of a variable frequency pulse train and directional signal from the microprocessor. A complete block diagram is shown in Fig. 28.27.

The position of the electrical stepping motor shaft will be a sharply defined *staircase* as shown in Fig. 28.28. The velocity of the shaft is, therefore, a function of the input pulse rate. The size of each step, both time- and position-wise, is determined by the pulse train and may also in practice have oscillatory overshoot if not well damped.

At frequencies exceeding 100 pulses/sec, the hydraulic valve in Fig. 28.26 does not have time to close. However, the number of pulses that the hydraulic motor lags behind the electric stepping motor is the amount of oil that should have been admitted into the hydraulic motor. This is remembered by the nut and lead screw summing point at the valve shaft. Each stored pulse threads the screw into the nut

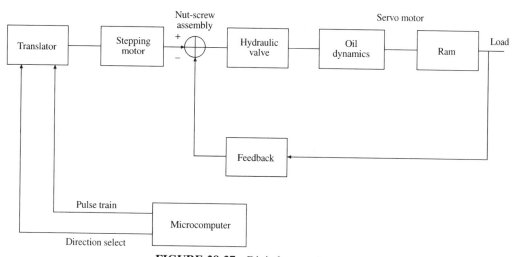

FIGURE 28.27 Digital control system.

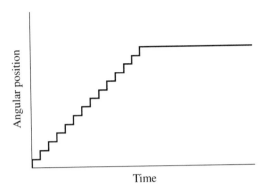

FIGURE 28.28 Electrical stepping motor output.

a specific distance. When each stored pulse is used, the screw threads out the nut a specified distance closing the 4-way hydraulic valve a similar distance. This delay is a function of motor speed and is analogous to loop gain in a conventional analog electrohydraulic servo.

Since 100 pulses/sec is normally faster than the valve can respond, it remains open and follows the hydraulic motor smoothly. Therefore, the steps in Fig. 28.28 are not transmitted through the system.

REFERENCES

Achariga, R., "Pitfalls the Inexperienced Fluid Power System Designer Should Avoid," *Hydraulic and Pneumatics*, Penton Publ., pp. 152–168, 1982.

"Aerospace Fluid Component Designer's Handbook," RPL-TDR-64-25, AF Rocket Propulsion Lab., Edwards, CA, 1964.

Anon., "Viscosity Considerations in the Selection of Hydraulic Fluids," *Hydraulics & Pneumatics*, Penton Press, pp. 23–26, 1980.

Benson, B., "Rotary and Linear Electrohydraulic Stepping Actuators," *Proc. 38th Nat. Conf. on Fluid Power*, Houston, TX, 1982.

BS 2917: 1977 Specification for Graphical Symbols Used on Diagrams for Fluid Power Systems and Components, British Standards Institute, London, 1977.

Collacott, R. A., *Mechanical Fault Diagnosis*, Chapman-Hall, 1977.

Dalibert, A., "Fire Resistant Fluids," Paper F.3, B.H.R.A., 2nd Fluid Power Conf., University of Surrey, 1971.

Davis, S. A. and Ledgerwood, B. K., *Electromechanical Components for Servomechanisms*, McGraw-Hill, New York, 1961.

Dummer, G. W., *Variable Resistors and Potentiometers*, Pitman, 1963.

Esposito, A., *Fluid Power with Applications*, Prentice-Hall, Englewood Cliffs, NJ, 1980.

Fitch, E. C., Bensch, L. E., and Tessmann, R. K., *Contamination Control for the Fluid Power Industry*, Pacific Scientific, Montclair, CA, 1978.

Fluid Power Designers Handbook, Parker-Hannifin Corp., Cleveland, OH.

Foord, B. A., "Specifying Fluid Filtration through Computer Simulation," *Proc. 34th Nat. Conf. on Fluid Power*, Philadelphia, PA, p. 197, 1978.

Forgeron, E. J. and McCormack, N., "Controlled Maintenance System for Hydraulics through Used Oil Analysis," *Proc. 34th Nat. Conf. on Fluid Power*, Philadelphia, PA, p. 187, 1978.

Harpur, N. F., "Some Design Considerations of Hydraulic Servos of the Jack Type," *Proc. Conf. on Hydraulic Seromechanisms*, *Int. Mech. Eng.*, London, 1953.

ISO 1219: 1976 Fluid Power Systems and Components—Graphical Symbols, International Standards Organization, 1976.

Keating, T. and Martin, H. R., "Mathematical Models for the Design of Hydraulic Actuators," *ISA Transactions*, Vol. 12, No. 2, 1973.

Keller, G., "Hydraulic System Analysis," *Hydraulics and Pneumatics*, Penton/IPC, Cleveland, OH, 1974.

Kelly, E. S., "Erosive Wear of Hydraulic Valves Operating in Fire Resistant Fluids," Paper F.4, B.H.R.A., 2nd Fluid Power Conf., University of Surrey, 1971.

Kern, J., *Hydrostatic Transmission Systems*, Intertext Books, London, 1969.

Louie, J., Burton, R. T., and Ukrainetz, P. R., "Fire Resistant Hydraulic Fluids—State of the Art Review," *Proc. 37th Nat. Conf. on Fluid Power*, Chicago, IL, p. 285, 1981.

Martin, H. R. and McCloy, D., *Control of Fluid Power*, 2nd ed., Ellis Horwood (John Wiley & Sons), New York, 1980.

Martin, H. R., *The Design of Hydraulic Components and Systems*, Ellis Horwood (John Wiley & Sons), New York, 1995.

Martin, H. R., "Effects of Pipes and Hoses on Hydraulic Circuit Noise and Performance," *37th Nat. Conf. Fluid Power*, Chicago, IL, pp. 71–76, 1981.

Merritt, H. E., *Hydraulic Control Systems*, John Wiley & Sons, New York, 1967.

Neubert, H., *Instrument Transducers*, Oxford University Press, 1963.

Samar, T., "Electrohydraulic Stepping Motor Drives Machine Tools Without Feedback," *Hydraulics & Pneumatics*, Penton Press, 1970.

Spencer, J., "Effective Contamination Control in Fluid Power Systems," Sperry Vickers Publication, Troy, MI, 1980.

Stanley, C. "Fire Resistant Fluids," Paper F.2, B.H.R.A., 2nd Fluid Power Conf., University of Surrey, 1971.

Tessman, R. K., "Ferrographic Measurements of Contaminant Wear In Gear Pumps," *Proc. 34th Nat. Conf. on Fluid Power*, Philadelphia, PA, p. 179, 1978.

Thoma, J., *Modern Oil Hydraulic Engineering*, Technical and Trade Press, 1970.

Turnball, D. E., *Fluid Power Engineering*, Newnes–Butterworths, London, 1976.

USA SY32-10-1967, *Standard Symbols for Fluid Power Diagrams*, American National Standards Institute, New York, 1967.

Wilson, W. E., "Positive Displacement Pumps and Fluid Motors," Pitman Publications (and ASME Paper 48-SA-14), 1950.

Yeaples, F. D., *Hydraulic and Pneumatic Power and Control*, McGraw-Hill, New York, 1966.

29 Compressed Air Systems

JOHN P. ROLLINS
Clarkson University
Potsdam, NY

CONTENTS

29.1 PURPOSES OF COMPRESSED AIR SYSTEMS

Compressed air systems are used widely throughout the manufacturing, construction, mining, agriculture, and food processing industries and many other fields. In manufacturing, compressed air operates hand tools, multiple air tools, hoists, clamps, automatic feed devices, painting equipment, and automatic controls. The relatively high power per unit weight of appliance accounts for many such applications including robot manipulators, where the smooth linear motion produced by an air cylinder with a parallel hydraulic damping circuit to control the motion is also important.

In the pharmaceutical and food industries, air supports biological processes producing foods and medicines. It also transfers fluidized solid foods such as grain. See Chapter 14 for a discussion of multiphase flow.

In the construction industries, dry cement is handled as a fluidized solid in pneumatic systems. Almost everyone has watched construction workers using demolition tools to break up old paving, as well as workers using spades, tampers, pneumatic wrenches, and many other air tools employed on new construction. The portable compressors supplying such systems are often rented during the off season to ski slope operators for use in producing manufactured snow. In mining systems, where

Handbook of Fluid Dynamics and Fluid Machinery, Edited by Joseph A. Schetz and Allen E. Fuhs
ISBN 0-471-12598-9 Copyright © 1996 John Wiley & Sons, Inc.

compressed air is universally used, the absence of electrical spark hazards in pneumatic systems is extremely important.

29.2 COMPRESSOR CAPACITY REQUIRED

The first step in designing a compressed air system is to assess existing and future demands. Many air tools operate at about 700 kPa (100 psig), and this is a widely used plant distribution pressure. Table 29.1 can serve as a guide in estimating air requirements for various tools, assuming that the pressure will have dropped to 630 kPa (90 psi) at the tools themselves. Figures from the table are only approximate and may vary considerably among tools of comparable designs from different manufacturers. Air capacities are usually stated in terms of *free air*, rather than actual volume at inlet conditions. Free air is defined as the volume which a given weight of air would have at atmospheric temperature and pressure. Since those conditions vary considerably from time to time, free air volume is subject to some variation. For this reason, standard conditions are often stated such as 16°C (60°F) and 101.3 kPa (14.7 psig), but the figures are not completely uniform among all manufacturers.

In estimating the combined air requirement of all the applications to be served by a system, one must remember that not all tools or other appliances will ordinarily be operated constantly at full capacity nor will they be in constant use, although some may be. The total air requirement will, therefore, not be the sum of the individual capacities. It is necessary to make the best estimate possible of load and time factors at each station and use those factors in determining the combined estimated average load. *Load factor* is that part of capacity at which the tool is run when it is in use, and the *time factor* is that part of the workday during which the tool is in actual use. It is also necessary to consider to what extent certain stations may be operated simultaneously, even though that is not the typical situation, in order to make a separate determination of the peak requirement.

Sand blasting is one of the operations which usually run steadily at full capacity. Others, such as riddle vibrators in the foundary are on full when running, but may have long off periods between applications. Still others, including hoists, and many air clamps and vises, as well as many air tools may not only be operated intermittently but may be used at varying throttle settings as well. Therefore, a study of the operating cycle at each station is often required. Previous experience with a similar system can be extremely useful.

Once the combined average and peak air requirements have been established, the compressor or compressors may be selected. If there is a considerable difference between average and peak requirements, the use of an air receiver to store air for the purpose of supplying the peak requirements can permit the selection of a smaller compressor. The compressor supplies air at somewhat above the average rate to recharge the receiver between peaks. A receiver may be placed near a point with highly varying load.

There is a tendency for new uses for compressed air to develop after a plant has been installed. The rate of increase will typically be between zero and 10% per annum, and the designer needs to use his best judgment in assessing the situation. To add a small compressor later may be the best way to provide this increase if it

TABLE 29.1 Air Requirements of Selected Tools

Tool	Free Air, cmm at 630 kPa, and 100% Load Factor
Grinders 6″ and 8″ wheels	1.42
Grinders, 2″ and 2½″ wheels	0.39–0.57
File and Burr machines	0.51
Rotary Sanders, 9″ pads	1.50
Rotary Sanders, 7″ pads	0.85
Sand Rammers and Tampers,	
1″ × 4″ cylinder	0.71
1¼″ × 5″ cylinder	0.79
1½″ × 6″ cylinder	1.10
Chipping Hammers, weighing 10–13 lbs.	0.79–0.85
heavy	1.10
weighing 2-4 lbs.	0.34
Nut Setters to $\frac{5}{16}$″ weighing 8 lbs.	0.57
Nut Setters ½″ to ¾″ weighing 18 lbs.	0.85
Sump Pumps, 145 gals. (at 50 foot head)	1.98
Paint Spray, average	0.198
varies from	.057–.566
Bushing Tools (Monument)	.425–.700
Carving Tools (Monument)	.283–.425
Plug Drills	1.13–1.42
Riveters, $\frac{3}{32}$″–1″ rivets	0.34
larger weighing 18–22 lbs.	0.99
Rivet busters	0.99–1.10
Wood Borers to 1″ diameter weighing 4 lbs.	1.13
2″ diameter weighing 26 lbs.	2.27
Steel Drills, Rotary Motors	
Capacity up to ¼″ weighing 1¼-4 lbs.	0.51–0.57
Capacity ¼″ to ⅜″ weighing 6–8 lbs.	0.57–1.13
Capacity ½″ to ¾″ weighing 9–14 lbs.	1.98
Capacity ⅞″ to 1″ weighing 25 lbs.	2.27
Capacity 1¼″ weighing 30 lbs.	2.69
Steel Drills, Piston Type	
Capacity ½″ to ¾″ weighing 13–15 lbs.	1.27
Capacity ⅞″ to 1¼″ weighing 25–30 lbs.	2.12–2.27
Capacity 1¼″ to 2″ weighing 40–50 lbs.	2.27–2.55
Capacity 2″ to 3″ weighing 55–75 lbs.	2.83–3.12

(Adapted with permission of Compressed Air and Gas Institute, *Compressed Air and Gas Handbook*, 4th ed., 1973.)

cannot be initially foreseen. Leakage also occurs, although good maintenance will, of course, minimize it. About ten percent may be added to the requirement to allow for leakage, according to the Compressed Air and Gas Handbook [Rollins (1973)], which also states that 20% more is normally added if water-cooled, double acting compressors are selected, and 40% more for single-acting air cooled compressors, to allow for other factors.

29.3 CENTRALIZED VS. DECENTRALIZED SYSTEMS

A centralized system will generally be more efficient than several smaller, local systems, but this is not necessarily always the case. If distribution pipes are too long, pressure drop in the lines may be excessive, and two or more separate systems may be more satisfactory and possibly cheaper as well. Alternatively, a single system with separate receivers near load points may be more economical.

A decentralized, multiple unit system represents the other extreme. An advantage of this choice is that, where future requirements are unknown, units may be purchased as new demands occur. The system is, therefore, more flexibile, but it has disadvantages too. Oveall costs will usually be greater in the end, and the maintenance of decentralized equipment may also be more difficult and costly. When compressors are located throughout a plant, one must remember the need for clean, cool inlet air. There may also be problems with noise. Management must weigh the various factors, including financing, in order to make the best decision. If a decentralized system is selected, a unitized compressor with integral receiver may be considered.

Special factors sometimes need to be considered. A plant may be closed over week-ends, for example, and it may be economical to provide a small compressor to meet the needs of maintenance and other applications that continue whether the plant is operating or not.

29.4 SPECIAL REQUIREMENTS

Some applications like so-called *clean rooms* require separate nonlubricated compressors because lubricant vapor cannot be tolerated in the air. Special attention to moisture removal by aftercoolers and separators may be required.

29.5 DISTRIBUTION PIPE SIZE

The air distribution system must provide air at adequate pressure at each station. In general, a pipe loop around the plant within each building will give best results. This gives a two way passage to each station with good pressure regardless where the load is heaviest. The pipe size should be selected so that the pressure drop will not be excessive at any station whichever direction the flow may take. All air pipes should be sloped downward away from the compressor about one quarter inch per foot toward a drop leg or trap for the removal of condensed moisture. The purpose of this is to prevent liquid water from reaching air tools or other air operated devices.

Example 29.1: The main line leading to an appliance finishing room is 36 m (120 ft) long, and the air requirement is 0.30 m^3/s (636 cfm), free air; allowances for leakage and future expansion are included in that figure. Find the size of steel pipe required for the main if the pressure drop is to be limited to 20 kPa (2.9 psi).

Assume average pressure and temperature in the main to be 700 kPag (101.6 psig) and 40 C (104 F). There are two elbows and three T's in the run.

(a) Assume as a first guess that a 2-inch pipe will be required:
(b) Reynold's number Re:
Air mass density:

$$\rho = \frac{pM}{RT}$$

$$\rho = \frac{(700000 + 101300 \text{ N/m}^2)(28.9 \text{ kg/kmol})}{(8315 \text{ J/kmol K})(40 + 273) \text{ K}}$$

$$\rho = 8.898 \text{ kg/m}^3$$

Dynamic viscosity is obtained from Chapter 2.

$$\mu = 1.9 \times 10^{-5} \text{ Ns/m}^2$$

A nominal 2-in. pipe has an ID of 2.067 in. which is 0.0525 m. The pipe cross sectional area, A, is

$$A = \pi d^2/4 = \pi(.0525)^2/4 = 0.002165 \text{ m}^2$$

The volume flow rate of air at free air conditions, Q_f, must be converted to the flow rate in the main, Q_m. Assuming free air conditions to be 101.3 kPa (14.7 psia) and 16 C (60 F)

$$Q_m = Q_f \left(\frac{P_f}{P_m}\right)\left(\frac{T_m}{T_f}\right)$$

$$Q_m = 0.30 \text{ m}^3/\text{s} \left(\frac{101.3 \text{ kPa}}{801.3 \text{ kPa}}\right)\left(\frac{40 + 273}{16 + 273}\right)$$

$$Q_m = 0.041 \text{ m}^3/\text{s}$$

The average velocity of air in the pipe is

$$V = Q_m/A = (0.041 \text{ m}^3/\text{s})/(0.002165 \text{ m}^2) = 18.97 \text{ m/s}$$

The Reynolds number is

$$\text{Re} = \frac{dV\rho}{\mu} = \frac{(.0525 \text{ m})(18.97 \text{ m/s})(9.898 \text{ kg/m}^3)}{(1.9 \times 10^{-5}) \text{ Ns/m}^2}$$

$$\text{Re} = 463000 \text{ (dimensionless)}$$

(c) The friction factor may be obtained from Fig. 29.1. In order to use the graph of friction factor as a function of Reynolds number, one needs the relative roughness ratio, ϵ/D. If it is assumed that the pipe is *commercial steel* or *wrought iron* and the upper graph in Fig 29.1 is selected, the following value

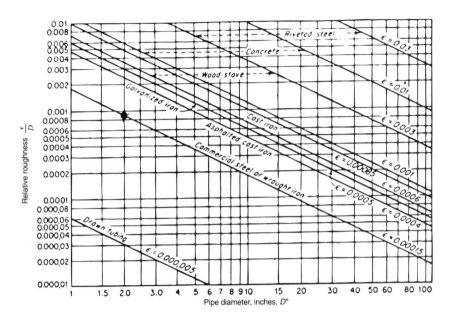

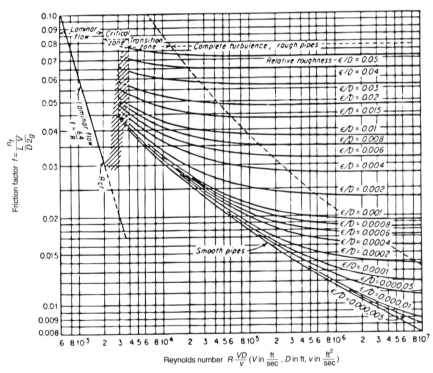

FIGURE 29.1 Pipe-friction factor based on relative roughness of various kinds and sizes of pipe. (Adapted from Figs. 1 and 8, Wright, D. K., Jr., "A New Friction Chart for Round Ducts," *Trans. ASHVE*, p. 303, 1945.)

of ϵ/D is found: $\epsilon/D = 0.0009$. Entering the lower graph in Fig. 29.1, f may be read: $f = 0.0195 \simeq 0.02$.

(d) Equivalent length of pipe l for elbows and T's; see Table 29.2

$$l = 120 + 2 \times 5.17 + 3 \times 2.67 = 136.52 \text{ ft} = 41.6 \text{ m}$$

(e) Pressure loss:

$$\Delta p = 0.48 \, \rho f \frac{L}{d} \, V^2$$

$$\Delta p = 0.48(8.898 \text{ kg/m}^3) \, (0.0195) \left(\frac{41.6 \text{ m}}{0.0525 \text{ m}} \right) (18.97 \text{ m/s})^2$$

$$\Delta p = 23.7 \text{ kPa}$$

The pressure loss is excessive. Since this is excessive, the calculation should be repeated for 2-1/2 in. dia. pipe. However, since pulsating conditions require larger pipe, a 3-in. line would be used. In this case, Δp will be 15.4 kPa (2.24 psi), which is satisfactory.

A long line, whether part of a loop or not, may require a receiver near its furthest point from the compressor in order to supply brief high demands at good pressure. A suitable drain to remove condensate from each receiver is necessary. Outlets should be provided near all points of application to minimize pressure losses in hoses.

29.5.1 Compressor Intake and Discharge Pipes

Inlet air should be taken from a clean, cool, dry location, with outdoors being usually best. It should be protected from rain and should be at least six feet above the ground. With many compressors, noise is a potential problem, and the inlet location should be selected with that in mind.

Steam lines or hot water lines should never be placed in the compressor intake. An increase from 16 C to 50 C in intake temperature reduces the compressor capacity by about 10%.

Reciprocating compressors have a pulsating intake flow which can rattle windows and shake building walls enough to cause serious structural damage. The pulses are low in frequency and are usually objectionable to personnel in the area and to nearby residents. For more details, the reader is referred to Sec. 23.2, Chapter 8, Compressible Flow, and Chapter 6, Waves in Fluids. The intake pipe should be carefully designed to minimize the problem. If the pulses are resonant with the natural frequency of the air which occupies the pipe volume, large vibrations can be induced. The length and diameter of the intake pipe can be selected so that the waves are not synchronized, remembering that the intake pulse frequency depends upon the number of intake strokes per revolution as well as the compressor speed. The intake should be located where damage to walls will not result, and enclosed courtyards even of well constructed buildings should be avoided. For large compressors, an inlet *snubber bottle* is often installed to reduce the inlet pressure variations thereby

TABLE 29.2 Loss of Pressure Through Screw Pipe Fittings-Steam, Air, Gas
[Given in equivalent lengths (meters) of straight pipe, schedule 40]

Nominal pipe size, in.	Actual inside diameter, cm	Gate valve	Long radius ell or on run of standard tee	Standard ell or on run of tee reduced in size 50 percent	Angle valve	Close return bend	Tee through side outlet	Globe valve
$\frac{1}{2}$	1.58	0.110	0.189	0.47	2.64	1.06	.94	5.3
$\frac{3}{4}$	2.09	0.146	0.250	0.63	3.47	1.40	1.26	7.0
1	2.66	0.186	0.320	0.80	4.45	1.77	1.60	8.9
$1\frac{1}{4}$	3.50	0.247	0.421	1.05	5.82	2.33	2.10	11.7
$1\frac{1}{2}$	4.09	0.287	0.491	1.23	6.93	2.73	2.45	13.6
2	5.25	0.369	0.631	1.58	8.75	3.51	3.14	17.5
$2\frac{1}{2}$	6.27	0.439	0.753	1.88	10.5	4.18	3.75	20.9
3	7.79	0.546	0.936	1.88	13.0	5.21	4.66	26.0
4	10.2	0.716	1.23	2.34	17.1	6.83	6.20	34.1
5	12.8	0.896	1.54	3.08	21.3	8.53	7.68	42.7
6	15.4	1.08	1.85	4.63	25.6	10.3	9.27	51.2
8	20.3	1.42	2.43	6.10	33.8	13.6	12.2	67.7
10	25.4	1.78	3.05	7.62	42.4	17.0	15.2	84.7
12	30.3	2.12	3.35	9.08	50.6	20.2	18.2	101.0

Reprinted with permission of McGraw-Hill Book Company, *Piping Handbook*, 4th ed., by Crocker, S., 1945.

protecting the surrounding structures and the compressor valves and to act as a silencer. Compressor valves may be slammed shut and broken if the intake pressure variations happen to be synchronized with the valve and valve spring natural motion.

Intake piping should be chemically cleaned and coated with suitable epoxy, neoprene or other special paint, and should be thoroughly inspected before startup to be sure that foreign materials are not present which could find their why into the compressor. A temporary inlet filter should also be used at that time. The filter must be strong enough to assure that it will not break and be carried into the intake itself. The number of joints should be sufficient to permit the intake line to be disassembled for cleaning.

The inlet pipe should be as short and direct as possible with long radius elbows if any are necessary. It should be well supported, and none of its weight should be born by the compressor.

An intake filter gives important protection to both compressors and appliances, especially the former. Such filters are divided into three general types: oil bath, dry, and viscous impingement. The first two of these types are generally available as filter silencers, which provide considerable sound reduction as well as air cleaning. Pressure drop in new filters is from 10 to 25 cm (4 to 10 in.) of water.

Oil bath filters draw the air through a reservoir followed by a fine mesh screen or other medium which together remove up to 98% of particles over 10 microns. Small amounts of oil are entrained in the air. Dry filters contain a filtering element to remove contaminants from the air without adding oil. Some of the dry elements may be cleaned by blowing with compressed air or with a detergent. Others are made of paper and are disposable. Such filters are especially useful for nonlubricated compressors. Centrifugal air compressors require special attention to filters, because even very small dust particles can accumulate unevenly on the rotor causing dynamic unbalance.

The discharge pipe from the compressor to the aftercooler and separator or to the receiver should be no smaller than the compressor outlet. It should be as short and direct as possible, with long radius elbows, if elbows are needed. It will be hot under running conditions and should not contact wood or other flammable materials. Gaskets, if any, should be of noncombustible material. If a low section of pipe forms a pocket, it should be provided with a drain valve or automatic trap to avoid accumulating oil or water in the low point. The pipe should drain toward the aftercooler or receiver. Any pipe through which hot air passes should have a removable section to permit inspection and cleanout of combustible buildup to avoid an otherwise potentially hazardous situation. The receiver should be off the floor to prevent rusting. The receiver should have a valve at the bottom to drain condensate. Long runs of vertical pipe in the discharge line adjacent to the compressor should not be used, because there is potential danger of oil collecting and running back into the compressor cylinder. There is danger of pressure pulsations in the discharge line as well as in the intake line. Certain pipe lengths should be avoided to prevent this problem. The compressor manufacturer can usually provide this information.

All pipe lines should be very well supported so that excessive dead weight does not load the compressor flanges, and so that expansion of the pipe due either to internal pressure or to temperature changes is allowed for. An expansion joint or an expansion loop is frequently the solution to such problems.

29.6 AFTERCOOLERS AND SEPARATORS

Humidity in the air taken into a compressor, except for that removed in an intercooler, remains in the compressed air discharged from the compressor. The air leaving the compressor is hot, but as it cools in the lines and receiver, the moisture is condensed. The liquid water thus produced can seriously interfere with the appliances, and, in the case of back drainage resulting from an incorrectly sloped discharge pipe, can even damage the compressor itself. Chilling accompanying the expansion of air through pneumatic devices will cause ice formation that clogs or partially clogs many such devices. Also, slugs of water entrained in the air pipes are thrown against fittings causing the familiar *water hammer* which is annoying. Such accumulations of water may sometimes freeze and burst the pipe. Also, some lubricating oil may be carried into the lines as vapor, condense there and cause problems such as buildup of deposits. Under unusual conditions, the condensed oil vapor can present an explosion hazard.

To remove water and oil vapors, an aftercooler and a separator are generally included in a compressed air system. The aftercooler is usually a tube-in-shell heat exchanger with air and vapor inside the tubes and cooling water outside. It should be placed between the compressor and the receiver, and should be located indoors to prevent freezing. The aftercooler should have a separator and a moisture trap, preferably automatic, for removal of water and oil. Moisture traps should also be placed in air lines where further cooling may take place, and the receiver should also have provisions for draining, preferably including an automatic trap.

REFERENCES

Crocker, S., *Piping Handbook*, 4th ed., McGraw-Hill, New York, 1945.

Rollins, J. P. (Ed.), *Compressed Air and Gas Handbook*, 4th ed., Compressed Air and Gas Institute, New York, 1973.

Wright, D. K., Jr., "A New Friction Chart for Round Ducts," *Trans. of Amer. Soc. Heating and Ventilating Engineers*, p. 303, 1945.

Index